Advanced Topics in Quantum Field Theory

Quantum field theory is the basis of our modern description of physical phenomena at the fundamental level. This systematic and comprehensive text emphasizes nonperturbative phenomena and supersymmetry. It includes a thorough discussion of various phases of gauge theories, extended objects and their quantization, and global supersymmetry from a modern perspective. This Second Edition is revised to include topics developed in the last decade, including higher-form global symmetries and their applications, anomalies in supersymmetric theories beyond Ferrara–Zumino, and non-Abelian supersymmetric vortex strings. A new final part is added, presenting detailed solutions to around 90 exercises, allowing students to check their understanding of the acquired knowledge and providing extra details to supplement the main text descriptions. This an indispensable book for graduate students and researchers in theoretical physics.

Mikhail Shifman is the Ida Cohen Fine Professor of Theoretical Physics at the University of Minnesota. He was awarded the 1999 Sakurai Prize for Theoretical Particle Physics, the 2006 Julius Edgar Lilienfeld Prize and the 2016 Dirac Medal and Prize for outstanding contributions to physics. In 2018, he was elected to the US National Academy of Sciences.

Advanced Topics in Quantum Field Theory

A Lecture Course

Second Edition

MIKHAIL SHIFMAN

University of Minnesota

CAMBRIDGE
UNIVERSITY PRESS

CAMBRIDGE
UNIVERSITY PRESS

University Printing House, Cambridge CB2 8BS, United Kingdom

One Liberty Plaza, 20th Floor, New York, NY 10006, USA

477 Williamstown Road, Port Melbourne, VIC 3207, Australia

314–321, 3rd Floor, Plot 3, Splendor Forum, Jasola District Centre, New Delhi – 110025, India

103 Penang Road, #05–06/07, Visioncrest Commercial, Singapore 238467

Cambridge University Press is part of the University of Cambridge.

It furthers the University's mission by disseminating knowledge in the pursuit of
education, learning, and research at the highest international levels of excellence.

www.cambridge.org
Information on this title: www.cambridge.org/9781108840422
DOI: 10.1017/9781108885911

First published 2012
Second edition published 2022

A catalogue record for this publication is available from the British Library.

Library of Congress Cataloging-in-Publication Data
Names: Shifman, Mikhail A., author.
Title: Advanced topics in quantum field theory : a lecture course /
Mikhail Shifman, University of Minnesota.
Cover design: Alexandra Rozenman, "Always Knock First"
Description: Second edition. | Cambridge, UK ; New York, NY : Cambridge
University Press, 2022. | Includes bibliographical references and index.
Identifiers: LCCN 2021029678 (print) | LCCN 2021029679 (ebook) |
ISBN 9781108840422 (hardback) | ISBN 9781108885911 (epub)
Subjects: LCSH: Quantum field theory.
Classification: LCC QC174.46 .S55 2021 (print) | LCC QC174.46 (ebook) |
DDC 530.14/3–dc23
LC record available at https://lccn.loc.gov/2021029678
LC ebook record available at https://lccn.loc.gov/2021029679

ISBN 978-1-108-84042-2 Hardback

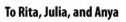

To Rita, Julia, and Anya

Contents

Preface to the Second Edition

Almost 10 years have elapsed since the first release of this textbook. Since then, quantum field theory has experienced significant developments – some of which are directly related to the topics covered in this text. They deserve to be known to today's students. In the new edition I add a number of sections devoted to advances in understanding anomalies in supersymmetric theories. Also added are newly discovered global anomalies for one- and higher-form symmetries. I tried to present them in the simplest way to make them accessible to the beginners. This insertion can be viewed as an introduction to higher-form symmetries and their global anomalies. Finally, I expand my soliton narrative by including more details regarding non-Abelian vortex strings. I revised many other sections and included extra exercises to make this lecture course even more pedagogical.

One of the most important changes is Part III, in which solutions to 87 exercises scattered across the text are collected. The solutions are not only intended to allow students to check their understanding of the acquired knowledge; their purpose is twofold. The curious reader will often find there extra details or information absent in the main body of the text.

Part III was prepared in collaboration with Gianni Tallarita, to whom I would like to say thank you. I am also grateful to my students Daniel Schubring and Chao-Hsiang Sheu.

Preface to the First Edition

Announcing the beginning of a Big Journey–Outlining the road map.

Quantum field theory remains the basis for the understanding and description of the fundamental phenomena in solid state physics and phase transitions, in high-energy physics, in astroparticle physics, and in nuclear physics multibody problems. It is taught in every university at the beginning of graduate studies. In American universities quantum field theory is usually offered in three sequential courses, over three or four semesters. Somewhat symbolically, these courses could be called Field Theory I, Field Theory II, and Field Theory III although the particular names may (and do) vary from university to university, and even in a given university, as time goes on.

Field Theory I treats relativistic quantum mechanics, spinors, and the Dirac equation and introduces the Hamiltonian formulation of quantum field theory and the canonical quantization procedure. Then basic field theories (scalar, Yukawa, quantum electrodynamics [QED], and Yang–Mills theories) are discussed and perturbation theory is worked out at the tree level. Field Theory I usually ends with a brief survey of the basic QED processes. Frequently used textbooks covering the above topics are F. Schwabl, *Advanced Quantum Mechanics* (Springer, 1997) and F. Mandl and G. Shaw, *Quantum Field Theory*, Second Edition (John Wiley and Sons, 2005).

Field Theory II begins with the path integral formulation of quantum field theory. Perturbation theory is generalized beyond tree level to include radiative corrections (loops). The renormalization procedure and renormalization group are thoroughly discussed, the asymptotic freedom of non-Abelian gauge theories is derived, and applications in quantum chromodynamics (QCD) and the Standard Model (SM) are considered. Sample higher-order corrections are worked out. The SM requires studies of the spontaneous breaking of the gauge symmetry (the Higgs phenomenon) to be included. A typical good modern text here is M. Peskin and D. Schroeder, *An Introduction to Quantum Field Theory* (Addison-Wesley, 1995). Some chapters from A. Zee, *Quantum Field Theory in a Nut Shell* (Princeton, 2003) and C. Itzykson and J.-B. Zuber, *Quantum Field Theory* (McGraw-Hill, 1980) can be used as a supplement.

Field Theory III has no canonical contents. Generically it is devoted to various advanced topics, but the choice of these advanced topics depends on the lecturer's taste and on whether one or two semesters are allocated. Sample courses that I have taught (or have witnessed in other universities) are (i) quantum field theory for solid state physicists (for critical phenomena conformal field theory is needed); (ii) supersymmetry; (iii) nonperturbative phenomena (broadly understood). In the first two categories some texts exist, but I would neither say that they are perfectly suitable for graduate students at the beginning of their career nor say that any single text could be used in class in isolation. Still, by and large one manages by combining existing textbooks.

In the third category, the set of books with pedagogical orientation is slim. Basically, it consists of Rubakov's text *Classical Theory of Gauge Fields* (Princeton, 2002), but, as can be seen from the title, this book covers a limited range of issues. A few topics are also discussed in R. Rajaraman, *Solitons and Instantons* (North-Holland, 1982).

I moved to the University of Minnesota in 1990. Since then, I have lectured on field theory many times. Field Theory III is my favorite. I choose topics based on my experience and personal judgment of what is important for students planning research at the front line in areas related to field theory. The two-semester lecture course goes on for 30 weeks. Lectures are given twice a week and last for 75 minutes per session. The audience is usually mixed, consisting of graduate students specializing in high-energy physics or in condensed-matter physics. This "two-phase" structure of the audience affects the topic selection process too, shifting the focus toward issues of general interest. The choice of topics in this course varies slightly from year to year, depending on the student class composition and their degree of curiosity, my current interests, and other factors.

Usually (but not always) I keep notes of my lectures. This book presents a compilation of these notes. The reader will find discussions of various advanced aspects of field theory spanning a wide range – from topological defects to supersymmetry, from quantum anomalies to false-vacuum decays.

A few words about other relevant textbooks are in order here. None covers the full spectrum of issues presented in this book. Some parts of my course do overlap to a certain extent with existing texts, in particular [1–15]; however, even in these instances the overlap is not complete. The chapters of this book are self-contained, so that any student familiar with introductory texts on field theory could start reading the book at any chapter. All appendices, as well as sections and exercises carrying an asterisk, can be omitted at a first reading, but the reader is advised to return to them later. A list of references can be found at the end of each chapter.

References

[1] M. Shifman, *ITEP Lectures on Particle Physics and Field Theory* (World Scientific, Singapore, 1999), Vols. 1 and 2.

[2] R. Rajaraman, *Solitons and Instantons* (North-Holland, Amsterdam, 1982).

[3] V. Rubakov, *Classical Theory of Gauge Fields* (Princeton University Press, 2002).

[4] Yu. Makeenko, *Methods of Contemporary Gauge Theory* (Cambridge University Press, 2002).

[5] A. Zee, *Quantum Field Theory in a Nutshell* (Princeton University Press, 2003).

[6] A. Vilenkin and E. P. S. Shellard, *Cosmic Strings and Other Topological Defects* (Cambridge University Press, 1994).

[7] N. Manton and P. Sutcliffe, *Topological Solitons* (Cambridge University Press, 2004).

[8] T. Vachaspati, *Kinks and Domain Walls* (Cambridge University Press, 2006).

[9] J. Wess and J. Bagger, *Supersymmetry and Supergravity*, Second Edition (Princeton University Press, 1992).

[10] J. Terning, *Modern Supersymmetry* (Clarendon Press, Oxford, 2006).

[11] M. Srednicki, *Quantum Field Theory* (Cambridge University Press, 2007).

[12] T. Banks, *Modern Quantum Field Theory: A Concise Introduction* (Cambridge University Press, 2008).

[13] Y. Frishman and J. Sonnenschein, *Non-Perturbative Field Theory* (Cambridge University Press, 2010).

[14] A. Smilga, *Lectures on Quantum Chromodynamics* (World Scientific, Singapore, 2001).

[15] A. S. Schwarz, *Topology for Physicists* (Springer-Verlag, Berlin, 1994).

Acknowledgments

This book was in the making for four years. I am grateful to many people who helped me en route. First and foremost I want to say thank you to Arkady Vainshtein and Alexei Yung, with whom I have shared the joy of explorations of various topics in modern field theory, some of which are described below. Not only have they shared with me their passion for physics, they have educated me in more ways than one. I would also like to thank my colleagues A. Armoni, A. Auzzi, S. Bolognesi, T. Dumitrescu, G. Dvali, A. Gorsky, Z. Komargodski, A. Losev, A. Nefediev, A. Ritz, S. Rudaz, N. Seiberg, E. Shuryak, M. Ünsal, G. Veneziano, and M. Voloshin, who offered generous advice. Dr Simon Capelin, the editorial director at Cambridge University Press, kindly guided me through the long process of polishing and preparing the manuscript. I am very grateful to Susan Parkinson, my copy editor, for careful and thoughtful reading of the manuscript and many useful suggestions.

I would like to thank Andrey Feldshteyn for the illustrations that can be seen at the beginning of each chapter. Alexandra Rozenman, a famous Boston artist, made her work available for the cover design. Thank you, Alya! Maxim Konyushikhin assisted me in typesetting this book in LATEX. He also prepared or improved certain plots and figures and checked crucial expressions. I am grateful to Sehar Tahir for help and advice on subtle aspects of LATEX. It is my pleasure to thank Ursula Becker, Marie Larson, and Laurence Perrin, who handled the financial aspects of this project. In the preparation I used funds kindly provided by William I. Fine Theoretical Physics Institute, University of Minnesota, and Chaires Internacionales de Recherche Blaise Pascal, France.

Without the encouragement I received from my wife, Rita, this book would have never been completed.

Useful General Formulas and Notation, Conventions, Abbreviations

Useful General Formulas and Notation

∂_L and ∂_R	2D chiral derivatives, p. 114
$\overrightarrow{\partial}(\overleftarrow{\partial})$	The partial derivative differentiates everything that stands to the right (left) of it.
$\overleftrightarrow{\partial} = \overrightarrow{\partial} - \overleftarrow{\partial}$	
$D_\alpha, \bar{D}_{\dot\alpha}$	Spinorial derivatives, pp. 439 and 472
$\mathcal{D}_\mu = \partial_\mu - i\, g\, A_\mu^a T^a$	
$\varepsilon^{\alpha\beta}, \varepsilon^{abc}$	2D and 3D Levi–Civita tensors, pp. 51, 421 (SUSY), 395
$\varepsilon^{\mu\nu\alpha\beta}$	Levi–Civita tensor in Minkowski space ($\varepsilon^{0123} = 1$) and Euclidean space ($\varepsilon_{1234} = 1$), pp. 184, 423
$\eta_{a\mu\nu}, \bar{\eta}_{a\mu\nu}$	't Hooft symbols, p. 187
$\bar{\eta}^{\dot\alpha}, \xi_\alpha$	Weyl spinors in 4D, p. 421
$F_{\mu\nu}$	Gauge field strength tensor, p. 422
$F_{\alpha\beta}, \bar{F}^{\dot\alpha\dot\beta}$	Gauge field strength tensor in spinorial notation, p. 423
$g^{\mu\nu} = \text{diag}\{+1, -1, -1, -1\}$	Metric in Minkowski space
γ^μ, γ^5	Dirac's 4D gamma matrices, p. 424
$\gamma^{0,1}$ or $\gamma^{t,z}$	2D gamma matrices, p. 426
$G_{\mu\nu}^a$	Gluon field strength tensor, pp. 22 and 23
$\sigma_1 = \begin{pmatrix} 0 & 1 \\ 1 & 0 \end{pmatrix},$ $\sigma_2 = \begin{pmatrix} 0 & -i \\ i & 0 \end{pmatrix},$ $\sigma_3 = \begin{pmatrix} 1 & 0 \\ 0 & -1 \end{pmatrix}$	Pauli matrices

$(\sigma^a)_{ij} (\sigma^a)_{pq} = 2\delta_{iq}\delta_{jp} - \delta_{ij}\delta_{pq}$ Completeness for the Pauli matrices

$\varepsilon_{abc} v^c (\sigma^a)_{ij} (\sigma^b)_{pq} = -i[(\vec{v}\vec{\sigma})_{ij}\,\delta_{pq} - 2(\vec{v}\vec{\sigma})_{iq}\,\delta_{jp} + (\vec{v}\vec{\sigma})_{pq}\,\delta_{ij}],$

Spatial vectors are denoted as $\vec{v}$ or $\boldsymbol{v}$

Useful relation for the Pauli matrices; $\vec{v}$ is an arbitrary 3-vector

$(\sigma^\mu)_{\alpha\dot\beta}$, $(\bar\sigma^\mu)^{\dot\beta\alpha}$	4D chiral σ matrices, p. 422
sign $p = \vartheta(p) - \vartheta(-p)$	Step function
$\tau_\mu^\pm$	Euclidean analogs of $(\sigma^\mu)_{\alpha\dot\alpha}$ and $(\bar\sigma^\mu)^{\dot\alpha\alpha}$, p. 187
$(\vec\tau)_{\alpha\beta}$, $(\vec\tau)_{\dot\alpha\dot\beta}$	Symmetric τ matrices for representations $(1,0)$ and $(0,1)$, p. 423
T^a	Generator of the gauge group; $C_2(R)$, $T(R)$, and T_G are defined on p. 484
T_G	Table 10.10, p. 550
W_α, $\bar W_{\dot\alpha}$	Supergeneralization of the gauge field strength tensor, p. 452
W_α^a	Non-Abelian superstrength tensor, generalizing $G_{\alpha\beta}^a$, p. 486
$\mathcal{W}$	Superpotential
$x_{L,R}^\mu$	Coordinates in the chiral superspaces, p. 437

Conventions

In this book I use the Lorentz–Heaviside system of units. In this system there are no 4π factors in the Maxwell equations, e.g.,

$$\mathrm{div}\vec E = \rho.$$

However, Coulomb's law takes the form

$$\vec E = \frac{1}{4\pi}\frac{q}{r^2}\,\vec n, \quad \vec n = \frac{\vec r}{r}.$$

In the Lorentz–Heaviside system $e^2 = 4\pi\alpha$ and the electromagnetic Hamiltonian $\mathcal{H} = \frac{1}{2}(\vec E^2 + \vec B^2)$ in the canonic normalization of the Lagrangian, or

$$\mathcal{H} = \frac{1}{2e^2}\left(\vec E^2 + \vec B^2\right)$$

in the noncanonic normalization in which e disappears from the interaction vertices. The flux of, say, electric field is defined as

$$\mathrm{Flux}_E = \int_{S_R} d^2 S_i E_i$$

in the canonic normalization. Then for the localized source the charge is expressible in terms of the flux, namely, $e = \mathrm{Flux}_E$.

Abbreviations

2D	two-dimensional
4D	four-dimensional
ADHM	Atiyah–Drinfel'd–Hitchin–Manin
ADS	Affleck–Dine–Seiberg
AF	asymptotic freedom
ANO	Abrikosov–Nielsen–Olesen
ASV	Armoni–Shifman–Veneziano
BPS	Bogomol'nyi–Prasad–Sommerfield
BPST	Belavin–Polyakov–Schwarz–Tyupkin
CC	central charge
CFIV	Cecotti–Fendley–Intriligator–Vafa
χSB	chiral symmetry breaking
CMS	curve(s) of the marginal stability
CP	CP-invariance; *also* complex projective space
DBI	Dirac–Born–Infeld
DR	dimensional regularization
FI	Fayet–Iliopoulos
GG	Georgi–Glashow
GUT	grand unified theory
GWS	Glashow–Weinberg–Salam
$I\!A$	instanton–anti-instanton
IR	infrared
LSP	lightest supersymmetric particle
NSVZ	Novikov–Shifman–Vainshtein–Zakharov
PV	Pauli–Villars
QCD	quantum chromodynamics
QED	quantum electrodynamics
QFT	quantum field theory
QM	quantum mechanics
SG	sine-Gordon
SM	standard model
SPM	superpolynomial model
SQCD	supersymmetric quantum chromodynamics, super-QCD
SQED	supersymmetric quantum electrodynamics, super-QED
SSG	super-sine-Gordon
SUSY	supersymmetry, supersymmetric
SYM	supersymmetric Yang–Mills (theory)
TWA	thin wall approximation
UV	ultraviolet
VEV	vacuum expectation value
WKB	Wentzel–Kramers–Brillouin
WZ	Wess–Zumino
WZNW	Wess–Zumino–Novikov–Witten

Be _tween_ St _andard_ & Mo _del_ St _rings_

Introduction

Presenting a brief review of the history of the subject. — The modern perspective.

Quantum field theory (QFT) was born as a consistent theory for a unified description of physical phenomena in which both quantum-mechanical aspects and relativistic aspects are important. In historical reviews it is always difficult to draw a line that would separate "before" and "after."[1] Nevertheless, it would be fair to say that QFT began to emerge when theorists first posed the question of how to describe the electromagnetic radiation in atoms in the framework of quantum mechanics. The pioneers in this subject were Max Born and Pascual Jordan, in 1925. In 1926 Max Born, Werner Heisenberg, and Pascual Jordan formulated a quantum theory of the electromagnetic field, neglecting polarization and sources to obtain what today would be called a free field theory. In order to quantize this theory they used the canonical quantization procedure. In 1927 Paul Dirac published his fundamental paper "The quantum theory of the emission and absorption of radiation." In this paper (which was communicated to the *Proceedings of the Royal Society* by Niels Bohr), Dirac gave the first complete and consistent treatment of the problem. Thus quantum field theory emerged inevitably, from the quantum treatment of the only known classical field, i.e. the electromagnetic field.

Dirac's paper in 1927 heralded a revolution in theoretical physics which he himself continued in 1928, extending relativistic theory to electrons. The Dirac equation replaced Schrödinger's equation for cases where electron energies and momenta were too high for a nonrelativistic treatment. The coupling of the quantized radiation field with the Dirac equation made it possible to calculate the interaction of light with relativistic electrons, paving the way to quantum electrodynamics (QED).

For a while the existence of the negative energy states in the Dirac equation seemed to be mysterious. At that time – it is hard to imagine – antiparticles were not yet known! It was Dirac himself who found a way out: he constructed a "Dirac sea" of negative-energy electron states and predicted antiparticles (positrons), which were seen as "holes" in this sea.

The hole theory enabled QFT to explore the notion of antiparticles and its consequences, which ensued shortly. In 1927 Jordan studied the canonical quantization of fields, coining the name "second quantization" for this procedure. In 1928 Jordan and Eugene Wigner found that the Pauli exclusion principle required the electron field to be expanded in plane waves with anticommuting creation and destruction operators.

---

[1] For a more detailed account of the first 50 years of quantum field theory see e.g. Victor Weisskopf's article [1] or the "Historical Introduction" in [2] and vivid personal recollections [3].

In the mid-1930s the struggle against infinities in QFT started and lasted for two decades, with a five-year interruption during World War II. While the infinities of the Dirac sea and the zero-point energy of the vacuum turned out to be relatively harmless, seemingly insurmountable difficulties appeared in QED when the coupling between the charged particles and the radiation field was considered at the level of quantum corrections. Robert Oppenheimer was the first to note that logarithmic infinities were a generic feature of quantum corrections. The best minds in theoretical physics at that time addressed the question how to interpret these infinities and how to get meaningful predictions in QFT beyond the lowest order. Now, when we know that every QFT requires an ultraviolet completion and, in fact, represents an effective theory, it is hard to imagine the degree of desperation among the theoretical physicists of that time. It is also hard to understand why the solution of the problem was evasive for so long. Landau used to say that this problem was beyond his comprehension and he had no hope of solving it [4]. Well ... times change. Today's students familiar with Kenneth Wilson's ideas will immediately answer that there are no actual infinities: all QFTs are formulated at a fixed short distance (corresponding to large Euclidean momenta) and then evolved to large distances (corresponding to small Euclidean momenta); the only difference between renormalizable and nonrenormalizable field theories is that the former are insensitive to ultraviolet data (which can be absorbed in a few low-energy parameters) while the latter depend on the details of the ultraviolet completion. But at that time theorists roamed in the dark. The discovery of the renormalization procedure by Richard Feynman, Julian Schwinger, and Sin-Itiro Tomonaga, which came around 1950, was a breakthrough, a ray of light. Crucial developments (in particular, due to Freeman Dyson) followed immediately. The triumph of quantum field theory became complete with the emergence of invariant perturbation theory, Feynman graphs, and the path integral representation for amplitudes,

$$\mathcal{A} = \int \prod_i \mathcal{D}\varphi_i e^{iS/\hbar}, \tag{0.1}$$

where the subscript i labels all relevant fields while S is the classical action of the theory calculated with appropriate boundary conditions.

In the mid-1950s Lev Landau, Alexei Abrikosov, and Isaac Khalatnikov discovered a feature of QED, the only respectable field theory of that time, that had a strong impact on all further developments in QFT. They found the phenomenon of zero charge (now usually referred to as infrared freedom): independently of the value of the bare coupling at the ultraviolet cut-off, the observed (renormalized) interaction between electric charges at "our" energies *must vanish* in the infinite cut-off limit. All other field theories known at that time were shown to have the same behavior. On the basis of this result, Landau pronounced quantum field theory dead [5] and called for theorists to seek alternative ways of dealing with relativistic quantum phenomena.[2] When I went to the theory department of ITEP[3] in 1970 to work on my Master's thesis, this attitude was still very much alive and studies of QFT were

---

[2] Of course, people "secretly" continued using field theory for orientation, e.g. for extracting analytic properties of the S-matrix amplitudes, but they did it with apologies, emphasizing that that was merely an auxiliary tool rather than the basic framework.

[3] The Institute of Theoretical and Experimental Physics in Moscow.

strongly discouraged, to put it mildly. Curiously, this was just a couple of years before the next QFT revolution.

The renaissance of quantum field theory, its second début, occurred in the early 1970s, when Gerhard't Hooft realized that non-Abelian gauge theories are renormalizable (including those in the Higgs regime) and, then, shortly after, David Gross, Frank Wilczek, and David Politzer discovered asymptotic freedom in such theories. Quantum chromodynamics (QCD) was born as *the* theory of strong interactions. Almost simultaneously, the standard model of fundamental interactions (SM) started taking shape. In the subsequent decade it was fully developed and was demonstrated, with triumph, to describe all known phenomenology to a record degree of precision. All fundamental interactions in nature fit into the framework of the standard model (with the exception of quantum gravity, of which I will say a few words later).

Thus, the gloomy prediction of the imminent *demise* of QFT – a wide spread opinion in the 1960s – turned out to be completely false. In the 1970s QFT underwent a conceptual revolution of the scale comparable with the development of renormalizable invariant perturbation theory in QED in the late 1940s and early 1950s. It became clear that the Lagrangian approach based on Eq. (0.1), while ideally suited for perturbation theory, is not necessarily the only (and sometimes, not even the best) way of describing relativistic quantum phenomena. For instance, the most efficient way of dealing with two-dimensional conformal field theories is algebraic. In fact, many different Lagrangians can lead to the same theory (according to Alexander Belavin, Alexander Polyakov, and Alexander Zamolodchikov, in 1981). This is an example of the QFT dualities, which occur not only in conformal theories and not only in two dimensions. Suffice it to mention that the sine-Gordon theory was shown long ago to be dual to the Thiring model. Even more striking were the extensions of duality to four dimensions. In 1994 Nathan Seiberg reported a remarkable finding: supersymmetric Yang–Mills theories with distinct gauge groups can be dual, leading to one and the same physics in the infrared limit!

Some QFTs were found to be integrable. Topological field theories were invented which led mathematical physicists to new horizons in mathematics, namely, in knot theory, Donaldson theory, and Morse theory.

Look through Introduction to Part II, Section 9.6.

The discovery of supersymmetric field theories in the early 1970s (which we will discuss later) was a milestone of enormous proportions, a gateway to a new world, described by QFTs of a novel type and with novel – and, quite often, – counterintuitive properties. In its impact on QFT, I can compare this discovery to that of the New World in 1492. People who ventured on a journey inside the new territory found treasures and exotic, and previously unknown, fruits: a richness of dynamical regimes in super-Yang–Mills theories, including a broad class of superconformal theories in four dimensions; exact results at strong coupling; hidden symmetries and cancellations; unexpected geometries and more.

Supersymmetric theories proved to be a powerful tool, allowing one to reveal intriguing aspects of gauge (color) dynamics at strong coupling. Continuing my analogy with Columbus's discovery of America in 1492, I can say that the expansion of QFT in the four decades that have elapsed, since 1970 has advanced us to the interior of a new continent. Our task is to reach, explore, and understand this continent and to try to open the ways to yet other continents. The reader should be warned that the very nature of the frontier explorations in QFT has changed considerably in comparison with what is found in older textbooks. A nice characterization of this change is given by an outstanding mathematical physicist, Andrey Losev, who writes [6]:

> In the good old days, theorizing was like sailing between islands of experimental evidence. And, if the trip was not in the vicinity of the shoreline (which was strongly recommended for safety reasons) sailors were continuously looking forward, hoping to see land – the sooner the better ...
>
> Nowadays, some theoretical physicists (let us call them sailors) [have] found a way to survive and navigate in the open sea of pure theoretical construction. Instead of the horizon they look at the stars,[4] which tell them exactly where they are. Sailors are aware of the fact that the stars will never tell them where the new land is, but they may tell them their position on the globe. In this way sailors – all together – are making a map that will at the end facilitate navigation in the sea and will help to discover new lands.
>
> Theoreticians become *sailors* simply because they just like it. Young people seduced by captains forming crews to go to a Nuevo El Dorado of Unified Quantum Field Theory or Quantum Gravity soon realize that they will spend all their life at sea. Those who do not like sailing desert the voyage, but for true potential sailors the sea becomes their passion. They will probably tell the alluring and frightening truth to their students – and the proper people will join their ranks.

Approximately at the same time as supersymmetry was born in the early-to-mid-1970s, a number of remarkable achievements occurred in uncovering the nonperturbative side of non-Abelian Yang–Mills theories: the discovery of extended objects such as monopoles (G. 'tHooft; A. Polyakov), domain walls, and flux tubes (H. Nielsen and P. Olesen) and, finally, tunneling trajectories (currently known as instantons) in Euclidean space–time (A. Polyakov and collaborators). A microscopic theory of magnetic monopoles was developed. It took people a few years to learn how to quantize magnetic monopoles and similar extended objects. The quasiclassical quantization of solitons was developed by Ludwig Faddeev and his school in St. Petersburg and, independently, by R. F. Dashen, B. Hasslacher, and A. Neveu. Then Y. Nambu, S. Mandelstam, and G. 't Hooft put forward (practically simultaneously but independently) the dual Meissner effect conjecture as the mechanism responsible for color confinement in QCD. It became absolutely clear that, unlike in QED, crucial physical phenomena go beyond perturbation theory and field theory *is* capable of describing them.

The phenomenon of color confinement can be summarized as follows. The spectrum of asymptotic states in QCD has no resemblance to the set of fields in the Lagrangian; at the Lagrangian level one deals with quarks and gluons while experimentalists detect pions, protons, glueballs, and other color singlet states – never quarks and gluons. Color confinement makes colored degrees of freedom inseparable. In a bid to understand this phenomenon Nambu, 't Hooft, and Mandelstam suggested a non-Abelian dual Meissner effect. According to their vision, non-Abelian monopoles condense in the vacuum, resulting in the formation of non-Abelian chromoelectric flux tubes between color charges, e.g. between a probe

---

[4] Here by "stars" he means aspects of the internal logic organizing the mathematical world rather than outstanding members of the community.

heavy quark and antiquark pair. Attempts to separate these probe quarks would lead to stretching of the flux tubes, so that the energy of the system grows linearly with separation. That is how linear confinement was visualized.

One may ask: where did these theorists get their inspiration? The Meissner effect, known for a long time and well understood theoretically, yielded a rather analogous picture. It answered the question: what happens if one immerses a magnetic charge and anticharge in a type-II superconductor?

If we place a probe magnetic charge and anticharge in empty space, the magnetic field they induce will spread throughout space, while the energy of the magnetic charge–anticharge configuration will obey the Coulomb $1/r$ law. The force will die off as $1/r^2$. Inside the superconductor, however, Cooper pairs condense, all electric charges are screened, and the photon acquires a mass; i.e., according to modern terminology the electromagnetic U(1) gauge symmetry is Higgsed. The magnetic field cannot be screened in this way; in fact, the magnetic flux is conserved. At the same time the superconducting medium cannot tolerate a magnetic field. This clash of contradictory requirements is solved through a compromise. A thin tube (known as an Abrikosov vortex) is formed between the magnetic charge and anticharge immersed in the superconducting medium. Within this tube superconductivity is destroyed – which allows the magnetic field to spread from the charge to the anticharge through the tube. The tube's transverse size is proportional to the inverse photon mass while its tension is proportional to the Cooper pair condensate. Increasing the distance between the probe magnetic charges (as long as they are within the superconductor) does not lead to their decoupling; rather, the magnetic flux tubes become longer, leading to linear growth in the energy of the system.

This physical phenomenon inspired Nambu, 't Hooft, and Mandelstam's idea of non-Abelian confinement as a dual Meissner effect. Many people tried to quantify this idea. The first breakthrough, instrumental in all later developments, came only 20 years later, in the form of the Seiberg–Witten solution of $\mathcal{N} = 2$ supersymmetric Yang–Mills theory. This theory has eight supercharges, which makes the dynamics quite "rigid" and helps one to find the full analytic solution at low energies. The theory bears a resemblance to quantum chromodynamics, sharing common family traits. By and large, one can characterize it as QCD's second cousin.

The problem of confinement in QCD *per se* (and in nonsupersymmetric theories in four dimensions in general) is not yet solved. Since this problem is of such paramount importance for the theory of strong interactions we will discuss at length instructive models of confinement in lower dimensions.

The topics listed above have become part of "operational" knowledge in the community of field theory practitioners. In fact, they transcend this community since many aspects reach out to string theorists, cosmologists, astroparticle physicists, and solid state theorists. My task is to present a coherent pedagogical introduction covering the basics of the above subjects in order to help prepare readers to undertake research of their own.

We will start from the Higgs effect in non-Abelian gauge theories. Then we will study the basic phases in which non-Abelian gauge theories can exist – Coulomb, conformal, Higgs, and so on. Some "exotic" phases discovered in the context of supersymmetric theories will not be discussed.

A significant part of this book will be devoted to topological solitons, that is, the topological defects occurring in various field theories. The term "soliton" was introduced in the 1960s, but scientific research on solitons had started much earlier, in the nineteenth century, when a Scottish engineer, John Scott-Russell, observed a large solitary wave in a canal near Edinburgh. Condensed matter systems in which topological defects play a crucial role have been well known for a long time: suffice it to mention the magnetic flux tubes in type II superconductors and the structure of ferromagnetic materials, with domain walls at the domain boundaries.

In 1961 Skyrme [7] was the first to introduce in particle physics a three-dimensional topological defect solution arising in a nonlinear field theory. Currently such solitons are known as Skyrmions. They provide a useful framework for the description of nucleons and other baryons in multicolor QCD (in the so-called 't Hooft limit, i.e. at $N_c \to \infty$ with $g^2 N_c$ fixed, where N_c is the number of colors and g^2 is the gauge coupling constant).

In general, in this book we will pay much attention to the broader aspects of multicolor gauge theories and the 't Hooft limit. We will see that a large-N expansion is equivalent to a topological expansion. Each term in a $1/N$ series is in one-to-one correspondence with a particular topology of Feynman graphs, e.g. planar graphs, those with one handle, and so on. Large-N analysis presents a very fruitful line of thought, allowing one to address and answer a number of the deepest questions in gauge theories.

As early as in 1965 Nambu anticipated the cosmological significance of topological defects [8]. He conjectured that the universe could have a kind of domain structure. Subsequently Weinberg noted the possibility of domain-wall formation at a phase transition in the early universe [9].

From the general theory of solitons we pass to a specific class of supersymmetric critical (or Bogomol'nyi–Prasad–Sommerfeld-saturated) solitons.

I will present a systematic and rather complete introduction to supersymmetry that is (almost) sufficient for bringing students to the cutting edge in this area.

Readers should be warned that nothing will be said on the quantum theory of gravity. There is no consistent theory of quantum gravity. Attempts to develop such a theory led people to the inception of critical string theory in the late 1970s. This theory builds on quantum field theory and, it is hoped, goes beyond it. It is believed that, after its completion, string theory will describe all fundamental interactions in nature, including quantum gravity. However, the completion of superstring theory seems to be in the distant future. Today neither is its mathematical structure clear nor its relevance to real-world phenomena established. A number of encouraging indications remain in disassociated fragments. If there is a definite lesson for us from string theory today, it is that the class of relativistic quantum phenomena to be considered must be expanded as far as possible and that we must explore, to the fullest extent, nonperturbative aspects in the hope of finding a path to quantum geometry, when the time is ripe, probably with many other interesting findings en route.

Finally, a few words on the history of supersymmetry are in order.[5] The history of supersymmetry is exceptional. All other major conceptual developments in physics

---

[5] For more details see [10].

have occurred because physicists were trying to understand or study some established aspect of nature or to solve some puzzle arising from data. The discovery in the early 1970s of supersymmetry, that is, invariance under the interchange of fermions and bosons, was a purely intellectual achievement, driven by the logic of theoretical development rather than by the pressure of existing data.

The discovery of supersymmetry presents a dramatic story. In 1970 Yuri Golfand and Evgeny Likhtman in Moscow found a superextension of Poincaré algebra and constructed the first four-dimensional field theory with supersymmetry, the (massive) quantum electrodynamics of spinors and scalars.[6] Within a year Dmitry Volkov and Vladimir Akulov in Kharkov suggested nonlinear realizations of supersymmetry and then Volkov and Soroka started developing the foundations of supergravity. Because of the Iron Curtain which existed between the then USSR and the rest of the world, these papers were hardly noticed. Supersymmetry took off after the breakthrough work of Julius Wess and Bruno Zumino in 1973. Their discovery opened to the rest of the community the gates to the Superworld. Their work on supersymmetry has become tightly woven into the fabric of contemporary theoretical physics.

Often students ask where the name "supersymmetry" comes from. The first paper of Wess and Zumino [11] was entitled "Supergauge transformations in four dimensions." A reference to supersymmetry (without any mention the word "gauge") appeared in one of Bruno Zumino's early talks [12]. In the published literature Salam and Strathdee were the first to coin the term *supersymmetry*. In the paper [13], in which these authors constructed supersymmetric Yang–Mills theory, super-symmetry (with a hyphen) was in the title, while in the body of the paper Salam and Strathdee used both the old terminology due to Wess and Zumino, "super-gauge symmetry," and the new one. This paper was received by the editorial office of *Physical Letters* on June 6, 1974, exactly eight months after that of Wess and Zumino [11]. An earlier paper, of Ferrara and Zumino [14] (received by the editorial office of *Nuclear Physics* on 27 May 1974),[7] where the same problem of super-Yang–Mills theory was addressed, mentions only supergauge invariance and supergauge transformations.

References for the Introduction

[1] V. Weisskopf, The development of field theory in the last 50 years, *Physics Today* **34**, 69 (1981).
[2] S. Weinberg, *The Quantum Theory of Fields* (Cambridge University Press, 1995), Vol. 1.
[3] S. Weinberg, Living with infinities [arXiv:0903.0568 [hep-th]].
[4] B. L. Ioffe, private communication.

---

[6] At approximately the same time, supersymmetry was observed as a world-sheet *two-dimensional* symmetry by string theory pioneers (Ramond, Neveu, Schwarz, Gervais, and Sakita). The realization that the very same superstring theory gave rise to supersymmetry in the target space came much later.
[7] The editorial note says it was received on May 27, 1973. This is certainly a misprint, otherwise the event would be acausal.

[5] L. Landau, in *Niels Bohr and the Development of Physics* (Pergamon Press, New York, 1955), p. 52.

[6] A. Losev, From Berezin integral to Batalin–Vilkovisky formalizm: a mathematical physicist's point of view, in M. Shifman (ed.), *Felix Berezin: Life and Death of the Mastermind of Supermathematics* (World Scientific, Singapore, 2007), p. 3.

[7] T. H. R. Skyrme, *Proc. Roy. Soc. A* **262**, 237 (1961) [reprinted in E. Brown (ed.), *Selected Papers, with Commentary, of Tony Hilton Royle Skyrme* (World Scientific, Singapore, 1994)].

[8] Y. Nambu, General discussion, in Y. Tanikawa (ed.), *Proc. Int. Conf. on Elementary Particles: In Commemoration of the Thirtieth Anniversary of Meson Theory*, Kyoto, September 1965, 327–333.

[9] S. Weinberg, *Phys. Rev. D* **9**, 3357 (1974) [reprinted in R. N. Mohapatra and C. H. Lai, (eds.), *Gauge Theories of Fundamental Interactions* (World Scientific, Singapore, 1981), pp. 581–602].

[10] G. Kane and M. Shifman (eds.), *The Supersymmetric World: The Beginnings of the Theory* (World Scientific, Singapore, 2000); S. Duplij, W. Siegel, and J. Bagger (eds.), *Concise Encyclopedia of Supersymmetry* (Kluwer Academic Publishers, 2004), pp. 1–28.

[11] J. Wess and B. Zumino, Supergauge transformations in four dimensions, *Nucl. Phys. B* **70**, 39 (1974).

[12] B. Zumino, Fermi–Bose supersymmetry (supergauge symmetry in four dimensions), in J. R. Smith (ed.), *Proc. 17th Int. Conf. High Energy Physics*, London, 1974, (Rutherford Lab., 1974).

[13] A. Salam and J. Strathdee, Super-symmetry and non-Abelian gauges, *Phys. Lett. B* **51**, 353 (1974).

[14] S. Ferrara and B. Zumino, Supergauge invariant Yang–Mills theories, *Nucl. Phys. B* **79**, 413 (1974).

PART I

BEFORE SUPERSYMMETRY

Spontaneous breaking of global and local symmetries. — The Higgs regime. — The Coulomb and infrared free phases. — Color confinement (closed and open strings). Does confinement imply chiral symmetry breaking? — Conformal regime. — Conformal window.

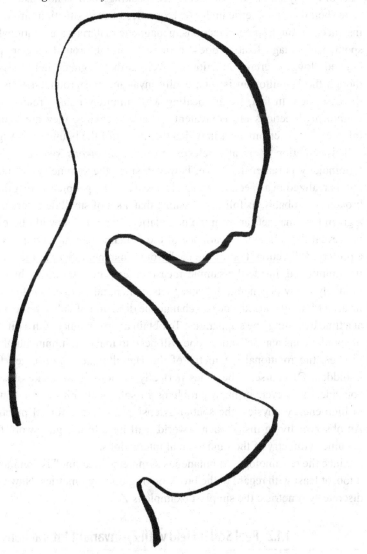

Illustration by Olga Kulakova: *Open string in nonperturbative regime*

1.1 Spontaneous Symmetry Breaking

1.1.1 Introduction

We will begin with a general survey of various patterns of spontaneous symmetry breaking in field theory. Our first task is to get acquainted with the breaking of global symmetries – at first discrete, then continuous. After that we will familiarize ourselves with the manifestations of spontaneous symmetry breaking.

Spontaneous symmetry breakdown: what does that mean?

Assume that a dynamical system under consideration is described by a Lagrangian $\mathcal{L}$ possessing a certain global symmetry $\mathcal{G}$. Assume that the ground state of this system is known. Generally speaking, there is no reason why the ground state should be symmetric under $\mathcal{G}$. Examples of such situations are well known. For instance, although spin interactions in magnetic materials are rotationally symmetric, spontaneous magnetization does occur: spins in the ground state are predominantly aligned along a certain direction, as well as the magnetic field they induce. Even though the Hamiltonian is rotationally invariant, the ground state is not. If this is the case then, in fact, we are dealing with infinitely many ground states, since all alignment directions are equivalent (strictly speaking, they are equivalent for an infinitely large ferromagnet in which the impact of the boundary is negligible).

This situation is usually referred to as *spontaneous symmetry breaking*. This terminology is rather deceptive, however, since the symmetry has not disappeared but, is realized in a special manner. The reason why people say that the symmetry is broken is, probably, as follows. Assume that a set of small detectors is placed inside a given ferromagnet far from the boundaries. Experiments with these detectors will not reveal the rotational invariance of the fundamental interactions because there is a preferred direction, that of the background magnetic field in the ferromagnet. For the uninitiated, inside-the-sample measurements give no direct hint that there are infinitely many degenerate ferromagnets, which, taken together, form a rotationally invariant family. Indeed, one can change the direction of only a finite number of spins at a time by tuning one's apparatus. To obtain a ferromagnet with a different direction of spontaneous magnetization, one will need to make an infinite number of steps.

A learned theoretician will be able to guess that the fundamental interaction is rotationally invariant from the presence of Goldstone bosons.

Thus, the rotational symmetry of the Hamiltonian, as observed from "inside," is hidden. Of course, it becomes perfectly obvious if we make observations from "outside." However, in many problems in solid state physics and in all problems in high-energy physics, the spatial extension is infinite for all practical purposes. An observer living inside such a world, will have to use guesswork to uncover the genuine symmetry of the fundamental interactions.

Since the terminology "spontaneous symmetry breaking" is common, we will use it too, at least with regard to the breaking of global symmetries. Now we will discuss discrete symmetries; the simplest example is Z_2.

1.1.2 Real Scalar Field with Z_2-Invariant Interactions

Let us consider a system with one real field $\phi(x)$ with action

$$S = \int d^D x \left[\frac{1}{2} \left(\partial_\mu \phi \right) \left(\partial^\mu \phi \right) - U(\phi) \right], \tag{1.1}$$

where $U(\phi)$ is the self-interaction (or potential energy) and D is the number of dimensions. In field theory one can consider three distinct cases, $D = 2$, $D = 3$, and $D = 4$. The first two cases may be relevant for both solid state and high-energy physics, while the third case refers only to high-energy physics.

The potential energy may be chosen in many different ways. In this subsection we will limit ourselves to the simplest choice, a quartic polynomial of the form

$$U(\phi) = \tfrac{1}{2}m^2\phi^2 + \tfrac{1}{4}g^2\phi^4, \tag{1.2}$$

where m^2 and g^2 are constants. We will assume that g^2 is small, so that a quasiclassical treatment applies.

It is obvious that the system described by Eqs. (1.1), (1.2) possesses a discrete Z_2 symmetry:

$$\phi(x) \longrightarrow -\phi(x). \tag{1.3}$$

The symmetry Z_2 as an example of the discrete global symmetry

Indeed, only even powers of ϕ enter the action. This is a global symmetry since the transformation (1.3) must be performed for all x simultaneously.

For the time being we will treat our theory purely classically but will use quantum-mechanical language. We will refer to the lowest energy state (the ground state) as the *vacuum*. To determine the vacuum states one should examine the Hamiltonian of the system,

$$\mathcal{H} = \int d^{D-1}x \left[\tfrac{1}{2}(\partial_0\phi)(\partial_0\phi) + \tfrac{1}{2}\left(\vec{\partial}\phi\right)\left(\vec{\partial}\phi\right) + U(\phi) \right]. \tag{1.4}$$

Since the kinetic term is positive definite, it is clear that the state of lowest energy is that for which the value of the field ϕ is constant, i.e., independent of the spatial and time coordinates. For a constant-field configuration the minimal energy is determined by the minimization of $U(\phi)$. We will refer to the corresponding value of ϕ as the vacuum value.

Within the given class of theories with the potential energy (1.2) we can find both dynamical scenarios: manifest Z_2 symmetry or spontaneously broken Z_2 symmetry, depending on the sign of the parameter m^2.

1.1.3 Symmetric Vacuum

Let us start from the case of positive m^2; see Fig. 1.1. The vacuum is achieved at

$$\phi = 0. \tag{1.5}$$

This solution is obviously invariant under the transformation (1.3). Thus the ground state of the system has the same Z_2 symmetry as the Hamiltonian. In this case we will say that *the vacuum does not break the symmetry spontaneously*. One can make one step further and consider small oscillations around the vacuum. Since the vacuum is at zero, small oscillations coincide with the field ϕ itself. In the quadratic approximation the action becomes

$$S_2 = \int d^D x \left[\tfrac{1}{2}\left(\partial_\mu\phi\right)\left(\partial^\mu\phi\right) - \tfrac{1}{2}m^2\phi^2 \right]. \tag{1.6}$$

We immediately recognize m as the mass of the ϕ particle. Moreover, from the quartic term $g^2\phi^4$ one can readily extract the interaction vertex and develop the

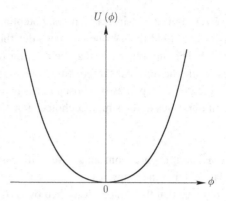

Fig. 1.1   The potential energy (1.2) at positive m^2.

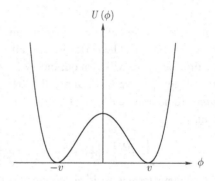

Fig. 1.2   The potential energy at negative m^2.

corresponding Feynman graph technique. The Z_2 symmetry of the interactions is apparent. Because of the invariance under (1.3), if in any scattering process the initial state has an odd number of particles then, so does the final state. Starting with any even number of particles in the initial state one can obtain only an even number of particles in the final state. Thus, a smart experimentalist, colliding two particles and never finding three, five, seven, and so on particles in his detectors, will deduce the Z_2 invariant nature of the theory.

1.1.4 Nonsymmetric Vacuum

Let us pass now to another case, that of negative m^2. To ease the notation we will introduce a positive parameter, $\mu^2 \equiv -m^2$. The new potential is shown in Fig. 1.2. Strictly speaking, I am cheating a little bit here; in fact, what is shown in Fig. 1.2 is *not* the potential (1.2). Rather, I have added a constant to this potential, $\Delta U = \mu^4/(4g^2)$, chosen in such a way as to adjust to zero the value of U at the minima. As you know, numerical additive constants in the Lagrangian are unobservable – they have no impact on the dynamics of the system.

The symmetric solution $\phi = 0$ is now at a maximum of the potential rather than a minimum. Small oscillations near this solution would be unstable; in fact, they would represent tachyonic objects rather than normal particles.

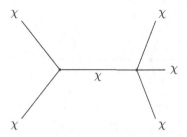

Fig. 1.3 The Feynman graph for the transition of two χ quanta into three in an asymmetric vacuum.

The true ground states are asymmetric with respect to (1.3),

$$\phi = \pm v, \qquad v = \frac{\mu}{g}. \tag{1.7}$$

The two-fold degeneracy of the vacuum follows from the Z_2 symmetry of the Lagrangian in (1.6). Indeed, under the action of (1.3) the positive vacuum goes into the negative vacuum, and vice versa.

In terms of v the potential takes the form

$$U(\phi) = \tfrac{1}{4}g^2 \left(\phi^2 - v^2\right)^2. \tag{1.8}$$

To investigate the physics near one of the two asymmetric vacua, let us define a new "shifted" field χ,

$$\phi = v + \chi, \tag{1.9}$$

which represents small oscillations, i.e., the particles of the theory. First let us examine the particle mass. To this end we substitute the decomposition (1.9) into the Lagrangian with a potential term given by Eq. (1.8). In this way we get

$$\mathcal{L} = \tfrac{1}{2} \left(\partial_\mu \chi\right) \left(\partial^\mu \chi\right) - \left(\mu^2 \chi^2 + \mu g \chi^3 + \tfrac{1}{4}g^2 \chi^4\right), \tag{1.10}$$

using Eq. (1.7) for v. By comparing the kinetic term with the term $\mu^2 \chi^2$ within the large parentheses we immediately conclude, for the mass of the χ quantum, that

$$m_\chi = \sqrt{2}\mu. \tag{1.11}$$

In the unbroken case of positive m^2 the particle's mass was m (see Eq. (1.6)). We see that changing the sign of m^2 leads to a factor of $\sqrt{2}$ difference in the particle mass.

The occurrence of the term cubic in χ in (1.10) is even more dramatic. Indeed this term, in conjunction with the quartic term, will generate amplitudes with an arbitrary number of quanta. For instance, the scattering amplitude for two quanta into three quanta is displayed in Fig. 1.3.[1]

The selection rule prohibiting the transition of an even number of particles into an odd number, as was the case for positive m^2 (a symmetric vacuum), is gone. Even for a smart physicist, doing scattering experiments, it would be rather hard now to discover the Z_2 symmetry of the original theory.

---

[1] Let us note parenthetically that there is an easy heuristic way to generate Feynman graphs in the asymmetric-vacuum theory from those of the symmetric theory. In the symmetric-vacuum theory, where all vertices are quartic, one starts for instance from the graph of Fig. 1.4a and replaces one external line by the vacuum expectation value of ϕ (Fig. 1.4b). Since ϕ_{vac} is just a number, one immediately arrives at the graph of Fig. 1.3.

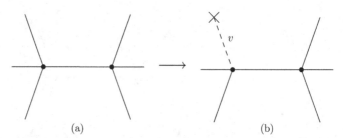

Fig. 1.4 Converting Feynman graphs in the symmetric theory (a) into those of the theory with asymmetric vacua (b). The cross on the broken line means that this line is replaced by the vacuum value of the field ϕ.

A trace of this symmetry remains in the broken phase, namely a relation between the cubic coupling constant in the Lagrangian $(-\mu g)$, the quartic constant $(-g^2/4)$, and the particle mass squared $(2\mu^2)$:

> *This relation does not hold for generic cubic and quartic interaction vertices in (1.2).*

$$\text{quartic constant} = -\frac{(\text{cubic constant})^2}{2m_\chi^2}. \tag{1.12}$$

A qualitative signature of the underlying spontaneously broken Z_2 symmetry is the existence of domain walls.

1.1.5 Equivalence of Asymmetric Vacua

Two questions remain to be discussed. Let us start with the simpler. What would happen if, instead of the vacuum at $\phi = v$, we (or, rather, nature) chose the second vacuum, at $\phi = -v$? The decomposition (1.9) would obviously be replaced by $\phi = -v + \chi$. This would change the sign of the cubic term in the Lagrangian, which, in turn, would entail the change in sign of all amplitudes with an odd number of external lines. We should remember, however, that it is not amplitudes but probabilities that are measurable. Since there is no interference between amplitudes with odd and even numbers of external lines, the sign is unobservable. *The physics in the two vacua is perfectly equivalent!*

This brings us to the second question: is there a *direct* manifestation of the fact that the underlying theory is Z_2 symmetric and the Z_2 symmetry is spontaneously broken by the choice of vacuum state? The answer is yes, at least in theory. We will discuss this phenomenon at length later (see Chapter 2).

1.1.6 Spontaneous Breaking of the Continuous Symmetry

To begin with, we will consider the simplest continuous symmetry, U(1). Consider a *complex* field $\phi(x)$ with action

$$S = \int d^D x \left[\left(\partial_\mu \phi \right)^* \left(\partial^\mu \phi \right) - U(\phi) \right], \tag{1.13}$$

where the potential energy $U(\phi)$ in fact depends only on $|\phi|$, for instance,

$$U(\phi) = m^2 |\phi|^2 + \tfrac{1}{2} g^2 |\phi|^4. \tag{1.14}$$

In this case the Lagrangian is invariant under a (global) phase rotation of the field ϕ:

$$\phi \rightarrow e^{i\alpha}\phi, \qquad \phi^* \rightarrow e^{-i\alpha}\phi^*. \tag{1.15}$$

If the mass parameter m^2 is positive, the minimum of the potential energy is achieved at $\phi = 0$. This is the unbroken phase. The vacuum is unique. There are two particles, that is, two elementary excitations, corresponding to $\operatorname{Re}\phi$ and $\operatorname{Im}\phi$. The mass of both these elementary excitations is m.

Changing the sign of m^2 from positive to negative drives one into the broken phase. The potential energy can be rewritten (after addition of an irrelevant constant) as

$$U(\phi) = \tfrac{1}{2}g^2 \left(|\phi|^2 - v^2\right)^2, \tag{1.16}$$

where

$$v^2 = \frac{\mu^2}{g^2} \equiv -\frac{m^2}{g^2}; \tag{1.17}$$

$U(\phi)$ has the form of a "Mexican hat," see Fig. 1.5. The degenerate minima in the potential energy are indicated by the black circle. An arbitrary point on this circle is a valid vacuum. Thus there is a continuous set of vacuum states, called the *vacuum manifold*. All these vacua are physically equivalent.

As an example let us consider the vacuum state at $\phi = v$. Near this vacuum the field ϕ can be represented as

$$\phi(x) = v + \frac{1}{\sqrt{2}}\varphi(x) + \frac{i}{\sqrt{2}}\chi(x), \tag{1.18}$$

where φ and χ are real fields. Then in terms of these fields

$$\mathcal{L} = \tfrac{1}{2}\left[(\partial_\mu\varphi)^2 + (\partial_\mu\chi)^2\right]$$
$$- \left[g^2v^2\varphi^2 + \frac{g^2v}{\sqrt{2}}\varphi(\varphi^2 + \chi^2) + \frac{g^2}{8}(\varphi^2 + \chi^2)^2\right]. \tag{1.19}$$

The mass of an elementary excitation of the φ field is $m_\varphi = \sqrt{2}gv = \sqrt{2}\mu$. A remarkable feature is that the mass of the χ quantum vanishes: the potential energy has no terms quadratic in χ in (1.19).

This is a general situation: the spontaneous breaking of continuous symmetries entails the occurrence of massless particles, which are referred to as Goldstone particles, or Goldstones for short.[2] In solid state physics they are also known as gapless excitations. For instance, in the example of the ferromagnet discussed at the beginning of this section such gapless excitations exist too; they are called magnons. Detecting magnons within the ferromagnet sample gives a clue that in fact one is dealing with an underlying symmetry that has been spontaneously broken.

In the problem at hand, that of a single complex field, the spontaneously broken symmetry is U(1). It has a single generator; hence the Goldstone boson, the phase of the order parameter, is unique.

> The Goldstone theorem, Section 6.5.1

---

[2] Sometimes the Goldstone bosons are referred to as the Nambu–Goldstone bosons. They were discussed first by Nambu in the context of the Bardeen–Cooper–Schrieffer superconductivity and independently by Vaks and Larkin who constructed a model now known as the Nambu–Jona-Lasinio model. [1]. In the context of high-energy physics they were discovered by Goldstone [2].

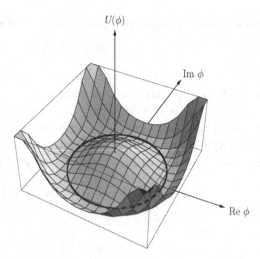

$U(\phi)$

Im ϕ

Re ϕ

Fig. 1.5 The potential energy (1.16). The black circle marks the minimum of the potential energy, the vacuum manifold.

To conclude this section we will consider another example, with a slightly more sophisticated pattern of symmetry breaking, which we will need in our study of monopoles (Section 4.1).

The model for analysis is a triplet of real fields ϕ_a ($a = 1, 2, 3$) with the Lagrangian

$$\mathcal{L} = \tfrac{1}{2}(\partial_\mu \vec{\phi})^2 - \left[-\tfrac{1}{2}\mu^2 \vec{\phi}^2 + \tfrac{1}{48}g^2 4(\vec{\phi}^2)^2\right], \qquad (1.20)$$

where $\vec{\phi} = \{\phi_1, \phi_2, \phi_3\}$ and $\mu^2 > 0$. It is obvious that this Lagrangian is O(3)-symmetric while the vacuum state is not. The minimum of the potential energy is achieved at $\vec{\phi}^2 = \mu^2/g^2$; thus $|\phi_{\text{vac}}| = \mu/g \equiv v$. The angular orientation of the vector of the vacuum field in the O(3) space ("isospace") is arbitrary. The vacuum manifold is a two-dimensional sphere of radius v. All points on this manifold are physically equivalent.

Suppose that we choose $\vec{\phi}_{\text{vac}} = \{0, 0, v\}$, i.e., we align the vacuum value of the field along the third axis in isospace. The original symmetry is broken down to U(1). The fact that there is a residual U(1) is quite transparent. Indeed, rotations in the isospace around the third axis do not change ϕ_{vac}. Thus, in this problem we are dealing with the following pattern of symmetry breaking:

$$O(3) \to U(1). \qquad (1.21)$$

Two out of three generators are broken; hence, we expect two Goldstone bosons. Let us see whether this expectation comes true.

Parametrizing the field $\vec{\phi}$ near this vacuum as $\vec{\phi}(x) = \{\varphi(x), \chi(x), v + \eta(x)\}$ and calculating $U(\varphi, \chi, \eta)$, it is easy to see that only one field, η, has a mass term, $m_\eta = \sqrt{2}\mu$, while the fields φ and χ have only cubic and quartic interactions and remain massless. The fields φ and χ present two Goldstone bosons in the problem at hand. The interaction depends on the combination $\varphi^2 + \chi^2$ and is invariant under the U(1) rotations

$$\varphi \to \varphi \cos \alpha + \chi \sin \alpha, \qquad \chi \to -\varphi \sin \alpha + \chi \cos \alpha, \qquad (1.22)$$

in full agreement with the existence of an unbroken U(1) symmetry.

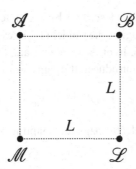

Fig. 1.6   Four towns on the map form a perfect square. Dotted lines are for orientation.

> *The signature of discrete symmetry breaking is the occurrence of domain walls (kinks).*

Summarizing, if continuous (global) symmetries are spontaneously broken then massless Goldstone bosons emerge, one such boson for each broken generator. The occurrence of Goldstones (gapless excitations) is the signature of spontaneous continuous symmetry breaking. A reservation must be added immediately: Goldstone bosons do not appear in $D = 1 + 1$ theories unless they are sterile. We will discuss this subtle aspect in more detail later (see Section 6.5.2).

The interactions of Goldstone bosons respect the unbroken symmetries of the theory. These symmetries are realized linearly; the broken part of the original symmetry is realized nonlinearly.

Exercise

1.1.1   Mayors of four towns located as shown in Fig. 1.6 decided to build a railroad connecting all four towns $\mathcal{A}$, $\mathcal{B}$, $\mathcal{L}$, and $\mathcal{M}$ with each other (possibly with connections). They also decided that its length must be minimal. The towns $\mathcal{A}$, $\mathcal{B}$, $\mathcal{L}$, $\mathcal{M}$ form a square on the map exhibiting a Z_4 symmetry – the symmetry with respect to rotations by $\pi/2$. What is the symmetry of the map with the minimal-length railroad?

1.2 Spontaneous Breaking of Gauge Symmetries

1.2.1 Abelian Theories

The simplest example of the spontaneous breaking of gauge symmetries is provided by the quantum electrodynamics (QED)[3] of a charged scalar field whose

---

[3] Strictly speaking, QED *per se* is under-defined at short distances, where the effective coupling grows and hits the Landau pole. Thus to make it consistent an ultraviolet completion is needed at short distances. For instance, one can embed QED into an asymptotically free theory. The Georgi–Glashow model, Section 4.1.1, gives an example of such an embedding. It is important to understand that different ultraviolet completions do not necessarily lead to the same physics in the infrared. For instance, Polyakov's confinement in three-dimensional QED illustrates this statement in a clear-cut manner; see Section 9.7.

self-interaction is described by the potential depicted in Fig. 1.5. This theory is obtained by gauging the model (1.13) with global U(1) symmetry that was studied in Section 1.1.6. In other words we add the photon field, whose interaction with the matter fields is introduced through a covariant derivative, giving

$$S = \int d^D x \left[-\tfrac{1}{4e^2} F_{\mu\nu} F^{\mu\nu} + \left(\mathcal{D}_\mu \phi \right)^* \left(\mathcal{D}^\mu \phi \right) - U(\phi) \right], \qquad (1.23)$$

where e is the electromagnetic coupling and the covariant derivative $\mathcal{D}$ is defined as

$$\mathcal{D}_\mu = \partial_\mu - i A_\mu. \qquad (1.24)$$

The kinetic term of the photon field is standard. Now the Lagrangian is invariant under the *local* U(1) transformation

$$\phi(x) \to e^{i\alpha(x)} \phi(x), \quad A_\mu(x) \to A_\mu(x) + \partial_\mu \alpha(x). \qquad (1.25)$$

If the potential has the form (1.16), the field ϕ develops an expectation value and the gauge U(1) symmetry is spontaneously broken.

I hasten to add that the terminology "spontaneously broken gauge symmetry," although widely accepted, is, in fact, rather sloppy and confusing.[4] What exactly does one mean by saying that the gauge symmetry is spontaneously broken? The gauge symmetry, in a sense, is not a symmetry at all. Rather, it is a description of x physical degrees of freedom in terms of $x + y$ variables, where y variables are redundant and the corresponding degrees of freedom are physically unobservable. Only those points in the field space that are given by gauge-nonequivalent configurations are to be treated as distinct.

If we decouple the photon by setting $e = 0$, the action (1.23) is invariant under global phase rotations. The condensation of the scalar field breaks this invariance, but the invariance of the "family of models" is not lost. Under this phase transformation one vacuum goes into another that is physically equivalent. Say, if we start from the vacuum characterized by a real value of the order parameter ϕ, then in the "rotated" vacuum the order parameter is complex. The spontaneous breaking of any global symmetry leads to a set of degenerate (and physically equivalent) vacua.

Switching on the electromagnetic interaction (i.e., setting $e \neq 0$), we lose the vacuum degeneracy – the degeneracy associated with the spontaneous breaking of the global symmetry. Indeed, all states related by phase rotation are gauge equivalent. They are represented by a single state in the Hilbert space of the theory. In other words, one can always *choose* the vacuum value of ϕ to be real. This is nothing other than the (unitary) gauge condition. Thus, the spontaneous breaking of the gauge symmetry does not imply, generally speaking, the existence of a degenerate set of vacua as is the case for the global symmetries. Then what does it mean, after all?

Unitary gauge, first appearance of the Higgs field

By inspecting the action (1.23) it is not difficult to see that if ϕ has a nonvanishing (and constant) value in the vacuum, the spectrum of the theory does not contain any massless vector particles. The photon acquires three polarizations and a mass $m_V = \sqrt{2} e v$, where v is a real parameter, $v = \langle \phi \rangle$. The remaining degree of freedom is a real (rather than complex) scalar field, the Higgs field, with mass $m_H = \sqrt{2} g v$.

---

[4] At present theorists tend to say that the theory is "Higgsed" when there is a spontaneously broken gauge symmetry.

This is seen from the decomposition (1.18), where χ must be set to zero because the field ϕ is real in the unitary gauge. The theoretical discovery of the Higgs phenomenon goes back to [3–5]. This regime is referred to as the *Higgs phase*. One massless scalar field is eaten up by the photon field in the process of the transition to the Higgs phase. In the Higgs phase the electric charge is screened by the vacuum condensates. Probe (static) electric charges will see the Coulomb potential $\sim 1/R$ at distances less than m_V^{-1} and the Yukawa potential $\sim \exp(-m_V R)/R$ at distances larger than m_V^{-1}. Moreover, the gauge coupling runs, according to the standard Landau formula, only at distances shorter than m_V^{-1} and becomes frozen at m_V^{-1}.

1.2.2 Phases of the Abelian Theory

Quantum electrodynamics was historically the first gauge theory studied in detail. This model is simple, with no mysteries. Nevertheless, it is nontrivial exhibiting three different types of behavior at large distances.

We have just identified the Higgs regime, in which all excitations are massive. At large distances there is no long-range interaction between charges.

Now we replace the scalar charged matter fields by spinor fields (electrons) with mass m. The same probe charges will experience a totally different interaction at large distances, the Coulomb interaction, with potential proportional to

$$V(R) \sim \frac{e^2(R)}{R},$$

where R is the distance between the probe charges. Classically e^2 is a constant. Quantum corrections due to virtual electron loops make e^2 run.

Its behavior is determined by the well-known Landau formula, which tells us that at large distances e^2 decreases logarithmically:

$$e^2(R) \sim \frac{1}{\ln R}. \tag{1.26}$$

If m is finite, the logarithmic fall-off is frozen at $R \sim m^{-1}$. The corresponding limiting value of e^2 is

$$e_*^2 = e^2(R = m^{-1}).$$

The potential between two distant static charges is

$$V(R) \sim \frac{e_*^2}{R}, \qquad R \to \infty. \tag{1.27}$$

The dynamical regime having this type of long-distance behavior is referred to as the *Coulomb phase*. In the case at hand we are dealing with the Abelian Coulomb phase.[5]

Now let us ask ourselves what happens if the electron mass vanishes. Unlike the massive case, where the running coupling constant is frozen at $R = m^{-1}$, in the theory with $m = 0$ the logarithmic fall-off (1.26) continues indefinitely: at asymptotically large R the effective coupling becomes arbitrarily small.

---

[5] Behavior like (1.27) can occur in non-Abelian gauge theories as well, as we will see later. Such non-Abelian gauge theories, with long-range potential (1.27), are said to be in the non-Abelian Coulomb phase.

Thus, in the asymptotic limit of massless spinor QED we have a free photon and a massless electron whose charge is completely screened. The theory has no localized asymptotic states and no mass shell, nor S matrix in the usual sense of this word. Still, it is well defined in, say, a finite volume.

This phase of the theory is referred to as an *infrared-free phase*. Sometimes it is also called *the Landau zero-charge phase*.

Summarizing, even in the simplest Abelian example we encounter three different phases, or dynamical regimes: the Coulomb phase, the Higgs phase, and the free (Landau) phase, depending on the details of the matter sector. All these regimes are attainable in non-Abelian models too.

The non-Abelian gauge theories are richer since they admit more dynamical regimes, to be discussed in Section 1.3.

1.2.3 Higgs Mechanism in Non-Abelian Theories

The Higgs mechanism in QED, considered in Section 1.2.1, extends straightforwardly to non-Abelian theories. The only difference is that U(1) is replaced by a non-Abelian group, which is then gauged. The essence of the construction remains the same.

Instead of the single complex field ϕ of QED (see Eq. (1.23)), we start with a multiplet of scalar fields ϕ_i belonging to a representation R of a non-Abelian group G. The representation R may be reducible; for simplicity, however, we will assume R to be irreducible for the time being. The generators of the group G in the representation R will be denoted T^a, where

$$\left[T^a, T^b\right] = i f^{abc} T^c, \qquad \mathrm{Tr}\left(T^a T^b\right) = T(R)\delta^{ab}, \qquad (1.28)$$

In the mathematical literature $T(R)$ is known as the Dynkin index.

and f^{abc} are the structure constants of the group G. In this book we will deal mostly with the unitary groups SU(N). Occasionally, the orthogonal groups O(N) will be involved.

Assume the self-interaction of the fields ϕ to be such that the lowest-energy state – the vacuum – breaks

$$G \to H, \qquad (1.29)$$

where H is a subgroup of G. A particular case is $H = 1$, corresponding to the complete breaking of G. In accordance with the general Goldstone theorem, the spontaneous breaking (1.29) entails the occurrence of dim G – dim H Goldstone bosons (here dim G is the dimension of the group, i.e., the number of its generators).

See Section 6.5.1.

Now, to gauge the theory, instead of the conventional derivative ∂_μ we introduce a covariant derivative

$$\mathcal{D}_\mu = \partial_\mu - iA_\mu, \qquad (1.30)$$

where

$$A_\mu \equiv A_\mu^a T^a \qquad (1.31)$$

and A_μ^a are the gauge fields. If $\phi(x)$ transforms as $\phi \to U(x)\phi$ for any $U(x) \in G$ then $\mathcal{D}_\mu\phi$ must transform in the same way:

$$\mathcal{D}_\mu\phi(x) \to U(x)\left(\mathcal{D}_\mu\phi(x)\right). \tag{1.32}$$

This requirement defines the transformation law of the gauge fields:

$$A_\mu \to U A_\mu U^{-1} + i U \partial_\mu U^{-1}. \tag{1.33}$$

The gauge field strength tensor (to be denoted by $G_{\mu\nu}$ rather than $F_{\mu\nu}$, to distinguish the non-Abelian and Abelian cases) is defined as[6]

$$G_{\mu\nu} \equiv i[\mathcal{D}_\mu, \mathcal{D}_\nu] = \partial_\mu A_\nu - \partial_\nu A_\mu - i[A_\mu, A_\nu]$$
$$= \left(\partial_\mu A_\nu^a - \partial_\nu A_\mu^a + f^{abc} A_\mu^b A_\nu^c\right) T^a \equiv G_{\mu\nu}^a T^a. \tag{1.34}$$

The kinetic term of the gauge field is

$$\mathcal{L}_{\text{YM}} = -\frac{1}{4g^2} G_{\mu\nu}^a G^{\mu\nu,a}, \tag{1.35}$$

while the scalar fields are described by the Lagrangian

$$\mathcal{L}_{\text{matter}} = \mathcal{D}_\mu\phi^*\left(\mathcal{D}^\mu\phi\right) - U(\phi) \tag{1.36}$$

where summation over the multiplet-R index is implied. In what follows we will use the notations $\mathcal{D}_\mu\phi^*$ and $\mathcal{D}_\mu\bar\phi$ indiscriminately.

Reminder: The canonical form of the Yangs–Mills Lagrangian (corresponding to canonically normalized kinetic term) is

$$\mathcal{L}_{\text{YM}} = -\frac{1}{4}G_{\mu\nu}^a G^{\mu\nu,a}, \quad G_{\mu\nu}^a = \partial_\mu A_\nu^a - \partial_\nu A_\mu^a + g f^{abc} T^2 A_\mu^b A_\nu^c, \quad \mathcal{D}_\mu = \partial_\mu - ig A_\mu.$$

In this form the coupling constant g appears in the interaction vertices. This convention is preferred in perturbation theory. In nonperturbative studies a more convenient convention is a (noncanonical) form obtained from the canonical one by the substitution $A_\mu \longrightarrow \frac{1}{g}A_\mu$. Then the coupling constant g disappears from all vertices, e.g.,

$$G_{\mu\nu}^a \longrightarrow \frac{1}{g}\left(\partial_\mu A_\nu^a - \partial_\nu A_\mu^a + f^{abc} A_\mu^b A_\nu^c\right).$$

The factor g^2 appears in the numerator of the gluon propagators. Equations (1.30), (1.34), and (1.35) refer to the noncanonical normalization. In this book I use both.

Now the dim G – dim H Goldstone bosons that existed before gauging are paired up with the gauge bosons to produce dim G – dim H three-component massive vector particles. In the unitary gauge one imposes dim G – dim H gauge conditions. If instead of $\langle \text{vac}|\phi|\text{vac}\rangle$ we use the shorthand ϕ_{vac} then $T^a\phi_{\text{vac}} = 0$, provided that $T^a \in H$. The corresponding dim H gauge bosons stay massless. The masses of the remaining dim G – dim H gauge bosons are obtained from the matrix

Mass formula for gauge bosons

$$m_{ab}^2 = 2g^2\left(\phi_{\text{vac}}^* T^a T^b \phi_{\text{vac}}\right), \qquad T^{a,b} \in G/H. \tag{1.37}$$

---

[6] It is obvious that the transformation law of $G_{\mu\nu}$ under the gauge transformation is

$$G_{\mu\nu} \to U G_{\mu\nu} U^{-1}.$$

Referring to [6] for a more detailed discussion of the generalities, in the remainder of this section we will focus on two examples of particular interest.

1.2.3.1 From SU(2)$_{\text{local}}$ to SU(2)$_{\text{global}}$

The model discussed in this subsection is the Glashow–Weinberg–Salam (GWS) model of electroweak interactions – *part* of the Standard Model (SM) of particle physics.[7]

The gauge group is SU(2). The structure constants are $f^{abc} = \varepsilon^{abc}$, where ε^{abc} is the Levi–Civita tensor $(a, b, c = 1, 2, 3)$. The matter sector consists of an SU(2) doublet of complex scalar fields ϕ^i, where $i = 1, 2$. In other words, the ϕ^i are the scalar quarks in the fundamental representation. The covariant derivative acts on ϕ^i as follows:

$$\mathcal{D}_\mu \phi(x) \equiv \left(\partial_\mu - i A^a_\mu T^a \right) \phi, \qquad T^a = \tfrac{1}{2} \tau^a, \tag{1.38}$$

where the τ^a are the Pauli matrices. We will choose the ϕ self-interaction potential to be in the form

$$U = \lambda \left(\bar{\phi} \phi - v^2 \right)^2. \tag{1.39}$$

Quite often it is said that this theory has just SU(2) gauge symmetry and nothing else. This is wrong. In fact, its symmetry is

$$\text{SU(2)}_{\text{gauge}} \times \text{SU(2)}_{\text{global}}. \tag{1.40}$$

One can prove this in a number of ways. Probably, the quickest proof is as follows. Let us introduce the 2×2 matrix

$$X = \begin{pmatrix} \phi^1 & -(\phi^2)^* \\ \phi^2 & (\phi^1)^* \end{pmatrix}. \tag{1.41}$$

The Lagrangian of the model rewritten in terms of X takes the form [7]

$$\mathcal{L} = -\frac{1}{4g^2} G^a_{\mu\nu} G^{\mu\nu, a} + \frac{1}{2} \text{Tr} \left(\mathcal{D}_\mu X \right)^\dagger \left(\mathcal{D}_\mu X \right) - \lambda \left(\frac{1}{2} \text{Tr} \, X^\dagger X - v^2 \right)^2. \tag{1.42}$$

Note that the generators T^a in the covariant derivative $\mathcal{D}$ act on the matrix X through matrix multiplication from the left. This Lagrangian is obviously invariant under the transformation

$$X(x) \to U(x) X(x) M^{-1}, \tag{1.43}$$

supplemented by (1.33), where M is an arbitrary x-independent matrix from SU(2)$_{\text{global}}$. The symmetry (1.40) is apparent. In the vacuum, $\frac{1}{2} \text{Tr} \, X^\dagger X = v^2$. Using gauge freedom (three gauge parameters), one can always choose the unitary gauge in which the vacuum value of X is

$$X_{\text{vac}} = v \begin{pmatrix} 1 & 0 \\ 0 & 1 \end{pmatrix}. \tag{1.44}$$

---

[7] The latter also includes QCD.

This vacuum expectation value breaks the $SU(2)_{gauge}$ and $SU(2)_{global}$ symmetries, but the diagonal global $SU(2)$ symmetry corresponding to $U = M$ remains unbroken. Thus, the symmetry-breaking pattern is

$$SU(2)_{gauge} \times SU(2)_{global} \to SU(2)_{diag}. \qquad (1.45)$$

Three would-be Goldstone bosons are eaten up by the gauge bosons, transforming them into massive W bosons belonging to the triplet (adjoint) representation of the unbroken $SU(2)_{diag}$ symmetry. There are no massless particles in this model. The physically observable excitations are three W bosons with mass $m_W = gv/\sqrt{2}$ and one Higgs particle (a singlet with respect to $SU(2)_{global}$) with mass $2\sqrt{\lambda}v$.

   This model will be discussed in more detail in Section 5.4.12 in the context of instanton calculus.

1.2.3.2 From $SU(2)_{local}$ to $U(1)_{local}$

Below, I will outline the Georgi–Glashow model [8]. If necessary, it can be easily generalized to $SU(N)$, with the gauge-symmetry-breaking pattern

$$SU(N) \to U(1)^{N-1}.$$

The Lagrangian of the model is

$$\mathcal{L} = -\frac{1}{4g^2}G^a_{\mu\nu}G^{\mu\nu,a} + \frac{1}{2}(\mathcal{D}_\mu \phi^a)(\mathcal{D}^\mu \phi^a) - \lambda(\phi^a\phi^a - v^2)^2, \qquad (1.46)$$

where ϕ^a is the triplet of real scalar fields in the adjoint representation; the covariant derivative in the adjoint acts as

$$\mathcal{D}_\mu \phi^a = \partial_\mu \phi^a + \varepsilon^{abc}A^b_\mu \phi^c. \qquad (1.47)$$

One can always choose a gauge (the unitary gauge) in which

| With matter |
| fields in the |
| adjoint rep- |
| resentation, |
| one can |
| say that |
| $O(3) \to O(2)$. |

$$\phi^1 = \phi^2 \equiv 0, \qquad \phi^3 \neq 0. \qquad (1.48)$$

The vacuum value of the field ϕ is

$$\phi^3_{vac} = v, \qquad (1.49)$$

which implies that the $SU(2)_{gauge}$ symmetry breaks down to $U(1)_{gauge}$. Since T^3 acts on ϕ_{vac} trivially, A^3_μ remains massless (a "photon"), while the two other gauge bosons become W bosons, acquiring mass $m_W = gv$, where

$$W^\pm = \frac{A^1_\mu \pm iA^2_\mu}{\sqrt{2}g}.$$

Besides the two W bosons and the photon there is another physical particle, the Higgs boson, with mass $m_H = 2\sqrt{2\lambda}v$. At distances much larger than m_W^{-1} the W bosons decouple and the theory reduces to QED.

   This model will be discussed in Chapter 4.

Exercise

1.2.1   Assume we have Yang–Mills theory with the gauge group SU(3) and the Higgs sector consisting of one real scalar field in the *adjoint* representation of SU(3). The latter develops a generic vacuum expectation value (large compared to Λ, the dynamical scale of the theory). Determine the most general pattern of Higgsing and the masses of all gauge bosons.

1.3 Phases of Yang–Mills Theories

The phase structure of non-Abelian gauge theories is richer than that of QED. In addition to the three regimes described in Section 1.2.2, which were known already in the 1960s, Yang–Mills theories can exhibit confining and conformal phases, phases with or without chiral symmetry breaking, and so on.

1.3.1 Confinement

We will start by discussing the confining phase. Consider pure Yang–Mills theory (1.35), where the gauge group is assumed to be SU(N) for arbitrary N. At short distances the running coupling constant falls off logarithmically [9],

$$\frac{\alpha(p)}{2\pi} = \frac{1}{\beta_0 \ln(p/\Lambda)}, \qquad \beta_0 = \frac{11N}{3}, \tag{1.50}$$

Asymptotic freedom

the interaction switches off, and one can detect – albeit indirectly – the gluon degrees of freedom as described by (1.35). The parameter Λ is the so-called *dynamical scale*.

At large distances we enter a strong coupling regime. The physically observed spectrum is drastically different from what we see in the Lagrangian. In the case at hand an experimentalist, if he or she could exist in the world of pure Yang–Mills theories, would observe a spectrum of glueballs that are, generally speaking, nondegenerate in mass. One can visualize the glueballs as a closed string (or, better, a tube), in a highly quantum state, i.e., a string-like field configuration which wildly oscillates, pulsates, and vibrates; see Fig. 1.7. If we add nondynamical (i.e., very heavy) quarks into the theory and set the quark and antiquark at a large distance from each other, such a string will stretch between them (as shown in the figure on the

Fig. 1.7   A quantum closed string as a glueball.

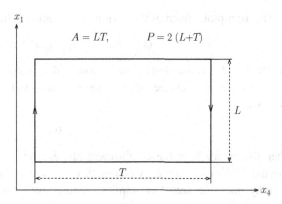

$$A = LT, \qquad P = 2\,(L+T)$$

Fig. 1.8 A Wilson contour C, with area A and perimeter P. The probe quark is dragged along this contour.

opening page of this chapter), connecting the pair of probe quarks[8] in an inseparable configuration. What is depicted in that figure is a highly quantum (presumably, nonperturbative) open string configuration with quarks attached at the ends. If we try to pull the quarks apart we just make the string longer, while the energy of the configuration grows linearly with separation.

This phase of the theory, whose existence was conjectured in 1973 [9], is referred to as color confinement. Although there is no analytic proof of color confinement that could be considered exhaustive, there is ample evidence that this regime does, indeed, occur. First, a version of color confinement was observed in certain supersymmetric Yang–Mills theories [10]. Second, the formation of tube-like configurations connecting heavy probe quarks was demonstrated numerically, in lattice simulations. I will not dwell on the dynamics leading to color confinement (this topic will be postponed until we have learned more of the underlying physics; see Chapters 3 and 9). It is worth noting, however, that there are distinct versions of confinement regimes, such as oblique confinement [11], Abelian and non-Abelian confinement, both of which are found in Yang–Mills theories, etc. Some examples will be considered in Chapter 9. The impatient and curious reader is directed to the original literature or the review paper [12].

Kenneth Wilson was the first to suggest [13] a very convenient criterion indicating whether a given gauge theory is in the confinement phase. Consider a gauge theory in Euclidean space–time. Introduce a closed contour, as shown in Fig. 1.8. Assume that $T \gg L \gg \Lambda^{-1}$, i.e., the contour is large.[9] Consider the *Wilson operator*

$$W(C) = \frac{1}{\dim_R} \mathrm{Tr}\, P \exp\left[i \oint_C A_\mu^a(x) T_R^a \, dx^\mu \right], \qquad (1.51)$$

where the subscript R indicates the representation of the gauge group to which the probe quark belongs (usually the fundamental representation).

---

[8] Probe quarks Q are those for which pair production in the vacuum can be ignored. This can be achieved by endowing them with a mass $m_Q \to \infty$. In contrast, dynamical quarks q are either massless or light, $m_q \ll \Lambda$, where Λ is the same scale parameter as in (1.50).

[9] Generally speaking the contour does not have to be rectangular, but for the rectangular contour the result is simpler to interpret.

The asymptotic form of the vacuum expectation value of $W(C)$ is

$$\langle W(C) \rangle_{\text{vac}} \propto \exp[-(\mu P + \sigma A)], \tag{1.52}$$

where $A = LT$ is the area of the contour and $P = 2(L + T)$ is the perimeter; μ and σ are numerical coefficients of dimension mass and mass squared, respectively. If we have

$$\sigma \neq 0 \tag{1.53}$$

then the theory is in the confinement phase, while at $\sigma = 0$ the theory does not confine.[10] We refer to these cases as the area law and the perimeter law, respectively.

Why does the area law implies confinement? The reason is that, on general grounds,

$$\langle W(C) \rangle_{\text{vac}} \propto \exp[-V(L)T] \tag{1.54}$$

if the contour is chosen as in Fig. 1.8. Hence, the area law means that the potential $V(L)$ between distant probe quarks Q and $\bar{Q}$ is $V(L) = \sigma L$ at $L \gg \Lambda^{-1}$. The coefficient σ is the string tension (in many publications it is denoted by T rather than σ).

1.3.2 Adding Massless Quarks

From pure Yang–Mills theory we pass to theories with matter. Considering N_f massless quarks in the fundamental representation is the first step. Each quark is described by a Dirac spinor and the overall number of Dirac spinors is N_f. At $N = 3$ and $N_f = 3$ we obtain quantum chromodynamics (QCD), the accepted theory of strong interactions in nature.

The most obvious impact of adding massless quarks is the change in β_0, the first coefficient in the Gell-Mann–Low function. Instead of the expression of β_0 in (1.50) we now have

$$\beta_0 = \tfrac{11}{3}N - \tfrac{2}{3}N_f. \tag{1.55}$$

If $N_f > \tfrac{11}{2}N$ then the coefficient changes sign, we lose asymptotic freedom, and the Landau regime sets in. The theory becomes infrared-free, much like QED with massless electrons. From a dynamics standpoint this is a rather uninteresting regime.

Let us assume that $N_f \leq \tfrac{11}{2}N$. Now we will address the question: what happens if N_f is only slightly less than the critical value $\tfrac{11}{2}N$? To answer this we need to know the two-loop coefficient in the β function.

1.3.3 Conformal Phase

See
Sections 1.4
and 8.4

The response of Yang–Mills theories to scale and conformal transformations is determined by the trace of the energy–momentum tensor

$$T^\mu_\mu \propto \beta(\alpha)G^a_{\mu\nu}G^{\mu\nu,a}, \tag{1.56}$$

---

[10] If $\sigma \neq 0$ the perimeter term is subleading. The parameter μ renormalizes the probe quark mass.

where $\beta(\alpha)$ is the Gell-Mann–Low function (also known as the β function). In SU(N) Yang–Mills theory with N_f quarks it has the form

$$\beta(\alpha) = \frac{\partial \alpha(\mu)}{\partial \ln \mu} = -\beta_0 \frac{\alpha^2}{2\pi} - \beta_1 \frac{\alpha^3}{4\pi^2} - \cdots , \qquad \alpha = \frac{g^2}{4\pi}, \tag{1.57}$$

where β_0 is given in (1.55) while

$$\beta_1 = \frac{17}{3} N^2 - \frac{N_f}{6N} \left(13 N^2 - 3\right). \tag{1.58}$$

At small α the term $\sim \beta_0$ in (1.57) dominates and so the β function is negative, implying asymptotic freedom at short distances. What is the large-distance behavior of the running coupling constant $\alpha(\mu)$?

Assume that

$$N_f = \tfrac{11}{2} N - \nu, \qquad 0 < \nu \ll \tfrac{11}{2} N. \tag{1.59}$$

Then the first coefficient, β_0, is anomalously small,

$$\beta_0 = \frac{2}{3}\nu. \tag{1.60}$$

> The ratio β_1/β_0 is negative.

At the same time the second coefficient is not suppressed; it is of a normal order of magnitude,

$$\beta_1 = -\tfrac{25}{4} N^2 + \tfrac{11}{4} + \tfrac{1}{6}\nu N^{-1} \left(13 N^2 - 3\right), \tag{1.61}$$

and *negative*!

As the scale μ decreases (at larger distances), the running gauge coupling constant grows and the second term in (1.57) eventually becomes important. Generally speaking, the second term takes over the first one at $N\alpha/\pi \sim 1$ (the strong coupling regime), when *all* terms in the α expansion of the β function are equally important and one cannot limit oneself to the first two terms. However, if N_f is only slightly less than $\tfrac{11}{2} N$ then the β function develops a zero at a value of α which is parametrically small,[11] namely, we have

> Position of IR fixed point

$$\frac{N\alpha_*}{2\pi} = \frac{N\beta_0}{-\beta_1} = \frac{8}{75} \frac{\nu}{N f(N,\nu)}, \tag{1.62}$$

where

$$f(N,\nu) = 1 - \frac{11}{25 N^2} - \frac{2\nu}{75 N^3} \left(13 N^2 - 3\right) \sim 1. \tag{1.63}$$

In other words, the second term catches up with the first one prematurely when $N\alpha/\pi \ll 1$. Hence we are at weak coupling and higher-order terms are inessential. The facts of the existence of this zero and its position are reliably established.

As an example, let me indicate that if $N = 3$ and $N_f = 15$ then

$$\frac{\alpha_*}{2\pi} = \frac{1}{44}. \tag{1.64}$$

The β function is shown in Fig. 1.9.

---

[11] By "parametrically" I mean that if, for instance, N is large while ν does not scale with N then $f(N,\nu) \to 1$, and $N\alpha_*/2\pi \to (8/75)(\nu/N)$.

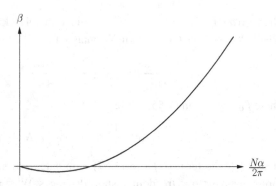

The β function at N_f slightly less than $\frac{11}{2}N$. The horizontal axis presents $N\alpha/2\pi$. The zero of the beta function is at $\frac{8}{75}v/N \ll 1$.

The zero of the β function depicted in Fig. 1.9 is nothing other than the infrared fixed point of the theory. If we start from the value of α lying between 0 and α_* and let α run then it will hit α_* in the infrared (remember, in the ultraviolet $\alpha(\mu)$ tends to 0).

Hence at large distances $\beta(\alpha) = \beta(\alpha^*) = 0$, implying that the trace of the energy–momentum tensor of the theory vanishes and so the theory is in the *conformal* phase. There are no localized particle-like states in the spectrum; rather, we are dealing with massless unconfined *interacting* quarks and gluons. All correlation functions at large distances exhibit a power-like behavior.[12] As long as α^* is small, the interactions of the massless quarks and gluons in the theory are weak at all distances, short and large, and thus amenable to the standard perturbative treatment. In particular, the potential between two probe, static, quarks at a large separation R will behave approximately as α^*/R, reminding us of conventional QED with massive electrons.

Since we are absolutely certain that, slightly below $N_f = \frac{11}{2}N$, we are in the conformal phase, on increasing v (i.e., decreasing N_f) we cannot leave this phase straight away. There should exist a critical value N_f^* of the number of flavors above which the theory is conformal in the infrared. The interval

| Conformal |
| window |

$$N_f^* \leq N_f \leq \tfrac{11}{2}N \qquad (1.65)$$

is referred to as a *conformal window*.[13] The exact value of N_f^* is unknown. From experiment we know that $N_f^* > 3$ at $N = 3$. On general grounds one can argue that $N_f^* \sim cN$, where c is a numerical constant of the order of unity. Of course, near the left-hand (lower) edge of the conformal window one should expect $N\alpha_*/2\pi \sim 1$ so that the theory, albeit conformal in the infrared, is strongly coupled. In particular, in this case there is no reason for the anomalous dimensions to be small.

Summarizing, if N_f lies in the interval (1.65) then the theory is in the conformal phase. For N_f close to the right-hand (upper) edge of the conformal window the theory is weakly coupled and all anomalous dimensions are calculable. Belavin and

---

[12] We will see in Chapter 8, Section 8.4, that the trace of the energy–momentum tensor in Yang–Mills theories with massless quarks is proportional to $\beta(\alpha)G_{\mu\nu}^a G^{\mu\nu,a}$. Basic data on conformal symmetry are collected in Appendix section 1.4. A more detailed discussion of the implications of conformal invariance in four and two dimensions can be found, e.g., in [14].

[13] This terminology was suggested in [12], and it took root.

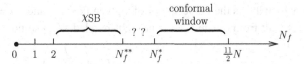

Fig. 1.10 Dynamical regimes change with the number of massless quarks N_f.

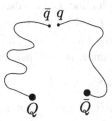

Fig. 1.11 The string between two probe quarks Q and $\bar{Q}$ can break through $\bar{q}q$ pair creation in Yang–Mills theories with dynamical quarks.

Migdal considered this model in the early 1970s [15]. Somewhat later, it was studied thoroughly by Banks and Zaks [16].

1.3.4 Chiral Symmetry Breaking

Next, in our journey along the N_f axis (Fig. 1.10) let us descend to $N_f = 1, 2, 3, \ldots$ Strictly speaking, dynamical quarks (in the fundamental representation) negate confinement understood in the sense of Wilson's criterion – the area law for the Wilson loop disappears. Indeed the string forming between the probe quarks can break, through $\bar{q}q$ pair creation, when the energy stored in the string becomes sufficient to produce such a pair (Fig. 1.11). As a result, sufficiently large Wilson loops obey the perimeter law rather than the area law. However, intuitively it is clear that, in essence this is the same confinement mechanism, although in the case at hand it is natural to call it *quark confinement*. The dynamical quarks are identifiable at short distances in a clear-cut manner and yet they never appear as asymptotic states. Experimentalists detect only color-singlet mesons of the type $\bar{q}q$ or baryons of the type qqq.

Theoretically, if necessary, one can suppress $\bar{q}q$ pair creation by sending N to ∞; see Chapter 9.

At $N_f \geq 2$ a new and interesting phenomenon shows up. The global symmetry of Yang–Mills theories with more than one massless quark flavor is

$$\mathrm{SU}(N_f)_L \times \mathrm{SU}(N_f)_R \times \mathrm{U}(1)_V. \tag{1.66}$$

The vectorial U(1) symmetry is simply the baryon number, while the axial U(1) is anomalous (see Chapter 8) and hence is not shown in (1.66). The origin of the chiral $\mathrm{SU}(N_f)_L \times \mathrm{SU}(N_f)_R$ symmetry is as follows. The quark part of the Lagrangian has the form

Massless quark sector

$$\mathcal{L}_{\mathrm{quark}} = \sum_f \bar{\Psi}_f i\,\slashed{D}\Psi^f, \tag{1.67}$$

where Ψ^f is the Dirac spinor of a given flavor f and $\mathcal{D} = \gamma^\mu \mathcal{D}_\mu$. Each Dirac spinor is built from one left- and one right-handed Weyl spinor,

$$\Psi^f_i = \begin{pmatrix} \xi^f_{\alpha,i} \\ \bar{\eta}^{\dot{\alpha},f}_i \end{pmatrix}, \tag{1.68}$$

Dirac spinor from two Weyl spinors

where i is the color index (i.e., the index of the fundamental representation of $SU(N)_{\text{color}}$) while f is the flavor index, $f = 1, 2, \ldots, N_f$. The left- and right-handed Weyl spinors in the kinetic term (1.67) totally decouple from each other. Hence, $\mathcal{L}_{\text{quark}}$ is invariant under the independent global rotations

$$\xi \to U\xi \quad \text{and} \quad \bar{\eta} \to U'\bar{\eta}, \qquad U \in SU(N_f)_L, \quad U' \in SU(N_f)_R. \tag{1.69}$$

Experimentally it is known that the chiral $SU(N_f)_L \times SU(N_f)_R$ symmetry is spontaneously broken at $N = 3$ and $N_f = 2, 3$, leaving the diagonal $SU(N_f)_V$ subgroup unbroken. $N_f^2 - 1$ massless Goldstone bosons – the pions – emerge as a result of this spontaneous breaking. This phenomenon bears the name *chiral symmetry breaking* (χSB). In Chapter 8 we will outline theoretical arguments demonstrating χSB in the limit $N \to \infty$ with N_f fixed.

There are qualitative arguments showing that in four-dimensional Yang–Mills theory χSB may be a consequence of quark confinement plus some general features of the quark–gluon interaction. In particular, a well-known picture is that of Casher [17] "explaining"[14] why in Yang–Mills theories with massless quarks (no scalar fields!) color confinement entails a Goldstone-mode realization of the global axial symmetry of the Lagrangian. A brief outline is as follows. If we deal with massless quarks, the left-handed quarks are decoupled from the right-handed quarks in the QCD Lagrangian. If spontaneous breaking of the chiral symmetry does not take place, this decoupling becomes an exact property of the theory: the quark chirality (helicity) is exactly conserved. Assume that we produce an energetic quark–antiquark pair in, say, e^+e^- annihilation. Let us place the origin at the annihilation point. If the quarks' energy is high then they can be treated quasiclassically. Let us say that in the given event the quark produced is right-handed and moves off in the positive z direction; the antiquark will then move off in the negative z direction. If the quark energy is high ($E \gg \Lambda$, where Λ is the QCD scale parameter) the distance L that the quark travels before confining effects become critical is large, $L \sim E/\Lambda^2$. Color confinement means that the quark cannot move indefinitely in the positive z direction; at a certain time $T \sim E/\Lambda^2$ it should turn back and start moving in the negative z direction. Let us consider this turning point in more detail. Before the turn, the quark's spin projection on the z axis is $+1/2$. Since by assumption the quark's helicity is conserved, after the turn, when p_z becomes negative, the quark's spin projection on the z axis must be $-1/2$ (Fig. 1.12). In other words, $\Delta S_z = -1$. The total angular momentum is conserved, consequently, $\Delta S_z = -1$ must be compensated. At the time of the turn, the quark is far from the antiquark and so they do not "know" what their respective partners are doing; conservation of angular momentum must be achieved locally. The only object that could be responsible for

Casher's argument

---

[14] I have used quotation marks since Casher's discussion could be said to be a little nebulous and imprecise.

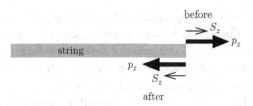

Fig. 1.12 Right-handed quark before and after the turning point.

compensating the quark ΔS_z is a QCD string that stretches in the z direction between the quark and the antiquark. The QCD string provides color confinement but it does not have L_z (more exactly, it is *presumed* to have no L_z) and, thus, cannot support the conservation of angular momentum in this picture. Thus, either the quark never turns (no confinement) or, if it does, chiral symmetry *must* be spontaneously broken.

The relation between quark confinement and χSB is a deep and intriguing dynamical question. Since I have nothing to add, let me summarize. There is a phase of QCD in which quark confinement and χSB coexist. On the N_f axis this phase starts at $N_f = 2$ and extends to some upper boundary $N_f = N_f^{**}$. We do not know whether N_f^{**} coincides with the left-hand edge of the conformal window N_f^*. It may happen that $N_f^{**} < N_f^*$, and the interval $N_f^{**} < N_f < N_f^*$ is populated by some other phase or phases (e.g., confinement without χSB) . . .

1.3.5 A Few Words on Other Regimes

Using various ingredients and mixing them in various proportions to construct a matter sector with the desired properties, one can reach other phases of Yang–Mills theories. For instance, by Higgsing the theory, as in Section 1.2.3.2, and breaking SU(N) down to U(1)$^{N-1}$ we can implement the Coulomb phase. Let us ask ourselves what happens if this Higgsing is implemented through the scalar fields in the fundamental representation, as in Section 1.2.3.1. If the vacuum expectation value $v \gg \Lambda$ then the theory is at weak coupling; it resembles the standard model. However, if $v \ll \Lambda$ then the theory is at strong coupling. Our intuition tells us that in this case it should resemble QCD, with a rich spectrum of composite color-singlet mesons having all possible quantum numbers.

There are convincing arguments [18] that there is no phase transition between these two regimes. Indeed, if the scalar fields are in the fundamental representation then the color-singlet interpolating operators that can be built from these fields and their covariant derivatives, and the gluon field strength tensor, span the space of physical (color-singlet or gauge-invariant) states in its *entirety*. All possible quantum numbers are covered. As the vacuum expectation value v changes from small to large, the strong coupling regime gives place smoothly to the weak coupling regime, possibly with a crossover in the middle. Each state existing at strong coupling is mapped onto its counterpart at weak coupling.

For instance, consider the operator

$$\mathrm{Tr}\left(\bar{X}i\overleftrightarrow{\mathcal{D}}_\mu X \tau^a\right). \tag{1.70}$$

At $v \ll \Lambda$ this operator produces a ρ meson and its excitations. The low-lying excitations could be seen as resonances. As v increases and becomes much larger than Λ the very same operator obviously reduces to $v^2 W_\mu$ plus small corrections. It produces a W boson from the vacuum. It produces excitations, too, but they are no longer resonances; rather, they are states that contain a number of W bosons and Higgs particles with the overall quantum numbers of a single W boson. Note that the global SU(2) symmetry of the model of Section 1.2.3.1 is respected in both regimes. All states appear in complete representations of SU(2), e.g., triplets, octets, and so on.

In the general case the following conjecture can be formulated (Fradkin and Shenker [18]):

Suppose that, in addition to gauge fields, a given non-Abelian theory contains a set of Higgs fields in the fundamental representation, which, by developing vacuum expectation values (VEVs) can "Higgs" the gauge group completely while the set of gauge-invariant operators built from the fields of the theory spans the space of all possible global quantum numbers (such as spin, isospin, and all other global symmetries of the Lagrangian). Then on decreasing all the above VEVs in proportion to each other from large to small values we do not pass through a Higgs-confinement phase transition. Rather, a crossover from weak to strong coupling takes place. If in addition there are massless fermions coupled to the gauge fields then there could be a phase transition separating the chirally symmetric and chirally asymmetric phases. This would be an example of χSB without confinement.[15] The opposite – confinement without χSB – is impossible in the absence of couplings between the fermion and scalar fields.

Contrived matter sectors can lead to more "exotic" phases. I have already mentioned oblique confinement. In supersymmetric Yang–Mills theories with matter in the adjoint representation a number of unconventional phases were found in [19]. We will not consider them here, as this aspect goes far beyond our scope in the present text.

Exercise

1.3.1   In QED with one massless Dirac fermion, identify the only one-loop diagram that determines charge renormalization. Calculate this diagram and show that the following relation holds for the running coupling constant:

$$\frac{1}{e^2(p)} = \frac{1}{e^2(\mu)} - \frac{1}{6\pi^2} \ln \frac{p}{\mu}.$$

Landau formula

Regardless of the value of $e^2(\mu)$, at $p \ll \mu$ (i.e., at large distances) we have $e^2(p) \rightarrow 0$. This phenomenon is known as the Landau zero-charge or infrared freedom. However, at large p namely, $p = \mu \exp[6\pi^2/e^2(\mu)]$, we hit the Landau pole in $e^2(p)$. When one approaches this pole from below, perturbation theory fails.

---

[15] Such examples are known in supersymmetric Yang–Mills theories.

1.4 Appendix: Basics of Conformal Invariance

Its general-ization, superconfor-mal symmetry, is briefly discussed in Section 10.19.3.

In this appendix we will review briefly some general features of conformal invariance. For a comprehensive consideration of conformal symmetry and its applications the reader is directed to [14, 20, 21].

In D-dimensional Minkowski space we have

$$ds^2 = g_{\mu\nu}(x)dx^\mu dx^\nu,$$

where for $D = 4$, for example,

$$g_{\mu\nu} = \text{diag}\{1, -1, -1, -1\} \equiv \eta_{\mu\nu}. \tag{1.71}$$

Under the general coordinate transformation

$$x \to x'$$

the original metric $g_{\mu\nu}$ is substituted by

$$g_{\mu\nu} \to g'_{\mu\nu}(x') = \frac{\partial x^\alpha}{\partial x'^\mu} \frac{\partial x^\beta}{\partial x'^\nu} g_{\alpha\beta}(x), \tag{1.72}$$

so that the interval ds^2 remains intact. Clearly, the general coordinate transformations form a very rich class that includes, as a subclass, transformations that change only the scale of the metric:

$$g'_{\mu\nu}(x') = \omega(x)g_{\mu\nu}(x). \tag{1.73}$$

All transformations belonging to this subclass form, by definition, the *conformal group*. It is obvious that, for instance, the global *scale* transformations

$$x \to x' = \lambda x, \quad \lambda \text{ is a number}, \tag{1.74}$$

is a conformal transformation. Moreover, the Poincaré group (of translations plus Lorentz rotations of flat space) is always a subgroup of the conformal group. The Minkowski metric (1.71) is invariant with respect to translations and Lorentz rotations.

In general, conformal algebra in four dimensions includes the following 15 generators:

P_μ (four translations);
K_μ (four special conformal transformations);
D (dilatation);
$M_{\mu\nu}$ (six Lorentz rotations).

Below, a few simple facts concerning the action of the conformal group in four dimensions are summarized. The set of 15 transformations given above forms a 15-parameter Lie group, the conformal group. This is a generalization of the 10-parameter Poincaré group, that is formed from 10 transformations generated by P_α and $M_{\alpha\beta}$. By considering the combined action of various infinitesimal transformations taken in a different order, the Lie algebra of the conformal group can be shown to be as follows:

$$i[P^\alpha, P^\beta] = 0,$$

$$i\left[M^{\alpha\beta}, P^\gamma\right] = g^{\alpha\gamma}P^\beta - g^{\beta\gamma}P^\alpha,$$

$$i\left[M^{\alpha\beta}, M^{\mu\nu}\right] = g^{\alpha\mu}M^{\beta\nu} - g^{\beta\mu}M^{\alpha\nu} + g^{\alpha\nu}M^{\mu\beta} - g^{\beta\nu}M^{\mu\alpha},$$

$$i[D, P^\alpha] = P^\alpha,$$

$$i[D, K^\alpha] = -K^\alpha,$$

$$i\left[M^{\alpha\beta}, K^\gamma\right] = g^{\alpha\gamma}K^\beta - g^{\beta\gamma}K^\alpha,$$

$$i\left[P^\alpha, K^\beta\right] = -2g^{\alpha\beta}D + 2M^{\alpha\beta},$$

$$i[D, D] = i[D, M^{\alpha\beta}] = i[K^\alpha, K^\beta] = 0. \qquad (1.75)$$

Conformal algebra

The first three commutators define the Lie algebra of the Poincaré group. The remaining commutators are specific to the conformal symmetry. If they were exact in our world this would mean, in particular, that

$$e^{i\alpha D}P^2 e^{-i\alpha D} = e^{2\alpha}P^2. \qquad (1.76)$$

The latter relation would imply, in turn, either that the mass spectrum is continuous or that all masses vanish. In neither case can one speak of the S matrix in the usual sense of this word. Instead of the on-shell scattering amplitudes, the appropriate objects for study in conformal theories are n-point correlation functions of the type

$$\langle O_1(x_1), \ldots, O_n(x_n) \rangle$$

whose dependence on $x_i - x_j$ is power-like. The powers, also known as critical exponents, depend on a particular choice of the operators O_i (and, certainly, on the theory under consideration).

Before establishing the conditions under which a given Lagrangian $\mathcal{L}$, which depends on the fields ϕ, is scale invariant or conformally invariant, we must decide how these fields ϕ transform under dilatation and conformal transformations. For translations and Lorentz transformations the rules are well known:

$$\delta_T^\alpha \phi(x) = -i\left[P^\alpha, \phi(x)\right] = \partial^\alpha \phi(x),$$

$$\delta_L^{\alpha\beta} \phi(x) = -i\left[M^{\alpha\beta}, \phi(x)\right] = \left(x^\alpha \partial^\beta - x^\beta \partial^\alpha + \Sigma^{\alpha\beta}\right)\phi(x), \qquad (1.77)$$

where $\Sigma^{\alpha\beta}$ is the spin operator. For the remaining five operations forming the conformal group, the following choice is consistent with (1.75):

$$\delta_D \phi(x) = (d + x\partial)\phi(x), \qquad (1.78)$$

$$\delta_C^\alpha \phi(x) = \left(2x^\alpha x^\nu - g^{\alpha\nu} x^2\right)\partial_\nu \phi(x) + 2x_\nu \left(g^{\nu\alpha}d - \Sigma^{\nu\alpha}\right)\phi(x), \qquad (1.79)$$

where d is a constant called the scale dimension of the field ϕ.

We can describe the generators of the conformal group in a slightly different language. Consider the infinitesimal coordinate transformation

$$x^\mu \to x'^\mu = x^\mu + \epsilon^\mu(x); \qquad (1.80)$$

then

$$\partial x^\beta / \partial x'^\rho = \delta_\rho^\beta - \partial \epsilon^\beta / \partial x^\rho$$

and to ensure that (1.73) holds one must take $\partial^\rho \varepsilon^\beta + \partial^\beta \varepsilon^\rho$ as being proportional to $\eta^{\beta\rho}$, namely, that

$$\partial^\rho \varepsilon^\beta + \partial^\beta \varepsilon^\rho = \frac{2}{D} (\partial \epsilon) \eta^{\beta\rho} \tag{1.81}$$

where $\eta^{\beta\rho}$ is the flat Minkowski metric. For $D > 2$ the maximal information one can extract from this relation is as follows:

(i) $\epsilon^\beta(x)$ is at most a quadratic function of x;
(ii) $\epsilon^\beta(x)$ can include a constant part

$$\epsilon^\beta = a^\beta$$

corresponding to ordinary x-independent translations;
(iii) the linear part can be of two types, either $\epsilon^\mu(x) = \lambda x^\mu$, where λ is a small number (dilatation), or $\epsilon^\mu(x) = \omega^\mu_\nu x^\nu$, where $\omega^{\mu\nu} = -\omega^{\nu\mu}$ (Lorentz rotations);
(iv) finally, the quadratic term satisfying Eq. (1.81) has the form

$$\epsilon^\mu(x) = b^\mu x^2 - 2x^\mu (bx), \tag{1.82}$$

where b^μ is a constant vector. Equation (1.82) corresponds to special conformal transformations. It is rather easy to see that the latter actually presents a combination of an inversion and a constant translation,

$$\frac{x'^\mu}{x'^2} = \frac{x^\mu}{x^2} + b^\mu. \tag{1.83}$$

Loosely speaking, in three or more dimensions conformal symmetry does not contain more information than Poincaré invariance plus scale invariance. If one is dealing with a local Lorentz- and scale-invariant Lagrangian, its conformal invariance will ensue.

A digression about the possible existence of "abnormal" theories

Caveat: The above assertion lacks the rigor of a mathematical theorem and, in fact, need not be true in subtle instances (such instances will *not* be considered in this book). In "normal" theories the scale and conformal currents are of the form [22]

$$S^\mu = x_\nu T^{\mu\nu}, \qquad C^\mu = \left[b_\nu x^2 - 2x_\nu (bx) \right] T^{\mu\nu}, \tag{1.84}$$

The vector b_ν is the same as in (1.82)

respectively. Here $T^{\mu\nu}$ is the conserved and symmetric energy–momentum tensor[16] that exists in any Poincaré-invariant theory and defines the energy–momentum operator of the theory:

$$P^\mu = \int d^{D-1}x\, T^{0\mu}, \qquad \dot{P}^\mu = 0. \tag{1.85}$$

Then the scale invariance implies that

$$\partial_\mu S^\mu = 0, \tag{1.86}$$

---

[16] Note that in some theories $T^{\mu\nu}$ is not unique. This allows for the so-called *improvements*, extra terms which are conserved by themselves and do not contribute to the spatial integral in (1.85). For instance, in the complex scalar field theory one can add

$$\Delta T^{\mu\nu} = \text{const} \times \left(g^{\mu\nu} \partial^2 - \partial^\mu \partial^\nu \right) \phi^\dagger \phi;$$

this improvement does not change P^μ but it does have an impact on the trace T^μ_μ.

which, in turn, entails[17]

$$T_\mu^\mu = 0. \tag{1.87}$$

Equation (1.87) then ensures that the conformal current is also conserved,

$$\partial_\mu C^\mu = 0. \tag{1.88}$$

Logically speaking, the representation (1.84) need not be valid in "abnormal" theories.[18]

For instance, Polchinski discusses [21] a more general extended representation in which[19]

$$S^\mu = x_\nu T^{\mu\nu} + \mathcal{S}^\mu, \tag{1.89}$$

where $\mathcal{S}^\mu$ is an appropriate local operator without an explicit dependence on x_ν. Then, (1.86) implies that

$$T_\mu^\mu = -\partial_\mu \mathcal{S}^\mu, \tag{1.90}$$

and the energy–momentum tensor is not traceless provided that $\partial_\mu \mathcal{S}^\mu \neq 0$. Generally speaking, the absence of a traceless energy–momentum tensor (possibly improved) is equivalent to the absence of *conformal* symmetry. Thus, "abnormal" scale-invariant theories need not be conformal.

After this digression, let us return to "normal" theories – those treated in this book. In such theories Eq. (1.84) is satisfied and scale invariance entails conformal invariance.

Applying the requirement of conformal invariance is practically equivalent to making all dimensional couplings in the Lagrangian vanish. In particular, all mass terms must be set to zero.

Warning: this last assertion is valid at the classical level and is, in fact, a necessary but not sufficient condition. Moreover classical conformal invariance may be (and typically is) broken at the quantum level owing to the scale anomaly; see Chapter 8. There are notable exceptions: for example $N = 4$ super-Yang–Mills theory (Section 10.18.3) is conformally invariant at the classical level. It remains conformally invariant at the quantum level too.

References for Chapter 1

[1] Y. Nambu, *Phys. Rev.* **117**, 648 (1960); V. G. Vaks ad A. I. Larkin, Sov. Phys. JETP, 13, 192 (1961).

[2] J. Goldstone, *Nuovo Cim.* **19**, 154 (1961).

---

[17] In theories in which improvements are possible one should analyze the set of all conserved and symmetric energy–momentum tensors to verify that there exists a traceless tensor in this set.

[18] The word "abnormal" is in quotation marks because so far we are unaware of explicit examples of local and Lorentz-invariant field theories of this type with not more than two derivatives. For an exotic example with four or more derivatives see [20]. Moreover, if the requirement of scale and Lorentz invariance in four dimensions is supplemented by *unitarity* the class of "abnormal" theories, in which scale invariance does not necessarily entail conformal invariance, is essentially empty. It is likely to be exhausted by theories reducible to free field theories of a special type [23].

[19] Here C^μ must also be extended compared to the expression in (1.84).

[3] F. Englert and R. Brout, *Phys. Rev. Lett.* **13**, 321 (1964).

[4] P. W. Higgs, *Phys. Rev. Lett.* **13**, 508 (1964).

[5] G. S. Guralnik, C. R. Hagen, and T. W. B. Kibble, *Phys. Rev. Lett.* **13**, 585 (1964).

[6] M. Peskin and D. Schroeder, *An Introduction to Quantum Field Theory* (Addison-Wesley, 1995), chapter 20.

[7] M. Shifman and A. Vainshtein, *Nucl. Phys. B* **362**, 21 (1991).

[8] H. Georgi and S. L. Glashow, *Phys. Rev. Lett.* **28**, 1494 (1972).

[9] D. J. Gross and F. Wilczek, *Phys. Rev. Lett.* **30**, 1343 (1973); H. D. Politzer, *Phys. Rev. Lett.* **30**, 1346 (1973).

[10] N. Seiberg and E. Witten, *Nucl. Phys. B* **426**, 19 (1994). Erratum: *ibid*. **430**, 485 (1994) [arXiv:hep-th/9407087].

[11] G. 't Hooft, *Nucl. Phys. B* **190**, 455 (1981).

[12] M. A. Shifman, *Prog. Part. Nucl. Phys.* **39**, 1 (1997) [arXiv:hep-th/9704114].

[13] K. G. Wilson, *Phys. Rev. D* **10**, 2445 (1974).

[14] Y. Frishman and J. Sonnenschein, *Non-Perturbative Field Theory* (Cambridge University Press, 2010).

[15] A. Belavin and A. Migdal, *JETP Lett.* **19**, 181 (1974); and Scale invariance and bootstrap in the non-Abelian gauge theories, Landau Institute Preprint-74-0894, 1974 (unpublished).

[16] T. Banks and Zaks, *Nucl. Phys. B* **196**, 189 (1982).

[17] A. Casher, *Phys. Lett. B* **83**, 395 (1979).

[18] K. Osterwalder and E. Seiler, *Ann. Phys.* **110**, 440 (1978); T. Banks and E. Rabinovici, *Nucl. Phys. B* **160**, 349 (1979); E. H. Fradkin and S. H. Shenker, *Phys. Rev. D* **19**, 3682 (1979).

[19] F. Cachazo, N. Seiberg, and E. Witten, *JHEP* **0302**, 042 (2003) [arXiv: hep-th/0301006].

[20] R. Jackiw, Field theoretic investigations in current algebra, section 7, in S. B. Treiman *et al.* (eds.), *Current Algebra and Anomalies* (Princeton University Press, 1985), p. 81.

[21] J. Polchinski, *Nucl. Phys. B* **303**, 226 (1988).

[22] J. Wess, *Nuov. Cim.* **18**, 1086 (1960).

[23] A. Dymarsky, Z. Komargodski, A. Schwimmer, and S. Theisen, *JHEP* **1510**, 171 (2015) [arXiv:1309.2921].

2 Kinks and Domain Walls

Separating in space degenerate vacua. — Quasiclassical treatment of kinks and domain walls. — Domain walls antigravitate. — Quantum corrections. — Can we observe fractional electric charges? Yes, we can, on domain walls.

2.1 Kinks and Domain Walls (at the Classical Level)

In this chapter we will consider a subclass of topological solitons. Let us assume that a field theory possesses a few (more than one) discrete degenerate vacuum states. A field configuration smoothly interpolating between a pair of distinct degenerate vacua is topologically stable. This subclass is rather narrow – for instance, it does not include vortices, a celebrated example of topological solitons. Vortices, flux tubes, monopoles, and so on will be discussed in subsequent chapters.

In nonsupersymmetric field theories the vacuum degeneracy requires the spontaneous breaking of some *global* symmetry – either discrete or continuous.[1] If there is no symmetry then, while a vacuum degeneracy may be present at the Lagrangian level for accidental reasons, it will be lifted by quantum corrections. For our current purposes we will focus on theories with spontaneously broken discrete symmetries. Then the set of vacua is, generally speaking, discrete.

We will start our studies from the simplest model,

$$S = \int d^D x \left[\frac{1}{2} \left(\partial_\mu \phi \right) \left(\partial^\mu \phi \right) - \frac{g^2}{4} \left(\phi^2 - v^2 \right)^2 \right], \tag{2.1}$$

with one real scalar field and Z_2 symmetry, $\phi \to -\phi$. Here v is a free parameter that is chosen to be positive. Then the Z_2 symmetry is spontaneously broken, since the lowest-energy state – the vacuum – is achieved at a nonvanishing value of ϕ. Since the global Z_2 is spontaneously broken, there are two degenerate vacua at $\phi_0 = \pm v$. The classical solution interpolating between these two vacua is the same for $D = 2, 3$, and 4 (where D is the number of space–time dimensions). In four dimensions we are dealing with a wall separating two domains. The wall's total energy is infinite because it has two longitudinal space dimensions and its energy is proportional to its surface area. At $D = 3$, the two domains are separated by a boundary line, with one longitudinal dimension. Hence the domain line energy is proportional to the length of the line. Finally, at $D = 2$ there are no longitudinal directions: the energy of the interpolating configuration is finite and localized in space. Thus, at $D = 2$ we are dealing with a particle of a special type called a *kink* (from the Dutch, meaning "a twist in a rope").

2.1.1 Domain Walls

Since we have two distinct vacua, we can imagine the following situation. Assume that on one side of the universe $\phi_{\text{vac}} = -v$ and on the other $\phi_{\text{vac}} = v$. As we know, the physics is the same on both sides. However, ϕ, being a continuous field, cannot change abruptly from $-v$ to v. A transition region must exist separating the domains $\phi_{\text{vac}} = -v$ and $\phi_{\text{vac}} = v$. This transition region is called the *domain wall* (Fig. 2.1). Needless to say, in the transition region the energy density is higher than it is far from the wall, in the vacua. The wall organizes itself in such a way that this energy excess is minimal. I will elucidate this point shortly. The very existence of the domain wall

---

[1] Note that in supersymmetric theories (theories where supersymmetry – SUSY – is unbroken) all vacua *must* have a vanishing energy density and are thus degenerate; see Part II.

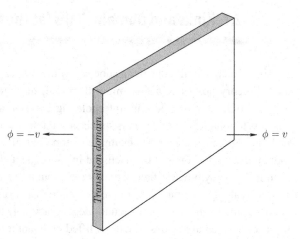

$\phi = -v \longleftarrow$ *Transition domain* $\longrightarrow \phi = v$

Fig. 2.1   The transition region between two degenerate vacua corresponding to the order parameters $\phi_{\text{vac}} = -v$ and $\phi_{\text{vac}} = v$ is a domain wall.

is due to the existence of two (in the case at hand) degenerate Z_2-asymmetric vacua. Thus, the domain wall is a *theoretical signature* of the spontaneous breaking of a discrete symmetry.

The above energy excess is of course proportional to the wall area A:

$$E_{\text{wall}} - E_{\text{vac}} = T_{\text{w}} A \tag{2.2}$$

| Wall tension | where the coefficient T_{w} is the wall *tension*.

Let us examine the case of $D = 4$, i.e., one time and three spatial dimensions x, y, z. First of all note that the domain wall, if it is unperturbed, is flat – it extends indefinitely in, say, the xy plane – and static, i.e., time-independent, in its rest frame. Thus, the field configuration describing the domain wall depends only on z,

$$\phi_{\text{wall}} = \phi_{\text{w}}(z), \qquad \phi_{\text{w}}(z \to -\infty) = -v, \qquad \phi_{\text{w}}(z \to \infty) = v. \tag{2.3}$$

The field configuration interpolating between $+v$ and $-v$ will be referred to as an *antiwall*. The flatness of the wall is obvious – it follows from the requirement of minimal energy. Equally obvious is its infinite extension – the boundary between two vacua cannot end (at least, not in the model under consideration; later on, in Section 2.2, we will consider more contrived models supporting wall junctions).

Needless to say, the choice of z as the axis perpendicular to the wall and x and y as axes parallel to the wall is a mere convention. One can always rotate the wall plane arbitrarily.

2.1.2 Visualizing the Wall Configuration

Let us try to visualize the phenomenon under discussion – the formation of two domains, in which the order parameter ϕ takes two distinct values, $\pm v$, separated by a transitional domain in which the field ϕ continuously interpolates between $\phi = -v$ and $\phi = v$. To this end it is convenient to consider $D = 1 + 1$ theories. The spatial dimension will be denoted by z. Thus, our theory is formulated on a line, $-\infty < z < \infty$. At each given point z we have a potential $U(\phi)$ with two degenerate

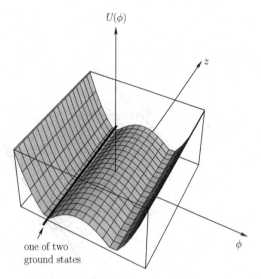

Fig. 2.2   A straight "rope" in the left trough represents one of two ground states.

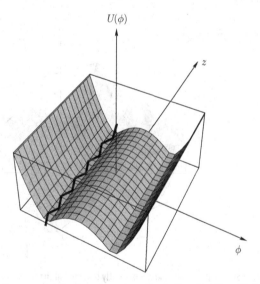

Fig. 2.3   Small oscillations near the ground state. The wave propagating in the z direction with time is interpreted as an elementary particle.

minima. One can imagine two parallel troughs separated by a barrier (Fig. 2.2), and an infinite rope that has to be placed on this profile in such a way as to minimize its energy. The ground state (i.e., the minimal energy state, the vacuum) is achieved when the rope, being perfectly straight, is placed either in one trough (the left-hand one in Fig. 2.2) or the other. Figure 2.3 depicts small oscillations around this ground state at a given moment of time. With advancing time the wave moves in the z direction. Upon quantization, such a wave is interpreted as an elementary particle.

In addition, there exists a topologically different class of stable static-field configurations. Imagine an (infinite) rope that starts in one trough and smoothly rolls over into the other trough (Fig. 2.4). The energy of this configuration is evidently *not*

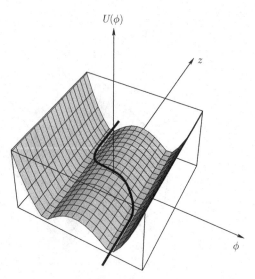

Fig. 2.4   A topologically distinct minimum of the energy functional. The "rope" crosses over from one trough to another.

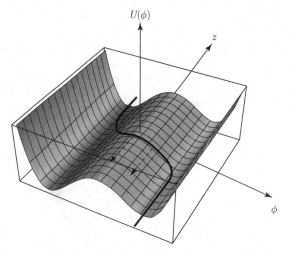

Fig. 2.5   If the Z_2 symmetry of the model is explicitly broken and the right-hand local minimum of $U(\phi)$ is slightly higher than the left-hand one then this configuration, with the "roll-over rope," is unstable. The position of the crossover will move with time towards the negative values of z in order to minimize the energy.

the absolute minimum. However, it does attain a minimum for the given boundary conditions $\phi(z \to -\infty) = -v$, $\phi(z \to \infty) = v$. Moreover, this configuration is obviously perfectly stable: one needs to invest infinite energy in order to "unroll" the rope back into one trough.

The situation changes if the two minima of the potential are not exactly degenerate. Consider, say, the potential $U(\phi)$ depicted in Fig. 2.5, where the right-hand minimum is slightly higher than the left-hand minimum. In this case, the Z_2 symmetry is *explicitly* broken from the very beginning; there is only one true vacuum, at $\phi = -v$. The second minimum of the potential at $\phi = v$ is a local minimum of energy and is not stable quantum-mechanically. Even if again you initially place the "rope" so that it crosses from one trough to the other, as in Fig. 2.5, it will start unrolling since it is energetically expedient for the length of the rope in the right-hand trough to be minimized and the length in the left-hand trough to be maximized. In this way we gain energy. The transition domain will keep moving in the direction of negative values of z; there is no static stable field configuration for the asymptotic behavior $\phi(z \to -\infty) = -v$, $\phi(z \to \infty) = v$ in this case.

> **Unstable (or quasistable) wall**

2.1.3 Classical Equation for the Domain Wall

The energy functional that determines the wall configuration follows from the action (2.1),

$$T_{\text{w}} \equiv \frac{\mathcal{H}}{A} = \int dz \left\{ \frac{1}{2} \left[\frac{\partial \phi_{\text{w}}(z)}{\partial z} \right]^2 + \frac{g^2}{4} \left[\phi_{\text{w}}^2(z) - v^2 \right]^2 \right\}, \qquad (2.4)$$

where $A = \int dx\, dy$ is the area of the wall, $A \to \infty$. The quantity T_{w} is called the wall *tension*; it measures the energy per unit area. For the time being we will discuss a purely classical domain-wall solution. We will consider quantum corrections later (see Section 2.4). Here let us note that neglecting quantum corrections is justified as long as the coupling constant is small, $g^2 \ll 1$. In this case the classical result is dominant – all quantum corrections are suppressed by powers of g^2.

The field configuration $\phi_{\text{w}}(z)$ minimizing the tension T_{w} (under the boundary conditions (2.3)) gives the wall solution. The condition of minimization leads to a certain equation for $\phi_{\text{w}}(z)$. To this end one slightly distorts the solution, so that $\phi_{\text{w}} \to \phi_{\text{w}} + \delta\phi$, then expands the functional T_{w} in $\delta\phi$, and requires the term linear in $\delta\phi$ to vanish. In this way one arrives at

> **Wall profile equation**

$$-\frac{d^2 \phi_{\text{w}}}{dz^2} + g^2 \phi_{\text{w}} \left(\phi_{\text{w}}^2 - v^2 \right) = 0. \qquad (2.5)$$

Of course, this is nothing other than the classical equation of motion in the model with action (2.1), restricted to the class of fields that depend only on one spatial coordinate, z.

The differential equation (2.5) is highly nonlinear. The domain-wall problem does not allow one to solve the equation by linearization, as is routinely done for small oscillations near the given vacuum (i.e., for particles). Such nondissipating localized solutions of nonlinear equations, whose very existence is due to nonlinearity, are generically referred to as *solitons*.

2.1.4 First-Order Equation

Although the solution of the particular nonlinear equation (2.5) can be found in every handbook on differential equations, it is instructive to imagine that this is not the case. Let us pretend that we are on an uninhabited island, with no handbooks or Internet, and play with Eq. (2.5), with the goal of maximally simplifying the search for a domain-wall solution.

As our imagination is ready to soar, we will replace the coordinate z by a fictitious time τ and ϕ_w by a fictitious coordinate X. Then, Eq. (2.5) takes the form

$$\ddot{X} = g^2 X (X^2 - v^2), \qquad (2.6)$$

where the "time" derivative is denoted by an overdot. One can immediately recognize the Newton equation of motion for a particle of mass $m = 1$ in the potential $-(g^2/4)(X^2 - v)^2$. In Newtonian motion, kinetic + potential energy is conserved; therefore

$$\frac{\dot{X}^2}{2} - \frac{g^2}{4}(X^2 - v)^2 = \text{const} = 0. \qquad (2.7)$$

The fact that the constant on the right-hand side vanishes follows from the boundary conditions (2.3). Indeed, in the infinite "past" and "future" both the kinetic and potential energies vanish.

Thus the existence of an integral of motion (the conserved energy) allows us to obtain the first-order differential equation

$$\dot{X} = \pm \frac{g}{\sqrt{2}}(X^2 - v^2), \qquad (2.8)$$

instead of the second-order equation (2.5).

Returning now to the original notation ϕ_w and z, we can rewrite Eq. (2.8) as follows:

$$\frac{d\phi_w}{dz} = -\frac{g}{\sqrt{2}}(\phi_w^2 - v^2). \qquad (2.9)$$

Here the sign ambiguity in Eq. (2.8) is resolved in the following way. Consider the field configuration interpolating between $-v$ at $z = -\infty$ and v at $z = +\infty$ (the wall, see Fig. 2.6). The left-hand side of Eq. (2.9) is positive and so is the right-hand side. In the case of the field configuration interpolating between $+v$ and $-v$ (the antiwall) one should choose the plus sign in Eqs. (2.8) and (2.9).

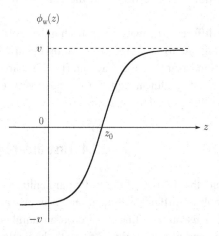

Fig. 2.6   The solution of Eq. (2.9) interpolating between $\phi_{vac} = -v$ and $\phi_{vac} = v$.

The first-order differential equation (2.9) is trivially integrable. Indeed,

$$z = -\frac{\sqrt{2}}{g} \int^{\phi_w} \frac{d\phi}{\phi^2 - v^2} = \frac{\sqrt{2}}{gv} \operatorname{arctanh}\left(\frac{\phi_w}{v}\right) + \text{const.} \tag{2.10}$$

Wall profile

We will denote the integration constant on the right-hand side by z_0, for reasons which will become clear soon. Then

$$\phi_w(z) = v \tanh\left[\frac{\mu}{\sqrt{2}}(z - z_0)\right], \tag{2.11}$$

The mass m of the ϕ quantum is

$$m = \sqrt{2}\mu,$$

where $\mu = gv$, as usual. The profile of this function is depicted in Fig. 2.6. The energy density (in other words, the Hamiltonian density) is defined as

$$\mathcal{E}(z) = \frac{1}{2}\left[\frac{d\phi_w(z)}{dz}\right]^2 + \frac{g^2}{4}\left[\phi_w^2(z) - v^2\right]^2, \tag{2.12}$$

cf. Eq. (2.4). If the tension T_w has dimension $D - 1$, the energy density $\mathcal{E}(z)$ has dimension D (in mass units). The plot of $\mathcal{E}(z)$ on the domain-wall configuration is presented in Fig. 2.7. Away from the vicinity of $z = z_0$ the energy density rapidly approaches zero, its vacuum value.

Comparing Figs. 2.6 and 2.7 we see that z_0 plays the role of the soliton center. In fact, instead of a single domain-wall solution we have found a whole family of solutions, labeled by a continuous parameter z_0. It is obvious that the tension T_w does not depend on z_0.

Soliton
moduli

The soliton center z_0 (or any other similar parameter occurring in a more complicated problem) is called the *collective coordinate* or *soliton modulus*. The existence of a family of wall solutions in the problem at hand is evident. Indeed, the original Lagrangian is invariant under arbitrary translations of the reference frame. At the same time, any given solution of the type (2.11), say, $v \tanh(\mu z/\sqrt{2})$, spontaneously breaks the translational invariance in the z direction. The existence of a family of solutions labeled by z_0 restores the translational symmetry.

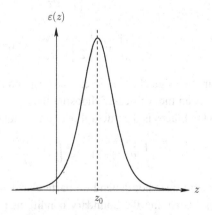

Fig. 2.7   The energy density $\mathcal{E}$ vs. z for the domain-wall solution (2.11).

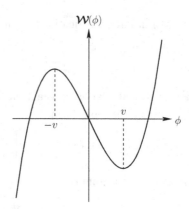

The superpotential $\mathcal{W}(\phi)$ vs. ϕ.

2.1.5 The Bogomol'nyi Bound

In this section we will discuss a more general derivation of the first-order equation for the domain-wall profile, which, among other things, will also reveal a topological aspect of the construction.

First, I will introduce an auxiliary function $\mathcal{W}(\phi)$ (it is called the *superpotential*, see Part II), where

$$\mathcal{W} = \frac{g}{\sqrt{2}} \left(\frac{1}{3}\phi^3 - v^2\phi \right). \tag{2.13}$$

Note that the extrema of the superpotential, i.e., the *critical* points where $\mathcal{W}' = 0$, correspond to the vacua of the model under consideration; see Fig. 2.8. The superpotential and the potential are related as follows:

$$U(\phi) = \frac{1}{2}\left(\frac{d\mathcal{W}}{d\phi} \right)^2.$$

Next, observe that the tension $T_{\rm w}$, see Eq. (2.4), can be rewritten as

$$T_{\rm w} = \int dz\, \frac{1}{2} \left\{ \left[\frac{d\phi(z)}{dz} \right]^2 + \left[\mathcal{W}'(\phi) \right]^2 \right\}$$

$$\equiv \int dz\, \left\{ \frac{1}{2}\left[\frac{d\phi(z)}{dz} \pm \mathcal{W}'(\phi) \right]^2 \mp \frac{d\phi}{dz}\mathcal{W}' \right\}. \tag{2.14}$$

There are two sign choices here: $\pm$ correspond to the wall and antiwall solutions. We will focus on the wall case, choosing the $+$ sign in the square brackets. The second term in the braces is the integral over a full (total) derivative:

The surface terms are topological. They are also referred to as topological charges.

$$\int dz\, \frac{d\phi}{dz}\mathcal{W}' \equiv \int dz\, \frac{d\mathcal{W}}{dz} \equiv \mathcal{W}(v) - \mathcal{W}(-v). \tag{2.15}$$

This term does not depend on particular details of the profile function $\phi(z)$: for *any* $\phi(z)$ satisfying the boundary conditions (2.3) it is the same. That is why it is called the *topological term*.

Combining Eqs. (2.14) and (2.15) one obtains

$$T_{\mathrm{w}} = -\Delta\mathcal{W} + \int dz \frac{1}{2} \left\{ \left[\frac{d\phi(z)}{dz} + \mathcal{W}'(\phi) \right]^2 \right\}, \qquad (2.16)$$

where

$$\Delta\mathcal{W} \equiv \mathcal{W}(v) - \mathcal{W}(-v).$$

Since the expression in the braces is positive definite, it is obvious that for *any* function interpolating between two vacua (see Eq. (2.3))

$$T_{\mathrm{w}} \geq -\Delta\mathcal{W} \equiv \frac{4}{3\sqrt{2}} \frac{\mu^3}{g^2}, \qquad (2.17)$$

i.e., the tension is larger than or equal to the topological charge. This is called the *Bogomol'nyi inequality* or *bound* [1]. It is saturated (i.e., it becomes an equality) if and only if the expression in the braces in Eq. (2.16) vanishes, i.e., the first-order equation

$$\frac{d\phi}{dz} = -\mathcal{W}'(\phi) \qquad (2.18)$$

holds (cf. Eq. (2.9)). Thus, the domain-wall profile minimizes the functional (2.16) in the class of field configurations with boundary conditions (2.3). In the case at hand the Bogomol'nyi bound is saturated, and the wall tension is

$$T_{\mathrm{w}} = \frac{4}{3\sqrt{2}} \frac{\mu^3}{g^2} = \frac{m^3}{3g^2}. \qquad (2.19)$$

Note the occurrence of the small parameter g^2 in the denominator. This is a general feature of solitons in the quasiclassical regime.

The above consideration can be readily generalized to a class of multifield models, with a set of fields $\phi_1, \phi_2, \ldots, \phi_n (n \geq 2)$, provided that the potential $U(\phi)$ reduces to

$$U(\phi) = \frac{1}{2} \sum_{\ell=1}^{n} \left(\frac{\partial\mathcal{W}}{\partial\phi_\ell} \right)^2, \qquad (2.20)$$

where $\mathcal{W}$ is a superpotential that depends, generally speaking, on all the ϕ_ℓ. In this case the vacua (the classical minima of energy) lie at the points where

$$\frac{\partial\mathcal{W}}{\partial\phi_\ell} = 0, \qquad \ell = 1, 2, \ldots, n, \qquad (2.21)$$

corresponding to the extrema of $\mathcal{W}$. Moreover, Eq. (2.16) takes the form

$$T_{\mathrm{w}} = -\Delta\mathcal{W} + \int dz \frac{1}{2} \left\{ \sum_{\ell=1}^{n} \left[\frac{d\phi_\ell(z)}{dz} + \frac{\partial\mathcal{W}}{\partial\phi_\ell} \right]^2 \right\}. \qquad (2.22)$$

The Bogomol'nyi bound is achieved provided that

$$\frac{d\phi_\ell}{dz} = -\frac{\partial\mathcal{W}}{\partial\phi_\ell} = 0, \qquad \ell = 1, 2, \ldots, n. \qquad (2.23)$$

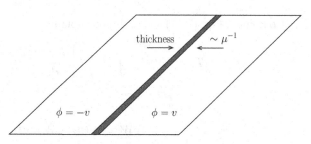

Fig. 2.9   Domain wall in $1 + 2$ dimensions.

2.1.6 Dimensions $D = 3$ and $D = 2$

In three dimensions (one time, two spatial) the domain wall is not really a wall. Rather, it is a line separating two distinct phases of the model on a plane (Fig. 2.9). The energy of the "wall" is equal to its tension times the length of the line.

In the $D = 2$ case, z is the only spatial dimension. At $D = 3$ and $D = 4$ the soliton under consideration is not localized in other spatial dimensions, i.e., x and/or y. At $D = 2$ the transverse spatial dimensions are absent; the action (2.1) contains no integrations over perpendicular coordinates. Thus, the soliton (2.11) presents a localized lump of energy – a particle. Correspondingly, the integrals (2.4) or (2.16) give the particle mass[2] rather than tension. It is instructive to check the dimensions of all relevant physical quantities. At $D = 2$ the parameter v is dimensionless, while g and μ have dimension of mass. Thus, the soliton mass is

$$M_k = \frac{4}{3\sqrt{2}} \frac{\mu^3}{g^2} \qquad \text{and} \qquad \dim(M_k) = \text{mass}. \qquad (2.24)$$

Kink mass | As mentioned earlier, for $D = 2$ the soliton (2.11) is often referred to as a *kink*. The size of the kink is of order μ^{-1} while its Compton wavelength is $M_k^{-1} \sim g^2/\mu^3$. At small g^2 (i.e., for $g/\mu \ll 1$) the Compton wavelength is much smaller than the kink size. In other words, for the kink under consideration the classical size is much larger than the quantum size. In essence the kink is a (quasi)classical object. As mentioned earlier, this is a general feature of all solitons at *weak coupling*.

The solution (2.11) refers to the kink rest frame. Since the kink mass is finite, one should be able to accelerate it. The kink solution corresponding to the motion of the particle with the velocity V is

$$\phi(t, z|V) = v \tanh\left(\frac{\mu}{\sqrt{2}} \frac{z - z_0 - Vt}{\sqrt{1 - V^2}}\right), \qquad (2.25)$$

where z_0 now plays the role of the kink center at $t = 0$ (see Exercise 2.1.4).

We will discuss quantum corrections to M_k in Section 2.4.

2.1.7 Topological Aspect

In this subsection we will consider in some depth the topological aspects of the problem, which have been mentioned already in Section 2.1.2. It is convenient to

---

[2] More exactly, they give the particle energy in the rest frame, see Exercise 2.1.4.

frame our discussion in terms of $D = 2$ theory, although all assertions can be reformulated with ease for $D = 3$ and $D = 4$.

Let us consider field configurations with finite energy (at $t = 0$). Finiteness of the energy implies that the field ϕ is a smooth function of z that tends to one of two vacua as $z \to \pm\infty$. (Otherwise, its potential energy would blow up.) Thus, in the problem at hand, we have two points at the spatial infinities $z = \pm\infty$, which are mapped by $\phi(z)$ onto two vacuum points. It is clear that there are four distinct classes of mappings:

$$
\begin{aligned}
&\{-\infty \to -v,\ +\infty \to -v\}, \\
&\{-\infty \to +v,\ +\infty \to +v\}, \\
&\{-\infty \to -v,\ +\infty \to +v\}, \\
&\{-\infty \to +v,\ +\infty \to -v\}.
\end{aligned}
\tag{2.26}
$$

It is impossible to leap from one class into another without passing en route a configuration with infinite energy. In particular, any time evolution caused by the dynamical equations of the model at hand or by local nonsingular sources will never take the field configuration from one class into another. One needs infinite energy (action) for such a jump. The class of a mapping is a *topological property*.

The first two classes are topologically trivial – they correspond to two vacua and oscillations over these vacua. These are the so-called vacuum sectors.

The kink sectors are topologically nontrivial. Kinks belong to the third class in Eq. (2.26), while antikinks belong to the fourth. The field configuration realizing the minimal energy in the kink sector is the soliton solution (2.11). Since its energy is minimal in the given sector (and field configurations do not leap from one sector to another) it is absolutely stable. Any other field configuration from the given class has a higher energy.

Summarizing, one can say that the existence of topologically stable solitons is due to the existence of nontrivial mappings of the spatial infinity onto the vacuum manifold of the theory. Let us remember this fact, as it has a general character.

One can go one step further and introduce a *topological current*. Equation (2.15) will prompt us to its form. Indeed, let us introduce a (pseudo)vector current

$$
J^\mu = -\varepsilon^{\mu\nu}\partial_\nu \mathcal{W}(\phi),
\tag{2.27}
$$

where $\varepsilon^{\mu\nu}$ is the absolutely antisymmetric tensor of the second rank, the Levi–Civita tensor with $\varepsilon^{01} = 1$ (remember, we are considering the $D = 2$ model). Unlike Noether currents, which are conserved only on equations of motion, the current J^μ is trivially conserved for *any* field configuration. The topological charge Q corresponding to the current J^μ is

$$
Q = \int dz J^0 = -\int dz(\partial \mathcal{W}/\partial z).
\tag{2.28}
$$

It is conserved too.

Topological stability

2.1.8 Low-Energy Excitations

So far we have considered a static, plane (unexcited) wall, i.e., a configuration with minimal energy in the given topological sector. Now let us discuss wall excitations. An excited wall is obtained from a static plane wall through the injection of a certain

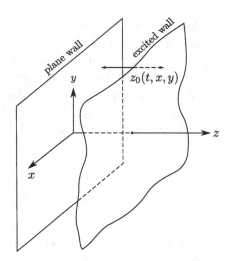

Fig. 2.10   The low-energy wall excitations are described by the effective world-sheet theory of the modulus field $z_0(t, x, y)$.

amount of energy, which perturbs the wall and destroys the time independence. If the energy injected is large enough (typically, larger than the inverse wall thickness μ), the corresponding perturbation changes the inner structure of the wall. For instance, its thickness becomes a time-dependent function of x and y. A wave of perturbation of the wall thickness propagates along the wall. At still higher energies the wall can emit quanta of the ϕ field from the wall surface into the three-dimensional bulk.

Here we will discuss briefly another type of perturbation, which neither changes the inner structure of the wall nor emits ϕ field quanta into the three-dimensional bulk. Assume that the energy injected into the wall is small, much less than the inverse thickness μ. From the standpoint of such excitations the wall can be viewed as infinitely thin, and the only process which can (and does) occur is a perturbation of the wall surface as a whole. In other words the wall center z_0 (see Section 2.1.4) becomes a slowly varying function of t, x, y describing waves propagating on the surface of the wall. This is depicted in Fig. 2.10. Fields such as $z_0(t, x, y)$, localized on the surface of the topological defect, are called *moduli fields*. In fact, they are the

| *Moduli fields* |

Goldstone fields associated with the spontaneous breaking of some symmetry on the topological defect under consideration.

In the case at hand, the bulk four-dimensional theory is translationally invariant. A given wall lying in the xy plane and centered at z_0 breaks translational invariance in the z direction. Hence, one should expect a Goldstone field to emerge. The peculiarity of this field is that it is localized on the wall (its "wave function" in the perpendicular direction is determined by the wall profile and falls off exponentially with the separation from the wall).

The low-energy oscillations of the wall surface are described by a low-energy effective theory of the moduli fields on the wall's world sheet. For brevity, people usually refer to such theories as *world-sheet theories*.

In the present case, the world-sheet theory can be derived trivially. Indeed, we start from the wall solution $\phi_w(z - z_0)$ and endow the field ϕ with a slow t, x, y-dependence coming only through the adiabatic dependence $z_0(t, x, y)$:

$$\phi_w(z - z_0) \rightarrow \phi_w(z - z_0(t, x, y)) \equiv \phi_w(z - z_0(x^P)),$$
$$x^P = \{t, x, y\}. \tag{2.29}$$

Here I have introduced three world-sheet coordinates x^P ($p = 0, 1, 2$) to distinguish them from the four coordinates x^μ ($\mu = 0, 1, 2, 3$) of the bulk theory. Then substituting (2.29) into the action (2.1) we get

$$S = \int d^4x \frac{1}{2} \left[\left(\frac{\partial \phi_w}{\partial z_0} \frac{\partial z_0}{\partial x^P} \right)^2 - \left(\frac{\partial \phi_w}{\partial z} \right)^2 - V(\phi_w) \right]$$

$$= -T_w \int dx\,dy + \frac{1}{2} \int dz \left(\frac{\partial \phi_w}{\partial z} \right)^2 \int d^3x \left(\frac{\partial z_0}{\partial x^P} \right)^2$$

$$= \text{const} + \frac{T_w}{2} \int d^3x \left(\frac{\partial z_0(x^P)}{\partial x^P} \right)^2. \tag{2.30}$$

World-sheet theory.

This is the action for a free field $z_0(x^P)$ on the wall's world sheet. There is no potential term – this obviously follows from the Goldstone nature of the field. The general form of the effective action in (2.30) is transparent and could have been obtained on symmetry grounds. Only the normalization factor $T_w/2$ requires a direct calculation.

2.1.9 Nambu–Goto and Dirac–Born–Infeld Actions

The world-sheet action in (2.30) captures only terms of the second order in derivatives. In fact, one should view it as the lowest-order term in the derivative expansion. If $\partial z_0/\partial x^P \sim 1$, i.e., the oscillations of the wall surface are not small, one must include in the world-sheet action terms of higher order in the derivatives. They cannot be obtained by the simple calculation described above. However, in the limit of infinitely thin walls (which are known as *branes*, or 2-branes, to be more exact) one can obtain the world-sheet action from a general argument. Indeed, in this limit the internal structure of the wall is irrelevant, and the action can depend only on the 3-volume swept by the brane in its evolution in space–time.

The action we are looking for was suggested, in connection with strings, by Nambu [2] and Goto [3]. We will need some facts from differential geometry going back to the nineteenth century. To ease the notation, in this section we will omit the subscript 0, so that the wall surface is parametrized by the function $z(t, x, y)$. The induced metric g_{pq} is defined as

$$g_{pq} = \partial_p X_\mu \partial_q X^\mu, \tag{2.31}$$

where

$$X^\mu = \{x^q, z(x^q)\}, \qquad \partial_p \equiv \partial/\partial x^P. \tag{2.32}$$

It is instructive to write down the explicit form for the induced metric:

$$g_{pq} = \begin{pmatrix} 1 - \dot{z}^2 & -\dot{z}\partial_x z & -\dot{z}\partial_y z \\ -\dot{z}\partial_x z & -1 - (\partial_x z)^2 & -\partial_x z\partial_y z \\ -\dot{z}\partial_y z & -\partial_x z\partial_y z & -1 - (\partial_y z)^2 \end{pmatrix}. \tag{2.33}$$

The world volume swept by the brane is $\int d^3x \sqrt{g(x^P)}$, where

$$g \equiv \det(g_{pq}). \tag{2.34}$$

The proportionality coefficient in the action can be readily established by expanding g up to the second order in derivatives and comparing with Eq. (2.30). In this way we arrive at

$$S_{\mathrm{NG}} = -T_{\mathrm{W}} \int d^3 x \sqrt{g(x^p)}, \qquad (2.35)$$

Nambu–Goto action

where the subscript NG stands for Nambu–Goto.

The Nambu–Goto action is not very convenient for practical calculations at $\partial z / \partial x^p \sim 1$ because of the square root in (2.35). Usually one replaces it by an equivalent Polyakov action [4], which we will not discuss here.

In certain problems (see, e.g., [5]), in addition to the translational modulus field $z_0(x^p)$ the domain walls possess a modulus φ of phase type; that is,

$$\varphi, \quad \varphi \pm 2\pi, \quad \varphi \pm 4\pi,$$

and so on are identified. On the brane (i.e., in $1 + 2$ dimensions), the massless field of phase type can be identified with a massless photon [6], namely,

Domain walls: geometry and electro-magnetism

$$F_{pq}(x^p) = \frac{e^2}{4\pi} \varepsilon_{pqp'}[\partial^{p'} \varphi(x^p)], \qquad (2.36)$$

where e is the electromagnetic coupling constant. This is discussed in detail in Section 9.7.3. The Nambu–Goto action can be generalized further to include electrodynamics on the brane's world sheet. This is done as follows:

$$S_{\mathrm{DBI}} = -T_{\mathrm{w}} \int d^3 x \sqrt{\det(g_{pq} + \alpha F_{pq})}, \qquad (2.37)$$

where the constant α is defined by

$$\alpha = \frac{1}{e\sqrt{T_{\mathrm{w}}}} \qquad (2.38)$$

and the subscript DBI stands for Dirac, Born, and Infeld, who were the first to construct this action. Expanding (2.37) in derivatives and keeping the quadratic terms, we get, in addition to (2.30), the standard action of the electromagnetic field:

$$S_{\mathrm{DBI}} \to -\frac{1}{4e^2} F_{pq} F^{pq}. \qquad (2.39)$$

2.1.10 Digression: Physical Analogies for the First-Order Equation

If z is interpreted as a "time" and ϕ as a "coordinate" then the set of the first-order equations

$$\dot{X}_\ell = -\frac{\partial \mathcal{W}(X)}{\partial X_\ell}, \qquad \ell = 1, 2, \ldots, n,$$

(cf. Eq. (2.23)) has a very transparent classical-mechanical analog. This is the equation describing the flow of a very viscous fluid (e.g., honey) on a "potential profile" $\mathcal{W}(X)$ at the given point. Indeed, $\dot{\vec{X}}$ is the fluid velocity while $-\vec{\nabla}\mathcal{W}(X)$ is the force acting on the fluid. In the limit of very high viscosity the term with acceleration can be neglected, and we are left with a limiting law: the velocity is proportional to the force.

Using this analog it is easy to guess, without actually solving the equation, whether the solution exists. For instance, a single glance at Fig. 2.8 tells us: yes, a droplet of honey placed at the point $-v$, at the maximum of the profile, will flow to the point v, the minimum. In fact, for the problem at hand, where we have a single variable, it is quite easy to find the analytic solution. This is not the case for problems with several fields. Then our intuition about viscous fluid flows is indispensable.

For cubic superpotentials Eq. (2.8) has another application. Let us shift the variable, so that $X + v \to X$, and rewrite (2.8) as

Logistic
equation

$$\dot{X} = \lambda(AX - X^2),$$

where λ and A are positive constants ($\lambda = g/\sqrt{2}$ and $A = 2v$). Written in this form, we have nothing other than the so-called *logistic equation*, describing, among other things, how new products saturate markets with time. Say, a new high-quality mass product appeared on the market on January 1, 2002. At first, it is manufactured by only one or two companies, not many instances of the product are in use, and most people do not know how good it is. The number of people who purchase the product at any given moment of time is proportional to the number of owners at this moment of time (it is supposed that the owners spread the word to their friends). Thus, at the beginning the number of owners X is small, and the number of purchases $\dot{X}$ is proportional to X. As time goes on, new companies start manufacturing this product and putting it on the market, and eventually almost everybody has one. This interferes destructively with new purchases. In a first approximation this destructive interference is described by a quadratic term with a negative coefficient.

We have already discussed the solution of this equation (see Fig. 2.6 and Eq. (2.11)). At the start the number of owners grows exponentially but then, quite abruptly, it reaches saturation (the market is full) and so the number of new purchases approaches zero in an exponential manner. This is the time for a new product to appear. If you are curious you can find the saturation time and the rate of growth at the initial stage in terms of parameters λ and A.

Exercises

2.1.1 As is well known, for a quantum-mechanical double-well potential, with Hamiltonian

$$\mathcal{H} = \frac{p^2}{2} + \frac{g^2}{4}\left(\phi^2 - v^2\right)^2, \qquad p = -i\frac{d}{d\phi}, \qquad (2.40)$$

the ground state is unique.[3] It is symmetric under $\phi \to -\phi$. The Z_2 symmetry present in the Hamiltonian is *not* broken in the ground state. Why, then, in a field theory treatment of the double-well potential does the ground state break Z_2 symmetry? What is the difference between quantum mechanics, (2.40), and field theory?

---

[3] See, e.g., [7], p. 183, problem 3.

2.1.2   Following Section 2.1.5 derive the Bogomol'nyi bound for the antiwall, i.e., the field configuration with minimal tension and the boundary conditions

$$\phi(z = -\infty) = v \quad \text{and} \quad \phi(z = \infty) = -v. \quad (2.41)$$

2.1.3   Find the thickness of the wall, i.e., the width of the energy distribution $\mathcal{E}(z)$ (see Eq. (2.12)). What is the asymptotic behavior of $\mathcal{E}(z)$ at $|z - z_0| \to \infty$? Express the result in terms of the mass of the elementary excitation.

2.1.4   Check that the moving-kink profile (2.25) is indeed the solution of the classical equation of motion

$$\left(\frac{\partial^2}{\partial t^2} - \frac{\partial^2}{\partial z^2} \right) \phi(t, z|V) + \frac{\partial U(\phi(t, z|V))}{\partial \phi} = 0. \quad (2.42)$$

At $V \neq 0$ does it satisfy a first-order differential equation? Calculate the classical energy E and (spatial) momentum P of the moving kink (2.25). Show that the standard relativistic relation $E^2 - P^2 = M_k^2$ holds.

2.1.5*   Consider a complexified version of the real model discussed earlier:

$$\mathcal{L} = \left(\partial_\mu \phi \right)^\dagger \partial^\mu \phi - |\mathcal{W}'(\phi)|^2, \quad (2.43)$$

where ϕ is a complex rather than a real field. The prime denotes differentiation with respect to ϕ while the dagger denotes complex conjugation. Assuming $\mathcal{W}$ to be a holomorphic function of ϕ, prove that the second-order equation of motion for a domain wall or kink following from (2.43) implies a first-order equation of the Bogomol'nyi type [8]. Note that the converse is of course trivially valid.

2.2 Higher Discrete Symmetries and Wall Junctions

2.2.1 Stable Wall Junctions: Generalities

So far we have considered isolated walls in the planar geometry. Everybody who has seen soap foam understands that, generally speaking, domain walls can intersect or join each other, forming complicated networks.

A wall junction is depicted in Fig. 2.11. It is a field configuration where three or more walls join along a line (in $D = 1 + 3$). The line along which the walls join is called the *wall junction*. As is clear from Fig. 2.11, all the fields in this wall junction configuration are z-independent: they depend only on x and y. Therefore for a static wall junction the problem is essentially two dimensional; see Fig. 2.12. The same picture as in Fig. 2.12 emerges for domain boundaries in $D = 1 + 2$. In this case the z coordinate does not appear at all. There is no analog of the junction configuration in $D = 1 + 1$.

In this section we will concentrate on wall junctions of the "hub and spokes" type, as in Fig. 2.12, which occur when a Z_n symmetry is spontaneously broken. We will orient the wall spokes in the xy plane as indicated in Fig. 2.12, namely, the hub is at the origin, the first spoke, say, runs along the x axis in the positive direction, the second runs at an angle $2\pi/n$, and so on. At the point P the theory

Conventions and definitions

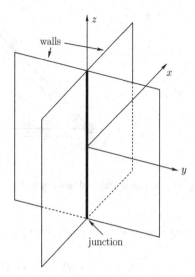

Fig. 2.11 The domain-wall junction. Here four domain walls join each other; the junction is oriented along the z axis.

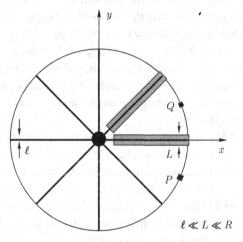

Fig. 2.12 The cross section of the domain-wall junction in the perpendicular plane. An eight-wall junction is shown.

"resides" in the first vacuum, at the point Q in the second, etc. This configuration is topologically stable.

First let us discuss general features of the tension associated with the wall junctions. In Fig. 2.12 the energy of the junction configuration (per unit length) is defined as the integral of the volume energy density over the area inside the circle, where it is assumed that the radius R of the circle tends to infinity:

$$E(R) = \frac{E_{\text{tot}}}{\text{length}} = \int_{|\vec{r}| \le R} \mathcal{E}(x, y)\, dx\, dy = T_1 R + T_2 + O(1/R), \qquad R \to \infty. \quad (2.44)$$

It is assumed that the parameters of the problem have been adjusted in such a way that the vacuum energy vanishes. This ensures that there is no R^2 term on the right-hand side of Eq. (2.44).

Defining the junction tension

It is intuitively clear that $T_1 = nT_{\text{w}}$, where T_{w}, is the tension of the isolated wall and n is the number of walls meeting at the junction. The quantity T_2 is the wall junction

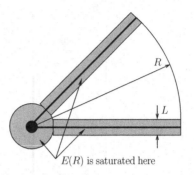

Fig. 2.13 A detail of Fig. 2.12. The wall junction and two neighboring walls are inside the shaded area.

tension. From now on it will be referred to as T_j, so that Eq. (2.44) takes the form

$$E(R) = nT_w R + T_j + O(1/R), \qquad R \to \infty. \tag{2.45}$$

A general proof of the fact that $T_1 = nT_w$, is quite straightforward. Of crucial importance is the fact that the wall thickness (i.e., the transverse dimension inside which the energy density is nonvanishing, while outside it vanishes with exponential accuracy) is R-independent at large R. This width is denoted by ℓ; see Fig. 2.12.

Figure 2.13 presents part of the junction configuration inside the circle $|\vec{r}| \leq R$. The rectangles around the spokes have width L, where L is an auxiliary parameter chosen to be much larger than the spoke width ℓ: $L \gg \ell$. In the limit $R \to \infty$ the width L stays fixed.

Outside the shaded areas the energy density $\mathcal{E}(x, y)$ vanishes, since the fields are at their vacuum values. The integral (2.44) is saturated within the near-hub circular domain of radius $\sim L$ and within the rectangles. Each rectangle obviously yields $T_w R$ plus terms that do not grow with R in the limit of large R. The latter are due to the fact that the expression $nT_w R$ does not correctly represent the circular domain of radius $\sim L$ around the hub (represented by the black circle in Fig. 2.12). This remark completes the proof of Eq. (2.45).

2.2.2 A Z_n Model with Wall Junctions

Now that we have discussed the definitions we can address the underlying dynamics. If the spontaneously broken discrete symmetry is Z_2, there are no stable static wall junctions (see Exercise 2.2.1 at the end of this section). They appear only for higher discrete symmetries, such as Z_n with $n \geq 3$. We will assume that the Z_n symmetry is realized through multiplication of (some of) the fields in the problem at hand by a phase, the simplest possibility.

In the theory of a single scalar field ϕ the Z_n symmetry with $n \geq 3$ can be realized as an invariance of the Lagrangian under multiplication by the phase $\exp(2\pi i k/n)$, where $k = 1, 2, \ldots, n$:

A sample model

$$\phi \to \exp\left(\frac{2\pi i k}{n}\right)\phi, \qquad k = 1, 2, \ldots, n. \tag{2.46}$$

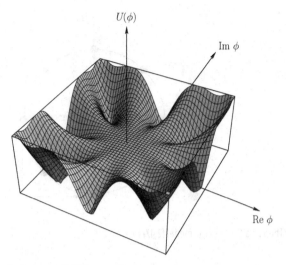

Fig. 2.14   The potential energy in the model (2.47) for $n = 8$.

Needless to say, it is necessary to have a *complex* field – a real field cannot do the job. The Z_n-symmetric Lagrangian with which we will deal is[4]

$$\mathcal{L} = \partial_\mu \bar{\phi} \partial^\mu \phi - U(\phi, \bar{\phi}), \qquad U(\phi, \bar{\phi}) = \mu^2 (1 - \nu \phi^n)(1 - \nu \bar{\phi}^n), \tag{2.47}$$

where the bar denotes complex conjugation and μ and ν are constants that can be chosen to be real and positive without loss of generality. The mass dimensions of μ and ν depend on D. In four dimensions the field ϕ has the dimension of mass; hence $\mu \propto [m]^2$ and $\nu \propto [m]^{-n}$. The potential (2.47) is depicted in Fig. 2.14.

The kinetic term in the Lagrangian (2.47) is in fact invariant under a larger symmetry, U(1), acting as $\phi \rightarrow \exp(i\alpha)\,\phi$ with arbitrary phase α. The potential term is invariant under the transformation (2.46).

In the vacuum the Z_n invariance of the Lagrangian is spontaneously broken; see Fig. 2.14. Correspondingly, there are n distinct vacuum states

$$\phi_{\text{vac}} = \nu^{-1/n} \exp\left(\frac{2\pi i\, k}{n}\right), \qquad k = 1, 2, \ldots, n, \tag{2.48}$$

where $\nu^{-1/n}$ is the arithmetic value of the root. The positions of the vacua in the complex ϕ plane are depicted in Fig. 2.15 by solid circles. At the positions of the circles $U(\phi)$ vanishes; at all other values of ϕ the potential $U(\phi)$ is strictly positive. As we already know, all n vacua are physically equivalent.

It is instructive to calculate the mass of an elementary excitation. To this end one must consider small oscillations near the vacuum value of ϕ. Since all the vacua are physically equivalent we can consider, for instance,

$$\phi = \nu^{-1/n} + \frac{\varphi + i\chi}{\sqrt{2}}, \tag{2.49}$$

where φ and χ are real fields.

---

[4] The model described by the Lagrangian (2.47) is by no means the most general possessing Z_n symmetry. At $n \geq 3$ it is nonrenormalizable. Since at the moment we are not interested in quantum corrections it will suit our purposes well.

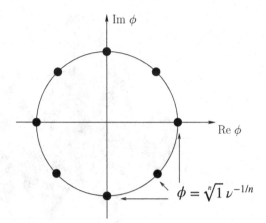

Fig. 2.15 The vacua (2.48) of the model (2.47) for $n = 8$.

Next, we follow a standard routine. Substitute Eq. (2.49) into Eq. (2.47) and expand the Lagrangian, keeping terms not higher than quadratic (the linear terms cancel). This quite straightforward calculation yields

$$m_\varphi = m_\chi = n\mu v^{1/n}. \tag{2.50}$$

Thus the mass of the two real scalars is degenerate. This is a special feature of the potential (2.47).

2.2.3 Elementary and Composite Walls

With n vacua one can have many distinct types of wall. For instance, one can have a wall separating the first vacuum from the second, the first from the third, and so on. A special role belongs to the so-called *elementary walls* – the walls separating two neighboring vacua. Note that nonelementary walls need not necessarily exist; they may turn out to be unstable. For example, in some models the wall separating the first vacuum from the third may decay into two elementary walls – the first–second and second–third – which experience mutual repulsion and eventually separate to infinity. The existence or nonexistence of nonelementary walls depends on the dynamical details of the model at hand. Elementary walls always exist. In Figs. 2.11 and 2.12 all the walls shown are elementary. In this case it is clear from the symmetry of the model that the minimal energy configuration is achieved if all relative angles between the walls are the same: $2\pi/n$.

2.2.4 Equation for the Wall Junction

The equation describing a wall-junction configuration is a two-dimensional reduction of the general classical equation of motion, which takes into account that the solution in which we are interested does not depend on time (i.e., it is static) or on z:

$$\left(\partial_x^2 + \partial_y^2\right)\phi = \frac{\partial U(\phi, \bar{\phi})}{\partial \bar{\phi}}. \tag{2.51}$$

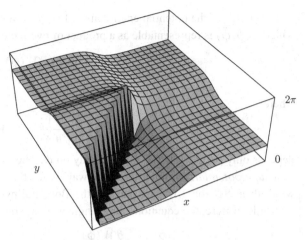

Fig. 2.16   Phase of the field ϕ for the domain-wall junction.

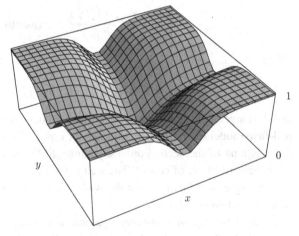

Fig. 2.17   Modulus of the field ϕ for the domain-wall junction.

The complex conjugate equation holds for $\bar{\phi}$. Moreover, appropriate boundary conditions must be imposed.

While Eq. (2.51) is general, the boundary conditions depend on the details of the model. In the model under consideration, where the vacuum pattern is fairly simple, see Eq. (2.48), the boundary conditions are obvious: (i) one should choose a solution $\phi(x, y)$ of Eq. (2.51) such that $\arg \phi(x, y)$ changes from 0 to 2π as we travel in the xy plane around a large circle centered at the origin (where the wall junction is assumed to lie); (ii) the solution must be symmetric under rotations in the xy plane by an angle $2\pi/n$. The first requirement, in conjunction with continuity of the solution, implies that

$$\phi(x, y) \to 0 \qquad \text{as} \qquad \sqrt{x^2 + y^2} \to 0.$$

Both features are clearly seen in Figs. 2.16 and 2.17, which display a numerical wall-junction solution of Eq. (2.51) for the model (2.47) with $n = 4$ and $v = 1$. The plots are taken from [9].

The choice of the potential energy in the Lagrangian (2.47) is the special case for which $U(\phi, \bar{\phi})$ is representable as a product of two factors:

$$U(\phi, \bar{\phi}) \equiv \frac{\partial \mathcal{W}(\phi)}{\partial \phi} \frac{\partial \bar{\mathcal{W}}(\bar{\phi})}{\partial \bar{\phi}}, \tag{2.52}$$

where

$$\mathcal{W}(\phi) = \mu \left(\phi - \frac{\nu}{n+1} \phi^{n+1} \right) \tag{2.53}$$

depends only on ϕ while $\bar{\mathcal{W}}$ depends only on $\bar{\phi}$. The function $\mathcal{W}(\phi)$ is referred to as a *superpotential*. In much the same way as for the real-field model with which we dealt in Section 2.1.5, we can use the Bogomol'nyi construction to derive the first-order differential equations for a single wall and the wall junction. Namely,

The BPS equations

$$\frac{\partial \phi}{\partial x} = e^{i\alpha} \frac{\partial \bar{\mathcal{W}}(\bar{\phi})}{\partial \bar{\phi}} \qquad \text{(single wall)}, \tag{2.54}$$

$$\frac{\partial \phi}{\partial \xi} = \frac{1}{2} e^{i\alpha} \frac{\partial \bar{\mathcal{W}}(\bar{\phi})}{\partial \bar{\phi}} \qquad \text{(junction)}, \tag{2.55}$$

where

$$\xi \equiv x + iy, \qquad \frac{\partial}{\partial \xi} \equiv \frac{1}{2} \left(\frac{\partial}{\partial x} - i \frac{\partial}{\partial y} \right),$$

and α is a phase to be determined by the boundary conditions, see below.[5] (In the real-field model $e^{i\alpha} = \pm 1$.) The solutions of Eqs. (2.54) and (2.55) are automatically the solutions of the second-order equation (2.51) for arbitrary α. The opposite is not necessarily true, of course. The wall solution of Eq. (2.54), $\phi(x)$, depends only on the single coordinate x. For the wall junction, Eq. (2.55), the solution $\phi(\xi, \bar{\xi})$ depends on two coordinates.

Let us show that the first-order equations above imply the second-order equation. We will do this exercise for, say, the wall-junction solution. To this end we differentiate both sides of Eq. (2.55) with respect to $\bar{\xi}$:

$$\frac{\partial}{\partial \bar{\xi}} \frac{\partial \phi}{\partial \xi} = \frac{1}{2} e^{i\alpha} \frac{\partial^2 \bar{\mathcal{W}}(\bar{\phi})}{\partial \bar{\phi}^2} \frac{\partial \bar{\phi}}{\partial \bar{\xi}} = \frac{1}{4} \frac{\partial^2 \bar{\mathcal{W}}(\bar{\phi})}{\partial \bar{\phi}^2} \frac{\partial \mathcal{W}(\phi)}{\partial \phi}, \tag{2.56}$$

where in the last formula on the right-hand side we have exploited the complex conjugate of (2.55). Using the definition (2.52) it is easy to see that Eq. (2.56) is equivalent to

$$4 \frac{\partial^2 \phi}{\partial \bar{\xi} \partial \xi} = \frac{\partial U(\phi, \bar{\phi})}{\partial \bar{\phi}}, \tag{2.57}$$

which is, in turn, equivalent to (2.51).

We will pause here to try to understand how the boundary conditions determine the value of the phase α in Eq. (2.54). This equation refers to complex ϕ and $\mathcal{W}$, therefore, even though the equation is first order, our intuition is not nearly as helpful in this case as it was in the real-field model. We have to rely on the mathematics. A conservation law that exists in this problem will help us. Consider the derivative

"An integral of motion"

---

[5] For complex scalar field models with potential energy (2.52) Eq. (2.55) was derived in [10].

$$\frac{\partial}{\partial x}\left(e^{-i\alpha}\mathcal{W} - e^{i\alpha}\bar{\mathcal{W}}\right) = e^{-i\alpha}\frac{\partial \mathcal{W}}{\partial \phi}\frac{\partial \phi}{\partial x} - e^{i\alpha}\frac{\partial \bar{\mathcal{W}}}{\partial \bar{\phi}}\frac{\partial \bar{\phi}}{\partial x}. \tag{2.58}$$

Now, using Eq. (2.54) and its complex conjugate we immediately conclude that the right-hand side vanishes. In other words,

$$\mathrm{Im}(e^{-i\alpha}\mathcal{W}) \tag{2.59}$$

is conserved on the wall solution, i.e., it is independent of x. Our task is to put this conservation law to work.

Assume for definiteness that the wall which we are going to construct interpolates between $\phi_{\mathrm{vac}} = v^{-1/n}$ and $\phi_{\mathrm{vac}} = v^{-1/n}\exp(2\pi i/n)$. Then

$$\mathcal{W}_{\mathrm{initial}} = \mu\phi\left(1 - \frac{v}{n+1}\phi^n\right)_{\phi = v^{-1/n}} = \frac{n}{n+1}\mu v^{-1/n}, \tag{2.60}$$

$$\mathcal{W}_{\mathrm{final}} = \mu\phi\left(1 - \frac{v}{n+1}\phi^n\right)_{\phi = v^{-1/n}\exp(2\pi i/n)}$$

$$= \frac{n}{n+1}\mu v^{-1/n}\exp\left(\frac{2\pi i}{n}\right).$$

Since $\mathrm{Im}\,(e^{-i\alpha}\mathcal{W})$ is conserved for the wall solution, comparing the initial and the final points we arrive at a condition on α, namely

$$\sin\alpha = \sin\left(\alpha - \frac{2\pi}{n}\right). \tag{2.61}$$

Its solution appropriate for our case is

$$\alpha = \frac{\pi}{2} + \frac{\pi}{n}; \tag{2.62}$$

| *Equation (2.62) is in agreement with (2.63).* |

see Fig. 2.18.

It is obvious that in the case at hand Eq. (2.62) is identical to

$$\alpha = \arg(\mathcal{W}_{\mathrm{final}} - \mathcal{W}_{\mathrm{initial}}). \tag{2.63}$$

In fact, this latter equation is universal: it is valid (i.e., it determines the phase α in Eq. (2.54)) in generic models with potential energy of the form (2.52).

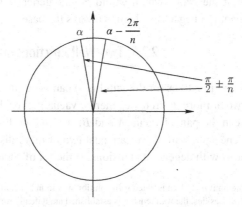

Fig. 2.18   Determination of the phase α in Eq. (2.54) from the boundary conditions.

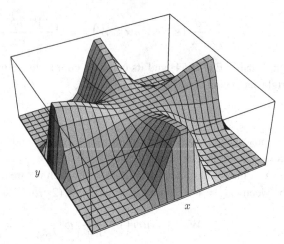

Fig. 2.19 Energy density of the domain-wall junction.

Unfortunately, in the model under consideration, analytic solutions are known neither for junctions nor even for isolated walls.[6] A few multifield models that admit analytic wall-junction solutions have been discussed in the literature (see, e.g., [13]). We will not consider them here because of their rather contrived structure. Instead, let us examine the energy density distribution for the wall-junction solution presented in Figs. 2.16 and 2.17. Figure 2.19 shows

$$\mathcal{E}(x, y) = U + \partial_x \bar{\phi} \partial_x \phi + \partial_y \bar{\phi} \partial_y \phi$$

as a function of x, y. It is clearly visible that four domain walls join each other in the junction, located at the origin, and that the energy density in the junction is lower than that in the core of the walls. This fact implies, in particular, that

$$T_{\mathrm{j}} < 0. \tag{2.64}$$

This negative tension of the wall junction is typical. For isolated objects, say, walls or strings, a negative tension cannot exist since then such objects would be unstable: they would crumple. The negativity of T_{j} does not necessarily lead to instability, however, since the wall junction does not exist in isolation; it is always attached to walls that have a positive tension. If the junction crumpled then so would the adjacent areas of the walls, which would be energetically disadvantageous provided that T_{j} were not too negative, which is always the case.

2.2.5 Two-Wall Junctions in $D = 1 + 3$

Two-wall junctions (or domain lines) can occur in theories of a special type supporting two or more discrete degenerate vacua. Let us discuss the wall interpolating, say, between the pair of vacua A and B. It turns out that in some theories there is more than one such wall: there can exist two or more distinct walls interpolating between A and B with degenerate tensions. If the set of such walls is discrete, they may form

---

[6] In the limit $n \gg 1$ an isolated wall solution was found [11] in the leading and the next-to-leading order in $1/n$. Besides, the wall tension is established analytically for any n while the junction tension only for $n \gg 1$; see [12].

stable two-wall junctions similar to those in Fig. 2.9 if you imagine that this figure is drawn in three spatial dimension rather than in two. The necessary condition for the theory to support two-wall junctions is a number of spontaneously broken discrete symmetries. Some of them will be responsible for the vacuum degeneracies. Some others, unbroken in the vacua, must be broken on the AB wall, which will guarantee that these walls with different inner structures still have one and the same tension.

Here I will briefly discuss only one simple example obtained from (2.1) by a straightforward generalization:

$$S = \int d^4x \left\{ (\partial_\mu \bar\phi)(\partial^\mu \phi) - \frac{g^2}{4} \left(\bar\phi\phi - v^2\right)^2 - \frac{\varkappa^2}{4}(2\bar\phi\phi - \phi^2 - \bar\phi^2) \right\}, \quad (2.65)$$

where ϕ is a complex scalar field and $\varkappa$ is a small parameter of dimension of mass,

$$\varkappa^2 \ll g^2 v^2. \quad (2.66)$$

Both parameters, $\varkappa$ and v are assumed to be real and positive.

If we put $\varkappa = 0$ the theory (2.65) has the phase U(1) global symmetry. With $\varkappa \neq 0$ the above U(1) is explicitly broken down to $Z_2 \times Z_2$ with regards to

$$\phi \to -\phi \quad \text{and} \quad \phi \leftrightarrow \bar\phi \quad (2.67)$$

transformations. The first Z_2 in (2.67) is spontaneously broken in the vacua,

$$\phi_{\text{vac}} = \bar\phi_{\text{vac}} = \pm v. \quad (2.68)$$

The above two vacua are degenerate; their energy density $\mathcal{E}_{\text{vac}} = 0$. Two distinct real degrees of freedom – two excitations over any of the two vacua – have masses $m_1 = gv$ and $m_2 = \varkappa$ (for the real and imaginary parts of ϕ, respectively).

What is the inner structure of the domain wall interpolating between $\phi_{\text{vac}} = v$ and $\phi_{\text{vac}} = -v$? Given the constraint (2.66) it is rather obvious that the wall solution sought has the form

$$\phi_{\text{w}}(z) = v \exp(i\alpha(z)) \quad (2.69)$$

with the boundary condition

$$\alpha(z = -\infty) = 0, \quad \alpha(z = \infty) = \pi. \quad (2.70)$$

The absolute value of ϕ does not change on the trajectory, hence heavy degrees of freedom (with mass m_1) are not excited inside the wall; see the second term in Eq. (2.65), which vanishes on the solution. This is energetically beneficial.

Then, the effective Lagrangian for the light degree of freedom α reduces to

$$\mathcal{L} = v^2 (\partial_\mu \alpha)(\partial^\mu \alpha) - \varkappa^2 v^2 (\sin \alpha)^2, \quad (2.71)$$

see (2.65) and (2.69). Using the same line of reasoning as in Sections 2.1.3 and 2.1.4 and taking into account the boundary conditions (2.70) we derive the first-order differential equation analogous to (2.9),

$$\frac{d\alpha_{\text{w}}}{dz} = \varkappa \sin \alpha_{\text{w}}. \quad (2.72)$$

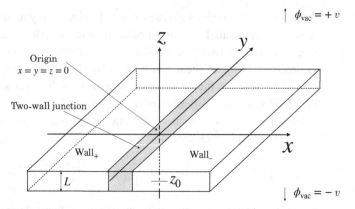

Fig. 2.20 Two distinct walls interpolatinng between two vacua (2.68) with a junction (domain line) between them

It is easy to find the solution to this equation with the given boundary conditions:[7]

$$\alpha_{\rm w} = 2\left[{\rm Arctan}\left(e^{\varkappa z - \varkappa z_0}\right)\right]. \tag{2.73}$$

Here z_0 is the z coordinate of the wall center; see Fig. 2.20. The plot of the function (2.73) is similar to that in Fig. 2.6 with a scale change on the vertical axis; namely, at $z = -\infty$ the curve asymptotically approaches zero while at $z = \infty$ it approaches π. Note that for $\forall z$ in Eq. (2.73) $\alpha_{\rm w} \geq 0$.

Now, let us return to the Z_2 invariance $\phi \leftrightarrow \bar{\phi}$ of our Lagrangian (2.65), which was not broken by the choice of our two vacuua in Eq. (2.68). The domain wall trajectory

$$\phi_{\rm w}(z) = v e^{i\alpha_{\rm w}(z)} \tag{2.74}$$

does break this Z_2. Therefore, another solution

$$\phi_{\rm w}(z) = v e^{-i\alpha_{\rm w}(z)} \tag{2.75}$$

must exist too. And it does! Clearly, the two solutions above have different inner structures. And yet, the tensions of both walls are the same; they can be readily obtained from the Bogomol'nyi completion,

$$T_{\rm w} = 4\varkappa v^2. \tag{2.76}$$

Their thickness in the z direction is $L \sim 1/\varkappa$ (Fig. 2.20).

Thus, we identified two distinct domain walls – both interpolating between $\phi_{\rm vac} = v$ and $\phi_{\rm vac} = -v$. The phase of ϕ inside the first wall is positive (wall$_+$) while in the second negative (wall$_-$). Their tensions are degenerate. Note that the following nonlocal parameter may be used to label these walls,[8] namely, walls$_\pm$,

$$\frac{1}{\pi}\frac{\varkappa}{v}\int_{-\infty}^{+\infty} dz\,{\rm Im}\,\phi(z) = \pm 1. \tag{2.77}$$

Due to the tension degeneracy, walls$_\pm$ can form a two-wall junction shown in Fig. 2.9 (a domain line). Assume this domain line is aligned along the y axis and its center is at $x = 0$. If we pierce the wall in the z direction at $x = 0$ and arbitrary

---

[7] The position of the wall center on the z axis is denoted by z_0. Its value can be arbitrary.
[8] As we will see below, in the presence of the wall junction one must calculate the integral in (2.77) at $|x| \gg \varkappa^{-1}$, see Fig. 2.20.

y the symmetry of the model tells us that the solution interpolating between the two vacua and passing through the middle of the junction must be a real function of z,

$$\text{Im}\,\phi(z)\Big|_{x=0} = 0, \quad \phi(z = -\infty) = v, \quad \phi(z = +\infty) = -v. \tag{2.78}$$

Equation (2.78) implies that the third term in (2.65) vanishes and the problem reduces to that discussed in Section 2.1.3. The only change we need to do is to rescale g^2 in (2.65), namely, $g^2 \to 2g^2$. Then the solution takes the form

$$\phi(z)\Big|_{x=0} = -v \tanh\left[gv\,(z - z_0)\right]. \tag{2.79}$$

Now it is obvious that in the middle of the junction (near $z = z_0$) heavy degrees of freedom (i.e., those of mass m_1) are excited. Restructuring of these degrees of freedom creates a hard core of the wall junction with the thickness $\sim (gv)^{-1}$. The restructuring mentioned above results in the restructuring of the light degrees of freedom that are attached as tails to the hard core in the x and z directions. The size of the tales $\sim \varkappa^{-1}$. Finally, the overall tension of the wall junction is

$$T_{\text{w junction}} \sim v^2. \tag{2.80}$$

Exercises

2.2.1 Explain why there are no stable wall junctions in the model (2.1) of Section 2.1, with the spontaneously broken Z_2 symmetry and doubly degenerate vacuum states.

2.2.2 Explain why for the wall junction, Eq. (2.55), the phase α can be eliminated.

2.2.3 Calculate the tension of the elementary wall in the model (2.47) in the limit $n \gg 1$, using the Bogomol'nyi construction. Find the condition on the parameters of the model under which $T_w/m_{\varphi,\chi}^3 \gg 1$. This is the condition of applicability of the quasiclassical approximation. The φ, χ fields are defined in Eq. (2.49).

2.2.4 Calculate the tension of the maximal wall junction for the model (2.47) in the limit $n \gg 1$. In the problem at hand we define the maximal wall junction as the junction with n spokes.

2.3 Domain Walls Antigravitate

So far we have ignored gravity. This is certainly an excellent approximation since gravity is extremely weak and usually cannot compete with other forces. However, if domain walls exist as cosmic objects in the universe, their gravitational interaction certainly cannot be neglected. In this section we will become acquainted with a remarkable facts: the gravitational field of domain walls in $D = 1 + 3$ is *repulsive* rather than attractive [14, 15]. This is the first example of *antigravity*, the dream of all science-fiction writers. Even though this observation will remain, most probably, a theoretical curiosity and will have no practical implications, it provides an interesting exercise, quite appropriate for this course.

2.3.1 Coupling the Energy–Momentum Tensor to a Graviton

To derive this "antigravity" result we will need to know a few facts from Einstein's relativity and from nonrelativistic quantum mechanics. I hasten to add that we do not assume a thorough knowledge of Einstein's relativity, just some basic notions: that gravity is mediated by gravitons, that gravitons are described by a massless spin-2 field $h_{\mu\nu}$, and that the interaction of $h_{\mu\nu}$ and matter occurs through the universal coupling of $h_{\mu\nu}$ to the energy–momentum tensor of the matter fields, $T^{\mu\nu}$:

$$\Delta\mathcal{L}_{\text{grav}} = \frac{1}{M_{\text{P}}} h_{\mu\nu} T^{\mu\nu}, \tag{2.81}$$

Linearized gravity is OK here.

where M_{P} is the Planck mass. The interaction (2.81) neglects nonlinear gravity effects. In the problem at hand we are assuming that $T_{\text{W}} \ll M_{\text{P}}^3$ since then there is no need to consider nonlinear effects.

The energy–momentum tensor is a symmetric conserved tensor. Its particular form depends on the model under consideration. The general rule for deriving $T^{\mu\nu}$ is as follows.

(i) Write the action of the model in general covariant form. For instance, in the real-field model of Section 1.1.2 we have

$$S = \int d^D x \sqrt{-g} \left[\frac{1}{2} g_{\mu\nu} (\partial^\mu \phi)(\partial^\nu \phi) - U(\phi) \right], \tag{2.82}$$

where $g_{\mu\nu}$ is the metric and $g \equiv \det(g_{\mu\nu})$. Note that in the actual world g is always negative, so that $\sqrt{-g}$ is defined unambiguously as the arithmetic value of the square root.

(ii) Differentiate the integrand in Eq. (2.82) with respect to $g_{\mu\nu}$ and set the metric to be flat after differentiation:

$$T^{\mu\nu} = 2 \frac{\delta \int \left[\sqrt{-g} \mathcal{L}(\phi, g_{\mu\nu}) \right]}{\delta g_{\mu\nu}} \Bigg|_{g_{\mu\nu} = \eta_{\mu\nu}}, \qquad \eta_{\mu\nu} = \text{diag}\{1, -1, -1, -1\}. \tag{2.83}$$

Alternatively, one could represent the metric $g_{\mu\nu}$ as $\eta_{\mu\nu}$ plus small fluctuations,

$$g_{\mu\nu} = \eta_{\mu\nu} + \frac{1}{M_{\text{P}}} h_{\mu\nu},$$

and linearize Eq. (2.82) with respect to $h_{\mu\nu}$.[9]

In this way we obtain that in the model at hand the energy–momentum tensor is

$$T^{\mu\nu} = (\partial^\mu \phi)(\partial^\nu \phi) - \eta^{\mu\nu} \left[\tfrac{1}{2} (\partial^\rho \phi)(\partial_\rho \phi) - U(\phi) \right]. \tag{2.84}$$

---

[9] In fact in the theory of scalar fields, the energy–momentum tensor is not unambiguously defined by the above procedure. So-called improvement terms are possible. The improvement terms are conserved by themselves, nondynamically, i.e., without the use of the equations of motion. Being full derivatives they do not change the energy–momentum operator P^μ. For instance, in the example under consideration one can add $g^{\mu\nu} \Box \phi^2 - \partial^\mu \phi \partial^\nu \phi$ to the energy–momentum tensor; cf. Eq. (10.470) and the subsequent discussion. This corresponds to the addition of $\int \sqrt{-g} R \phi^2$ to the action, where R is the scalar curvature. Improvement terms would not affect our derivation.

Note that $\sqrt{-g} = 1 + \frac{1}{2}M_{\mathrm{P}}^{-1}h_{\mu\nu}\eta^{\mu\nu}$ plus (irrelevant) terms that are quadratic or high-order in h. This expression is obviously symmetric. It is instructive to check directly the conservation of $T^{\mu\nu}$. Let us calculate the divergence:

$$\partial_\mu T^{\mu\nu} = (\Box\phi)(\partial^\nu\phi) + (\partial^\mu\phi)(\partial_\mu\partial^\nu\phi) - (\partial^\rho\phi)(\partial_\rho\partial^\nu\phi) + \frac{\partial U(\phi)}{\partial\phi}(\partial^\nu\phi)$$

$$= \left[\Box\phi + \frac{\partial U(\phi)}{\partial\phi}\right](\partial^\nu\phi) = 0, \tag{2.85}$$

where the second line vanishes because it is proportional to the equation of motion.

2.3.2 The Domain-Wall Energy–Momentum Tensor Is Unusual

The next step in our antigravity calculation is to find the energy–momentum tensor for the domain wall and for an isolated localized object (say, a metal ball or a planet). In what follows it will be assumed that the measurement of gravity is done at distances much larger than the typical sizes of the gravitating bodies; see Fig. 2.21.

For a domain wall at rest we have

$$T_{\mathrm{w}}^{\mu\nu} = T_{\mathrm{w}}\,\mathrm{diag}\{1, -1, -1, 0\}, \tag{2.86}$$

where T_{w} is the wall tension and it is assumed that the wall lies in the xy plane. For an isolated localized *nonrelativistic* body (a particle, a ball, or a planet)

$$T_{\mathrm{body}}^{\mu\nu} = M\,\mathrm{diag}\{1, 0, 0, 0\}, \tag{2.87}$$

where M is the total mass of the body.

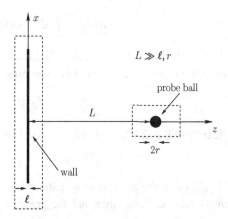

Fig. 2.21 The gravitational interaction between a domain wall and a distant localized body. The broken rectangles denote the integration domains for determination of the corresponding energy–momentum tensors. The distance L is assumed to be much larger than l and r.

A few comments on the derivation of the above expressions are in order. Let us start with the domain wall. In the rest frame the domain-wall profile ϕ_w depends only on z. Therefore, only the z derivatives survive in Eq. (2.84):

$$
T_w^{\mu\nu} = \begin{cases} 0 & \text{if } \mu \neq \nu, \\ \left[\frac{1}{2}(\partial_z\phi)^2 + U(\phi)\right] & \text{if } \mu = \nu = 0, \\ -\left[\frac{1}{2}(\partial_z\phi)^2 + U(\phi)\right] & \text{if } \mu = \nu = x \text{ or } y, \\ \left[\frac{1}{2}(\partial_z\phi)^2 - U(\phi)\right] & \text{if } \mu = \nu = z. \end{cases}
\tag{2.88}
$$

Since we are supposing that the effect under investigation is measured far from the wall, we should integrate over z for the domain where the wall is located (see Fig. 2.21). To this end we observe that on the one hand

$$
\int dz\, \tfrac{1}{2}(\partial_z\phi_w)^2 = \int dz\, U(\phi_w) = \tfrac{1}{2}T_w,
\tag{2.89}
$$

which immediately leads to Eq. (2.86).

On the other hand, for an isolated on-mass-shell particle with momentum p and mass M,

$$
\langle p|T^{\mu\nu}|p\rangle = \frac{1}{2E}2p^\mu p^\nu, \qquad E \equiv p^0.
\tag{2.90}
$$

In the rest frame this is the same as Eq. (2.87)

2.3.3 Repulsion from the Walls

To see that a probe body experiences repulsion in the gravitational field generated by the wall, we will compare the interaction of two probe bodies with that between the wall and a probe body. As is well known, the structure of the potential can be inferred from the Born graph describing the scattering of two interacting bodies (Fig. 2.22). According to the Born formula,[10] the scattering amplitude $A(\vec{q})$ is proportional to the Fourier transform of the potential,

$$
A(\vec{q}) \propto \int dx\, V(\vec{x})e^{-i\vec{q}\vec{x}},
\tag{2.91}
$$

where $V(\vec{x})$ is the interaction potential. The inverse of this formula gives the potential in terms of the Fourier transform of the scattering amplitude,

$$
V(\vec{x}) \propto \int dq\, A(\vec{q})\, e^{i\vec{q}\vec{x}}.
\tag{2.92}
$$

As is clear from Fig. 2.22, the Born scattering amplitude is proportional to

$$
T^{(1)\mu\nu} D_{\mu\nu,\alpha\beta}(q)\, T^{(2)\,\alpha\beta},
\tag{2.93}
$$

where $D_{\mu\nu,\alpha\beta}(q)$ is the graviton propagator, which in turn is proportional to the graviton density matrix. Remember that the graviton is described by a massless spin-2 field,

$$
D_{\mu\nu,\alpha\beta}(q) \propto \frac{1}{q^2}\left(\eta_{\mu\alpha}\eta_{\nu\beta} + \eta_{\mu\beta}\eta_{\nu\alpha} - \eta_{\mu\nu}\eta_{\alpha\beta} + \text{longitudinal terms}\right),
\tag{2.94}
$$

---

[10] See [7], section 126.

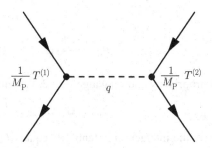

Fig. 2.22 The Born graph for the scattering of two bodies due to one-graviton exchange. The broken line denotes the graviton propagator.

where the longitudinal terms contain the momentum q. These longitudinal terms (which are gauge dependent) are irrelevant since they drop out upon multiplication by $T^{(1)\mu\nu}$ or $T^{(2)\,\alpha\beta}$, because of the transversality of the energy–momentum tensor.

One-graviton exchange

Combining Eqs. (2.93) and (2.94) we arrive at the conclusion that the interaction potential $V(\vec{x})$ can be written as

$$V(\vec{x}) \propto \frac{1}{M_{\mathrm{P}}^2} \left(2T^{(1)\,\mu\nu} T^{(2)}_{\mu\nu} - T^{(1)\,\mu}_{\mu} T^{(2)\,\nu}_{\nu} \right) \times (\text{Fourier transform of } - 1/\vec{q}^2). \quad (2.95)$$

The expression in parentheses determines the sign of the interaction between the two bodies. Let us calculate it for three distinct cases:

$$2T^{(1)\mu\nu} T^{(2)}_{\mu\nu} - T^{(1)\mu}_{\mu} T^{(2)\nu}_{\nu} = \begin{cases} T^{(1)00}T^{(2)00} = M^{(1)}M^{(2)} & \text{(ball–ball)}, \\ -T^{(1)00}T^{(2)00} = -T^{(1)}M^{(2)} & \text{(wall–ball)}, \quad (2.96) \\ -3T^{(1)00}T^{(2)00} = -3T^{(1)}T^{(2)} & \text{(wall–wall)}, \end{cases}$$

where $T^{(1,2)}$ are the wall tensions. To ease the notation I have dropped the subscript w. It is worth noting that we have assumed the walls to be parallel to each other in the case of the wall–wall interaction. We see that if the gravitational interaction between two localized probe bodies (balls at rest) is attractive – which is certainly the case – then the gravitational interaction between two distant walls and between a wall and a ball is repulsive. Note that the corrections due to the motion of the probe bodies relative to the walls (which are taken to be at rest) are proportional to powers of their velocity v, a small parameter in the nonrelativistic limit. Equation (2.96) reproduces Newton's well-known law according to which the gravitational potential of two distant nonrelativistic bodies is proportional to the product of their masses. For the walls it is their tension that enters.

Instead of determining the interaction from the Born scattering amplitudes, one could follow a more traditional route and solve the Einstein equations for a source term generated by the wall,

$$R_{\mu\nu} - \tfrac{1}{2} g_{\mu\nu} R = \frac{1}{M_{\mathrm{P}}^2} T_{\mathrm{w},\mu\nu}, \quad (2.97)$$

where $R_{\mu\nu}$ is the Ricci tensor and R is the scalar curvature [16]. Convoluting both sides with $g^{\mu\nu}$ one finds that the scalar curvature is given by

$$R = -M_{\mathrm{P}}^{-2} T^{\alpha}_{\mathrm{w},\alpha}$$

and, hence,

$$R_{\mu\nu} = \frac{1}{M_{\rm P}^2}(T_{{\rm w},\mu\nu} - \tfrac{1}{2}g_{\mu\nu}T_{{\rm w},\alpha}^\alpha). \tag{2.98}$$

In an appropriately chosen gauge Eq. (2.98) implies that

$$h_{\mu\nu} \propto \Box^{-1}(T_{{\rm w},\mu\nu} - \tfrac{1}{2}g_{\mu\nu}T_{{\rm w},\alpha}^\alpha). \tag{2.99}$$

Of course, the interaction potential V equals $T_{\mu\nu}^{(2)}h_{\mu\nu}$. This returns us to Eq. (2.96) and simultaneously confirms the formula for the graviton density matrix given in Eq. (2.94).

Suppose that we are interested not only in the sign of the gravity interaction but also in its functional form, i.e., the dependence on the distance between two gravitating bodies. As follows from Eq. (2.95), to find this dependence one has to perform the Fourier transform of $1/\vec{q}^2$ in various numbers of dimensions δ: 1, 2, or 3. One encounters similar Fourier transformations in numerous other problems. It makes sense to derive here a general formula:

The Fourier transform formula

$$\int d^\delta \vec{q}\, e^{i\vec{x}\vec{q}}\left(\frac{1}{\vec{q}^2}\right)^n = 2^{\delta/2}\,\pi^{\delta/2}\,x^{1-\delta/2}\int dq\, q^{\delta/2-2n}J_{\delta/2-1}(qx)$$

$$= 2^{\delta-2n}\,\pi^{\delta/2}\frac{\Gamma(\delta/2-n)}{\Gamma(n)}x^{2n-\delta}, \tag{2.100}$$

where

$$q \equiv |\vec{q}|, \qquad x \equiv |\vec{x}|,$$

$J_{\delta/2-1}$ is a Bessel function and δ and n are treated as arbitrary integers such that the integral (2.100) exists. The first line in Eq. (2.100) is obtained upon integration over the angle between $\vec{x}$ and $\vec{q}$ and the second line presents the result of integration over $|\vec{q}|$. A few important particular cases are as follows:

$$-\int d^\delta \vec{q}\, e^{i\vec{x}\vec{q}}\frac{1}{\vec{q}^2} = \begin{cases} 2\pi^2\left(-\frac{1}{|\vec{x}|}\right), & \delta = 3, \\ \pi|\vec{x}|, & \delta = 1. \end{cases} \tag{2.101}$$

The first expression gives the gravitational interaction of two localized bodies (we recover the familiar $1/r^2$ Newtonian force) and the second the wall–ball interaction. Here the force is distance-independent, in full accord with intuition.

Exercises

2.3.1   The Lagrangian of a free photon field is

$$\mathcal{L} = -\tfrac{1}{4}F_{\mu\nu}F^{\mu\nu},$$

where $F_{\mu\nu}$ is the photon field strength tensor. (This is called the Maxwell Lagrangian). Find the energy–momentum tensor of the photon and show that it is (i) conserved; (ii) traceless. Do the same for the three-dimensional free Maxwell theory. Does the trace of the energy–momentum change in this case? Does the canonical energy–momentum tensor allow for improvement terms in this problem?

2.3.2* Explain what happens in Eq. (2.100) if $n = 1$ and $\delta = 2$. What does one get for $V(x)$ in this case?

2.4 Quantization of Solitons (Kink Mass at One Loop)

We will discuss the quasiclassical quantization of solitons, which is applicable to solitons in weakly coupled theories. Although this procedure is conceptually similar to that of the canonical quantization of fields, it was not worked out until the early 1970s [17, 18], when theorists first addressed in earnest various soliton problems in field theory.

There is a super-symmetric analog in Section 11.2.

We will consider the simplest example, the calculation of the mass of a kink appearing in the two-dimensional theory of one real field. To simplify our task further we will find the one-loop correction to the classical expression in the logarithmic approximation. Nonlogarithmic terms will be discarded. This formulation of the problem provides a pedagogical environment that is free of excessive technicalities and, at the same time, exhibits the essential features of the procedure.[11]

2.4.1 Why the Classical Expression for the Kink Mass Has to Be Renormalized

The model we will deal with is described by the action

$$S = \int d^2x \left[\tfrac{1}{2} \left(\partial_\mu \phi \right)^2 - V(\phi) \right], \tag{2.102}$$

where

$$V(\phi) = \tfrac{1}{2} \left[\mathcal{W}'(\phi) \right]^2,$$

$$\mathcal{W} = \frac{g}{\sqrt{2}} \left(\frac{\phi^3}{3} - v^2 \phi \right). \tag{2.103}$$

Here ϕ is a real field and χ in Fig. 2.23 is defined as $\phi = v + \chi$. This theory is renormalizable. A kink in two dimensions is a particle; its mass is finite and is

The classical kink mass

determined by the bare parameters in Eq. (2.103). Namely,

$$M_{\text{k}} = \frac{m^3}{3g^2} \tag{2.104}$$

where m is the mass of the elementary excitation in either of the two vacua; $m = gv\sqrt{2}$. The kink mass is a physical parameter and as such must be expressible in terms of the renormalized quantities. While g^2 is not logarithmically renormalized in two dimensions, the elementary excitation mass is renormalized. This renormalization in the log approximation is described by Fig. 2.23 and a similar graph for renormalization of v (see Exercise 2.4.1).

One-loop mass renormalization in 2D

Calculation of this diagram is straightforward and leads to the following relation between the renormalized and the bare mass parameters:

$$m_{\text{R}}^2 = m^2 - \frac{3g^2}{2\pi} \ln \frac{M_{\text{uv}}^2}{m^2}, \tag{2.105}$$

---

[11] For a detailed list references relevant to this calculation see [19] (the references span two decades).

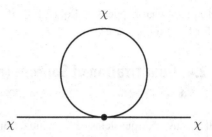

Fig. 2.23 One of the diagrams for mass renormalization. The field χ is defined as in Eq. (1.9), and in Exercise 2.4.1 (it should not be confused with $\chi(t, z)$ in Eq. (2.106) and the equations that follow it). At the classical level the mass m of the elementary excitations is $m_\chi = \sqrt{2}gv$. In calculating the one-loop correction to m one should also take into account renormalization of v. For further details see Exercise 2.4.1. Its solution is on page 647.

where M_{uv} is the ultraviolet cutoff (see also Exercise 2.4.1 at the end of this section). From the renormalizability of the theory under consideration it is clear that M_k must be renormalized in such a way that m^3 in Eq. (2.104) is replaced by m_R^3. Our task is to see how this happens and extract general lessons from this calculation of the kink mass renormalization.

$\chi(t, z)$ in (2.106) and the $\chi_n(z)$ below are not to be confused with the field χ in Section 1.1.4.

2.4.2 Mode Decomposition

The principles of the quasiclassical quantization of solitons are the same as in the standard canonical quantization procedure. There are important nuances, however, that are specific to the soliton problem.

Our starting point is the field decomposition

$$\phi(t, z) = \phi_k(z) + \chi(t, z), \tag{2.106}$$

where $\phi_k(z)$ is the kink solution (see Eq. (2.11)), which is a large classical background field, while $\chi(t, z)$ describes small fluctuations in this background, to be quantized. On general grounds one can represent $\chi(t, z)$ as

$$\chi(t, z) = \sum_n a_n(t) \chi_n(z), \tag{2.107}$$

where the basis set of functions $\{\chi_n(z)\}$ must be complete and orthonormal. The functions $\chi_n(z)$ must also satisfy appropriate boundary conditions, which we will discuss shortly. Generally speaking, one can use any complete and orthonormal set of functions. One set will prove to be the most convenient for the above decomposition.

To see that this is indeed the case let us substitute (2.107) into the action (2.102) and expand the action in the quantum field χ. Since the background field ϕ_k is the solution to the classical equation of motion, the term linear in χ vanishes and we arrive at

$$S[\phi] = S[\phi_k] + \int dt\, dz \left[\tfrac{1}{2}[\dot{\chi}(t, z)]^2 - \tfrac{1}{2}\chi(t, z)L_2\,\chi(t, z)\right] + \cdots, \tag{2.108}$$

where the ellipses indicate terms cubic in χ and higher, which are not needed at one loop. In deriving this equation we have integrated by parts and used the boundary conditions $\chi(\pm\infty) = 0$; see below. Moreover, L_2 is a linear differential operator of the second order,

$$L_2 = \left[-\frac{\partial^2}{\partial z^2} + (\mathcal{W}'')^2 + \mathcal{W}'\mathcal{W}'''\right]_{\phi=\phi_k(z)}. \tag{2.109}$$

Using

$$\phi_k(z) = \frac{m}{\sqrt{2}\,g}\tanh\frac{mz}{2} \tag{2.110}$$

and Eq. (2.103) we obtain the Hamiltonian for the quantum part of the dynamical system in question, in the form

$$\mathcal{H} = \int dz\left[\tfrac{1}{2}[\dot{\chi}(t,z)]^2 + \tfrac{1}{2}\chi(t,z)L_2\chi(t,z)\right], \tag{2.111}$$

where

$$L_2 = -\partial_z^2 + m^2\left[1 - \tfrac{3}{2}\left(\cosh\tfrac{1}{2}mz\right)^{-2}\right]. \tag{2.112}$$

The form of the Hamiltonian (2.111) prompts us to the most natural way of mode decomposition. Indeed, L_2 is a Hermitian operator whose eigenfunctions constitute a complete basis, which can be made orthonormal. Let us define $\chi_n(z)$ by

$$L_2\,\chi_n(z) = \omega_n^2\chi_n(z) \tag{2.113}$$

and impose appropriate normalization conditions,

$$\int dz\,\chi_n(z)\chi_k(z) = \delta_{nk}. \tag{2.114}$$

Using $\chi_n(z)$ as a basis in Eq. (2.107) and substituting this decomposition into Eq. (2.111) we arrive at

$$\mathcal{H} = \sum_n\left(\frac{1}{2}\dot{a}_n^2 + \frac{\omega_n^2}{2}a_n^2\right). \tag{2.115}$$

This is the sum of the Hamiltonian for decoupled harmonic oscillators. This decoupling is the result of our using the L_2 modes in the mode decomposition. As usual, the canonical quantization procedure requires us to treat a_n and $\dot{a}_n$ as operators, rather than c-numbers, satisfying the commutation relations

$$[a_n, \dot{a}_m] = i\delta_{mn}. \tag{2.116}$$

An unexcited kink corresponds to all oscillators being in the ground state. The sum of the zero-point energies for an infinite number of oscillators represents a quantum correction to the kink mass, $\delta M_k = \sum_n \tfrac{1}{2}\omega_n$.

Before discussing the quantization procedure in more detail, and in particular how to make the above formal expression for δM_k meaningful, I will pause to make a few crucial remarks.

Equation (2.113) can be interpreted as the Schrödinger equation corresponding to the potential depicted in Fig. 2.24. As we will see shortly, this potential has two discrete levels with $\omega^2 < m^2$; the levels with $\omega^2 > m^2$ form a continuous spectrum. To make the sum over n well defined, we must discretize the spectrum. To this end let us introduce a "large box," i.e., impose certain confining boundary conditions at

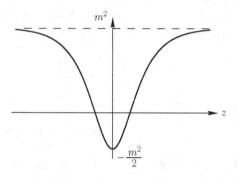

Fig. 2.24   The potential in L_2.

$z = \pm L/2$ where L is an auxiliary large parameter that we will allow to tend to infinity at the end of our calculation.

The particular choice of boundary conditions is not important as long as we apply them consistently. Needless to say, the final physical results should not depend on this choice. The simplest choice is to require that

$$\chi_n(z) = 0 \qquad \text{at } z = \pm \frac{L}{2}. \tag{2.117}$$

Note that the two eigenfunctions with $\omega^2 < m^2$ satisfy Eq. (2.117) automatically at $L \to \infty$. For eigenfunctions with $\omega^2 > m^2$ the boundary conditions (2.117) discretize the spectrum.

To say that each mode in the mode decomposition gives rise to a (decoupled) harmonic oscillator is not quite accurate; it is true for all modes with *positive* eigenvalues. However, in the problem at hand one mode is special. Its eigenvalue vanishes.[12] Such modes are referred to as *zero modes* and must be treated separately, because the fluctuations in the functional space along the "direction" of the zero modes are not small.

Zero modes

The occurrence of zero modes (a single zero mode in the case at hand) can be understood from a general argument. The solution (2.110) represents a kink centered at the origin. This particular solution breaks the translational invariance of the problem. The breaking is spontaneous, which means that, in fact, there must exist a family of solutions centered at every point on the z axis – translational invariance is restored by this family. The latter is parametrized by a collective coordinate z_0, the kink center:

$$\phi_k(z - z_0) = \frac{m}{\sqrt{2}g} \tanh \frac{m(z - z_0)}{2}. \tag{2.118}$$

Two solutions, $\phi_k(z - z_0)$ and $\phi_k(z - z_0 - \delta z_0)$, where δz_0 is a small variation of the kink center, have the same mass. Therefore, it is clear that the zero mode χ_0 is proportional to the derivative of ϕ_k:

$$\chi_0(z - z_0) \sim \frac{\partial}{\partial z_0} \phi_k(z - z_0). \tag{2.119}$$

---

[12] In stable systems there are no modes with negative eigenvalues.

Normalizing to unity we get

$$\chi_0(z) = \frac{1}{\sqrt{M_k}} \frac{\partial \phi_k(z)}{\partial z} = \sqrt{\frac{3m}{8}} \frac{1}{[\cosh(mz/2)]^2}. \tag{2.120}$$

This result – the proportionality of the zero modes to the derivatives of the classical solution with respect to the appropriate collective coordinates – is general. In the case at hand there is a single collective coordinate and a single zero mode. In other problems classical solutions can be described by a number of collective coordinates (moduli). The number of zero modes always matches the number of collective coordinates.

The sums in Eqs. (2.107) and (2.115) run over $n \neq 0$. For a discussion of the second discrete level see Section 2.4.5.

2.4.3 Dynamics of the Collective Coordinates

In two dimensions a kink is a particle. If we consider a kink in the ground state, none of the oscillator modes is excited. The kink dynamics are described by a single variable, z_0. We will carry out the quantization of this variable (usually referred to as the translational modulus) in the adiabatic approximation.

In this approximation we assume that the kink moves very slowly, so that the time dependence of the kink solution enters only through the time dependence of its center:

$$\phi(t, z) = \phi_k(z - z_0(t)). \tag{2.121}$$

Substituting the *ansatz* (2.121) into Eq. (2.102) we arrive at

$$S = \int dz dt \left\{ -\left[\tfrac{1}{2}\phi_k'^2 + V(\phi_k) \right] + \tfrac{1}{2}(\phi_k')^2 \dot{z}_0^2 \right\} = \int dt \left(-M_k + \tfrac{1}{2}M_k \dot{z}_0^2 \right). \tag{2.122}$$

Here we have used Eq. (2.120) and the fact that the expression in the square brackets is the kink mass. The corresponding Hamiltonian is

$$\mathcal{H} = M_k + \frac{M_k}{2} \dot{z}_0^2 = M_k + \frac{p_{z_0}^2}{2M_k}, \tag{2.123}$$

where p_{z_0} is the canonical momentum:

$$[p_{z_0}, z_0] = -i. \tag{2.124}$$

There is no potential term in Eq. (2.123). The reason is clear: z_0 reflects the translational invariance of the original field theory and hence the kink energy cannot depend on z_0 *per se*, only on the kink velocity $\dot{z}_0$. Equations (2.123) and (2.124) represent the first-quantized description of a freely moving particle characterized by a single degree of freedom, its position. Equation (2.123) prompts us to how one can generalize the Hamiltonian to go, if necessary, beyond the assumption $\dot{z}_0^2 \ll 1$, namely: $\mathcal{H} \to \sqrt{M_k^2 + p_{z_0}^2}$.

2.4.4 Nonzero Modes

Let us return to the expression for the Hamiltonian,

$$\mathcal{H} = M_k + \frac{M_k}{2}\dot{z}_0^2 + \sum_{n \neq 0}\left(\frac{1}{2}\dot{a}_n^2 + \frac{\omega_n^2}{2}a_n^2\right). \tag{2.125}$$

Quantum fluctuations in the "direction" of nonzero modes are described by the last term. To specify the quantum state of the kink we must specify the quantum state of each harmonic oscillator in the sum. Let us consider the situation when the kink is in the ground state. All oscillators then are in the ground state too, which obviously implies that

$$M_k^{\text{one-loop}} = M_k + \sum_{n \neq 0}\frac{\omega_n}{2}. \tag{2.126}$$

To calculate the sum over the zero-point energies we must know the spectrum of the operator (2.112). Fortunately, the Schrödinger equation (2.113) has been very well studied in the literature.[13] The potential in this equation is a special case – it is called "reflectionless" – and we will use this fact below.

The spectrum has two discrete eigenvalues, $\omega_0^2 = 0$ and $\omega_1^2 = 3m^2/4$. All other eigen-values lie above m^2. This part of the spectrum would be continuous if it were not for the "large box" boundary conditions (2.117). Let us forget about these boundary conditions for a moment. The general solution of (2.113) is given in [7]; however, we do not need its explicit form. It is sufficient to know the following.

First, the solutions with $\omega^2 > m^2$ are labeled by a continuous index p. This index is related to the eigenvalue ω_p^2 by

$$p = \sqrt{\omega_p^2 - m^2} \tag{2.127}$$

and spans the interval $(0, \infty)$.

Second, there is no reflection in the potential (2.112). In other words, choosing one of two linearly independent solutions in such a way that

$$\chi_p(z) = e^{ipz} \quad \text{at } z \to +\infty, \qquad (p > 0) \tag{2.128}$$

(i.e., choosing the right-moving wave) we have the same exponential in the other asymptotic region:

$$\chi_p(z) = e^{ipz + i\delta_p} \text{ at } \quad z \to -\infty. \tag{2.129}$$

The left-moving wave, e^{-ipz}, which appears at $z \to -\infty$ in generic potentials does not appear in the problem at hand. The only impact of the potential is a phase shift δ_p where

$$e^{i\delta_p} = \left(\frac{1 + ip/m}{1 - ip/m}\right)\left(\frac{1 + 2ip/m}{1 - 2ip/m}\right). \tag{2.130}$$

[13] See [7], pp. 73, 80.

The second, linearly independent, solution with the same eigenvalue can be chosen as $\chi_p(-z)$. The general solution then has the form

$$A\chi_p(z) + B\chi_p(-z), \tag{2.131}$$

where A and B are arbitrary constants.

Now let us discretize the spectrum imposing the boundary conditions (2.117). From Eq. (2.131) we get two relations for A and B,

$$A\,\chi_p\left(\frac{L}{2}\right) + B\,\chi_p\left(-\frac{L}{2}\right) = 0, \qquad A\,\chi_p\left(-\frac{L}{2}\right) + B\,\chi_p\left(\frac{L}{2}\right) = 0. \tag{2.132}$$

A nontrivial solution exists if and only if

$$\frac{\chi_p(L/2)}{\chi_p(-L/2)} = \pm 1. \tag{2.133}$$

This constraint in conjunction with (2.128)–(2.130) gives us the following equation for p:

$$e^{ipL - i\delta_p} = \pm 1, \tag{2.134}$$

or, equivalently,

$$pL - \delta_p = \pi n, \qquad n = 0, 1, \ldots \tag{2.135}$$

Let us denote the nth solution of the last equation by $\tilde{p}_n$. For what follows I note that for an "empty" vacuum (no kink) the corresponding equation would be

$$pL = \pi n \tag{2.136}$$

and the nth solution would be

$$p_n = \frac{\pi n}{L}. \tag{2.137}$$

We need to calculate the sum $\sum_{n \neq 0} \omega_n/2$. At large p the eigenvalues grow as p, and the sum is quadratically divergent. Should we be surprised? No.

The high-lying modes do not "notice" the kink background; they are the same as for the "empty" vacuum, whose energy density is indeed quadratically divergent.

<table>
<tr><td>

*Subtracting
the vacuum
fluctuations*

</td><td>

When we measure the kink mass we perform the measurement relative to the vacuum energy. Thus the vacuum energy must be subtracted from the sum $\sum \omega_n/2$, which becomes

</td></tr>
</table>

$$\delta M_{\rm k} = \sum \left(\frac{\omega_n}{2} - \frac{\omega_{{\rm vac},n}}{2}\right)$$
$$= \frac{1}{2} \sum \left(\sqrt{m^2 + \tilde{p}_n^2} - \sqrt{m^2 + p_n^2}\right) + \text{second bound-state energy}. \tag{2.138}$$

The need to subtract the vacuum energy is a *general rule* in this range of problems.

Since our task is the calculation of $\delta M_{\rm k}$ with logarithmic accuracy we will omit from the sum (2.138) the contribution of the second bound state (with $\omega_1^2 = 3m^2/4$). For any preassigned n, the difference $\sqrt{m^2 + \tilde{p}_n^2} - \sqrt{m^2 + p_n^2}$ is arbitrarily close to zero at $L \to \infty$. Only summing over a large number of terms with $n \sim mL$ gives a logarithmic effect. Under these conditions we can write

$$\delta M_k = \frac{1}{2} \sum_n \frac{\tilde{p}_n^2 - p_n^2}{2\sqrt{m^2 + p_n^2}} = \frac{1}{2} \sum_n \frac{p_n \delta_{p_n}}{L\sqrt{m^2 + p_n^2}}, \tag{2.139}$$

where Eqs. (2.135) and (2.136) have been used. Above, $\delta_{p_n} = \tilde{p}_n - p_n$. Keeping in mind the limit $L \to \infty$ we can replace summation over n by integration over p:

$$\sum_n \longrightarrow \int_0^\infty \frac{dp\, L}{\pi}. \tag{2.140}$$

Then we get

$$\delta M_k = -\frac{1}{2\pi} \int_0^\infty dp\, \frac{d\delta_p}{dp} \left(m^2 + p^2\right)^{1/2}. \tag{2.141}$$

Here we have integrated by parts and used $\delta_0 = \delta_\infty = 0$. The derivative of the phase δ_p is readily calculable from Eq. (2.130),

$$\frac{d\delta_p}{dp} = \frac{2}{m} \left(\frac{1}{1 + y^2} + \frac{2}{1 + 4y^2}\right), \qquad y \equiv \frac{p}{m}. \tag{2.142}$$

Substituting this expression into (2.141) and discarding nonlogarithmic contributions we get

$$\delta M_k = -\frac{3m}{2\pi} \int dy/y. \tag{2.143}$$

This integral is logarithmic. The divergence at small y (small p) is an artifact of the approximation we have used. In fact, comparing Eqs. (2.142) and (2.143) we see that at the lower limit of integration the logarithmic integral is cut off at $y \sim 1$.

The divergence at large y (large p) is a genuine ultraviolet divergence, typical of renormalizable field theories. To regularize this divergence we must introduce an ultraviolet cutoff M_{uv}. Then at the upper limit of integration the logarithmic integral (2.143) has a cutoff at $y = M_{uv}/m$.

As a result, we finally arrive at

$$\begin{aligned} M_k^{\text{one-loop}} &= M_k - \frac{3m}{4\pi} \ln\left(\frac{M_{uv}}{m}\right)^2 \\ &= \frac{m^3}{3g^2} - \frac{3m}{4\pi} \ln\left(\frac{M_{uv}}{m}\right)^2. \end{aligned} \tag{2.144}$$

| (2.144) and (2.145) match! |

Let us compare this result with the expression for the mass parameter m renormalized at one loop, see Eq. (2.105). We observe, with satisfaction, that the logarithmically divergent term is completely absorbed in the renormalized mass m_R,

$$M_k^{\text{one-loop}} = \frac{m_R^3}{3g^2}. \tag{2.145}$$

Note that the coupling constant g is not logarithmically renormalized in the present model.

In our simplified analysis we have ignored nonlogarithmic (finite) renormalizations of M_k and m at one loop. These were first calculated in a pioneering paper (see the second paper in [18]). The result after incorporating them is

$$M_k^{\text{one-loop}} = \frac{m_R^3}{3g^2} - m_R \left(\frac{3}{2\pi} - \frac{\sqrt{3}}{12} \right).$$ (2.146)

2.4.5 Kink Excitations

If we excite any oscillator corresponding to the $n \neq 0$ modes we get an excited kink state. Of particular interest is the mode with eigenvalue $\omega_1^2 = 3m^2/4$. It corresponds to an eigenvibration of the kink that dies off exponentially at $|z - z_0| \gg m^{-1}$. The state with ℓ quanta in this oscillator will have energy $\sqrt{3}m\ell/2$. The $\ell = 1$ state is below the continuum, i.e., it lies below the threshold of the two-particle states "kink + elementary excitation." Therefore it is stable. The higher-ℓ states decay into the ground-state kink and one or more elementary excitations.

Exercises

2.4.1   Derive the renormalization formula in Eq. (2.105).

2.4.2   Prove by direct calculation that $\chi_0(z)$ satisfies the equation

$$L_2 \, \chi_0(z) = 0.$$ (2.147)

2.4.3   Using the explicit form of the zero mode, prove that the Schrödinger equation

$$L_2 \, \chi_n(z) = \omega_n^2 \chi_n(z)$$ (2.148)

has no negative eigenvalues.

2.5 Charge Fractionalization

In this section we will become acquainted with fermions in the context of soliton physics. Fermions are unavoidable in supersymmetric models. However, they can appear in nonsupersymmetric models too. In some ways, dealing with fermions in nonsupersymmetric models is a simpler task. Once fermions have been introduced we encounter, quite frequently, interesting and counterintuitive effects in the soliton background. Charge fractionalization is one such phenomenon. We will discuss other spectacular effects due to fermions in topologically nontrivial backgrounds in subsequent sections.

Let us remember that at weak coupling, when the quasiclassical treatment is applicable, the soliton background field is strong. Since the fermions present purely quantum effects, in the leading approximation we can first construct the soliton, ignoring the presence of fermions altogether, and then consider fermion-induced effects in the given background; the impact of fermions on the background field reveals itself at higher orders.

2.5.1 Kinks in Two Dimensions and Dirac Fermions

This model with Majorana fermions is in Section 11.2.

The simplest model in which the presence of fermions leads to interesting phenomena is the model with kinks discussed in Section 2.1. The bosonic sector of the model includes one *real* scalar field ϕ. Here we will couple it to a *Dirac* fermion.

The Lagrangian of the model can be chosen, for instance, as follows:

$$\mathcal{L} = \tfrac{1}{2}\partial_\mu \phi \partial^\mu \phi - \tfrac{1}{4}g^2 \left(\phi^2 - v^2\right)^2 + \bar{\psi}i\,\partial\!\!\!/\,\psi + \lambda\phi\bar{\psi}\psi, \qquad (2.149)$$

where ϕ is a real scalar field, g and λ are positive coupling constants, and ψ is the Dirac (complex two-component) spinor,

$$\psi = \left(\begin{array}{c} \psi_1 \\ \psi_2 \end{array} \right). \qquad (2.150)$$

Warning: these γ matrices are "nonstandard," cf. Section 10.2.2.

Here, convenient choice of gamma matrices is

$$\gamma^0 = \sigma_2, \qquad \gamma^1 = i\sigma_3, \qquad \gamma^5 = \gamma^0\gamma^1 = -\sigma_1, \qquad (2.151)$$

where $\sigma_{1,2,3}$ are the Pauli matrices. The bosonic part of the Lagrangian (the first two terms in Eq. (2.149)) is the same as in Section 2.1, with the very same kinks. Therefore we will bypass this part of the construction, focusing on the fermion part represented by the second two terms in Eq. (2.149).

In our model there exists a global U(1) symmetry,

$$\psi \to e^{i\alpha}\psi, \qquad \bar{\psi} \to \bar{\psi}e^{-i\alpha}. \qquad (2.152)$$

This symmetry has an obvious interpretation: it relates to the fermion charge. The fermion current

$$j^\mu = \bar{\psi}\gamma^\mu\psi \qquad (2.153)$$

has no divergence:

$$\partial_\mu j^\mu = 0. \qquad (2.154)$$

Equation (2.154) is an immediate consequence of the equations of motion. It implies, in turn, that the fermion charge, defined as

$$Q = \int dz\, j^0(z), \qquad (2.155)$$

is conserved.

Besides its global U(1) symmetry this model possesses a Z_2 symmetry:

$$\phi \to -\phi, \qquad \psi \to \gamma^5\psi, \qquad \bar{\psi} \to -\bar{\psi}\gamma^5. \qquad (2.156)$$

This Z_2 symmetry is spontaneously broken in the vacuum. There are two vacuum states, at $\phi = \pm v$. In both vacua the mass of the elementary fermion excitations is equal,

$$m \equiv m_\psi = \lambda v, \qquad (2.157)$$

see Eq. (2.149). Note that the *sign of the mass term* in the Lagrangian changes when one passes from one vacuum state, at $\phi = -v$, to the other, at $\phi = v$. The kink solution interpolates between the two vacua. The mass term vanishes at the center of the kink solution. The fact that the mass term changes sign on the kink will play a crucial role in what follows.

The canonical quantization of the field ψ in the given vacuum is straightforward. Let us consider for definiteness the vacuum at $\phi = -v$. The free fermion field Lagrangian is

$$\mathcal{L}_\psi = \bar{\psi} i \, \partial\!\!\!/ \, \psi - m \bar{\psi} \psi, \qquad (2.158)$$

Mode decomposition: plane waves

where m is given by Eq. (2.157). The field ψ can be decomposed into plane waves. Then the standard procedure of quantization of the field ψ in a box of size L yields

$$\psi = \sum_p \frac{1}{\sqrt{2EL}} \left(a_p u_p e^{-i(Et-pz)} + b_p^\dagger v_p e^{i(Et-pz)} \right), \qquad (2.159)$$

where $p \equiv p_z$ and $E(p) = \sqrt{p^2 + m^2}$. This expression describes fermion annihilation and antifermion creation: a_p and b_p^+ are the corresponding annihilation and creation operators. With our choice of gamma matrices the spinors u_p and v_p can be defined as follows:

$$u_p = \begin{pmatrix} \sqrt{E} \\ (-p + im)/\sqrt{E} \end{pmatrix}, \qquad v_p = \begin{pmatrix} \sqrt{E} \\ (-p - im)/\sqrt{E} \end{pmatrix} \qquad (2.160)$$

The standard anticommutation relations are implied for the creation and annihilation operators:

$$\{a_p, a_{p'}^\dagger\} = \delta_{pp'}, \qquad \{b_p, b_{p'}^\dagger\} = \delta_{pp'}; \qquad (2.161)$$

all other anticommutators vanish. It is not difficult to check that Eq. (2.161) entails the proper anticommutation relation for the field ψ, namely

$$\{\psi_\alpha(t, z), \psi_\beta^\dagger(t, z')\} = \delta_{\alpha\beta} \delta(z - z'). \qquad (2.162)$$

As usual the vacuum state of the theory must be defined as the state that is annihilated by all the operators a_p and b_p:

$$a_p |\text{vac}\rangle = b_p |\text{vac}\rangle = 0. \qquad (2.163)$$

Then the state $a_p^\dagger |\text{vac}\rangle$ describes a fermion elementary excitation i.e., a fermion with momentum p, while $b_p^\dagger |\text{vac}\rangle$ describes an antifermion. Furthermore, if one uses the decomposition (2.159) in the expression for the fermion charge (2.155), one obtains

$$Q = \sum_p \left[a_p^\dagger a_p - \left(b_p^\dagger b_p - 1 \right) \right]. \qquad (2.164)$$

It should be clear that a definite fermion charge can be assigned to each elementary excitation of the theory. Equation (2.164) implies that the charge of the fermion is unity and that of the antifermion is minus unity, while the charge of the bosonic elementary excitation vanishes. At the same time, Eq. (2.164) reveals a drawback in our definition of the fermion charge. Namely, if we try to calculate the fermion charge of the vacuum state (2.163) then we will find that it is positive and infinite. This additive infinite constant has no impact on the charges of the excitations – that is why usually one just ignores it.

As we will see shortly, when we come to the soliton fermion charge we have to use a more careful definition preserving the neutrality of the vacuum state. Fortunately, it is very easy to amend the fermion current (2.153) using its C invariance (charge conjugation). To this end let us introduce the charge-conjugated fermion field ψ^c and the fermion current for this field. The charge-conjugated field ψ^c must depend linearly on ψ^* and must satisfy the same equation as ψ, thus

$$i\,\partial\!\!\!/\,\psi + \lambda\phi\psi = 0, \qquad i\,\partial\!\!\!/\,\psi^c + \lambda\phi\psi^c = 0, \tag{2.165}$$

where we have taken into account that the ϕ field is C-even. Since our $\gamma_{0,1}$ matrices are purely imaginary, it is obvious that $\psi^c \equiv \psi^*$.

| Now, if we introduce the fermion current as

$$j^\mu = \tfrac{1}{2}\left(\bar\psi\gamma^\mu\psi - \overline{\psi^c}\gamma^\mu\psi^c\right), \tag{2.166}$$

Amended fermion charge

instead of Eq. (2.153), it is still conserved, while the expression for the fermion charge becomes

$$Q = \sum_p \{a_p^\dagger a_p - b_p^\dagger b_p\}. \tag{2.167}$$

The amended definition (2.166) is *identical* to that presented in Eq. (2.153) up to a constant – the infinite additive constant in the vacuum charge mentioned above. Now the vacuum is neutral, as it should. The charges of all elementary excitations stay the same; for any finite number n, any ensemble of n quanta has integer fermion charge.

For future comparison I give here the second-quantized expression for (the fermion part of) the Hamiltonian,

$$H = \sum_p E(p)\left(a_p^\dagger a_p + b_p^\dagger b_p\right), \tag{2.168}$$

where an infinite additive constant has been omitted.

Now, after this rather extended digression on canonical quantization, we are ready to address the kink problem in the presence of fermions. Since the kink solution is static, one can still use the Hamiltonian (canonical) quantization. The decomposition in plane waves (2.159) is no longer appropriate, however. In the kink background a plane wave is no longer a solution of the equation of motion. If the kink center is fixed, translational invariance is lost: p_z does not commute with the Hamiltonian. The decomposition (2.159) will *not* diagonalize the Hamiltonian.

Let us write the equations for the eigenvalues of the Dirac operator in the presence of a kink. To this end we take the classical equation of motion

$$(i\,\partial\!\!\!/\, + \lambda\phi_k)\psi = \begin{pmatrix} -\partial_z + \lambda\phi_k & \partial_t \\ -\partial_t & \partial_z + \lambda\phi_k \end{pmatrix}\psi = 0 \tag{2.169}$$

and substitute there

$$\psi = e^{-i\omega t}\begin{pmatrix} \tilde\chi \\ \chi \end{pmatrix}. \tag{2.170}$$

In the kink background, with the kink center fixed at the origin, a discrete Z_2 symmetry survives corresponding to the transformation $z \to -z$. The eigenfunctions of the corresponding operator can be classified according to this Z_2 symmetry: under $z \to -z$ they are either even or odd.

To be more specific, let us introduce two conjugated operators,

$$P = \partial_z + \lambda\phi_k, \qquad P^\dagger = -\partial_z + \lambda\phi_k, \tag{2.171}$$

where ϕ here stands for the kink solution (2.11),

$$\phi = v \tanh\left(\frac{gvz}{\sqrt{2}}\right). \tag{2.172}$$

The precise form of the kink solution is not important at this stage. What is important, though, is that the kink solution is topologically nontrivial: at $z \to -\infty$ the field ϕ_k tends to a negative constant and at $z \to \infty$ to a positive constant.

From the pair of the conjugate operators P and $P^\dagger$ one can construct two *Hermitian* operators, namely

$$\begin{aligned} L_2 &= P^\dagger P = -\partial_z^2 + \lambda^2 \phi_k^2 - \lambda \phi_k', \\ \tilde{L}_2 &= PP^\dagger = -\partial_z^2 + \lambda^2 \phi_k^2 + \lambda \phi_k', \end{aligned} \tag{2.173}$$

where $\phi_k' = \partial_z \phi_k$.

We need to choose Hermitian operators since only for such operators does the set of eigenfunctions present a complete orthonormal system suitable for decomposition.

It is clear that all eigenvalues of L_2 and $\tilde{L}_2$ are non-negative (and the eigenfunctions are real). In fact, the spectra of both operators are the same, with the exception of the zero mode. To be able to discuss this point more carefully we need to discretize the spectrum, in much the same way as a "large box" discretizes the spectrum of p_z in the canonical quantization near the trivial vacuum (see above). Convenient boundary conditions in the present case are as follows. At $z = \pm L/2$ the eigenfunctions $\tilde{\chi}_n$ of $\tilde{L}_2$ satisfy the constraint

$$\tilde{\chi}_n(z = \pm L/2) = 0, \tag{2.174}$$

where

$$\tilde{L}_2 \tilde{\chi}_n(z) = \omega_n^2 \tilde{\chi}_n(z). \tag{2.175}$$

Convenient boundary conditions

For the eigenfunctions χ_n of L_2 we impose the boundary conditions

$$P \chi_n(z = \pm L/2) = 0, \tag{2.176}$$

where

$$L_2 \chi_n(z) = \omega_n^2 \chi_n(z). \tag{2.177}$$

Equations (2.175) and (2.177) reflect the fact (mentioned above) that the spectra of L_2 and $\tilde{L}_2$ are degenerate under these boundary conditions.

Indeed, let $\chi_n(z)$ be a normalized eigenfunction of the operator L_2. Then

$$\tilde{\chi}_n = \frac{1}{\omega_n} P \chi_n \tag{2.178}$$

is the normalized eigenfunction of $\tilde{L}_2$ having the same eigenvalue. The converse is also true. If $\tilde{\chi}_n$ is a normalized eigenfunction of $\tilde{L}_2$ then

$$\chi_n = \frac{1}{\omega_n} P^\dagger \tilde{\chi}_n \tag{2.179}$$

is the normalized eigenfunction of L_2 having the same eigenvalue.

The only subtlety occurs for the zero mode. The operator L_2 has a zero mode, $L_2 \chi_0 = 0$, while $\tilde{L}_2$ does not. Why is this?

For a zero mode to occur in L_2 it is necessary that $P \chi_0 = 0$. This equation has a *normalizable* solution,

$$\chi_0 \propto \exp\left(-\lambda \int_0^z \phi_k dz\right). \tag{2.180}$$

If λ is positive (which I am assuming) and $\phi(z)$ has the asymptotic behavior specified after Eq. (2.172) then the zero mode (2.180) is normalizable.

For a zero mode to occur in $\tilde{L}_2$ it would be necessary that $P^\dagger \chi_0 = 0$, which would require that

$$\chi_0 \propto \exp\left(\lambda \int_0^z \phi_k dz\right),$$

which is non-normalizable. This fact – that only one of these two operators has a zero mode – will have far-reaching consequences.

Now, if we use the eigenfunctions of the operators L_2 and $\tilde{L}_2$ for the decomposition of the fermion field ψ, the fermion part of the Hamiltonian will be diagonalized. The second-quantized expression for the fermion field takes the form

$$\psi(t, z) = a_0 \begin{pmatrix} 0 \\ \chi_0(z) \end{pmatrix} + \sum_{n \neq 0} \left[e^{-i\omega_n t} \frac{a_n}{\sqrt{2}} \begin{pmatrix} \tilde{\chi}_n(z) \\ -i\chi_n(z) \end{pmatrix} + e^{i\omega_n t} \frac{b_n^\dagger}{\sqrt{2}} \begin{pmatrix} \tilde{\chi}_n(z) \\ i\chi_n(z) \end{pmatrix} \right], \tag{2.181}$$

with a similar expression for $\psi^\dagger$. The operators a_n and $b_n^\dagger$ are interpreted respectively as annihilation and creation operators, with the standard anticommutation relations

$$\{a_n a_{n'}^\dagger\} = \delta_{nn'}, \qquad \{b_n b_{n'}^\dagger\} = \delta_{nn'}. \tag{2.182}$$

Using the completeness of both sets of eigenfunctions, $\chi_n(z)$ and $\tilde{\chi}_n(z)$, it is not difficult to check that the basic anticommutation relation (2.162) is satisfied (in the limit $L \to \infty$).

In the kink background, the fermion part of the Hamiltonian reduces to

$$H = \int dz \, \psi^\dagger(t, z) \left[-\gamma^0 \left(i\gamma^1 \partial_z + \lambda\phi \right) \right] \psi(t, z)$$

$$= \int dz \begin{pmatrix} \psi_1^\dagger \\ \psi_2^\dagger \end{pmatrix}^T \begin{pmatrix} 0 & iP \\ -iP^\dagger & 0 \end{pmatrix} \begin{pmatrix} \psi_1 \\ \psi_2 \end{pmatrix}$$

$$= \sum_{n \neq 0} \omega_n \left(a_n^\dagger a_n + b_n^\dagger b_n \right), \tag{2.183}$$

where I have dropped an additive (infinite) constant in the last line. Note that the operators $a_0, a_0^\dagger$ relating to the zero mode do not enter the second-quantized Hamiltonian (2.183) which looks essentially identical to that in Eq. (2.168).

Now our task is to build the lowest-energy state, the ground-state kink, which is an analog of the vacuum state in the case of the trivial solution $\phi = \pm v$. It is no surprise that there are two such states for a given kink. The fermion level associated with the zero mode may or may not be filled – both options lead to the same energy.

Indeed, as is obvious from Eq. (2.183), the minimum energy in the fermion sector is achieved when all levels with $n \neq 0$ are empty, i.e.,

$$a_n|\text{kink}\rangle = b_n|\text{kink}\rangle = 0, \qquad n \neq 0. \tag{2.184}$$

Here |kink⟩ denotes the ground-state kink. Since a_0 does not enter the Hamiltonian, the condition a_0|kink⟩ $= 0$ is not mandatory. Let us first assume that this condition is imposed,

$$a_0|\text{kink}\rangle = 0. \tag{2.185}$$

This is the condition that this level is empty. One can build another state, let us call it |kink'⟩, such that

$$|\text{kink}'\rangle = a_0^\dagger|\text{kink}\rangle. \tag{2.186}$$

This is the state with a filled zero level. Both states, |kink⟩ and |kink'⟩, have the same energy,

$$\langle\text{kink}|H|\text{kink}\rangle = \langle\text{kink}'|H|\text{kink}'\rangle. \tag{2.187}$$

The reason for this is obvious: since this fermion level has zero energy, whether or not it is filled does not matter.

2.5.2 What Is the Fermion Charge of the Kink?

Now we are able to deduce the fermion charge of the kink. As explained earlier, we should measure the charge using the amended current (2.166). Substituting the decomposition (2.181) into (2.166) one obtains

$$Q_{\text{kink}} = \tfrac{1}{2}\left(a_0^\dagger a_0 - a_0 a_0^\dagger\right) + \tfrac{1}{2}\sum_{n\neq 0}\left(a_n^\dagger a_n + b_n b_n^\dagger - b_n^\dagger b_n - a_n a_n^\dagger\right). \tag{2.188}$$

There is no ambiguous additive constant here – the expression for the current has been adjusted already in such a way that the trivial vacuum $\phi = \pm v$ carries zero charge. In order to find the fermion charge of the kink we sandwich Eq. (2.188) between |kink⟩ or |kink'⟩, using the conditions (2.184) and (2.185) and the definition (2.186):

$$\langle\text{kink}|Q_{\text{kink}}|\text{kink}\rangle = -\tfrac{1}{2}, \qquad \langle\text{kink}'|Q_{\text{kink}}|\text{kink}'\rangle = \tfrac{1}{2}. \tag{2.189}$$

The result is remarkable! There are two kink ground states, and both have fractional charge. Remember that any finite number of elementary excitations in the trivial vacua can only produce an integer-charge state. Technically, the occurrence of the fermion charge $\pm 1/2$ is due to the existence of a *single* fermion zero mode in the kink background.

Other models, with an odd number of fermion zero modes on solitons, are known. *In all such problems the fermion charge of the soliton is fractional.*

When I say that there is one fermion zero mode on the kink, I need to qualify this. The zero mode represented by the first term in (2.181) is complex. Consider the equation on ψ and that on $\psi^\dagger$ in the kink background (see Eq. (2.165)). Both have a solution. Since we are dealing with Dirac (complex) fermions, even though the functional form for the solution is the same (proportional to χ_0) these are two distinct zero modes. The corresponding moduli parameter is complex – we have a_0 and $a_0^\dagger$, which are independent.

Were we dealing with the Majorana fermion, we would get only one modulus. This situation is also referred to as the one-fermion zero mode. One encounters such

See Part II for many important examples.

an example in supersymmetry (see Chapter 11 Section 11.2). In problems where the Dirac fermion has one zero mode we end up with fermion charge fractionalization. An even more unusual phenomenon occurs when the Majorana fermion has one zero mode – the very distinction between bosons and fermions is lost in this case.

Our derivation of the fact that the fermion charge of the kink is $\pm 1/2$ is completely sound, albeit rather technical. This fact is so counterintuitive that the curious reader may be left unsatisfied in a search for the underlying physics. Without delving into details, we will say only that the missing half of the fermion charge does not totally disappear. It "delocalizes," i.e., it leaves the soliton and attaches itself to a boundary of the "large box." In no local experiments (performed in the vicinity of the kink) can one observe the "missing 1/2." An experimentalist investigating the kink states will simply detect $\pm 1/2$.

Exercise

2.5.1   Look through later chapters and identify other examples of charge fractionalization.

References for Chapter 2

[1] E. B. Bogomol'nyi, *Stability of Classical Solutions, Sov. J. Nucl. Phys.* **24**, 449 (1976) [reprinted in C. Rebbi and G. Soliani (eds.), *Solitons and Particles* (World Scientific, Singapore, 1984) pp. 389–394].

[2] Y. Nambu, Quark Model and Factorization of the Veneziano Amplitude, in *Lectures at the Copenhagen Symposium on Symmetries and Quark Models* (Gordon and Breach, New York, 1970), p. 269.

[3] T. Goto, *Prog. Theor. Phys.* **46**, 1560 (1971).

[4] A. M. Polyakov, *Phys. Lett. B* **103**, 207 (1981).

[5] M. Shifman and A. Yung, *Phys. Rev. D* **67**, 125 007 (2003) [arXiv:hep-th/0212293].

[6] A. M. Polyakov, *Nucl. Phys. B* **120**, 429 (1977).

[7] L.D. Landau and E.M. Lifshitz, *Quantum Mechanics*, Third Edition (Pergamon Press, 1977).

[8] D. Bazeia, J. Menezes, and M. M. Santos, *Phys. Lett. B* **521**, 418 (2001) [arXiv:hep-th/0110111].

[9] D. Binosi and T. ter Veldhuis, *Phys. Lett. B* **476**, 124 (2000) [hep-th/9912081].

[10] B. Chibisov and M. A. Shifman, *Phys. Rev. D* **56**, 7990 (1997). Erratum: *ibid.* **58**, 109901 (1998) [arXiv:hep-th/9706141]; see also [13].

[11] G. R. Dvali and Z. Kakushadze, *Nucl. Phys. B* **537**, 297 (1999) [hep-th/9807140].

[12] A. Gorsky and M. A. Shifman, *Phys. Rev. D* **61**, 085001 (2000) [hep-th/9909015]; D. Binosi and T. ter Veldhuis, *Phys. Lett. B* **476**, 124 (2000), [arXiv:hep-th/9912081.

[13] H. Oda, K. Ito, M. Naganuma, and N. Sakai, *Phys. Lett. B* **471**, 140 (1999) [hep-th/9910095]; M. A. Shifman and T. ter Veldhuis, *Phys. Rev. D* **62**, 065004 (2000) [hep-th/9912162].

[14] A. Vilenkin, *Phys. Lett. B* **133**, 177 (1983).

[15] J. Ipser and P. Sikivie, *Phys. Rev. D* **30**, 712 (1984).

[16] L. D. Landau and E. M. Lifshitz, *The Classical Theory of Fields* (Pergamon Press, 1979), Sections 91–95.

[17] L. D. Faddeev and L. Takhtajan, Particles for the sine-Gordon Equation, in *Sessions of the I. G. Petrovskii Seminar, Usp. Mat. Nauk.* **28**, 249 (1974); V. Korepin and L. D. Faddeev, Quantization of solitons, *Theor. Math. Phys.* **25**, 1039 (1975); L. D. Faddeev and V. E. Korepin, *Phys. Rept.* **42**, 1 (1978).

[18] R. F. Dashen, B. Hasslacher, and A. Neveu, Nonperturbative methods and extended hadron models in field theory. 1. Semiclassical functional methods, *Phys. Rev. D* **10**, 4114 (1974); Nonperturbative methods and extended hadron models in field theory. 2. Two-dimensional models and extended hadrons, *Phys. Rev. D* **10**, 4130 (1974) [reprinted in C. Rebbi and G. Soliani (eds.), *Solitons and Particles* (World Scientific, Singapore, 1984), pp. 297–305].

[19] M. A. Shifman, A. I. Vainshtein, and M. B. Voloshin, *Phys. Rev. D* **59**, 045 016 (1999) [arXiv:hep-th/9810068].

3 Vortices and Flux Tubes (Strings)

Global, local, and (in passing) semilocal vortices. — Abelian and non-Abelian strings. — How they gravitate. — Index theorem. — Fermion zero modes on the string.

3.1 Vortices and Strings

In field theory solitons of a "curly type" are called vortices, for a good reason. They are close relatives of tornadoes and of the vortices on a water surface that are a matter of every-day experience. Vortices can develop in field theories with spontaneously broken continuous symmetries in which vacuum manifolds have a circular structure. The simplest example can be found in models with gauge U(1) in the Higgs phase, with which we will start. This example was found long ago: in 1957 it was discussed by Abrikosov [1] in the context of superconductivity; in 1973 Nielsen and Olesen [2] considered relativistic vortices after the advent of the Higgs model in high-energy physics. After we have become acquainted with Abrikosov–Nielsen–Olesen (ANO) vortices we will discuss some generalizations.

Topological defects of the vortex type can be considered in $1 + 2$ and $1 + 3$ dimensions. In the latter case they represent flux tubes (strings). In the former case we are dealing with vortices *per se*.

In passing from the classical vortex solution in $1 + 2$ dimensions to the flux-tube solution in $1 + 3$ dimensions, the form of the solution *per se* does not change. In $1 + 3$ dimensions we will always assume that the flux tube under consideration is parallel to the z axis. Then the static flux-tube solution depends only on x and y and coincides with the static vortex solution in $1 + 2$ dimensions. With this convention the magnetic field inside the flux tube is aligned in the z direction, i.e., $\vec{B} = \{0, 0, B_3\}$. The vortex magnetic field is a scalar quantity under spatial rotations: in $1 + 2$ dimensions the photon field strength tensor $F_{\mu\nu}$ has a single spatial component F_{12}, which transforms as the time component of a 3-vector.

Vortices in $1 + 2$ dimensions are particles and are characterized by their mass. Strings in $1 + 3$ dimensions are extended objects. They are characterized by their energy per unit length, the string *tension*.

Even though the classical solutions in $1 + 2$ and $1 + 3$ dimensions coincide, the determination of quantum corrections to masses or tensions depends critically on the number of dimensions, since the quantum corrections "know" of the presence of the z direction. Thus they should be treated separately in these two cases.

3.1.1 Global Vortices

| *U(1) is not gauged here.* |

The simplest vortex that one can imagine emerges in the theory of a single complex scalar field with U(1) symmetry, for which

$$\mathcal{L} = |\partial_\mu \phi|^2 - U(\phi) \tag{3.1}$$

where

$$U(\phi) = \lambda \left(|\phi|^2 - v^2 \right)^2 . \tag{3.2}$$

In the vacuum $|\phi| = v$, but the phase of the field ϕ may rotate. Imagine a point on the xy plane and a contour C which encircles this point (Fig. 3.1). Imagine that, as

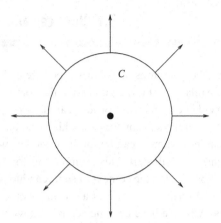

Fig. 3.1 The vortex of the ϕ field. The arrows show the value and phase of the complex field ϕ at given points on a contour that encircles the origin (the vortex center).

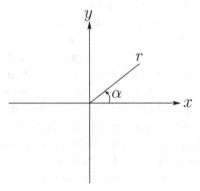

Fig. 3.2 Polar coordinates in the xy plane, $r = \sqrt{x^2 + y^2}$.

we travel along this contour, the phase of the field ϕ increases from 0 to 2π, or from 0 to 4π, and so on; ϕ is said to "wind." In other words,

$$\phi(r, \alpha) \to v e^{in\alpha} \quad \text{at } r \to \infty, \tag{3.3}$$

where we are using polar coordinates: α is the angle in the xy plane, r is the radius (Fig. 3.2), and n is an integer. Such a field configuration is called a vortex. It is clear, on topological grounds, that the winding of the field ϕ cannot be unwound by any continuous field deformation. Mathematically this is expressed as follows. The vacuum manifold in the case at hand is a circle. We map this abstract circle onto a spatial circle as depicted in Fig. 3.1. Such maps are categorized by topologically distinct classes, labeled by integers that are positive, negative or zero:

$$\pi_1(U(1)) = \mathbb{Z}.$$

Topological formula for the first homotopy group.

The integer labeling a class counts how many times we wind around the vacuum-manifold circle when we sweep the spatial circle once. The map is orientable: by sweeping the vacuum manifold clockwise we can wind around the spatial circle clockwise or anticlockwise.

Although such *global* vortices may play a role if their spatial dimensions are assumed to be finite, their energy diverges (logarithmically) in the limit of infinite sample size. Indeed,

$$\partial_i \phi \sim inv\partial_i \alpha = -ine\varepsilon_{ij}\frac{x_j}{r^2} \text{ as } r \to \infty \quad (i,j = 1,2), \tag{3.4}$$

which implies that

$$E = \int d^2x \left\{\partial_i\bar{\phi}\partial_i\phi + U(\phi)\right\} \xrightarrow{\phi=ve^{in\alpha}} 2\pi v^2 n^2 \int \frac{dr}{r} \to \infty. \tag{3.5}$$

Thus, the global vortex mass (the flux-tube tension in $D = 4$ dimensions) diverges logarithmically both at large and small r. The small-r divergence can be cured if we let $\phi \to 0$ in the vicinity of the vortex center. To cure the large-r divergence we have to introduce a gauge field.

3.1.2 The Abrikosov–Nielsen–Olesen Vortex (or String)

A way out allowing one to make the vortex energy finite is well known.[1] To this end one needs to gauge the U(1) symmetry. The Abrikosov–Nielsen–Olesen (ANO) vortex is a soliton in the gauge theory with a charged scalar field whose vacuum expectation value breaks U(1) spontaneously. The model is described by the Lagrangian

U(1) is gauged.

$$\mathcal{L} = -\frac{1}{4e^2}F_{\mu\nu}^2 + |\mathcal{D}^\mu\phi|^2 - U(\phi), \tag{3.6}$$

where $F_{\mu\nu}$ is the photon field strength tensor,

$$F_{\mu\nu} = \partial_\mu A_\nu - \partial_\nu A_\mu,$$

and the covariant derivative is defined by

$$\mathcal{D}_\mu\phi = (\partial_\mu - in_e A_\mu)\phi, \qquad (\mathcal{D}_\mu\phi)^\dagger = (\partial_\mu + in_e A_\mu)\phi^\dagger \tag{3.7}$$

where n_e is the electric charge of the field ϕ (in the units of e, for instance, $n_e = \pm 1/2, \pm 1, \ldots$).

The potential energy $U(\phi)$ is chosen in such a way as to guarantee that the Higgs mechanism does take place. Equation (3.2) achieves this. As usual, the constants λ and e are assumed to be small, so that a quasiclassical treatment is justified.

This model is invariant under the U(1) gauge transformations

$$\phi \to e^{i\beta(x)}\phi, \qquad A_\mu \to A_\mu + \frac{1}{n_e}\partial_\mu\beta. \tag{3.8}$$

Usually the gauge is chosen in such a way that, in the vacuum,

$$A_\mu = 0, \qquad \phi = v. \tag{3.9}$$

---

[1] Since the transverse size of the ANO string is of order $m_{V,H}^{-1}$, see below, and the energy density is well localized, some people refer to the ANO string as local. Strings occupying an intermediate position between the global strings of Section 3.1.1 and the ANO strings, whose transverse size can be arbitrary while their tension is finite, go under the name of *semilocal*. For a review see [3]. An example of a semilocal string is the CP(1) instanton provided that one elevates the CP(1) model to four dimensions. Semilocal strings will not be considered in this text.

*Unitary
gauge*

This is called the unitary gauge. The phase of v can be chosen arbitrarily; usually it is assumed that v is *real*. It is obvious that Eq. (3.9) corresponds to the minimal energy, the vacuum. In the unitary gauge the scalar field in the vacuum is coordinate independent.

Owing to the Higgs mechanism the vector field acquires a mass

$$m_V = \sqrt{2}en_ev; \qquad (3.10)$$

Im ϕ is eaten by the Higgs mechanism, so that $\phi(x) = v + \eta(x)/\sqrt{2}$. The surviving real scalar field $\eta(x)$, which is not eaten up by the vector field, is called the Higgs field. Its mass is

$$m_H = 2\sqrt{\lambda}v. \qquad (3.11)$$

In order to see that the soliton finite-energy solution does exist in this model, and to find it, let us first consider all nonsingular field configurations that are static (time-independent) in the gauge $A_0 = 0$. Imposing the gauge $A_0 = 0$, we still have the freedom of doing time-independent (but space-dependent) gauge transformations. We will keep this freedom in reserve for the time being. The only requirement that we impose now is the finiteness of the energy:

$$\mathcal{E}[\vec{A}(\vec{x}), \phi(\vec{x})] = \int d^2x \left[\frac{1}{4e^2}F_{ij}F_{ij} + |\mathcal{D}_i\phi|^2 + U(\phi) \right] < \infty. \qquad (3.12)$$

To ensure that the energy is finite it is necessary (but not sufficient) that $U(\phi) \to 0$ at $|\vec{x}| \to \infty$, i.e.,

$$|\phi| \to v \qquad \text{as } |\vec{x}| \to \infty. \qquad (3.13)$$

Let us choose a circle of large radius R (eventually we will let $R \to \infty$) centered at the origin. The absolute value of ϕ on this circle must be v; however, the phase of the field ϕ is not fixed by the condition $\int d^2x \, U(\vec{x}) < \infty$. Thus, one can choose

$$\phi = ve^{if(\alpha)} \qquad (3.14)$$

*Winding
number*

on the large circle. The *winding number* does not depend on details of the function $f(\alpha)$, but only on its global (topological) properties. An example of $f(\alpha)$ that belongs to the class $n = 1$ (i.e., a single winding) is $f(\alpha) = \alpha$. By performing a "small" time-independent gauge transformation we can always transform $f(\alpha)$ into any other function from the $n = 1$ class; see Fig. 3.3.

The same is true with regard to the phase functions $f(\alpha)$ chosen to belong to other classes, with integer (positive or negative) $n \neq 1$. Any continuous function satisfying the boundary conditions $f(0) = 0$ and $f(2\pi) = 2n\pi$ can be transformed into $f(\alpha) = n\alpha$. This is an analog of the unitary gauge in the topologically trivial class $n = 0$.

The condition

$$\phi(x) \to ve^{in\alpha} \qquad (3.15)$$

at large r is necessary but not sufficient to ensure the finiteness of the energy functional (3.12). Indeed, assume that $\vec{A} \to 0$ at $|x| \to \infty$. Then we have

$$\int d^2x |\mathcal{D}_i\phi|^2 \to \int d^2x |\partial_i\phi|^2 \to 2\pi n^2 v^2 \int dr \frac{1}{r}.$$

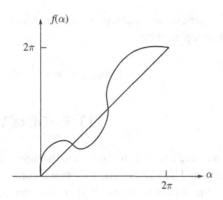

Fig. 3.3 The phase functions (3.14) from the $n = 1$ class. This class is defined by the boundary conditions $f(0) = 0$ and $f(2\pi) = 2\pi$.

The last integral diverges logarithmically at large r, as in Eq. (3.5).

This divergence, due to the winding of ϕ, can be eliminated. Indeed, $\partial_i \phi$ is not the correct measure of the variation in ϕ, since it is the covariant derivative that counts. One can try to introduce the gauge potential $\vec{A}$ in such a way that (i) at $|x| \to \infty$ it is pure gauge and no field strength tensor F_{ij} is generated (otherwise, there would be a divergence owing to the F_{ij}^2 term); (ii) $\mathcal{D}_i \phi \to 0$ fast enough that there is no divergence in the $\int d^2x |\mathcal{D}_i \phi|^2$ term.

Using Eqs. (3.4) and (3.7) it is not difficult to see that to meet the above requirements we must switch on the gauge potential in such a way that asymptotically, at large r, it tends to

$$A_i = \frac{n}{n_e} \partial_i \alpha = -\frac{n}{n_e} \varepsilon_{ij} \frac{x_j}{r^2}, \qquad i, j = 1, 2, \tag{3.16}$$

where ε_{ij} is the two-dimensional Levi–Civita tensor. It is clear that then both $\mathcal{D}_i \phi$ and F_{ij} fall off at infinity faster than $1/r^2$ (in fact, they fall off exponentially fast), and the energy integral converges.

The form of the gauge potential (3.16) is in one-to-one correspondence with the form of the phase in the asymptotics of ϕ; see Eq. (3.15). One can write an integral representation for the *winding number*:

The winding number is the flux of the magnetic field in the string's core, in units $n_e/(2\pi)$.

$$n = \frac{n_e}{2\pi} \oint_{|x|=R \to \infty} dx^i A_i = \frac{n_e}{2\pi} \int d^2x\, B, \tag{3.17}$$

where B is the magnetic field,

$$B = \tfrac{1}{2} F_{ij} \varepsilon^{ij} = F_{12}. \tag{3.18}$$

The second equality on the right-hand side is due to Stokes' theorem, which allows one to transform the contour integral into a surface integral over $F_{ij}\varepsilon^{ij}$. We see that the winding number is proportional to the flux of the magnetic field carried by the string in its core. If we define the magnetic flux as $n_e \int d^2x F_{12}$ (see the definition of the covariant derivative in (3.7)), then for the *minimal winding* ANO string we have

$$\text{Flux} = n_e \int d^2x F_{12} = 2\pi. \tag{3.19}$$

If the Lagrangian we deal with is *canonically* normalized, then the the above relation should be replaced by

$$\text{Flux}_{\text{canon}} = n_e e \int d^2x F_{12} = 2\pi. \tag{3.20}$$

3.1.3 The Critical Vortex

*Super-
symmetric
counterpart
in
Section 11.5*

So far we have focused on two issues: the topological stability of the U(1) vortex and how gauging U(1) allows one to obtain a vortex of finite energy. Neither the precise form of the soliton solution nor its mass were addressed. Now it is time to discuss these issues. We will consider a special limiting case, the critical, or Bogomol'nyi–Prasad–Sommerfield (BPS), vortex.

For generic values of the scalar coupling λ (see Eq. (3.2)), the scalar-field mass (also called the Higgs-field mass) is distinct from that of the photon. The ratio of the vector-field mass and the Higgs mass is an important parameter in the theory of superconductivity since it characterizes the superconductor type; see, e.g., [4]. Namely, for $m_H < m_V$ we have a type I superconductor (the vortices attract each other), while for $m_H > m_V$ we have a type II superconductor (the vortices repel each other). This is related to the fact that the scalar field produces an attraction between two vortices, while the electromagnetic field produces a repulsion.

*Super-
conductors
of the I and
II kind*

The boundary separating type I and type II superconductors corresponds to the special case $m_H = m_V$, i.e., to a special value of the quartic coupling λ given by

$$\lambda = \frac{n_e^2}{2} e^2; \tag{3.21}$$

see Eqs. (3.10) and (3.11). In this case the vortices do not interact.

It is well known that the vanishing of the interaction between two parallel strings at the special point $m_H = m_V$ can be explained by a criticality (i.e., BPS saturation) of the Abrikosov–Nielsen–Olesen vortex. At this point the vortex satisfies the first-order equations and saturates the Bogomol'nyi bound.

*Bogomol'nyi
completion
in the vortex
problem*

This bound follows from the following representation for the vortex mass (string tension) T:

$$T = \int d^2x \left[\frac{1}{4e^2} F_{ij}^2 + |\mathcal{D}_i\phi|^2 + \frac{n_e^2}{2} e^2 \left(|\phi|^2 - v^2 \right)^2 \right]$$

$$= \int d^2x \left\{ \frac{1}{2} \left[\frac{1}{e} B + n_e e \left(|\phi|^2 - v^2 \right) \right]^2 + |(\mathcal{D}_1 + i\mathcal{D}_2)\phi|^2 \right\} + 2\pi v^2 n. \tag{3.22}$$

The representation (3.22) is known as the *Bogomol'nyi completion*. It is not difficult to see that the first and second lines in Eq. (3.22) are identical up to the boundary term. The difference between them reduces to

$$-\int d^2x \left[n_e B \left(|\phi|^2 - v^2 \right) + i\bar{\phi}[\mathcal{D}_2\mathcal{D}_1]\phi \right] \tag{3.23}$$

plus an integral over a total derivative that vanishes. The terms proportional to $|\phi|^2$ cancel each other; the remainder is the flux times v^2.

The minimal value of the tension is reached when both terms in the integrand of Eq. (3.22) vanish,

$$B + n_e e^2 \left(|\phi|^2 - v^2\right) = 0, \qquad (\mathcal{D}_1 + i\mathcal{D}_2)\phi = 0. \qquad (3.24)$$

(Let me note parenthetically that within the Landau–Ginzburg approach to super-conductivity the same system of first-order differential equations was derived by G. Sarma in the early 1960s; see [4].)

String
tension

If Eqs. (3.24) are satisfied, the vortex mass (string tension) is

$$T = 2\pi v^2 n, \qquad (3.25)$$

where the winding number n counts the quantized magnetic flux. The linear dependence of the n-vortex mass on n implies the absence of interactions between the vortices.

To solve Eqs. (3.24) one must find an appropriate *ansatz*. For the elementary, $n = 1$, vortex it is convenient to introduce two profile functions $\varphi(r)$ and $f(r)$, as follows:

$$\phi(x) = v\varphi(r)e^{i\alpha}, \qquad A_i(x) = -\frac{1}{n_e}\varepsilon_{ij}\frac{x_j}{r^2}[1 - f(r)], \qquad (3.26)$$

where $r = \sqrt{x^2 + y^2}$ is the distance and α is the polar angle; see Fig. 3.2. Moreover, it is convenient to introduce a dimensionless distance ρ, where

$$\rho = n_e e\, vr. \qquad (3.27)$$

A remarkable fact: the *ansatz* (3.26) is compatible with the set of equations (3.24) and, upon substitution in (3.24), results in the following two equations for the profile functions:

$$-\frac{1}{\rho}\frac{df}{d\rho} + \varphi^2 - 1 = 0, \qquad \rho\frac{d\varphi}{d\rho} - f\varphi = 0. \qquad (3.28)$$

The boundary conditions for the profile functions are rather obvious from the form of the *ansatz* (3.26) and from our previous discussion. At large distances we have

$$\varphi(\infty) = 1, \qquad f(\infty) = 0. \qquad (3.29)$$

At the same time, at the origin the smoothness of the field configuration under consideration (i.e., the absence of singularities) requires that

$$\varphi(0) = 0, \quad f(0) = 1. \qquad (3.30)$$

These boundary conditions are such that the scalar field reaches its vacuum value at infinity. Equations (3.28) with the above boundary conditions lead to a unique solution for the profile functions, although its analytic form is not known. A numerical solution is presented in Fig. 3.4. At large r the asymptotic behavior of the profile functions is

$$1 - \varphi(r) \sim \exp(-m_V r), \qquad f(r) \sim \exp(-m_V r). \qquad (3.31)$$

The ANO vortex breaks the translational invariance. It is characterized by two collective coordinates (or moduli) x_0 and y_0, which indicate the position of the string center.

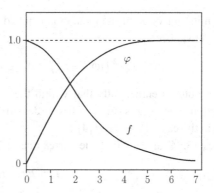

Fig. 3.4 Profile functions of the string as functions of the dimensionless variable $m_V\, r$. The gauge and scalar profile functions are given by f and φ, respectively.

3.1.4 Noncritical Vortex or String

If $m_H \neq m_V$ then Bogomol'nyi completion does not work. One has to solve the second-order equations of motion which follow from minimization of the energy functional in Eq. (3.12) with $U(\phi)$ given in Eq. (3.2). The *ansatz* (3.26) remains to be applicable. It goes through the second-order equations of motion and yields

$$\frac{d}{dr}\left(\frac{1}{r}\frac{df}{dr}\right) - 2n_e^2 e^2 v^2 \frac{\varphi^2}{r} f = 0,$$

$$-\frac{d}{dr}\left(r\frac{d\varphi}{dr}\right) + 2\lambda v^2 r\varphi\left(\varphi^2 - 1\right) + \frac{\varphi}{r}f^2 = 0. \tag{3.32}$$

These equations must be supplemented by the boundary conditions (3.29) and (3.30). One can then solve them numerically.

In the limiting case of small m_V (i.e., $m_H/m_V \gg 1$) one can quite easily find the vortex mass or string tension with logarithmic accuracy. This was first done in Abrikosov's original paper, in 1957. Let us linearize Eqs. (3.32) at large r using the boundary conditions (3.29) and (3.30). Then we get

$$r\frac{d}{dr}\left(\frac{1}{r}\frac{df}{dr}\right) - m_V^2 f = 0,$$

$$\frac{1}{r}\frac{d}{dr}\left[r\frac{d(1-\varphi)}{dr}\right] - m_H^2(1-\varphi) = 0, \tag{3.33}$$

implying the following asymptotic behavior:

$$f(r) \sim \sqrt{r}\,\exp(-m_V r), \qquad 1 - \varphi \sim \left(\frac{1}{\sqrt{r}}\right)\exp(-m_H r). \tag{3.34}$$

At the origin both these profile functions, f and $1 - \varphi$, tend to unity. Away from the origin they monotically decrease: $1 - \varphi$ becomes exponentially small at distances $r \sim m_H^{-1}$ while f does so at much larger distances, $r \sim m_V^{-1}$. At distances $r \ll m_V^{-1}$, we have effectively, a global vortex with logarithmically divergent mass since the vector field has not yet developed. The logarithmic divergence is cut off from below at $r \sim m_H^{-1}$. Thus, in the limit $m_H/m_V \gg 1$ we have for the vortex mass or string tension

Extreme type-II superconductor

$$T \to 2\pi v^2 \ln\left(m_H/m_V\right). \tag{3.35}$$

The opposite limit, $m_V/m_H \gg 1$, is also of interest. In this limit we have [5]

$$T \to 2\pi v^2 \frac{1}{\ln(m_V/m_H)}. \tag{3.36}$$

The light Higgs limit was studied only quite recently, in 1999, by A. Yung because the limit $m_V/m_H \gg 1$ is attainable only in supersymmetric theories. In nonsupersymmetric theories, even if one fine-tunes the Higgs mass to be small at the tree level, radiative corrections shift it to larger values. In fact, the Higgs mass is constrained from below [6]:

$$m_H^2 \gtrsim \frac{e^2}{4\pi^2} m_V^2. \tag{3.37}$$

3.1.5 Translational Moduli

The solution discussed above describes a vortex centered at the origin. To obtain a solution when the center is at the point $(\vec{x}_0)_\perp \equiv \{x_0, y_0\}$ in the perpendicular plane, one must perform the substitution

$$\vec{x}_\perp \to \vec{x}_\perp - (\vec{x}_0)_\perp \tag{3.38}$$

everywhere in the above solution. Equation (3.38) is, of course, equivalent to $x \to x - x_0$ and $y \to y - y_0$. The two parameters x_0 and y_0 are the translational moduli of the vortex (or string) solution.

Exercise

3.1.1 Prove that the gauge potential with the asymptotics (3.16) is pure gauge.

3.2 Non-Abelian Vortices or Strings

In this section we will discuss the simplest example of non-Abelian vortices or strings. What does this mean? As we already know, the U(1) gauge theories in the Higgs regime support ANO strings. Needless to say, non-Abelian strings emerge in non-Abelian gauge theories with a judiciously chosen matter sector [8]. Not every flux-tube solution in non-Abelian theories is a non-Abelian string. To fall into this class, the flux-tube solution must have the possibility of arbitrary rotations in the "internal" non-Abelian group space.[2]

To explain this in more detail let us recall that the non-Abelian magnetic field B_i^a has two indices, the geometric index i characterizing its orientation in space

---

[2] Some authors, especially in the literature of 1980s and 1990s, called "non-Abelian" any string appearing in non-Abelian field theories. This was rather unfortunate, since the magnetic field orientation in these strings was rigidly fixed by the choice of gauge-symmetry-breaking pattern. I suggest that this dated terminology be abandoned. "Non-Abelian" should be reserved for those flux tubes that have orientational moduli in the internal space.

and the color index $a(a = 1, 2, 3$ for SU(2)). If the string axis is directed in the z direction, only the $i = 3$ component of B_i^a is nonvanishing; $B_i^a = 0$ for $i = 1, 2$.

Orientational moduli

The third component, B_3^a, is still a three-component vector in SU(2). In non-Abelian strings its orientation in SU(2) can be arbitrary. The solution must have two internal "orientational" moduli, which parametrize the direction of B_3^a in SU(2), in addition to two translational moduli x_0 and y_0. The ANO string has only the translational moduli. The orientational moduli possess a nontrivial interaction which reflects the structure of the gauge and flavor symmetries of the model under consideration.

A basic model

As a conceptual prototype, let us consider a model (to be generalized shortly) with Lagrangian

$$\mathcal{L} = -\frac{1}{4g_2^2}\left(F_{\mu\nu}^a\right)^2 - \frac{1}{4g_1^2}\left(F_{\mu\nu}\right)^2 + \left(\mathcal{D}_\mu\phi^A\right)^*\left(\mathcal{D}_\mu\phi^A\right)$$
$$- \frac{g_2^2}{2}\left(\phi_A^*\frac{\tau^a}{2}\phi^A\right)^2 - \frac{g_1^2}{8}\left[\left(\phi_A^*\right)\left(\phi^A\right) - 2v^2\right]^2. \tag{3.39}$$

It describes two gauge bosons, SU(2) and U(1). The corresponding coupling constants are denoted by g_2 and g_1, respectively. The matter sector consists of two scalar fields ($A = 1, 2$), each in the doublet representation of SU(2)$_\text{gauge}$. Note that the coupling constants governing the scalar-field self-interactions coincide with the gauge coupling constants. This special choice is made to ensure the equality of the Higgs and gauge boson masses, which, as we already know, leads to BPS saturation of the string solutions (i.e., the reduction of the second-order equations of motion to the first-order Bogomol'nyi equations).

The covariant derivative is defined as

$$\mathcal{D}_\mu\phi = \partial_\mu\phi - \frac{i}{2}A_\mu\phi - \frac{i}{2}A_\mu^a\tau^a\phi. \tag{3.40}$$

As is obvious from this definition, the U(1) charges of the fields ϕ^A, $A = 1, 2$, are $\frac{1}{2}$. This choice is convenient; it simplifies many expressions to be presented below. To keep the theory at weak coupling we consider large values of the parameter v^2 in (3.39), i.e., $v \gg \Lambda$.

Besides the gauge symmetry SU(2) × U(1), the Lagrangian (3.39) has a global flavor SU(2) symmetry. To see this in an explicit way it is convenient to introduce a 2×2 matrix of the fields ϕ,

$$\Phi = \begin{pmatrix} \phi^{11} & \phi^{12} \\ \phi^{21} & \phi^{22} \end{pmatrix}, \tag{3.41}$$

Matter fields in matrix form

where the first superscript refers to the SU(2)$_\text{gauge}$ group and the second to the flavor group (i.e., $A = 1, 2$). In terms of Φ the matter part of the Lagrangian (3.39) takes the form

$$\mathcal{L}_\text{matter} = \text{Tr}\,(\mathcal{D}_\mu\Phi)^\dagger(\mathcal{D}_\mu\Phi) - U(\Phi, \Phi^\dagger), \tag{3.42}$$

where

$$U(\Phi, \Phi^\dagger) = \frac{g_2^2}{2}\text{Tr}\left(\Phi^\dagger\frac{\tau^a}{2}\Phi\right)\text{Tr}\left(\Phi^\dagger\frac{\tau^a}{2}\Phi\right) + \frac{g_1^2}{8}\left[\text{Tr}\left(\Phi^\dagger\Phi\right) - 2v^2\right]^2. \tag{3.43}$$

The flavor transformation has the following effect on Φ:

$$\Phi \rightarrow \Phi U \tag{3.44}$$

while the color transformation acts as follows:

$$\Phi \to \tilde{U}\Phi, \tag{3.45}$$

where U and $\tilde{U}$ are arbitrary matrices from the groups $SU(2)_{\text{flavor}}$ and $SU(2)_{\text{color}}$, respectively.

The flavor $SU(2)$ symmetry of (3.42) and (3.43) is obvious. To verify the color $SU(2)$ symmetry of $U(\Phi, \Phi^\dagger)$ one can use, for instance, the identity

$$\text{Tr}\left(\Phi^\dagger \frac{\tau^a}{2}\Phi\right)\text{Tr}\left(\Phi^\dagger \frac{\tau^a}{2}\Phi\right) = -\frac{1}{4}\text{Tr}\left(\Phi^\dagger\Phi\right)\text{Tr}\left(\Phi^\dagger\Phi\right) + \frac{1}{2}\text{Tr}\left(\Phi^\dagger\Phi\Phi^\dagger\Phi\right)$$

following from the Fierz transformation for the Pauli matrices.

3.2.1 Symmetries and the Vacuum Structure of the Model

One may ask oneself why the interaction potential of the fields Φ given in Eq. (3.43) is chosen in such a special way. This is done on purpose: we want to ensure a special symmetry-breaking pattern in the vacuum of the theory.

Let us have a closer look at Eq. (3.43). It consists of two non-negative terms. The absolute minimum of the potential is obviously $U = 0$. To achieve this minimum each of the two terms must vanish. The vanishing of the second term requires that $\Phi \neq 0$ and $\Phi \propto v$. Then, to make the first term vanish one can choose Φ to be proportional to the unit matrix, since $\text{Tr }\tau^a = 0$ for all a.

After these remarks the vacuum field configuration is obvious:

$$\Phi_{\text{vac}} = v\begin{pmatrix} 1 & 0 \\ 0 & 1 \end{pmatrix}, \qquad (A^a_\mu)_{\text{vac}} = 0. \tag{3.46}$$

| *Color–flavor locking* |

Of course, any field configuration that is gauge equivalent to (3.46) presents the (same) vacuum solution.

We see that the vacuum of the model is invariant under a combined color–flavor *global* $SU(2)$:

$$\Phi \to U^\dagger\Phi U. \tag{3.47}$$

This feature will ensure occurrence of the orientational moduli in the string solution, making it non-Abelian.

The phenomenon described above is usually referred to as *color–flavor locking*. This mechanism for color–flavor locking in models with an equal number of colors and flavors was devised in 1972 [7].

The masses of the (Higgsed) gauge bosons are

$$m^2_{V_{U(1)}} = g_1^2 v^2,$$
$$m^2_{V_{SU(2)}} = g_2^2 v^2. \tag{3.48}$$

3.2.2 Abrikosov–Nielsen–Olesen versus "Elementary" (1,0) and (0,1) Strings

Even if we ignore the $SU(2)$ gauge bosons altogether, the model that we are discussing still supports the conventional Abrikosov–Nielsen–Olesen strings. The existence of the ANO string is due to the fact that $\pi_1(U(1)) = \mathbb{Z}$, ensuring its

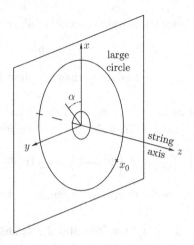

Fig. 3.5   Geometry of a string.

topological stability. For this solution one can discard the $SU(2)_{\text{gauge}}$ part of the action, putting $A_\mu^a = 0$. Correspondingly, there will be no SU(2) winding of Φ. A nontrivial topology is realized through a U(1) winding of Φ,

$$\Phi(x) = v e^{i\alpha(x)}, \qquad |x| \to \infty, \tag{3.49}$$

and

$$A_i = -2\varepsilon_{ij}\frac{x_j}{r^2}, \qquad i,j = 1,2, \tag{3.50}$$

where α is the angle in the perpendicular plane (Fig. 3.5) and r is the distance from the string axis in the perpendicular plane. Equations (3.49) and (3.50) refer to a minimal ANO string with a minimal winding. The factor 2 in Eq. (3.50) is due to the fact that the U(1) charge of the matter fields is 1/2. Needless to say, the tension of the ANO string is given by the standard formula

$$T_{\text{ANO}} = 4\pi v^2, \tag{3.51}$$

where the factor 4π instead of the 2π in Eq. (3.25) appears due to the two flavors.

This is not the string in which we are interested here, however – in fact, in the problem at hand there are "more elementary" strings with half the above tension, so that the ANO string can be viewed as a bound state of two elementary strings. Where do they come from? Since $\pi_1(SU(2))$ is trivial, at first sight it might seem that in the $SU(2) \times U(1)$ theory there are no new options. This conclusion is wrong, however; one can combine the Z_2 center of SU(2) with the element $-1 \in U(1)$ to get a topologically stable string-like solution, possessing both windings, i.e., in SU(2) and U(1), of the following type:

$$\Phi(x) = v \exp\left[i\alpha(x)\frac{1 \pm \tau^3}{2}\right], \qquad |x| \to \infty,$$

$$A_i = -\varepsilon_{ij}\frac{x_j}{r^2}, \qquad A_i^3 = \mp\varepsilon_{ij}\frac{x_j}{r^2}, \qquad i,j = 1,2. \tag{3.52}$$

In this *ansatz* only one of the two flavors winds around the string axis. Correspondingly, the U(1) magnetic flux is half that in the ANO case. To see that this is so it is sufficient to perform a Bogomol'nyi completion of the energy functional, obtaining

$$\mathcal{E} = \int d^2x \left\{ \frac{1}{2g_2^2} \left[F_{12}^a + \frac{g_2^2}{2} \text{Tr} \left(\Phi^\dagger \tau^a \Phi \right) \right]^2 + \frac{1}{2g_1^2} \left[F_{12} + \frac{g_1^2}{2} \text{Tr} \left(\Phi^\dagger \Phi - v^2 \right) \right]^2 \right.$$

$$\left. + \left[(\mathcal{D}_1 + i\mathcal{D}_2)\phi^A \right]^* \left[(\mathcal{D}_1 + i\mathcal{D}_2)\phi^A \right] + v^2 F_{12} \right\}. \tag{3.53}$$

Here we have omitted a (vanishing) surface term. Equation (3.53) shows that for a BPS-saturated string its tension is determined exclusively by the flux of the U(1) field,

$$\tau_\pm = v^2 \int d^2x \, F_{12} = v^2 \oint_{\text{large circle}} \vec{A} \, d\vec{r} = 2\pi v^2. \tag{3.54}$$

The $\pm$ subscript corresponds to two types of elementary string in which either only ϕ^1 or only ϕ^2 is topologically nontrivial; see the boundary conditions (3.52).

We will refer to the strings corresponding to the boundary conditions (3.52) as $(1,0)$ and $(0,1)$. It is instructive to reiterate the reason for their topological stability. The SU(2) group space is a sphere. The homotopy group $\pi_1(SU(2))$ is trivial. However, if we map half the large circle (encircling the string in the perpendicular plane) onto this sphere, fixing the beginning and the end at the north and south poles and the remaining half on half the U(1) circle, in such a way that the mapping starts and ends at the same north and south poles, this mapping will be noncontractable to a trivial mapping. Of course, we are relying on the fact that -1 and 1 are elements of both the SU(2) sphere (the center elements) and the U(1) circle. Note that the boundary conditions (3.52) break the Z_2 invariance of the theory under consideration:

> *These strings are also known as Z_2 strings.*

$$\Phi \rightarrow \tau^1 \Phi \tau^1, \qquad (A^a \tau^a) \rightarrow \tau^1 (A^a \tau^a) \tau^1. \tag{3.55}$$

Under this Z_2 symmetry the strings $(1,0)$ and $(0,1)$ interchange. This explains the degeneracy of the tensions.

3.2.3 First-Order Equations for Elementary Strings

Now let us study elementary strings. The first-order equations for the BPS strings following from the energy functional (3.53) are

$$\tilde{F}_3^a + \frac{g_2^2}{2} \left(\bar{\phi}_A \tau^a \phi^A \right) = 0, \qquad a = 1, 2, 3,$$

$$\tilde{F}_3 + \frac{g_1^2}{2} \left(|\phi^A|^2 - 2v^2 \right) = 0, \tag{3.56}$$

$$(\mathcal{D}_1 + i\mathcal{D}_2)\phi^A = 0,$$

where

$$\tilde{F}_m = \frac{1}{2} \varepsilon_{mnk} F_{nk}, \qquad m, n, k = 1, 2, 3. \tag{3.57}$$

To construct the $(0,1)$ and $(1,0)$ strings we further restrict the gauge field A_μ^a to a single color component, namely A_μ^3, by setting $A_\mu^1 = A_\mu^2 = 0$; then we consider the Φ fields of 2×2 color–flavor diagonal form,

$$\Phi^{kA}(x) \neq 0 \qquad \text{for } k = A = 1, 2. \tag{3.58}$$

The off-diagonal components of the matrix Φ are set to zero.

The $(1, 0)$ string arises when the first flavor has unit winding number and the second flavor does not wind at all. And, vice versa, the $(0, 1)$ string arises when the second flavor has unit winding number and the first flavor does not wind. Consider for definiteness the $(1, 0)$ string. (The $(0, 1)$ string solution is easy to obtain through (3.55).) The solutions of the first-order equations (3.56) can be sought using the following *ansatz* [8]:

$$\Phi(x) = v \begin{pmatrix} e^{i\alpha}\varphi_1(r) & 0 \\ 0 & \varphi_2(r) \end{pmatrix},$$

$$A_i^3(x) = -\varepsilon_{ij}\frac{x_j}{r^2}[1 - f_3(r)], \tag{3.59}$$

$$A_i(x) = -\varepsilon_{ij}\frac{x_j}{r^2}[1 - f(r)],$$

| *Non-Abelian*
| *string ansatz* |

where the profile functions φ_1, φ_2 for the scalar fields and f_3, f for the gauge fields depend only on r ($i, j = 1, 2$). Applying this *ansatz* one can rearrange the first-order equations (3.46) in the form

$$r\frac{d}{dr}\varphi_1 - \frac{1}{2}(f + f_3)\,\varphi_1 = 0,$$

$$r\frac{d}{dr}\varphi_2 - \frac{1}{2}(f - f_3)\,\varphi_2 = 0,$$

$$-\frac{1}{r}\frac{d}{dr}f + \frac{g_1^2 v^2}{2}\left(\varphi_1^2 + \varphi_2^2 - 2\right) = 0, \tag{3.60}$$

$$-\frac{1}{r}\frac{d}{dr}f_3 + \frac{g_2^2 v^2}{2}\left(\varphi_1^2 - \varphi_2^2\right) = 0.$$

Furthermore, one needs to specify the boundary conditions that would determine the profile functions in these equations, namely,

$$f_3(0) = 1, \qquad f(0) = 1,$$

$$f_3(\infty) = 0, \qquad f(\infty) = 0 \tag{3.61}$$

for the gauge fields, while the boundary conditions for the Higgs fields are

$$\varphi_1(\infty) = 1, \qquad \varphi_2(\infty) = 1, \qquad \varphi_1(0) = 0. \tag{3.62}$$

Note that, since the field φ_2 does not wind, it need not vanish at the origin and it does not. Numerical solutions of the Bogomol'nyi equations (3.60) for the $(0, 1)$ and $(1, 0)$ strings were found in [8], from which Figs. 3.6 and 3.7 are taken.

3.2.4   Making Non-Abelian Strings from Elementary Strings: Non-Abelian Moduli

The theory under consideration preserves global SU(2) symmetry, a diagonal subgroup of $SU(2)_{gauge}$ and $SU(2)_{flavor}$. At the same time, a straightforward inspection of the asymptotics (3.52) shows that both elementary strings break this global SU(2) symmetry down to the U(1) subgroup corresponding to rotations around the third axis in SU(2) space. This means that there should exist a general family of solutions [8] described by non-Abelian moduli. Their role is to propagate two "elementary"

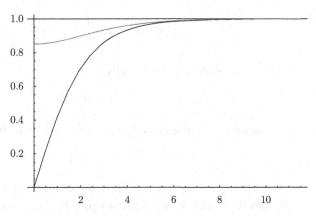

Fig. 3.6 Vortex profile functions $\varphi_1(r)$ and $\varphi_2(r)$ of the $(1,0)$ string. Note that $\varphi_1(0) = 0$.

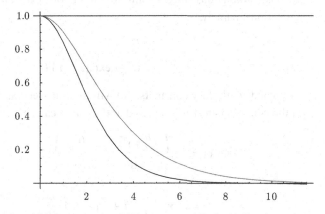

Fig. 3.7 The profile functions $f_3(r)$ (lower curve) and $f(r)$ (upper curve) for the $(1,0)$ string.

solutions inside SU(2) space. The $(1,0)$ and $(0,1)$ strings discussed above are just two representatives of this continuous family.

Let us elucidate the above assertion. While the vacuum field $\Phi_{\mathrm{vac}} = vI$ (here I is a 2×2 unit matrix) is invariant under the global $SU(2)_{C+F}$ symmetry,

$$\Phi \to U\Phi U^{-1}, \tag{3.63}$$

the string configuration (3.59) is not. Therefore, if there is a single solution of the form (3.59) then there must be in fact a whole family of solutions, obtained by combined global gauge–flavor rotations. Say, for the Φ fields,

Rotating the Z_2 string in group space

$$\Phi(x) \to e^{i\vec{\omega}\vec{\tau}/2}\Phi(x)e^{-i\vec{\omega}\vec{\tau}/2}. \tag{3.64}$$

Thus, applying an SU(2) transformation to an elementary string we "rotate" it in SU(2), producing a different embedding. In fact, we are dealing here with the coset SU(2)/U(1), as should be clear from Eq. (3.59): rotations around the third axis in SU(2) space leave the solution (3.59) intact.

Thus, introduction of the moduli matrix U allows us to obtain a generic solution for the non-Abelian string Bogomol'nyi equation having the following asymptotics at $|x| \to \infty$:

$$\Phi(x) = v \exp\left(i\alpha(x)\frac{1+\vec{S}\vec{\tau}}{2}\right), \tag{3.65}$$

where $\vec{S}$ is a moduli vector defined by

$$\vec{S}\vec{\tau} = U\tau^3 U^{-1}. \tag{3.66}$$

The unitarity of U implies that the vector $\vec{S}$ is subject to the following constraint:

$$\vec{S}^2 = 1. \tag{3.67}$$

At $\vec{S} = (0,0,\pm 1)$ we get the field configurations of Eq. (3.52). Every given matrix U defines the moduli vector $\vec{S}$ unambiguously. The inverse is not true, however. If we consider the left-hand side of Eq. (3.66) as given, then the solution for U is obviously ambiguous since for any solution U one can construct two "gauge orbits" of solutions, namely,

$$\begin{aligned} U &\to U \exp(i\beta\tau_3),\\ U &\to \exp\left(i\gamma\vec{S}\vec{\tau}\right) U, \end{aligned} \tag{3.68}$$

with β and γ arbitrary constants. We will use this freedom in what follows. At finite $|x|$ the non-Abelian string centered at the origin can be written as [8]

$$\begin{aligned} \Phi(x) &= Uv\begin{pmatrix} e^{i\alpha}\varphi_1(r) & 0 \\ 0 & \varphi_2(r) \end{pmatrix} U^{-1}\\ &= v\exp\left[\frac{i}{2}\alpha(1+\vec{S}\vec{\tau})\right]U\begin{pmatrix} \varphi_1(r) & 0 \\ 0 & \varphi_2(r) \end{pmatrix} U^{-1}, \end{aligned} \tag{3.69}$$

$$A_i^a(x) = -S^a \varepsilon_{ij}\frac{x_j}{r^2}[1-f_3(r)],$$

$$A_i(x) = -\varepsilon_{ij}\frac{x_j}{r^2}[1-f(r)],$$

where the profile functions are the solutions to Eq. (3.60). Note that

$$U\begin{pmatrix} \varphi_1 & 0 \\ 0 & \varphi_2 \end{pmatrix} U^{-1} = \frac{\varphi_1 + \varphi_2}{2} + \vec{S}\vec{\tau}\frac{\varphi_1 - \varphi_2}{2}. \tag{3.70}$$

It is now clear that this solution smoothly interpolates between the $(1,0)$ and $(0,1)$ strings as we go from $\vec{S} = (0,0,1)$ to $\vec{S} = (0,0,-1)$.

Since the SU(2)$_{C+F}$ symmetry is not broken by the vacuum expectation values, it is physical and has nothing to do with the gauge rotations "eaten" by the Higgs mechanism. The orientational moduli $\vec{S}$ are not gauge artifacts. Rather, they parametrize the coset SU(2)/U(1) = S$_2$. To see this, we can construct gauge-invariant operators that have an explicit $\vec{S}$-dependence. This procedure is instructive.

As an example, let us define a "non-Abelian" field strength (denoted by boldface type),

$$\widetilde{\boldsymbol{F}}_3^a = \frac{1}{v^2}\mathrm{Tr}\left(\Phi^\dagger\widetilde{F}_3^b\frac{\tau^b}{2}\Phi\tau^a\right), \tag{3.71}$$

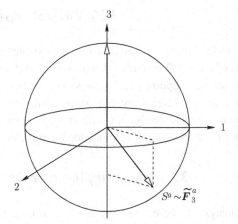

Fig. 3.8   The bosonic moduli S^a introduced in (3.66) describe the orientation of the color-magnetic flux for the rotated (0,1) and (1,0) strings in the O(3)-group space, Eq. (3.72).

where the subscript 3 labels the z axis, the direction of the string (Fig. 3.8). From the very definition it is clear that this field is *gauge invariant*.[3] Moreover, Eq. (3.69) implies that

$$\widetilde{\boldsymbol{F}}_3^a = -S^a \frac{(\varphi_1^2 + \varphi_2^2)}{2} \frac{1}{r} \frac{df_3}{dr}. \tag{3.72}$$

From this formula we readily infer the physical meaning of the moduli $\vec{S}$: the flux of the *color*-magnetic field[4] in the flux tube is directed along $\vec{S}$ (Fig. 3.8). For the strings in Eq. (3.59), see also Eq. (3.52), the color-magnetic flux is directed along the third axis in the O(3)-group space, either upward or downward (i.e., toward either the north or the south pole). These are the north and south poles of the coset $SU(2)/U(1) = S_2$.

Singular gauge, or combing the hedgehog

To conclude this section, I present the non-Abelian string solution (3.69) in the singular gauge in which the Φ fields at $|x| \to \infty$ tend to fixed vacuum expectation values (VEVs) and do not wind (i.e., do not depend on the polar angle α as $|x| \to \infty$). In the singular gauge we have

$$\Phi = v\, U \begin{pmatrix} \varphi_1(r) & 0 \\ 0 & \varphi_2(r) \end{pmatrix} U^{-1},$$

$$A_i^a(x) = S^a\, \varepsilon_{ij} \frac{x_j}{r^2} f_3(r), \tag{3.73}$$

$$A_i(x) = \varepsilon_{ij} \frac{x_j}{r^2} f(r).$$

In this gauge the spatial components of A_μ fall off fast at large distances. If the color-magnetic flux is defined as the circulation of A_i over a circle encompassing the string axis, the flux will be saturated by an integral coming from the small circle around the (singular) string origin.

---

[3] In the vacuum, where the matrix Φ is that of vacuum expectation values, $\widetilde{F}_3^a$ and $\widetilde{F}_3^a$ coincide.
[4] Defined in a gauge-invariant way; see Eq. (3.71).

3.2.5 Vortex Strings in SM

It is curious to note that vortex strings exist in the Standard Model. They combine elements of the ANO and non-Abelian strings and, in addition, a Dirac type monopole at the string endpoint [9]. These vortex strings, usually referred to as Z strings, are not quite stable. They can break (nonperturbatively) through the monopole–antimonopole pair production with an exponentially suppressed probability.

I briefly discuss them in Appendix 3.6 at the end of this chapter (page 120).

3.2.6 Low-Energy Theory on the String World Sheet

The analysis to be carried out below is similar to that of Section 2.1.8. The non-Abelian string solution under consideration is characterized by four moduli – two translational moduli x_0 and y_0 parametrizing the position of the string center in the perpendicular plane, plus two orientational moduli described by the vector $\vec{S}$ subject to the constraint $\vec{S}\vec{S} = 1$. To obtain the world-sheet theory we promote these moduli to be moduli fields

$$x_0(t, z), \quad y_0(t, z), \quad \vec{S}(t, z), \tag{3.74}$$

depending on t and z adiabatically. The coordinates $\{t, z\}$ on the string world sheet can be combined into a two-dimensional coordinate x^p ($p = 0, 3$). The fields (3.74) are Goldstone bosons localized on the string. The first two fields are due to the spontaneous breaking of translational invariance in the directions x and y, while the second two are due to the breaking of the global SU(2) symmetry of the bulk theory down to U(1) on the string solution.

As in Section 2.1.8 we start from the static z-independent string solution (3.73) parametrized by two translational moduli, as explained in Section 3.1.5, e.g.,

$$r = |\vec{x}_\perp - (\vec{x}_0)_\perp| \qquad \text{where } \vec{x}_\perp = \{x, y\} \equiv \{x^j\}, \tag{3.75}$$

and so on. Then we substitute the "shifted" solution into the four-dimensional Lagrangian (3.39), assuming that the moduli fields $(\vec{x}_0)_\perp$ depend on $x^p \equiv \{t, z\}$ ($p = 0, 3$). Finally, we integrate over $d^2 x_\perp$. There is no potential in the effective two-dimensional action obtained in this way. The kinetic terms of the moduli fields (they are of the second order in the derivatives) are obtained from the kinetic terms in (3.39). Their structure is obvious on symmetry grounds:

World-sheet theory, $x^p \equiv \{t, z\}$.

$$S^{(1+1)} = \int dt\, dz \left[\frac{T}{2} \left(\frac{\partial \vec{x}_\perp}{\partial x^p} \right)^2 + \frac{\beta}{2} \left(\frac{\partial \vec{S}}{\partial x^p} \right)^2 \right], \qquad \vec{S}^2 = 1, \tag{3.76}$$

where T is the string tension and β is a constant. The orientational part of the world-sheet action is the famous O(3) sigma model, which will be discussed in detail in Chapter 6.

The coefficient $T/2$ in front of the first term in the world-sheet action (3.76) (the translational part of the action) is universal and can be established in just the same way as in Section 2.1.8. To derive the coefficient β in terms of the parameters of the four-dimensional theory (3.39) one has to carry out an actual calculation which, although straightforward, is rather cumbersome. For curious readers this calculation is presented in Appendix section 3.5, at the end of this chapter. Here I just quote the answer,

$$\beta = \frac{2\pi}{g_2^2}. \tag{3.77}$$

Exercises

3.2.1 Calculate the masses of the elementary excitations of the fields ϕ in the vacuum (3.46).

3.2.2 The vector $\vec{\omega}$ in (3.64) consists of a set of three constant parameters, $\omega_{1,2,3}$. Which of these parameters lead to nontrivial rotations of the Z_2 string solutions in $SU(2)_{C+F}$? Which act trivially?

3.3 Fermion Zero Modes

In this section we will add fermions and explore the impact they produce on strings. For simplicity we will limit our consideration to ANO strings. The generalization to non-Abelian strings is straightforward. Some fermion-induced effects in non-Abelian strings will be discussed in Part II, which is devoted to supersymmetry.

We will start from the bosonic model described in Section 3.1.2. To ease the notation we will set $n_e = 1$, i.e., we will assume the U(1) charge of the field ϕ to be unity. In addition to the photon and ϕ fields we introduce a Dirac (four-component) field Ψ, which is composed of two Weyl spinors, ξ_α and $\bar{\eta}^{\dot{\alpha}}$, according to (1.68).[5] Instead of the conventional fermion mass term of the type $\bar{\Psi}\Psi$, we introduce a "Higgs" mass term through the Yukawa coupling of the fermions with the ϕ fields. Since the U(1) charge of ϕ is unity, the only allowed Yukawa term is of the type $\overline{\Psi^C}\Psi\phi$, where the superscript C stands for charge conjugation,

$$\Psi^C = \gamma^2 \Psi^*, \tag{3.78}$$

while the U(1) charge of Ψ (as well as that of $\overline{\Psi^C}$) must be $-1/2$, i.e., under the U(1) transformation we have

$$\Psi \to e^{-i\beta/2}\Psi. \tag{3.79}$$

Then the covariant derivative acting on Ψ is

$$\mathcal{D}_\mu \Psi = \left(\partial_\mu + \frac{i}{2}A_\mu\right)\Psi. \tag{3.80}$$

With all these conventions, the fermion part of the Lagrangian takes the form

$$\mathcal{L}_\Psi = \bar{\Psi}i\,\slashed{\mathcal{D}}\,\Psi + \frac{h}{2}\overline{\Psi^C}\Psi\phi + \frac{h^*}{2}\bar{\Psi}\Psi^C\,\bar{\phi}, \tag{3.81}$$

[5] For more details see the beginning of Part II, Section 10.2.1.

Table 3.1. The U(1) charge of the fields ξ and η			
ξ	$\bar{\xi}$	η	$\bar{\eta}$
$-\frac{1}{2}$	$\frac{1}{2}$	$\frac{1}{2}$	$-\frac{1}{2}$

where the Yukawa coupling h can always be chosen to be real and positive,[6] by an appropriate rotation of the field ϕ. The gauge field A_μ is defined in Eq. (3.26), while the string's geometry is depicted in Fig. 3.5.

$\mathcal{L}_\Psi$ in spinorial notation

For further analysis it is convenient to rewrite Eq. (3.81) in two-component form,

$$\mathcal{L}_\Psi = \bar{\xi}_{\dot{\alpha}}(\bar{\sigma}^\mu)^{\dot{\alpha}\alpha} i\mathcal{D}_\mu \xi_\alpha + \eta^\alpha (\sigma^\mu)_{\alpha\dot{\alpha}} i\mathcal{D}_\mu \bar{\eta}^{\dot{\alpha}}$$

$$+ \frac{ih}{2}\left[\phi\left(\xi^2 + \bar{\eta}^2\right) - \bar{\phi}\left(\eta^2 + \bar{\xi}^2\right)\right] \tag{3.82}$$

where we use the spinoral notation explained in Section 10.2 at the beginning of Part II. The U(1) charges of the ξ, η fields are shown in Table 3.1.

The gauge U(1) symmetry is broken in the vacuum $\phi = v$ (where, as usual, we assume v to be real and positive), and, as a result, the fields ξ and η acquire masses

$$m_F = hv. \tag{3.83}$$

However, in the core of the string $\phi \to 0$; hence, the fermions are massless inside the flux tube and therefore one may expect the occurrence of localized zero modes.

Our task is to determine the fermion zero modes in the two-dimensional Dirac operator[7] in the string background. Why is this important? If such modes exist – and they do[8] – the fermion dynamics on the string world sheet is that of the free fermion theory, with no mass gap, i.e., the world-sheet fermions are massless and can travel freely along the string. Witten suggested [11] using this property to construct (with the introduction of yet another U(1) gauge field, which remains un-Higgsed) superconducting cosmic strings. We will not go into details of this astrophysical topic, but the interested reader is referred to the textbook [12].

Fermions and cosmic strings

Before calculating the fermion zero modes let us discuss a general strategy allowing one to find out a priori, without direct calculation, whether such modes exist in a given model with a given background. This strategy is based on the index of the Dirac operator and is applicable for generic fermion sectors.

3.3.1 Index Theorems

Assume that we have an abstract Dirac operator $i\mathcal{D}$ acting on some spinor ψ and that γ^j matrices in this operator are such that there exists an analog of the conventional γ^5, i.e., a Hermitian matrix such that $\gamma^5\gamma^j = -\gamma^j\gamma^5$ for all j and $(\gamma^5)^2 = 1$. Define the eigenmode of this operator by

$$i\mathcal{D}\psi = E\psi, \tag{3.84}$$

---

[6] Assuming that h is real and positive, hereafter the asterisk will be omitted.
[7] The Dirac operator in the transverse xy plane.
[8] They were found originally by Jackiw and Rossi [10].

where E is the eigenvalue. It is real provided that $i\slashed{D}$ is Hermitian. All modes must be normalizable (we will follow the standard convention of the unit norm). For all nonvanishing eigenvalues the eigenmodes are paired in the following sense: assume that ψ is a solution of (3.84). Then $\tilde{\psi} = \gamma^5\psi$ is the solution of the equation $i\slashed{D}\tilde{\psi} = -E\tilde{\psi}$, i.e., $\gamma^5\psi$ is the eigenmode of the same Dirac operator having eigenvalue $-E$. For this reason, for each nonzero mode $\int \psi^\dagger\gamma^5\psi = 0$. This fact will be exploited below.

This does not have to be the case for zero modes. If $E = 0$ then $\tilde{\psi}$ must coincide with ψ up to a phase factor,[9] which must be either $+1$ or -1 because $(\gamma^5)^2 = 1$. Let us call the mode "left-handed" if $\gamma^5\psi = \psi$ and "right-handed" if $\gamma^5\psi = -\psi$. Then the number of left-handed zero modes n_L minus the number of right-handed zero modes n_R is an *index*, a quantity that does not depend on continuous deformations of the background field in the expression for the Dirac operator.

Here is a brief outline of the proof [13] (all subtleties are omitted). We start from an axial current

$$a^\mu = \psi^\dagger\gamma^\mu\gamma^5\psi. \tag{3.85}$$

To regularize the Green's function of the Dirac operator in (3.84) we must endow it with a small mass m:

$$i\slashed{D} \rightarrow i\slashed{D}_{\text{reg}} = i\slashed{D} - im, \tag{3.86}$$

where m is set to zero at the very end. The corresponding Lagrangian takes the form

$$\mathcal{L} = \psi^\dagger i\slashed{D}_{\text{reg}}\psi. \tag{3.87}$$

The Green's function for the operator (3.86) is

$$G(x,y) = \sum_{\forall \text{ modes}} \frac{\psi_\ell(x)\psi_\ell^\dagger(y)}{E_\ell - im}, \tag{3.88}$$

where ψ_ℓ and $\psi_\ell^\dagger$ are the eigenmodes of the operator $i\slashed{D}$ with eigenvalues E_ℓ. Now, the divergence of the axial current $\partial_\mu a^\mu$ following from (3.87) can be written as

$$\partial_\mu a^\mu = -2m\,\psi^\dagger\gamma^5\psi = 2m\,\text{Tr}\left[\gamma^5 iG(x,x)\right]. \tag{3.89}$$

Substituting Eq. (3.88) into (3.89), integrating over x, taking account of the mode normalization and taking the limit $m \rightarrow 0$ we get

$$\int \partial_\mu a^\mu = -2\left(n_L - n_R\right) \equiv -\left[n_L(\psi) + n_L(\psi^\dagger) - n_R(\psi) - n_R(\psi^\dagger)\right]. \tag{3.90}$$

| Index theorem |

This is the desired result: the integral $\int \partial_\mu a^\mu$ counts the number of zero modes of the Dirac operator $i\slashed{D}$, or, to be exact, the difference between the numbers of zero modes of opposite chiralities. If, from some additional arguments we know that, say, $n_R = 0$ then the integral $\int \partial_\mu a^\mu$ predicts n_L.

Why is this number an index? The left-hand side of (3.90) is an integral over a full derivative. Hence it depends only on the behavior at the boundaries and does not change in response to local variations in the background field. If $\int \partial_\mu a^\mu$ does not

---

[9] If the number of zero modes ψ_0 is larger than 1 then these modes can be diagonalized with respect to the action of γ^5, $\gamma^5\psi_0 = \pm\psi_0$.

vanish – and this is the case in topologically nontrivial backgrounds – zero modes of the operator $i\,\slashed{D}$ must exist.

3.3.2 Fermion Zero Modes for the ANO String

Now, it is time to return to the model (3.82). Here we will specify the general analysis of Section 3.3.1 and discuss an index theorem establishing the number of fermion zero modes on a string [14]. Then we will find these modes explicitly and present the string world-sheet theory for fermions.

The sum of the Lagrangians in (3.6) and (3.82) defines a four-dimensional theory with fermions which supports ANO strings. The string solution depends on only *two* coordinates $x_i(i = 1, 2)$, and the fermion zero modes sought for are those of the two-dimensional Dirac operator. Hence, we need to calculate the index for the two-dimensional, rather than the four-dimensional, theory. If $h \neq 0$, the four-dimensional theory (3.6), (3.82) has no global chiral symmetry at all.

However, after reduction of the theory (3.82) to two dimensions a global chirality does emerge.[10] To see that this is the case, observe the following. In two dimensions there is no distinction between the dotted and undotted indices (see Section 10.2). Moreover, we can eliminate the upper indices altogether, expressing the two-dimensional reduced Lagrangian in terms of two decoupled spinors ξ_α and η_α (in what follows, we will write ξ and η for short) which enter symmetrically in the Yukawa part of the Lagrangian but have opposite-sign couplings to the photon field (see Table 3.1). The Yukawa part of the Lagrangian is proportional to $\phi\xi_1\xi_2 - \bar{\phi}\eta_1\eta_2+$ Hermitian conjugate (H.c.), where ξ_1 and η_1 are left-handed components (in the two-dimensional sense) while ξ_2 and η_2 are right-handed. The term $\phi\xi_1\xi_2 - \bar{\phi}\eta_1\eta_2$, as well as all the other terms, stay invariant under the two independent global rotations

$$\begin{aligned} \xi_1 &\to e^{i\gamma}\xi_1, & \xi_2 &\to e^{-i\gamma}\xi_2, \\ \eta_1 &\to e^{i\tilde{\gamma}}\eta_1, & \eta_2 &\to e^{-i\tilde{\gamma}}\eta_2. \end{aligned} \tag{3.91}$$

Now we define two-dimensional gamma matrices relevant to the problem at hand:

$$\gamma_1 = -\sigma_1, \quad \gamma_2 = -\sigma_2, \quad \gamma^5 = \sigma_3. \tag{3.92}$$

The invariance of the Lagrangian under (3.91) implies that the two axial currents

$$a_i = \xi^\dagger\gamma_i\gamma^5\xi, \qquad \tilde{a}_i = \eta^\dagger\gamma_i\gamma^5\eta \tag{3.93}$$

are conserved. Their conservation is broken at the quantum level, owing to anomalies, which will be discussed in detail in Chapter 8. Taking into account the fact that the couplings of ξ, η to the photon field are $\pm\frac{1}{2}$, we obtain

$$\partial_i a_i = \frac{1}{4\pi}\varepsilon^{ij}F_{ij}, \qquad \partial_i\tilde{a}_i = -\frac{1}{4\pi}\varepsilon^{ij}F_{ij}. \tag{3.94}$$

Compare Eq. (3.94) with the winding number (3.17). We see that in the string background

$$\int \partial_i a_i d^2x = 1, \qquad \int \partial_i\tilde{a}_i d^2x = -1, \tag{3.95}$$

---

[10] By "chirality" I mean here the two-dimensional chirality.

which entails in turn that

$$n_R(\xi) + n_R(\xi^\dagger) - n_L(\xi) - n_L(\xi^\dagger) = 1,$$

$$n_L(\eta) + n_L(\eta^\dagger) - n_R(\eta) - n_R(\eta^\dagger) = 1. \tag{3.96}$$

The implication of Eq. (3.96) is that ξ has one (real) zero mode in ξ_R (i.e., ξ_2) while η has one (real) zero mode in η_L (i.e., η_1).

It is not difficult to calculate the zero modes explicitly. For instance, for the ξ field the equations to be solved are

$$-(\mathcal{D}_1 - i\mathcal{D}_2)\xi_2 - h\bar{\phi}\xi_2^\dagger = 0,$$

$$-(\mathcal{D}_1 + i\mathcal{D}_2)\xi_1 + h\bar{\phi}\xi_1^\dagger = 0. \tag{3.97}$$

Using Eq. (3.26) and the geometrical definitions from Fig. 3.5 we can rewrite the covariant derivatives as

$$\mathcal{D}_1 - i\mathcal{D}_2 = e^{-i\alpha}\frac{\partial}{\partial r} - \frac{i}{r}e^{-i\alpha}\frac{\partial}{\partial\alpha} + \frac{1-f}{2r}e^{-i\alpha},$$

$$\mathcal{D}_1 + i\mathcal{D}_2 = e^{i\alpha}\frac{\partial}{\partial r} + \frac{i}{r}e^{i\alpha}\frac{\partial}{\partial\alpha} - \frac{1-f}{2r}e^{i\alpha}. \tag{3.98}$$

In addition, $\bar{\phi}(x) = v\varphi(r)\exp(-i\alpha)$. The boundary conditions are as follows: (i) at infinity the solution must decay as $e^{-m_F r}$; at the origin it must be regular, which implies that if $\xi(0) \neq 0$, then the solution must have no winding (winding is possible only if $\xi(0) = 0$). Comparing and examining Eqs. (3.98) and (3.99) one readily concludes that only the equation for ξ_2 has a solution satisfying the above boundary conditions:

Constructing
zero modes

$$\xi_2 = \zeta\exp\left[-\int_0^r dr\left(hv\varphi + \frac{1-f}{2r}\right)\right], \tag{3.99}$$

where ζ is a real Grassmann constant and $\xi_1 = 0$. The large-r asymptotics of (3.99) is $r^{-1/2}e^{-m_F r}$.

As for the η field, the zero-mode equations have exactly the same form as (3.98) after the substitution

$$\xi_\alpha \to \eta^{\alpha\dagger}. \tag{3.100}$$

Thus, the solution satisfying the appropriate boundary conditions exists only for $\eta^{2\dagger}$. Since η^2 is the same as η_1 one can write

$$\eta_1 = v\exp\left[-\int_0^r dr\left(hv\varphi + \frac{1-f}{2r}\right)\right], \tag{3.101}$$

where v is a real Grassmann number.

Fermion
moduli fields
on the ANO
string

On the string world sheet ζ and v acquire a (slow) dependence on t and z and become two-dimensional fermion fields. Clearly, one can combine them in a two-dimensional Majorana field ψ,

$$\psi = \begin{pmatrix} v(t,z) \\ \zeta(t,z) \end{pmatrix}, \tag{3.102}$$

with action

$$S = \int dt\, dz\, i\bar{\psi}\left(\gamma^0 \frac{\partial}{\partial t} + \gamma^z \frac{\partial}{\partial z}\right)\psi, \qquad \gamma^0 \gamma^z = -\sigma_3. \tag{3.103}$$

This action emerges as a result of the substitution of the zero-mode solutions found above into Eq. (3.82).

3.3.3 A Brief Digression: Left- and Right-Handed Fermions in Four and Two Dimensions

The matrices $(\bar{\sigma}^\mu)^{\dot{a}a}$ are defined in the beginning of Part II.

In four dimensions (three spatial dimensions) there exists the spin operator $\frac{1}{2}\vec{\sigma}$. The helicity is defined as the projection of the spin onto a particle's momentum. A particle with negative helicity is referred to as left-handed and one with positive helicity as right-handed. Thus, the four-dimensional left-handed spinor satisfies the equation

$$i\partial_\mu\, (\bar{\sigma}^\mu)^{\dot{a}a}\, \xi_a = 0, \tag{3.104}$$

which is equivalent to

$$\vec{n}\vec{\sigma}\xi = -\xi, \qquad \vec{n} \equiv \frac{\vec{p}}{p_0}. \tag{3.105}$$

Alternatively, one can define the four-dimensional left-handed Dirac spinor as the spinor satisfying the condition $\gamma^5 \Psi = \Psi$, where γ^5 is the four-dimensional γ^5 matrix.

In two dimensions (one spatial dimension), and thus in the absence of spatial rotations, spin does not exist. The above left-handed spinor ξ becomes the *Dirac spinor* in two dimensions. It satisfies the same equation, (3.105):

$$n_3\sigma_3\, \xi = -\xi, \qquad n_3 = \pm 1, \tag{3.106}$$

In two dimensions $\gamma^5 = -\sigma_3$.

with $n_{1,2}$ set to zero. However, σ_3 no longer represents the spin operator. Instead, in two dimensions, it plays the role of $-\gamma^5$ (see Section 10.2.2). For the left-handed spinors $\sigma_3\xi = 1$, which entails that n_3 is negative and the particle moves to the left along the z axis, in the literal sense; see Fig. 3.9. In the context of two-dimensional field theory, such particles are called *left-movers*. For the right-handed spinors $\sigma_3\xi = -1$, implying that n_3 is positive and the particle in question is a *right-mover*. In the coordinate space, the equation[11]

$$i\left(\frac{\partial}{\partial t} - \frac{\partial}{\partial z}\sigma_3\right)\xi = 0 \tag{3.107}$$

Useful definitions

implies that the left-movers depend on $t + z$ while the right-movers depend on $t - z$. This is sometimes expressed by the following equations:

$$\partial_L\, \xi_R = 0, \qquad \partial_R\, \xi_L = 0, \tag{3.108}$$

where

$$\partial_L \equiv \frac{\partial}{\partial t} + \frac{\partial}{\partial z}, \qquad \partial_R \equiv \frac{\partial}{\partial t} - \frac{\partial}{\partial z}. \tag{3.109}$$

---

[11] This is the reduced version of (3.104).

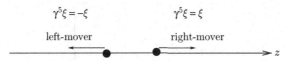

Left- and right-handed spinors in two dimensions.

3.4 String-Induced Gravity

In Section 2.3 we considered the gravitational interaction of a probe body with a domain wall in $1 + 3$ dimensions and found, to our surprise, that the domain wall antigravitates. Now we will discuss the gravity induced by a flux tube (string). The finding that awaits us is no less remarkable. It turns out that locally, at any given spatial point away from the string, the string exerts no gravity at all. However, an experimenter traveling around such a string in a plane perpendicular to its axis will discover, after performing a full rotation, that the full angle α swept is less than 2π, namely, that

$$\alpha = 2\pi - 8\pi GT_{\text{str}}, \tag{3.110}$$

where G is Newton's constant; we assume here that $GT_{\text{str}} \ll 1$. Thus, the geometry of the $1 + 3$ dimensional space with a string at the origin is conical (Fig. 3.10).

It is convenient to divide our analysis of the problem into two steps. First we will prove, on very general grounds, that the curvature tensor vanishes identically everywhere except at the string itself (the z axis, see Fig. 3.10). Then we will find the angle deficit.

A brief inspection of Fig. 3.10 tells us that the problem at hand is essentially $1 + 2$ dimensional. This means that the static solution we are looking for is t, z independent. The metric can be chosen as follows:[12]

$$g_{tt} = -g_{zz} = 1, \qquad g_{zt} = g_{zx} = g_{zy} = g_{tx} = g_{ty} = 0, \tag{3.111}$$

while all components of the metric tensor $g_{\alpha\beta}$ with $\alpha, \beta = x, y$ depend only on x and y. Under these circumstances all components of the Riemann curvature tensor $R_{\mu\nu\alpha\beta}$ with at least one index z vanish. This tensor is then defined by the same expressions as in $1 + 2$ dimensions.

Now let us calculate the number of independent components of $R_{\mu\nu\alpha\beta}$ in $1 + 2$ dimensions. This calculation can be found in a number of textbooks, e.g., in the section "Properties of the curvature tensor" of [15]. Let us start with those components which have only two different indices, i.e., $R_{\mu\nu\mu\nu}$ (note that there is no summation over μ and ν here). A pair of values for μ and ν can be chosen from the triplet $0, 1, 2$ in three distinct ways. Owing to the fact that

$$R_{\mu\nu\alpha\beta} = -R_{\nu\mu\alpha\beta}, \qquad R_{\mu\nu\alpha\beta} = -R_{\mu\nu\beta\alpha}, \tag{3.112}$$

each selected pair of μ and ν gives only one independent component. Therefore, we have three independent components of the type $R_{\mu\nu\mu\nu}$.

---

[12] The $t, x, y,$ and z coordinates will also be denoted by sub- or superscripts $0, 1, 2, 3$.

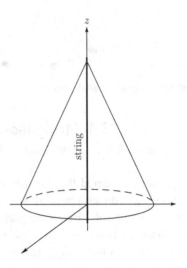

Fig. 3.10 Flux tube (string) geometry.

In addition,

$$R_{\mu\nu\alpha\beta} = R_{\alpha\beta\mu\nu};\tag{3.113}$$

thus, there are three independent components with three distinct sets of indices,

$$R_{0102}, \quad R_{1012}, \quad \text{and} \quad R_{2120}.\tag{3.114}$$

All other components are reducible to (3.114) by virtue of the symmetry properties of the curvature tensor. We conclude that in $1 + 2$ dimensions the curvature tensor has six independent components. The (symmetric) Ricci tensor $R_{\mu\nu}$ has exactly the same number of components. This means that the six linear equations defining the Ricci tensor,

$$g^{\alpha\beta} R_{\beta\mu\alpha\nu} = R_{\mu\nu},\tag{3.115}$$

represent a solvable set where the $R_{\beta\mu\alpha\nu}$ are to be treated as unknowns while the $g^{\alpha\beta}$ are given coefficients. This system of equations can be solved algebraically. Thus, in $1 + 2$ dimensions all components of the curvature tensor are algebraically expressible in terms of the components of the Ricci tensor.

Moreover, the Einstein equation

$$R_{\mu\nu} - \tfrac{1}{2} R g_{\mu\nu} = 8\pi G T_{\mu\nu}\tag{3.116}$$

tells us that in empty space (i.e., away from the string), where $T_{\mu\nu} = 0$, the Ricci tensor vanishes. Since the system (3.115) is algebraically solvable, the fact that $R_{\mu\nu} = 0$ implies the vanishing of all components of the curvature tensor $R_{\mu\nu\alpha\beta}$ everywhere in space except along the z axis.

Now, let us pass to the second stage. First we need to establish the general structure of the energy–momentum tensor for the string solution. In the Abelian model discussed in Section 3.1.2 the energy–momentum tensor takes the form

The string energy– momentum tensor

$$T^{\mu\nu} = -\frac{1}{e^2} \left(F^{\mu\alpha} F^{\nu\beta} g_{\alpha\beta} - \tfrac{1}{4} g^{\mu\nu} F^{\alpha\beta} F_{\alpha\beta} \right)$$
$$+ \mathcal{D}^{\mu}\phi^* \mathcal{D}^{\nu}\phi + \mathcal{D}^{\nu}\phi^* \mathcal{D}^{\mu}\phi - g^{\mu\nu} \left[\mathcal{D}^{\alpha}\phi^* \mathcal{D}_{\alpha}\phi - U(\phi) \right].\tag{3.117}$$

Using the properties of the flux-tube solution one can readily derive that, for a straight string oriented along the z axis,

$$T^{\mu\nu} = T_{\text{str}} \, \text{diag}\{1,0,0,-1\}\delta^{(2)}(x_\perp), \tag{3.118}$$

cf. Eq. (2.86). In fact, this is the general expression for the energy–momentum tensor of a straight infinitely thin string; it does not depend on the underlying microscopic model.

Assuming the gravitational field to be weak (i.e., $GT_{\text{str}} \ll 1$), the metric can be linearized around the Minkowski metric, so that

$$g_{\mu\nu} = \eta_{\mu\nu} + h_{\mu\nu}, \qquad \eta_{\mu\nu} = \text{diag}\{1,-1,-1,-1\}. \tag{3.119}$$

If we impose the harmonic gauge,

$$\partial_\nu(h^\nu_\mu - \tfrac{1}{2}\delta^\nu_\mu h^\sigma_\sigma) = 0, \tag{3.120}$$

the linearized Einstein equation takes the following simple form:

$$\Box h_{\mu\nu} = -16\pi G(T_{\mu\nu} - \tfrac{1}{2}\eta_{\mu\nu}T^\sigma_\sigma), \tag{3.121}$$

where the indices have been raised and lowered here using the Minkowski metric $\eta_{\mu\nu}$.

Substituting Eq. (3.118) into (3.121) and using the fact that $h_{\mu\nu}$ depends only on x and y, we readily find the solution for the metric:

$$h_{00} = h_{33} = 0, \qquad h_{11} = h_{22} \equiv h = 8\,GT_{\text{str}}\ln\frac{r}{r_0}, \tag{3.122}$$

where all other components vanish, $r = (x^2 + y^2)^{1/2}$, and r_0 is an integration constant.[13] Needless to say, our solution (3.122) confirms the *ansatz* (3.111).

To understand the physical meaning of the metric we have just derived, it is convenient to write down an expression for the interval in cylindrical coordinates:

$$ds^2 = dt^2 - dz^2 - (1-h)(dr^2 + r^2 d\theta^2). \tag{3.123}$$

It is not difficult to check that if we introduce new radial and angular coordinates $\tilde{r}, \tilde{\theta}$, where

$$\left(1 - 8GT_{\text{str}}\ln\frac{r}{r_0}\right)r^2 = (1 - 8GT_{\text{str}})\tilde{r}^2,$$
$$\tilde{\theta} = (1 - 4GT_{\text{str}})\theta \tag{3.124}$$

(in deriving the above equation we have kept only terms of first order in GT_{str}), then in the new coordinates the interval (3.123) takes the form

$$ds^2 = dt^2 - dz^2 - d\tilde{r}^2 - \tilde{r}^2 d\tilde{\theta}^2. \tag{3.125}$$

This last result confirms our previous conclusion that the geometry around a straight string is locally identical to that of flat space. There is no global equivalence, however, since the angle $\tilde{\theta}$ varies in the interval

$$0 \le \tilde{\theta} < 2\pi(1 - 4G\,T_{\text{str}}). \tag{3.126}$$

---

[13] Formally, $h_{\mu\nu}$ becomes large at exponentially large distances from the string. This is an artifact of the given coordinate choice.

The angle deficit is

$$\Delta\alpha = 8\pi G\, T_{\mathrm{str}}, \tag{3.127}$$

as we saw in Eq. (3.110) at the start of this section. This result was first obtained by A. Vilenkin [16].

Exercise

3.4.1   Use Eq. (3.122) to calculate the Riemann curvature tensor directly, in order to confirm that it vanishes at $r \neq 0$. Remember that (3.122) is obtained to the first order in GT_{str}.

3.5 Appendix: Calculation of the Orientational Part of the World-Sheet Action for Non-Abelian Strings

Here I present some details of the derivation in [17] of the world-sheet action for non-Abelian strings. The general strategy was outlined in Section 3.2.6.

Because of the Goldstone nature of the moduli fields their world-sheet interaction has no potential term. To obtain the kinetic term (more exactly, the part relevant to the orientational moduli fields), we substitute the solution (3.73), with its adiabatic dependence on x^p through $\vec{S}(t, z)$, into the action (3.39). In doing so we immediately observe that we must modify the solution (3.73).

Indeed, Eq. (3.73) is obtained as a *global* SU(2) rotation of the elementary $(1, 0)$ string. Now we will make this transformation local (i.e., now $\vec{S}$ will depend on t and z). Because of this, the t and z components of the gauge potential no longer vanish. They must be added to (3.73).

The following *ansatz* for these components (to be checked *a posteriori*) is fairly obvious:

$$A_p = -i(\partial_p U)U^{-1}\rho(r), \qquad p = 0, 3, \tag{3.128}$$

where $\rho(r)$ is a new profile function.

As was mentioned after Eq. (3.67), the parametrization of the matrix U is ambiguous. Consequently, if we introduce

$$\alpha_p \equiv -i\left(\partial_p U\right)U^{-1}, \qquad \alpha_p \equiv \alpha_p^a\left(\frac{\tau^a}{2}\right), \tag{3.129}$$

then the functions α_p^a are defined modulo the two gauge transformations following from Eq. (3.68). Equation (3.66) implies that

$$\alpha_p^a - S^a\left(S^b\alpha_p^b\right) = -\varepsilon^{abc}\, S^b\, \partial_p S^c, \tag{3.130}$$

and we can impose the condition $S^b\alpha_p^b = 0$. Then

$$\alpha_p^a = -\varepsilon^{abc}\, S^b\partial_p S^c, \qquad -i\left(\partial_p U\right)U^{-1} = -\tfrac{1}{2}\tau^a\,\varepsilon^{abc}\,S^b\partial_p S^c. \tag{3.131}$$

The function $\rho(r)$ in Eq. (3.128) is determined through a minimization procedure that generates an equation of motion for $\rho(r)$. Note that it must vanish at infinity:

$$\rho(\infty) = 0. \tag{3.132}$$

The boundary condition at $r = 0$ will be determined shortly.

The kinetic term for $\vec{S}$ comes from the gauge and Φ kinetic terms in Eq. (3.39). Using (3.73) and (3.128) to calculate the SU(2) gauge field strength we find that

$$F_{pi} = \frac{1}{2} \left(\partial_p S^a\right) \tau^a \varepsilon_{ij} \frac{x_j^\perp}{r^2} f_3[1 - \rho(r)] + i \left(\partial_p U\right) U^{-1} \frac{x_i^\perp}{r} \frac{d\rho(r)}{dr}. \tag{3.133}$$

For $\mathrm{Tr}\, F_{pi}^2$ to give a finite contribution to the action one must require that

$$\rho(0) = 1. \tag{3.134}$$

Substituting the field strength (3.133) into the action (3.39) and including, in addition, the kinetic term of the Φ fields, we arrive at

$$S^{(1+1)} = \frac{\beta}{2} \int dt\, dz (\partial_p S^a)^2, \tag{3.135}$$

where the coupling constant β is given by an integral, as follows:

$$\beta = \frac{2\pi}{g_2^2} \int_0^\infty r\, dr \left\{ \left(\frac{d}{dr}\rho(r)\right)^2 + \frac{1}{r^2} f_3^2 (1 - \rho)^2 \right.$$
$$\left. + g_2^2 \left[\frac{\rho^2}{2} \left(\phi_1^2 + \phi_2^2\right) + (1 - \rho)(\phi_1 - \phi_2)^2 \right] \right\}. \tag{3.136}$$

The above functional must be minimized with respect to ρ, with boundary conditions given by (3.132) and (3.134). Varying (3.136) with respect to ρ, one readily obtains a second-order equation for $\rho(r)$:

$$-\frac{d^2}{dr^2}\rho - \frac{1}{r}\frac{d}{dr}\rho - \frac{1}{r^2} f_3^2 (1 - \rho) + \frac{g_2^2}{2} \left(\phi_1^2 + \phi_2^2\right) \rho - \frac{g_2^2}{2}(\phi_1 - \phi_2)^2 = 0. \tag{3.137}$$

<table>
<tr><td>Deriving the coupling constant on the string world sheet</td><td>After some algebra and extensive use of the first-order equations (3.60) one can show that the solution to (3.137) satisfying the boundary conditions (3.132) and (3.134) is as follows:</td></tr>
</table>

$$\rho = 1 - \frac{\phi_1}{\phi_2}. \tag{3.138}$$

Substituting this solution back into the expression for the sigma model coupling constant (3.136) one can check that the integral in (3.136) reduces to a total derivative and that it is given by $f_3(0) = 1$. Namely,

$$I \equiv \int_0^\infty r\, dr \left\{ \left(\frac{d}{dr}\rho(r)\right)^2 + \frac{1}{r^2} f_3^2 (1 - \rho)^2 \right.$$
$$\left. + g_2^2 \left[\frac{\rho^2}{2} \left(\phi_1^2 + \phi_2^2\right) + (1 - \rho)(\phi_1 - \phi_2)^2 \right] \right\}$$
$$= \int_0^\infty dr \left(-\frac{d}{dr} f_3\right) = 1, \tag{3.139}$$

where I have used the first-order equations (3.60) for the profile functions of the string. We conclude that the two-dimensional sigma model coupling β is determined by the four-dimensional non-Abelian coupling as follows:

$$\beta = \frac{2\pi}{g_2^2}. \tag{3.140}$$

3.6 Appendix: Two-String Junctions and Z Strings in SM

Warning: This appendix contains material from Chapters 4, 7, and 11. In this section I adhere to the canonical normalization, as is standard for the SM Lagrangian.

In Sections 3.2.2–3.2.4 we discussed the simplest example of non-Abelian flux tubes (vortex strings) supported by a U(2) gauge theory with a judiciously designed Higgs sector. At the classical level we found a continuous set of such strings labeled by a three-component vector $\vec{S}$ with the constraint $\vec{S}^2 = 1$. All these vortex strings have one and the same tension, $2\pi v^2$. At the quantum level this tension degeneracy is lifted. However, in the supersymmetric version of the model a Z_2 degeneracy survives – we have two distinct strings with one and the same tension. I will first discuss an elegant phenomenon – the so-called two-string junction – occurring in this case. Then we will consider the Z strings.

3.6.1 Two-String Junctions

To consider this phenomenon it is crucial to have distinct strings with degenerate tensions. I will use "elementary" $(1,0)$ and $(0,1)$ vortex strings. To lift the continuous degeneracy we will add an $\vec{S}$ dependent potential in (3.76), a mass term perturbation

$$\Delta \mathcal{L}_m = -m^2 \left\{ 1 - (S_3)^2 \right\}. \tag{3.141}$$

This will allow us to consider the phenomenon of the two-string junction at the classical level. Equation (3.76) guarantees that two vortex-string solutions with the minimal tension are $\vec{S} = \{0, 0, \pm 1\}$ corresponding to the elementary $(1,0)$ and $(0,1)$ vortex strings. They can form a junction as shown in Fig. 3.11.

The internal structure of the vortex string to the left of M is obviously different from that to the right: in the $(1,0)$ string the magnetic fluxes of the Abelian field B_μ and the non-Abelian field A_μ^3 are parallel, while in the $(0,1)$ string they are antiparallel. Their magnetic fluxes through the vortex strings are

$$\text{Flux}_{B,A^3} = \pm \frac{2\pi}{g}. \tag{3.142}$$

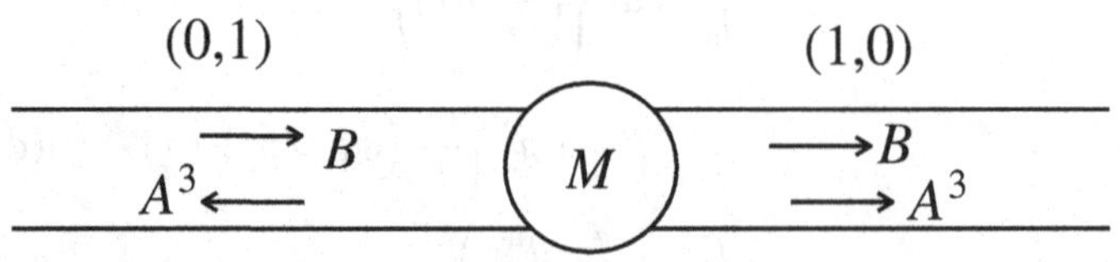

Fig. 3.11　Two-string junction.

This result takes into account the fact that the coupling appearing in the covariant derivative (in canonic normalization) is $g/2$.

A restructuring of internal degrees of freedom needed to pass from (1,0) to (0,1) occurs in the transitional domain M. The magnetic flux of B_μ incoming the blob M from the left is equal to that leaving M to the right. On the other hand, the total magnetic flux of A_μ^3 emanating from the blob is $4\pi/g$, the sum of the fluxes outgoing in the right and left directions. This is exactly equal to that of the magnetic monopole of the Dirac type. Thus, the transition domain M can be interpreted as a magnetic monopole permanently attached to two non-Abelian strings.

This string-junction construction is topologically stable – unbreakable. More details can be found in the original papers [17].

3.6.2 Z Strings

A somewhat similar phenomenon exists in the Standard Model (SM), where it is known as the Z string attached to a monopole of the Dirac type. We will focus only on the gauge and Higgs sector of the Glashow–Weinberg–Salam (GWS) model ignoring fermions – quarks and leptons.

To emphasize the conceptual side of the Z string construction, I will simplify the GWS model using a single coupling constant g instead of three SM constants g, g' and λ, namely,

$$\mathcal{L} = -\frac{1}{4}\left(F_{\mu\nu}^a\right)^2 - \frac{1}{4}\left(F_{\mu\nu}\right)^2 + |D_\mu\phi|^2 - \frac{g^2}{4}\left(\phi^*\phi - \frac{v^2}{2}\right)^2,$$

$$\mathcal{D}_\mu\phi = \left(\partial_\mu - ig A_\mu^a \frac{\tau^a}{2} - \frac{i}{2}g B_\mu\right). \tag{3.143}$$

The Higgs sector ϕ consists of a single SU(2) doublet built of two complex fields,

$$\phi = \begin{pmatrix} \phi^1 \\ \phi^2 \end{pmatrix}. \tag{3.144}$$

The standard convention for the vacuum expectation value of the Higgs field is

$$\phi_{\text{vac}} = \frac{1}{\sqrt{2}}\begin{pmatrix} 0 \\ v \end{pmatrix}. \tag{3.145}$$

Under this condition, as we will see later, the electric charge of the field ϕ^2 (which includes the physical Higgs field h) is zero. Conservation of the electric charge is guaranteed in this model.

Putting $\phi^1 = 0$ can be viewed as a gauge condition (unitary gauge). Note the factor $1/\sqrt{2}$ in Eq. (3.145), a historical convention in the Standard Model. It is absent in the theory of non-Abelian strings discussed in Chapter 3.

With the preceding conventions we arrive at

$$m_{\mathcal{A}} = 0, \quad m_{W^\pm} = \frac{gv}{2}, \quad m_Z = \frac{gv}{\sqrt{2}}, \quad m_h = \frac{gv}{\sqrt{2}}, \tag{3.146}$$

where the photon $\mathcal{A}$ and Z-boson fields are defined as

$$\mathcal{A}_\mu = \frac{1}{\sqrt{2}}\left[A_\mu^3 + B_\mu\right], \quad Z_\mu = \frac{1}{\sqrt{2}}\left[A_\mu^3 - B_\mu\right], \tag{3.147}$$

while the $W^\pm$ bosons are related to $A_\mu^{1,2}$,

$$W_\mu^\pm = \frac{1}{\sqrt{2}} \left(A_\mu^1 \mp i A_\mu^2 \right). \tag{3.148}$$

Now, we are ready to build the Z string. At this point, the Nambu construction significantly deviates from that of non-Abelian strings. Indeed, in the latter case both components ϕ_{vac}^1 and ϕ_{vac}^2 do not vanish (because of the presence of *two* Higgs doublets rather than one in SM). Therefore, two distinct windings

$$\phi_{\text{large } R} = \exp\left[i\alpha(x)\frac{1 \pm \tau^3}{2} \right] \phi_{\text{vac}} \tag{3.149}$$

act nontrivially giving rise to the $(1,0)$ and $(0,1)$ elementary strings. In the Standard Model, however, Eq. (3.145) implies that the winding

$$\exp\left[i\alpha(x)\frac{1 + \tau^3}{2} \right] \phi,$$

does not affect ϕ^1 because $\phi^1 = 0$, nor does it affect the lower component of ϕ because $(1 + \tau^3)_{22}$ vanishes for the lower component. This is natural – the vacuum expectation value ϕ_{vac}^2 does not break the electric charge conservation (discussed earlier). There is no analogue of $(1,0)$ vortex string.

At the same time, the opposite rotation

$$\phi_{\text{large } R} = \frac{1}{\sqrt{2}} \begin{pmatrix} 0 \\ v e^{i\alpha} \end{pmatrix} = \exp\left[i\alpha(x)\frac{1 - \tau^3}{2} \right] \phi_{\text{vac}} \tag{3.150}$$

does wind ϕ_{vac}, giving rise to the gauge tail of the Z field. The corresponding flux tube carries the Z field magnetic flux,

$$\int d^2x (\partial_1 Z_2 - \partial_2 Z_1) = \text{Flux}_Z \neq 0. \tag{3.151}$$

This is what people call the Z string. In the terminology of non-Abelian strings it is akin to the $(0,1)$ elementary string. In essence, it is an Abelian string of the ANO type. The $\mathcal{A}$ and $W^\pm$ fields are not excited in the core of the Z string. As was mentioned, unlike the non-Abelian string, the Z string is stable only perturbatively. Beyond perturbation theory it can break through the monopole–antimonopole pair production.

First, we will forget about the Z string breaking and calculate its tension. To this end we ignore $\mathcal{A}_\mu$, $W_\mu^\pm$, ϕ^1 for the time being. Then we can write a reduced Lagrangian,

$$\mathcal{L}_{\text{red}} = -\frac{1}{4} \left(F_{\mu\nu}^Z \right)^2 + \left| D_\mu \phi^2 \right|^2 - \frac{g^2}{4} \left[\left(\phi^2 \right)^* \left(\phi^2 \right) - \frac{v^2}{2} \right]^2,$$

$$D_\mu \phi^2 \to \left(\partial_\mu + ig\frac{1}{\sqrt{2}} Z_\mu \right) \phi^2, \tag{3.152}$$

which allows one to perform the Bogomolny completion. The theory (3.152) is Abelian, and so is the Z string. The Z-field flux following from this theory is

$$\text{Flux}_Z = -\frac{2\sqrt{2}\,\pi}{g}, \tag{3.153}$$

(cf. (3.152) and Sec. 3.1.2). After performing the Bogomol'nyi completion in Eq. (3.152) we arrive at

$$
\mathcal{H} = \frac{1}{2}\left(F_{12}^Z\right)^2 + \frac{1}{2}\sum_{i=1,2}|D_i\varphi|^2 + \frac{g^2}{16}\left(|\varphi|^2 - v^2\right)^2
$$

$$
= \frac{1}{2}\left\{\left[F_{12}^Z + \frac{g}{2\sqrt{2}}\left(|\varphi|^2 - v^2\right)\right]^2\right.
$$

$$
\left. + \left(D_1 + iD_2\right)\varphi^*\left(D_1 - iD_2\right)\varphi + \frac{g}{\sqrt{2}}v^2 F_{12}^Z\right\}. \tag{3.154}
$$

In the preceding equation, to ease the notation I introduced φ,

$$
\phi^2 = \frac{\varphi}{\sqrt{2}}. \tag{3.155}
$$

Taking into account BPS saturation, Eq. (3.154) implies for the Z-string tension

$$
T_Z = \frac{1}{2}\frac{g}{\sqrt{2}}v^2 \, \mathrm{Flux}_Z = \pi v^2. \tag{3.156}
$$

3.6.3 Nambu Monopoles

As was mentioned, the Z string is stable in perturbation theory, i.e., no emission of any *finite number* of $W^\pm$, Z bosons, or photons can occur if the vortex string is in the ground state at rest. However, the Z string can break by virtue of under-the-barrier production of the monopole–antimonopole pairs [18]. In Section 7.2.3 it is demonstrated that the breaking probability per unit time and unit length is (see Eq. (7.50))

$$
\Gamma_{\text{breaking}} \sim v^2 e^{-S_*}, \quad S_* = \frac{\pi M_M^2}{T}, \tag{3.157}
$$

where M_M is the monopole mass and T is the string tension; see Eq. (3.156). For the estimate of the monopole mass we will take that of the 't Hooft-Polyakov monopole,

$$
M_M \sim \frac{4\pi v}{g}, \tag{3.158}
$$

see Eq. (4.26). Then,

$$
S_* = c\,\frac{8\pi^2}{g^2}, \tag{3.159}
$$

Nonperturbatively, Nambu monopoles can decay

where c stands for a constant of order 1. The critical action (3.159) is of the same order of magnitude as the instanton action. Thus, Γ_{breaking} is extremely small.

Let us assume, however, that such a breaking occurred and study one of the two fragments, depicted in Fig. 3.12. In this figure we see a part of the Z string with the magnetic monopole attached to it.[14]

As in the case of non-Abelian strings, the B field flux goes through M untouched while the flux associated with A_μ^3 (it is a part of the flux (3.160)) changes its

---

[14] The endpoint domain M is not described by the Lagrangian (3.152) since the monopole inner structure entangles all gauge fields, including $W_\mu^\pm$, as well as the ϕ fields.

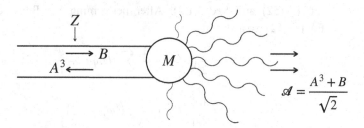

Fig. 3.12   The Nambu monopole with the Z string attached to it.

direction and, combined with the Abelian flux on the right-hand side of M, creates the dispersed flux of the photon-related magnetic field from the monopole,

$$\int_{S_R} d^2 S \, (\partial_1 \mathcal{A}_2 - \partial_2 \mathcal{A}_1) = \text{Flux}_{\mathcal{A}} = \frac{2\sqrt{2}\,\pi}{g} = \frac{2\pi}{e} \qquad (3.160)$$

which spreads out to infinity. Here integration runs over the surface of a large sphere S. In the last equality in (3.160) I used the fact that under my (unrealistic) SM conventions,

$$e = \frac{g}{\sqrt{2}}. \qquad (3.161)$$

The fluxes due to A_μ^3 emanated from the monopole to the right and to the left in Fig. 3.12 are the same as in the non-Abelian strings; see Fig. 3.11. The total monopole magnetic flux due to A_μ^3 is $4\pi/g$ both for the confined monopoles in Figs. 3.11 and 3.12.

The magnetic flux of the 't Hooft–Polyakov monopole (Section 4.1) is $4\pi/g = 2\pi(\frac{g}{2})^{-1}$. Comparison with Eq. (3.160) is subtle – one should take into account the difference in the normalization of the coupling constants. However, the quantization conditions $Q_M e_{\min} = 2\pi$ are the same in both cases if $e_{\min}$ is understood as the *minimal* nonvanishing electric charge existing in SM and the Georgi–Glashow model.

References for Chapter 3

[1] A. A. Abrikosov, *ZhETF* **32**, 1442 (1957) [Engl. transl. *Sov. Phys. JETP* **5**, 1174 (1957); reprinted in C. Rebbi and G. Soliani (eds.), *Solitons and Particles* (World Scientific, Singapore, 1984), p. 356].

[2] H. B. Nielsen and P. Olesen, *Nucl. Phys. B* **61**, 45 (1973) [Reprinted in C. Rebbi and G. Soliani (eds.), *Solitons and Particles* (World Scientific, Singapore, 1984), p. 365].

[3] A. Achucarro and T. Vachaspati, *Phys. Rept.* **327**, 347 (2000) [arXiv:hep-ph/9904229].

[4] P. G. De Gennes, *Superconductivity of Metals and Alloys* (Benjamin, New York, 1966).

[5] A. Yung, *Nucl. Phys. B* **562**, 191 (1999) [hep-th/9906243].

[6] A. Linde, *JETP Lett.* **23**, 64 (1976); *Phys. Lett.* **70B**, 306 (1977); S. Weinberg, *Phys. Rev. Lett.* **36**, 294 (1976).

[7] K. Bardakci and M. B. Halpern, *Phys. Rev. D* **6**, 696 (1972).

[8] R. Auzzi, S. Bolognesi, J. Evslin, K. Konishi, and A. Yung, *Nucl. Phys. B* **673**, 187 (2003) [hep-th/0307287].

[9] Y. Nambu, *Nucl. Phys. B* **130**, 505 (1977).

[10] R. Jackiw and P. Rossi, *Nucl. Phys. B* **190**, 681 (1981).

[11] E. Witten, *Nucl. Phys. B* **249**, 557 (1985).

[12] A. Vilenkin and E. P. S. Shellard, *Cosmic Strings and Other Topological Defects* (Cambridge University Press, 1994).

[13] A. S. Schwarz, *Phys. Lett. B* **67**, 172 (1977); L. S. Brown, R. D. Carlitz, and C. K. Lee, *Phys. Rev. D* **16**, 417 (1977); S. Coleman, The uses of instantons, in S. Coleman (ed.), *Aspects of Symmetry* (Cambridge University Press, 1985), p. 265.

[14] J. E. Kiskis, *Phys. Rev. D* **15**, 2329 (1977); M. M. Ansourian, *Phys. Lett. B* **70**, 301 (1977); N. K. Nielsen and B. Schroer, *Nucl. Phys. B* **120**, 62 (1977); E. J. Weinberg, *Phys. Rev. D* **24**, 2669 (1981).

[15] L. D. Landau and E. M. Lifshitz, *The Classical Theory of Fields* (Pergamon Press, Oxford, 1979).

[16] A. Vilenkin, *Phys. Rev. D* **23**, 852 (1981).

[17] R. Auzzi, S. Bolognesi, J. Evslin, K. Konishi and A. Yung, *Nucl. Phys. B* **673**, 187 (2003) [hep-th/0307287]; M. Shifman and A. Yung, *Phys. Rev. D* **70**, 045004 (2004) [arXiv:hep-th/0403149].

[18] John Preskill and Alexander Vilenkin, *Phys. Rev. D* **47**, 2324, Section VIII.

Becoming acquainted with 't Hooft – Polyakov monopoles. — Nontriviality of the second homotopy group, or why these monopoles are topologically stable. — The Bogomol'nyi limit. — Quantizing the monopole moduli. — Dyons. — Skyrmions in multicolor QCD. — Nontriviality of the third homotopy group. — Interpreting Skyrmions as baryons. — Exotic Skyrmions.

4.1 Magnetic Monopoles

Now we will discuss magnetic monopoles – very interesting particles which carry a magnetic charge. They emerge in non-Abelian gauge theories in which the gauge symmetry is spontaneously broken down to an Abelian subgroup. The simplest example was found by 't Hooft [1] and Polyakov [2]. The model with which they worked had been devised by Georgi and Glashow [3] for a different purpose. As it often happens, the Georgi–Glashow model turned out to be more valuable than the original purpose; this is long forgotten while the model itself is alive and well and is in constant use by theorists.

4.1.1 The Georgi–Glashow Model

I begin with a brief description of the Georgi–Glashow (GG) model. The gauge group is SU(2) and the matter sector consists of one real scalar field ϕ^a in the adjoint representation (i.e., an SU(2) triplet). The Lagrangian of the GG model is

$$\mathcal{L} = -\frac{1}{4g^2} G^a_{\mu\nu} G^{\mu\nu,a} + \frac{1}{2} (\mathcal{D}_\mu \phi^a)(\mathcal{D}^\mu \phi^a) - \lambda (\phi^a \phi^a - v^2)^2, \qquad (4.1)$$

where

$$G^a_{\mu\nu} = \partial_\mu A^a_\nu - \partial_\nu A^a_\mu + \varepsilon^{abc} A^b_\mu A^c_\nu \qquad (4.2)$$

and the covariant derivative in the adjoint acts as follows:

$$\mathcal{D}_\mu \phi^a = \partial_\mu \phi^a + \varepsilon^{abc} A^b_\mu \phi^c. \qquad (4.3)$$

We will also use matrix notation for the field ϕ^a, writing

$$\phi = \phi^a \frac{\tau^a}{2} \qquad (4.4)$$

where the τ^a are the Pauli matrices. Below we focus on a special limit of critical (or BPS) monopoles. This limit corresponds to a vanishing scalar coupling, $\lambda \to 0$. The only role of the last term in Eq. (4.1) is to provide a boundary condition for the scalar field.

One can speak of magnetic charges only in those theories that support a long-range (Coulomb) magnetic field. Therefore, the pattern of the symmetry breaking should be such that some gauge bosons remain massless. In the Georgi–Glashow model (4.1) the pattern is as follows:

$$\text{SU}(2) \to \text{U}(1). \qquad (4.5)$$

To see that this is indeed the case let us note that the ϕ^a self-interaction term (the last term in Eq. (4.1)) forces ϕ^a to develop a vacuum expectation value

$$\phi^a_{\text{vac}} = v\delta^{3a}, \qquad \phi_{\text{vac}} = v\frac{\tau_3}{2}. \qquad (4.6)$$

Unitary gauge condition.

The direction of the vector ϕ^a in SU(2) space (hereafter to be referred to as the color space) can be chosen arbitrarily. One can always reduce it to the form (4.6) by a global color rotation. Thus, Eq. (4.6) can be viewed as a (unitary) gauge condition on the field ϕ.

This gauge is very convenient for discussing the particle content of the theory (for the present we mean elementary excitations rather than solitons). A color rotation around the third axis does not change the vacuum expectation value of ϕ^a,

$$\exp\left(i\alpha\frac{\tau_3}{2}\right)\phi_{\text{vac}}\exp\left(-i\alpha\frac{\tau_3}{2}\right) = \phi_{\text{vac}}. \tag{4.7}$$

Thus the third component of the gauge field remains massless, and we will refer to it as a "photon":

$$A_\mu^3 \equiv A_\mu, \qquad F_{\mu\nu} = \partial_\mu A_\nu - \partial_\nu A_\mu. \tag{4.8}$$

The first and the second components form massive vector bosons (W bosons for short)

$$W_\mu^\pm = \frac{1}{\sqrt{2}\,g}\left(A_\mu^1 \pm iA_\mu^2\right). \tag{4.9}$$

As usual in the Higgs mechanism, the massive vector bosons eat up the first and second components of the scalar field ϕ^a. The third component, the physical Higgs field, can be parametrized as

$$\phi^3 = v + \varphi, \tag{4.10}$$

where φ is the physical Higgs field. In terms of these fields the Lagrangian (4.1) can be readily rewritten as

| GG Lagrangian |

$$\begin{aligned}
\mathcal{L} = {}&-\frac{1}{4g^2}F_{\mu\nu}F_{\mu\nu} + \frac{1}{2}(\partial_\mu\varphi)^2 \\
&-\left(\mathcal{D}_\nu W_\mu^+\right)\left(\mathcal{D}_\nu W_\mu^-\right) + \left(\mathcal{D}_\mu W_\mu^+\right)\left(\mathcal{D}_\nu W_\nu^-\right) + g^2(v+\varphi)^2 W_\mu^+ W_\mu^- \\
&-2iW_\mu^+ F_{\mu\nu}W_\nu^- + \frac{g^2}{4}\left(W_\mu^+ W_\nu^- - W_\nu^+ W_\mu^-\right)^2,
\end{aligned} \tag{4.11}$$

where we have used integration by parts. The covariant derivative now includes only the photon field:

$$\mathcal{D}_\mu W^\pm = \left(\partial_\mu \pm iA_\mu\right)W^\pm. \tag{4.12}$$

The last line in (4.11) presents the magnetic moment of the charged (massive) vector bosons and their self-interaction. In the limit $\lambda \to 0$, which is assumed in (4.11), the physical Higgs field is massless. The monopole obtained in the $\lambda = 0$ limit is called

| Critical monopole |

critical. Noncritical monopole masses are discussed in Section 4.1.6.

The mass of the $W^\pm$ bosons is

$$m_W = gv. \tag{4.13}$$

4.1.2 Monopoles – Topological Argument

After this brief introduction to the Georgi–Glashow model in the perturbative sector, in other words a discussion of elementary excitations, I will explain why this model predicts a topologically stable soliton. Assume that the monopole's center is at the

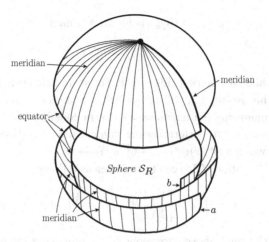

meridian

meridian

equator

Sphere $\mathcal{S}_R$

$b \rightarrow$

$\leftarrow a$

meridian

Fig. 4.1   Illustration of how a sphere $\mathcal{S}_G$ can be wrapped twice around another 2-sphere (i.e., $n = 2$). The white sphere in the middle is $\mathcal{S}_R$. The covering surface is indicated by meridians. The edges a and b should be identified.

origin and consider a large sphere $\mathcal{S}_R$ of radius R also with its center at the origin. Since the mass of the monopole is finite, by definition, $\phi^a \phi^a = v^2$ on this sphere.

As we recall, ϕ^a is a three-component vector in isospace subject to the constraint $\phi^a \phi^a = v^2$, which gives us a two-dimensional sphere $\mathcal{S}_G$. Thus, we are dealing here with mappings of $\mathcal{S}_R$ into $\mathcal{S}_G$. Such mappings split into distinct classes labeled by an integer n, counting how many times the sphere $\mathcal{S}_G$ is swept when the sphere $\mathcal{S}_R$ is swept once (see Fig. 4.1). The topologically trivial mapping corresponds to $n = 0$; for topologically nontrivial mappings $n = \pm 1, \pm 2, \ldots$ Mathematically, the above topological considerations are concisely expressed by the formula

Topological formula for the second homotopy group

$$\pi_2(SU(2)/U(1)) = \mathbb{Z}. \tag{4.14}$$

Here π_2 represents a maping of the coordinate-space (two-dimensional) sphere $\mathcal{S}_R$ onto the group space $SU(2)/U(1)$ relevant to the monopole problem. The group $SU(2)$ is divided by $U(1)$ because for each given vector ϕ^a there is a $U(1)$ subgroup that does *not* rotate it. The $SU(2)$ group space is a three-dimensional sphere while that of $SU(2)/U(1)$ is a two-dimensional sphere. As we will see shortly, the one-monopole field configuration corresponds to a mapping with $n = 1$. Since it is impossible to deform it continuously to the topologically trivial mapping, the monopoles are topologically stable.

4.1.3 Mass and Magnetic Charge

Classically the monopole mass is given by the energy functional following from the Lagrangian (4.1),

$$E = \int d^3x \left[\frac{1}{2g^2} B_i^a B_i^a + \frac{1}{2} \left(\mathcal{D}_i \phi^a \right) \left(\mathcal{D}_i \phi^a \right) \right], \tag{4.15}$$

$$B_i^a = -\frac{1}{2} \varepsilon_{ijk} G_{jk}^a. \tag{4.16}$$

The magnetic and Higgs fields are assumed to be time independent,

$$B_i^a = B_i^a(\vec{x}), \qquad \phi^a = \phi^a(\vec{x}),$$

while all electric fields vanish. For static fields it is natural to assume that $A_0^a = 0$. This assumption will be verified *a posteriori*, after we find the field configuration minimizing the functional (4.15). In equation (4.15) we assume the limit $\lambda \to 0$. However, in performing the minimization we should keep in mind the boundary condition $\phi^a(\vec{x})\phi^a(\vec{x}) \to v^2$ at $|\vec{x}| \to \infty$.

Equation (4.15) can be rewritten as follows:

$$E = \int d^3x \left[\frac{1}{2} \left(\frac{1}{g} B_i^a - \mathcal{D}_i\phi^a \right) \left(\frac{1}{g} B_i^a - \mathcal{D}_i\phi^a \right) + \frac{1}{g} B_i^a \mathcal{D}_i\phi^a \right]. \tag{4.17}$$

(The signs above correspond to the monopole solution; one must change the signs for the *antimonopole* solution.) It is easy to show that the last term on the right-hand side is a full derivative. Indeed, after integrating by parts and using the equation of motion $\mathcal{D}_i B_i^a = 0$ we get

$$\int d^3x \left(\frac{1}{g} B_i^a \mathcal{D}_i\phi^a \right) = \frac{1}{g} \int d^3x \, \partial_i \left(B_i^a \phi^a \right)$$

$$= \frac{1}{g} \int_{S_R} d^2 S_i \left(B_i^a \phi^a \right). \tag{4.18}$$

In the last line we have made use of Gauss' theorem and passed from volume integration to that over the surface of a large sphere. Thus, the last term in Eq. (4.17) is topological.

The combination $B_i^a \phi^a$ can be viewed as a gauge-invariant definition of the magnetic field $\vec{\mathcal{B}}$. More exactly,

$$\mathcal{B}_i = \frac{1}{v} B_i^a \phi^a. \tag{4.19}$$

| Singular gauge |

Indeed, far from the monopole core one can always assume ϕ^a to be aligned in the same way as in the vacuum (in the singular gauge, cf. Section 3.2.4), i.e., $\phi^a = v\delta^{3a}$. Then $\mathcal{B}_i = B_i^3$. The advantage of the definition (4.19) is that it is gauge independent.

Furthermore, the magnetic charge Q_M inside a sphere S_R can be defined through the flux of the magnetic field through the surface of the sphere,

$$Q_M = \int_{S_R} d^2 S_i \frac{1}{g} \mathcal{B}_i. \tag{4.20}$$

Using the boundary conditions (4.27), (4.33) and (4.31), to be derived below, we see that

$$\mathcal{B}_i \equiv \frac{1}{v} B_i^a \phi^a \quad \to \quad n^i \frac{1}{r^2} \quad \text{at } r \to \infty \tag{4.21}$$

and calculate the 't Hooft-Polyakov monopole magnetic flux

$$\text{Magn. Flux} = \int_{S_R} d^2 S_i \, \mathcal{B}_i = 4\pi. \tag{4.22}$$

Therefore, the monopole magnetic charge is [1]

$$Q_M = \int_{S_R} d^2 S_i \, \frac{1}{g} \, \mathcal{B}_i = \frac{4\pi}{g}. \tag{4.23}$$

Combining Eqs. (4.20), (4.19), and (4.18) we conclude that

$$E = v Q_M + \int d^3 x \left[\frac{1}{2} \left(\frac{1}{g} B_i^a - \mathcal{D}_i \phi^a \right) \left(\frac{1}{g} B_i^a - \mathcal{D}_i \phi^a \right) \right]. \tag{4.24}$$

The minimum of the energy functional is attained at

$$\frac{1}{g} B_i^a - \mathcal{D}_i \phi^a = 0. \tag{4.25}$$

Bogomol'nyi bound. The mass of the field configuration realizing this minimum – the monopole mass – is obviously given by

$$M_M = \frac{4\pi v}{g}. \tag{4.26}$$

Thus, the mass of the critical monopole is in one-to-one relationship with the magnetic charge of the monopole. Equation (4.25) is nothing other than the Bogomol'nyi equation in the monopole problem. If it is satisfied, the second-order differential equations of motion are satisfied too.

4.1.4 Solution of the Bogomol'nyi Equation for Monopoles

To solve the Bogomol'nyi equation we need to guess the dependence of the relevant fields on the gauge and Lorentz indices. The topological arguments of Section 4.1.2 prompt us to an appropriate *ansatz* for ϕ^a. Indeed, as one sweeps S_R the vector ϕ^a must sweep the group space sphere. The simplest choice is to identify these two spheres point-by-point,

$$\phi^a = v \frac{x^a}{r} = v n^a, \qquad r \to \infty, \tag{4.27}$$

where $n^i \equiv x^i / r$. This field configuration obviously belongs to the class of topologically nontrivial mappings with $n = 1$. The SU(2) group index a has become entangled with the coordinate $\vec{x}$. Polyakov proposed to refer to such fields as "hedgehogs."

Next, observe that finiteness of the monopole energy requires the covariant derivative $\mathcal{D}_i \phi^a$ to fall off faster than $r^{-3/2}$ at large r; cf. Eq. (4.15). Since

$$\partial_i \phi^a = v \frac{1}{r} \left(\delta^{ai} - n^a n^i \right) \sim \frac{1}{r}, \qquad r \to \infty, \tag{4.28}$$

---

[1] A remark: the conventions for charge normalization used in different books and papers may vary. In his original paper on the magnetic monopole [4], Dirac used the convention $e^2 = \alpha$ and the electromagnetic Hamiltonian $\mathcal{H} = (8\pi)^{-1}(\vec{E}^2 + \vec{B}^2)$. Then, the electric charge is defined through the flux of the electric field as $e = (4\pi)^{-1} \int_{S_R} d^2 S_i E_i$, and an analogous definition holds for the magnetic charge. We are using the convention according to which $e^2 = 4\pi\alpha$ and the electromagnetic Hamiltonian $\mathcal{H} = (2g^2)^{-1}(\vec{E}^2 + \vec{B}^2)$. Then $e = g^{-1} \int_{S_R} d^2 S_i E_i$ while $Q_M = g^{-1} \int_{S_R} d^2 S_i B_i$. See also page xvii.

one must choose A_i^b in such a way as to cancel (4.28). It is not difficult to see that this requires

$$A_i^a = \varepsilon^{aij}\frac{1}{r}n^j, \qquad r \to \infty. \tag{4.29}$$

Then the term $1/r$ in $\mathcal{D}_i\phi^a$ is canceled.

The monopole profile functions.

   Equations (4.27) and (4.29) determine the index structure of the field configuration with which we are going to deal. The appropriate *ansatz* is perfectly clear now:

$$\phi^a = vn^a H(r), \qquad A_i^a = \varepsilon^{aij}\frac{1}{r}n^j F(r), \tag{4.30}$$

where H and F are functions of r with boundary conditions

$$H(r) \to 1, \qquad F(r) \to 1 \qquad \text{at } r \to \infty \tag{4.31}$$

and

$$H(r) \to 0, \qquad F(r) \to 0 \qquad \text{at } r \to 0. \tag{4.32}$$

The boundary condition (4.31) is equivalent to Eqs. (4.27) and (4.29), while the boundary condition (4.32) guarantees that our solution is nonsingular at $r \to 0$. The absence of singularity at $r \to 0$ is a necessary feature of admissible solutions.

   After some straightforward algebra we get

$$B_i^a = \left(\delta^{ai} - n^a n^i\right)\frac{1}{r}F' + n^a n^i\frac{1}{r^2}\left(2F - F^2\right),$$

$$\mathcal{D}_i\phi^a = v\left[\left(\delta^{ai} - n^a n^i\right)\frac{1}{r}H(1 - F) + n^a n^i H'\right], \tag{4.33}$$

where a prime denotes differentiation with respect to r.

   Let us return now to the Bogomol'nyi equation (4.25). This comprises a set of nine first-order differential equations. Our *ansatz* has only two unknown functions. The fact that the *ansatz* goes through and we get two scalar equations on two unknown functions from the Bogomol'nyi equations is a highly nontrivial check. Comparing Eqs. (4.25) and (4.33) we get

$$F' = gvH(1 - F),$$

$$H' = \frac{1}{gv}\frac{1}{r^2}\left(2F - F^2\right). \tag{4.34}$$

The functions H and F are dimensionless; it is convenient to make the radius r dimensionless too. It is obvious that a natural unit of length in the problem at hand is $(gv)^{-1}$. From now on we will measure r in these units, so that

From now on F and H are functions of ρ and the prime indicates d/dρ.

$$\rho = gvr. \tag{4.35}$$

The functions H and F are to be considered as functions of ρ while below the prime will denote differentiation over ρ. Then the system (4.34) takes the form

$$F' = H(1 - F),$$

$$H' = \frac{1}{\rho^2}\left(2F - F^2\right). \tag{4.36}$$

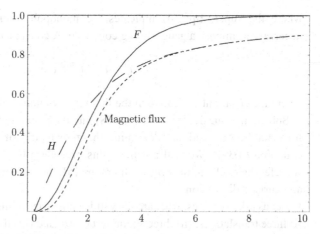

The functions F (solid line) and H (long-broken line) in the critical monopole solution, vs. ρ. The short-broken line shows the flux of the magnetic field $\mathcal{B}_i$ (in units 4π) through the sphere of radius ρ. The figure was drawn by Richard Morris.

These equations are obviously nonlinear. Nevertheless, they have the following known analytical solutions (quite a rarity in the world of nonlinear differential equations!):

$$F = 1 - \frac{\rho}{\sinh \rho},$$

$$H = \frac{\cosh \rho}{\sinh \rho} - \frac{1}{\rho}. \tag{4.37}$$

At large ρ, F tends to unity exponentially fast while $1 - H$ tends to 0 as ρ^{-1}. This is due to the masslessness of the Higgs particle in the limit $\lambda = 0$.

At large ρ the functions H and F tend to unity (cf. Eq. (4.31)) while at $\rho \to 0$ we have

$$F \sim \rho^2, \qquad H \sim \rho.$$

They are plotted in Fig. 4.2. Calculating the flux of the magnetic field through the large sphere we verify that, for the solution at hand, $Q_M = 4\pi/g$.

4.1.5 Collective Coordinates (Moduli)

The monopole solution presented in the previous subsection breaks a number of valid symmetries of the theory, for instance, translational invariance. As usual, the symmetries are restored after the introduction of collective coordinates (moduli), which convert a given solution into a family of solutions.

Our first task is to count the number of moduli in the monopole problem. A straightforward way to arrive at this number is to count the linearly independent zero modes. To this end, one represents the fields A_μ and ϕ as a sum of the monopole background plus small deviations,

$$A_\mu^a = A_\mu^{a\,(0)} + a_\mu^a, \qquad \phi^a = \phi^{a\,(0)} + \delta\phi^a, \tag{4.38}$$

where the superscript (0) indicates the monopole solution. At this point it is necessary to impose a gauge-fixing condition. A convenient such condition is

$$\frac{1}{g}\mathcal{D}_i a_i^a + \varepsilon^{abc}\phi^{b(0)}\delta\phi^c = 0, \tag{4.39}$$

where the covariant derivative in the first term contains only the background field.

This is generalization of the background gauge.

Substituting the decomposition (4.38) into the Lagrangian one can find a quadratic form for $\{a, \delta\phi\}$ and can determine the zero modes of this form (subject to the condition (4.39)). We will not track this procedure in detail; for such an account we refer the reader to the original literature [5]. Instead, cutting corners, we will give an heuristic discussion.

Let us ask ourselves: what are the valid symmetries of the model at hand? They are (i) three translations, (ii) three spatial rotations, and (iii) three rotations in the SU(2) group. Not all these symmetries are independent. It is not difficult to check that the spatial rotations are equivalent to the SU(2) group rotations for the monopole solution; thus, we should not count them independently. This leaves us with six symmetry transformations.

One should not forget, however, that two of those six act nontrivially in the "trivial vacuum." Indeed, the latter is characterized by the condensate (4.6). While rotations around the third axis in the isospace leave the condensate intact (see Eq. (4.7)), rotations around the first and second axes do not. Such rotations should not be taken into account, as the vacuum is assumed to be chosen in a particular (and unique) way. Thus the number of moduli in the monopole problem is $6 - 2 = 4$. These four collective coordinates have a very transparent physical interpretation. Three of them correspond to translations. They are introduced into the solution through the substitution

$$\vec{x} \to \vec{x} - \vec{x}_0. \tag{4.40}$$

The vector $\vec{x}_0$ now plays the role of the monopole center. Needless to say, the unit vector $\vec{n}$ is now defined as $\vec{n} = (\vec{x} - \vec{x}_0)/|\vec{x} - \vec{x}_0|$.

The fourth collective coordinate is related to the unbroken U(1) symmetry of the model. This is the rotation around the direction of alignment of the field ϕ. In the trivial vacuum ϕ^a is aligned along the third axis in color space. The monopole generalization of Eq. (4.7) is

$$A^{(0)} \to U A^{(0)} U^{-1} + iU\left(\partial U^{-1}\right),$$
$$\phi^{(0)} \to U \phi^{(0)} U^{-1} = \phi^{(0)}, \tag{4.41}$$
$$U = \exp\left(i\alpha\phi^{(0)}/v\right),$$

The α modulus

where the fields $A^{(0)}$ and $\phi^{(0)}$ are understood here to be in matrix form,

$$A^{(0)} = A^{a(0)}\frac{\tau^a}{2}, \qquad \phi^{(0)} = \phi^{a(0)}\frac{\tau^a}{2}.$$

Unlike the trivial vacuum, which is not changed under (4.7), the monopole solution for the vector field does change its form. The change looks like a gauge transformation. Note, however, that the gauge matrix U does not tend to unity at $r \to \infty$. Thus, this transformation is in fact a global U(1) rotation. The physical meaning of the collective coordinate α will become clear shortly. Now let us note that: (i) for small α, Eq. (4.41) reduces to

$$\delta \vec{A}^a = \alpha \frac{1}{v} \left(\vec{\nabla} \phi^{(0)} \right)^a, \qquad \delta \phi = 0, \qquad (4.42)$$

and this is compatible with the gauge condition (4.39); (ii) the variable α is compact, since the points α and $\alpha + 2\pi$ can be identified (the transformed $A^{(0)}$ is identically the same[2] for α and $\alpha + 2\pi$). In other words, α is an angle variable.

Having identified all four moduli relevant to the problem we can proceed to the quasi-classical quantization. The task is to obtain quantum mechanics of the moduli. Let us start from the monopole center coordinate $\vec{x}_0$. To this end, as usual, we assume that $\vec{x}_0$ weakly depends on time t, so that the only time dependence of the solution enters through $\vec{x}_0(t)$. The time dependence is important only in time derivatives, so that the quantum-mechanical Lagrangian of these moduli can be obtained from the following expression:

$$\begin{aligned}
\mathcal{L}_{\text{QM}} &= -M_M + \int d^3 x \left\{ \frac{1}{2g^2} G^a_{0i} G^a_{0i} + \frac{1}{2} (\nabla_0 \phi^a)^2 \right\} \\
&= -M_M + \int d^3 x \left\{ \frac{1}{2g^2} \dot{A}^a_i \dot{A}^a_i + \frac{1}{2} \dot{\phi}^a \dot{\phi}^a \right\} \\
&\stackrel{?}{=} -M_M + \frac{1}{2} (\dot{x}_0)_k (\dot{x}_0)_j \int d^3 x \left\{ \left[\frac{1}{g} \frac{\partial A^{a(0)}_i}{\partial (x_0)_k} \right] \right. \\
&\qquad\qquad \left. \times \left[\frac{1}{g} \frac{\partial A^{a(0)}_i}{\partial (x_0)_j} \right] + \left[\frac{\partial \phi^{a(0)}}{\partial (x_0)_k} \right] \left[\frac{\partial \phi^{a(0)}}{\partial (x_0)_j} \right] \right\},
\end{aligned}$$

$$(4.43)$$

where the subscript QM stands for quantum mechanics. The question mark above the third equals indicates that the subsequent transition, although formally correct, is not quite accurate. The square brackets in Eq. (4.43) represent (unnormalized) zero modes of the corresponding fields. If it were not for the gauge invariance, the (unnormalized) zero modes would indeed be obtained by differentiating the solution with respect to the collective coordinates:

$$\begin{aligned}
a^{a,\text{zm}}_{i(k)} &= \frac{1}{g} \frac{\partial A^{a(0)}_i}{\partial (x_0)_k} = -\frac{1}{g} \partial_k A^{a(0)}_i, \\
\delta \phi^{a,\text{zm}}_{(k)} &= \frac{\partial \phi^{a(0)}}{\partial (x_0)_k} = -\partial_k \phi^{a(0)},
\end{aligned} \qquad (4.44)$$

where the superscript zm indicates a zero mode while the subscript (k) indicates the kth zero mode. We have used the fact that the solution depends on $\vec{x}_0$ only through the combination $\vec{x} - \vec{x}_0$.

We note that Eq. (4.44) is incomplete. Because of the gauge freedom, differentiation over the collective coordinates can be supplemented by a gauge transformation.

---

[2] More accurately, this statement refers to the spatial infinity, where $\phi^{(0)}$ has magnitude v. At finite distances $A^{(0)}$ is gauge-transformed. But for gauge-invariant physical states the action of a gauge transformation depends only on the behavior of the transformation at the spatial infinity. If it equals 1 at infinity, it leaves the states invariant.

As a matter of fact we *must* do a gauge transformation, since the zero modes (4.44) do not satisfy the gauge condition (4.39). It is not difficult to guess the gauge transformation that must be made:

"Perfected" zero modes

$$a_{i(k)}^{a,\mathrm{zm}} = -\frac{1}{g}\left(\partial_k A_i^{a(0)} - \mathcal{D}_i A_k^{a(0)}\right) = -\frac{1}{g} G_{ki}^{a(0)},$$

$$\delta\phi_{(k)}^{a,\mathrm{zm}} = -\mathcal{D}_k \phi^{a(0)}.$$

(4.45)

For the kth zero mode, the phase of the gauge matrix U is proportional to $A_k^{a(0)}$. With these expressions for the zero modes the gauge condition (4.39) is satisfied since it reduces to the original (second-order) equations of motion.

Now, the expressions on the right-hand side of Eq. (4.45) replace those in the square brackets in Eq. (4.43), and we arrive at

$$\mathcal{L}_{\mathrm{QM}} = -M_M + \frac{1}{2}(\dot{x}_0)_k(\dot{x}_0)_j \int d^3x \left\{\left[\frac{1}{g}G_{ik}^{a(0)}\right]\left[\frac{1}{g}G_{ij}^{a(0)}\right] + \left[\mathcal{D}_k\phi^{a(0)}\right]\left[\mathcal{D}_j\phi^{a(0)}\right]\right\}.$$

(4.46)

Averaging over the angular orientations of $\vec{x}$ yields

$$\mathcal{L}_{\mathrm{QM}} = -M_M + \frac{1}{2}(\dot{\vec{x}}_0)^2 \int d^3x \left\{\frac{2}{3}\frac{1}{g^2}B_i^{a(0)}B_i^{a(0)} + \frac{1}{3}\mathcal{D}_i\phi^{a(0)}\mathcal{D}_i\phi^{a(0)}\right\}$$

$$= -M_M + \frac{M_M}{2}(\dot{\vec{x}}_0)^2.$$

(4.47)

Quantum mechanics of moduli

This last result readily follows when one combines Eqs. (4.15) and (4.25). Of course, it could have been guessed from the very beginning since it is nothing other than the Lagrangian describing the free nonrelativistic motion of a particle of mass M_M endowed with the coordinate $\vec{x}_0$.

Having tested the method in a case where the answer was obvious, let us apply it to the fourth collective coordinate, α. In this case we get

$$\mathcal{L}_{[\alpha]} = \frac{1}{2}\frac{M_M}{m_W^2}\dot{\alpha}^2.$$

(4.48)

The starting point is the second line in Eq. (4.43). We then use the fact that $\phi^{(0)}$ does not depend on α; the only source of α dependence is that of $A_i^{(0)}$, as indicated in Eq. (4.41). Since α is a modulus of the angular type it is *a priori* clear that $\mathcal{L}_{[\alpha]}$ must be invariant under shifts $\alpha \to \alpha + \mathrm{const}$, implying that $\mathcal{L}_{[\alpha]}$ can contain only $\dot{\alpha}$, not α itself.[3] If this is the case then we can calculate $\mathcal{L}_{[\alpha]}$ at small α using Eq. (4.42). Combining Eqs. (4.15) and (4.25) we arrive at (4.48).

The canonical momentum conjugate to the angular variable α is

$$\pi_{[\alpha]} = \frac{\delta\mathcal{L}_{[\alpha]}}{\delta\dot{\alpha}} = \frac{M_M}{m_W^2}\dot{\alpha} \to -i\frac{d}{d\alpha}$$

(4.49)

or, equivalently,

$$\dot{\alpha} = \frac{m_W^2}{M_M}\pi_{[\alpha]},$$

---

[3] This fact can be readily checked by an explicit calculation.

resulting in the following contribution to the full Hamiltonian:

$$\mathcal{H}_{[\alpha]} = \frac{1}{2}\frac{m_W^2}{M_M}\pi_{[\alpha]}^2, \tag{4.50}$$

where $\mathcal{H}_{[\alpha]}$ is the part of the Hamiltonian relevant to the angular variable α. The full quantum-mechanical Hamiltonian describing the moduli dynamics is thus

$$\mathcal{H} = M_M + \frac{\vec{p}^2}{2M_M} + \frac{1}{2}\frac{m_W^2}{M_M}\pi_{[\alpha]}^2, \tag{4.51}$$

where the momentum operator $\vec{p}$ is given by

$$\vec{p} \equiv -i\frac{d}{d\,\vec{x}_0}.$$

This Hamiltonian describes the free motion of a spinless particle endowed with an internal (compact) variable α.

A monopole is a spinless particle in this model.

While the spatial part of $\mathcal{H}$ does not raise any questions, the α dynamics deserves an additional discussion. The α motion is free, but one should not forget that α is an angle. Because of the 2π periodicity, the corresponding wave functions must have the form

$$\Psi(\alpha) = e^{ik\alpha}, \tag{4.52}$$

where $k = 0, \pm 1, \pm 2, \ldots$ Strictly speaking, only the ground state, $k = 0$, describes the monopole – a particle with magnetic charge $4\pi/g$ and vanishing electric charge. Excitations with $k \neq 0$ correspond to a particle with magnetic charge $4\pi/g$ *and* electric charge $e = kg$, the so-called *dyon*.

Meet the dyons.

To see that this is indeed the case, let us note that for $k \neq 0$ the expectation value of $\pi_{[\alpha]}$ is k. Hence, the expectation value of

$$\dot{\alpha} = \frac{m_W^2}{M_M}\pi_{[\alpha]} \quad \text{is} \quad \frac{m_W^2}{M_M}k. \tag{4.53}$$

Now, let us define a gauge-invariant electric field $\mathcal{E}_i$ (analogous to $\mathcal{B}_i$ in Eq. (4.19)), as follows:

$$\mathcal{E}_i \equiv \frac{1}{v}E_i^a\phi^a = \frac{1}{v}\phi^{a(0)}\dot{A}_i^{a(0)} = \frac{1}{v^2}\dot{\alpha}\phi^{a(0)}(\mathcal{D}_i\phi^{a(0)}). \tag{4.54}$$

The last equality follows from Eq. (4.42). Since for the critical monopole $\mathcal{D}_i\phi^{a(0)} = (1/g)B_i^{a(0)}$, we see that

$$\mathcal{E}_i = \dot{\alpha}\frac{1}{m_W}\mathcal{B}_i \tag{4.55}$$

Electric flux

and the flux of the gauge-invariant electric field through the large sphere is

$$\frac{1}{g}\int_{S_R}d^2S_i\mathcal{E}_i = \frac{m_W^2 k}{M_M}\frac{1}{m_W}\frac{1}{g}\int_{S_R}d^2S_i\mathcal{B}_i = m_W k\frac{Q_M}{M_M}, \tag{4.56}$$

where we have replaced $\dot{\alpha}$ by its expectation value. Thus, the flux of the electric field reduces to

$$Q_E = \frac{1}{g}\int_{S_R}d^2S_i\,\mathcal{E}_i = kg. \tag{4.57}$$

I did not plan
to discuss
dyons. They
popped out
after
quantization
of modulus
α.

which proves the above assertion that the electric charge of the dyon under consideration is kg. In deriving (4.57) we used Eqs. (4.23) and (4.26).

It is interesting to note that the mass of the dyon can be written as

$$M_D = M_M + \frac{1}{2}\frac{m_W^2}{M_M}k^2 \approx \sqrt{M_M^2 + m_W^2 k^2} = v\sqrt{Q_M^2 + Q_E^2}. \qquad (4.58)$$

In supersymmetric theories, for critical dyons (Section 11.6) the last formula will be exact.

Magnetic monopoles were introduced into the theory by Dirac in 1931 [4]. He considered macroscopic electrodynamics and derived a self-consistency condition for the product of the magnetic charge of the monopole Q_M and the elementary electric charge e,[4]

$$Q_M e = 2\pi. \qquad (4.59)$$

This is known as the Dirac quantization condition. For the 't Hooft–Polyakov monopole we have just derived that $Q_M g = 4\pi$, twice as large as in the Dirac quantization condition (cf. Eq. (4.23)). Note, however, that g is the electric charge of the W bosons. It is *not* the minimal possible electric charge that can be present in the theory at hand. If quarks in the fundamental (doublet) representation of SU(2) were introduced into the Georgi–Glashow model, their U(1) charge would be $e = g/2$, and the Dirac quantization condition would be satisfied for these elementary charges.

4.1.6 Noncritical Monopole

The topological argument tells us that stable monopoles do exist even if $\lambda \neq 0$. However, in this case they cannot be obtained from the first-order differential equations. One has to solve the second-order equations of motion. The *ansatz* (4.30) remains valid. Although the profile functions F and H are not known analytically, they can be found numerically. The formula for the monopole mass becomes

$$M_M = \frac{4\pi v}{g} f\left(\frac{\lambda}{g^2}\right), \qquad (4.60)$$

where f is a dimensionless function normalized to unity, $f(0) = 1$. To see that this is indeed the case, one takes the *ansatz* (4.30), (4.33) and substitutes it into the energy functional, passing to the dimensionless spatial variables

$$\vec{\rho} = gv\vec{x}, \qquad \rho = |\vec{\rho}|. \qquad (4.61)$$

Then

$$E = \frac{4\pi v}{g}\int_0^\infty d\rho\, \rho^2 \left[\frac{(F')^2}{\rho^2} + \frac{(2F - F^2)^2}{2\rho^4}\right.$$
$$\left. + \frac{H^2(1-F)^2}{\rho^2} + \frac{(H')^2}{2} + \frac{\lambda}{g^2}\left(H^2 - 1\right)^2\right], \qquad (4.62)$$

where F and H are functions of the dimensionless variable ρ, the prime denotes differentiation over ρ, and Eq. (4.62) corresponds to three distinct terms in the Hamiltonian: B^2, $(\mathcal{D}\phi)^2$, and $\lambda(\phi^2 - v^2)^2$. The overall factor v/g sets the scale

---

[4] In Dirac's original convention the charge quantization condition is, in fact, $Q_M e = \frac{1}{2}$.

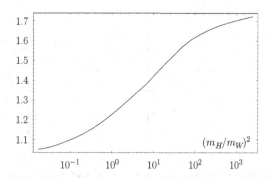

Fig. 4.3 The monopole mass (in units of $4\pi v/g$) as a function of the ratio $m_H^2/m_W^2 \equiv \lambda/g^2$ (from [6]). As $m_H/m_W \to 0$ the mass tends to unity while in the opposite limit, $m_H/m_W \to \infty$, the monopole mass ≈ 1.79.

of the monopole mass, while it becomes obvious that the λ-dependence enters only through the ratio λ/g^2. Physically this is nothing other than the ratio m_H^2/m_W^2, where m_H is the mass of the Higgs particle. The function f in Eq. (4.60) varies smoothly [6] from 1 to ≈ 1.79 as m_H^2/m_W^2 changes from 0 to ∞ (see Fig. 4.3).

4.1.7 Singular Gauge, or How to Comb a Hedgehog

The *ansatz* (4.30) for the monopole solution that we have used so far is very convenient for revealing the nontrivial topology lying behind this solution, i.e., that SU(2)/U(1) = S_2 in the group space is mapped onto the spatial S_2. However, it is often useful to gauge-transform the monopole solution in such a way that the scalar field becomes oriented along the third axis in color space, $\phi^a \sim \delta^{3a}$, in all space (i.e., at all $\vec{x}$), repeating the pattern of the "plane" vacuum (4.6). Polyakov suggested that one should refer to this gauge transformation as "combing the hedgehog." Comparison of Figs. 4.4a and 4.4b shows that this gauge transformation cannot be nonsingular. Indeed, the matrix U which combs the hedgehog

$$U^\dagger (n^a \tau^a) U = \tau^3, \tag{4.63}$$

has the form

$$U = \frac{1}{\sqrt{2}} \left(\sqrt{1 + n^3} + i\frac{n^2 \tau^1 - n^1 \tau^2}{\sqrt{1 + n^3}} \right), \tag{4.64}$$

where $\vec{n}$ is the unit vector in the direction of $\vec{x}$. The matrix U is obviously singular at $n^3 = -1$ (see Fig. 4.4). This is a gauge artifact since all physically measurable quantities are nonsingular and well defined. In the "old," Dirac, description of the monopole [7] (see also [8]) the singularity of U at $n^3 = -1$ would correspond to the Dirac string.

In the singular gauge the monopole magnetic field B_i at large $|\vec{x}|$ takes a "color-combed" form:

$$B_i \to \frac{\tau^3}{2}\frac{n^i}{r^2} = 4\pi\frac{\tau^3}{2}\frac{n^i}{4\pi r^2}. \tag{4.65}$$

The latter equation demonstrates the same magnetic charge, $Q_M = 4\pi/g$, as was derived in Section 4.1.3.

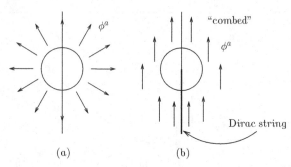

(a)   (b)

Fig. 4.4 Transition from the radial to singular gauge: "combing the hedgehog." (a) Radial gauge; (b) singular gauge. Note that a Dirac string is created by this transition.

4.1.8 Monopoles in SU(N)

Let us return to critical monopoles and extend the construction presented above from SU(2) to SU(N). The starting Lagrangian is the same as in Eq. (4.1), but with the replacement of the structure constants ε^{abc} of SU(2) by the SU(N) structure constants f^{abc}. The potential of the scalar-field self-interaction can be of a more general form than that of Eq. (4.1). The details of this potential are unimportant for our purposes since in the critical limit it tends to zero; its only role is to fix the vacuum value of the field ϕ at infinity.

Below we will need to use the elements of group theory (Lie algebras). Some of these will be reviewed *en route*, while other useful formulas are collected in Appendix section 4.3 at the end of this chapter. The reader is also referred to the books on group theory written by Howard Georgi and Pierre Ramond [9].

Elements of Lie algebra

Recall that all generators of the Lie algebra can be divided into two groups: the Cartan generators H_i, which commute with each other, and a set of raising (lowering) operators E_α (labeled by the root vectors; see below),

$$E_\alpha^\dagger = E_{-\alpha}. \tag{4.66}$$

For SU(N) – on which we are focusing here – there are $N - 1$ Cartan generators, which can be chosen as follows:

$$H^1 = \frac{1}{2}\,\mathrm{diag}\{1, -1, 0, \ldots, 0\},$$

$$H^2 = \frac{1}{2\sqrt{3}}\,\mathrm{diag}\{1, 1, -2, 0, \ldots, 0\},$$

$$\vdots$$

$$H^m = \frac{1}{\sqrt{2m(m+1)}}\,\mathrm{diag}\{1, 1, 1, \ldots, -m, \ldots, 0\}, \tag{4.67}$$

$$\vdots$$

$$H^{N-1} = \frac{1}{\sqrt{2N(N-1)}}\,\mathrm{diag}\{1, 1, 1, \ldots, 1, -(N-1)\}.$$

There are also $N(N-1)/2$ raising generators E_α and $N(N-1)/2$ lowering generators $E_{-\alpha}$. The Cartan generators are analogs of $\tau_3/2$ of SU(2) while the $E_{\pm\alpha}$ are analogs of $\tau_{\pm}/2$. Moreover, the $N(N-1)$ vectors $\boldsymbol{\alpha}, -\boldsymbol{\alpha}$ are called root vectors. They have $(N-1)$ components:

$$\boldsymbol{\alpha} = \{\alpha_1, \alpha_2, \ldots, \alpha_{N-1}\}. \tag{4.68}$$

By making an appropriate choice of the basis, any element of SU(N) algebra can be brought into a Cartan subalgebra. Correspondingly, the vacuum value of the (matrix) field $\phi \equiv \phi^a T^a$ can always be chosen to be of the form

$$\phi_{\text{vac}} = \boldsymbol{hH}, \tag{4.69}$$

where $\boldsymbol{h}$ is an $(N-1)$-component vector:

$$\boldsymbol{h} = \{h_1, h_2, \ldots, h_{N-1}\}. \tag{4.70}$$

For simplicity we will assume that, for all simple roots $\boldsymbol{\gamma}$ (see Appendix section 4.3) $\boldsymbol{h\gamma} > 0$ (otherwise, we would just change the condition defining positive roots in order to meet this constraint).

Depending on the form of the self-interaction potential, distinct patterns of gauge symmetry breaking can take place. We will discuss only the case when the gauge symmetry is maximally broken,

$$\text{SU}(N) \rightarrow \text{U}(1)^{N-1}. \tag{4.71}$$

The unbroken subgroup is Abelian. This situation is general. In special cases, when $\boldsymbol{h}$ is orthogonal to $\boldsymbol{\alpha}^m$, for some m (or a set of m) the unbroken subgroup will contain non-Abelian factors, as will be explained below. These cases will not be considered here.

The topological argument proving the existence of a variety of topologically stable monopoles in the above set-up parallels that of Section 4.1.2, except that Eq. (4.14) is replaced by

$$\pi_2\left(\text{SU}(N)/\text{U}(1)^{N-1}\right) = \pi_1\left(\text{U}(1)^{N-1}\right) = \mathbb{Z}^{N-1}. \tag{4.72}$$

| Topological formula for the second homotopy group |

There are $N-1$ independent windings in the SU(N) case.

In the matrix form, $A_\mu \equiv A_\mu^a T^a$, where the T^a are the matrices of the SU(N) generators in the fundamental representation, normalized as

$$\text{Tr}\left(T^a T^b\right) = \frac{1}{2}\delta^{ab}, \tag{4.73}$$

the gauge field A_μ can be represented as

$$A_\mu^a T^a = \sum_{m=1}^{N-1} A_\mu^m H^m + \sum_\alpha A_\mu^\alpha E_\alpha. \tag{4.74}$$

In (4.74) the A_μ^m $(m = 1, \ldots, N-1)$ can be viewed as photons and the A_μ^α as W bosons. The mass terms are obtained from the term

$$\text{Tr}\left(\left[A_\mu, \phi\right]\right)^2$$

in the Lagrangian. Substituting here Eqs. (4.69), (4.74), and (4.195) it is easy to see that the W-boson masses are

$$(m_W)_\alpha = g\boldsymbol{h\alpha}. \tag{4.75}$$

For each α the set of $N - 1$ "electric charges" of the W bosons is given by $N - 1$ components of α.

A special role belongs to the $N - 1$ massive bosons corresponding to the simple roots γ (see Appendix section 4.3): they can be thought of as *fundamental*, in the sense that the quantum numbers and masses of all other W bosons can be obtained as linear combinations (with non-negative integer coefficients) of those of the fundamental W bosons. With regard to the masses this is immediately seen from Eq. (4.75) in conjunction with

$$\alpha = \sum_\gamma k_\gamma \, \gamma. \tag{4.76}$$

The construction of SU(N) monopoles reduces, in essence, to that of an SU(2) monopole followed by various embeddings of SU(2) in SU(N). Note that each simple root γ defines an SU(2) subgroup[5] of SU(N) with the following three generators:

$$\begin{aligned}
t^1 &= \frac{1}{\sqrt{2}}(E_\gamma + E_{-\gamma}), \\
t^2 &= \frac{1}{\sqrt{2}\,i}(E_\gamma - E_{-\gamma}), \\
t^3 &= \gamma H,
\end{aligned} \tag{4.77}$$

with the standard algebra $[t^i, t^j] = i\varepsilon^{ijk}t^k$.[6] If the basic SU(2) monopole solution corresponding to the Higgs vacuum expectation value v is denoted as $\{\phi^a(\boldsymbol{r}; v), A_i^a(\boldsymbol{r}; v)\}$, see Eq. (4.30), the construction of a specific SU(N) monopole proceeds in three steps: (i) a simple root γ is chosen; (ii) the vector $\boldsymbol{h}$ is decomposed into two components, parallel and perpendicular with respect to γ, so that

$$\begin{aligned}
\boldsymbol{h} &= \boldsymbol{h}_\parallel + \boldsymbol{h}\perp, \\
\boldsymbol{h}_\parallel &= \tilde{v}\gamma, \qquad \boldsymbol{h}_\perp \gamma = 0, \\
\tilde{v} &\equiv \gamma \boldsymbol{h} > 0;
\end{aligned} \tag{4.78}$$

(iii) $A_i^a(\boldsymbol{r}; v)$ is replaced by $A_i^a(\boldsymbol{r}; \tilde{v})$ and a covariantly constant term is added to the field $\phi^a(\boldsymbol{r}; \tilde{v})$ to ensure that at $r \to \infty$ it has the correct asymptotic behavior, namely, $2\mathrm{Tr}\phi^2 = \boldsymbol{h}^2$.

Algebraically the SU(N) monopole solution takes the form

$$\phi = \phi^a(\boldsymbol{r}; \tilde{v})t^a + \boldsymbol{h}_\perp H, \qquad A_i = A_i^a(\boldsymbol{r}; \tilde{v})t^a. \tag{4.79}$$

Note that the mass of the corresponding W boson is $(m_W)_\gamma = g\tilde{v}$, fully in parallel with the SU(2) monopole.

It is instructive to verify that (4.79) satisfies the BPS equation (4.25). To this end it is sufficient to note that $[\boldsymbol{h}_\perp H, A_i] = 0$, which in turn implies that

$$\nabla_i(\boldsymbol{h}_\perp H) = 0.$$

What remains to be done? We must analyze the magnetic charges of the SU(N) monopoles and their masses. In the singular gauge (Section 4.1.7) the Higgs field is

---

[5] Generally speaking, each root α defines an SU(2) subalgebra according to Eq. (4.77), but we will deal only with the simple roots for reasons which will become clear shortly.

[6] Simple roots for SU(N) are normalized as $\gamma^2 = 1$.

aligned in the Cartan subalgebra, $\phi \sim \boldsymbol{h}\boldsymbol{H}$. The magnetic field at large distances from the monopole core, being commutative with ϕ, also lies in the Cartan subalgebra. In fact, from Eq. (4.77) we infer that the combing of the SU(N) monopole implies that

$$B_i \rightarrow 4\pi\gamma\boldsymbol{H}\frac{n^i}{4\pi r^2}, \tag{4.80}$$

which in turn implies that the set of $N - 1$ magnetic charges of the SU(N) monopole is given by the components of the $(N - 1)$-vector

$$\boldsymbol{Q}_M = \frac{4\pi}{g}\gamma. \tag{4.81}$$

Of course, the very same result is obtained in a gauge-invariant manner from the defining formula:

$$2\operatorname{Tr}(B_i\phi) \longrightarrow (\boldsymbol{Q}_M\boldsymbol{h})\frac{g}{4\pi}\frac{n_i}{r^2} \qquad \text{as } r \rightarrow \infty. \tag{4.82}$$

Equation (4.17) implies that the mass of this monopole is

$$(M_M)_\gamma = \boldsymbol{Q}_M\boldsymbol{h} = \frac{4\pi\tilde{v}}{g}, \tag{4.83}$$

which may be compared with the mass of the corresponding W bosons,

$$(m_W)_\gamma = g\gamma\boldsymbol{h} = g\tilde{v}, \tag{4.84}$$

in perfect parallel with the SU(2) monopole results of Section 4.1.3. The Dirac quantization condition is replaced by the general magnetic charge quantization condition

$$\exp\left(ig\boldsymbol{Q}_M\boldsymbol{H}\right) = 1, \tag{4.85}$$

| *Composite* |
| *monopoles* |

valid for all SU(N) groups.

Let us ask ourselves what happens if one builds a monopole on a nonsimple root. Such a solution is in fact composite: it is a combination of basic "simple-root" monopoles whose mass and quantum number (magnetic charge) are obtained by summing up the masses and quantum numbers of the basic monopoles according to Eq. (4.76).

4.1.9 The SU(3) Example

There are two simple roots in SU(3) and, consequently, two basic monopoles. The third root is nonsimple. The corresponding monopole is composite. The roots are two-component vectors; therefore, they can be drawn in a plane (Fig. 4.5).

To begin with, let us assume that the vector $\boldsymbol{h}$ belongs to sector A (see Fig. 4.5a). Then the simple roots can be chosen in the standard form, namely,

$$\gamma = \begin{cases} (1, 0), \\ \left(-\frac{1}{2}, \frac{\sqrt{3}}{2}\right), \end{cases} \tag{4.86}$$

(see the root vectors γ_1 and γ_2 in Fig. 4.5a), while the Cartan generators $H_{1,2}$ are given by

$$H_1 = \frac{1}{2}\operatorname{diag}(1, -1, 0), \qquad H_2 = \frac{1}{2\sqrt{3}}\operatorname{diag}(1, 1, -2). \tag{4.87}$$

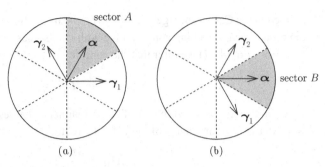

Fig. 4.5   The SU(3) root vectors.

As a result, for the two basic monopoles we have

$$g\, \boldsymbol{Q}_M \boldsymbol{H} = 2\pi \times \begin{cases} \text{diag}\,(1, -1, 0), \\ \text{diag}\,(0, 1, -1), \end{cases} \tag{4.88}$$

Masses of two SU(3) monopoles

and the quantization condition (4.85) is satisfied. The masses of the basic monopoles are $(4\pi/g)h_1$ and $(2\pi/g)(\sqrt{3}h_2 - h_1)$, respectively. Here $h_{1,2}$ stand for the two components of $\boldsymbol{h}$.

Let us now consider the composite monopole corresponding to the positive nonsimple root $\boldsymbol{\alpha} = \boldsymbol{\gamma}_1 + \boldsymbol{\gamma}_2$. Its mass is given by the formula

$$(M_M)_\alpha = \frac{4\pi}{g}(\boldsymbol{\alpha}\boldsymbol{h}) = \frac{4\pi}{g}\left(\frac{h_1}{2} + \frac{\sqrt{3}}{2}h_2\right), \tag{4.89}$$

which is the sum of the masses of two basic monopoles.

Finally, let us ask ourselves what happens if the vector $\boldsymbol{h}$ does not belong to sector A? For instance, in Fig. 4.5b this vector lies in sector B. In this case we must adjust the definition of a positive root appropriately. Remember, that the way in which the six root vectors of SU(3) are divided into two classes, positive and negative, is merely a convention. If $\boldsymbol{h}$ belongs to sector B, the positive roots must be chosen as shown in Fig. 4.5b; the horizontally oriented root vector is nonsimple but the other two are simple. The masses of the basic monopoles in this case are

$$\frac{4\pi}{g}\left(\frac{h_1}{2} \pm \frac{\sqrt{3}}{2}h_2\right). \tag{4.90}$$

4.1.10 The θ Term Induces a Fractional Electric Charge on the Monopole (Witten's Effect)

Witten noted [10] that in *CP*-nonconserving theories the dyon electric charge need not be integral. In non-Abelian gauge theories a crucial source of *CP*-nonconservation is due to the topologically nontrivial vacuum structure. It is known as the *vacuum angle* or θ *term*. (See also Chapter 8 below or Section 5 in [11].) This phenomenon, the occurrence of the vacuum angle, is not seen in perturbation theory and was discovered in the 1970s [12].

The θ term, which can be added to the Yang–Mills Lagrangian (4.1) without spoiling its renormalizability is

$$\mathcal{L}_\theta = \frac{\theta}{32\pi^2} G^a_{\mu\nu} \widetilde{G}^{a\mu\nu}, \qquad \widetilde{G}^{a\mu\nu} = \frac{1}{2} \varepsilon^{\mu\nu\alpha\beta} G^a_{\alpha\beta}. \tag{4.91}$$

This interaction violates P and CP but not C.

As is well known, the θ term can be represented as a full derivative. The corresponding action reduces to a surface term (and hence does not affect the classical equations of motion or the perturbation theory. A θ-dependence emerges, however, in instanton-induced effects (Chapter 5). This was observed shortly after the advent of instantons in Yang–Mills theories. Later, Witten realized that in the presence of magnetic monopoles the vacuum angle θ produces nontrivial effects too. Namely, $\theta \neq 0$ shifts the allowed values of the electric charge in the monopole sector of the theory.

In terms of the electric and magnetic fields the θ term takes the form

$$\mathcal{L}_\theta = -\frac{\theta}{8\pi^2} \vec{E}^a \vec{B}^a = \frac{\theta}{8\pi^2} \dot{\vec{A}}^a \vec{B}^a \tag{4.92}$$

(remembering that we are in the $A_0 = 0$ gauge). Combining this expression with Eqs. (4.42), (4.25), (4.15), and (4.26) to calculate the θ-term contribution in the action, we get an extra term in the quantum-mechanical Lagrangian describing the moduli dynamics. Namely, Eq. (4.48) gets modified as follows:

$$\mathcal{L}_{[\alpha]} = \frac{1}{2} \frac{M_M}{m_W^2} \dot{\alpha}^2 + \frac{\theta}{2\pi} \dot{\alpha}. \tag{4.93}$$

The last term – the additional contribution – is a full derivative with respect to time. This was certainly expected. Integrals of full derivatives in the action have no impact on the equations of motion. Since the extra term is linear in $\dot{\alpha}$, the quantum-mechanical Hamiltonian of the system, being expressed in terms of $\dot{\alpha}$, is the same as in Section 4.1.5. What does change, however, is the expression for the canonical momentum. Equation (4.93) implies that

$$\pi_{[\alpha]} = \frac{M_M}{m_W^2} \dot{\alpha} + \frac{\theta}{2\pi}. \tag{4.94}$$

The quantum-mechanical Hamiltonian is

$$\mathcal{H}_{[\alpha]} = \frac{1}{2} \frac{m_W^2}{M_M} \left(\pi_{[\alpha]} - \frac{\theta}{2\pi} \right)^2. \tag{4.95}$$

The wave functions remain the same as in Eq. (4.52) while Eq. (4.53) becomes

$$\dot{\alpha} = \frac{m_W^2}{M_M} \left(\pi_{[\alpha]} - \frac{\theta}{2\pi} \right) \qquad \rightarrow \qquad \frac{m_W^2}{M_M} \left(k - \frac{\theta}{2\pi} \right), \tag{4.96}$$

Electric charge is no longer restricted to integer values. It can even be irrational!

where $k = 0, \pm 1, \pm 2, \ldots$ Repeating the derivation following Eq. (4.53) we find that the dyon electric charge in the presence of the θ term is [10]

$$Q_E = g \left(k - \frac{\theta}{2\pi} \right), \qquad k = 0, \pm 1, \pm 2, \ldots \tag{4.97}$$

In fact, we could have dropped the condition $k = 0, \pm 1, \pm 2, \ldots$ provided that, simultaneously, we allowed θ to vary from $-\infty$ to ∞ rather than restricting it to the

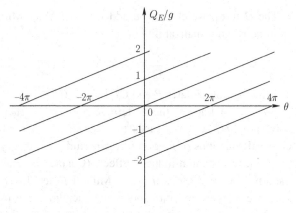

Fig. 4.6   Evolution of the monopole or dyon electric charges vs. θ.

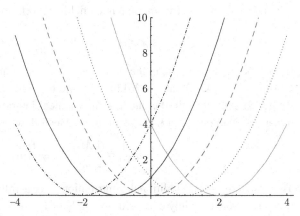

Fig. 4.7   The difference $M_D - M_M$ in units $m_W^2/(2M_M)$ vs. $\theta/(2\pi)$.

interval $0 \leq \theta \leq 2\pi$; see Fig. 4.6. Note that at, say, $\theta = 2\pi$ the $Q_E = -1$ dyon becomes the monopole while the monopole becomes the $Q_E = 1$ dyon, and so on. Varying θ intertwines the monopole and dyon states.

In the absence of θ, the dyon states with positive and negative values of k (i.e., positive and negative values of Q_E) were doubly degenerate for all $|k| > 0$, in full accord with the mass formula (4.58). Now, at generic values of θ, the mass formula (4.58) gets modified to

$$M_D = M_M + \frac{1}{2}\frac{m_W^2}{M_M}\left(k - \frac{\theta}{2\pi}\right)^2. \tag{4.98}$$

The degeneracy has been lifted. A restructuring of levels takes place at $\theta = \pm\pi$, $\pm 3\pi, \ldots$ (Fig. 4.7).

Note that the dyon mass formula, written in the form

$$M_D = v\sqrt{Q_M^2 + Q_E^2}, \tag{4.99}$$

remains valid at $\theta \neq 0$ for fractional electric charges. This is a remarkable fact!

4.1.11 Monopoles and Fermions

Fascinating monopole-induced effects ensue as soon as we introduce fermions. The most spectacular example is the fact that protons in the standard model[7] decay into baryon-charge-0 states with unsuppressed probability in the vicinity of a magnetic monopole. This phenomenon is known as the *monopole catalysis* of proton decay [13]. In this text we will not go into details of monopole catalysis phenomenology;[8] a couple of brief remarks can be found in Section 5.7. Instead, we will focus on the purely theoretical side, or, to be more exact, on just one aspect: the analysis of the fermion zero modes in the monopole background, which can be viewed as a preparation for the topic of monopole catalysis. We will consider one Dirac fermion either in the adjoint or in the fundamental representation of the SU(2) gauge group.

4.1.11.1 Zero Modes for Adjoint Fermions

Spinorial notation is discussed at length in the beginning of Part II.

One Dirac spinor is equivalent to two Weyl spinors, to be denoted by λ and ψ, respectively. The fermion part of the Lagrangian to be considered below is

$$\mathcal{L}_{\text{adj}f} = \lambda^{\alpha,a} i \mathcal{D}_{\alpha\dot\alpha} \bar\lambda^{\dot\alpha,a} + \psi^{\alpha,a} i \mathcal{D}_{\alpha\dot\alpha} \bar\psi^{\dot\alpha,a}$$
$$- \sqrt{2}\varepsilon_{abc}\left(\phi^a \lambda^{\alpha,b}\psi^c_\alpha + \phi^a \bar\lambda^b_{\dot\alpha}\bar\psi^{\dot\alpha,c}\right). \tag{4.100}$$

Equations for the fermion zero modes can be readily derived from the Lagrangian (4.100):

$$i\mathcal{D}_{\alpha\dot\alpha}\lambda^{\alpha,c} - \sqrt{2}\varepsilon_{abc}\phi^a \bar\psi^b_{\dot\alpha} = 0,$$
$$i\mathcal{D}_{\alpha\dot\alpha}\psi^{\alpha,c} + \sqrt{2}\varepsilon_{abc}\phi^a \bar\lambda^b_{\dot\alpha} = 0, \tag{4.101}$$

plus the Hermitian conjugates. After a brief consideration we conclude that this corresponds to two complex, or equivalently four real, zero modes.[9] Two of the modes are obtained if we substitute into (4.101)

$$\lambda^\alpha = F^{\alpha\beta}, \qquad \bar\psi_{\dot\alpha} = \sqrt{2}\mathcal{D}_{\alpha\dot\alpha}\phi. \tag{4.102}$$

The other two solutions correspond to the following substitution:

$$\psi^\alpha = F^{\alpha\beta}, \qquad \bar\lambda_{\dot\alpha} = \sqrt{2}\mathcal{D}_{\alpha\dot\alpha}\phi. \tag{4.103}$$

With four real fermion collective coordinates, the monopole supermultiplet is four dimensional: it includes two bosonic states and two fermionic. (This counting refers just to the monopole, without its antimonopole partner. The antimonopole supermultiplet also includes two bosonic and two fermionic states.)

---

[7] Also referred to as the SM. Beyond any doubts, SM is *the* theory of our world.

[8] At this stage I always suggest to my students a problem (formulated below) with the assurance that whoever comes up with the correct solution will immediately get the highest grade and will be allowed to skip the remainder of the course. So far no solution has been offered. The problem: how many kilograms of magnetic monopoles one must find in the depths of the universe and bring back to Earth in order to meet all energy needs of humankind for the next three centuries?

[9] This means that a monopole is described by two complex or four real fermion collective coordinates.

4.1.11.2 Dirac Fermion in the Fundamental Representation

The spectacular effects mentioned at the beginning of Section 4.1.11 are caused by color-doublet Dirac fermions in the monopole field. Let us introduce a Dirac fermion composed of two Weyl fermions ξ_α and $\bar{\eta}^{\dot{\alpha}}$, each of which transforms in the fundamental representation of the $SU(2)_{\text{gauge}}$ group. The fermion mass is generated through the coupling to the Higgs field ϕ. The fermions become massive once ϕ develops a vacuum expectation value.

Then two things happen. First, the half-integer color spin of the fermion (sometimes referred to as the *isospin*) converts itself into the regular spatial spin, so that the overall angular momentum of the monopole + fermion system is integer rather than semi-integer. This is rather counterintuitive.

Second, the system under consideration exhibits the interesting phenomenon of charge fractionalization, very similar to that occurring when kinks are coupled to fermions, as in Section 2.5.2. In the case at hand, we will have one *complex* fermion zero mode [14] (two real fermion moduli), implying that the degenerate monopole multiplet includes one state with fermion charge $\frac{1}{2}$ and another with fermion charge $-\frac{1}{2}$. Correspondingly, the monopoles acquire fractional electric charges, $\pm\frac{1}{2}$ of that of the elementary fermion [14, 15]. Thus, in the presence of the fundamental fermions the monopoles become dyons even at $\theta = 0$. This aspect is also similar to what we discussed in Section 2.5.2.

The fermion part of the Lagrangian is

$$\mathcal{L}_{\text{fund}\,f} = \bar{\xi}_{\dot{\alpha}} i \mathcal{D}^{\dot{\alpha}\alpha} \xi_\alpha + \eta^\alpha i \mathcal{D}_{\alpha\dot{\alpha}} \bar{\eta}^{\dot{\alpha}} - h\eta^\alpha \phi \xi_\alpha - h\bar{\xi}_{\dot{\alpha}} \phi \bar{\eta}^{\dot{\alpha}}, \qquad (4.104)$$

where the Yukawa coupling can always be chosen to be real and positive. The fermion equations of motion following from (4.104) are

$$i\mathcal{D}^{\dot{\alpha}\alpha} \xi_\alpha - h\phi \bar{\eta}^{\dot{\alpha}} = 0,$$

$$i\mathcal{D}_{\alpha\dot{\alpha}} \bar{\eta}^{\dot{\alpha}} - h\phi \xi_\alpha = 0. \qquad (4.105)$$

Examining Eqs. (4.105) in the "empty" vacuum (i.e., without monopoles) we readily obtain that the fermion mass terms are $\pm hv/2$, implying that the fermion mass is

$$m_f = \frac{hv}{2}. \qquad (4.106)$$

The fermion charge of the elementary fermion excitation is ± 1, while the electromagnetic $U(1)$ charge is $\pm\frac{1}{2}$.

Now we will move on to address the monopole background problem. The monopole solution is given in Eq. (4.30). For clarity we will denote the spatial matrices acting on spinorial indices of ξ and $\bar{\eta}$ as σ^i and the $SU(2)$ color matrices by τ^i, although both are in fact the Pauli matrices. The distinction is that the τ_i act on the color indices of ξ and $\bar{\eta}$. Our considerations will simplify if we adopt the following convention: the action of the color generators $\vec{\tau}$ on ξ and $\bar{\eta}$ (say, $\vec{\tau}\xi$) will be written in the form $\xi(\vec{\tau})^T$ and $\bar{\eta}(\vec{\tau})^T$, where the superscript T stands for transposition. We are assuming that if ξ and $\bar{\eta}$ are regarded as two-by-two matrices then their spatial index comes first and their gauge $SU(2)$ index comes second.

The monopole background field is time-independent, and so are the fermion zero modes. They can depend *only on the three spatial coordinates* x_i. Thus Eq. (4.105) can be rewritten in the form of two decoupled equations

$$\mathcal{D}_i \sigma^i (\xi + i\bar{\eta}) - h\phi(\xi + i\bar{\eta}) = 0,$$
$$\mathcal{D}_i \sigma^i (\xi - i\bar{\eta}) + h\phi(\xi - i\bar{\eta}) = 0. \tag{4.107}$$

In three dimensions we cannot use index theorems of the type discussed in Section 3.3.1 because no three-dimensional γ^5 matrix exists. Instead, one should turn to the Callias theorem [16, 17], which relates the difference between the numbers of the zero modes for the operators $L_- = \mathcal{D}_i \sigma^i - h\phi$ and $L_+ = \mathcal{D}_i \sigma^i + h\phi$ to the topological charge of the background field.[10]

The derivation of Callias' theorem involves a number of cumbersome details which we will not discuss here. The mathematically oriented reader is directed to [16, 17]. A consequence from Callias' theorem is that the above-mentioned difference is 1 in the monopole field (4.30). We will see below that the first equation in (4.107) has a single solution while the second has none.

The spinors ξ and η can be considered as 2×2 matrices: the first index is spinorial, the second refers to color. A simple inspection of Eqs. (4.30) and (4.107) prompts us to the form of *ansatz* that will satisfy Eqs. (4.107),

$$\xi + i\bar{\eta} = \tau_2 X(r), \qquad \xi - i\bar{\eta} = \tau_2 \tilde{X}(r), \tag{4.108}$$

where X and $\tilde{X}$ are some functions of r to be determined from (4.107). We should remember that, say,

$$h\phi(\xi + i\bar{\eta}) = \frac{h}{2}\phi^a(\xi + i\bar{\eta})(\tau^a)^T = m_f n^a H(r)\tau_2 X(r)(\tau^a)^T$$
$$= -(\vec{n}\vec{\tau}\tau_2)m_f HX. \tag{4.109}$$

Master equation for zero modes.

With the *ansatz* (4.108), the structure $(\vec{n}\vec{\tau}\tau_2)$ emerges in all the terms in Eq. (4.107). Therefore it cancels out, leaving us with equations with no indices,

$$X' + \frac{1}{r}XF + m_f XH = 0,$$
$$\tilde{X}' + \frac{1}{r}\tilde{X}F - m_f \tilde{X}H = 0. \tag{4.110}$$

Given the asymptotics of the functions H and F indicated in Eq. (4.31) we may conclude that the first equation has a normalizable solution,

$$X = \text{const} \times \exp\left[-\int_0^r dr \left(m_f H + \frac{F}{r}\right)\right], \tag{4.111}$$

while the solution for $\tilde{X}$ (regular at the origin) grows at infinity. The large-r behavior of X is

$$X \to \frac{e^{-m_f r}}{r}, \qquad r \to \infty. \tag{4.112}$$

---

[10] Note that the operators $L_\pm$ are not complex conjugates.

Exercises

4.1.1   Verify that Eq. (4.45) is consistent with the gauge condition (4.39).

4.1.2   For N_f Dirac fermions in the doublet representation of SU(2), one finds N_f complex zero modes in the monopole background. The corresponding fermion moduli can be written in terms of creation and annihilation operators a_0^i and $a_0^{i\dagger}$ ($i = 1, 2, \ldots, N_f$) obeying the anticommutation relations

$$\{a_0^i, a_0^j\} = \{a_0^{i\dagger}, a_0^{j\dagger}\} = 0,$$
$$\{a_0^i, a_0^{j\dagger}\} = \delta^{ij}. \tag{4.113}$$

     (a) Construct operators obeying the Lie algebra of SU(N_f) in terms of the operators a_0^i and $a_0^{j\dagger}$.

     (b) Show that the monopole ground state has multiplicity 2^{N_f}. To which representations of SU(N_f) does it belong?

4.1.3   Present an explicit proof of the fact that the monopole solution (4.30) stays intact under the combined action $\vec{L} + \vec{T}$, where $\vec{L}$ and $\vec{T}$ denote the generators of the spatial and SU(2) color rotations, respectively.

4.2 Skyrmions

This section is devoted to the studies of the Skyrmion model for baryons which treats baryons as quasiclassical solitons in the chiral theory. This is parametrically justified in the 't Hooft limit, i.e., in the limit

$$N \to \infty, \qquad g^2 N \text{ fixed}, \tag{4.114}$$

where g is the gauge coupling constant; see Section 9.2. As will become clear shortly, the Skyrmion model does *not* represent the exact solution of QCD in the baryon sector. However, it has its virtues. Arguably it captures all regularities of the baryon world (see Section 9.2.7 and the three following subsections). In some well-defined instances the Skyrmion model predictions are expected to be quite precise while in other instances they are expected to be valid only semi-quantitatively.

From the microscopic standpoint (i.e., that of QCD), the low-lying baryons such as protons and neutrons at large N have the color structure

$$q^{i_1} q^{i_2} q^{i_3} \ldots q^{i_N} \, \varepsilon_{i_1 i_2 i_3 \ldots i_N}, \tag{4.115}$$

where q is the quark field. The above baryon consists of N quarks fully antisymmetrized in the color space – then the spatial wave function can be fully symmetrized. In other words, all quarks can be viewed as S-wave constituents. As we will see in Section 4.2.9, this circumstance is of paramount importance for the success of the Skyrmion model.

4.2.1 Preamble: Global Symmetries of QCD

To begin with, let us recall some basic facts regarding quantum chromodynamics (QCD).

At low energies QCD can be described as Yang–Mills theory with two or three light Dirac fermions q in the fundamental representation of SU(N), where the

QCD Lagrangian in the chiral limit, n flavors

number of colors $N = 3$ in the actual world. To a good approximation we can consider the light quarks to be massless. Then the QCD Lagrangian takes the form

$$\mathcal{L} = -\frac{1}{4} G^a_{\mu\nu} G^{\mu\nu a} + \sum_{f=1}^{n} \bar{q}_f \gamma^\mu i \mathcal{D}_\mu q^f, \tag{4.116}$$

where $G^a_{\mu\nu}$ is the gluon field strength tensor, and n is the number of the massless flavors (two or three in the actual world). The global symmetry of the above Lagrangian is well known:[11]

$$\mathrm{SU}(n)_L \times \mathrm{SU}(n)_R \times \mathrm{U}(1)_V. \tag{4.117}$$

The vectorial U(1) symmetry, the last factor in Eq. (4.117), is responsible for baryon number conservation. The *baryon current* is

$$J^B_\mu = \frac{1}{N} \sum_{f=1}^{n} \bar{q}_f \gamma_\mu q^f. \tag{4.118}$$

The chiral part of (4.117) describes the invariance of the QCD Lagrangian under independent SU(n) rotations of the left- and right-handed quarks, $q_{L,R} = (1 \mp \gamma_5)q/2$,

$$q_L^f \to L^f_g q^g_L, \qquad q_R^{\bar{f}} \to R^{\bar{f}}_{\bar{g}} q^{\bar{g}}_R, \tag{4.119}$$

Global flavor rotations

where L and R are the SU(n)$_{L,R}$ matrices. To emphasize their independence we use barred and unbarred flavor right- and left-handed indices, respectively.

Let us make a brief excursion into a fancy world in which the chiral symmetry of the Lagrangian would be linearly implemented in the physical spectrum. We hasten to add that this is *not* our world; see Section 8.3.2. Nevertheless, this sci-fi digression may teach us something useful.

The SU(n)$_L \times$ SU(n)$_R$ chiral symmetry is conveniently represented in terms of the Weyl spinors

$$[q_L]^{if}_\alpha, \qquad [q_R]^{i\bar{f}}_{\dot{\alpha}}, \tag{4.120}$$

where $\alpha, \dot{\alpha} = 1, 2$ are spinorial indices of the Lorentz group, $i = 1, \ldots, N$ is the color index and $f, \bar{f} = 1, \ldots, n$ are "subflavor" indices of two independent, left and right, SU(n) groups. The reader should note that in this section we use square brackets to emphasize the matrix nature of a quantity.

The interpolating fields for colorless hadrons can be constructed from the quark fields. For instance, the spin-zero mesons are described by the meson matrix M,

$$M^f_{\bar{f}} = [\bar{q}_R]^\alpha_{i\bar{f}} [q_L]^{if}_\alpha = \bar{q}_{\bar{f}} \frac{1 - \gamma_5}{2} q^f. \tag{4.121}$$

---

[11] To refresh one's memory one could look through Sections 12 and 14 in [11].

The baryon charge of M clearly vanishes. The matrix M realizes the $\{n, n\}$ representation of $SU(n)_L \times SU(n)_R$ and contains $2n^2$ real fields. The mirror reflection of the space coordinates, the P-parity operation, which transforms $q_{L\alpha}^{if}$ to $q_{R\dot{\alpha}}^{if}$ and vice versa, acts on the matrix $M_{\bar{f}}^f$ as follows:

$$PM = M^\dagger. \tag{4.122}$$

It means that the Hermitian part of M describes n^2 scalars while the anti-Hermitian part describes n^2 pseudoscalars. In terms of the diagonal $SU(n)_V$ symmetry (when $L = R$) these n^2 fields form an adjoint representation plus a singlet.

Starting from spin 1, there exist interpolating $q\bar{q}$ operators of a different chiral structure. In the case of spin-1 mesons one can introduce

$$\left[V_\mu^L\right]_g^f = \sigma_\mu^{\alpha\dot{\alpha}} [\bar{q}_L]_{\dot{\alpha}ig} [q_L]_\alpha^{if} = \bar{q}_g \gamma_\mu \frac{1 - \gamma_5}{2} q^f, \tag{4.123}$$

where $\sigma^\mu = \{1, \vec{\sigma}\}$.

Subtracting the trace we get the $(n^2 - 1, 1)$ representation, while the trace part is the $(1, 1)$ representation of $SU(n)_L \times SU(n)_R$. The matrix V_μ^L is Hermitian; therefore it represents n^2 fields of spin 1. These fields are singlets of $SU(n)_R$ and adjoints or singlets of $SU(n)_L$ (as well as of $SU(n)_V$). Under the parity transformation V_μ^L goes to

$$\left[V_\mu^R\right]_{\bar{g}}^{\bar{f}} = \sigma_\mu^{\alpha\dot{\alpha}} [\bar{q}_R]_{\alpha i \bar{g}} [q_R]_{\dot{\alpha}}^{i\bar{f}} = \bar{q}_{\bar{g}} \gamma_\mu \frac{1 + \gamma_5}{2} q^{\bar{f}}. \tag{4.124}$$

The vector and axial-vector particles are described respectively by the sum and the difference of V_μ^L and V_μ^R.

Let us note in passing that spin-1 mesons can also be described by an antisymmetric tensor field transforming in the $(1, 1)$ representation of the Lorentz group, instead of the $(\frac{1}{2}, \frac{1}{2})$ representation displayed above:

$$\left[H_{\mu\nu}\right]_{\bar{f}}^f = [\sigma_\mu, \bar{\sigma}_\nu]^{\alpha\beta} \left\{[\bar{q}_R]_{\alpha i \bar{f}} [q_L]_\beta^{if} + (\alpha \leftrightarrow \beta)\right\} = \bar{q}_{\bar{f}} \sigma_{\mu\nu} \frac{1 - \gamma_5}{2} q^f, \tag{4.125}$$

where $\bar{\sigma}^\mu = \{1, -\vec{\sigma}\}$. The chiral features of this tensor current are different from $[V_\mu^L]_g^f$ but the same as those of the spin-0 fields $M_{\bar{f}}^f$, Eq. (4.121). Moreover, by applying the total derivative we see that the tensor current $[H_{\mu\nu}]_{\bar{f}}^f$ is equivalent to $[M_\mu]_{\bar{f}}^f$. Indeed,

$$\partial^\nu \left[H_{\mu\nu}\right]_{\bar{f}}^f = -i\bar{q}_{\bar{f}} \overset{\leftrightarrow}{\mathcal{D}}_\mu \frac{1 + \gamma_5}{2} q^f. \tag{4.126}$$

The QCD Lagrangian (4.116) has another (classical) symmetry, $U(1)_A$, corresponding to the following rotations of the left- and right-handed fields in opposite directions,

$$q_L \to e^{i\eta} q_L, \qquad q_R \to e^{-i\eta} q_R, \tag{4.127}$$

This axial $U(1)_A$ is anomalous at the quantum level (Chapter 8).

The anomaly is suppressed by $1/N$ in the 't Hooft limit $N \to \infty$ with $g^2 N$ fixed [18], Section 9.2. Thus, in the 't Hooft limit, the $U(1)_A$ charge becomes a good quantum number. Note that the $U(1)_A$ charge of the meson matrices M and H in Eqs. (4.121) and (4.125) respectively is 2, while that of V in Eqs. (4.123) and (4.124) is zero.

Thus, were the chiral symmetry realized linearly, for any given spin we would have two types of chiral multiplets, charged and neutral with respect to $U(1)_A$. Each multiplet would contain n^2 states of each parity.[12]

Nambu–Goldstone realization of the chiral symmetry

The above introduction to the theory of representations of chiral symmetry is needed in order to say that we see nothing of the kind in nature. We do not observe $2 \times n^2$ degenerate multiplets in the meson spectrum. The reason is that chiral symmetry is realized *nonlinearly*, in the *Nambu–Goldstone mode* [19].

4.2.2 Massless Pions and the Chiral Lagrangian

As is well known, the chiral $SU(n)_L \times SU(n)_R$ symmetry is spontaneously broken down to the diagonal $SU(n)_V$ symmetry. Only the vectorial $SU(n)_V$ symmetry is realized linearly in QCD and is seen in the spectrum. The above spontaneous breaking implies the existence of $n^2 - 1$ Goldstone bosons, massless pions. Below we will mostly focus on the case of two massless flavors, $n = 2$.

In this case there are three pion fields $\pi^a(x)(a = 1, 2, 3)$. The pion dynamics is concisely described by an $SU(2)$ matrix field $U(x)$,

$$U(x) = \exp\left(\frac{i}{F_\pi}\tau^a\pi^a(x)\right), \qquad U \in SU(2), \tag{4.128}$$

where the τ^a are the Pauli matrices and

$$F_\pi \approx 93\,\text{MeV}$$

is a so-called pion constant.[13] Under an $SU(2)$ transformation by unitary matrices L and R, U transforms as

$$U \to LUR^\dagger. \tag{4.129}$$

The Lagrangian (usually referred to as the *chiral* Lagrangian) must be invariant under both transformations, while the vacuum state must respect only the diagonal combination $L = R$. The Lagrangian must be expandable in powers of derivatives.

Chiral Lagrangian

The lowest-order term has two derivatives and can be written as

$$\mathcal{L}^{(2)} = \frac{F_\pi^2}{4}\text{Tr}\left(\partial_\mu U \partial^\mu U^\dagger\right). \tag{4.130}$$

---

[12] The case $n = 2$ is special. Owing to the quasireality of the fundamental representation of $SU(2)$, the eight-dimensional representation of $SU(2)_L \times SU(2)_R$ given by the 2×2 matrix $M_{\tilde{f}}^f$ becomes reducible and can be split into two four-dimensional representations. This can be done by imposing the group-invariant conditions

$$\tau_2 M_\pm^* \tau_2 = \pm M_\pm.$$

Then

$$M_+ = \sigma - i\vec{\tau}\vec{\pi}, \qquad M_- = i\eta + \vec{\tau}\vec{\sigma},$$

where all fields are real. The quadruplet M_+ contains the isosinglet scalar σ and the isotriplet of pseudoscalars $\vec{\pi}$ while in M_- the pseudoscalar η is isosinglet and scalars form the isotriplet $\vec{\sigma}$. Switching on the large-N axial $U(1)_A$, we observe that the $U(1)_A$ transformations mix M_+ and M_-, thus restoring an eight-dimensional representation.

[13] The constant F_π is related to the constant f_π that determines the $\pi \to \mu\nu$ decay rate; $F_\pi = f_\pi/\sqrt{2}$, see Section 8.3.3. This aspect need not concern us for the time being.

It dates back to the work of Gell-Mann and Lévy [20]. The invariance of this term under the global transformation (4.129) is obvious. In what follows it will be important that F_π^2 is proportional to the number of colors N.

In the fourth order in derivatives one can write in the chiral Lagrangian many terms that are invariant under (4.129); they are classified in [21]. We will not dwell on this classification. For our purposes it suffices to limit ourselves to one of these terms,

The Skyrme term

$$\mathcal{L}^{(4)} = \frac{1}{32e^2} \mathrm{Tr}\left[\left(\partial_\mu U\right) U^\dagger, (\partial_\nu U) U^\dagger\right]^2. \qquad (4.131)$$

This operator, which goes under the name of the *Skyrme term*, is of special importance; it is singled out because it is of the second order in the time derivative. As we will see shortly, this allows us to apply a Hamiltonian description. The constant e^2 in Eq. (4.131) is a dimensionless parameter, $e \sim 4.8$. Note that $1/e^2$ is also proportional to N.

The chiral Lagrangian we will deal with is the sum of the two terms (4.130) and (4.131),

$$\mathcal{L} = \mathcal{L}^{(2)} + \mathcal{L}^{(4)}. \qquad (4.132)$$

The vacuum mainfold

Any constant (x-independent) matrix U represents the lowest-energy state, the vacuum of the theory. Each matrix U represents a point in the space of vacua, which is usually referred to as the *vacuum manifold*. Performing a generic chiral transformation, we move from one point of the vacuum manifold to another. However, some chiral transformations, applied to a given vacuum, leave it intact. It is not difficult to understand that all vacua of the theory are invariant under the diagonal $\mathrm{SU}(n)_V$ symmetry operation of the chiral $\mathrm{SU}(n)_L \times \mathrm{SU}(n)_R$ group. The easiest way to see this is to consider the vacuum $U = 1$. It is obviously invariant under (4.129) provided that $R = L$. Thus, the vacuum manifold is the coset

$$\{\mathrm{SU}(n)_L \times \mathrm{SU}(n)_R\}/\mathrm{SU}(n)_V. \qquad (4.133)$$

The chiral Lagrangian (4.132) describes a $\{\mathrm{SU}(n)_L \times \mathrm{SU}(n)_R\}/\mathrm{SU}(n)_V$ *sigma model*. The coset (4.133) is referred to as the *target space* of the sigma model.

The chiral transformations (4.129) generate flavor-nonsinglet currents. As we know from the microscopic theory, there is another conserved current, the baryon current (4.118). What happens with the baryon current in the chiral theory (4.132)?

Needless to say, the baryon charge vanishes identically in the meson sector. Thus, if there is a "projection" of the baryon current (4.118) in the chiral theory, its expression in terms of U must obey the following property: it must vanish identically for all fields presenting small oscillations of U around its vacuum value.

Baryon current

Such a conserved current does exist,

$$J^{B\mu} = -\frac{\varepsilon^{\mu\nu\alpha\beta}}{24\pi^2} \mathrm{Tr}\left(U^\dagger \partial_\nu U\right)\left(U^\dagger \partial_\alpha U\right)\left(U^\dagger \partial_\beta U\right), \qquad (4.134)$$

and the baryon charge B takes the form

$$B = -\frac{\varepsilon^{ijk}}{24\pi^2} \int d^3x\, \mathrm{Tr}\left(U^\dagger \partial_i U\right)\left(U^\dagger \partial_j U\right)\left(U^\dagger \partial_k U\right). \qquad (4.135)$$

We leave it as an exercise to prove that the current (4.134) is conserved "topologically" (i.e., one does not need to use equations of motion in the proof) and that $B \equiv 0$

order by order in the expansion of (4.128) in the fields π, assuming that $|\pi| \ll 1$ and $\pi(x) \to 0$ as $|\vec{x}| \to \infty$.

We will see shortly that for topologically nontrivial configurations of the field U – i.e., solitons – the baryon charge B can take any *integer* value, positive or negative. The coefficient $24\pi^2$ in the denominator is chosen to make the baryon charge of the lightest soliton unity. In fact, (4.135) presents the topological charge of the field configuration $U(x)$.

If the number massless flavors is three or larger, the so-called Wess–Zumino–Novikov–Witten (WZNW) term [22] must be added to the chiral Lagrangian (4.132). Although we will not discuss it in detail, a few remarks about it will be made below. For two flavors the Wess–Zumino–Novikov–Witten term vanishes identically.

4.2.3 Baryons as Topologically Stable Solitons

The idea that baryons, such as nucleons or delta particles, might be solitons in the model (4.132) has a long history. The first suggestion had been made by Skyrme [23] long before QCD was born. Then Finkelstein and Rubinstein showed that such solitons, being made of pions, could nevertheless obey Fermi statistics [24]. With the advent of QCD, Skyrme's idea was essentially forgotten. In the 1980s it was revived, however, by Witten whose two papers [25] and subsequent research [26] gave impetus to a new direction, which can be called the Skyrme phenomenology.[14]

What guided Witten in his arguments in favor of the baryon interpretation of Skyrmions? In the 't Hooft limit, QCD reduces to the theory of an infinite number of stable mesons whose interactions are governed by $1/N$, where N is the number of colors (Section 9.2). This parameter plays the role of a coupling constant in an effective meson theory. We can see this regularity clearly in the Lagrangian (4.132) provided that we use Eq. (4.128) and expand the Lagrangian in powers of π, remembering that $F_\pi^2 \sim N$. Then we can readily convince ourselves that the kinetic term is $O(N^0)$, the term quartic in π is $O(N^{-1})$, and so on.

Baryons, being composite states of N quarks, must have masses proportional to N, or, in other words, to the inverse coupling constant. As we know from previous chapters of this book, this behavior is typical of solitons in the quasiclassical approximation.

Why do topologically stable static solitons exist in the sigma model (4.132)? Assume that we are considering a t-independent field configuration $U(\vec{x})$. For its energy to be finite, $U(\vec{x})$ must approach a constant at the spatial infinity. This means that in mapping our three-dimensional space onto the space of unitary matrices U we are compactifying the three-dimensional space, making it topologically equivalent to a three-dimensional sphere. If so, any mapping $U(\vec{x})$ can be viewed as an element in the third homotopy group $\pi_3(\mathrm{SU}(2))$. Since

Topological formula for the third homotopy group

$$\pi_3(\mathrm{SU}(2)) = \mathbb{Z}, \tag{4.136}$$

all continuous mappings fall into distinct classes characterized by integers that count how many times S_3, the group space of $\mathrm{SU}(2)$, is swept when the coordinate three-dimensional sphere is swept just once. Since these mappings are orientable, the above

---

[14] To a certain extent, these papers were motivated by earlier studies of Balachandran *et al.* [27].

integers can be positive or negative. The corresponding topological charge is given in Eq. (4.135). For topologically trivial mappings, $B = 0$. It is natural to expect that topologically nontrivial solitons with $B = 1$ correspond to baryons. Note that the topological classification based on (4.136) will be essential in Chapter 5, devoted to instantons.

The topological argument above is general and does not specify a particular choice of the sigma model with target space (4.133) which dynamically supports topologically stable solitons. Why is the kinetic term (4.130) not sufficient for our purposes and why is it necessary to add the Skyrme term? The so-called Derrick theorem [28] answers these questions. It tells us that the energy functional

Derrick's
theorem

derived from $\mathcal{L}^{(2)}$ has no minimum, or, more exactly, its minimum is reached only asymptotically in the (singular) limit of zero-radius functions.

Indeed, assume $U_0(\vec{x})$ to be a solution to the classical (static) equation of motion following from $\mathcal{L}^{(2)}$. The corresponding energy functional is

$$E^{(2)}\left[U_0(\vec{x})\right] = \int d^3x \mathcal{H}^{(2)}\left[U_0(\vec{x})\right], \qquad (4.137)$$

where $\mathcal{H}^{(2)}$ is the part of the Hamiltonian density that is quadratic in the spatial derivatives; the superscript 2 will remind us of this fact.

Consider now a trial function $U_0(\lambda\vec{x})$, where λ is a numerical factor, substitute this function in (4.137), change the integration variable $\vec{x} \to \lambda\vec{x}$, and perform the integration. We immediately arrive at

$$E^{(2)}\left[U_0(\lambda\vec{x})\right] = \frac{1}{\lambda}E^{(2)}\left[U_0(\vec{x})\right], \qquad (4.138)$$

which is lower than $E^{(2)}\left[U_0(\vec{x})\right]$ provided that $\lambda > 1$, in contradiction with the assumption. The energy functional gets lower as the support of the function $U_0(\vec{x})$ shrinks to zero.

Now, let us switch on the Skyrme term. Following the same line of reasoning we get

$$E\left[U_0(\lambda\vec{x})\right] \equiv E^{(2)}\left[U_0(\lambda\vec{x})\right] + E^{(4)}\left[U_0(\lambda\vec{x})\right]$$

$$= \frac{1}{\lambda}E^{(2)}\left[U_0(\vec{x})\right] + \lambda E^{(4)}\left[U_0(\vec{x})\right], \qquad (4.139)$$

where the superscript 4 labels those contributions that come from the four-derivative term $\mathcal{L}^{(4)}$.[15] Now we can satisfy the initial assumption, that $E[U_0(\vec{x})]$ is the minimum of the energy functional, provided that

$$E^{(2)}\left[U_0(\vec{x})\right] = E^{(4)}\left[U_0(\vec{x})\right].$$

Before passing to a detailed analysis of the Skyrmion solution let us ask (and

Topological
formula for
the fourth
homotopy
group

answer) the following question: how can one show that the topologically stable solitons in the model at hand are fermions?

The fact from topology that $\pi_4(\mathrm{SU}(2)) = \mathbb{Z}_2$ is crucial. If we consider space–time dependent mappings $U(t, \vec{x})$ with boundary condition

$$U(t, \vec{x}) \to \mathrm{const} \qquad \text{as } t \to \pm\infty, |\vec{x}| \to \infty,$$

---

[15] In deriving $\mathcal{H}^{(4)}$ it is essential that the Skyrme term does not contain more than two *time* derivatives.

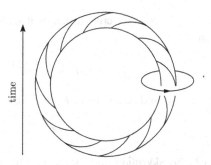

time

A soliton–antisoliton pair is created from the vacuum; the soliton is rotated by a 2π angle; the pair is then annihilated. This represents the nontrivial homotopy class in $\pi_4(\mathrm{SU}(2))$.

all such mappings fall into two topological classes: trivial (i.e., continuously contractible to 1) and nontrivial. An explicit field configuration $U(t, \vec{x})$ which tends to unity at the space–time infinity and represents the nontrivial class in $\pi_4(\mathrm{SU}(2))$ can be described as follows. At $t = -\infty$ we start from $U = 1$. As we move forward in time, we gradually create a soliton–antisoliton pair and separate them by a spatial interval; then we rotate, say, the soliton by 2π without touching the antisoliton; then we bring them together and annihilate them (see Fig. 4.8). Clearly, this 2π-rotated field configuration is topologically nontrivial – i.e., noncontractible to unity. If we assign to it a weight factor -1 (and to the topologically trivial configuration with no soliton rotation a weight factor $+1$) then we are quantizing the soliton as a fermion. That this is possible was first noted in [24]. Witten took a step further and showed, by analyzing the WZNW term for three flavors, that in fact it is necessary: the soliton *must* be a fermion if and only if N is odd, in full agreement with the quark picture of baryons as composite states of N quarks. We will return to this issue in Section 4.2.7.

4.2.4 The Skyrmion Solution

To begin with, we note that the energy functional following from (4.132) admits the Bogomol'nyi representation. Indeed,

$$
\begin{aligned}
E[U(\vec{x})] &= \int d^3x \left(\frac{F_\pi^2}{2} I_i^a I_i^a + \frac{1}{4e^2} \varepsilon^{abc} \varepsilon^{\tilde{a}bc} I_i^a I_j^b I_i^{\tilde{a}} I_j^{\tilde{b}} \right) \\
&= \frac{1}{2} \int d^3x \left[\left(F_\pi I_i^a + \frac{1}{2e} \varepsilon^{abc} \varepsilon^{ijk} I_j^b I_k^c \right)^2 - \frac{F_\pi}{e} \varepsilon^{abc} \varepsilon^{ijk} I_i^a I_j^b I_k^c \right] \\
&= 6\pi^2 \frac{F_\pi}{e} B + \frac{1}{2} \int d^3x \left(F_\pi I_i^a + \frac{1}{2e} \varepsilon^{abc} \varepsilon^{ijk} I_j^b I_k^c \right)^2,
\end{aligned}
\tag{4.140}
$$

where

$$
I_i = (\partial_i U) U^\dagger = i I_i^a \tau^a
\tag{4.141}
$$

and we have used the definition of the baryon charge B in Eq. (4.135) and assumed that it is positive (otherwise, we would have changed the relative sign in the parentheses).

If the Skyrmion were critical, i.e., if it were the baryon charge-1 solution to the equation

$$F_\pi I_i^a = -\frac{1}{2e} \varepsilon^{abc} \varepsilon^{ijk} I_j^b I_k^c, \tag{4.142}$$

then its mass would be related to F_π as follows:

$$M_{sk} = 6\pi^2 \frac{F_\pi}{e}. \tag{4.143}$$

Example of a problem in which the Bogomol'nyi bound exists but is not saturated

In fact, the Skyrmion does not satisfy the BPS equation (4.142). This equation has no solutions with appropriate boundary conditions. The Skyrmion satisfies the second-order equation of motion, and its mass is $\sim 23\%$ higher then the lower bound (4.143) Nevertheless, this bound sets a natural scale for the Skyrmion mass.

The classical (static) equations of motion following from the Lagrangian (4.132) contain the second and fourth orders in spatial derivatives and are highly nonlinear. It is not difficult to derive them. We will take a simpler route, however, and derive the Skyrme equation directly for an appropriate *ansatz*,

$$U_0(\vec{x}) = \exp\left(iF(r)\frac{\tau^j x_j}{r}\right), \qquad r = |\vec{x}|, \tag{4.144}$$

where the dimensionless function $F(r)$ parametrizes the Skyrmion profile. This is a hedgehog *ansatz* of the Polyakov type. For the function (4.144) to be regular at the origin and tend to a constant at the spatial infinity (which guarantees finite energy) we must impose the conditions

$$F(r) = \pi \times \text{ integer at } r = 0 \quad \text{and} \quad F(r) = \pi \times \text{ integer at } r \to \infty. \tag{4.145}$$

Substituting (4.144) into the definition of the baryon (and topological) charge (4.135), after some straightforward algebra we reduce the integrand to a full derivative and find

$$B = -\frac{1}{\pi}\left[F(r) - \frac{1}{2}\sin 2F(r)\right]_0^\infty. \tag{4.146}$$

Given Eq. (4.145), the second term can be omitted. Thus, if we are interested in the baryon charge-1 solution we can set the following boundary conditions:

Boundary conditions for the Skyrmion profile function

$$F(0) = \pi, \quad F(\infty) = 0. \tag{4.147}$$

Now we can substitute the *ansatz* (4.144) into the energy functional. In this way we arrive at

$$M_{sk} = 4\pi \int_0^\infty r^2 dr \left\{ \frac{F_\pi^2}{2}\left[\left(\frac{\partial F}{\partial r}\right)^2 + 2\frac{\sin^2 F}{r^2}\right] + \frac{1}{2e^2}\frac{\sin^2 F}{r^2}\left[\frac{\sin^2 F}{r^2} + 2\left(\frac{\partial F}{\partial r}\right)^2\right]\right\}$$

$$= \frac{2\pi F_\pi}{e} \int_0^\infty d\rho \left[\rho^2 F'^2 + 2\sin^2 F + \sin^2 F\left(2F'^2 + \frac{\sin^2 F}{\rho^2}\right)\right], \tag{4.148}$$

Skyrmion profile function and mass

where ρ is a dimensionless variable,

$$\rho = eF_\pi r, \tag{4.149}$$

and the prime indicates differentiation over ρ. Note that eF_π scales as N^0.

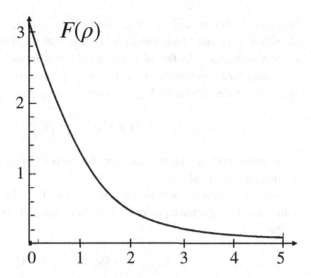

Fig. 4.9
The Skyrme profile function vs. ρ defined in (4.149).

The Skyrme profile function F minimizes the above energy functional, with constraints following from the boundary conditions (4.147). The variational equation in F following from (4.148) is

$$\left(\rho^2 F' + 2F' \sin^2 F\right)' - \sin 2F \left(1 + F'^2 + \frac{1}{\rho^2} \sin^2 F\right) = 0. \qquad (4.150)$$

It was solved numerically in [26]. The plot of $F(\rho)$ is depicted in Fig. 4.9. The corresponding value of the Skyrmion mass is

$$M_{\text{sk}} = 6\pi^2 \frac{F_\pi}{e} \times 1.23 \approx 73 \frac{F_\pi}{e}. \qquad (4.151)$$

4.2.5 Skyrmion Quantization

The complete and consistent calculation of quantum corrections to the classical results described above is impossible because the chiral theory is nonrenormalizable. It is effectively a low-energy theory. In the ultraviolet, quantum corrections are governed by the microscopic theory, QCD.

Despite this we can (and must) quantize Skyrmion moduli (collective coordinates). The classical Skyrmion solution presented in Section 4.2.4 has no definite spin or flavor quantum numbers. One can determine them only upon moduli quantization.

The Skyrmion moduli are in one-to-one correspondence with the zero modes in the Skyrmion background. That such zero modes do exist follows from the symmetries of the theory spontaneously broken by the *ansatz* (4.144). The Skyrmion solution in this particular *ansatz* implies that in fact there is a large family of solutions with shifted origins, or rotated spatial or flavor coordinates.

To perform a spatial translation one makes the replacement $\vec{x} \rightarrow \vec{x} - \vec{x}_0$, where $\vec{x}_0$ plays the role of the soliton center. In the quasiclassical quantization we let $\vec{x}_0$

depend (slowly) on t. Then we substitute $U_0(\vec{x} - \vec{x}_0(t))$ into the expression for the Hamiltonian of our chiral model, assuming that the only time dependence is that coming from $\vec{x}_0(t)$. In the adiabatic approximation we assume $\dot{\vec{x}}_0$ to be small and keep only terms quadratic in $\dot{\vec{x}}_0$. In this way we arrive at a quantum-mechanical Hamiltonian describing the Skyrmion's motion in space,

$$\mathcal{H} = M_{\text{sk}} + \frac{M_{\text{sk}}}{2} \left(\dot{\vec{x}}_0 \right)^2. \tag{4.152}$$

This corresponds to the free motion of a particle in three-dimensional space, and quantization is trivial.

Now let us turn to rotations of the *ansatz* (4.144). First, one can obtain another solution of the Skyrme equation (4.150) by rotating the spatial coordinates in $U_0(\vec{x})$, so that

$$x_i \rightarrow x_i' = O_{ij} x_j, \qquad O \in O(3), \tag{4.153}$$

where O_{ij} is an arbitrary 3×3 orthogonal matrix. Second, one can rotate this field configuration in the SU(2) flavor space (remember, we have two flavors in the case at hand), by applying an arbitrary unitary matrix, so that

$$U_0 \rightarrow A^\dagger U_0 A, \qquad A \in \text{SU}(2), \tag{4.154}$$

where A acts on the indices in the fundamental representation. Each of the matrices O and A involves three parameters. These parameters are not independent, however. Indeed, the hedgehog *ansatz* (4.144) entangles the spatial variables with the flavor variables (through the product $\vec{\tau}\vec{x}$). Therefore, each flavor rotation is equivalent to a spatial rotation. Indeed, each orthogonal (real) matrix O_{ij} can be represented as

$$O_{ij} = \frac{1}{2} \text{Tr} \left(\tau_i \mathcal{B}^\dagger \tau_j \mathcal{B} \right), \tag{4.155}$$

where $\mathcal{B}$ is some unitary 2×2 matrix. If we combine the rotations (4.153) and (4.154) we get

$$A \left(\tau_i \vec{x}_i' \right) A^\dagger = A \tau_i A^\dagger \frac{1}{2} \text{Tr} \left(\tau_i \mathcal{B}^\dagger \tau_j \mathcal{B} \right) x_j = A \mathcal{B}^\dagger \tau_j \mathcal{B} A^\dagger x_j = \tau_j x_j, \tag{4.156}$$

provided that $\mathcal{B} = A$. The "rotated" coordinate $\vec{x}'$ is defined in (4.153). In the above expression we used a formula from page xvi,

$$(\tau^i)_{mn} (\tau^i)_{pq} = 2 \delta_{mq} \delta_{np} - \delta_{mn} \delta_{pq}.$$

Equation (4.155) implies

$$A U_0(O\vec{x}) A^\dagger = U_0(\vec{x}).$$

Thus, the hedgehog *ansatz* is invariant under rotations generated by $\vec{J} + \vec{T}$, where $\vec{J}$ is the spatial rotation generator, while $\vec{T}$ generates rotations in flavor space. In the present case one has three rotational moduli; they can be introduced as three parameters in the matrix A.

Following the standard quasiclassical quantization procedure, we introduce time-dependent collective coordinates $A(t)$ into the solution, i.e., we set

$$U(\vec{x}, t) = A(t)U_0(\vec{x})A^\dagger(t), \tag{4.157}$$

and substitute (4.157) into the Hamiltonian of the chiral model (4.132). The algebra that follows is rather tedious but straightforward. Omitting the intermediate stages we present here the quantum-mechanical Hamiltonian, which includes the rotational degrees of freedom and replaces Eq. (4.152):

$$\mathcal{H} = M_{\text{sk}} + \frac{M_{\text{sk}}}{2}\left(\dot{\vec{x}}_0\right)^2 + \frac{I_{\text{sk}}}{2}\vec{\omega}^2, \tag{4.158}$$

> **QM Hamiltonian for Skyrmions**

where $\vec{\omega}$ is the angular velocity of the Skyrmion,

$$\omega_i = -i\,\text{Tr}\left(\tau^i A^\dagger \dot{A}\right), \qquad A^\dagger \dot{A} = i\omega_j\frac{\tau^j}{2}, \tag{4.159}$$

and I_{sk} is the moment of inertia,

$$I_{\text{sk}} = \frac{\pi}{3}\frac{1}{(eF_\pi)e^2}\lambda,$$

$$\lambda = 8\int_0^\infty d\rho(\sin F)^2\left\{\rho^2 + \left[4\rho^2 F'^2 + (\sin F)^2\right]\right\} \sim 51. \tag{4.160}$$

The rotational part of the Hamiltonian (4.158) is that of a spherical quantum top. The quantization of quantum tops is considered in detail in books on quantum mechanics; see, e.g., [29]. Owing to the fact that the flavor rotations of the Skyrmion are identical to those in space, upon quantization we get only states whose spin J is equal to the isospin T. The rotational energy is

$$E_{\text{rot}} = \frac{J(J+1)}{2I_{\text{sk}}} = \frac{T(T+1)}{2I_{\text{sk}}}. \tag{4.161}$$

> **The factor $1/e^2$ scales as N while $\lambda = O(N^0)$.**

Note that the moment of inertia I_{sk} scales as N, implying that the rotational energies are proportional to $1/N$. The ratio of the rotational energy of the Skyrmion to its mass is $O(1/N^2)$. It is parametrically small at large N, where the (quasiclassical) description of baryons as Skyrmions is valid.

I will outline one possible way of deriving Eq. (4.161) [26]. Any unitary 2×2 matrix can be parametrized as

$$A = a_0 + i\vec{a}\vec{\tau}, \qquad a_0^2 + \vec{a}^2 = 1. \tag{4.162}$$

In terms of the a_i the rotational part of the Hamiltonian (4.158) becomes

$$\mathcal{H}_{\text{rot}} = 2I_{\text{sk}}\sum_{i=0}^{3}(\dot{a}_i)^2. \tag{4.163}$$

To carry out the quantization we express the Hamiltonian in terms of the conjugate momenta $p_i = 4I_{\text{sk}}\dot{a}_i$, make the replacement $p_i \rightarrow -i\partial/\partial a_i$ (which guarantees the appropriate commutation relation $[p_i, a_j] = -i\delta_{ij}$), and obtain

$$\mathcal{H}_{\text{rot}} \overset{?}{=} \frac{1}{8I_{\text{sk}}}\sum_{i=0}^{3}\left(-\frac{\partial^2}{\partial a_i^2}\right). \tag{4.164}$$

The question mark over the equality sign warns us that, because of the constraint

$$a_0^2 + \vec{a}^2 = 1, \tag{4.165}$$

the expression $\sum \partial^2 / \partial a_i^2$ in (4.164) is a symbolic shorthand. In fact, the operator $\sum \partial^2 / \partial a_i^2$ must be understood as the Laplacian on the 3-sphere of a unit radius, $\vec{\nabla}_{S_3}^2$, or, in other words, the angular part of the four-dimensional Laplacian, which can be written as

$$-\vec{\nabla}_{S_3}^2 = -\left\{ \frac{\partial^2}{\partial \theta_1^2} + 2 \cot \theta_1 \frac{\partial}{\partial \theta_1} + \frac{1}{\sin^2 \theta_1} \left[\frac{\partial^2}{\partial \theta_2^2} + \cot \theta_2 \frac{\partial}{\partial \theta_2} \right] \right.$$
$$\left. + \frac{1}{\sin^2 \theta_1 \sin^2 \theta_2} \frac{\partial^2}{\partial \theta_3^2} \right\}. \tag{4.166}$$

The angle variables θ_i ($i = 1, 2, 3$) are defined by

$$a_0 = \cos \theta_1, \qquad a_1 = \cos \theta_2 \sin \theta_1, \qquad a_2 = \cos \theta_3 \sin \theta_1 \sin \theta_2,$$
$$a_3 = \sin \theta_1 \sin \theta_2 \sin \theta_3, \qquad 0 \le \theta_{1,2} \le \pi, \qquad 0 \le \theta_3 \le 2\pi. \tag{4.167}$$

The eigenfunctions of the operator in (4.166) are components of homogeneous traceless polynomials of the type $a_i a_j \cdots a_k$ – traces. If the order of the polynomial is ℓ then the eigenvalue is $\ell(\ell + 2)$, which can be checked by a straightforward inspection.

The rotational collective coordinates are introduced in the Skyrme *ansatz* via $U_0 \to U = A U_0 A^\dagger$. If we change the sign of A (or, equivalently, set $a_i \to -a_i$) we get the same U. Naively, one might expect that only symmetric eigenfunctions, $\psi(a_i) = \psi(-a_i)$, should be kept. Actually, Finkelstein and Rubinstein demonstrated [24] that there are two consistent ways to quantize the given soliton: either symmetric eigenfunctions $\psi(a_i) = \psi(-a_i)$ are required for all solitons or antisymmetric eigenfunctions $\psi(a_i) = -\psi(-a_i)$ are required for all solitons. In the former case ℓ is even, and the soliton is quantized as a boson. In the latter case ℓ is odd, and the soliton is quantized as a fermion.

Combining Eqs. (4.164) and (4.166) we conclude that

$$\mathcal{H}_{\text{rot}} = \frac{1}{8 I_{\text{sk}}} \left(-\vec{\nabla}_{S_3}^2 \right) \longrightarrow \frac{1}{2 I_{\text{sk}}} \frac{\ell}{2} \left(\frac{\ell}{2} + 1 \right). \tag{4.168}$$

This coincides with Eq. (4.161) provided that we set

Remember that $I_{\text{sk}} \sim N$ at large N. If $N \to \infty$, all states $J = T = \frac{1}{2}, \frac{3}{2}, \ldots$ are degenerate; cf. Section 9.2.10.

$$\frac{\ell}{2} = J = T. \tag{4.169}$$

The mass splitting between the states $J = T = 1/2$ (nucleons) and $J = T = 3/2 (\Delta s)$ is

$$\Delta M = \frac{3}{2 I_{\text{sk}}}. \tag{4.170}$$

4.2.6 Some Numerical Results

Cf. Section 9.2.10.

Some numerical results will be presented here. The reader should be warned that we do not expect too precise an agreement with the data. The reason is obvious. The parameter justifying our quasiclassical treatment is $1/N$. For $N = 3$ one can expect that dimensionless expressions of the order of $1/N$ are ~ 0.3 and those of the order of $1/N^2$ are ~ 0.1. As we will see soon, the ratio $\Delta M/M_{sk}$, which is theoretically of the order of $1/N^2$, is in fact ~ 0.3.

Experimentally the Δ-proton mass difference is ~ 290 MeV. Substituting this number into Eq. (4.170) and using Eq. (4.160) and $F_\pi \sim 92$ MeV we get

$$e \sim 4.8. \qquad (4.171)$$

Equation (4.151) now implies that $M_{sk} \sim 1.46$ GeV, to be compared with $M_{p,n} \sim 0.94$ GeV.

We see that the numbers come out reasonably, although the agreement is not precise. Some other quantities, such as the charge radii and magnetic moments, were calculated and analyzed in [26] following the same lines of reasoning. Qualitatively the description of baryons as Skyrmions comes out correctly, although some theoretical numbers deviate from their experimental counterparts by ~30% or 40%. Discrepancies of this order of magnitude are to be expected.

4.2.7 The WZNW Term

Now let us pass to the more general case of three or more massless flavors. To explain the problem occurring at $n \geq 3$ it is sufficient to consider $n = 3$. It turns out that in this case the effective chiral Lagrangian (4.132), including the two terms (4.130) and (4.131), cannot be complete since it misses an important part of the interaction intrinsic to the Goldstone bosons in the problem at hand.

Expanding the above chiral Lagrangian in powers of the Goldstone fields, it is not difficult to see that the set of amplitudes that this Lagrangian generates conserves the number of bosons modulo 2, i.e., two bosons in the initial state can produce only two, four, six, and so on bosons in the final state. This is certainly true if we apply the low-energy chiral description to the pion triplet. However, with three massless flavors, when the Goldstone bosons form an SU(3)$_{flavor}$ octet rather than a triplet, this is no longer true. The simplest example violating the above rule is the allowed process $K^+ K^- \to \pi^+ \pi^- \pi^0$. What must be done in order to amend the Lagrangian (4.132) appropriately?

Here I will only outline the answer to this question without delving into a number of very interesting aspects, which we do not have the space to discuss here.

It turns out that the terms we need to add, when written as four-dimensional local operators, form an infinite series. To sum them and express the result in a compact form as a single operator, one needs to leave four dimensions and pass to a five-dimensional space [22]. Let us imagine our space–time as a very large four-dimensional sphere $\mathcal{M}$. A given field configuration U represents a mapping of $\mathcal{M}$ into the group manifold of SU(3) (remember that in the case under discussion U is a 3×3 unitary matrix belonging to SU(3)) (Fig. 4.10a).

Topological formula for the fourth homotopy group

Since $\pi_4(\mathrm{SU}(3)) = 0$, the 4-sphere in SU(3) defined by the mapping $U(x)$ is the boundary of a five-dimensional disc Q (Fig. 4.10b).

On the SU(3) manifold there is a unique fifth-rank antisymmetric tensor ω_{ijklm}, which is invariant under SU(3)$_L \times$ SU(3)$_R$ [22],

$$\omega_{ijklm} = -\frac{i}{240\pi^2} \mathrm{Tr}\left(U^\dagger \frac{\partial U}{\partial y^i} U^\dagger \frac{\partial U}{\partial y^j} U^\dagger \frac{\partial U}{\partial y^k} U^\dagger \frac{\partial U}{\partial y^l} U^\dagger \frac{\partial U}{\partial y^m} \right), \qquad (4.172)$$

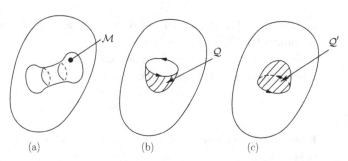

Fig. 4.10   Space–time, imagined as a 4-sphere, is mapped into the SU(3) manifold. In part (a), space–time is symbolically denoted as a 2-sphere. In parts (b) and (c), space–time is reduced to a circle that bounds the discs Q and Q'. The SU(3) manifold is symbolized by the interior of the region represented by the large oval.

where the y^i ($i = 1, 2, \ldots, 5$) are coordinates on the disc Q. The normalization factor $-i/(240\pi^2)$ is derived as follows.

Define the functional

$$\Gamma = \int_Q \omega_{ijklm} d\,\Sigma^{ijklm}, \tag{4.173}$$

where $d\Sigma^{ijklm}$ is an element of the disc area, with the intention of including $i\Gamma$ in the action of the chiral model, i.e., using $\exp(i\Gamma)$ as an additional weight factor in the Feynman path integrals defining the amplitudes of the chiral model.

It is clear that the disc Q is not unique. The mapping of the four-sphere $\mathcal{M}$ is also the boundary of another five-dimensional disc Q' (Fig. 4.10c). If we introduce[16]

$$\Gamma' = -\int_{Q'} \omega_{ijklm} d\,\Sigma^{ijklm} \tag{4.174}$$

then we must require that

$$e^{i\Gamma} = e^{i\Gamma'}, \tag{4.175}$$

implying that

$$\int_{Q+Q'} \omega_{ijklm} d\,\Sigma^{ijklm} = 2\pi \times \text{integer}. \tag{4.176}$$

Equation (4.176) must be valid for an integral taken over any five-dimensional sphere in the eight-dimensional SU(3) manifold, since $Q + Q'$ is in fact a closed five-dimensional sphere (Fig. 4.10).

The topological classification of mappings of the five-dimensional sphere into SU(3) is based on the fact that

Topological formula for the fifth homotopy group	$$\pi_5(\text{SU}(3)) = \mathbb{Z}. \tag{4.177}$$

There is a trivial mapping and also a mapping in which, if a five-sphere is swept once, its image in SU(3) is also swept once (a basic topologically nontrivial

---

[16] The minus sign in Eq. (4.174) is due to the fact that now the orientation of the boundary is opposite to that in Eq. (4.173).

mapping). The coefficient in Eq. (4.172) was chosen in such a way that, for the basic mapping,

$$\int_{S_0} \omega_{ijklm} d\Sigma^{ijklm} = 2\pi. \tag{4.178}$$

The action of the chiral model takes the form

$$S = \int d^4x \left(\mathcal{L}^{(2)} + \mathcal{L}^{(4)} \right) + \nu\Gamma. \tag{4.179}$$

The last term is referred to as the WZNW term; the coefficient ν at this level is an arbitrary integer number. In Section 4.2.8 we will see, after establishing contact with QCD, that $\nu = N$, where N is the number of colors.

In SU(3) the matrix field U is parametrized as

$$U(x) = \exp\left(\frac{i}{F_\pi} \pi(x) \right) \equiv \exp\left(\frac{i}{F_\pi} \pi^a(x) \lambda^a \right), \qquad U \in \text{SU(3)}, \tag{4.180}$$

where the λ^a are the Gell-Mann matrices. Then $U^\dagger \partial_i U = (i/F_\pi)\partial_i \pi + O(\pi^2)$ and

$$\begin{aligned}
\omega_{ijklm} d\Sigma^{ijklm} &= \frac{1}{240\pi^2 F_\pi^5} d\Sigma^{ijklm} \, \text{Tr}\left[\partial_i \pi \partial_j \pi \partial_k \pi \partial_l \pi \partial_m \pi + O(\pi^6) \right] \\
&= \frac{1}{240\pi^2 F_\pi^5} d\Sigma^{ijklm} \, \text{Tr}\left[\partial_i \left(\pi \partial_j \pi \partial_k \pi \partial_l \pi \partial_m \pi \right) + O(\pi^6) \right]. \tag{4.181}
\end{aligned}$$

The WZNW term is an integral over a full derivative. Equation (4.181) demonstrates this only to order $O(\pi^5)$, but in fact it is valid at higher orders also. Then by Stokes' theorem the WZNW term can be expressed as an integral over the boundary of Q. This boundary is our four-dimensional space–time, by construction,

$$\Gamma = \frac{1}{240\pi^2 F_\pi^5} \int d^4x \, \varepsilon^{\mu\nu\alpha\beta} \, \text{Tr}\left[\pi \partial_\mu \pi \partial_\nu \pi \partial_\alpha \pi \partial_\beta \pi + O(\pi^6) \right]. \tag{4.182}$$

We see that the WZNW term reduces to an infinite series of local four-dimensional operators, as mentioned above.

Now, assuming that $\nu = N$ let us determine whether the soliton is a boson or a fermion. To this end, following Witten [25] we will compare the amplitudes for two processes. First we consider a soliton sitting at rest, at a certain point in space from time 0 until time T, where T is a very large parameter (at the very end we can let $T \to \infty$). Second, we consider a process in which the soliton is adiabatically rotated by 2π during the same time interval. The first amplitude is obviously $\exp(-iM_{\text{sk}}T)$. To determine the second amplitude it is worth noting that in the limit $T \to \infty$ neither $\mathcal{L}^{(2)}$ nor $\mathcal{L}^{(4)}$ contribute to this amplitude, because these terms in the chiral Lagrangian are second order in the time derivative, while integration of the action produces only the first power of T. However, the WZNW term is of first order in the time derivative. Therefore it distinguishes between a soliton sitting at rest and a soliton adiabatically rotated by 2π. Obviously, for the soliton at rest $\Gamma = 0$ while for the adiabatically rotated soliton $\Gamma = \pi$ [25]. Thus, the corresponding amplitude is

$$e^{-iM_{\text{sk}}T + iN\Gamma} = (-1)^N e^{-iM_{\text{sk}}T}, \tag{4.183}$$

implying that the Skyrmion is of necessity a fermion for all odd N (in particular, $N = 3$).

4.2.8 Determining ν

Our task in this section is to prove that the integer ν in the WZNW term in (4.179) coincides with the number of colors in the underlying microscopic theory, QCD. To this end we will step aside, to generalize the WZNW term to include electromagnetic interactions. Thus, we will derive a low-energy effective Lagrangian that describes not only Goldstone boson interactions but also those involving photons.

We start by introducing a 3×3 charge matrix Q of quarks:

$$Q = \begin{pmatrix} \frac{2}{3} & 0 & 0 \\ 0 & -\frac{1}{3} & 0 \\ 0 & 0 & -\frac{1}{3} \end{pmatrix} \tag{4.184}$$

with Q acting on the quark flavor triplet as

$$Q \begin{pmatrix} u \\ d \\ s \end{pmatrix} = \begin{pmatrix} \frac{2}{3}u \\ -\frac{1}{3}d \\ -\frac{1}{3}s \end{pmatrix}.$$

The action of Q on the matrix U must be defined as the commutator $[Q, U]$. Then automatically the charges of the charged mesons are ± 1 and those of neutral mesons are 0.

It is not difficult to check that the action (4.179) is invariant under the *global* charge rotation $U \rightarrow \exp(i \epsilon Q) U \exp(-i \epsilon Q)$, which for small rotations takes the form

$$U \rightarrow U + i\epsilon [Q, U], \tag{4.185}$$

where ϵ is a constant rotation parameter. We need to promote the above global symmetry to a gauge U(1) symmetry also described by (4.185) but where the parameter ϵ is an arbitrary function of x,

$$\epsilon \rightarrow \epsilon(x).$$

To this end we introduce into the theory the photon field A_μ, which is coupled to the matrix U through the covariant derivative

$$i\partial_\mu U \rightarrow i\mathcal{D}_\mu U \equiv i\partial_\mu U + eA_\mu[Q, U] \tag{4.186}$$

and transforms under the gauge transformation as [17]

$$A_\mu \rightarrow A_\mu + \frac{1}{e}\partial_\mu \epsilon. \tag{4.187}$$

Upon the replacement $\partial_\mu \rightarrow \mathcal{D}_\mu$, the $\mathcal{L}^{(2)} + \mathcal{L}^{(4)}$ part of the action (4.179) becomes gauge invariant. However, this does not work for the WZNW term Γ. Under the local gauge rotation (4.185) we have

---

[17] The covariant derivative in Eqs. (4.186), (4.187) is introduced with the sign in front of e opposite to that in Witten's paper [25]. This changes the sign in front of e in (4.189).

$$\Gamma \rightarrow \Gamma - \int d^4x (\partial_\mu \epsilon) J^\mu,$$

$$J^\mu = \frac{1}{48\pi^2} \varepsilon^{\mu\nu\alpha\beta} \operatorname{Tr} \left[Q(\partial_\nu UU^\dagger) \left(\partial_\alpha UU^\dagger \right) \left(\partial_\beta UU^\dagger \right) \right.$$
$$\left. + Q \left(U^\dagger \partial_\nu U \right) \left(U^\dagger \partial_\alpha U \right) \left(U^\dagger \partial_\beta U \right) \right].$$ (4.188)

Using this transformation law one can check that the functional

$$\tilde{\Gamma}(U, A_\mu) = \Gamma(U) + e \int d^4x A_\mu J^\mu + \frac{ie^2}{24\pi^2} \int d^4x \varepsilon^{\mu\nu\alpha\beta} \left(\partial_\mu A_\nu \right) A_\alpha$$
$$\times \operatorname{Tr} \left[Q^2 (\partial_\beta U) U^\dagger + Q^2 U^\dagger \left(\partial_\beta U \right) + QU Q U^\dagger \left(\partial_\beta U \right) U^\dagger \right] \quad (4.189)$$

is gauge invariant.

Thus, replacing (4.179) by

$$\tilde{S} = \int d^4x \left\{ \frac{F_\pi^2}{4} \operatorname{Tr} \left(\mathcal{D}_\mu U \mathcal{D}^\mu U^\dagger \right) + \frac{1}{32e^2} \operatorname{Tr} \left[\left(\mathcal{D}_\mu U \right) U^\dagger, (\mathcal{D}_\nu U) U^\dagger \right]^2 \right\} + v\tilde{\Gamma}$$
(4.190)

we get an effective low-energy action that includes electromagnetism.

How does this help to establish the value of v? It does so in a rather simple way. Indeed the term $v\tilde{\Gamma}$, among others, contains the $\pi^0 \rightarrow \gamma\gamma$ amplitude, which can be obtained by expanding U to first order in $\pi(x)$ and integrating by parts; the result is

$$\mathcal{A}(\pi^0 \rightarrow \gamma\gamma) = \frac{ve^2}{96\pi^2 F_\pi} \pi^0 \varepsilon^{\mu\nu\alpha\beta} F_{\mu\nu} F_{\alpha\beta}. \quad (4.191)$$

However, the famous calculation of this decay from the quark triangle anomaly [30] yields the same amplitude but with $v = N$. This calculation is discussed in detail in Chapter 8; cf. Eqs. (8.83) and (8.86). One must take into account the fact that $F_\pi = f_\pi/\sqrt{2}$.

Returning to the decay $K^+ K^- \rightarrow \pi^+\pi^-\pi^0$ (Section 4.2.7), which was the primary motivation for the introduction of the WZNW term, it is rather curious to note that its amplitude must be (in units given in Eq. (4.182)) an integer number equal to the number of colors in QCD.

Chiral theory + photons

For those who may have forgotten: π^0 is the Goldstone (pseudoscalar) boson with quark content $\bar{u}u - \bar{d}d$.

4.2.9 Beyond the Conventional

The quasiclassical treatment of Skyrmions is parametrically justified in the limit $N \gg 1$. Nevertheless, in our actual world $N = 3$; in our world QCD is the theory of quarks in the fundamental representation of SU(3) interacting through the octet of non-Abelian gauge bosons. One may ask whether analytic continuation from $N = 3$ to large N is unique. The answer to this question is negative.

The standard procedure [31], well known under the name of the *'t Hooft large-N limit* (Section 9.2), is as follows. The gauge group SU(3) is replaced by SU(N), the quarks are assigned to the fundamental representations of SU(N), and N is sent to infinity while the 't Hooft coupling $\lambda \equiv g^2 N$ is kept fixed.

Instead, however, one can consider an alternative limit in which the quarks $Q^{[\alpha\beta]}$ are assumed to be in the two-index antisymmetric representation of SU(N)$_{\text{gauge}}$.

At $N = 3$ the two-index antisymmetric quark is identical to the quark in the (anti)fundamental representation:

$$Q^{[\alpha\beta]} \sim \varepsilon^{\alpha\beta\gamma} q_\gamma.$$

At $N > 3$ the above relation between the two-index antisymmetric and fundamental representations no longer holds, and we arrive[18] at a different large-N limit [32]. Unlike the 't Hooft limit, it does not discard fermion loops.

Assume we have a few quark flavors (say, two or three) in the two-index antisymmetric representation of color.[19] Since the fermion fields are Dirac and in the complex representation of the gauge group, the theory has the same chiral symmetry as QCD for n flavors of fundamental quarks, namely, $\mathrm{SU}(n)_L \times \mathrm{SU}(n)_R$, and it is spontaneously broken in the same way,[20]

$$\mathrm{SU}(n)_L \times \mathrm{SU}(n)_R \to \mathrm{SU}(n)_V. \tag{4.192}$$

Therefore the low-energy chiral Lagrangian must have the same structure as that discussed earlier in this section, including the WZNW term at $n = 3$. In particular, it supports topologically stable solitons, i.e., Skyrmions, which are already very familiar to us. There is an important parametric distinction, however.

In the 't Hooft large-N limit the constants F_π and $1/e$ in the chiral Lagrangian scale as $\sqrt{N}$ but now, for the two-index antisymmetric quarks, they scale as N. Moreover, the coefficient in front of the WZNW term also changes. Previously $\nu = N$, but now one can readily convince oneself that[21]

$$\nu_{Q^{[\alpha\beta]}} = \frac{N(N-1)}{2}. \tag{4.193}$$

Under these circumstances the Skyrmions will have a mass scaling as $M_{\mathrm{sk}} \sim N^2$, and their statistics will be determined by the factor $(-1)^{N(N-1)/2}$ [34]. If we identify them with baryons in this model, the relation between Skyrmions and the quark picture of baryons becomes questionable.

Indeed, the simplest color-singlet composite hadron of the baryon type can be built of $N/2$ quarks as follows:

$$\varepsilon_{\alpha_1\alpha_2...\alpha_N} Q^{[\alpha_1\alpha_2]} \cdots Q^{[\alpha_{N-1}\alpha_N]}. \tag{4.194}$$

Here we limit ourselves to one of four possible cases, namely, that with N even and $N/2$ odd. The other cases can be considered in a similar manner. If N is even and $N/2$ is odd then $N(N-1)/2$ is odd too. The smallest value of N falling into this class is $N = 6$.

Upon inspecting (4.194) one might conclude that the baryon mass must be proportional to $N/2$. This is a wrong conclusion, however.

---

[18] In [32] it was suggested that one should refer to this limit as the *orientifold* large-N limit, for reasons which need not concern us here.

[19] The number of such quarks n cannot be larger than five since at $n > 5$ asymptotic freedom of the theory is lost.

[20] Arguments in favor of this pattern of chiral symmetry breaking can be found in [33].

[21] One can obtain this equality using the same derivation as that of Section 4.2.8. Only the last step is different: in the triangle anomaly responsible for $\pi^0 \to \gamma\gamma$ one must replace N by $N(N-1)/2$.

For quarks in the the fundamental representation of SU(N) the color wave function is antisymmetric, which allows all these to be in the S wave in coordinate space. For antisymmetric two-index spinor fields the color wave function (4.194) is symmetric, which requires the spinors to occupy "orbits" with angular momentum up to $\sim N/2$. The ground state of such a hadron is a degenerate Fermi gas; it is obtained by filling all the lowest energy states up to the Fermi surface. This is a more complicated dynamical situation in which one may expect that the quasiclassical Skyrme description fails even despite the fact that the pattern of the chiral symmetry breaking is the same as in QCD with fundamental quarks. If the number of flavors of the $Q^{[\alpha_i \alpha_j]}$ quarks could be $\sim N$ this would save the Skyrme model for the baryons fully antisymmetric in flavor. Alas ... As I have mentioned previously, the maximal number of flavors is five – otherwise we lose asymptotic freedom.

I should add that "Skyrmions" are cutrrently widely discussed in condensed matter physics, especially in magnetic phenomena. Unlike Skyrmions in high energy physics, in condensed matter all solitons whose stability is based on $\pi_n(S^n) = \mathbb{Z}$ are called Skyrmyons.

Exercises

4.2.1 Prove that the current (4.134) is conserved topologically (i.e., one does not need to use equations of motion in the proof) and that $B \equiv 0$ order by order in the expansion of (4.128) in the fields π, assuming that $|\pi| \ll 1$ and $\pi(x) \to 0$ as $|\vec{x}| \to \infty$.

4.2.2 Prove Eq. (4.155).

4.2.3 Prove the gauge invariance of the functional (4.189).

4.2.4 Derive Eq. (4.146).

4.3 Appendix: Elements of Group Theory for SU(*N*)

The topic to be discussed below is covered in the physicist-oriented texts on group theory cited in [9].

The $(N-1)$-component root vectors $\boldsymbol{\alpha} = \{\alpha_1, \alpha_2, \ldots, \alpha_{N-1}\}$ and $-\boldsymbol{\alpha}$ are defined by

$$[H_i, E_\alpha] = \alpha_i E_\alpha, \qquad \left[H_i, E_\alpha^\dagger\right] = -\alpha_i E_\alpha^\dagger, \qquad (4.195)$$

where the Hermitian conjugate of E_α is defined by

$$E_\alpha^\dagger = E_{-\alpha}.$$

It is customary to normalize the generators in the following way:

$$\mathrm{Tr}\left(H_i\, H_j\right) = \tfrac{1}{2}\delta_{ij}, \qquad \mathrm{Tr}\left(E_\alpha^\dagger E_\beta\right) = \tfrac{1}{2}\delta_{\alpha\beta}. \qquad (4.196)$$

Then all the root vectors, the total number of which is $N(N - 1)$, are normalized to unity:

$$\alpha^2 = 1. \tag{4.197}$$

Moreover, one can show that for each α

$$[E_\alpha, E_{-\alpha}] = \alpha_i H_i. \tag{4.198}$$

If $\alpha + \beta \neq 0$ but $\alpha + \beta$ is a root then

$$\left[E_\alpha, E_\beta\right] \propto E_{\alpha+\beta}; \tag{4.199}$$

otherwise $\left[E_\alpha, E_\beta\right] = 0$.

Positive vs. negative roots. Simple roots

It is convenient to divide all the roots into two halves, positive and negative. For instance, one can define the positive roots as the set of root vectors such that the *first* nonzero component of every vector is positive. Alternatively, one can choose to call a root positive if its *last* nonzero component is positive. This gives an arbitrary division of the space into two halves. It is important that every root is either positive or negative. In our notation the αs are positive roots and the $-\alpha$ are negative.

In addition, the notion that we need here is that of *simple roots*. A simple root is a positive root which cannot be written as the sum of two positive roots. There are $N - 1$ simple roots in SU(N) – let us call them γ – and they are linearly independent. Any positive root α can be written as a sum of simple roots γ with *non-negative integer* coefficients k_γ,

$$\alpha = \sum_\gamma k_\gamma \gamma. \tag{4.200}$$

Needless to say, not all possible combinations $\sum k_\gamma \gamma$ with non-negative integer coefficients are roots (we have $N(N - 1)/2$ positive roots in SU(N)). A possible set of simple roots in SU(N) is

$$\gamma^1 = \{1, 0, 0, 0, \ldots, 0\},$$

$$\gamma^2 = \left\{-\frac{1}{2}, \frac{\sqrt{3}}{2}, 0, 0, \ldots, 0\right\},$$

$$\gamma^3 = \left\{0, -\frac{\sqrt{1}}{\sqrt{3}}, \sqrt{\frac{2}{3}}, 0, \ldots, 0\right\},$$

$$\vdots$$

$$\gamma^m = \left\{0, 0, \ldots, -\sqrt{\frac{m-1}{2m}}, \sqrt{\frac{m+1}{2m}}, \ldots, 0\right\},$$

$$\vdots$$

$$\gamma^{N-1} = \left\{0, 0, \ldots, -\sqrt{\frac{N-2}{2(N-1)}}, \sqrt{\frac{N}{2(N-1)}}\right\}. \tag{4.201}$$

The angle between all neighboring simple-root vectors is 120°, while non-neighboring simple-root vectors are perpendicular. This is indicated in the Dynkin diagram in Fig. 4.11.

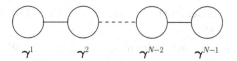

$\gamma^1 \qquad \gamma^2 \qquad \gamma^{N-2} \qquad \gamma^{N-1}$

Fig. 4.11   The Dynkin diagram for SU(N).

References for Chapter 4

[1] G. 't Hooft, *Nucl. Phys. B* **79**, 276 (1974).

[2] A. M. Polyakov, *Pisma Zh. Eksp. Teor. Fiz.* **20**, 430 (1974) [Engl. transl. *JETP Lett.* **20**, 194 (1974), reprinted in C. Rebbi and G. Soliani (eds.), *Solitons and Particles* (World Scientific, Singapore, 1984), p. 522].

[3] H. Georgi and S. L. Glashow, *Phys. Rev. Lett.* **28**, 1494 (1972).

[4] P. A. M. Dirac, *Proc. Roy. Soc. A* **133**, 60 (1931).

[5] E. Mottola, *Phys. Lett. B* **79**, 242 (1978) (E). Erratum: *ibid.* **80**, 433 (1979).

[6] E. Bogomol'nyi and M. Marinov, *Sov. J. Nucl. Phys.* **23**, 355 (1976). Usually people refer to T. W. Kirkman and C. K. Zachos, *Phys. Rev. D* **24**, 999 (1981), but the work of Bogomol'nyi and Marinov was much earlier.

[7] The classical work on the Dirac monopole is T. T. Wu and C. M. Yang, *Phys. Rev. D* **12**, 3845 (1975).

[8] S. R. Coleman, The magnetic monopole: fifty years later, in A. Zichichi (ed.), *The Unity of the Fundamental Interactions, Proc. 1981 Int. School of Subnuclear Physics*, Erice, Italy (Plenum Press, New York, 1983).

[9] H. Georgi, *Lie Algebras in Particle Physics* (Benjamin/Cummings, Menlo Park, 1982); Second Edition (Westview Press, 1999); P. Ramond, *Group Theory* (Cambridge University Press, 2010).

[10] E. Witten, *Phys. Lett. B* **86**, 283 (1979).

[11] A. Smilga, *Lectures on Quantum Chromodynamics* (World Scientific, Singapore, 2001).

[12] R. Jackiw and C. Rebbi, *Phys. Rev. Lett.* **37**, 172 (1976); C. G. Callan, R. F. Dashen, and D. J. Gross, *Phys. Lett. B* **63**, 334 (1976) [reprinted in M. Shifman (ed.), *Instantons in Gauge Theories* (World Scientific, Singapore, 1994)].

[13] V. A. Rubakov, *Nucl. Phys. B* **203**, 311 (1982); C. G. Callan, *Nucl. Phys. B* **212**, 391 (1983); C. G. Callan and E. Witten, *Nucl. Phys. B* **239**, 161 (1984).

[14] R. Jackiw and C. Rebbi, *Phys. Rev. D* **13**, 3398 (1976) [reprinted in C. Rebbi and G. Soliani (eds.), *Solitons and Particles* (World Scientific, Singapore, 1984), p. 331].

[15] J. A. Harvey, *Phys. Lett. B* **131**, 104 (1983).

[16] C. Callias, *Commun. Math. Phys.* **62**, 213 (1978).

[17] E. Poppitz and M. Ünsal, *JHEP* **0903**, 027 (2009) [arXiv:0812.2085 [hep-th]].

[18] G. 't Hooft, *Nucl. Phys. B* **72**, 461 (1974).

[19] Y. Nambu and G. Jona-Lasinio, *Phys. Rev.* **122**, 345 (1961); *Phys. Rev.* **124**, 246 (1961); V. G. Vaks and A. I. Larkin, *Sov. Phys. JETP*, **13**, 192 (1961).

[20] M. Gell-Mann and M. Levy, *Nuovo Cim.* **16**, 705 (1960).

[21] J. Gasser and H. Leutwyler, *Ann. Phys.* **158**, 142 (1984); *Nucl. Phys. B* **250**, 465 (1985).

[22] J. Wess and B. Zumino, *Phys. Lett. B* **37**, 95 (1971); S. P. Novikov, *Dokl. Akad. Nauk SSSR Section Matem.* **260** 31 (1981) [*Sov. Math. Doklady*, **24**, 22 (1981)]; E. Witten, *Nucl. Phys. B* **223**, 422 (1983).

[23] T. H. R. Skyrme, *Proc. Roy. Soc. Lond. A* **260**, 127 (1961); *Nucl. Phys.* **31**, 556 (1962); *J. Math. Phys.* **12**, 1735 (1971); *Int. J. Mod. Phys. A* **3**, 2745 (1988).

[24] D. Finkelstein and J. Rubinstein, *J. Math. Phys.* **9**, 1762 (1968).

[25] E. Witten, *Nucl. Phys. B* **223**, 422 (1983); *Nucl. Phys. B* **223**, 433 (1983) [Reprinted in C. Rebbi and G. Soliani (eds.), *Solitons and Particles* (World Scientific, Singapore, 1984), p. 617].

[26] G. S. Adkins, C. R. Nappi, and E. Witten, *Nucl. Phys. B* **228**, 552 (1983).

[27] A. P. Balachandran, V. P. Nair, S. G. Rajeev, and A. Stern, *Phys. Rev. Lett.* **49**, 1124 (1982). Erratum: *ibid*. **50**, 1630 (1983); *Phys. Rev. D* **27**, 1153 (1983). Erratum: *ibid*. **27**, 2772 (1983).

[28] G. H. Derrick, *J. Math. Phys.* **5**, 1252 (1964).

[29] L. D. Landau and E. M. Lifshitz, *Quantum Mechanics: Non-Relativistic Theory*, Third Edition (Butterworth–Heinemann, Oxford, 1981).

[30] S. L. Adler, *Phys. Rev.* **177**, 2426 (1969); J. S. Bell and R. Jackiw, *Nuovo Cim. A* **60**, 47 (1969); W. A. Bardeen, *Phys. Rev.* **184**, 1848 (1969); see also the book S. B. Treiman, E. Witten, R. Jackiw, and B. Zumino, *Current Algebra and Anomalies* (World Scientific, Singapore, 1985).

[31] G. 't Hooft, *Nucl. Phys. B* **72**, 461 (1974).

[32] A. Armoni, M. Shifman, and G. Veneziano, *Phys. Rev. Lett.* **91**, 191601 (2003) [arXiv: hep-th/0307097].

[33] S. Dimopoulos, *Nucl. Phys. B* **168**, 69 (1980); M. E. Peskin, *Nucl. Phys. B* **175**, 197 (1980); Y. I. Kogan, M. A. Shifman, and M. I. Vysotsky, *Sov. J. Nucl. Phys.* **42**, 318 (1985); J. J. Verbaarschot, *Phys. Rev. Lett.* **72**, 2531 (1994) [hep-th/9401059]; A. Smilga and J. J. Verbaarschot, *Phys. Rev. D* **51**, 829 (1995) [hep-th/9404031]; M. A. Halasz and J. J. Verbaarschot, *Phys. Rev. D* **52**, 2563 (1995) [hep-th/9502096].

[34] A. Armoni and M. Shifman, *Nucl. Phys. B* **670**, 148 (2003) [arXiv:hep-th/0303109].

5 Instantons

Dealing with tunneling processes in field theory. — Transition to the Euclidean space-time. — Nontriviality of the third homotopy group in Yang–Mills. — Everything you need to know about the Belavin–Polyakov–Schwartz–Tyupkin instanton. — Instanton-induced baryon number violation in the standard model. — What is the holy grail function?

In previous chapters we advanced along the road of increasing codimensions: from codimension 1, for domain walls, to codimension 3 for monopoles and Skyrmions. These objects were considered in the static limit. Now we will pass to objects with codimension 4: instantons [1]. It is clear that in four-dimensional space–time static objects cannot have codimension 4. Thus instanton solutions depend on time (albeit Euclidean time). Physical phenomena whose understanding requires instantons are drastically different from those discussed previously. Instantons appear in problems in which there is *tunneling* between (energy-degenerate) field-theoretic states separated by a barrier [2–4]. Such problems are common in quantum mechanics (e.g., the famous double-well potential), where they can be solved in a number of different ways. In four-dimensional field theories, instanton calculus becomes essentially the only feasible method applicable. What are the physical implications of instantons?

Instantons describe tunneling in quasiclassical approximation.

First and foremost, instantons reveal a nontrivial vacuum structure in non-Abelian gauge theories, i.e., the existence of a vacuum angle θ and of the so-called θ vacuum. In Yang–Mills theories with massless fermions (quarks), instantons explain the nonconservation of the flavor-singlet axial current. This nonconservation was a great mystery in QCD before the discovery of instantons [5]. And, finally, in theories with chiral fermions such as the standard model, tunneling in the θ vacuum described by instantons gives rise to baryon number violation [6]. The baryon-number-violating processes due to instantons possess a remarkable property: their cross sections grow exponentially with energy [7]. How high can the exponential enhancement factor grow? In a bid to answer this question an interesting phenomenon was discovered [8–10] referred to as "premature unitarization." All these topics will be discussed in this chapter. We will not consider instanton-based models of the QCD vacuum (such as the instanton liquid model, which is thoroughly presented in [11]). Crucial instanton-induced effects in some supersymmetric theories will be covered in Part II. Two very detailed introductory articles on instantons [12, 13] can be recommended[1] to those readers who want to familiarize themselves further with the related ideas, techniques, and developments.

5.1 Tunneling in Non-Abelian Yang–Mills Theory

Instantons are localized objects in four-dimensional (Euclidean) space–time. Originally Polyakov suggested the name "pseudoparticles," which did not take root, however, and now is used rather rarely. The term "instantons" was suggested by 't Hooft. The physical role of instantons is as follows. In the quasiclassical approximation they describe the least-action trajectory (in Euclidean time) that connects two distinct energy-degenerate states in the space of fields. The initial point of the instanton trajectory at $t = -\infty$ is one such state, while the final point at $t = \infty$ is another such state. Naturally, instantons are present only in those theories in which energy-degenerate states in the space of fields exist. They minimize the (Euclidean) action, under the given boundary conditions. Therefore, instantons present classical

---

[1] In fact, a significant part of this chapter is an adaptation of several sections from [13]. For superinstanton calculus see Section 10.19.

solutions of the Euclidean equations of motion. In fact, as we will see shortly, they are Bogomol'nyri–Prasad–Sommerfield (BPS) objects satisfying the so-called duality equations [5]. In non-Abelian gauge theories they were discovered by Belavin, Polyakov, Schwarz, and Tyupkin [5] and are usually referred to as BPST instantons.

First we will consider pure Yang–Mills theory for the gauge group SU(N). For pedagogical reasons we will mostly focus on SU(2). In QCD the gauge group is SU(3). The fermion fields (quarks) will be incorporated later. At that stage we will pass from SU(2) to SU(3).

5.1.1 Nontrivial Topology in the Space of Fields in Yang–Mills Theories

The Yang–Mills Lagrangian has the form[2]

$$\mathcal{L} = -\frac{1}{4}G^a_{\mu\nu}G^a_{\mu\nu} \tag{5.1}$$

where $G_{\mu\nu}$ is the gluon field strength tensor,

$$G^a_{\mu\nu} = \partial_\mu A^a_\nu - \partial_\nu A^a_\mu + g\,f^{abc}A^b_\mu A^c_\nu, \tag{5.2}$$

g is the gauge coupling constant, and f^{abc} is a structure constant of the gauge group. For SU(2),

$$f^{abc} = \varepsilon^{abc}, \qquad a,b,c = 1,2,3.$$

The issue to be discussed in this section is independent of the particular choice of gauge group.

The first question to be asked is, from where to where does the system of the Yang–Mills fields tunnel?

At first glance it is not obvious at all that the Lagrangian (5.1) has a discrete set of degenerate classical minima.[3] But it does!

The space of fields in field theories is infinite dimensional. Most of these field-theoretical degrees of freedom are oscillator-like and thus, having just a single ground state, present no interest for our current purposes. However, we will demonstrate that in Yang–Mills theories there exists one composite degree of freedom, a direction in the infinite-dimensional space of fields along which the Yang–Mills system can tunnel. If we forget for a while about the other degrees of freedom and focus on this chosen degree of freedom, we will see degenerate states connected by "under-the-barrier" trajectories.

A close analogy that one can keep in mind while analyzing Yang–Mills theories in the context of tunneling is the quantum mechanics of a particle living on a vertically oriented circle and subject to a constant gravitational force (Fig. 5.1). Classically the particle with the lowest possible energy (i.e., in the ground state of the system) just stays at rest at the bottom of the circle. Quantum-mechanically, zero-point oscillations come into play. Within a perturbative treatment we will deal exclusively with small oscillations near the equilibrium point at the bottom of the circle. For such

---

[2] Note that the normalization of the Yang–Mills fields in this chapter is different from that in the previous chapters.

[3] We will call them *pre-vacua* for reasons that will become clear later.

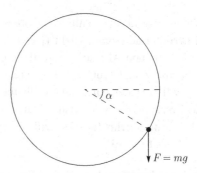

Fig. 5.1   A particle on a one-dimensional topologically nontrivial manifold, the circle.

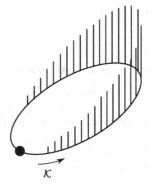

Fig. 5.2   Nontrivial topology in the space of gauge fields in the $\mathcal{K}$ direction. The circumference of the circle is 1. The vertical lines indicate the strength of the potential acting on the effective degree of freedom living on the circle.

small oscillations, the existence of the upper part of the circle plays no role. It could be eliminated altogether with no impact on the zero-point oscillations.

From studies in quantum mechanics we know, however, that the genuine ground-state wave function is different. The particle oscillating near the origin "feels" that it could wind around the circle on which it belongs, by tunneling through the potential barrier it experiences at the top of the circle (the barrier is similar to that shown in Fig. 5.2).

To single out the relevant degree of freedom in the infinite-dimensional space of the gluon fields, it is necessary to proceed to the Hamiltonian formulation of Yang–Mills theory. This implies, of course, that the time component of the four-potential A_μ has to be gauged away, $A_0 = 0$. Then,

$$\mathcal{H} = \tfrac{1}{2} \int d^3x \left(E_i^a E_i^a + B_i^a B_i^a \right), \tag{5.3}$$

where $\mathcal{H}$ is the Hamiltonian and the $E_i^a = \dot{A}_i^a$ are to be treated as canonical momenta.

Two subtle points should be mentioned in connection with this Hamiltonian. First, the equation div $\boldsymbol{E}^a = \rho^a$, intrinsic to the original Yang–Mills theory, does not stem from this Hamiltonian *per se*. This equation must be imposed by hand, as a constraint on the states from the Hilbert space. Second, the gauge freedom is not fully eliminated. Gauge transformations which depend on x but not t are still allowed. This freedom is reflected in the fact that, instead of two transverse degrees of freedom

$A_{\perp}^a$, the Hamiltonian above has three (the three components of A^a). Imposing, say, the Coulomb gauge condition,

$$\partial_i A_i^a = 0, \tag{5.4}$$

we could get rid of the "superfluous" degree of freedom, a procedure quite standard in perturbation theory (in the Coulomb gauge). Alas! If we want to keep and reveal the topologically nontrivial structure of the space of Yang–Mills fields, the Coulomb gauge condition cannot be imposed. We have to work, with certain care, with an "undergauged" Hamiltonian.

Quasiclassically, the state of the system described by the Hamiltonian (5.3) at any given moment of time is characterized by the field configuration $A_i^a(x)$; x indicates a set of three spatial coordinates. Since we are interested in the zero-energy states – classically, they are obviously the states with minimal possible energy – the corresponding gauge field A_i must be pure gauge,

$$A_i(x)|_{\text{vac}} = iU(x)\partial_i U^\dagger(x), \tag{5.5}$$

Matrix
notation

where U is a matrix belonging to SU(2) that depends on the spatial components x of the 4-coordinates. We have also introduced the matrix notation

$$A_\mu = g A_\mu^a \frac{\tau^a}{2}. \tag{5.6}$$

Moreover, we are interested only in those zero-energy states that may be connected with each other by tunneling transitions, i.e., the corresponding classical action must be finite. The latter requirement results in the following boundary condition:[4]

$$U(x) \to 1, \qquad |x| \to \infty, \tag{5.7}$$

or $U(x)$ tends to any other constant matrix U_0 that is independent of the direction in the three-dimensional space along which x tends to infinity. This boundary condition *compactifies* our three-dimensional space, which thus becomes topologically equivalent to the three-dimensional sphere S_3. The group space of SU(2) is also a three-dimensional sphere, however. Indeed, any matrix belonging to SU(2) can be parametrized as

$$M = A + iB\tau, \qquad M \in \text{SU}(2). \tag{5.8}$$

Here A and B comprise four real parameters; τ are the Pauli matrices. The conditions $M^+M = 1$ and $\det M = 1$ are both met provided that

$$A^2 + B^2 = 1. \tag{5.9}$$

Topological
formula for
the third
homotopy
group, cf.
(4.136)

Since $U(x)$ is a matrix from SU(2) and the space of all coordinates x is topologically equivalent to a three-dimensional sphere (after the compactification $U(x) \to 1$ at $|x| \to \infty$), the function $U(x)$ realizes a mapping of the sphere in coordinate space onto a sphere in the group space. Intuitively it is obvious that all continuous mappings $S_3 \to S_3$ are classified according to the number of coverings, which is the number of times the group-space sphere S_3 is swept when the coordinate x sweeps the sphere in coordinate space once. The number of coverings can be zero

---

[4] If (5.7) is not satisfied then $G_{0i} \sim \dot{A}_i$ will scale at large fixed t as $1/|x|$ and the integral $\int d^3x G_{0i}^2$ will be divergent, implying an infinite action. See a remark in Section 5.3.1 and/or the discussion in [14].

(a topologically trivial mapping), one, two, and so on (see Fig. 4.1). The number of coverings can be negative, too, since the mappings $S_3 \rightarrow S_3$ are orientable [15]. Mathematically this is expressed by the formula

$$\pi_3(S_3) = \mathbb{Z}. \tag{5.10}$$

In other words, the matrices $U(x)$ can be sorted into distinct classes $U_n(x)$, labeled by an integer $n = 0, \pm 1, \pm 2, \ldots$, referred to as the *winding number*. All matrices belonging to a given class $U_n(x)$ are reducible to each other by a continuous x-dependent gauge transformation. At the same time, no continuous gauge transformation can transform $U_n(x)$ into $U_{n'}(x)$ if $n \neq n'$. The unit matrix represents the class $U_0(x)$. For $n = 1$ one can take, for instance,[5]

$$U_1(x) = \exp\left[-i\pi \frac{x\,\tau}{\left(x^2 + \rho^2\right)^{1/2}}\right], \tag{5.11}$$

where ρ is an arbitrary parameter. An example of a matrix from U_n is $[U_1(x)]^n$.

Any field configuration $A_i(x)|_{\mathrm{vac}} = iU_n(x)\partial_i U_n^\dagger(x)$, being pure gauge, corresponds to the lowest possible energy – zero energy. As a matter of fact, the set of points $\{U_n\}$ in the space of fields consists simply of the gauge images of the same physical point (which is analogous to the bottom of the circle in Fig. 5.1). The fact that the matrices U_n from different classes are not continuously transformable to each other indicates the existence of a "hole" in the space of fields, with noncontractible loops winding around this "hole."

We are finally ready to identify the degree of freedom corresponding to motion around this circle. Let us consider the vector

Chern–
Simons
current.

$$K^\mu = 2\varepsilon^{\mu\nu\alpha\beta}\left(A_\nu^a \partial_\alpha A_\beta^a + \frac{g}{3}\varepsilon^{abc}A_\nu^a A_\alpha^b A_\beta^c\right), \qquad \varepsilon^{0123} = 1. \tag{5.12}$$

The vector K^μ is called the *Chern–Simons current*; it plays an important role in instanton calculus. We will encounter it more than once in what follows. Now, define the charge $\mathcal{K}$ corresponding to the Chern–Simons current,

$$\mathcal{K} = \frac{g^2}{32\pi^2}\int K_0(x)d^3x. \tag{5.13}$$

It is not difficult to show that for any pure gauge field $A_i^a(x)$ the Chern–Simons charge $\mathcal{K}$ measures the winding number: for any field of the type (5.5) we have (cf. Eqs. (4.135) and (4.146))

$$\mathcal{K} = n. \tag{5.14}$$

Summarizing, moving in the "direction of $\mathcal{K}$" in the space of Yang–Mills fields we observe that this particular direction has the topology of a circle. The points $\mathcal{K}$,

Compare
with
Section 8.1.

$\mathcal{K} \pm 1$, $\mathcal{K} \pm 2$, and so on, are physically one and the same point. The integer values of $\mathcal{K}$ correspond to the bottom of the circle in Fig. 5.1.

It is convenient to visualize the dynamics of the Yang–Mills system in the "direction of $\mathcal{K}$" as in Fig. 5.2. The vertical lines indicate the potential energy – the higher the line the larger the potential energy. It is well known (see, e.g., the

---

[5] Let us note in passing that exactly the same topological classification is the basis of the theory of Skyrmions; see Section 4.2.

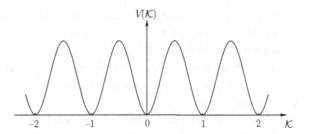

Fig. 5.3   If we unwind the circle of Fig. 5.2 onto a line we get a periodic potential.

textbooks [16]) that the only consistent way of treating quantum-mechanical systems living on a circle (i.e., those with angle-type degrees of freedom) is to cut the circle and map it many times onto a straight line. In other words, we pretend that the variable $\mathcal{K}$ lives on the line (Fig. 5.3). Any integer value of $\mathcal{K}$ in Fig. 5.3 corresponds to a pure gauge configuration with zero energy. If $\mathcal{K}$ is not an integer, however, the field strength tensor is nonvanishing and the energy of the field configuration is positive. Viewed as a function on the line, the potential energy $V(\mathcal{K})$ is, of course, periodic – with unit period.

To take into account the fact that the original problem is formulated on the circle, we impose a (quasi)periodic Bloch boundary condition on the wave functions Ψ,

$$\Psi(\mathcal{K}+1) = e^{i\theta}\Psi(\mathcal{K}). \tag{5.15}$$

The phase θ, $0 \leq \theta \leq 2\pi$, appearing in the Bloch quasiperiodic boundary condition is a hidden parameter, the *vacuum angle*. The boundary condition (5.15) must be the same for the wave functions of all states. We will return to the issue of the vacuum angle later on. The classical minima of the potential in Fig. 5.3 can be called *pre-vacua*. The correct wave function of the quantum-mechanical vacuum state of Bloch form is a linear combination of these pre-vacua.

Introducing the vacuum angle

We would like to emphasize here a subtle point that in many presentations remains unclear. It might seem that the systems depicted in Figs. 5.2 and 5.3 (a particle on a circle and that in a periodic potential) are physically identical. This is not quite the case. In periodic potentials, say in crystals, one can always introduce impurities that would slightly violate periodicity. For a system on the circle this cannot be done. Thus the correct analog system for Yang–Mills theories, where the gauge invariance is a sacred principle, is that of Fig. 5.2.

Assume that at $t = -\infty$ and at $t = +\infty$ our system is at one of the classical minima (zero-energy states) depicted in Fig. 5.3, but that the minimum in the past is different from that in the future. Assume that at $t = -\infty$ the winding number $\mathcal{K} = n$ while at $t = +\infty$ the winding number $\mathcal{K} = n \pm 1$. In Fig. 5.2 this means that our system tunnels from the point marked by the small solid circle under the hump of the potential and back to the same point.

Consider now a field configuration $A_\mu(t, \boldsymbol{x})$ continuously interpolating (with minimal action) between these two states in Euclidean time, i.e., the least-action tunneling trajectory. This is the BPST instanton.[6]

---

[6] An illuminating discussion of the tunneling interpretation in the *Minkowski* space is presented in [17].

The analysis outlined above (based on the Hamiltonian formulation) is convenient for establishing the existence of a nontrivial topology and nonequivalent (pre-) vacuum states and, hence, the existence of nontrivial interpolating field configurations corresponding to tunneling. In practice, however, the Hamiltonian gauge $A_0 = 0$ is rarely used in constructing the instanton solutions. It is inconvenient for this purpose.

Below we will describe a standard procedure based on a specific *ansatz* for $A_\mu(x)$ in which all four Lorentz components of A_μ are nonvanishing. This *ansatz* entangles the color and Lorentz indices; the field configurations emerging in this way are, following Polyakov, generically referred to as "hedgehogs," as mentioned earlier.

5.1.2 Theta Vacuum and θ Term

Compare with Section 8.1.4.

The existence of a noncontractible loop in the space of fields A_μ leads to drastic consequences for the vacuum structure in non-Abelian gauge theories. Let us take a closer look at the potential of Fig. 5.3. The argument presented below is formulated in quasiclassical language. One should keep in mind, however, that the general conclusion is valid, even though the quasiclassical approximation is inappropriate, in quantum chromodynamics, where the coupling constant becomes large at large distances.

Classically, the lowest-energy state of the system depicted in Fig. 5.3, occurs when the system is in a minimum of the potential. Quantum-mechanically, zero-point oscillations arise. The wave function[7] corresponding to oscillations near the nth zero-energy state, Ψ_n, is localized near the corresponding potential minimum. The genuine wave function is delocalized, however, and takes the form

$$\Psi_\theta = \sum_{n=0,\pm1,\pm2,\dots} e^{in\theta} \Psi_n \tag{5.16}$$

where θ is a parameter,

$$0 \le \theta \le 2\pi, \tag{5.17}$$

Here θ is the vacuum angle mentioned after (5.5).

analogous to the quasimomentum in the physics of crystals [16]. If the nth term in the sum is the nth "pre-vacuum," the total sum represents the θ vacuum. The vacuum angle θ is a global fundamental constant characterizing the boundary condition on the wave function. It does not make sense to say that in one part of the space θ takes some value while in another part it takes a different value or depends on time. Once this parameter is set we stay in the world corresponding to the given θ vacuum forever. Worlds with different values of θ have orthogonal wave functions; for any operator O acting on the Hilbert space of physical states

$$\langle \Psi_\theta | O | \Psi_{\theta'} \rangle = 0 \qquad \text{if } \theta \neq \theta'. \tag{5.18}$$

Superselection rule

This property is referred to as the *superselection rule*.

The energy of Ψ_θ can (and does) depend on θ, generally speaking, and so do other physically measurable quantities. From the definition of the vacuum angle it is clear

---

[7] In application to Yang–Mills theories and QCD we should rather use the term wave functional; nevertheless, we will continue referring to the wave function.

that the θ-dependence of all physical observables, including the vacuum energy, must be periodic with period 2π.

Since all pre-vacua states Ψ_n are degenerate in energy, the question is often raised of why one should form a linear combination, the θ vacuum. Is it possible to take, say, Ψ_0 as the vacuum wave function?

The answer is negative and can be explained at different levels. Purely theoretically, if we want to implement the full gauge invariance of the theory, including invariance under "large" gauge transformations, we must pass from Ψ_n to Ψ_θ.

At a more pragmatic level one can say that the introduction of Ψ_θ is necessary to maintain the property of cluster decomposition, which must take place in any sensible field theory. What is cluster decomposition? This property means that the vacuum expectation value of the T product of any two operators, $O_1(x_1)$ and $O_2(x_2)$, at large separations $|x_1 - x_2| \to \infty$ must tend to $\langle O_1 \rangle \langle O_2 \rangle$. If the vacuum wave function were chosen to be Ψ_n, this property would not be valid (see, for example, the text below Eq. (8.39)).

Finally, by proceeding to Ψ_θ we ensure that the vacuum state is stable under small perturbations. This would not be the case if the vacuum wave function were Ψ_n. For instance, a small mass term of the quark fields could then cause a drastic restructuring of the vacuum wave function.

Although the physical meaning of the parameter θ is absolutely transparent within the Hamiltonian formulation, when we speak of instantons in field theory, usually, we have in mind a Lagrangian formulation based on path integrals. In the Lagrangian

| *The θ term* | formalism the vacuum angle is introduced as a θ term in the Lagrangian,

$$\mathcal{L} = -\tfrac{1}{4}G_{\mu\nu}^a G^{a,\mu\nu} + \mathcal{L}_\theta, \qquad \mathcal{L}_\theta = \theta\frac{g^2}{32\pi^2}G^{a,\mu\nu}\widetilde{G}_{\mu\nu}^a, \tag{5.19}$$

where

$$\widetilde{G}_{\mu\nu}^a = \tfrac{1}{2}\varepsilon_{\mu\nu\alpha\beta}G^{a,\alpha\beta}, \qquad \varepsilon^{0123} = 1. \tag{5.20}$$

Note that if $\theta \neq 0$ or π, the θ term violates P and T invariance.

Before the discovery of instantons it was believed that QCD naturally conserves P and CP. Indeed, the only gauge-invariant Lorentz scalar operator that can be constructed from the A_μ fields of dimension 4 violating P and T is $G\widetilde{G}$. This operator, however, presents a full derivative: $G\widetilde{G} = \partial_\mu K_\mu$, where K_μ is the Chern–Simons current (5.12). It was believed that such a full derivative can have no impact on the action.

In the instanton field, however, the integral over $G\widetilde{G}$ does *not* vanish. The reasons for this will be explained below. What is important for us now is the fact that by adding the θ term to the QCD Lagrangian we break P and CP for strong interactions if $\theta \neq 0$ or π. Since it is known experimentally that P and CP symmetries are conserved for strong interactions to a very high degree of accuracy, this means that in nature the vacuum angle is fine-tuned and is very close to zero.[8] Estimates show that $\theta \lesssim 10^{-9}$ [18, 19].

Thus, with the advent of instantons the naturalness of QCD is gone. Can this fine-tuning be naturally explained? There exist several suggestions of how one could

---

[8] The second solution, with $\theta = \pi$, is incompatible with the experimental data, for subtle reasons.

solve the problem of P and CP conservation in QCD in a natural way. One of the most popular is the axion conjecture [20]. This topic, however, lies outside our scope. Interested readers are referred to [19] for a pedagogical review. We will simply assume that $\theta = 0$ although theoretically, in a hypothetical world, it could take any value from the interval $[0, 2\pi]$.

In Minkowski space the θ term (5.19) is real. It becomes purely imaginary on passing to Euclidean space. Certainly, this does not mean any loss of unitarity. So why do we need to pass to Euclidean space?

The reason is not hard to find: the classical solutions describing the tunneling trajectories are those of the Euclidean equations of motion. In order to pass to Euclidean time one can choose two alternative routes. In pure Yang–Mills theory with no fermions, it is advantageous to formulate a Euclidean version of the theory from the very beginning and to work only with this version. The Euclidean formulation can also be developed in the presence of fermions, provided that all fermions in the theory are described by Dirac fields, i.e., are nonchiral. This is what we will do in this chapter.

This approach does not work, however, for chiral fermions, or for many supersymmetric field theories. For such problems one must choose the second route, which will be discussed in Part II.

Exercise

5.1.1   Using Eq. (5.5) for the pure gauge field together with the matrix $U_1(x)$ from Eq. (5.11) corresponding to a unit winding, show that $\mathcal{K} = 1$. Show that $\mathcal{K} = n$ for the winding-n matrices $U_n(x)$.

5.2 Euclidean Formulation of QCD

First we will discuss the passage from Minkowski to Euclidean time. Then we will describe the gauge-boson fields in Euclidean space. Finally, anticipating the uses of instantons in QCD, we will consider the Euclidean version of Dirac fermions.[9]

Warning!

Note: In this section a caret is used to denote a quantity in Euclidean space. The Greek letters $\mu, \nu, \ldots$ denote indices running from 0 to 3 for Minkowskian quantities; for Euclidean quantities (with a caret) they run from 1 to 4. The Latin letters i, j take the values 1,2,3.

In Minkowski space one distinguishes between contravariant and covariant vectors, written as v^μ and v_μ, respectively. The spatial vector v coincides with the spatial components of the contravariant four-vector v^μ,

---

[9] I would like to emphasize that a full Euclidean formulation of the theory is not necessary for the instanton studies; see Part II. The only necessary element is the transition from Minkowski to Euclidean time. Nevertheless, below we will construct a complete Euclidean version of Yang–Mills theories because this formulation [6, 13] will be convenient for practical purposes.

$$v = \{v^1, v^2, v^3\}.$$

In Euclidean space the distinction between the lower and upper vectorial indices is immaterial; we consider just one vector $\hat{v}_\mu (\mu = 1, 2, 3, 4)$.

In passing to Euclidean space, the spatial coordinates are not changed, $\hat{x}_i = x^i$. For the time coordinate x_0 we make the substitution

$$x_0 = -i\hat{x}_4. \tag{5.21}$$

Clearly, when x_0 is continued to imaginary values the zeroth component of the vector potential A_μ also becomes imaginary.

We define the Euclidean vector potential $\hat{A}_\mu$ as follows:

$$A^i = -\hat{A}_i, \qquad A_0 = i\hat{A}_4. \tag{5.22}$$

With this definition, the quantities $\hat{A}_\mu (\mu = 1, 2, 3, 4)$ form a Euclidean vector. The difference between (5.22) and the corresponding relations for the vector x^μ see (5.21) is introduced for convenience in the subsequent expressions.[10]

Thus, for the operator of covariant differentiation,

$$\mathcal{D}_\mu = \partial_\mu - ig A_\mu^a T^a, \tag{5.23}$$

where the T^a are matrices of the generators in the representation being considered, we obtain

$$\mathcal{D}^i = -\hat{\mathcal{D}}_i, \qquad \mathcal{D}_0 = i\hat{\mathcal{D}}_4,$$
$$\hat{\mathcal{D}}_\mu = \frac{\partial}{\partial \hat{x}_\mu} - ig \hat{A}_\mu^a T^a. \tag{5.24}$$

For the field strength tensor $G_{\mu\nu}$ we get

$$G_{ij}^a = \hat{G}_{ij}^a, \qquad G_{0j}^a = i\hat{G}_{4j}^a, \tag{5.25}$$

where the Euclidean field strength tensor $\hat{G}_{\mu\nu}^a$ is defined as follows:

$$\hat{G}_{\mu\nu}^a = \frac{\partial}{\partial \hat{x}_\mu} \hat{A}_\nu^a - \frac{\partial}{\partial \hat{x}_\nu} \hat{A}_\mu^a + g f^{abc} \hat{A}_\mu^b \hat{A}_\nu^c. \tag{5.26}$$

It is expressed in terms of $\hat{A}_\mu$ and $\partial/\partial \hat{x}_\mu$ in just the same way as the Minkowskian $G_{\mu\nu}^a$ is expressed in terms of A_μ and $\partial/\partial x_\mu$.

This concludes the bosonic part of the transition. To complete the transition to Euclidean space, what remains to be done is to derive similar expressions for the Dirac spinor fields. We begin with the definition of the four Hermitian γ-matrices $\hat{\gamma}_\mu$:

$$\hat{\gamma}_4 = \gamma_0, \quad \hat{\gamma}_i = -i\gamma^i,$$
$$\{\hat{\gamma}_\mu, \hat{\gamma}_\nu\} = 2\delta_{\mu\nu}, \tag{5.27}$$

Euclidean
gamma
matrices

where γ_0 and γ^i are the conventional Dirac matrices.

---

[10] If we use the definition $\hat{A}_i = A^i$ $(i = 1, 2, 3)$ then in all the following connection formulas it is necessary to make the substitution $g \to -g$.

In Euclidean space the fields ψ and $\bar{\psi}$ over which we integrate in the path integral must be regarded as independent anticommuting variables. It is convenient to define the variables $\hat{\psi}$ and $\hat{\bar{\psi}}$ as follows:

$$\psi = \hat{\psi}, \qquad \bar{\psi} = -i\hat{\bar{\psi}}. \tag{5.28}$$

Under rotations of the pseudo-Euclidean (Minkowski) space, $\bar{\psi}$ transforms as $\psi^\dagger \gamma_0$. In Euclidean space $\hat{\bar{\psi}}$ transforms as $\hat{\psi}^\dagger$. Indeed, under infinitesimal rotations in Minkowski space characterized by the parameters $\omega_{\mu\nu}$, the spinor ψ varies as follows:

$$\delta\psi = -\tfrac{1}{4}(\gamma_\mu\gamma_\nu - \gamma_\nu\gamma_\mu)\omega^{\mu\nu}\psi. \tag{5.29}$$

One can readily deduce from Eq. (5.29) the variation in $\bar{\psi} = \psi^\dagger\gamma_0$:

$$\begin{aligned}
\delta\left(\psi^\dagger\gamma_0\right) &= -\tfrac{1}{4}\psi^\dagger\gamma_0\gamma_0\left(\gamma_\nu^\dagger\gamma_\mu^\dagger - \gamma_\mu^\dagger\gamma_\nu^\dagger\right)\gamma_0\omega^{\mu\nu} \\
&= \tfrac{1}{4}\left(\psi^\dagger\gamma_0\right)(\gamma_\mu\gamma_\nu - \gamma_\nu\gamma_\mu)\omega^{\mu\nu};
\end{aligned} \tag{5.30}$$

as a result, $\psi_1^\dagger\gamma_0\psi_2$ is a scalar and $\psi_1^\dagger\gamma_0\gamma_\mu\psi_2$ a vector.

During the transition to Euclidean space the parameters ω_{ij} do not change, while $\omega_{0j} = i\omega_{4j}$. For the variations in $\hat{\psi}$ and $\hat{\psi}^\dagger$ under rotations, we then obtain

$$\delta\hat{\psi} = \tfrac{1}{4}\left(\hat{\gamma}_\mu\hat{\gamma}_\nu - \hat{\gamma}_\nu\hat{\gamma}_\mu\right)\hat{\omega}_{\mu\nu}\hat{\psi}, \qquad \delta\hat{\psi}^\dagger = -\tfrac{1}{4}\psi^\dagger\left(\hat{\gamma}_\mu\hat{\gamma}_\nu - \hat{\gamma}_\nu\hat{\gamma}_\mu\right)\hat{\omega}_{\mu\nu}, \tag{5.31}$$

so that $\hat{\psi}_1^\dagger\hat{\psi}_2$ and $\hat{\psi}_1^\dagger\hat{\gamma}_\mu\hat{\psi}_2$ are a scalar and a vector, respectively.

Finally, we can write down the Euclidean action of QCD,

$$iS = -\hat{S},$$

$$S = \int d^4x\left[-\frac{1}{4}G^a_{\mu\nu}G^{a\mu\nu} + \bar{\psi}\left(i\gamma^\mu D_\mu - m\right)\psi + \theta\frac{g^2}{32\pi^2}G^a_{\mu\nu}\widetilde{G}^{a\mu\nu}\right], \tag{5.32}$$

$$\hat{S} = \int d^4\hat{x}\left[\frac{1}{4}\hat{G}^a_{\mu\nu}\hat{G}^a_{\mu\nu} + \hat{\bar{\psi}}\left(-i\hat{\gamma}_\mu\hat{D}_\mu - im\right)\hat{\psi} + i\theta\frac{g^2}{32\pi^2}\hat{G}^a_{\mu\nu}\hat{\widetilde{G}}^a_{\mu\nu}\right],$$

Levi–Civita
tensor in
Euclidean
space

where it is assumed that $\hat{\psi}$ is a column vector in the space of flavors (with a triplet color index, suppressed in (5.32)) and m is a mass matrix in this space. Note that in Euclidean space the Levi–Civita tensor $\varepsilon_{\mu\nu\alpha\beta}$ is defined in such a way that $\varepsilon_{1234} = 1$. The mass matrix can always be chosen to be diagonal.

The Minkowskian weight factor $\exp(iS)$ in the path integral becomes $\exp(-\hat{S})$ in Euclidean space.

Below, in this chapter, we will use the Euclidean formulation while omitting the carets. The expressions given above make it possible to relate relevant quantities in the pseudo-Euclidean and Euclidean spaces.

To conclude this section we note that if we are considering quantities such as the vacuum expectation values of time-ordered products of currents for space-like external momenta, in the case when the sources do not produce real hadrons from the vacuum, theEuclidean-space formulation is not only merely possible but in fact is more adequate than the pseudo-Euclidean. The region of time-like momenta, where there are singularities, can be reached by means of analytic continuation.

Exercise

5.2.1 Find the transformation law for the following fermion bilinear combination:
$\hat{\psi}_1^\dagger \frac{1}{2} \left(\hat{\gamma}_\mu \hat{\gamma}_\nu - \hat{\gamma}_\nu \hat{\gamma}_\mu \right) \hat{\psi}_2$.

5.3 BPST Instantons: General Properties

5.3.1 Finiteness of the Action and the Topological Charge

In Section 5.1 we learned that the initial and final states between which the instanton interpolates are characterized by the winding number (5.13). Now we will consider an interpolating trajectory $A_\mu(x)$, not necessarily an instanton but any trajectory with a *finite* action. Here x is a Euclidean coordinate, a four-dimensional space–time vector. Our task is to demonstrate that all such trajectories $A_\mu(x)$ fall into distinct classes characterized by the topological charge Q, which can take any integer value. Here

$$Q = \mathcal{K}(x_4 = +\infty) - \mathcal{K}(x_4 = -\infty) \equiv \Delta\mathcal{K}. \tag{5.33}$$

The BPST instanton has $Q = \pm 1$.

Equations (5.12) and (5.13) imply that a gauge-invariant local representation exists for the topological density of the charge Q (making unnecessary the transition to the $A_0 = 0$ gauge), namely, the topological density is $(g^2/32\pi^2)G^a_{\mu\nu}\widetilde{G}^a_{\mu\nu}$ and

$$Q = \frac{g^2}{32\pi^2} \int d^4x\, G^a_{\mu\nu} \widetilde{G}^a_{\mu\nu}, \tag{5.34}$$

Topological charge

where

$$\widetilde{G}^a_{\mu\nu} = \tfrac{1}{2}\varepsilon_{\mu\nu\alpha\beta}G^a_{\alpha\beta}, \qquad \varepsilon_{1234} = 1. \tag{5.35}$$

The statement that (5.33) and (5.34) coincide can be verified by representing $G^a_{\mu\nu}\widetilde{G}^a_{\mu\nu}$ in the form of a total derivative,

$$G_{\mu\nu}\widetilde{G}_{\mu\nu} = \partial_\mu K_\mu, \tag{5.36}$$

where the Chern–Simons current K_μ can be found from (5.12). Next, invoking the Gauss formula

$$\int d^3x\, \partial_i K_i = \int_{\text{surface } S_2} K_i\, dS_i \to 0,$$

we transform the volume integral (5.34) into an integral of K_0 over the three-dimensional space presenting the boundary of the Euclidean space–time at $t \to \pm\infty$, cf. (5.13).

Let us pose the question: what must be the behavior of the vector fields A^a_μ as $x \to \infty$ if the Yang–Mills (Euclidean) action proportional to $\int d^4x G^2_{\mu\nu}$ is to be finite? From Eq. (5.32) it is clear that the field strength tensor $G^a_{\mu\nu}$ must decrease at infinity faster than $1/x^2$. This requires A^a_μ to be pure gauge in the limit $x \to \infty$:

$$A_\mu \equiv \tfrac{1}{2}g A^a_\mu \tau^a \to i S \partial_\mu S^\dagger, \qquad x \to \infty, \qquad (5.37)$$

where S is a unitary unimodular matrix. As long as the expression (5.37) holds, the field strength tensor $G^a_{\mu\nu}$ vanishes and the total action is finite.

Thus, the behavior of A^a_μ at large x is determined by the matrix S at large distances from the instanton center, i.e., on the three-dimensional "boundary" S_3 of four-dimensional space. As a result, the problem of classifying the fields A^a_μ that give a finite action reduces to the topological classification of the SU(2) matrices S in terms of their dependence on points on S_3, the hypersphere in Euclidean space. For classifying continuous mappings from S_3 onto the group space SU(2) the following topological formula is relevant:[11]

$$\pi_3(\mathrm{SU}(2)) = \mathbb{Z}, \qquad (5.38)$$

which is exactly the same as in our previous analysis of distinct pre-vacua in QCD, see (5.10). By the way, this is an independent confirmation of the boundary condition (5.5). Equation (5.38) proves the existence of distinct classes, labeled by integers, of interpolating trajectories connecting distinct pre-vacua.

The simplest example of a nontrivial (not reducible to 1) matrix S is

$$S_1 = \frac{x_4 + i\boldsymbol{x}\boldsymbol{\tau}}{\sqrt{x^2}}. \qquad (5.39)$$

It corresponds to the unit topological charge. For a topological charge n we can take, for instance, a matrix of the form

$$S_n = (S_1)^n, \qquad n = 0, \pm 1, \pm 2, \ldots \qquad (5.40)$$

Of course, one could choose a different form of the matrix S_n corresponding to charge n, but the difference between any alternative choice and S_n in Eq. (5.40) must reduce to a topologically trivial gauge transformation.

Warning: Equation (5.39) does not correspond to the $A_4 = 0$ gauge.

For the careful reader it should be clear already that there exist two related, but not identical, topological arguments. The first argument, discussed in detail in Section 5.1, reveals the existence of distinct topologically nonequivalent zero-energy states characterized by winding numbers. Outlined here is a four-dimensional topological view; it refers to the topology of the trajectories connecting (in Euclidean space–time) the distinct zero-energy states discussed in Section 5.1.

The field configuration $A_\mu(x_4, \boldsymbol{x})$ satisfying Eq. (5.37) with $S = S_1$ interpolates between the state with winding number $\mathcal{K}$ and that with winding number $\mathcal{K} + 1$. To see that this is indeed the case we must, of course, transform the instanton into the $A_4 = 0$ gauge, which we will do in Section 5.4.4.

For $S = S_2$ we are dealing with the trajectory $A_\mu(x_4, \boldsymbol{x})$ connecting $\mathcal{K}$ and $\mathcal{K} + 2$, etc. For arbitrary n the topological charge Q of any field configuration $A_\mu(x_4, \boldsymbol{x})$ satisfying Eq. (5.37) is given by Eq. (5.33).

---

[11] Below, in Section 5.4.7, we will also use the fact that the homotopy group $\pi_3(\mathrm{SU}(N)) = \mathbb{Z}$ for all N.

5.3.2 Entanglement of the Color and Lorentz Indices

Yang–Mills theories are invariant under both global color notations and Lorentz rotations. The instanton solution, to be discussed below, spontaneously breaks both these symmetries. However, a diagonal combination remains unbroken. This is therefore a typical Polyakov hedgehog. I will comment briefly on the entanglement of the instanton color and Lorentz indices, with the intention of returning to this issue later on when we discuss the instanton's collective coordinates (moduli).

On the other hand, under a global rotation in color space the matrix S in Eqs. (5.37) and (5.39) transforms as follows:

$$S \to U^\dagger S, \qquad (5.41)$$

where U is a *constant* matrix from SU(2).

On the other hand, the group of rotations in four-dimensional Euclidean space is well known to be SO(4) = SU(2) × SU(2). The generators of the two SU(2) subgroups have the forms

$$I_1^a = \frac{1}{4}\eta_{a\mu\nu}M_{\mu\nu}, \quad \begin{pmatrix} a = 1,2,3 \\ \mu,\nu = 1,2,3,4 \end{pmatrix}, \qquad (5.42)$$
$$I_2^a = \frac{1}{4}\bar{\eta}_{a\mu\nu}M_{\mu\nu}$$

where $M_{\mu\nu} = -ix_\mu\partial/\partial x_\nu + ix_\nu\partial/\partial x_\mu +$ spin part are the operators generating infinitesimal rotations in the $\mu\nu$ plane and the $\eta_{a\mu\nu}$ are numerical symbols given by

| 't Hooft symbols |

$$\eta_{a\mu\nu} = \begin{cases} \varepsilon_{a\mu\nu}, & \mu,\nu = 1,2,3 \\ -\delta_{a\nu} & \mu = 4, \\ \delta_{a\mu}, & \nu = 4, \\ 0 & \mu = \nu = 4. \end{cases} \qquad (5.43)$$

The symbols $\bar{\eta}_{a\mu\nu}$ in (5.42) differ from η by a change in the sign in front of δ. The sets of parameters η and $\bar{\eta}$ are called the *'t Hooft symbols*. The coordinate vector x_μ transforms in the representation $(\frac{1}{2},\frac{1}{2})$ of SU(2) × SU(2) . This is conveniently seen by considering transformations of the matrix[12]

$$x_4 + i\mathbf{x}\boldsymbol{\tau} = i\tau_\mu^+ x_\mu, \qquad (5.44)$$

The $\tau_\mu^\pm$ are Euclidean analogs of Minkowski σ^μ and $\bar{\sigma}^\mu$, Section 10.2.1.

which determines the numerator in Eq. (5.39). Here we introduce the notation

$$\tau_\mu^\pm = (\boldsymbol{\tau}, \mp i). \qquad (5.45)$$

For $\tau_\mu^\pm$ we have

$$\tau_\mu^+\tau_\nu^- = \delta_{\mu\nu} + i\eta_{a\mu\nu}\tau^a, \qquad \tau_\mu^-\tau_\nu^+ = \delta_{\mu\nu} + i\bar{\eta}_{a\mu\nu}\tau^a. \qquad (5.46)$$

It is not difficult to find the transformation law for the matrix (5.44):

$$\exp(i\varphi_1^a I_1^a + i\varphi_2^a I_2^a)\, i\tau_\mu^+ x_\mu = \exp[-i\varphi_1^a(\tau^a/2)]i\tau_\mu^+ x_\mu \exp[i\varphi_2^a(\tau^a/2)], \qquad (5.47)$$

---

[12] These $\tau_\mu^\pm$ matrices are Euclidean analogs of the Minkowski matrices $(\sigma^\mu)_{\alpha\dot\alpha}$ and $(\bar{\sigma}^\mu)^{\dot\alpha\alpha}$, Section 10.2.1: $\tau^+ \leftrightarrow \bar{\sigma}$, $\tau^- \leftrightarrow \sigma$.

where φ_1^a and φ_2^a are the parameters of four-dimensional rotations. In other words, a four-dimensional rotation of x_μ is equivalent to multiplication by unitary unimodular matrices from the left and also from the right, corresponding to two SU(2) subgroups of SO(4). Thus, if we rotate the coordinates according to (5.47) with $\varphi_2^a = 0$ and then perform a compensating *global* color rotation with $U = \exp[-i\varphi_1^a(\tau^a/2)]$ then the asymptotics (5.39) of the instanton solution remains intact. Shortly we will see that the same statement applies to the instanton solution *per se*, not just to its asymptotics. In other words if, instead of the generators of the SU(2) subgroup of the four-dimensional SO(4) rotations I_1^a, we introduce $I_1^a + T^a$ (where T^a generates the global color rotations) as the "angular momentum operators" then the instanton has spin zero with regard to this combined "angular momentum."

Instanton = hedgehog.

The SU(2) gauge group is distinguished (as compared with other non-Abelian gauge groups) by the dimension of the coordinate space and the fact that SO(4) = SU(2) × SU(2). Further clarifying remarks about why the SU(2) group is singled out are presented in Section 5.4.5.

5.3.3 Bogomol'nyi Completion and the Instanton Action

Although we do not yet have the explicit form of the instanton solution, we can nevertheless calculate the value of its action. Indeed, for positive values of the topological charge Q, the Euclidean action can be rewritten in the form

$$S = \int d^4x \frac{1}{4} G_{\mu\nu}^a G_{\mu\nu}^a = \int d^4x \left[\frac{1}{4} G_{\mu\nu}^a \widetilde{G}_{\mu\nu}^a + \frac{1}{8} \left(G_{\mu\nu}^a - \widetilde{G}_{\mu\nu}^a \right)^2 \right]$$
$$= Q \frac{8\pi^2}{g^2} + \frac{1}{8} \int d^4x \left(G_{\mu\nu}^a - \widetilde{G}_{\mu\nu}^a \right)^2 . \tag{5.48}$$

This is the Bogomol'nyi completion. It is clear from the relation (5.48) that in the class of functions with a given positive Q the minimum of the action is attained for

$$G_{\mu\nu}^a = \widetilde{G}_{\mu\nu}^a, \tag{5.49}$$

which is known as the self-duality equation. The Q-instanton action S_Q is equal to $8\pi^2 Q/g^2$. Functions with different Q values cannot be related by a continuous deformation if the action is to remain finite. Therefore, minimization of the action can be carried out separately in each class of functions having a given Q. The BPST

Instanton action

instanton has $Q = 1$. The instanton action is

$$S_1 = \frac{8\pi^2}{g^2}. \tag{5.50}$$

Since the instanton trajectory minimizes the action, it represents an extremum in the functional integral over the gauge fields.

The case of negative Q is obtained from (5.48) by the reflection $x_{1,2,3} \to -x_{1,2,3}$, under which $G_{\mu\nu}\widetilde{G}_{\mu\nu} \to -G_{\mu\nu}\widetilde{G}_{\mu\nu}$ and accordingly $Q \to -Q$. Thus, the minimum of the action for negative Q is $(8\pi^2/g^2)|Q|$. It is attained when

$$G_{\mu\nu}^a = -\widetilde{G}_{\mu\nu}^a. \tag{5.51}$$

The latter equation is referred to as the anti-self-duality equation. Its minimal topologically nontrivial solution is anti-instanton, with $Q = -1$. The anti-instanton action is the same as that for the instanton.

As can be seen from this discussion, the self-duality and anti-self-duality conditions $G_{\mu\nu}^a = \pm \widetilde{G}_{\mu\nu}^a$ automatically lead to the fulfillment of the equations of motion $\mathcal{D}_\mu G_{\mu\nu} = 0$. This can also be seen directly; indeed for, say, a self-dual field we have

$$
\begin{aligned}
\mathcal{D}_\mu G_{\mu\nu}^a = \mathcal{D}_\mu \widetilde{G}_{\mu\nu}^a &= \tfrac{1}{2}\varepsilon_{\mu\nu\gamma\delta}\mathcal{D}_\mu G_{\gamma\delta}^a \\
&= \tfrac{1}{6}\varepsilon_{\mu\nu\gamma\delta}\left(\mathcal{D}_\mu G_{\gamma\delta}^a + \mathcal{D}_\gamma G_{\delta\mu}^a + \mathcal{D}_\delta G_{\mu\gamma}^a\right) = 0,
\end{aligned} \tag{5.52}
$$

> Bianchi
> identity

where we have used the Bianchi identity

$$
\mathcal{D}_\mu G_{\gamma\delta}^a + \mathcal{D}_\gamma G_{\delta\mu}^a + \mathcal{D}_\delta G_{\mu\gamma}^a = 0. \tag{5.53}
$$

Not every solution of the classical Yang–Mills equations motion is (anti-)self-dual. However, it was proved that for $|Q| = 1$ every solution of the classical equations of motion is (anti-)self-dual.

5.4 Explicit Form of the BPST Instanton

5.4.1 Solution with $Q = 1$

As discussed in the previous section, the asymptotic behavior of A_μ^a for the solution with $Q = 1$ is

$$
\begin{aligned}
g A_\mu^a \frac{\tau^a}{2} &\to i S_1 \partial_\mu S_1^\dagger, \qquad x \to \infty, \\
S_1 &= \frac{i\tau_\mu^+ x_\mu}{\sqrt{x^2}},
\end{aligned} \tag{5.54}
$$

where the matrices $\tau_\mu^\pm$ were defined in (5.45). We will also use the 't Hooft symbols $\eta_{a\mu\nu}$ and $\bar{\eta}_{a\mu\nu}$ defined in Eqs. (5.43). Some useful relations for $\eta_{a\mu\nu}$ are given below in Section 5.4.3.

The expression for the asymptotic behavior of A_μ^a can be rewritten in terms of the 't Hooft symbols as follows:

$$
A_\mu^a \to \frac{2}{g}\eta_{a\mu\nu}\frac{x_\nu}{x^2}, \qquad x \to \infty. \tag{5.55}
$$

For an instanton with its center at the point $x = 0$, it is natural to assume the same angular dependence of the field for all x, i.e., to seek a solution in the form

$$
A_\mu^a \to \frac{2}{g}\eta_{a\mu\nu}\frac{x_\nu}{x^2}f(x^2), \tag{5.56}
$$

where

$$
\begin{aligned}
f(x^2) &\to 1, & x^2 &\to \infty, \\
f(x^2) &\to \text{const} \times x^2, & x^2 &\to 0.
\end{aligned} \tag{5.57}
$$

The last condition corresponds to the absence of a singularity at the origin (in fact, the power of x is determined from the general solution (5.61)). The *a posteriori* justification for the *ansatz* (5.56) will be the construction of a self-dual expression for $G_{\mu\nu}^a$. From (5.56) we obtain

$$G_{\mu\nu}^a = -\frac{4}{g}\left\{\eta_{a\mu\nu}\frac{f(1-f)}{x^2} + \frac{x_\mu\eta_{a\nu\gamma}x_\gamma - x_\nu\eta_{a\mu\gamma}x_\gamma}{x^4}\left[f(1-f) - x^2 f'\right]\right\}. \quad (5.58)$$

$G_{\mu\nu}^a$ and $\widetilde{G}_{\mu\nu}^a$ in terms of the profile function

Here the prime denotes differentiation with respect to x^2. In deriving (5.58), we have used the relation for $\varepsilon^{abc} \times \eta_{b\mu\gamma}\eta_{c\nu\delta}$ from the list of formulas in Section 5.4.3 below. Using the formula for $\varepsilon_{\mu\nu\gamma\delta}\eta_{a\delta\rho}$ from the same list, we obtain for $\widetilde{G}_{\mu\nu}^a$ the expression

$$\widetilde{G}_{\mu\nu}^a = -\frac{4}{g}\left\{\eta_{a\mu\nu}f' - \frac{x_\mu\eta_{a\nu\gamma}x_\gamma - x_\nu\eta_{a\mu\gamma}x_\gamma}{x^4}\left[f(1-f) - x^2 f'\right]\right\}. \quad (5.59)$$

The condition for self-duality, $G_{\mu\nu}^a = \widetilde{G}_{\mu\nu}^a$, implies the equation

$$f(1-f) - x^2 f' = 0, \quad (5.60)$$

which determines the function f:

$$f(x^2) = \frac{x^2}{x^2 + \rho^2}, \quad (5.61)$$

where ρ^2 is a constant of integration and ρ is called the *instanton size* or the *instanton radius*. Given the solution (5.61), translational invariance guarantees the existence of a whole family of instanton solutions whose centers are at an arbitrary point x_0. To obtain this family it is necessary to replace x by $x - x_0$. We will discuss ρ, x_0 and other

Translational and size moduli of the BPST instanton

collective coordinates (moduli) in more detail later. Note that if $f - \frac{1}{2}$ is denoted as X and if $x^2 = e^z$ then the equation for f becomes identical to the first-order differential equation obtained in the kink problem

$$\dot{X} = \frac{1}{4} - X^2$$

(see Chapter 2). Summarizing, the final expression for an instanton with its center at the point x_0 and with size ρ has the form

$$A_\mu^a = \frac{2}{g}\eta_{a\mu\nu}\frac{(x - x_0)_\nu}{(x - x_0)^2 + \rho^2}, \quad (5.62)$$

$$G_{\mu\nu}^a = -\frac{4}{g}\eta_{a\mu\nu}\frac{\rho^2}{\left[(x - x_0)^2 + \rho^2\right]^2}.$$

It can now be verified that the instanton action is $8\pi^2/g^2$, as was shown in the general form. The anti-instanton (anti-self-dual) solution is obtained from (5.62) by the substitution $\eta_{a\mu\nu} \to \bar{\eta}_{a\mu\nu}$. Note that A_μ^a falls off at infinity slowly, as $1/x$.

5.4.2 Singular Gauge. The 't Hooft Multi-Instanton *Ansatz*

It is often convenient to use the expression for A_μ^a in the so-called *singular gauge*, when the "bad" behavior of A_μ^a is transferred from infinity to the instanton center. Such a transfer can be realized by a gauge transformation[13] by a matrix $U(x)$

---

[13] More precisely, this transformation should be called a quasigauge transformation, since at the point where $U(x)$ has a singularity (and there must be such a singularity) this transformation changes the gauge-invariant quantities, for example, $G_{\mu\nu}^a G_{\mu\nu}^a$. To use such transformations it is necessary to consider a space–time with punctured singular points. This we will do, remembering that physical quantities remain nonsingular at the singular points.

which becomes identical with the matrix $S(x)$ from (5.37) at $x \to \infty$. The gauge transformation has the conventional form

$$g \frac{\tau^a}{2} \bar{A}_\mu^a = U^\dagger g \frac{\tau^a}{2} A_\mu^a U + iU^\dagger \partial_\mu U,$$

$$g \frac{\tau^a}{2} \bar{G}_{\mu\nu}^a = U^\dagger g \frac{\tau^a}{2} G_{\mu\nu}^a U, \quad (5.63)$$

where the overbars label the fields in the singular gauge. For an instanton centered at x_0 we take the following gauge matrix:

$$U = \frac{i\tau_\mu^+ (x - x_0)_\mu}{\sqrt{(x - x_0)^2}}. \quad (5.64)$$

Then for the potential $\bar{A}_\mu^a$ and the field strength tensor $\bar{G}_{\mu\nu}^a$ in the singular gauge we obtain

$$\bar{A}_\mu^a = \frac{2}{g} \bar{\eta}_{a\mu\nu} (x - x_0)_\nu \frac{\rho^2}{(x - x_0)^2 \left[(x - x_0)^2 + \rho^2 \right]},$$

$$\bar{G}_{\mu\nu}^a = -\frac{8}{g} \left[\frac{(x - x_0)_\mu (x - x_0)_\rho}{(x - x_0)^2} - \frac{1}{4} \delta_{\mu\rho} \right] \bar{\eta}_{a\nu\rho} \frac{\rho^2}{\left[(x - x_0)^2 + \rho^2 \right]^2} \quad (5.65)$$

$$- (\mu \leftrightarrow \nu),$$

where the bar indicates (only in this section) that the fields we are dealing with are in the singular gauge. It is obvious that the quantities $G_{\mu\nu}^a G_{\gamma\delta}^a$ are invariants of the gauge transformation (see, however, footnote 13 at the beginning of this subsection). Note also that (5.65) contains the symbols $\bar{\eta}_{a\mu\nu}$ but not the $\eta_{a\mu\nu}$. This difference is due to the fact that in the singular gauge the topological charge (5.34) is saturated in the neighborhood of $x = x_0$ and not at infinity.[14] The expression (5.65) for $\bar{A}_\mu^a$ can be rewritten in the form

$$\bar{A}_\mu^a = -\frac{1}{g} \bar{\eta}_{a\mu\nu} \partial_\nu \ln \left[1 + \frac{\rho^2}{(x - x_0)^2} \right]. \quad (5.66)$$

| 't Hooft multi- instanton solution |

As was noted by 't Hooft [21], this expression can be generalized to topological charges Q greater than unity. Indeed, if

$$A_\mu^a = -\frac{1}{g} \bar{\eta}_{a\mu\nu} \partial_\nu \ln W(x) \quad (5.67)$$

then for $G_{\mu\nu}^a - \widetilde{G}_{\mu\nu}^a$ we obtain

$$G_{\mu\nu}^a - \widetilde{G}_{\mu\nu}^a = \frac{1}{g} \bar{\eta}_{a\mu\nu} \frac{\partial_\rho \partial_\rho W}{W} \quad (5.68)$$

(see again the properties of the η symbols in Section 5.4.3). The self-duality of $G_{\mu\nu}^a$ requires fulfillment of the harmonic equation

$$\partial_\rho \partial_\rho W = 0. \quad (5.69)$$

---

[14] Something to memorize: in the regular gauge the instanton field is proportional to $\eta_{a\mu\nu}$ while the anti-instanton field it is proportional to $\bar{\eta}_{a\mu\nu}$. In the singular gauge, however, the instanton field is proportional to $\bar{\eta}_{a\mu\nu}$ and that of the *anti*-instanton to $\eta_{a\mu\nu}$.

The solution with topological charge Q has the form

$$W = 1 + \sum_{i=1}^{n} \frac{\rho_i^2}{(x - x_i)^2}, \qquad (5.70)$$

i.e., it describes instantons with their centers at points x_i. The effective scale of an instanton whose center is at the point x_i is obviously

$$\rho_i^{\text{eff}} = \rho_i \left[1 + \sum_{k \neq i} \frac{\rho_k^2}{(x_k - x_i)^2} \right]^{-1/2}. \qquad (5.71)$$

It should be noted that the choice of A_μ^a in the form (5.67) does not give the most general solution for topological charge Q, since all Q-instantons described by (5.67) have the same orientation in color space. The general Q-instanton solution (the so-called Atiyah–Drinfel'd–Hitchin–Manin construction [22], ADHM for short) attributes eight moduli parameters per instanton (in the SU(2) case; in the general case there are $4N$ moduli per instanton and so $4N|Q|$ moduli altogether). We will not describe the general constriction here. However, we will establish the number of moduli per instanton in a generic multi-instanton configuration in Section 5.4.5.

5.4.3 Relations for the η Symbols

Here we give a list of relations for the symbols $\eta_{a\mu\nu}$ and $\bar{\eta}_{a\mu\nu}$, defined by Eqs. (5.43):

$$\eta_{a\mu\nu} = \frac{1}{2}\varepsilon_{\mu\nu\alpha\beta}\eta_{a\alpha\beta},$$
$$\eta_{a\mu\nu} = -\eta_{a\nu\mu}, \quad \eta_{a\mu\nu}\eta_{b\mu\nu} = 4\delta_{ab},$$
$$\eta_{a\mu\nu}\eta_{a\mu\lambda} = 3\delta_{\nu\lambda}, \quad \eta_{a\mu\nu}\eta_{a\mu\nu} = 12,$$
$$\eta_{a\mu\nu}\eta_{a\gamma\lambda} = \delta_{\mu\gamma}\delta_{\nu\lambda} - \delta_{\mu\lambda}\delta_{\nu\gamma} + \varepsilon_{\mu\nu\gamma\lambda},$$
$$\varepsilon_{\mu\nu\lambda\sigma}\eta_{a\gamma\sigma} = \delta_{\gamma\mu}\eta_{a\nu\lambda} - \delta_{\gamma\nu}\eta_{a\mu\lambda} + \delta_{\gamma\lambda}\eta_{a\mu\nu}, \qquad (5.72)$$
$$\eta_{a\mu\nu}\eta_{b\mu\lambda} = \delta_{ab}\delta_{\nu\lambda} + \varepsilon_{abc}\eta_{c\nu\lambda},$$
$$\varepsilon_{abc}\eta_{b\mu\nu}\eta_{c\gamma\lambda} = \delta_{\mu\gamma}\eta_{a\nu\lambda} - \delta_{\mu\lambda}\eta_{a\nu\gamma} - \delta_{\nu\gamma}\eta_{a\mu\lambda} + \delta_{\nu\lambda}\eta_{a\mu\gamma},$$
$$\eta_{a\mu\nu}\bar{\eta}_{b\mu\nu} = 0, \quad \eta_{a\gamma\mu}\bar{\eta}_{b\gamma\lambda} = \eta_{a\gamma\lambda}\bar{\eta}_{b\gamma\mu}.$$

To pass from the relations for $\eta_{a\mu\nu}$ to those for $\bar{\eta}_{a\mu\nu}$ it is necessary to make the following substitutions:

$$\eta_{a\mu\nu} \to \bar{\eta}_{a\mu\nu}, \qquad \varepsilon_{\mu\nu\gamma\delta} \to -\varepsilon_{\mu\nu\gamma\delta}. \qquad (5.73)$$

5.4.4 Instanton in the $A_0 = 0$ Gauge

Topological charge vs. winding numbers

In Section 5.3.1 I mentioned that the relation between the instanton topological charge Q and the winding numbers of the zero-energy states in the distant past and distant future, between which it interpolates,

$$Q = \mathcal{K}' - \mathcal{K}, \qquad (5.74)$$

is most transparently seen in the $A_0 = 0$ gauge.[15] Now we can explicitly demonstrate this relation.

Equations (5.56) and (5.61) imply that the instanton field is given by

$$A_\mu = \frac{x^2}{x^2 + \rho^2} i S_1 \partial_\mu S_1^\dagger, \tag{5.75}$$

where $A_\mu = g A_\mu^a (\tau^a / 2)$ and the matrix S_1 is defined in Eq. (5.54). Let us now impose the condition that the time component of the gauge-transformed field A_μ vanishes identically,

$$U^\dagger A_4 U + i U^\dagger \partial_4 U = 0. \tag{5.76}$$

Substituting the expression for the instanton field we get the following equation for the gauge matrix U transforming the BPST instanton to the $A_0 = 0$ gauge,

$$\dot{U} + \frac{x^2}{x^2 + \rho^2} \left(S_1 \dot{S}_1^\dagger \right) U = 0, \tag{5.77}$$

where

$$S_1 \dot{S}_1^\dagger = i \frac{x\tau}{x^2} \tag{5.78}$$

and the dot denotes differentiation with respect to the time coordinate $x_4 = \tau$. The reader should be careful not to confuse the Pauli matrices τ with the time coordinate τ! The solution of (5.78) is obvious:

$$U(\tau, x) = \left[\exp \int_{-\infty}^{\tau} \left(\frac{-i x \tau}{x^2 + \rho^2} \right) d\tau \right] U(\tau = -\infty, x). \tag{5.79}$$

The instanton field in the $A_0 = 0$ gauge takes the form

$$A_i(\tau, x) = i \left[\frac{x^2}{x^2 + \rho^2} U^\dagger(\tau, x) \left(S_1 \partial_i S_1^\dagger \right) U(\tau, x) + U^\dagger(\tau, x) \partial_i U(\tau, x) \right]. \tag{5.80}$$

In the distant past and distant future

$$A_i \to i \left(U^\dagger S_1 \right) \partial_i \left(U^\dagger S_1 \right)^{-1}. \tag{5.81}$$

Moreover, $S_1 \to 1$ at $\tau \to \pm \infty$. For $U(t = +\infty)$, we have

$$U(\tau = +\infty, x) = \left[\exp \int_{-\infty}^{+\infty} \left(-\frac{i x \tau}{x^2 + \rho^2} d\tau \right) \right] U(\tau = -\infty, x)$$

$$= \exp \left[\left(-\frac{i \pi x \tau}{\sqrt{x^2 + \rho^2}} \right) \right] U(\tau = -\infty, x). \tag{5.82}$$

Correspondingly, the instanton solution in the $A_0 = 0$ gauge is

$$A_i(\tau, x_i) = \begin{cases} 0, & \text{at } \tau = -\infty, \\ i U^\dagger(x) \partial_i U(x) & \text{at } \tau = \infty, \end{cases} \tag{5.83}$$

---

[15] This is generally accepted physicists' jargon. Since we are in Euclidean space–time now, it would be more exact to speak of the $A_4 = 0$ gauge.

where

$$U(\boldsymbol{x}) = U(\tau = +\infty, \boldsymbol{x}) \equiv \exp\left(-i\pi \frac{\boldsymbol{x}\tau}{\sqrt{\boldsymbol{x}^2 + \rho^2}}\right). \tag{5.84}$$

Using (5.12) and (5.13) we write

$$\mathcal{K} = \int d^3x \, \frac{1}{8\pi^2} \, \varepsilon^{ijk} \, \mathrm{Tr}\left(A_i \partial_j A_k - \frac{2i}{3} A_i A_j A_k\right), \quad A_i \equiv g \frac{\tau^a}{2} A_i^a. \tag{5.85}$$

By virtue of the identity $\partial_j U^\dagger = -U^\dagger \left(\partial_j U\right) U^\dagger$ Eq. (5.85) reduces to

$$\mathcal{K} = \int d^3x \, \frac{1}{24\pi^2} \, \varepsilon^{ijk} \, \mathrm{Tr}\left[\left(U\partial_i U^\dagger\right)\left(U\partial_j U^\dagger\right)\left(U\partial_k U^\dagger\right)\right], \tag{5.86}$$

where

$$U(\tau = +\infty, \vec{x}) \equiv U(\vec{x}) = \exp\left(-i\pi \frac{\boldsymbol{x}\tau}{\sqrt{\boldsymbol{x}^2 + \rho^2}}\right). \tag{5.87}$$

This confirms the statement that $\mathcal{K} = 1$, cf. Section 4.2.4 and Exercise on page 169 (see also page 660).

5.4.5 Instanton Collective Coordinates (Moduli)

The instanton solution presented in Eq. (5.62) has the following collective coordinates: the instanton size ρ (associated with dilatations) and four parameters represented by the instanton center x_0 (associated with translations). The issue of collective coordinates is important, since each gives rise to a zero mode and the latter play a special role in calculating the instanton determinant, and, eventually, the instanton measure. Thus it is imperative to establish a complete set of collective coordinates. In this section we will analyze the set of collective coordinates for the SU(2) instantons.

The action in pure Yang–Mills theory, Eq. (5.48), has no dimensional parameters and is conformally invariant at the classical level. Since the instanton is the solution of the classical equations of motion (which are naturally conformally invariant too), the set of collective coordinates appearing in the generic instanton solution is determined by the conformal group. Each given instanton solution breaks (spontaneously) some invariances. Conformal symmetry is restored only upon consideration of the family of the solution as a whole. Those symmetry transformations that act on the instanton solution nontrivially generate another solution belonging to the same family, with "shifted" values of the collective coordinates. Thus each symmetry transformation from the conformal group which does not leave the instanton solution intact requires a separate collective coordinate.

The conformal group in four dimensions includes 15 transformations (it is briefly reviewed in Appendix section 1.4 see also, e.g., [23]), comprising four translations, six Lorentz rotations (in Euclidean space it is more appropriate to speak of six SO(4) rotations), four proper conformal transformations, and one dilatation. Moreover, the Yang–Mills action is gauge invariant. We do not need to consider (small) gauge transformations of the instanton, since they produce just the same solution in a different gauge. Global rotations in color space have to be considered, however. In SU(2) theory there are three global rotations. Thus, *a priori* one could expect the generic instanton solution to depend on 18 collective coordinates. So far, we have only seen five. Where are the remaining collective coordinates?

The proper conformal transformations can be represented as a combination of translations and inversion. Under inversion

$$x_\mu \to x'_\mu = \frac{x_\mu}{x^2}, \qquad A_\mu(x) \to x'^2 A_\mu(x'). \tag{5.88}$$

Translations are already represented by the corresponding collective coordinate, x_0. Now, if we start from the original BPST instanton with unit radius and make an inversion, we will obviously get an *anti-instanton* in the singular gauge,

$$\frac{2}{g}\eta_{a\mu\nu}\frac{x_\nu}{x^2+1} \xrightarrow{\text{inversion}} \frac{2}{g}\eta_{a\mu\nu}\frac{x_\nu}{x^2(x^2+1)} \tag{5.89}$$

(see Eqs. (5.62) and (5.65)). Thus, no new collective coordinates are associated with the proper conformal transformations.

What remains to be discussed? We must consider the six rotations in Euclidean space and the three global color rotations. An heuristic argument was given in Section 5.3.2. Here we will show, in a more comprehensive manner, that only three linear combinations of these nine generators act on the instanton solution nontrivially; the result is three extra collective coordinates, which will be defined explicitly.

To this end it is convenient to pass to a spinorial formalism (described in detail in Section 10.2 in the context Minkowski space). This formalism becomes practically indispensable in dealing with chiral fermions. To facilitate a comparison with Section 10.19 we will focus here on the *anti-instanton* solution.

Let us start from the anti-instanton solution that follows from (5.62)

$$g A_\mu^a \tau^a = 2\bar{\eta}_{a\mu\nu} x_\nu \tau^a \left(x^2 + \rho^2\right)^{-1}, \tag{5.90}$$

where the gauge field is treated as a matrix in the color space. Nothing interesting happens with the denominator, so we will forget about it for a short while and concentrate on the numerator,

$$N_{ij,\mu} \equiv 2\bar{\eta}_{a\mu\nu} x_\nu \left(\tau^a\right)_{ij}. \tag{5.91}$$

To pass from the vectorial to the spinorial formalism we multiply $N_{ij,\mu}$ by $(\tau_\mu^-)_{p\dot{q}}$. The matrix τ_μ^- was defined in Eq. (5.45). To distinguish the two SU(2) subgroups of O(4) we will use the dotted index for SU(2)$_R$ and undotted for SU(2)$_L$. Then

$$N_{ij,\mu} \to N_{ij,p\dot{q}} \equiv N_{ij,\mu}\left(\tau_\mu^-\right)_{p\dot{q}} = 2\bar{\eta}_{a\mu\nu} x_\nu \left(\tau^a\right)_{ij}\left(\tau_\mu^-\right)_{p\dot{q}}. \tag{5.92}$$

Using the definition of $\bar{\eta}_{a\mu\nu}$ from Eq. (5.43) and various completeness conditions for the Pauli matrices, we obtain, after some algebra,

$$N_{ij,p\dot{q}} = 2i\left[\delta_{pj}\left(x\tau^-\right)_{i\dot{q}} - \varepsilon_{ip}\varepsilon_{js}\left(x\tau^-\right)_{s\dot{q}}\right], \tag{5.93}$$

where $(x\tau^-)$ is a shorthand for $(x_\mu \tau_\mu^-)$. Thus, the anti-instanton field takes the form given by

$$g A_\mu^a \left(\tau^a\right)_{ij}\left(\tau_\mu^-\right)_{p\dot{q}} = \frac{N_{ij,p\dot{q}}}{x^2+\rho^2}. \tag{5.94}$$

*Anti-
instanton in
the spinorial
notation*

The dotted index of the $SU(2)_R$ subgroup goes from the left- to the right-hand side intact, while the index p of $SU(2)_L$ becomes entangled with the color indices. A remark in passing: in what follows it is instructive to rewrite (5.93) in terms of $\tilde{N}_{ij,p\dot{q}}$:

$$\tilde{N}_{ij,p\dot{q}} \equiv \left(\tau^2\right)_{ik} N_{kj,p\dot{q}} = 2i \left[\delta_{ip}\left(x\tau^2\tau^-\right)_{j\dot{q}} + \delta_{jp}\left(x\tau^2\tau^-\right)_{i\dot{q}}\right]. \tag{5.95}$$

This expression is slightly neater than (5.93). The reason why will become clear in Section 10.19.1.

In the instanton solution the entanglement pattern is different, namely, the undotted index of $SU(2)_L$ goes through, while the dotted index of $SU(2)_R$ becomes entangled with the color indices (see Exercise 5.4.1). In both cases, in spinorial notation the 't Hooft symbols are traded for the Pauli matrices.

Now we are ready to discuss what happens with the (anti-)instanton under Lorentz and/or color rotations. Transformations from $SU(2)_R$ (which act on the dotted indices) rotate x and A in the same way. In other words, the form of the anti-instanton solution (5.93) does not change at all; no collective coordinates corresponding to the $SU(2)_R$ rotations emerge in the anti-instanton solution.

We are left with the color rotations and Lorentz transformations from $SU(2)_L$. It is easy to see that they are not independent. Color transformations are equivalent to transformations from $SU(2)_L$. Indeed, the global color rotation acts on the 4-potential A as $A \rightarrow MAM^{\dagger}$ while the Lorentz rotation acts as $A \rightarrow LA$, where M and L are $SU(2)$ matrices. We obtain for the transformed 4-potential

$$\frac{2i}{x^2 + \rho^2}\left[(LM^{\dagger}) \otimes (Mx\tau^-) + \left(M\tau^2\tilde{L}\right) \otimes \left(\tilde{M}^{\dagger}\tau^2x\tau^-\right)\right]. \tag{5.96}$$

where the tildes indicate transposed matrices and to ease the notation all indices are omitted. Their convolution in (5.96) is evident from (5.93). Now we use

$$\tau^2\tilde{L} = L^{\dagger}\tau^2, \qquad M^*\tau^2 = \tau^2M$$

and impose the condition

$$M = L. \tag{5.97}$$

Under this condition the transformed 4-potential expressed in terms of the transformed x looks exactly like the original 4-potential expressed in terms of the original x.

*Cf. Section
10.19.8.*

This means that out of six transformations (three global color rotations and three $SU(2)_L$ rotations) only three are independent, giving rise to three moduli. We can choose them to be associated either with the global color rotations (as is usually assumed) or with those from $SU(2)_L$. If we follow the first route then, the three orientational moduli emerge from the matrix M,

$$A \rightarrow MAM^{\dagger}.$$

In the conventional formalism the orientational moduli are usually parametrized by an orthogonal matrix O_{ab}:

$$\eta_{a\mu\nu} \rightarrow O_{ab}\eta_{b\mu\nu}, \qquad \bar{\eta}_{a\mu\nu} \rightarrow O_{ab}\bar{\eta}_{b\mu\nu}. \tag{5.98}$$

The relation between O_{ab} and M is as follows:

$$O_{ab} = \tfrac{1}{2}\mathrm{Tr}\left(M\tau^a M^\dagger \tau^b\right). \tag{5.99}$$

The advantage of the spinorial formalism is obvious – there is no need to introduce the 't Hooft symbols and the hedgehog nature of the instanton is transparent.

Summarizing, eight collective coordinates characterize the SU(2) instanton. Correspondingly, we will observe eight zero modes. For higher gauge groups the number of collective coordinates corresponding to global color rotations increases. Altogether, in the group SU(N) the BPST instanton has $4N$ collective coordinates. This counting was first carried out in [24]. We will return to the discussion of the SU(N) instanton in Section 5.4.7.

5.4.6 SU(2) Instanton Measure

The *instanton measure* is defined as a weight factor in the functional integral associated with a given saddle point, the instanton saddle point in the case at hand. The exponential part of the weight factor, $\exp(-S_0)$, where S_0 is given by Eq. (5.50), is obvious. Therefore, when one speaks of the calculation of instanton measure one is implying, in fact, calculation of the pre-exponential factor. A full calculation, which was first carried out in [6][16] is tedious albeit straightforward. We will not dwell on this here. Instead, I will make a few observations that will allow us to establish the instanton measure $d\mu_{\text{inst}}$ up to an overall numerical constant.

The calculation of quantum corrections in $d\mu_{\text{inst}}$ (in one loop) amounts to integrating over small fluctuations of the gauge fields near the instanton solution in the quadratic approximation. Thus, we represent the field A_μ^a in the form

$$A_\mu^{a\,\text{inst}} + a_\mu^a \tag{5.100}$$

and expand the Yang–Mills action functional $S[A]$ with respect to the deviation field a_μ^a. In the quadratic approximation we obtain

$$S[A] = \frac{8\pi^2}{g^2} - \frac{1}{2}\int d^4x\, a_\mu^a \left[L_{\mu\nu}^{ab}\left(A_\mu^{a\,\text{inst}}\right)\right] a_\nu^b, \tag{5.101}$$

where

$$L_{\mu\nu}^{ab}\left(A_\mu^{a\,\text{inst}}\right) = \left(\mathcal{D}^2\delta_{\mu\nu} - \mathcal{D}_\mu\mathcal{D}_\nu\right)\delta^{ab} - g\varepsilon^{abc}G_{\mu\nu}^c \tag{5.102}$$

Quadratic expansion of the action near the instanton solution

and the fields G and A in (5.102) are those of the instanton. Path integration over $a_\mu^a(x)$ gives $(\det L)^{-1/2}$ in the instanton measure.

The latter statement is symbolic, for many reasons. First, we must fix the gauge and – a necessary consequence – introduce corresponding ghost fields, which result in a ghost operator determinant in addition to $(\det L)^{-1/2}$. Second, the operator L has zero modes. Formal substitution of the zero eigenvalues into $(\det L)^{-1/2}$ would lead to infinities. This was expected, and how to deal with them is well known: the zero modes must be excluded from $(\det L)^{-1/2}$. They reappear, however, in the form of integrals over all collective coordinates in $d\mu_{\text{inst}}$. Finally, the product

---

[16] The reprinted version of this paper takes account of the corrections summarized in the erratum to [6] and some other corrections.

of nonzero eigenvalues in $\det L$ diverges in the ultraviolet and so requires an ultraviolet regularization. Most often used for this purpose is the Pauli–Villars (PV) regularization, which prescribes that $\det L$ should be replaced as follows:

$$\det L \to (\det L)^{\text{reg}} = \frac{\det L}{\det(L + M_{\text{uv}}^2)}, \qquad (5.103)$$

where M_{uv} is the PV regulator mass (the ultraviolet cutoff).

The most labor- and time-consuming aspect is the treatment of the nonzero modes. As we will soon see, the impact of the nonzero modes on $d\mu_{\text{inst}}$ can be guessed without difficulty, taking into account the renormalizability of Yang–Mills theory.

Let us focus first on the zero modes, which are excluded from $(\det L)^{-1/2}$. Each zero mode gives rise to an integral over the corresponding modulus times a Jacobian due to the transition to integration over the moduli (which produces $\sqrt{S_0}$ per collective coordinate). The factor M_{uv} per zero mode comes from the ultraviolet regularization of $(\det L)^{-1/2}$, see Eq. (5.103). As we already know (see Section 5.4.5), the SU(2) instanton has eight collective coordinates: x_0 (the position of its center), ρ (its size), and three Euler angles, θ, φ, and ψ, which specify the orientation of the instanton in one of two SU(2) groups: either that of the color space or the (dotted) SU(2)$_R$ of the Lorentz group SO(4) = SU(2) × SU(2). Assembling all these zero-mode contributions, we arrive at

$$d\mu_{\text{inst}}^{\text{zm}} = \text{const} \times e^{-S_0} \left(M_{\text{uv}} S_0^{1/2} \right)^8 \int d^4x_0 \sin\theta \, d\theta \, d\varphi \, d\psi \, \rho^3 d\rho. \qquad (5.104)$$

The measure on the right-hand side is obviously invariant under translations and global SU(2) rotations. The factor ρ^3 in the integrand arises from the Jacobian associated with the transition to integration over θ, φ, and ψ; it is readily established on dimensional grounds.

Performing integration over the Euler angles θ, φ, and ψ and parametrizing the nonzero mode contribution in $d\mu_{\text{inst}}$ by a function Φ_1 in the exponent, we can rewrite Eq. (5.104) as follows:

$$d\mu_{\text{inst}} = \text{const} \times \left(\frac{8\pi^2}{g^2} \right)^4 \int \frac{d^4x_0 d\rho}{\rho^5} \exp\left[-\frac{8\pi^2}{g^2} + 8\ln(M_{\text{uv}}\rho) + \Phi_1 \right]. \qquad (5.105)$$

Needless to say, because the theory in question is renormalizable, only the renormalized coupling constant can appear in the instanton measure. To distinguish between these two couplings let us endow (temporarily) the bare coupling constant with a subscript 0. Then the expression in the exponent becomes

$$\frac{8\pi^2}{g_0^2} - 8\ln(M_{\text{uv}}\rho) \stackrel{?}{=} \frac{8\pi^2}{g^2(\rho)}, \qquad (5.106)$$

where for the moment I will ignore Φ_1. I denote by $g^2(\rho)$ the running coupling constant renormalized at the scale ρ^{-1}. The question mark over the equality sign warns us that it is not quite correct. To make it fully correct the factor 8 in front of the logarithm on the left-hand side of (5.106) must be replaced by b_0, the first coefficient in the Gell-Mann–Low function (also known as the β function), which governs the running law of the effective (renormalized) coupling constant. In the Yang–Mills theory for the gauge group SU(2),

The first cofficinet in the β function for SU(2) Yang–Mills

$$b_0 = \frac{22}{3} \equiv 8 - \frac{2}{3}. \tag{5.107}$$

Now it is quite evident that if we performed an honest calculation of Φ_1, collecting all nonzero mode contributions, we would obtain

$$\Phi_1 = -\frac{2}{3}\ln(M_{\mathrm{uv}}\rho) + \mathrm{const.} \tag{5.108}$$

The constant term renormalizes the overall constant in Eq. (5.105), which we will not be calculating anyway, while the logarithmic term corrects the coefficient in front of the logarithm in (5.105), (5.106), reducing the factor 8 to 22/3. The result is

$$\frac{8\pi^2}{g_0^2} - \frac{22}{3}\ln(M_{\mathrm{uv}}\rho) = \frac{8\pi^2}{g^2(\rho)}. \tag{5.109}$$

The factor g^{-8} in the pre-exponent in (5.105) remains unrenormalized in this approximation. To see its renormalization explicitly one should perform a two-loop calculation in the instanton background, a task which goes beyond the scope of the present text.

In summary, switching on one-loop quantum corrections we get the instanton measure in the form

Instanton density

$$d\mu_{\mathrm{inst}} \equiv \int \frac{d^4x_0\,d\rho}{\rho^5}d(\rho), \tag{5.110}$$

$$d(\rho) = \mathrm{const} \times \left(\frac{8\pi^2}{g^2}\right)^4 \exp\left(-\frac{8\pi^2}{g^2(\rho)}\right), \tag{5.111}$$

where the function $d(\rho)$ is referred to as the *instanton density*.

Here we will digress and return to the decomposition for the first coefficient in the Yang–Mills β function b_0 given in Eq. (5.107). In our instanton calculation the first term in (5.107), +8, comes from the zero modes while the second term, $-2/3$, which has the opposite sign and is much smaller in absolute value, comes from the nonzero modes in $(\det L)^{-1/2}$. The fact that these two contributions have distinct physical origins can be detected in conventional perturbative calculations (in ghost-free gauges) of gauge coupling renormalization. The negative term, $-2/3$, represents a "normal" screening of the "bare charge" at large distances. This is the only contribution whose analog survives in Abelian gauge theories. The positive term, +8, represents the antiscreening that is characteristic only to non-Abelian gauge theories. We discuss this issue in more detail in Appendix section 5.8.1.

5.4.7 Instantons in $SU(N)$

So far, we have discussed instantons in SU(2) Yang–Mills theories. Since the gauge group of QCD is SU(3) we should address the question of instantons in higher gauge groups. Now we will consider instantons in SU(N) with $N \geq 3$. The very fact of the existence of BPST instantons in SU(N) is due to the nontriviality of the relevant homotopy group,

Topological formula for the third homotopy group

$$\pi_3(\mathrm{SU}(N)) = \mathbb{Z} \tag{5.112}$$

for all N.

To construct an SU(N) instanton we simply embed the SU(2) instanton solution (5.62) or (5.65) in SU(N). This embedding is not unambiguous, as we will see shortly. The so-called *minimal embedding* (the most conventional) is carried out as follows. We select an SU(2) subgroup of SU(N) and choose generators in the fundamental representation. As a particular example, we choose the first three generators as the following $N \times N$ matrices

$$T^1 = \frac{1}{2}\left(\begin{array}{cc|c} 0 & 1 & \cdot \\ 1 & 0 & \cdot \\ \hline \cdot & \cdot & \cdot \end{array}\right), \quad T^2 = \frac{1}{2}\left(\begin{array}{cc|c} 0 & -i & \cdot \\ i & 0 & \cdot \\ \hline \cdot & \cdot & \cdot \end{array}\right), \quad T^3 = \frac{1}{2}\left(\begin{array}{cc|c} 1 & 0 & \cdot \\ 0 & -1 & \cdot \\ \hline \cdot & \cdot & \cdot \end{array}\right), \quad (5.113)$$

where the dots in the above definition represent zero matrices of appropriate dimensions. For instance, for SU(3) the three generators are

$$T^a = \tfrac{1}{2}\lambda^a, \qquad a = 1,2,3, \tag{5.114}$$

where the λ^a are the Gell-Mann matrices.

Next, we define the SU(N) instanton field using this matrix notation, as follows:

$$A_\mu^{\mathrm{SU}(N)\mathrm{inst}} = \sum_{a=1,23} A_\mu^a T^a, \tag{5.115}$$

where A_μ^a is given in Eqs. (5.62) and (5.65) for nonsingular and singular gauges, respectively. Equation (5.115) thus implies that

$$G_{\mu\nu}^{\mathrm{SU}(N)\mathrm{inst}} = \sum_{a=1,2,3} G_{\mu\nu}^a T^a. \tag{5.116}$$

Using the general definitions it is not difficult to see that the above SU(N) instanton solution is (i) self-dual, (ii) has unit topological charge, and (iii) has minimal (nontrivial) action $8\pi^2/g^2$. This embedding procedure is standard, and the instanton thus obtained is referred to as the SU(N) BPST instanton. Of course, in order to generate a full family of solutions we must include additional collective coordinates corresponding to global rotations of the given SU(2) subgroup within SU(N). This aspect will be discussed in Section 5.4.8.

A brief discussion is in order here regarding alternative embeddings. Long ago Wilczek noted [25] that if $T^{1,2,3}$ satisfy the SU(2) algebra and form *any* representation of SU(2) then the 4-potential (5.115) will give a self-dual field strength tensor, which thereby satisfies the classical equations of motion. For instance, in the physically interesting case of SU(3) we might choose three 3×3 Hermitian traceless matrices

| Wilczek's instanton, topological charge 4 |

$$\hat{T}^1 = \frac{1}{2}\begin{pmatrix} 0 & \sqrt{2} & 0 \\ \sqrt{2} & 0 & \sqrt{2} \\ 0 & \sqrt{2} & 0 \end{pmatrix}, \qquad \hat{T}^2 = \frac{1}{2}\begin{pmatrix} 0 & -i\sqrt{2} & 0 \\ i\sqrt{2} & 0 & -i\sqrt{2} \\ 0 & i\sqrt{2} & 0 \end{pmatrix},$$

$$\hat{T}^3 = \begin{pmatrix} 1 & 0 & 0 \\ 0 & 0 & 0 \\ 0 & 0 & -1 \end{pmatrix}, \tag{5.117}$$

satisfying the SU(2) algebra

$$[\hat{T}^i, \hat{T}^j] = i\varepsilon^{ijk}\hat{T}^k. \tag{5.118}$$

Now we can place this instanton in SU(3).

The general expression for the topological charge replacing (5.34) is

$$Q = \frac{g^2}{16\pi^2} \int d^4x \, \mathrm{Tr} \left(G_{\mu\nu} \widetilde{G}_{\mu\nu} \right). \tag{5.119}$$

For the generators (5.113) this reduces to (5.34), yielding $Q = 1$, while for those in Eq. (5.117) the topological charge is four times larger because $\mathrm{Tr}\,\hat{T}^i\hat{T}^j = 2\delta^{ij}$, to be compared with $\mathrm{Tr}\,T^iT^j = \frac{1}{2}\delta^{ij}$ in the fundamental representation. Correspondingly, the action of the Wilczek instanton is four times larger than that of the minimal instanton. From the standpoint of the latter the Wilczek solution presents a particular limiting case of a generic four-instanton solution, which can be obtained by bringing together four separated BPST instantons, each with unit topological charge.

5.4.8 The SU(N) Instanton Measure

Here I will briefly outline a calculation of the SU(N) instanton measure and its density $d(\rho)$. The relation between $d\mu_{\mathrm{inst}}$ and $d(\rho)$ for SU(N) instantons is the same as for SU(2); see (5.110).

The first question is how does the number of the instanton zero modes change in SU(N)? We already know that the instanton field uses only the SU(2) subgroup of the complete group. Suppose for definiteness that this subgroup occupies the top left-hand corner in the $N \times N$ matrix of generators (Fig. 5.4). It is clear that the five zero modes associated with translations and dilatations remain the same as in SU(2). Only the modes associated with the group rotations are changed. In SU(2) there were three rotational zero modes and, correspondingly, three rotational moduli residing in the matrix O_{ab} in Eq. (5.98). In SU(N) these three modes correspond to the three generators (5.113) at the top left in Fig. 5.4.

Those of the remaining SU(N) generators that lie in the $(N-2) \times (N-2)$ matrix at the bottom right of Fig. 5.4 obviously do not rotate this particular instanton field. Thus, to the three SU(2) rotations only $4(N-2)$ additional unitary rotations are added. They lie in two strips that overlap the SU(2) corner in Fig. 5.4.

The total number of zero modes of the BPST instanton is

$$5 + 3 + 4(N - 2) = 4N.$$

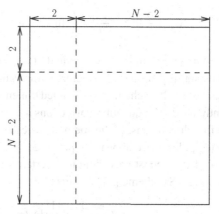

Fig. 5.4   Counting the generators of the group rotations in SU(N).

The first coefficient of the β function for SU(N) Yang–Mills theory

Of course, knowing what we already know, we can immediately say that this number, $4N$, is in one-to-one correspondence with the coefficient of the "antiscreening" logarithm in the the formula for running $g^2(\rho)$ in SU(N). Indeed, in this case the first coefficient of the β function can be written as

$$(b_0)_{\mathrm{SU}(N)} = \frac{11N}{3} \equiv 4N - \frac{N}{3}, \tag{5.120}$$

where the terms $4N$ and $-N/3$ come from the antiscreening and screening contributions (Figs. 5.23 and 5.22 in Appendix section 5.8), respectively.

Since SU(N) is a compact group and the SU(N) group space is finite, we can integrate explicitly over the collective coordinates associated with the instanton orientation in the SU(N) group space. The algebraic manipulations are rather tedious; here we limit ourselves to a few remarks regarding the final answer for the SU(N) instanton density $d(\rho)$,

SU(N) instanton density

$$d(\rho) = \frac{C_1}{(N-1)!\,(N-2)!} \left(\frac{8\pi^2}{g^2}\right)^{2N} e^{-8\pi^2/g^2(\rho) - C_2 N}, \tag{5.121}$$

where $g^2(\rho)$ is expressed in terms of the bare charge g_0^2 as follows:

$$\frac{8\pi^2}{g_0^2} - \frac{11N}{3}\ln(M_{\mathrm{uv}}\rho) = \frac{8\pi^2}{g^2(\rho)}. \tag{5.122}$$

The constants C_1 and C_2 can be found by a certain modification [26] of 't Hooft's calculations [6]. Compared with the SU(2) case it is necessary to take into account the additional $4(N-2)$ vector fields with color indices belonging to the two strips in Fig. 5.4. These "extra" fields contribute both through the zero and the nonzero modes.

This is not the end of the story, however, if we want to establish the values of both numerical constants, C_1 and C_2, in Eq. (5.121). To this end we need to find the embedding volume of SU(2) in SU(N), a rather complicated problem (see [26]). A factor $[(N-1)!\,(N-2)!\,]^{-1}$ is associated with this the embedding volume. I will just quote the final results for C_1 and C_2

$$C_1 = \frac{2e^{5/6}}{\pi^2} \approx 0.466, \tag{5.123}$$

$$C_2 = \frac{5}{3}\ln 2 - \frac{17}{36} + \frac{1}{3}(\ln 2\pi + \gamma) + \frac{2}{\pi^2}\sum_{s=1}^{\infty}\frac{\ln s}{s^2} \approx 1.296.$$

Connect with dimensional regularization. See Appendix section 5.8.2.

The constant C_2 depends on the method of regularization, which actually defines the bare constant. Equation (5.123) refers to the Pauli–Villars (PV) regularization.

Instead of the PV scheme the so-called dimensional regularization (DR) scheme is frequently used. The quantum corrections are calculated in $4 - \epsilon$ dimensions rather than in four dimensions. In this method, instead of logarithms of the ultraviolet cutoff parameter, poles in $1/\epsilon$ appear. To proceed from PV to DR we make the replacement $\ln M \to (1/\epsilon) + \mathrm{const}$ according to a certain rule. For instance, using the minimal subtraction (MS) scheme [27] one gets

$$C_{2\mathrm{MS}} = C_2 - \frac{1}{6} - \frac{11}{6}(\ln 4\pi - \gamma) = C_2 - 3.888. \tag{5.124}$$

Needless to say, simultaneously one must use $8\pi^2/g_{\mathrm{MS}}^2(\rho)$ in the exponent. Of course the relations between the *observable* amplitudes do not depend on the particular choice of regularization scheme. The instanton density *per se* is not observable. It is an element of a theoretical construction.

For further details about the passage from the PV scheme to those used in perturbation theory the reader is referred to Appendix section 5.8.2.

It is worth noting that, for a given N, the main ρ-dependence of the instanton density is determined by the running coupling $g^2(\rho)$ in the exponent. Substituting Eq. (5.122) into (5.121) we observe that $d(\rho)$ is a very steep function of ρ,

$$d(\rho) \sim \rho^{11N/3}, \tag{5.125}$$

i.e., it grows as a rather high power of ρ at large ρ. Thus, any ensemble of instantons will be dominated by the large-ρ instantons unless the instanton density is somehow cut off (e.g., through Higgsing the theory). At large ρ the gauge coupling constant becomes strong, and we completely lose theoretical control; quasiclassical methods are no longer applicable. This is the reason why instantons turn out to be rather powerless in solving the confinement problem in QCD despite the high expectations they originally raised [1]. Nevertheless, BPST instantons constitute an important element of the theoretical toolkit in other applications.

5.4.9 Instanton-Induced Interaction of Gluons

In this section we will discuss gluon scattering amplitudes induced by an instanton of fixed size ρ. In the leading approximation the set of these amplitudes is summarized by the effective Lagrangian

$$\mathcal{L}_\rho(x_0) = d(\rho)\rho^{-5}d\rho \left[\exp\left(\frac{2\pi^2\rho^2}{g}O^{ab}\bar{\eta}_{b\mu\nu}G_{\mu\nu}^a(x_0)\right)\right] + (\bar{\eta} \to \eta), \tag{5.126}$$

where O^{ab} is a global color rotation matrix containing three moduli (parametrizing three rotation angles). To find a multigluon scattering amplitude one must expand (5.126) up to an appropriate order in the field G. Note that the instanton-induced effective Lagrangian contains the $\bar{\eta}$ symbols in the exponent while that for the anti-instanton is obtained by the substitution $\bar{\eta} \to \eta$. The effective Lagrangian (5.126) has a number of parallels and a number of uses. For instance, it allows one readily to obtain the instanton–anti-instanton (IA) interaction, a crucial component of instanton-based models of the QCD vacuum [11]. While we will not discuss these models, some other applications will be considered, for instance, a three-dimensional analog of (5.126), in Section 9.7 and the exponential growth of instanton-induced cross sections, in Section 5.6.

Now let us derive the Lagrangian (5.126). The problem is formulated as follows [29]. Assume that one has a number of gluons with momenta $\Lambda \ll |p_i| \ll \rho^{-1}$. These gluons scatter in the "vacuum," where, by construction, we place an instanton of a size ρ that is much smaller than the wavelengths of the gluons involved. From the gluon point of view such an instanton presents a point-like vertex, which we want to find in the leading approximation.

To this end we will calculate in the given approximation, the transition amplitude between the vacuum and n gluons in two distinct ways and then compare the answers.

First, we will obtain this amplitude directly from instanton calculus and then from the effective Lagrangian. This will fix the form of the effective Lagrangian.

The reduction formula (e.g., [30]) for the amplitude of interest can be written as

$$\langle n \text{ gluons}|0\rangle = \left\langle 0 \left| i^n \prod_{k=1}^{n} \int dx_k e^{ip_k x_k} \epsilon_{\mu_k}^{a_k} p_k^2 A_{\mu_k}^{a_k}(x_k) \right| 0 \right\rangle, \tag{5.127}$$

Reduction formulas can be found in old texts; see, e.g., Bjorken and Drell.

where p_k and $\epsilon_{\mu_k}^{a_k}$ are the 4-momentum and the polarization vector of the kth gluon and $A_{\mu_k}^{a_k}(x_k)$ is the operator for the gluon field. To find the one-instanton contribution to (5.127) we follow a standard procedure consisting of a few steps. First we proceed to Euclidean space. Then, in the leading approximation, we replace the gluon field operator $A_{\mu_k}^{a_k}(x_k)$ by the classical instanton expression $\bar{A}_{\mu_k}^{a_k}(x_k - x_0)$ given in Eq. (5.65). The *singular gauge* is used because the reduction formula (5.127) is valid only for those fields which fall off fast enough as $|x_k - x_0| \to \infty$. In the nonsingular gauge we would have to replace the inverse propagator p_k^2 for each gluon in (5.127) by a more complicated expression.

Finally, we multiply the result by $d(\rho)\rho^{-5}d\rho$ and arrive at

$$\langle n \text{ gluons}|0\rangle = d(\rho)\rho^{-5}d\rho\, e^{-ix_0\Sigma p_k} \prod_{k=1}^{n} \left[\int dx_k e^{-ip_k x_k} (-p_k^2)\epsilon_{\mu_k}^{a_k} \bar{A}_{\mu_k}^{a_k}(x_k) \right], \tag{5.128}$$

where all quantities on the right-hand side are Euclidean. It is not difficult to find the Fourier transform of the instanton solution, which we need only in the limit $p\rho \to 0$:

$$\int dx e^{-ipx}(-p^2)\,\bar{A}_\mu^a(x) = \frac{4i\pi^2}{g}\bar{\eta}_{a\mu\nu}\, p_\nu \rho^2, \qquad p\rho \ll 1. \tag{5.129}$$

Substituting (5.129) into (5.128) we get

$$\langle n \text{ gluons}|0\rangle = d(\rho)\rho^{-5}d\rho\, e^{-ix_0\Sigma p_k} \prod_{k=1}^{n} \left[\frac{4i\pi^2}{g}\bar{\eta}_{a\mu_k\nu_k}\epsilon_{\mu_k}^{a_k}(p_k)_{\nu_k}\rho^2 \right]. \tag{5.130}$$

Exactly the same formula is obtained, in the leading approximation,[17] from the effective Lagrangian (5.126) with gauge field

$$A_\mu^a(x) = \sum_k \epsilon_\mu^a(p_k)e^{-ip_k x}, \tag{5.131}$$

which completes the proof. The factorials that occur in the expansion of the exponential cancel against the combinatorial coefficients.

The 't Hooft symbols in Minkowski space

To transform the instanton-induced effective Lagrangian to Minkowski space it is sufficient to replace $\bar{\eta}_{a\mu\nu}$ in Eq. (5.126) by $\bar{\eta}_{a\mu\nu}^M$, where

$$\bar{\eta}_{a\mu\nu}^M = \begin{cases} \bar{\eta}_{aij}, & \mu = i, \nu = j; i,j = 1,2,3, \\ -i\bar{\eta}_{a4j}, & \mu = 0, \nu = j; j = 1,2,3. \end{cases} \tag{5.132}$$

By the same token, $\eta_{a\mu\nu} \to \eta_{a\mu\nu}^M$.

---

[17] By the leading approximation we mean that corresponding to the highest possible power of $1/g$ and the lowest power in $p\rho$. Beyond the leading approximation, the exponent in (5.126) will contain other operators with, say, derivatives $\mathcal{D}_a G_{\mu\nu}$ or two or more Gs, along with a series in g.

The master formula (5.126) allows us easily to find the leading term in the IA interaction at large distances, the so-called dipole–dipole interaction. Indeed, Eq. (5.126), which was originally derived to describe the gluon scattering amplitudes is valid for any "background" field. In particular, this field can be caused by a distant anti-instanton of size ρ_A. If we substitute into Eq. (5.126) the value of the gluon field strength tensor induced by the anti-instanton centered at y_0 (assuming that $|x_0 - y_0|\rho_{I,A} \gg 1$) then we will get a formula [14] describing the instanton–anti-instanton interaction at large separation:[18]

Dipole–dipole IA interaction

$$\mathscr{A}_{IA} \sim \exp\left[-\frac{16\pi^2}{g^2} - \frac{32\pi^2}{g^2}\rho_I^2\rho_A^2\,\eta_{a\lambda\mu}\bar{\eta}_{b\lambda\nu}O^{ab}\frac{(x_0 - y_0)_\mu (x_0 - y_0)_\nu}{(x_0 - y_0)^6}\right]. \quad (5.133)$$

The anti-instanton centered at y_0 should be taken in the singular gauge; see Eq. (5.65), where $\bar{\eta}$ must be substituted by η. The interaction term obviously depends on the relative orientation of the IA pseudoparticles in color space. Setting

$$x_0 - y_0 \equiv R,$$

one can rewrite the relative orientation factor as

$$\eta_{a\lambda\mu}\bar{\eta}_{b\lambda\nu}O^{ab}R_\mu R_\nu = -\left[4(\hat{v}R)^2 - \hat{v}^2 R^2\right], \quad (5.134)$$

where the unit vector $\hat{v}_\mu$ is defined by $i\hat{v}_\mu\tau_\mu^- \equiv M$; see the end of Section 5.4.5. If you have difficulty in deriving Eq. (5.134), look at the solution of Exercise 5.4.2.

Let us rewrite Eq. (5.133) as follows:

$$\mathscr{A}_{IA} \sim \exp\left[-\left(\frac{16\pi^2}{g^2} + S_{\text{int}}\right)\right]. \quad (5.135)$$

This defines the interaction "energy" S_{int} of the IA system. Thus

$$S_{\text{int}} = \frac{32\pi^2}{g^2}\rho_I^2\rho_A^2\,\eta_{a\lambda\mu}\bar{\eta}_{b\lambda\nu}\,O^{ab}\frac{R_\mu R_\nu}{R^6}. \quad (5.136)$$

Note that if the instanton and anti-instanton are aligned in color space, i.e., $\hat{v}$ and R are parallel, then S_{int} is negative ($-S_{\text{int}}$ is positive) and maximal in its absolute value, reaching $96\pi^2/(g^2 R^4)$. The IA system is attractive in this case. This should be intuitively clear. For other relative orientations the IA interaction can be repulsive.

In this way one determines the IA interaction as a systematic double expansion, in the ratio $\rho/|x_0 - y_0|$ and also in the coupling constant.

For pedagogical reasons we will consider here a somewhat different (and less known) derivation of the IA interaction, which does not use the language of classical fields. It allows one to connect the classical problem of the IA interaction energy with the quantum problem of instanton-induced cross sections, on which we will focus in Section 5.6.1. In the present section we will apply this method to reproduce the dipole–dipole IA interaction (5.133).

The graphs relevant to this calculations are depicted in Fig. 5.5. An instanton with size ρ_I is placed at x and an anti-instanton with size ρ_A at the origin; $|x| \gg \rho_{I,A}$ is required.

---

[18] Note that two instantons or two anti-instantons do not interact, since both configurations are exact solutions of the (anti-)self-duality equations and saturate the bound $S \geq Q(8\pi^2/g^2)$. The action for two instantons is exactly equal to $16\pi^2/g^2$ and is independent of their separation.

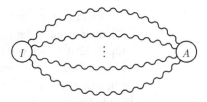

(a) One-gluon exchange   (b) Multigluon exchange

Fig. 5.5   The *IA* interaction from the instanton-induced effective Lagrangian (5.126). The instanton is at the point x while the anti-instanton is at the origin. The vertices in diagrams (a) and (b) are generated by expanding the exponent in Eq. (5.126) and keeping only the linear part of each $G_{\mu\nu}$ operator appearing in the expansion.

Figure 5.5b is an iteration of Fig. 5.5a; we will start from the one-gluon exchange between the instanton and anti-instanton presented in Fig. 5.5a.

First we expand the exponential in Eq. (5.126) and a similar one for the anti-instanton; we keep the terms linear in $G_{\mu\nu}^a$ in these expansions and contract $G(x)$ and $G(0)$ to get

$$\left(\frac{4\pi^4}{g^2}\rho_I^2\rho_A^2\right)O_I^{ab}\bar{\eta}_{b\mu\nu}O_A^{cd}\eta_{d\alpha\beta}\left\langle G_{\mu\nu}^a(x)G_{\alpha\beta}^c(0)\right\rangle, \tag{5.137}$$

where $\left\langle G_{\mu\nu}^a(x)G_{\alpha\beta}^c(0)\right\rangle$ is the free Green's function for the gauge field. Moreover, in the Green's function $\left\langle A_\mu(x)A_\nu(0)\right\rangle$ only the $\delta_{\mu\nu}$ part is retained, since the part $x_\mu x_\nu$ drops out. Then

$$\left\langle G_{\mu\nu}^a(x)G_{\alpha\beta}^c(0)\right\rangle = \frac{2\delta^{ac}}{\pi^2}\left(\delta_{\nu\alpha}x_\mu x_\beta + \delta_{\mu\beta}x_\nu x_\alpha\right.$$
$$\left.-\delta_{\nu\beta}x_\mu x_\alpha - \delta_{\mu\alpha}x_\nu x_\beta\right)\frac{1}{x^6} + \cdots, \tag{5.138}$$

where the ellipses indicate terms that do not contribute since

$$\eta_{a\mu\nu}\bar{\eta}_{b\mu\nu} = 0.$$

Next, it is not difficult to see that Fig. 5.5b just exponentiates the expression (5.137). Indeed, the factor $1/(n!)^2$ from the expansion of the effective Lagrangians is supplemented by the factor $n!$ coming from the combinatorics. In this way we immediately reproduce $-S_{\text{int}}$ in the form given in Eq. (5.136) (with the replacement $x_0 - y_0 \to x$).

5.4.10 Switching on the Light (Massless) Quarks

So far we have considered instantons in pure Yang–Mills theory. In our long journey through instanton calculus it is now time to turn to fermions; light or massless fermions play a very important role and drastically change some aspects of instanton-related physics. As an example I will mention the fact that in Yang–Mills theories with *massless* quarks and nonvanishing vacuum angle the θ-dependence of physical observables disappears.

The most suitable formalism for the treatment of light (massless) fermions is that of chiral spinors. However, chiral spinors *per se* cannot be continued in Euclidean space in a straightforward way. This will be discussed in Part II of this book (which is devoted to supersymmetry). For the time being we will limit ourselves to Dirac spinors. The corresponding theory can be formulated directly in Euclidean space–time (see Section 5.2). We will focus on fermions in the fundamental representation of the gauge group (i.e., quarks), which, for simplicity, will be assumed to be SU(2). There are no conceptual difficulties in generalizing the results to other groups, e.g., SU(N) with arbitrary N. At first we will keep a nonvanishing quark mass m in our analysis, but assuming that $m\rho \ll 1$; then we will let m tend to 0 and will observe, with surprise, the emergence of a new and very interesting physical phenomenon.

Thus our task is to calculate the path integral over the Fermi fields in the presence of a fixed-size instanton. In the Euclidean action, a fermion with mass m adds a term of the form (see (5.32))

$$S_F^{(E)} = \int d^4x\, \bar{\psi}\left(-i\gamma_\mu \mathcal{D}_\mu - im\right)\psi. \tag{5.139}$$

Keeping in mind that ψ and $\bar{\psi}$ are anticommuting fields, integrating them out yields

$$\mu_F = \int D\psi D\bar{\psi}\, e^{-S_F} = \det\left(i\gamma_\mu \mathcal{D}_\mu + im\right). \tag{5.140}$$

As usual the determinant can be understood as a product of the eigenvalues of the corresponding operator,

$$\det\left(i\gamma_\mu \mathcal{D}_\mu + im\right) = \prod_n (\lambda_n + im), \tag{5.141}$$

where the *real numbers* λ_n are the eigenvalues of the Hermitian operator $i\gamma_\mu \mathcal{D}_\mu$ having eigenfunctions u_n:

$$i\gamma_\mu \mathcal{D}_\mu u_n(x) = \lambda_n u_n(x), \tag{5.142}$$

with appropriate boundary conditions. These are imposed at a large but finite distance R from the instanton center to make the eigenfunctions $u_n(x)$ normalizable.

For any $\lambda_n \neq 0$ there exists a companion eigenvalue $-\lambda_n$. Indeed, let us define an eigenfunction $\tilde{u}_n(x) = \gamma_5 u_n(x)$. Then it is easy to see that $\tilde{u}_n$ satisfies the equation $i\gamma_\mu \mathcal{D}_\mu \tilde{u}_n(x) = -\lambda_n \tilde{u}_n(x)$. The only exception is in the case of the zero modes for which $\tilde{u}_n = \pm u_n$ and $\lambda_n = 0$. They do not have to be doubled.

Leaving aside the possible zero modes for a short while, we can say that

$$\det\left(i\gamma_\mu \mathcal{D}_\mu + im\right) \longrightarrow \prod_{n=0}^{\infty} \left(\lambda_n^2 + m^2\right) \tag{5.143}$$

up to an irrelevant overall factor (which is canceled by the same factor, coming from a regulator determinant). Thus, in Euclidean space the determinant arising from integrating out the Dirac fermions is positive definite. This is an important property, which makes lattice gauge theories with Dirac fermions relatively simple in comparison with theories with chiral fermions.

The occurrence of a zero mode in (5.142) will force the determinant to vanish in the limit $m = 0$. As we will see shortly, this will have far-reaching consequences. But first we will establish the existence of two zero modes per Dirac fermion, one in ψ and another in $\bar{\psi}$. We recall that ψ and $\bar{\psi}$ are to be treated as independent fields in Euclidean space–time.

Let us show that, in the instanton field background, Eq. (5.142) has one and only one normalizable solution with $\lambda = 0$; (5.142) then becomes

$$i\gamma_\mu \mathcal{D}_\mu u_0 = 0. \tag{5.144}$$

To find the above solution we pass to two-component spinors $\chi_{L,R}$ using the Weyl representation for the γ matrices,

$$\gamma_\mu = \begin{pmatrix} 0 & -i\sigma_\mu^- \\ i\sigma_\mu^+ & 0 \end{pmatrix}, \qquad \{\gamma_\mu, \gamma_\nu\} = 2\delta_{\mu\nu}, \tag{5.145}$$

$$u_0 = \begin{pmatrix} \chi_L \\ \chi_R \end{pmatrix}, \qquad \sigma_\mu^+ \mathcal{D}_\mu \chi_L = 0, \quad \sigma_\mu^- \mathcal{D}_\mu \chi_R = 0, \tag{5.146}$$

where[19]

$$\sigma_\mu^\pm = (\sigma, \mp i). \tag{5.147}$$

(Compare with (5.45).) Both σ and τ denote the Pauli matrices; we use τ in connection with the color indices and σ in connection with the Lorentz indices. Of course, when these indices get entangled, the distinction becomes blurred.)

Using the relations (5.46), the commutator

$$\left[\mathcal{D}_\mu, \mathcal{D}_\nu\right] = -ig\, G^a_{\mu\nu}\, (\tau^a/2),$$

and the explicit form of the gluon field strength tensor $G^a_{\mu\nu}$ from Eq. (5.62), we obtain the following equations for $\chi_{L,R}$ in the nonsingular gauge:

$$-\mathcal{D}_\mu^2 \chi_L = 0, \qquad \left\{-\mathcal{D}_\mu^2 + 4\sigma\tau \frac{\rho^2}{[(x-x_0)^2 + \rho^2]^2}\right\} \chi_R = 0. \tag{5.148}$$

The operator $-\mathcal{D}_\mu^2$ is a sum of the squares Hermitian operators: $-\mathcal{D}^2 = \left(i\mathcal{D}_\mu\right)^2$, i.e., it is positive definite. Therefore it has no vanishing eigenvalues and thus, $\chi_L = 0$.

In the equation for χ_R, we use a basis in the space of spinor and color indices that diagonalizes the matrix $\sigma\tau$. We recall that σ acts on the spinor indices while τ acts on the color indices. This basis corresponds to the addition of the ordinary spin and color spin to a total "angular momentum" equal to zero (when $\sigma\tau = -3$) or unity (when $\sigma\tau = +1$). It again follows from the positive definiteness of $-\mathcal{D}_\mu^2$ that the only suitable case for us, the only hope for obtaining a zero mode, occurs when the total "angular momentum" is equal to zero, which implies that $\sigma\tau = -3$ and completely determines the dependence of χ_R on the indices:

| Spin and color are entangled. |

$$(\sigma + \tau)\chi_R = 0, \qquad (\chi_R)_{\alpha k} \sim \varepsilon^{\alpha k}, \tag{5.149}$$

where $\alpha = 1, 2$ and $k = 1, 2$ are the spin and color indices, respectively. Their entanglement is obvious.

---

[19] At this point it is in order to make a comparison with the Minkowski formalism presented in Section 10.2.1. First, we note that the Euclidean "left- and right-handed" spinors are identified as $\chi_L \leftrightarrow \xi_\alpha$ and $\chi_R \leftrightarrow \bar{\eta}^{\dot\alpha}$, which is natural. Furthermore, $\sigma_\mu^+ \equiv \tau_\mu^+$ must be identified with $(\bar\sigma^\mu)^{\dot\alpha\alpha}$ and $\sigma_\mu^- \equiv \tau_\mu^-$ with $-(\sigma^\mu)_{\alpha\dot\alpha}$; cf. Eq. (10.40). We already know about the last identification $\sigma_\mu^- \leftrightarrow -(\sigma^\mu)_{\alpha\dot\alpha}$, from Section 5.4.5.

The dependence of χ_R on the coordinates can be readily found from the explicit form of $\mathcal{D}_\mu^2$. After a simple, albeit somewhat lengthy calculation we arrive at the final result for the zero mode:

$$u_0(x) = \frac{1}{\pi}\frac{\rho}{(x^2+\rho^2)^{3/2}}\begin{pmatrix}0\\1\end{pmatrix}\varphi, \qquad \varphi_{\alpha k} = \varepsilon^{\alpha k}. \qquad (5.150)$$

Here the normalizing condition

$$\int u^\dagger u\, d^4x = 1$$

has already been taken into account.

In concluding this section it is worth presenting the expression for the zero mode $u_0^{\text{sing}}(x)$ in the singular gauge, which we will need later:

$$u_0^{\text{sing}}(x) = \frac{1}{\pi}\frac{\rho}{(x^2+\rho^2)^{3/2}}\frac{x_\mu\gamma_\mu}{\sqrt{x^2}}\begin{pmatrix}1\\0\end{pmatrix}\varphi. \qquad (5.151)$$

To perform the transition to the singular gauge we multiply (5.150) by the gauge transformation matrix (5.64). At large x both expressions fall off as $1/x^3$.

5.4.11 Tunneling Interpretation in the Presence of Massless Fermions. The Index Theorem

Since the instanton contribution is proportional to $\det(i\gamma_\mu\mathcal{D}_\mu + im)$, and since the operator $i\gamma_\mu\mathcal{D}_\mu$ has a zero mode in the instanton field, it is tempting to conclude that in the massless limit the instanton contribution vanishes. How can one reconcile this result with the tunneling interpretation?

The introduction of fermions certainly does not affect the nontrivial topology in the space of the gauge fields. The existence of a noncontractible loop remains intact, and with this loop comes the necessity of considering a wave function of the Bloch type (Section 5.1). The instanton trajectory connects Ψ_n and Ψ_{n+1} under the barrier and is related to the probability of tunneling. If this probability were to vanish, what could have gone wrong with this picture?

To answer this question we must expand the picture of "tunneling in the $\mathcal{K}$ direction" by coupling the variable $\mathcal{K}$ to (an infinite number of) the fermion degrees of freedom. In order to make the situation more transparent we will slightly distort some details. We will assume that the motion of the system in the $\mathcal{K}$ direction is slow while the fermion degrees of freedom are fast, so that an approximation of the Born–Oppenheimer type is applicable. In this approximation the motion in the $\mathcal{K}$ direction is treated adiabatically. We first freeze $\mathcal{K}$, then consider the dynamics of the fermion degrees of freedom, integrate them out, and, at the last stage, return to the evolution of $\mathcal{K}$.[20] Certainly, in QCD all degrees of freedom are equally fast and no Born–Oppenheimer approximation can be developed. The general feature of the underlying dynamics in which we are interested does not depend, however, on this approximation.

---

[20] This picture becomes exact in the two-dimensional Schwinger model considered in Section 8.1. The essence of the phenomenon is the same in both theories.

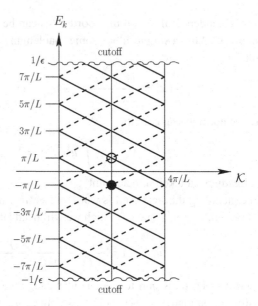

Fig. 5.6   The fermion energy levels v. $\mathcal{K}$.

Thus for each given value of $\mathcal{K}$ we must determine the fermion component of the

The Dirac sea in the instanton transition

wave function. This is done by building the Dirac sea in the fermion sector with $\mathcal{K}$ frozen. The structure of the Dirac sea depends on the value of $\mathcal{K}$.

When $\mathcal{K}$ varies adiabatically, the energy of the fermion levels evolves continuously. The points $\mathcal{K} = n$ and $\mathcal{K} = n + 1$, being gauge copies of each other, are physically identical. This means that the set of energy levels of the Dirac sea at $\mathcal{K} = n$ is identical to the set at $\mathcal{K} = n + 1$.

This does not mean, however, that the individual levels do not move. When $\mathcal{K}$ changes by one unit, some fermion levels with positive energy can dive into the negative-energy sea while those from this sea, with negative energies, can appear as levels with positive energies. As a whole the set will be intact but some levels interchange their positions (Fig. 5.6).

For each value of $\mathcal{K}$ we build the Dirac sea by filling all negative energy states and leaving all positive states unfilled. Let us say that at $\mathcal{K} = n$ we have built it in this way. If in the process of motion in the $\mathcal{K}$ direction, at $\mathcal{K} = n + 1/2$, say, one level dives into the sea and one jumps out, this must be interpreted as fermion production, since the state we end up with at $\mathcal{K} = n + 1$ is an excited state with respect to the filled Dirac sea at $\mathcal{K} = n$. Indeed, it has one filled positive-energy level and one hole. Thus, the tunneling trajectory connects the states $\Psi_n \Phi_n^{\text{ferm}}$ and $\Psi_{n+1} \Phi_{n+1}^{\text{ferm}}$, where the fermion components Φ_n^{ferm} and Φ_{n+1}^{ferm} differ by the quantum numbers of the fermion sector. In Section 5.4.10 we calculated the probability of the tunneling transition when there is no change in the fermion state, and we got zero in the limit $m \to 0$. We can now understand that we should not be discouraged: this zero value could have been expected since the tunnelings occur in such a way that the fermion quantum numbers are forced to change in the tunneling process.

The argument presented above is exact for the two-dimensional Schwinger model (or two-dimensional spinor electrodynamics); here, instead of $\mathcal{K}$, one considers the fermion level evolution as a function of A_1; see Chapter 8. In QCD the situation is complicated by the presence of infinitely many degrees of freedom but if we focus on $\mathcal{K}$, disregarding the other degrees of freedom the overall picture is the same. An

Impact of chiral anomaly, see Section 8.2

argument demonstrating the validity of this picture in QCD is based on the chiral (triangle) anomaly. Assume that we have one massless quark, q. At the classical level both the vector and axial currents

$$J_\mu^V = \bar{q}\gamma_\mu q, \qquad J_\mu^A = \bar{q}\gamma_\mu\gamma_5 q \tag{5.152}$$

are conserved:

$$\partial^\mu J_\mu^V = 0, \quad \partial^\mu J_\mu^A = 0. \tag{5.153}$$

The second equation implies $m = 0$. At the quantum level the axial current is anomalous,

$$\partial^\mu J_\mu^A = \frac{g^2}{16\pi^2} G^{a,\mu\nu}\widetilde{G}_{\mu\nu}^a. \tag{5.154}$$

Let us now integrate over x and evaluate this equation in the instanton field. On the left-hand side we first integrate over the spatial variables. Then the left-hand side reduces to

$$\int_{-\infty}^{\infty} dt\,\partial_0 \int J_0^A d^3x = Q_5(t = \infty) - Q_5(t = -\infty). \tag{5.155}$$

The right-hand side of (5.154) becomes

$$\frac{g^2}{16\pi^2}\int d^4x \left(G_{\mu\nu}^a \widetilde{G}_{\mu\nu}^a\right)_{\text{inst}} = 2Q = 2[\mathcal{K}(t = \infty) - \mathcal{K}(t = -\infty)]. \tag{5.156}$$

We see that in the theory with one massless quark, in the instanton transition the chiral charge (i.e., Q_5) is forced to change by two units: say, a left-handed quark is converted into a right-handed quark with unit probability. If we want to obtain a nonvanishing tunneling probability we have to incorporate this feature.

The change in the chiral charge, $\Delta Q_5 \neq 0$, in the tunneling transition is in one-to-one correspondence with the occurrence of the zero modes in the Dirac equation for the self-dual fields. The number of fermion zero modes is related to the topological charge of the gauge field by the famous Atiyah–Singer (or index) theorem [31],

Atiyah–Singer index theorem

which was derived in the instanton context in [32] (see also [12, 33, 34]). Specifically, if the number of the normalizable zero modes of positive (negative) chirality is $n_+(n_-)$ then

$$n_+ - n_- = Q \tag{5.157}$$

for each Dirac fermion field Ψ in the fundamental representation ($\bar{\Psi}$ is counted as an independent field). A brief but illuminating discussion of the derivation of Eq. (5.157) can be found in an article by Coleman [12]. As a matter of fact this theorem is equivalent to the triangle anomaly in the axial vector current presented above.

Summarizing the contents of this subsection and those of Section 5.4.10 we can say that each instanton (or anti-instanton) emits or absorbs two Weyl fermions of the same chirality per massless quark flavor.[21] In the theory with N_f massless

---

[21] The reader is invited to consider how this is compatible with the statement after (5.156) that a left-handed quark is converted into a right-handed quark.

flavors (Dirac spinor fields in the fundamental representation) every instanton or anti-instanton generates a vertex with $2N_f$ fermion lines, known as the 't Hooft vertex [6].

Let us note in passing that the presence of massless fermions, combined with the triangle anomaly in $\partial^\mu J_\mu^A$, results in another drastic consequence: the θ term becomes unobservable even if $\theta \neq 0$. Indeed, one can rewrite $\mathcal{L}_\theta$ from (5.19) as

$$\mathcal{L}_\theta = \frac{\theta}{2}\partial^\mu J_\mu^A, \tag{5.158}$$

i.e., a full derivative of the gauge-invariant quantity. Such full derivatives drop out of the action. This is in sharp distinction with the full derivative of the Chern–Simons current from (5.12), which, as we know, gives a nonvanishing contribution in the action once we switch on the instanton field. The Chern–Simons current is not gauge invariant.

This argument implies that in a theory with light quarks all θ-dependent effects must be proportional to the quark mass.

5.4.12 Instantons in the Higgs Regime

Quantum chromodynamics was the original testing ground for instanton calculus. Soon it became clear that small-size instantons are suppressed in QCD while large-size instantons dominate, because of the instanton measure. Analyses of the large-size instantons lie well beyond the scope of this book, because the theory becomes strongly coupled and our quasiclassical approximation fails. Quantum chromodynamics is not the only gauge theory of practical importance. Are there any other theories in which large-size instantons are suppressed?

The standard model for electroweak interactions is a gauge theory too. A drastic difference between the dynamics of QCD and that of the standard model arises because the non-Abelian gauge group is spontaneously broken in the standard model owing to the Higgs mechanism, and the coupling constant is frozen at values of momenta of the order of the W boson mass; it never becomes strong. Correspondingly, the color confinement and other peculiar phenomena of QCD do not take place. Since the nontrivial topology in the space of the gauge fields is not affected by the Higgs phenomenon, instantons (as tunneling trajectories) exist in the standard model too, leading to certain nonperturbative effects. The one that has been under the most intense scrutiny is baryon number violation at high energies [7] (for reviews see [35]). We will focus on this in Section 5.5. Here we concentrate on the theoretical issues. As a matter of fact, as we will see below, the consideration of instantons in the Higgs regime even has certain advantages over the QCD instantons. Since the coupling constant never becomes large, quasiclassical approximation is always justified in a description of the tunneling phenomena based on instantons. This is in contradistinction with QCD, where the instanton contribution is dominated by large-size instantons; these are obviously outside the scope of applicability of quasiclassical methods.

The first encounter with a truncated standard mode; χ is the Higgs field.

In what follows we will limit ourselves to a truncated standard model, the SU(2) gauge group, with minimal Higgs sector consisting of one complex Higgs doublet $\chi^i (i = 1, 2)$. The U(1) subgroup, as well as the fermions, present in the standard model are discarded.

It is convenient for our purposes to write the Lagrangian in the Higgs sector in a slightly unconventional form. The model at hand has a global SU(2) symmetry, because the doublet χ^i can be rotated into the conjugated doublet $\varepsilon^{ij}(\chi^j)^*$. This global symmetry is responsible for the fact that all three W, Z bosons are degenerate if the U(1) interaction is switched off in the standard model. The SU(2) symmetry of the χ sector becomes explicit if we introduce a matrix field

Look through Section 1.2.3.

$$X = \begin{pmatrix} \chi^1 & -\left(\chi^2\right)^* \\ \chi^2 & \left(\chi^1\right)^* \end{pmatrix}, \tag{5.159}$$

and rewrite the standard Higgs Lagrangian in terms of this matrix:

$$\Delta\mathcal{L}_\chi = \tfrac{1}{2}\operatorname{Tr}(\mathcal{D}_\mu X)^\dagger \mathcal{D}_\mu X - \lambda \left(\tfrac{1}{2}\operatorname{Tr} X^\dagger X - v^2\right)^2, \tag{5.160}$$

where $\mathcal{D}_\mu X = (\partial_\mu - igA_\mu)X$. The complex doublet field χ^i develops a vacuum expectation value v. This parameter can be arbitrary. If $v \gg \Lambda$ we are at weak coupling.

Because the Higgs field is in the fundamental representation of the color group, there is no clear-cut distinction between the confinement phase and the Higgs phase. As the vacuum expectation value (VEV) of the Higgs field χ changes continuously from large values to smaller values, we flow continuously from the weak coupling regime to the strong coupling regime. The spectra of all physical states, and all other measurable quantities, change smoothly [36].

One can argue that this is the case in many different ways. Perhaps the most straightforward line of reasoning is as follows. Using the Higgs field in the fundamental representation one can build gauge-invariant interpolating operators for *all* possible physical states. The Källen–Lehmann spectral functions corresponding to these operators, which carry complete information on the spectrum, depend smoothly on v. When the latter parameter is large the Higgs description is more convenient; when it is small it is more convenient to think in terms of bound states. There is no sharp boundary; we are dealing with a single Higgs–confining phase [36]. For a more detailed discussion see Section 1.3.5.

As v changes from large to small values, no phase transition is expected to occur.

All physical states form representations of the global SU(2) group. Consider, for instance, the SU(2) triplets produced from the vacuum by the operators

$$W_\mu^a = -\tfrac{1}{2}i \operatorname{Tr}\left(X^\dagger \overset{\leftrightarrow}{\mathcal{D}}_\mu X \tau^a\right), \qquad a = 1, 2, 3. \tag{5.161}$$

The lowest-lying states produced by these operators in the weak coupling regime (i.e., when $v \gg \Lambda$) coincide with the conventional W bosons of the Higgs picture, up to a normalization constant. The mass of the W bosons is $\sim gv$. If $v \lesssim \Lambda$, however, it is more appropriate to consider the bound states of the χ "quarks" as forming a vector meson triplet with respect to the global SU(2) symmetry ("ρ mesons"). Their mass is $\sim \Lambda$. The continuous evolution of v results in the continuous evolution of the mass of the corresponding states. It is easy to check that the complete set of gauge-invariant operators that one can build in this model spans the whole Hilbert space of physical states.

Now we will focus on two problems: calculation of the instanton action in the Higgs regime and of the height of the barrier in Fig. 5.3.

5.4.12.1 Instanton Action

Strictly speaking, if the scalar field develops a vacuum expectation value then the only exact solution of the classical equation of motion is the zero-size instanton. For each given value of ρ one can make the action of the tunneling trajectory smaller by choosing a smaller value of ρ, so that $8\pi^2/g^2$ is achieved asymptotically (see below). Since the nontrivial topology remains intact (one direction in the space of fields forms a circle), for proper understanding of the tunneling phenomena one cannot disregard the trajectories connecting the zero-energy gauge copies (pre-vacua) in Euclidean time, even though they are not exact solutions any more. Following 't Hooft [6], we will consider *constrained instantons* – trajectories that minimize the action under the condition that the value of ρ is fixed. Our analysis will be somewhat heuristic. The construction is described more rigorously in, for example, Ref. [37].

Technically the procedure can be summarized as follows. First we find the solution of the classical (Euclidean) equations of motion for the gauge field, ignoring the scalar field altogether. The solution is of course the familiar instanton. Then we look for a solution of the equations of motion for the χ field in the given instanton background. This solution minimizes the Higgs part of the action. A nonvanishing scalar field, in turn, induces a source term in the equation for the gauge field, which can be neglected. This source term will push the instanton towards smaller sizes, in particular, by cutting off the tails of the A_μ field at large distances (where they should become exponentially small). The distance at which this occurs is of order $1/(gv)$. If we are interested in distances of order $1/v$ – and the instanton contributions are indeed saturated at such distances – then we can neglect this effect and continue to disregard the back reaction of the scalar field in considering instantons whose sizes are fixed by hand.

To keep our analysis as simple as possible we will assume further that the scalar self-coupling $\lambda \to 0$. The only role of the scalar self-interaction then is to provide the boundary condition at large distances,

$$\frac{1}{2}\mathrm{Tr}\left(X^\dagger X\right) \to v^2. \tag{5.162}$$

The equation of motion of the scalar field is completely determined by the kinetic term in the Lagrangian,

$$\mathcal{D}_\mu^2 X = 0. \tag{5.163}$$

If the instanton field is written as in Eq. (5.75), i.e.,

$$A_\mu = \frac{x^2}{x^2 + \rho^2} i S_1 \partial_\mu S_1^+ \tag{5.164}$$

Scalar (Higgs) field in instanton background (for the anti-instanton, $S_1 \leftrightarrow S_1^\dagger$; the matrix S_1 is defined in Eq. (5.54)), it is not difficult to check that the solution of the equation $\mathcal{D}_\mu^2 X = 0$ takes the form

$$X = v \left(\frac{x^2}{x^2 + \rho^2}\right)^{1/2} S_1. \tag{5.165}$$

Asymptotically the modulus of the Higgs field approaches its value in the "empty" vacuum, while the "phase" part of the scalar field has a hedgehog winding. At small x the vacuum expectation value is suppressed.

Using the fact that the Higgs field satisfies the equation of motion, we can readily rewrite the contribution of the Higgs kinetic term in the action as

$$\int d^4x \partial_\mu \tfrac{1}{2} \operatorname{Tr}\left(X^\dagger \mathcal{D}_\mu X\right). \tag{5.166}$$

Moreover,

$$X^\dagger \mathcal{D}_\mu X = v^2 \rho^2 \frac{x_\mu}{(x^2 + \rho^2)^2} + v^2 \rho^2 \frac{x^2}{(x^2 + \rho^2)^2}\left(S_1^\dagger \partial_\mu S_1\right). \tag{5.167}$$

The contents of the last parentheses, being an element of the algebra, are proportional to τ^a and hence vanish when the color trace is taken. Therefore, the trace is determined entirely by the first term. Now exploiting the Gauss theorem and rewriting the volume integral as that over the surface of a large sphere with area element dS_μ, we arrive at

$$\int d^4x\, \partial_\mu \tfrac{1}{2} \operatorname{Tr}\left(X^+ \mathcal{D}_\mu X\right) = \int dS_\mu v^2 \rho^2 \frac{x_\mu}{(x^2 + \rho^2)^2} = 2\pi^2 v^2 \rho^2. \tag{5.168}$$

Summarizing, the extra term in the action induced by a nonvanishing vacuum expectation value of the Higgs field has the form[22]

$$\Delta S = 2\pi^2 v^2 \rho^2. \tag{5.169}$$

> *The 't Hooft interaction; cf. Section 10.19.9.*

This term is called the 't Hooft interaction, since 't Hooft was the first to calculate it [6]. It explicitly exhibits the feature we anticipated earlier – the smaller the instanton size ρ the smaller is the instanton action. It is clear that the instanton contribution to physical quantities is determined by an integral over ρ. Following the derivation leading to Eqs. (5.110) and (5.111) we can readily obtain the instanton measure in the problem at hand,

$$d\mu_{\text{inst}} = \text{const} \times d^4x_0 \frac{d\rho}{\rho^5} \exp\left\{-\left[\frac{8\pi^2}{g^2(\rho)} + 2\pi^2 v^2 \rho^2\right]\right\}. \tag{5.170}$$

The effective coupling $g^2(\rho)$ is given by a formula similar to Eq. (5.109) but with a slightly different coefficient:

> *Including the extra scalar field in the β function*

$$\frac{22}{3} \rightarrow \frac{22}{3} - \frac{1}{6}.$$

The exponential suppression, $\exp(-2\pi^2 v^2 \rho^2)$, of the instanton density at large ρ due to the 't Hooft term guarantees that $\rho \sim v^{-1}$. This, in turn, justifies the approximations made: the back reaction of the scalar field on the gauge field becomes important at much larger distances, $x \sim (gv)^{-1}$.

In the SU(2) theory the 't Hooft interaction depends on only one collective coordinate, ρ. In more complicated examples it may acquire dependences on other collective coordinates. For instance, if we consider an SU(3) model with one Higgs triplet (breaking SU(3) down to SU(2)) then the 't Hooft interaction will depend, roughly speaking, on the orientation of the instanton in the color space relative to the direction of the Higgs VEV. The 't Hooft term becomes $2\pi^2 v^2 \rho^2 \cos^2(\alpha/2)$, where α is an angle. If the instanton under consideration resides in the corner of SU(3)

---

[22] It is instructive to compare this calculation with that in Section 10.19.9.

corresponding to the unbroken $SU(2)$ then $\alpha = \pi$ and the 't Hooft term vanishes. Further details can be found in [38].

5.4.12.2 *IA* Interaction Due to the Higgs Field

In Section 5.4.9 we analyzed the IA interaction in pure Yang–Mills theory, in the leading (dipole) approximation. We found that at large separations this interaction falls off as $1/|R|^4$, where $R = x_0 - y_0$. Are there changes in the Higgs regime?

The answer is positive. First, the $1/|R|^4$ law is no longer valid at separations larger than $1/(gv)$. Because of the Higgsing of the theory, at such large distances all interactions fall off exponentially.

Second, even at distances $\rho < |R| < (gv)^{-1}$, before the onset of the exponential falloff there is an extra contribution to the IA interaction that is directly connected with the Higgs field, see Fig. 5.7. The graph in Fig. 5.7a presents the mass insertion in the gluon propagator (corresponding to the expansion $(p^2 - m_W^2)^{-1} \rightarrow p^{-2} + m_W^2 p^{-4}$). The W-boson mass is due to the Higgs field expectation value. This insertion is proportional to v^2 and can be obtained from Eq. (5.138). To this end the leading term in the gluon propagator, proportional to R^{-4}, must be supplemented by the next-to-leading term, proportional to $m_W^2 R^{-2}$. The angular dependence of this part of S_{int} is obviously the same as in Eq. (5.136).

Another contribution, with the same R^{-2} dependence on the IA separation, comes from the diagram in Fig. 5.7b, which describes the Higgs field exchange between the pseudoparticles. Conceptually, the corresponding calculation parallels that of Section 5.4.9.

The fastest way to proceed is as follows. Start from Eq. (5.170). Although this expression for S_{int} in the exponent on the right-hand side was obtained under the assumption that v is constant, it remains valid for all slowly varying background fields. Therefore, it follows that

$$S_{\text{int}}^H = 2\pi^2 \rho^2 v^2 \rightarrow \pi^2 \rho_I^2 \, \text{Tr}(X^+ X); \tag{5.171}$$

cf. Eq. (5.162). Now, one can substitute $\frac{1}{2} \text{Tr}(X^+ X)$ by the expression for the operator X in the *anti-instanton* background, see (5.165), namely,

$$\tfrac{1}{2}\text{Tr}\,(X^+ X) = v^2 \left(1 - \frac{\rho_A^2}{R^2 + \rho^2}\right). \tag{5.172}$$

The unit term on the right-hand side must be discarded, as it has nothing to do with the IA interaction. Then the part of the IA interaction due to Higgs exchange takes the form (at $R \gg \rho$)

$$S_{\text{int}}^H = -2\pi^2 v^2 \frac{\rho_I^2 \rho_A^2}{R^2}. \tag{5.173}$$

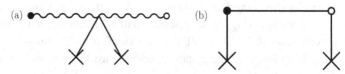

Fig. 5.7 An additional contribution to the *IA* interaction due to the Higgs field. The crosses denote the vacuum expectation value of the Higgs field. (a) Mass term of the gauge boson; (b) Higgs exchange.

5.4.13 Instanton Gas

In many instances, physical quantities are saturated by an ensemble of instantons rather than by a single instanton. The number of instantons and anti-instantons in the ensemble can be arbitrary. In addition to integrating over the instanton measure one must sum over the number of instantons and anti-instantons. If the pseudoparticles[23] in this ensemble are well separated and unaffected by each other's presence, one can speak of an instanton gas. The instanton gas approximation was introduced by Callan, Dashen, and Gross [14] in a bid to solve the problem of confinement in QCD. This attempt was unsuccessful; the instanton gas is not a good approximation in QCD even at the qualitative level. This is due to the fact that the ρ integrations diverge at large ρ: each instanton tends to become large and overlap with the others.

This is not the case in the Higgs regime. The 't Hooft term renders all ρ integrations convergent. The instantons are well isolated and so their interactions are negligible. Indeed, while the typical instanton size $\rho \sim v^{-1}$, the average density of pseudoparticles (i.e., their number per unit four-dimensional volume),

$$\frac{1}{V_4} \int_{V_4} d\mu_{\text{inst}} \sim v^4 \exp\left\{-\left[\frac{8\pi^2}{g^2(v)}\right]\right\}, \tag{5.174}$$

is exponentially suppressed. The average distance between two pseudoparticles,

$$\bar{R} \sim v^{-1} \exp\left[\frac{2\pi^2}{g^2(v)}\right]$$

is exponentially large, implying the exponential suppression of interactions. We are dealing with an extremely rarefied instanton gas here.

As is well known, the properties of the given ensemble in the gas approximation are determined by the characteristics of a single "molecule," an instanton or anti-instanton in the case at hand. Assume that we want to calculate the instanton contribution in the partition function

$$Z = e^{-E_{\text{vac}} V_4}\Big|_{V_4 \to \infty}. \tag{5.175}$$

Since the pseudoparticles do not interact, we can write

$$Z = \sum_{n+,n-} \prod \int d\mu_{\text{inst}} \int d\mu_{\text{anti-inst}}$$

$$= \sum_{n+,n-} \frac{1}{n_+!} \left(v^4 V_4\right)^{n_+} \exp\left\{-\left[\frac{8\pi^2 n_+}{g^2(v)}\right]\right\}$$

$$\times \frac{1}{n_-!} \left(v^4 V_4\right)^{n_-} \exp\left\{-\left[\frac{8\pi^2 n_-}{g^2(v)}\right]\right\}, \tag{5.176}$$

where $n_+ (n_-)$ is the number of instantons (anti-instantons); the overall measure is obtained as a product of the measures of all the pseudoparticles. We have neglected a pre-exponential constant in the instanton measure (cf. Eq. (5.121)) and set the vacuum angle θ equal to 0. Performing the summation we arrive at

$$Z = \exp\left(2v^4 V_4 e^{-8\pi^2/g^2(v)}\right), \tag{5.177}$$

---

[23] We will use the collective term "pseudoparticle" for instantons and anti-instantons, as suggested by Polyakov.

which implies, in turn, that

$$\delta_{\text{inst}} E_{\text{vac}} = -2v^4 e^{-8\pi^2/g^2(v)}. \tag{5.178}$$

The vacuum energy in the gas approximation is given by that of one instanton and one anti-instanton. The multi-instanton sum (5.176) exponentiates the one-instanton contribution. Note that the instanton contribution in E_{vac} is negative, in full accordance with the general statement that in switching on tunneling one lowers the ground-state energy (see, e.g., the famous quantum-mechanical double-well-potential problem [12, 13, 39]).

5.4.14 The Height of the Barrier. Sphalerons

Let us return to the tunneling interpretation of instantons discussed in Section 5.1 and ask the question: what is the height of the barrier in Figs. 5.2 or 5.3, through which the tunneling described by the instanton trajectory takes place? This issue is not as simple as it might seem at first sight.

Indeed, the QCD Lagrangian at the classical level contains no dimensional parameters. Since the instantons are solutions of the classical equations of motion (in Euclidean space), they do not carry dimensional constants other than the instanton size, which is a variable parameter. The height of the barrier must have the dimension of mass. Therefore the smaller the instanton size ρ, the higher the barrier seen by an instanton of size ρ, so that the classical action stays constant at $8\pi^2/g^2$. This is possible, of course, since the space of fields is actually infinite dimensional. The one-dimensional plot depicted in Fig. 5.3 is purely symbolic. Since the infrared limit of QCD is not tractable quasiclassically, it is impossible to determine the lowest possible height of the barrier under which the system tunnels. All we can say is that it is of order Λ_{QCD}.

The situation changes drastically in the Higgs regime considered in Section 5.4.12. The vacuum expectation value of the Higgs field provides masses for all gauge bosons. If the vacuum expectation value is much larger than Λ_{QCD} then the coupling constant stays small and the quasiclassical picture is fully applicable. Under these circumstances the question of what is the minimal height of the barrier in Fig. 5.3 becomes amenable to quantitative analysis. From this figure it is clear that when the system is at a position of maximum potential energy of the barrier, this is a solution of the *static* equations of motion since the position at the top of the barrier is an equilibrium. It is also clear that the equilibrium is unstable since it corresponds to a maximum of energy rather than a minimum.

Thus, we will look for the solution of the static equations of motion following from the Yang–Mills–Higgs Lagrangian in the $A_0 = 0$ gauge. By inspecting the structure of these equations it is easy to guess an *ansatz* that untangles the color and Lorentz indices,

$$A_i^a = \frac{1}{g} \varepsilon^{iak} \frac{x_k}{r} f(r), \qquad X = \frac{\tau x}{r} h(r), \tag{5.179}$$

where $r = \sqrt{x^2}$ and f, h are profile functions to be determined from the equations. The boundary conditions are obvious: at $r \to 0$ both functions must tend to zero to avoid singularities; at $r \to \infty$ the function h tends to v while $f(r) \to -2/r$. The latter condition is necessary to ensure that $A_i^a(r)$ becomes pure gauge at infinity; then the energy density of the gauge field will vanish at large r. Simultaneously the energy density of the scalar field also vanishes, in spite of the winding of the field X. The overall energy of the field configuration under consideration can be expected to be finite if both conditions are met.

Technically, instead of solving the equations of motion it is more convenient to write out the energy functional and minimize it with respect to f and h under the given boundary conditions. Substituting our *ansatz* into the Lagrangian (in Minkowski space) presented in Eq. (5.160), we readily obtain in the $\lambda \to 0$ limit[24]

$$\mathcal{H} = 4\pi \int_0^\infty r^2 dr \left[\frac{1}{g^2} \left(f'^2 + \frac{2}{r^2} f^2 + \frac{2}{r} f^3 + \frac{1}{2} f^4 \right) + h'^2 + 2h^2 \left(\frac{1}{r} + \frac{f}{2} \right)^2 \right].$$

(5.180)

The contents of the first pair of parentheses are from the gauge part (integration by parts has been carried out for one term). The second and the third terms in the square brackets represent the Higgs part. Since all terms in $\mathcal{H}$ are positive definite it is clear that a minimum of this functional exists; it can be found numerically.

Before minimization it is convenient to rescale the fields and the variable r to make them dimensionless. We set

$$f = gvF, \qquad h = vH, \qquad r = R(gv)^{-1}.$$

(5.181)

In terms of the rescaled fields the energy functional takes the form[25]

$$\mathcal{H} = 4\pi \frac{v}{g} \int_0^\infty R^2 dR \left[\left(F'^2 + \frac{2}{R^2} F^2 + \frac{2}{R} F^3 + \frac{1}{2} F^4 \right) + H'^2 + 2H^2 \left(\frac{1}{R} + \frac{F}{2} \right)^2 \right],$$

(5.182)

> Minimizing this functional we find sphalerons.

where the prime denotes differentiation over R. The expression in square brackets contains no parameters and neither do the boundary conditions for the dimensionless fields F, H; at $R \to \infty$ the function H approaches unity and the function F tends to $-2/R$. Numerical minimization of the integral in (5.182) is straightforward and is achieved on the profile functions F and H depicted in Fig. 5.8. The only parameter of the problem, v/g, is an overall factor. This means that the energy of the solution obtained by minimizing the energy functional $\mathcal{H}$ is

$$E \equiv \mathcal{H}_{\min} = \text{const} \times \frac{v}{g},$$

(5.183)

where the constant is of order unity. Its exact numerical value is not important for our illustrative purposes. It can be found in the original papers; see, e.g., [40].

---

[24] Compare the expressions (5.180) and (5.182) with the corresponding expression in Eq. (4.62) given in the context of monopole calculus. Caution: the notation is different!

[25] One may want to check that the first term in Eq. (5.182) vanishes for pure gauge (i.e., for $F = -2/R$). For this check to be carried out one should remember that the integrand contains a full derivative term not presented in (5.182), namely, $(R F^2)'$. Incorporating it we add in the first term (inside the brackets) an additional term,

$$\frac{F^2}{R^2} + \frac{2FF'}{R}.$$

Then it becomes immediately clear that the integrand vanishes upon substitution $F \to -2/R$.

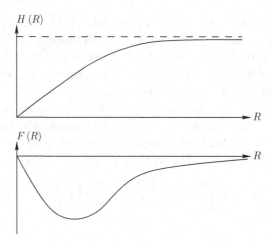

Fig. 5.8   The solutions for the sphaleron profile functions.

The static solution outlined above, corresponding to the top of the barrier, is called a *sphaleron*, from the Greek adjective *sphaleros* meaning unstable, ready to fall. It was found in [41] in SU(2) theory and rediscovered later in the context of the standard model by Klinkhamer and Manton [40], who were the first to interpret the sphaleron energy

$$M_{\text{sph}} = C \times \frac{v}{g} \qquad (5.184)$$

> *The sphaleron mass. In SU(2) theory $C = 2\sqrt{2}\pi^2$.*

as the height of the barrier separating distinct pre-vacua of the Yang–Mills theory in the Higgs regime (see also [42]). It is instructive to examine the position of the sphaleron on the plot of Fig. 5.3 directly, by calculating the winding number of the corresponding gauge field. Note that at large distances

$$(A_i)_{\text{sph}} \to iU\partial_i U^+, \qquad U = \frac{\tau x}{r}. \qquad (5.185)$$

The matrix U takes different values as we approach infinity from different directions. Thus the condition of compactification, which we impose on the vacuum gauge field, does not hold for the sphaleron. Correspondingly, the winding number $\mathcal{K}\left[(A_i)_{\text{sph}}\right]$ need not be integer. A direct calculation (which I leave as an exercise for the reader) readily yields

$$\mathcal{K}\left[(A_i)_{\text{sph}}\right] = \tfrac{1}{2}, \qquad (5.186)$$

demonstrating that the sphaleron sits right in the middle between two classical minima,[26] with $\mathcal{K} = 0$ and $\mathcal{K} = 1$.

To give a well-defined quantitative meaning to the height of the barrier in the absence of the Higgs field we must regularize the Yang–Mills theory in the infrared domain. A possible regularization was suggested in [44], where the Yang–Mills fields

> *Sphaleron in Yang–Mills on a sphere*

were put on a three-dimensional sphere of finite radius instead of the flat space of conventional QCD. The radius of the sphere plays essentially the same role as $(gv)^{-1}$ in the Higgs picture. If this radius is small, the quasiclassical consideration becomes

---

[26] The sphaleron field configuration is unstable with regard to decay into either of the two adjacent minima. The decay (explosion) process is discussed in, e.g., [43].

closed and one discovers analogs of the sphaleron solution in a natural way. The advantage of this regularization over the Higgs field regularization is the existence of analytic expressions. Both the sphaleron field configuration and its energy can be found analytically [44]. In particular, the sphaleron energy turns out to be $3\pi^2/g^2$ times the inverse radius of the sphere.

5.4.15 Global Anomaly

In Section 5.4.10 we discussed the massless quark effects in the presence of instantons. In particular, a formula for counting the number of fermion zero modes was derived. An inspection of this formula leads one to a perplexing question. Indeed let us assume that, instead of QCD, we are dealing with an SU(2) theory with one massless left-handed Weyl fermion transforming as a doublet with respect to SU(2). So far only Dirac fermions have been considered; one Dirac fermion is equivalent to *two* Weyl fermions. Now we want to consider a chiral theory. Before the advent of instantons this theory was believed to be perfectly well defined. It has no internal anomalies; see Chapter 8. Moreover, in perturbation theory, taken order by order, one

Witten's SU(2) anomaly

encounters no reasons to make the theory suspect. And yet this theory is pathological. Analysis of the instanton-induced effects helps to reveal the pathology.

Indeed, following the line of reasoning in Section 5.4.10, in the SU(2) theory for a *single* massless left-handed Weyl fermion we would immediately discover that an instanton-induced fermion vertex of the 't Hooft type must be *linear* in the fermion field. Indeed, for an instanton transition with one Dirac fermion we have $\Delta Q_5 = 2$,[27] but since the Weyl fermion is only half the Dirac fermion, $\Delta Q_5 = 1$!

This contradiction made it obvious to many that something was unusual in this theory. The intuitive feeling of pathology was formalized by Witten, who showed [45] that the theory is ill defined because of the *global anomaly*. Such a theory is mathematically inconsistent. It simply cannot exist.

One proof of the global anomaly is based on fermion level restructuring in the instanton transition. The key elements are the following: (i) the vacuum-to-vacuum amplitude in the theory with one Weyl fermion is proportional to $\sqrt{\det(i\slashed{D})}$; (ii) only one fermion level changes its position with regard to the Dirac sea zero when $\mathcal{K} = n$ passes to $\mathcal{K} = n + 1$, as opposed to one *pair* in the case of the Dirac fermion, Fig. 5.6. This forces the partition function to vanish, making all correlation functions ill defined. For further details see [45].

Exercises

5.4.1 Generalize our derivation of the anti-instanton field in the spinorial notation, see Eq. (5.94), to instantons. Hint: Treat the indices of the color matrix as dotted.

5.4.2 Prove Eq. (5.134) through a direct calculation using definitions and results presented in Sections 5.3.2, 5.4.3, and 5.4.5.

---

[27] The Weyl fermion's contribution to the chiral anomaly is half of that of the Dirac fermion.

5.4.3   Calculate the integral in (5.79) explicitly. Find the instanton field in the $A_0 = 0$ gauge for arbitrary values of τ.

5.4.4   Verify that the expression (5.165) is indeed a solution of Eq. (5.163).

5.4.5   Verify Eq. (5.186).

5.5 Applications: Baryon Number Nonconservation at High Energy

5.5.1 Where Baryon Number Violation Comes From

In the previous sections of this chapter we have become acquainted with instanton calculus. Now it is time to discuss practical applications. The question of baryon number violation at high energies, i.e., those approaching the sphaleron mass, is one of the most interesting applications. The essence of this remarkable phenomenon, to be considered in some detail below, is as follows. As is well-known, the baryon number is not conserved in the standard model (SM). This nonconservation is caused by instantons and is suppressed (in the transition rate) by the square of the instanton factor, $[\exp(-2\pi/\alpha)]^2$, making proton decay unobservable. It turns out [35] that in scattering processes at high energies baryon number violation is exponentially enhanced. At energies below the sphaleron mass the cross sections of such processes grow *exponentially*; they level off only at $E \sim M_{\text{sph}}$. The reason for the exponential growth is the multiple production of W bosons and Higgs particles. At $E \sim M_{\text{sph}}$ the number of particles produced approaches $1/\alpha$ and a finite fraction of the suppressing exponent $4\pi/\alpha$ has gone. The result is a gigantic enhancement. However, in spite of this gigantic enhancement, a residual suppression of the type $\exp(-c\pi/\alpha)$ apparently still persists, c being a numerical factor strictly less than 4. As a result, baryon number violating processes remain unobservable even at high energies, albeit many orders of magnitude "less unobservable" than at low energies.

To understand how all this works we should remember the basic lessons we learned from the previous instanton studies:

(i) The vacuum in Yang–Mills theories has a complex structure. The vacuum wave function is a linear superposition of an infinite set of pre-vacua labeled by the winding (or Chern–Simons) number $\mathcal{K} = 0, \pm 1, \pm 2$, etc.

(ii) The instanton is the tunneling trajectory connecting these pre-vacua. The instanton contributions are well defined and exponentially suppressed in the Higgs regime.

(iii) The introduction of massless Dirac fermions leads to a new phenomenon, nonconservation of the axial charge: $\Delta Q_5 = 2$ per flavor in an instanton transition with $\Delta \mathcal{K} = 1$.

Now we will expand our explorations and study instanton-induced effects in the fermion sector of *chiral theories*.

5.5.1.1 Chiral Theory: What Is It?

Important:
defining
chiral
theories

Assume that we have a non-Abelian gauge theory. (For definiteness one can think of SU(N) as its gauge group, but this is not necessary.) Fermions in various representations of the gauge group comprise the fermion sector. If the set of fermions is such that no Lorentz-invariant and *gauge-invariant* mass term, bilinear in the fermion fields, is possible then the theory is said to be *chiral*.

Yang–Mills theories with any number of Dirac fermions are obviously nonchiral. Thus, the class of chiral theories is limited to those in which the fermion sector consists of Weyl fermions in complex (i.e., nonreal) representations of the gauge group. Not any set of the Weyl fermions makes a given gauge theory chiral, for the following reasons. First, some sets are reducible to a number of Dirac fermions. For instance, one Weyl (left-handed) fermion in the fundamental representation plus one Weyl (left-handed) fermion in the antifundamental representation comprise one

Internal
chiral
anomaly,
Fig. 5.9, cf.
Section 8.2.

Dirac spinor. Second, the choice of fermion sector is subject to a very rigid constraint: the set of Weyl fermions in question should not induce the *internal* chiral anomaly (in four dimensions it is also known as the triangle anomaly). The dangerous triangle diagrams appear as a one-loop correction to the three-gauge-boson vertex. They are depicted in Fig. 5.9. The anomalous part of these graphs violates gauge symmetry, and the theory becomes inconsistent at the quantum level. Such a theory simply does not exist. The anomalous part of the triangle graph in Fig. 5.9, with gauge bosons A_μ^a, A_ν^b, and A_λ^c, is proportional to

$$\sum_R \left[\sum_{\text{left}} \text{Tr}_R \left(T^a \left\{ T^b, T^c \right\} \right) - \sum_{\text{right}} \text{Tr}_R \left(T^a \left\{ T^b, T^c \right\} \right) \right], \qquad (5.187)$$

where $T^{a,b,c}$ denote the generators of the gauge group in the representation R to which a given fermion belongs, the sums run over all left-handed and right-handed fermions, respectively, and over all representations, and Tr_R denotes the trace in the representation R. Finally, the braces $\{\cdots\}$ stand for an anticommutator. The anticommutator emerges from the sum of two triangle diagrams in which the fermions circle in opposite directions. Note that if T^a is a generator in the representation R, the generator in the representation $\bar{R}$ is $-\tilde{T}^a$, where the tilde means transposition.

Equation (5.187) is very restrictive. Only very special sets of chiral fermions satisfy this constraint. Let us give some examples.

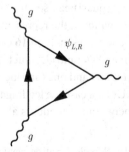

Fig. 5.9   Triangle graph which can lead to internal anomalies in chiral theories.

The simplest example would be the SU(2) gauge theory with one Weyl, say, left-handed, fermion in the fundamental (doublet) representation. Equation (5.187) is trivially satisfied since the anticommutator of two SU(2) generators, $\{\tau^b/2, \tau^c/2\}$, equals $\delta^{bc}/2$. This implies in turn that the trace in (5.187) vanishes trivially.

Furthermore, in this theory it is impossible to write down the mass term. Indeed, if the fermion field is denoted by ψ^i_α, where α, is the Lorentz index while i is the color SU(2) index, the only appropriate mass term would be $\psi^i_\alpha \psi^j_\beta \varepsilon^{\alpha\beta} \varepsilon_{ij}$. However, this expression vanishes identically since ψ is an anticommuting variable.

Thus, at first sight one-doublet SU(2) theory seems to be a good model to represent the class of chiral theories. Unfortunately, this theory has a global anomaly (see Section 5.4.15) and, because of this, cannot exist.

Next, if ψ is in the two-index symmetric representation of SU(2), it is equivalent to ψ in the adjoint representation, which is real. Thus, this theory is nonchiral.

The simplest chiral theory is obtained when ψ is in the three-index symmetric representation, SU(2)-spin 3/2. This theory has no internal anomalies (nor global anomaly) and no Lorentz- and gauge-invariant mass term is possible, for the same reason as in the case of one fundamental fermion.

Another well-known example of a chiral theory is the SU(5) theory with k decuplets $\psi^{[ij]}$ (as usual the square brackets around the indices denote antisymmetrization) and k antiquintets χ_i of left-handed fermions. Finally, one could mention the so-called *quiver* theories in which the gauge group is a product

$$\mathrm{SU}(N)_1 \times \mathrm{SU}(N)_2 \times \cdots \times \mathrm{SU}(N)_k \qquad (5.188)$$

and the set of left-handed fermions consists of k *bifundamentals*

$$\psi^{i_1}_{j_2}, \quad \psi^{i_2}_{j_3}, \quad \cdots, \quad \psi^{i_{k-1}}_{j_k}, \quad \psi^{i_k}_{j_1}.$$

The $k = 2$ quiver is nonchiral. Each fermion transforms in the fundamental representation with regard to one SU(N) and in the antifundamental representation with regard to its nearest neighbor SU(N) (the one to the right). At $k = 2$ the quiver theory is nonchiral since a gauge-invariant mass term can be built. If $k \geq 3$ then the quiver theory is chiral.

An easy way to build a chiral theory that will be internally anomaly-free is to start from a larger anomaly-free theory and pretend that the gauge symmetry of the original model is somehow spontaneously broken down to a smaller group. The gauge bosons corresponding to the broken generators are frozen out. The matter fields that are singlets with respect to the unbroken subgroup can be discarded. The remaining matter sector may well be chiral, but there will be no internal anomalies. For instance, to obtain the SU(5) theory we may start from SO(10), where all representations are (quasi)real, so this theory is automatically anomaly-free. Assume that we introduce matter in the representation **16** of SO(10). Now, we break SO(10) down to SU(5). The representation **16** can be decomposed with respect to SU(5) as a singlet, a quintet, and an (anti)decuplet. Drop the singlet. We are left with the SU(5) model with one quintet and one (anti)decuplet. Further, we can break SU(5) down to SU(3)×SU(2)×U(1), a cascade leading to SM and GWS.[28] The fermion sector of the Glashow–Weinberg–Salam model is also chiral. For obvious reasons it is of special

---

[28] SM = the Standard Model of particle physics; GWS = the Glashow–Weinberg–Salam model of electroweak interactions.

importance. Below, we will discuss baryon number nonconservation in the context of this model. We will consider mostly the conceptual aspects, leaving technicalities aside. For those who are interested in the technicalities we can recommend the review papers [35].

5.5.1.2 Simplifying the Standard Model

We do not have to consider the standard model in full. Inessential details would just overshadow the essence of the phenomenon that we would like to study. Therefore, at the start we will simplify the model, somewhat distorting it in comparison with the genuine SM. We will keep, however, those features which are relevant to baryon number nonconservation.

First, it is sufficient to consider just one generation. Second, we will set the Weinberg angle θ_W equal to 0. This decouples the photon field, which, becoming sterile, can be safely discarded at $\theta_W = 0$; then the gauge group becomes $SU(3)_{\text{strong}} \times SU(2)_{\text{weak}}$ and the masses m_W of all three "weak" gauge bosons W^a become equal. Finally, we will disregard the Higgs couplings to fermions. In this approximation the fermion fields remain massless. The coupling of the Higgs doublet to the gauge fields is fixed by the gauge coupling. We will use the convention

$$\mathcal{L}_H = \left| \left(\partial_\mu - i g_2 A^a_\mu \frac{\tau^a}{2} \right) H \right|^2 - \lambda \left(|H|^2 - v^2 \right)^2, \tag{5.189}$$

where g_2 and λ are coupling constants (the subscript 2 emphasizes that g_2 is the gauge coupling of $SU(2)_{\text{weak}}$), H is the Higgs doublet, and v is its vacuum expectation value.[29] In this convention the W-boson mass is given by

$$m_W = \frac{g_2 v}{\sqrt{2}}.$$

Finally, we will assume that $\lambda \ll g_2^2$. This allows us to ignore the Higgs particle self-interaction.

The fermion sector of the simplified model is as follows. We have three doublets of left-handed (colored) quarks,

$$q^{i,a} = \begin{pmatrix} u_L^a & i = 1 \\ d_L^a & i = 2 \end{pmatrix} \tag{5.190}$$

where $a = 1, 2, 3$ is the color index, and one doublet of left-handed leptons,

$$\ell^i = \begin{pmatrix} \nu_L \\ e_L \end{pmatrix}. \tag{5.191}$$

The right-handed components u_R^a, d_R^a, and e_R are singlets with respect to $SU(2)_{\text{weak}}$. Thus these fields do not participate in weak interactions.

The above simplifications do not distort the essence of the phenomenon. The results will remain valid in SM: the Higgs coupling to fermions does not change the anomalies (5.194), to be considered below, nor is the inclusion of the $U(1)_Y$ gauge field (i.e., the switching on of $\sin^2 \theta_W \neq 0$) crucial. The $U(1)_Y$ gauge field is not

---

[29] In many textbooks the normalization of the vacuum expectation value differs by $1/\sqrt{2}$, so that then $m_W = g_2 v/2$.

involved in SU(2) instantons. The effects due to this field on the SU(2) instanton measure are negligible. The sphaleron mass $M_{\mathrm{sph}} \sim v/g$ is slightly different in our simplified model compared to its value[30] in the full SM where $\sin^2 \theta_W \approx 0.23$ and $\lambda \gtrsim g_2^2$. This change is numerically small [40].

5.5.1.3 Anomalous and Nonanomalous Global Symmetries

The above theory is free of internal anomalies. Let us now discuss global symmetries. The baryon and lepton charges are defined as

$$Q_B = \int d^3x \, J_B^0, \quad Q_L = \int d^3x \, J_L^0, \tag{5.192}$$

where

$$J_B^\mu = \tfrac{1}{3} \left[\bar{q}_{i,a} \gamma^\mu q^{i,a} + (\bar{u}_R)_a \gamma^\mu u_R^a + (\bar{d}_R)_a \gamma^\mu d_R^a \right],$$
$$J_L^\mu = \bar{\ell}_i \gamma^\mu \ell^i + \bar{e}_R \gamma^\mu e_R. \tag{5.193}$$

In any Feynman graph Q_B and Q_L are conserved separately: the number of incoming quark lines is equal to the number of outgoing lines and the same is true for the lepton lines. This is illustrated in Fig. 5.10, where we have two incoming q lines and one ℓ line, and exactly the same numbers of outgoing lines. However, both currents (5.193) have anomalies with respect to the gauge bosons of SU(2)$_{\mathrm{weak}}$. Now we will calculate them. Note that all terms in Eq. (5.193) with the right-handed fermions are irrelevant since the right-handed fields, being SU(2)$_{\mathrm{weak}}$ singlets, do not interact with the W bosons.

> *Some advice: consult Section 8.2.*

The anomalies are determined by the triangle diagram of Fig. 5.11. Taking into account the normalization of the baryon current and the fact that the left-handed quark doublet q is repeated three times because of the three colors, we conclude that

$$\partial_\mu J_B^\mu = \partial_\mu J_L^\mu = \frac{g_2^2}{16\pi^2} \frac{1}{2} F_{\mu\nu}^a \widetilde{F}^{\mu\nu a}, \tag{5.194}$$

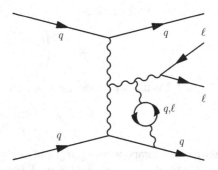

Fig. 5.10   In perturbation theory any Feynman graph conserves Q_B and Q_L separately.

---

[30] The sphaleron mass in SU(2) theory was evaluated in Section 5.4.14. There I omitted the numerical factors. Reinstating these numerical factors we have [40]

$$M_{\mathrm{sph}} = \pi \frac{m_W}{\alpha_2} = 2\sqrt{2}\pi^2 \frac{v}{g^2}.$$

Numerically the expression above is close to 7 TeV.

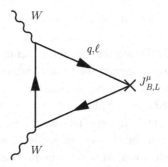

Triangle anomaly in the divergence of J_B^μ and J_L^μ. In contradistinction to Fig. 5.9, in the given triangle only two vertices are due to gauge bosons; the third vertex is due to the *external* current (5.193).

where $F_{\mu\nu}^a$ is the W-boson field strength tensor,

$$F_{\mu\nu}^a = \partial_\mu W_\nu^a - \partial_\nu W_\mu^a + g_2 \varepsilon^{abc} W_\mu^b W_\nu^c, \qquad a, b, c = 1, 2, 3, \tag{5.195}$$

and the factor $\frac{1}{2}$ in Eq. (5.194) is due to the fact that it is the left-handed Weyl fermion rather than the Dirac fermion that propagates in the triangle loop. Recall that in Minkowski space $F_{\mu\nu}^a \tilde{F}^{\mu\nu a} = -2\vec{E}^a \vec{B}^a$.

Equation (5.194) obviously implies that (i) the baryon and lepton charges are not separately conserved because the right-hand side can be nonvanishing, generally speaking; (ii) $Q_B - Q_L$ is a conserved quantum number. Integrating Eq. (5.194) over d^4x, we can express the nonconservation of $Q_{B,L}$ as follows:

$$\Delta Q_B = \Delta Q_L = \frac{g_2^2}{32\pi^2} \int d^4x F_{\mu\nu}^a \tilde{F}^{\mu\nu a}$$

$$= \frac{g_2^2}{32\pi^2} \int d^4x \partial_\mu K^\mu = \Delta\mathcal{K}, \tag{5.196}$$

where the Chern–Simons current K^μ and the Chern–Simons charge $\mathcal{K}$ were discussed in Section 5.4.11. The integral in the second line vanishes in perturbation theory, which explains the baryon number conservation and lepton number conservation in perturbation theory. However, if the gauge field fluctuations are strong (nonperturbative), so that $F_{\mu\nu}^a \sim 1/g_2$, the right-hand side of Eq. (5.196) is not necessarily zero. In particular, for the instanton field $\Delta\mathcal{K} = 1$.

5.5.1.4 Instanton-Induced Effects

The one-instanton contribution in the theory with massless quarks is described by an effective multifermion vertex, also known as the 't Hooft vertex; see Section 5.4.11. The number of fermion lines at this vertex is equal to the number of fermion zero modes. There is one zero mode per Weyl fermion in the instanton field; therefore, in the model at hand,

$$S_{tH} = \int \frac{d^4x_0 d\rho}{\rho^5} (qqq\ell) \times \exp\left[-\frac{2\pi}{\alpha_2(\rho)} - 2\pi^2 v^2 \rho^2\right], \tag{5.197}$$

where I have omitted the SU(2) and the SU(3) indices of the quark fields q and the SU(2) indices of ℓ. I have also omitted the orientational moduli in the measure associated with rotations of the instanton within SU(2), as well as the pre-factors in Eq. (5.197).

First let us discuss the fermion structure in Eq. (5.197). It describes the annihilation of three q quanta into one $\bar{\ell}$ quantum. Three quarks comprise a proton or neutron. Therefore, one can say that this vertex is responsible for proton decay into e^+ (accompanied by, say, a photon or π^0-meson emission necessary to maintain energy–momentum conservation). The baryon and lepton charges of the initial and final states are $(1, 0)$ and $(0, -1)$, respectively, so that the conservation law $\Delta Q_B = \Delta Q_L$ is explicit. Moreover, $\Delta Q_B = -1$ in this transition.

The amplitude $\mathcal{A}_{qqq\ell}$ of this transition is determined by the integral over ρ in Eq. (5.197). This integral is obviously saturated at $\rho \sim 1/v$; hence,

$$\mathcal{A}_{qqq\ell} \sim \exp\left[-\frac{2\pi}{\alpha_2(v)}\right]. \tag{5.198}$$

> The notation $Q\!\!\!/_B$ means that baryon number is not conserved.

The fact that the values of ρ are typically of order $1/v$ justifies our use of the undistorted instanton solution, because distortions of the solution due to the W-boson mass occur at much larger distances $\sim 1/m_W \sim 1/(g_2 v)$. Since $g_2^2(v)$ is small, the probability of baryon-number-violating instanton-induced decays is exponentially suppressed:

$$\Gamma_{Q\!\!\!/_B} \sim \exp\left[-\frac{4\pi}{\alpha_2(v)}\right] \sim 10^{-170}, \tag{5.199}$$

where we have used the experimental value of $\alpha_2(v)$ obtained in the framework of SM,

$$\alpha_2 = \frac{g_2^2(v)}{4\pi} = \frac{\alpha}{\sin^2\theta_W} \approx \frac{1}{31}. \tag{5.200}$$

Needless to say, the decay rate (5.199) is unobservable.

The question is: can one enhance this rate by changing the experimental conditions to make it observable?

5.5.1.5 High Temperature

It is clear that to overcome (5.199) we must get rid (at least, in part) of the exponential suppression typical of all tunneling processes. The incredibly large suppression (5.199) is due to the fact that, for baryon number nonconservation to show up, the system at hand must tunnel under a very high barrier (depicted in Fig. 5.3).

At the same time, if we heat the system up to high temperatures, of the order of the barrier height (5.183), transitions between adjacent pre-vacua, with $\Delta\mathcal{K} \neq 0$, can proceed via thermal jumps over the barrier [46–48]. Once the system jumps up to the saddle point (i.e., the sphaleron is thermally created) it will fall back into a different pre-vacuum, with, roughly speaking, unit probability. These transitions are described by classical statistical mechanics, and so their rate is governed by the Boltzmann factor

$$\Gamma_{Q\!\!\!/_B} \sim \exp\left(-\frac{M_{\mathrm{sph}}}{T}\right), \tag{5.201}$$

rather than by the instanton exponent (5.199). At $T \sim M_{\text{sph}}$ the Q_B-violating processes are unsuppressed. The sphaleron mass is

$$M_{\text{sph}} = \pi \frac{m_W}{\alpha_2} = 2\sqrt{2}\pi^2 \frac{v}{g_2} \sim 7\,\text{TeV}. \qquad (5.202)$$

Temperature dependence of the sphaleron mass

However, this is the zero-temperature value. In fact, the loss of the exponential suppression does occur at lower temperatures since the vacuum expectation value of the Higgs field and, hence, m_W and the sphaleron mass are temperature dependent. The vacuum expectation value vanishes at and above the $SU(2)_{\text{gauge}}$-restoring phase transition, which takes place at $T \gtrsim 100\,\text{GeV}$. At this point the barrier disappears and $\Delta \mathcal{K} \neq 0$ transitions occur all the time.

In hot Big Bang cosmology there was a time in the past when the temperature was $T \gtrsim 100\,\text{GeV}$ or higher. At that time Q_B and Q_L were strongly violated. There are models [49] of baryon asymmetry generation in which the only source of baryon number violation is the mechanism discussed above. Unfortunately, temperatures $T \gtrsim 100\,\text{GeV}$ are not attainable in controllable terrestrial conditions.

5.6 Instantons at High Energies

In our search for baryon number violations that could be tested in laboratories it is natural to pose the following question. *Can high energies play the same role as a heat bath in facilitating $\Delta \mathcal{K} \neq 0$ jumps?* In other words, can baryon-number-violating transitions occur in collisions of energetic particles at energies $E \sim M_{\text{sph}}$ with an unsuppressed (or, at least, a less suppressed) rate?

To answer this question we need to find out how to calculate, or at least estimate, instanton-induced cross sections at high energies. This will be the subject of this section. We will see that although the rate of baryon-number-violating transitions grows exponentially with energy (below the sphaleron mass), only a finite fraction of the exponential suppression in (5.199) can be eliminated.

5.6.1 Cross Section of Instanton-Induced Processes

This subsection is intended to give a general idea of the phenomenon of exponential growth with energy, thus allowing us to understand the origins of the exponential growth of the instanton-induced cross section. We will systematically omit numerical factors that might overshadow the general picture. A more concise and technically efficient method of calculation will be presented in Section 5.6.3.

Consider the process

$$qqq\ell \rightarrow \underbrace{WW\ldots W}_{n}, \qquad (5.203)$$

describing proton–electron annihilation into an arbitrary number n of W bosons. Now, our task is to show that the total cross section $\sigma(qqq\ell \rightarrow W$ bosons$)$ grows exponentially with energy [35]. The instanton-induced Q_B vertex contains the pre-exponential operator $qqq\ell$ and the exponential factors given in Eqs. (5.111) and (5.126). Since for the time being we are interested in the behavior of the exponent

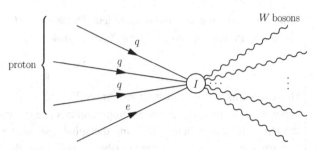

Fig. 5.12 Instanton-induced $p - e$ annihilation into an arbitrary number n of W bosons. In the initial state $qqq\ell$, $Q_B = 1$ and $Q_L = 1$ while in the final state $Q_B = Q_L = 0$.

we will omit the pre-exponential factors. The amplitude $\mathcal{A}(qqq\ell \to W$ bosons$)$ is depicted in Fig. 5.12 and can be obtained by expanding the exponent in Eq. (5.126):

$$\mathcal{A}\left(qqq\ell \to \underbrace{WW\cdots W}_{n}\right) \sim \left(\frac{\rho^2 E}{n\,g_2}\right)^n \exp\left(-\frac{2\pi}{\alpha_2}\right), \qquad (5.204)$$

where the factor $1/n!$ from the expansion is canceled in passing from the operator G^n to the n-boson amplitude because of the combinatorics. We will assume the W bosons to be relativistic (this assumption will be justified *a posteriori*) and will not differentiate between their momenta; the average momentum of each W boson is taken to be E/n, where E is the total energy (in the center-of-mass frame). This rather rough approximation is sufficient to establish the energy dependence of the exponent.

Squaring the amplitude and integrating over the n-particle phase space [50],

Phase space for n massless particles

$$V_n \sim \frac{1}{E^4} \frac{(\mathrm{const} \times E^2)^n}{(n-1)!\,(n-2)!} \qquad (5.205)$$

we get

$$\sigma(qqq\ell \to W \text{ bosons}) \sim \exp\left(-\frac{4\pi}{\alpha_2}\right) \sum_n \int d\rho_1^2 d\rho_2^2 \exp\left[-2\pi^2 v^2(\rho_1^2 + \rho_2^2)\right]$$

$$\times \left[\frac{1}{n!\,(n-1)!\,(n-2)!}\left(\frac{\rho_1^2 \rho_2^2 E^4}{n^2 g_2^2}\right)^n\right], \qquad (5.206)$$

where the extra factor $1/n!$ on the right-hand side comes from the Bose nature of the final particles. Next, we integrate over ρ_1^2 and ρ_2^2 using the stationary-point approximation; this yields

$$\sigma(qqq\ell \to W \text{ bosons}) \sim \exp\left(-\frac{4\pi}{\alpha_2}\right) \sum_n \frac{1}{(n!)^3}\left(\frac{\mathrm{const} \times E^4}{v^4 g_2^2}\right)^n. \qquad (5.207)$$

The stationary-point value of the instanton size is

$$\rho_*^2 \sim \frac{n}{v^2}. \qquad (5.208)$$

(Here and below an asterisk subscript denotes the stationary-point value of the parameter in question.) The summation over n can also be carried out using the stationary-point approximation, which finally leads us to

$$n_* = \text{const} \times \left(\frac{E^4}{v^4 g_2^2} \right)^{1/3}. \tag{5.209}$$

Therefore we arrive at

$$\sigma(qq q\ell \to W \text{ bosons}) \sim \exp\left[-\frac{4\pi}{\alpha_2} \left(1 - \text{const} \times \frac{E^{4/3} g_2^{4/3}}{v^{4/3}} \right) \right]. \tag{5.210}$$

All the constants in these expressions are positive numbers that are calculable; see below. Substituting the stationary-point value of n from Eq. (5.209) into Eq. (5.208), we obtain the characteristic value of the instanton size,

$$\rho_*^2 \sim \text{const} \times \frac{1}{v^2 g_2^2} \frac{E^{4/3} g_2^{4/3}}{v^{4/3}}. \tag{5.211}$$

Before continuing, it is instructive to express all the parameters found above in terms of the natural energy scale for the problem at hand, M_{sph}; see Eq. (5.202). To this end we will introduce a "dimensionless energy"

$$\mathcal{E} = E/M_{\text{sph}}. \tag{5.212}$$

The $\mathcal{Q}_B$ cross section is due to multi-W-boson production.

In terms of this dimensionless energy the values of n and ρ that saturate the summed integral (5.206) scale as follows:

$$n_* = \text{const} \times \frac{\mathcal{E}^{4/3}}{\alpha_2}, \quad \rho_* = \text{const} \times \frac{\mathcal{E}^{2/3}}{m_W}, \tag{5.213}$$

while the total $\mathcal{Q}_B$ cross section itself takes the form

$$\sigma(qq q\ell \to W \text{ bosons}) \sim \exp\left[-\frac{4\pi}{\alpha_2} \left(1 - \text{const} \times \mathcal{E}^{4/3} \right) \right]. \tag{5.214}$$

We can see that as the energy grows the baryon-number-violating cross section grows exponentially, $\sim \exp(\mathcal{E}^{4/3}/\alpha_2)$, which "eats up" part of the suppressing exponent $4\pi/\alpha_2$ in (5.199). Our calculation is trustworthy as long as $\rho_* \ll 1/m_W$. This condition translates into

$$\mathcal{E} \ll 1, \tag{5.215}$$

i.e., we must stay well below the sphaleron mass. The very same condition justifies our neglect of the W-boson masses and the use of the relativistic approximation in phase space. Indeed, the average energy of each W boson produced scales as $m_W \mathcal{E}^{-1/3}$. If we could extrapolate to $\mathcal{E} \sim 1$, formally we would get an unsuppressed cross section. Of course, near the sphaleron mass all our approximations break down. We will discuss later what stops the exponential growth in (5.214).

5.6.2 The Holy Grail Function

The particular dependence on $\mathcal{E}$ of the $\mathcal{Q}_B$ cross section, presented in (5.214), is just the lowest-order term in the expansion of the general formula [51]

$$\sigma(qq q\ell \to W \text{ bosons}) \sim \exp\left[-\frac{4\pi}{\alpha_2} (1 - F(\mathcal{E})) \right]. \tag{5.216}$$

Michael Mattis suggested the name *holy grail function* for $F(\mathcal{E})$. The name took root, and we will use it consistently in what follows. The boundary condition for $F(\mathcal{E})$ is $F(0) = 0$. Equation (5.216) exactly describes the one-instanton contribution in the limit[31]

$$\alpha_2 \to 0, \qquad \mathcal{E} \equiv \frac{E}{M_{\mathrm{sph}}} \to \text{fixed value}, \qquad \mathcal{E} \ll 1. \qquad (5.217)$$

Intuitive as it is, this general formula can be derived on essentially dimensional grounds [51]. Its emergence will become clear after we familiarize ourselves with a more advanced method of calculation in Section 5.6.3.

5.6.3 Total Cross Section via Dispersion Relation

The exponential growth of the $\mathcal{Q}_B$ cross section was discovered [35] through squaring the instanton-induced amplitude and integrating over the multiparticle phase space, as outlined in Section 5.6.1. In this subsection we will discuss a faster and more efficient way of obtaining Eq. (5.210): through unitarity [52]. Using this trick we will be able to derive Eq. (5.210) and similar expressions for other contributions with ease, by relating them to the instanton–anti-instanton interaction. The total cross section that we have just calculated can be pictorially represented as in Fig. 5.13. As is well known, unitarity relates this cross section to the imaginary part of the forward

> *This is the optical theorem; see [39].*

scattering amplitude depicted in Fig. 5.14. The latter (in the center-of-mass frame where the total momentum $P_\mu = \{E, 0, 0, 0\}$) is proportional to

$$\mathcal{A}(qqq\ell \to qqq\ell) \sim \exp\left(-\frac{4\pi}{\alpha_2}\right) \int d^4x_0 \, e^{iEt_0} \int dO \exp[-S_{IA}(x_0)], \qquad (5.218)$$

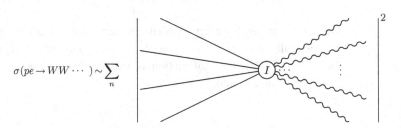

$$\sigma(pe \to WW \cdots) \sim \sum_n$$

Fig. 5.13   The cross section of *pe* annihilation into an arbitrary number of *W* bosons.

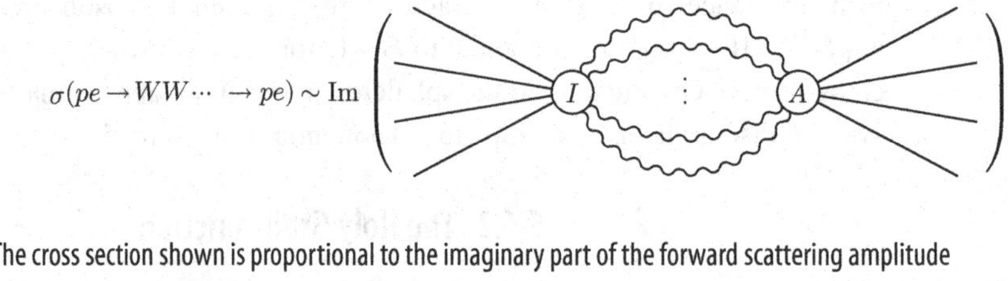

$$\sigma(pe \to WW \cdots \to pe) \sim \mathrm{Im}$$

Fig. 5.14   The cross section shown is proportional to the imaginary part of the forward scattering amplitude $qqq\ell \to qqq\ell$, depicted here.

---

[31] In practice, it is sufficient to consider α_2 as a small parameter imposing the constraint $\alpha_2 \ll \mathcal{E}$. Note that the holy grail function in (5.216) slightly differs from the definition in Mattis review [35].

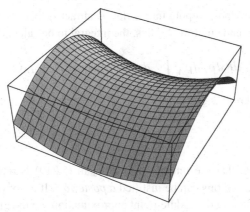

Fig. 5.15   The stationary point of $S_{IA}(R, \rho)$ is a saddle point.

where the instanton center is at x_0, the anti-instanton center is at the origin, $\int dO$ stands for the integral over the relative orientation of the pseudoparticles, and S_{IA} is the instanton–anti-instanton interaction; see, e.g., Eq. (5.133). Finally, t_0 is the time component of x_0. Irrelevant pre-exponential factors are omitted.

Of course, since all the above quantities belong to Euclidean space, the amplitude (5.218) has no imaginary part. Upon the analytic continuation of E to Minkowski space, namely, $E_E \to -iE_M$, it acquires an imaginary part

$$\sigma(qq q\ell \to W \text{ bosons}) \sim \exp\left(-\frac{4\pi}{\alpha_2}\right) \text{Im} \int d^4 x_0\, e^{Et_0} \int dO \exp[-S_{IA}(x_0)].$$
(5.219)

The integral over x_0 in Eq. (5.219) is ill defined since it diverges at both small and large $|x_0|$. However, its imaginary part is well defined and can be found using the steepest-descent method. The extremum of the exponent is a saddle point rather than a maximum or minimum (Fig. 5.15). To make the integral convergent we must rotate the contour in the complex plane (at $R = R_*$, see Eq. (5.226) below), aligning it along the imaginary axis. This yields the desired imaginary unit in front of the integral [52].

5.6.3.1 W Bosons

The IA interaction due to W

The instanton–anti-instanton interaction due to gauge field exchanges that was derived in Section 5.4.9 can be rewritten as follows:

$$-S_{IA} = \frac{32\pi^2}{g_2^2}\left[4\frac{(\hat{v}R)^2}{R^2} - 1\right]\frac{\rho_I^2 \rho_A^2}{R^4},$$
(5.220)

where R is the IA separation, $R_\mu \equiv (x_0)_\mu$, and the unit vector v_μ parametrizes the relative orientation of the pseudoparticles,

$$\hat{v}_\mu \tau_\mu^- \equiv M_A^\dagger M_I, \qquad \hat{v}^2 = 1.$$
(5.221)

The dipole–dipole interaction is attractive if $\hat{v}$ is parallel to R and repulsive if it is perpendicular to R. Thus, the integral to be calculated by steepest descent is

$$\int dR \, d\rho_I^2 \, d\rho_A^2 \int_0^\pi d\gamma \sin^2 \gamma \exp\left[ER - S_{IA}(R, \gamma) - 2\pi^2 v^2 \left(\rho_I^2 + \rho_A^2 \right) \right], \quad (5.222)$$

where γ is given by

$$\cos \gamma = \frac{\hat{v}_\mu R_\mu}{|R|}. \quad (5.223)$$

Note that the integral over $d^4 x_0$ in (5.219) is replaced by an integral over dR in (5.222); this can be justified *a posteriori*. It is convenient to integrate over the angle γ first. In the saddle-point approximation the integral is saturated at

$$\cos^2 \gamma = 1;$$

therefore the dominant interaction is attractive, as expected, and reduces to

$$-S_{IA} = \frac{96\pi^2}{g_2^2} \frac{\rho_I^2 \rho_A^2}{R^4}. \quad (5.224)$$

The saddle-point value of the other integration parameters is determined by

$$\frac{\rho_I^2}{g_2^2} \frac{1}{R^4} = \frac{\rho_A^2}{g_2^2} \frac{1}{R^4} = \frac{v^2}{48}, \qquad E = 384\pi^2 \frac{\rho_I^2 \rho_A^2}{g_2^2} \frac{1}{R^5}, \quad (5.225)$$

leading to

$$\left(\rho_{I,A} \right)_*^2 = \text{const} \times \frac{1}{m_W^2} \mathcal{E}^{4/3}, \qquad R_* = \text{const} \times \frac{1}{m_W} \mathcal{E}^{1/3}, \quad (5.226)$$

which, in turn, implies that at the stationary point

$$F_G(\mathcal{E}) = g_2^2 \left[ER + \frac{\rho_I^2 \rho_A^2}{g_2^2} \frac{1}{R^4} - 2\pi^2 v^2 \left(\rho_I^2 + \rho_A^2 \right) \right]_* = C\mathcal{E}^{4/3}, \quad (5.227)$$

where F_G denotes the gauge-boson part of the holy grail function and $C \sim 1$. Equation (5.227) should be compared with Eq. (5.214) and (5.216).

5.6.3.2 Higgs Particles

The instanton-induced vertex for Higgs particle emission and the corresponding instanton–anti-instanton interaction was derived in Section 5.4.12.2, for convenience we reproduce it here:

> *The IA interaction action due to Higgs particles*

$$S_{IA}^H = -2\pi^2 v^2 \frac{\rho_I^2 \rho_A^2}{R^2}. \quad (5.228)$$

The dependence on the IA separation differs from that in Eq. (5.220); hence, one can expect a different $\mathcal{E}$ dependence of the Higgs part of the holy grail function. As was explained in Section 5.4.12.2, S_{IA}^H is not the only source of R^{-2} in the instanton–anti-instanton interaction. There is another contribution [53] in the IA interaction that is proportional to R^{-2}, namely, that associated with the diagram in Fig. 5.7a. This graph gives the m_W^2 correction to the dipole–dipole interaction discussed above. At large R it scales as

$$\frac{\rho^4}{g_2^2} \frac{1}{R^4} \left(m_W^2 R^2 \right) \sim \rho^4 v^2 \frac{1}{R^2}. \tag{5.229}$$

The functional dependence is the same as in (5.228) while the overall coefficient and the color structure (omitted here) are different, of course. Unlike (5.228), the m_W^2 correction to the dipole–dipole interaction (5.229) depends on the relative $I A$ orientation. Including (5.228) and (5.229) in S_{IA} and using the saddle-point values of the parameters, we arrive at

$$F_H(\mathcal{E}) \sim g_2^2 \left[v^2 \frac{\rho^4}{R^2} \right]_* = \text{const} \times \mathcal{E}^2. \tag{5.230}$$

| Holy grail function | For completeness I reproduce here the $O(\mathcal{E}^{4/3})$ and $O(\mathcal{E}^2)$ terms in the holy grail function with their numerical coefficients [52]: |

$$F(\mathcal{E}) = \left(\frac{9}{16\sqrt{2}} \right)^{2/3} \mathcal{E}^{4/3} - \frac{3}{32} \mathcal{E}^2. \tag{5.231}$$

The next term,

$$O\left(\mathcal{E}^{8/3} \ln \frac{1}{\mathcal{E}} \right),$$

was found in [54]. It enters in $F(\mathcal{E})$ with a positive coefficient. Thus $F(\mathcal{E})$ presents a series in $\mathcal{E}^{2/3}$, possibly with logarithms at high orders.

5.6.3.3 Deriving the General Formula

Now we have acquired enough experience to substantiate the general formula (5.216). Indeed, the general form of the instanton–anti-instanton interaction is

$$S_{IA} = \frac{1}{g_2^2} \frac{\rho^4}{R^4} f_1 \left(\frac{\rho^2}{R^2} \right) + v^2 \rho^2 f_2 \left(\frac{\rho^2}{R^2} \right), \tag{5.232}$$

where $f_{1,2}$ are functions of the dimensionless variable ρ^2/R^2. Moreover, the saddle-point values of ρ and R are

$$\rho_* = \frac{\mathcal{F}_1(\mathcal{E})}{m_W}, \qquad R_* = \frac{\mathcal{F}_2(\mathcal{E})}{m_W}, \tag{5.233}$$

where $\mathcal{F}_{1,2}$ are some other functions; $\mathcal{F}_{1,2}(\mathcal{E}) \ll 1$ at $\mathcal{E} \ll 1$. Equation (5.233) follows, in essence, from dimensional analysis. Combining (5.232) and (5.233) we arrive at (5.216). Moreover, if we invoke, in addition, Eq. (5.227), we will see that the expansion of $F(\mathcal{E})$ runs in powers of $\mathcal{E}^{2/3}$.

5.6.4 Premature Unitarization

Thus, we have established that the behavior of the instanton-induced Q_B cross section is as follows:

$$\sigma(pe \to W + \text{Higg particles}) \sim \exp\left\{-\frac{4\pi}{\alpha_2}\left[1 - \left(\frac{9}{16\sqrt{2}}\right)^{2/3}\mathcal{E}^{4/3} + \frac{3}{32}\mathcal{E}^2 + \dots\right]\right\},$$

(5.234)

where the ellipses represent higher-order terms in the expansion of the holy grail function. The expansion is valid at $\mathcal{E} \ll 1$. If we take Eq. (5.234) at its face value and formally extrapolate it up to $\mathcal{E} \sim 1$, we will see that at $\mathcal{E} \approx 2$ the holy grail function reaches unity and the exponent in (5.234) vanishes. The vanishing of the exponent would mean that at $E \approx 2M_{\text{sph}}$ the exponential suppression of the baryon-number-violating cross section is totally eliminated and this cross section reaches its maximal possible value,[32] $\sigma_{Q_B} \sim 1/M_{\text{sph}}^2$. Of course, formal extrapolation is by no means justified, and one could say that the higher-order terms in the expansion of $F(\mathcal{E})$ omitted in (5.234) are such that at $\mathcal{E} \gtrsim 1 \sim$ the holy grail function levels off at a positive value strictly less than unity, say $F = 1/3$. Then a finite part of the suppressing exponent will be eliminated but the exponential suppression will persist.

However, some estimates of higher-order terms suggest that $F(\mathcal{E})$ does indeed cross unity.[33] Does this mean that at energies $\mathcal{E} > 1$ the baryon number violation becomes unsuppressed?

The answer to this question is negative. The mechanism that cuts off exponential growth is known as *premature unitarization*. It was suggested in [9]; see also [8, 10]. What is unitarization? If, say, an S-wave scattering amplitude grows with energy and reaches its unitary limit (full saturation of the corresponding scattering phase), the very same interaction automatically screens off further growth, preventing the cross section from exceeding its unitary limit, which scales as $1/s$ where $\sqrt{s}$ is the total energy. The screening occurs through rescattering. This is illustrated in Fig. 5.16, which presents a two-body scattering process. Assume that the point-like vertex is λs, where λ is a constant (Fig. 5.16a). At large s this amplitude violates unitarity. However, the sum of all iterations (Fig. 5.16b) is, roughly,

$$\frac{\lambda s}{1 + \text{const} \times \lambda s} \to \text{const}$$

(5.235)

at large s. This mechanism has been well known since the early days of scattering theory in quantum mechanics.

A peculiarity of instanton-induced cross sections is that the growth in the point-like vertex is exponentially fast while the vertex itself is exponentially small. In this case, as we will explain shortly, unitarization occurs prematurely, i.e., the amplitude does not "wait" until it reaches its unitary limit for the iterations to become important; they produce screening long before that.

---

[32] The cross section $\sigma_{Q_B} \sim 1/M_{\text{sph}}^2$ is the maximum for any scattering process at $E = \sqrt{s} \sim M_{\text{sph}}$ occurring if we have a single wave or just a few waves are involved, say, S- and P-waves. It is not difficult to see that our saddle-point calculation predicts that all particles produced in the collision under consideration are found in the S- wave.

[33] It is unclear whether one can actually define $F(\mathcal{E})$ beyond a perturbative expansion, whether or not it is convergent (as opposed to asymptotic), and so on.

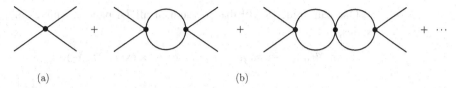

Fig. 5.16   (a) Point-like two-body scattering and (b) its iterations.

Fig. 5.17   A helix-like curve represents instanton–anti-instanton interactions due to multiple W-boson and Higgs particle exchanges.

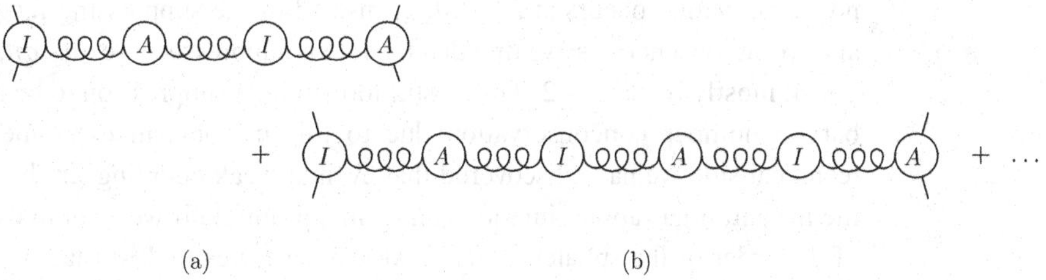

Fig. 5.18   The multi-instanton contribution due to iterations of the one-instanton mechanism. Each term has successively more and more IA pairs arranged in a chain.

Let us redraw the graph in Fig. 5.14 symbolically, as shown in Fig. 5.17. Each of the two blobs (vertices) carries a factor $\exp(-2\pi/\alpha_2)$, while the link between them is $\exp[4\pi F(\mathcal{E})/\alpha_2]$. Using chemical terminology we can refer to the links as bonds; this is quite appropriate since they represent the instanton–anti-instanton interaction. Then the amplitude depicted in Fig. 5.17 is

$$\exp\{-4\pi[1 - F(\mathcal{E})]/\alpha_2\}, \tag{5.236}$$

while that in Fig. 5.18a is

$$\exp\{-8\pi[1 - \tfrac{3}{2}F(\mathcal{E})]/\alpha_2\}. \tag{5.237}$$

The simple observation is that in (5.237) the factor in braces vanishes while that in (5.236) is still $4\pi/(3\alpha_2)$. In fact, iterating the same bond function (i.e., including in the chain of Fig. 5.18 an arbitrary number of IA pairs), it is easy to see that the chain reaches unity when the one-instanton result for the amplitude is $\exp(-2\pi/\alpha_2)$ – the geometric mean of the results with the original suppression and with no suppression. This argument is independent of one's choice of bond function as long as the latter grows with $\mathcal{E}$. In fact, this argument implies that the one-instanton approximation breaks down for the $\mathcal{Q}_B$ cross sections at energies below the sphaleron mass. Multi-instantons are instrumental in premature unitarization.

One can argue [8–10] that the sum of all IA pairs assembles into a geometric series,

$$\text{Im}\mathcal{A}(qqq\ell \to qqq\ell) \sim \sigma_{\mathcal{Q}_B} = \text{const} \times \exp(-2\pi/\alpha_2)$$

$$\times \sum_{k=1,3,\dots}^{\infty} (-1)^k \exp\left[\left(-\frac{2\pi}{\alpha_2} + \frac{4\pi F}{\alpha_2}\right)k\right]$$

$$= \text{const} \times \exp(-2\pi/\alpha_2)\left\{\cosh[(2\pi - 4\pi F)/\alpha_2]\right\}^{-1}$$

$$< \text{const} \times \exp(-2\pi/\alpha2). \tag{5.238}$$

Summation over multiple IA pairs

The alternating signs in (5.238) come from counting a negative mode Gaussian factor i in the multi-instanton configuration of Fig. 5.18.

Summarizing, the exponential growth of the instanton-induced $\mathcal{Q}_B$ cross section at high energies, up to the sphaleron mass, is firmly established. Maximum baryon nonconservation occurs at $E \sim M_{\text{sph}}$. Instead of the suppressing factor (5.199) that appears at low energies, we find that the rate is suppressed as $\exp(-c\pi/\alpha_2(v))$, where $c < 4$; mostly likely $c \sim 2$. This is still too strong a suppression to be able to observe baryon number nonconservation due to this mechanism experimentally. On the technical side we have discovered that even at weak coupling (in the Higgs regime) the instanton-gas approximation can be inappropriate if we want to explore energies of the order of the sphaleron mass. At such energies multi-instanton configurations with $S_{IA} \sim S_{\text{inst}}$, i.e., strongly interacting IA pairs, are important.

5.7 Other Ideas Concerning Baryon Number Violation

If it should turn out to be the case that the standard model is a part of a grand unified theory (GUT), then, in addition to the anomaly and associated relation (5.196), there is another mechanism of baryon number nonconservation, namely, through the superheavy (leptoquark) gauge bosons X and Y;[34] see Fig. 5.19, which presents the amplitude

$$e^- + d^{i_0} \to X \to \varepsilon^{i_0 j k} \bar{u}_j \bar{u}_k. \tag{5.239}$$

The proton decay rate associated with this mechanism (for a review see, e.g., [60]) is easy to estimate:

$$\Gamma_{\text{proton}} \sim \alpha^2 m_{\text{proton}} \left(\frac{m_{\text{proton}}}{M_X}\right)^4, \tag{5.240}$$

where α is the common value of the three gauge couplings at the unification scale. Since $M_X \sim 10^{16}$ GeV, the suppression in (5.240), compared with the typical hadronic width, is "only" ~ 66 orders of the magnitude (cf. Eq. (5.199)).

This section could have been entitled "Are there ways to enhance $\mathcal{Q}_B$ processes other than heating the system up to temperatures exceeding the sphaleron mass?" A remarkable alternative was suggested by Rubakov [61] (see also [62]), who noted that the suppression disappears in the presence of a magnetic monopole: magnetic monopoles catalyze proton decays.

---

[34] For a pedagogical introduction to GUTs see, e.g., [59].

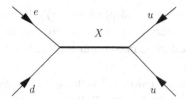

Fig. 5.19 One of the diagrams responsible for proton decay in GUTs.

Again, if the standard model is part of a grand unified theory, then magnetic monopoles should exist in nature since GUTs support them as topologically stable solitons. Their core size is determined by M_X^{-1}. Given the fact that the scale of grand unification is very large, $M_X \sim 10^{16}$ GeV, in processes at "our" energies one can view GUT monopoles as point-like sources of a strong magnetic field. Essentially, one can treat them as the Dirac monopoles of the era before 't Hooft and Polyakov. The masses of the GUT monopoles are even higher than M_X, namely, $M_M \sim M_X/\alpha$.

In the previous sections we saw that exponential suppression is intrinsic to $\mathcal{Q}_B$ transitions as long as they proceed through under-the-barrier tunneling. The corresponding action is very large. The suppression in (5.240) is due to the fact that the range of the $\mathcal{Q}_B$ interaction is very short. Both circumstances would be negated if someone having a monopole in their possession brought it to "our" world and placed it in the vicinity of protons [61, 62].

Let us first discuss the changes in the mechanism (5.196) in the presence of a magnetic monopole. Now one can generate a virtual electric field (in addition to the already existing magnetic field of the monopole) such that (i) the overall action stays the *same as that of the magnetic monopole* and, simultaneously, (ii) the integral over $\vec{B}\vec{E}$ on the right-hand side of (5.196) (remember that in Minkowski space $F_{\mu\nu}^a \tilde{F}^{\mu\nu a} = -2\vec{E}^a\vec{B}^a$) is saturated, $\Delta\mathcal{K} = 1$, implying that $\Delta\mathcal{Q}_B = 1$. It is important that the monopole size $r_M \sim M_X^{-1}$ *per se* is absolutely irrelevant, since it plays no role in the analysis, and can be set to zero.

To see that this is indeed the case, let us consider the following trial field configuration. Let us draw a sphere of radius R with the monopole at the center. At time $-T/2$ we (adiabatically) switch on an electric field

$$\vec{E} = C\frac{\vec{n}}{RT}, \tag{5.241}$$

inside this sphere (in (5.241) C is a constant, R and T are arbitrary parameters, and $\vec{n}$ is the unit vector $\vec{r}/r$). At time $T/2$ we switch it off. To avoid singularities the trial electric field must vanish in the near vicinity of the origin. The additional contribution to the energy due to $\vec{E} \neq 0$ is $\int d^3\vec{r}\vec{E}^2 \sim R/T$ and is arbitrarily small in the limit $R/T \to 0$. However, $\int d^3\vec{r}\, dt\, \vec{B}\vec{E}$, the contribution to the right-hand side of (5.196) is independent of R and T, namely $\int d^3\vec{r}\, dt\, \vec{B}\vec{E} = O(1)$. This argument illustrates that the cross section $\sigma(p + M \to M + e^+ + \text{pions})$ is expected to be of a typical hadronic scale.

We would come to the same conclusion if we discussed the mechanism of Fig. 5.19. The existence of fermion zero modes in the monopole background, in the limit $r_M \to 0$, is crucial. The spectator monopole captures one of the proton composites, say, the d^{i_0} quark (with color index i_0) onto the S-wave orbit with a probability that is independent of r_M. Because of (5.239) the monopole *per se* has no definite baryon (or, equivalently, lepton) number. As a result the captured d^{i_0} quark is converted into an anti-u diquark, $\varepsilon^{i_0 jk} \bar{u}_j \bar{u}_k$, plus a positron, with probability $O(1)$ [62].

In a bid to understand better the hadronic aspect of the monopole catalysis of proton decay, Callan and Witten suggested [63] that one should treat the proton at hand as a Skyrmion. They demonstrated that the Dirac monopole "unwinds" the Skyrmion. Neither the GUT scale nor the weak scale are relevant to this unwinding, in full agreement with the above arguments.

Exercise

5.7.1   Prove that the expression for the winding number $\mathcal{K}$ in this chapter and for the baryon number in Section 4.2 are in one-to-one correspondence.

5.8 Appendices

5.8.1 Gauge Coupling Renormalization in Gauge Theories. Screening versus Antiscreening

It is instructive to consider gauge coupling (charge) renormalization in the ghost-free Coulomb gauge. Our task is to compare Abelian and non-Abelian gauge theories.[35]

Let us start from quantum electrodynamics (QED). Assume that we have two heavy charged bodies (probes), with charges $\pm e_0$ where $\pm e_0$ is a bare electric charge appearing in the Lagrangian. One can measure the charge through the Coulomb interaction of the probe bodies. The corresponding Feynman diagram is shown in Fig. 5.20, where the wavy line depicts photon exchange. The heavy probe bodies are (almost) at rest; the photon 4-momentum q is assumed to tend to zero. We will choose a reference frame in which $q^\mu = \{q_0, 0, 0, q^3\}$. If $\Gamma_\mu^{(1)}$ and $\Gamma_\mu^{(2)}$ are the vertices for the first and second probe bodies, the amplitude $\mathcal{A}$ corresponding to Fig. 5.20 can be written as

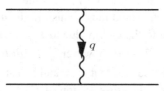

Fig. 5.20   Scattering of two heavy probe charges (denoted by thick lines) in QED, in the tree approximation. The photon exchange is denoted by a wavy line. The momentum transfer is q.

---

[35] In this appendix I follow [64].

$$\mathcal{A}_0 = \frac{e_0^2}{q^2} \Gamma_\mu^{(1)} \Gamma_\nu^{(2)} g^{\mu\nu}, \tag{5.242}$$

where we have taken into account the transversality of the vertices,

$$q^\mu \Gamma_\mu^{(1)} = q^\mu \Gamma_\mu^{(2)} = 0, \tag{5.243}$$

and the subscript 0 (in $\mathcal{A}_0$) means that we are dealing with a tree diagram (no loops).

The very same transversality implies that $q_0 \Gamma_0^{(1,2)} = q_3 \Gamma_3^{(1,2)}$. Using these conditions in Eq. (5.242), we arrive at

$$\mathcal{A}_0 = \frac{e_0^2}{q^2} \left[\Gamma_0^{(1)} \Gamma_0^{(2)} \left(1 - \frac{q_0^2}{q_3^2} \right) - \sum_{\ell=1,2} \Gamma_\ell^{(1)} \Gamma_\ell^{(2)} \right]$$

$$= -e_0^2 \left(\frac{1}{q_3^2} \Gamma_0^{(1)} \Gamma_0^{(2)} + \frac{1}{q^2} \sum_{\ell=1,2} \Gamma_\ell^{(1)} \Gamma_\ell^{(2)} \right). \tag{5.244}$$

The first term in the second line describes the instantaneous Coulomb interaction (this is obvious upon performing a Fourier transformation and passing to coordinate space). The second term has a pole at $q^2 = 0$. It describes a (retarded) propagation of an electromagnetic wave with two possible transverse polarizations. We can determine the charge through measurement of the Coulomb interaction. Thus, for our purposes the second term can be omitted.

The one-loop correction to the Coulomb interaction (5.244) in QED is given by the diagram in Fig. 5.21 with the electron in a loop. A straightforward calculation gives

$$(\mathcal{A}_0 + \mathcal{A}_1)_{\text{QED}} = -\frac{e_0^2}{q_3^2} \Gamma_0^{(1)} \Gamma_0^{(2)} \left(1 - \frac{e_0^2}{12\pi^2} \ln \frac{M_{\text{uv}}^2}{-q^2} \right) + \cdots \tag{5.245}$$

where we have omitted irrelevant terms and assumed that $|q^2| \gg m_e^2$. Thus, the effective (renormalized) coupling constant in QED, which measures the strength of interaction at the scale q^2 (note, that in the process at hand $-q^2 > 0$) is

$$e^2(q^2) = e_0^2 \left(1 - \frac{e_0^2}{12\pi^2} \ln \frac{M_{\text{uv}}^2}{-q^2} \right) \rightarrow e_0^2 \left(1 + \frac{e_0^2}{12\pi^2} \ln \frac{M_{\text{uv}}^2}{-q^2} \right)^{-1} \tag{5.246}$$

Landau formula

where the first relation presents the one-loop expression while the second relation is the result of summing up all leading logarithms (the summation can be performed using, e.g., the renormalization group). At the scale q^2 the effective charge is smaller than the bare charge. This is natural. The reason is obvious: the bare charge is

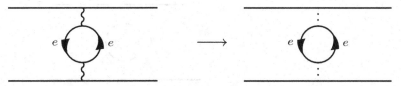

Fig. 5.21 One-loop correction to the Coulomb interaction in QED. The Coulomb part of the photon propagator D_{00} is denoted by the dotted lines.

screened. Indeed, the bare charge is defined at the shortest distances $\sim M_{uv}^{-1}$. Assume for definiteness that the probe charge is positive. Electron–positron pairs created in the vacuum as a result of field-theoretic fluctuations polarize the vacuum. The probe charge attracts negatively charged electrons while positively charged positrons are repelled. Thus a cloud of virtual electrons screens the original positive probe charge. An effective charge seen at some distance from the probe charge is smaller than the bare charge and the further we go, the smaller is the screened charge.

Bare charge screening is a rather general phenomenon. It takes place in all four-dimensional renormalizable field theories except non-Abelian Yang–Mills theories. Formally, bare charge screening is in one-to-one correspondence with the fact that the imaginary part of the diagram in Fig. 5.21 (the discontinuity in q^2 at positive q^2) is always positive, owing to unitarity.

One can ask then: what miracle happens in passing from QED to non-Abelian Yang–Mills theories? In QED the photons are coupled to the electrons and do not interact with each other directly. In non-Abelian Yang–Mills theories gluons themselves are the sources for gluons (gauge bosons). There are three- and four-gluon vertices. Moreover, as we know from Eq. (5.244), the gluon "quanta" can be Coulomb (their propagation is described by the component D_{00} of the Green's function) or physical transversal (their propagation is described by the component $D_{\ell\ell'}$ of the Green's function with $\ell, \ell' = 1, 2$). The diagram depicted in Fig. 5.22 is qualitatively similar to the one-loop correction in QED in Fig. 5.21b. It produces screening of the bare charge. The only difference from Eq. (5.244) is insignificant: the coefficient $-e_0^2/12\pi^2$ is replaced by $-g_0^2/24\pi^2$ in the SU(2) Yang–Mills theory.

A qualitative difference arises due to the diagram in Fig. 5.23. This graph depicts the transition of the Coulomb quantum (described by D_{00}) into a pair that is "transverse plus Coulomb." We should remember that the Coulomb interaction is instantaneous: D_{00} depends on q_3^2 rather than on q^2, see (5.244). This means that the contribution of this diagram does not have an imaginary part (there is no discontinuity in q^2 at positive q^2). Unitarity no longer determines the sign of this correction. An explicit calculation shows that it has the opposite sign; the graph in Fig. 5.23 produces antiscreening,

$$(\mathcal{A}_0 + \mathcal{A}_1)_{\mathrm{SU(2)\,YM}} = \mathcal{A}_0 \left(1 - \frac{g_0^2}{16\pi^2} \frac{2}{3} \ln \frac{M_{uv}^2}{-q^2} + \frac{g_0^2}{16\pi^2} 8 \ln \frac{M_{uv}^2}{q_3^2} \right)$$

$$\rightarrow \mathcal{A}_0 \left[1 - \frac{g_0^2}{16\pi^2} \ln \frac{M_{uv}^2}{-q^2} \left(8 - \frac{2}{3} \right) \ln \frac{M_{uv}^2}{q_3^2} \right]^{-1}, \qquad (5.247)$$

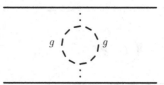

Fig. 5.22   One-loop correction to the Coulomb interaction in Yang–Mills theory. The transverse (physical) gluons are denoted by the broken lines. This diagram is similar to that in Fig. 5.21.

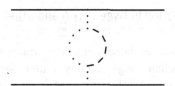

Fig. 5.23   One-loop correction to the Coulomb interaction, specific to non-Abelian Yang-Mills theories.

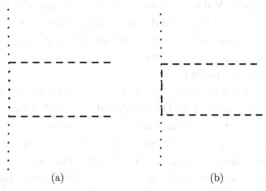

(a)   (b)

Fig. 5.24   Comparison of the loops in Figs. 5.22 and 5.23. The interaction proceeds via the exchange of (a) Coulomb and (b) transverse quanta.

where the first and second corrections in the parentheses in the first line are due to Figs. 5.22 and 5.23, respectively.[36] Now the bare charge is smaller than that seen at a distance!

One can give an heuristic argument why these two diagrams produce effects of opposite signs. To this end let us compare the loops in these graphs, as in Fig. 5.24, where I have cut one transverse gluon line in order to make clearer the analogy with QED to be presented shortly. Figure 5.24a contains an exchange of a Coulomb quantum and Fig. 5.24b an exchange of a transverse gluon quantum. The effect of the Coulomb quanta is repulsion of charges of the same sign, while the exchange of transverse quanta leads to an attraction of parallel currents (the Biot–Savart law).

The only circumstance that remains unexplained by the above arguments is that the antiscreening effect, represented by the coefficient 8 in Eq. (5.247), is numerically much stronger in Yang–Mills than the screening effect, represented by $-2/3$. For us, this is a lucky circumstance since the numerical dominance of antiscreening over screening makes non-Abelian Yang–Mills theories asymptotically free.

It is remarkable that the same binary fission of the one-loop quantum correction, eight as against $-2/3$, is clearly seen in the instanton calculation, cf. (5.105) and (5.108), where the distinction is associated with zero as against nonzero modes.

---

[36] The result presented in the first line, in precisely this form, was obtained by I. Khriplovich [55] before the discovery of asymptotic freedom and the advent of QCD. A curious story of the "pre-observation" of asymptotic freedom is recounted in [56].

5.8.2 Relation between Λ_{PV} and Other Λs Used in Perturbative Calculations

The standard regularization in performing loop calculations in perturbation theory in non-Abelian gauge theories is dimensional regularization (DR), supplemented by, say, minimal subtraction (MS) renormalization. Determinations of the scale parameter Λ in QCD that can be found in the literature refer to this scheme. At the same time, in calculating nonperturbative effects (e.g., instantons) one routinely uses the Pauli–Villars (PV) regularization scheme. The reason is that the instanton field is (anti-)self-dual, and this notion cannot be continued to $4 - \epsilon$ dimensions. The PV regularization and renormalization scheme in this context was suggested by 't Hooft [6]; it was further advanced in supersymmetric instanton calculus in [57]; see also the review paper [58].

In his classic work [6] 't Hooft himself demonstrated how one can proceed from the PV scheme to DR combined with MS. However, even now the situation in question is somewhat confusing. The one-loop calculation of α_{PV} in terms of α_{MS} was carried out in Section XIII of [6]. Unfortunately, the key expression (13.7) in that article contained an error.[37] The error was noted and corrected by Hasenfratz and Hasenfratz [65], see also [66]. What was confusing was that a later reprint of 't Hooft's paper (see the second reference in [6]) presented the corrected derivation, and the updated result (13.9), without mentioning that the required corrections had been made.[38]

In this textbook we are using appropriately corrected expressions.

In an associated regularization scheme [28], known as the modified minimal subtraction ($\overline{\text{MS}}$) scheme, one obtains

$$C_{2\overline{\text{MS}}} = C_2 - \frac{5}{16} \approx 1.54,$$

$$\frac{8}{g_{\overline{\text{MS}}}^2} = \frac{8\pi^2}{g_{\text{MS}}^2} - \frac{11}{6} N (\ln 4\pi - \gamma). \tag{5.248}$$

References for Chapter 5

[1] A. M. Polyakov, *Phys. Lett. B* **59**, 82 (1975) [reprinted in M. Shifman (ed.), *Instantons in Gauge Theories* (World Scientific, Singapore, 1994), p. 19].

[2] V. N. Gribov, 1976, unpublished.

[3] R. Jackiw and C. Rebbi, *Phys. Rev. Lett.* **37**, 172 (1976) [reprinted in M. Shifman (ed.), *Instantons in Gauge Theories* (World Scientific, Singapore, 1994), p. 25].

---

[37] Numerically, the error is rather insignificant. Nevertheless, it was unfortunate that this error propagated even in reviews, e.g., [13].

[38] Equation (13.9) of the reprinted article still contains a typo: -1 on the right-hand side should be replaced by $-1/2$. This misprint has no impact on subsequent expressions in the reprinted article.

[4] C. G. Callan, R. F. Dashen, and D. J. Gross, *Phys. Lett. B* **63**, 334 (1976) [reprinted in M. Shifman (ed.), *Instantons in Gauge Theories* (World Scientific, Singapore, 1994), p. 29].

[5] A. A. Belavin, A. M. Polyakov, A. S. Schwarz, and Yu. S. Tyupkin, *Phys. Lett. B* **59**, 85 (1975) [reprinted in M. Shifman (ed.), *Instantons in Gauge Theories* (World Scientific, Singapore, 1994), p. 22].

[6] G. 't Hooft, *Phys. Rev. D* **14**, 3432 (1976). Erratum: *ibid.* 18, 2199 (1978) [reprinted in M. Shifman (ed.), *Instantons in Gauge Theories* (World Scientific, Singapore, 1994), p. 70].

[7] A. Ringwald, *Nucl. Phys. B* **330**, 1 (1990); O. Espinosa, *Nucl. Phys. B* **343**, 310 (1990); L. D. McLerran, A. I. Vainshtein, and M. B. Voloshin, *Phys. Rev. D* **42**, 171 (1990).

[8] V. I. Zakharov, *Nucl. Phys. B* **353**, 683 (1991).

[9] M. Maggiore and M. A. Shifman, *Nucl. Phys. B* **371**, 177 (1992); *Phys. Rev. D* **46**, 3550 (1992).

[10] G. Veneziano, *Mod. Phys. Lett. A* **7**, 1661 (1992).

[11] Edward V. Shuryak, *The QCD Vacuum, Hadrons and the Superdense Matter* (World Scientific, Singapore, 2003).

[12] S. Coleman, The uses of instantons, in S. Coleman (ed.), *Aspects of Symmetry* (Cambridge University Press, 1985), p. 265.

[13] V. Novikov, M. Shifman, A. Vainshtein, and V. Zakharov, ABC of instantons, in M. Shifman (ed.), *ITEP Lectures on Particle Physics and Field Theory* (World Scientific, Singapore, 1999), Vol. 1, p. 201.

[14] C. G. Callan, R. F. Dashen, and D. J. Gross, *Phys. Rev. D* **17**, 2717 (1978); *Phys. Rev. D* **19**, 1826 (1979).

[15] A. Schwartz, *Topology for Physicists* (Springer, Berlin, 1994).

[16] S. Flügge, *Practical Quantum Mechanics* (Springer, Berlin, 1971), Vol. 1, Problem 28; C. Kittel, *Quantum Theory of Solids* (Wiley & Sons, New York, 1963), Chapter 9; N. Ashcroft and N. Mermin, *Solid State Physics* (Sounders College, Philadelphia, 1976), Chapter 8.

[17] K. M. Bitar and S. J. Chang, *Phys. Rev. D* **17**, 486 (1978).

[18] R. J. Crewther, P. Di Vecchia, G. Veneziano, and E. Witten, *Phys. Lett. B* **88**, 123 (1979). Erratum: *ibid.* **91**, 487 (1980); M. A. Shifman, A. I. Vainshtein, and V. I. Zakharov, *Nucl. Phys. B* **166**, 493 (1980).

[19] J. E. Kim and G. Carosi, *Axions and the Strong CP Problem* [arXiv:0807.3125 [hep-ph]].

[20] S. Weinberg, *Phys. Rev. Lett.* **40**, 223 (1978); F. Wilczek, *Phys. Rev. Lett.* **40**, 279 (1978).

[21] G. 't Hooft, unpublished. The 't Hooft solution is presented and discussed in R. Jackiw, C. Nohl, and C. Rebbi, *Phys. Rev. D* **15**, 1642 (1977).

[22] M. F. Atiyah, N. J. Hitchin, V. G. Drinfeld, and Yu. I. Manin, *Phys. Lett. A* **65**, 185 (1978) [reprinted in M. Shifman (ed.), *Instantons in Gauge Theories* (World Scientific, Singapore, 1994), p. 133]; V. G. Drinfeld and Yu. I. Manin, *Commun. Math. Phys.* **63**, 177 (1978).

[23] R. Jackiw, Field theoretic investigations in current algebra, Section 7, in S. Treiman, R. Jackiw, B. Zumino, and E. Witten (eds.), *Current Algebra*

Anomalies (Princeton University Press, 1985), p. 81; P. Fayet and S. Ferrara, *Phys. Rept.* **32**, 249 (1977).

[24] R. Jackiw and C. Rebbi, *Phys. Lett. B* **67**, 189 (1977); C. W. Bernard, N. H. Christ, A. H. Guth, and E. J. Weinberg, *Phys. Rev. D* **16**, 2967 (1977) [reprinted in M. Shifman (ed.), *Instantons in Gauge Theories* (World Scientific, Singapore, 1994), pp. 149–153].

[25] F. Wilczek, *Phys. Lett. B* **65**, 160 (1976) [reprinted in M. Shifman (ed.), *Instantons in Gauge Theories* (World Scientific, Singapore, 1994), p. 116].

[26] C. W. Bernard, *Phys. Rev. D* **19**, 3013 (1979) [reprinted in M. Shifman (ed.), *Instantons in Gauge Theories* (World Scientific, Singapore, 1994), p. 109].

[27] G. 't Hooft and M. J. G. Veltman, *Nucl. Phys. B* **44**, 189 (1972); G. 't Hooft, *Nucl. Phys. B* **62** 444 (1973).

[28] W. A. Bardeen, A. J. Buras, D. W. Duke, and T. Muta, *Phys. Rev. D* **18**, 3998 (1978).

[29] M. A. Shifman, A. I. Vainshtein, and V. I. Zakharov, *Nucl. Phys. B* **165**, 45 (1980).

[30] J. D. Bjorken and S. D. Drell, *Relativistic Quantum Fields* (McGraw-Hill, New York, 1965).

[31] M. F. Atiyah and I. M. Singer, *Ann. Math.* **87**, 484 (1968); **87**, 546 (1968); **93**, 119 (1971).

[32] A. S. Schwarz, *Phys. Lett. B* **67**, 172 (1977).

[33] L. S. Brown, R. D. Carlitz, and C. K. Lee, *Phys. Rev. D* **16**, 417 (1977).

[34] D. Friedan and P. Windey, *Nucl. Phys. B* **235**, 395 (1984) [reprinted in S. Ferrara (ed.), *Supersymmetry* (North-Holland/World Scientific, 1987), p. 572].

[35] A. Ringwald, Nucl. Phys. B **330**, 1 (1990); O. Espinosa, Nucl. Phys. B **343**, 310 (1990); M. P. Mattis, *Phys. Rept.* **214**, 159 (1992); V. A. Rubakov and M. E. Shaposhnikov, *Phys. Usp.* **39**, 461 (1996) [arXiv:hep-ph/9603208].

[36] K. Osterwalder and E. Seiler, *Ann. Phys.* **110**, 440 (1978); T. Banks and E. Rabinovici, *Nucl. Phys. B* **160**, 349 (1979); E. H. Fradkin and S. H. Shenker, *Phys. Rev. D* **19**, 3682 (1979).

[37] The idea of the constrained instanton was first put forward in Y. Frishman and S. Yankielowicz, *Phys. Rev. D* **19**, 540 (1979); I. Affleck, *Nucl. Phys. B* **191**, 429 (1981) [reprinted in M. Shifman (ed.), *Instantons in Gauge Theories* (World Scientific, Singapore, 1994), p. 247].

[38] M. A. Shifman and A. I. Vainshtein, *Nucl. Phys. B* **362**, 21 (1991) [reprinted in M. Shifman (ed.), *Instantons in Gauge Theories* (World Scientific, Singapore, 1994), p. 97].

[39] L. D. Landau and E. M. Lifshitz, *Quantum Mechanics*, Third Edition (Elsevier, Amsterdam, 1977), Section 50 (Problems).

[40] F. R. Klinkhamer and N. S. Manton, *Phys. Rev. D* **30**, 2212 (1984); F. R. Klinkhamer and R. Laterveer, *Z. Phys. C* **53**, 247 (1992); Y. Brihaye and J. Kunz, *Phys. Rev. D* **47**, 4789 (1993).

[41] R. F. Dashen, B. Hasslacher, and A. Neveu, *Phys. Rev. D* **10**, 4138 (1974).

[42] L. G. Yaffe, *Phys. Rev. D* **40**, 3463 (1989).

[43] D. M. Ostrovsky, G. W. Carter, and E. V. Shuryak, *Phys. Rev. D* **66**, 036004 (2002) [arXiv: hep-ph/0204224].

[44] A. V. Smilga, *Nucl. Phys. B* **459**, 263 (1996) [arXiv:hep-th/9504117].

[45] E. Witten, *Phys. Lett. B* **117**, 324 (1982) [reprinted in S. Treiman, R. Jackiw, B. Zumino, and E. Witten (eds.), *Current Algebra and Anomalies* (Princeton University Press, 1985) p. 429].

[46] A. Polyakov, Models and mechanisms in gauge theory, in *Proc. 9th Int. Symp. on Lepton and Photon Interactions at High Energy*, Batavia, Illinois, August 1979, eds. T. B. W. Kirk and H. D. I. Abarbanel (Batavia, Fermilab., 1980), p. 521.

[47] V. A. Kuzmin, V. A. Rubakov, and M. E. Shaposhnikov, *Phys. Lett. B* **155**, 36 (1985).

[48] P. Arnold and L. D. McLerran, *Phys. Rev. D* **36**, 581 (1987); *Phys. Rev. D* **37**, 1020 (1988).

[49] L. D. McLerran, *Phys. Rev. Lett.* **62**, 1075 (1989); B. H. Liu, L. D. McLerran, and N. Turok, *Phys. Rev. D* **46**, 2668 (1992).

[50] G. I. Kopylov, *Fundamentals of the Kinematics of Resonances* (Nauka, Moscow, 1970), in Russian; E. Byckling and K. Kajantie, *Particle Kinematics* (John Wiley & Sons, 1973).

[51] S. Y. Khlebnikov, V. A. Rubakov, and P. G. Tinyakov, *Nucl. Phys. B* **350**, 441 (1991).

[52] V. V. Khoze and A. Ringwald, *Nucl. Phys. B* **355**, 351 (1991).

[53] A. V. Yung, *Instanton induced effective Lagrangian in the gauge Higgs theory*, Report SISSA-181-90-EP, 1990.

[54] D. Diakonov and M. V. Polyakov, *Nucl. Phys. B* **389**, 109 (1993); I. Balitsky and A. Schafer, *Nucl. Phys. B* **404**, 639 (1993) [arXiv:hep-ph/9304261].

[55] I. B. Khriplovich, *Sov. J. Nucl. Phys.* **10**, 235 (1969).

[56] M. Shifman, Historical curiosity: how asymptotic freedom of the Yang–Mills theory could have been discovered three times before Gross, Wilczek and Politzer, but was not, in M. Shifman (ed.), *At the Frontier of Particle Physics* (World Scientific, Singapore, 2000), Vol. 1 p. 126.

[57] V. A. Novikov, M. A. Shifman, A. I. Vainshtein, and V. I. Zakharov, *Nucl. Phys. B* **260**, 157 (1985).

[58] M. A. Shifman and A. I. Vainshtein, Instantons versus supersymmetry: fifteen years later, in M. Shifman (ed.), *ITEP Lectures on Particle Physics and Field Theory* (World Scientific, Singapore, 1999) Vol. 2, pp. 485–647 [hep-th/9902018].

[59] R. N. Mohapatra, *Unification and Supersymmetry*, Third Edition (Springer, 2002); L. B. Okun, *Leptons and Quarks* (Elsevier, 1985).

[60] P. Nath and P. Fileviez Pérez, *Phys. Rept.* **441**, 191 (2007) [arXiv:hep-ph/0601023].

[61] V. A. Rubakov, *Nucl. Phys. B* **203**, 311 (1982).

[62] C. Callan, *Nucl. Phys. B* **212**, 391 (1983).

[63] C. Callan and E. Witten, *Nucl. Phys. B* **239**, 161 (1984).

[64] V. A. Novikov, L. B. Okun, M. A. Shifman, A. I. Vainshtein, M. B. Voloshin, and V. I. Zakharov, *Phys. Rept.* **41**, 1 (1978).

[65] A. Hasenfratz and P. Hasenfratz, *Nucl. Phys. B* **193**, 210 (1981).

[66] G. M. Shore, *Ann. Phys.* **122**, 321 (1979).

It all started with the Heisenberg O(3) *sigma model. — A geometric representation, or from* O(3) *to* CP(1). *— Generalization to* CP($N - 1$) *models. — Gauged formulation. — Calculation of the Gell–Mann–Low or β function. — Continuous symmetries cannot be spontaneously broken in two dimensions.*

6.1 O(3) Sigma Model

6.1.1 The S Field and O(3) Model

The O(3) sigma model is a representative of a huge class of models which are generically referred to as sigma models. This model is a showcase example because it exhibits a plethora of interesting phenomena, some them of a very general nature. Besides, it is of practical importance in both solid state and high-energy physics.

The O(3) sigma model can be traced back to Heisenberg's model of antiferromagnets formulated in the 1930s, which was designed for the description of interacting spins. The O(3) sigma model is a double limit of Heisenberg's model – when the spin is large (i.e., the classical limit) and the distance between the spin sites is small (i.e., the continuous limit).

In this case the model can be formulated in terms of a triplet of fields $\vec{S}(x)$, where $\vec{S} = \{S^a\}$, $a = 1, 2, 3$, is a 3-vector in an internal space (referred to as the *target space*), constrained by the condition

$$[\vec{S}(x)]^2 \equiv S^a(x)S^a(x) \equiv 1. \tag{6.1}$$

Thus the absolute value of $\vec{S}$ is fixed; only the angular variables are dynamical. In other words, one can say that the fields $\{S^a\}$ live on a two-dimensional sphere S_2 of unit radius; see Fig. 6.1. Thus, in the case at hand the target space is S_2. In generic sigma models, the target spaces can be much more complicated than a sphere.

The dynamics is described by the Lagrangian

$$\mathcal{L} = \frac{1}{2g^2}(\partial_\mu \vec{S})(\partial^\mu \vec{S}), \tag{6.2}$$

where g^2 is the coupling constant and the factor 2 in the denominator is introduced for convenience. The invariance under global (x-independent) rotations of the vector $\vec{S}$ (Fig. 6.1) is explicit in Eq. (6.2). This is the only term of second order in derivatives compatible with this invariance. Usually in sigma model studies people limit themselves to such terms. In principle, one could add terms with higher derivatives on the right-hand side of Eq. (6.2) that are compatible with the

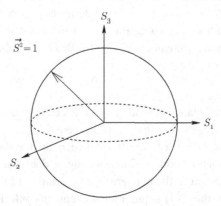

Fig. 6.1 The target space of the O(3) sigma model is S_2.

symmetries of the model under consideration. For instance, in Section 4.2 it turned out necessary to include quartic in addition to quadratic terms. In this section we will limit ourselves to quadratic terms.

Thus, let us focus on the Lagrangian (6.2) *per se*. The corresponding action is

$$S = \frac{1}{2g^2} \int d^D x (\partial_\mu \vec{S})(\partial^\mu \vec{S}). \tag{6.3}$$

Classically the model is well defined for any D. However, only at $D = 2$ is the model renormalizable. The fact that for $D = 2$ the coupling constant is dimensionless hints at the renormalizability of the model. At $D = 4$, say, the coupling constant $1/(2g^2)$ has dimension $(\text{mass})^2$, and quantum corrections proliferate in much the same way as in quantum gravity. At $D = 2$ the O(3) sigma model considered in Euclidean space has a nontrivial topology and, hence, instantons (see Section 6.4) – this is another reason why we should concentrate on this case.

To say that $(\partial \vec{S})^2$ is the only term of second order in derivatives compatible with O(3) symmetry is not quite accurate. In two dimensions, and only in two dimensions, one can add another term,

Topological term

$$\mathcal{L}_\theta = \frac{\theta}{8\pi} S^a (\partial_\mu S^b)(\partial_\nu S^c) \varepsilon^{\mu\nu} \varepsilon_{abc}, \tag{6.4}$$

where θ is a dimensionless parameter, the *vacuum angle* (in solid state physics it is called the quasimomentum). Furthermore, $\varepsilon^{\mu\nu}$ and ε_{abc} are Levi–Civita tensors acting in the configurational and target spaces, respectively ($\mu, \nu = 1, 2$ and $a, b, c = 1, 2, 3$). The additional term, presented in Eq. (6.4), is called the θ term or topological term. It has no impact whatsoever in perturbation theory. To see that this is the case, it is enough to show that $\mathcal{L}_\theta$ does not change the equations of motion. Indeed, let us find the variation in ΔS under the change $\vec{S} \to \vec{S} + \delta \vec{S}$, in the linear approximation,

$$\delta \left(\int d^2 x \, \mathcal{L}_\theta \right) = \frac{\theta}{8\pi} \int d^2 x \, \varepsilon^{\mu\nu} \varepsilon_{abc} \left[(\delta S^a)(\partial_\mu S^b)(\partial_\nu S^c) + 2 S^a (\partial_\mu \delta S^b)(\partial_\nu S^c) \right]$$

$$= \frac{\theta}{8\pi} \int d^2 x \, \varepsilon^{\mu\nu} \varepsilon_{abc} \left\{ 2 \partial_\mu \left[S^a (\delta S^b) \partial_\nu S^c \right] + 3 (\delta S^a)(\partial_\mu S^b)(\partial_\nu S^c) \right\}. \tag{6.5}$$

The first term in the second line is a full derivative. Since we are assuming, as usual, that $\delta \vec{S} \to 0$ as $|x| \to \infty$, this term drops out. Let us examine the last term in the second line. The constraint $\vec{S}^2 = 1$ implies that $S^a \delta S^a = 0$. The same is valid with respect to $\partial_\mu \vec{S}$: namely, $S^a \partial_\mu S^a = 0$. Thus, all three vectors involved,

The topological term is a full derivative; see Exercise 6.2.2.

$$\delta \vec{S}, \quad \partial_1 \vec{S}, \quad \text{and} \quad \partial_2 \vec{S},$$

lie in the plane perpendicular to $\vec{S}$, i.e., they are coplanar. The convolution of three 3-planar vectors with ε_{abc} yields zero.

Thus, $\delta(\int d^2 x \mathcal{L}_\theta) = 0$ and, hence, the topological term (6.4) does not affect the equations of motion. Consequently, it does not show up in perturbation theory. It is important in the nonperturbative solution of the model, however. In particular, at $\theta = \pi$ the O(3) sigma model becomes conformal while at $\theta \neq \pi$ a mass gap is generated. Unfortunately, I cannot dwell on this aspect in this text.

6.1.2 Representation in Complex Fields: CP(1) Model

As already mentioned, the target space in the O(3) sigma model is S_2. The two-dimensional sphere is a very special manifold, it is a representative of a class of spaces called *Kähler* spaces. The Kähler manifolds admit the introduction of complex coordinates, much in the same way as one can parametrize a two-dimensional plane by a complex number z. Now we will show how a complex field $\phi(x)$ can be introduced on S_2. If the original field $\vec{S}(x)$ is constrained, see Eq. (6.1), the field $\phi(x)$ has two components,

$$\phi_1(x) \equiv \mathrm{Re}\,\phi(x), \qquad \phi_2(x) \equiv \mathrm{Im}\,\phi(x), \tag{6.6}$$

which are unconstrained. This is convenient for the construction perturbation theory.

The complex coordinates on S_2 can be introduced by virtue of *stereographic* projection; see Fig. 6.2. This figure displays the target space sphere (with unit radius) on which $\vec{S}$ lives, and a ϕ plane which touches the sphere at the north pole. This plane admits the introduction of the complex coordinate ϕ in a standard manner: if ϕ_1 and

Stereographic projection of a sphere onto a plane

ϕ_2 are Cartesian coordinates, $\phi = \phi_1 + i\phi_2$. A ray of light is emitted from the south pole; it pierces the sphere and the plane at the points denoted by small crosses. We then map these points onto each other:

$$S^1 = \frac{2\phi_1}{1 + \phi_1^2 + \phi_2^2}, \qquad S^2 = \frac{2\phi_2}{1 + \phi_1^2 + \phi_2^2}, \qquad S^3 = \frac{1 - \phi_1^2 - \phi_2^2}{1 + \phi_1^2 + \phi_2^2}. \tag{6.7}$$

The inverse transformation has the form

$$\phi = \frac{S^1 + iS^2}{1 + S^3}. \tag{6.8}$$

It is clear that this is a one-to-one correspondence. The only point which deserves a comment is the south pole ($S^1 = S^2 = 0$, $S^3 = -1$); it is mapped onto infinity. Since physically this is a non-singular point on the target space, the only functions of ϕ that are allowed for consideration are those that have a well-defined limit at

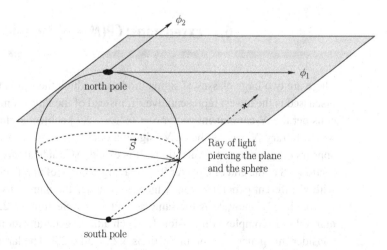

Fig. 6.2 Introduction of complex coordinates on S_2 through stereographic projection. The ϕ plane touches the sphere at its north pole.

$|\phi| \to \infty$, irrespective of the direction in the ϕ plane. After a few simple but rather tedious algebraic transformations, one obtains the action of the O(3) sigma model in terms of ϕ:

$$S = \int d^2x \frac{1}{(1 + \bar{\phi}\phi)^2} \left(\frac{2}{g^2} \partial_\mu \bar{\phi} \, \partial^\mu \phi + \frac{\theta}{2\pi i} \varepsilon^{\mu\nu} \partial_\mu \bar{\phi} \, \partial_\nu \phi \right). \tag{6.9}$$

Geometric representation; the metric of the sphere is in Fubini–Study form; see Section 6.2.2.

Let me note in passing that the expression $(1 + \bar{\phi}\phi)^{-2}$ in front of the parentheses is nothing other than the metric of the target space sphere in the given parametrization.

In this representation the sigma model (6.9) is usually referred to as the CP(1) model, where CP stands for complex projective. This model opens a class of CP($N - 1$) models, about which we will say a few words shortly. The target spaces of CP($N - 1$) models are complex projective spaces of higher dimension (see Section 6.2.2).

Where have the symmetries of the original Lagrangian (6.2) gone? Only one global U(1) symmetry is apparent in Eq. (6.9):

$$\phi \to e^{i\alpha}\phi, \qquad \bar{\phi} \to e^{-i\alpha}\bar{\phi}. \tag{6.10}$$

Two other symmetries are realized nonlinearly:

$$\phi \to \phi + \epsilon + \bar{\epsilon}\phi^2, \qquad \bar{\phi} \to \bar{\phi} + \bar{\epsilon} + \epsilon\bar{\phi}^2, \tag{6.11}$$

where ϵ is a small complex parameter. To verify invariance under (6.11) we observe that

$$\begin{aligned} \delta(1 + \bar{\phi}\phi)^{-2} &= -2(1 + \bar{\phi}\phi)^{-2}(\epsilon\bar{\phi} + \bar{\epsilon}\phi), \\ \delta(\partial_\mu \bar{\phi} \, \partial_\nu \phi) &= (\partial_\mu \bar{\phi} \, \partial_\nu \phi)2(\epsilon\bar{\phi} + \bar{\epsilon}\phi). \end{aligned} \tag{6.12}$$

6.2 Extensions: CP($N - 1$) Models

There are two large classes of sigma models of which the sigma model we have just discussed is the lowest representative. If, instead of the three-component vector $\vec{S}$ we consider the N-component vector $\vec{S} = \{S^1, \ldots, S^N\}$ subject to the constraint $\vec{S}^2 = 1$, with arbitrary N, we get the O(N) sigma model with target space S_{N-1}. This model appears in a number of applications. It is exactly solvable at large N (see Appendix section 9.8 at the end of Chapter 9). For a larger range of applications one has to deal with a different generalization, which goes under the name of CP($N - 1$) models. As we already know, S_2 is the same as CP(1); see Section 6.1.2. CP(1) is a Kähler manifold of (complex) dimension 1, which admits generalization to any N. We will consider this generalization in Sections 6.2.2 and 6.2.3. The large-N solution of the CP($N - 1$) model is presented in Section 9.4. Finally, supersymmetric sigma models are discussed in Part II.

6.2.1 Kähler Spaces: Generalities

Before passing to the analysis of CP($N-1$) models I will summarize general elements of the Kählerian geometry. The defining element is the so-called *Kähler potential*

$$\mathcal{K}(\phi^i, \bar{\phi}^j). \tag{6.13}$$

The Kähler space metric $G_{i\bar{j}}$ carries one barred (antiholomorphic) and one unbarred (holomorphic) index, i and $\bar{j}$, respectively, and must be expressible (at least locally) as the second derivative of the Kähler potential,

> **Kähler metric and potential**

$$G_{i\bar{j}} = \frac{\partial^2}{\partial \phi^i \partial \bar{\phi}^j} \mathcal{K} \equiv \partial_k \partial_{\bar{j}} \mathcal{K}. \tag{6.14}$$

Here and in what follows, I will use the following shorthand:

$$\partial_i \equiv \frac{\partial}{\partial \phi^i}, \quad \partial_{\bar{j}} \equiv \frac{\partial}{\partial \bar{\phi}^j}. \tag{6.15}$$

From the preceding definition, it is obvious that the metric is symmetric, $G_{i\bar{j}} = G_{\bar{j}i}$. Also it is obvious that

$$\partial_j G_{k\bar{n}} = \partial_k G_{j\bar{n}}, \quad \partial_{\bar{m}} G_{k\bar{n}} = \partial_{\bar{n}} G_{k\bar{m}}, \tag{6.16}$$

and $G_{k\bar{n}} = (G_{\bar{k}n})^\dagger$. The metric with the upper indices is defined through the relation

$$G_{j\bar{m}} G^{\bar{m}i} = \delta_j^i. \tag{6.17}$$

The nonvanishing Christoffel symbols carry either all nonbarred indices or all barred,

$$\Gamma_{jk}^i = G^{\bar{m}i}\, \partial_j G_{k\bar{m}}, \quad \Gamma_{\bar{j}\bar{k}}^{\bar{i}} = G^{\bar{i}m}\, \partial_{\bar{j}} G_{\bar{k}m}. \tag{6.18}$$

It is obvious that $\Gamma_{ij}^k = \Gamma_{ji}^k$ (see (6.16)). Given the Christoffel symbols, we can define the Riemann curvature,

$$R^{\bar{i}}{}_{\bar{j}k\bar{n}} = \partial_k \Gamma_{\bar{j}\bar{n}}^{\bar{i}}, \quad R_{i\bar{j}k\bar{n}} = G_{i\bar{m}} \partial_k \Gamma_{\bar{j}\bar{n}}^{\bar{m}}, \tag{6.19}$$

and its complex conjugate. The Riemann tensor $R_{i\bar{j}k\bar{n}}$ thus defined is symmetric under interchanges $i \leftrightarrow k$ or $\bar{j} \leftrightarrow \bar{n}$. An alternative representation of the Riemann tensor on the CP($N-1$) spaces is given in (10.399).

Note that Eq. (6.19) defines the Riemann tensor $R_{i\bar{j}k\bar{n}}$ with the alternating holomorphic and antiholomorphic indices. In the literature, people also use the curvature tensor with a different ordering of indices, namely, with two neighboring unbarred indices $R_{\bar{i}jk\bar{n}}$, which can be defined as

$$R_{\bar{i}jk\bar{n}} = G_{i\bar{m}}\left(-\partial_k \Gamma_{j\bar{n}}^{\bar{m}}\right) \tag{6.20}$$

(see [1]). Note the extra minus sign in (6.20) compared to (6.19). In this book, I use only the curvature tensor $R_{ijk\bar{\ell}}$ defined in (6.19).

Finally, for the Ricci tensor and scalar curvature, we have

$$R_{m\bar{n}} = -G^{\bar{j}i} R_{i\bar{j}m\bar{n}} = -R^{\bar{j}}{}_{\bar{j}m\bar{n}}, \quad \mathcal{R} = 2\, G^{\bar{n}m} R_{m\bar{n}}. \tag{6.21}$$

There is also another useful formula specific for Kähler geometry, namely,

$$R_{m\bar{n}} = -\partial_m \partial_{\bar{n}} \log\left(\text{Det}\{G\}\right). \tag{6.22}$$

where Det $\{G\}$ in the right-hand side is the determinant of the metric $G_{i\bar{j}}$. Equation (6.22) explicitly demonstrates that $R_{m\bar{n}} = R_{\bar{n}m}$.

6.2.2 CP($N-1$) Models

The two-dimensional sphere, the target space of the CP(1) sigma model, is arguably the simplest symmetric Kähler space with a nontrivial metric. It is not difficult to generalize this model to cover the case of multiple fields ϕ^i and $\bar{\phi}^{\bar{j}}$, where the holomorphic and antiholomorphic indices i and $\bar{j}$ run over the values $1, 2, \ldots, N-1$.[1] The class of ($N-1$)-dimensional symmetric Kähler spaces to which CP(1) belongs as a particular (and the most simple) case is called CP($N-1$).[2]

The Kähler potential of the CP($N-1$) model can be chosen as:

$$\mathcal{K} = \frac{2}{g^2} \log\left(1 + \sum_{i\bar{j}}^{N-1} \bar{\phi}^{\bar{j}} \delta_{\bar{j}i} \phi^i\right). \tag{6.23}$$

Note the coupling constant g^2 that appears in the definition of $\mathcal{K}$. Its normalization is consistent with that in Eq. (6.3). In the O(3) model discussed in Section 6.1.1, it is obvious from its symmetries that a single coupling constant g fully characterizes the target space S^2, namely, $1/g$ is the radius of this sphere. Upon renormalization, S^2 remains the same sphere; it is only its radius that can "run." In CP($N-1$) models with $N > 2$, it is not a priori obvious. And nevertheless, it is a valid statement – a single constant fully describes the renormalization flow of this target space. Why is this the case?

The CP($N-1$) manifold is not only compact and *Kählerian*, but also symmetric and, most importantly, *Einsteinian*. The latter means that all relevant two-index tensors are proportional to the metric, for instance,

$$R_{i\bar{j}} = \frac{g^2}{2} N G_{i\bar{j}}, \qquad R_{i\bar{j}k\bar{n}} R_{\bar{m}}^{\bar{j}k\bar{n}} = -g^2 R_{i\bar{m}}, \ldots \tag{6.24}$$

The inclusion of $2/g^2$ in (6.23) has no impact on the Christoffel symbols, nor on the curvature components $R^i_{\bar{j}k\bar{n}}$, nor on the Ricci tensor $R_{i\bar{j}}$. It shows up, however, in (6.24) and (6.27). The ensuing metric

$$G_{k\bar{\ell}} = \frac{2}{g^2}\left[\frac{\delta_{k\bar{\ell}}}{1 + \sum_{i\bar{j}}^{N-1} \bar{\phi}^{\bar{j}} \delta_{\bar{j}i} \phi^i} - \frac{\phi^k \bar{\phi}^{\bar{\ell}}}{\left(1 + \sum_{i\bar{j}}^{N-1} \bar{\phi}^{\bar{j}} \delta_{\bar{j}i} \phi^i\right)^2}\right] \tag{6.25}$$

is called the Fubini–Study metric. Moreover, for CP($N-1$) a simple expression exists expressing the curvature tensor $R_{i\bar{j}k\bar{\ell}}$ in terms of a combination of metrics, namely,

$$R_{i\bar{j}k\bar{\ell}} = -\frac{g^2}{2}\left(G_{i\bar{j}}G_{k\bar{\ell}} + G_{i\bar{\ell}}G_{k\bar{j}}\right). \tag{6.26}$$

---

[1] The antiholomorphic index $\bar{j}$ is the index of the complex-conjugate field: $\bar{\phi}^{\bar{j}}$ is the same as $\overline{\phi^j}$.

[2] Here we are counting complex dimensions. For instance, the complex dimension of CP(1) is 1, while its real dimension is 2.

Finally, using the general expression for the scalar curvature (6.21), we obtain

$$\mathcal{R} = g^2 N(N - 1). \tag{6.27}$$

The Lagrangian of the CP(N − 1) model can be written as

$$\mathcal{L} = G_{i\bar{j}} \left(\partial_\mu \bar{\phi}^{\bar{j}}\right) \left(\partial^\mu \phi^i\right) + \frac{\theta}{2\pi i} \varepsilon^{\mu\nu} \left(\frac{g^2}{2} G_{i\bar{j}}\right) \left(\partial_\mu \bar{\phi}^{\bar{j}}\right) \left(\partial_\nu \phi^i\right). \tag{6.28}$$

Here, I added a topological term.

For $N = 2$, when we have a single field ϕ (and its complex conjugate), we return to the CP(1) model. For those who want to know more about the various mathematical aspects of the Kählerian sigma models, I can recommend two books [1, 2] and the review papers of Perelomov [3].

6.2.3 An Alternative Formulation of CP(N - 1) Models

The formulation of the CP(N − 1) model discussed in Section 6.2.2 is based on an explicit geometric description of the target space. In fact, many people refer to it as the *geometric formulation*. Now we will acquaint ourselves with an alternative formulation known as the *gauged formulation*.

In constructing the Lagrangian we start from an N-plet of complex "elementary" fields n^i, where $i = 1, 2, \ldots, N$. The fields n^i are scalar (i.e., spin-0) and transform in the fundamental representation of SU(N). These fields are subject to the single constraint

$$\bar{n}_i n^i = 1, \tag{6.29}$$

where the bar stands for complex conjugation. Thus, we have $2N$ real fields with one constraint, which leaves us with $2N − 1$ real degrees of freedom. From Section 6.2.2 we know that the CP(N − 1) model has $2N − 2$ real degrees of freedom. Thus, we must eliminate one more degree of freedom. This is achieved through a U(1) gauging. We introduce an auxiliary U(1) gauge field A_μ, with no kinetic term, to make the Lagrangian locally U(1) invariant. The possibility of imposing a gauge condition reduces the number of degrees of freedom to $2N − 2$.

Gauged formulation: in the literature g^{-2} is often denoted as β

Concretely, we specify the Lagrangian in the following way [4]:

$$\mathcal{L} = \frac{2}{g^2} \left|\mathcal{D}_\mu n^i\right|^2, \tag{6.30}$$

where the covariant derivative $\mathcal{D}_\mu$ is defined as

$$\mathcal{D}_\mu n^i \equiv \left(\partial_\mu + iA_\mu\right) n^i. \tag{6.31}$$

In terms of these fields the θ term takes the form

$$\mathcal{L}_\theta = \frac{\theta}{2\pi} \varepsilon^{\mu\nu} \partial_\mu A_\nu. \tag{6.32}$$

The fact that the θ term is a full derivative is explicit in this expression, as is the local U(1) invariance of the model at hand.

Since the field A_μ enters $\mathcal{L}$ without derivatives, one can eliminate it by virtue of the equations of motion,

$$A_\mu = \frac{i}{2} \left(\bar{n}_i \overleftrightarrow{\partial}_\mu n^i\right), \tag{6.33}$$

where the constraint (6.29) has been used. If we insert the above expression into the Lagrangian we then obtain

$$\mathcal{L} = \frac{2}{g^2} \left[\left(\partial_\mu \bar{n}_i \right) \left(\partial^\mu n^i \right) + \left(\bar{n}_i \partial_\mu n^i \right)^2 \right],   \tag{6.34}$$

$$\int d^2x \mathcal{L}_\theta = \frac{\theta}{2\pi i} \int d^2x \, \varepsilon^{\mu\nu} \left(\partial_\mu \bar{n}_i \right) \left(\partial_\nu n^i \right).   \tag{6.35}$$

In this form the fact that only $2N - 2$ real degrees of freedom are independent is not so obvious.

The fields ϕ^i of the geometric formulation are related to n^i as follows: we single out one component of n^i, say, n^N, and define

$$\phi^i = \frac{n^i}{n^N}, \quad i = 1, 2, \ldots, N - 1.   \tag{6.36}$$

| *How to pass* |
| *from* CP(1) |
| *to* O(3) |

For $N = 2$, when we deal with the CP(1) model, equivalent to O(3), it is helpful to have handy expressions relating the $\vec{S}$ fields to the n fields. Given the fact that in this case the n^i are spinors of SU(2) while $\vec{S}$ is the O(3) vector, it is not difficult to guess these expressions, namely,[3]

$$S^a = \bar{n} \tau^a n, \qquad a = 1, 2, 3,   \tag{6.37}$$

where the τ^a are the Pauli matrices. In the subsequent derivation we will need the Fierz transformation for the Pauli matrices (see, e.g., [5] or page xvi),

$$\vec{\tau}_{\alpha\beta} \, \vec{\tau}_{\delta\gamma} = 2\delta_{\alpha\gamma} \, \delta_{\delta\beta} - \delta_{\alpha\beta} \, \delta_{\delta\gamma}.   \tag{6.38}$$

Making use of this transformation we conclude that

$$\vec{S}^2 = (\bar{n}n)^2 = 1,   \tag{6.39}$$

provided that $\bar{n}n = 1$ and

$$\left(\partial_\mu \vec{S} \right)^2 = 4 \left[\left(\partial_\mu \bar{n}_i \right) \left(\partial^\mu n^i \right) + \left(\bar{n}_i \partial_\mu n^i \right)^2 \right].   \tag{6.40}$$

This establishes the equivalence of (6.3) and $\mathcal{L}$ in (6.34). The equivalence of the θ term representations in (6.4) and (6.35) must and can be verified too. The easiest way is to choose a reference frame in the target space in such a way that (at a given point x) one has $n^1 = 1$, $n^2 = 0$, and $\partial_\mu n^1 = 0$, while $\partial_\mu n^2 \neq 0$. This is consistent with (6.29) and can always be achieved.

The large-N solution of the model (6.34) is discussed in Chapter 9.

Exercises

6.2.1   Given the CP(1) Kähler metric

$$G = \frac{2}{g^2} \frac{1}{(1 + \phi\bar{\phi})^2}$$

find the Christoffel symbols, the curvature tensor, the Ricci tensor, and the scalar curvature.

---

[3] The relation $S^a = -\bar{n} \tau^a n$, with the opposite sign, is possible too. If we choose this relation, the sign of the θ term in Eq. (6.35) must be reversed.

6.2.2 Prove that the topological term in Eq. (6.9) can be represented as a full derivative,

$$\frac{\varepsilon^{\mu\nu}\left(\partial_\mu\bar{\phi}\right)(\partial_\nu\phi)}{(1+\bar{\phi}\phi)^2} = \partial_\mu K^\mu, \qquad K^\mu = \varepsilon^{\mu\nu}\left(\frac{\bar{\phi}\partial_\nu\phi}{1+\bar{\phi}\phi}\right).$$

Is K^μ invariant under the O(3) transformations (6.10), (6.11)? What is the maximal symmetry of the expression in parentheses on the right-hand side of the second formula?

6.2.3 Start from the Lagrangian (6.28) with $\theta = 0$ and $N = 3$, i.e., consider the CP(2) model. Calculate nonvanishing Christoffel symbols, components of the Riemann tensor, the Ricci tensor, and the scalar curvature.

6.3 Asymptotic Freedom in the O(3) Sigma Model

Asymptotic freedom in the O(3) *model discovered by Polyakov*

The content of this section is rather technical. Unlike many other sections where the physical meaning is emphasized, here I stress the computational aspects. This is reasonable since theoretical physicists sometimes have to do rather cumbersome calculations. My task is two-fold: (i) I want to demonstrate the power and elegance of the background field method; (ii) I want to find the coupling constant renormalization and show that the model at hand is asymptotically free (AF), i.e., the interaction becomes weak in the ultraviolet domain and strong in the infrared.

Since tasks of a technical nature are unavoidable, one should learn how to do them in a fun way, making the technical work as enjoyable as possible. To illustrate this, we will derive the law for the running of the coupling constant in the O(3) model following two distinct routes: first using a standard roadmap and then, later, via a shortcut for more experienced drivers.

6.3.1 Goldstone Fields in Perturbation Theory

First I will demonstrate an important aspect of the model – the perturbative spontaneous breaking of global symmetries (which, as we will see later on, is absent in a nonperturbative treatment). One begins by observing the most salient feature of the model at hand: the global O(3), symmetry which is spontaneously broken by the choice of the vacuum state. The vacuum manifold is depicted in Fig. 6.1. One can quantize the theory near any value of $\vec{S}$. All physical results will be equivalent for any choice of $\vec{S}_{\text{vac}} \equiv \langle\vec{S}\rangle$. A convenient choice is the north pole, i.e., $S^1 = S^2 = 0$, $S^3 = 1$. Near this vacuum, the fluctuations in S^1 and S^2 are small, and one can expand the Lagrangian (6.2) in these fields, treating them as quantum fluctuations,

$$\mathcal{L} = \frac{1}{2g^2}\left\{\left[(\partial_\mu S^1)^2 + (\partial_\mu S^2)^2\right]\right.$$
$$\left. + (S^1\partial_\mu S^1)^2 + (S^2\partial_\mu S^2)^2 + 2S^1 S^2(\partial_\mu S^1)(\partial_\mu S^2) + \cdots\right\}, \qquad (6.41)$$

where we have replaced S^3 as follows:

$$S^3 = \sqrt{1-(S^1)^2-(S^2)^2} = 1 - \frac{1}{2}\left[(S^1)^2 + (S^2)^2\right] + \cdots \qquad (6.42)$$

The ellipses in Eqs. (6.41) and (6.42) denote higher powers of the S fields. As was mentioned earlier, the topological term plays no role in perturbation theory.

The terms in the square brackets in Eq. (6.41) determine the propagator of the quantum fields $S^1(x)$ and $S^2(x)$; the other terms present interactions. Note the absence of mass terms for the fields $S^1(x)$ and $S^2(x)$. In our vacuum the original O(3) symmetry is broken down to O(2) – rotations around the axis connecting the north and south poles leave the vacuum intact. Since two symmetry generators are broken in the vacuum, one should expect the occurrence of two massless bosons, in accordance with the Goldstone theorem. They do occur – the fields $S^1(x)$ and $S^2(x)$ are massless.[4]

6.3.2 Perturbation Theory and Background Field Method

While Eq. (6.42) demonstrates, in a very nice manner, the Goldstone phenomenon (at a perturbative level), it would be unwise to use this representation of the model to calculate the coupling constant renormalization at one loop. It is certainly possible but would be rather awkward.

Instead, we will turn to the complex field (geometrical) representation; see Eq. (6.9) which gives the Lagrangian of the CP(1) model. We will exploit the *background field method*, which can be summarized as follows. The field $\phi(x)$ is decomposed into two components,

$$\phi(x) = \phi_0(x) + g_0 q(x), \tag{6.43}$$

Expansion in the quantum fields: quadratic terms

where $\phi_0(x)$ is a background c-number field while $q(x)$ is the quantum field propagating in loops. From now on we will denote the bare coupling in the original Lagrangian by g_0 rather than g, to distinguish it from the renormalized coupling. Upon substituting Eq. (6.43) into the Lagrangian (6.9) one obtains

$$\mathcal{L}[\phi(x)] = \frac{2}{g_0^2} \frac{1}{(1 + \bar{\phi}_0\phi_0)^2} \partial_\mu \bar{\phi}_0 \, \partial^\mu \phi_0 + q \times \text{(equation of motion)}$$

$$+ 2\frac{\partial_\mu \bar{q}\partial^\mu q}{(1 + \bar{\phi}_0\phi_0)^2} - 4\left(\partial_\mu \bar{q}\partial^\mu \phi_0 + \partial_\mu q \, \partial^\mu \bar{\phi}_0\right)\frac{\bar{\phi}_0 q + \phi_0\bar{q}}{(1 + \bar{\phi}_0\phi_0)^3}$$

$$+ 2\partial_\mu \bar{\phi}_0 \, \partial^\mu \phi_0 \left[\frac{3(\bar{\phi}_0 q + \phi_0\bar{q})^2}{(1 + \bar{\phi}_0\phi_0)^2} - \frac{2\bar{q}q}{1 + \bar{\phi}_0\phi_0}\right]\frac{1}{(1 + \bar{\phi}_0\phi_0)^2} + \cdots, \tag{6.44}$$

where the ellipses denote cubic and higher-order terms in q, which are relevant for two or more loops.

A few comments are in order concerning the first line in the above equation. The first term, the background Lagrangian, is the original Lagrangian from which we started. Our task is to calculate the one-loop correction to this Lagrangian. Then, this one-loop correction combined with $(2/g_0^2)(1 + \bar{\phi}_0\phi_0)^{-2}\partial_\mu \bar{\phi}_0 \, \partial^\mu \phi_0$ will yield an

---

[4] I hasten to add that a consideration going beyond perturbation theory will restore the full O(3) symmetry of the vacuum and eliminate the Goldstone bosons. This is a very special feature of $D = 1 + 1$ dimensional models, which has no analog at $D \geq 3$.

effective one-loop Lagrangian, from which we will determine the coupling constant renormalization at one loop.

The second term in the first line of Eq. (6.44) is linear in $q(x)$. Besides the equation of motion it contains full derivatives that drop out in the action. Within the background field method one *must* set this term equal to zero. If the background field $\phi_0(x)$ satisfies the equation of motion (as is the case in many instances) then the term linear in $q(x)$ vanishes automatically. One advantage of the background field method is the possibility of choosing the background field $\phi_0(x)$ arbitrarily, in such a way as to maximally facilitate the calculation we have to perform. The choice depends, generally speaking, on the particular problem under consideration. If $\phi_0(x)$ is chosen in such a way that the original equation of motion is not satisfied, we must add source terms to our theory to make the chosen $\phi_0(x)$ satisfy the equation of motion, now including the source terms. This is always possible to achieve. Then the expansion of the Lagrangian in the quantum field $q(x)$ will contain no terms linear in q, thus ensuring that the quantization procedure for $q(x)$ oscillating near zero is stable. The presence of a linear term would force the theory to slide away from $q \sim 0$.

After the field $\phi(x)$ has been split into the background and quantum parts, the nonlinear invariance transformation (6.11) is linearized for $q(x)$. Namely, the quadratic part of the Lagrangian (6.44) is invariant under the following transformations performed simultaneously:

Target space invariance

$$\phi_0 \to \phi_0 + \epsilon + \bar{\epsilon}\phi_0^2, \qquad \bar{\phi} \to \bar{\phi}_0 + \bar{\epsilon} + \epsilon\bar{\phi}_0^2,$$

$$q \to q + 2\bar{\epsilon}\phi_0 q, \qquad \bar{q} \to \bar{q} + 2\epsilon\bar{\phi}_0\bar{q}. \tag{6.45}$$

Here I will point out another advantage of the background field method. In the original formulation of the theory, which was in terms of the fields $\vec{S}$ or ϕ, it was impossible to introduce a mass term for those fields without destroying the full symmetry of the model. This would make the infrared and ultraviolet regularization of the one-loop correction a tricky task. In the background field method the symmetry transformation for $q(x)$ is linear, see Eq. (6.45). This fact enables one to introduce a mass term for the q field of the type $\mu^2\bar{q}q$ without violating the symmetry of the model. Hence, our regularization, both in the infrared and ultraviolet, will be compatible with all symmetries.

With this information in hand let us rewrite the quadratic part of the quantum Lagrangian for $q(x)$:

$$\mathcal{L}^{(2)}[q(x)] = 2\frac{\partial_\mu\bar{q}\,\partial^\mu q - \mu^2\bar{q}q}{(1+\bar{\phi}_0\phi_0)^2} - 4(\partial_\mu\bar{q}\,\partial^\mu\phi_0 + \partial_\mu q\partial^\mu\bar{\phi}_0)\frac{\bar{\phi}_0 q + \phi_0\bar{q}}{(1+\bar{\phi}_0\phi_0)^3}$$

$$+ 2\partial_\mu\bar{\phi}_0\partial^\mu\phi_0\left[\frac{3(\bar{\phi}_0 q + \phi_0\bar{q})^2}{(1+\bar{\phi}_0\phi_0)^2} - \frac{2\bar{q}q}{1+\bar{\phi}_0\phi_0}\right]\frac{1}{(1+\bar{\phi}_0\phi_0)^2} + \ldots, \tag{6.46}$$

$\mu^2\bar{q}q$ is added for IR regularization.

where we have added (in the numerator of the first term) a mass term for the purpose of regularization.

So far, the background field $\phi_0(x)$ has not been specified. In principle we could proceed further, making no assumptions regarding $\phi_0(x)$. However, one can immensely simplify the calculation by making a wise choice of background field.

In the case at hand a good choice is, for instance, a plane wave background,

$$\phi_0(x) = f e^{ikx}, \qquad \bar{\phi}_0(x) = \bar{f} e^{-ikx}, \tag{6.47}$$

where f is a dimensionless constant. The value of f is arbitrary and the wave vector k is assumed to be small. This means that one cannot expand in f; however, one can expand in k. As we will see shortly we will need to keep terms quadratic in k; cubic and higher-order terms are irrelevant.

The background field should be chosen in such a way that the operator whose renormalization is under investigation does not vanish. The plane wave background satisfies this (necessary) condition since

$$\frac{1}{(1 + \bar{\phi}_0 \phi_0)^2} \partial_\mu \bar{\phi}_0 \partial^\mu \phi_0 = \frac{k^2 |f|^2}{(1 + |f|^2)^2} \neq 0. \tag{6.48}$$

Why is the choice (6.47) good? Simplifications occur due to the fact that $\bar{\phi}_0 \phi_0$ reduces to a constant. One cannot choose ϕ_0 itself to be constant (the simplest choice) since this would violate the necessary condition above. So, we settle for the second best choice.

The first term on the right-hand side of Eq. (6.46) is of zeroth order in k, the second is linear in k, while the third is quadratic. This establishes a hierarchy: the first term is "large" while the other two are "small" and can be treated as a perturbation. Thus, we will determine the propagator of the field q from the first term; the second and third terms will determine the interaction "vertices." I hasten to add that the propagator and vertices with which we are dealing with here have nothing to do with the propagator and interaction vertices in the vacuum. The background field propagator (Fig. 6.3) could include, say, any number of interactions of the quantum field q with the background field $\bar{\phi}_0 \phi_0$. Moreover, the interaction "vertices" are quadratic in q, so the word "vertices" applies here in a Pickwick sense (a way that is not immediately obvious).

Since, with our choice of the background field, $\bar{\phi}_0 \phi_0$ is just a constant,

$$\bar{\phi}_0 \phi_0 = \bar{f} f,$$

it is very easy to obtain the propagator of the q field,

$$D(p) = \frac{(1 + \bar{f} f)^2}{2} \frac{i}{p^2 - \mu^2}, \tag{6.49}$$

where p is the momentum flowing through the q line, see Fig. 6.3.

Now let us turn our attention to the "vertices." We will start from a simple "vertex," that in the second line of Eq. (6.46). It is explicitly proportional to $\partial_\mu \bar{\phi}_0 \partial^\mu \phi_0 \propto k^2$. Since we do not need to keep terms higher than k^2, we can neglect the k-dependence of the ϕ_0 field in the square brackets, making the replacements $\phi_0 \to f$, $\bar{\phi}_0 \to \bar{f}$. In this way we arrive at (Fig. 6.4a)

$$\diamond = i \, 2k^2 \frac{\bar{f} f}{(1 + \bar{f} f)^2} \left[\frac{3(\bar{f} q + f \bar{q})^2}{(1 + \bar{f} f)^2} - \frac{2 \bar{q} q}{1 + \bar{f} f} \right]$$

$$\to i \, 2k^2 \frac{\bar{f} f}{(1 + \bar{f} f)^2} \left[\frac{6 \bar{f} f}{(1 + \bar{f} f)^2} - \frac{2}{1 + \bar{f} f} \right] \bar{q} q, \tag{6.50}$$

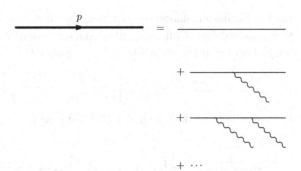

Fig. 6.3   The propagator of the quantum field q (thick solid line) in the background field $\bar{\phi}_0 \phi_0$. This propagator sums up all insertions of $\bar{\phi}_0 \phi_0$, denoted by wavy lines.

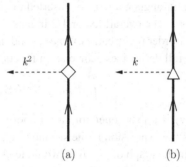

Fig. 6.4   Interaction "vertices" proportional to (a) k^2 and (b) k.

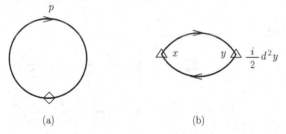

Fig. 6.5   One-loop correction to the effective Lagrangian due to (6.50) and (6.54). The interaction "vertex" (6.54), shown in (b), should be inserted twice, while that of (6.50), shown in (a), need not be iterated.

where in the second line we retain only the term with one incoming q line and one outgoing; only this term is relevant for our calculation. Since the "vertex" (6.50) is proportional to k^2, there is no need to iterate it. The corresponding contribution to the effective one-loop Lagrangian is depicted in Fig. 6.5a.

The calculation of this graph is straightforward. The tadpole loop is proportional to

$$\int \frac{d^2 p}{(2\pi)^2} D(p) = \int \frac{d^2 p}{(2\pi)^2} \frac{(1 + \bar{f}f)^2}{2} \frac{i}{p^2 - \mu^2}$$

$$\xrightarrow{\text{Euclid}} \frac{(1 + \bar{f}f)^2}{2} \int \frac{dp^2}{4\pi} \frac{1}{p^2 + \mu^2}, \tag{6.51}$$

where a Euclidean rotation $p_0 \to ip_0$ and angular integration are performed in passing to the second line. Collecting all pre-factors from Eq. (6.50) and cutting off the integral over p^2 in the ultraviolet at M_{uv}^2, we obtain

$$\mathcal{L}_a^{\text{one-loop}} = k^2 \bar{f} f \left[\frac{6\bar{f}f}{(1+\bar{f}f)^2} - \frac{2}{1+\bar{f}f} \right] \frac{1}{4\pi} \ln \frac{M_{\mathrm{uv}}^2}{\mu^2}. \tag{6.52}$$

It is time now to deal with the $O(k)$ interaction "vertex," see the first line in Eq. (6.46):

$$\Delta = \frac{4k^\mu}{(1+\bar{f}f)^3} \left\{ \bar{f}f \left[q\partial_\mu\bar{q} - \bar{q}\partial_\mu q \right] + \left[\frac{f^2}{2}(\partial_\mu\bar{q}^2)e^{2ikx} - \frac{\bar{f}^2}{2}(\partial_\mu q^2)e^{-2ikx} \right] \right\}. \tag{6.53}$$

Since this "vertex" is proportional to k, and we are looking for terms $O(k^2)$ in the one-loop Lagrangian, it must be inserted twice. The corresponding graph is depicted in Fig. 6.5b. (The overall factor $1/2$ indicated in this figure reflects the fact that this is a second-order perturbation.) The second line in Eq. (6.53) contains two operators which are full derivatives. Correlation functions of the type

$$\int d^2 y \, e^{2ik(x-y)} \langle \partial_\mu \bar{q}^2(x), O(y) \rangle,$$

where $O(y)$ is a local operator, are proportional to the first or higher powers of k. Therefore, the expression in the second line of Eq. (6.53) can be safely omitted – its insertion in the graph of Fig. 6.5b would lead to terms in the one-loop action that are cubic in k or higher. Thus, for our purposes we can make the replacements

$$\Delta \to \frac{4\bar{f}fk^\mu}{(1+\bar{f}f)^3} \left[q\partial_\mu\bar{q} - \bar{q}\partial_\mu q \right] \tag{6.54}$$

and

$$\mathcal{L}_b^{\text{one-loop}} = \left[\frac{4\bar{f}f}{(1+\bar{f}f)^3} \right]^2 \left[\frac{(1+\bar{f}f)^2}{2} \right]^2 k^\mu k^\nu (-2i)$$

$$\times \int \frac{d^2p}{(2\pi)^2} \frac{p_\mu}{p^2-\mu^2} \frac{p_\nu}{p^2-\mu^2}, \tag{6.55}$$

where we have taken into account the fact that there are four terms in the correlation function $\langle \Delta, \Delta \rangle$ and they are equal to each other.

Performing the integral in Eq. (6.55) is trivial. Owing to the Lorentz symmetry the product $p_\mu p_\nu$ can be replaced by $(1/2)g_{\mu\nu}p^2$. Performing the Euclidean rotation and cutting off the p^2 integration at M_{uv}^2 in the ultraviolet, as we have done previously, we arrive at

$$\mathcal{L}_b^{\text{one-loop}} = -4k^2 \frac{(\bar{f}f)^2}{(1+\bar{f}f)^2} \frac{1}{4\pi} \ln \frac{M_{\mathrm{uv}}^2}{\mu^2}. \tag{6.56}$$

Combining this result with Eq. (6.52) we obtain

$$\mathcal{L}^{\text{one-loop}} = -2k^2 \frac{\bar{f}f}{(1+\bar{f}f)^2} \frac{1}{4\pi} \ln \frac{M_{\mathrm{uv}}^2}{\mu^2}. \tag{6.57}$$

The last step is to interpret the result of our calculation. Recall that the background field method operates in a way such that at no stage is the symmetry (6.45) violated.

This means, that after integration over the quantum field $q(x)$, the expression for the effective Lagrangian as a function of $\phi_0(x)$ must be invariant under

$$\phi_0 \to \phi_0 + \epsilon + \bar{\epsilon}\phi_0^2, \qquad \bar{\phi}_0 \to \bar{\phi}_0 + \bar{\epsilon} + \epsilon\bar{\phi}_0^2. \tag{6.58}$$

The *only* structure satisfying this requirement and containing not more than two derivatives is $(1 + \bar{\phi}_0\phi_0)^{-2}\partial_\mu\bar{\phi}_0\partial^\mu\phi_0$. This is perfectly consistent with Eq. (6.57). Moreover, upon inspecting Eq. (6.57) we immediately conclude that

$$\mathcal{L}^{\text{one-loop}} = \frac{1}{(1 + \bar{\phi}_0\phi_0)^2}\partial_\mu\bar{\phi}_0\,\partial^\mu\phi_0\left(-\frac{1}{2\pi}\right)\ln\frac{M_{\text{uv}}^2}{\mu^2}. \tag{6.59}$$

Assembling $\mathcal{L}^{(0)}$ from (6.44) and $\mathcal{L}^{\text{one-loop}}$ we arrive at

$$\mathcal{L} = \mathcal{L}^{(0)} + \mathcal{L}^{\text{one-loop}} = \frac{2}{g^2(\mu)}\frac{1}{(1 + \bar{\phi}\phi)^2}\partial_\mu\bar{\phi}\,\partial^\mu\phi, \tag{6.60}$$

where the running constant $g^2(\mu)$ is expressed in terms of the bare constant and the logarithm of the ultraviolet cutoff,

$$\frac{1}{g^2(\mu)} = \frac{1}{g_0^2} - \frac{1}{4\pi}\ln\frac{M_{\text{uv}}^2}{\mu^2}. \tag{6.61}$$

Coupling constant renormalization exhibits AF.

The minus sign in front of the logarithm gives the celebrated asymptotic freedom. Indeed, the β function obtained by differentiating Eq. (6.61) over $\ln\mu$ is negative,

$$\beta(g^2) \equiv \frac{\partial g^2}{\partial\ln\mu} = -\frac{g^4}{2\pi}. \tag{6.62}$$

Deep in the ultraviolet domain, as μ^2 grows, $g^2(\mu)$ decreases. However, in the infrared domain, with μ^2 decreasing, $g^2(\mu)$ grows and eventually blows off at

$$\frac{g_0^2}{4\pi}\ln\frac{M_{\text{uv}}^2}{\mu^2} \sim 1.$$

No matter how small g_0^2 is, one can always find a μ^2 such that the running constant is of order 1. This is the (infrared) domain of strong coupling, where perturbation theory in the coupling constant fails, and other methods for solution of the theory should be sought (for instance, expansion in $1/N$ in CP$(N-1)$; see Section 9.4).

Equation (6.61) can be rewritten as

$$g^2(\mu) = g_0^2\left(1 - \frac{g_0^2}{4\pi}\ln\frac{M_{\text{uv}}^2}{\mu^2}\right)^{-1}, \tag{6.63}$$

or

$$g^2(\mu) = \frac{4\pi}{\ln(\mu^2/\Lambda^2)}, \qquad \Lambda^2 = M_{\text{uv}}^2 e^{-4\pi/g_0^2}. \tag{6.64}$$

Dynamical scale parameter Λ through dimensional transmutation

We see that neither the bare coupling constant nor the ultraviolet cutoff appear separately in the running coupling constant; rather, they enter through a very particular combination, Λ, which is usually referred to as the *scale parameter* of the theory. This feature – the emergence of a particular combination of g_0^2 and M_{uv}^2 in the running coupling – is due to the renormalizability of the model under consideration. Unlike g_0^2, which is dimensionless, Λ has the dimension of mass. Trading the dimensionless bare coupling for the scale parameter is called *dimensional transmutation*. All physically nontrivial phenomena occur in the domain $\mu \sim \Lambda$. At $\mu \gg \Lambda$ the theory is at weak coupling, and its dynamics is quite transparent and amenable to perturbative treatment.

The asymptotic freedom of the O(3) sigma model was discovered in the 1970s by Polyakov [6].

In this section I used the simplest version of the background field method and the simplest choice of the background field (6.47). Its advantage is that the metric for the given background field is x independent. The disadvantage is that this version is not applicable to arbitrary σ models. Rather, it works only in the Kähler σ sigma models. In the next section, I explain why this is the case in a few words.

6.3.3 Why Kähler σ Models are special

Let us analyze the calculation of the tadpole diagram in Fig. 6.5a in more general terms. The diamond vertex in the tadpole graph is in fact the second derivative

$$\diamond = (\partial_\mu \phi_0^i)(\partial^\mu \bar{\phi}_0^{\bar{j}})(\partial_k \partial_{\bar{m}} G_{i\bar{j}})(\delta \phi^k)(\delta \bar{\phi}^{\bar{m}}), \tag{6.65}$$

where $\delta \phi = gq$ in my previous notation. Substituting it in the tadpole graph and multiplying by the Green function we arrive at

$$(\partial_\mu \phi_0^i)(\partial^\mu \bar{\phi}_0^{\bar{j}}) [G^{\bar{m}k}] [\partial_k \partial_{\bar{m}} G_{i\bar{j}}] \frac{1}{2\pi} \log \frac{M_0}{\mu}$$

$$= (\partial_\mu \phi_0^i)(\partial^\mu \bar{\phi}_0^{\bar{j}}) \left[-R_{i\bar{j}} - \left(\partial_i G^{\bar{m}k} \right) \left(\partial_{\bar{j}} G_{k\bar{m}} \right) \right] \frac{1}{2\pi} \log \frac{M_0}{\mu}, \tag{6.66}$$

where I used various general identities from Section 6.2.2. All expressions in the square brackets are to be evaluated at $\phi = \phi_0$. It is not difficult to check that the graph in Figure 6.5b cancels the term proportional to $(\partial_i G^{\bar{m}k})(\partial_{\bar{j}} G_{k\bar{m}})$, and, therefore, the one-loop correction to the Lagrangian is proportional to $R_{i\bar{j}} \sim G_{i\bar{j}}$ as it should be on general grounds [7].[5]

6.3.4 Shortcut (or What You Can Do with Experience)

Now we will see how to do the same calculation in a trice. Being confident that the background field method preserves the full symmetry of the model, see Eq. (6.58), we will not waste our time checking that the sum of all one-loop graphs does indeed yield the required structure $|\partial_\mu \phi|^2/(1 + |\phi|^2)^2$. We will take this for granted. Then it is sufficient to find the coefficient in front of $|\partial_\mu \phi|^2 = k^2| f|^2$ to get the coupling constant renormalization. In other words, assuming that both background parameters, k and $| f|$, are small we will expand not only in k but also in f and keep only the terms $O(| f|^2, k^2)$.

Then the Lagrangian for the quantum field $q(x)$ given in Eq. (6.46) simplifies and takes the form

$$\mathcal{L}^{(2)}[q(x)] = 2\partial_\mu \bar{q}\, \partial^\mu q - 4k^2| f|^2 \bar{q}q, \tag{6.67}$$

---

[5] Note also that the product $\left(\partial_i G^{\bar{m}k} \right) \left(\partial_{\bar{j}} G_{k\bar{m}} \right)$ is suppressed by $\bar{f}f$ compared to $R_{i\bar{j}}$, cf. Section 6.3.4.

which at one loop produces the renormalization

$$-4k^2 |f|^2 \langle \bar{q}q \rangle. \tag{6.68}$$

The only surviving diagram is the tadpole graph of Fig. 6.5a with the standard p^{-2} propagator for the q field. A straightforward (and very simple) calculation of this tadpole leads to the replacement of the bare coupling as follows:

$$\frac{1}{g_0^2} \to \frac{1}{g_0^2} - 2\langle \bar{q}q \rangle = \frac{1}{g_0^2} - \frac{1}{4\pi} \int \frac{dp^2}{p^2}, \tag{6.69}$$

which is equivalent to Eq. (6.61).

6.3.5 The β Functions of CP($N-1$)

Equation (6.61) determines the one-loop β function (6.62) of the CP(1) model. The lesson we learned from Section 6.3.4 allows us to immediately extend this result to the general case of arbitrary N. Starting from (6.23) and assuming that

$$\phi_0^i = \begin{cases} f e^{ikx}, & i = 1, \\ 0, & i = 2, \dots, N-1, \end{cases} \tag{6.70}$$

we get the Lagrangian that must replace (6.67), in the form

$$\mathcal{L}^{(2)}[q(x)] = 2\partial_\mu \bar{q}^i \, \partial^\mu q^i - 2k^2 |f|^2 \left(2\bar{q}^1 q^1 + \sum_{i=2}^{N-1} \bar{q}^i q^i \right). \tag{6.71}$$

Correspondingly, we get N tadpoles in the CP($N-1$) model, each of which is half the CP(1) tadpole. As a result,

$$\beta(g^2)_{\text{one-loop}} = \frac{\partial g^2}{\partial \ln \mu} = -\frac{g^4 N}{4\pi}. \tag{6.72}$$

One- and two-loop β functions in CP($N-1$)

This N-dependence is in agreement with the general analysis [8].

The two-loop β function can be calculated as well. Although straightforward, the procedure is quite tedious and time-consuming owing to the large number of two-loop graphs involved. It is an instructive exercise for mastering the background field technique, but I would recommend it only to the most courageous and advanced readers. The result is

$$\beta(g^2)_{\text{two-loop}} = -\frac{g^4 N}{4\pi} \left(1 + \frac{g^2}{2\pi} \right). \tag{6.73}$$

A traditional calculation of the two-loop β function in the CP($N-1$) model can be found in [10]. In the large-N 't Hooft limit (Chapter 9) the coupling constant g^2 scales as $1/N$. This scaling implies that $\beta(g^2)_{\text{one-loop}}$ survives in the limit $N \to \infty$ while the two-loop term in (6.73) is subleading in $1/N$ and drops out at large N. This is consistent with the large-N solution of the CP($N-1$) model presented in Chapter 9, which is exhausted by one loop.

For the advanced reader one can suggest an alternative route of derivation of the second term in (6.73).[6] In Part II (Section 10.12.3.4) we will study a supersymmetric

---

[6] In fact, this problem is recommended to readers who intend to master Part II, devoted to supersymmetry; after studying supersymmetry, such readers should return to this section and do this exercise.

extension of the CP(N − 1) model. We will learn that, on general grounds (i) the β function in this model is exhausted by one loop [8] and (ii) fermions contribute to the β function only at the second and higher loops (they do not show up at one loop). This implies that in the nonsupersymmetric CP(N−1) model under discussion here, the two-loop coefficient in $\beta(g^2)_{\text{two-loop}}$ is equal to minus one times the fermion contribution. The advantage of this indirect calculation is that there exists a single fermion diagram that contributes to $\beta(g^2)_{\text{two-loop}}$; see [9] and Fig. S.7.

Exercises

6.3.1 Given the Lagrangian in (6.9) find the equation of motion for the ϕ field. Do the same in the S representation, starting from the action (6.3) and the constraint (6.1).

6.3.2 Identify the two-loop diagram presenting the fermion contribution mentioned in the last paragraph of Section 6.3.5.

6.3.3 Calculate the running coupling constant in the O(N) sigma model at one loop using the background field technique. If problems arise, see Appendix section 9.8 in Chapter 9.

6.4 Instantons in CP(1)

In this section we will use Euclidean notation similar to that of Section 5.2. The Euclidean Levi-Civita tensor is defined as $\varepsilon_{12} = -\varepsilon_{21} = 1$.

Instantons in the CP(N − 1) model (first found in the pioneering work [6]) are remarkably simple. This is the reason why they serve as an excellent theoretical laboratory and present a basis for a large number of various investigations. A seminal paper in this range of ideas is [11].

As we know from Chapter 5, the first thing to do in instanton studies is to pass to Euclidean space–time. The Euclidean action formally looks as in (6.9), although the space–time metric is now diag{1, 1} rather than diag{1, −1}:

$$S_{\text{E}} = \int d^2x \frac{2}{g^2} \frac{\partial_\mu \bar{\phi} \partial_\mu \phi}{(1 + \bar{\phi}\phi)^2}. \tag{6.74}$$

For simplicity we have omitted the θ term; it can be easily reintroduced if necessary. The Bogomol'nyi completion takes the form

$$S_{\text{E}} = \int d^2x \frac{1}{g^2} \left[\left(\partial_\mu \bar{\phi} \mp i\varepsilon_{\mu\nu} \partial_\nu \bar{\phi} \right) \left(\partial_\mu \phi \pm i\varepsilon_{\mu\rho} \partial_\rho \phi \right) \mp 2i\, \varepsilon_{\mu\nu} \partial_\mu \bar{\phi} \partial_\nu \phi \right] (1 + \bar{\phi}\phi)^{-2}. \tag{6.75}$$

Euclidean action

The last term in the square brackets presents an integral over a full derivative (see Exercise 6.2.2) and thus reduces to the topological term. The minimal action is achieved if

$$\partial_\mu \phi \pm i\varepsilon_{\mu\nu} \partial_\nu \phi = 0. \tag{6.76}$$

This is the (anti-)self-duality equation. For definiteness, let us take the upper sign in Eqs. (6.75) and (6.76). Moreover, instead of two real coordinates $x_{1,2}$ let us introduce complex coordinates

$$z = x_1 + ix_2, \qquad \bar{z} = x_1 - ix_2;$$

$$\frac{\partial}{\partial z} = \frac{1}{2}\left(\frac{\partial}{\partial x_1} - i\frac{\partial}{\partial x_2}\right), \quad \frac{\partial}{\partial \bar{z}} = \frac{1}{2}\left(\frac{\partial}{\partial x_1} + i\frac{\partial}{\partial x_2}\right). \tag{6.77}$$

In terms of these complex coordinates the self-duality equation (6.76) takes the form [12]

$$\frac{\partial \phi}{\partial \bar{z}} = 0. \tag{6.78}$$

Remembering that ϕ is the coordinate on the target space sphere S_2 (with south pole corresponding to $\phi \to \infty$) we can assert that the solution of (6.78) is given by any meromorphic function of z. Why meromorphic? As usual, we require the action to be finite. This means that if at a certain point $z = z_*$ the function $\phi(z)$ is singular then the limit $\phi(z \to z_*)$ should be such as to guarantee the convergence of (6.74). This leaves us with only poles. A similar situation occurs at $|z| \to \infty$. The limit $\phi(|z| \to \infty)$ must be independent of the angular direction, for the same reason. Thus, the two-dimensional space–time is compactified and is topologically equivalent to the sphere S_2. The target space is S_2.[7] The topological stability of the instanton solution is due to the fact that

$$\pi_2(\text{SU}(2)/\text{U}(1)) = \pi_1(\text{U}(1)) = \mathbb{Z}. \tag{6.79}$$

Thus, the CP(1) instanton solutions can have topological charges $\pm 1, \pm 2, \pm 3, \ldots$ in much the same way as in Yang–Mills theories (Chapter 5). In terms of the complex variables $z, \bar{z}$ the Euclidean expression for the topological charge is

$$Q_E = \frac{1}{\pi} \int d^2x \left(\frac{\partial \bar{\phi}}{\partial \bar{z}} \frac{\partial \phi}{\partial z} - \frac{\partial \bar{\phi}}{\partial z} \frac{\partial \phi}{\partial \bar{z}}\right)(1 + \bar{\phi}\phi)^{-2} \tag{6.80}$$

$$= -\frac{i}{2\pi} \int d^2x \, \varepsilon_{\mu\nu} \, \partial_\mu \bar{\phi} \, \partial_\nu \phi \, (1 + \bar{\phi}\phi)^{-2}. \tag{6.81}$$

A single instanton is represented by a single pole in $\phi(z)$,

$$\phi(z) = \frac{a}{z - b}, \tag{6.82}$$

and has unit topological charge. Choosing the upper sign in (6.75) we rewrite the Bogomol'nyi representation as follows:

$$S_E = \int d^2x \frac{1}{g^2} \left(\partial_\mu \bar{\phi} - i\varepsilon_{\mu\nu}\partial_\nu \bar{\phi}\right)\left(\partial_\mu \phi + i\varepsilon_{\mu\rho}\partial_\rho \phi\right) + \frac{4\pi Q_E}{g^2}, \tag{6.83}$$

implying that the instanton action is

$$S_0 = \frac{4\pi}{g^2}. \tag{6.84}$$

The complex numbers a and b in (6.82) are instanton moduli. It is obvious that b represents two (real) translational moduli. In other words, b is the instanton center.

Instanton action and moduli

---

[7] This is equivalent to the coset SU(2)/U(1).

As far as a is concerned, its interpretation requires some work, which I leave as an exercise to the reader. Let me just formulate the answer. Assume that a is represented as

$$a = \rho e^{i\alpha}, \qquad \rho \equiv |a|. \tag{6.85}$$

Then ρ plays the role of the instanton size, in much the same way as in Yang–Mills theories. To understand the meaning of α we should remember that at weak coupling the SU(2) symmetry of the model at hand is spontaneously broken down to U(1) by a particular choice of the vacuum state. We have made this choice implicitly, choosing the vacuum at the north pole of the target space sphere. At large separations from the center the instanton solution must tend to the vacuum value. In (6.82), $\phi(z)$ tends to zero as $z \to \infty$, which is exactly the north pole on the target space sphere. While the vacuum is invariant under rotations around the vertical axis in the target space, this U(1) symmetry is explicitly broken on every given instanton solution. This explains the occurrence of the angular modulus α.

The general solution with k instantons is quite simple too and has the form

$$\phi(z) = \sum_{j=1}^{k} \frac{a_j}{z - b_j}. \tag{6.86}$$

The overall number of moduli is $4k$ in this case.

To obtain anti-instantons rather than instantons we must return to Eqs. (6.75) and (6.76) and choose the lower signs. For anti-instantons, ϕ is a meromorphic function of $\bar{z}$ rather than z and the topological charges are negative.

To conclude this section, I will make a few remarks regarding the instanton measure in CP(1). In fact, with information already at our disposal, we can reconstruct it, up to an overall constant, without direct calculation.[8] We will follow the same line of reasoning as in Section 5.4.6.

In the case at hand we have four zero modes; hence the part of the measure due to these zero modes is

$$d\mu_{\text{inst}}^{\text{zm}} = \text{const} \times \left[\exp\left(-\frac{4\pi}{g_0^2} \right) \right] \left(M_{\text{uv}}^2 \frac{4\pi}{g_0^2} \right)^2 d^2 b \, d\rho \, d\alpha. \tag{6.87}$$

The right-hand side unambiguously emerges from consideration of (6.84) plus symmetry and dimensional arguments. Now, as in the case of the Yang–Mills instantons (Section 5.4.6), the nonzero modes additionally contribute the logarithmic term $-2\ln(M_{\text{uv}}\rho)$ in the exponent. This follows from Eq. (6.64). Thus, using Eq. (6.64) we can write the one-instanton measure in terms of $g^2 \equiv g^2(\rho)$ or in terms of Λ:

Instanton measure in CP(1)

$$d\mu_{\text{inst}} = \text{const} \times \Lambda^2 d^2 b \frac{d\rho}{\rho}. \tag{6.88}$$

The fact that the measure is divergent at large ρ is not surprising – we witnessed the same phenomenon for the Yang–Mills instantons – it means only that the one-instanton approximation (as well as the instanton gas) becomes invalid at large

---

[8] A multipage direct calculation can be found in, e.g., [13]. If you want, you can compare it with the subsequent paragraph. It is true, however, that the overall constant, which remains undetermined in Eqs. (6.87) and (6.88), is unambiguously found in a straightforward direct calculation [13].

sizes.[9] What was, perhaps, unexpected, is that there is a logarithmic *ultraviolet* divergence of the instanton measure at $\rho \to 0$. We will not dwell on this issue, referring the interested reader to [14], where nonperturbative UV infinities in various models are discussed in some detail. What is important is that nonperturbative UV divergences do not require extra (i.e., new) renormalization constants in observable physical quantities. The instanton measure by itself is unobservable.

Exercise

6.4.1   Calculate the topological charge for the k-instanton solution.

6.5 The Goldstone Theorem in Two Dimensions

6.5.1 The Goldstone Theorem

This is a very simple but very powerful and general theorem, which states [15]:

> Assume that there is a *global continuous* symmetry in the field theory under consideration. If this symmetry is spontaneously broken [then] the particle spectrum *must* contain a massless boson (the Goldstone boson) coupled to the broken generator. A Goldstone boson corresponds to each broken generator, so that the number of the Goldstone bosons is equal to the number of the broken generators.

The proof is quite straightforward. Given a global continuous symmetry of the Lagrangian, one can always construct a Noether current $J^\mu(x)$ that is conserved:

$$\partial_\mu J^\mu = 0. \tag{6.89}$$

For the time being we will assume that the current J is Hermitian, $J^\dagger = J$. This assumption can easily be lifted.

The corresponding charge Q is obtained from J^0 by integrating over space:

$$Q = \int d^{D-1}x\, J^0(x), \qquad \dot{Q} = 0. \tag{6.90}$$

Assume that there is a local field $\phi(x)$ (it may be composite) such that

$$\chi(x) = [Q, \phi(x)], \tag{6.91}$$

where $\chi(x)$ is another field (which may also be composite) Then χ is an order parameter for the given symmetry. If χ develops a nonvanishing vacuum expectation value then the vacuum state is asymmetric; the symmetry generated by Q is spontaneously broken. Indeed,

$$\langle \text{vac}|\chi(x)|\text{vac}\rangle \equiv v \neq 0 \tag{6.92}$$

---

[9] A remark for curious readers: an instanton melting at large densities was demonstrated in [11] in a clear-cut manner. This derivation became possible owing to the fact that it is much easier to treat two-dimensional models than four-dimensional models.

implies that

$$\langle \text{vac}|Q\phi(x) - \phi(x)Q|\text{vac}\rangle = v \neq 0, \tag{6.93}$$

which implies, in turn, that

$$Q|\text{vac}\rangle \neq 0. \tag{6.94}$$

The vacuum state is not annihilated by Q. Hence, it is asymmetric. The symmetry of the Lagrangian is spontaneously broken by the vacuum state.

If the symmetry were not broken,

$$e^{i\alpha Q}|\text{vac}\rangle = |\text{vac}\rangle \qquad \text{or} \qquad Q|\text{vac}\rangle = 0, \tag{6.95}$$

resulting in a vanishing expectation value of the order parameter χ.

Now, where are the Goldstone bosons? In order to see them consider the correlation function

$$\Pi^{\mu}(q) = -i \int e^{iqx} d^D x \langle \text{vac}|T\{J^{\mu}(x), \phi(0)\}|\text{vac}\rangle. \tag{6.96}$$

Multiply Π^{μ} by q_{μ} and let $q_{\mu} \to 0$. We obtain

$$
\begin{aligned}
q_{\mu}\Pi^{\mu} &= -\int \left(\frac{\partial}{\partial x^{\mu}} e^{iqx}\right) d^D x \langle \text{vac}|T\cdot\left\{J^{\mu}(x), \phi(0)\right\}|\text{vac}\rangle \\
&= \int_{q\to 0} e^{iqx} d^D x \frac{\partial}{\partial x^{\mu}} \langle \text{vac}|T\left\{J^{\mu}(x), \phi(0)\right\}|\text{vac}\rangle \\
&= \langle \text{vac}|\chi|\text{vac}\rangle, \tag{6.97}
\end{aligned}
$$

where we have used Eqs. (6.89)–(6.91). Since the right-hand side does not vanish, neither does the left-hand side and this implies that

| Goldstone theorem |

$$\Pi^{\mu}(q) = v\frac{q^{\mu}}{q^2} \qquad \text{as } q \to 0. \tag{6.98}$$

The pole in Π^{μ} at $q = 0$ demonstrates the inevitability of a massless boson coupled both to ϕ and J^{μ}.

6.5.2 Why Does This Argument Not Work in Two Dimensions?

| No Goldstones in two dimensions! |

This subsection could have been entitled "Coleman versus Goldstone." Coleman noted [16] that it is virtually impossible to have spontaneously broken continuous global symmetries in two-dimensional field theory. Even if at the classical level an order parameter is set to be nonvanishing, quantum fluctuations are always strong enough to screen it completely. No matter how small the original coupling constants are, interactions grow in the infrared domain and eventually become strong enough to restore the full symmetry of the Lagrangian. All continuous global symmetries of the Lagrangian are thus linearly realized in the particle spectrum in two-dimensional field theory.

The above statement does not apply to models in which interactions switch off in the infrared domain, so that all particles become sterile. For instance, in the 't Hooft model [17] (two-dimensional multicolor QCD) the quark condensate vanishes at any finite number of colors N; however, a nonvanishing quark condensate does develop

[18] in the limit $N \to \infty$, in which all mesons in the spectrum become sterile. The axial symmetry of the model is spontaneously broken at $N = \infty$, and a pion emerges.

The Coleman theorem does not apply to the spontaneous breaking of gauge symmetries (the Higgs mechanism). If a global symmetry is gauged, then the would-be Goldstone boson is eaten up by the gauge field, which becomes massive. The Higgs regime is attainable in two dimensions, although it must be added that the Higgs phase in $1 + 1$ dimensions is somewhat peculiar and does not exactly coincide with that in $1 + 2$ or $1 + 3$ dimensions; see Section 9.3.

The original proof of the Coleman theorem [16] is rather formal; it is based on the fact that infrared divergences associated with massless particles in two dimensions invalidate a certain postulate of the axiomatic field theory with regard to the expectation value

$$\langle \text{vac}|\phi(x)\phi(0)|\text{vac}\rangle, \tag{6.99}$$

where ϕ is the same field as in Eq. (6.91). (Note the absence of a T product.) In fact, the essence of the phenomenon is clear from the physical standpoint. Assuming that massless (Goldstone) bosons exist, the behavior of the corresponding Green's functions is pathological at large distances. Indeed, if in the momentum space

$$\int d^2x\, e^{ipx}\langle \text{vac}|T\{\phi(x), \phi(0)\}|\text{vac}\rangle = \frac{i}{p^2} \tag{6.100}$$

then the massless particle Green's function in the coordinate space has the form

$$\langle \text{vac}|T\{\phi(x), \phi(0)\}|\text{vac}\rangle = \int \frac{d^2p}{(2\pi)^2} e^{-ipx} \frac{i}{p^2}. \tag{6.101}$$

The integral on the right-hand side is divergent at small p, in the infrared domain. If we regularize it somehow (e.g., by giving the particle a small mass which will be put to zero at the end) then we discover that the Green's function

$$\langle \text{vac}|T\{\phi(x), \phi(0)\}|\text{vac}\rangle = -\frac{1}{4\pi} \ln x^2 + C \tag{6.102}$$

grows at large distances! Moreover, the constant C on the right-hand side is ill defined: it can take arbitrary values depending on the regularization. One can derive the sameexpression for the Green's function, bypassing the momentum space representation, by directly solving the defining equation

Equation for Green's function

$$\partial^2 G(x) = -i\delta^{(2)}(x). \tag{6.103}$$

The logarithmic growth in the massless particle propagator at large distances is a specific feature of two dimensions. In higher dimensions, $G(x)$ falls off at large $|x|$. If logarithmic growth did indeed take place then the signal produced by a ϕ quantum emitter at the origin would be detected, amplified, in a distant ϕ quantum absorber (placed at a point x). In a well-defined theory this cannot happen.

There are two ways out. If the would-be Goldstone particles interact, their interaction becomes strong in the infrared domain and a mass gap is dynamically generated. Then all particles in the spectrum become massive. In the absence of massless Goldstone bosons, all generators of the global symmetries must annihilate the vacuum. The full global symmetry of the Lagrangian is then realized linearly (i.e., there is no spontaneous symmetry breaking). We will consider in detail an example of such a solution in Appendix section 9.8.

Another way out, which keeps massless *noninteracting* particles in the spectrum, is to make sure that all physically attainable emitters and absorbers are of a special form such that their correlation functions fall off at large distances in spite of Eq. (6.102). Typically, this happens when the theory under consideration has global U(1) symmetries. The spontaneous breaking of a U(1) symmetry would produce a single Goldstone boson, call it α, with Lagrangian

$$\mathcal{L} = \frac{F^2}{2} \partial_\mu \alpha \, \partial^\mu \alpha, \tag{6.104}$$

where F is a dimensionless constant and α, being a phase variable, is defined mod 2π: α, $\alpha + 2\pi$, $\alpha + 4\pi$, and so on are identified.

In this case all physically measurable operators must be periodic in α, with period 2π. Only such operators belong to the physical Hilbert space; the others are unphysical. For instance, the correlation function

$$\langle \text{vac} | T \left\{ e^{i\alpha(x)}, e^{-i\alpha(0)} \right\} | \text{vac} \rangle \tag{6.105}$$

is physically measurable while $\langle T\{\alpha(x), \alpha(0)\} \rangle$ is not.

Let us calculate the correlation function (6.105). To this end we will expand both exponents and observe that, upon averaging over the vacuum state, only terms with equal powers of α will produce a nonvanishing contribution, namely,

$$\sum_{k=0}^{\infty} \frac{1}{(k!)^2} \langle \text{vac} | T\{[i\alpha(x)]^k, [-i\alpha(0)]^k\} | \text{vac} \rangle$$

$$= \sum_{k=0}^{\infty} \frac{1}{k!} \langle \text{vac} | T\{\alpha(x), \alpha(0)\} | \text{vac} \rangle^k$$

$$= \exp[\langle \text{vac} | T\{\alpha(x), \alpha(0)\} | \text{vac} \rangle]. \tag{6.106}$$

Using Eq. (6.102) for the Green's function we arrive at

$$\langle \text{vac} | T\{ e^{i\alpha(x)}, e^{-i\alpha(0)} \} | \text{vac} \rangle = \exp\left(-\frac{1}{4\pi F^2} \ln x^2 + C \right)$$

$$\propto \left(\frac{1}{x^2} \right)^{1/(4\pi F^2)}. \tag{6.107}$$

Thus this correlation function decays at large distances, as it should.

If the U(1) symmetry was spontaneously broken then one would expect the order parameter $\langle \text{vac} | e^{i\alpha} | \text{vac} \rangle$ to be nonvanishing, say,

$$\langle \text{vac} | e^{i\alpha} | \text{vac} \rangle = e^{i\alpha_0}. \tag{6.108}$$

However, the right-hand side of Eq. (6.107) vanishes at $|x| \to \infty$, albeit in a power-like manner, implying (through cluster decomposition) that

$$\langle \text{vac} | e^{i\alpha} | \text{vac} \rangle = 0. \tag{6.109}$$

The order parameter in Eq. (6.108) is averaged over all α_0, and the original U(1) symmetry is *not* broken although the mass gap is not generated either. The above statement is rather subtle and requires extra comments.

(i) If $F \to \infty$ the fall-off in (6.107) flattens out and at the point $F = \infty$ is replaced by an x independent constant. Hence, $\langle \text{vac} | e^{i\alpha} | \text{vac} \rangle \neq 0$. This is exactly what happens in the 't Hooft model at $N = \infty$ when all interactions are switched off; see Section 9.5 and Exercise 9.5.1. Equation (6.109) is valid at large but finite F. In this window a massless boson is still present in the spectrum. Its coupling to the operator $e^{i\alpha}$ vanishes simultaneously with the vanishing of the order parameter.

Berezinskii–Kosterlitz–Thouless phase transition

(ii) As F decreases, nonperturbative effects due to proliferation of vortices in the Euclidean version of this theory become important, and a mass gap is generated, leading to an exponential fall-off of the two-point function (6.107). This is the so-called Berezinskii–Kosterlitz–Thouless phase transition. Its discovery was reported in [19].

See Section 10.10.

It is worth making one last remark, in conclusion. Supersymmetry is definitely a continuous symmetry, yet its spontaneous breaking in two dimensions is not forbidden by the Coleman theorem. This is due to the fact that the Goldstone particle occurring in this case – a Goldstino – is a spin-1/2 fermion. The massless fermion Green's function in two dimensions falls off with distance as $1/x$.

Exercise

6.5.1 Prove the Goldstone theorem, assuming that the conserved current J_μ is non-Hermitian. Then J_μ has a partner, $J_\mu^\dagger$, which is also conserved:

$$\partial^\mu J_\mu = \partial^\mu J_\mu^\dagger = 0.$$

References for Chapter 6

[1] M. Nakahara, *Geometry, Topology and Physics*, Second Edition (IOP Pubishers, Bristol, 2003).

[2] Sigurdur Helgason, *Differential Geometry, Lie Groups, and Symmetric Spaces*, (American Mathematical Society, 2001).

[3] A. M. Perelomov, *Phys. Rept.* **146**, 135 (1987); *Phys. Rept.* **174**, 229 (1989).

[4] E. Witten, *Nucl. Phys. B* **149**, 285 (1979).

[5] V. Berestetskii, E. Lifshitz, and L. Pitaevskii, *Quantum Electrodynamics* (Pergamon, 1980), Section 17.

[6] A. M. Polyakov, *Phys. Lett.* **59**, 79 (1975).

[7] S. V. Ketov, *Quantum Non-linear Sigma-models*, (Springer-Verlag: Berlin, 2001).

[8] A. Y. Morozov, A. M. Perelomov, and M. A. Shifman, *Nucl. Phys. B* **248**, 279 (1984).

[9] X. Cui and M. Shifman, *Phys. Rev. D* **85**, 045004 (2012), see Eq. (69) with $N_f = -1$.

[10] D. Friedan, *Phys. Rev. Lett.* **45**, 1057 (1980) and (in an expanded form) S. J. Graham, *Phys. Lett. B* **197**, 543-547 (1987).

[11] V. A. Fateev, I. V. Frolov, and A. S. Schwarz, *Sov. J Nucl. Phys.* **30**, 590 (1979); *Nucl. Phys. B* **154**, 1 (1979).

[12] A. M. Perelomov, *Commun. Math. Phys.* **63**, 237 (1978).

[13] A. Jevicki, *Nucl. Phys. B* **127**, 125 (1977); A. M. Din, P. Di Vecchia, and W. J. Zakrzewski, *Nucl. Phys. B* **155**, 447 (1979).

[14] T. Banks and N. Seiberg, *Nucl. Phys. B* **273**, 157 (1986).

[15] V. G. Vaks and A. I. Larkin, *Sov. Phys. JETP* **13**, 192 (1961); Y. Nambu, *Phys. Rev. Lett.* **4**, 380 (1960); Y. Nambu and G. Jona-Lasinio, *Phys. Rev.* **122**, 345 (1961); *Phys. Rev.* **124**, 246 (1961); J. Goldstone, *Nuov. Cim.* **19**, 154 (1961); J. Goldstone, A. Salam, and S. Weinberg, *Phys. Rev.* **127**, 965 (1962). For a historic review see D. V. Shirkov, *Mod. Phys. Lett. A* **24**, 2802 (2009) [arXiv:0903.3194 [physics.hist-ph]].

[16] S. R. Coleman, *Commun. Math. Phys.* **31**, 259 (1973).

[17] G. 't Hooft, *Nucl. Phys. B* **75**, 461 (1974).

[18] A. R. Zhitnitsky, *Phys. Lett. B* **165**, 405 (1985); *Yad. Fiz.* **43**, 1553 (1986) [*Sov. J. Nucl. Phys.* **43**, 999 (1986)].

[19] V. Berezinskii, *Sov. Phys. –JETP* **32**, 493 (1971); J. M. Kosterlitz and D. J. Thouless, *Journal of Physics* **C6**, 1181 1973.

False-Vacuum Decay and Related Topics

False-vacuum decay: what does it mean? — Under-the-barrier tunneling. — Bounces and their generalizations. — Metastable string breaking and domain wall fusion can be treated as false-vacuum decay too.

7.1 False-Vacuum Decay

This section could have been entitled "How water starts to boil," or "How the universe could have been destroyed," or in a dozen similar ways. We will consider the problem of false-vacuum decay, which finds a large number of applications in cosmology, high-energy physics, and solid state physics. Later we will discuss some interesting applications, for instance, the decay rate of metastable strings through monopole pair creation.

Metastable states emerge when the potential energy of a system has more than one minimum, say, one global minimum and one local minimum, separated by a barrier. The simplest model allowing one to study the phenomenon is a model of a real scalar field having the potential presented in Fig. 7.1.

This is a deformation of the Z_2-symmetric model considered in Chapter 2. We break the Z_2 symmetry by a small linear perturbation, so that

$$L = \frac{1}{2}\left(\partial_\mu \phi\right)^2 - V(\phi),$$

$$V(\phi) = \frac{\lambda}{4}\left(\phi^2 - v^2\right)^2 + \frac{\mathcal{E}}{2v}\phi + \text{const},$$

(7.1)

> *It is technically convenient to impose the condition $V(\phi_+) = 0$.*

where $\mathcal{E}$ is assumed to be a small parameter and the constant on the right-hand side is adjusted in such a way that in the right-hand minimum $V(\phi_+) = 0$. This is certainly not necessary (the overall constant is unobservable), but it is very convenient. If $\mathcal{E} = 0$, we return to the Z_2-symmetric model with two degenerate vacua at $\phi = \pm v$. As $\mathcal{E} > 0$, only the vacuum at $\phi_- \approx -v$ is genuine; the vacuum at $\phi_+ \approx v$ becomes metastable. The difference between the energy densities in the metastable and true vacua is $\mathcal{E}$. If our system originally resides in the false vacuum and $\mathcal{E}$ is small, it will live there for a long time before, eventually, the false vacuum will decay into the true vacuum. The decay is similar to the nucleation processes of statistical physics, such as the crystallization of a supersaturated solution or the boiling of a superheated liquid. In the latter, the system goes through bubble creation. The false vacuum corresponds to the superheated phase and the true vacuum to the vapor phase.

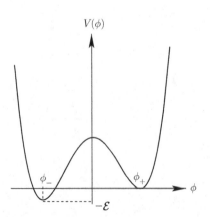

Fig. 7.1   A two-minimum potential, with a genuine vacuum at ϕ_- and a false vacuum at ϕ_+.

The bubbles cannot be too small since then the gain in volume energy would not be enough to compensate for the loss due to bubble surface energy. Thus physical bubbles can be only of a critical size or larger. Subcritical bubbles "exist" under the barrier.

Our task here is to analyze the problem in $D = 1 + 1$, $1 + 2$, and $1 + 3$ dimensions. We will do this from two different perspectives: (i) that of the Euclidean tunneling picture and (ii) that of the dynamics of "true vacuum bubbles" in Minkowski time.

The theory of false-vacuum decay was worked out in the 1970s by Kobzarev, Okun, and Voloshin [1] and by Coleman [2]. In this section we will follow closely two excellent reviews [3, 4].

7.1.1 Euclidean Tunneling

Thus, we will start from a system placed in the false vacuum. It is obvious that to any finite order in perturbation theory the instability will never reveal itself, since perturbation theory describes small oscillations near the equilibrium position. However, occasionally a large fluctuation of the field ϕ may occur, so that it spills from the false vacuum to the true one. It is natural to expect that the action corresponding to large field fluctuations will be large, so that the problem will be tractable quasiclassically. I will confirm this expectation a posteriori.

The process with which we are dealing is a tunneling process. Small bubbles are classically forbidden. As is well known, an appropriate description of tunneling is provided by passing to a Euclidean formulation (see Section 2.1). I will outline this approach now and then we will discuss its relation to the bubble dynamics in real time.

| Euclidean action |

After Euclidean rotation the action takes the form

$$S = \int d^D x \left[\frac{1}{2} \left(\partial_\mu \phi \right)^2 + V(\phi) \right]. \tag{7.2}$$

The standard strategy for solving tunneling problems in the leading quasiclassical approximation is straightforward. We look for a field configuration that (i) approaches the false vacuum ϕ_+ in the distant (Euclidean) past and in the distant future and approaches the true vacuum at intermediate times; (ii) extremizes the action. Extremizing the action means that we must find a solution of the classical Euclidean equation of motion,

| The bounce solution gives decay probability rather than amplitude. Maximal action. |

$$\partial^2 \phi - V'(\phi) = 0. \tag{7.3}$$

This corresponds to the motion of the system in the potential $-V$, starting from ϕ_+, moving towards ϕ_- and then bouncing back to ϕ_+. Such a solution is called a *bounce* (Fig. 7.2). It should be intuitively clear that the bounce solution is O(4) symmetric (for $D = 1 + 3$) (Fig. 7.3). In fact, this assumption can be proved rigorously [5].

Let us place the origin at the center of the bounce solution. Then the O(4) symmetry reduces Eq. (7.3) to

$$\frac{d^2}{dr^2} \phi(r) + \frac{3}{r} \frac{d}{dr} \phi(r) = V'(\phi), \qquad r = \sqrt{x_\mu^2}. \tag{7.4}$$

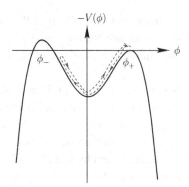

Fig. 7.2 The trajectory $\phi_b(\tau, \vec{x} = 0)$ (broken line). Here ϕ_b stands for the bounce solution and τ is Euclidean time.

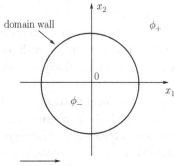

The arrow of Euclidean time

Fig. 7.3 Geometry of the bounce solution. The perpendicular coordinates $x_{3,4}$ are not shown.

The boundary conditions corresponding to Fig. 7.2 are

$$\phi(r \to \infty) = \phi_+, \tag{7.5}$$

$$\phi(r = 0) \approx \phi_-, \tag{7.6}$$

$$\left.\frac{d\phi}{dr}\right|_{r=0} = 0. \tag{7.7}$$

Boundary conditions

The last condition guarantees that the bounce solution is nonsingular at the origin. From the mathematical standpoint, only Eqs. (7.5) and (7.7) are valid boundary conditions while Eq. (7.6) is superfluous. It is admittedly vague (because of the approximate rather than exact equality). Physically it expresses the fact that the final point of the tunneling trajectory is the true vacuum. The approximate equality becomes exact in the limit $\mathcal{E} \to 0$ (see below).

Let us show, at the qualitative level, that a bounce solution with the above properties exists. To this end it is convenient to reinterpret Eq. (7.4) as describing the mechanical motion of a particle with mass $m = 1$ and coordinate ϕ, which depends on the "time" r, in the potential $-V(\phi)$ (see Fig. 7.2) and subject to a viscous damping force (friction) with coefficient inversely proportional to the time. The particle is released at time zero (with vanishing velocity) somewhere close to ϕ_-; it must reach ϕ_+ at infinite time.

It is clear that on the one hand if the particle is released sufficiently far to the right of ϕ_- then it will never climb all the way up to ϕ_+. It will undershoot. This situation is depicted in Fig. 7.2. On the other hand, if it is released too close to ϕ_- then it will reach ϕ_+ with a nonvanishing velocity and will overshoot. Indeed, by choosing $\phi(0)$ arbitrarily close to ϕ_- we can always ensure that the nonlinear terms in Eq. (7.4) are negligibly small for at least some time. Then this equation can be linearized; we obtain

$$\left(\frac{d^2}{dr^2} + \frac{3}{r}\frac{d}{dr} - \mu^2\right)(\phi(r) - \phi_-) = 0, \qquad \mu^2 \equiv V''(\phi = \phi_-) > 0. \tag{7.8}$$

The solution of the linearized equation is

$$\phi(r) - \phi_- = 2[\phi(0) - \phi_-]I_1(\mu r)(\mu r)^{-1}, \tag{7.9}$$

where $I_1(\mu r)$ is a modified Bessel function. By choosing $\phi(0) - \phi_-$ positive and small, one guarantees that $\phi(r) - \phi_-$ is small for arbitrarily large r. However, for sufficiently large r the friction term becomes arbitrarily small. Neglecting the friction term leaves us with the equation

$$\frac{d^2}{dr^2}\phi(r) = V'(\phi), \tag{7.10}$$

for which "energy" is conserved. If at the moment of time when Eq. (7.10) becomes a good approximation,

$$-V(\phi(r)) > -V(\phi_+),$$

then the particle will overshoot. Thus, there should exist a starting point in the vicinity of ϕ_- that yields the trajectory we need: when released at this point at time zero with vanishing velocity, the particle reaches ϕ_+ at infinite time with vanishing velocity.

Having established the existence of a (Euclidean) field configuration, relevant to tunneling from the false to the ture vacuum, that extremizes the action we can now verify the fact that this solution yields a maximum of the action rather than a minimum. This implies, in turn, the existence of a negative mode in the bounce background. (Below we will see that the negative mode corresponds to a change in the radius of the bubble in Fig. 7.3.) The existence of the negative mode is vitally important. Indeed, false-vacuum decay manifests itself in the occurrence of an imaginary part of the vacuum energy of the false vacuum. Thus, the bounce contribution to the vacuum energy density *must be purely imaginary*. The i factor emerges from $\mathrm{Det}^{-1/2}$ accounting for small fluctuations near the classical bounce solution provided that there is one and only one negative mode.

Let $\phi_b(x)$ denote the bounce solution of the classical equations (7.4). Consider a family of functions $\phi(x; \nu) \equiv \phi_b(x/\nu)$, where ν is a positive parameter. The action for this family is

$$S[\phi(x; \nu)] = \tfrac{1}{2}\nu^2 \int d^4x(\partial_\mu\phi_b)^2 + \nu^4 \int d^4x\, V(\phi_b). \tag{7.11}$$

Since $\phi_b(x)$ extremizes the action we have

$$\left.\frac{\partial S[\phi(x; \nu)]}{\partial \nu}\right|_{\nu=1} = 0,$$

Bounce
action

implying that

$$\int d^4x \, (\partial_\mu \phi_b)^2 = -4 \int d^4x \, V(\phi_b). \tag{7.12}$$

In deriving Eq. (7.11) we have relied on the convergence (finiteness) of the action integral. Using Eq. (7.12) one can represent the second derivative over v at $v = 1$ as follows:

$$\left. \frac{\partial^2 S[\phi(x; v)]}{\partial v^2} \right|_{v=1} = -2 \int d^4x \, \left(\partial_\mu \phi_b\right)^2 < 0. \tag{7.13}$$

> *The negative mode resides in the bounce size.*

This shows that inflating or deflating the bounce decreases the action.

The analytical solution of Eq. (7.4) is not known. However, we have a pretty thorough idea of its properties, and this will allow us to find the decay rate at small $\mathcal{E}$ (to leading order in $\mathcal{E}$). Indeed, the thickness of the transitional domain where the field ϕ changes its value from ϕ_+ to ϕ_- is determined by the mass of the elementary excitation, $V''(\phi_+)$ or $V''(\phi_-)$. At the same time the radius R of the bubble depends on $\mathcal{E}$; the smaller is $\mathcal{E}$, the larger is the radius. At sufficiently small $\mathcal{E}$ the radius R becomes parametrically larger than the bubble wall thickness. This is called the *thin wall approximation* (TWA). If $R \gg m^{-1}$ then we can (i) neglect the curvature of the bubble, treating the bubble wall as a flat domain wall; (ii) approximate the field outside the bubble by $\phi = \phi_+$ and inside the bubble by $\phi = \phi_-$.

Then the action integral (7.2) can be decomposed into three parts: an integral outside the bubble, an integral inside it, and an integral over the transitional domain (the wall). The first integral obviously vanishes (see the marginal remark after Eq. (7.1)), the second yields the bubble volume times $-\mathcal{E}$, while the third reduces to the bubble wall surface times T, where T is the tension of the flat wall:

$$S = T \times \left\{ \begin{array}{l} 2\pi^2 R^3 \\ 4\pi R^2 \\ 2\pi R \end{array} \right\} - \mathcal{E} \times \left\{ \begin{array}{l} \frac{1}{2}\pi^2 R^4 \\ \frac{4}{3}\pi R^3 \\ \pi R^2 \end{array} \right\}, \quad \begin{array}{l} D = 4, \\ D = 3, \\ D = 2. \end{array} \tag{7.14}$$

Recall that $T = m^3/(3\lambda)$. So far R is a free parameter. To find the bounce action we have to extremize (7.14) with respect to R. The critical value of R is

$$R_* = (D - 1)\frac{T}{\mathcal{E}}. \tag{7.15}$$

> *Critical action.*

It is seen that the extremum of the action is indeed a maximum, and R_* becomes arbitrarily large at small $\mathcal{E}$. This justifies the TWA. The value of the action at the extremum is

$$S_* = \left\{ \begin{array}{ll} \frac{27}{2}\pi^2 T^4/\mathcal{E}^3, & D = 4, \\ \frac{16}{3}\pi T^3/\mathcal{E}^2, & D = 3, \\ \pi T^2/\mathcal{E}, & D = 2. \end{array} \right. \tag{7.16}$$

This concludes our calculation. The false-vacuum decay rate (per unit time and unit volume) is

$$d\Gamma_{\text{false-vac}} \sim e^{-S_*}. \tag{7.17}$$

7.1.2 False-Vacuum Decay in Minkowski Space–Time

So far we have carried out a rather formal calculation of false-vacuum decay by considering Euclidean bubbles and calculating the action of an extremal Euclidean bubble. To get a better idea of the underlying physics it is instructive to consider the same process directly in Minkowski space–time.

As already mentioned, this decay occurs through bubble nucleation. This time we speak of *real bubbles in Minkowski space–time*. If, say, the original problem was four-dimensional then the bubbles of which we are speaking are three-dimensional while their surface presents S_2, a two-dimensional sphere. The Euclidean bubble in this case was four-dimensional while its surface is S_3.

The surface S_2 is made of a domain wall separating two phases with $\phi = \phi_+$ and $\phi = \phi_-$. Like any surface it is characterized by its tension T, which can be readily calculated in the microscopic theory.

It is clear that, as discussed above, classically the existence of such a bubble is possible only provided its radius is larger than a certain minimal radius. Indeed, the volume energy of the interior of the bubble is negative with respect to the outer false vacuum (this is our gain):

$$E_{\text{vol}} = -\mathcal{E}V, \tag{7.18}$$

where

$$V = \begin{cases} \frac{4}{3}\pi r^3, & D = 1 + 3, \\ \pi r^2, & D = 1 + 2, \\ 2r, & D = 1 + 1. \end{cases} \tag{7.19}$$

and r is the radius of the (Minkowskian) bubble. (At $D = 1+1$ we are dealing not with a bubble but, rather, with an interval of size $2r$.) Besides, there is a positive energy associated with the surface tension and its motion (if the bubble is expanding). This is our loss. The surface of the minimal-size bubble is at rest. Therefore, the positive energy associated with the surface is

$$E_{\text{surf}} = TA, \tag{7.20}$$

where T is the tension and A is the surface area,

$$A = \begin{cases} 4\pi r^2, & D = 1 + 3, \\ 2\pi r, & D = 1 + 2, \\ 2, & D = 1 + 1. \end{cases} \tag{7.21}$$

Since the total energy of the spontaneously nucleated bubble with respect to the initial phase must vanish, one concludes that the minimal radius of the classical bubble is

$$r_* = (D - 1)\frac{T}{\mathcal{E}}, \tag{7.22}$$

the same as the extremal size of the Euclidean bubble; see Eq. (7.15). This is certainly no accident. We will return to this point later. Bubbles of smaller sizes occur "under the barrier."

In developing the macroscopic theory of the bubble we have assumed, as previously, that the bubble radius is much larger than the wall thickness. In this case the separation of the volume and surface energies has a clear-cut meaning, and, moreover, one can neglect the bubble's curvature and treat the tension effect as that for a flat wall.

Before attempting the calculation of the probability of quantum nucleation, let us discuss the classical dynamics of (spherical) bubbles. If the TWA is valid – which is the case at small $\mathcal{E}$ – the bubble can be described by a single dynamical variable r, the bubble radius. The relativistic Lagrangian for an expanding bubble consists of two terms: (i) the kinetic term[1] describing the motion of the surface, whose mass is $4\pi r^2 T$ and (ii) the potential part describing the negative volume energy inside the bubble, $-\frac{4}{3}\pi r^3 \mathcal{E}$. (We assume here that the number of spatial dimensions is three. For two spatial dimensions and for one, the formulas for the bubble surface and volume must be changed accordingly.) The total Lagrangian has the form

Relativistic Lagrangian describing (Minkowskian) dynamics of r

$$L = -4\pi r^2 T\sqrt{1 - \dot{r}^2} + \tfrac{4}{3}\pi r^3 \mathcal{E}, \tag{7.23}$$

where

$$\dot{r} = \frac{dr}{dt}$$

is the speed of the (expanding) wall. The canonical momentum following from this Lagrangian is

$$p = \frac{\delta L}{\delta \dot{r}} = 4\pi r^2 T \frac{\dot{r}}{\sqrt{1 - \dot{r}^2}}, \tag{7.24}$$

which implies in turn that the Hamiltonian H is given by

$$H = p\dot{r} - L = 4\pi r^2 T \frac{1}{\sqrt{1 - \dot{r}^2}} - \frac{4\pi}{3} r^3 \mathcal{E}. \tag{7.25}$$

Combining Eqs. (7.25) and (7.24) we find

$$\left(H + \frac{4\pi}{3} r^3 \mathcal{E}\right)^2 - p^2 = \left(4\pi r^2 T\right)^2. \tag{7.26}$$

As already mentioned, the energy of a spontaneously nucleated bubble vanishes. Replacing H in Eq. (7.26) by zero we arrive at the following relation:

$$p = 4\pi r^2 T \sqrt{\frac{r^2}{r_*^2} - 1}, \tag{7.27}$$

where $r_* = 3T/\mathcal{E}$ is the minimal radius of a classical bubble; see Eq. (7.22) with $D = 4$.

Comparing (7.24) and (7.27) it is easy to see that

$$\dot{r} = \sqrt{1 - \left(\frac{r_*}{r}\right)^2}. \tag{7.28}$$

Clearly, the classical description applies only provided $r > r_*$, so that the expression under the square root is positive. The solution of this equation is

---

[1] The kinetic term for a relativistic particle is $-m(1 - v^2)^{1/2}$; see, e.g., [6].

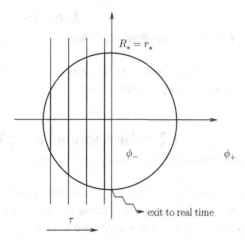

Fig. 7.4 Time slices of the Euclidean (four-dimensional) bubble represent evolution of the subcritical Minkowskian (three-dimensional) bubble under the barrier.

$$r = \sqrt{r_*^2 + t^2} \qquad \text{or} \qquad r^2 - t^2 = r_*^2. \tag{7.29}$$

The last expression is explicitly Lorentz invariant: the bubble wall trajectory lies on an invariant hyperboloid in space–time. This means that the center of the expanding bubble is at rest in any inertial frame, a rather surprising result.

The domain $r < r_*$ is classically forbidden. The bubble dynamics in this domain corresponds to under-the-barrier tunneling. We have already discussed this process from the Euclidean standpoint. Remember that the critical radius of the Euclidean bubble, (7.15), matches the minimal radius of the classical bubble, (7.22). Under the barrier, the bubble evolves in imaginary time (which corresponds to consecutive slices of Euclidean four-dimensional bubble at various values of the Euclidean time). When the bubble radius reaches R_*, which is also the minimal classically allowed value, it goes classical, expanding further in real time (Fig. 7.4).

The tunneling probability can be calculated in a more conventional way, using the well-known WKB formula [7],

$$\Gamma \sim \exp\left(-2 \int_0^{r_*} dr \, |p(r)|\right), \tag{7.30}$$

where $p(r)$ is obtained from the classical expression (7.27) by analytic continuation in the classically forbidden domain $r < r_*$. In this way we get

$$\Gamma \sim \exp\left(-2 \int_0^{r_*} dr \, 4\pi r^2 T \sqrt{1 - \frac{r^2}{r_*^2}}\right)$$

$$= \exp\left(-\frac{\pi^2}{2} T r_*^3\right) = \exp\left(-\frac{27}{2}\pi^2 \frac{T^4}{\mathcal{E}^3}\right). \tag{7.31}$$

This coincides identically with the result of the Euclidean treatment; see Eq. (7.16) for $D = 4$. The derivation changes in a minimal way for $D = 3$ and $D = 2$ – one has to use appropriate expressions for the bubble volume and surface in Eq. (7.25). The two other results in Eq. (7.16) are then recovered.

Exercise

7.1.1   Give an argument to explain why spontaneously nucleated bubbles (in Minkowski space–time) have a spherical form.

7.2 False-Vacuum Decay: Applications

In this section we will consider some important applications. It turns out that the ideas presented above are applicable in a number of problems which – at first sight – look different and seemingly have little to do with the false-vacua problem. In fact, the examples to be analyzed below are akin to each other and can indeed be interpreted in terms of false-vacuum decays. We will start from metastable string decays; for this particular problem we will also discuss the underlying microscopic physics (see Section 7.2.3).

7.2.1 Decay of Metastable Strings

Metastable string-like configurations (or flux tubes) appear in various contexts in high-energy physics. For instance, one can embed a $U(1)$ gauge theory supporting an Abrikosov–Nielsen–Olesen (ANO) string into a non-Abelian theory with the matter sector constructed in a special way. Then, such a string can break by the creation of a monopole–antimonopole pair at the endpoints of the two broken pieces [8]. In QCD-like theories one can consider so-called symmetric 2-strings, which can decay into antisymmetric 2-strings having a smaller tension through the creation of a pair of gluelumps [9]. This is shown in Fig. 7.5.

The strings have one spatial dimension; their world sheet is two dimensional. The probability of the processes depicted in this figure (per unit length per unit time) is exponentially small provided that $\mathcal{E} \ll \mu^2$, where

$$\mathcal{E} = T_1, \qquad \mu \text{ is the monopole mass in Fig. 7.5(a)},$$
$$\mathcal{E} = T_1 - T_2, \qquad \mu \text{ is the gluelump mass in Fig. 7.5(b)}. \qquad (7.32)$$

Then one can calculate the exponent in the decay rate in exactly the same way as for the false-vacuum decay in two dimensions, using the TWA. Calculation of the pre-exponent is more subtle since quantum fluctuations around the extremal field configuration playing the role of the bounce "know" that they are occurring in four rather than two dimensions. But this task is achievable too [10].

The strings at the top of Figs. 7.5a, b are excited states (false vacua). Those at the bottom are ground states (true vacua). In Euclidean time the processes proceed through the formation of bubbles of the genuine ground states (either no string for the process in Fig. 7.5a or a smaller-tension stable string in the process in Fig. 7.5b), as shown in Fig. 7.6. Given the definitions (7.32), one can write the Euclidean bubble action responsible for the tunneling processes under consideration as follows:

Bubble action

$$S_{\text{bubble}} = 2\pi r \mu - \pi r^2 \mathcal{E}, \qquad (7.33)$$

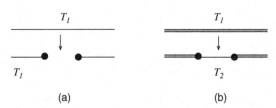

Fig. 7.5   A metastable string can break (a) through monopole–antimonopole pair creation; (b) a metastable string with tension T_1 can decay into a string with a smaller tension T_2 through gluelump pair creation. The symbol ● denotes (anti)monopoles in (a) and gluelumps in (b). The double lines in (b) denote the string with the larger tension.

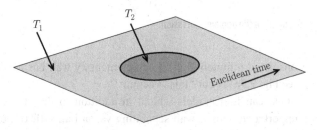

Fig. 7.6   The bounce configuration describing a semiclassical tunneling trajectory in Euclidean time.

where $\mathcal{E}$ is defined in (7.32) At this stage we need to invoke results from Section 7.1. Compare (7.33) with the last line in Eq. (7.14). The critical action is given in the last line of (7.16). Equation (7.33) immediately leads us to a decay rate (per unit length) [8]

$$d\Gamma_{\text{breaking}}/dL = C \exp\left(-\frac{\pi\mu^2}{\mathcal{E}}\right), \tag{7.34}$$

where C is a pre-exponential factor. As already mentioned, this pre-exponential factor was calculated in one loop in [10]; the result was

$$C = \frac{\mathcal{E}}{2\pi} F^2, \tag{7.35}$$

where

$$F = \begin{cases} \dfrac{e}{2\sqrt{\pi}}, & e = 2.71828\ldots, \quad \text{for Fig. 7.5a,} \\[2ex] \dfrac{1}{\sqrt{2\pi}}\sqrt{\dfrac{1}{\kappa+1}}\Gamma(\kappa+1)\left(\dfrac{e}{\kappa}\right)^{\kappa}, & \kappa \equiv \dfrac{T_1+T_2}{T_1-T_2} \quad \text{for Fig. 7.5b.} \end{cases} \tag{7.36}$$

7.2.2 Domain-Wall Fusion

Now we will consider another problem related to false-vacuum decay. Assume that we have two parallel domain walls: the first wall separates vacua I and II while the second separates vacua II and III. We will refer to these walls as elementary. All vacua are degenerate; therefore, at large distances from each other the elementary walls are stable. Assume that they are nailed in space at a distance d from each other, d being much larger than the wall thickness (then the walls can be

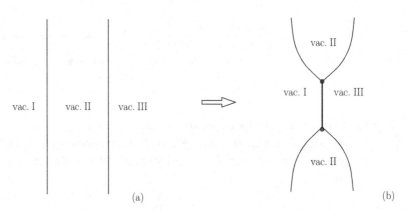

Fig. 7.7   Geometry of the domain wall fusion.

considered infinitely thin). Each elementary wall has tension T_1. This configuration (see Fig. 7.7a) is our "false vacuum."

One can find models which, in addition to the above two "elementary" walls, support a composite wall separating vacua I and III. If the tension of the composite wall is T_2, the fact that two elementary walls are bound into one composite implies that

$$T_2 < 2T_1. \tag{7.37}$$

For simplicity we will assume weak binding, i.e.,

$$\frac{T_2 - 2T_1}{T_1} \ll 1. \tag{7.38}$$

It is clear that fusing two elementary walls into one composite is energetically expedient. However, the walls cannot fuse in their entirety in one quantum leap, since this "global" fusion would require infinite action. As in Section 7.1, fusion will occur through a bubble (a patch) of the composite wall. Figure 7.7b shows the geometry of the fused configuration. This is our "true vacuum." If you looked in the perpendicular direction you would see a domain of fused walls that has a circular shape, with radius r, which cannot be smaller than a critical value. Indeed, the gain in energy due to fusion is accompanied by a loss in energy in the boundary region near the junction. In this boundary region the elementary walls are warped, which increases their area and, hence, the energy.[2] At the critical radius r_* the gain in energy due to wall fusion in the central domain (the fused patch) is exactly compensated by the loss in energy due to the necessary warping of the elementary wall. At $r < r_*$ the system tunnels under the barrier. At $r > r_*$ the fused patch expands classically. To find the fusion rate (with exponential accuracy) we will pass to Euclidean time and find the solution of the classical equations for the bounce field configuration.

The plane parallel to the elementary walls is parametrized by coordinates x and y, while the perpendicular coordinate is z. The Euclidean time is τ. The first elementary wall is at $z = d/2$ and the second is at $z = -d/2$. The x and y coordinates are chosen in such a way that the center of the fused patch lies at $x = y = 0$.

---

[2] We will neglect the mass of the wall junction *per se*. The wall junction is represented by small solid circles in Fig. 7.7b.

As in Section 7.1, the bounce configuration is spherically symmetric in Euclidean time. This means that the solution depends on the coordinate

$$r = (x^2 + y^2 + \tau^2)^{1/2}, \tag{7.39}$$

rather than on x, y, and τ separately. The boundary conditions are

$$z(r) \to \pm\frac{d}{2} \qquad \text{as } r \to \infty. \tag{7.40}$$

In weak binding, see (7.38), the curvature of the fused-wall configuration at $r > r_*$ is small (see Eq. (7.47) below), and the walls (we will parametrize them as $z(r)$) can be described by the linearized equation

$$\Delta z = 0, \tag{7.41}$$

everywhere except at the junction line.

The solutions of the above equation for the top and bottom warped elementary walls are

$$z_1(x, y, \tau) = -\frac{A}{r} + \frac{d}{2}, \quad z_2(x, y, \tau) = \frac{A}{r} - \frac{d}{2}, \quad r > r_*, \tag{7.42}$$

where A is a constant to be determined below. Figure 7.7b shows that the two elementary walls meet at $z = 0$ and $r = r_*$. Then Eq. (7.42) implies that

$$r_* = \frac{2A}{d}. \tag{7.43}$$

It is obvious that r_* is the radius of the world volume of the composite wall at the moment it leaves Euclidean space and enters Minkowski space ($\tau = 0$).

The total Euclidean action is the sum of two contributions:

Euclidean action for wall fusion

$$S = (T_2 - 2T_1)\frac{4\pi}{3}r_*^3 + 2T_1 \int_{r_*}^{\infty} 4\pi r^2 \, dr \, \frac{z'^2}{2}$$

$$= -(2T_1 - T_2)\frac{4\pi}{3}r_*^3 + T_1\pi r_* d^2, \tag{7.44}$$

where the first term comes from the composite wall in the middle while the second term comes from the two warped regions of the elementary walls. The action (7.44) is regularized: the contribution of the two parallel undistorted walls (Fig. 7.7a) is subtracted. In deriving this action we have used Eq. (7.43). The first term in Eq. (7.44) is negative and is dominant at large r_*. The second term is positive and is dominant at small r_*. Somewhere in between, there lies a maximum of the action. The bounce solution is at the tip of this hill; it can be obtained by extremizing Eq. (7.44) with respect to r_*:

$$r_* = \frac{d}{2}\sqrt{\frac{T_1}{2T_1 - T_2}}, \tag{7.45}$$

$$S_* = \frac{\pi}{3}T_1 d^3 \sqrt{\frac{T_1}{2T_1 - T_2}}.$$

Critical radius and action

The probability of wall fusion per unit time and unit area is proportional to [11]

$$d\Gamma_{\text{fusion}} \sim e^{-S_*} \sim \exp\left(-\frac{\pi}{3}T_1 d^3 \sqrt{\frac{T_1}{2T_1 - T_2}}\right). \tag{7.46}$$

*Before
starting this
subsection
the reader is
invited to
review the
sections on
monopoles,
strings, and
false-vacuum
decay.*

Finally, we must check that the linearization approximation is valid. The necessary condition is $|z'| \ll 1$, which is equivalent to $A/r_*^2 \ll 1$. Equations (7.43) and (7.45) imply that

$$\frac{A}{r_*^2} \sim \frac{d}{r_*} \sim \sqrt{\frac{2T_1 - T_2}{T_1}}, \tag{7.47}$$

and the condition $A/r_*^2 \ll 1$ is met at weak binding; see Eq. (7.38). Note that the interwall distance d must be large enough to ensure that $S_* \gg 1$.

7.2.3 Breaking Flux Tubes through Monopole Pair Production: The Microscopic Physics

Above we carried out a macroscopic investigation of string breaking and calculated the corresponding decay rate in the quasiclassical approximation. This section is devoted to the conceptual aspects. We will discuss why and how ANO-like strings can break. We will turn to the microscopic physics underlying string decays [12] and explain why monopole pair creation is crucial.

The 't Hooft–Polyakov monopoles appear as solitons in the Georgi–Glashow model (see Section 4.1). The Georgi–Glashow model *per se* does not support stable ANO flux tubes, as is obvious on topological grounds. Indeed, in this model the gauge group G is SU(2). It has a trivial first homotopy group, $\pi_1(\mathrm{SU}(2)) = 0$. This is illustrated in Fig. 7.8. As we saw in Chapter 3, the necessary condition for the existence of topologically stable flux tubes is the nontriviality of $\pi_1(G)$. However, we can generalize the GG model to make possible *quasistable strings*. Assume that we add an extra matter field in the fundamental representation of SU(2), a doublet, to be referred to as the "quark field." We will arrange the (self-)interaction of the scalar fields, adjoint and fundamental, in a special way. Namely, we will choose relevant parameters to make the adjoint scalar develop a very large vacuum expectation value (VEV),

$$V \gg \Lambda, \tag{7.48}$$

where Λ is the dynamical scale of the SU(2) theory. This VEV of the adjoint field breaks the SU(2) gauge group down to U(1) and ensures that the theory at hand is weakly coupled. Below the scale V one is left with the quantum electrodynamics of two charged fields, descendants of the quark doublet. The charged quark fields are then forced (through an appropriate choice of potential) to develop a small VEV v:

*Two-scale-
gauge
symmetry
breaking
(Higgsing)*

$$v \ll V. \tag{7.49}$$

In low-energy U(1) theory we can forget about the heavy adjoint field as well as the super-heavy monopoles. (The monopole mass is very large indeed, $M_M \sim V/g$.) The low-energy U(1) theory is scalar QED, with a charged field developing a vacuum expectation value. This is a classical set-up for ANO flux tubes. In the low-energy theory *per se* these flux tubes are topologically stable, since $\pi_1(\mathrm{U}(1)) = Z$.

However, in the full SU(2) theory there are no stable strings. Therefore, the strings of the low-energy theory will become unstable in the full theory. There is a way of "unwinding" the ANO string winding on the SU(2) group manifold (Fig. 7.8). Dynamically, this unwinding is an under-the-barrier process, the corresponding

Fig. 7.8   The first homotopy group of SU(2)/U(1) is trivial. (This illustration is from Wikipedia.)

action being very large in the limit $v \ll V$. As we will see shortly, the physical interpretation of this tunneling process is that of monopole–antimonopole pair creation accompanied by the annihilation of a segment of the string. Our task in this section is illustrative: to present an analytic *ansatz* which explicitly "unwinds"

Probability of metastable string decay

the U(1) string in the full SU(2) theory. The metastable string decay rate (the probability per unit time per unit length that the string will decay) was calculated in Section 7.2.1. For convenience, I reproduce it here in a more explicit notation,

$$\Gamma_{\text{breaking}} \sim v^2 \exp\left(-\frac{\pi M_M^2}{T_{\text{ANO}}}\right),\tag{7.50}$$

where M_M is the monopole mass and T_{ANO} is the string tension. Recall that $M_M^2 \sim V^2/g^2$ while $T_{\text{ANO}} \sim v^2$, so that the decay rate is exponentially suppressed,

$$-\ln \Gamma_{\text{breaking}} \sim \frac{V^2}{v^2 g^2} \gg 1.$$

This is in full accord with the physics of the string-breaking process. Indeed, the energy needed to produce the monopole–antimonopole pair is huge; a very long string segment must annihilate to release this energy. This is a tunneling process with highly suppressed probability.

7.2.3.1 Formulation of the Extended Model

Euclidean

Consider an SU(2) gauge theory with (Euclidean) action [12]

$$S = \int d^4x \left[\frac{1}{4g^2} F_{\mu\nu}^a F^{\mu\nu a} + \frac{1}{2}(\mathcal{D}_\mu \phi^a)^2 + |\mathcal{D}_\mu q|^2 + V(q,\phi) \right],\tag{7.51}$$

Extension of GG model

where ϕ^a, $a = 1, 2, 3$, is a real scalar field in the adjoint representation of SU(2) while q_k, $k = 1, 2$, is a complex scalar field in the fundamental representation. I will call q_1 and q_2 *scalar quarks*. The quantity $V(q,\phi)$ is a scalar self-interaction potential. We will use both matrix and vector notation for the adjoint fields, writing, say,

$$\phi \equiv \frac{\tau^a}{2} \phi^a.$$

The covariant derivative $\mathcal{D}_\mu$ acts in the adjoint and fundamental representations according to standard rules. The simplest form of the potential $V(q,\phi)$ that will serve our purpose is

$$V(q,\phi) = \lambda \left(|q|^2 - v^2\right)^2 + \tilde{\lambda} \left(\phi^a \phi^a - V^2\right)^2 + \gamma \left| \left(\phi - \frac{V}{2}\right) q \right|^2,\tag{7.52}$$

where v and V are parameters having the dimension of mass and $\lambda, \tilde{\lambda}$, and γ are dimensionless coupling constants.

As usual, we will limit ourselves to the case of weak coupling, when all four coupling constants g^2, λ, $\tilde{\lambda}$, and γ, are small (this requires, in particular, that $V \gg \Lambda$). Then a quasiclassical treatment applies. To arrange the desired double-scale (hierarchical) pattern of symmetry breaking, we must ensure a hierarchy of the vacuum expectation values. Namely, the breaking SU(2) $\rightarrow$ U(1) occurs at a high scale while U(1) $\rightarrow$ nothing occurs at a much lower scale,

$$v \ll V. \tag{7.53}$$

At the first stage the adjoint field ϕ develops a VEV that can always be aligned along the third axis in isospace,

$$\langle \phi^a \rangle = \delta^{a3} V. \tag{7.54}$$

This breaks the gauge SU(2) group down to U(1) and gives mass to the $W^\pm$ bosons and to one real adjoint scalar ϕ^3:

$$m_{W^\pm} = gV, \qquad m_{\text{adj}} \equiv m_a = 2\sqrt{2\tilde{\lambda}}\, V, \tag{7.55}$$

Mass spectrum of the theory: elementary excitations

while the two other adjoint scalars, ϕ^1 and ϕ^2, are "eaten up" by the Higgs mechanism. Note that simultaneously the second component of the quark field, q_2, acquires a large mass,

$$M_{q_2} = \sqrt{\gamma}\, V, \tag{7.56}$$

due to the last term in the potential (7.52).

What is left below the scales (7.55) and (7.56), in the low-energy U(1) theory? We are left with the U(1) gauge field A^3_μ, interacting with one complex scalar quark q_1. The Euclidean action is

Covariant derivative in the low-energy action

$$S_{\text{QED}} = \int d^4x \left[\frac{1}{4g^2} F^3_{\mu\nu} F^{\mu\nu 3} + \left| \mathcal{D}_\mu q_1 \right|^2 + \lambda \left(|q_1|^2 - v^2 \right)^2 \right]. \tag{7.57}$$

Note that the covariant derivative in the low-energy action acts on q_1 as follows:

$$\mathcal{D}_\mu q_1 = \left(\partial_\mu - \frac{i}{2} A^3_\mu \right) q_1.$$

The U(1) charge of q_1 is 1/2.

At this second stage the charged field q_1 develops a VEV and the low-energy U(1) theory finds itself in the Higgs regime,

$$\langle q_1 \rangle = v \quad (\text{while } \langle q_2 \rangle = 0). \tag{7.58}$$

At this stage the gauge symmetry is completely broken. The breaking of U(1) gives a mass to the photon field A^3_μ, namely,

$$m_\gamma = \frac{1}{\sqrt{2}} gv, \tag{7.59}$$

while the mass of the light component of the quark field, q_1, is

$$m_{q_1} = 2\sqrt{\lambda}\, v. \tag{7.60}$$

In the low-energy U(1) theory one can forget about the heavy quark field q_2. The only place where q_2 surfaces again is in the "unwinding" *ansatz*, Eq. (7.66) below, at $\theta \neq 0$, see Fig. 7.10. For the time being, to ease the notation, we will drop the subscript 1 in mentioning the scalar quark field, setting $m_q \equiv m_{q_1} = 2\sqrt{\lambda}v$.

The theory (7.57) is an Abelian Higgs model which supports the standard ANO strings (Section 3.1). For generic values of λ in Eq. (7.57) the quark mass m_{q_1} (the inverse correlation length) and the photon mass m_γ (the inverse penetration depth) are distinct. Their ratio is an important parameter in the theory of superconductivity, characterizing superconductor type. Namely, for $m_{q_1} < m_\gamma$ one is dealing with a type I superconductor, in which two strings at large separations attract each other. For $m_{q_1} > m_\gamma$, however, the superconductor is of type II, in which two strings at large separations repel each other. This behavior is related to the fact that the scalar field generates attraction between two vortices while the electromagnetic field generates repulsion. The boundary separating superconductors of types I and II corresponds to $m_{q_1} = m_\gamma$, i.e., to a special value of the quartic coupling λ, namely, $\lambda = g^2/8$. Then the vortices do not interact (BPS saturation). I hasten to add that the above relation will not be maintained; the ratio λ/g^2 will be treated as an arbitrary parameter.

Supercon-ductors of types I and II

7.2.3.2 A Brief Review of ANO Strings

More details in Section 3.1

The classical field equations for an ANO string with unit winding number are solved by the standard *ansatz*

$$q_1(x) = q(r)e^{-i\alpha},$$
$$A_0^3 \equiv 0, \tag{7.61}$$
$$A_i^3(x) = 2\varepsilon_{ij}\frac{x_j}{r^2}[1 - f(r)].$$

Here

$$r = \sqrt{\sum_{i=1,2} x_j^2}$$

is the distance from the vortex center while α is the azimuthal angle in the 12 plane transverse to the vortex axis (the subscripts $i, j = 1, 2$ denote coordinates x and y in this plane). Moreover, $q(r)$ and $f(r)$ are profile functions. Note that $\partial_i\alpha = -\varepsilon_{ij}x_j/r^2$, see Fig. 7.9. The electric charge of the q field is $1/2$.

The profile functions q and f in Eq. (7.61) are real and satisfy the second-order differential equations

$$q'' + \frac{1}{r}q' - \frac{1}{r^2}f^2q - m_q^2\frac{q(q^2 - v^2)}{2v^2} = 0,$$
$$f'' - \frac{1}{r}f' - \frac{m_\gamma^2}{v^2}q^2f = 0, \tag{7.62}$$

for generic values of λ (a prime stands here for a derivative with respect to r), plus the boundary conditions

$$q(0) = 0, \quad f(0) = 1,$$
$$q(\infty) = v, \quad f(\infty) = 0, \tag{7.63}$$

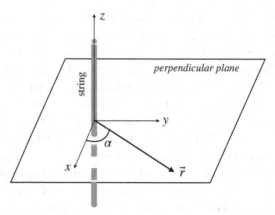

Fig. 7.9 Geometry of the string aligned along the z axis. The magnetic flux of the string can be calculated as the circulation of the photon four-potential along a large circle in the x, y plane, see (7.65).

which ensure that the scalar field reaches its VEV ($q_1 = v$) at infinity and that the vortex at hand carries one unit of magnetic flux.

The expression for the tension T (the energy per unit length) for an ANO string in terms of the profile functions (7.61) has the form

$$T_{\text{ANO}} = 2\pi \int r\,dr \left[\frac{2}{g^2} \frac{f'^2}{r^2} + q'^2 + \frac{f^2}{r^2} q^2 + \lambda (q^2 - v^2)^2 \right]. \tag{7.64}$$

Magnetic flux

The magnetic field flux $\int d^2x\, B^3$ can be written as

$$\frac{1}{2} \int B^3 \, dx\, dy \equiv \frac{1}{2} \oint A_i^3 dx_i = 2\pi. \tag{7.65}$$

The overall factor $\frac{1}{2}$ is due to our convention on the electric charge of q, see the expression for the covariant derivative after (7.57).

7.2.3.3 Decaying Strings: An Unwinding Configuration

To visualize how decay could be possible, note that the winding in (7.61) runs along the "equator" of the SU(2) group space (which is S_3) and, therefore, can be shrunk to zero by contracting the loop towards the south or north pole (Fig. 7.10).

It is not difficult to engineer an *ansatz* demonstrating the possibility of unwinding the field configuration (7.61) through the loop shrinkage in SU(2) group space [12]. The *ansatz* that does this is parametrized by an angle θ:

$$\begin{pmatrix} q_1 \\ q_2 \end{pmatrix} = U \begin{pmatrix} 1 \\ 0 \end{pmatrix} q_\theta(r),$$

$$A_0 \equiv 0, \qquad A_3 \equiv 0,$$

$$A_j(x) = iU\partial_j U^{-1}[1 - f_\theta(r)], \qquad j = 1, 2, \tag{7.66}$$

$$\phi = VU\frac{\tau_3}{2}U^{-1} + \Delta\phi,$$

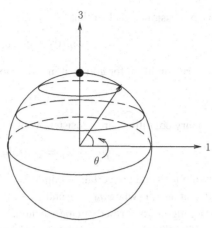

SU(2) group space

Fig. 7.10 Unwinding the ANO *ansatz*. The SU(2) group space is a three-dimensional sphere. The contour spun by the trajectory $U = \exp(-i\alpha\tau_3)$ ($\alpha \in [0, 2\pi]$) is the equator of the sphere. Our task is to contract the contour continuously up to the north pole.

"*Unwinding*" U | where the "unwinding" matrix is given by

$$U = e^{-i\alpha\tau_3} \cos\theta + i\tau_1 \sin\theta,$$

$$U \in SU(2), \text{ i.e., } UU^\dagger = 1, \quad \det U = 1. \tag{7.67}$$

Here α is the azimuthal angle in the coordinate space, see Fig. 7.10. (Eventually, upon quantization, θ becomes a slowly varying function of z and t, i.e., a field $\theta(t, z)$.)

The gauge and quark fields in (7.66) are parametrized by profile functions $f_\theta(r)$ and $q_\theta(r)$ depending on the parameter θ, which varies in the interval $[0, \pi/2]$. They satisfy the same boundary conditions,

$$q_\theta(0) = 0, \quad f_\theta(0) = 1,$$

$$q_\theta(\infty) = v, \quad f_\theta(\infty) = 0, \tag{7.68}$$

as the ANO *ansatz* in the low-energy U(1) theory; see Eq. (7.63). The boundary conditions at zero are chosen to ensure the absence of singularities of the "unwinding" field configuration at $r = 0$.

The term $\Delta\phi$ in the last line of Eq. (7.66) is needed to make sure that there is no singularity in ϕ at $r \to 0$. For an axially symmetric string the function $\Delta\phi$ can be chosen in the form

$$\Delta\phi = \varphi_\theta(r)\left(\frac{\tau_1}{2}\sin\alpha - \frac{\tau_2}{2}\cos\alpha\right), \tag{7.69}$$

where we have assumed that the component of $\Delta\phi$ along τ_3 is zero, while $\varphi_\theta(r)$ is an extra profile function that depends on θ as a parameter. The $a = 1, 2$ components of $\Delta\phi$ cannot be set equal to zero. To see this, substitute Eqs. (7.67) and (7.69) into the last line in Eq. (7.66). Then we obtain

$$\phi = \frac{\tau_3}{2}V\cos 2\theta - \left(\frac{\tau_1}{2}\sin\alpha - \frac{\tau_2}{2}\cos\alpha\right)[V\sin 2\theta - \varphi_\theta(r)]. \tag{7.70}$$

From this expression it is clear that ϕ has no singularity at $r = 0$ provided that

$$\varphi_\theta(0) = V \sin 2\theta. \tag{7.71}$$

The boundary condition for $\varphi_\theta(r)$ at infinity should be chosen as follows:

$$\varphi_\theta(\infty) = 0. \tag{7.72}$$

Both boundary conditions are consistent with the initial and final conditions

$$\varphi_\theta(r)|_{\theta=0} = \varphi_\theta(r)|_{\theta=\pi/2} = 0, \tag{7.73}$$

which are obvious and are certainly implied.

Note that at large r, when $q_\theta \to v$ and $f_\theta \to 0$ and $\varphi_\theta \to 0$, our field configuration presents a gauge-transformed "plane vacuum." This ensures that, at every given θ, the energy functional converges at large r. The convergence of the energy functional at small r is guaranteed by the boundary conditions $q_\theta(0) = 0$, $f_\theta(0) = 1$, and (7.71).

Now let us have a closer look at our unwinding *ansatz*. At $\theta = 0$ it is identical to the ANO string *ansatz*. The heavy field ϕ is strictly aligned along the third axis in the SU(2) space. The heavy "W bosons" $A_\mu^{1,2}$ are not excited; only the photon field A_μ^3 is involved in addition to the light quark field q^1. Now we continuously deform θ from 0 to $\pi/2$. At $\theta > 0$ we climb up (and then down) a huge potential energy hill. Indeed, at $\theta \neq 0$ the heavy "W bosons" $A_\mu^{1,2}$ are excited, as well as the heavy quark components, as is readily seen from Eqs. (7.66) and (7.67). As θ evolves from 0 to larger values, the field orientation in the SU(2) (or *isotopic*) space changes. For instance, the "W bosons" are excited while the form of A_μ^3 at large r varies as follows:

$$2\varepsilon_{ij}\frac{x_j}{r^2} \to \left(2\varepsilon_{ij}\frac{x_j}{r^2}\right)\cos^2\theta, \tag{7.74}$$

see (7.66). As for the scalar adjoint field ϕ^a, its $a = 3$ component decreases, at $\theta = \frac{\pi}{4}$ vanishes and then changes sign.

Finally, one last remark regarding our *ansatz* (7.66). The string magnetic flux calculated for a given θ in the interval $[0, \pi/2]$, takes the form

$$2\pi \cos^2\theta. \tag{7.75}$$

It changes from 2π at $\theta = 0$ to zero at $\theta = \pi/2$ in complete agreement with the unwinding interpretation.

For each given θ one can calculate the tension of the "distorted" string $T(\theta)$, provided that all relevant profile functions are found through minimization. Although this calculation is possible, in fact it is not advisable: we will need only the gross features of $T(\theta)$, which can be inferred without any calculations.

As θ approaches $\pi/2$, the unwinding of the ANO string is complete. Indeed, at $\theta = \pi/2$ we find ourselves in the empty vacuum: the gauge matrix U becomes $U = i\tau_1$, all components of the gauge field vanish, and ϕ becomes aligned along the third axis again, so that the ϕ quanta are not excited either. Note the change of sign of ϕ^3 at $\theta = \pi/2$ compared to the value of ϕ^3 at $\theta = 0$. This change of sign implies that it is the q^2 field that is light in this vacuum, rather than q^1. The q^1 degrees of freedom are not excited, in full accord with Eq. (7.66), while q^2 reduces to its vacuum value. The energy density of the empty vacuum vanishes. The potential energy of the unwinding field configuration is depicted in Fig. 7.11.

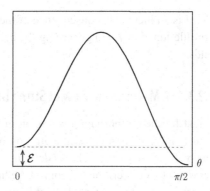

Fig. 7.11 The potential energy $T(\theta)$. Note that $T(0) \equiv \mathcal{E} - T_{\text{ANO}}$.

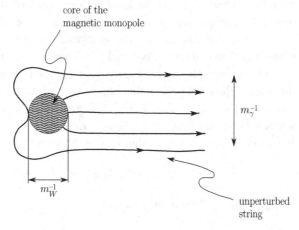

Fig. 7.12 The right-hand half of the broken string as obtained from Eq. (7.66). One unit of the magnetic charge is produced in the shaded area. The arrows indicate the magnetic field flux.

Indeed, imagine a perpendicular (to the string axis) plane sliding in Fig. 7.12 from right to left. When it is far to the right of the shaded area – the monopole – the total magnetic flux through this plane is 2π. However as we approach the monopole and then cross it, the flux diminishes and approaches zero when the plane is far to the left.

Above, I mentioned that calculating $T(\theta)$ by starting from Eq. (7.66) and minimizing the profile functions was not advisable. Why? The point is that though our unwinding *ansatz*, being relatively simple, is perfect for illustrative purposes, it is too restrictive to be fully realistic. While the *ansatz* (7.66) does describe the production of a magnetic charge at the end of a broken string, this magnetic change is in fact a highly excited monopole-like state rather than a 't Hooft–Polyakov monopole. To see this, let us inspect the picture of the magnetic flux corresponding to Eq. (7.66). This picture is presented in Fig. 7.12. The magnetic flux emerges in a bulge near the left end point. The longitudinal dimension of the bulge is $\sim m_W^{-1}$, a size typical of the monopole core. At the same time its transverse dimension (in a plane perpendicular to the string axis) is of order m_γ^{-1}. This is considerably larger than the monopole core size. The stretching of the core in the perpendicular direction is the reason why this lump is in fact (logarithmically) heavier than the 't Hooft–Polyakov monopole; it is an inevitable consequence of the fact that the *ansatz* (7.66) contains a single profile function $f_\theta(r)$ governing the behavior of both the photon and the W boson fields.

7.2.3.4 A Macroscopic View of String Breaking through Tunneling

We have just finished a thorough discussion of the unwinding *ansatz*. It represents a family of field configurations depending on one parameter, θ which can change continuously from 0 to $\pi/2$. The underlying theory describing tunneling in the parameter θ is a two-dimensional theory of the field $\theta(t, z)$, where z is the coordinate along the string.[3] What would we do next if our *ansatz* was perfect?

At $\theta = 0$ we have the ANO flux tube, at $\theta = \pi/2$ an empty vacuum. We start from $\theta = 0$ (more exactly, we let θ perform small oscillations near 0). This is our metastable state, a "false vacuum." The breaking of the tube occurs through tunneling to $\theta = \pi/2$. The state at $\theta = \pi/2$ is a "true vacuum." When tunneling occurs the string is broken into two parts – each part ending with a monopole or antimonopole.

For tunneling to happen, a large segment of the tube must be annihilated. Indeed, the mass of the monopole–antimonopole pair created is $\sim V/g$, and this mass has to come from the energy of the annihilated segment of the flux tube. If the length of this segment is L, the energy is $\sim LT_{\text{ANO}} \sim Lv^2$, where T_{ANO} is the tension of the ANO flux tube. Thus, the energy balance takes the form

$$Lv^2 \sim V/g \quad \text{or} \quad L \sim \frac{V}{gv^2}. \tag{7.76}$$

Compare this with the monopole size

$$\ell_M \sim \frac{1}{gV}. \tag{7.77}$$

> *Discussing the perfect unwinding ansatz*

We see that indeed $L/\ell_M \sim V^2/v^2 \gg 1$.

In a perfect *ansatz*, the endpoint domain of the broken string would be roughly a sphere with radius $\sim m_W^{-1}$ presenting the core of a practically unperturbed 't Hooft–Polyakov monopole, since at distances of order m_W^{-1} the effect of (magnetic charge) confinement is negligible; it comes into play only at distances $\sim m_\gamma^{-1}$. Thus, the mass of the endpoint bulge in the perfect *ansatz* must be $M_M + O(v/g)$. The $O(v/g)$ correction reflects the distortion of the 't Hooft–Polyakov monopole at distances $\sim m_\gamma^{-1}$.

The distorted endpoint domain of the broken string has a much smaller size than the length of the annihilated segment (the true vacuum). This justifies the use of the theory of false-vacuum decay in the thin wall approximation. In this approximation only two parameters are relevant: the difference between the energy densities in the false vacuum and in the true vacuum (this difference is T_{ANO}, the string tension) and the surface energy of the bubble whose creation describes the tunneling. This surface energy is fully determined by the monopole mass M_M. The ratio of the bubble wall thickness and the bubble size is $\sim 1/(Lm_\gamma) \sim v/V \ll 1$.

---

[3] We ignore all possible nonbreaking deformations of the string and focus on a single variable $\theta(t, z)$ responsible for the string annihilation.

Returning to the field $\theta(t, z)$ we observe that, indeed, it has two classical equilibrium positions, at $\theta = 0$ and $\theta = \pi/2$. To find the tension of the bubble wall we have (in the thin wall approximation) to ignore a small nondegeneracy of the true- and false-vacuum energies. Assuming that these two classical equilibrium positions are degenerate, we have to find a kink corresponding to interpolation between $\theta = \pi/2$ at $z = -\infty$ and $\theta = 0$ at $z = \infty$. The kink's mass is that of a (distorted) monopole.

Equation (7.50) is obtained in this way.

Exercise

7.2.1 Modify the microscopic model considered in Section 7.2.3.1 as follows: discard the quark field in the fundamental representation of SU(2) and introduce, instead, a "second" (light) adjoint matter field χ^a, $a = 1, 2, 3$. The pattern of the symmetry breaking remains hierarchical: first the heavy field ϕ develops a vacuum expectation value V that breaks SU(2) down to U(1). Then the light field χ develops a (small) vacuum expectation value v that breaks U(1).

Repeat the string breaking analysis, introducing appropriate changes where necessary, and calculate the decay rate.

References for Chapter 7

[1] I. Y. Kobzarev, L. B. Okun, and M. B. Voloshin, *Sov. J. Nucl. Phys.* **20**, 644 (1975).

[2] S. R. Coleman, *Phys. Rev. D* **15**, 2929 (1977). Erratum: *ibid.* **16**, 1248 (1977); C. G. Callan and S. R. Coleman, *Phys. Rev. D* **16**, 1762 (1977).

[3] M. Voloshin, in A. Zichichi (ed.), *Vacuum and Vacua: The Physics of Nothing* (World Scientific, Singapore, 1996), p. 88.

[4] S. Coleman, *Aspects of Symmetry* (Cambridge University Press, 1985), p. 327.

[5] S. R. Coleman, V. Glaser, and A. Martin, *Commun. Math. Phys.* **58**, 211 (1978).

[6] L. D. Landau and E. M. Lifshitz, *The Classical Theory of Fields* (Pergamon Press, 1987), Chapter 2, Eq. (8.2).

[7] L. D. Landau and E. M. Lifshitz, *Quantum Mechanics* (Pergamon Press, 1989), Section VII.50, Eq. (50.5).

[8] A. Vilenkin, *Nucl. Phys. B* **196**, 240 (1982); J. Preskill and A. Vilenkin, *Phys. Rev. D* **47**, 2324 (1993).

[9] A. Armoni and M. Shifman, *Nucl. Phys. B* **671**, 67 (2003) [arXiv:hep-th/0307020].

[10] A. Monin and M. B. Voloshin, *Phys. Rev. D* **78**, 065048 (2008) [arXiv:0808.1693 [hep-th]].

[11] S. Bolognesi, M. Shifman, and M. B. Voloshin, *Phys. Rev. D* **80**, 045 010 (2009) [arXiv:0905.1664 [hep-th]].

[12] M. Shifman and A. Yung, *Phys. Rev. D* **66**, 045 012 (2002) [hep-th/0205025].

8 Chiral and Other Anomalies

A clash between global chiral symmetries and gauge symmetry leads to anomalies. — External and internal anomalies. — Two faces of the anomaly. — The power of the 't Hooft matching. — A brief encounter with the scale anomaly. — One-form anomalies.

8.1 Chiral Anomaly in the Schwinger Model

Our first encounter with the chiral anomalies in gauge theories occurred in Chapter 5. We have invoked them, in a pragmatic way, more than once. The current chapter is designed to explain the conceptual issues behind the anomalies. The questions to be asked are "Why do they appear?" and "What do they imply?". Here we will address these questions on a more systematic basis.

This topic is important, since anomalies play a role in a number of subtle aspects of gauge dynamics. Our first task will be to understand the physical meaning of the phenomenon. This is best done in a simple example [1], that of a two-dimensional model which can be treated at weak coupling – the Schwinger model on a spatial circle. This example clearly demonstrates that (i) anomalies appear when two contradictory requirements clash and so we have to choose one of them as "sacred" (usually gauge invariance); (ii) anomalies have two faces, infrared and ultraviolet; and (iii) the infinite number of degrees of freedom in field theory is crucial. The chiral anomaly involves fermions. There is another anomaly in gauge theories, the scale anomaly. It occurs even in pure Yang–Mills theory, with no quarks. We will familiarize ourselves with a number of methods allowing us to derive both these anomalies and then pass to the implications. We will discuss the 't Hooft matching condition, one of the few tools that are applicable to non-Abelian theories at strong coupling, and we will prove that the chiral symmetry of QCD must be spontaneously broken, at least at large N. As an illustrative example of the usefulness of a proper understanding of the anomalies we will calculate the $\pi^0 \to \gamma\gamma$ decay rate. Many more applications are known; they would be found in a good textbook on particle theory. With regret, I have to leave them aside in this general field theory text.

Finally, I will briefly discuss a (relatively) recently discovered class of anomalies which are referred to as "generalized global anomalies." The simplest example is the one-form anomaly in the Schwinger model on a spatial circle; it will be considered in Section 8.5. In various (sometimes exotic) field-theoretic models one can encounter higher-form anomalies and some other generalized global anomalies. Unfortunately, I cannot include the latter in my pedagogical narrative.

A few words on terminology is in order here. One of the topics of Chapters 8 listed on page 298 is titled "External and internal anomalies." The latter are the anomalies inside the given theories which, if not canceled, ruin the theory as such. For instance, if in the Standard Model we remove, say, the electron field, this model becomes internally inconsistent.

The external anomalies may emerge after we gauge some global symmetries of the theory under consideration. The gauge fields introduced in the process of gauging are non-dynamic external (or background) fields. Being auxiliary, they are our helpers, allowing us to reveal subtle aspects of the original theory which may not be seen at strong coupling. The external anomalies are instrumental in 't Hooft matching, see Section 8.3.

8.1.1 Schwinger Model on a Circle

Two-dimensional QED for a massless Dirac fermion seems to be the simplest gauge model. The Lagrangian is

$$\mathcal{L} = -\frac{1}{4e_0^2} F_{\mu\nu} F^{\mu\nu} + \bar{\psi} i \mathcal{D} \psi, \tag{8.1}$$

where $F_{\mu\nu}$ is the photon field strength tensor,

Defining the covariant derivative in the Schwinger model.

$$F_{\mu\nu} = \partial_\mu A_\nu - \partial_\nu A_\mu, \tag{8.2}$$

and e_0 is the gauge coupling constant, having the dimension of mass for $D = 2$. Moreover, $\mathcal{D}_\mu$ is the covariant derivative, given by

$$i\mathcal{D}_\mu = i\partial_\mu + A_\mu, \tag{8.3}$$

Consult Sections 3.3.3 and 10.2.2.

and ψ is the two-component spinor field. The gamma matrices in Minkowski space can be chosen in the following way:

$$\gamma^0 = \sigma_2, \qquad \gamma^1 = -i\sigma_1, \qquad \gamma^5 = -\sigma_3. \tag{8.4}$$

The spinor $\psi_L = \begin{pmatrix} \psi_1 \\ 0 \end{pmatrix}$ will be called left-handed ($\gamma^5 \psi_L = -\psi_L$) and the spinor $\psi_R = \begin{pmatrix} 0 \\ \psi_2 \end{pmatrix}$ will be called right-handed ($\gamma^5 \psi_R = \psi_R$). Note also that $\bar{\psi} = \psi^\dagger \gamma^0$.

In spite of considerable simplifications compared with four-dimensional QED, the dynamics of the model (8.1) is still too complicated for our purposes. Indeed, the set of asymptotic states in this model drastically differs from the fields in the Lagrangian. In the two-dimensional theory the photon, as is well known, has no transverse degrees of freedom and essentially reduces to the Coulomb interaction.[1] The latter, however, grows linearly with distance. This linear growth of the Coulomb potential results in confinement of the charged fermions in the Schwinger model irrespective of the value of the coupling constant e_0. The model (8.1) was used as a prototype for describing color confinement in QCD (see, e.g., [2] and Section 9.5).

In order to simplify the situation further let us do the following. Consider the system described by the Lagrangian (8.1) on a finite spatial domain of length L. If L is small, $e_0 L \ll 1$, the Coulomb interaction never becomes strong and one can actually treat it as a small perturbation; in particular, in a first approximation its effect can be neglected altogether. We will impose periodic boundary conditions on the field A_μ and antiperiodic ones on ψ. Thus, the problem to be considered below is the Schwinger model on the circle. Notice that the antiperiodic boundary condition is imposed on the fermion field for convenience only. As will be seen, any other boundary condition (periodic, for instance) would do as well; nothing would change except minor technical details. Thus,

Boundary conditions

$$A_\mu(t, x = -L/2) = A_\mu(t, x = L/2),$$

$$\psi(t, x = -L/2) = -\psi(t, x = L/2). \tag{8.5}$$

Equations (8.5) imply that the fields A_μ and ψ can be expanded in Fourier modes, $\exp\left(ikx\frac{2\pi}{L}\right)$ for bosons and $\exp\left[i(k + \frac{1}{2})x\frac{2\pi}{L}\right]$ for fermions ($k = 0, \pm 1, \pm 2, \ldots$).

---

[1] It is instructive to compare this assertion with those in Section 9.5.

Now, let us recall that the Lagrangian (8.1) is invariant under the local gauge transformations

$$\psi \to e^{i\alpha(t,x)}\psi, \qquad A_\mu \to A_\mu + \partial_\mu \alpha(t,x). \tag{8.6}$$

It is evident that all modes for the field A_1 except the zero mode (i.e., $k = 0$) can be gauged away. Indeed, the term of the type $a(t) \sin\left(kx\frac{2\pi}{L}\right)$ in A_1 is gauged away by virtue of the gauge function

$$\alpha(t,x) = L(2\pi k)^{-1} a(t) \cos\left(kx\frac{2\pi}{L}\right).$$

The latter is periodic on the circle and does not violate the conditions (8.5), as required. Thus, in the most general case we can treat A_1 as an x-independent constant.

This is not the end of the story, however, since the possibilities provided by gauge invariance are not yet exhausted. There exists another class of admissible gauge transformations – sometimes, they are referred to as "large" gauge transformations – with a gauge function that is not periodic in x,

| Large gauge |
| transforma- |
| tions. |

$$\alpha = \frac{2\pi}{L} nx, \qquad n = \pm 1, \pm 2, \ldots, \tag{8.7}$$

where n is an integer. In spite of its nonperiodicity, such a choice of gauge function is also compatible with the conditions (8.5). This is readily verifiable: since $\partial\alpha/\partial x =$ const and $\partial\alpha/\partial t = 0$ the periodicity for A_μ is not violated. An analogous assertion is also valid for the phase fact $e^{i\alpha}$: the difference in the phases at the endpoints of the interval $x \in [-L/2, L/2]$ is equal to $2\pi n$.

As a result, we arrive at the conclusion that the variable A_1 (remember that it has no x-dependence; it depends only on time) should not be considered on the whole interval $(-\infty, \infty)$; the points

$$A_1, \quad A_1 = \pm\frac{2\pi}{L}, \quad A_1 = \pm\frac{4\pi}{L}, \quad \ldots$$

are gauge equivalent and must be identified. In other words, the variable A_1 is an independent variable only on the interval $[0, \frac{2\pi}{L}]$. Going beyond these limits we find ourselves in a gauge image of the original interval. Following the commonly accepted terminology, we say that A_1 lives on a circle of circumference $\frac{2\pi}{L}$.

| A_1 is an |
| angle-type |
| variable. |

It is well known that the gauge invariance of electrodynamics is closely interrelated with the conservation of electric charge. Indeed, the Lagrangian (8.1) (for finite as well as infinite L) admits multiplication of the fermion field by a constant phase,

$$\psi \to e^{i\alpha}\psi, \qquad \psi^\dagger \to \psi^\dagger e^{-i\alpha}.$$

Using a standard line of reasoning one easily derives from this phase invariance the conservation of the electric current:

$$j^\mu = \overline{\psi}\gamma^\mu\psi, \qquad \dot{Q}(t) = 0, \qquad Q = \int dx\, j^0(x,t).$$

The vanishing of the divergence $\partial_\mu j^\mu$ follows from the equations of motion.

The classical Lagrangian (8.1) exhibits the second conservation law. Observe that (8.1) is invariant under another phase rotation, the global axial transformation

$$\psi \to e^{-i\alpha\gamma^5}\psi, \qquad \psi^\dagger \to \psi^\dagger e^{i\alpha\gamma^5},$$

which multiplies the left- and right-handed fermions by opposite phases (remembering that $\gamma^5 = -\sigma_3$). At the classical level the axial current

$$j^{\mu 5} = \overline{\psi} \gamma^\mu \gamma^5 \psi$$

is conserved in just the same way as the electromagnetic current. One can readily check, using the equations of motion, that $\partial_\mu j^{\mu 5} = 0$. If the axial charge of the left-handed fermions is $Q_5 = +1$ then for the right-handed fermions $Q_5 = -1$. The conservation of Q and Q_5 is equivalent to the conservation of the numbers of the left-handed and right-handed fermions separately. This fact is obvious for any Born (tree) graph. Indeed, in all such graphs the fermion lines are continuous, photon emission does not change their chirality, and the number of ingoing fermion legs is equal to that of the outgoing legs. In the exact answer including all quantum effects, however, only the sum of the chiral charges is conserved, i.e., only one of the two classical symmetries survives quantization of the theory.

As will be seen below, the characteristic excitation frequencies for A_1 are of order e_0 while those associated with the fermionic degrees of freedom are of order L^{-1}. Since $e_0 L \ll 1$ the variable A_1 is adiabatic with respect to the fermionic degrees of freedom. Consequently, the Born–Oppenheimer approximation is justified in our case. In the next subsection we will analyze in more detail the fermion sector, assuming temporarily that A_1 is a fixed (time-independent) quantity. From Eqs. (8.10) and (8.11) below it is evident that the fermionic frequencies are indeed of order L^{-1}. Calculation of the A_1 frequencies will be carried out later, see (8.31).

For our pedagogical purposes we can confine ourselves to the study of the limit $e_0 L \ll 1$. Those readers who would like to know about the solution of the Schwinger model on a circle with arbitrary L should turn to the original publications (e.g., [3]).

8.1.2 Dirac Sea: The Vacuum Wave Function

Following the standard prescription of the adiabatic approximation we will freeze the time dependence of the photon field A_μ and consider it as "external." Regarding the $\mu = 0$ component of the photon field, it is responsible for the Coulomb interaction between charges; the corresponding effect is of the order $e_0 L \ll 1$ and does not show up in the leading approximation to which we will limit ourselves in the present section. Thus, we can put $A_0 = 0$. The difference between these two components lies in the fact that the fluctuations in A_0 are small, while this is not the case for A_1. The wave function is not localized in A_1 in the vicinity of $A_1 = 0$. It is just this phenomenon – delocalization of the A_1 wave function and the possibility of penetration to large values of A_1 – that will lead to observable manifestations of the chiral anomaly.

In two-dimensional electrodynamics the Dirac equation determining the fermion energy levels has the form

$$\left[i\frac{\partial}{\partial t} - \sigma_3 \left(i\frac{\partial}{\partial x} + A_1 \right) \right] \psi = 0. \tag{8.8}$$

For the kth stationary state, $\psi \sim \exp(-iE_k t)\psi_k(x)$ and its energy is given by

$$E_k \psi_k(x) = \sigma_3 \left(i\frac{\partial}{\partial x} + A_1 \right) \psi_k(x). \tag{8.9}$$

Furthermore, the eigenfunctions are proportional to

$$\psi_k \sim \exp\left[i\left(k + \frac{1}{2}\right)\frac{2\pi}{L}x\right], \qquad k = 0, \pm1, \pm2, \ldots \tag{8.10}$$

The extra term $\frac{1}{2}\frac{2\pi}{L}x$ in the exponent ensures the antiperiodic boundary conditions; see Eqs. (8.5). As a result, we conclude that the energy of the kth level for left-handed fermions is

$$E_{k(L)} = -\left(k + \frac{1}{2}\right)\frac{2\pi}{L} + A_1, \tag{8.11a}$$

*Level flow.
Rearrange-
ment of
levels in
gauge
equivalent
point*

while for the right-handed fermions

$$E_{k(R)} = \left(k + \frac{1}{2}\right)\frac{2\pi}{L} - A_1. \tag{8.11b}$$

The energy-level dependence on A_1 is displayed in Fig. 8.1. The broken lines show the behavior of $E_{k(L)}$ and the solid lines show $E_{k(R)}$. At $A_1 = 0$ the energy levels for the left-handed and right-handed fermions are degenerate. As A_1 increases, the degeneracy is lifted and the levels split. At the point $A_1 = 2\pi/L$ the overall structure of the energy levels is precisely the same as for $A_1 = 0$; degeneracy occurs again. The identity of the points $A_1 = 0$ and $A_1 = 2\pi/L$ is a remnant of the gauge invariance of the original theory (see the discussion in Section 8.1.1).

We note that this identity is achieved in a nontrivial way; in passing from $A_1 = 0$ to $A_1 = 2\pi/L$ a restructuring of the fermion levels takes place. The left-handed levels are shifted upwards by one interval while the right-handed levels are shifted downwards by one interval. This phenomenon, the restructuring of the fermion levels, is the essence of the chiral anomaly as will become clear shortly.

Let us proceed from the one-particle Dirac equation to field theory. Our first task is the construction of the ground state, the vacuum. To this end, following the well-known Dirac prescription we fill up all levels lying in the Dirac sea, leaving all positive-energy levels empty. The notation $|1_{L,R}, k\rangle$ and $|0_{L,R}, k\rangle$, respectively, will

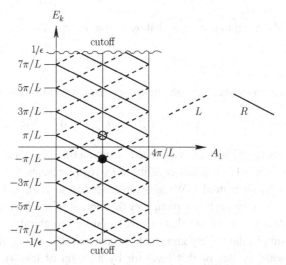

Fig. 8.1   Fermion energy levels as a function of A_1.

be used below for full and empty levels with a given k. The subscript L (R) indicates that we are dealing with the left-handed (right-handed) fermions.

Recall that A_1 is a slowly varying adiabatic variable; the corresponding quantum mechanics will be considered later. At first, the value of A_1 is fixed in the vicinity of zero, $A_1 \approx 0$. Then the fermion wave function of the vacuum, as seen from Fig. 8.1, reduces to

$$\Psi_{\text{ferm. vac.}} = \left(\prod_{k=0,1,2,\ldots} |1_L, k\rangle \right) \left(\prod_{k=-1,-2,\ldots} |0_L, k\rangle \right)$$

$$\times \left(\prod_{k=-1,-2,\ldots} |1_R, k\rangle \right) \left(\prod_{k=0,1,2,\ldots} 0_R, k\rangle \right). \tag{8.12}$$

The Dirac sea, consisting of the negative-energy levels, is completely filled. Now let A_1 increase adiabatically from 0 to $\frac{2\pi}{L}$. The same figure shows that at $A_1 = \frac{2\pi}{L}$ the wave function (8.12) describes a state that, from the standpoint of the normally filled Dirac sea, contains one left-handed particle and one right-handed hole (the small circles in Fig. 8.1).

Do the *quantum numbers* of the fermion sea change in the process of the transition from $A_1 = 0$ to $A_1 = 2\pi/L$? Answering this question, we would say that the appearance of a particle and a hole does not change the electric charge since the electric charges of the particle and the hole are obviously opposite. In other words, the electromagnetic current is conserved. However, the axial charges of the left-handed particle and the right-handed hole are the same ($Q_5 = -1$) and, hence, for the transition at hand,

$$\Delta Q_5 = -2. \tag{8.13}$$

A more formal analysis, to be carried out shortly, will confirm this assertion.

Equation (8.13) can be rewritten as $\Delta Q_5 = -(L/\pi)\Delta A_1$. Dividing by Δt, the transition time, we get

$$\dot{Q}_5 = -\frac{L}{\pi}\dot{A}_1, \tag{8.14}$$

which implies, in turn, that the conserved quantity has the form

$$\int dx \left(j^{05} + \frac{1}{\pi}A_1 \right). \tag{8.15}$$

Anomaly in the axial current derived from the level flow

The current corresponding to the charge (8.15) is obviously

$$\widetilde{j}^{\mu 5} = j^{\mu 5} + \frac{1}{\pi}\varepsilon^{\mu\nu}A_\nu, \qquad \partial_\mu \widetilde{j}^{\mu 5} = 0, \qquad \partial_\mu j^{\mu 5} = -\frac{1}{2\pi}\varepsilon^{\mu\nu}F_{\mu\nu}, \tag{8.16}$$

where $\varepsilon^{\mu\nu}$ is the Levi–Civita antisymmetric tensor and $\varepsilon^{01} = -\varepsilon^{10} = 1$. (Notice that $\varepsilon_{01} = -1$.) The last equality in (8.16) represents the famous axial anomaly in the Schwinger model. We have succeeded in deriving it by "hand-waving" arguments, i.e., by inspecting a picture of the motion of the fermion levels in the external field $A_1(t)$. It turns out that in this language the chiral anomaly presents an extremely simple and widely known phenomenon: the crossing of the zero point in the energy scale by this or that level (or by a group of levels). The presence of an infinite number of levels and the Dirac "multiparticle" interpretation, according to which the emergence of a filled level from the sea is equivalent to the appearance of a particle while the submergence of an empty level into the sea is equivalent to the production of a hole – an antiparticle – constitute the essential elements of the construction. With a finite number of levels there is no place for such an interpretation and there can be no quantum anomaly.

I would like to draw the reader's attention to a somewhat different, although intimately related with the previous, aspect of the picture. The fermion levels move parallel to each other through the bulk of the Dirac sea. Therefore, the disappearance of the levels beyond the zero-energy mark occurs simultaneously with the disappearance of their "copies" beyond the ultraviolet cutoff, which is always implicitly present in field theory; below, we will introduce this cutoff explicitly. Because of this, the heuristic derivation of the anomaly given in this section and a more standard treatment based on ultraviolet regularization are actually one and the same. Often it turns out to be more convenient just to trace the crossing of the ultraviolet cutoff by the levels from the Dirac sea. Beyond toy models, in QCD-like theories, the latter approach becomes an absolute necessity, not a question of convenience, due to the notorious "infrared slavery." The connection between the ultraviolet and infrared interpretations of the anomaly is discussed in more detail in Sections 8.1.3 and 8.1.7. The interested reader is referred to the original work [4], where the subtle points are thoroughly analyzed.

8.1.3 Ultraviolet Regularization

In spite of the transparent character of this heuristic derivation, almost all the "evident" points above could be questioned by the careful reader. Indeed, why is the wave function (8.12) the appropriate choice? In what sense is the energy of this state minimal, taking into account the fact that, according to (8.11),

$$E \sim - \sum_{k=0}^{\infty} \left(k + \frac{1}{2} \right) \frac{2\pi}{L}$$

and the sum is ill defined (the series is divergent)? Moreover, it is usually asserted that the quantum anomalies are due to the necessity for ultraviolet regularization of the theory. If so, why speak of the Dirac sea and the crossing of the zero-energy point by the fermion levels?

Surprisingly, all these questions are connected with each other. It may be instructive to start with the last. I want to explain that ultraviolet regularization, mentioned in passing in Section 8.1.2, is actually the key element. More than that, the derivation sketched above tacitly assumes a quite specific regularization.

The fermion levels stretch in the energy scale up to indefinitely large energies, positive or negative. The wave function (8.12) describing the fermion sector at $A_1 \approx 0$ contains, in particular, the direct product of an infinitely large number of filled states $|1_R, k\rangle, |1_L, k\rangle$ with negative energy. It is clear that such an object – an infinite product – is ill defined, and one cannot avoid some regularization in calculating physical quantities. The contribution corresponding to large energies (momenta) should be somehow cut off.

At first sight, it would seem sufficient simply to discard the terms with $|k| > |k|_{max} (|k|_{max}$ is a fixed number independent of A_1). This *is* a regularization,

Making the cutoff in a gauge-invariant manner

of course, but, clearly enough, the prescription will lead to a violation of gauge invariance and to electric charge nonconservation. Indeed, in gauge theories the momentum p always appears only in the combination $p + A$, not simply as p (or, equivalently, k).

In order to preserve gauge invariance, it is possible and convenient to use a regularization called in the literature the Schwinger, or ϵ, splitting. This regularization will provide a solid mathematical basis for the heuristic derivation presented above. Instead of the original currents

$$j^\mu = \bar{\psi}(t, x)\gamma^\mu \psi(t, x), \qquad j^{\mu 5} = \bar{\psi}(t, x)\gamma^\mu \gamma^5 \psi(t, x), \tag{8.17}$$

we introduce the regularized objects

$$j^\mu_{\text{reg}} = \bar{\psi}(t, x + \epsilon)\gamma^\mu \psi(t, x) \exp\left(i \int_x^{x+\epsilon} A_1 \, dx\right),$$

$$j^{\mu 5}_{\text{reg}} = \bar{\psi}(t, x + \epsilon)\gamma^\mu \gamma^5 \psi(t, x) \exp\left(i \int_x^{x+\epsilon} A_1 dx\right). \tag{8.18}$$

It is implied that $\epsilon \to 0$ in the final answer for the physical quantities. At the intermediate stages, however, all computations are performed with fixed ϵ. The exponential factor in (8.18) ensures the gauge invariance of the "split" currents. Without this factor, multiplying $\psi(t, x)$ by an x-dependent phase to obtain $\exp[i\alpha(x)]\psi(t, x)$, yields

$$\psi^\dagger_\alpha(t, x + \epsilon)\psi_\beta(t, x) \to \exp[-i\alpha(x + \epsilon) + i\alpha(x)]\psi^\dagger_\alpha(t, x + \epsilon)\psi_\beta(t, x). \tag{8.19}$$

Applying the gauge transformation (8.6) to A_1 compensates for the phase factor in Eq. (8.19).

Now, there appears to be no difficulty in calculating the electric and axial charges of the state (8.12) in a well-defined manner. If

$$Q = \int dx \, j^0_{\text{reg}}(t, x), \qquad Q_5 = \int dx \, j^{05}_{\text{reg}}(t, x) \tag{8.20}$$

then for the vacuum wave function we evidently obtain

$$Q = Q_L + Q_R, \qquad Q_5 = -Q_L + Q_R, \tag{8.21}$$

$$Q_L = \sum_k \exp\left\{-i\epsilon\left[\left(k + \frac{1}{2}\right)\frac{2\pi}{L} - A_1\right]\right\},$$

$$Q_R = \sum_{k'} \exp\left\{-i\epsilon\left[\left(k' + \frac{1}{2}\right)\frac{2\pi}{L} - A_1\right]\right\}, \tag{8.22}$$

where k and k' run over all the filled levels. In the limit $\epsilon \to 0$, the charges Q_L and Q_R both turn into a sum of unities, each unity representing one energy level from the Dirac sea. Equations (8.22) once again demonstrate the gauge invariance of the Schwinger regularization. Indeed, the cutoff suppresses the states with $|p + A_1| \gtrsim \epsilon^{-1}$.

The phase factor in Eqs. (8.18) ensures that the suppressing function contains the desired combination, $p + A$.

I hasten to add here that although superficially Eqs. (8.22) do not differ from each other, actually they do not coincide because the summations run over different values of k. The particular values are easy to establish from Fig. 8.1.[2] Let $|A_1| < \pi/L$. Then in a "left-handed" sea the filled levels have $k = 0, 1, 2, \ldots$ In a "right-handed" sea the filled levels correspond to $k = -1, -2, \ldots$ Thus, if $|A_1| < \pi/L$ we have

$$Q_L = \sum_{k=0}^{\infty} \exp\left(i\epsilon E_{k(L)}\right),$$

$$Q_R = \sum_{k=-1}^{-\infty} \exp\left(-i\epsilon E_{k(R)}\right). \tag{8.23}$$

Performing summation and expanding in ϵ we arrive at

$$(Q_L)_{\text{vac}} = -(Q_R)_{\text{vac}} = \frac{e^{i\epsilon A_1}}{2i\,\sin(\epsilon\pi/L)}$$

$$= \frac{L}{2\pi i \epsilon} + \frac{L}{2\pi} A_1 + O(\epsilon), \tag{8.24}$$

We pause here to summarize our results. Equation (8.24) shows that under our choice of the vacuum wave function (8.12) the charge of the vacuum vanishes, $Q = Q_L + Q_R = 0$. Moreover, there is no time dependence: charge is conserved. The axial charge consists of two terms: the first term represents an infinitely large *constant* and the second gives a linear A_1-dependence. In the transition $(A_1 \approx 0) \rightarrow (A_1 \approx 2\pi/L)$ the axial charge changes by minus two units (see Eq. (8.21)).

These conclusions are not new for us. We found just the same from the illustrative picture described in Section 8.1.2 in which the electric and axial charges of the Dirac sea were determined intuitively. Now we have learned how to sum up the infinite series $\sum_k 1$, the charges of the "left-handed" and "right-handed" seas, by virtue of a well-defined procedure that automatically cuts off the levels with $|p + A_1| \gtrsim \epsilon^{-1}$.

The procedure suggests an alternative language for describing axial charge nonconservation in the transition $(A_1 \approx 0) \rightarrow (A_1 \approx 2\pi/L)$. Previously we thought that the nonconservation was due to the level crossing of the zero-energy point. It is equally correct – as we see now – to say that the nonconservation can be explained as follows: one right-handed level leaves the sea via the lower boundary (the cutoff $-\epsilon^{-1}$) and one new left-handed level appears in the sea through the same boundary (Fig. 8.1). Both phenomena – the crossing of the zero-energy point and the departure (arrival) of the levels via the ultraviolet cutoff – occur simultaneously, though, and represent two different facets of the same anomaly, which admits both the infrared and the ultraviolet interpretation.

Gauge invariance should be maintained by all means!

One last remark concerning the axial charge is in order. Instead of Eqs. (8.18) one could regularize the axial charge in a different way, so that $\partial_\mu j^{\mu 5} = 0$ and $\Delta Q_5 = 0$. (A nice exercise for the reader!) Under such a regularization, however, the expression for the axial current would not be gauge invariant. Specifically, the conserved axial current, apart from Eqs. (8.18), would include an extra term $\frac{1}{\pi} \epsilon^{\mu\nu} A_\nu$, cf. Eqs. (8.16). As already mentioned, there is no regularization ensuring simultaneous gauge invariance and conservation of $j^{\mu 5}$.

---

[2] See also Eq. (8.12).

8.1.4 The Theta Vacuum

Compare
with
Section 5.1.2.

Now, we will leave the issue of charges and proceed to the calculation of the fermion sea energy, a problem that could not be solved at the naive level, without regularization. Fortunately, all the necessary elements are now in place.

The fermion part of the Hamiltonian, cf. Eqs. (8.9),

$$H = \int_{-L/2}^{L/2} dx \, \psi^\dagger(t, x) \sigma_3 \left(i \frac{\partial}{\partial x} + A_1 \right) \psi(t, x), \tag{8.25}$$

reduces after the ϵ splitting to

$$H_{\mathrm{reg}} = \int_{-L/2}^{L/2} dx \, \psi^\dagger(t, x + \epsilon) \, \sigma_3 \left(i \frac{\partial}{\partial x} + A_1 \right) \psi(t, x) \exp \left(i \int_x^{x+\epsilon} A_1 \, dx \right). \tag{8.26}$$

This formula implies, in turn, the following regularized expression for the energies of the "left-handed" and "right-handed" seas:

$$E_L = \sum_{k=0}^{\infty} E_{k(L)} \exp(i\epsilon \, E_{k(L)}), \qquad E_R = \sum_{k=-1}^{-\infty} E_{k(R)} \exp(-i\epsilon \, E_{k(R)}), \tag{8.27}$$

where the energies of the individual levels $E_{k(L,R)}$ are given in (8.11) and the summation runs over all levels having a negative energy. The values of the summation indices in Eqs. (8.27) correspond to $|A_1| < \pi/L$. Expressions (8.27) have an obvious meaning: in the limit $\epsilon \to 0$ they simply reduce to the sum of the energies of all filled fermion levels from the Dirac sea. The additional exponential factors guarantee the convergence of the sums.

Dirac sea
energy

Furthermore, we notice that E_L and E_R can be obtained by differentiating the expressions (8.23) and (8.24) for $Q_{L,R}$ with respect to ϵ. (Equation (8.23) presents geometrical progressions that are trivially summable.) Expanding in ϵ we get

$$E_{\mathrm{sea}} = E_L + E_R = \frac{L}{2\pi} \left(A_1{}^2 - \frac{\pi^2}{L^2} \right) + \text{a constant independent of } A_1. \tag{8.28}$$

In the expression above we will omit the infinite A_1-independent constant term (the last term in (8.28)). Then the term in the parentheses vanishes at the points $A_1 = \pm\pi/L$ (see Fig. 8.2).

I promised in
Section 8.1.1
to do this
check.

Two remarks are in order here. First, it is instructive to check that the Born–Oppenheimer approximation, which we have assumed from the very beginning, is indeed justified. In other words, let us verify that the dynamics of the variable A_1 is slow in the scale characteristic of the fermion sector. The effective Lagrangian determining the quantum mechanics of A_1 is

$$\mathcal{L} = \frac{L}{2e_0^2} \dot{A}_1^2 - \frac{L}{2\pi} A_1^2. \tag{8.29}$$

This describes a harmonic oscillator, with ground-state wave function

$$\Psi_0(A_1) = \left(\frac{L}{e_0 \pi^{3/2}} \right)^{1/4} \exp \left(-\frac{L A_1^2}{2e_0 \sqrt{\pi}} \right) \tag{8.30}$$

and level splitting

$$\omega_A = \frac{e_0}{\sqrt{\pi}}. \tag{8.31}$$

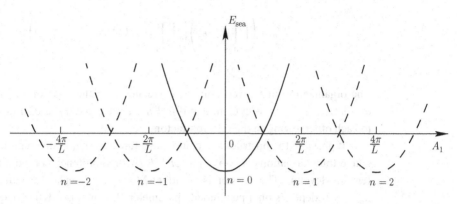

Fig. 8.2   Energy of the Dirac sea in the Schwinger model on a circle. The solid line corresponds to Eq. (8.12). The broken lines reflect the restructuring of the Dirac sea that is necessary if $|A_1| > \frac{\pi}{L}$.

The characteristic frequencies in the fermion sector are $\omega_{\text{ferm}} \sim L^{-1}$. Hence,

$$\frac{\omega_A}{\omega_{\text{ferm}}} \sim e_0 L \ll 1. \tag{8.32}$$

The second remark concerns the structure of the total vacuum wave function. We have convinced ourselves that

$$\Psi_{\text{vac}} = \Psi_{\text{ferm vac}} \Psi_0(A_1) \tag{8.33}$$

is an eigenstate of the Hamiltonian of the Schwinger model on the circle in the Born–Oppenheimer approximation. The wave function (8.33) is satisfactory from the point of view of "small" gauge transformations, i.e., those continuously deformable to the trivial (unit) transformation. (More exactly, Eq. (8.33) refers to the specific gauge in which the gauge degrees of freedom associated with A_1 are eliminated and A_1 is independent of x.) This wave function, however, is not invariant under "large" gauge transformations $A_1 \to A_1 + 2\pi k/L$, where $k = \pm 1, \pm 2, \ldots$

The essence of the situation becomes clear if we return to Fig. 8.1. When A_1 performs small and slow oscillations in the vicinity of zero, the Dirac sea is filled in the way shown in Eq. (8.12). But A_1 can oscillate in the vicinity of the gauge equivalent point $A_1 = 2\pi/L$ as well. In this case, if we do *not* restructure the fermion sector and leave it just as in Eq. (8.12) then the configuration of Eq. (8.12) is obviously *not* the vacuum – it corresponds to one particle plus one hole. This assertion is confirmed, in particular, by a plot showing the Dirac sea energy as a function of A_1 (Fig. 8.2). In order to construct the configuration of lowest energy in the vicinity of $A_1 = 2\pi/L$ it is necessary to fill the fermion levels as follows:

$$\prod_{k=1,2,3,\ldots} |1_L, k\rangle \prod_{k=0,-1,-2,\ldots} |1_R, k\rangle;$$

the empty levels are not shown explicitly, cf. Eq. (8.12).

Thus, the Hilbert space splits naturally into distinct sectors corresponding to different structures of the fermion sea. The wave function of the ground state in the nth sector has the form

The nth pre-vacuum

$$\Psi_n = \left(\prod_{k=n}^{\infty} |1_L, k\rangle\right)\left(\prod_{k=n-1}^{-\infty} |1_R, k\rangle\right)\Psi_0\left(A_1 - \frac{2\pi}{L}n\right), \tag{8.34}$$

$$n = 0, \pm 1, \pm 2, \ldots$$

The organization of the fermion sea correlates with the position of the "center of oscillation" of A_1. It is evident that if $n \neq n'$ then Ψ_n and $\Psi_{n'}$ are strictly orthogonal to each other, owing to the fermion factors.

Is it possible to construct a vacuum wave function that is invariant under "large" gauge transformations $A_1 \to A_1 + 2\pi k/L$ (with simultaneous renumbering of the fermion levels)? The answer is positive. Moreover, such a wave function is not unique. It depends on a new hidden parameter θ, which is often called the vacuum angle in the literature. Consider the linear combination

$$\Psi_{\theta\,\text{vac}} = \sum_n e^{i n \theta}\Psi_n. \tag{8.35}$$

This linear combination is also an eigenfunction of the Hamiltonian having the lowest energy, in just the same way as Ψ_n. But, unlike Ψ_n, these "large" gauge transformations leave $\Psi_{\theta\,\text{vac}}$ essentially intact. More exactly, under $A_1 \to A_1 + 2\pi/L$ the wave function (8.35) is multiplied by $e^{i\theta}$. This overall phase of the wave function is unobservable; all physical quantities resulting from averaging over the θ vacuum are invariant under gauge transformations.

Summarizing, we have now become acquainted with another model in which

Previously we discussed the θ vacuum in Chapter 5.

the notions of the vacuum angle θ and the θ vacuum are absolutely transparent: the Schwinger model on the spatial circle. The presence of the vacuum angle θ in the wave function is imitated in Lagrangian language by adding a so-called topological density to the Lagrangian. In the Schwinger model the topological density is

$$\Delta\mathcal{L}_\theta = \frac{\theta}{4\pi}\varepsilon^{\mu\nu}F_{\mu\nu}. \tag{8.36}$$

This extra term in the action is an integral over the full derivative; it does not affect the equations of motion and gives a vanishing contribution for any topologically trivial configuration $A_\mu(t, x)$. The topological density $\Delta\mathcal{L}_\theta$ shows up only if

$$\int_{-L/2}^{L/2} dx\, [A_1(t = +\infty, x) - A_1(t = -\infty, x)] = 2\pi k, \qquad |k| = 1, 2, \ldots \tag{8.37}$$

8.1.5 Topological Aspects

It is not by chance that here I am drawing the reader's attention to topological properties. It is very instructive to discuss topological aspects of the theoretical construction under consideration in more detail; this parallels a similar discussion in Chapter 5, where we exploited the path integral formulation of Yang–Mills theory using the Lagrangian formalism. At the same time, in the Schwinger model so far (Section 8.1.4) we have used Hamiltonian language in establishing the existence of the θ vacuum.

The Schwinger model possesses U(1) gauge invariance. An element of the U(1) group, as it is well known, can be written as $e^{i\alpha}$. Using gauge freedom one can reduce the fields $A_1(t, x)$ or $\psi(t, x)$ at a given moment of time to a standard form,

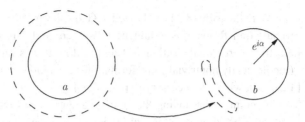

Fig. 8.3 Mapping of circle a in coordinate space into U(1). The broken-line contour near circle b shows a topologically trivial mapping.

by choosing an appropriate gauge function $\alpha(t, x)$. The standard form of A_1 is $A_1 = $ const, which can vary between, say, zero and $2\pi/L$. The gauge-equivalent points $A_1 = 0, \pm 2\pi/L, \pm 4\pi/L, \dots$ are connected by "large" (topologically nontrivial) gauge transformations.

Moreover, under our boundary conditions the variable x represents a circle of length L and, consequently, we are dealing here with the (continuous) mappings of the circle in configuration space into the gauge group U(1). The set of the mappings can be divided into classes. The mathematical formula expressing this fact is

The same topology as in the case of ANO strings

$$\pi_1(U(1)) = \mathbb{Z}. \tag{8.38}$$

The meaning of Eq. (8.38) is very simple. Within each class all mappings, by definition, can be reduced to each other by continuous deformations. However, there are no continuous deformations transforming mappings from one class into those in another class.

When the mappings of a circle onto U(1) are considered, the difference between the classes is especially transparent (see Fig. 8.3). Assume that we start from a certain point, go around circle a (following the path indicated by the broken line) once, and return to the starting point. In doing so, we have simultaneously gone around the b circle $0, \pm 1, \pm 2$, etc. times. (The negative sign corresponds to circulation in the opposite direction.) The number of windings around circle b labels a class of the mapping. It is clear that all mappings with a given winding number are continuously deformable into each other. Conversely, different winding numbers guarantee that a continuous deformation is impossible. The letter $\mathbb{Z}$ in Eq. (8.38) denotes the set of integers and shows that the set of different mapping classes is isomorphic to the set of integers; each class is characterized by an integer having the meaning of the winding number. The mappings corresponding to the winding number zero are called topologically trivial; the others are topologically nontrivial.

This information is sufficient to establish the existence of vacuum sectors labeled by n ($n = 0, \pm 1, \pm 2, \dots$), for which $(A_\mu)_{\text{vac}} \sim \partial_\mu \alpha_{(n)}$, without any explicit construction such as (8.34) ($\alpha_{(n)}$ belongs to the nth class). The necessity of introducing the vacuum angle θ also stems from the same information.

8.1.6 The Necessity of the θ Vacuum

The last issue to be discussed in connection with the Schwinger model is as follows. Sometimes the question is raised as to why the vacuum wave function cannot be chosen in the form (8.34) with fixed n. Gauge invariance under "small" (topologically trivial) transformations would be preserved, and this would automatically imply electric charge conservation. What would be lost is only invariance under "large" (topologically nontrivial) transformations; it would seem that there is nothing bad in that.[3] So, why is it necessary to pass to $\Psi_{\theta\,\text{vac}} = \sum_n e^{in\theta}\Psi_n$?

> *Cluster decomposition and stability with regards to, e.g., mass deformations*

The point is that taking Ψ_n as the vacuum wave function would violate clusterization, a basic property in field theory, which can be traced back to the causality and unitarity of the theory. The following is meant by clusterization: the vacuum expectation value of the product of several local operators at causally independent points must be reducible to the product of the vacuum expectation values for each operator; for example,

$$\langle O_1 O_2 \rangle = \langle O_1 \rangle \langle O_2 \rangle. \tag{8.39}$$

The violation of this clusterization can be demonstrated explicitly. Consider the two-point function

$$\mathcal{A}(t) = \langle \Psi_n | T\{O^\dagger(t), O(0)\} | \Psi_n \rangle,$$
$$O(t) = \int \bar\psi(t,x)(1+\gamma^5)\psi(t,x)\,dx. \tag{8.40}$$

The operator O changes the axial charge of the state by two units (it adds a particle and a hole to the Dirac sea) and $O^\dagger$ returns it back, and, as a result, $\mathcal{A}(t) \neq 0$. Moreover, if $t \to \infty$ in the Euclidean domain then $\mathcal{A}(t) \to$ const. (For a concrete calculation based on the bosonization method, see, e.g., [2]. In [2] the limit $L \to \infty$ is considered but all relevant expressions can be readily rewritten for finite L.) The fact that $\mathcal{A}(t)$ tends to a nonvanishing constant at $t \to \infty$ means, according to clusterization, that the operators $\bar\psi(1 \pm \gamma^5)\psi$ acquire a nonvanishing vacuum expectation value.

However, if $|\text{vac}\rangle = |\Psi_n\rangle$ then $\langle \bar\psi(1 \pm \gamma^5)\psi \rangle = 0$, for a trivial reason: the operator $\bar\psi(1 \pm \gamma^5)\psi$ acting on Ψ_n produces an electron and a hole, and the corresponding state is obviously orthogonal to Ψ_n itself.

The clusterization property restores itself if one passes to the θ vacuum (8.35). In this case there emerges a nondiagonal expectation value,

$$\langle \Psi_{n+1} | \bar\psi(1 \pm \gamma^5)\psi | \Psi_n \rangle \sim L^{-1} \exp\left(-\frac{\pi^{3/2}}{e_0 L}\right). \tag{8.41}$$

If the line of reasoning based on clusterization seems too academic to the reader, it might be instructive to consider another argument, connected with Eqs. (8.40) and the subsequent discussion. Let us ask the question: what will happen if instead of the massless Schwinger model we consider a model with a small mass, i.e., we introduce an extra mass term $\Delta\mathcal{L}_m = -m\bar\psi\psi$ into the Lagrangian (8.1)? Naturally, all physical quantities obtained in the massless model will be shifted. It is equally natural to require, however, the shifts to be small for small m, so that there is no change in the limit $m \to 0$. Otherwise, we would encounter an unstable situation while in fact we would like to have the mass term as a small perturbation. For more details see Section 9.6.3.

---

[3] The contents of this subsection should be compared with Section 5.1.2. For a discussion of the subtle and contrived modifications which are possible but will not concern us here, see [5, 6].

In the presence of the degenerate states (and the states Ψ_n with different n are degenerate), however, any perturbation is potentially dangerous and can lead to large effects. Just such a disaster occurs, in particular, if $\Delta\mathcal{L}_m$, acting on the vacuum, is nondiagonal.

If we prescribe states like Ψ_n to be the vacuum then $\Delta\mathcal{L}_m$ will by no means be diagonal, as follows from the discussion after Eqs. (8.40). This we cannot accept. However, the mass term is certainly diagonalized in a basis consisting of the wave functions (8.35):

$$\langle\Psi_{\theta'\,\text{vac}}|\Delta\mathcal{L}_m|\Psi_{\theta\,\text{vac}}\rangle = 0 \qquad \text{if } \theta' \neq \theta. \tag{8.42}$$

8.1.7 Two Faces of the Chiral Anomaly*

In concluding this section, it will be extremely useful to discuss the connection between the picture presented above and the more standard derivation of the chiral anomaly in the Schwinger model. This discussion will represent a bridge between the physical picture described above and the standard approach to anomalies.

We have already emphasized the double nature of the anomaly, which shows up as an infrared effect in the current and an ultraviolet effect in the divergence of the current. The line of reasoning used thus far has put more emphasis on the infrared aspect of the problem – the finite "box" served as a natural infrared regularization. The same result for $\partial_\mu j^{\mu 5}$ as in Eqs. (8.16) could be obtained with no reference to infrared regularization, however.

The conventional treatment of the issue is based on the standard Feynman diagram technique. The usual explanation, to be found in numerous textbooks, connects the anomalies to the ultraviolet divergence of certain Feynman graphs. The assertion of ultraviolet divergence is valid if one is dealing directly with $\partial_\mu j^{\mu 5}$. Thus, the emphasis is shifted to the ultraviolet aspect of the anomaly.

Below, I first sketch the standard derivation. Then I show that, as a rule, the diagrammatic language used, for the analysis of $\partial_\mu j^{\mu 5}$ from the point of view of ultraviolet regularization can be successfully used for an "infrared" derivation of the anomaly. The fact that the anomalies reveal themselves in the infrared behavior of Feynman graphs is rarely mentioned in the literature, and, hence, deserves a more detailed discussion. The pragmatically oriented reader *can omit this subsection at first reading*.

Thus, we would like to demonstrate that

$$\partial_\mu j^{\mu 5} = -\frac{1}{2\pi}\varepsilon^{\mu\nu}F_{\mu\nu}, \tag{8.43}$$

by considering directly $\partial_\mu j^{\mu 5}$, not $j^{\mu 5}$ as previously. Then we need only ultraviolet regularization; in particular, the theory can be considered in an infinite space since the finiteness of L does not affect the result at short distances.

A convenient method of ultraviolet regularization is due to Pauli and Villars. In the model at hand it reduces to the following. In addition to the original massless fermions in the Lagrangian, heavy regulator fermions are introduced with mass $M_0(M_0 \to \infty)$ and the opposite metric. The latter means that each loop of the regulator fermions is supplied with an extra minus sign relative to the normal fermion loop. The interaction of the regulator fermions with the photons is assumed to be just the same as for the original fermions, the only difference being the mass. Then the role of the Pauli–Villars fermions in low-energy processes ($E \ll M_0$) is to provide an ultraviolet cutoff in the formally divergent integrals with fermion loops. Clearly, such a regularization procedure automatically guarantees gauge invariance and electromagnetic current conservation.

In a model regularized according to Pauli and Villars the axial current has the form

$$j^{\mu 5} = \bar{\psi}\gamma^\mu\gamma^5\psi + \bar{R}\gamma^\mu\gamma^5 R, \tag{8.44}$$

where R is the fermion regulator. In calculating the divergence of the regularized current the naive equations of motion can be used. Then

$$\partial_\mu j^{\mu 5} = 2iM_0\bar{R}\gamma^5 R.$$

The divergence does not vanish (the axial current is not conserved!), but, as expected, $\partial_\mu j^{\mu 5}$ contains only the regulator's anomalous term.

The last step is contraction of the regulator fields in the loop in order to convert $M_0\bar{R}\gamma^5 R$ into the "normal" light fields in the limit $M_0 \to \infty$. The relevant diagrams are displayed in Figs. 8.4a,b,c, where the solid lines denote the standard heavy fermion propagator $i(\not{p} - M_0)^{-1}$. Graph (a) does not depend on the external field. The corresponding contribution to $\partial_\mu j^{\mu 5}$ represents a number that can be set equal to zero. Graph (c), with two photon legs, and all others having more legs die off in the limit $M_0 \to \infty$. The only surviving graph is (b). Calculation of this diagram is trivial:

$$2iM_0\bar{R}\gamma^5 R \to -\frac{1}{2\pi}\varepsilon^{\mu\nu}F_{\mu\nu}. \tag{8.45}$$

(Do not forget that there is an extra minus sign in Pauli–Villars fermion loops.) We have reproduced the anomalous relation (8.43) obtained previously by a different method.

The easiest method allowing one to check Eq. (8.45) in another way is, probably, the so-called background field technique. I will not enlarge on its details here because these would lead us far astray. The interested reader is referred to the review [7], where all relevant nuances are fully discussed. We will limit ourselves to the

| *Background* |
| *field formula* |

intuitively obvious features and use self-evident notation. Thus

$$2iM_0\bar{R}\gamma^5 R = -2M_0 \operatorname{Tr}\left[\gamma^5(\not{P} - M_0)^{-1}\right], \tag{8.46}$$

where $P_\mu = i\mathcal{D}_\mu = i\partial_\mu + A_\mu$ is the generalized momentum operator, and we have taken into account the fact that the minus sign in the fermion loop does not appear for the regulator fields.

Moreover,

$$(\not{P} - M_0)^{-1} = (\not{P} + M_0)\left(P^2 + \tfrac{1}{2}i\varepsilon^{\mu\nu}F_{\mu\nu}\gamma^5 - M_0^2\right)^{-1}. \tag{8.47}$$

Now, since $M_0 \to \infty$ the contents of the trace in Eq. (8.46) can be expanded in inverse powers of M_0:

$$\operatorname{Tr}\left[\gamma^5(\not{P} - M_0)^{-1}\right]$$

$$= \operatorname{Tr}\left[\gamma^5(\not{P} + M_0)\left(\frac{1}{P^2 - M_0^2} - \frac{1}{P^2 - M_0^2}\tfrac{1}{2}i\varepsilon^{\mu\nu}F_{\mu\nu}\gamma^5\frac{1}{P^2 - M_0^2} + \cdots\right)\right]. \tag{8.48}$$

The first term in the expansion vanishes after the trace of the γ matrices has been taken. The third and all other terms denoted by the ellipses are irrelevant because they vanish in the limit $M_0 \to \infty$. The only relevant term is the second, in which we can substitute the operator P_μ by the momentum p_μ since the result is explicitly proportional to the background field $F_{\mu\nu}$, and the chiral anomaly in the Schwinger model is linear in $F_{\mu\nu}$. Then

$$2iM_0\bar{R}\gamma_5 R = -2M_0^2 \int \frac{d^2p}{(2\pi)^2} \frac{i}{(p^2 - M_0^2)^2} \varepsilon^{\mu\nu} F_{\mu\nu}.$$

Upon performing Wick rotation and integrating over p we arrive at Eq. (8.45).

This computation completes the standard derivation of the anomaly. One needs a rather rich imagination to be able to see in these formal manipulations the simple physical nature of the phenomenon described above (the restructuring of the fermion sea and the level crossing). Nevertheless, it is the same phenomenon viewed from a different angle – less transparent but more economic since we can get the final result very quickly using the well-developed machinery of the diagram technique, familiar to everybody.

Let us ask the question: what is the infrared connection (or infrared face, if you wish) of the anomaly in diagram language? To extract the infrared aspect from the Feynman graphs it is necessary to turn back to a consideration of the current $j^{\mu 5}$. Our aim is to calculate the matrix element of the current $j^{\mu 5}$ in the background photon field. Unlike $\partial_\mu j^{\mu 5}$ the matrix element $\langle j^{\mu 5} \rangle$ contains an infrared contribution. Because of this, it is impossible to consider $\langle j^{\mu 5} \rangle$ for an on-mass-shell photon, with momentum $k^2 = 0$. We are forced to introduce "off-shellness" to ensure infrared regularization (a substitute for finite L, see above). Thus, we will consider the photon field A_μ, which does not obey the equations of motion.

General arguments (such as gauge invariance) imply the following expression for the matrix element $\langle j^{\mu 5} \rangle$ stemming from diagram (d) of Fig. 8.4:

$$\langle j^{\mu 5} \rangle = \text{const} \times \frac{k^\mu}{k^2} \varepsilon^{\alpha\beta} F_{\alpha\beta}, \tag{8.49}$$

where the constant on the right-hand side can be determined by explicit computation of the graph. In principle, there is one more structure with the appropriate dimension

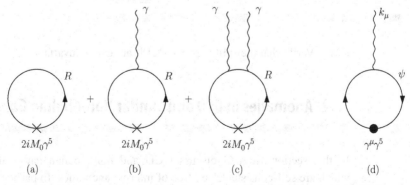

Fig. 8.4   Diagrammatic representation of the anomaly in the axial current in the Schwinger model. (a), (b), (c): Heavy regulator fields in the divergence of the current. (d): Infrared anomalous contribution in $\bar{\psi}\gamma^\mu\gamma^5\psi$.

and quantum numbers, namely $\varepsilon^{\mu\nu}A_\nu$, but it cannot appear by itself if gauge invariance is to be maintained. In other words, one can say that the *local* structur $\varepsilon^{\mu\nu}A_\nu$ can always be eliminated by subtraction of an ultraviolet counterterm.

It is worth noting that, purely kinematically,

$$k^\mu \varepsilon^{\alpha\beta} F_{\alpha\beta} = -2i\varepsilon^{\mu\nu}[k^2 A_\nu - k_\nu(k^\rho A_\rho)]. \tag{8.50}$$

It can be seen that, in order to distinguish an infrared singular term proportional to k^{-2} from the local term depending on ultraviolet regularization, it is necessary to assume that $k^\rho A_\rho \neq 0$. The infrared singular term is fixed unambiguously by diagram (d) of Fig. 8.4. The easiest way to obtain it is to compute this graph in a straightforward way:

$$\langle j^{\mu 5}\rangle = (-1)\int \frac{d^2p}{(2\pi)^2} \mathrm{Tr}\left[\gamma^\mu \gamma^5 \frac{i\not{p}}{p^2} i\gamma^\rho \frac{i(\not{p}+\not{k})}{(p+k)^2}\right] A_\rho. \tag{8.51}$$

Performing the p integration and disregarding terms that are nonsingular in k^2, we get

$$\int \frac{p^\alpha}{p^2} \frac{(p+k)^\beta}{(p+k)^2} \frac{d^2p}{(2\pi)^2} \to \frac{i}{4\pi} \frac{k^\alpha k^\beta}{k^2},$$

which implies, in turn, that

$$\langle j^{\mu 5}\rangle_{\text{singular}} = -\frac{1}{4\pi k^2}\mathrm{Tr}(\gamma^\mu \gamma^5 \not{k} \gamma^\rho \not{k})A_\rho \to \frac{1}{\pi k^2}\varepsilon^{\mu\nu}k_\nu(k^\rho A_\rho).$$

Anomaly from the IR side

Now, inserting the local term in order to restore gauge invariance and using Eq. (8.50) we arrive at

$$\langle j^{\mu 5}\rangle = -\frac{i}{2\pi}\frac{k^\mu}{k^2}\varepsilon^{\alpha\beta}F_{\alpha\beta}. \tag{8.52}$$

Taking the divergence is equivalent to multiplying the right-hand side by $-ik_\mu$, and so we have reproduced, now for the third time, the anomalous relations (8.43).

Let us draw the reader's attention to the pole k^{-2} in Eq. (8.52). The emergence of this pole is the manifestation of the infrared nature of the anomaly. We see that it can be derived from this side with the familiar Feynman technique.

Exercise

8.1.1   Verify that the split currents (8.18) are gauge invariant.

8.2 Anomalies in QCD and Similar Non-Abelian Gauge Theories

In this section we will discuss QCD and non-Abelian gauge theories at large which are self-consistent, i.e., free of internal anomalies. In particular, dealing with chiral theories we should follow strict rules in constructing the matter sector (see Section 5.5.1.1). Nevertheless, these theories have external anomalies: the scale anomaly and those in the divergence of external axial currents.[4] The latter are also referred to as chiral (or triangle, or Adler–Bell–Jackiw [8]) anomalies. We will analyze and derive the chiral and scale anomalies using QCD as a showcase. More exactly, we will assume that the theory under consideration has the gauge group SU(N) and contains N_f massless quarks (Dirac fields in the fundamental representation). In this section it will be convenient to write the action in the canonical normalization,

$$S = \int d^4x \left(-\tfrac{1}{4} G^a_{\mu\nu} G^{\mu\nu a} + \sum_{f=1}^{N_f} \bar{\psi}_f i\mathcal{D}\psi^f \right). \tag{8.53}$$

We will start by examining the classical symmetries of the above action.

In addition to the scale invariance (implying, in fact, full conformal invariance) of the action, which we will discuss later, (8.53) has the following symmetry:

Global symmetries of QCD

$$U(1)_V \times U(1)_A \times SU(N_f)_L \times SU(N_f)_R \tag{8.54}$$

acting in the matter sector. The vector U(1) corresponds to the baryon number conservation, with current

$$j^B_\mu = \tfrac{1}{3} \bar{\psi}_f \gamma_\mu \psi^f. \tag{8.55}$$

The axial U(1) symmetry corresponds to the overall chiral phase rotation

$$\psi^f_L \to e^{i\alpha}\psi^f_L, \qquad \psi^f_R \to e^{-i\alpha}\psi^f_R, \qquad \psi_{L,R} = \tfrac{1}{2}(1 \mp \gamma^5)\psi. \tag{8.56}$$

The axial current generated by (8.56) is

$$j^\mu_A = \bar{\psi}_f \gamma^\mu \gamma^5 \psi^f. \tag{8.57}$$

Singlet and nonsinglet axial currents

Finally, the last two factors in (8.54) reflect the invariance of the action with regard to the chiral flavor rotations

$$\psi^f_L \to U^f_g \psi^g_L, \qquad \psi^f_R \to \tilde{U}^f_g \psi^g_R, \tag{8.58}$$

where U and $\tilde{U}$ are arbitrary (independent) matrices from SU(N_f). Equation (8.58) implies conservation of the following vector and axial currents:

$$j^a_\mu = \bar{\psi} \gamma_\mu T^a \psi, \qquad j^{5a}_\mu = \bar{\psi} \gamma^\mu \gamma^5 T^a \psi. \tag{8.59}$$

Here the T^a are the generators of the *flavor* SU(N_f) group in the fundamental representation. These generators act in the flavor space, i.e., ψ is a column of the ψ^f while the matrices T^a act on this column.

At the quantum level (i.e., including loops with a regularization) the fate of the above symmetries is different. The vector U(1) invariance generated by (8.55) remains a valid anomaly-free symmetry at the quantum level.[5] The same is true with regard to the vector SU(N_f) currents: they are conserved. The axial currents are anomalous. One should distinguish, though, between the singlet current (8.57)

---

[4] By external I mean currents that are not coupled to the gauge fields of the theory under consideration. We start from global symmetries as in (8.54) and "pretend" to gauge them by introducing non-dynamical gauge fields.

[5] I hasten to make a reservation. This statement is valid in vector-like theories. As we already know from Section 5.6, this is not true in chiral models such as the standard model, but for the time being we are discussing QCD.

and the SU(N_f) currents $j_\mu^{5a} = \bar\psi_f \gamma^\mu \gamma^5 T^a \psi^f$. The former is anomalous in QCD *per se*. The latter become anomalous only upon the introduction of appropriate external vector currents. As we will see later, this circumstance is in one-to-one correspondence with the spontaneous breaking of the axial SU(N_f) symmetry in QCD, which is accompanied by the emergence of $N_f^2 - 1$ Goldstone bosons. The vector SU(N_f) symmetry is realized linearly.

In the weakly coupled Schwinger model considered in Section 8.1.1 we could take both the infrared and ultraviolet routes (and we actually did so) to derive the chiral anomaly. The first route is closed in QCD, since this theory is strongly coupled in the infrared domain and this invalidates any conclusions based on Feynman graph calculations. Neither quarks nor gluons are relevant in the infrared. However, the second route is open and we will take it in the following subsections. We will limit ourselves to a one-loop analysis. Higher loops, where present, generally speaking, lie outside the scope of this book. The only exception is a class of supersymmetric gauge theories, to be considered in Part II (Section 10.16).

8.2.1 Chiral Anomaly in the Singlet Axial Current

Differentiating (8.57) naively, we get $\partial_\mu j_A^\mu = \bar\psi_f \overleftarrow{\mathcal{D}} \gamma^5 \psi^f - \bar\psi_f \gamma^5 \mathcal{D} \psi^f = 0$ by virtue of the equation of motion $\mathcal{D}\psi^f = 0$. Experience gained from the Schwinger model teaches us, however, that the axial current conservation will not hold when we switch to a gauge-symmetry-respecting regularization. To make the calculation of the anomaly reliable we must exploit only Green's functions at short distances. This means that we must focus directly on $\partial_\mu j_A^\mu$ and use an appropriate ultraviolet regularization. The following demonstration will be based on the Schwinger and Pauli–Villars regularizations.[6]

8.2.1.1 The Schwinger Regularization

In this regularization we ε-split the current,

$$j_\mu^{A,R}(x) = \bar\psi_f(x+\varepsilon)\gamma_\mu\gamma^5 \left\{ \exp\left[\int_{x-\varepsilon}^{x+\varepsilon} ig\, A_\rho(y)\, dy^\rho\right]\right\} \psi^f(x-\varepsilon). \qquad (8.60)$$

Here the superscript R indicates that the current has been regularized, while $A_\rho \equiv A_\rho^a T^a$. The parameter ε must be set to zero at the very end. The exponent is necessary to ensure the gauge invariance of the regularized current $j_\mu^{A,R}$ after the split

Look back
through
Section 8.1.3.

$$\bar\psi_f(x+\varepsilon)\psi^f(x-\varepsilon). \qquad (8.61)$$

Next, we differentiate with respect to x using the equations of motion above. Expanding in ε and keeping $O(\varepsilon)$ terms we arrive at

$$\partial^\mu j_\mu^{A,R} = \bar\psi_f(x+\varepsilon)\Big[-ig\slashed{A}(x+\varepsilon)\,\gamma^5 - \gamma^5\, ig\slashed{A}(x-\varepsilon)$$
$$+ ig\,\gamma^\mu\gamma^5\varepsilon^\beta\, G_{\mu\beta}(x)\Big]\psi^f(x-\varepsilon). \qquad (8.62)$$

---

[6] The widely used dimensional regularization is awkward and inappropriate in problems in which γ^5 is involved.

The third term in the square brackets in (8.62) contains the gluon field strength tensor and results from differentiation of the exponential factor. The gluon 4-potential A_μ and the field strength tensor $G_{\mu\beta}$ are treated as background fields. For convenience we impose the Fock–Schwinger gauge condition on the background field, settting $y^\mu A_\mu(y) = 0$ (for a pedagogical course on this gauge and its uses see [7]).[7] In this gauge $A_\mu(y) = \frac{1}{2} y^\rho G_{\rho\mu}(0) + \cdots$ Now, we contract the quark lines (8.61) to form the quark Green's function $S(x - \varepsilon, x + \varepsilon)$ in the background field,[8]

$$\partial^\mu j_\mu^{A,\,R} = -ig N_f \, \mathrm{Tr}_{C,L} \left[-2i \varepsilon^\rho G_{\rho\mu}(x) \gamma^\mu \gamma^5 S(x - \varepsilon, x + \varepsilon) \right]$$

$$= -N_f \frac{g^2}{2} G_{\rho\mu}(0)^a \tilde{G}_{\alpha\phi}(x)^a \frac{\varepsilon^\rho \varepsilon^\alpha}{\varepsilon^2} \frac{1}{8\pi^2} \, \mathrm{Tr}_L \left(\gamma^\mu \gamma^5 \gamma^\phi \gamma^5 \right)$$

$$= \frac{N_f g^2}{16\pi^2} \left(G^{\alpha\beta a} \tilde{G}_{\alpha\beta}^a \right)_{\text{background}}, \tag{8.63}$$

where

$$\tilde{G}_{\alpha\beta} = \tfrac{1}{2} \varepsilon_{\alpha\beta\rho\mu} G^{\rho\mu} \tag{8.64}$$

and the subscripts C and L indicate traces over the color and Lorentz indices, respectively. The most crucial point is that the Green's function $S(x - \varepsilon, x + \varepsilon)$ is used *only* at very short distances $\varepsilon \to 0$, where it is reliably known in the form of an expansion in the background field. We need only the first nontrivial term in this expansion (the Fock–Schwinger gauge),

$$S(x, y) = \frac{1}{2\pi^2} \frac{r_\beta \gamma^\beta}{(r^2)^2} - \frac{1}{8\pi^2} \frac{r^\alpha}{r^2} g \tilde{G}_{\alpha\phi}(x) \gamma^\phi \gamma^5 + \cdots, \qquad r = x - y. \tag{8.65}$$

In passing from the second to the third line in Eq. (8.63) we have averaged over the angular orientations of the 4-vector ε.

8.2.1.2 Pauli–Villars Regularization

Paralleling our two-dimensional studies in Section 8.1.7, we will introduce the Pauli–Villars fermion regulators R with mass M_R, to be sent to infinity at the very end. Then the regularized singlet axial current takes the form

$$j_\mu^{A,\,R} = \bar{\psi}_f \gamma_\mu \gamma^5 \psi^f + \bar{R}_f \gamma_\mu \gamma^5 R^f. \tag{8.66}$$

Since the current is now regularized, its divergence can be calculated according to the equations of motion:

$$\partial^\mu j_\mu^{A,\,R} = 2i M_R \, \bar{R}_f \gamma^5 R^f. \tag{8.67}$$

As expected, the result contains only the regulator term. Our next task is to project it onto "our" sector of the theory in the limit $M_R \to \infty$. In this limit only the two-gluon operator will survive, as depicted in the triangle diagram of Fig. 8.5. This

---

[7] This gauge condition is not obligatory, of course. Although it is convenient, one can work in any other gauge; the final result is gauge independent.

[8] A step-by-step derivation of (8.63) can be found on p. 609 in [7].

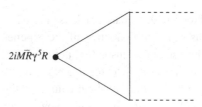

Fig. 8.5 Diagrammatic representation of the triangle anomaly. The solid and broken lines denote the regulator and gluon fields, respectively.

diagram can be calculated either by the standard Feynman graph technique or using the background field method [7], which is quite straightforward in the case at hand,

$$2iM_R \bar{R}_f \gamma^5 R^f \to 2iM_R N_f \operatorname{Tr}_{C,L} \left(\gamma^5 \frac{i}{i\, \mathcal{D} - M_R} \right)$$

$$\to -2M_R N_f \operatorname{Tr}_{C,L} \left[\gamma^5 \frac{1}{(i\mathcal{D})^2 - M_R^2 + \frac{1}{2} ig G_{\mu\nu}\sigma^{\mu\nu}} \, (i\, \mathcal{D} + M_R) \right].$$

$$\tag{8.68}$$

Here I have omitted the extra minus sign that would have been necessary if it were an ordinary fermion loop, but, given that the triangle loop in Fig. 8.5 applies to regulator fields, the extra minus sign must not be inserted. The term $i\, \mathcal{D}$ in the final parentheses can be dropped because of the factor γ^5 in the trace. Remembering that $M_R \to \infty$, one can expand the denominator in $G\sigma$. The zeroth-order term in this expansion vanishes for the same reason. The term $O(G\sigma)$ vanishes after taking the color trace. The term $O((G\sigma)^2)$ does not vanish, but all higher-order terms are suppressed by positive powers of $1/M_R$ and disappear in the limit $M_R \to \infty$. In this way

$$2iM_R \bar{R}_f \gamma^5 R^f \to \frac{M_R^2 g^2}{2} N_f \operatorname{Tr}_C(G_{\mu\nu} G_{\alpha\beta}) \operatorname{Tr}_L \left(\gamma^5 \sigma^{\mu\nu} \sigma^{\alpha\beta} \right)$$

$$\times \int \frac{d^4 p}{(2\pi)^4} \frac{1}{(p^2 - M_R^2)^3}, \tag{8.69}$$

which, in turn, implies that

$$\partial^\mu j_\mu^A = N_f \frac{g^2}{16\pi^2} G^{\alpha\beta a} \tilde{G}_{\alpha\beta}^a, \tag{8.70}$$

Chiral anomaly

in full accord with the result (8.63) obtained in the Schwinger regularization. The characteristic distances saturating the triangle loop in Fig. 8.5 are of order $M_R^{-1} \to 0$ at $M_R \to \infty$.

8.2.1.3 The Chiral Anomaly for Generic Fermions

What changes occur in the chiral anomaly if instead of the fundamental representation we consider fermions in some other representation R? The answer to this question is simple. If we inspect the derivations in Sections 8.2.1.1 and 8.2.1.2 we will observe that the result for the anomalous divergence of the U(1) axial current is proportional to the color trace $\operatorname{Tr} T^a T^b$. For the fundamental representation in SU(N),

$$\operatorname{Tr} T^a T^b = \tfrac{1}{2}\delta^{ab}.$$

In the general case,

$$\mathrm{Tr}\, T^a T^b = T(R)\delta^{ab},$$

See
Eq. (10.346)
and
Table 10.3.

where $T(R)$ is one-half the Dynkin index for the given representation. Thus, if we have N_f massless Dirac fermions in the representation R then Eq. (8.70) must be replaced by the following formula:

$$\partial^\mu(\bar{\psi}_f \gamma_\mu \gamma^5 \psi_f) = N_f \frac{T(R)g^2}{8\pi^2} G^{\alpha\beta a}\tilde{G}^a_{\alpha\beta}. \tag{8.71}$$

For instance, for the adjoint representation in SU(N) one has $T(\mathrm{adj}) = N$. Note that, for real representations such as the adjoint, one can consider not only Dirac fermions but Majorana fermions as well. For each Majorana fermion we have $N_f = \frac{1}{2}$. The same is true with regard to the Weyl fermions with which one deals in chiral Yang–Mills theories.

8.2.2 Introducing External Currents

What does this mean? Assume that we are studying QCD. Then our dynamical gauge bosons are gluons. However, typically we have a number of color-singlet conserved *vector* currents that can be "gauged" too. These vector currents correspond to *global* symmetries. One can couple these currents to "external" *nondynamical* gauge bosons, thinking of them as gauge bosons of a weakly coupled theory whose dynamics can be ignored. Axial currents that are initially anomaly-free can (and typically will) acquire anomalies with regard to these external nondynamical gauge bosons.

For example, the currents j_μ^a given in Eq. (8.59) are conserved. Gauging the global SU(N_f)$_V$ symmetry, we introduce auxiliary vector bosons $A^{\mu a}$ with coupling $j_\mu^a A^{\mu a}$. Now, the divergence of $j_\mu^{5,a}$, which was anomaly-free in QCD *per se*, will acquire an $F\tilde{F}$ term, with Fs built from the above auxiliary vector bosons $A^{\mu a}$.

To illustrate further this point in a graphic way, let us assume $N_f = 2$. Then ψ is a two-component column in flavor space, while the three generator matrices are in fact the Pauli matrices (up to a normalizing factor $\frac{1}{2}$). The background gauge fields are $A^{\mu 1,2,3}$ or, alternatively, $A^{\mu 3}$ and $A^{\mu\pm}$. The current j_μ^B in (8.55) is conserved too. Therefore, we can also introduce an external field A_μ with coupling $A_\mu \bar{\psi}_f \gamma^\mu \psi_f$. Another possible alternative is to gauge the electromagnetic interaction in addition to $A^{\mu a}$. Then we will have a photon (which is an external gauge boson with regard to QCD) interacting with the current $\frac{2}{3}\bar{u}\gamma_\mu u - \frac{1}{3}\bar{d}\gamma_\mu d$. The latter current is a linear combination of the isotriplet and isosinglet,

$$j_\mu^{\mathrm{em}} = \frac{2}{3}\bar{u}\gamma_\mu u - \frac{1}{3}\bar{d}\gamma_\mu d = \frac{1}{2}(\bar{u}\gamma_\mu u - \bar{d}\gamma_\mu d) + \frac{1}{6}(\bar{u}\gamma_\mu u + \bar{d}\gamma_\mu d). \tag{8.72}$$

To distinguish the photon field from other external gauge bosons, temporarily (in this subsection) we will denote it by $\mathcal{A}^\mu$. Then the interaction takes the form $e\mathcal{A}^\mu j_\mu^{\mathrm{em}}$.

It is instructive to study this simple example further and to derive the anomaly in the $j_\mu^{5,a}$ currents. Keeping in mind a particularly important application, to be discussed shortly, we will limit ourselves to the neutral component, which we denote by a^μ:

$$a_\mu \equiv j_\mu^{5(a=3)} = \tfrac{1}{2} \left(\bar{u}\gamma_\mu\gamma^5 u - \bar{d}\gamma_\mu\gamma^5 d \right). \tag{8.73}$$

Third component (in the isospace) of the flavor axial current defined in (8.59)

We will have to analyze the same graph as previously (Fig. 8.5), with regulator fields for the u and d quarks. They carry exactly the same quantum numbers as those of the u and d quarks. The only difference is that the regulator loop, as usual, has the opposite sign.[9] It is obvious that the current a^μ is anomaly-free in QCD *per se* since the triangle loops for the u and d quark regulators exactly cancel each other. Including the external photons with the interaction $e\mathcal{A}^\mu j_\mu^{\rm em}$, which obviously distinguishes between u and d, ruins the cancelation.

In fact, we do not have to repeat the full computation. All we have to do is to reevaluate the diagram in Fig 8.5 with the external gluons replaced by photons. Starting from Eq. (8.70), derived in Section 8.2.1.2, we must take into account the difference in the vertex factors in this triangle graph. First, we will deal with the color factors. While, in (8.70), for the gluon background field we used $\mathrm{Tr}_C(T^a T^b) = \tfrac{1}{2}\delta^{ab}$, in the case of the photon background field we replace this by $\mathrm{Tr}_C 1 = N = 3$. Next, in the u loop we make the replacement $g \to Q_u e$ and, in the d loop, $g \to Q_d e$. (Here $Q_u = \tfrac{2}{3}$ and $Q_d = -\tfrac{1}{3}$.) As a result,

$$N_f g^2 \to \frac{1}{2}(Q_u^2 - Q_d^2)e^2, \tag{8.74}$$

where the factor $\tfrac{1}{2}$ in (8.74) is due to the factor $\tfrac{1}{2}$ in the definition (8.73). Assembling all the factors, we arrive at

$$\partial_\mu a^\mu = \frac{\alpha}{4\pi} N(Q_u^2 - Q_d^2)\mathcal{F}_{\mu\nu}\tilde{\mathcal{F}}^{\mu\nu}, \tag{8.75}$$

where $\mathcal{F}_{\mu\nu} = \partial_\mu\mathcal{A}_\nu - \partial_\nu\mathcal{A}_\mu$. Generalization to other external currents is straightforward.

Studying anomalies in the presence of external currents provides us with a powerful tool for uncovering subtle aspects of strong dynamics at large distances, as we will see shortly.

8.2.3 Longitudinal Part of the Current

Under certain circumstances one can reconstruct from (8.75) the longitudinal part of the current [9, 10]. Let us separate the longitudinal and transverse parts of a^μ:

$$a^\mu \equiv a_\parallel^\mu + a_\perp^\mu, \qquad \partial_\mu a_\perp^\mu = 0. \tag{8.76}$$

It is clear that (8.75), viewed as an equation for the current, says nothing about $a_\perp^\mu$. However, it imposes a constraint on $a_\parallel^\mu$, which allows one to determine $a_\parallel^\mu$ unambiguously under appropriate kinematical conditions. Namely, assume that the photons in (8.75) are produced with momenta $k^{(1)}$ and $k^{(2)}$ and are on the mass shell, i.e.,

$$(k^{(1)})^2 = 0, \qquad (k^{(2)})^2 = 0. \tag{8.77}$$

The total momentum transferred from the current a^μ to the pair of photons is $q_\mu = k_\mu^{(1)} + k_\mu^{(2)}$ (Fig. 8.6). Then

---

[9] This is in addition to the requirement of taking the regulator masses in the limit $M_R = \infty$ at the very end.

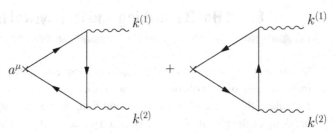

Fig. 8.6   Anomaly in a^μ.

$$\mathcal{F}_{\mu\nu}\tilde{\mathcal{F}}^{\mu\nu} \longrightarrow -2 \times 2 \times \varepsilon^{\mu\nu\alpha\beta} k_\mu^{(1)} \epsilon_\nu^{(1)} k_\alpha^{(2)} \epsilon_\beta^{(2)}. \tag{8.78}$$

Here $\epsilon_\mu^{(1,2)}$ is the polarization vector of the first or second photon. The first factor 2 in (8.78) comes from combinatorics: one can produce the first photon either from the first $\mathcal{F}_{\mu\nu}$ tensor or the second. Gauge invariance with regard to the external photons is built into our regularization.

The statement resulting from (8.75) and (8.77) is as follows [9, 10]: for on-mass-shell photons the two-photon matrix element of $a_{||}^\mu$ is determined unambiguously:

$$\langle 0|a_{||}^\mu|2\gamma\rangle = i\frac{q^\mu}{q^2} \frac{\alpha}{\pi} N(Q_u^2 - Q_d^2)\varepsilon^{\mu\nu\alpha\beta} k_\mu^{(1)} \epsilon_\nu^{(1)} k_\alpha^{(2)} \epsilon_\beta^{(2)}. \tag{8.79}$$

This result is exact and is valid for any value of q^2, in particular, at $q^2 \to 0$. The emergence of the pole $1/q^2$, with far-reaching physical consequences, should be emphasized. Note that the gluon anomaly in the singlet axial current (see Eq. (8.71)) does not imply the existence of a pole in $a_{||}^\mu$ at $q^2 \to 0$, because one cannot make gluons on-shell – the condition (8.77), which is crucial for the derivation of (8.79), cannot be met.

That (8.79) is the solution to (8.75) is obvious. That it is the *only* possible solution is less obvious. The reader is referred to [9, 10] for a comprehensive proof.

Exercise

8.2.1   Consider the two-dimensional CP(1) model with fermions presented in Section 10.12.3.4. Find the anomaly in the divergence of the axial current $\bar{\psi}\gamma^\mu\gamma^5\psi$. Can it be called the triangle anomaly?

8.2.2   (a) Prove that if $F_{\mu\nu}$ is x-independent one can always choose a gauge in which

$$A_\mu(x) = \frac{1}{2}x^\rho F_{\rho\mu}.$$

Hint: start from the Taylor expansion for A_μ in arbitrary gauge.

(b) The Green function $S(x, y)$ is defined as $-i\langle T\{\psi(x)\bar{\psi}(y)\}\rangle$ and satisfies the equation

$$i\mathcal{D} S(x, y) = \delta^{(4)}(x - y). \tag{8.80}$$

From this equation find the first term in (8.65). Then, using $A_\mu(x) = \frac{1}{2}x^\rho F_{\rho\mu}$ as a perturbation in (8.80), find the second term in (8.65).

8.3 't Hooft Matching and Its Physical Implications

In this section we will turn to physical consequences. We will start from a general interpretation of the pole in (8.79) and similar anomalous relations for other currents, formulate the 't Hooft matching condition, prove (at large N) the spontaneous breaking of the global $SU(N_f)_A$ symmetry, and, finally, calculate the $\pi^0 \to 2\gamma$ decay width.

8.3.1 Infrared Matching

Poles do not appear in physical amplitudes for no reason. In fact, the only way an amplitude can acquire a pole is through the coupling of massless particles in the spectrum of the theory to the external currents under consideration. There are two possible scenarios: (i) spontaneous breaking of the global axial symmetry (it would be more exact to say that it is realized nonlinearly); (ii) linear realization with massless spin-$\frac{1}{2}$ fermions.

In the first case massless Goldstone bosons appear in the physical spectrum. They must be coupled to $j_\mu^{5,a}$ and external vector gauge bosons. Equation (8.79), or a similar equation for other currents, presents a constraint on the product of the Goldstone boson couplings that can always be met.

The second scenario is more subtle and, apparently, is rather exotic. It is true that the triangle loop (Fig. 8.6) with massless spin-$\frac{1}{2}$ fermions yields q^μ/q^2 in the longitudinal part $a_{\parallel}^\mu$ of the axial current [9, 10]. However, not only is the kinematic factor q^μ/q^2 exactly predicted by the anomaly; the coefficient in front of this factor is known *exactly* too. For instance, in the example of Section 8.2.3, this coefficient is $(\alpha/\pi)N(Q_u^2 - Q_d^2)$. For the chiral symmetry to remain unbroken, the massless spin-$\frac{1}{2}$ (composite) fermions that might be potential contributors to the triangle loop must reproduce this coefficient exactly, which, generally speaking, is a highly nontrivial requirement. The search for massless spin-$\frac{1}{2}$ fermions that could match the coefficient in front of q^μ/q^2 in $a_{\parallel}^\mu$ constitutes the celebrated 't Hooft matching procedure [10].

Needless to say, if free massless N-colored quarks existed in the spectrum of asymptotic states then they would automatically provide the required matching.[10] Alas ... quark confinement implies the absence of quarks in the physical spectrum. The only spin-$\frac{1}{2}$ fermions we deal with in QCD are composite baryons.

8.3.2 Spontaneous Breaking of the Axial Symmetry

Let us see whether we can match (8.79) with the baryon contribution. We will put $N = 3$, as in our world, and consider first $N_f = 2$. Then the lowest-lying spin-1/2

---

[10] In all theories that are strongly coupled in the infrared the only proper way of obtaining $a_{\parallel}^\mu$ in the form (8.79) is an ultraviolet derivation through the external anomaly. However, if we pretended to forget all the correct things about QCD and just blindly calculated the triangle loop of Fig. 8.6 with *noninteracting* massless quarks, we would get exactly the same formula. I hasten to add that this coincidence acquires a meaning only in the context of 't Hooft matching. Feynman diagrams, in particular that in Fig. 8.6, which are saturated by quarks in the infrared have no meaning in QCD-like theories with confinement.

baryons are the proton and neutron (p and n), with electric charges charges $Q_p = 1$ and $Q_n = 0$, respectively. Hence, only p contributes in the triangle loop in Fig. 8.6. If it were massless, it would generate a formula repeating (8.79) but with the substitution

$$N \left(Q_u^2 - Q_d^2 \right) \rightarrow Q_p^2. \tag{8.81}$$

The right- and left-hand sides in Eq. (8.81) are equal! Thus, in this particular case, 't Hooft matching does not rule out a linearly realized axial SU(2) symmetry for the massless baryons p and n. This could be merely a coincidence, though. Therefore, let us not jump to conclusions. We will examine the stability of the above matching.

To this end we add the third quark, s, keeping intact the axial current to be analyzed; see (8.73). The electromagnetic current (8.72) acquires an additional term $-\frac{1}{3}\bar{s}\gamma_\mu s$. The anomaly-based prediction (8.79) remains intact.

In the theory with u, d, and s quarks the lowest-lying spin-$\frac{1}{2}$ baryons form the baryon octet

$$B = (p, n, \Sigma^\pm, \Lambda, \Sigma^0, \Xi^-, \Xi^0). \tag{8.82}$$

If both the vector and axial SU(3) flavor symmetries are realized linearly, the baryon–baryon–photon coupling constants and the constants $\langle B|a^\mu|B\rangle$ at zero momentum transfer are unambiguously determined from the baryon quantum numbers (for instance, $\langle \Sigma^+|a^\mu|\Sigma^+\rangle = \bar{\Sigma}\gamma^\mu\gamma^5\Sigma$). Calculating the triangle diagram of Fig. 8.6 (or, more exactly, its longitudinal part) we find that the baryon octet does *not* contribute there owing to cancelations: the proton contribution (the quark content uud) is canceled by that of Ξ^- (the quark content ssd) while the Σ^- contribution (the quark content dds) is canceled by Σ^+ (the quark content uus). Other baryons from (8.82) are neutral and decouple from the photon. Seemingly, the absence of matching tells us that global SU(3)$_A$ symmetry must be spontaneously broken.

Although the above argument is suggestive, it is still inconclusive. It tacitly assumes that baryons with other quantum numbers, e.g., $J^P = \frac{1}{2}^-$, are irrelevant in the calculation of $a_{||}^\mu$, which need not be the case. How can one prove that the combined contribution of all baryons cannot be equal to (8.79)?

To answer this question let us explore the N-dependence in Eq. (8.79). An anomaly-based calculation naturally produces the factor N on the right-hand side. At the same time, the linear dependence on N cannot be obtained by saturating the triangle loop by baryons at large N [11]: each baryon loop is suppressed exponentially, as e^{-N}, since each baryon consists of N quarks. This observation proves that the global SU(N_f)$_A$ symmetry must be spontaneously broken, at least in the multicolor limit. As a result, $N_f^2 - 1$ massless Goldstone bosons (pions) emerge in the spectrum. Note that this argument is inapplicable to the singlet axial current (see the remark at the end of Section 8.2.3); the singlet pseudoscalar meson need not be massless.

Consult Section 9.2.

Caveat: To my mind, the above assertion of exponential suppression of the baryon loops has the status of a "physical proof" rather than a mathematical theorem. It is intuitively natural, indeed. However, in the absence of a full dynamical solution of Yang–Mills theories at strong coupling, one cannot completely rule out exotic scenarios in which the loop expansion in $1/N$ (implying e^{-N} for baryons) is invalid; see [12]. I do think that this expansion is valid in QCD per se. Doubts remain *concerning models with more contrived fermion sectors. Note that in two dimensions examples of baryons defying the formal 1/N expansion are known.*

8.3.3 Predicting the $\pi^0 \to 2\gamma$ Decay Rate

If the global $SU(N_f)_A$ symmetry is realized nonlinearly, through the Goldstone bosons (which in the case of two flavors are called pions), saturation of the anomaly-based formula (8.79) is trivial (Fig. 8.7). The pole in $a_{\parallel}^\mu$ is due to the pion contribution. The constraint (8.79) provides us with a relation between the $a^\mu \to \pi^0$ amplitude and the $\pi^0 \to 2\gamma$ coupling constant. The result has been known since the 1960s. For completeness I will recall its derivation.

The $\pi^0 \to \gamma\gamma$ amplitude can be parametrized as

$$A(\pi^0 \to 2\gamma) = F_{\pi 2\gamma} \mathcal{F}_{\mu\nu} \tilde{\mathcal{F}}^{\mu\nu} \to -4F_{\pi 2\gamma} k_\mu^{(1)} \epsilon_\nu^{(1)} k_\alpha^{(2)} \epsilon_\beta^{(2)} \varepsilon^{\mu\nu\alpha\beta}, \qquad (8.83)$$

where we use the same notation as in Sections 8.2.2 and 8.2.3. Moreover, the amplitude $\langle 0|a^\mu|\pi^0\rangle$ is parametrized by the constant f_π playing the central role in pion physics,

> *Pion constant f_π.*

$$\langle 0|a^\mu|\pi^0\rangle = \frac{1}{\sqrt{2}} i f_\pi q_\mu, \qquad f_\pi \approx 130 \text{ MeV}. \qquad (8.84)$$

Then the pion contribution to the matrix element on the left-hand side of Eq. (8.79) is

$$\langle 0|a_{\parallel}^\mu|2\gamma\rangle = i\frac{q^\mu}{q^2} \frac{f_\pi}{\sqrt{2}} 4F_{\pi 2\gamma} \, \varepsilon^{\mu\nu\alpha\beta} k_\mu^{(1)} \epsilon_\nu^{(1)} k_\alpha^{(2)} \epsilon_\beta^{(2)}. \qquad (8.85)$$

Comparing with (8.79) we arrive at the following formula:

$$F_{\pi 2\gamma} = \frac{N}{2\sqrt{2}} \frac{1}{f_\pi} \frac{\alpha}{\pi} \left(Q_u^2 - Q_d^2\right) \to \frac{1}{2\sqrt{2} f_\pi} \frac{\alpha}{\pi}. \qquad (8.86)$$

This is in good agreement with experiment.

Before the advent of QCD people did not know about color, so naturally the factor $N = 3$ was omitted from the prediction (8.86). In fact, analysis of the $\pi^0 \to \gamma\gamma$ decay led to one of the very few quantitative proofs of the existence of the color in the early 1970s.

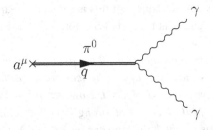

Fig. 8.7   Pion saturation of the anomaly.

Exercise

8.3.1   Assume the number of colors to be large, and try to saturate the triangle graph
in Fig. 8.6 by baryons. What N dependence would you expect?

8.4 Scale Anomaly

In this section we will briefly discuss the scale anomaly in Yang–Mills theories. For
simplicity we will limit ourselves to pure Yang–Mills theories, i.e., those without
matter, for which

$$S = \int d^4x \left(\frac{-1}{4g_0^2}\right) G_{\mu\nu}^a G^{\mu\nu a}, \tag{8.87}$$

where the subscript 0 indicates the bare coupling constant. At the classical level the
action (8.87) is obviously invariant under the scale transformations

$$x \to \lambda^{-1}x, \qquad A_\mu^a \to \lambda A_\mu^a, \tag{8.88}$$

where λ is an arbitrary real number. Barring subtleties (see Appendix section 1.4
at the end of Chapter 1), the scale invariance of the theory with any local Lorentz-
invariant Lagrangian implies the full conformal symmetry [13]. Roughly speaking,
scale-invariant theories contain only *dimensionless* constants in the Lagrangian
(otherwise, the action would not be invariant under the scale transformations). Thus,
the conformal invariance of the action is quite clear, at least at the intuitive level.

<table>
<tr><td>Look
through
Appendix
section 1.4.</td></tr>
</table>

The scale transformations are generated by the current [13]

$$j_\nu^D = x^\mu \theta_{\mu\nu}, \tag{8.89}$$

where $\theta_{\mu\nu}$ is the symmetric and conserved energy–momentum tensor of the theory
under consideration. For instance, in pure Yang–Mills theory, (8.87),

$$\theta_{\mu\nu} = -\frac{1}{g^2}\left(G_{\mu\alpha}^a G_\nu^{\alpha a} - \frac{1}{4}g_{\mu\nu}G_{\alpha\beta}^a G^{\alpha\beta a}\right). \tag{8.90}$$

The classical scale invariance of (8.87) implies that the current j_ν^D is conserved,
$\partial^\nu j_\nu^D = 0$. Indeed,

$$\partial^\nu j_\nu^D = \theta_\mu^\mu \tag{8.91}$$

and the trace of the energy–momentum tensor (8.90) obviously vanishes, $\theta_\mu^\mu = 0$.

The vanishing of θ_μ^μ is valid only at the *classical* level. At the quantum level θ_μ^μ
acquires an anomalous part. We will derive this (scale) anomaly at one loop. Unlike
the chiral anomaly, we do not have to deal with γ^5 here; therefore, the simplest
derivation is based on dimensional regularization. Namely, instead of considering
the action (8.87) in four dimensions we will consider it in $4 - \epsilon$ dimensions, where
$\epsilon \to 0$ at the very end. In $4 - \epsilon$ dimensions $\int d^{4-\epsilon}x\, G_{\mu\nu}^2$ is not scale invariant.
The change in $\int d^{4-\epsilon}x\, G_{\mu\nu}^2$ under the scale transformation is proportional to ϵ. One
should not forget, however, that $1/g_0^2$, being expressed in terms of the renormalized

coupling, also depends on ϵ; this latter dependence contains $1/\epsilon$. As a result, in the limit $\epsilon \to 0$, a finite term giving us the noninvariance of (8.87) remains.

Concretely,

$$\delta S = \int d^{4-\epsilon}x \left[-\frac{1}{4}\left(\frac{1}{g^2} + \frac{\beta_0}{8\pi^2}\frac{1}{\epsilon}\right)(\lambda^\epsilon - 1)\, G^a_{\mu\nu}G^{\mu\nu a} \right]$$

$$\to \int d^4x \ln\lambda \left(-\frac{\beta_0}{32\pi^2} G^a_{\mu\nu}G^{\mu\nu a} \right) \qquad (8.92)$$

where $\beta_0 = 11\,N/3$ is the first coefficient of the β function; cf. Eq. (1.57). Equation (8.92) immediately leads to the conclusion that [14]

$$\theta^\mu_\mu = -\frac{\beta_0}{32\pi^2}\, G^a_{\mu\nu}G^{\mu\nu a}. \qquad (8.93)$$

Anomaly in
θ^μ_μ

This expression for θ^μ_μ remains valid even in the presence of massless fermions, although the value of β_0 changes, of course.

The scale anomaly formula (8.93) expresses the fact that, although the classical Yang–Mills action contains only dimensionless constants, a dynamical scale parameter Λ of dimension of mass is generated at the quantum level; this phenomenon is referred to as *dimensional transmutation*. All hadronic masses are proportional to Λ. The expectation value of $G^2_{\mu\nu}$ over a given hadron is proportional to the mass of this hadron [15] (in the chiral limit).

8.5 One-Form Anomalies

In this section we will address a relatively new class of anomalies discovered in 2014 [16–19], the so-called 1-form anomalies.[11] They are relevant to problems in which one of possible symmetries is described by a nonlocal parameter: 1-form, 2-form, or higher forms. In this textbook I consider only the simplest case of 1-forms.

*Discrete
higher-form
symmetries*

The 1-form symmetry we will consider below is a discrete symmetry, which turns out to be in conflict with a discrete chiral symmetry. Only one of them can be maintained. The concept of two conflicting symmetries with one of them sacrificed at the quantum level is already familiar to us from Sections 8.2 and 8.3.

The anomalies that were discussed in these sections – i.e., the "old anomalies" – are referred to as 0-form anomalies in the current context. The most well-known example of the 0-form anomaly is the divergence of the axial current – the Adler–Bell–Jackiw anomaly. The axial current conservation is in conflict with gauge invariance. The discrete global symmetries instrumental in [16–19] are related to special transformations – "large" gauge transformations of the type we encountered in Chapter 5 and Section 8.1.

*0-form
anomalies,
1-form
anomalies,
2-form
anomalies,
etc.*

After these preliminary remarks we turn to the simplest example – the Schwinger model at weak coupling (see Section 8.1) – for explanation of the concept of the phenomenon. As we know, at weak coupling this model is solvable, therefore no additional information emerges from the study of two conflicting global symmetries. However, anomalies survive in the regime of strong coupling and then the study of

---

[11] The class of generalized global symmetries includes also the so-called non-invertible symmetries [20, 21] which will not be considered in this text. In this section I follow in part Ref. [19].

anomalies becomes useful and highly informative; see Section 8.3. The Schwinger model is ideal for pedagogical purposes.

In Section 8.1.1 I introduced an x independent parameter A_1 defined on the circle, i.e., for all integer k the values of $A_1 + 2k\frac{\pi}{L}$ must be identified. Here I will introduce essentially the same parameter in a different form and notation,[12] explicitly demonstrating its nonlocality, namely the Polyakov line (the notation $\langle \cdots \rangle$ means averaging over the ground state),

| Polyakov line |

$$P = \exp\left(i \int_0^L dx A_1(x,t)\right), \quad P_0 = \left\langle \exp\left(i \int_0^L dx A_1(x,t)\right)\right\rangle. \tag{8.94}$$

The integral in the exponent (mathematicians call it 1-form) runs along the compactified direction. Since the points $x = 0$ and $x = L$ in the coordinate space are identified the integration path is in fact closed; we deal here with a particular case of the Wilson loop (see Eq. (1.51) on page 27).

The Schwinger model has a U(1) gauge symmetry. The Polyakov line is gauge invariant.[13] It is obviously a nonlocal operator. Its physical meaning is as follows: assume one drags an (infinitely) heavy probe charge-1 particle along the closed integration contour. The impact of the probe quark is the phase factor in (8.94).

8.5.1 $\mathbb{Z}_N$ 1-Form Symmetry or "Center" Symmetry

In the Schwinger model the gauge symmetry is U(1) – this group is Abelian, and every element of U(1) is also its center element. By definition, the center elements of the group commute with all other elements.[14] That is why I used quotation marks in this title of this subsection. In the case at hand, it would be more appropriate to say $\mathbb{Z}_N$ 1-form symmetry. The origin of "center symmetry" is historical. In discussions of Yang-Mills theory it was common to refer to its $\mathbb{Z}_N$ 1-form symmetry as "center" since it is related to the center elements of SU(N). Currently, more accurate terminology becomes dominant.

Let us turn to the model (8.1), (8.36). For the time being we ignore fermions and limit our analysis to pure photodynamics on the spatial circle,[15]

$$S = \int dt\, dx \left[-\frac{1}{4e^2} F_{\mu\nu} F^{\mu\nu} + \frac{\theta}{4\pi} \varepsilon^{\mu\nu} F_{\mu\nu}\right] \tag{8.95}$$

$$\to S = \int dt \left[\frac{L}{2e^2}(\dot{A}_1)^2 + \frac{\theta}{2\pi} \dot{A}_1\right]. \tag{8.96}$$

We observe an A-shift symmetry of (8.96),

$$A_1 \to A_1 + c, \tag{8.97}$$

with any constant c.

For the time being let us assume that $\theta = 0$. Examining the general condition for the wave function on the circle,

---

[12] A_1 in Section 8.1 reduces to $(-i/L)\log P$, where P is the Polyakov line (8.94).

[13] On page 27 we treat non-Abelian theories. In this case the definition of the Polyakov line includes trace and path ordering in the preexponent, irrelevant for the Abelian theory.

[14] This is not the case in non-Abelian groups. For instance, in SU(2) the only center elements are 1 and $\exp(i\pi\tau_3) = -1$.

[15] In the absence of interactions we can set $A_0 \equiv 0$. The x-dependent (time-independent) gauge transformations then allow us to reduce $A_1(t,x) \to A_1(t)$.

$$\Psi\left(A_1 + \frac{2\pi}{L}\right) = e^{i\theta}\,\Psi(A_1), \tag{8.98}$$

we conclude that the ground state wave function $\Psi(A_1)$ is flat in the fundamental interval $\left[0, \frac{2\pi}{L}\right]$. This implies in turn that classically P defined in Eq. (8.94) can have any phase. Then P_0 vanishes after averaging. By definition, the vanishing of P_0 is the *center symmetry*. Hence, the model under consideration is classically center-symmetric.

Once we switch on the charge-1 fermion (Sections 8.1.1–8.1.6) we will see that the center symmetry is explicitly broken. The phase of P no longer spreads evenly all over the fundamental interval $\left[0, \frac{2\pi}{L}\right]$. The Polyakov line no longer vanishes. From Fig. 8.2 on page 308 we see that $P_0 = 1$. The question to be addressed now is as follows: Are there situations in which at least a discrete part of the center symmetry survives?

Dynamical charge-2 fermion and nondynamical charge-1 field as a probe

The answer is yes. To see this, we will slightly generalize the model considered in Section 8.1 – we will replace the charge-1 *massless* Dirac fermion with that of charge-2. The charge-1 matter will appear in our model as a very heavy fermion field (with mass $M_h \gg e$).

8.5.2 Massless Charge 2 and Heavy Charge 1

Let us turn to the beginning of Section 8.1. Passing to charge-2 massless fermion ψ (instead of charge-1) one must replace

$$A_1 \rightarrow 2A_1 \tag{8.99}$$

in all expressions in Section 8.1.2. Above, the factor 2 on the right-hand side represents the charge of the Dirac fermion. In principle, we could have chosen any integer value of q but for our purposes it is sufficient to limit ourselves to $q = 2$. Were it not for the charge-1 heavy field Ψ_h, we could just rename $A_1(t) \rightarrow 2A_1(t)$ and return to the already solved problem. "New" A_1 will be defined on the interval $A_1 \in [0, \pi/L]$. The vacuum state will be unique. However, the presence of probe charge-1 objects, even very heavy, does not allow us to do that.

Under the large gauge transformations the interval of periodicity in A_1 cannot be reduced to $\left[0, \frac{\pi}{L}\right]$ by redefinition of A_1. It is still $\left[0, \frac{2\pi}{L}\right]$ due to the presence of the charge-1 field. The latter does not allow us to gauge-identify the points $A_1 = 0$ and $A_1 = \frac{\pi}{L}$. The nearest gauge images are the points $A_1 = 0$ and $A_1 = \frac{2\pi}{L}$. The interval $0 \leq A_1 \leq 2\pi/L$ is the *fundamental domain or fundamental interval* under the large gauge transformation.

An immediate consequence is the change in Eq. (8.13). Now, in the course of the A_1 adiabatic evolution from 0 to $2\pi/L$ four levels in the Dirac sea have to be repopulated (see Fig. 8.8), rather than two levels shown in Fig. 8.1. In other words,

$$\Delta Q_5 = -4. \tag{8.100}$$

Correspondingly, the discrete chiral symmetry of this model is $Z_4 = Z_2 \times Z_2$. One of the Z_2 factors above (say, the last one) is unobservable because it corresponds to the sign change of the fermion field while all observables are bilinear in the fermion fields. The first Z_2 factor is detectable.

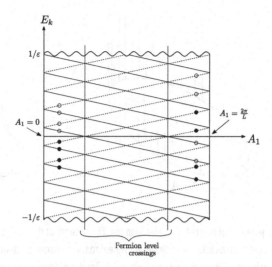

The same as in Fig. 8.1 but for charge-2 fermions. Now there are two vertical lines of level crossings in the fundamental domain. The Dirac sea must be rearranged twice.

What changes occur in the narrative and results obtained in Section 8.1 if we replace charge-1 massless fermion by charge-2? The answer to this question can be guessed from Fig. 8.1. Because of $q = 2$ (see Eq. (8.99)) the slope of the fermion levels as a function of A_1 doubles, and hence, as was mentioned, the fermion levels cross the horizontal axis twice inside the fundamental interval: first at $A_1 = \pi/2L$ and then at $A_1 = 3\pi/2L$. (For brevity I will refer to this phenomenon as *level crossing*.) Inside the fundamental interval two levels from the Dirac sea (i.e., filled) appear above the sea level, and two unpopulated levels sink in the sea. We have to repopulate four fermion levels, according to (8.100) instead of Eq. (8.13). The effective potential looks similar to that in Fig. 8.2, except that the scale on the horizontal axis has to be changed, implying that there are two minima in the fundamental domain: one at $A_1 = 0$ (or $2\pi/L$, which is the same) and the second minimum at $A_1 = \pi/L$. This phenomenon is illustrated in Fig. 8.9b.

Strictly speaking, I was not quite accurate. Let us remember about the heavy charge-1 fermion field. Its impact is small, suppressed by powers of $1/M_h$, but it still could be detected at $M_h \neq \infty$ since it changes the potential,

$$V(A_1) \rightarrow V(A_1) + \delta V(A_1). \tag{8.101}$$

Here $V(A_1)$ is generated by massless charge-2 fermion and is periodic with the period π/L while $\delta V(A_1)$ is a small correction generated by heavy charge-1 fermion. The period of $\delta V(A_1)$ is $2\pi/L$. If M_h is large but $M_h \neq \infty$ the degeneracy between two minima in Fig. 8.9b is lifted: the one at $A_1 = \pi/L$ becomes slightly higher than the minimum at $A_1 = 0$. As a result, the Z_2 shift symmetry by π/L becomes exact only in the limit of $M_h \rightarrow \infty$. Therefore, it cannot be viewed as an honest-to-god large gauge transformation. This Z_2 symmetry is asymptotic; in what follows I will neglect its violation.

The Polyakov lines for the two minima we see in Fig. 8.9b are

$$P_0 = 1 \text{ and } P_0 = -1, \tag{8.102}$$

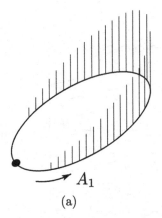
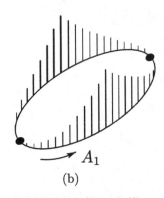

(a)   (b)

Fig. 8.9   The potential generated by the fermion sea. The strength of the potential is depicted by vertical lines growing from a circle topologically equivalent to the fundamental domain of A_1. The circumference of this circle is $2\pi/L$. (Remember $A_1 = 0$ and $A_1 = 2\pi/L$ are identified because of large gauge transformations.) In Fig. 8.9a we see one minimum that corresponds to Fig. 8.2. In Fig. 8.9b we see the emergence of the second minimum due to a global Z_2 symmetry.

respectively. Note that there is no tunneling between the two minima. Usually tunnelings do occur in quantum mechanics, resulting in a smeared wave function. In the case at hand, however, the fermion components of the corresponding wave functions are orthogonal, cf. Eq. (8.34). Because of the repopulation of the fermion sea no tunneling between these two vacua occurs.

We are left with two logical possibilities: first, we choose one of the two vacua in (8.102) as the genuine ground state. This means that in the vacuum P_0 is either 1 or -1, and the center symmetry is spontaneously broken. The chiral Z_4 remains unbroken, however, as is seen from Fig. 8.8. Indeed, Fig. 8.8 implies that starting from the minimum at $A_1 = 0$ and shifting to its gauge copy at $A_1 = 2\frac{\pi}{L}$ we produce two holes above the level of the Dirac sea, and simultaneously two populated levels above the sea sink in the sea. Hence, the vacuum condensate is four-fermion,

$$\langle \bar{\psi}\psi\,\bar{\psi}\psi \rangle \neq 0, \quad \langle \bar{\psi}\psi \rangle = 0. \tag{8.103}$$

The second logical possibility is to take two minima and form linear combinations of the wave functions corresponding to two minima (see Eq. (8.34)), namely,

$$\Psi_{\pm} = \frac{1}{\sqrt{2}}\left\{ \Psi_{\mathrm{Dirac}\,0}\,\Psi_0(A_1) \pm \Psi_{\mathrm{Dirac}\,1}\,\Psi_0\left(A_1 - \frac{\pi}{L}\right)\right\}, \tag{8.104}$$

where the Dirac sea wave functions $\Psi_{\mathrm{Dirac}\,0}$ and its counterpart $\Psi_{\mathrm{Dirac}\,1}$ can be read off from (8.34). Then it is obvious that for both vacua $\Psi_{\pm}$ the Polyakov line

$$(P_0)_{\pm} = \langle \Psi_{\pm} | P | \Psi_{\pm} \rangle = \frac{1}{2}\left\{ \langle \Psi_{\mathrm{Dirac}\,0}\,\Psi_0(A_1) | P | \Psi_{\mathrm{Dirac}\,0}\,\Psi_0(A_1) \rangle \right.$$

$$\left. + \left\langle \Psi_{\mathrm{Dirac}\,1}\,\Psi_0\left(A_1 - \frac{\pi}{L}\right) | P | \Psi_{\mathrm{Dirac}\,1}\,\Psi_0\left(A_1 - \frac{\pi}{L}\right)\right\rangle \right\}$$

$$= \frac{1}{2}(1 - 1) = 0. \tag{8.105}$$

Hence, the center symmetry is unbroken in this case. (The operator P is defined in (8.94).) Simultaneously, the operator $\bar{\psi}\psi$ acquires a vacuum expectation value. Indeed

$$\langle\bar{\psi}\psi\rangle = \pm\left\{\langle\Psi_{\text{Dirac}\,0}\,\Psi_0(A_1)\,|\bar{\psi}\psi|\,\Psi_{\text{Dirac}\,1}\,\Psi_0\left(A_1 - \frac{\pi}{L}\right)\rangle + \text{H.c.}\right\}. \qquad (8.106)$$

The right-hand side does not vanish; in fact, it was calculated in Section 8.1; see (8.41). The $\bar{\psi}\psi$ condensate above is an unambiguous signature of the spontaneous breaking of the chiral Z_4 symmetry down to Z_2.

8.5.3 Why Only One of Two Discrete Symmetries Can Survive

Thus our calculation leads us to one of two possible scenarios. Could we have anticipated this outcome without the actual calculation, on purely algebraic basis?

In the following I will prove that the two Z_2 symmetries (shift and chiral) are incompatible in the ground state. We start with Eq. (8.100) which tells us that for charge-2 fermions the conserved charge $\tilde{Q}_5$ is

$$\tilde{Q}_5 \equiv Q_5 + \frac{2LA_1}{\pi}; \qquad (8.107)$$

cf. Eq. (8.14). Remember that A_1 varies from 0 to its gauge copy, $\frac{2\pi}{L}$.

For the subsequent proof we will need the generator of the chiral Z_4 transformation corresponding to (8.107). It has the form

$$G(Z_4) = \exp\left(\frac{i\pi\tilde{Q}_5}{2}\right). \qquad (8.108)$$

The generator of the Z_2 center-symmetry transformation acting on the Polyakov line (8.94) must shift $A_1 L$ in the exponent by π. This generator is expressed through the canonic momentum corresponding to A_1 as follows:

$$G(Z_2\,\text{center}) = \exp\left(\frac{\pi}{L}\frac{\partial}{\partial A_1}\right). \qquad (8.109)$$

Both $G(Z_4)$ and $G(Z_2\,\text{center})$ commute with the effective Hamiltonian but do *not* commute with each other. Indeed, commuting (8.108) and (8.109) using (8.107) we easily obtain

$$G(Z_2\,\text{center})\,G(Z_4) = e^{i\pi}\,G(Z_4)\,G(Z_2\,\text{center})$$

$$= -G(Z_4)\,G(Z_2\,\text{center}). \qquad (8.110)$$

Since the generators of the symmetries do not commute at most one of them can survive in the ground state.

8.5.4 What Scenario Should We Choose

We are allowed to introduce perturbations that can explicitly break any of the preceding discrete symmetries. For instance, by keeping M_h large but finite we lift the degeneracy of two minima in Fig. 8.9b. One of them becomes false vacuum. If the split is large enough the linear combination in (8.104) is irrelevant, the true vacuum is unique and therefore the center symmetry is necessarily broken. This is the first scenario.

On the other hand, we can introduce the Z_4 breaking mass term $\Delta\mathcal{L}_m = m\bar{\psi}\psi$ as a dominant deformation, tending $M_h \to \infty$. If we want the theory to be smooth in m we should choose the wave function (8.104). This is the second scenario.

Exercise

8.5.1   Assume we have a particle living on a circular ring of radius 1 in a certain potential.[16]

If the coordinate of the particle on the ring is denoted by the angle a then the Lagrangian of the system takes the form

$$\mathcal{L} = \frac{1}{2}\dot{a}^2 + \frac{\theta}{2\pi}\dot{a} - V(a), \tag{8.111}$$

where $V(a)$ is a potential and the second term is the topological θ term. Choose a specific potential $V(a)$,[17]

$$V(a) = 1 - \cos(2a). \tag{8.112}$$

The coordinate a obviously varies in the interval $[0, 2\pi)$. The angle $\theta = 2\pi$ is the same as $\theta = 0$, while $\theta = \pi$ is the same as $\theta = -\pi$, as is evident from Eq. (8.113). The Bloch theory teaches us that the boundary conditions on the wave function $\Psi(a)$ are quasiperiodic, rather than periodic, namely,

$$\Psi(a + 2\pi) = e^{i\theta}\,\Psi(a), \tag{8.113}$$

where the phase θ on the right-hand side coincides with the θ angle in (8.111). In this exercise let us assume that $\theta = \pi$. Then the boundary condition (8.113) becomes *antiperiodic*.

The Lagrangian (8.111), (8.112) has two discrete symmetries: C-symmetry[18] corresponding to $a \to -a$ and the shift (S_{shift}) symmetry: $a \to a + \pi$. Prove that the above two symmetries do not commute with each other, i.e.,

$$S_{\text{shift}}C - CS_{\text{shift}} \neq 0. \tag{8.114}$$

What is the implication of (8.114) for the ground state?

---

[16] For instance, if the plane of the ring is oriented vertically, the gravity force acting on the particle will induce a potential whose value at the lowest point of the ring can be set to zero, then at the highest point of the ring the potential reaches its maximal value. This is symbolically depicted in Figs. 5.1 and 5.2 (page 176).

[17] This is *not* the potential emerging from Fig. 5.1 because of the cosine argument. Rather, it is the potential qualitatively describing Fig. 8.9b.

[18] The transformation $a \to -a$ must be supplemented by the sign change of θ in the topological term in (8.111). However, was mentioned earlier, $\theta = \pi$ and $\theta = -\pi$ are equivalent, see (8.113).

References for Chapter 8

[1] M. A. Shifman, *Phys. Rept.* **209**, 341 (1991) [see also M. Shifman (ed.), *Vacuum Structure and QCD Sum Rules* (North-Holland, Amsterdam, 1992)].

[2] A. Casher, J. B. Kogut, and L. Susskind, *Phys. Rev. Lett.* **31**, 792 (1973). J. B. Kogut and L. Susskind, *Phys. Rev. D* **11**, 3594 (1975).

[3] J. E. Hetrick and Y. Hosotani, *Phys. Rev. D* **38**, 2621 (1988).

[4] V. N. Gribov, *Gauge Theories and Quark Confinement* (Phasis, Moscow, 2002), p. 271.

[5] M. A. Shifman and A. V. Smilga, *Phys. Rev. D* **50**, 7659 (1994) [arXiv:hep-th/9407007].

[6] N. Seiberg, *JHEP* **1007**, 070 (2010) [arXiv:1005.0002 [hep-th]].

[7] V. A. Novikov, M. A. Shifman, A. I. Vainshtein, and V. I. Zakharov, *Fortsch. Phys.* **32**, 585 (1984) [see also M. Shifman (ed.), *Vacuum Structure and QCD Sum Rules* (North-Holland, Amsterdam, 1992)].

[8] S. L. Adler, *Phys. Rev.* **177**, 2426 (1969); J. S. Bell and R. Jackiw, *Nuovo Cim. A* **60**, 47 (1969).

[9] A. D. Dolgov and V. I. Zakharov, *Nucl. Phys. B* **27**, 525 (1971).

[10] G. 't Hooft, Naturalness, chiral symmetry, and spontaneous chiral symmetry breaking, in G. 't Hooft, C. Itzykson, A. Jaffe, *et al.* (eds.), *Recent Developments in Gauge Theories* (Plenum Press, New York, 1980) [reprinted in E. Farhi *et al.* (eds.), *Dynamical Symmetry Breaking* (World Scientific, Singapore, 1982), p. 345, and in G. 't Hooft, *Under the Spell of the Gauge Principle* (World Scientific, Singapore, 1994), p. 352].

[11] S. R. Coleman and E. Witten, *Phys. Rev. Lett.* **45**, 100 (1980).

[12] D. Amati and E. Rabinovici, *Phys. Lett. B* **101**, 407 (1981).

[13] S. B. Treiman, E. Witten, R. Jackiw, and B. Zumino, *Current Algebra and Anomalies* (World Scientific, Singapore, 1985).

[14] J. C. Collins, A. Duncan, and S. D. Joglekar, *Phys. Rev. D* **16**, 438 (1977).

[15] M. A. Shifman, A. I. Vainshtein, and V. I. Zakharov, *Phys. Lett. B* **78**, 443 (1978).

[16] A. Kapustin and N. Seiberg, *JHEP* **1404**, 001 (2014) [arXiv:1401.0740 [hep-th]].

[17] D. Gaiotto, A. Kapustin, N. Seiberg, and B. Willett, *JHEP* **1502**, 172 (2015) [arXiv:1412.5148 [hep-th]].

[18] D. Gaiotto, A. Kapustin, Z. Komargodski, and N. Seiberg, *JHEP* **1705**, 091 (2017) [arXiv:1703.00501 [hep-th]].

[19] M. Anber and E. Poppitz, *JHEP* **1809**, 076 (2018) [arXiv:1807.00093 [hep-th]].

[20] T. Rudelius and S. H. Shao, JHEP **12**, 172 (2020) [arXiv:2006.10052].

[21] M. Nguyen, Y. Tanizaki and M. Ünsal, JHEP **03**, 238 (2021) [arXiv:2101.02227].

Confinement in 4D Gauge Theories and Models in Lower Dimensions

The confinement phase as a physical phenomenon. — General ideas on confinement in four dimensions. — The dual Meissner effect: what is it? — The 't Hooft large-N limit and a "stringy" picture. — Known examples of confining behavior in lower dimensions.

9.1 Confinement in Non-Abelian Gauge Theories: Dual Meissner Effect

Look through Section 1.3.1.

The most salient feature of pure Yang–Mills theory is linear confinement. If one takes a heavy probe quark and antiquark separated by a large distance, the force between them does not fall off with distance; the potential energy grows linearly. This is the explanation of the empirical fact that quarks and gluons (the microscopic degrees of freedom in QCD) never appear as asymptotic states. The physically observed spectrum consists of color-singlet mesons and baryons. This phenomenon is known as *color confinement* or, in a more narrow sense, *quark confinement*. In the early days of QCD it was also referred to as *infrared slavery*.

Quantum chromodynamics (QCD) and Yang–Mills theories at strong coupling in general are not yet analytically solved. Therefore, it is reasonable to ask the following questions.

Are there physical phenomena in which the interaction energy between two interacting bodies grows with distance at large distances? Do we understand the underlying mechanism?

Superconductors and Abrikosov vortices

The answer to these questions is positive. The phenomenon of a linearly growing potential was predicted by Abrikosov [1] in superconductors of the second type, which, in turn, were predicted by Abrikosov [2] and discovered experimentally in the 1960s. The corresponding set-up is shown in Fig. 9.1. In the center region of this figure we see a superconducting sample, with two very long magnets attached to it. A superconducting medium does not tolerate a magnetic field; however, the flux of the magnetic field must be conserved. Therefore, the magnetic field lines emanating from the north pole of one magnet find their way to the south pole of the other magnet, through the medium, by the formation of a flux tube. Inside the flux tube the Cooper pair condensate vanishes and the superconductivity is destroyed. The flux tube has a fixed tension, implying a constant force between the magnetic poles as long as they are within the superconducting sample. The phenomenon described above is sometimes referred to as the Meissner effect.

Meissner effect

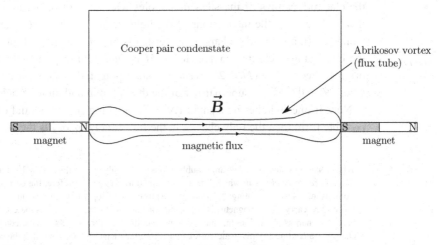

Cooper pair condenstate

Abrikosov vortex (flux tube)

$$\vec{B}$$

S N   S N

magnet   magnetic flux   magnet

Fig. 9.1 The Meissner effect in QED, in a superconductor of the second kind.

Of course, the Meissner effect of Abrikosov type occurs in an Abelian theory, QED: the flux tube that forms in this case is Abelian. In Yang–Mills theories we are interested in non-Abelian analogs of the Abrikosov vortices. Moreover, while in the Abrikosov case the flux tube is that of the magnetic field, in QCD and QCD-like theories the confined objects are quarks; therefore, the flux tubes must be "chromoelectric" rather than chromomagnetic. In the mid-1970s, Nambu, 't Hooft, and Mandelstam (independently) put forward the idea [3] of a "dual Meissner effect" as the underlying mechanism for color confinement.[1] Within their conjecture, in chromoelectric theories "monopoles" condense, leading to the formation of "non-Abelian flux tubes" between the probe quarks. At this time the Nambu– 't Hooft–Mandelstam paradigm was not even a physical scenario, rather a dream, since people had no clue as to the main building blocks, such as non-Abelian flux tubes. After the Nambu–'t Hooft–Mandelstam conjecture had been formulated, however, many works were published on this subject.

Super-Yang–Mills theories are considered in Part II.

A milestone in this range of ideas was the Seiberg–Witten solution [4] of $N = 2$ super-Yang–Mills theory slightly deformed by a superpotential breaking $N = 2$ down to $N = 1$. In the $N = 2$ limit, the theory has a moduli space. If the gauge group is SU(2), on the moduli space the SU(2)$_{\text{gauge}}$ symmetry is spontaneously broken down to U(1). Therefore, the theory possesses 't Hooft–Polyakov monopoles [5] (Sections 4.1.1 and 4.1.2). Two special points on the moduli space were found [4] (they are called the *monopole and dyon points*), in which the monopoles (dyons) become massless. In these points the scale of the gauge symmetry breaking

$$\text{SU}(2) \rightarrow \text{U}(1) \tag{9.1}$$

is determined by the dynamical parameter Λ of the $N = 2$ super-Yang–Mills theory.

All physical states can be classified with regard to the unbroken U(1) symmetry. It is natural to refer to the U(1) gauge boson as a photon. In addition to the photon all its superpartners, being neutral, remain massless at this stage while all other states, with non-vanishing "electric" charges, acquire masses of the order of Λ. In particular, the two gauge bosons corresponding to SU(2)/U(1) – it is natural to call them $W^{\pm}$ – have masses $\sim \Lambda$. All such states are "heavy" and can be integrated out.

In the low-energy limit, near the monopole and dyon points, one is dealing with the electrodynamics of massless monopoles. One can formulate an effective local theory describing the interactions of the light states. This is a U(1) gauge theory in which the (magnetically) charged matter fields M, $\tilde{M}$ are those of monopoles while

Dual QED

the U(1) gauge field that couples to M, $\tilde{M}$ is *dual* with respect to the photon of the original theory. The ($N = 2$)-preserving superpotential has the form $\mathcal{W} = \mathcal{A} M \tilde{M}$, where $\mathcal{A}$ is the $N = 2$ superpartner of the dual photon and photino fields.

Now, if one switches on a small ($N = 2$)-breaking superpotential of the simplest possible form then the only change in the low-energy theory is the emergence of an

---

[1] While Nambu's and Mandelstam's publications are easily accessible, it is hard to find the EPS *Conference Proceedings* in which 't Hooft presented his vision. Therefore, the corresponding passage from his talk is worth quoting: "...[monopoles] turn to develop a non-zero vacuum expectation value. Since they carry color-magnetic charges, the vacuum will behave like a superconductor for color-magnetic charges. What does that mean? Remember that in ordinary electric superconductors, magnetic charges are confined by magnetic vortex lines... We now have the opposite: it is the color charges that are confined by electric flux tubes."

extra term $m^2 \mathcal{A}$ in the superpotential ($m \ll \Lambda$). Its impact is crucial: it triggers the monopole condensation, $\langle M \rangle = \langle \tilde{M} \rangle = m$, which implies, in turn, that the dual U(1) symmetry is spontaneously broken and the dual photon acquires a mass $\sim m$. As a consequence, Abrikosov flux tubes are formed. Viewed within the dual theory, they carry the magnetic field flux. With regards to the original microscopic theory these are the electric field fluxes.

Thus Seiberg and Witten demonstrated, for the first time, the existence of the dual Meissner effect in a judiciously chosen non-Abelian gauge field theory. If one injects a (very heavy) probe quark and antiquark into this theory, with necessity, a flux tube forms between them, leading to linear confinement.

The flux tubes in the Seiberg–Witten solution were investigated in detail in 1995–1997 as described in [6]. These flux tubes are Abelian, and so is the confinement caused by their formation. What does that mean? At the scale of distances at which the flux tube is formed (the inverse mass of the Higgsed U(1) photon) the gauge group that is operative is Abelian. In the Seiberg–Witten analysis this is the dual U(1) symmetry. The off-diagonal (charged) gauge bosons are very heavy in this scale and play no direct role in the flux tube formation and confinement that ensues. Naturally, the spectrum of composite objects in this case turns out to be richer than that in QCD and similar theories with non-Abelian confinement. By non-Abelian confinement I mean a dynamical regime such that at distances at which flux tube formation occurs all gauge bosons are equally important.

> Abelian vs.
> non-Abelian
> confinement

Moreover, the string's topological stability is based on $\pi_1(\mathrm{U}(1)) = \mathbb{Z}$. Therefore, N strings do not annihilate as they should in QCD-like theories.

The two-stage symmetry-breaking pattern, with SU(2) $\rightarrow$ U(1) occurring at a high scale while at a much lower scale we have U(1) $\rightarrow$ nothing, has no place in QCD-like theories, as we know from experiment. In such theories, presumably all non-Abelian gauge degrees of freedom take part in string formation and are operative at the scale at which the strings are formed. Although it is believed that the strings in the Seiberg–Witten solution belong to the same universality class as those in QCD-like theories, the status of this statement is conjectural. An analytic theory of color confinement in QCD remains elusive.

Why then do people think that the Nambu–'t Hooft–Mandelstam picture is correct, i.e., that a version of a dual Meissner effect is responsible for quark confinement in QCD? Qualitative evidence in favor of a string-like picture behind confinement in QCD comes from consideration of the 't Hooft large-N limit and from various models in lower dimensions. We will discuss these two aspects one by one.

9.2 The 't Hooft Limit and 1/*N* Expansion

9.2.1 Introduction

In asymptotically free gauge theories in the confining phase, the gauge coupling g^2 is not in fact an expansion parameter. Through dimensional transmutation it sets the scale of physical phenomena,

$$\Lambda = M_{\text{uv}} \exp\left(-\frac{8\pi^2}{\beta_0 g_0^2} + \cdots\right), \tag{9.2}$$

Dimensional transmutation

where M_{uv} is the ultraviolet cutoff, g_0 is the bare coupling at the cutoff, β_0 is the first coefficient in the Gell-Mann–Low function, and the ellipses stand for higher-order terms. The hadron masses are of order Λ, the charge radii of order Λ^{-1}, and so on. The incredible variety of the hadronic world is explained by a variety of numerical coefficients – all, generally speaking, of order 1.

Well, the above statement is not true or, better to say, it is not the whole truth. A hidden expansion parameter was found by 't Hooft [7]. In the actual world, quantum chromodynamics is based on the gauge group SU(3). If, instead, we consider the gauge group SU(N) then a smooth limit can be attained at $N \to \infty$ provided that the gauge coupling scales as follows:

't Hooft limit

$$g^2 N = \text{const.} \tag{9.3}$$

This limit is referred to as the *'t Hooft limit*.

Statement: great simplifications occur in the 't Hooft limit. As we will see shortly, only planar diagrams survive. Thus it is also known as the *planar limit*. Moreover, the $1/N$ expansion is in one-to-one correspondence with the topology of the surface on which the corresponding Feynman graphs can be drawn.

The planar graphs can be drawn on a plane (with identified infinite points, so that topologically we must deal with a sphere). In pure Yang–Mills theory the next-to-leading term is suppressed as $1/N^2$; it is associated with a surface with one handle (the toric topology). The $O(1/N^4)$ term comes from the two-handle topology, and so on. The combination $g^2 N$ in Eq. (9.3) is referred to as the *'t Hooft coupling*,

't Hooft coupling

$$\lambda \equiv g^2 N. \tag{9.4}$$

Planar diagrams can contain any power of the 't Hooft coupling. The first coefficient in the β function in Yang–Mills theory is

$$\beta_0 = \tfrac{11}{3} N,$$

therefore it is the 't Hooft coupling that appears in the dynamically generated scale (9.2).

Passing to QCD, i.e., Yang–Mills theory with quarks, we start from the observation that in the actual world quarks belong to the fundamental representation of SU(3). If we assume that this assignment stays intact in multicolor QCD then each extra quark loop is suppressed by $1/N$ (see below). Therefore, in the 't Hooft limit each process is dominated by contributions with the minimal possible number of quark loops. Below we will derive these results and outline some consequences.

9.2.2 *N*-Counting and Topology

Let us examine the combinatorics of Feynman diagrams in the large-N limit. For large N there are many colors and therefore many possible intermediate states, so that the sum over these intermediate states gives rise to N factors. It is convenient to think of the gluon field as an $N \times N$ matrix $A^j_{\mu\,i}$, with an upper fundamental index and a lower antifundamental index, which gives us N^2 components. More exactly this

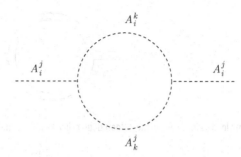

Fig. 9.2 One-loop gluon contribution to the gluon vacuum polarization. In this and subsequent graphs gluons are denoted by broken lines.

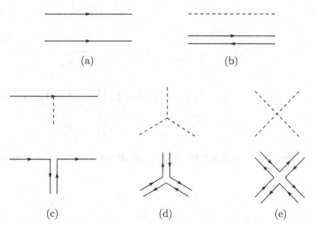

Fig. 9.3 't Hooft double line notation. The lower diagram shows each QCD propagator or interaction vertex in the double-line notation.

> **'t Hooft graphs**

matrix is traceless, so that the number of components is $N^2 - 1$ but the difference between N^2 and $N^2 - 1$ can be neglected at large N. Note that the quark and antiquark fields ψ^i and $\bar{\psi}_j$ carry a fundamental and an antifundamental index, respectively. Thus, for keeping track of color factors (and for this purpose only) one can represent the gluon field as a quark–antiquark pair. This circumstance will be used shortly to construct the 't Hooft double-line graphs, which encode all information on color loops in a very transparent manner.

Let us consider a typical Feynman diagram, for instance, the gluon contribution to the gluon vacuum polarization, Fig. 9.2. Let us specify the color indices of the incoming and outgoing lines as i, j. Then the pair of gluons propagating in the loop is A_k^i and A_j^k, summation over k being implied. Thus, this diagram is in fact $O(g^2 N)$ and is of leading order in the $1/N$ expansion.

An easy way to see how the N factor appears is to redraw the graph in Fig. 9.2 in the double-line language. If a quark or antiquark is represented in a Feynman diagram as a single line with an arrow, the direction of the arrow distinguishing quark from antiquark, we should represent the gluon as a double line, with opposite arrows on the two lines, representing the corresponding color flow, as in Fig. 9.3. In the double-line representation each closed loop gives a factor N. For instance, Fig. 9.4 represents Fig. 9.2 in the double-line language. The occurrence of N is trivially seen in this language.

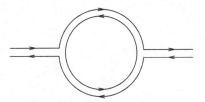

Fig. 9.4   The same loop as in Fig. 9.2 in the 't Hooft double-line notation. The arrows denote color flow.

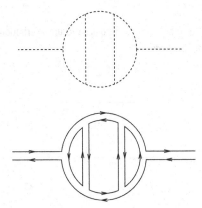

Fig. 9.5   Three-loop gluon contribution to the gluon vacuum polarization. This graph is planar.

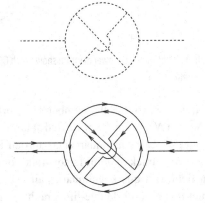

Fig. 9.6   A nonplanar three-loop gluon contribution to the gluon vacuum polarization.

An example of a more complicated planar three-loop graph is presented in Fig. 9.5, in the standard and 't Hooft notation. One can immediately convince oneself that this graph is $O((g^2N)^3)$. As was mentioned, nonplanar graphs do not survive in the 't Hooft limit. For instance, a three-loop graph that does not survive is indicated in Fig. 9.6. It is impossible to draw this diagram on a plane without line crossings (at points where there are no interaction vertices). This diagram has six interaction vertices, but only one large and tangled color loop which gives us

$$g^6N \sim \frac{1}{N^2}(g^2N)^3.$$

In other words, we get $1/N^2$ suppression compared to its planar counterpart in Fig. 9.5. By experimenting with other examples it is not difficult to guess that this

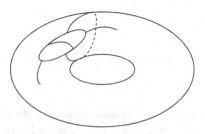

Fig. 9.7 The non-planar graph of Fig. 9.6 drawn on a torus.

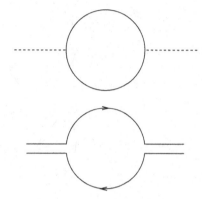

Fig. 9.8 One-loop quark contribution to the gluon vacuum polarization. In this and subsequent graphs quarks are denoted by solid lines.

conclusion must be general: nonplanar Feynman diagrams with gluons always vanish at least as $1/N^2$ for large N. Note that the diagram of Fig. 9.6 can be drawn without self-intersections on a torus; see Fig. 9.7.

As far as the quark loops are concerned, the fact that for large N there are N^2 gluon states and only N quark states suggests that all internal quark loops are suppressed by $1/N$. Indeed, let us consider the one-quark loop contribution to the gluon propagator (Fig. 9.8). Inspecting the double-line representation we note that the closed color loop that appeared in the gluon graph in Fig. 9.4, is absent in the quark graph. The reason is that the quark propagator corresponds to a single color line, not two. As a result, the contribution of Fig. 9.8 is proportional to

$$g^2 \sim \frac{1}{N}(g^2 N).$$

This conclusion is also general: any internal quark loop is suppressed by $1/N$. Therefore, in the 't Hooft limit one should consider only planar graphs with the minimal number of quark loops.

It is not always possible to get rid of the quark loops altogether. For instance, if one is considering the photon polarization operator, the photon, being coupled only to quarks, necessarily creates a quark–antiquark pair; see Fig. 9.9. The same is true for n-point functions induced by quark bilinear operators $\bar{\psi}\psi$, $\bar{\psi}\gamma^5\psi$, and so on. The free-quark diagram depicted in Fig. 9.9 is of order N, corresponding to the color sum for the quark running around the loop. One can make arbitrary gluon insertions without changing this N-dependence, as long as planarity is conserved. For instance, the diagram of Fig. 9.10 is of order

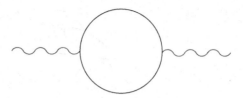

Fig. 9.9 One-quark loop in the photon polarization operator. In this and subsequent graphs the wavy lines denote photons or other external sources that are bilinear in the quark fields.

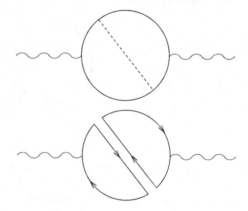

Fig. 9.10 Two-loop contribution to the photon polarization operator.

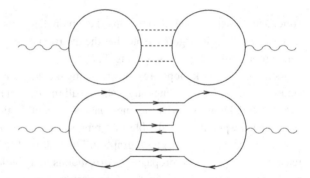

Fig. 9.11 Disconnected quark loops in the photon polarization operator.

$$g^2 N^2 \sim N(g^2 N).$$

However, the diagrams of Figs. 9.11 and 9.12, with an extra quark loop, are of order $\left(g^2 N\right)^3$, i.e., they carry the relative suppression factor $1/N$.

The above two rules for survival in the 't Hooft limit – planarity and the minimal number of the quark loops – must be supplemented by a third rule, which applies if the quark loop is coupled to external sources, as in Fig. 9.9. The three-loop diagram of Fig. 9.13 is drawn on the plane. However, expressing it in double-line language, Fig. 9.14, it is not difficult to see that it has only a single closed color loop and so is of order $g^4 N$, i.e., it carries the relative suppression factor $1/N^2$. This diagram differs from the previous examples in that the gluon lines are attached on both sides of the fermion loop. Thus, the third rule can be formulated as follows: the leading

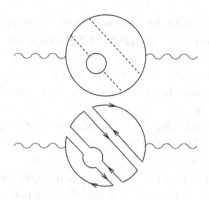

Fig. 9.12   Quark insertion in the gluon propagator in the photon polarization operator.

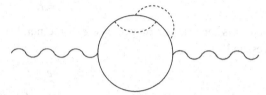

Fig. 9.13   A suppressed planar graph with gluon lines on both sides of the quark propagator.

Fig. 9.14   The 't Hooft double-line representation for the diagram of Fig. 9.13.

contributions in the n-point functions induced by the quark bilinear operators are planar diagrams with *quarks at the edge*.

9.2.3 The 't Hooft Limit and String Theory

Let us first limit ourselves to pure Yang–Mills theory and consider the set of diagrams for the vacuum energy (i.e., without external lines). One can think of each double-line graph as a surface obtained by gluing polygons together at the double lines. Since each line has an arrow on it, and double lines have oppositely directed arrows, one can only construct orientable polygons.

General derivation of the 't Hooft counting rules

To compute the N-dependence one needs to count the powers of N from sums over closed color-index loops, as well as factors $1/\sqrt{N}$ from the explicit N-dependence in the coupling constants. It is convenient to use a rescaled Lagrangian to "mechanize" the derivation of N-counting. To this end we define a QCD Lagrangian as follows:

$$\mathcal{L} = N\left(-\frac{1}{2\lambda}\,\mathrm{Tr}\,G_{\mu\nu}G^{\mu\nu} + \sum \bar{\psi}_f\,i\,\mathcal{D}\psi^f\right). \tag{9.5}$$

This Lagrangian has an overall factor N; nevertheless, the theory does not reduce to a classical theory of quarks and gluons in the $N \to \infty$ limit because the numbers of components of ψ and A_μ grow with N as N and N^2, respectively. The coupling λ is defined in Eq. (9.4). The sum in (9.5) runs over all quark flavors, which are assumed to be massless for simplicity. The number of flavors does not scale with N, by assumption.

One can readily determine the powers of N in any Feynman graph using Eq. (9.5) and the 't Hooft notation. Every vertex contributes a factor N, and every propagator contributes a factor $1/N$. In addition, every color loop gives a factor N. In the double-line notation, where Feynman graphs correspond to glued polygons that form surfaces, each color loop is the edge of a polygon and, in addition, defines a face of the surface. As a result, any connected vacuum graph scales with N as

$$N^{v-e+f} = N^\chi, \tag{9.6}$$

where v is the number of vertices, e is the number of edges, f is the number of faces, and

$$\chi \equiv v - e + f \tag{9.7}$$

Euler character

is a topological invariant of two-dimensional surfaces known as the *Euler character*. For any connected orientable surface we have

$$\chi = 2 - 2h - b, \tag{9.8}$$

where h is the number of handles and b is the number of boundaries (or holes). For a sphere, $h = 0$, $b = 0$, $\chi = 2$; for a torus, $h = 1$, $b = 0$, $\chi = 0$, and so on. The Euler character is related to the genus g of the surface as follows:

$$\chi = 2 - 2g. \tag{9.9}$$

Here g stands for genus.

The maximum power of N is 2, from diagrams with $h = b = 0$.

To illustrate the above analysis we can inspect the planar diagram in Fig. 9.15. It has three color loops and two vertices. After drawing it on a sphere according to the rules specified above, we can identify three edges, three surfaces, and two vertices.

Now let us switch on quarks. A quark is represented by a single line; therefore, a closed quark loop is a boundary. Compared with the surfaces one obtains in pure Yang–Mills theory, for each quark loop one must remove one polygon. For instance, in planar graphs one obtains a sphere with one hole. Correspondingly, $b = 1$ and, instead of N^2, now one obtains N.

Summarizing, large-N diagrams in QCD look like two-dimensional surfaces. For example, the leading diagram in the pure-glue sector has the topology of a sphere and the leading diagram in the quark sector is a surface with the quark as the

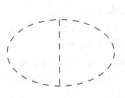
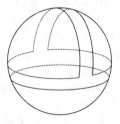

outermost edge. One can imagine all possible planar gluon exchanges as filling out the surface into a two-dimensional world sheet. It has been conjectured that this is the way in which large-N QCD might be connected with string theory, planar diagrams representing the leading-order string theory diagrams [7]. *The topological counting rule for the $1/N$ suppression factors in QCD is the same as that for the string coupling constant in the string loop expansion* (see, e.g., [8]).

Take, for instance, a toric surface in closed-string theory. If we depict it as lying on the horizontal plane and slice it by a vertical plane moving from left to right, we will see that it describes the propagation of a closed string, with a subsequent split into two closed strings, which then reassemble themselves. This process is of order g_s^2, where g_s is the string coupling constant. Compared with the spherical surface the process is suppressed by g_s^2. At the same time, in pure Yang–Mills theory, according to (9.6) and (9.8), the same suppression is $1/N^2$. Thus the string coupling g_s must indeed be identified with $1/N$. The processes with quark loops are related to open strings.

> *String coupling $g_s \leftrightarrow 1/N$*

As we will see in Section 9.2.4, the $1/N$ expansion, by and large, is supported by the known phenomenology of hadron physics. This is the reason why, starting with 't Hooft, people have believed that QCD has an underlying string representation and the $1/N$ expansion in QCD is related to a topological expansion in string dynamics. Unfortunately, the connection between large-N QCD and string theory has never been made precise.

In a sense, now a logical circle is closed: on the theoretical side, as we saw in Section 9.1, in some supersymmetric Yang–Mills theories flux tubes emerge, providing a natural basis for color confinement through the dual Meissner effect. Going in the opposite direction, through phenomenology, we learn about the attractiveness of the $1/N$ expansion in QCD and how it hints at an underlying string representation of QCD, which, when established, will describe confining dynamics.

9.2.4 Implications of the 1/N Expansion in Mesons (in Brief)

Let us see how well the $1/N$ expansion reproduces basic regularities of the hadronic world. Assuming confinement and using $1/N$ one can deduce the following in the 't Hooft limit.

> *Regularities of the hadronic world*

(i) Mesons – quarkonia and glueballs – are stable and noninteracting at $N = \infty$. The meson masses scale as N^0 (with the exception of η'). The number of meson states for given J^{PC} and flavor quantum numbers is infinite.

(ii) The amplitudes for quarkonia decays of the type $a \to bc$ are suppressed as $1/\sqrt{N}$ while the $ab \to cd$ scattering amplitudes are suppressed as $1/N$, and so on. For glueballs the corresponding suppression factors are $1/N$ and $1/N^2$, respectively. The widths of quarkonium mesons scale as $1/N$ while those of glueballs as $1/N^2$.

(iii) As $N \to \infty$ an effective QCD Lagrangian presents an infinite number of terms corresponding to an infinite number of stable mesons, which are described by tree interaction amplitudes suppressed by powers of $1/N$.

(iv) The multibody decays of excited mesons are dominated by resonant two-body final states, whenever these states are available.

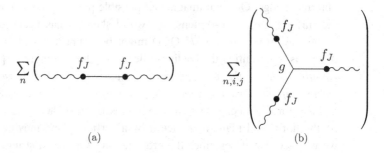

Fig. 9.16 The phenomenological representation of two- and three-point functions of quark bilinears. The sum runs over an infinite number of quarkonium mesons with appropriate quantum numbers. The coupling f_J is determined from the amplitude $\langle \text{vac}|J|\text{meson}\rangle$, while g stands for the tri-meson coupling.

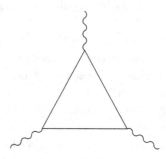

Fig. 9.17 The quark loop diagram for the three-point function $\langle J(x) J(y) J(0) \rangle$.

    (v) Flavor-singlet quarkonia do not mix among themselves, nor do they mix with glueballs at $N \to \infty$.

  (vi) The $\bar{q}q\bar{q}q$ exotics are absent in the limit $N = \infty$;

 (vii) Scattering processes in strong interactions (e.g., $\pi\pi$ scattering) can be described in terms of tree diagrams with the exchange of physical mesons (hence, Regge phenomenology is justified at high energies).

The pattern summarized above, following from the $1/N$ expansion, is indeed observed in hadronic phenomenology. No other explanation of all these regularities that is as universal as $1/N$ has been found. This is a strong evidence that the $1/N$ expansion is a good approximation in our world, in which the underlying theory of strong interactions is quantum chromodynamics.

We will not derive most of the above results; an excellent pedagogical presentation can be found in [9]. For illustration I will show how one can establish the validity of (ii). All other statements can be verified in a similar way.

We start from the two-point function $\langle J(x)J(0)\rangle$, where $J = \bar{\psi}\Gamma\psi$ is a bifermion operator, for instance the vector current (in which case $\Gamma = \gamma^\mu$). On the theoretical side this correlation function is described by the graphs depicted in Figs. 9.9 and 9.10, and similar diagrams. As we already know, in the 't Hooft limit all these diagrams scale as N^1. On the phenomenological side the correlation function under consideration is presented by an infinite sum of mesonic poles; see Fig. 9.16a. Each pole enters with a weight $|f_J|^2$, where f_J is the coupling constant of the nth meson in the given channel. This fact implies that $f_J \sim \sqrt{N}$.

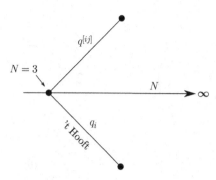

Fig. 9.18   The quark fields of three-color QCD can be generalized to the multicolor case in two different ways, since at $N = 3$ the two-index antisymmetric field $q^{[ij]}$ is the same as the (anti)fundamental field q_j.

Next, we must establish the N-dependence of the three-point function $\langle J(x)J(y) J(0)\rangle$. The simplest Feynman diagram for this three-point function is shown in Fig. 9.17. Needless to say, all planar diagrams must be summed up. The result scales as N^1. Let us compare it with the phenomenological "mesonic" representation; see Fig. 9.16b. Each term in the mesonic sum is proportional to $f_J^3 g$, where g is the decay constant. Thus $f_J^3 g \sim N$, implying that $g \sim 1/\sqrt{N}$. This completes the proof of point (ii) above.

9.2.5 Alternative Large-N Expansion

As we learned in Section 9.2.1, the 't Hooft large-N expansion is based on the assumption that, independently of the value of N, the quark fields are in the fundamental representation of SU(N). The $1/N$ suppression of each quark loop ensues. This is not the only possible choice, however. Indeed, consider a Dirac fermion field $\psi^{[ij]}$ in the two-index antisymmetric representation of SU(N). At $N = 3$ this field is identical to that in the antifundamental representation ψ_i. Indeed, for SU(3), $\bar{\psi}_{[ij]}\varepsilon^{ijk} \sim \psi^k$. In other words, at $N = 3$ it describes the standard quark. However, the continuation to larger values of N is totally different (see Fig. 9.18). The field $\psi^{[ij]}$ has $\frac{1}{2}N(N - 1)$ color components. At large N the number of color degrees of freedom in $\psi^{[ij]}$ is $\frac{1}{2}N^2$ rather than N. Needless to say, there can be more than one flavor of antisymmetric quarks. Thus, it is obvious that extrapolation to large N, with the subsequent $1/N$ expansion, can go via distinct routes with the same starting point: (i) quarks in the *fundamental* representation; (ii) quarks in the *two-index antisymmetric* representation.

Quark loops are not suppressed in ASV.

The first option gives rise to the standard 't Hooft $1/N$ expansion [7] while the second leads to an alternative Armoni–Shifman–Veneziano (ASV) expansion [10, 11].[2] The relation between $1/N$ and the graph topology remains the same.

---

[2] Corrigan and Ramond suggested as early as 1979 [12] replacing the 't Hooft model by a model with one two-index antisymmetric quark $\psi_{[ij]}$ and two fundamental quarks $q^i_{1,2}$. Their motivation originated from some awkwardness in the treatment of baryons in the 't Hooft model, where all baryons, being composed of N quarks, have masses scaling as N and thus disappear from the spectrum at $N = \infty$. If the fermion sector contains $\psi_{[ij]}$ and $q^i_{1,2}$ then, even at large N, there are three-quark baryons of the type $\psi_{[ij]}q^i_1 q^j_2$. However, the symmetry between all quarks comprising baryons is lost, an obvious drawback.

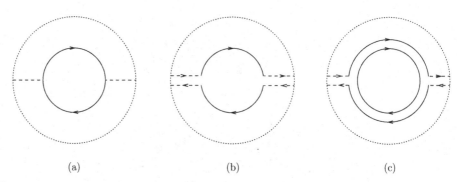

                    (a)   (b)   (c)

Fig. 9.19   (a) A typical contribution to the vacuum energy. (b) The planar contribution in 't Hooft large-N expansion. (c) The ASV large-N expansion. The dotted circle represents a sphere, so that every line hitting the dotted circle gets connected "on the other side."

Therefore, all consequences from the planar graph dominance at $N \to \infty$ remain intact. However, quark loops are no longer suppressed.

The 't Hooft expansion has enjoyed a significant success in phenomenology. It has provided a qualitative explanation for the well-known regularities of the hadronic world. Although the standard large-N ideology definitely captures the basic regularities, it gives rise to certain puzzles as far as the subtle details are concerned. Indeed, in the 't Hooft expansion, the width of $q\bar{q}$ mesons scales as $1/N$ while that of glueballs scales as $1/N^2$. In other words the latter are expected to be narrower than quarkonia, which is hardly the case in reality. Moreover, the Zweig rule[3] is not universally valid in actuality. It is known to be badly violated for scalar and pseudoscalar mesons.

Thus, the 't Hooft expansion seemingly underestimates the role of quarks, at least in some cases. The ASV large-N expansion eliminates the quark loop suppression. It opens the way for a large-N phenomenology in which quark loops (i.e., dynamical quarks) do play a non-negligible role. An additional bonus is that in the ASV large-N expansion, one-flavor QCD connects with supersymmetric Yang–Mills theory (Sections 9.2.6 and 10.13), via *planar equivalence*.

To illustrate the difference between the 't Hooft and ASV large-N expansions, I exhibit in Fig. 9.19 a planar contribution to the vacuum energy in two expansions. Mentioning a few important distinctions between these two expansions in meson phenomenology, we note that (i) the decay widths of *both* glueballs and quarkonia scale with N in a similar manner, as $1/N^2$; this can be deduced by analyzing the appropriate diagrams with quark loops of the type displayed in Figs. 9.19b, c; (ii) the unquenching of quarks in the vacuum gives rise to quark-induced effects that are not suppressed by $1/N$; in particular, the vacuum energy density becomes quark-mass dependent at the leading order in $1/N$. In baryon phenomenology, the predictions of the 't Hooft and ASV large-N expansions were compared in [14]. Both large-N limits generate an emergent spin–flavor symmetry (Section 9.2.10) that leads to the

---

[3] The Zweig or Okubo–Zweig–Iizuka rule, states that any QCD process describeable by Feynman graphs that can be cut into two pieces by cutting *only* internal gluon lines is suppressed. The default example of a Zweig-suppressed decay is $\phi \to \pi^+ \pi^- \pi^0$. For a review of this rule and the fascinating story of its discovery, see [13].

vanishing of particular linear combinations of baryon masses at specific orders in the expansions. Experimental evidence shows that these relations hold at the expected orders regardless of which large-N limit one uses, suggesting the validity of either limit in the study of baryons.

9.2.6 Planar Equivalence

In this section I will show that two confining Yang–Mills theories with obviously different fermion contents can be equivalent to each other in the $N \rightarrow \infty$ limit for a judiciously chosen set of correlation functions. In other words, there is a sector of these theories, usually referred to as the *common sector*, in which they are indistinguishable from each other at $N = \infty$. First, I will establish the existence of planar-equivalent pairs of theories. Then we will discuss how we can benefit from this.

Consider two SU(N) Yang–Mills theories. In the simplest case [15] one of the theories to be compared has a Weyl spinor in the adjoint representation of SU(N). Let us call this theory the *parent*. As we will learn in Part II (Section 10.14), the parent theory is nothing other than $\mathcal{N} = 1$ super-Yang–Mills. The fermion field is that of a gluino, with the standard notation λ^a where a is the color index of the adjoint representation.

The second theory (a *daughter theory*), to be compared with the first, has a single Dirac fermion in the two-index antisymmetric representation. This is the theory that we discussed in Section 9.2.5, with one flavor. Both theories have the same gauge group and the same gauge coupling.

See Section 10.14.

The gluino field λ^a can also be written as $\lambda^i_j \equiv \lambda^a (T^a)^i_j$, with one upper and one lower color index (i.e., a fundamental and an antifundamental index), the T^a being generators of the gauge group. To pass from the parent to the daughter theory we replace λ^i_j by two Weyl spinors $\eta_{[ij]}$ and $\xi^{[ij]}$, with two antisymmetrized indices. We can combine the Weyl spinors into one Dirac spinor, either $\psi^{[ij]} \sim (\xi, \bar{\eta})$ or $\psi_{[ij]} \sim (\eta, \bar{\xi})$. Note that the number of fermion degrees of freedom in $\psi_{[ij]}$ is $N^2 - N$, while in the parent theory it is $N^2 - 1$, i.e., the same as in the large-N limit.

The hadronic (color-singlet) sectors of the parent and daughter theories are different, generally speaking. Thus, in the parent theory, composite fermions with mass scaling as N^0 exist and, moreover, they are degenerate with their bosonic superpartners. In the daughter theory any interpolating color-singlet current with fermion quantum numbers contains a number of constituents growing with N. Hence, at $N = \infty$ the spectrum contains only bosons.

Classically the parent theory has a single global symmetry – an R symmetry corresponding to the chiral rotations of the gluino field. In fact, the corresponding current is axial-vector. Instantons break this symmetry down to Z_{2N}, through the chiral anomaly discussed in Section 8.2.1. The daughter theory has, in addition, the conserved anomaly-free current

$$\bar{\eta}_{\dot\alpha}\eta_\alpha - \bar{\xi}_{\dot\alpha}\xi_\alpha. \tag{9.10}$$

Definition of the common sector

In terms of the Dirac spinor this is the vector current $\bar{\psi}\gamma_\mu\psi$. From the fact of the existence of (9.10) in the daughter theory it is clear that even in the bosonic sector the spectra of these two theories are different. The common sector of both theories

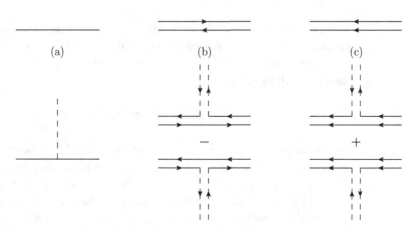

Fig. 9.20   (a) A fermion propagator and a fermion–fermion–gluon vertex; (b) the parent theory, $\mathcal{N} = 1$ super-Yang–Mills; (c) the daughter theory.

is defined as follows: any given interpolating (color-singlet) operator of the parent theory belonging to the common sector must have a projection onto the daughter theory, and vice versa. In particular, all glueballs belong to the common sector. In both theories the Z_{2N} symmetry is spontaneously broken down to Z_2 by bifermion condensates $\langle \lambda \lambda \rangle$ and $\langle \bar{\psi} \psi \rangle$, respectively, implying the existence of N degenerate vacua[4] in both cases.

Now I will explain, using broad brush strokes, why planar equivalence occurs. For details of a proof valid at the perturbative and nonperturbative levels the reader is referred to [15]. The Feynman rules in both theories in the 't Hooft double-line notation are shown in Fig. 9.20. The difference is that the arrows on the fermionic lines point in the same direction in the daughter theory, since the fermion is in the antisymmetric two-index representation, in contrast with the supersymmetric theory where the gluino is in the adjoint representation and hence the arrows point in opposite directions. This difference between the two theories does not affect planar graphs, provided that each gaugino line is replaced by the sum of $\eta_{[\cdot\cdot]}$ and $\xi^{[\cdot\cdot]}$.

There is a one-to-one correspondence between the planar graphs of the two theories. Diagrammatically this works as follows; see, for example, Fig. 9.21. Consider any planar diagram of the parent $\mathcal{N} = 1$ theory: by the definition of planarity it can be drawn on a sphere. The fermionic propagators form closed, nonintersecting, loops that divide the sphere into regions. Each time we cross a fermionic line the orientation of the color-index loops (each producing a factor N) changes from clockwise to counterclockwise, and vice versa, as can be seen in Fig. 9.21b. Thus, the fermionic loops allow one to attribute to each of the above regions a binary label (say, ± 1), according to whether the color loops go clockwise or counterclockwise in the given region. Imagine now that one cuts out all the regions with label -1 and glues them back onto the sphere, after having flipped them upside down. We then obtain a planar diagram of the daughter theory in which all color loops go, by convention, clockwise. The overall number associated with both diagrams will be the same since

---

[4] At finite N the parent theory has N vacua, while the daughter theory discussed in this section has $N-2$ vacua.

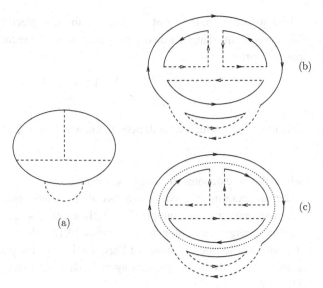

Fig. 9.21 (a) A typical planar contribution to the vacuum energy. The same in 't Hooft notation for (b) the parent theory; (c) the daughter.

the diagrams within each region always contain an even number of powers of g, so that the relative minus signs of Fig. 9.20 do not matter.

In fact, in the above argument, we have ignored certain subtleties, so that the careful reader might get somewhat worried. For instance, in the parent theory gluinos are Weyl fermions, while in the daughter theory fermions are Dirac. Therefore, an explanatory remark is in order here.

First, let us replace the Weyl gluino of $\mathcal{N} = 1$ super-Yang–Mills theory by a Dirac spinor ψ^i_j. Each fermion loop in the parent theory is then obtained from the Dirac loop by multiplying the latter by $\frac{1}{2}$. Let us keep this factor $\frac{1}{2}$ in mind.

In the daughter theory, instead of considering the antisymmetric spinor $\psi_{[ij]}$ we will consider a Dirac spinor in the *reducible* two-index representation ψ_{ij}, without imposing any (anti)symmetry conditions on i, j. Thus, this reducible two-index representation is a sum of two irreducible representations symmetric and antisymmetric. It is rather obvious that at $N \to \infty$ any loop of $\psi_{[ij]}$ yields the same result as the very same loop with $\psi_{\{ij\}}$, which implies in turn that, to get the fermion loop in the antisymmetric daughter, one can take the Dirac fermion loop in the above reducible representation and multiply it by $\frac{1}{2}$.

Given that there is the same factor $\frac{1}{2}$ on the side of the parent and daughter theories, what remains to be done is to prove that the Dirac fermion loops for ψ^i_j and ψ_{ij} are identical at $N \to \infty$. To this end, from now on we will focus on the color factors.

Equivalence proof in more detail: counting color factors on both sides

Let the generator of SU(N) in the fundamental representation be T^a and that in the antifundamental be $\bar{T}^a$:

$$T^a = T^a_N, \qquad \bar{T}^a = T^a_{\bar{N}}. \tag{9.11}$$

Then the generator in the adjoint representation is

$$T^a_{\text{adj}} \sim T^a_{N \otimes \bar{N}} = T^a_N \otimes 1 + 1 \otimes T^a_{\bar{N}}$$

$$\equiv T^a \otimes 1 + 1 \otimes \bar{T}^a, \tag{9.12}$$

where we have made use of the large-N limit, neglecting the singlet (trace) part. Moreover, in the daughter theory the generator of the reducible $N \otimes N$ representation can be written as

$$T^a_{\text{two-index}} = T^a_N \otimes 1 + 1 \otimes T^a_N \equiv T^a \otimes 1 + 1 \otimes T^a$$
$$\text{or} \quad \bar{T}^a \otimes 1 + 1 \otimes \bar{T}^a. \tag{9.13}$$

One more thing which we will need to know is that (e.g., [16])

$$\bar{T} = -\tilde{T} = -T^*, \tag{9.14}$$

where the tilde denotes the transposed matrix.

Let us examine the color structure of a generic planar diagram for a gauge-invariant quantity. For example, Fig. 9.21a exhibits a four-loop planar graph for the vacuum energy. The color decomposition (9.12), (9.13) is equivalent to using the 't Hooft double-line notation, see Figs. 9.21b, c. In the parent theory each fermion–gluon vertex contains T^a_{adj}; in passing to the daughter theory we make the replacement $T^a_{\text{adj}} \rightarrow T^a_{\text{two-index}}$.

Upon substitution of Eqs. (9.12) and (9.13) the graph at hand splits into two (disconnected!) parts:[5] an inner part (inside the dotted ellipse in Fig. 9.21c) and an outer part (outside the dotted ellipse in Fig. 9.21c). These two parts do not communicate, because of planarity (i.e., in the large-N limit). The outer parts in Figs. 9.21b, c are the same. They are proportional to the trace of the product of two Ts in the two cases

$$\text{Tr} \, \bar{T}^a \bar{T}^a.$$

This is the first factor. The second comes from the inner part of Figs. 9.21b, c. In the parent theory the inner factor is built from six Ts, one in each fermion-gluon vertex, and three Ts in the three-gluon vertex $\text{Tr}([A_\mu A_\nu]\partial_\mu A_\nu)$, where $A_\mu \equiv A^a_\mu T^a$. In the daughter theory the inner factor is obtained from that in the parent theory by replacing all Ts by $\bar{T}$s. According to Eq. (9.14), $\bar{T} = -\tilde{T}$ (remember that a tilde denotes the transposed matrix). This fact implies that the only difference between the inner blocks in Figs. 9.21b, c is the reversal in the direction of color flow on each 't Hooft line. Since the inner part is a color singlet by itself, the above reversal has no impact on the color factor – the color factors are identical in the parent and daughter theories.

It may be instructive to illustrate how this works using a more conventional notation. For the inner part of the graph in Fig. 9.21b we have a color factor $\text{Tr}(T^a T^b T^c) \, f^{abc}$, while in the daughter theory we have $\text{Tr}(\bar{T}^a \bar{T}^b \bar{T}^c) \, f^{abc}$. Using

$$[T^a, T^b] = i f^{abc} T^c \qquad \text{and} \qquad [\bar{T}^a, \bar{T}^b] = i f^{abc} \bar{T}^c,$$

we immediately come to the conclusion that the above two color factors coincide.

Now we will consider the benefits that one can extract from planar equivalence. At $N \rightarrow \infty$ all results applicable in one theory can be copied into the other.[6] In particular, all predictions (in the common sector) obtained in $\mathcal{N} = 1$ super-Yang–Mills

---

[5] More exactly, what is meant here is the color structure of the graph.
[6] This refers only to the common sector.

theory stay valid in the daughter theory. For example, we can assert that the β function of the daughter theory is

$$\beta(\alpha) = -\frac{1}{2\pi} \frac{3N\alpha^2}{1 - N\alpha/(2\pi)} \left[1 + O\left(\frac{1}{N}\right)\right], \qquad \alpha = \frac{g^2}{4\pi} \qquad (9.15)$$

(cf. Section 10.21). Note that the corrections are $1/N$ rather than $1/N^2$. For instance, the exact first coefficient of the β function is $-3N - \frac{4}{3}$ as against $-3N$ in the parent theory.

The same equivalence applies to the vacuum states of both theories: their vacuum structure is identical at $N \to \infty$, up to $1/N$ corrections.

9.2.7 Baryons in the 't Hooft Limit

Large-N QCD can be treated as a weakly coupled field theory of mesons. It is a theory of effective local meson fields, with effective local interactions, in which the three-meson coupling scales as $1/\sqrt{N}$, the four-meson as $1/N$, and so on. At large N all coupling constants are weak. As we know already, many weakly coupled field theories possess, in addition to elementary excitations, heavy solitonic states whose masses diverge at weak coupling as the inverse of the coupling. Are there such states in QCD and its effective mesonic counterpart? The answer is positive. In QCD we have N-quark states – baryons – whose mass is proportional to N. As a reflection of this fact, the low-energy mesonic theory must have solitons with nonvanishing baryon numbers and masses scaling as N. These are the Skyrmions, considered in Section 4.2. By and large, the Skyrmion model results give a satisfactory description of the low-lying baryons. And yet, a model is just a model ... It turns out, however, that some implications of the Skyrmion model are model-independent; they follow from QCD in the 't Hooft limit without the invoking of particular details of the Skyrme model *per se*. Here we will focus on such general aspects of the baryon theory in multicolor QCD [17].

Look through Section 4.2 to refresh you knowledge.

Baryons are color-singlet hadrons composed of N quarks in the fundamental representation. N is the minimal number of the baryon constituents since the SU(N) invariant Levi–Civita tensor (the ε symbol) has N indices,

$$B \sim \varepsilon_{i_1 \cdots i_N} q^{i_1} \cdots q^{i_N}. \qquad (9.16)$$

The ε symbol is fully antisymmetric in color. Since quarks obey Fermi statistics, the baryon must be completely symmetric in other quantum numbers such as spin and flavor.

The number of quarks in baryons grows with N, so one might think that extrapolation from $N = 3$ to the large-N limit is not a good procedure for baryons. However, we will see that for baryons, as for mesons, the expansion parameter is $1/N$ and that one can compute baryonic properties in a systematic semiclassical expansion in $1/N$. The results are in good agreement with experiment and shed light on the spin–flavor structure of baryons. In fact, the main achievement of large-N analysis in the baryon sector is the realization that there is a deep connection between QCD and two popular models of baryons: the quark model and the Skyrme model. Some seemingly naive results of the quark model get a solid theoretical justification.

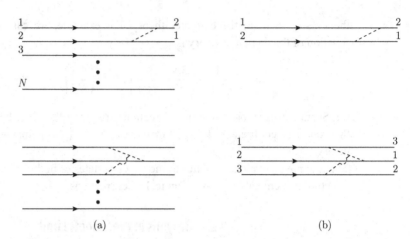

Fig. 9.22 (a) Two-and three-quark interactions in baryons (upper and lower panels on the left) and (b) the corresponding connected components. The numbers labeling the quark lines indicate color.

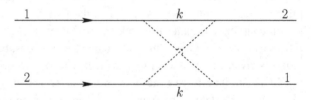

Fig. 9.23 An example of a "planar" two-body baryon graph. The gluon lines do not intersect.

9.2.8 The N-Counting Rules for Baryons

Let us start by deriving N-counting rules for baryon graphs. To this end we draw the incoming baryon as N quarks, with the colors arranged in order, $1, \ldots, N$. The colors of the outgoing quark lines are then a permutation of $1, \ldots, N$. The two- and three-quark interactions are depicted in Fig. 9.22a. The connected parts are presented in Fig. 9.22b. A connected part that contains n quark lines will be referred to as an n-body interaction. The colors on the outgoing quarks in the n-body interaction are a permutation of the colors on the incoming quarks, and the colors are distinct. Each outgoing line can be identified with an incoming line of the same color in a unique way.

Let us start with the two-body interaction, with the color assignments given in Fig. 9.22b. It has an explicit g^2 factor in addition to the combinatorial factor $\frac{1}{2}N(N-1)$ reflecting the number of ways in which one can choose two lines out of N. Thus, this contribution scales as N. The double gluon exchange depicted in Fig. 9.23 does not look planar at first sight. However, if we take into account the color loop corresponding to summation over the color index k we will conclude that this graph is proportional to $g^4 \times N$ times the combinatorial factor $\frac{1}{2}N(N-1)$, i.e., it scales as N too.[7]

---

[7] Baryon graphs in the double-line notation can have color index lines crossing each other owing to fermion line "twists."

*Baryon
n-body
interactions
scale as N
for all n.*

Moreover, the same scaling law applies to three-body interactions, as is clearly seen from the three-body contribution in Fig. 9.22b, which is proportional to g^4 times the combinatorial factor $\frac{1}{6}N(N-1)(N-2)$. A similar examination gives us the N-counting rules for all n-body interactions in baryons: the kernel itself scales as N^{1-n} but there are $O(N^n)$ ways of choosing n quarks from an N-quark baryon. Thus the net effect of n-body interactions is of order N, independently of the value of n.

If the quarks are relativistic, it is difficult to get a closed-form equation, such as in Section 9.5 below. For our purposes it will be sufficient to consider [9] the (unrealistic) case of N heavy quarks, with masses m such that $m \gg \Lambda$. The interactions of such quarks in a baryon can be described by a nonrelativistic Hamiltonian,

$$H = Nm + \sum_i \frac{p_i^2}{2m} + \frac{1}{N} \sum_{i \neq j} V_2(x_i - x_j) + \frac{1}{N^2} \sum_{i \neq j \neq k} V_3(x_i - x_j, x_i - x_k) + \cdots$$

(9.17)

where the ellipses represent four-body, five-body, etc. terms. The contribution of each term to the total energy scales as N. The interaction terms in the Hamiltonian (9.17) are the sum of many small contributions, so fluctuations are small and each quark can be considered to move in an average background potential. Consequently, the Hartree–Fock approximation (see, e.g., [18]) is exact in the large-N limit. The ground state wave function can be written as [9]

$$\Psi_0(x_1, \ldots, x_N) = \prod_{i=1}^N \Phi_0(x_i).$$

(9.18)

Using the representation (9.18) and applying the Hamiltonian (9.17) one obtains for $\Phi_0(x)$ an N-independent eigenvalue equation of the Hartree–Fock type. Hence, the spatial wave function $\Phi_0(x)$ is N-independent, so the baryon size is fixed in the $N \to \infty$ limit; it does not scale with N. This conclusion has far-reaching consequences. Needless to say, the baryon mass is proportional to N, as was expected.

The N-counting rules can be extended to baryon matrix elements of color-singlet operators. Consider a one-body operator such as $\bar{q}q$. The baryon matrix element $\langle B|\bar{q}q|B \rangle$ has N terms, since the operator can be inserted on any of the quark lines. (I assume here that the baryons in the initial and final states have the same momenta; for instance, they could be at rest.) At first sight one could conclude that this matrix element scales as N. In fact, this is the upper bound, generally speaking, because there can be cancelations between the N possible insertions. Such cancelations are crucial in unraveling the structure of baryons. Similarly, N^2 is the upper bound on two-body-operator matrix elements such as $\langle B|\bar{q}q\,\bar{q}q|B \rangle$, since there are N^2 ways of inserting the operator $\bar{q}q\,\bar{q}q$ in a baryon (see Fig. 9.24), while cancelations are possible.

9.2.9 Meson–Baryon Couplings and Scattering Amplitudes

The baryon–meson coupling constant g_{MBB} is proportional to $\sqrt{N}$. This can be seen from Fig. 9.25, which shows the matrix element of a fermion bilinear in a baryon and implies that

$$f_M\, g_{MBB} \sim N.$$

(9.19)

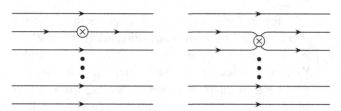

Fig. 9.24   Baryon matrix elements of a one-body operator such as $\bar{q}q$ and a two-body operator such as $\bar{q}q\,\bar{q}q$. The operator insertion is denoted by $\otimes$.

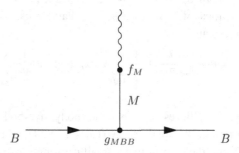

Fig. 9.25   Meson saturation of the $\langle B|\bar{q}\Gamma q|B\rangle$ matrix element.

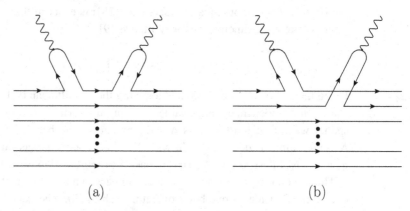

Fig. 9.26   Diagrams for baryon–meson scattering.

Given that $f_M \sim \sqrt{N}$ we obtain

$$g_{MBB} \sim \sqrt{N}. \tag{9.20}$$

The baryon–meson scattering amplitude is $O(1)$. Two contributions to the scattering amplitude are depicted in Fig. 9.26. Figure 9.26a has N possible insertions of the fermion bilinear form, and two meson f_M factors that are each $\sqrt{N}$, so the net scattering amplitude is $O(1)$. The two bilinears must be inserted on the same quark line to conserve energy – the incoming meson injects energy into the quark line, which must be removed by the outgoing meson to reproduce the original baryon. If the bilinears are inserted on different quark lines, as in Fig. 9.26b, an additional gluon exchange is needed to transfer energy between the two quark lines. The number of ways of choosing two quarks is N^2, the meson f_M couplings are $1/\sqrt{N}$ each and the gluon exchange gives an extra $g^2 \sim 1/N$, so the total $MB \to MB$ amplitude is indeed $O(1)$. (More exactly, this is the upper bound on $MB \to MB$, since it is assumed that no cancelation takes place in the estimate of $MB \to MB$; see above.)

Summarizing, the amplitude $B \to BM$ is of order $\sqrt{N}$, and that for $MB \to MB$ is of order unity. One can similarly show that the amplitude for $B + M \to B + M + M$ is of order $1/\sqrt{N}$, etc. As in the case of purely mesonic amplitudes, each additional meson gives a factor $1/\sqrt{N}$ suppression.

One can also investigate, in a similar fashion, the amplitudes for transitions of the type ground state baryon + meson $\to$ excited baryon. We will not enlarge upon this issue, instead referring the interested reader to the review [17].

9.2.10 Spin–Flavor Symmetry for Baryons

The large-N counting rules for baryons imply some highly nontrivial constraints among baryon couplings. The simplest to derive are relations between pion–baryon couplings or, equivalently, baryon–axial-current matrix elements. Related results also hold for ρ–baryon couplings, etc. To derive the axial-current relations, consider pion–nucleon scattering at fixed energy in the $N \to \infty$ limit. The argument simplifies in the chiral limit, where the pion is massless, but this assumption is not necessary. The two assumptions required are that the baryon mass and g_A (the axial– nucleon coupling) are both of order N. We have seen that the N-counting rules imply that g_A is of order N unless there is a cancelation among the leading terms. In the nonrelativistic quark model, $g_A = \frac{1}{3}(N+2)$ so such a cancelation does not occur. It is reasonable to accept that g_A is of order N in QCD even though, generally speaking, it need not have exactly its nonrelativistic value, $\frac{1}{3}(N + 2)$.

The standard form of the pion–nucleon vertex is

$$\frac{\partial_\mu \pi^a}{f_\pi} \left\langle B \left| \bar{q}\gamma^\mu \gamma^5 \frac{\tau^a}{2} q \right| B \right\rangle, \tag{9.21}$$

where the $\tau^a/2$ are the generators of the SU(2)$_{\text{flavor}}$ group. In Section 9.2.9 we learned that this amplitude is of order $\sqrt{N}$. Recoil effects are of order $1/N$, since the baryon mass is of order N and the pion energy is of order unity and can be neglected. This allows one to simplify the expression for the nucleon axial current. The time component of the axial current between two nucleons at rest vanishes, i.e., $\langle B|\bar{q}\gamma^0 \gamma^5 \frac{\tau^a}{2} q|B \rangle = 0$. The space components of the axial current between nucleons at rest can be written as

$$\left\langle B \left| \bar{q}\gamma^i \gamma^5 \frac{\tau^a}{2} q \right| B \right\rangle = gN\langle B|X^{ia}|B \rangle, \tag{9.22}$$

where X^{ia} is a set of matrices acting in the flavor and spin spaces; this set comprises nine matrices since $a = 1, 2, 3$ and $i = 1, 2, 3$. The coupling constant g is $O(1)$, and so are the matrix elements of X^{ia}; g has been factored out so that the normalization of X^{ia} can be chosen so as to simplify future expressions. For instance, for nucleons X^{ia} is a 4×4 matrix defined on the nucleon states $|p \uparrow\rangle, |p \downarrow\rangle, |n \uparrow\rangle$, and $|n \downarrow\rangle$, and each matrix from the set X^{ia} has a finite $N \to \infty$ limit.

The leading contribution to pion-nucleon scattering is from the pole graphs depicted in Fig. 9.27, which contribute at order E provided that the intermediate state is degenerate with the initial and final states. Otherwise, the pole graph contribution

Fig. 9.27 Pion–nucleon scattering diagrams of order E, where E is the pion energy. The third diagram is $1/N$ suppressed in the large-N limit.

is of order E^2, cf. Eq. (9.21). In the large-N limit, the pole graphs are of order N, since each pion–nucleon vertex is of order $\sqrt{N}$. There is also a direct two-pion–nucleon coupling, which contributes at order E and is of order $1/N$ in the large-N limit and so can be neglected.

With this information we can write the pion–nucleon scattering amplitude for $\pi^a(q) + B(k) \to \pi^b(q') + B(k')$ following from the pole graphs in Fig 9.27 as

$$-iq^i q'^j \frac{N^2 g^2}{f_\pi^2} \left(\frac{1}{q^0} X^{jb} X^{ia} - \frac{1}{q'^0} X^{ia} X^{jb} \right) \pi^a \pi^b; \qquad (9.23)$$

Amplitude for πB forward scattering

the amplitude (9.23) is written in matrix form, e.g., $X^{jb} X^{ia}$ is the product of two 4×4 matrices and is itself a 4×4 matrix, acting on the spin and isospin indices of the initial and final nucleons or, equivalently, on the spin and flavor quantum numbers of nucleons. Both initial and final nucleons are on-shell, so $q^0 = q'^0$. Since $f_\pi \sim \sqrt{N}$ the overall amplitude is of order N, which violates unitarity at fixed energy and also contradicts large-N counting (Section 9.2.9).

We observed this degeneracy in the Skyrme model, Section 4.2.

Thus, a large-N effective theory of baryons which includes only the interactions of the $J = T = \frac{1}{2}$ nucleon multiplet with pions is inconsistent. There must be other states degenerate with nucleons (which show up as intermediate states in Fig. 9.27) that cancel the order-N amplitude in Eq. (9.23), so that the total amplitude is of order unity, consistent with unitarity.

This means that one must generalize X^{ia} to be an operator acting on this degenerate set of baryons rather than a 4×4 matrix. As we will see shortly the set of degenerate baryons is, in fact, infinite at $N = \infty$, and so is the dimension of X. With this generalization the form of Eq. (9.23) is unchanged but, in addition, we must impose the consistency condition [19, 20],

$$\left[X^{ia}, X^{jb} \right] = 0 \qquad \text{for all } a, b, i, j. \qquad (9.24)$$

This consistency condition implies that the baryon axial currents are represented by a set of operators X^{ia} which commute in the large-N limit. In addition, there are obviously extra commutation relations,

$$\begin{aligned}
\left[J^i, X^{jb} \right] &= i\varepsilon^{ijk} X^{kb}, \\
\left[T^a, X^{jb} \right] &= i\varepsilon^{abc} X^{jc},
\end{aligned} \qquad (9.25)$$

following from the fact that X^{ia} has spin 1 and isospin 1. Here the J^i are spin generators while the T^a are isospin generators.

The algebra presented in Eqs. (9.24) and (9.25) is a so-called *contracted* SU($2N_f$) algebra, where $N_f = 2$ is the number of quark flavors. To see this, consider the algebra of operators in the nonrelativistic quark model, which has an SU(4) symmetry. The operators are

$$J^i = q^\dagger \frac{\sigma^i}{2} q, \qquad T^a = q^\dagger \frac{\tau^a}{2} q, \qquad G^{ia} = q^\dagger \frac{\sigma^i}{2} \frac{\tau^a}{2} q, \tag{9.26}$$

where the G^{ia} are spin–flavor generators. The commutation relations involving the G^{ia} are as follows:

$$\left[G^{ia}, G^{jb} \right] = \frac{i}{2} \varepsilon^{ijk} \delta^{ab} J^k + \frac{i}{2} \varepsilon^{abc} \delta^{ij} T^c,$$
$$\left[J^i, G^{jb} \right] = i \varepsilon^{ijk} G^{kb}, \tag{9.27}$$
$$\left[T^a, G^{jb} \right] = i \varepsilon^{abc} G^{jc}.$$

The algebra (9.24) and (9.25) for large-N baryons is obtained from (9.27) by taking the limit

$$X^{ia} \equiv \lim_{N \to \infty} \left(\frac{1}{N} G^{ia} \right). \tag{9.28}$$

Then the SU(4) commutation relations (9.27) turn into the commutation relations (9.24) and (9.25). The limiting process (9.28) is known as a Lie algebra contraction.

Lie algebra
contraction
for SU(4)

Thus, we conclude that in QCD with two flavors the large-N limit has a contracted SU(4) spin–flavor symmetry in the baryon sector. This is the symmetry of the constituent quark model for baryons too (see [21]). This circumstance explains why the naive quark model turned out to be successful in describing baryons. For instance, from the 1960s this model has been known to give $-\frac{3}{2}$ for the ratio of the proton and neutron magnetic moments and $\frac{3}{2}$ for the ratio of the couplings $g_{\pi N\Delta}$ and $g_{\pi NN}$. At the same time the large-N analysis with its solid theoretical basis, outlined above, yields [17]

$$\mu_p / \mu_n = -\tfrac{3}{2} + O(N^{-2}), \qquad g_{\pi N\Delta} / g_{\pi NN} = \tfrac{3}{2} + O(N^{-2}). \tag{9.29}$$

The unitary irreducible representations of the contracted Lie algebra can be obtained using the theory of induced representations and can be shown to be *infinite dimensional*. This means that the X^{ia} must be treated as infinite-dimensional matrices or, equivalently, as operators acting in a Fock space. That is what we will do from now on. The simplest irreducible representation for two flavors is a tower of states with $J = T = \frac{1}{2}, \frac{3}{2}, \frac{5}{2}$, etc. For $\frac{1}{2}$ we have two spin and isospin states, for $\frac{3}{2}$ we have four spin and isospin states, and so on. At $N = \infty$ all these states are degenerate. The spectrum splits only at the level of $1/N$ corrections, namely, the baryon mass splitting is proportional to $J^2/N = j(j+1)/N$.[8] In particular,

$$M_\Delta - M_N \sim \frac{1}{N}, \tag{9.30}$$

while, at the same time, $M_{N,\Delta} \sim N$.

The degenerate set $J = T = \frac{1}{2}, \frac{3}{2}, \frac{5}{2}$, etc. is exactly the set of states of the Skyrme model (Section 4.2.5), which is also endowed with the same algebra. The same is true with regard to the large-N generalization of the nonrelativistic quark model. This statement explains why the predictions for the dimensionless ratios in these models are more general than the models themselves. In fact, all such predictions can be obtained, in a model-independent way, from the large-N analysis of baryons.

---

[8] This formula is not valid for values of j that are too high, i.e., the values of j that scale with N as a positive power of N.

More precisely, in the large-N limit the leading-order predictions for the pion–baryon coupling ratios, magnetic moment ratios, mass splitting ratios, and so on are the same as those obtained in the Skyrme model or in the nonrelativistic quark model [22], because both these models also have a contracted SU(4) spin–flavor symmetry in this limit.

The operators X^{ia} can be completely determined (up to an overall normalization g), since they constitute the generators of the SU(4)$_{contracted}$ algebra. It is useful to have an explicit $N \rightarrow \infty$ realization of this algebra. To this end one can use, as a possible option, the realization provided by the Skyrme model. The Skyrmion solution is characterized by the rotational moduli matrix $A(t)$, which is parametrized by the quantum-mechanical variables $\vec{\omega}$ quantized via the canonic commutation relation $[\dot{\omega}^i, \omega^j] \sim \delta^{ij}$ (Section 4.2.5). In terms of this moduli matrix we have

$$X^{ia} \sim \text{Tr}\left(A\tau^i A^\dagger \tau^a\right). \tag{9.31}$$

Since the X operators contain A but not $\dot{A}$, they commute. It is clear that their spin and isospin rotation properties are exactly those in (9.25).

For finite N the contracted SU(4) group is no longer the symmetry of the baryon sector of multicolor QCD. Nevertheless, many results obtained in the naive quark model can be rederived in QCD using SU(4)$_{contracted}$ in the leading approximation and then calculating $1/N$ corrections one by one [22–24].

From two flavors to three

In nature there are three light quarks: u, d, s. If for a moment we neglect the s-quark mass then the spin–flavor symmetry, exact in the $N = \infty$ limit, is SU(6) rather than SU(4).[9] Now, to obtain predictions for actual baryons one must include not only $1/N$ corrections but (where necessary) also those due to $m_s \neq 0$, i.e., SU(3)$_{flavor}$-breaking corrections. One of the most successful predictions obtained in this way is a mass formula for the baryons from the decuplet (see, e.g., [17]):

$$\tfrac{1}{4}M(\Delta) + \tfrac{3}{4}M(\Xi^*) = \tfrac{1}{4}M(\Omega) + \tfrac{3}{4}M(\Sigma^*) + O(\epsilon^3/N^2), \tag{9.32}$$

where ϵ is an SU(3)-breaking parameter proportional to m_s. Experimentally, the accuracy of this mass formula is 0.9×10^{-3}.

9.3 Abelian Higgs Model in 1 + 1 Dimensions

The Coleman theorem discussed in Chapter 6, Section 6.5, tells us that continuous global symmetries cannot be spontaneously broken in two-dimensional theories. It is natural to raise a question whether gauge symmetries can be Higgsed in two dimensions. The answer is no so straightforward. As they say, it depends …

In this section we will consider the Abelian Higgs model in 1 + 1 dimension in the regime which in 1 + 2 or 1 + 3 dimensions would be the standard Higgs regime. We will see that, instead (under certain constraints on the coupling constants), we obtain confinement whose origin is associated with an instanton gas [25]; see also [26].

---

[9] Algebraically, one can identify the spin–flavor symmetry with SU(6) of the nonrelativistic quark model [21].

At the same time the Higgs boson is eaten up by the gauge field, just as in the standard Higgs mechanism.[10]

9.3.1 The Model

We have already dealt with the Abelian Higgs model in Chapter 3, devoted to flux tubes. For convenience I reproduce here the action of the model in Euclidean space,

$$S = \int d^2x \left[\frac{1}{4e^2} F_{\mu\nu}^2 + \left| \mathcal{D}_\mu \phi \right|^2 + \lambda \left(|\phi|^2 - v^2 \right)^2 \right], \tag{9.33}$$

where ϕ is a complex scalar field (with charge 1), the covariant derivative is defined by

$$\mathcal{D}_\mu \phi = (\partial_\mu - iA_\mu)\phi, \tag{9.34}$$

v is a dimensionless constant, and λ is a positive constant with dimension mass squared. We will assume that $e^2 \ll \lambda |v^2|$ and $|v^2| \gg 1$, which ensures weak coupling.[11]

The action (9.33) admits an extension: we could add a θ term. In Euclidean space it has the form

$$\Delta S_\theta = -i \frac{\theta}{4\pi} \int d^2x \, \varepsilon_{\mu\nu} F_{\mu\nu}, \tag{9.35}$$

> *The vacuum angle θ is chosen to vanish.*

where $\varepsilon_{\mu\nu}$ is the Euclidean Levi–Civita tensor, with $\varepsilon_{12} = 1$ (do not confuse it with the Minkowski Levi–Civita tensor). For simplicity, at first we will set $\theta = 0$.

The model (9.33) is almost superrenormalizable. I write "almost" because the dimensionless parameter v^{-2} still experiences logarithmic running, in much the same way as g^2 in four-dimensional Yang–Mills theory. It plays the role of the asymptotically free coupling constant in loop expansion. The only diagram contributing to logarithmic renormalization of v^2 is the tadpole graphs of the type depicted in Fig. 9.29 (p. 369) with the ϕ field propagating in the loop. Its calculation is trivial and yields

$$v^2(\mu) = v_0^2 - \frac{1}{\pi} \log \frac{M_0}{\mu} \equiv \frac{1}{\pi} \log \frac{\mu}{\Lambda}, \tag{9.36}$$

where μ is the running renormalization point, M_0 is the ultraviolet cutoff and Λ in the last equality above defines the dynamical scale (analogous to that in QCD). If we choose the physical masses m (which are $m_\gamma \sim ev$ or $m_\phi \sim \sqrt{\lambda}v$) to be much larger than Λ we guarantee that the theory at hand is weakly coupled. This requirement is easy to meet provided that $v^2 \gg 1$ and $m/\Lambda \gg 1$. The running stops at $\mu \sim m$.

---

[10] There is a curious story associated with the discovery of this phenomenon. Here is a quotation from Sidney Coleman's lecture *The Uses of Instantons* [25]: "The fact that the Abelian Higgs model in two dimensions does not display the Higgs phenomenon was discovered independently by two of my graduate students, Frank De Luccia and Paul Steinhardt. They did not write up their results because I did not believe them. I take this occasion to apologize for my stupidity. – SC."

[11] These two constraints do not preclude one from choosing $\lambda = e^2/2$, which would correspond to the Bogomol'nyi limit. Such a choice is convenient although not crucial for what follows.

9.3.2 Negative vs. Positive v^2

Cf. Section 9.5.4 and Eq. (9.91).

If v^2 is negative then the model (9.33) describes the electrodynamics of charged particles. In $1 + 1$ dimensions the massless photon has no transverse (propagating) degrees of freedom. However, the photon field induces the Coulomb interaction between charged particles. The Coulomb potential in $1+1$ dimensions grows linearly with distance. Hence, isolated charged particles do not exist and two opposite charges are in fact confined. All we can see in "experiments" are neutral bound states.

A much less trivial dynamical situation takes place at positive v^2. In three or four dimensions, choosing v^2 to be positive would trigger the Higgs phenomenon and we would end up with a massive photon, $m_\gamma = \sqrt{2}ev$, and screened electric charges. The interaction of probe charges at distances $L \gg 1/(ev)$ would be exponentially small. This is not the whole story in two dimensions: although the photon gets a mass, nonperturbative effects give rise to a long-range force, which corresponds to a linear potential at very large distances, and charge confinement. The slope of this potential is not proportional to e^2 as in the case of negative v^2 but is exponentially small.

This dynamical pattern is due to instantons. At large v^2 the model (9.33) can be treated quasiclassically. Instantons are solutions of the classical equations of motion that technically coincide with the static vortex solutions in three dimensions (or flux-tube solutions in four dimensions) studied in Chapter 3. Therefore, all we learned there can be directly applied here. The vortex mass must be reinterpreted as the instanton action S_{inst}. I recall that $S_{\text{inst}} \geq 2\pi n v^2$, where n is the topological charge, given by the integral

$$n = \frac{1}{4\pi} \int d^2x\, \varepsilon_{\mu\nu} F_{\mu\nu}. \tag{9.37}$$

The equality $S_{\text{inst}} = 2\pi n v^2$ is achieved in the Bogomol'nyi limit, i.e., at $\lambda/e^2 \to 1/2$, see Section 3.1.3. If $n = 1$,

$$S_{\text{inst}} = c\,(\omega)\,2\pi v^2, \quad \omega \equiv \lambda/e^2,$$

where the coefficient c in front of $2\pi v^2$ is dimensionless and depends on the ratio λ/e^2; moreover, c is always ≥ 1 and $c \sim 1$ if the ratio of the coupling constants $\lambda/e^2 \sim 1$. As we will see later, c grows logarithmically when the ratio λ/e^2 becomes $\gg 1$.

Unlike the QCD instantons, in the model at hand the instanton size ρ is not a modulus. It is determined by the inverse mass of the Higgsed photon: $\rho \sim 1/(ev)$. There are two moduli, the two coordinates of the instanton center on the plane. Thus, the instanton measure takes the form

$$d\mu_{\text{inst}} = \mu^2 d^2 x_0\, e^{-S_{\text{inst}}}, \tag{9.38}$$

Wilson loop criterion: area vs. perimeter law

where x_0 denotes the coordinates of the instanton center and μ^2 is the pre-exponential factor in the instanton measure. Its precise value is unimportant for our purposes.

The quickest way to infer charge confinement in the model at hand is to calculate the Wilson loop

$$\langle W \rangle = \left\langle \exp\left(iq \oint_C A_\mu dx_\mu\right)\right\rangle, \tag{9.39}$$

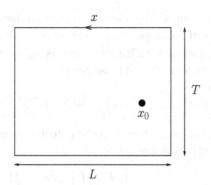

Fig. 9.28 The contour C in Eq. (9.39) representing the Euclidean trajectory of the probe particle. The instanton (anti-instanton) is shown by the solid circle. The size of the contour is large: $T, L \gg 1/(ev)$, see also Eq. (9.49).

describing an infinitely heavy probe particle of charge q making a loop along the closed contour C depicted in Fig. 9.28.

One-instanton action
$S_{\text{inst}} \geq 2\pi v^2$

Let us start from the one-instanton contribution. Expanding the exponent in a Taylor series, we obtain

$$\exp\left(iq \oint_C A_\mu dx_\mu\right)_{\text{inst}} = \int d^2 x_0 \left(\mu^2 e^{-S_{\text{inst}}}\right) iq \oint_C A_\mu^{\text{inst}} dx_\mu$$
$$+ \int d^2 x_0 \left(\mu^2 e^{-S_{\text{inst}}}\right) \frac{(iq)^2}{2!} \left(\oint_C A_\mu^{\text{inst}} dx_\mu\right)^2 + \cdots , \quad (9.40)$$

where the integral over A_μ^{inst} runs over the contour depicted in Fig. 9.28. Now, we can apply Stokes' theorem:

$$\oint_C A_\mu \, dx_\mu = \int d^2 x \, F_{12}. \quad (9.41)$$

It is obvious that $\int d^2 x \, F_{12}$ vanishes if the instanton is outside the contour C. However, if it is inside the contour the integral reduces to

$$\int d^2 x \, F_{12} = 2\pi \quad (9.42)$$

for all x_0 except those which are within a distance $\sim 1/(ev)$ from the contour (remember, the instanton solution falls off exponentially at distances $\gtrsim 1/(ev)$ from the instanton center and at the same time $L, T \to \infty$). Thus, Eq. (9.40) takes the form

$$\exp\left(iq \oint_C A_\mu \, dx_\mu\right)_{\text{inst}} = LT \left(\mu^2 e^{-S_{\text{inst}}}\right) \left(e^{2\pi iq} - 1\right). \quad (9.43)$$

Now we add the anti-instanton contribution, which at $\theta = 0$ differs only in sign:

$$\left(\oint_C A_\mu \, dx_\mu\right)_{\text{anti-inst}} = -\left(\oint_C A_\mu \, dx_\mu\right)_{\text{inst}} = -2\pi. \quad (9.44)$$

This concludes our calculation of the Wilson loop in the one-instanton approximation:

$$\exp\left(iq \oint_C A_\mu \, dx_\mu\right)_{\text{inst+anti-inst}} = -LT \left(2\mu^2 e^{-S_{\text{inst}}}\right) [1 - \cos(2\pi q)]. \quad (9.45)$$

Next, we must sum over the instanton–anti-instanton ensemble with arbitrary numbers of pseudoparticles in the vacuum, which can be treated in the instanton gas approximation (see Chapter 5). In this approximation, summing over the ensemble exponentiates the result presented in Eq. (9.45):

$$\left\langle \exp\left(iq \oint_C A_\mu \, dx_\mu\right)\right\rangle = \exp\left\{-LT\left(2\mu^2 e^{-S_{\text{inst}}}\right)[1 - \cos(2\pi q)]\right\}, \qquad (9.46)$$

which implies, in turn, that the potential energy of two probe charges q and $-q$ separated by a distance L is

$$V(L) = L\left(2\mu^2 e^{-S_{\text{inst}}}\right)[1 - \cos(2\pi q)]. \qquad (9.47)$$

We see that the model at hand (being classically in the Higgs regime) in fact generates linear confinement for all probe charges $q \neq 1, 2, \ldots$ Why is there no confinement at $q = 1, 2, \ldots$? One should remember that the model has dynamical fields ϕ of charge unity, which screen the probe charges if they are integer. Fractional charges remain unscreened. It is remarkable that the linear potential $V(L)$ depends on q periodically, as $1 - \cos(2\pi q)$, rather than q^2. Thus, linear confinement is not due to one-photon exchange as is the case for negative v^2. Qualitatively there is not much difference between these two cases, of negative and positive v^2. Quantitatively, however, there is a huge difference since in the latter case the potential, being linear, is exponentially weak since it is proportional to $e^{-S_{\text{inst}}}$.

> Instanton-generated linear potential is exponentially weak.

We will defer the study of the θ-dependence – an interesting issue in this model [25] – till Section 9.3.3 below.

The result presented in (9.47) was obtained by virtue of a Euclidean calculation. It is worth discussing the corresponding physical picture in Minkowski space. Assume we place the positive probe charge eq at the origin of the spatial axis x, while the negative probe charge $-eq$ is placed far away to the right, at $x = L$. Let us ask ourselves what is the potential $V(x)$ as a function of x in the interval $x \in [0, L]$.

Since the photon mass does not vanish, as we move from the origin to the right, at first at $x \gtrsim m_\gamma^{-1}$ we see screening, i.e., the standard exponential falloff of the potential. The screening length ℓ_{scr} can be estimated from the following equation:

$$e^{-m_\gamma \ell_{\text{scr}}} \sim e^{-c\, 2\pi v^2}, \qquad (9.48)$$

implying

$$\ell_{\text{scr}} \sim \frac{c\, 2\pi v^2}{m_\gamma} \gg \frac{1}{m_\gamma}. \qquad (9.49)$$

When we reach $x \sim \ell_{\text{scr}}$ and continue our journey toward $x = L$ the coherent effect of the instanton gas becomes strong enough to shatter screening, and the exponential falloff smoothly passes into linear growth regime with the exponentially suppressed coefficient.

It should be noted, however, that $c \sim 1$ only under the condition $m_\phi \sim m_\gamma$ i.e $\sqrt{\lambda} \sim e$. If the Higgs field is much heavier than the Higgsed photon, we are deeply inside type II superconductivity and in calculating S_{inst} one must use the Abrikosov formula containing Abrikosov's logarithm,

$$S_{\text{inst}} = 2\pi v^2 \log \frac{m_\phi}{m_\gamma} \qquad (9.50)$$

or $c = \log m_\phi/m_\gamma$. A special situation occurs if

$$\frac{m_\phi}{m_\gamma} \to \infty.$$

Then $S_{\text{inst}} \to \infty$ and the instanton gas disappears. At $c \to \infty$ the screening length in Eq. (9.49) tends to infinity. Then screening is never replaced by linear confinement. The probe charges q are not confined.

9.3.3 $\theta \neq 0$

We conclude this section with a remark on the θ dependence. The θ dependence of the instanton contribution is

$$E_{\text{vac}}(\theta) \sim \exp\left(-S_{\text{inst}} + i\theta\right) + \text{H.c}, \qquad (9.51)$$

which explicitly exhibits 2π periodicity.

In two dimensions the θ term (9.35) in the action is equivalent to a spatially constant background electric field sourced by charges Q

$$\frac{Q}{e} = \pm\frac{\theta}{2\pi} \qquad (9.52)$$

at both spatial infinities (see, e.g., Section 9.6.1 or [26]). If we add probe charges eq at the points $x = 0$ and $x = L$, inside this interval the energy density (measured from the density at $q = 0$) becomes

$$\Delta\mathcal{E} = \mu^2 \exp\left(-S_{\text{inst}}\right)\left[\cos\theta - \cos(\theta + 2\pi q)\right], \qquad (9.53)$$

while the total extra energy in this interval is

$$\Delta E = L\Delta\mathcal{E}, \qquad (9.54)$$

provided that $L \gg \ell_{\text{scr}}$; see Eq. (9.49). Equations (9.53) and (9.54) are to be compared with (9.47).

A brief remark is in order regarding Eqs. (9.50) and (9.51). In the limit $m_\phi/m_\gamma \to \infty$ the instanton action becomes infinite and θ dependence disappears. Gone with it is linear confinement. And conversely, whenever the vacuum energy density has a nontrivial θ dependence, introducing probe charges qe in two-dimensional theory we will see the Coulomb regime confining the probe charges.

9.4 CP(N − 1) at Large N

The two-dimensional model that we are going to consider was introduced and discussed in Chapter 6. In the standard normalization the Lagrangian of the model is

$$\mathcal{L} = \frac{2}{g^2}\left[(\partial_\mu \bar{n}_i)(\partial^\mu n^i) + (\bar{n}_i\partial_\mu n^i)^2\right], \qquad (9.55)$$

where n^i is the SU(N) N-plet and is subject to the constraint

$$\bar{n}_i n^i = 1. \qquad (9.56)$$

Confinement in CP(N − 1). Supersymmetry destroys it; see Appendix section 10.27.1.

Below we will solve the model at large N and demonstrate that the n^i quanta are *confined*, i.e., they do not exist in the spectrum of the theory as asymptotic states. Instead, all asymptotic states are bound states of the type $\bar{n}n$. In solving the model we will follow [27, 28].

More convenient for our purposes is a linear gauged realization in which an auxiliary U(1) gauge field A_μ (with no kinetic term) is introduced. We will see that because of quantum corrections a kinetic term for A_μ is generated, which guarantees the confinement of the n^i quanta in this two-dimensional model.[12] The constraint (9.56) will be taken into account through introduction of the Lagrange multiplier field $\sigma(x)$ with a term $\sigma(\bar{n}_i n^i - 1)$ in the Lagrangian. In addition, we will replace the coupling g^2 by a 't Hooft coupling λ that does not scale with N at large N:

$$\lambda \equiv \frac{g^2 N}{2}, \qquad \lambda \ll 1. \tag{9.57}$$

As a result, from (9.55) we obtain the Lagrangian with which we will work,

$$\mathcal{L} = \frac{N}{\lambda}(\partial_\mu - iA_\mu)\bar{n}_i(\partial^\mu + iA^\mu)n^i - \sigma(\bar{n}_i n^i - 1), \tag{9.58}$$

while the partition function is

$$Z = \int D\bar{n}\, Dn\, DA\, D\sigma \exp\left[i \int d^2x\, \mathcal{L}(\bar{n}, n, A, \sigma)\right]. \tag{9.59}$$

In this form the U(1) gauge invariance of the model is explicit.

Let us ask ourselves how many independent degrees of freedom are incorporated in (9.58). The number of complex fields n is N. The real constraint (9.56) eliminates one real degree of freedom. Another real degree of freedom is eliminated because of the U(1) gauge invariance. Altogether, we are left with $N-1$ complex degrees of freedom. This is precisely the number of independent degrees of freedom in CP($N - 1$); see Section 6.2.2.

The Lagrangian (9.58) is bilinear in the n fields; therefore, one can perform the path integral over these fields exactly. However, the subsequent integral over A and σ cannot be done exactly. We will use the fact that at large N the action is large and, hence, a stationary phase (saddle point) approximation is applicable.

9.4.1 Vacuum Structure

First let us determine the vacuum of the model. Integration over $\bar{n}$ and n in (9.59) yields

$$Z = \int DA\, D\sigma \exp\left\{-N \operatorname{Tr}\, \ln\left[-(\partial_\mu + iA_\mu)^2 - \frac{\lambda\sigma}{N}\right] + i \int d^2x\, \sigma\right\}. \tag{9.60}$$

The Lorentz invariance of the theory tells us that if the saddle point exists then it must be achieved at an x-independent value of σ. Hence for the purpose of vacuum determination we can assume σ to be constant and then vary (9.60) with respect to

---

[12] Recall that in this case the Coulomb potential grows linearly with separation; see below.

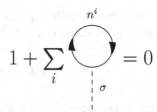

$$1 + \sum_i \quad = 0$$

Fig. 9.29 The vanishing of the linear in σ term (the tadpole term) in the effective Lagrangian.

σ and require the result to vanish. The same Lorentz invariance tells us that at the saddle point $A_\mu = 0$. In this way we arrive at the following equation:

$$i + \lambda \int \frac{d^2k}{(2\pi)^2} \frac{1}{k^2 - \lambda\sigma/N + i\epsilon} = 0, \quad \epsilon \to +0. \tag{9.61}$$

Diagrammatically this equation is depicted in Fig. 9.29. The integral in (9.61) is logarithmic and diverges in the ultraviolet; therefore we will cut it off at M_{uv}^2. In this way, starting from (9.61), we arrive at the equation

$$\frac{\lambda\sigma}{N} = M_{\text{uv}}^2 \, e^{-4\pi/\lambda}, \tag{9.62}$$

implying that the vacuum value of σ is

$$\sigma_{\text{vac}} = N M_{\text{uv}}^2 \frac{1}{\lambda} e^{-4\pi/\lambda}. \tag{9.63}$$

Thus, the assumption of the existence of a saddle point is confirmed *a posteriori*: a solution with $\sigma_{\text{vac}} > 0$ does exist. Examining the original Lagrangian (9.58) one sees that a positive vacuum expectation value $\lambda N^{-1} \langle \sigma \rangle$ is simply a mass term of the n field. The n-field mass,

---

The subscript n labels the parameters of the n field, e.g., e_n

---

$$M_n^2 \equiv M_{\text{uv}}^2 \, e^{-4\pi/\lambda}, \tag{9.64}$$

is dynamically generated. The same expression can be used to define a dynamical scale analogous to the dynamical scale in QCD,

$$\Lambda \equiv M_{\text{uv}} \, e^{-2\pi/\lambda} = M_n. \tag{9.65}$$

In this section I will use the n-field mass M_n and the scale parameter Λ indiscriminately.

Let us pause here to make two comments regarding Eq. ((9.64). First, it is obvious that M is N independent (i.e., it does not scale with N). Second, the renormalization-group invariance of the right-hand side allows one to obtain the β function governing

---

$M_n \equiv \Lambda$

---

the running law of the coupling constant λ, namely,

$$M_{\text{uv}} \frac{\partial\lambda}{\partial M_{\text{uv}}} = -\frac{\lambda^2}{2\pi}, \tag{9.66}$$

implying that the β function for $\alpha = g^2/4\pi = \lambda/(2\pi N)$ is

$$\beta(\alpha) = -N\alpha^2. \tag{9.67}$$

Cf.
Eq. (6.72).

This should be compared with the expression for the β function obtained in Chapter 6 through a standard perturbative calculation.

9.4.2 Spectrum

Next, to determine the spectrum of the theory, let us examine the fluctuations of σ and A around their vacuum values. (To consider the σ fluctuations one must perform the shift $\sigma \to \sigma - \sigma_{\text{vac}}$.)

Expanding the effective action (9.60) around the saddle point, one can easily check that the cubic and higher orders in σ and A are suppressed by powers of $1/N$. The linear term of the expansion vanishes. This is the essence of Eq. (9.61). Therefore, we need to focus only on the quadratic terms of the expansion.

The quadratic term in σ can be readily found; it does not vanish but plays little role in the dynamical confinement mechanism under discussion. In this discussion we can just replace $\sigma \to \sigma_{\text{vac}}$, use (9.64) for M, and forget about the σ fluctuations.

It is not difficult to check that the cross term of σA type also vanishes (see Fig. 9.30). Therefore, we need only consider the terms quadratic in A. To this end one must calculate two graphs depicted in Fig. 9.31. A straightforward computation yields for the sum of these diagrams

$$\frac{N}{12\pi\Lambda^2}(-g_{\mu\nu}k^2 + k_\mu k_\nu)[1 + O(k^2/\Lambda^2)].\tag{9.68}$$

This expression is automatically transversal, as expected given the U(1) gauge invariance of (9.58).

Photon
kinetic term
generation

Observe that the $O(k^4)$ terms in (9.68) are irrelevant for our spectrum exploration. What is relevant is the $O(k^2)$ term, which, in fact, represents the standard kinetic term $-\frac{1}{4}F_{\mu\nu}^2$ of the photon field. We see that, indeed, the one-loop corrections generate a kinetic term for the A_μ field, which, originally, was introduced as auxiliary.

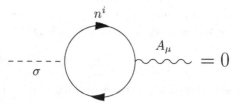

Fig. 9.30   The vanishing of the σA mixing term in the effective Lagrangian.

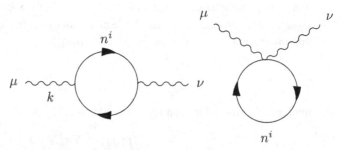

Fig. 9.31 $O(A^2)$ terms in the effective Lagrangian.

To summarize our achievements, in the large-N limit, to leading order, we have derived two important facts: (i) the generation of a mass term for the n quanta; (ii) the generation of the kinetic term[13]

$$-\frac{N}{48\pi\Lambda^2}F_{\mu\nu}F^{\mu\nu}$$

(9.69)

for the photon field.

In what follows it is convenient to rescale the A field to make its kinetic term (9.69) canonically normalized. Upon this rescaling the effective Lagrangian takes the form

$$\mathcal{L}_{\text{eff}} = -\frac{1}{4}F_{\mu\nu}^2 + (\partial_\mu - ie_n A_\mu)\bar{n}_i(\partial^\mu + ie_n A^\mu)n^i - M^2\bar{n}_i n^i,$$

(9.70)

where to ease notation I dropped the subscript n in the mass, $M_n \to M$, the electric charge of the quanta of the n field is

$$e_n \equiv \Lambda\sqrt{\frac{12\pi}{N}}.$$

(9.71)

It has dimension of mass, which is correct for the electric charge in two-dimensional theories. Moreover, one should stress that at large N the electric charge becomes small, $e_n/M \ll 1$, which implies, in turn, weak coupling.

Recall that the only impact of the massless gauge field (the photon) in two dimensions is the Coulomb interaction. The Coulomb potential energy grows linearly with separation:

$$V(x, y) = \frac{12\pi M^2}{N}r, \qquad r = |x - y|.$$

(9.72)

The above growth leads to permanent confinement for $\bar{n}n$ pairs. That is why Witten referred to the n quanta as "quarks" transforming in the fundamental representation of the (global) SU(N) group.

Given the fact that the slope in (9.72) is small for large N, the conventional nonrelativistic Schrödinger equation with Hamiltonian

$$H = 2M - \frac{1}{M}\frac{d^2}{dr^2} + \frac{12\pi M^2}{N}r$$

(9.73)

is applicable for low-lying bound states. If the excitation number $k \ll N$ then the mass of the kth bound state is

$$M_k = 2M + \text{const} \times M\left(\frac{k}{N}\right)^{2/3}.$$

(9.74)

The constant in the above equation is evaluated in the solution to Exercise 9.4.1; see Eq. (PS.342) on page 689. It turns out to be rather large, around 32. As k approaches N one should abandon the nonrelativistic description in favor of an appropriate relativistic equation. There are $\sim N^{2/3}$ nonrelativistic levels.

I have just mentioned that the n "quarks" form N-plets with regard to global SU(N). Thus, the $\bar{n}n$ "mesons" can belong either to the adjoint or to the trivial (singlet) representation of SU(N). At large N the adjoint and singlet mesons are degenerate, as can be seen from, e.g., (9.73). This degeneracy is not a consequence

---

[13] An extra factor $\frac{1}{2}$ in (9.69) compared to (9.68) comes in passing from Feynman graphs to the effective Lagrangian.

of any symmetry and, in fact, is lifted at finite N. Indeed, for $N = 2$ the model at hand is just the O(3) model[14] considered in Chapter 6. The spectrum of excitations in this model is known from the exact solution [29]. It consists of one triplet; there are no singlets. This can be understood only if, with N decreasing, the number of stable bound states decreases too, the higher excitations becoming unstable. The lowest-lying adjoint mesons have nowhere to decay and must be stable. The singlet mesons must split from the adjoint mesons, become heavier, and decay at $N = 2$.

9.4.3 Comparison with Chapter 6

Lagrangian (9.58) describes the CP($N - 1$) model in the gauged linear formulation (gauged linear sigma model, GLSM). In Chapter 6 we mostly dealt with the geometric formulation obtained after integrating out all auxiliary field, i.e., A_μ, σ, and one of the n fields. With this geometric formulation we pass to a nonlinear sigma model (NLSM). In the geometric formulation one of the components of n^i acquires a vacuum expectation value, and the theory is in the Higgs regime, with $N - 1$ rather than N dynamic n fields. The flavor symmetry is broken

Target space geometry. NLSM vs. GLSM

$$SU(N) \rightarrow SU(N - 1) \times U(1).$$

As we see now from the large-N solution, treating NLSM as we did in Chapter 6 – i.e., converting GLSM to NLSM – we are actually moving away from the solution of the model in the infrared domain. NLSM is very convenient for perturbation theory and quasi-classical analysis (instantons). On the other hand, using NLSM as a starting point we miss crucial features of the IR solution in particular, confinement.

Indeed, from the large-N solution we infer that the number of the dynamical n fields is N rather than $N - 1$. The photon is *not* Higgsed, hence, the theory is in the confining regime. Correspondingly, the original flavor SU(N) symmetry is fully restored. That is why we must have a full N-plet of n_i "quarks" in the fundamental representation of SU(N). The "microscopic" n^i fields should be viewed as quarks that are confined, so that the physical spectrum consists of SU(N) adjoint and (generally speaking) singlet bound states, "mesons".

9.4.4 An Alternative Perspective

The fact that the n-field quanta do not exist as asymptotic states in the spectrum of the CP($N - 1$) model, only as $\bar{n}n$ mesons, can be inferred from a different point of view, which will enrich our understanding of the issue. To explain this alternative interpretation [30], it is instructive first to compare the solution presented in Section 9.4.2 with that of the supersymmetric CP($N - 1$) model to be studied in Part II. In the supersymmetric version there is no confinement and the n quarks (belonging to the fundamental and antifundamental representations of SU(N)) exist as asymptotic states in the physical spectrum. This is in one-to-one correspondence with the fact that in the supersymmetric model the photon acquires a mass, and what would have been a Coulomb interaction falls off exponentially at large distances.

See, in particular, Appendix section 10.27.1.

---

[14] The large-N solution of the O(N) sigma model is interesting in itself although unremarkable from the standpoint of the confinement problem. We will consider it in Appendix section 9.8.

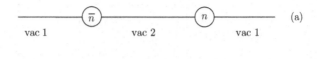

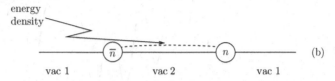

A kink–antikink state in (a) the supersymmetric and (b) the nonsupersymmetric CP($N-1$) models. In the supersymmetric case both vacua, 1 and 2, have the same (vanishing) energy density. In the nonsupersymmetric case, vacuum 2 is a quasivacuum, whose energy density is slightly higher than that of the genuine vacuum, vacuum 1.

In the supersymmetric CP($N-1$) model there are N distinct degenerate vacua, all with vanishing energy density (see [31] and Section 10.22). As we already know, all theories with discrete degenerate vacua support kinks, which interpolate between the different vacua. In the late 1970s Witten demonstrated [28] that the fields n^i in CP($N-1$) in fact represent kinks interpolating between a given vacuum and its neighbor. The multiplicity of such kinks is N [32]: they form an N-plet. This is the origin of the superscript i in n^i. I will not justify the above statements here since their proof would lead us far astray. Let us just accept them and see what happens. A kink–antikink configuration in one spatial dimension is shown in Fig. 9.32, where the supersymmetric CP($N-1$) case is displayed at the top (Fig. 9.32a). It is clear that the energy of this configuration does not depend on the distance between n and $\bar{n}$, so that these "quarks" are free to travel to the corresponding spatial infinities and, thus, are unconfined.

Now, let us pass to the nonsupersymmetric CP($N-1$) model, Fig. 9.32b. In this model the genuine vacuum is unique. In the 1990s Witten proved [33] that at large N there are, in fact, of order N quasivacua, which lie higher in energy than the genuine vacuum but become stable in the limit $N \rightarrow \infty$ (Fig. 9.33). This is due to the fact that the energy split between two neighboring (quasi)vacua is $O(1/N)$. The kink interpretation of n and $\bar{n}$ remains valid. Assume that the $\bar{n}$ in Fig. 9.32b interpolates between the genuine vacuum (vacuum 1) and the first quasivacuum (vacuum 2), while n returns us to the genuine vacuum. Owing to the energy split between vacuum 1 and vacuum 2, the energy of this configuration will contain a term $\Delta\mathcal{E} L$ where $\Delta\mathcal{E}$ is a (positive) excess of energy and L is the distance between the "quarks" n and $\bar{n}$. It is obvious that the energy separation cannot become infinite since this would require an infinite amount of energy to be pumped into the system. This is typical linear confinement, with $\bar{n}n$ "mesons" in the physical spectrum.

A lesson we should learn from this alternative interpretation is that the mechanism of linear confinement in the CP($N-1$) model is specific to two dimensions and cannot be lifted to four dimensions. Complete duality between the two alternative pictures presented in Sections 9.4.2 and 9.4.4, respectively, takes place only because the (massless) photon has no propagating degrees of freedom in two dimensions. Its impact is completely equivalent to that of the energy split between two neighboring vacua in Figs. 9.32b and 9.33.

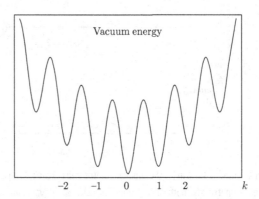

Fig. 9.33 The vacuum structure of the nonsupersymmetric CP($N - 1$) model at large N and $\theta = 0$. The genuine vacuum is labeled by $k = 0$. All minima with $k \neq 0$ are quasivacua, which become stable at $N = \infty$.

Exercises

9.4.1   Derive the $(k/N)^{2/3}$ law in Eq. (9.74).

9.4.2   Consider the CP($N - 1$) model at large N, assuming that the theory resides in the first quasivacuum (i.e., $k = 1$ in Fig. 9.33). Using the technology developed in Chapter 7 and the results of the present section find the decay rate of this false vacuum into the genuine ground state (i.e., $k = 0$ in Fig. 9.33).

9.5 The 't Hooft Model

9.5.1 Introduction

It turns out that combining planarity with the suppression of the fermion loops provides us with enough power to solve QCD in two dimensions. By solving QCD I mean not only establishing the fact that the physical spectrum comprises color singlets (color confinement) but, in fact, calculating the whole spectrum and understanding all the basic regularities. We will try to trace the relation between the spontaneous breaking of the chiral symmetry and color confinement. Two-dimensional multicolor QCD is usually referred to as the 't Hooft model [34].

9.5.2 The 't Hooft Model

The model we will study is two-dimensional QCD with the Lagrangian

$$\mathcal{L} = -\tfrac{1}{4} F_{\mu\nu}^a F^{a\,\mu\nu} + \sum_i \bar{\psi}_i (i \, \slashed{D} - m_i)\psi_i, \qquad (9.75)$$

where

$$i\mathcal{D}_\mu = i\partial_\mu + g A_\mu^a T^a \qquad (9.76)$$

The kinetic term in (9.75) is normalized canonically; g is in the covariant derivative (9.76).

and

$$F_{\mu\nu}^a = \partial_\mu A_\nu^a - \partial_\nu A_\mu^a + g\, f^{abc} A_\mu^b A_\nu^c. \tag{9.77}$$

The gauge group is SU(N), with generators T^a in the fundamental representation,

$$T^a T^a = \tfrac{1}{2}\left(N - N^{-1}\right). \tag{9.78}$$

The summation over i in Eq. (9.75) runs over quark flavors and m_i is the mass of the ith quark. The fermions are described by Dirac spinors, which in two dimensions are two-component complex spinors. In this section two-dimensional gamma matrices will be chosen as usual:[15]

$$\gamma^0 = \sigma_2, \qquad \gamma^1 = -i\sigma_1, \qquad \gamma^5 = \gamma^0\gamma^1 = -\sigma_3. \tag{9.79}$$

If some quarks are massless then the Lagrangian (9.75) possesses a chiral symmetry, in much the same way as four-dimensional QCD. In what follows we will limit ourselves to a single massless quark (plus an infinitely heavy antiquark playing the role of the force center). This will be sufficient for our purposes. Then the corresponding chiral symmetry is U(1)$_L$×U(1)$_R$, or, equivalently, U(1)$_V$×U(1)$_A$. There are two conserved currents, namely, the vector current (the quark number current) $V^\mu = \bar\psi\gamma^\mu\psi$ and the axial current $A^\mu = \bar\psi\gamma^\mu\gamma^5\psi$:

$$\partial_\mu V^\mu = \partial_\mu A^\mu = 0. \tag{9.80}$$

Here ψ is the field of the massless quark. Note that, unlike its four-dimensional counterpart, the two-dimensional axial current is anomaly-free. Note also that in two dimensions V^μ and A^μ are algebraically related to each other,

$$V^\mu = -\varepsilon^{\mu\nu} A_\nu, \tag{9.81}$$

It two dimensions there are no propagating gluons.

as follows from Eq. (9.79).

In reducing four-dimensional QCD to two dimensions we gain a crucial simplification. In two dimensions the gluon field has no physical transverse degrees of freedom. In fact, what remains is just the Coulomb interaction, which is characterized in two dimensions by a linearly growing potential. This is the physical reason lying behind color confinement in the 't Hooft model. Of course, this mechanism of color confinement is much more primitive than the mechanism which is presumed to act in the real world of four dimensions. Still, the model is not completely trivial; it is in the strong coupling regime and one can draw from it some instructive lessons.

Formally, the triviality of the gluon sector is best seen in a judiciously chosen gauge. We will consistently use the axial gauge, in which

$$A_1^a \equiv 0. \tag{9.82}$$

Then only A_0 survives, and the only component of the field strength tensor present in two dimensions, F_{01}, is linear in A_0:

$$F_{01}^a = -\partial_1 A_0^a. \tag{9.83}$$

---

[15] Warning: The choice (9.79) differs from that popular in the literature; see, e.g., [35].

Note that, as usual, no time derivative of A_0 is present in the Lagrangian. The gluon part of the Lagrangian reduces to

$$\mathcal{L}_{\text{gluon}} = \tfrac{1}{2}(\partial_1 A_0^a)^2. \tag{9.84}$$

The second crucial simplification is due to the fact that there are no quark loop insertions in the 't Hooft limit, $N \to \infty$ with $g^2 N$ fixed: each internal quark loop is suppressed by $1/N$. This property is not specific to two dimensions (Section 9.2). The solvability of the model at hand is the combined effect of two crucial properties: the absence of gluon "branchings" and the absence of internal quark loops.

I will define the 't Hooft coupling as[16]

$$\lambda \equiv \frac{g^2}{4\pi} N. \tag{9.85}$$

The action is

$$S = \int dt\, dz\, \mathcal{L}, \tag{9.86}$$

The coupling λ has dimension $[m^2]$.

where t stands for time and z is the spatial coordinate. Where there is no likelihood of confusion, x will denote collectively the space and time coordinates: $x^\mu = \{t, z\}$.

9.5.3 The Gluon Green's Function

In the axial gauge the only surviving component of the gluon Green's function is

$$D_{00}^{ab}(t, z) = \left\langle T\left\{A_0^a(t, z)\, A_0^b(0)\right\}\right\rangle. \tag{9.87}$$

The absence of a time derivative in Eq. (9.84) implies that $D_{00}^{ab}(t, z)$ is local in time,

$$D_{00}^{ab}(t, z) \sim \delta(t). \tag{9.88}$$

Thus, the gluon-exchange-mediated interaction is instantaneous in the model at hand. In momentum space,

$$D_{00}^{ab}(p) \equiv \int d^2x\, e^{ip_\mu x^\mu} D_{00}^{ab}(x) = \frac{i}{p^2}\delta^{ab}, \tag{9.89}$$

Gluon propagator in the axial gauge; p is the spatial momentum.

where $p^\mu \equiv \{p^0, p\}$ and $D(p)$ is p^0-independent. (Henceforth we will omit the Lorentz and color indices where there is no danger of confusion.) The spatial dependence of $D(t, z)$ can be obtained either from the Fourier transform of (9.89), of which I will say more later, or directly from the equation

$$-\partial_z^2 D(t, Z) = i\delta(t)\delta(z). \tag{9.90}$$

The solution to this equation is obvious,

$$D(t, z) = -i\delta(t)\left(\tfrac{1}{2}|z| + C\right), \tag{9.91}$$

where C is an arbitrary constant. The occurrence of an arbitrary constant is physically transparent. If $|z|$ is the confining potential, C shifts the origin on the energy scale. One could fix the value of C by an appropriate additional requirement.

---

[16] My normalization of g is standard. It differs, however, by $\sqrt{2}$ from that adopted in the pioneering paper [7] and in many following publications.

It is instructive to derive Eq. (9.91) through the Fourier transform of (9.89). Since $D(p)$ is p^0-independent, the integral over $dp^0/(2\pi)$ is trivial and immediately produces $\delta(t)$. The spatial integral over $dp/(2\pi)$ is divergent in the infrared domain, at $p = 0$, and must be regularized. It is quite common to regularize it according to the following prescription:

$$\fint_{-\infty}^{\infty} \frac{dp}{2\pi} \frac{1}{p^2} F(p) \equiv \lim_{\varepsilon \to 0} \frac{1}{2} \int_{-\infty}^{\infty} \frac{dp}{2\pi} \left(\frac{1}{p^2 - i\varepsilon} + \frac{1}{p^2 + i\varepsilon} \right) F(p), \qquad (9.92)$$

where ε is a positive infinitesimal parameter and $F(p)$ is an arbitrary nonsingular function with an appropriate fall-off at infinities. It is straightforward to check that under this regularization $\fint_{-\infty}^{\infty} dp/(2\pi p^2)$, vanishes, while

$$\fint_{-\infty}^{\infty} \frac{dp}{2\pi} e^{ipz} \frac{1}{p^2} = -\frac{1}{2} |z|. \qquad (9.93)$$

Sometimes I will omit the regularization sign. Where necessary an appropriate infrared regularization is implied. Where we need to emphasize the standard principal value we will preface an integral with P.V.[17]

In other words, the infrared regularization in Eq. (9.92) leads to $C = 0$. To get a nonvanishing C one could add, for instance, a term proportional $\delta(p)$ in the parentheses in Eq. (9.92). Clearly, this ambiguity must cancel in all equations for physical quantities. Once a regularization procedure is specified, it is important to adhere to it in all calculations until the final values for the physical observables are obtained. Using the fact that

$$\fint_{-\infty}^{\infty} \frac{dp}{2\pi} \frac{1}{p^2} = 0$$

in the regularization of Eq. (9.92), one can rewrite (9.93) in terms of the conventionally (and unambiguously) defined principal value,

$$-\frac{1}{2} |z| = \fint \frac{dp}{2\pi} (e^{ipz} - 1) \frac{1}{p^2}.$$

9.5.4 Equation for Heavy–Light Mesons

Now I am ready to explain (in gross terms) quark confinement in this model and address the issue of quark–antiquark bound states, i.e., mesons. Originally, the spectral problem was solved [34] in the infinite-momentum frame.[18] The corresponding equation is known as the 't Hooft equation. Although this equation has significant computational advantages, the underlying physics is hidden in rather obscure boundary conditions. In addition, the phenomenon of chiral symmetry breaking and its relation to color confinement remains unclear. To make this transparent it is convenient to formulate the problem in a different way.

---

[17] The conventional definition of the principal value is as follows:

$$\text{P.V.} \int_{-\infty}^{\infty} \frac{dp}{p} F(p) \equiv -\fint_{-\infty}^{\infty} \frac{dp}{p} F(p) \equiv \lim_{\varepsilon \to 0} \left[\int_{-\infty}^{-\varepsilon} \frac{dp}{p} F(p) + \int_{\varepsilon}^{\infty} \frac{dp}{p} F(p) \right].$$

[18] An excellent pedagogical discussion of both the derivation of the 't Hooft equation, with appropriate boundary conditions, and the numerical results can be found in the 176-page Ph.D. thesis of K. Hombostel [36] (the KEK scanned version).

The meson to be considered below is built from an infinitely heavy antiquark at rest at the origin and a dynamical quark with mass m, which may or may not vanish.

We will study $q\bar{Q}$ mesons.
We will refer to the dynamical quark as the light quark. The heavy antiquark is the source of the Coulomb field in which the light quark moves. Since the (infinitely) heavy quark has no dynamics, the light-quark Lagrangian can be written separately from the heavy (anti)quark part, namely,

$$\mathcal{L}_{\text{light}} = \bar{\psi} \left[\gamma^0(i\partial_0 + g A_0) + i\gamma^1 \partial_1 - m - \Sigma \right] \psi, \tag{9.94}$$

where A_0 is a t-independent confining potential and Σ is the light quark self-energy, to be considered in Section 9.5.5. This Lagrangian takes into account all gluon exchanges between the static force center and the light quark. Planarity allows one to perform a complete summation over the color degrees of freedom. The color indices in Eq. (9.94) are implicit in ψ and A_0. (The light-quark self-energy is diagonal in color.) Therefore, for the *color-singlet quark–antiquark* state one can replace $g A_0$ by an effective Abelian combination:

$$g A_0 \rightarrow -V = -g^2 \frac{\text{Tr}\,(T^a T^a)}{N} \frac{1}{2}|z| = -2\pi\lambda \frac{1}{2}|z| = \lambda \oint dk \frac{1}{k^2} e^{ikz}, \tag{9.95}$$

simultaneously omitting all color indices elsewhere (i.e., on ψ, etc.). Here we have used Eq. (9.93) to pass to the Fourier representation.

The light-quark Lagrangian (9.94) implies the following equation of motion:

Dirac equation for the light quark
$$\mathcal{E}\psi(z) = \left[\pi\lambda|z| - i\gamma^5\partial_z + \gamma^0(m + \Sigma) \right] \psi(z), \tag{9.96}$$

where the time dependence of the wave function

$$\psi \sim e^{-i\mathcal{E}t}$$

is explicitly accounted for, so that $\psi(z)$ in Eq. (9.96) depends only on z. To begin with, let us assume that the light quark is not particularly light, namely that $m \gg \sqrt{\lambda}$.

See Section 9.5.5.
In this case one can neglect Σ in Eq. (9.96). Then, for the low-lying levels,

$$\mathcal{E} = m + E, \qquad E \ll m, \tag{9.97}$$

we find that Eq. (9.96) reduces to the standard nonrelativistic Schrödinger equation for the (one-component) wave function $\Psi(z)$,

$$\left(-\frac{1}{2m}\partial_z^2 + \pi\lambda|z| \right) \Psi(z) = E\Psi(z), \tag{9.98}$$

Qualitative discussion of highly excited bound states of a quark with mass $m \gg \sqrt{\lambda}$ and an infinitely heavy antiquark at the origin
with a linearly growing potential. Needless to say, the spectrum of this problem is discrete, and all bound states are localized. The energy level splitting is $O\left[\sqrt{\lambda}(\sqrt{\lambda}/m)^{1/3} \right]$.

More interesting are the highly excited states, $\mathcal{E} \gg m$. Now the nonrelativistic approximation is inadequate, of course. We must return to the version of Eq. (9.96) with Σ omitted,

$$(V - i\gamma^5\partial_z + \gamma^0 m)\psi(z) = \mathcal{E}\psi(z), \qquad V = \pi\lambda|z|, \tag{9.99}$$

for a relativistic treatment. I recall that $\psi(z)$ has two components:

$$\psi = \begin{pmatrix} \psi_1 \\ \psi_2 \end{pmatrix}. \tag{9.100}$$

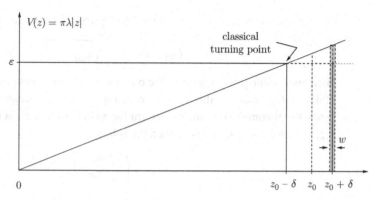

Fig. 9.34 Analyzing Eq. (9.99) in different domains. In this figure $z_0 = \mathcal{E}/(\pi\lambda)$ and $\delta = m/(\pi\lambda)$. We are assuming that $\mathcal{E} \gg m \gg \sqrt{\lambda}$. The width of the shaded domain $w = O(1/\sqrt{\lambda})$.

Like any Dirac equation, (9.99) has solutions with positive and negative energies. The latter should be ignored since they represent (dynamical) antiquarks, which cannot form finite-energy bound states with an infinitely heavy antiquark at the origin.

Our task here is to understand the qualitative features of the solution of Eq. (9.99), rather than actually to solve the spectral problem in this way (this is much more easily done in an infinite-momentum frame [34, 36]). To develop a qualitative picture we can rely on the fact that at $\mathcal{E} \gg m \gg \sqrt{\lambda}$ the classically allowed domain of the quark motion in the potential $\pi\lambda|z|$ is huge, and the potential itself changes very slowly: if $\Delta z \sim 1/\sqrt{\lambda}$ then the variation in the potential $\Delta V \sim \sqrt{\lambda}$, to be compared with $\mathcal{E}$. Under these circumstances one can consider Eq. (9.99) piecewise. Relevant intervals on the z axis are depicted in Fig. 9.34, which displays only positive values of z (for negative z it must be mirror-reflected). If we neglect m then at $z = z_0 \equiv \mathcal{E}/(\pi\lambda)$ the total energy $\mathcal{E}$ becomes equal to the potential energy, so that classically this would be a turning point. Taking account of $m \neq 0$ shifts the turning point to $z_0 - \delta$, where $\delta \equiv m/(\pi\lambda)$.

In the domain $|z| < z_0 - \delta$, the quark is fast, its mass can be neglected, and its motion is quasiclassical.[19] The solution is given by left- and right-moving waves with slowly varying wavelengths

$$\psi_\ell = \left(\begin{array}{c} \exp\{-i\int [\mathcal{E} - V(z)]\,dz\} \\ 0 \end{array} \right), \quad \psi_r = \left(\begin{array}{c} 0 \\ \exp\left\{i\int [\mathcal{E} - V(z)]\,dz\right\} \end{array} \right),$$
(9.101)

Quasiclassical solutions for highly excited states

where m is set to zero. Any linear combination of ψ_ℓ and ψ_r is a solution too. To balance the momentum, we can close $\psi(z) = \psi_\ell \pm \psi_r$. Two different signs in this expression reflect the Z_2 symmetry of the problem: invariance under $z \to -z$, $\psi_1 \to \psi_2$, and $\psi_2 \to -\psi_1$. Thus, the combinations $\psi(z) = \psi_\ell \pm \psi_r$ correspond, in a sense, to symmetric and antisymmetric wave functions.

---

[19] To be more exact this is true if $|z| \ll z_0$, when the quark moves essentially as a free ultrarelativistic on-mass-shell particle. Qualitatively, we can extend this domain up to $|z| < z_0 - \delta$. The quark mass is non-negligible at $|z - z_0| \sim \delta$ but since $\delta \ll z_0$ the corresponding error is small. At $|z| > z_0 - \delta$ the quark momentum becomes purely imaginary; the quark goes off-shell.

In the interval $z_0 - \delta < |z| < z_0 + \delta$ the quark effective momentum, defined as

$$\mathcal{E}_{\text{eff}} = \mathcal{E} - V = \sqrt{p_{\text{eff}}^2 + m^2},$$

becomes purely imaginary and the oscillating regime gives place to an exponential fall-off. One can see this readily from Eq. (9.99). For simplicity, we can neglect $|\mathcal{E} - V(z)|$ compared to the mass term for z close to z_0 (i.e., in the interval $z_0 - \delta < |z| < z_0 + \delta$). Then, in this domain the solution is

$$\psi = \begin{pmatrix} e^{-mz} \\ -e^{-mz} \end{pmatrix}, \tag{9.102}$$

where the explicit form of the γ matrices in Eq. (9.79) is used. The quark "tunnels" under the barrier and, as we move from the left-hand to the right-hand edge of this interval, the wave function is suppressed by $\exp(-m^2/\lambda)$, an enormous exponential suppression.

In the shaded domain of thickness $w = O(1/\sqrt{\lambda})$ near $z = z_0 + \delta$, Eq. (9.99) ceases to be applicable since the neglect of Σ in (9.96) in passing to (9.99) is now unjustified. At $|z| = z_0 + \delta$ the dynamical quark becomes effectively massless; the gap between quarks and antiquarks disappears. At still larger $|z|$, Eq. (9.99) no longer describes dynamical quarks since the effective energy becomes negative.

The last thing to do is to match Eqs. (9.101) and (9.102) at $z = z_0 - \delta$. In fact, within the accuracy of the approximations made above, we can replace the matching at $z = z_0 - \delta$ by that at $z = z_0$. Accounting for both signs in $\psi_\ell \pm \psi_r$ the matching condition can be written as

$$\int_0^{z_0} [\mathcal{E} - V(z)]\, dz = \frac{\pi}{2} n, \qquad z_0 = \frac{\mathcal{E}}{\pi \lambda}, \tag{9.103}$$

where n is the excitation number,

$$n \gg \frac{m}{\sqrt{\lambda}}.$$

Equation (9.103) implies the following quantization of energy:

$$\mathcal{E} = \pi \sqrt{\lambda} \sqrt{n}. \tag{9.104}$$

As expected, $\mathcal{E}^2$ is linear in n. The energy-level splitting is

$$\Delta \mathcal{E} = \frac{\pi}{2} \frac{\sqrt{\lambda}}{\sqrt{n}}. \tag{9.105}$$

The limit of the massless dynamical quarks, $m = 0$, is most important but complicated. The integral Bethe–Salpeter equation emerges here.

This is much smaller than in the nonrelativistic case (see the estimate after Eq. (9.98)).

Now let us pass to the most difficult case, that of massless dynamical quarks. We must return to Eq. (9.96), put $m = 0$ and keep the quark self-energy Σ, which will play a crucial role:

$$\mathcal{E}\psi(z) = (\pi \lambda |z| - i\gamma^5 \partial_z + \gamma^0 \Sigma)\psi(z). \tag{9.106}$$

In fact this equation is symbolic, since Σ is a nonlocal function of z. As we will see in Section 9.5.5, it is local in momentum space; there is a closed-form exact equation for $\Sigma(p)$. Therefore, it is convenient to pass to wave functions in momentum space, $\psi(p)$:

$$\psi(z) = \int \frac{dp}{2\pi} e^{ipz} \psi(p). \tag{9.107}$$

Before we will be ready to rewrite Eq. (9.106) in momentum space, untangling en route the positive- and negative-energy solutions and discarding the latter, we will need to carry out a more thorough investigation of the quark self-energy.

9.5.5 The Quark Green's Function

In this section we will use the large-N limit to calculate the quark self-energy and the quark propagator exactly. An exact calculation becomes possible because only the planar diagrams survive, and these can be readily summed over. To warm up and get some experience we will start, however, from the one-loop graph presented in Fig. 9.35. We will denote the quark self-energy by $-i\Sigma$, so that the quark Green's function is

$$G_{ij}(p_0, p) = \int d^2x \, e^{ip_\mu x^\mu} \langle T\{\psi_i(x), \bar{\psi}_j(0)\}\rangle = \frac{i}{\slashed{p} - m - \Sigma} \delta_{ij}, \tag{9.108}$$

where, as usual, $\slashed{p} = p_\mu \gamma^\mu$ and we use the fact, to be confirmed below, that the quark self-energy is diagonal in color space (i.e., it is proportional to δ_{ij}). In the $A_1 = 0$ gauge, which was chosen once and for all, Σ depends only on the spatial components of the quark momentum p, not on p^0. This will be seen shortly. It is not difficult to calculate the graph in Fig. 9.35 although one has to deal with rather cumbersome expressions. We benefit from the fact that only D_{00} is nonvanishing and perform the integral over the time component of the loop momentum using residues. In this way we arrive at

$$\Sigma(p) = \frac{\lambda}{2} \left\{ -2\gamma^1 \left[\frac{p}{m^2 + p^2} + \frac{m^2}{2(m^2 + p^2)^{3/2}} \ln \frac{\sqrt{m^2 + p^2} + p}{\sqrt{m^2 + p^2} - p} \right] \right.$$

$$\left. -m \left[\frac{2}{m^2 + p^2} - \frac{p}{(m^2 + p^2)^{3/2}} \ln \frac{\sqrt{m^2 + p^2} + p}{\sqrt{m^2 + p^2} - p} \right] \right\}. \tag{9.109}$$

p is the spatial component of the two-momentum p^μ with upper index.

From this exercise one should extract three lessons, which are not limited to the one-loop case. They are of a general nature: (i) the loop expansion parameter is $\lambda/(m^2 + p^2)$ and explodes at m and $|p| < \sqrt{\lambda}$, so that summation of an infinite series

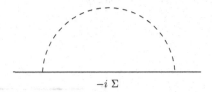

$-i\,\Sigma$

Fig. 9.35   The quark self-energy $-i\Sigma$ at one loop. The solid and broken lines represent quarks and gluons, respectively.

is necessary; (ii) in the $A_1 = 0$ gauge Σ depends only on the spatial component of momentum; (iii) its general Lorentz structure is

$$\Sigma(p) = A(p) + B(p)\,\gamma^1, \tag{9.110}$$

where A and B are some real functions of p (for real p). From Eq. (9.108) we see that the combination that will appear in the quark Green's function is (see Eq. (9.113))

$$p^0\gamma^0 - [m + p\,\gamma^1 + A(p) + B(p)\,\gamma^1]. \tag{9.111}$$

It is customary to exchange A and B for two other functions, E_p and θ_p which have a clear-cut physical meaning, and parametrize the quark Green's function in a more convenient way. Namely,

$$E_p \equiv \sqrt{(m + A)^2 + (p + B)^2},$$

$$m + A = E_p \cos\theta_p, \qquad p + B = E_p \sin\theta_p, \tag{9.112}$$

| Definition of E_p; the first appearance of the Bogoliubov (or chiral) angle. |

where for consistency we require that E_p be positive for all real p. The angle θ_p is referred to as the Bogoliubov angle or, more commonly, the chiral angle. The exact quark Green's function now can be rewritten as

$$G = i\,\frac{p^0\gamma^0 - E_p \sin\theta_p\gamma^1 + E_p \cos\theta_p}{p_0^2 - E_p^2 + i\varepsilon}. \tag{9.113}$$

Closed-form exact equations can be obtained for E_p and θ_p. This is due to the fact that in the 't Hooft limit the quark self-energy is saturated by "rainbow graphs." An example of a rainbow graph is given in Fig. 9.36. Intersections of gluon lines and insertions of internal quark loops are forbidden, and so are gluon lines on the other side of the quark line, see Section 9.2.2. This diagrammatic structure implies the equation depicted in Fig. 9.37, where the bold solid line denotes the exact Green's function (9.113). Algebraically,

$$\Sigma(p) = -ig^2 T^a T^a \int \frac{d^2k}{(2\pi)^2}\,\gamma^0 G(k)\gamma^0 D_{00}(p - k). \tag{9.114}$$

It is easy to see that this equation sums an infinite sequence of rainbow graphs.

Fig. 9.36   An example of the rainbow graph in $\Sigma(p)$.

$$\Sigma =$$

Fig. 9.37   Exact equation for $\Sigma(p)$, summing all rainbow graphs. The bold solid line is the exact quark propagator (9.113).

Using Eq. (9.89) for the photon Green's function in conjunction with (9.113) and performing integration over k^0, the time component of the loop momentum, by virtue of residues it is not difficult to obtain

$$\Sigma(p) = \frac{\lambda}{2} \oint dk \left[\gamma^1 \sin\theta_k \frac{1}{(p-k)^2} + \cos\theta_k \frac{1}{(p-k)^2} \right],$$ (9.115)

which implies, in turn,

$$E_p \cos\theta_p - m = \frac{\lambda}{2} \oint dk \cos\theta_k \frac{1}{(p-k)^2},$$

$$E_p \sin\theta_p - p = \frac{\lambda}{2} \oint dk \sin\theta_k \frac{1}{(p-k)^2},$$ (9.116)

with boundary conditions

$$\theta_p \to \begin{cases} \dfrac{\pi}{2} & \text{as } p \to \infty, \\ -\dfrac{\pi}{2} & \text{as } p \to -\infty, \end{cases}$$ (9.117)

determined by the free-quark limit (Section 9.5.6). This set of equations was first obtained by Bars and Green [35]. Multiplying the first equation by $\sin\theta_p$ and the second by $\cos\theta_p$ and subtracting one from the other one gets an integral equation for the chiral angle, namely,

$$p \cos\theta_p - m \sin\theta_p = \frac{\lambda}{2} \oint dk \sin(\theta_p - \theta_k) \frac{1}{(p-k)^2}.$$ (9.118)

Assuming that the chiral angle is found one then can get E_p from the equation

$$E_p = m \cos\theta_p + p \sin\theta_p + \frac{\lambda}{2} \oint dk \cos(\theta_p - \theta_k) \frac{1}{(p-k)^2}.$$ (9.119)

Note that for heavy dynamical quarks ($m \gg \sqrt{\lambda}$) and $|p| \sim \sqrt{\lambda}$ Eq. (9.118) reduces to $m \sin\theta_p = 0$, with the trivial solution

$$\theta_p = 0$$ (9.120)

up to terms $O(\lambda/m^2)$. Equation (9.120) is equivalent to the statement that in this limit $E_p \to m$ up to corrections $O(\lambda/m)$.[20]

9.5.6 Chiral Symmetry Breaking. Solutions for Σ

Our primary goal now is to discuss the 't Hooft model in the limit $m = 0$, when its Lagrangian is symmetric under the chiral rotations,

$$\psi_R \to e^{i\alpha} \psi_R, \qquad \psi_L \to e^{-i\alpha} \psi_L, \qquad \psi_{R,L} = \tfrac{1}{2}\left(1 \pm \gamma^5\right)\psi.$$ (9.121)

---

[20] In fact, depending on the regularization of the infrared divergences, a correction of the order of $(m)^0$ could appear. For instance, one could introduce an extra term of the form $2\kappa\delta(p)$ in the parentheses in Eq. (9.92), where κ is a constant. This will have an impact on Eqs. (9.119), (9.123), and (9.95). What is important is that our final equation, (9.140), being unambiguously defined in terms of the P.V., remains valid in any regularization. For simplicity, in intermediate derivations, we stick to a regularization with no $O((m)^0)$ shift in the dynamical quark mass m. Let us note parenthetically that the boundary conditions (9.117) will be satisfied for $|p| \gg m$.

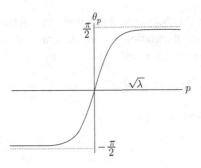

Fig. 9.38   A stable solution for the chiral angle θ_p as a function of p.

<table>
<tr><td>

The spontaneous breaking of the axial symmetry will be explicitly seen in the solution to be given below.

</td><td>

In other words, the axial current of massless quarks is conserved; see Eq. (9.80). Jumping ahead of myself, let me say that in the 't Hooft model the chiral symmetry *is spontaneously broken.*

Let us rewrite Eqs. (9.118) and (9.119) for the chiral angle and E_p in the limit $m = 0$:

</td></tr>
</table>

$$p \cos \theta_p = \frac{\lambda}{2} \!\!\!\!\!\!\diagup\!\!\!\!\int dk\ \sin(\theta_p - \theta_k) \frac{1}{(p-k)^2}, \tag{9.122}$$

$$E_p = p \sin \theta_p + \frac{\lambda}{2} \!\!\!\!\!\!\diagup\!\!\!\!\int dk\ \cos(\theta_p - \theta_k) \frac{1}{(p-k)^2}. \tag{9.123}$$

An immediate consequence is that θ_p is an odd function of p, while $E(p)$ is even.

Upon examining Eq. (9.122) it is not difficult to guess an analytic solution,

$$\theta_p = \frac{\pi}{2} \operatorname{sign} p, \tag{9.124}$$

where sign p is the sign function,

$$\operatorname{sign} p = \vartheta(p) - \vartheta(-p).$$

Substituting this solution into Eq. (9.123) one obtains

$$E_p = |p| - \frac{\lambda}{|p|}. \tag{9.125}$$

Alas … this analytic solution is unphysical. This is obvious from the fact that $E(p)$ becomes negative for $|p| < \sqrt{\lambda}$. This feature of the solution (9.125) – negativity at small $|p|$ – cannot be amended by a change in the infrared regularization. In fact, one can show [37] that (9.125) does *not* correspond to the minimum of the vacuum energy.

A stable solution has the form depicted in Fig. 9.38. It was (numerically) obtained in [38]. It is smooth everywhere. For $|p| \ll \sqrt{\lambda}$ it is linear in p. Its asymptotic approach to $\pm\pi/2$ at $|p| \gg \sqrt{\lambda}$ will be discussed later.

Now, let us calculate the chiral condensate, the vacuum expectation value $\langle \bar{\psi}\psi \rangle$,

$$\langle \bar{\psi}\psi \rangle = -\operatorname{Tr} \int \frac{d^2 p}{(2\pi)^2} G(p_0, p), \tag{9.126}$$

where Tr stands for the trace with respect to both the color and the Lorentz indices, and the quark Green's function $G(p_0, p)$ is defined in Eq. (9.113). Taking the trace and performing the p_0 integration we arrive at

$$\langle \bar{\psi}\psi \rangle = -N \int \frac{dp}{2\pi} \cos \theta_p. \qquad (9.127)$$

For the singular solution (9.124) the above quark condensate vanishes, since $\cos \theta_p \equiv 0$. However, for the physical smooth solution depicted in Fig. 9.38 the quark condensate does not vanish. In fact, $\langle \bar{\psi}\psi \rangle$ was calculated analytically (as a self-consistency condition) in [39] using methods going beyond the scope of the present section. The result was

$$\langle \bar{\psi}\psi \rangle = -\frac{N}{\sqrt{6}}\sqrt{\lambda}. \qquad (9.128)$$

A comparison of Eqs. (9.127) and (9.128) provides us with a constraint on the integral over $\cos \theta_p$. Moreover, these two expressions, in conjunction with Eq. (9.122), allow us to determine the leading pre-asymptotic correction in θ_p at $|p| \gg \sqrt{\lambda}$. Indeed, in this limit the right-hand side of Eq. (9.122) reduces to (for $p > 0$)

$$\frac{\lambda}{2p^2} \int dk \, \sin\left(\frac{\pi}{2} - \theta_k\right) = \frac{\lambda}{2p^2} \int dk \, \cos \theta_k, \qquad (9.129)$$

while for the left-hand side we have

$$p \sin\left(\frac{\pi}{2} - \theta_p\right) \rightarrow p\left(\frac{\pi}{2} - \theta_p\right). \qquad (9.130)$$

This implies, in turn, that

$$\theta_p = \frac{\pi}{2}\text{sign}\, p - \frac{\pi}{\sqrt{6}}\left(\frac{\sqrt{\lambda}}{p}\right)^3 + \cdots, \quad |p| \gg \sqrt{\lambda}. \qquad (9.131)$$

At the same time, from Eq. (9.123) we deduce that there is no p^{-3} correction in $E/|p|$; the leading correction is of order λ^3/p^6.

Let us pause here to discuss the phase of the condensate (9.128). The quark condensate is not invariant under the transformation (9.121), implying the existence of a continuous family of degenerate vacua and a massless Goldstone boson, a "pion" (see Section 6.5.1). Under the circumstances, the phase of $\langle \bar{\psi}_R \psi_L \rangle$ is ambiguous and depends on the way in which a given vacuum is picked up. If a small mass term

$$-\left(m\bar{\psi}_R \psi_L + \text{H.c.}\right)$$

is added to the Lagrangian for infrared regularization, it lifts the degeneracy, forcing the theory to choose a particular vacuum. Equation (9.116) with the asymptotics (9.117) and the result quoted in (9.128) correspond to the limit $m \rightarrow 0$ with the mass parameter m real and positive. This is the standard convention.

In the conclusion of this subsection, let us ask ourselves the physical meaning of the chiral angle θ_p introduced through the quark Green's function; see (9.112). To answer this question, let us have a closer look at the free-quark Dirac equation,

$$\mathcal{E}\psi(p) = (\gamma^5 p + \gamma^0 m)\psi(p) \qquad (9.132)$$

Defining the phase of the quark condensate

where ψ is the two-component spinor (9.100). For any given value of p one solution has positive energy, while the other has negative energy and must be discarded. To diagonalize the Hamiltonian one can make the following unitary transformation:

$$\psi(p) \rightarrow \exp\left[\frac{\gamma^1}{2}\left(\frac{\pi}{2} - \alpha\right)\right]\psi(p),$$

$$\sin\alpha = \frac{p}{\sqrt{p^2 + m^2}}, \qquad \cos\alpha = \frac{m}{\sqrt{p^2 + m^2}}. \tag{9.133}$$

The angle $\alpha \rightarrow \frac{\pi}{2}$ sign p as $|p| \rightarrow \infty$. Then the Hamiltonian (9.132) takes the form

$$H \rightarrow \exp\left[-\frac{\gamma^1}{2}\left(\frac{\pi}{2} - \alpha\right)\right] H \exp\left[\frac{\gamma^1}{2}\left(\frac{\pi}{2} - \alpha\right)\right],$$

$$= \gamma^5\sqrt{p^2 + m^2}, \tag{9.134}$$

i.e., it is diagonal. This implies that

$$\exp\left[\frac{\gamma^1}{2}\left(\frac{\pi}{2} - \alpha\right)\right]\begin{pmatrix} \chi \\ \phi \end{pmatrix} \tag{9.135}$$

is the eigenfunction of the original Hamiltonian, with positive energy if $\chi = 0$, $\phi \neq 0$ and negative energy if $\chi \neq 0$, $\phi = 0$.

What changes when we switch on the Coulomb interaction and pass to Eq. (9.96) from the free-quark equation (9.132)? I claim that if α is replaced by θ_p, so that

$$\psi(p) \equiv \exp\left[\frac{\gamma^1}{2}\left(\frac{\pi}{2} - \theta_p\right)\right]\begin{pmatrix} 0 \\ \phi(p) \end{pmatrix}, \tag{9.136}$$

then the equation for $\phi(p)$ obtained from (9.96) will describe light-quark propagation in the Coulomb potential induced by the infinitely heavy antiquark at the origin. Keeping χ instead of ϕ in Eq. (9.135) would lead us, instead, to a system containing a heavy antiquark plus a light antiquark. This system has no bound states. This separation of positive and negative energy solutions is known as the *Foldy–Wouthuysen transformation* (see, e.g., [40]).

9.5.7 The Bethe–Salpeter Equation

Thus, we start from Eq. (9.96) and make the Fourier transformation (9.107). Then Eq. (9.96) takes the form

$$\mathcal{E}\psi(p) = -\lambda \oint dk \frac{1}{(p-k)^2}\psi(k) + (\gamma^5 E_p \sin\theta_p + \gamma^0 E_p \cos\theta_p)\psi(p), \tag{9.137}$$

where we have used Eqs. (9.95), (9.111), and (9.112). The substitution of Eq. (9.136) into (9.137) is straightforward. In the first term on the right-hand side we get

$$\exp\left[-\frac{\gamma^1}{2}\left(\frac{\pi}{2} - \theta_p\right)\right]\exp\left[\frac{\gamma^1}{2}\left(\frac{\pi}{2} - \theta_k\right)\right] = \cos\frac{\theta_p - \theta_k}{2} + \gamma^1\sin\frac{\theta_p - \theta_k}{2}. \tag{9.138}$$

The term with $\sin[(\theta_p - \theta_k)/2]$ drops out in the integral because it is odd under $p \to -p$, $k \to -k$. The second term is treated analogously to Eq. (9.134). In this way we arrive at the *Bethe–Salpeter equation* for the function $\phi(p)$ describing the bound states:

$$\mathcal{E}\phi(p) = -\lambda \int \frac{dk}{(p-k)^2} \cos \frac{\theta_p - \theta_k}{2} \phi(k) + E_p\phi(p), \tag{9.139}$$

where we assume, as usual, that E_p is positive.

It is convenient to make an extra step: substitute Eq. (9.123) into (9.139). In this way we arrive at the equivalent Beth–Salpeter equation:

$$\mathcal{E}\phi(p) = p \sin \theta_p \, \phi(p)$$
$$- \lambda \int \frac{dk}{(p-k)^2} \left[\cos \frac{\theta_p - \theta_k}{2} \phi(k) - \left(\cos \frac{\theta_p - \theta_k}{2} \right)^2 \phi(p) \right]. \tag{9.140}$$

The advantage of this equation over (9.139) is that now the integral on the right-hand side can be regularized by virtue of the standard principal value prescription; see footnote 17, at the end of Section 9.5.3.

It is not difficult to derive the boundary conditions on $\phi(p)$ and some properties of the wave function; as follows.

(i) It can be taken as real, nonsingular, and either symmetric or antisymmetric under $p \to -p$,

$$\phi(-p) = \pm\phi(p),$$

and

(ii) at large $|p|$,

$$\phi(p) \sim \begin{cases} \dfrac{1}{|p|^3} & \text{symmetric levels,} \\[2mm] \dfrac{1}{p^4} & \text{antisymmetric levels.} \end{cases} \tag{9.141}$$

This asymptotic behavior is necessary to guarantee the cancelation of the leading term (at large p) on the right-hand side of Eq. (9.139).

Analytic solutions of Eq. (9.122) and the spectral Bethe–Salpeter equation (9.140) are not known. However, they can be solved numerically (see, e.g., [50]).

Exercises

9.5.1 The quark condensate (9.128) is the order parameter for the (continuous) axial symmetry. The fact that $\langle \bar{\psi}\psi \rangle \neq 0$ implies the spontaneous breaking of this symmetry in the 't Hooft model and the occurrence of the massless pion. Why does this not contradict the Coleman theorem (Section 6.5)?

9.5.2 Derive the nonrelativistic limit of Eq. (9.96), i.e., Eq. (9.98). Find the relation between $\Psi(z)$ and $\psi_{1,2}(z)$.

9.6 Schwinger Model

Julian Schwinger was the first to consider quantum electrodynamics (QED) with charged fermions in $1 + 1$ dimensions [51]. We discussed the *massless* Schwinger model in Section 8.1 in connection with the chiral anomaly. In Section 8.1, the spatial dimension x was compactified on a circle S_1 of circumference L such that $eL \ll 1$. The latter condition guaranteed that the theory was at weak coupling. This consideration was sufficient for anomaly analysis.

In this section, from the compactified space-time $R_1 \times S_1$ we will return to infinite space dimension, i.e., R_2 space-time, as in Section 8.1.7. Our task is to analyze dynamical features of the Schwinger model at strong coupling, i.e., on R_2. We will limit ourselves to the simplest case of a single Dirac fermion with charge 1 (in the units of e) and consider both massless and massive limits. The fermion mass is denoted by m. To further simplify the subsequent explanations I will use the word "quark" to refer to the ψ field in Eq. (9.142) and below.

As we will see, independently of the value of m there are no charged states in the physical spectrum (no charged asymptotic states). This is by itself a significant distinction between the Schwinger model and 4D QED. However, this is not all – two more dynamical features emerge that are absent in 4D QED. Their impact depends on m.

First, we will prove that for $m = 0$ dynamical Higgsing takes place, the photon field acquires a nonvanishing mass term m_γ, and all electric charges are screened. The only asymptotic state in the physical spectrum is a spin-0 boson of mass $e/\sqrt{\pi}$, a remnant of the photon that "swallowed" an electron–positron pair as a "composite Higgs state." Non-propagating massless photon transformed itself into a propagating physical degree of freedom. Second, the vacuum θ angle appears in the Schwinger model with necessity, in contradistinction with 4D QED. At $m = 0$ it is unobservable. However, if $m \neq 0$ physics depends on θ in the most direct way.

At $m \neq 0$, the physical spectrum in the Schwinger model contains a few *neutral* bound states of the quark–antiquark type. The number of such stable states depends on the ratio m/e. The binding is due to linear potential characteristic of 2D Coulomb interaction. In this case we have a confining regime similar to that in QCD.

For the reader's convenience I reproduce here the Lagrangian of the Schwinger model from Section 8.1.1 (in canonic normalization),

$$\mathcal{L}_{\text{Sch}} = -\frac{1}{4} F_{\mu\nu} F^{\mu\nu} + \bar{\psi}\, \gamma^\mu \left(i\, \partial_\mu + e A_\mu \right) \psi - m \bar{\psi}\psi + \frac{\theta}{4\pi} \varepsilon^{\mu\nu} \left(e F_{\mu\nu} \right). \qquad (9.142)$$

9.6.1 Emergence of θ

In this section I explain in heuristic terms [52] why in $1 + 1$ QED the θ angle is necessary for the description of the ground state and why, in fact, it can be viewed as a (spatially constant) background electric field.

In classical physics a constant electric field is the solution of static Maxwell's equation

$$\text{div}\vec{E} = \rho \qquad (9.143)$$

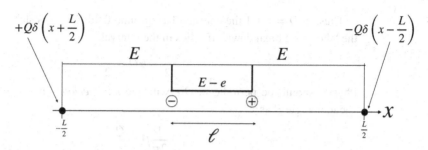

Trial "quark" pair creation in the constant electric field in 2D QED. Closed circles at $x = \pm L/2$ denote infinitely heavy particles at spatial infinities (i.e., it is assumed that at the end $L \to \infty$). The "quark–antiquark" separation is ℓ. The parameters Q, E, and e are positive in this figure.

inside a capacitor (here ρ is the electric charge density vanishing inside the capacitor). In the 2D model described by $\mathcal{L}_{\text{Sch}}$ the afore mentioned capacitor reduces to a pair of charges $Q = Ce$ (one positive, one negative) nailed at both spatial infinities; see Fig. 9.39. They can be interpreted as infinitely heavy nondynamical particles. The parameter C is a nonnegative constant not necessarily integer.

Generally speaking, the described configuration is unstable under quantum "dielectric breakdown" – production of "quark" pairs that eventually screen the capacitor and nullify the constant electric field. If the number of spatial dimensions is larger than one the nullification will be complete (see the remark below Eq. (9.4.7)). In one spatial dimension, as we will see, only sufficiently strong background fields will be screened.

Let us examine a trial configuration, a "quark" pair produced in the background field E as shown in Fig. 9.39. The energy density in the domain between the "quark" pair components is

$$\mathcal{E} = \frac{1}{2}(E - e)^2 \tag{9.144}$$

instead of the original $E^2/2$. The resulting energy change is

$$\frac{E^2 \ell}{2}\left[\left(1 - \frac{e}{E}\right)^2 - 1\right]. \tag{9.145}$$

Here E is the electric field (which in the case at hand has only one component), the distance ℓ defined in Fig. 9.39 can be arbitrarily large. The charge parameters Q and e are assumed to be positive. Equation (9.143) then implies that E is positive too since $E = Q$. The careful reader will notice that in (9.145) I omitted the term $(-2m)$, the energy loss due to the quark masses. This is always justified provided ℓ is sufficiently large.

From (9.145) it is obvious that generally speaking the quark pair creation lowers the energy of the configuration in Fig. 9.39. Therefore, the pair creation resulting in screening will continue until the $\pm Q$ charges and the original value of E hit the bound $Q = E = \frac{e}{2}$ at which point the energy in (9.145) hits zero. Further quark pair creation is energetically inexpedient.

If we swap positive and negative charges at infinities and the quark charge signs the electric field will become negative. Then the corresponding bound for the pair creation is $E = -\frac{e}{2}$.

Thus, in $D = 1 + 1$ the constant background field $E = Q$ is stable with regards to the "dielectric breakdown" if it lies in the interval

$$-\frac{e}{2} \leq \pm Q \leq \frac{e}{2}. \tag{9.146}$$

Physics should be periodic in Q/e with the *unit period*. Then one can define the vacuum angle θ as [21]

$$\frac{1}{2\pi}\theta = \frac{Q}{e}, \tag{9.147}$$

cf. Eq. (9.52). In other words, in the 2D Schwinger model the θ angle can be interpreted as a background electric field.

With three spatial dimensions, even if one could construct a gigantic capacitor at macroscopically large distances the constant field would be unstable – it would be always screened to zero by pair production.

Since the θ term in (9.142) is a full derivative, it is unobservable in perturbative calculations. In much the same way as in QCD, the θ-related effects show up only nonperturbatively, at strong coupling.

9.6.2 Photon Mass in the Massless Schwinger Model

In this section we will drop the "quark" mass parameter in (9.142), i.e., put $m = 0$. We will see that a finite photon mass is generated because of a peculiar IR dynamics, i.e., at strong coupling, see Eq. (9.151). In fact, Eq. (8.29) gives a hint that the photon mass $m_\gamma \neq 0$ emerges at one loop. Let us trace its genesis in more detail.

Let us start from the one-loop graph for the photon polarization depicted in Fig. 9.40a. For pedagogical purposes I will analyze this graph in the coordinate space rather than in momentum. For massless fields this is the fastest way of calculation of one-loop and some other diagrams.

Indeed, in two dimensions the massless fermion Green function is [22]

$$G(x, 0) = -\frac{1}{2\pi} \frac{i\left(x_\alpha \gamma^\alpha\right)}{x^2}. \tag{9.148}$$

Then the polarization operator

$$i\,\Pi^{\mu\nu}(x, 0) = \underbrace{(-1)}_{\text{ferm. in loop}} \underbrace{(ie)^2}_{\text{two vert.}} \text{Tr}\left(\gamma^\mu G(x, 0)\gamma^\nu G(0, x)\right)$$

$$= -\frac{e^2}{2\pi^2}\left(\frac{g^{\mu\nu}}{x^2} - \frac{2x^\mu x^\nu}{x^4}\right) \rightarrow i\frac{e^2}{\pi}\left(g^{\mu\nu} - \frac{p^\mu p^\nu}{p^2}\right), \tag{9.149}$$

where in the second line in (9.149) I performed the Fourier transformation. A subtlety is hidden in this transformation. Indeed, one can readily check that the coordinate expression in the second line is transversal, i.e., vanishes upon differentiating over, say, ∂_μ. However, the Fourier transform recovers the singular part proportional to p^{-2}, but not a contact p independent term proportional to $g^{\mu\nu}$ in the last expression. This contact term can be recovered either by a more careful calculation (e.g., in $2 - \epsilon$ dimensions) or added by hand to restore transversality in the momentum space.

---

[21] See also Eqs. (8.35), (8.36), and Fig. 8.2.
[22] See, e.g., [53]. Equation (9.148) is the Fourier transform of $\frac{ip}{p^2}$.

Photon polarization operator (a) and summation of 1-particle reducible Feynman graphs for the photon propagator (b).

Now, let us sum the chain of diagrams shown in Fig. 9.40b. In the photon propagator we arrive at a progression that sums up to

$$\frac{g^{\mu\nu}}{p^2} \to \frac{g^{\mu\nu}}{p^2 - (e^2/\pi)}. \tag{9.150}$$

Schwinger mechanism of dynamical Higgsing

The longitudinal terms in the photon propagator proportional to $p^\mu p^\nu$ are irrelevant in this calculation. As was announced in the beginning of Section 9.4, the photon mass shifts from zero to $\frac{e^2}{\pi}$. I should stress that the graph in Fig. 9.40a is the only one-particle irreducible graph contributing to the photon mass. This is due to the following two-dimensional γ matrix identities: $\gamma^\mu \gamma^\nu = g^{\mu\nu} + \varepsilon^{\mu\nu}\gamma^5$ and $\gamma^\alpha \gamma^\mu \gamma_\alpha = 0$. The relation

$$m_\gamma^2 = \frac{e^2}{\pi} \tag{9.151}$$

is exact even beyond perturbation theory.[23]

Alternatively, one can use the last expression in (9.149) to derive the mass term in the effective Lagrangian

$$\Delta\mathcal{L} = \frac{1}{2}\frac{e^2}{\pi}(A^\mu)^2, \tag{9.152}$$

where the term containing $p_\mu A^\mu$ is omitted. Equation (9.152) is in full correspondence with (8.29).

The massive gauge particle in 1+1 dimension has one physical degree of freedom that can be treated as spin-zero particle. This is the only degree of freedom in the physical spectrum of the massless Schwinger model. There is no long-range interaction in the limit $m = 0$. All "quarks" with nonvanishing electric charges are completely screened.

9.6.3 Adding a Mass Term $m \neq 0$ to the Fermion Field

Adding a nonzero mass m term to the fermion field ψ,

$$\Delta\mathcal{L}_\psi = -m\bar{\psi}\psi, \tag{9.153}$$

see (9.142), results in a drastic change in dynamics. The photon no longer acquires $m_\gamma \neq 0$ through dynamical Higgsing; it remains massless. Electric charges are no longer fully screened. In fact, they form bound states similar to QCD "mesons" due

---

[23] This can be proven by exactly solving the massless Schwinger model by virtue of *bosonization*, see, e.g., the review volume [54].

to a residual Coulomb interaction. Insight into the contrast between the massless and the massive models can be obtained from comparing this section with Section 9.3.3.

In perturbation theory the expansion parameter is e/m. If $m \ll e$ we are at strong coupling. However, in this regime one can expand in the inverse parameter, m/e. In the Schwinger model on R_2 instantons and other quasiclassical methods do not work. However, we can still prove to the order $O(m^1)$ that two *probe* charges $\pm eq$ separated by distance ℓ experience a long-range interaction similar to (9.53) with a different (albeit also suppressed) coefficient. To this end we should include the effect due to the mass term (9.153).

As was discussed in Section 9.3.3, the q dependence introduced by the probe charges goes hand-in-hand with the θ dependence, as in Eq. (9.53). From Section 8.1.6 we know that the fermion vacuum expectation value $\langle \bar{\psi}\psi \rangle \neq 0$ and depends on θ as $\cos\theta$. The coefficient can be extracted from (8.41) by extrapolating $L \rightarrow e^{-1}$. Therefore, the corresponding energy density is

$$\mathcal{E}_{\bar{\psi}\psi} = c\, me \cos\theta, \qquad c \sim 1. \tag{9.154}$$

The above coefficient in front of cosine was verified by virtue of the exact solution through bosonization. The rule (9.53) then provides us with the answer for the interaction between the probe charges $\pm qe$, cf. Eq. (9.53). Let us first put $\theta = 0$. Then

$$V_{\text{int}} = c\, me\, \ell \left[1 - \cos(2\pi q) \right]. \tag{9.155}$$

If we expand this formula in q assuming $q \ll 1$ we arrive at the linear confinement

$$V_{\text{int}} \sim \frac{m}{e}\, (qe)^2 \ell \tag{9.156}$$

with the coefficient suppressed as m/e. This fact shows that a residual screening of the charge qe takes place in the interval of distances up to

$$\ell_{\text{scr}} \sim \frac{1}{2m} \log \frac{e}{m} \gg \frac{1}{2m},$$

to be compared with Eq. (9.49). Moreover, in Eq. (9.156) $\ell \gg \ell_{\text{scr}}$.

The probe quarks are confined for all noninteger values of q. The vanishing of the coefficient in front of ℓ for integer values of q means that the e^+e^- pair creation breaks the "string" in pieces, producing "meson" states, much in the same way as in QCD.

If the fermions are heavy, i.e., $m \gg e$, the e^+e^- pair creation is highly suppressed and we are left with a simple problem: a nonrelativistic positive charge at the point x and a negative one at the point y bound by the Coulomb potential

$$V_{\text{Coul}} \sim |e|^2 |x - y|. \tag{9.157}$$

Most of the e^+e^- bound states are unstable; a few of the lowest can be stable.

Exercise

9.6.1   Calculate the polarization operator (9.149) directly in the momentum space using dimensional regularization. Assume that you have one Dirac spinor with mass $m \neq 0$ and electric charge 1 (in the units of e). Do not integrate over the Feynman parameter. First show that the result is automatically transversal before integration over the Feynman parameter. In the two limiting cases,

$$(a)\, p^2 \gg m^2; \text{ and } (b)\, |p^2| \ll m^2,$$

perform this integration and find the final answers.

9.7 Polyakov's Confinement in 2 + 1 Dimensions

Polyakov's model of color confinement [41] was historically the first gauge model where confinement was analytically established in 2 + 1 dimensions. Polyakov's formula for three-dimensional confinement is concise: "compact electrodynamics confines electric charges in 2 + 1 dimensions." In this section we will elaborate on this formula.

At energies relevant to Polyakov's confinement the W bosons decouple, and the GG model reduces to compact electrodynamics.

Unfortunately, the mechanism leading to color confinement in this case, as we will see shortly, is specifically three-dimensional. It cannot be generalized to four dimensions. Still, Polyakov's model remains a useful theoretical laboratory. Its advantages are: (i) the emergence of "strings" attached to color charges and (ii) the calculability of the string tension. Its main disadvantage, besides the above-mentioned limitation to three dimensions, is that the color confinement taking place in this model is essentially Abelian. Attempts to apply Polyakov's results in hadronic physics are described in the papers [42, 43].[24]

9.7.1 Theoretical Setup

Polyakov's confinement emerges in the Georgi–Glashow (GG) model [44] in 2 + 1 dimensions, where monopoles are reinterpreted as instantons in the Euclidean version of the model. Both the Georgi–Glashow model and monopoles were discussed in detail in Chapter 4. Here we briefly review this model, limiting ourselves to the SU(2) case and making adjustments appropriate to a Euclidean formulation.

The Lagrangian of the Georgi–Glashow model in 2 + 1 dimensions includes gauge fields and a real scalar field, both in the adjoint representation of SU(2). In passing to the Euclidean (2 + 1)-dimensional space we will use the following definitions (the Euclidean quantities are marked by carets, which will be dropped shortly, after the transition to the Euclidean space is complete):

The reader is advised to consult Section 5.2.

$$x^i = \hat{x}_i, \qquad i = 1, 2,$$
$$x^0 = -i\hat{x}_3;$$
$$A^m = -\hat{A}_m, \qquad m = 1, 2,$$
$$A^0 = i\hat{A}_3. \tag{9.158}$$

---

[24] The second of these references analyzes four-dimensional Yang–Mills theory with massless quarks in various representations of the gauge group compactified on $R_3 \times S_1$. When the radius of S_1 becomes much greater than Λ^{-1} we return to the four-dimensional theory.

The Lagrangian of the model is obtained from $3 + 1$ dimensions by reducing one coordinate and the corresponding component of the vector field. In Euclidean space

$$\mathcal{L} = \frac{1}{4g^2} G^a_{\mu\nu} G^a_{\mu\nu} + \frac{1}{2}(\nabla_\mu \phi^a)(\nabla_\mu \phi^a) - \lambda(\phi^a \phi^a - v^2)^2, \tag{9.159}$$

where $\mu, \nu = 1, 2, 3$. The covariant derivative in the adjoint acts according to

$$\nabla_\mu \phi^a = \partial_\mu \phi^a + \varepsilon^{abc} A^b_\mu \phi^c, \tag{9.160}$$

and the Euclidean metric is $g_{\mu\nu} = \text{diag}\{+1, +1, +1\}$.

As previously, we will work in the critical (or BPS) limit of vanishing scalar coupling, $\lambda \to 0$. The only role of the last term in Eq. (9.159) is to provide a boundary condition for the scalar field,

$$\left(\phi^a \phi^a\right)_{\text{vac}} = v^2, \tag{9.161}$$

where v is a real positive parameter. One can always choose the gauge in such a way that

$$\phi^{1,2} = 0, \qquad \phi^3 = v. \tag{9.162}$$

Then the third component of A_μ (i.e., A^3_μ) remains massless. It can be referred to as a "photon." At the same time, the $A^\pm_\mu = \frac{1}{\sqrt{2}g}(A^1_\mu \mp A^2_\mu)$ components become W bosons and acquire mass gv; see Section 4.1.

The classical equations of motion which follow from Eq. (9.159) are differential equations of second order. In the BPS limit they can be replaced, however, by first-order "duality" equations

$$-\frac{1}{2g} \varepsilon_{\mu\nu\rho} G^a_{\nu\rho} = \pm \nabla_\mu \phi^a, \tag{9.163}$$

in much the same way as for monopoles, cf. Eq. (4.25). Formally this is exactly the same equation as that for the static monopoles in $3 + 1$ dimensions considered in the A_0 gauge. Hence it has the same functional solution, albeit the interpretation is different. What used to be the monopole mass becomes the instanton action:

$$S_{\text{inst}} = 4\pi \frac{v}{g} = 4\pi \frac{m_W}{g^2}, \tag{9.164}$$

> *In three dimensions g^2 has the dimension of mass.*

where m_W is the mass of the W boson in the model at hand. In $2 + 1$ dimensions the coupling g^2 has the dimension of mass, so that S_{inst} is dimensionless, as it should be. This is to be compared with the $(3 + 1)$-dimensional GG model, where formally the same expression gives the monopole mass.

As we will see shortly, the energy scale of the phenomenon in which we are interested is much lower than m_W. The only source of m_W-dependence is the instanton measure. At energies much lower than m_W the model at hand contains only photons (plus nondynamical probe electric charges). This is why it is sometimes referred to as *compact electrodynamics*.

9.7.2 Instanton Measure

As usual the instanton measure is trivially determined by the instanton action, the zero modes in the instanton background, and the renormalizability of the theory (see Chapter 5 for details), up to an overall numerical factor:

$$d\mu_{\text{inst}} = \text{const} \times S_{\text{inst}}^2 \, M_{\text{uv}}^4 \, \frac{1}{m_W} d^3 x_0 \exp\left(-S_{\text{inst}} + 4\ln\frac{m_W}{M_{\text{uv}}}\right), \qquad (9.165)$$

where M_{uv} is an ultraviolet parameter appearing in the Pauli–Villars regularization (the only one suitable for instanton calculations). Various factors in Eq. (9.165) have distinct origins. First, $\exp(-S_{\text{inst}})$ is the classical instanton exponent. Furthermore, the factor $(\sqrt{S_{\text{inst}}}M_{\text{uv}})^4$ in the pre-exponent arises because there are four zero modes in the instanton background – three translational (manifesting themselves in d^3x_0 in the measure, where x_0 is the instanton center), plus an additional zero mode associated with the unbroken U(1) symmetry of the model. The corresponding collective coordinate α is of angular type. Equation (9.165) assumes that the integration over α is done. The norm of this rotational zero mode is $S_{\text{inst}}^{1/2}m_W^{-1}$ whereas it was $S_{\text{inst}}^{1/2}$ for the translational modes; this explains the factor $1/m_W$ in $d\mu_{\text{inst}}$. Finally, the logarithm in the exponent must come from modes other than the zero modes, i.e., the nonzero modes. We have not calculated it, but we know that it must be there because the ultraviolet parameter M_{uv} cannot be present in the overall answer for $d\mu_{\text{inst}}$: it must cancel out. Indeed, the model at hand is super-renormalizable in $2 + 1$ dimensions: neither m_W nor g^2 receive logarithmically divergent corrections. Therefore the occurrence of $\exp(-4\ln M_{\text{uv}})$ from nonzero modes is unavoidable.

The only remaining question is: which infrared parameter is available to make the argument of the logarithm dimensionless? The answer is that the only relevant infrared parameter at our disposal is m_W. This concludes our derivation of the instanton measure up to a numerical coefficient.

Assembling all factors together we get

$$d\mu_{\text{inst}} = \text{const} \times m_W^5 g^{-4} d^3 x_0 \exp(-S_{\text{inst}}). \qquad (9.166)$$

The validity of the quasiclassical approximation implies that $v \gg g$ and, hence, $S_{\text{inst}} \gg 1$. As a result, the instanton measure is exponentially suppressed.[25]

9.7.3 Low-Energy Limit. Dual Representation

In the low-energy range $E \ll m_W$, the presence of the W bosons in the spectrum of the model is irrelevant, and we can focus on the U(1) field, to be referred to as a *photon*, which has not been Higgsed and hence must be considered as massless for the time being. Later on I will show that in fact it does acquire a tiny mass, but this mass is associated with nonperturbative instanton effects and is exponentially suppressed as in Eq. (9.166).

Now we will return, for a while, to Minkowski space. It is obvious that in $2 + 1$ dimensions the photon field has only one physical (transverse) polarization. This means that it must have a dual description in terms of one scalar field φ, namely [41],

Polyakov's duality

$$F_{\mu\nu} = k\varepsilon_{\mu\nu\rho}\partial^\rho\varphi, \qquad (9.167)$$

where k is a numerical coefficient to be determined below and $\varepsilon_{\mu\nu\rho}$ is the completely antisymmetric unit tensor of the third rank ($\varepsilon^{012} = \varepsilon_{012} = 1$). Equation (9.167)

---

[25] I hasten to warn the reader that the pre-exponent in Eq. (9.166) does not quite match the corresponding expression presented in [41], which was later copied into [42].

defines the field φ in terms of $F_{\mu\nu}$ in a nonlocal way. At the same time, $F_{\mu\nu}$ is related to φ locally.

Polyakov's duality equation (9.167) can be illustrated as follows.

Let us consider a three-dimensional QED in the absence of source (probe) charges. One can always use the so-called first-order formalism, in which the gauge field strength tensor $F_{\mu\nu}$ is treated as an independent variable field while the condition

$$F_{\mu\nu} = \partial_\mu A_\nu - \partial_\nu A_\mu \tag{9.168}$$

is implemented through a Lagrange multiplier field φ

$$\Delta S_\varphi = -\int d^3x\, C\, \varphi(x) \left(\partial_\mu \epsilon^{\mu\alpha\beta} F_{\alpha\beta}\right)$$

$$= \int d^3x\, C \left(\partial_\mu \varphi\right) \epsilon^{\mu\alpha\beta} F_{\alpha\beta}, \tag{9.169}$$

enforcing the identity $\partial_\mu \left(\epsilon^{\mu\alpha\beta} F_{\alpha\beta}\right) = 0$ for all x, which, in turn, entails (9.168), at least locally. Here C is an arbitrary numerical constant.

The total action takes the form

$$S = \int d^3x \left[-\frac{1}{4e^2} F_{\mu\nu} F^{\mu\nu} + C \left(\partial_\mu \varphi\right) \epsilon^{\mu\alpha\beta} F_{\alpha\beta}\right]. \tag{9.170}$$

Now, varying with respect to $F^{\mu\nu}$ we arrive at

$$F^{\mu\nu} - 2\, e^2\, C\, \epsilon^{\mu\nu\alpha} \partial_\alpha \varphi = 0, \tag{9.171}$$

cf. Eq. (9.167). If in this equation we chose $\mu = 0$ and $\nu = \rho = 1, 2$ then $F_{0i} = E_i$, where $\vec{E}$ is the electric field,[26] and

$$E_i = k\varepsilon_{ij}\partial^j \varphi, \qquad i, j = 1, 2. \tag{9.172}$$

Here ε_{ij} is the completely antisymmetric unit tensor of the second rank, $\varepsilon_{12} = \varepsilon^{12} = 1$.

Our next task is to determine the coefficient k. To this end let us place a heavy probe charge at the origin, as shown in Fig. 9.41. Usually, the charge is not quantized in the U(1) theory. However, our theory is in fact *compact* electrodynamics; the minimal U(1) charge in this model is $\frac{1}{2}$. Indeed, the probe particle must belong to a representation of SU(2). If it belongs to the doublet representation, it has the U(1) charge $\pm\frac{1}{2}$. We will assume that the charge of the probe particle at the origin in Fig. 9.41 is $\frac{1}{2}$.

The electric field induced by the probe particle is radial. A brief inspection of Eq. (9.172) shows the the radial orientation of $\vec{E}$ requires the scalar function φ to be r-independent. Moreover, it should depend on the polar angle α as const $\times\alpha$. The normalization constant that we have just introduced can always be included in the coefficient k. Thus, we can write that for the minimum probe charge

$$\varphi = \alpha. \tag{9.173}$$

---

[26] For what follows it is useful to note that $F_{12} = -B$, where B is the magnetic field. In $2 + 1$ dimensions B is a Lorentz scalar.

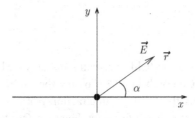

Fig. 9.41   Probe heavy charge $\frac{1}{2}$ at the origin.

Needless to say, Eq. (9.173) implies that the scalar field φ is compact and defined mod 2π; the points

$$\varphi, \quad \varphi \pm 2\pi, \quad \varphi \pm 4\pi, \quad \ldots \tag{9.174}$$

are identified. Thus, Polyakov's observation that in 2 + 1 dimensions the photon field is dual to a real scalar field needs an additional specification: the real scalar field at hand is compact; it is defined on a circle of circumference 2π.

The static equation that determines the electric field of the probe charge is

$$\text{div } \vec{E} = -\tfrac{1}{2} g^2 \delta^{(2)}(\vec{r}), \tag{9.175}$$

where $\frac{1}{2}$ represents the minimal charge. Its solution is obvious,

$$\vec{E} = -\frac{g^2}{4\pi} \frac{\vec{r}}{r^2}. \tag{9.176}$$

Let us compare this expression with $\varepsilon_{ij}\partial^j \varphi$ (see Eq. (9.172)), substituting as φ the solution (9.173). Then we have

$$\varepsilon_{ij}\left(\partial^j \varphi\right) \rightarrow -\frac{\vec{r}}{r^2}. \tag{9.177}$$

Thus, we conclude that

$$k = \frac{g^2}{4\pi}. \tag{9.178}$$

In the model at hand the dimension of $F_{\mu\nu}$ is m^2, the dimension of g^2 is m, and the dimension of φ is m^0.

For the sake of convenience we can summarize the result of our derivation in the form

$$F_{\mu\nu} = \frac{g^2}{4\pi} \varepsilon_{\mu\nu\rho}\partial^\rho \varphi. \tag{9.179}$$

The energy of the $\{\vec{E}, B\}$ field configuration in the original (compact) electrodynamics and in the dual description has the form

$$\mathcal{E} = \frac{1}{2g^2} \int d^2 x \left(\vec{E}^2 + B^2\right) = \frac{g^2}{32\pi^2} \int d^2 x \left[\left(\vec{\nabla}\varphi\right)^2 + \dot{\varphi}^2\right]. \tag{9.180}$$

Observe that the canonical momentum part of the original theory gets transformed into the canonical coordinate part of the dual theory (i.e., $\vec{E}^2 \rightarrow \left(\vec{\nabla}\varphi\right)^2$), and vice versa ($B^2 \rightarrow \dot{\varphi}^2$). Finally, Eq. (9.180) implies that the Lagrangian of the dual model is

$$\mathcal{L}_{\text{dual}} = \frac{\kappa^2}{2} \left[\dot{\varphi}^2 - \left(\vec{\nabla}\varphi\right)^2\right] = \frac{\kappa^2}{2} \left(\partial_\mu \varphi\right) \left(\partial^\mu \varphi\right), \tag{9.181}$$

where

$$\kappa = \frac{g}{4\pi}. \tag{9.182}$$

In 2 + 1 dimensions the Coulomb force falls off as $1/r$, and the Coulomb potential grows as $\ln r$.

At this level the field φ remains massless, which is in one-to-one correspondence with the Coulomb law (9.176) for the probe electric charges. Note that in $2 + 1$ dimensions the Coulomb interaction *per se* confines the probe charges. This is a weak logarithmic confinement, however. Our task is to arrive at a linear confinement of the type that takes place in QCD. This is a far less trivial task, but we will achieve it shortly. To this end we will need to show that instantons do generate a potential term for the dual field φ. Although suppressed by $\exp(-S_{\text{inst}})$, this term will lead to a qualitative restructuring of the theory at large distances.

I conclude this section with the following side remark on the literature, intended for the curious reader who would like to learn more about compact electrodynamics. In [46] Polyakov explored the origin of the duality relation (9.167) within a discretized approach, in the spirit of statistical mechanics on lattices (see Section 4.3 of Polyakov's book). From a mathematical standpoint the same question was discussed in [47].

9.7.4 Instanton-Induced Interaction

Now we return to our instanton theme. We have already calculated the measure, see Eq. (9.166), which we will rewrite here in the form

$$d\mu_{\text{inst}} = \tfrac{1}{2}\mu^3 \, d^3 \, x_0, \tag{9.183}$$

where μ is a parameter having the dimension of mass,

$$\mu^3 = \text{const} \times m_W^5 \, g^{-4} \, \exp(-S_{\text{inst}}). \tag{9.184}$$

Since $S_{\text{inst}} \gg 1$, μ is exponentially suppressed:

$$\mu \ll m_W. \tag{9.185}$$

We will show that the instanton-induced interaction (at distances $\gg m_W^{-1}$) can be written as

$$\mathcal{L}_{\text{inst}} = \tfrac{1}{2}\mu^3 \, \exp(\pm i\varphi), \tag{9.186}$$

where the plus sign in the exponent represents an instanton and the minus sign an anti-instanton.

Is it difficult to construct an effective instanton-induced Lagrangian? Not at all. The procedure is described in Sections 5.4.9 and 5.4.12.2. In the problem at hand it can be implemented as follows. Let us consider the correlation function

$$\left\langle 0 \left| T \left\{ \mathcal{B}_{\gamma_1}(x_1), \mathcal{B}_{\gamma_2}(x_2), \dots, \mathcal{B}_{\gamma_n}(x_n) \right\} \right| 0 \right\rangle_{\text{one-inst}}, \tag{9.187}$$

where[27]

$$\mathcal{B}^\gamma(x) \equiv -\tfrac{1}{2}\varepsilon^{\mu\nu\gamma} F_{\mu\nu}(x), \tag{9.188}$$

---

[27] The careful reader may have observed that $\mathcal{B}^\gamma$ coincides with the gauge-invariant magnetic field (4.19) in $(3+1)$-dimensional theory. This observation allows one easily to copy Eq. (4.21) (which was actually calculated in a nonsingular gauge) into Eq. (9.189).

n is arbitrary, $x_1, x_2, \ldots, x_n$ are arbitrary coordinates, and $F_{\mu\nu}(x)$ is the electromagnetic field strength tensor of the compact electrodynamics under consideration. For definiteness the instanton is placed at the origin. All distances x_k are assumed to be large, $|x_k| \gg m_W^{-1}$, $k = 1, 2, \ldots, n$.

To obtain the one-instanton contribution (9.187) in the leading approximation, one passes to Euclidean space and then substitutes each operator $\mathcal{B}_{\gamma_k}(x_k)$ by its classical value in the instanton field (the latter is taken in the limit of large distances from the instanton center),

$$\mathcal{B}_{\gamma_k}(x_k) \to \frac{n_{\gamma_k}}{(x_k)^2}, \qquad n_{\gamma_k} \equiv \frac{(x_k)_{\gamma_k}}{\sqrt{(x_k)^2}}. \tag{9.189}$$

The n-point function (9.187) reduces to, on the one hand,

$$\frac{1}{2}\mu^3 \prod_{k=1}^{n} \frac{(x_k)_{\gamma_k}}{(x_k)^2 \sqrt{(x_k)^2}}. \tag{9.190}$$

On the other hand, by construction (or, equivalently, by definition of the effective Lagrangian), it is possible to express the same n-point function as

$$(-ik)^n \left\langle 0 \left| T \left\{ [\partial_{\gamma_1}\varphi(x_1), \partial_{\gamma_2}\varphi(x_2), \ldots, \partial_{\gamma_n}\varphi(x_n)] \times \mathcal{L}_{\text{inst}}(0) \right\} \right| 0 \right\rangle. \tag{9.191}$$

Note that Eq. (9.179) implies that in Euclidean space

$$\mathcal{B}_\gamma(x) = -ik \left[\partial_\gamma \varphi(x) \right]; \tag{9.192}$$

see Section 5.2.

Now we are finally ready to verify that Eq. (9.186) solves the problem. Indeed, expanding $\mathcal{L}_{\text{inst}}$ in φ we observe that the relevant term in the expansion of $\mathcal{L}_{\text{inst}}$ saturating the n-point function under consideration is $[i\varphi(0)]^n/n!$. Furthermore, the factor $n!$ in the denominator is canceled by a combinatorial factor, the number of possible contractions in Eq. (9.191). Therefore, Eq. (9.191) reduces to

> Instanton Lagrangian in 3D is presented in (9.186)

$$\frac{1}{2}\mu^3 \, k^n \prod_{k=1}^{n} \partial_{\gamma_k} D(x_k, 0), \tag{9.193}$$

where $D(x_k, 0)$ is the Green's function (in Euclidean space–time), which is determined from Eq. (9.181),

$$D(x, 0) = -\frac{\kappa^{-2}}{4\pi} \frac{1}{\sqrt{x^2}} = -\frac{4\pi}{g^2} \frac{1}{\sqrt{x^2}} = -\frac{1}{g\kappa} \frac{1}{\sqrt{x^2}}, \tag{9.194}$$

see Fig. 9.42. Substituting this expression into (9.193) and differentiating $D(x, 0)$ we observe, with satisfaction,[28] perfect coincidence with Eq. (9.190), which confirms the exponential *ansatz* for $\mathcal{L}_{\text{inst}}$. Additional indirect confirmation comes from the fact that $\mathcal{L}_{\text{inst}}$ in Eq. (9.186) is 2π-periodic in φ. The requirement of 2π-periodicity of the effective interaction is equivalent to requiring the compactness of φ, a result derived above from independent arguments.

---

[28] It is crucial that $k = g^2(4\pi)^{-1}$, an independent consequence of Polyakov's identification of the $(2+1)$-dimensional photon with a real scalar field according to Eq. (9.167).

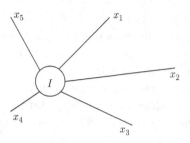

Fig. 9.42 One-instanton saturation of the n-point functions (9.187) and (9.191). In this example $n = 5$. The instanton is at the origin.

Assembling the instanton and anti-instanton contributions, we arrive at the following effective Lagrangian for the field φ:

$$\mathcal{L}_{\text{dual}} = \frac{\kappa^2}{2} \left(\partial_\mu \varphi\right) \left(\partial^\mu \varphi\right) + \mu^3 \cos \varphi. \tag{9.195}$$

Besides the kinetic term established previously, it contains an exponentially suppressed potential term generated at the nonperturbative level. This is the Lagrangian of the *sine-Gordon* model.

9.7.5 Domain "Line" in 2 + 1 Dimensions as a String

Now, we calculate the dual photon mass. To this end we expand $\cos \varphi$ and compare the coefficient in front of φ^2 with that in the kinetic term. In this way we obtain

$$m_\varphi = \mu^{3/2} \kappa^{-1}. \tag{9.196}$$

The typical transverse size of the confining string will be determined by the length scale m_φ^{-1}. Needless to say m_W/m_φ is exponentially large, which justifies our approximation – compact electrodynamics with no W-boson excitations.

Now, what is a string in the case at hand? Rather paradoxically, Polyakov's string in the Georgi–Glashow model is a "domain wall." More precisely it is a "domain line," since in two spatial dimensions the transitional domain separating two vacua is a line (Fig. 9.43). Moreover, the two vacua just mentioned, $\varphi_{\text{vac}} = 0$ and $\varphi_{\text{vac}} = 2\pi$ (or, in general, $\varphi_{\text{vac}} = 2\pi n$ and $\varphi_{\text{vac}} = 2\pi(n + 1)$, where n is an arbitrary integer) represent one and the same vacuum since φ and $\varphi + 2\pi n$ are identified because of the compactness of the field φ. This is crucial because otherwise the domain line could not be interpreted as a string. Indeed, the necessary conditions for a topological defect to be a string are: (i) the one-dimensional nature of the defect; (ii) that when one travels away from the string in the transverse direction, at large distances one should find oneself in the same vacuum no matter in which direction one goes. While the first requirement is obviously satisfied for the domain line, the second usually is not because usually the vacua on the two sides of the line are physically distinct. For compact fields we can have the same vacuum on both sides of a topological defect of domain-wall or domain-line type. There is a well-known example of this

Domain-line solution, cf. Section 11.2.1

phenomenon in 3 + 1 dimensions: axion walls. The axion field is compact too.

The domain-line (or domain-wall) solution in the sine-Gordon model is well known. Repeating the procedure of Chapter 2 we can write

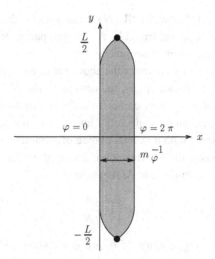

Fig. 9.43 "Domain wall" in 2 + 1 dimensions. The solid circles represent probe charges $\pm\frac{1}{2}$. It is assumed that $L \gg m_\varphi^{-1}$. The transitional domain, which is a domain line and a string simultaneously, is shaded.

$$\varphi = 2\left[\text{arcsin}\,\text{tanh}\left(m_\varphi y\right) + \frac{\pi}{2}\right], \tag{9.197}$$

interpolating between $\varphi_{\text{vac}} = 0$ at $y = -\infty$ and $\varphi_{\text{vac}} = 2\pi$ at $y = \infty$; see Fig. 9.43. The tension of the Polyakov string is

$$T = 8\mu^{3/2}\kappa = 8m_\varphi \kappa^2 = \frac{2km_\varphi}{\pi}. \tag{9.198}$$

Note that this tension is much larger than m_φ^2.

For this string to develop between two probe charges the distance L between the charges must be $L \gg m_\varphi^{-1}$. At distances $\lesssim m_\varphi^{-1}$ each charge is surrounded by an essentially (two-dimensional) Coulomb field, with force lines spreading homogeneously in all directions. At distances $\sim m_\varphi^{-1}$ the "flux tube" starts forming. If the probe charges have opposite signs, and $L \gg m_\varphi^{-1}$, they will be connected by the "flux tube" and the energy of the configuration will grow as TL. At very large distances $L \gg m_\varphi^{-1}$, linear confinement sets in.

9.7.6 Polyakov's Confinement in SU(N)

Conceptually the confinement mechanism in the SU(N) generalization of the Georgi–Glashow model remains the same. There are technical differences [48], however, which we will briefly discuss below.

The adjoint field $\phi \equiv \phi^a T^a$ in the Lagrangian (9.159) now has $N^2 - 1$ components. Its vacuum expectation value (VEV) can always be chosen as

$$\phi \equiv \phi^a T^a = \text{diag}\{v_1, v_2, \dots, v_N\}, \tag{9.199}$$

where

$$\sum_{k=1}^{N} v_k = 0 \tag{9.200}$$

and the T^a denote the SU(N) generators (in the fundamental representation). Thus, its VEV is parametrized by $N - 1$ free parameters. Moreover, the theory has $N - 1$ distinct instanton-monopoles.

For a generic choice of the above parameters, all $N(N-1)$ W bosons have different masses; correspondingly, the actions of the $N - 1$ instantons are different too. The dominant effect will come from the instanton-monopole with the minimal action. The others can be neglected. Thus we return essentially to the SU(2) case.

However, with a special choice of parameters one can achieve the degeneracy of all W boson masses (instanton actions). In this case all $N - 1$ instanton-monopoles are equally important. Assume that

$$\phi \equiv \phi^a T^a = v \, \text{diag} \left\{ 1 - \frac{1}{N}, 1 - \frac{3}{N}, \dots, -\left(1 - \frac{3}{N}\right), -\left(1 - \frac{1}{N}\right) \right\}. \quad (9.201)$$

Then the eigenvalues of ϕ are equidistant and symmetric with respect to the C transformation $v \to -v$ (with subsequent reordering of the eigenvalues). Moreover, $\exp(2\pi i \phi / v)$ takes the form $\sqrt[N]{\pm 1}$. (To be more exact, we have $\sqrt[N]{1}$ for odd N and $\sqrt[N]{-1}$ for even N.) The vector $\boldsymbol{h}$ defined in Eq. (4.69) is given by

$$\boldsymbol{h} = \frac{v\sqrt{2}}{N} \left\{ \sqrt{1 \times 2}, \sqrt{2 \times 3}, \dots, \sqrt{m(m+1)}, \dots, \sqrt{(N-1)N} \right\}. \quad (9.202)$$

It is easy now to calculate the masses of the $N - 1$ lightest W bosons:

$$(m_W)_\gamma = g\gamma \boldsymbol{h} = \frac{2gv}{N}, \quad (9.203)$$

cf. Eq. (4.75). Here γ stands for a simple root vector; see Appendix section 4.3 in Chapter 4. The actions of all $N - 1$ instanton-monopoles are the same:

$$S_{\text{SU}(N) \text{ inst}} = \frac{4\pi}{g} \gamma \boldsymbol{h} = \frac{8\pi v}{gN} = \frac{4\pi}{g^2} (m_W)_\gamma. \quad (9.204)$$

Next, it is not difficult to derive the effective Lagrangian for $N - 1$ dual "photons," an analog of Eq. (9.195). This is a good exercise. Instead of a single dual field φ we have an $(N - 1)$-component vector,

$$\boldsymbol{\varphi} = \{\varphi_1, \varphi_2, \dots, \varphi_{N-1}\}. \quad (9.205)$$

The energy functional for the dual fields takes the form

$$\mathcal{E}_{\text{dual}}^{\text{SU}(N)} = \int d^2 x \left[\frac{\kappa^2}{2} (\partial_k \boldsymbol{\varphi})(\partial_k \boldsymbol{\varphi}) - \mu^3 \sum_{i=1}^{N-1} \cos \boldsymbol{\varphi} \gamma_i \right]. \quad (9.206)$$

Here μ^3 is the same as in Eq. (9.184) but with the substitution $S_{\text{inst}} \to S_{\text{SU}(N) \text{ inst}}$.

For generic N we obtain a rather complicated system of coupled sine-Gordon models. For pedagogical purposes it is instructive to consider the SU(3) example, the next in complexity after SU(2). Then we have two dual photons, φ_1 and φ_2, and the corresponding energy functional takes the form

$$\mathcal{E}_{\text{dual}}^{\text{SU}(3)} = \int d^2 x \left[\frac{\kappa^2}{2} \sum_{i=1}^{2} (\partial_k \varphi_i)(\partial_k \varphi_i) - \mu^3 \left(\cos \varphi_1 + \cos \frac{\varphi_1 - \sqrt{3}\varphi_2}{2} \right) \right]. \quad (9.207)$$

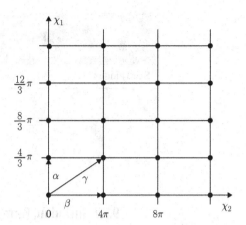

Fig. 9.44   Periodicity on the $\chi_1\chi_2$ plane. The solid circles denote the vacuum configuration.

The instanton action is $8\pi v/(3g)$. The mass eigenvalues of the fields $\varphi_{1,2}$ are

$$m_1^2 = \frac{3\mu^3}{2\kappa^2}, \qquad m_2^2 = \frac{\mu^3}{2\kappa^2}. \tag{9.208}$$

The diagonal combinations can be readily found, too:

$$\chi_1 = \frac{\sqrt{3}\varphi_1 - \varphi_2}{2}, \qquad \chi_2 = \frac{\varphi_1 + \sqrt{3}\varphi_2}{2}. \tag{9.209}$$

In terms of these diagonal fields the energy functional reduces to a simple formula,

$$\mathcal{E}_{\text{dual}}^{\text{SU}(3)} = \int d^2x \left[\frac{\kappa^2}{2} \sum_{i=1}^{2} (\partial_k \chi_i)(\partial_k \chi_i) - 2\mu^3 \cos\frac{\sqrt{3}\chi_1}{2} \cos\frac{\chi_2}{2} \right]. \tag{9.210}$$

The periodicity on the χ_1, χ_2 plane is shown in Fig. 9.44. As we already know, strings are domain lines interpolating between vacuum configurations. The solutions are static and depend on one spatial coordinate, y. It is not easy to find solutions for generic strings (of the type γ in Fig. 9.44). Two particular solutions, that satisfy the classical equations of motion with the required boundary conditions, are fairly obvious. They correspond to the interpolations α and β in Fig. 9.44. Of special interest is the α trajectory, $\chi_2 = 0$. In terms of the original dual photons it corresponds to

> *For* SU(N) *with $N > 2$ one has a variety of strings.*

$$\varphi_1 = -\sqrt{3}\varphi_2. \tag{9.211}$$

At the endpoints of the string represented by the domain wall α are the fundamental probe quark and antiquark.

What does that mean? A probe quark in the fundamental representation (the SU(3) triplet) has the charges with respect to the third and eighth photons shown in Table 9.1. The string α connects Q_2 and $\overline{Q_2}$. Its tension is

$$T = \frac{16\sqrt{2}}{\sqrt{3}} \mu^{3/2}\kappa. \tag{9.212}$$

The β string is composite; it connects $Q_1^2 Q_2$ and $\overline{Q_1^2 Q_2}$. Its tension is larger than (9.212) by a factor $\sqrt{3}$.

Table 9.1. The U(1) charges of the probe fundamental quark Q_i (SU(3) indices 1, 2, and 3) with respect to the third and eighth photons

SU(3) index	q_3	q_8
1	$\frac{1}{2}$	$\frac{1}{2\sqrt{3}}$
2	$-\frac{1}{2}$	$\frac{1}{2\sqrt{3}}$
3	0	$-\frac{2}{2\sqrt{3}}$

9.7.7 Including Fermions

Let us ask ourselves what happens to Polyakov's confinement if, in addition to gauge fields and the adjoint scalar field ϕ, we add fermions. If they are coupled to ϕ as in Eq. (11.206) or Eq. (4.104) the instanton-monopoles have fermion zero modes; see Section 4.1.11. This means that the contribution of the instanton-monopoles to the potential vanishes. Instead of generating a sine-Gordon interaction of the type (9.195) or (9.207) they generate fermion condensates. Polyakov's confinement is destroyed.[29]

9.8 Appendix: Solving the O(N) Model at Large N

Previously we considered the O(3) sigma model in perturbation theory and found that the global O(3) is spontaneously broken down to O(2); correspondingly two massless Goldstone bosons emerge (see Sections 6.1 and 6.3). My present task is to show that, beyond perturbation theory, in the full solution, a mass gap is generated, and the full symmetry of the Lagrangian is *restored* in the O(N) model for arbitrary N.[30]

Look through Section 6.5.2.

Instead of the three S fields of the O(3) model let us consider N fields $S^a (a = 1, 2, \ldots, N)$ defined on the unit sphere,

$$S^a(x)S^a(x) \equiv 1. \tag{9.213}$$

In what follows $1/N$ will play the role of the expansion parameter.

The Lagrangian is similar to that of the O(3) model,

$$\mathcal{L} = \frac{1}{2g_0^2}(\partial_\mu S^a)(\partial_\mu S^a), \qquad a = 1, 2, \ldots, N. \tag{9.214}$$

The O(N) invariance under global (x-independent) rotations of the N component vector S^a is explicit in Eq. (9.214). The model is known as the O(N) sigma model.

---

[29] It is instructive to note that in four-dimensional Yang–Mills theory with massless fermions compactified on $R_3 \times S_1$, Polyakov's confinement does take place due to Euclidean configurations – bions – which are more complicated than the instanton-monopoles. This topic lies beyond the scope of the present textbook. The curious reader is directed to [43].

[30] This appendix is based on Section 2 from [49].

In much the same way as in the O(3) model, the O(N) model in perturbation theory gives rise to the spontaneous symmetry breaking O(N) $\rightarrow$ O($N-1$), which leads in turn to $N-1$ massless interacting Goldstones. At one loop, Eq. (6.63) for the running coupling constant of the O(3) model is replaced by

$$g^2(\mu) = g_0^2 \left[1 - \frac{g_0^2}{4\pi}(N-2) \ln \frac{M_{\text{uv}}^2}{\mu^2} \right]^{-1}, \tag{9.215}$$

which implies that the one-loop β function of the O(N) model is

$$\beta(g^2) \equiv \mu \frac{\partial}{\partial \mu} g^2(\mu) = -\frac{N-2}{2\pi} g^4. \tag{9.216}$$

Note that, for $N = 3$, Eq. (9.215) reduces to (6.63) as of course it should.

As an easy warm up, let us derive Eq. (9.215). To avoid cumbersome expressions with subscripts en route, the calculation we will carry out below will be at $N = 4$ (i.e., for an S_3 target space). The generalization to arbitrary N is easy. In polar coordinates, S_3 with unit radius is parametrized by three angle variables, ξ, θ, and φ and the metric

Calculating the β function in the O(N) model

$$g_{ab} = \text{diag}\{1, \ \sin^2 \xi, \ \sin^2 \xi \ \sin^2 \theta\}. \tag{9.217}$$

In polar coordinates the Lagrangian takes the form

$$\mathcal{L} = \frac{1}{2g_0^2}(\partial_\mu \xi \, \partial^\mu \xi + \sin^2 \xi \partial_\mu \theta \, \partial^\mu \theta + \sin^2 \xi \sin^2 \theta \, \partial_\mu \varphi \, \partial^\mu \varphi). \tag{9.218}$$

The most convenient choice of background field is

$$\xi_0 = \frac{\pi}{2}, \qquad \theta_0 = \frac{\pi}{2}, \qquad \varphi_0 = \tilde{\varphi}, \tag{9.219}$$

where $\tilde{\varphi}(x)$ is a slowly varying function of x.

Splitting all fields into background and quantum parts we arrive at

$$\mathcal{L} = \mathcal{L}_0 + \frac{1}{2g_0^2}(\partial_\mu \xi_{\text{qu}} \partial^\mu \xi_{\text{qu}} + \partial_\mu \theta_{\text{qu}} \partial^\mu \theta_{\text{qu}} + \partial_\mu \varphi_{\text{qu}} \partial^\mu \varphi_{\text{qu}})$$

$$- \frac{1}{2g_0^2} \left(\xi_{\text{qu}}^2 + \theta_{\text{qu}}^2 \right) \partial_\mu \tilde{\varphi} \partial^\mu \tilde{\varphi},$$

$$\mathcal{L}_0 = \frac{1}{2g_0^2} \partial_\mu \tilde{\varphi} \partial^\mu \tilde{\varphi}, \tag{9.220}$$

in an approximation quadratic in the quantum fields. Now we are ready to perform the desired one-loop calculation. The terms $\partial_\mu \xi_{\text{qu}} \partial^\mu \xi_{\text{qu}}$ and so on in the first line determine the quantum field propagators, while the term in the second line is the vertex to be evaluated in the one-loop approximation. We must calculate two trivial and identical tadpoles: one for the ξ_{qu} field and the other for the θ_{qu} field. In the general case the number of tadpoles is obviously $N - 2$. In this way we are able to reproduce Eq. (9.215).

Now let us turn to the main task of this section – the solution of the O(N) model using the $1/N$ expansion. The interaction of the Goldstone bosons emerges from the constraint (9.213). If it were not for this constraint, the theory would be free. It is clear that it is inconvenient to deal with constraints of such type; therefore, we will account for the constraint (9.213) by means of a Lagrange multiplier $\alpha(x)$. The Euclidean action can be rewritten as

$$S[S(x), \alpha(x)] = \frac{1}{2} \int d^2x \left[\left(\partial_\mu S^a \right) \left(\partial_\mu S^a \right) + \frac{\alpha(x)}{\sqrt{N}} \left(S^a S^a - \frac{N}{f_0} \right) \right], \qquad (9.221)$$

where I have introduced a new constant f_0:

$$\frac{1}{g_0^2} \equiv \frac{N}{f_0}, \qquad (9.222)$$

and changed the normalization of the S field to make the kinetic term canonically normalized. The factor $1/\sqrt{N}$ in front of $\alpha(x)$ is chosen for convenience. As we will see shortly, in order to get sensible results one must assume that the product $g_0^2 N \equiv f_0$ stays fixed at large N while the coupling constant g_0^2 itself scales as N^{-1}.

Since the field $\alpha(x)$ enters the action without derivatives it can be eliminated, resulting in the equation of motion

$$S^a(x) S^a(x) = \frac{N}{f_0}, \qquad (9.223)$$

which is equivalent to the constraint (9.213). Our task is to solve the theory (9.221) by assuming that the parameter N is large and expanding in $1/N$.

We will find the propagator of the S field in the saddle-point approximation, and then we will check that this approximation is perfectly justified at large N. In fact, the expansion around the saddle point is equivalent to the expansion in $1/N$.

To determine the propagator of the S field let us add the source term $\int d^2x\, J^a(x) S^a(x)$ to the action and then calculate the generating functional $Z[J^a(x)]$, where

$$Z[J^a(x)] = \int \mathcal{D}\alpha(x) \mathcal{D}S^a(x) \exp\left\{ -S[S(x), \alpha(x)] + \int d^2x\, J^a(x) S^a(x) \right\}, \qquad (9.224)$$

where on the right-hand side we have the path integral over all S^a fields and α. Since the action (9.221) is linear in α, integration over α returns us to the original action (9.214) plus the constraint on the S fields. However, we will integrate in the order indicated in Eq. (9.224) – first over S^a and then over α. Since the action (9.221) is bilinear in S, the integral over S is Gaussian and is easily found. To warm up, let us first put $J^a(x) = 0$. Then

$$Z \equiv \int \mathcal{D}\alpha(x) \mathcal{D}S^a(x) \exp\{ -S[S(x), \alpha(x)] \}$$

$$= \int \mathcal{D}\alpha(x) \exp\{ -S_{\text{eff}}[\alpha(x)] \}, \qquad (9.225)$$

where

$$S_{\text{eff}} = \frac{N}{2} \operatorname{Tr} \ln\left[-\partial^2 + \frac{\alpha(x)}{\sqrt{N}} \right] - \int d^2x \frac{\sqrt{N}}{2f_0} \alpha(x). \qquad (9.226)$$

The factor N in front of the trace of the logarithm in S_{eff} appears because there are N fields S^a and they are decoupled from each other in Eq. (9.221). Note that the trace of the logarithm is identically equal to

$$\ln \mathrm{Det}\left[-\partial^2 + \frac{\alpha(x)}{\sqrt{N}}\right].$$

The existence of a sharp stationary point in the integral over $\alpha(x)$ at $\alpha \neq 0$ is crucial in what follows. This will allow us to do the path integral over α using the saddle-point technique. As we will see shortly, this is equivalent to the $1/N$ expansion.

First, we note that the stationary value of α, if it exists, must be x-independent. This is due to the Lorentz symmetry. Let us denote this constant by $\sqrt{N}m^2$, where m^2 does not scale with N (we will confirm this scaling law later). Then

$$\alpha(x) = \sqrt{N}\, m^2 + \alpha_{\mathrm{qu}}(x), \tag{9.227}$$

where $\alpha_{\mathrm{qu}}(x)$ describes deviations from the stationary point $\sqrt{N}\, m^2$. In fact, $\alpha_{\mathrm{qu}}(x)$ will turn out to describe quantum fluctuations of the α field. We will expand S_{eff} in α_{qu} assuming the fluctuations to be small and then we will check that this is indeed the case.

The effective α action as a functional of α_{qu} takes the form

$$
\begin{aligned}
S_{\mathrm{eff}} = {} & \frac{N}{2}\,\mathrm{Tr}\,\ln(-\partial^2 + m^2) - \int d^2 x\, \frac{N}{2f_0}m^2 \\
& - \int d^2 x\, \frac{\sqrt{N}}{2f_0}\alpha_{\mathrm{qu}}(x) + \frac{\sqrt{N}}{2}\,\mathrm{Tr}\left(\frac{1}{-\partial^2 + m^2}\alpha_{\mathrm{qu}}\right) \\
& - \frac{1}{4}\,\mathrm{Tr}\left[\left(\frac{1}{-\partial^2 + m^2}\right)\alpha_{\mathrm{qu}}\left(\frac{1}{-\partial^2 + m^2}\right)\alpha_{\mathrm{qu}}\right] + \cdots,
\end{aligned}
\tag{9.228}
$$

where the ellipses denote terms cubic in α_{qu} and higher. The two terms on the first line are inessential constants (they affect only the overall normalization of Z, in which we are not interested). The two terms on the second line are linear in α_{qu}. If our conjecture of the existence of the stationary point at $\alpha(x) = \sqrt{N}\, m^2$ is valid, the sum of these two linear terms must vanish identically. Let us have a closer look at this condition. The functional trace of $(-\partial^2 + m^2)^{-1}\alpha_{\mathrm{qu}}$ can be identically transformed as

$$
\begin{aligned}
\frac{\sqrt{N}}{2}\,\mathrm{Tr}\left(\frac{1}{-\partial^2 + m^2}\alpha_{\mathrm{qu}}\right) &\overset{\mathrm{def}}{=} \frac{\sqrt{N}}{2}\int d^2 x\, \left\langle x\left|\frac{1}{-\partial^2 + m^2}\right|x\right\rangle \alpha_{\mathrm{qu}}(x) \\
&= \frac{\sqrt{N}}{2}\left\langle 0\left|\frac{1}{-\partial^2 + m^2}\right|0\right\rangle \int d^2 x\, \alpha_{\mathrm{qu}}(x) \\
&= \int \frac{d^2 p}{(2\pi)^2}\frac{1}{p^2 + m^2}\frac{\sqrt{N}}{2}\int d^2 x\, \alpha_{\mathrm{qu}}(x),
\end{aligned}
\tag{9.229}
$$

where we used the fact that $\langle x|(-\partial^2 + m^2)^{-1}|x\rangle$ is translationally invariant and therefore x-independent. We see that this term is identically canceled by another term linear in α_{qu} provided that

$$\frac{1}{f_0} = \int \frac{d^2 p}{(2\pi)^2}\frac{1}{p^2 + m^2} = \frac{1}{4\pi}\ln\frac{M_{\mathrm{uv}}^2}{m^2}. \tag{9.230}$$

> *The parameter m is the mass of the S quanta N-plet; see Eq. (9.233).*

The relation between the bare coupling, the ultraviolet cutoff, and the parameter m^2 is a self-consistency condition. The stationary point in the α integration exists if and only if Eq. (9.230) has a solution. Such a solution is easy to find:[31]

$$m^2 = M_{uv}^2 \exp\left(-\frac{4\pi}{f_0}\right). \tag{9.231}$$

If f_0 is N-independent then so is m^2, as was anticipated. Equation (9.231) implies, of course, that the O(N) model is asymptotically free, like O(3). Indeed, for fixed m and $M_{uv} \to \infty$, the bare coupling constant vanishes, $f_0 \to 0$. Moreover it demonstrates, in a transparent manner, dimensional transmutation.

Our next task is to understand the physical meaning of the parameter m^2. To this end we return to Eq. (9.224), with the source term switched on. Repeating step by step the analysis carried out above we now obtain

$$Z[J^a(x)] = \int \mathcal{D}\alpha(x) \exp\left\{ -S_{\text{eff}}[\alpha(x)] + \frac{1}{2} \int d^2x \, J^a \left[\frac{1}{-\partial^2 + \alpha(x)/\sqrt{N}} J^a\right](x) \right\}$$

$$= \text{const} \times \exp\left[\frac{1}{2} \int d^2x \, J^a(x) \left(\frac{1}{-\partial^2 + m^2} J^a\right)(x)\right]$$

$$+ \; 1/N \text{ corrections.} \tag{9.232}$$

By expanding $Z[J^a(x)]$ in $J^a(x)$ and examining the terms quadratic in $J^a(x)$ we observe that the Green's function of the fields S^a has the form (in the leading order in $1/N$)

$$\langle S^a, S^b \rangle \to \frac{\delta^{ab}}{p^2 + m^2}, \qquad a, b = 1, \ldots, N. \tag{9.233}$$

This is a remarkable formula. It shows that all N fields are on an equal footing; in fact, they form an N-plet of O(N). The symmetry of the Lagrangian, O(N), is *not* spontaneously broken. All N fields are massive, with the same mass m, rather than massless. This is another manifestation of the fact that the global O(N) symmetry, which was broken in perturbation theory, is restored at the nonperturbative level so that there are no massless Goldstones.

Of course, strictly speaking the results obtained above refer to the large-N limit. By themselves they tell us nothing about what happens at $N = 3$. A special analysis is needed in order to check that in decreasing N from ∞ down to 3 one encounters no singularities, i.e., that there is no qualitative difference between the large-N model and O(3). Such an analysis has been carried out in the literature. The conclusions perfectly agree with the exact solution of the O(3) model [29], which also demonstrates that there is no spontaneous symmetry breaking and there are no Goldstones.

In the course of our discussion, I have mentioned, more than once, that the expansion (9.227) around the saddle point is equivalent to a $1/N$ expansion. It would

---

[31] For finite N the expression for m^2 is

$$m^2 = M_{uv}^2 \exp\{-4\pi N/[f_0(N-2)]\}.$$

The exponent tends to that of Eq. (9.231) in the limit $N \to \infty$. The above expression is easy to check by comparing the expression for Λ^2 for the O(3) model with the β function in the O(N) model, Eq. (9.216).

be in order now to prove this assertion. As already explained, the term linear in α_{qu} vanishes. The bilinear term is given in the third line of Eq. (9.228),

$$S_{eff}^{(2)} = -\frac{1}{4} \text{Tr}\left[\left(\frac{1}{-\partial^2 + m^2}\right)\alpha_{qu}\left(\frac{1}{-\partial^2 + m^2}\right)\alpha_{qu}\right]$$

$$\equiv -\frac{1}{4}\int d^2x\, d^2y\, \alpha_{qu}(x)\Gamma(x-y)\alpha_{qu}(y), \tag{9.234}$$

where Γ is the inverse propagator of the α "particle,"

$$D^{(\alpha)}(p) = -\frac{2}{\Gamma(p)}. \tag{9.235}$$

Here $D^{(\alpha)}$ is the α propagator and $\Gamma(p)$ is the Fourier transform of $\Gamma(x)$:

$$\Gamma(p) = \int \frac{d^2q}{(2\pi)^2} \frac{1}{(q^2+m^2)[(p+q)^2+m^2]}$$

$$= \frac{1}{2\pi}\frac{1}{\sqrt{p^2(p^2+4m^2)}}\ln\frac{\sqrt{(p^2+4m^2)}+\sqrt{p^2}}{\sqrt{(p^2+4m^2)}-\sqrt{p^2}}. \tag{9.236}$$

Note that the propagator $D^{(\alpha)}$ has no poles in p^2, only a cut starting at $p^2 = -4m^2$. This means that the α field is not a real particle; rather, it is a resonance-like state.

Knowing the propagator $D^{(\alpha)}$ one can readily calculate the higher-order corrections due to deviations from the saddle point (i.e., the loops generated by α_{qu} exchanges). The most convenient way to formulate the result in a concise manner is using a new perturbation theory in terms of new Feynman graphs (Fig. 9.45). Note that this perturbation theory has nothing to do with that in the coupling constant g^2. In fact, g^2 does not show up explicitly at all; it only enters in the new Feynman graphs through the parameter m^2. The expansion parameter of the new perturbation theory is $1/N$.

The Feynman rules in Fig. 9.45 describe the propagation of N massive particles with Green's function $[D^{ab}(p)] = \delta^{ab}(p^2+m^2)^{-1}$ (Fig. 9.45a), the propagation of the α "particle" with Green's function $D^{(\alpha)}(p) = -2/\Gamma(p)$ (Fig. 9.45b), and their interaction, given by the vertex $\Gamma_{ab} = -(1/\sqrt{N})\delta^{ab}$ (Fig. 9.45c). The graphs shown in Fig. 9.46 are accounted for in $D^{(\alpha)}(p)$. They should not be included again, to avoid double counting. The same applies to the graphs of tadpole type (Fig. 9.47).

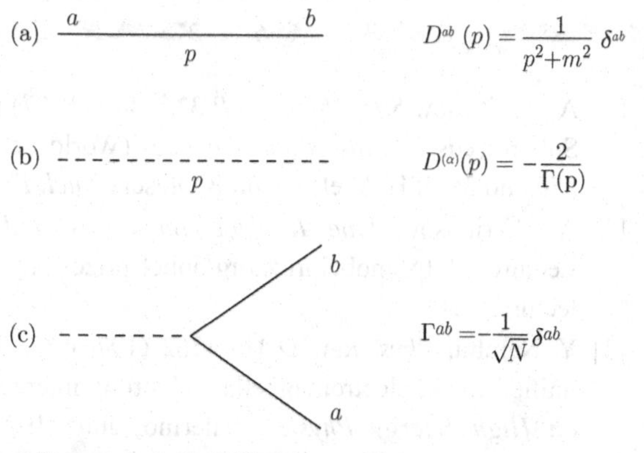

(a) $\quad a \underline{\hspace{4cm}} b \qquad\qquad D^{ab}(p) = \frac{1}{p^2+m^2}\delta^{ab}$
$\qquad\qquad\qquad p$

(b) $\quad \text{-----------} \qquad\qquad D^{(\alpha)}(p) = -\frac{2}{\Gamma(p)}$
$\qquad\qquad\qquad p$

(c) $\qquad\qquad\qquad\qquad \Gamma_{ab} = -\frac{1}{\sqrt{N}}\delta^{ab}$

Fig. 9.45   Feynman rules for the 1/N expansion in the O(N) sigma model.

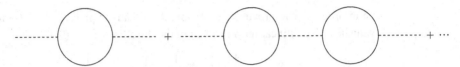

The one-particle reducible graphs included in the α propagator, to be discarded in perturbation theory following from Fig. 9.45.

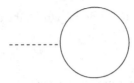

Tadpole graphs vanish under the condition (9.230). They are not to be included in perturbation theory, summarized in Fig. 9.45.

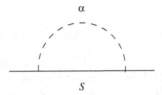

The leading correction to the S mass.

They vanish because the condition (9.230) is satisfied, and there are no linear in α_{qu} terms in S_{eff}.

I would like to reiterate that the $1/N$ perturbation theory, presented in Fig. 9.45, drastically differs from that in the coupling constant g^2. First and foremost, it explicitly incorporates the crucial nonperturbative effects: symmetry restoration and mass generation. The number of S particles is N not $N - 1$, from the very beginning. Second, the structure of the $1/N$ expansion becomes transparent: each αSS vertex introduces a factor $1/\sqrt{N}$. For instance, the leading correction to the S mass is due to the graph in Fig. 9.48; it is proportional to $1/N$.

References for Chapter 9

[1] A. Abrikosov, *Sov. Phys. JETP* **32**, 1442 (1957) [reprinted in C. Rebbi and G. Soliani (eds.), *Solitons and Particles* (World Scientific, Singapore, 1984), pp. 356 and 365]. H. Nielsen and P. Olesen, *Nucl. Phys. B* **61**, 45 (1973).

[2] A. Abrikosov, *Type II Superconductors and the Vortex Lattice*, Nobel Lecture [http://nobelprize.org/nobel_prizes/physics/laureates/2003/abrikosov-lecture.pdf]].

[3] Y. Nambu, *Phys. Rev. D* **10**, 4262 (1974); G. 't Hooft, Gauge theories with unified weak, electromagnetic and strong interactions, in *Proc. EPS Int. Conf. on High Energy Physics*, Palermo, June 1975, ed. A. Zichichi (Editrice Compositori, Bologna, 1976); S. Mandelstam, *Phys. Rept.* **23**, 245 (1976).

[4] N. Seiberg and E. Witten, *Nucl. Phys. B* **426**, 19 (1994); Erratum, *ibid*. **430**, 485 (1994) [hep-th/9407087]; *Nucl. Phys. B* **431**, 484 (1994) [hep-th/9408099].

[5] G. 't Hooft, *Nucl. Phys. B* **79**, 276 (1974); A. M. Polyakov, *JETP Lett*. **20**, 194 (1974).

[6] M. R. Douglas and S. H. Shenker, *Nucl. Phys. B* **447**, 271 (1995) [hep-th/9503163]; A. Hanany, M. J. Strassler, and A. Zaffaroni, *Nucl. Phys. B* **513**, 87 (1998) [hep-th/9707244].

[7] G. 't Hooft, *Nucl. Phys. B* **72**, 461 (1974); see also Planar diagram field theories, in G. 't Hooft, *Under the Spell of the Gauge Principle* (World Scientific, Singapore, 1994), p. 378.

[8] B. Zwiebach, *A First Course in String Theory* (Cambridge University Press, 2004).

[9] E. Witten, *Nucl. Phys. B* **160**, 57 (1979).

[10] A. Armoni, M. Shifman, and G. Veneziano, *Phys. Rev. Lett.* **91**, 191601 (2003) [hep-th/0307097]; *Phys. Lett. B* **579**, 384 (2004) [hep-th/0309013].

[11] A. Armoni, M. Shifman, and G. Veneziano, From super-Yang–Mills theory to QCD: planar equivalence and its implications [arXiv:hep-th/0403071] in M. Shifman *et al.* (eds.) *From Fields to Strings: Circumnavigating Theoretical Physics* (World Scientific, Singapore, 2004), Vol. 1, p. 353.

[12] E. Corrigan and P. Ramond, *Phys. Lett. B* **87**, 73 (1979).

[13] G. Zweig, *Int. J. Mod. Phys. A* **25**, 3863 (2010) [arXiv:1007.0494 [physics.hist-ph]].

[14] A. Cherman, T. D. Cohen, and R. F. Lebed, *Phys. Rev. D* **80** (2009) [arXiv:0906.2400 [hep-ph]].

[15] A. Armoni, M. Shifman, and G. Veneziano, *Nucl. Phys. B* **667**, 170 (2003) [arXiv:hep-th/0302163]; *Phys. Rev. D* **71**, 045 015 (2005) [arXiv:hep-th/0412203].

[16] M. Peskin and D. Schroeder, *An introduction to Quantum Field Theory* (Addison-Wesley, 1995).

[17] A. V. Manohar, Large-N QCD [arXiv:hep-ph/9802419] in R. Gupta, A. Morel, E. de Rafael, and F. David (eds.), *Probing the Standard Model of Particle Interactions* (North Holland, Amsterdam, 1999) p. 1091.

[18] H. Lipkin, *Quantum Mechanics* (North-Holland, Amsterdam, 1973); F. Schwabl, *Advanced Quantum Mechanics* (Springer, Berlin, 1997).

[19] J.-L. Gervais and B. Sakita, *Phys. Rev. Lett.* **52**, 87 (1984); *Phys. Rev. D* **30**, 1795 (1984).

[20] R. Dashen and A.V. Manohar, *Phys. Lett. B* **315**, 425; 438 (1993).

[21] F. Kokkedee, *Quark theory* (Benjamin, New York, 1969).

[22] A.V. Manohar, *Nucl. Phys. B* **248**, 19 (1984).

[23] R. Dashen, E. Jenkins, and A.V. Manohar, *Phys. Rev. D* **49**, 4713 (1994).

[24] R. Dashen, E. Jenkins, and A.V. Manohar, *Phys. Rev. D* **51**, 3697 (1995).

[25] F. De Luccia and P. Steinhardt, unpublished; C. G. Callan, R. F. Dashen, and D. J. Gross, *Phys. Lett. B* **66**, 375 (1977); S. Coleman, The uses of instantons, in *Aspects of Symmetry* (Cambridge University Press, 1985).

[26] David Tong, *Lectures on Gauge Theory* (see Chapter 7, "Quantum Field Theory in Two Dimensions," www.damtp.cam.ac.uk/user/tong/gaugetheory .html).

[27] A. D'Adda, M. Lüscher, and P. Di Vecchia, *Nucl. Phys. B* **146**, 63 (1978); A. D'Adda, P. Di Vecchia, and M. Lüscher, *Nucl. Phys. B* **152**, 125 (1979).

[28] E. Witten, *Nucl. Phys. B* **149**, 285 (1979).

[29] A. B. Zamolodchikov and Al. B. Zamolodchikov, *Ann. Phys.* **120**, 253 (1979).

[30] A. Gorsky, M. Shifman, and A. Yung, *Phys. Rev. D* **71**, 045 010 (2005) [arXiv:hep-th/0412082].

[31] E. Witten, *Nucl. Phys. B* **202** (1982) 253 [reprinted in S. Ferrara (ed.) *Supersymmetry* (North Holland/World Scientific, Amsterdam–Singapore, 1987), Vol. 1, p. 490]; *J. High Energy Phys.* **12**, 019 1998 (see Appendix).

[32] B. S. Acharya and C. Vafa, On domain walls of $N = 1$ supersymmetric Yang–Mills in four dimensions [hep-th/0103011].

[33] E. Witten, *Phys. Rev. Lett.* **81**, 2862 (1998) [hep-th/9807109].

[34] G. 't Hooft, *Nucl. Phys. B* **75**, 461 (1974) [reprinted in G. 't Hooft, *Under the Spell of the Gauge Principle* (World Scientific, Singapore, 1994), p. 461].

[35] I. Bars and M. B. Green, *Phys. Rev. D* **17**, 537 (1978).

[36] K. Hornbostel, *The application of light cone quantization to quantum chromodynamics in (1+1)-dimensions*, SLAC PhD thesis, 1988.

[37] Y. S. Kalashnikova and A. V. Nefediev, *Phys. Usp.* **45**, 347 (2002) [hep-ph/0111225].

[38] F. Lenz, M. Thies, S. Levit, and K. Yazaki, *Ann. Phys.* **208**, 1 (1991).

[39] A. Zhitnitsky, *Phys. Lett. B* **165**, 405 (1985); *Sov. J. Nucl. Phys.* **43**, 999 (1986); *ibid.* **44**, 139 (1986).

[40] F. Schwabl, *Advanced Quantum Mechanics* (Springer, 1997), Chapter 9.

[41] A. M. Polyakov, *Nucl. Phys. B* **120**, 429 (1977).

[42] A. Kovner, Confinement, magnetic $Z(N)$ symmetry and low-energy effective theory of gluodynamics, in M. Shifman (ed.), *At the Frontier of Particle Physics: Hand-book of QCD* (World Scientific, Singapore, 2001), Vol. 3, p. 1777 [hep-ph/0009138]; I. I. Kogan and A. Kovner, Monopoles, vortices and strings: confinement and deconfinement in 2+1 dimensions at weak coupling, in M. Shifman (ed.), *At the Frontier of Particle Physics* (World Scientific, Singapore, 2001), Vol. 4, p. 2335 [hep-th/0205026].

[43] M. Shifman and M. Ünsal, *Phys. Rev. D* **78**, 065 004 (2008) [arXiv:0802.1232 [hep-th]]; *Phys. Rev. D* **79**, 105 010 (2009) [arXiv:0808.2485 [hep-th]].

[44] H. Georgi and S. L. Glashow, *Phys. Rev. Lett.* **28**, 1494 (1972).

[45] P. Sikivie, *Nucl. Phys. Proc. Suppl.* **87**, 41 (2000) [arXiv:hep-ph/0002154].

[46] A. M. Polyakov, *Gauge fields and Strings* (Harwood Academic Press, Newark, 1987).

[47] E. Witten, Dynamics of quantum fields. Lecture 8. Abelian duality, in P. Deligne *et al.* (eds.), *Quantum Fields and Strings. A Course for Mathematicians* (AMS, 1999), Vol. 2, p. 1119.

[48] N. J. Snyderman, *Nucl. Phys. B* **218**, 381 (1983).

[49] V. A. Novikov, M. A. Shifman, A. I. Vainshtein, and V. I. Zakharov, *Phys. Rept.* **116**, 103 (1984).

[50] L. Y. Glozman, V. K. Sazonov, M. Shifman, and R. F. Wagenbrunn, *Phys. Rev. D* **85**, 094030 (2012) [arXiv:1201.5814 [hep-th]].

[51] J. Schwinger, *Phys. Rev.* **128**, 2425, 1962. Schwinger's solution was completed in J. Lowenstein and J. A. Swieca, *Annals of Phys.* **68**, 172 (1971).

[52] S. R. Coleman, *Annals Phys.* **101**, 239 (1976).

[53] M. Shifman, *Quantum Field Theory II* (World Scientific, Singapore, 2019), pp. 96–99, 272–273.

[54] M. Stone (ed.), *Bosonization*, (World Scientific, Singapore, 1994).

PART II

INTRODUCTION TO SUPERSYMMETRY

Basics of Supersymmetry with Emphasis on Gauge Theories

Extending the Poincaré algebra. — Quantum dimensions of space–time. — Superfields. — Simplest models in four and two dimensions. — Becoming acquainted with supergauge invariance. — Supersymmetric Yang–Mills theories and super-Higgs mechanism. — Hypercurrent. — Exact results, or the power of supersymmetry.

10.1 Introduction

Global symmetries play a crucial role in explorations of fundamental interactions, both from the standpoint of phenomenology and from the point of view of dynamical studies. Supersymmetry is arguably the most beautiful invention in theoretical physics in the twentieth century. It is the *supreme* symmetry: it extends the set of geometric symmetries to its limit. There are very few geometric symmetries in nature and they are precious. They are expressed by energy–momentum conservation and Lorentz invariance – consequences of the homogeneity of space–time. For some time it was believed that that was the end of the story. The Coleman–Mandula theorem, which we will consider shortly (in Section 10.3) tells us that no other exact symmetry can be geometric; all other conserved quantities, such as electric charge, are of an internal nature and can only be Lorentz scalars. The spinorial extension of Poincaré algebra was discovered as a "loophole" in the above Coleman–Mandula theorem. It turns out that conserved quantities with spinorial indices can exist. They are called *supercharges*. The anticommutator of two supercharges is proportional to the energy–momentum operator.

Supersymmetry is unique because supersymmetry transformations connect bosons and fermions – particles of different spins, combining them in common supermultiplets and relating their masses and other properties. Mathematically it can be expressed as the emergence of extra quantum (Grassmannian) dimensions in addition to our conventional space–time. As we will see below, many miraculous consequences ensue from the Grassmannian dimensions.

"Supersymmetry, if it holds in nature, is part of the quantum structure of space and time. In everyday life we measure space and time by numbers, 'It is now three o'clock, the elevation is two hundred meters above the sea level,' and so on. Numbers are classical concepts, known to humans since long before quantum mechanics was developed in the early twentieth century. The discovery of quantum mechanics changed our understanding of almost everything in physics, but our basic way of thinking about space and time has not yet been affected. Showing that nature is supersymmetric would change that, by revealing a quantum dimension of space and time, not measurable by ordinary numbers ..." [1].

Since its inception in the early 1970s supersymmetric field theory has experienced unprecedented development. Supersymmetry has become one of the most powerful tools in the uncovering of the subtle and long-standing mysteries of gauge dynamics at strong coupling. The first ever analytical proof that the dual Meissner effect is *the* mechanism of color confinement was obtained [2] in a supersymmetric extension of Yang–Mills theory. Supersymmetric field theory is an excellent testing ground for a number of ideas that are hardly accessible to analytical study by other methods. That is why the knowledge of nonperturbative supersymmetry is essential for those whose research interests are in high-energy physics.

In this chapter we will uncover the foundations of supersymmetric field theory while aiming to avoid excessive technicalities. We will limit ourselves to global supersymmetry. Local supersymmetry (supergravity) is beyond the scope of the present book. My task is to acquaint the reader with a set of basic concepts and adequate formalism in preparation for his/her own "supersymmetry odyssey" in this vast and still growing area.

A few words on the history of the "superdiscovery" are in order. I quote here Julius Wess [3], one of the founding fathers of supersymmetry:

It started with the work of Golfand and Likhtman [4]. They thought about adding spinorial generators to the Poincaré algebra, in that way enlarging the algebra. This was about 1970, and they were really on the track of supersymmetry. [...] I think that this is the right question: can we enlarge the algebra, the concept of symmetry, by new algebraic concepts in order to get new types of symmetries?

Then in 1972 there was a paper by Volkov and Akulov [5] who argued along the following lines. We know that with spontaneously broken symmetries there are Goldstone particles, supposed to be massless. In nature we know spin-$\frac{1}{2}$ particles that have, if any, a very small mass, these are the neutrinos. Could these fermions be Goldstone particles of a broken symmetry? Volkov and Akulov constructed a Lagrangian, a non-linear one, that turned out to be supersymmetric. [...]

Another path to supersymmetry came from two-dimensional dual models. Neveu and Schwarz [6][1] [...] had constructed models which had spinorial currents related to supergauge transformations that transform scalar fields into spinor fields. The algebra of the transformation, however, only closed on [the] mass shell. The spinorial currents were called supercurrents and that is where the name "supersymmetry" comes from.

In 1974 Bruno Zumino and I published a paper [8] where we established supersymmetry in four dimensions, constructed renormalizable Lagrangians and exhibited nonrenormalization properties at the one-loop level ...

The paper of Wess and Zumino started an explosive development and showed, to the entire theoretical community, a new way – a way to supersymmetry.

The standard classic text in this field is the textbook by Wess and Bagger [9]. A generation of theorists has used the Wess–Bagger notation. We will follow the same tradition, with one exception. The choice of the metric tensor in [9] is $g^{\mu\nu} = $ diag $\{-1, 1, 1, 1\}$. I find it more convenient, however, to work with the standard Minkowski metric $g^{\mu\nu} = $ diag $\{1, -1, -1, -1\}$. Accordingly, some formulas must be modified but these modifications are minimal.

A number of special topics, such as the supergraph technique and topics related to supersymmetric phenomenology, will not be covered here.[2] The interested reader is referred to the textbooks quoted in [10–21]. For mathematically oriented students I can recommend [22]. An excellent compilation of the groundbreaking original papers of the 1970s and early 1980s can be found in [23].

---

[1] See also Gervais and Sakita [7].

[2] The reason for the omission of supersymmetric phenomenology should be fairly obvious. This area is in rapid change and its discussion is likely to become obsolete by the time this textbook is issued. So far, by the end of 2021, no traces of supersymmetry are seen at LHC. The superdiagram technique is an indispensable element of general supersymmetry. We will skip this topic, with regret, owing to space and time limitations and to its availability in many textbooks.

10.2 Spinors and Spinorial Notation

Supersymmetry unifies bosons with fermions. The conserved supercharges are spinors. Therefore our first task is to recall the spinorial formalism in four, three, and two dimensions.

10.2.1 Four-Dimensional Spinors and Related Topics

Let us start with four dimensions (see, e.g., [24]). Four-dimensional spinors realize an irreducible representation of the Lorentz group that has six generators: three spatial rotations and three Lorentz boosts. There are two types of spinor right-handed and left-handed, indicated by dotted and undotted indices, respectively, as follows:[3]

$$\text{right-handed:} \quad \bar{\eta}^{\dot{\alpha}}, \quad \dot{\alpha} = 1, 2, \tag{10.1}$$

$$\text{left-handed:} \quad \xi_{\alpha}, \quad \alpha = 1, 2. \tag{10.2}$$

Let us write the transformation law for the undotted spinors as

$$\tilde{\xi}_{\alpha} = U_{\alpha}{}^{\beta} \xi_{\beta}, \tag{10.3}$$

where for spatial rotations

$$U_{\text{rot}} = \exp\left(i \frac{\theta}{2} \vec{n}\vec{\sigma}\right), \quad \det U = 1, \quad U^{\dagger}U = 1, \tag{10.4}$$

θ is the rotation angle, and $\vec{n}$ is the rotation axis. For the Lorentz boosts,

$$U_{\text{boost}} = \exp\left(-\frac{\phi}{2} \vec{n}'\vec{\sigma}\right), \quad \det U = 1, \quad U^{\dagger}U \neq 1. \tag{10.5}$$

Here $\tanh \phi = v$, where v is the 3-velocity while $\vec{n}'$ is its direction. Moreover, $\vec{\sigma}$ represents the Pauli matrices. As is seen from Eq. (10.5) the transformation matrix for the Lorentz boost is not unitary, because the Lorentz group is the noncompact $O(1,3)$ rather than the compact $O(4)$. If we passed from Minkowski to Euclidean space then the Euclidean "Lorentz group" would be $O(4)$. By definition,

$$\bar{\eta}_{\dot{\alpha}} \sim (\xi_{\alpha})^{*}, \tag{10.6}$$

where the sign $\sim$ means "transforms as." Therefore, for the dotted spinors the Lorentz transformation requires the complex-conjugate matrix:

$$\tilde{\bar{\eta}}_{\dot{\alpha}} = (U^{*})_{\dot{\alpha}}^{\dot{\beta}} \bar{\eta}_{\dot{\beta}} \quad \text{or} \quad \tilde{\bar{\eta}}^{\dot{\alpha}} = \begin{cases} (U_{\text{rot}})_{\dot{\beta}}^{\dot{\alpha}} \bar{\eta}^{\dot{\beta}}, & \text{for rotations,} \\ \left(U_{\text{boost}}^{-1}\right)_{\dot{\beta}}^{\dot{\alpha}} \bar{\eta}^{\dot{\beta}}, & \text{for boosts.} \end{cases} \tag{10.7}$$

---

[3] This convention is standard in supersymmetry but is opposite to that accepted in the textbook [24], where the left-handed spinor is dotted. Sometimes we will omit spinorial indices. Then, in order to differentiate between left- and right-handed spinors, we will indicate the latter by overbars, e.g., $\bar{\eta}$ is a shorthand for $\bar{\eta}^{\dot{\alpha}}$.

If the three generators of the spatial rotations are denoted by L^i and the three Lorentz boost generators by N^i, it is obvious[4] that $L^i + iN^i$ does not act on ξ_α while $L^i - iN^i$ does not act on $\bar{\eta}^{\dot{\alpha}}(i = 1, 2, 3)$. The spinors ξ_α and $\bar{\eta}^{\dot{\alpha}}$ are referred to as the *chiral* or *Weyl* spinors. In four dimensions one chiral spinor is equivalent to one Majorana spinor, while two chiral spinors – one dotted and one undotted – comprise one Dirac spinor (see below).

Weyl and Majorana spinors

In order to be invariant, every spinor equation must have on each side the same number of undotted and dotted indices, since otherwise the equation becomes invalid under a change of reference frame. We must remember, however, that taking the complex conjugate implies interchanging the dotted and undotted indices. For instance, the relation $\bar{\eta}^{\dot{\alpha}\beta} = \left(\xi^{\alpha\beta}\right)^*$ is invariant.

To build Lorentz scalars we must convolute either the undotted or the dotted spinors (separately). For instance, the products

$$\chi^\alpha \xi_\alpha \quad \text{or} \quad \bar{\psi}_{\dot{\beta}} \, \bar{\eta}^{\dot{\beta}} \tag{10.8}$$

are invariant under Lorentz transformation. The lowering and raising of the spinorial indices is achieved by applying the invariant Levi–Civita tensor *from the left*:[5]

$$\chi^\alpha = \varepsilon^{\alpha\beta}\chi_\beta, \qquad \chi_\alpha = \varepsilon_{\alpha\beta}\chi^\beta, \tag{10.9}$$

Two-dimensional Levi–Civita tensor

and the same applies for the dotted indices. The two-index Lorentz-invariant Levi–Civita tensor is defined as follows:

$$\begin{aligned} \varepsilon^{\alpha\beta} &= -\varepsilon^{\beta\alpha}, & \varepsilon^{12} &= -\varepsilon_{12} = 1, \\ \varepsilon^{\dot{\alpha}\dot{\beta}} &= -\varepsilon^{\dot{\beta}\dot{\alpha}} & \varepsilon^{\dot{1}\dot{2}} &= -\varepsilon_{\dot{1}\dot{2}} = 1. \end{aligned} \tag{10.10}$$

We will follow a standard shorthand notation:

$$\eta\chi \equiv \eta^\alpha \chi_\alpha, \qquad \bar{\eta}\bar{\chi} \equiv \bar{\eta}_{\dot{\alpha}}\bar{\chi}^{\dot{\alpha}}. \tag{10.11}$$

Note that this convention acts differently for left- and right-handed spinors. It is very convenient because

$$(\eta\chi)^\dagger = (\eta^\alpha \chi_\alpha)^\dagger = (\chi_\alpha)^*(\eta^\alpha)^* = \bar{\chi}\bar{\eta}, \tag{10.12}$$

where

$$\bar{\chi}_{\dot{\alpha}} \equiv (\chi_\alpha)^*, \qquad \bar{\eta}^{\dot{\alpha}} \equiv (\eta^\alpha)^*. \tag{10.13}$$

Moreover, using the properties (10.10) of the Levi–Civita tensor and the Grassmannian nature of the fermion variables, we get

$$\begin{aligned} \chi^\alpha \chi^\beta &= -\tfrac{1}{2}\varepsilon^{\alpha\beta}\chi^2, & \chi_\alpha \chi_\beta &= \tfrac{1}{2}\varepsilon_{\alpha\beta}\chi^2, \\ \bar{\chi}^{\dot{\alpha}}\bar{\chi}^{\dot{\beta}} &= \tfrac{1}{2}\varepsilon^{\dot{\alpha}\dot{\beta}}\bar{\chi}^2, & \bar{\chi}_{\dot{\alpha}}\bar{\chi}_{\dot{\beta}} &= -\tfrac{1}{2}\varepsilon_{\dot{\alpha}\dot{\beta}}\bar{\chi}^2. \end{aligned} \tag{10.14}$$

---

[4] For the left-handed states $\vec{L} = \tfrac{1}{2}\vec{\sigma}$, $\vec{N} = \tfrac{i}{2}\vec{\sigma}$. The algebra of these generators is as follows: $[L^i, L^j] = i\varepsilon^{ijk}L^k$, $[L^i, N^j] = i\varepsilon^{ijk}N^k$, and $[N^i, N^j] = -i\varepsilon^{ijk}L^k$, implying that $[L^i - iN^i, L^j + iN^j] = 0$. Note that under spatial rotations ξ_α and $\bar{\eta}^{\dot{\alpha}}$ transform in the same way. This is not the case for Lorentz boosts.

[5] The same rule, multiplying by the Levi–Civita tensor from the left, applies to quantities with several spinorial indices: dotted, undotted, or mixed.

Vector quantities (the $(\frac{1}{2},\frac{1}{2})$ representation of the Lorentz group) are obtained in the spinorial notation by convoluting a given vector with the matrix

$$(\sigma^\mu)_{\alpha\dot\beta} = \{1, \vec\sigma\}_{\alpha\dot\beta}. \tag{10.15}$$

For instance,

$$A_{\alpha\dot\beta} = A_\mu (\sigma^\mu)_{\alpha\dot\beta}, \qquad A^\mu = \tfrac{1}{2} A_{\alpha\dot\beta} (\bar\sigma^\mu)^{\dot\beta\alpha}, \tag{10.16}$$

where

$$A_\mu \equiv (A^0, -\vec{A}) \equiv (A^0, -A^1, -A^2, -A^3) \equiv (A_t, -A_x, -A_y, -A_z). \tag{10.17}$$

The convolution of two 4-vectors is then

$$A_\mu B^\mu = \frac{1}{2} A_{\alpha\dot\beta} B^{\dot\beta\alpha}, \qquad A_{\alpha\dot\beta} A^{\dot\beta\gamma} = \delta_\alpha^\gamma A_\mu A^\mu. \tag{10.18}$$

The square of a 4-vector is understood as follows:

$$A^2 \equiv A_\mu A^\mu = \frac{1}{2} A_{\alpha\dot\beta} A^{\dot\beta\gamma}. \tag{10.19}$$

If the matrix $(\sigma^\mu)_{\alpha\dot\beta}$ is "right-handed" then it is convenient to introduce its "left-handed" counterpart,

$$(\bar\sigma^\mu)^{\dot\beta\alpha} = \{1, -\vec\sigma\}_{\dot\beta\alpha}. \tag{10.20}$$

> The matrices $(\sigma^\mu)_{\alpha\dot\beta}$ and $(\bar\sigma^\mu)^{\dot\beta\alpha}$ are defined in (10.15) and (10.20).

To obtain the matrix $(\bar\sigma^\mu)^{\dot\beta\alpha}$ from $(\sigma^\mu)_{\alpha\dot\beta}$ we raise the indices of the latter according to the rule (10.9) and then transpose the dotted and undotted indices. It should be remembered that

$$(\sigma^\mu)_{\alpha\dot\beta}(\bar\sigma^\nu)^{\dot\beta\gamma} + (\sigma^\nu)_{\alpha\dot\beta}(\bar\sigma^\mu)^{\dot\beta\gamma} = 2g^{\mu\nu}\delta_\alpha^\gamma. \tag{10.21}$$

An immediate consequence is

$$(\sigma^{\{\mu})_{\alpha\dot\beta}(\sigma^{\nu\}})_\gamma{}^{\dot\beta} \equiv (\sigma^\mu)_{\alpha\dot\beta}(\sigma^\nu)_\gamma{}^{\dot\beta} + (\sigma^\nu)_{\alpha\dot\beta}(\sigma^\mu)_\gamma{}^{\dot\beta} = -2\, g^{\mu\nu}\, \varepsilon_{\alpha\gamma},$$

$$(\sigma^{[\mu})_{\alpha\dot\beta}(\sigma^{\nu]})_\gamma{}^{\dot\beta} \equiv (\sigma^\mu)_{\alpha\dot\beta}(\sigma^\nu)_\gamma{}^{\dot\beta} - (\sigma^\nu)_{\alpha\dot\beta}(\sigma^\mu)_\gamma{}^{\dot\beta}$$

$$= -2 \left\{ \begin{array}{l} (\tau^i)_{\alpha\gamma} \text{ if } \mu = 0,\ \nu = i, \\ \varepsilon^{jk\ell}(i\tau^\ell)_{\alpha\gamma} \text{ if } \mu = j,\ \nu = k, \end{array} \right\}, \qquad (\vec\tau)_{\alpha\gamma} = \{-\sigma_z, i\mathbf{1}, \sigma_x\}_{\alpha\gamma}. \tag{10.22}$$

Remember that if you see τ^i matrices sandwiched between the parentheses they are *not* the Pauli matrices. Yet another consequence is

$$(\sigma^\mu)_{\gamma\dot\beta}(\sigma_\mu)_{\alpha\dot\alpha} = 2\varepsilon_{\gamma\alpha}\varepsilon_{\dot\beta\dot\alpha}, \qquad (\sigma^\mu)_{\gamma\dot\delta}(\bar\sigma_\mu)^{\dot\beta\alpha} = 2\delta_{\dot\delta}^{\dot\beta}\delta_\gamma^\alpha. \tag{10.23}$$

Two-index antisymmetric Lorentz tensors have six components and can be expressed in terms of two 3-vectors. The most well-known example is the electromagnetic field tensor

$$F^{\mu\nu} = \begin{pmatrix} 0 & -E_x & -E_y & -E_z \\ E_x & 0 & -B_z & B_y \\ E_y & B_z & 0 & -B_x \\ E_z & -B_y & B_x & 0 \end{pmatrix}, \quad F_{\mu\nu} = \begin{pmatrix} 0 & E_x & E_y & E_z \\ -E_x & 0 & -B_z & B_y \\ -E_y & B_z & 0 & -B_x \\ -E_z & -B_y & B_x & 0 \end{pmatrix}, \tag{10.24}$$

where $\vec{E}$ and $\vec{B}$ are the electric and magnetic fields, respectively. A standard shorthand for two-index antisymmetric tensors is as follows:

$$F^{\mu\nu} = (-\vec{E}, \vec{B}), \qquad F_{\mu\nu} = (\vec{E}, \vec{B}). \tag{10.25}$$

Two-index antisymmetric tensors such as those in (10.24) realize the representations $(0, 1) + (1, 0)$ of the Lorentz group. The passage from vectorial to spinorial notation in this case proceeds according to the general rules, namely,

$$\begin{aligned}
F_{\alpha\beta} &= -\tfrac{1}{2} F_{\mu\nu} (\sigma^\mu)_{\alpha\dot\gamma} (\sigma^\nu)^{\dot\gamma}_\beta \equiv (\vec{E} - i\vec{B})(\vec{\tau})_{\alpha\beta}, \\
F^{\dot\alpha\dot\beta} &= \tfrac{1}{2} F_{\mu\nu} (\bar\sigma^\mu)^{\dot\alpha\gamma} (\bar\sigma^\nu)^{\dot\beta}_\gamma \equiv (\vec{E} + i\vec{B})(\vec{\tau})^{\dot\alpha\dot\beta},
\end{aligned} \tag{10.26}$$

where $(\vec{\tau})_{\alpha\beta}$ and $(\vec{\tau})^{\dot\alpha\dot\beta}$ are two triplets of matrices:

$$\begin{aligned}
(\vec{\tau})_{\alpha\beta} &= \{-\sigma_z, i\mathbf{1}, \sigma_x\}_{\alpha\beta}, \qquad (\vec{\tau})^{\alpha\beta} = \{\sigma_z, i\mathbf{1}, -\sigma_x\}_{\alpha\beta}, \\
(\vec{\tau})^{\dot\alpha\dot\beta} &= \{\sigma_z, -i\mathbf{1}, -\sigma_x\}_{\dot\alpha\dot\beta}, \qquad (\vec{\tau})_{\dot\alpha\dot\beta} = \{-\sigma_z, -i\mathbf{1}, \sigma_x\}_{\dot\alpha\dot\beta};
\end{aligned} \tag{10.27}$$

the indices on the right-hand sides are understood as regular matrix indices, for instance $\mathbf{1}_{\alpha\beta} = \delta_{\alpha\beta}$. Note that both the sets in (10.27) are symmetric with respect to the interchanges $\alpha \leftrightarrow \beta$ and $\dot\alpha \leftrightarrow \dot\beta$, implying that $F_{\alpha\beta} = F_{\beta\alpha}$ and $\bar{F}^{\dot\alpha\dot\beta} = \bar{F}^{\dot\beta\dot\alpha}$. This property expresses the fact that $F_{\alpha\beta}$ belongs to the irreducible representation $(1, 0)$ and $\bar{F}^{\dot\alpha\dot\beta}$ to the irreducible representation $(0, 1)$. Furthermore,

$$\begin{aligned}
F_{\alpha\beta} F^{\alpha\beta} &= 2(\vec{B}^2 - \vec{E}^2 + 2i\vec{E}\vec{B}) = F_{\mu\nu} F^{\mu\nu} - i F_{\mu\nu} \tilde{F}^{\mu\nu}, \\
\bar{F}_{\dot\alpha\dot\beta} \bar{F}^{\dot\alpha\dot\beta} &= 2(\vec{B}^2 - \vec{E}^2 - 2i\vec{E}\vec{B}) = F_{\mu\nu} F^{\mu\nu} + i F_{\mu\nu} \tilde{F}^{\mu\nu},
\end{aligned} \tag{10.28}$$

where $\tilde{F}^{\mu\nu}$ is the dual field tensor,

$$\tilde{F}^{\mu\nu} = \frac{1}{2} \varepsilon^{\mu\nu\rho\sigma} F_{\rho\sigma}, \tag{10.29}$$

4D Levi–Civita tensor

and $\varepsilon^{\mu\nu\rho\sigma}$ is the four-index Levi–Civita tensor,

$$\varepsilon^{0123} = 1, \qquad \varepsilon_{0123} = -1, \qquad \varepsilon^{\mu\nu\rho\sigma} \varepsilon_{\mu\nu\rho\sigma} = -24. \tag{10.30}$$

With this definition,

$$\tilde{F}^{\mu\nu} = (-\vec{B}, -\vec{E}), \tag{10.31}$$

i.e., the duality transformation acts as

$$\vec{E} \to \vec{B}, \qquad \vec{B} \to -\vec{E}. \tag{10.32}$$

Note that in Minkowski space the dual of the dual field is not the original field; rather

$$\widetilde{\tilde{F}^{\mu\nu}} = \tfrac{1}{2} \varepsilon^{\mu\nu\rho\sigma} \tilde{F}_{\rho\sigma} = -F^{\mu\nu}. \tag{10.33}$$

It is instructive to check our rules of passage of the gauge fields to the Euclidean space formulated in Section 5.2. This transition is $\vec{E} \to -i\hat{\vec{E}}, \ \vec{B} \to \hat{\vec{B}}$. (Here I mark by caret the Euclidean quantities as in Section 5.2). Then the combination $\vec{E} - i\vec{B}$ in Eq. (10.26) takes the form

$$-i\left(\hat{\vec{E}} + \hat{\vec{B}} \right),$$

implying that $F_{\alpha\beta}$ does *not* vanish in the antiselfdual field while $\bar{F}^{\dot\alpha\dot\beta}$ does not vanish in the selfdual field (but vanishes in the antiselfdual field). See also page 522.

We pause here to define two other matrices that are useful in discussing the transformation laws of Weyl spinors with respect to Lorentz rotations. One can combine (10.4) and (10.5) in a unified formula (see Exercise 10.2.1 on page 427) if one introduces[6]

$$(\sigma^{\mu\nu})_\alpha{}^\beta \equiv \frac{1}{4}\left\{(\sigma^\mu)_{\alpha\dot\beta}(\bar\sigma^\nu)^{\dot\beta\beta} - (\sigma^\nu)_{\alpha\dot\beta}(\bar\sigma^\mu)^{\dot\beta\beta}\right\} = \left\{-\frac{1}{2}\vec\sigma, \frac{i}{2}\vec\sigma\right\}_{\alpha\beta},$$

$$(\bar\sigma^{\mu\nu})^{\dot\beta}{}_{\dot\alpha} \equiv \frac{1}{4}\left\{(\bar\sigma^\mu)^{\dot\beta\alpha}(\sigma^\nu)_{\alpha\dot\alpha} - (\bar\sigma^\nu)^{\dot\beta\alpha}(\sigma^\mu)_{\alpha\dot\alpha}\right\} = \left\{\frac{1}{2}\vec\sigma, \frac{i}{2}\vec\sigma\right\}_{\dot\beta\dot\alpha}. \tag{10.34}$$

Note that $\sigma^{\mu\nu}$ must act on left-handed spinors with lower indices, while $\bar\sigma^{\mu\nu}$ acts on right-handed spinors, with upper indices.

Let us now return to the question of constructing Dirac and Majorana spinors from the Weyl spinors. Dirac spinors, also known as bispinors, naturally appear in theories with extended supersymmetry (i.e., those in which the number of conserved supercharges is larger than the minimal number). They can be obtained as follows:

$$\Psi = \begin{pmatrix} \xi_\alpha \\ \bar\eta^{\dot\alpha} \end{pmatrix}. \tag{10.35}$$

More exactly, this is the Dirac bispinor in the Weyl (or spinor) representation. Bispinors in the *standard or Dirac*, and *Majorana* representations will be discussed at the end of this section. Each Dirac spinor requires one left- and one right-handed Weyl spinor. Sometimes, instead of (10.35), the following notation is used:

$$\Psi \equiv \Psi_L + \Psi_R, \qquad \Psi_L = \begin{pmatrix} \xi_\alpha \\ 0 \end{pmatrix}, \qquad \Psi_R = \begin{pmatrix} 0 \\ \bar\eta^{\dot\alpha} \end{pmatrix}. \tag{10.36}$$

The kinetic term for these Weyl spinors,[7]

$$\mathcal{L}_{\text{kin}} = i\bar\xi_{\dot\beta}(\bar\sigma^\mu)^{\dot\beta\alpha}\partial_\mu\xi_\alpha + i\eta^\alpha(\sigma^\mu)_{\alpha\dot\beta}\partial_\mu\bar\eta^{\dot\beta} \tag{10.37}$$

can be rewritten in terms of the Dirac spinor as

$$\mathcal{L}_{\text{kin}} = i\bar\Psi\gamma^\mu\partial_\mu\Psi, \tag{10.38}$$

where

$$\bar\Psi = \Psi^\dagger\gamma^0 \tag{10.39}$$

and

$$\gamma^\mu = \begin{pmatrix} 0 & \sigma^\mu \\ \bar\sigma^\mu & 0 \end{pmatrix} \tag{10.40}$$

> γ *matrices realize Clifford algebra.*

are the Dirac matrices (in the spinor, or Weyl, representation).[8] It is obvious that they satisfy the basic anticommutation relation (the Clifford algebra)

$$\gamma^\mu\gamma^\nu + \gamma^\nu\gamma^\mu = 2g^{\mu\nu} \tag{10.41}$$

---

[6] Remember the definition in Eq. (10.25).

[7] Remember that $\bar\xi_{\dot\beta} = (\xi_\beta)^*$ and $\eta^\alpha = \left(\bar\eta^{\dot\alpha}\right)^*$.

[8] Some signs in our definition differ from those in the popular textbook [24] but coincide with those in M. Peskin and D. Schroeder, *An Introduction to Quantum Field Theory*, [25].

and, in addition, that

$$(\gamma^0)^\dagger = \gamma^0, \qquad (\gamma^i)^\dagger = -\gamma^i, \quad i = 1, 2, 3. \tag{10.42}$$

Often used in applications are the so-called $\Sigma^{\mu\nu}$ matrices,

$$\Sigma^{\mu\nu} = \frac{1}{4}(\gamma^\mu\gamma^\nu - \gamma^\nu\gamma^\mu) = \begin{pmatrix} \sigma^{\mu\nu} & 0 \\ 0 & \bar{\sigma}^{\mu\nu} \end{pmatrix} \tag{10.43}$$

representing the Dirac bispinor analog of the 2×2 matrices in Eq. (10.34). In the Dirac formalism there is a special combination of the γ matrices which plays an important role, the so-called chiral projector. Namely, let us introduce the γ_5 matrix, which anticommutes with all Dirac matrices, i.e., $\gamma_5\gamma^\mu = -\gamma^\mu\gamma_5$, as follows:

$$\gamma_5 \equiv i\gamma^0\gamma^1\gamma^2\gamma^3 = \begin{pmatrix} -1 & 0 \\ 0 & 1 \end{pmatrix}, \qquad (\gamma_5)^2 = 1, \qquad (\gamma_5)^\dagger = \gamma_5. \tag{10.44}$$

The chiral projector is $\frac{1}{2}(1 \pm \gamma_5)$. Indeed,

$$\Psi_L = \tfrac{1}{2}(1 - \gamma_5)\Psi, \qquad \Psi_R = \tfrac{1}{2}(1 + \gamma_5)\Psi. \tag{10.45}$$

If the Dirac spinor Ψ is defined as in Eq. (10.35) then the charge-conjugated Dirac spinor Ψ^C can be defined as [11, 24]

$$\Psi^C = \begin{pmatrix} \eta_\alpha \\ \bar{\xi}^{\dot\alpha} \end{pmatrix}. \tag{10.46}$$

Also widely used are bispinors in the standard (or Dirac) representation and in the Majorana representation. If we denote the former by Ψ_D the transition from Ψ defined in (10.35) to Ψ_D is as follows:

$$\Psi_D = U_{\text{D}\leftarrow\text{W}}\,\Psi, \qquad U_{\text{D}\leftarrow\text{W}} = \frac{1}{\sqrt{2}}\begin{pmatrix} 1 & 1 \\ 1 & -1 \end{pmatrix}. \tag{10.47}$$

It is obvious that the transfer matrix above is unitary. Thus, Ψ_D expressed in terms of ξ and $\bar{\eta}$ (both with the *lower* indices) takes the form

$$\Psi_D = \frac{1}{\sqrt{2}}\begin{pmatrix} \xi + i\sigma^2\bar{\eta} \\ \xi - i\sigma^2\bar{\eta} \end{pmatrix}. \tag{10.48}$$

| Dirac and Majorana bispinors |

The Majorana bispinor in the Weyl representation reduces to

$$\lambda = \begin{pmatrix} \eta_\alpha \\ \bar{\eta}^{\dot\alpha} \end{pmatrix}. \tag{10.49}$$

We see that the Majorana bispinor describes two degrees of freedom and is equivalent to the Weyl two-component spinor. Thus, given a Weyl spinor we can always construct a Majorana bispinor and vice versa. If the Weyl spinor corresponds to a right-handed particle and a left-handed antiparticle, the Majorana bispinor describes a "neutral" particle of both polarizations, coinciding with its antiparticle. Both formalisms, Weyl and Majorana, are used in supersymmetric field theory, though the Weyl formalism is most common. Using Eq. (10.49) it is easy to derive

$$\bar{\lambda}\lambda = \left(\bar{\eta}_{\dot\beta}\bar{\eta}^{\dot\beta} + \eta^\beta\eta_\beta\right), \qquad \bar{\lambda}\gamma_5\lambda = \left(\bar{\eta}_{\dot\beta}\bar{\eta}^{\dot\beta} - \eta^\beta\eta_\beta\right), \tag{10.50}$$

and

$$\bar{\lambda}\gamma^\mu\lambda = 0, \qquad \tfrac{1}{2}\bar{\lambda}\gamma^\mu\gamma^5\lambda = \eta^\alpha(\sigma^\mu)_{\alpha\dot{\beta}}\bar{\eta}^{\dot{\beta}}. \qquad (10.51)$$

The Majorana representation is defined by the requirement that complex conjugated of the Majorana bispinor coincides with itself, $\lambda_M^* = \lambda_M$, while all γ matrices are purely imaginary. If we introduce two auxiliary two-component spinors,[9]

$$a = (1+i)(1+\sigma^2)\eta, \quad b = -(1+i)(1-\sigma^2)\eta, \qquad (10.52)$$

then

$$\lambda_M = \frac{1}{2\sqrt{2}}\begin{pmatrix} a + a^* \\ b + b^* \end{pmatrix}, \quad U_{M\leftarrow D} = \frac{1}{\sqrt{2}}e^{i\pi/4}\begin{pmatrix} 1 & \sigma^2 \\ \sigma^2 & -1 \end{pmatrix}. \qquad (10.53)$$

It is obvious that λ_M is real. For further details see Exercises 10.2.3 and 10.2.4 on page 427 and the solution on pages 691–692.

10.2.2 Two and Three Dimensions

In two dimensions, we have no spatial rotations and only one Lorentz boost. The Dirac spinor is a two-component complex spinor,[10]

$$(\Psi_D)_\alpha = \begin{pmatrix} \Psi_1 \\ \Psi_2 \end{pmatrix}, \qquad \Psi_{1,2}^* \neq \Psi_{1,2}. \qquad (10.54)$$

It is convenient to choose 2×2 γ matrices as follows:

$$(2D) \qquad \gamma^0 = \sigma_2, \qquad \gamma^1 = -i\sigma_1. \qquad (10.55)$$

The chiral projector is the same as in Eq. (10.45), with γ_5 matrix given by

$$\gamma_5 \equiv \gamma^0\gamma^1 = \begin{pmatrix} -1 & 0 \\ 0 & 1 \end{pmatrix}, \qquad (\gamma_5)^2 = 1, \qquad (\gamma_5)^\dagger = \gamma_5. \qquad (10.56)$$

Obviously, chiral spinors in two dimensions have one complex component. Since both γ matrices in Eq. (10.55) are purely imaginary, the Majorana spinors exist too. They have two real components,

$$(\chi_M)_\alpha = \begin{pmatrix} \chi_1 \\ \chi_2 \end{pmatrix}, \qquad \chi_{1,2}^* = \chi_{1,2}. \qquad (10.57)$$

In three dimensions chirality does not exist, as there is no analog of the γ_5 matrix.[11] Three γ matrices with the Clifford algebra can be chosen as follows:

$$(3D) \qquad \gamma^0 = \sigma_y, \qquad \gamma^1 = -i\sigma_x, \qquad \gamma^2 = -i\sigma_z. \qquad (10.58)$$

Thus, in three dimensions the Dirac spinor has two complex components, in much the same way as in two dimensions. Since all three γ matrices in Eq. (10.58) are purely imaginary, one can define a Majorana spinor. Again, as in two dimensions, it has two real components.

---

[9] In Eq. (10.52) η is taken with the lower index, in much the same way as $\bar{\eta}$ in (10.48).
[10] With the conventions to be presented, Ψ_1 is a left-mover while Ψ_2 is a right-mover.
[11] The product $\gamma^0\gamma^1\gamma^2$ reduces to unity. The same statement is valid for any odd number of dimensions.

Exercises

10.2.1 Using the matrices (10.34) write the transformation laws (10.4) and (10.5) in a unified way, setting $U = \exp\left(-\frac{1}{2}\omega_{\mu\nu}\sigma^{\mu\nu}\right)$ and $(U^{-1})^* = \exp\left(-\frac{1}{2}\omega_{\mu\nu}\bar{\sigma}^{\mu\nu}\right)$. Identify the transformation parameters $\omega_{\mu\nu}$ in terms of the parameters in Eqs. (10.4) and (10.5).

10.2.2 Derive a set of 4×4 matrices $\Sigma^{\mu\nu}$ that are analogs of (10.34) when applied to the Dirac spinor, i.e., they should realize six Lorentz rotations

$$\tilde{\Psi} = \exp\left(-\frac{1}{2}\omega_{\mu\nu}\Sigma^{\mu\nu}\right)\Psi, \qquad \Sigma^{\mu\nu} = \begin{pmatrix} \sigma^{\mu\nu} & 0 \\ 0 & \bar{\sigma}^{\mu\nu} \end{pmatrix}. \tag{10.59}$$

10.2.3 Show that the existence of Majorana spinors is in one-to-one correspondence with the fact that it is possible to choose γ matrices obeying the Clifford algebra such that these γ matrices are purely imaginary. Starting from the expression

$$\mathcal{L}_{\text{kin}} = i\bar{\eta}_{\dot{\beta}}(\bar{\sigma}^{\mu})^{\dot{\beta}\alpha}\partial_{\mu}\eta_{\alpha} - \frac{m}{2}\left(\eta^{\alpha}\eta_{\alpha} + \bar{\eta}_{\dot{\alpha}}\bar{\eta}^{\dot{\alpha}}\right) \tag{10.60}$$

and the definition of the Majorana spinor given earlier, find the γ matrices in the Majorana representation corresponding to Eq. (10.53).

10.2.4 Compare three sets of γ matrices (in the Weyl or chiral, Dirac, and Majorana representations) and show that all three sets are unitary equivalent to each other.

10.3 The Coleman–Mandula Theorem

The Coleman–Mandula theorem singles out supersymmetry as the only possible geometric extension of the Poincaré invariance in four-dimensional field theory (it is applicable also in three dimensions but does not apply in two dimensions, as will become clear shortly). In fact, the theorem as originally formulated [26] states that in dynamically nontrivial theories, i.e., those with a nontrivial S matrix, no geometric extensions of the Poincaré algebra are possible. In other words, besides the already known conserved quantities carrying Lorentz indices (the energy–momentum operator P_{μ} and the six Lorentz transformations $M_{\mu\nu}$) no such new conserved quantities can appear. According to the theorem, the only additional conserved charges that are allowed must be Lorentz scalars such as the electromagnetic charge. In 1970 Golfand and Likhtman found [4] a loophole in this theorem: the implicit assumption that all Lorentz indices must be vectorial. This paper of Golfand and Likhtman was entitled "Extension of the algebra of Poincaré group generators and violation of P invariance." They were the first to obtain what is now known as the super-Poincaré algebra in four dimensions.

The essence of the proof of the Coleman–Mandula theorem is simple. Since the original argumentation [26] is not quite transparent, in my presentation I will follow Witten's rendition [27] of the proof.[12]

Let us start from a free field theory. Such a theory can have, besides the energy–momentum tensor, other conserved Lorentz tensors with three or more vectorial indices. For instance, it is easy to check that, for two real fields φ_1 and φ_2 with Lagrangian

$$\mathcal{L} = \partial_\mu \varphi_1 \partial^\mu \varphi_1 + \partial_\mu \varphi_2 \partial^\mu \varphi_2 - m^2 \left(\varphi_1^2 + \varphi_2^2 \right), \tag{10.61}$$

the three-index tensor

$$J_{\mu\rho\sigma} = \partial_\rho \partial_\sigma \varphi_1 \partial_\mu \varphi_2 - \varphi_2 \partial_\rho \partial_\sigma \partial_\mu \varphi_1 \tag{10.62}$$

is transversal with regard to μ, implying the conservation of $Q_{\rho\sigma}$:

$$\dot{Q}_{\rho\sigma} = 0, \qquad Q_{\rho\sigma} = \int d^3x \, J_{0\rho\sigma}.$$

However, there are no Lorentz-invariant interactions which can be added that would preserve this conservation. The basic idea is that the conservation of P_μ and $M_{\mu\nu}$ leaves only the scattering angle unknown in an elastic two-body collision. Additional exotic conservation laws would fix the scattering angle completely, leaving only a discrete set of possible angles. Since we are assuming that the scattering amplitude is an analytic function of angle (assumption number 1) it then must vanish for all angles.

Let us consider a particular example. Assume that we have a conserved *traceless symmetric* tensor $Q_{\mu\nu}$, i.e., $\dot{Q}_{\mu\nu} = 0$. By Lorentz invariance, its matrix element in a one-particle state of momentum p_μ and spin zero is

$$\langle p|Q_{\mu\nu}|p\rangle = \text{const} \times (p_\mu p_\nu - \tfrac{1}{4} g_{\mu\nu} p^2). \tag{10.63}$$

Apply this to an elastic two-body collision of identical particles with incoming momenta p_1, p_2, and outgoing momenta q_1, q_2, assuming that before and after scattering the initial and final particles 1 and 2 are widely separated (assumption number 2: no long-range forces). The matrix element of $Q^{\mu\nu}$ in the two-particle state $|p_1 p_2\rangle$ is then the sum of the matrix elements in the states $|p_1\rangle$ and $|p_2\rangle$. Conservation of the symmetric traceless charge $Q^{\mu\nu}$ together with energy–momentum conservation would yield

$$p_1^\mu + p_2^\mu = q_1^\mu + q_2^\mu,$$
$$p_1^\mu p_1^\nu + p_2^\mu p_2^\nu = q_1^\mu q_1^\nu + q_2^\mu q_2^\nu. \tag{10.64}$$

This would imply, in turn, that the scattering angle vanishes. For the extension of this argument to nonidentical particles, particles with spin, and inelastic collisions, see the original paper [26]. The theorem does not go through in two dimensions because in two dimensions (one time, one space) there are no scattering angles.

As already mentioned, the Coleman–Mandula theorem does not apply to spinorial conserved charges. To elucidate the point let us start from a free theory of a complex scalar and a free two-component (Weyl) fermion,

---

[12] A more technical and thoroughly detailed discussion can be found in Weinberg's textbook [21], pp. 13–22.

$$\mathcal{L} = \partial_\mu \bar{\varphi} \partial^\mu \varphi + i \bar{\psi}_{\dot{\beta}} (\bar{\sigma}^\mu)^{\dot{\beta}\alpha} \partial_\mu \psi_\alpha. \tag{10.65}$$

Note that in supersymmetric field theories, in instances where there is no danger of confusion, the bar symbol is conventionally used to mark Hermitian conjugated fields, i.e., $\bar{\varphi} \equiv \varphi^*$ and $\bar{\psi}_{\dot{\beta}} \equiv (\psi_\beta)^*$. As in the free bosonic case, in this theory one can write a number of spin-3/2, spin-5/2, etc. conserved operators,[13] for instance,

$$J_\alpha^\mu = \partial_\rho \bar{\varphi} (\sigma^\rho \bar{\sigma}^\mu \psi)_\alpha, \tag{10.66}$$

$$\tilde{J}_\alpha^{\mu\nu} = \partial_\rho \bar{\varphi} (\sigma^\rho \bar{\sigma}^\mu \partial^\nu \psi)_\alpha, \tag{10.67}$$

and so on. None of these currents survives the inclusion of nontrivial interactions, except J_α^μ. As we will see later, one can add to Eq. (10.66) appropriate corrections $O(g)$ in such a way that J_α^μ continues to be conserved, say, in a theory with Lagrangian

$$\mathcal{L} = \partial_\mu \bar{\varphi} \partial^\mu \varphi + i \bar{\psi}_{\dot{\beta}} (\bar{\sigma}^\mu)^{\dot{\beta}\alpha} \partial_\mu \psi_\alpha - V, \tag{10.68}$$

$$V = g^2 (\bar{\varphi}\varphi)^2 + \left(g\varphi \psi^\alpha \psi_\alpha + g\bar{\varphi} \bar{\psi}_{\dot{\alpha}} \bar{\psi}^{\dot{\alpha}} \right), \tag{10.69}$$

where g is the coupling constant, which is assumed to be real. At the same time, $\tilde{J}_\alpha^{\mu\nu}$ and higher currents cannot be amended to maintain conservation in the presence of interactions. The conserved supercharges are

$$Q_\alpha \equiv \int d^3x \, J_\alpha^0, \qquad \bar{Q}_{\dot{\alpha}} \equiv \int d^3x \, \bar{J}_{\dot{\alpha}}^0, \qquad \alpha, \dot{\alpha} = 1, 2. \tag{10.70}$$

It can be seen that there are four of these in the case at hand. The loophole in the original Coleman–Mandula theorem is as follows: unlike the conserved bosonic currents, say, $J_{\mu\rho\sigma}$ in Eq. (10.62), the conservation of Q_α and $\bar{Q}_{\dot{\alpha}}$ does not impose constrains on particles' momenta in the scattering processes; rather it relates the various amplitudes for bosons and fermions and, in particular, makes equal the masses of boson–fermion superpartners.

Let us prove that the conservation of higher spinorial currents, such as $\tilde{J}_\alpha^{\mu\nu}$ in Eq. (10.67), is ruled out in nontrivial theories. Unlike bosonic generators, the fermion generators enter in superalgebra with anticommutators rather than commutators. Consider the anticommutator $\{\tilde{Q}_\alpha^\nu, \bar{\tilde{Q}}_{\dot{\alpha}}^\gamma\}$. It cannot vanish since $\tilde{Q}_\alpha^\nu$ is not identically zero and, since $\tilde{Q}_\alpha^\nu$ has components of spin up to 3/2, the above anticommutator has components of spin up to 3. Since the anticommutator is conserved if $\tilde{Q}_\alpha^\nu$ is conserved, and since the Coleman–Mandula theorem does not permit the conservation of any bosonic operator of spin 3 in any interacting theory, $\tilde{Q}_\alpha^\nu$ cannot be conserved.

> *No conserved supercharges beyond spin 1/2*

10.4 Superextension of the Poincaré Algebra

10.4.1 The Poincaré Algebra

The Poincaré algebra includes 10 generators: four components of the energy–momentum operator P_μ and six generators of the Lorentz transformations. Above

---

[13] To be more exact, this family of currents is transversal with regard to μ, namely, $\partial_\mu \tilde{J}^{\mu\nu\cdots} = 0$.

we denoted the generators of spatial rotations as $\vec{L}$ and those of Lorentz boosts as $\vec{N}$. These two triplets can be combined together in a two-index antisymmetric tensor $M^{\mu\nu}$,

$$M^{\mu\nu} = (-\vec{N}, -\vec{L}),\tag{10.71}$$

cf. (10.25). On the left-handed spinors $M^{\mu\nu}$ acts as $M^{\mu\nu} = i\sigma^{\mu\nu}$ while on the right-handed spinors $M^{\mu\nu} = i\bar{\sigma}^{\mu\nu}$. The Poincaré algebra has the form

$$[P_\mu, P_\nu] = 0,$$
$$[M_{\mu\nu}, P_\lambda] = i\left(g_{\nu\lambda}P_\mu - g_{\mu\lambda}P_\nu\right),$$
$$[M_{\mu\nu}, M_{\rho\sigma}] = i\left(g_{\nu\rho}M_{\mu\sigma} + g_{\mu\sigma}M_{\nu\rho} - g_{\mu\rho}M_{\nu\sigma} - g_{\nu\sigma}M_{\mu\rho}\right).\tag{10.72}$$

The generators of the Lorentz transformations contain, generally speaking, two terms: an orbital part and a spin part.

10.4.2 Superextension of the Poincaré Algebra

Let us discuss the simplest supersymmetric extension of (10.72) in four dimensions. I have already mentioned that the minimum number of supergenerators is four. They can be written in the Weyl or Majorana representations. In the Weyl representation we are dealing with supercharges Q_α and $\bar{Q}_{\dot\alpha}$. Since Q_α is a Weyl spinor its transformation properties with respect to the Poincaré group are known,

$$[P_\mu, Q_\alpha] = [P_\mu, \bar{Q}^{\dot\alpha}] = 0,$$
$$[M^{\mu\nu}, Q_\alpha] = i(\sigma^{\mu\nu})_\alpha{}^\beta Q_\beta, \qquad [M^{\mu\nu}, \bar{Q}^{\dot\alpha}] = i(\bar{\sigma}^{\mu\nu})^{\dot\alpha}{}_{\dot\beta}\bar{Q}^{\dot\beta}.\tag{10.73}$$

The matrices $\sigma^{\mu\nu}$ and $\bar{\sigma}^{\mu\nu}$ were defined in (10.34). To close the algebra we need to specify the anticommutators $\{Q_\alpha, \bar{Q}_{\dot\beta}\}$ and $\{Q_\alpha, Q_\beta\}$. Needless to say, for spinorial generators, because of their fermion nature, we must consider anticommutators rather than commutators.

The first anticommutator above can only be proportional to $P_{\alpha\dot\beta}$, since the latter is the only conserved operator with the appropriate Lorentz indices. The standard normalization is as follows:

$$\{Q_\alpha, \bar{Q}_{\dot\beta}\} = 2P_\mu \left(\sigma^\mu\right)_{\alpha\dot\beta} = 2P_{\alpha\dot\beta}.\tag{10.74}$$

Regarding $\{Q_\alpha, Q_\beta\}$, the simplest choice allowed by the Jacobi identities is

$$\{Q_\alpha, Q_\beta\} = \{\bar{Q}_{\dot\alpha}, \bar{Q}_{\dot\beta}\} = 0.\tag{10.75}$$

This is the super-Poincaré algebra first obtained by Golfand and Likhtman [4].

Possible further extensions of the Golfand–Likhtman superalgebra were investigated by Haag, Łopuszański, and Sohnius [28]. They demonstrated that, besides the minimal supersymmetry with four supercharges, one can construct extended supersymmetries, with up to 16 supercharges in four dimensions. The minimal supersymmetry[14] is referred to as $\mathcal{N} = 1$. Correspondingly, one can consider $\mathcal{N} = 2$

---

[14] The very definition of $\mathcal{N} = 1$ depends on the number of dimensions. For instance, in three dimensions the $\mathcal{N} = 1$ supersymmetry has two supercharges rather than four.

> *Thorough discussion is in Chapter 11.*

(eight supercharges) or $N = 4$ (16 supercharges).[15] We will briefly discuss some extended supersymmetries later.

Haag, Łopuszański, and Sohnius also indicated another way of extending the super-Poincaré algebra (10.74) and (10.75), namely, by the inclusion of central charges – elements of the superalgebra commuting with all other generators.[16] The central charges act as numbers whose values depend on the sector of the theory under consideration. They reflect the possible existence of conserved topological currents and topological charges [32]. For instance, if a theory under consideration supports topologically stable domain walls, the right-hand side of (10.75) can be modified as follows:

$$\{Q_\alpha, Q_\beta\} = C_{\alpha\beta}, \tag{10.76}$$

where $C_{\alpha\beta}$ is a triplet of central charges (the number of components in the set is three because $C_{\alpha\beta}$ is obviously symmetric in α, β).[17] Such superalgebras are referred to as *centrally extended*. We will return to studies of centrally extended superalgebras in Sections 10.25, 11.1, 11.3, 11.5.1.1, and 11.6.1.2. Now let us discuss some fundamental consequences of (10.74).

10.4.3 Vanishing of the Vacuum Energy

A basic property discovered at the very early stage of the supersymmetry saga was that in any theory with unbroken supersymmetry the vacuum energy density vanishes. Indeed, let us start from Eq. (10.74) and consider the sum

$$\tfrac{1}{4} \sum_\alpha \left(Q_\alpha (Q_\alpha)^\dagger + (Q_\alpha)^\dagger Q_\alpha \right) = P^0. \tag{10.77}$$

Sandwiching both sides of this equation between the vacuum state we get

$$E_{\text{vac}} = \tfrac{1}{4} \sum_\alpha \left\langle 0 \left| Q_\alpha (Q_\alpha)^\dagger + (Q_\alpha)^\dagger Q_\alpha \right| 0 \right\rangle. \tag{10.78}$$

If the supersymmetry is unbroken then the vacuum is annihilated by supercharges,

$$Q_\alpha |0\rangle = (Q_\alpha)^\dagger |0\rangle = 0, \tag{10.79}$$

implying that $E_{\text{vac}} = 0$. If the supersymmetry is spontaneously broken then $Q_\alpha |0\rangle \neq 0$ or $Q_\alpha^\dagger |0\rangle \neq 0$ and from Eq. (10.78) it is obvious that $E_{\text{vac}} > 0$. Thus, in supersymmetric theories the vacuum energy density is positive definite; the vanishing of the vacuum energy is the *necessary and sufficient* condition for supersymmetry to be valid.

---

[15] In two and three dimensions extended supersymmetries other than $N = 2$ and $N = 4$ exist; see, e.g., [29–31].

[16] For a pedagogical discussion see Section 3 of [10]. In that textbook one can also find super-Lie algebras extensively used in the mathematics and superconformal and super de Sitter algebras appearing in some problems in field theory. An application of superconformal algebra will be discussed in Section 10.19.3. For a detailed consideration of general graded Lie algebras, including super-Jacobi identities, the reader is referred to [21], Section 25.1.

[17] Strictly speaking, the Lorentz transformations act non-trivially on the set of numbers $C_{\alpha\beta}$, in accordance with their Lorentz indices. Therefore, referring to them as "central charges" is a physics jargon. Sometimes they are called brane charges in the literature. Honest-to-god central charges have no Lorentz indices. In what follows I will ignore this subtlety.

10.4.4 Bose–Fermi Degeneracy

In supersymmetric theories, if there is a boson of mass $m > 0$, then a fermion with the same mass must exist too. The degeneracy $m_B = m_F$ follows from the fact that the boson states $|B\rangle$ and fermion states $|F\rangle$ are related through $|F\rangle \sim Q|B\rangle$ or $|F\rangle \sim \bar{Q}|B\rangle$, and the Hamiltonian H commutes with the supercharges Q; hence, if $H|B\rangle = m|B\rangle$ then $H|F\rangle = m|F\rangle$.

10.4.5 Equal Numbers of Bosonic and Fermionic Degrees of Freedom in Every Supermultiplet

To prove that there are equal numbers of bosonic and fermionic degrees of freedom in every supermultiplet, let us note that $P^2 = P_\mu P^\mu$, which is a Casimir operator of the Poincaré algebra, is also the Casimir operator of the superalgebra because

$$[P^2, Q_\alpha] = [P^2, \bar{Q}_{\dot{\alpha}}] = 0. \tag{10.80}$$

Another Poincaré-group Casimir operator can be obtained from the Pauli–Lubanski spin vector W^μ,

Pauli–Lubanski pseudovector

$$W^\mu = \tfrac{1}{2} \varepsilon^{\mu\nu\rho\sigma} P_\nu M_{\rho\sigma}, \tag{10.81}$$

namely

$$W^2 = W_\mu W^\mu = -m^2 \vec{J}^2, \tag{10.82}$$

where m^2 is the mass squared (the eigenvalue of the operator P^2) and the eigenvalue of the angular momentum operator $\vec{J}^2$ is $j(j + 1)$. However, W^2 does not commute with the supercharges, $[W^2, Q_\alpha] \neq 0$, as follows from Eq. (10.73). Thus, massive irreducible superalgebra representations must contain different spins.[18] In four dimensions, in $\mathcal{N} = 1$ theories, we will deal with the two or three subsequent spin values.

To see that the number of bosonic and fermionic states in supermultiplets is equal we observe that Q_α and $\bar{Q}_{\dot{\alpha}}$ each change the fermion number by one unit. Thus, $Q_\alpha|B\rangle = |F\rangle$, $Q_\alpha|F\rangle = |B\rangle$, and the same holds for $\bar{Q}_{\dot{\alpha}}$. Hence the anticommutator $\{Q_\alpha, \bar{Q}_{\dot{\alpha}}\}$ maps the fermionic sector into itself, and the bosonic sector into itself. Owing to (10.74) the same mapping is accomplished by P^μ, which is a one-to-one operator. It follows then that Q_α and $\bar{Q}_{\dot{\alpha}}$ are also one-to-one operators and, hence, the bosonic and fermionic sectors have the same dimensions.

A somewhat more formal proof proceeds as follows. Since Q_α changes the fermion number by one unit, we may write

$$(-1)^{N_f} Q_\alpha = -Q_\alpha (-1)^{N_f}, \tag{10.83}$$

where N_f is the fermion number operator. Now, consider a finite-dimensional supermultiplet R. Then

---

[18] For massless particles $P^2 = 0$ and $W^2 = 0$. Then, instead of spin we must consider helicity; see below. Massless irreducible representations must contain different helicities.

$$\mathrm{Tr}\left((-1)^{N_f}\{Q_\alpha, \bar{Q}_{\dot\alpha}\}\right) = \mathrm{Tr}\left[-Q_\alpha(-1)^{N_f}\bar{Q}_{\dot\alpha} + (-1)^{N_f}\bar{Q}_{\dot\alpha}Q_\alpha\right]$$
$$= \mathrm{Tr}\left[-Q_\alpha(-1)^{N_f}\bar{Q}_{\dot\alpha} + Q_\alpha(-1)^{N_f}\bar{Q}_{\dot\alpha}\right]$$
$$= 0, \tag{10.84}$$

where the cyclic property of the trace is used. Now using the basic anticommutator (10.74), we conclude that

$$\mathrm{Tr}\left[(-1)^{N_f}P_\mu\right] = 0. \tag{10.85}$$

Thus, for the states in the supermultiplet in which the value of P_μ is fixed to be nonvanishing (and one and the same for the given supermultiplet),

$$\mathrm{Tr}\left[(-1)^{N_f}\right] = 0. \tag{10.86}$$

Since $(-1)^{N_f}$ is $+1$ for a bosonic state and -1 for a fermionic state, Eq. (10.86) implies that, for each irreducible supermultiplet,

Fermion–boson degeneracy

$$n_F = n_B. \tag{10.87}$$

This property is very important for understanding why the vacuum energy density vanishes in supersymmetric theories. Indeed, let us consider a free field theory. As is well known, even in a free field theory bosons and fermions contribute to the vacuum energy owing to the zero-point oscillations. The bosonic contribution is

$$\sum_B \sum_{\vec{p}} \sqrt{m_B^2 + \vec{p}^2}, \tag{10.88}$$

where the (divergent) sum runs over all bosonic degrees of freedom and over all spatial momenta $\vec{p}$ and m_B is the mass of a given bosonic mode. The fermionic contribution is

$$-\sum_F \sum_{\vec{p}} \sqrt{m_F^2 + \vec{p}^2}, \tag{10.89}$$

where the sum runs over all fermionic degrees of freedom, and the extra minus sign is associated with the fermion loop. The vanishing vacuum energy requires cancelation, which is only possible if $m_B = m_F$ and the number of degrees of freedom matches inside each supermultiplet. We already know about the mass degeneracy for Bose–Fermi pairs. The argument at the beginning of this subsection proves the match (10.87). It is noteworthy that the cancelation of the vacuum energy density under these conditions was mentioned as early as the 1940s by Pauli [33].

10.4.6 Building Supermultiplets

Since the group corresponding to the Poincaré algebra is not compact, all its unitary representations (except the trivial representation) are infinite dimensional. This infinite dimensionality simply corresponds to the familiar fact that particle states are labeled by the continuous parameters P^μ, their 4-momenta. Finite dimensional representations can be organized using a trick invented by Wigner, the so-called *little group*; this is the group of (usually compact) transformations remaining after "freezing out" some of the noncompact transformations in a certain conventional way.

In the present case, the noncompact part of the Poincaré group is comprised of boosts and translations. For massive particles, we can use a boost to pass to a frame in which the particle is at rest,

$$P_\mu = (m, 0, 0, 0). \tag{10.90}$$

The little group in this case is just those Lorentz transformations that preserve the 4-vector P^μ, namely the group of spatial rotations SO(3). Thus massive particles belong to representations of SO(3) labeled by the spin j, which can be either integer (for bosons) or half-integer (for fermions). Any given spin-j representation is $(2j+1)$ dimensional with states $|j, j_z\rangle$ labeled by j_z, where

$$j_z = -j, -j + 1, \ldots, j - 1, j. \tag{10.91}$$

Massless states can be classified in a similar manner except that now, instead of the rest frame, we choose a frame in which

$$P_\mu = (E, 0, 0, E) \tag{10.92}$$

with a given (and fixed) value of E. This choice leaves the freedom of SO(2) rotations in the xy plane. All representations of SO(2) are one dimensional and are labeled by a single eigenvalue, the helicity λ, which measures the projection of the angular momentum onto the direction of motion (the z axis in the present case). As we know, λ is constrained. Since the helicity is the eigenvalue of the generator of rotations around the z axis, a rotation by an angle φ around that axis produces a phase $e^{i\lambda\varphi}$. The full 2π rotation results in $e^{2\pi i\lambda}$. This phase must reduce to 1 for bosons and -1 for fermions, implying that λ is integer for bosons and half-integer for fermions.

Now we will establish the particle content of supermultiplets. Let us start with a massive particle state $|a\rangle$ in its rest frame (10.90). For this state, the supersymmetry algebra (10.74) becomes (remembering that $\bar{Q}_{\dot\beta} = (Q_\beta)^\dagger$)

$$\{Q_\alpha, (Q_\beta)^\dagger\} = 2m\delta_{\alpha\beta}, \qquad \{Q_\alpha, Q_\beta\} = 0, \qquad \{\bar{Q}_{\dot\alpha}, \bar{Q}_{\dot\beta}\} = 0, \tag{10.93}$$

where $\alpha, \beta = 1, 2$. Representations of this algebra are easy to construct, since essentially it is the algebra of two creation and annihilation operators (up to a rescaling of Q by $\sqrt{2m}$). If we assume that Q_α annihilates a state $|a\rangle$, i.e., $Q_\alpha|a\rangle = 0$, then we find the following four-dimensional representation:[19]

$$|a\rangle, \qquad (Q_1)^\dagger|a\rangle, \qquad (Q_2)^\dagger|a\rangle, \qquad (Q_1)^\dagger(Q_2)^\dagger|a\rangle. \tag{10.94}$$

Suppose that $|a\rangle$ is a spin-j particle. The $(Q_\beta)^\dagger$ operators are doublets with respect to the right-handed rotation representing spin $\frac{1}{2}$. Thus, the states $(Q_\beta)^\dagger|a\rangle$, by the rule for the addition of angular momenta, have spins $j+\frac{1}{2}$ and $j-\frac{1}{2}$ if $j \neq 0$ while for $j = 0$ they have only spin $\frac{1}{2}$. The operator $(Q_1)^\dagger(Q_2)^\dagger$ transforms as a singlet of the right-handed rotations (ie. $j = 0$). Therefore, the state $(Q_1)^\dagger(Q_2)^\dagger|a\rangle$ has the same spin j as $|a\rangle$. Thus, if we start from a spinless particle, the corresponding supermultiplet

---

[19] Generally speaking, the last term in (10.94) could have been written as $(Q_\alpha)^\dagger(Q_\beta)^\dagger|a\rangle$. However, the combination symmetric in the spinorial indices vanishes because of (10.93). The antisymmetric spin-0 combination survives and reduces to $(Q_1)^\dagger(Q_2)^\dagger|a\rangle$. The product of three Qs is always reducible, by virtue of (10.93), to a linear combination of Qs.

contains two spin-0 bosons and one spin-$\frac{1}{2}$ (Weyl) fermion. If we start from the $j \neq 0$ bosonic state $|a\rangle$, the corresponding supermultiplet has $2(2j + 1)$ bosonic states and the following numbers of Weyl-fermionic states:

$$2\left(j - \frac{1}{2}\right) + 1 \quad \text{and} \quad 2\left(j + \frac{1}{2}\right) + 1. \tag{10.95}$$

If we start from the $j \neq 0$ fermionic state $|a\rangle$, the corresponding supermultiplet has $2(2j + 1)$ Weyl-fermionic states while the structure of the bosonic states is the same as in (10.95). As anticipated, the total number of boson degrees of freedom always matches that of the fermion degrees of freedom.

I pause here to give two examples that will be used frequently in what follows. For massive particles we can have (i) the massive chiral multiplet with spins $j = \{0, 0, \frac{1}{2}\}$ corresponding to massive complex scalar and Weyl fermion fields $\{\phi, \psi_\alpha\}$ and (ii) the massive vector multiplet with $j = \{0, \frac{1}{2}, \frac{1}{2}, 1\}$ with massive field content $\{h, \psi_\alpha, \lambda_\alpha, A_\mu\}$, where h is a real scalar field. In terms of degrees of freedom, it is clear that the massive vector multiplet has the same number as a massless chiral multiplet plus a massless vector multiplet (see below). This is indeed the case dynamically: massive vector multiplets arise as a supersymmetric analog of the Higgs mechanism.

The super-Higgs mechanism is discussed in Section 10.9.

For massless particles we choose the reference frame (10.92). The superalgebra (10.74) then reduces to

$$\{Q_\alpha, (Q_\beta)^\dagger\} = 4E \begin{pmatrix} 1 & 0 \\ 0 & 0 \end{pmatrix}. \tag{10.96}$$

This implies that Q_2 and $(Q_2)^\dagger$ vanish for all massless representations. Let us denote by $|b\rangle$ the initial state annihilated by Q_1. Then it is readily seen that the massless supermultiplets are just two dimensional, containing

$$|b\rangle \quad \text{and} \quad (Q_1)^\dagger|b\rangle. \tag{10.97}$$

If $|b\rangle$ has helicity λ then $(Q_1)^\dagger|b\rangle$ has helicity $\lambda + \frac{1}{2}$.

By *CPT* invariance, such a multiplet will always appear in field theory with its opposite helicity multiplet $\{-\lambda, -\lambda - \frac{1}{2}\}$.

For massless particles, we will be interested in the chiral supermultiplet with helicities given by

$$\lambda = \{-\tfrac{1}{2}, 0, 0, \tfrac{1}{2}\}. \tag{10.98}$$

The corresponding degrees of freedom are associated with a complex scalar and a Weyl (or Majorana) fermion. We will be interested also in the vector multiplet with helicities

$$\lambda = \{-1, -\tfrac{1}{2}, \tfrac{1}{2}, 1\}. \tag{10.99}$$

Here the corresponding degrees of freedom are associated with a vector gauge boson and a Majorana fermion.

Other massless supersymmetry multiplets contain fields with spin $\frac{3}{2}$ or greater and are relevant in supergravity, a theory which will not be considered here.

Chiral multiplets are the supersymmetric analogs of matter fields, while vector multiplets are analogs of gauge fields. The conventional terminology is as follows: the fermions in the chiral multiplets are referred to as *quarks* and their scalar superpartners as *squarks*; the fermionic superpartners of gauge bosons are termed *gauginos*.

Exercises

10.4.1 Using the Jacobi identities show that, for instance, $[P^\mu, Q_\alpha]$ cannot be proportional to $(\sigma^\mu)_{\alpha\dot\beta} \bar{Q}^{\dot\beta}$; it must vanish.
Hint. Consider the Jacobi identities for P_μ, $M_{\nu\lambda}$, and Q_α.

10.4.2 Rewrite the four supercharges and the above superalgebra in the Majorana representation.

10.5 Superspace and Superfields

10.5.1 Superspace

Field theory presents a conventional formalism for describing the relativistic quantum mechanics of an (infinitely) large number of degrees of freedom. The basic building blocks of this formalism are fields of spin 0, $\frac{1}{2}$, and 1 that depend locally on the space–time point x^μ. With supersymmetry it is very natural to expand the concept of space–time to the concept of *superspace*. The energy–momentum operator generates translations in four-dimensional space–time, so it is natural that anticommuting supercharges should generate "super" translations in an anticommuting space. This breakthrough idea was pioneered by Salam and Strathdee [34].

Thus, a linear realization of supersymmetry is achieved by enlarging space–time to include four anticommuting variables θ^α and $\bar\theta^{\dot\alpha}$ representing the "quantum" or "fermionic" dimensions of superspace. The advantages of this formalism are immediately obvious: superspace allows a simple and explicit description of the action of supersymmetry on the component fields and provides a very efficient method of constructing superinvariant Lagrangians.

A finite element of the group corresponding to the $\mathcal{N} = 1$ superalgebra (10.74), (10.75) can be written as

$$G(x^\mu, \theta, \bar\theta) = \exp\left[i\left(\theta Q + \bar\theta\bar{Q} - x^\mu P_\mu\right)\right], \tag{10.100}$$

where θ^α and $\bar\theta^{\dot\beta} \equiv (\theta^\beta)^*$ are Grassmann variables,[20]

$$\{\theta^\alpha, \theta^\beta\} = \{\bar\theta^{\dot\alpha}, \bar\theta^{\dot\beta}\} = \{\theta^\alpha, \bar\theta^{\dot\beta}\} = 0,$$

$$\left\{\frac{\partial}{\partial\theta^\alpha}, \frac{\partial}{\partial\theta^\beta}\right\} = \left\{\frac{\partial}{\partial\bar\theta^{\dot\alpha}}, \frac{\partial}{\partial\bar\theta^{\dot\beta}}\right\} = \left\{\frac{\partial}{\partial\theta^\alpha}, \frac{\partial}{\partial\bar\theta^{\dot\beta}}\right\} = 0. \tag{10.101}$$

---

[20] The Leibniz rule for the Grassmann derivative is

$$(\partial/\partial\theta^\alpha)\,\theta^\beta\theta^\gamma = (\partial\theta^\beta/\partial\theta^\alpha)\theta^\gamma - \theta^\beta\,(\partial\theta^\gamma/\partial\theta^\alpha).$$

We want to construct a linear representation of the group whose elements are parametrized in Eq. (10.100). This can be done by considering the action of the group elements (10.100) on the superspace

$$\{x^\mu, \theta^\alpha, \bar{\theta}^{\dot{\alpha}}\} \tag{10.102}$$

in the following way. It is not difficult to show that

$$G(x^\mu, \theta, \bar{\theta})\, G(a^\mu, \epsilon, \bar{\epsilon}) = G(x^\mu + a^\mu + i\epsilon\sigma^\mu\bar{\theta} - i\theta\sigma^\mu\bar{\epsilon}, \theta + \epsilon, \bar{\theta} + \bar{\epsilon}). \tag{10.103}$$

To prove this equality we can use the Hausdorff formula

$$e^A e^B = (\exp A + B + \tfrac{1}{2}[A, B] + \cdots) \tag{10.104}$$

and take into account the fact that the series on the right-hand side terminates at the first commutator for the group elements considered here. Thus, the (super)coordinate transformations

$$\begin{aligned} &\{x^\mu, \theta^\alpha, \bar{\theta}^{\dot{\alpha}}\} \longrightarrow \{x^\mu + \delta x^\mu, \theta^\alpha + \delta\theta^\alpha, \bar{\theta}^{\dot{\alpha}} + \delta\bar{\theta}^{\dot{\alpha}}\}, \\ &\delta\theta^\alpha = \epsilon^\alpha, \qquad \delta\bar{\theta}^{\dot{\alpha}} = \bar{\epsilon}^{\dot{\alpha}}, \qquad \delta x_{\alpha\dot{\alpha}} = -2i\theta_\alpha\bar{\epsilon}_{\dot{\alpha}} - 2i\bar{\theta}_{\dot{\alpha}}\epsilon_\alpha \end{aligned} \tag{10.105}$$

add supersymmetry to the translational and Lorentz transformations.[21]

Two invariant subspaces, $\{x_L^\mu, \theta^\alpha\}$ and $\{x_R^\mu, \bar{\theta}^{\dot{\alpha}}\}$, are spanned by half of the Grassmann coordinates:

Two invariant (chiral) subspaces of the superspace	$$\begin{aligned} \{x_L^\mu, \theta^\alpha\}: \qquad & \delta\theta^\alpha = \epsilon^\alpha, \qquad \delta(x_L)_{\alpha\dot{\alpha}} = -4i\theta_\alpha\bar{\epsilon}_{\dot{\alpha}}, \\ \{x_R^\mu, \bar{\theta}^{\dot{\alpha}}\}: \qquad & \delta\bar{\theta}^{\dot{\alpha}} = \bar{\epsilon}^{\dot{\alpha}}, \qquad \delta(x_R)_{\alpha\dot{\alpha}} = -4i\bar{\theta}_{\dot{\alpha}}\epsilon_\alpha, \end{aligned} \tag{10.106}$$

where

$$\begin{aligned} (x_L)_{\alpha\dot{\alpha}} &= x_{\alpha\dot{\alpha}} - 2i\theta_\alpha\bar{\theta}_{\dot{\alpha}}, \\ (x_R)_{\alpha\dot{\alpha}} &= x_{\alpha\dot{\alpha}} + 2i\theta_\alpha\bar{\theta}_{\dot{\alpha}}. \end{aligned} \tag{10.107}$$

Sometimes it is more convenient to use vectorial notation. Then

$$x_L^\mu = x^\mu - i\theta^\alpha(\sigma^\mu)_{\alpha\dot{\alpha}}\bar{\theta}^{\dot{\alpha}}, \qquad x_R^\mu = x^\mu + i\theta^\alpha(\sigma^\mu)_{\alpha\dot{\alpha}}\bar{\theta}^{\dot{\alpha}}. \tag{10.108}$$

Readers with a more advanced mathematical background might like to note the following. Ordinary space–time can be defined as the coset space obtained as

(Poincaré group)/(Lorentz group).

By the same token superspace can be defined as the coset space

(super-Poincaré group)/(Lorentz group).

The points of the latter are orbits obtained by the action of the Lorentz group in the super-Poincaré group. If we choose a certain point as the origin then the superspace can be parametrized by (10.100).

10.5.2 Superfields

In conventional field theory we are dealing with *fields*, that are scalar, spinor, or vector functions of the coordinates x^μ. In supersymmetric theories we are

---

[21] To derive the last equation in (10.105), use the definition (10.16).

dealing, rather, with *superfields* [34, 35], which are functions of the coordinates on superspace. Expanding the superfields in powers of the supervariables θ^α and $\bar\theta^{\dot\alpha}$, we get a set of regular fields. This set is finite since the square of a given Grassmann parameter vanishes. Thus the highest term in the expansion in Grassmann parameters is $\theta^2\bar\theta^2 \equiv \theta^\alpha\theta_\alpha\bar\theta_{\dot\alpha}\bar\theta^{\dot\alpha}$.

The most general superfield with no external indices is

$$S(x,\theta,\bar\theta) = \phi + \theta\psi + \bar\theta\bar\chi + \theta^2 F + \bar\theta^2 G + \theta^\alpha A_{\alpha\dot\beta}\bar\theta^{\dot\beta}$$
$$+ \theta^2(\bar\theta\bar\lambda) + \bar\theta^2(\theta\rho) + \theta^2\bar\theta^2 D, \tag{10.109}$$

where $\phi, \psi, \bar\chi, \ldots, D$ depend only on x^μ and are referred to as the *component fields*.

Superfields form linear representations of superalgebra. In general, however, these representations are highly reducible. We need to eliminate extra component fields by imposing covariant constraints. In other words, superfields shift the problem of finding supersymmetry representations to that of finding appropriate constraints. Note that we must reduce superfields without restricting their x-dependence, for instance using differential equations in x space.

As an example let us inspect Eq. (10.109). It is easy to see that it gives a reducible representation of the supersymmetry algebra. If all the fields in (10.109) were propagating and ϕ had spin j (assuming it to be massive) then there would be component fields with spins $j, j \pm \frac{1}{2}$, and $j \pm 1$, which is larger than the irreducible supermultiplets found in Section 10.4.6. To get an irreducible field representation we must impose a constraint on the superfield that (anti)commutes with the supersymmetry algebra. One such constraint is simply the reality condition $S^\dagger = S$, which leads to a vector superfield that can be parametrized as follows:

| *Reducible vs. irreducible representations* |

$$V(x,\theta,\bar\theta) = C + i\theta\chi - i\bar\theta\bar\chi + \frac{i}{\sqrt{2}}\theta^2 M - \frac{i}{\sqrt{2}}\bar\theta^2\bar M$$
$$- 2\theta^\alpha\bar\theta^{\dot\alpha}v_{\alpha\dot\alpha} + \left[2i\theta^2\bar\theta_{\dot\alpha}\left(\bar\lambda^{\dot\alpha} - \frac{i}{4}\partial^{\dot\alpha\alpha}\chi_\alpha\right) + \text{H.c.}\right]$$
$$+ \theta^2\bar\theta^2\left(D - \frac{1}{4}\partial^2 C\right), \tag{10.110}$$

where

$$\partial^{\dot\alpha\alpha} = (\bar\sigma^\mu)^{\dot\alpha\alpha}\,\partial_\mu. \tag{10.111}$$

The superfield V is real, $V = V^\dagger$, implying that the bosonic fields C, D, and $v^\mu = \frac{1}{2}(\bar\sigma^\mu)^{\dot\alpha\alpha}v_{\alpha\dot\alpha}$ are real. The other fields are complex, and the bar denotes, as usual, complex conjugation. As we will see in Section 10.6.8, (super)gauge freedom will eliminate the unwanted components $C, \chi, \bar\chi, M$, and $\bar M$, reducing the physical content of V to that of (10.99), namely, $V(x,\theta,\bar\theta) \rightarrow -2\theta^\alpha\bar\theta^{\dot\alpha}v_{\alpha\dot\alpha} + \left\{2i\theta^2\bar\theta\bar\lambda - 2i\bar\theta^2\theta\lambda\right\} + \theta^2\bar\theta^2 D$. This will allow us to use a vector superfield in constructing supersymmetric gauge theories.

At first sight, the parametrization (10.110) might seem contrived. Why not drop $\partial^2 C$ and $\partial\chi$ in the last and last but one terms? This is always possible by redefining D and $\bar\lambda$. The reason behind this particular parametrization of the vector superfield will become clear in Section 10.6.8, however.

The transformations (10.105) generate supertransformations of the fields, which can be written as

$$\delta S = i \left(\epsilon Q + \bar{\epsilon} \bar{Q} \right) S, \tag{10.112}$$

(cf. Eq. (10.100)), where S is a generic superfield, which can be a vector superfield, or a *chiral superfield*; see below. In this way the supercharges Q and $\bar{Q}$ can be defined as differential operators acting in superspace,[22]

$$Q_\alpha = -i\frac{\partial}{\partial\theta^\alpha} + \bar{\theta}^{\dot\alpha}\partial_{\alpha\dot\alpha}, \qquad \bar{Q}_{\dot\alpha} = i\frac{\partial}{\partial\bar{\theta}^{\dot\alpha}} - \theta^\alpha\partial_{\alpha\dot\alpha}, \qquad \left\{Q_\alpha, \bar{Q}_{\dot\alpha}\right\} = 2i\partial_{\alpha\dot\alpha}. \tag{10.113}$$

These differential operators give an explicit realization of the supersymmetry algebra, Eqs. (10.74), (10.75), and (10.73), where $P_{\alpha\dot\alpha} = i\partial_{\alpha\dot\alpha}$.

It is also possible to introduce *superderivatives*. They are defined as differential operators anticommuting with Q_α and $\bar{Q}_{\dot\alpha}$,

$$D_\alpha = \frac{\partial}{\partial\theta^\alpha} - i\bar{\theta}^{\dot\alpha}\partial_{\alpha\dot\alpha}, \qquad \bar{D}_{\dot\alpha} = -\frac{\partial}{\partial\bar{\theta}^{\dot\alpha}} + i\theta^\alpha\partial_{\alpha\dot\alpha}, \qquad \left\{D_\alpha, \bar{D}_{\dot\alpha}\right\} = 2i\partial_{\alpha\dot\alpha}. \tag{10.114}$$

Superderivatives allow us to impose constraints on superfields. Instead of the reality condition $S^\dagger = S$ leading to the vector superfield V we can impose so-called *chiral* (or *antichiral*) superfield constraints [36],

$$\bar{D}_{\dot\alpha}\Phi = 0 \qquad \text{or} \qquad D_\alpha\bar{\Phi} = 0. \tag{10.115}$$

The definitions of the covariant superderivatives above and the (anti)chiral coordinates (10.107) and (10.108) are not independent. In fact,

$$\bar{D}_{\dot\alpha}x^\mu_L = 0, \qquad D_\alpha x^\mu_R = 0. \tag{10.116}$$

Moreover, in the chiral subspace $\{x^\mu_L, \theta^\alpha\}$ the superderivatives $\bar{D}_{\dot\alpha}$ and D_α are realized as

$$\bar{D}_{\dot\alpha} = -\frac{\partial}{\partial\bar{\theta}^{\dot\alpha}}, \qquad D_\alpha = \frac{\partial}{\partial\theta^\alpha} - 2i\bar{\theta}^{\dot\alpha}\partial_{\alpha\dot\alpha}; \tag{10.117}$$

similar expressions are valid in the second subspace, $\{x^\mu_R, \bar{\theta}^{\dot\beta}\}$. This immediately leads us to solutions of the superfield constraints (10.115). For example, the chiral superfield (in the chiral basis) does not depend on $\bar{\theta}^{\dot\alpha}$:

$$\Phi(x_L, \theta) = \phi(x_L) + \sqrt{2}\theta^\alpha\psi_\alpha(x_L) + \theta^2 F(x_L). \tag{10.118}$$

Thereby the chiral superfield Φ (or antichiral $\bar{\Phi}$) describes the minimal supermultiplet which includes one complex scalar field $\phi(x)$ (two bosonic states) and one complex Weyl spinor $\psi^\alpha(x)$, $\alpha = 1, 2$ (two fermionic states). The F term is an auxiliary component since the F field is nonpropagating. As we will see shortly, this field will appear in Lagrangians without a kinetic term. Chiral superfields are used for constructing the matter sectors of various theories.

It is not difficult to see that the constraints (10.115) are self-consistent and give rise to irreducible representations of the superalgebra. The consistency of (10.115) is

---

[22] Note that I have introduced, in accordance with (10.16), the derivative $\partial_{\alpha\dot\alpha} = (\sigma^\mu)_{\alpha\dot\alpha}\,\partial_\mu \equiv 2\partial/\partial x^{\alpha\dot\alpha}$.

explained by the fact that the operators D_α and $\bar{D}_{\dot\alpha}$ anticommute with the generators Q and $\bar{Q}$ of the supersymmetry algebra. Therefore D_α and $\bar{D}_{\dot\alpha}$ commute with the combination $\epsilon Q + \bar{\epsilon}\bar{Q}$ appearing in supertransformations.

10.5.3 Properties of Superfields

It is easy to verify that linear combinations of superfields are themselves superfields. Similarly, products of superfields are superfields because Q and $\bar{Q}$ can be viewed as linear differential operators. Given a superfield, we can use the space–time derivatives to generate a new one. At the same time, the Grassmann derivatives $\partial/\partial\theta^\alpha$ and $\partial/\partial\bar{\theta}^{\dot\alpha}$, when applied to a superfield, do not produce a superfield. We can use the covariant superderivatives D_α and $\bar{D}_{\dot\alpha}$ to construct irreducible representations of the supersymmetry (new superfields); for instance, $\bar{D}^2 D_\alpha V$ is a chiral superfield that will play an important role below. Note, however, that $D_\alpha\Phi$ (with Φ chiral) is *not* a chiral superfield. Indeed,

$$\bar{D}_{\dot\alpha}(D_\alpha\Phi) = \left\{\bar{D}_{\dot\alpha}, D_\alpha\right\}\Phi = 2i\partial_{\alpha\dot\alpha}\Phi \neq 0.$$

However, $D^2\Phi$ is an *anti*chiral superfield since $D_\alpha(D^2\Phi) = 0$.

Acting with $\bar{D}^2$ or D^2 on a generic superfield, we get a chiral or antichiral superfield, respectively.

10.5.4 Supertransformations of the Component Fields

Let us start from the component expansion (10.118) of the chiral superfield. The supertransformations (10.106) imply that

$$\begin{aligned}
\Phi + \delta\Phi &= \phi(x_L + \delta x_L) + \sqrt{2}(\theta^\alpha + \delta\theta^\alpha)\psi_\alpha(x_L + \delta x_L) \\
&\quad + (\theta^\alpha + \delta\theta^\alpha)\,(\theta_\alpha + \delta\theta_\alpha)F(x_L + \delta x_L) \\
&= \phi(x_L) + \left[\partial_\mu\phi(x_L)\right] 2i\bar{\epsilon}_{\dot\alpha}\,(\bar{\sigma}^\mu)^{\dot\alpha\alpha}\,\theta_\alpha \\
&\quad + \sqrt{2}\theta^\alpha\psi_\alpha(x_L) + \sqrt{2}\epsilon^\alpha\psi_\alpha(x_L) + \sqrt{2}\theta^\alpha\left[\partial_\mu\psi_\alpha(x_L)\right]2i\bar{\epsilon}_{\dot\alpha}\,(\bar{\sigma}^\mu)^{\dot\alpha\beta}\,\theta_\beta \\
&\quad + \theta^2 F(x_L) + 2\theta^\alpha\epsilon_\alpha F(x_L),
\end{aligned} \tag{10.119}$$

Supertrans-
formations of
the
component
fields from
the chiral
superfield

where we have kept only terms linear in the supertransformation parameters ϵ and $\bar{\epsilon}$.

Let us have a closer look at the above decomposition. Comparing the terms with the same powers of θ, we arrive at the following supertransformations for the component fields:

$$\begin{aligned}
\delta\phi &= \sqrt{2}\epsilon^\alpha\psi_\alpha, \\
\delta\psi_\alpha &= -\sqrt{2}i\partial_{\alpha\dot\alpha}\phi\bar{\epsilon}^{\dot\alpha} + \sqrt{2}\epsilon_\alpha F, \\
\delta F &= i\sqrt{2}\left(\partial_{\alpha\dot\alpha}\psi^\alpha\right)\bar{\epsilon}^{\dot\alpha}.
\end{aligned} \tag{10.120}$$

Here we have used the identity $\bar{\epsilon}\bar{\sigma}^\mu\theta = -\theta\sigma^\mu\bar{\epsilon}$ and the standard convention for spinorial index convolution; see Eq. (10.11).

Needless to say, the transformation laws for the component fields of $\bar{\Phi}$ follow from (10.120) by Hermitian conjugation. Note that the last component of Φ transforms through a total derivative, $\delta F \sim \partial\psi$. This property is of paramount importance for the construction of supersymmetric theories.

Table 10.1. Lorentz spins of component field	
Lorents spin	Component field
$(0,0)$	scalar ϕ
$(\frac{1}{2},0)$	spinor ψ_α
$(0,\frac{1}{2})$	spinor $\bar{\psi}_{\dot{\alpha}}$
$(\frac{1}{2},\frac{1}{2})$	vector $A_{\alpha\dot{\alpha}}$
$(1,0)$	tensor $F_{\alpha\beta} \sim F^{\mu\nu} - i\tilde{F}^{\mu\nu}$ i.e. $\vec{E} - i\vec{B}$
$(0,1)$	tensor $F^{\dot{\alpha}\dot{\beta}} \sim F^{\mu\nu} + i\tilde{F}^{\mu\nu}$ i.e. $\vec{E} + i\vec{B}$

The above procedure can be repeated for the vector superfield (10.110). We will not do it here; the corresponding algebra is rather cumbersome (see Exercise 10.5.3 at the end of this section). Vector superfields will be used below for the description of gauge fields. As was already mentioned, a judicious supergauge choice (the so-called Wess–Zumino gauge, see Section 10.6.8 below), allows one to eliminate completely the components $C, \chi, \bar{\chi}, M$, and $\bar{M}$, of the vector superfield. Only the supertransformation for the last component of V will be of importance for us now, namely

$$\delta D = \epsilon^\alpha \left(\partial_{\alpha\dot{\beta}} \bar{\lambda}^{\dot{\beta}} \right) + \left(\lambda^\beta \overleftarrow{\partial}_{\beta\dot{\alpha}} \right) \bar{\epsilon}^{\dot{\alpha}}. \tag{10.121}$$

As in the case of the chiral superfield, the last component of the vector superfield is transformed through a total derivative. It is clear now that this is a general property. Let us remember this general feature. We will return to it when constructing superinvariant actions in subsequent sections, for instance, in Section 10.6.

For convenience, I list in Table 10.1 all the component fields with which we will be dealing in what follows.

Exercises

10.5.1 Prove the equalities in (10.116). Find $\bar{D}_{\dot{\alpha}} x_R$ and $D_\alpha x_L$.

10.5.2 Show that

$$\{D_\alpha, Q_\beta\} = \{D_\alpha, \bar{Q}_{\dot{\beta}}\} = \{\bar{D}_{\dot{\alpha}}, Q_\beta\} = \{\bar{D}_{\dot{\alpha}}, \bar{Q}_{\dot{\beta}}\} = 0. \tag{10.122}$$

10.5.3 Write the supersymmetry transformations (in components) for the vector superfield (10.110).

10.6 Superinvariant Actions

In this section I will explain how, using the superfield formalism, one can construct superinvariant actions describing all the variety of supersymmetric models.

10.6.1 Rules of Grassmann Integration

It is very easy to tabulate all possible integrals over the Grassmann variables, also known as Berezin integrals [37]. Assume that we have a set of Grassmann variables $\theta_i (i = 1, 2, \ldots)$. Then

$$\int d\theta_i = 0, \qquad \int d\theta_i\, \theta_j = \delta_{ij}. \tag{10.123}$$

Normalization of the Grassmann integrals

A two-fold integral is to be understood as a product of integrals, etc. Usually we work with integrals over all Grassmann variables in the given superspace (or its invariant subspace), for instance

$$\int d^4\theta \equiv \int d^2\theta\, d^2\bar{\theta} \to \int d\theta_1\, d\theta_2\, d\bar{\theta}_{\dot{1}}\, d\bar{\theta}_{\dot{2}}. \tag{10.124}$$

We will normalize the integral $\int d^2\theta\, d^2\bar{\theta}$ in such a way that

$$\int \theta^2 \bar{\theta}^2\, d^2\theta\, d^2\bar{\theta} = 1. \tag{10.125}$$

Integrals over the chiral subspaces will be normalized as follows:

$$\int \theta^2\, d^2\theta = 1, \qquad \int \bar{\theta}^2\, d^2\bar{\theta} = 1. \tag{10.126}$$

While the Grassmann variables θ and $\bar{\theta}$ have dimension $[\text{length}]^{1/2}$, the differentials $d\theta$ and $d\bar{\theta}$ have dimension $[\text{length}]^{-1/2}$. If c is a number, then $d(c\theta) = c^{-1} d\theta$. This follows from the second equation in (10.123).

10.6.2 Kinetic Terms for Matter Fields

If we have a vector superfield V, its last component D is the coefficient in front of $\theta^2\bar{\theta}^2$. Equations (10.123) and (10.125) then imply that $D = \int d^4\theta V$. Since the change in D under supertransformations is a full derivative, see Eq. (10.121), the action

$$S = \int d^4x\, d^4\theta\, V(x, \theta, \bar{\theta}) \tag{10.127}$$

is superinvariant. The Lagrangian

$$\mathcal{L} = \int d^4\theta\, V(x, \theta, \bar{\theta}) \tag{10.128}$$

is superinvariant up to a total derivative. Let us see how one can exploit this to construct the kinetic terms of the matter fields. If Φ and $\bar{\Phi}$ are chiral and antichiral superfields, respectively, their product is a vector superfield. As this is our first encounter with a product superfield it will be helpful to write out the components of this product:

$$\bar{\Phi}\Phi = \left[\bar{\phi} + i\theta^\alpha(\partial_{\alpha\dot\alpha}\bar{\phi})\bar{\theta}^{\dot\alpha} - \tfrac{1}{4}\theta^2\bar{\theta}^2\partial^2\bar{\phi} + \sqrt{2}\bar{\theta}\bar{\psi}\tfrac{1}{\sqrt{2}}i\bar{\theta}^2(\theta^\alpha\partial_{\alpha\dot\alpha}\bar{\psi}^{\dot\alpha}) + \bar{\theta}^2\bar{F} \right]$$

$$\times \left[\phi - i\theta^\alpha(\partial_{\alpha\dot\alpha}\phi)\bar{\theta}^{\dot\alpha} - \tfrac{1}{4}\theta^2\bar{\theta}^2\partial^2\phi + \sqrt{2}\theta\psi + \tfrac{1}{\sqrt{2}}i\theta^2(\psi^\alpha\overleftarrow{\partial}_{\alpha\dot\alpha}\bar{\theta}^{\dot\alpha}) + \theta^2 F \right]$$

$$= \bar{\phi}\phi + \sqrt{2}\theta\psi\bar{\phi} + \sqrt{2}\bar{\theta}\bar{\psi}\phi$$

$$+ i\theta^\alpha(\partial_{\alpha\dot\alpha}\bar{\phi})\bar{\theta}^{\dot\alpha}\phi - i\theta^\alpha(\partial_{\alpha\dot\alpha}\phi)\bar{\theta}^{\dot\alpha}\bar{\phi} + 2(\theta\psi)(\bar{\theta}\bar{\psi})$$

$$- \tfrac{i}{\sqrt{2}}\bar{\theta}^2\theta^\alpha(\bar{\phi}\overleftrightarrow{\partial}_{\alpha\dot\alpha}\psi^{\dot\alpha}) - \tfrac{i}{\sqrt{2}}\theta^2(\psi^\alpha\overleftrightarrow{\partial}_{\alpha\dot\alpha}\bar{\phi})\bar{\theta}^{\dot\alpha} + \sqrt{2}\theta^2\bar{\theta}\bar{\psi}F + \sqrt{2}\bar{\theta}^2\theta\psi\bar{F}$$

$$+ \theta^2\bar{\theta}^2\left(\tfrac{1}{2}\partial_\mu\bar{\phi}\partial^\mu\phi - \tfrac{1}{4}\bar{\phi}\partial^2\phi - \tfrac{1}{4}\phi\partial^2\bar{\phi} + \tfrac{i}{2}\psi^\alpha\overleftrightarrow{\partial}_{\alpha\dot\alpha}\bar{\psi}^{\dot\alpha} + \bar{F}F \right), \qquad (10.129)$$

where all component fields depend on the space–time point x and

$$\overleftrightarrow{\partial} \equiv \overrightarrow{\partial} - \overleftarrow{\partial}.$$

> $\overleftrightarrow{\partial} \equiv \overrightarrow{\partial} - \overleftarrow{\partial}.$

It is evident that the superinvariant action

$$S_{\text{kin}} = \int d^4x\, d^4\theta\, \bar{\Phi}\Phi = \int d^4x \left(\partial_\mu\bar{\phi}\,\partial^\mu\phi + i\bar{\psi}_{\dot\alpha}\partial^{\dot\alpha\alpha}\psi_\alpha + \bar{F}F \right) \qquad (10.130)$$

(I have dropped the full derivatives in the integrand) presents the kinetic terms for the matter fields ϕ and ψ. Here

$$\bar{\partial}^{\dot\alpha\alpha} \equiv \partial_\mu\,(\bar{\sigma}^\mu)^{\dot\alpha\alpha}. \qquad (10.131)$$

As previously stated, the F component appears in the Lagrangian without derivatives and can be eliminated by virtue of the equations of motion. It does not represent any physical (propagating) degrees of freedom.

10.6.3 Potential Terms of the Matter Fields

By definition, the potential terms in the Lagrangian are those that enter with no derivatives and, generally speaking, are quadratic or of higher order in the component fields. For instance, the mass terms are quadratic both in the boson and fermion fields. In search of such terms in the superinvariant actions we should focus our attention on integrals over the chiral superspaces. Indeed, let us consider a function $\mathcal{W}(\Phi)$ of the chiral field that is termed a *superpotential*. Most commonly the superpotential is assumed to be a polynomial function of Φ. If we want to limit ourselves to renormalizable field theories in four dimensions, $\mathcal{W}(\Phi)$ must be at most cubic in Φ (see Section 10.6.4 below).

> Super-
> potentials

Since Φ is a chiral superfield, so is $\mathcal{W}(\Phi)$. We already know that the change in the last component of the chiral superfield under supertransformations (i.e., the component proportional to θ^2) is a total derivative. To project out the last component we must integrate over $d^2\theta$. Consequently, the action

$$S_{\text{pot}} = \int d^2\theta\, d^4x_L\, \mathcal{W}(\Phi(x_L, \theta)) + \text{H.c.}$$

$$= \int d^4x\, d^2\theta\, \mathcal{W}(\Phi(x, \theta)) + \text{H.c.} \qquad (10.132)$$

is superinvariant. The corresponding Lagrangian is invariant up to a total derivative. Note that the superpotential has dimension [mass]^3.

As a warm-up exercise let us consider a quadratic function,[23]

$$\mathcal{W}(\Phi) = \frac{m}{2}\Phi^2, \tag{10.133}$$

where m is a (complex!) mass parameter, and show that the corresponding superpotential term in conjunction with the kinetic term (10.130) generates masses for the fields ϕ and ψ. To this end let us first calculate $[\Phi(x,\theta)]^2$. Using Eq. (10.118), we arrive at

$$\Phi^2 = \phi^2 + 2\sqrt{2}\phi\theta^\alpha\psi_\alpha - \theta^2\psi^2 + 2\theta^2\phi F. \tag{10.134}$$

Now it is obvious that with this quadratic superpotential we get

$$\begin{aligned} S_{\text{pot}} &= \int d^4x\, d^2\theta \left(\frac{m}{2}\Phi^2\right) + \text{H.c.} \\ &= \int d^4x \left(m\phi F + \bar{m}\bar{\phi}\bar{F} - \frac{m}{2}\psi^2 - \frac{\bar{m}}{2}\bar{\psi}^2\right), \end{aligned} \tag{10.135}$$

to be added to Eq. (10.130). Next we combine all terms containing F in the Lagrangian:

$$\mathcal{L}_F = F\bar{F} + m\phi F + \bar{m}\bar{\phi}\bar{F}. \tag{10.136}$$

The equations of motion for F and $\bar{F}$ imply that

$$\bar{F} = -m\phi, \qquad F = -\bar{m}\bar{\phi}. \tag{10.137}$$

Substituting these back into $\mathcal{L}_F$ we obtain $\mathcal{L}_F = -|m|^2|\phi|^2$. Assembling all the elements, we conclude that the supersymmetric (noninteracting) Lagrangian that is built from one chiral superfield is

$$\mathcal{L} = \partial_\mu\bar{\phi}\partial^\mu\phi - |m|^2|\phi|^2 + i\bar{\psi}_{\dot\alpha}\partial^{\dot\alpha\alpha}\psi_\alpha - \frac{m}{2}\psi^2 - \frac{\bar{m}}{2}\bar{\psi}^2. \tag{10.138}$$

Needless to say, the masses of the scalar and spinor particles are equal and are given by the parameter $|m|$.

10.6.4 The Wess–Zumino Model

Now we will include interactions. We start from the simplest version with a single chiral superfield, with the intention of generalizing it later to the case of an arbitrary number of chiral superfields and a nonminimal kinetic term.

Thus, the model (it was invented by Wess and Zumino [36] and bears their name) contains one chiral superfield $\Phi(x_L,\theta)$ and its complex conjugate $\bar{\Phi}(x_R,\bar{\theta})$, which is antichiral. The action for the model is

$$S = \int d^4x\, d^4\theta\, \Phi\bar{\Phi} + \int d^4x\, d^2\theta\, \mathcal{W}(\Phi) + \int d^4x\, d^2\bar{\theta}\, \bar{\mathcal{W}}(\bar{\Phi}). \tag{10.139}$$

Note that the first term is an integral over the full superspace, while the second and the third run over the chiral subspaces. The *holomorphic* function $\mathcal{W}(\Phi)$ must be

---

[23] Quadratic expressions in the action give rise to terms corresponding to free (noninteracting) fields, just as in nonsupersymmetric theories.

viewed as a generic superpotential. In terms of components, the Lagrangian has the form

$$\mathcal{L} = (\partial^\mu \bar{\phi})(\partial_\mu \phi) + i\bar{\psi}_{\dot{\alpha}} \partial^{\dot{\alpha}\alpha} \psi_\alpha + \bar{F}F + \left[F\mathcal{W}'(\phi) - \tfrac{1}{2}\mathcal{W}''(\phi)\psi^2 + \text{H.c.} \right]. \tag{10.140}$$

From Eq. (10.140) it is obvious that F can be eliminated by virtue of the classical equation of motion

$$\bar{F} = -\frac{\partial \mathcal{W}(\phi)}{\partial \phi}, \tag{10.141}$$

so that the *scalar potential* describing the self-interaction of the field ϕ is

$$V(\phi, \bar{\phi}) = \left| \frac{\partial \mathcal{W}(\phi)}{\partial \phi} \right|^2. \tag{10.142}$$

Remark: in supersymmetric theories it is customary to denote the chiral superfield and its lowest (bosonic) component by the same letter, making no distinction between capital and small ϕ. Usually it is clear from the context what is meant in each particular case.

If one limits oneself to renormalizable theories, the superpotential $\mathcal{W}$ must be a polynomial function of Φ of power not higher than 3. In the model at hand, with one chiral superfield, the generic superpotential can always be reduced to the following "standard" form:

$$\mathcal{W}(\Phi) = \frac{m}{2}\Phi^2 - \frac{\lambda}{3}\Phi^3; \tag{10.143}$$

If one wishes, the quadratic term can be eliminated by a c-numerical shift of the field Φ,

$$\mathcal{W}(\Phi) = \frac{m^2}{4\lambda}\Phi - \frac{\lambda}{3}\Phi^3; \tag{10.144}$$

c-numerical terms in $\mathcal{W}$ can be omitted. Moreover, by using R symmetries (Section 10.7), one can choose the phases of the constants m and λ at will; we will choose them to be real and positive.

10.6.5 Vacuum Degeneracy

Typically, in supersymmetric field theories, the vacuum is not unique. In nonsupersymmetric theories this happens only if some *global* symmetry is spontaneously broken. In supersymmetric theories, vacuum degeneracy (even a continuous degeneracy) can take place without spontaneous breaking of any global symmetry. This feature is of paramount importance for practical applications. Therefore, the study of any given supersymmetric model should begin with the analysis of its *vacuum manifold*.

Let us study the set of classical vacua for the very simple Wess–Zumino model introduced in Section 10.6.4. In the case of a vanishing superpotential, $\mathcal{W} = 0$, any coordinate-independent field $\Phi_{\text{vac}} = \phi_0$ can serve as a vacuum. The vacuum manifold is then the one-dimensional (complex) manifold $C^1 = \{\phi_0\}$. The continuous degeneracy is due to the absence of potential energy, while the kinetic energy vanishes for any constant ϕ_0.

This continuous degeneracy is lifted by the superpotential. In particular, the superpotential (10.144) implies two degenerate classical vacua,

$$\phi_{\text{vac}} = \pm \frac{m}{2\lambda}. \tag{10.145}$$

Thus, the continuous manifold of vacua C^1 reduces to two points. Both vacua are physically equivalent. This equivalence can be explained by the spontaneous breaking of Z_2 symmetry, $\Phi \to -\Phi$, present in the superpotential (10.144). (One should remember that the overall phase of the superpotential is unobservable; in the Lagrangian (10.140) with superpotential (10.144) the above Z_2 symmetry is implemented as $\phi \to -\phi$ and $\psi \to i\psi$, so that $\psi^2 \to -\psi^2$.)

In the general case, Eq. (10.142) implies that the potential energy is positive definite. It vanishes only at critical points of $\partial \mathcal{W}/\partial \Phi$, where the F terms vanish, i.e., at

$$\frac{\partial \mathcal{W}(\phi)}{\partial \phi} = 0. \tag{10.146}$$

If $\mathcal{W}(\phi)$ is a polynomial of nth order, this equation has $n-1$ solutions. At some values of parameters the critical points can coalesce; for instance, if $m \to 0$ then the two solutions (10.145) coincide. However, if the theory is well defined at the quantum level, we will still see two vacuum states.

10.6.6 Hypercurrent in the Wess–Zumino Model. Generalities*

This section can be omitted at first reading. The reader could return to it after Sections 10.6.8, 10.7, or 10.16.

Let us consider an operator superfield $\mathcal{J}_{\alpha\dot\alpha}$ transforming in the representation $\left\{\frac{1}{2}, \frac{1}{2}\right\}$ of the Lorentz group,

$$\mathcal{J}_{\alpha\dot\alpha} = -\frac{1}{3}\left(D_{\dot\alpha}\bar\Phi\right)(D_\alpha\Phi) + \frac{2}{3}i\bar\Phi\overleftrightarrow{\partial}_{\dot\alpha\alpha}\Phi, \qquad \mathcal{J}^\mu \equiv \frac{1}{2}\left(\bar\sigma^\mu\right)^{\dot\beta\beta}\mathcal{J}_{\beta\dot\beta}. \tag{10.147}$$

One can call $\mathcal{J}_{\alpha\dot\alpha}$ a *hypercurrent* since the various components of this operator are related to a U(1) current, the supercurrent, and the energy–momentum tensor of the Wess–Zumino model, respectively. The above supermultiplet of conserved operators was first derived by Ferrara and Zumino; see Section 10.16. Sometimes in the literature people refer to this superfield as the *Ferrara–Zumino multiplet*. The term *hypercurrent* seems preferrable since it is more general. The Ferrara–Zumino multiplet in the Wess–Zumino model exemplifies a generic hypercurrent construction.

General formula

The hypercurrent defined above is obviously real. This is a general feature valid in all models. Using the equations of motion one can calculate its superderivative, obtaining the general formula

$$D^\alpha \mathcal{J}_{\alpha\dot\alpha} = \bar D_{\dot\alpha}\bar X, \tag{10.148}$$

where $\bar X$ is an antichiral superfield. In the Wess–Zumino model at hand

$$\bar X = 2\left(\bar{\mathcal{W}} - \frac{1}{3}\bar\Phi\bar{\mathcal{W}}'\right). \tag{10.149}$$

The easiest way to check Eq. (10.148) is to compare the lowest components in the left- and right-hand sides of the relation

$$\partial^{\dot\alpha\alpha}\mathcal{J}_{\alpha\dot\alpha} = \frac{1}{2i}\left(D^2 X - \bar D^2 \bar X\right), \tag{10.150}$$

which follows from (10.148) (one can take into account Eq. (10.151) in this comparison).

Equation (10.148) is *generic*: it applies in all the supersymmetric models to be considered below, with a single exception.[24] Equation (10.149) is specific to the Wess–Zumino model. Note that, for purely cubic superpotentials, $\bar{X} = 0$, implying that $D^\alpha \mathcal{J}_{\alpha\dot\alpha} = 0$. Taking the superderivative $\bar{D}^{\dot\alpha}$ of $D^\alpha \mathcal{J}_{\alpha\dot\alpha}$ and then doing the same in the reverse order, using $\{\bar{D}^{\dot\alpha}, D^\alpha\} = 2i\partial^{\dot\alpha\alpha}$, cf. Eq. (10.114), we conclude that in this case $\partial^{\dot\alpha\alpha} \mathcal{J}_{\alpha\dot\alpha} = 0$.

The lowest component of $\mathcal{J}^\mu$ is

$$R^\mu \equiv \tfrac{1}{2}(\bar{\sigma}^\mu)^{\dot\beta\beta} R_{\beta\dot\beta} = -\tfrac{1}{3}\bar{\psi}_{\dot\alpha}(\bar{\sigma}^\mu)^{\dot\alpha\alpha}\psi_\alpha + \tfrac{2}{3}\bar{\phi}i\overleftrightarrow{\partial}^\mu\phi. \tag{10.151}$$

The U(1) charge corresponding to this current generates phase rotations

$$\phi \to \exp(\tfrac{2}{3}i\alpha)\phi, \qquad \psi \to \exp(-\tfrac{1}{3}i\alpha)\psi. \tag{10.152}$$

For cubic superpotentials in (10.139) this current is obviously conserved. The corresponding U(1) symmetry of the Wess–Zumino model with a cubic superpotential is referred to as the R symmetry (see Section 10.7). The commutator of the R current with the supercharges then produces a conserved spin-$\tfrac{3}{2}$ operator. The only such operator is the supercurrent. It resides in the θ (or $\bar\theta$) component of the hypercurrent $\mathcal{J}_{\alpha\dot\alpha}$. The subsequent commutator produces a spin-2 conserved operator. The only nontrivial operator of this type[25] is the energy–momentum tensor, which appears in the $\theta\bar\theta$ component of $\mathcal{J}_{\alpha\dot\alpha}$. All higher components are conserved trivially, in much the same way as $\varepsilon^{\mu\nu\alpha\beta}\partial_\alpha R_\beta$. They will not concern us here.

Now let us consider the precise composition of the higher components of the hypercurrent $\mathcal{J}_{\alpha\dot\alpha}$ for generic superpotentials (i.e., components higher than the lowest component, (10.151)). As mentioned above, the θ component is associated with the supercurrent,

$$J_{\alpha\beta\dot\beta} = 2\sqrt{2}\,i\left\{\left[(\partial_{\alpha\dot\beta}\bar\phi)\psi_\beta - i\varepsilon_{\beta\alpha}F\bar\psi_{\dot\beta}\right]\right.$$
$$\left. - \tfrac{1}{6}\left[\partial_{\alpha\dot\beta}(\psi_\beta\bar\phi) + \partial_{\beta\dot\beta}(\psi_\alpha\bar\phi) - 3\varepsilon_{\beta\alpha}\partial^\gamma_{\dot\beta}(\psi_\gamma\bar\phi)\right]\right\}, \tag{10.153}$$

| *Supercurrent, with an "improve-ment"* |

which, in mixed spinorial–vectorial notation, can be written as

$$J^\mu_\alpha = \tfrac{1}{2}(\bar\sigma^\mu)^{\dot\beta\beta} J_{\alpha\beta\dot\beta}, \qquad \varepsilon^{\beta\alpha} J_{\alpha\beta\dot\beta} = J^{\mu,\alpha}\left(\sigma_\mu\right)_{\alpha\dot\beta}. \tag{10.154}$$

The second line in Eq. (10.153) is a full derivative, and so can be shown to produce no contribution to the supercharge. This term is conserved separately, and so is the term in the first line. The second line in Eq. (10.153) is the so-called *improvement*. In nonsupersymmetric formulations we could have perfectly well omitted the second line. However, the general supersymmetric formula (10.148) tells us that the supertrace $\varepsilon^{\beta\alpha} J_{\alpha\beta\dot\beta}$ must be directly reducible to the equations of motion, and the combination in (10.153) is the only one satisfying this requirement. Indeed,

---

[24] This exception is the class of theories with the Fayet–Iliopoulos term, Section 10.6.9. See [38] for a dramatic account of this finding. A sequel, which could have been entitled "Two-dimensional theories with four supercharges" is presented in [39].

[25] There is also a trivially conserved spin-2 operator $\varepsilon^{\mu\nu\alpha\beta}\partial_\alpha R_\beta$. Unlike the energy–momentum tensor, it is antisymmetric in μ, ν.

$$\varepsilon^{\beta\alpha} J_{\alpha\beta\dot{\beta}} = 2\sqrt{2}\left(-\bar{\phi}\partial_{\gamma\dot{\beta}}\psi^{\gamma} + 2iF\bar{\psi}_{\dot{\beta}}\right). \tag{10.155}$$

Now we can assert that

$$\mathcal{J}_{\alpha\dot{\alpha}} = R_{\alpha\dot{\alpha}} - \left[i\theta^{\beta}\left(J_{\beta\alpha\dot{\alpha}} - \tfrac{2}{3}\varepsilon_{\beta\alpha}\,\varepsilon^{\gamma\delta} J_{\delta\gamma\dot{\alpha}}\right) + \text{H.c.}\right] + \dots, \tag{10.156}$$

General
formula

where the ellipses stand for terms of higher orders in θ, to which we will turn shortly. To verify the above composition of the θ component of $\mathcal{J}_{\alpha\dot{\alpha}}$ we apply the superderivative D^{α} from the left, obtaining

$$D^{\alpha}\mathcal{J}_{\alpha\dot{\alpha}}\big|_{\theta=\bar{\theta}=0} = \tfrac{1}{3}i\,\varepsilon^{\gamma\delta} J_{\delta\gamma\dot{\alpha}}$$
$$= \tfrac{2\sqrt{2}}{3}i\left(-\bar{\phi}\,\partial_{\gamma\dot{\alpha}}\psi^{\gamma} + 2iF\bar{\psi}_{\dot{\alpha}}\right). \tag{10.157}$$

The lowest
component
of $\bar{D}_{\dot{\alpha}}\bar{X}$ is
$\tfrac{1}{3}i\varepsilon^{\gamma\delta} J_{\delta\gamma\dot{\alpha}}$.

Then we use the equations of motion for ψ and F and compare the result with the lowest component of $\bar{D}_{\dot{\alpha}}2\left(\bar{W} - \tfrac{1}{3}\bar{\Phi}\bar{W}'\right)$. Noting, with satisfaction, a perfect coincidence, I hasten to add that Eq. (10.156) is more general than its derivation in the Wess–Zumino model would suggest. It is valid in all models with (10.148).

The last calculation to be done in this subsection is that of the $\theta\bar{\theta}$ component of $\mathcal{J}_{\alpha\dot{\alpha}}$ and the components of $\bar{D}_{\dot{\alpha}}\bar{X}$ that are linear in θ (or $\bar{\theta}$). In this case, vectorial notation turns out to be more concise than spinorial notation. In this notation the supercurrent takes the form

$$\mathcal{J}_{\mu} = R_{\mu} + \left[\bar{\theta}_{\dot{\alpha}}(\bar{\sigma}^{\nu})^{\dot{\alpha}\alpha}\theta_{\alpha}\right]\left(2T_{\nu\mu} - \tfrac{2}{3}g_{\nu\mu}T_{\varphi}^{\varphi} - \tfrac{1}{2}\varepsilon_{\nu\mu\rho\sigma}\partial^{\rho}R^{\sigma}\right) + \cdots, \tag{10.158}$$

General
formula

where the ellipses stand for irrelevant powers of θ and $\bar{\theta}$. Here

$$T^{\mu\nu} = T_b^{\mu\nu} + T_f^{\mu\nu}, \tag{10.159}$$

where

$$T_b^{\mu\nu} = \partial^{\mu}\bar{\phi}\partial^{\nu}\phi + \partial^{\nu}\bar{\phi}\partial^{\mu}\phi - g^{\mu\nu}\left(\partial^{\chi}\bar{\phi}\partial_{\chi}\phi - F\bar{F}\right)$$
$$+ \tfrac{1}{3}\left(g^{\mu\nu}\partial^2 - \partial^{\mu}\partial^{\nu}\right)\phi\bar{\phi} \tag{10.160}$$

is the boson part of the energy–momentum tensor operator and

$$T_f^{\mu\nu} = \tfrac{1}{4}(\bar{\psi}\bar{\sigma}^{\mu}i\overset{\leftrightarrow}{\partial^{\nu}}\psi + \bar{\psi}\bar{\sigma}^{\nu}i\overset{\leftrightarrow}{\partial^{\mu}}\psi)$$
$$- g^{\mu\nu}\left(\tfrac{1}{2}\bar{\psi}\bar{\sigma}^{\rho}i\overset{\leftrightarrow}{\partial_{\rho}}\psi - \tfrac{1}{2}W''\psi^2 - \tfrac{1}{2}\overline{W}''\bar{\psi}^2\right) \tag{10.161}$$

is the corresponding fermion part. The second line in (10.160) presents the improvement term, which is analogous to that in the second line of (10.153). It is separately conserved and gives no contribution to the energy–momentum operator P^{μ}. It plays the same role as in (10.153), i.e., it ensures that the trace of the energy–momentum tensor reduces to the equations of motion. Indeed, with this term included,

$$(T_b)_{\chi}^{\chi} = \bar{\phi}\partial^2\phi + \phi\,\partial^2\bar{\phi} + 4\bar{F}F. \tag{10.162}$$

Note that the second line in (10.161) vanishes on the equations of motion.

Equation (10.158) is general in much the same way as Eq. (10.156), although our particular derivation, implying (10.160) and (10.161), was carried out for the Wess–Zumino model. It is instructive to check that Eq. (10.148) is valid for the $\bar{\theta}$ component too. To this end, starting from (10.158) we calculate the $\bar{\theta}$ term in $D^{\alpha}\mathcal{J}_{\alpha\dot{\alpha}}$,

$$D^{\alpha}\mathcal{J}_{\alpha\dot{\alpha}}\big|_{\bar{\theta}} = \bar{\theta}_{\dot{\alpha}}\left(i\,\partial_{\mu}R^{\mu} - \tfrac{2}{3}T_{\mu}^{\mu}\right). \tag{10.163}$$

Next, we use the equations of motion to calculate $\partial_\mu R^\mu$ and T^μ_μ on the one hand and

$$\bar{D}_{\dot\alpha}\left(2\bar{W} - \tfrac{2}{3}\bar{\Phi}\bar{W}'\right)\Big|_{\bar\theta}$$

on the other. Comparing the latter with the right-hand side of (10.163), we observe perfect agreement.[26]

The hypercurrent satisfying Eq. (10.148), whose component expansion is given by (10.156) and (10.158), is referred to as the Ferrara–Zumino hypercurrent [40]. We will discuss hypercurrents in more detail in Section 10.16.

> *The $\bar\theta$ component of $\bar{D}_{\dot\alpha}\bar{X}$ is $i\partial_\mu R^\mu - \left(\tfrac{2}{3}\right)T^\mu_\mu$.*

10.6.7 Generalized Wess–Zumino Models*

> *The section can be omitted at first reading. The reader could return to it after Sections 10.12.3.4 and 10.12.4.*
>
> *This action is also known as the Landau–Ginzburg action.*

The generalized Wess–Zumino model describes the interactions of an arbitrary number of chiral superfields, with more general kinetic terms of the type appearing in sigma models. Sigma model Lagrangians describe fields whose interactions derive from the fact that they are constrained and belong to certain manifolds. The latter are referred to as *target spaces*.

In many instances generalized Wess–Zumino models emerge as effective theories describing the low-energy behavior of "fundamental" gauge theories, in much the same way as the pion chiral Lagrangian presents a low-energy limit of QCD. In this case the models need not be renormalizable, the superpotential need not be polynomial, and the kinetic term need not be canonical. The most general action compatible with four supercharges and containing not more than two space–time derivatives ∂_μ is

$$S = \int d^4x\, d^4\theta\, \mathcal{K}(\Phi^i, \bar{\Phi}^{\bar{j}}) + \left[\int d^4x\, d^2\theta\, \mathcal{W}(\Phi^i) + \text{H.c.}\right], \qquad (10.164)$$

where $\Phi^i(i = 1, 2, \ldots, n)$ is a set of chiral superfields and $\bar{\Phi}^{\bar{j}} = (\Phi^j)^*$; the superpotential $\mathcal{W}$ is an analytic function of the chiral variables Φ^i while the kinetic term is determined by the function $\mathcal{K}$, which depends on both the chiral, Φ^i, and antichiral, $\bar{\Phi}^{\bar{j}}$, fields. Usually $\mathcal{K}$ is referred to as the *Kähler potential* (or the Kähler function). The Kähler potential is real.

In components, the Lagrangian takes the form

$$\mathcal{L} = G_{i\bar{j}}\partial_\mu\phi^i\partial^\mu\bar{\phi}^{\bar{j}} - G^{i\bar{j}}\frac{\partial \mathcal{W}}{\partial\phi^i}\frac{\partial\bar{\mathcal{W}}}{\partial\bar{\phi}^{\bar{j}}}$$
$$+ G_{i\bar{j}}i\bar{\psi}^{\bar{j}}\bar{\sigma}^\mu D_\mu\psi^i + \frac{1}{4}R_{i\bar{j}k\bar{l}}(\psi^i\psi^k)(\bar{\psi}^{\bar{j}}\bar{\psi}^{\bar{l}})$$
$$- \frac{1}{2}\left[\left(\frac{\partial^2\mathcal{W}}{\partial\phi^j\partial\phi^k} - \Gamma^i_{jk}\frac{\partial\mathcal{W}}{\partial\phi^i}\right)\psi^j\psi^k + \text{H.c.}\right], \qquad (10.165)$$

where

$$G_{i\bar{j}} = \frac{\partial^2\mathcal{K}}{\partial\phi^i\partial\bar{\phi}^{\bar{j}}} \qquad (10.166)$$

---

[26] The details of this comparison are left as an instructive exercise for the reader.

plays the role of the metric in the space of fields (the target space) and $G^{i\bar{j}}$ is the inverse metric,

$$G^{i\bar{j}}G_{k\bar{j}} = \delta^i_k, \qquad G^{i\bar{j}}G_{i\bar{k}} = \delta^{\bar{i}}_{\bar{k}}. \tag{10.167}$$

Moreover,

$$D_\mu \psi^i = \partial_\mu \psi^i + \Gamma^i_{kl}\partial_\mu \phi^k \psi^l \tag{10.168}$$

> **Kähler geometry**

is the (target space) covariant derivative, Γ^i_{kl} are the (target space) Christoffel symbols,

$$\Gamma^i_{kl} = G^{i\bar{m}}\frac{\partial G_{k\bar{m}}}{\partial \phi^l}, \qquad \bar{\Gamma}^{\bar{i}}_{\bar{k}\bar{l}} = G^{m\bar{i}}\frac{\partial G_{m\bar{k}}}{\partial \bar{\phi}^{\bar{l}}}, \tag{10.169}$$

and $R_{i\bar{j}k\bar{l}}$ is the (target space) Riemann tensor,

$$R_{i\bar{j}k\bar{l}} = \frac{\partial^2 G_{i\bar{j}}}{\partial \phi^k \partial \bar{\phi}^{\bar{l}}} - \Gamma^m_{ik}\bar{\Gamma}^{\bar{m}}_{\bar{j}\bar{l}}G_{m\bar{m}}. \tag{10.170}$$

Note that the covariant derivative defined above acting on the metric produces zero, $D_\mu G_{i\bar{j}} = 0$.

The metric (10.166) defines a Kähler manifold. By definition this is a manifold that allows one to introduce complex (instead of real) coordinates. Therefore, the real dimension of Kähler manifolds is always even. However, not every space with an even number of real coordinates is Kähler. The two-dimensional plane and the two-dimensional sphere are Kähler manifolds, while the four-dimensional sphere is not.

What is the vacuum manifold in the model (10.164)? In the absence of a superpotential, i.e., for $\mathcal{W} = 0$, any set ϕ^0_i of constant fields is a possible vacuum. Thus, the vacuum manifold is the Kähler manifold of the complex dimension n and the metric $G_{i\bar{j}}$ defined in Eq. (10.166). If $\mathcal{W} \neq 0$ (this is only possible for noncompact Kähler manifolds), the conditions of *F-flatness*,

$$\frac{\partial \mathcal{W}}{\partial \phi^i} = 0, \qquad i = 1, 2, \ldots, n, \tag{10.171}$$

single out some submanifold of the original Kähler manifold. This submanifold may be continuous or discrete. If no solution of the above equations exists, the supersymmetry is spontaneously broken. We will address the issue of the spontaneous breaking of supersymmetry in due course (Section 10.10).

10.6.8 Abelian Gauge-Invariant Interactions

As well known, in nonsupersymmetric gauge theories the matter fields transform under a gauge transformation as

$$\phi(x) \to e^{i\alpha(x)}\phi(x), \qquad \bar{\phi}(x) \to e^{-i\alpha(x)}\bar{\phi}(x), \tag{10.172}$$

while, for the gauge field,

$$A_\mu(x) \to A_\mu(x) + \partial_\mu \alpha(x) \tag{10.173}$$

where $\alpha(x)$ is an arbitrary function of x. To maintain gauge invariance, the partial derivatives acting on the matter fields must be replaced by covariant derivatives; for instance,

$$\partial_\mu \phi \, \partial^\mu \bar{\phi} \to \mathcal{D}_\mu \phi \mathcal{D}^\mu \bar{\phi}, \tag{10.174}$$

where

$$\mathcal{D}_\mu \phi = (\partial_\mu - iA_\mu)\phi, \qquad \mathcal{D}_\mu \bar{\phi} = (\partial_\mu + iA_\mu)\bar{\phi}. \tag{10.175}$$

Equations (10.172) prompt us as to how to extend the (Abelian) local gauge invariance to supersymmetric theories. Indeed, in the latter the matter sector is described by chiral superfields replacing the scalar fields in Eq. (10.172). Therefore, the x-dependent phase in (10.172) must be promoted to a chiral superfield Λ as follows:

$$\Phi(x_L, \theta) \to e^{i\Lambda(x_L,\theta)}\Phi(x_L, \theta), \qquad \bar{\Phi}(x_R, \bar{\theta}) \to e^{-i\bar{\Lambda}(x_R,\bar{\theta})}\bar{\Phi}(x_R, \bar{\theta}). \tag{10.176}$$

Note that, unlike in the nonsupersymmetric transformation (10.172), Λ and $\bar{\Lambda}$ are different: the first is the chiral superfield and the second the antichiral superfield; the first depends on x_L and θ while the second depends on x_R and $\bar{\theta}$. Hence $\Lambda - \bar{\Lambda} \neq 0$.

How can one generalize Eq. (10.173) to construct a gauge-invariant kinetic term? The gauge field is a component of the vector superfield V. Let us try the following supertransformation:

$$V(x, \theta, \bar{\theta}) \to V(x, \theta, \bar{\theta}) - i \left[\Lambda(x_L, \theta) - \bar{\Lambda}(x_R, \bar{\theta}) \right]. \tag{10.177}$$

It is obvious that the combination $\bar{\Phi}e^V\Phi$ is gauge invariant, i.e., it is invariant under the simultaneous action of (10.176) and (10.177). Consequently, it transforms $\partial_\mu \phi \, \partial^\mu \bar{\phi}$ into $\mathcal{D}_\mu \phi \mathcal{D}^\mu \bar{\phi}$. The same happens with the fermion kinetic term; the partial derivative in $i\bar{\psi}_{\dot{\alpha}} \bar{\partial}^{\dot{\alpha}\alpha} \psi_\alpha$ becomes covariant.[27]

Since this is a crucial point it is instructive to have a closer look at the above procedure in terms of components. If we parametrize $\Lambda(x_L, \theta)$ as

$$\Lambda \equiv \varphi + \sqrt{2}\theta\eta + \theta^2 \mathcal{F}, \tag{10.178}$$

then, under (10.177), we have

$$C \to C - i(\varphi - \bar{\varphi}), \qquad \chi \to \chi - \sqrt{2}\eta, \qquad M \to M - \sqrt{2}\mathcal{F},$$
$$v_{\alpha\dot{\beta}} \to v_{\alpha\dot{\beta}} + \tfrac{1}{2}\partial_{\alpha\dot{\beta}}(\varphi + \bar{\varphi}), \qquad \lambda \to \lambda, \qquad D \to D. \tag{10.179}$$

Here we have used Eqs. (10.110) and (10.178) in calculating $\Lambda - \bar{\Lambda}$. If we require $C - i(\varphi - \bar{\varphi})$ to vanish, the lowest component of the vector superfield vanishes and simultaneously the last component reduces to $\theta^2 \bar{\theta}^2 D$. This explains the peculiar choice of parametrization (10.110).

We see that the C, χ, and M components of the vector superfield can be gauged away, and thus[28]

$$V = -2\theta^\alpha \bar{\theta}^{\dot{\alpha}} A_{\alpha\dot{\alpha}} - 2i\bar{\theta}^2(\theta\lambda) + 2i\theta^2(\bar{\theta}\bar{\lambda}) + \theta^2 \bar{\theta}^2 D. \tag{10.180}$$

---

[27] Supersymmetrization of the gauge transformations (10.176), (10.177) was the path that led Wess and Zumino to the discovery of supersymmetric theories.

[28] To make contact with the standard notation we will denote by $A_{\alpha\dot{\alpha}}$ the shifted vector component field $v_{\alpha\dot{\alpha}} + \tfrac{1}{2}\partial_{\alpha\dot{\alpha}}(\varphi + \bar{\varphi})$.

The Wess–Zumino gauge is the most commonly used.

This is called the *Wess–Zumino gauge*. This gauge, bearing the name of those who devised it, is routinely imposed when the component formalism is used. However, imposing the Wess–Zumino gauge condition in supersymmetric theories does not fix the gauge completely. The component Lagrangian at which one arrives in the Wess–Zumino gauge still possesses gauge freedom with respect to nonsupersymmetric (old-fashioned) gauge transformations.

10.6.9 Supersymmetric QED

Supersymmetric quantum electrodynamics (SQED) is the simplest and, historically, the first [4] supersymmetric gauge theory. This model supersymmetrizes QED. In QED the electron is described by a Dirac field. One Dirac field is equivalent to two chiral (Weyl) fields: one left-handed and one right-handed, both with electric charge 1. Alternatively, one can decompose the Dirac field as two left-handed fields, one with charge $+1$, the other with charge -1. Each Weyl field is accompanied in SQED by a complex scalar field, known as a selectron. Thus, we conclude that we need to introduce two chiral superfields, Q and $\tilde{Q}$, of opposite electric charge.

Apart from the matter sector there exists the gauge sector, which includes the photon and photino. As explained above, these are represented by a vector superfield V. The SQED Lagrangian is

$$\mathcal{L} = \left(\frac{1}{4e^2} \int d^2\theta \, W^2 + \text{H.c.} \right) + \int d^4\theta \left(\bar{Q} e^V Q + \bar{\tilde{Q}} e^{-V} \tilde{Q} \right)$$
$$+ \left(m \int d^2\theta \, Q\tilde{Q} + \text{H.c.} \right), \tag{10.181}$$

where e is the electric charge, m is the electron or selectron mass, and the *chiral* superfield $W_\alpha(x_L, \theta)$ is the supergeneralization of the photon field strength tensor,

Definition of W_α in the Abelian case

$$W_\alpha \equiv \tfrac{1}{8} \bar{D}^2 D_\alpha V = i \left(\lambda_\alpha + i\theta_\alpha D - \theta^\beta F_{\alpha\beta} - i\theta^2 \partial_{\alpha\dot\alpha} \bar{\lambda}^{\dot\alpha} \right). \tag{10.182}$$

In the units of e the charge of Q is $+1$ and that of $\tilde{Q}$ is -1; see Eq. (10.175).

The chiral "superphoton" field strength W and $\bar{W}$ have mass dimension $\tfrac{3}{2}$. They are gauge invariant in the Abelian theory and satisfy the additional constraint equation (a supergeneralization of the Bianchi identity)

Super-Bianchi identity

$$D^\alpha W_\alpha = \bar{D}_{\dot\alpha} \bar{W}^{\dot\alpha}. \tag{10.183}$$

The lowest component of this constraint expresses the fact that D is real. Equation (10.183) is also the superspace version of the Bianchi identity, which in nonsupersymmetric QED has the form $\partial^\mu \tilde{F}_{\mu\nu} = 0$. The above Bianchi identity is equivalent to

$$\partial_{\dot\beta}^\beta F_{\alpha\beta} = \partial_\alpha^{\dot\alpha} \bar{F}_{\dot\alpha\dot\beta} \tag{10.184}$$

in spinorial formalism. The component field supertransformations following from Eq. (10.182) are

$$\delta\lambda_\alpha = iD\epsilon_\alpha - F_{\alpha\beta}\epsilon^\beta, \qquad \delta D = \left(\lambda^\alpha \overleftarrow{\partial}_{\alpha\beta} \right) \bar{\epsilon}^{\dot\beta} + \epsilon^\beta \partial_{\beta\dot\alpha} \bar{\lambda}^{\dot\alpha},$$
$$\delta F_{\alpha\beta} = -i \left(\partial_{\beta\dot\beta}\lambda_\alpha + \partial_{\alpha\dot\beta}\lambda_\beta \right) \bar{\epsilon}^{\dot\beta}. \tag{10.185}$$

The form of the Lagrangian (10.181) is uniquely fixed by the supergauge invariance

$$Q \to e^{i\Lambda}Q, \quad \bar{Q} \to e^{-i\bar{\Lambda}}\bar{Q}, \quad \tilde{Q} \to e^{-i\Lambda}\tilde{Q}, \quad \bar{\tilde{Q}} \to e^{i\bar{\Lambda}}\bar{\tilde{Q}},$$
$$V \to V - i\left[\Lambda - \bar{\Lambda}\right], \quad W_\alpha \to W_\alpha, \quad \bar{W}_{\dot{\alpha}} \to \bar{W}_{\dot{\alpha}}. \tag{10.186}$$

Integration over $d^2\theta$ singles out the θ^2 component of the chiral superfields W^2 and $Q\tilde{Q}$, i.e., the F terms, while the $d^2\theta d^2\bar{\theta}$ integration singles out the $\theta^2\bar{\theta}^2$ component of the real superfields $\bar{Q}e^V Q$ and $\bar{\tilde{Q}}e^{-V}\tilde{Q}$, i.e., the D terms. The fact that the electric charges of Q and $\tilde{Q}$ are opposite is explicit in Eq. (10.181). The theory describes the conventional electrodynamics of one Dirac and two complex scalar fields. In addition, it includes photino–electron–selectron couplings and the self-interaction of the selectron fields, which has a special form, to be discussed below; see Eq. (10.191).

In Abelian gauge theories one may add another term to the Lagrangian, the *Fayet–Iliopoulos* term [41] (also known as the ξ term),

$$\Delta\mathcal{L}_\xi = -\xi \int d^2\theta \, d^2\bar{\theta} \, V(x, \theta, \bar{\theta}) \equiv -\xi D. \tag{10.187}$$

| Fayet–Iliopoulos term |

It plays an important role in the dynamics of some gauge models.

The D component of V is an auxiliary field (like F); it enters the Lagrangian as follows:

$$\mathcal{L}_D = \frac{1}{2e^2}D^2 + D(\bar{q}q - \bar{\tilde{q}}\tilde{q}) - \xi D + \cdots, \tag{10.188}$$

where the ellipses denote D-independent terms and ξ will be assumed to be positive hereafter. Eliminating D by substituting the classical equation of motion we get the so-called D potential describing the self-interaction of selectrons:

$$V_D = \frac{1}{2e^2}D^2, \quad D = -e^2(\bar{q}q - \bar{\tilde{q}}\tilde{q} - \xi). \tag{10.189}$$

This is only part of the scalar potential. The full scalar potential $V(q, \tilde{q})$ is obtained by adding the part generated by the F terms of the matter fields, see Eq. (10.142) with $\mathcal{W}$ replaced by $mQ\tilde{Q}$:

$$V(q, \tilde{q}) = \frac{e^2}{2}(\bar{q}q - \bar{\tilde{q}}\tilde{q} - \xi)^2 + |mq|^2 + |m\tilde{q}|^2. \tag{10.190}$$

In components, the Lagrangian (10.181) of supersymmetric QED has the form

$$\mathcal{L} = \frac{1}{e^2}\left(-\frac{1}{4}F_{\mu\nu}F^{\mu\nu} + \bar{\lambda}_{\dot{\alpha}}i\partial^{\dot{\alpha}\alpha}\lambda_\alpha\right)$$
$$+ \left(\mathcal{D}^\mu \bar{q}\,\mathcal{D}_\mu q + \bar{\psi}_{\dot{\alpha}}i\mathcal{D}^{\dot{\alpha}\alpha}\psi_\alpha\right) + \left(\mathcal{D}^\mu \bar{\tilde{q}}\,\mathcal{D}_\mu\tilde{q} + \bar{\tilde{\psi}}_{\dot{\alpha}}i\mathcal{D}^{\dot{\alpha}\alpha}\tilde{\psi}_\alpha\right)$$
$$+ \left[i\sqrt{2}(\lambda\psi)\bar{q} + \text{H.c.}\right] + \left[-i\sqrt{2}\left(\lambda\tilde{\psi}\right)\bar{\tilde{q}} + \text{H.c.}\right]$$
$$- m\left(\psi\tilde{\psi} + \text{H.c.}\right) - V(q, \tilde{q}). \tag{10.191}$$

Here λ is the photino field, q and $\tilde{q}$ are the scalar fields (selectrons), i.e., the lowest components of the superfields Q and $\tilde{Q}$, respectively, and ψ and $\tilde{\psi}$ are the fermion components of Q and $\tilde{Q}$. The scalar potential $V(q, \tilde{q})$ is given in Eq. (10.190). One should not forget that the electric charges of Q and $\tilde{Q}$ are opposite; therefore,

$$iD_\mu q = \left(i\partial_\mu + A_\mu\right) q, \qquad iD_\mu \tilde{q} = \left(i\partial_\mu - A_\mu\right) \tilde{q},$$
$$iD_\mu \psi = \left(i\partial_\mu + A_\mu\right) \psi, \qquad iD_\mu \tilde{\psi} = \left(i\partial_\mu - A_\mu\right) \tilde{\psi}. \tag{10.192}$$

In deriving the component form of the Lagrangian for supersymmetric QED we used

| W^2 in QED |

the identity

$$W^2(x_L, \theta) = -\lambda^2 - 2i(\lambda\theta)D + 2\lambda^\alpha F_{\alpha\beta}\theta^\beta$$
$$+ \theta^2 \left(D^2 - \tfrac{1}{2}F^{\alpha\beta}F_{\alpha\beta}\right) + 2i\theta^2 \bar{\lambda}_{\dot\alpha}\partial^{\dot\alpha\alpha}\lambda_\alpha. \tag{10.193}$$

In nonsupersymmetric field theory the terms in the third line of Eq. (10.191) would be referred to as the Yukawa terms. This is not the case in supersymmetric theories, where these terms represent a supergeneralization of the gauge interaction. It is the cubic part of the superpotential that is referred to as the super-Yukawa term.

10.6.10 Flat Directions

As already mentioned, in the study of each supersymmetric model one starts by establishing the vacuum manifold. Equation (10.190) allows us to examine the structure of the vacuum manifold in supersymmetric QED with the Fayet–Iliopoulos term.

The energy of any field configuration in supersymmetric theory is positive definite. Thus, any configuration with vanishing energy is automatically a vacuum, i.e., the vacuum manifold is determined by the condition $V(q, \tilde{q}) = 0$. Assume first that the mass term and the ξ term are absent, $m = \xi = 0$, i.e., we are dealing with massless supersymmetric QED. Then the equation to solve is

$$V(q, \tilde{q}) = \frac{e^2}{2} \left(\bar{q}q - \bar{\tilde{q}}\tilde{q}\right)^2 \equiv 0. \tag{10.194}$$

This equation does not have a unique solution; rather, it has a continuous complex noncompact manifold of solutions of the type

$$q = \varphi, \qquad \tilde{q} = \varphi \tag{10.195}$$

(modulo a gauge transformation), where φ is a complex parameter. One can think of the potential $V(q, \tilde{q})$ as a mountain ridge; the *flat direction* (a D-flat direction in the present case) then presents the flat bottom of a valley. This explains the origin of the term *vacuum valleys*, which is sometimes used to denote the flat directions. The (classical) vacuum manifold in the present case is a one-dimensional complex line C_1, parametrized by φ. Each point of this manifold can be viewed as the vacuum of a particular theory. If $\varphi \neq 0$ in the vacuum, the theory is in the Higgs regime; the photon and its superpartners become massive. The photon field "eats up" one of the real scalar fields residing in $Q, \tilde{Q}$ and so acquires a mass; another real scalar field acquires the very same mass. The photino teams up with a linear combination of two Weyl spinors in $Q, \tilde{Q}$ and becomes a massive Dirac field, with the same mass as the photon. One Weyl spinor and one complex scalar remain massless. This phenomenon – the super-Higgs mechanism – will be discussed in more detail in Section 10.9. The flat direction (vacuum valley) in which the gauge symmetry is

| Higgs branch |

realized in the Higgs mode is referred to as the *Higgs branch*. Supersymmetric gauge theories with flat directions are abundant.

In the model at hand, on the Higgs branch the set of massless degrees of freedom consists of the field φ that describes excitations along the flat direction and its superpartner ψ. These two fields can be assembled into a single chiral superfield $\Phi(x_L, \theta) = \varphi(x_L) + \sqrt{2}\,\theta\psi(x_L) + \theta^2\mathcal{F}$, which is described by the massless Wess–Zumino model with the Kähler potential $\bar{\Phi}\Phi$, i.e., the flat metric.[29]

The above discussion applied to the Wess–Zumino gauge. The gauge-invariant parametrization of the vacuum manifold is given by the product of the chiral superfields $Q\tilde{Q}$. This product is also a chiral superfield, of zero charge; therefore it is obviously *(super)gauge invariant*. Neutral combinations such as $Q\tilde{Q}$ are referred to as *chiral invariants*. In the model under consideration there exists only one chiral invariant. Generally speaking, in supersymmetric gauge theories with nontrivial matter sectors one can construct several chiral invariants. The problem of establishing flat directions then reduces to the analysis of all chiral invariants and all possible constraints between them. In general vacuum manifolds are parametrized by chiral invariants.

In supersymmetric QED with $\xi = m = 0$, every point of the flat direction is in one-to-one correspondence with the value of $Q\tilde{Q} = \Phi^2$. The superfield Φ is also known as the *moduli field*. Theories with a flat direction are said to have a moduli space.

What happens if ξ and/or $m \neq 0$? If $\xi \neq 0$ while m still vanishes then a one-dimensional complex vacuum manifold (the Higgs branch) survives, although it ceases to be flat. Indeed, now

$$V(q, \tilde{q}) = \frac{e^2}{2}\left(\bar{q}q - \bar{\tilde{q}}\tilde{q} - \xi\right)^2. \tag{10.196}$$

The D-flatness condition is

$$\bar{q}q - \bar{\tilde{q}}\tilde{q} - \xi = 0. \tag{10.197}$$

The solution of the above D-flatness equation can be presented as follows (see, e.g., [42, 43]):

$$q = \sqrt{\xi}\,e^{i\alpha}\cosh\rho, \qquad \tilde{q} = \sqrt{\xi}\,e^{i\alpha}\sinh\rho, \tag{10.198}$$

(modulo a gauge transformation). The chiral invariant $Q\tilde{Q}$ then takes the form

$$Q\tilde{Q}\big|_{\theta=0} = q\tilde{q} = \frac{\xi}{2}e^{2i\alpha}\sinh 2\rho \equiv \frac{\xi}{2}\varphi, \tag{10.199}$$

where the right-hand side defines a new chiral field φ, the lowest component of the moduli superfield[30]

$$\Phi = \frac{2}{\xi}Q\tilde{Q}. \tag{10.200}$$

For this lowest component we have

$$\partial_\mu\bar{\varphi}\,\partial^\mu\varphi = 4(\cosh 2\rho)^2\left[(\partial_\mu\rho\,\partial^\mu\rho) + (\tanh 2\rho)^2(\partial_\mu\alpha\partial^\mu\alpha)\right]. \tag{10.201}$$

---

[29] Warning: the word "flat" is used in this range of questions in two distinct meanings, not to be confused with each other. First, we talk about a flat direction, implying a continuous manifold (in the space of fields) of degenerate vacua at zero energy. Second, the word "flat" can refer to the Kähler geometry of the vacuum manifold, whose Kähler metric in general may or may not be flat.

[30] The superfield Φ defined in Eq. (10.200) and considered below is unrelated to the superfield Φ in the first half of this section, where we dealt with the $\xi = 0$ case.

Now, let us derive the metric on the target space and, hence, the Kähler potential. The parametrization (10.198) must be substituted into the appropriate part of the Lagrangian (10.191), i.e., the second line. Besides the regular derivatives acting on the fields $q, \tilde{q}$ we should take into account the photon field. In the present case the latter reduces to

$$A_\mu = -\frac{i}{2} \frac{\bar{q}\overset{\leftrightarrow}{\partial}_\mu q - \bar{\tilde{q}}\overset{\leftrightarrow}{\partial}_\mu \tilde{q}}{\bar{q}q + \bar{\tilde{q}}\tilde{q}} = \frac{\partial_\mu \alpha}{\cosh 2\rho}, \tag{10.202}$$

in the limit when all degrees of freedom except those residing in Φ become very heavy. Then the bosonic kinetic term following from (10.191) is

$$\mathcal{L}(\rho, \alpha) = \xi(\cosh 2\rho) \left[(\partial_\mu \rho \, \partial^\mu \rho) + (\tanh 2\rho)^2 \left(\partial_\mu \alpha \, \partial^\mu \alpha \right) \right]$$
$$= \frac{\xi}{4} \frac{1}{\sqrt{1 + \bar{\varphi}\varphi}} \partial_\mu \bar{\varphi} \, \partial^\mu \varphi. \tag{10.203}$$

This result implies, in turn, that the metric G is given by

$$G = \frac{\xi}{4} \frac{1}{\sqrt{1 + \bar{\varphi}\varphi}} \tag{10.204}$$

and the corresponding Kähler potential has the form

$$K(\Phi, \bar{\Phi}) = \xi \left(\sqrt{1 + \bar{\Phi}\Phi} - \operatorname{arctanh} \sqrt{1 + \bar{\Phi}\Phi} \right). \tag{10.205}$$

The dynamics of the moduli fields is described by a supersymmetric sigma model (i.e., a generalized Wess–Zumino model with vanishing superpotential) with Kähler potential (10.205). For a more detailed consideration of this problem the reader is referred to Appendix section 10.27.2 on page 564.

Introducing the mass term $m \neq 0$ and setting $\xi = 0$ one lifts the vacuum degeneracy, making the bottom of the valley (10.194) nonflat. The vanishing of the F terms $F_Q = -\bar{m}\bar{\tilde{q}}$ and $F_{\tilde{Q}} = -\bar{m}\bar{q}$ implies that

$$q = \tilde{q} = 0. \tag{10.206}$$

The mass term pushes the theory towards the origin of the D-flat direction. The Higgs branch disappears and the vacuum becomes unique.

In the general case, $\xi \neq 0$ and $m \neq 0$. Then the condition (10.206) of vanishing F terms is inconsistent with the vanishing of the D term, Eq. (10.197). Thus the theory has no zero-energy state. Hence, the supersymmetry is spontaneously broken (see Section 10.10.2).

The occurrence of flat directions is the most crucial feature of supersymmetric gauge theories regarding the dynamics of supersymmetry breaking.

10.6.11 Complexification of the Coupling Constants

The coupling constants in supersymmetric theories appear in the action in the F terms, i.e., in the integrals over chiral subspaces. For instance, all the coupling constants in the superpotential appear in this way, through the integral $\int d^2\theta W$. This is also the case for the inverse gauge coupling constant, which appears as a coefficient in front of $\int d^2\theta W^2$; see Eq. (10.181). There is a crucial consequence which will be repeatedly exploited in what follows.

All such coupling constants must be viewed as complex numbers, i.e., complex chiral or antichiral parameters. The dependence of the F terms and chiral superfields on the chiral complex parameters must be holomorphic. (The dependence of the $\bar{F}$ terms and antichiral superfields on the antichiral complex parameters must be holomorphic too.)

The proof of the above assertion is simple. Indeed, let us promote the above coupling constants to the rank of (auxiliary nondynamic) chiral superfields. In other words, one can treat them as the lowest components of the appropriate chiral (antichiral) superfields, for instance the mass parameter m in Eq. (10.181) gives rise to a chiral superfield $M(X_L, \theta) = m + \ldots$ Assuming the *lowest* component of these superfields to develop vacuum expectation values (which do not break supersymmetry), we find that we return to the original action upon substituting the auxiliary chiral superfields by their expectation values.

The only role of the auxiliary chiral superfields is to develop expectation values for their lowest components. All degrees of freedom residing there are those of infinitely heavy "particles" that are nondynamical.

It is clear that any calculation of integrals over the chiral subspace,

$$\int d^2\theta\, f(x_L, \theta),$$

or calculation of chiral superfields must produce results that depend only on the above auxiliary chiral superfields; they *cannot* depend on the antichiral superfields. This concludes the proof of holomorphic dependence.

It is instructive to discuss the physical meaning of the complexified gauge coupling in Eq. (10.181). Let us parametrize $1/e^2$ as follows:[31]

| Complexified coupling |

$$\frac{1}{e^2} = \frac{1}{\tilde{e}^2} - i\frac{\theta}{8\pi^2}, \qquad (10.207)$$

where the tilde (temporarily) marks the real part of $1/e^2$, while $-\theta/(8\pi^2)$ is the imaginary part. Assembling Eqs. (10.28), (10.181), (10.193), and (10.207) we arrive at the following kinetic terms for the photon and photino:

$$\Delta\mathcal{L}_{\gamma,\tilde{\gamma}} = -\frac{1}{4\tilde{e}^2}F^{\mu\nu}F_{\mu\nu} + \frac{\theta}{32\pi^2}F^{\mu\nu}\tilde{F}_{\mu\nu} + \frac{1}{\tilde{e}^2}\bar{\lambda}_{\dot{\alpha}}i\partial^{\dot{\alpha}\alpha}\lambda_\alpha + \frac{1}{2\tilde{e}^2}D^2, \qquad (10.208)$$

where I have omitted a full derivative term of the type $\partial^{\dot{\alpha}\alpha}\left(\bar{\lambda}_{\dot{\alpha}}\lambda_\alpha\right)$. Equation (10.208) demonstrates in a clear-cut manner that the imaginary part of the complexified gauge coupling constant plays the role of the θ angle. Note that there is a "wrong" positive sign in front of D^2. This sign would not be allowed for a dynamical field.

---

[31] In the literature one quite often encounters a different normalization of the holomorphic variable associated with the gauge coupling, namely,

$$\tau \equiv i\frac{4\pi}{\tilde{e}^2} + \frac{\theta}{2\pi} = i\frac{4\pi}{e^2}.$$

Exercises

10.6.1 Prove the assertion following Eq. (10.138). Hint: Pass to the Majorana representation for the spinor fields.

10.6.2 Obtain Eq. (10.165) by a straightforward algebraic derivation from Eq. (10.164) using the component decomposition of the chiral superfields and the definitions of the target space geometry given in Section 10.6.7.

Hint: The expression for the F term following from the corresponding equation of motion is

$$F^i = \frac{1}{2}\Gamma^i_{jk}\,\psi^j\psi^k - G^{i\bar{j}}\frac{\partial\bar{W}}{\partial\bar{\phi}^{\bar{j}}}.$$

10.6.3 Explain why the supertransformation laws in the first line in Eq. (10.185) differ from those in the Exercise 10.5.3. Is it a mistake?

10.6.4 Show that the target space with the metric (10.204), which is the vacuum manifold for supersymmetric QED with the Fayet–Iliopoulos term, is a two-dimensional hyperboloid up to small corrections dying off at $|\varphi| \to 0$ and $|\varphi| \to \infty$.

10.7 *R* Symmetries

The Coleman–Mandula theorem states that all global symmetries must commute with the generators of the Poincaré group. However, it is not necessary for them to commute with all generators of the super-Poincaré group.

The associativity of the super-Poincaré algebra implies that there can exist at most one (independent) Hermitian U(1) generator R that does not commute with the supercharges:

$$[R, Q_\alpha] = -Q_\alpha, \qquad [R, \bar{Q}_{\dot\alpha}] = +\bar{Q}_{\dot\alpha}. \tag{10.209}$$

This single U(1) symmetry, if it exists in the given model, is called the R *symmetry*. Since the R symmetry does not commute with supersymmetry, the component fields of the chiral superfields do not all carry the same R charge. Let us call the R charge of the lowest component field of the given superfield the R charge of the superfield.

To see in more detail how this works we will now focus on a chiral superfield Φ with superpotential

$$\mathcal{W} = \Phi^3. \tag{10.210}$$

The first encounter was at the beginning of Section 10.6.6.

The R transformations that we will assign to the component fields are as follows:

$$\phi(x_L) \to \phi(x_L)\exp\left(\tfrac{2}{3}i\alpha\right), \qquad \psi(x_L) \to \psi(x_L)\exp\left[\left(\tfrac{2}{3}-1\right)i\alpha\right],$$
$$F \to F\exp\left[\left(\tfrac{2}{3}-2\right)i\alpha\right], \tag{10.211}$$

where α is a constant phase. The above expressions *define* the R charge $r(\Phi)$ of the superfield Φ to be 2/3:

$$\Phi(x_L, \theta) \overset{\text{def}}{\to} e^{2i\alpha/3}\Phi(x_L, e^{-i\alpha}\theta). \tag{10.212}$$

Now, in the superinvariant actions we will make the following changes in the Grassmann parameters θ and $\bar{\theta}$:

$$\theta \to e^{i\alpha}\theta, \qquad \bar{\theta} \to e^{-i\alpha}\bar{\theta}. \tag{10.213}$$

Thus, we assign an R charge $+1$ to θ and an R charge -1 to $\bar{\theta}$. According to the rules of Grassmann (Berezin) integration (Section 10.6.1), simultaneously,

$$d^2\theta \to e^{-2i\alpha}\,d^2\theta, \qquad d^2\bar{\theta} \to e^{2i\alpha}\,d^2\bar{\theta}. \tag{10.214}$$

Combining Eqs. (10.210), (10.212), (10.213), and (10.214) we conclude that

$$\int d^2\theta\,\mathcal{W}(x,\theta) \to \int \left(e^{-2i\alpha}\,d^2\theta\right)\left[\mathcal{W}(x,\theta)e^{2i\alpha}\right], \tag{10.215}$$

i.e., the integral stays invariant under the transformations (10.211). One can check this statement explicitly by inspecting the component Lagrangian (10.140) of the Wess–Zumino model, setting $m = 0$ in (10.143).

The general lesson is that a given supersymmetric theory is R invariant provided that the R charge of the superpotential is $+2$.

Supersymmetric QED is also R invariant at $m = 0$, i.e., at vanishing superpotential. The R charge of the superfield V is zero while that of the chiral superfield W_α is $+1$. In other words,

$$r(\lambda) = 1, \qquad r(D) = 0, \qquad r(F_{\alpha\beta}) = 0, \qquad r(\bar{\lambda}) = -1. \tag{10.216}$$

The R charges of Q and $\tilde{Q}$ can be taken to be $\frac{2}{3}$, for instance,

$$r(q) = r(\tilde{q}) = \tfrac{2}{3}, \qquad r(\psi) = r(\tilde{\psi}) = -\tfrac{1}{3}, \tag{10.217}$$

and so on.

The above R charges are sometimes referred to as the canonical R charges and the corresponding R current as the *geometric R current*. In fact, the actual situation with the conserved R current is more complicated. As a rule the conserved R current, if it exists, is a combination of the geometric R current and the flavor currents. We will have more encounters with the R symmetry and R parity in what follows (e.g., Section 10.16.7.1).

| Geometric R current |

10.8 Nonrenormalization Theorem for F Terms

When we speak of renormalization, we are implying the calculation of an effective Lagrangian, starting from a bare Lagrangian formulated at a high ultraviolet scale M_0 that must be viewed as the scale of the ultraviolet cutoff. We calculate this effective Lagrangian at a scale μ, assuming that $\mu \ll M_0$. This calculation can be carried out either in perturbation theory, loop by loop, or including nonperturbative effects.

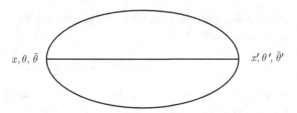

$x, \theta, \bar{\theta}$ ... $x', \theta', \bar{\theta}'$

Fig. 10.1   A typical two-loop supergraph for the vacuum energy.

Originally, the theorem stating the nonrenormalization of F terms in the effective Lagrangian was proved [20, 44] in perturbation theory. That is why, as we will see later, it can be violated nonperturbatively in certain theories. An extension of the theorem covering nonperturbative effects in some other theories was worked out in [45]. We will review this too.

At this stage I face a rather peculiar situation. A discussion of the Feynman graph calculations using supergraph formalism is beyond the scope of the present course.[32] And yet, I would like to explain that this formalism implies the vanishing of the loop corrections for the F terms. To this end, of necessity I will have to invoke heuristic arguments.

The nonrenormalization theorem derives from the observation [44] that in supergraph perturbation theory any radiative (loop) correction to the effective action can always be written as a full superspace integral $\int d^4\theta$, with an integrand that is a local function of superfields. The F terms are integrals over chiral subspaces and therefore cannot receive quantum corrections.

I will try to illustrate the above statement in a somewhat more quantitative manner. To warm up let us start from the vacuum energy, which, as we already know, vanishes in all theories with unbroken supersymmetry: consider the typical two-loop vacuum (super)graph shown in Fig. 10.1.

Each line on the graph represents a Green's function of some superfield. We do not need to know these Green's functions explicitly. The crucial point is that (if one works in the coordinate representation) each interaction vertex can be written as an integral over $d^4x\, d^2\theta d^2\bar{\theta}$. Assume that we substitute explicit expressions for Green's functions and vertices in the integrand and carry out integration over the (super)coordinates of the second vertex, keeping the first vertex fixed. As a result, we will arrive at an expression of the form

$$E_{\text{vac}} = \int d^4x\, d^2\theta d^2\bar{\theta} \times \left(\text{a function of } x_\mu, \theta_\alpha, \bar{\theta}_{\dot{\alpha}}\right). \tag{10.218}$$

Since superspace is homogeneous (there are no points that are singled out, we can freely make supertranslations since any point in the superspace is equivalent to any other point) the integrand in Eq. (10.218) can only be a constant. If so, the result vanishes because of the Berezin rules of integration over the Grassmann variables θ and $\bar{\theta}$.

What remains to be demonstrated is that the one-loop vacuum (super)graph, not representable in this form, also vanishes. The one-loop (super)graph, however, is the

---

[32] The reader interested in this formalism is referred to [9, 10, 13, 20].

same as for free particles and we know already that for free particles $E_{\text{vac}} = 0$, see Eqs. (10.88) and (10.89), thanks to the balance between the bosonic and fermionic degrees of freedom.

This concludes the proof of the fact that if the vacuum energy is zero at the classical level it remains zero to any finite order – there is no renormalization. What changes if, instead of the vacuum energy, we consider the renormalization of the F terms?

The proof presented above can be readily modified to include this case as well. Technically, instead of vacuum loops we must consider now loop (super)graphs in a background field.

Shifman–Vainshtein proof

The basic idea is as follows. In any supersymmetric theory there are several – at least four – supercharge generators. In a generic background all supersymmetries are broken since the background field is *not* invariant under supertransformations, generally speaking. One can select a "magic" background field, however, which leaves part of the supertransformations as valid symmetries. For this specific background field some terms in the effective action will vanish and others will not. (Typically, the F terms do not vanish.) The nonrenormalization theorems refer to those terms which do not vanish in the background field chosen.

Consider, for definiteness, the Wess–Zumino model discussed in Section 10.6.4. An appropriate choice of background field in this case is

$$\bar{\phi}_0 = 0, \qquad \phi_0 = C_1 + C_2^\alpha \theta_\alpha + C_3 \theta^2, \qquad (10.219)$$

where $C_{1,2,3}$ are c-numerical constants and the subscript 0 indicates the background field. In making this choice we are assuming that ϕ_0 and $\bar{\phi}_0$ are treated as independent background fields, that are not connected by complex conjugation (i.e., we have in mind a kind of analytic continuation). The x-independent chiral field (10.219) is invariant under the action of $\bar{Q}_{\dot\alpha}$, i.e., under the following transformations:

$$\delta\theta_\alpha = 0, \qquad \delta\bar{\theta}_{\dot\alpha} = \bar{\varepsilon}_{\dot\alpha}, \qquad \delta x_{\alpha\dot\alpha} = -2i\theta_\alpha \bar{\varepsilon}_{\dot\alpha}. \qquad (10.220)$$

Next, to calculate the effective action we decompose the superfields as follows:

$$\Phi = \phi_0 + \Phi_{\text{qu}}, \qquad \bar{\Phi} = \bar{\phi}_0 + \bar{\Phi}_{\text{qu}}, \qquad (10.221)$$

where the subscript "qu" denotes the quantum part of the superfield, Then we expand the action in Φ_{qu} and $\bar{\Phi}_{\text{qu}}$, dropping the linear terms, and treat the remainder as the action for the quantum fields. Next we integrate out the quantum fields, order by order, keeping the background field fixed. The crucial point is that in the given background field (i) the integral $\int d^2x\, \mathcal{W}$ does *not* vanish and (ii) there exists an exact supersymmetry under $\bar{Q}$-generated supertransformations.

This means that boson–fermion degeneracy holds just as in the "empty" vacuum. All lines in the graph in Fig. 10.1 must be treated now as Green's functions in the background field (10.219). After substituting these Green's functions and integrating over all vertices except the first, we arrive at an expression of the type

$$\int d^4x\, d^2\bar{\theta} \times \left(a\ \bar{\theta}_{\dot\alpha}\text{-independent function}\right) = 0. \qquad (10.222)$$

The $\bar{\theta}$-independence follows from the fact that our superspace is homogeneous in the $\bar{\theta}$ direction even in the presence of the background field (10.219). This completes the proof [46] of F-term nonrenormalization.

The kinetic term $\int d^4\theta\,\bar{\Phi}\Phi$ vanishes in the background (10.219), so nothing can be said about its renormalization from the above argument; explicit calculation tells us that this term gets renormalized in loops, of course.

Remark: following a similar line of reasoning it is not difficult to prove [46] (see the footnote on p. 481 of [46]; see also [38, 47]) that the Fayet–Iliopoulos term is not renormalized at two and higher loops. In addition, ξ is not renormalized at one loop if the matter sector is nonchiral with regard to the given U(1), i.e., if all chiral superfields enter in pairs with the opposite electric charges, as, for example, in Section 10.6.9 where the U(1) charge of Q is +1 while that of $\tilde{Q}$ is −1.

| *The Fayet–Iliopoulos term is not renormalized.* |

Now we will discuss the F-term nonrenormalization theorem from another perspective, suggested by Seiberg [45], which, in certain instances, allows one to go beyond perturbation theory. Consider the coupling constants that appear in the superpotential (the masses, Yukawa couplings, etc.) as classical background chiral superfields. It then follows that these couplings can only appear in the effective superpotential holomorphically, i.e., if λ is a coupling then only λ and not $\bar{\lambda}$ can appear in any quantum corrections to the superpotential since the superpotential $\mathcal{W}$ is a function only of the chiral superfields. This simple observation allows one in many instances to prove the nonrenormalization theorem at the nonperturbative level.

| *Seiberg's proof* |

Consider as an example the Wess–Zumino model of Section 10.6.4 with superpotential (10.143). By holomorphy, the effective superpotential is

$$\mathcal{W}_{\text{eff}} = f(\Phi, m, \lambda), \tag{10.223}$$

i.e., it is a function of Φ, m, and λ and *not of* their complex conjugates. The theory under consideration has no global symmetries other than supersymmetry. At $m \neq 0$ the R symmetry is explicitly broken.

However, one can restore it if, simultaneously with the rotations (10.211) corresponding to $r(\Phi) = \frac{2}{3}$, one rotates m with $r(m) = \frac{2}{3}$. Then both terms in the superpotential $\mathcal{W}(\Phi) = (m/2)\Phi^2 - (\lambda/3)\Phi^3$ (see Eq. (10.143)) transform in the same way and R symmetry is recovered.

In the same way one can add another U(1) symmetry, with the charges $Q_{\text{U(1)}}$ given in Table 10.2.

The above symmetries (plus dimensional arguments) imply that the effective superpotential must take the following form:

Table 10.2. The $Q_{\text{U(1)}}$ and R charges

Fields or parameters	U(1) charges	R charges
Φ	+1	$\frac{2}{3}$
m	−2	$\frac{2}{3}$
λ	−3	0

$$\mathcal{W}_{\text{eff}} = m\Phi^2 \times f\left(\frac{\lambda\Phi}{m}\right)$$

$$= \sum_{n=-\infty}^{\infty} c_n \lambda^n m^{1-n} \Phi^{n+2}, \tag{10.224}$$

where f is a function and the c_n are numerical coefficients in its Laurent expansion.

Next we observe that at $\lambda = 0$ the theory is free, which requires all coefficients c_n with negative n to vanish. Moreover, the Wilsonian effective action cannot be singular in m in the limit $m \to 0$. This is due to the fact that by definition the Wilsonian action[33] contains no contributions from virtual momenta below μ. This excludes $n > 1$, leaving us with only two terms in the second line of Eq. (10.224), namely, $n = 0$ and 1; this implies in turn that $\mathcal{W}_{\text{eff}}$ coincides with the bare superpotential. There is no renormalization.

That the complex parameters in $\mathcal{W}_{\text{eff}}$ are not renormalized does not mean that no physical amplitudes proportional to powers of λ receive quantum corrections from loops. Renormalization comes from the kinetic term:

$$\int d^4\theta\, \bar{\Phi}\Phi \to Z \int d^4\theta\, \bar{\Phi}\Phi, \tag{10.225}$$

where Z is the field renormalization factor, implying that

$$m_{\text{r}} = \frac{m}{Z}, \qquad \lambda_{\text{r}} = \frac{\lambda}{Z^{3/2}}. \tag{10.226}$$

The subscript r indicates the renormalized mass and coupling constant. The scattering amplitudes contain m_{r} and λ_{r}. The field renormalization Z factor drops out if we consider the ratio

$$\frac{m_{\text{r}}^3}{\lambda_{\text{r}}^2} = \frac{m^3}{\lambda^2}. \tag{10.227}$$

This ratio presents an example of a physically measurable quantity – a domain-wall tension in the Wess–Zumino model – that receives no quantum corrections.

The nonrenormalization theorem for F terms and its possible nonperturbative violations are crucial in two practical problems of paramount importance: the mass hierarchy problem and the related issue of dynamical supersymmetry breaking.

10.9 Super-Higgs Mechanism

When a charged chiral superfield acquires a nonvanishing expectation value, the gauge symmetry is spontaneously broken. In the usual Higgs mechanism, gauge bosons "eat" scalars and become massive. In supersymmetry they will "eat" chiral superfields. We will first familiarize ourselves with this phenomenon as it occurs in supersymmetric QED (see Section 10.6.9) and then generalize it to non-Abelian theories.

---

[33] By construction, the effective action does not include one-particle-reducible diagrams. Analyzing the expansion in (10.224), one can observe that its structure is exactly that of a tree diagram.

As we know from Section 10.4.6 (see Eq. (10.95) with $j = 1/2$) the *massive* vector superfield contains four fermionic states (one Dirac fermion) and four bosonic states (one vector particle with three polarizations plus one real scalar particle). However, a *massless* vector superfield has only two bosonic and two fermionic states. Thus, to become massive, it has to "swallow" two bosonic and two fermionic states, which is exactly the content of a chiral superfield.

Let us examine the super-Higgs mechanism at work in the simplest example of U(1) gauge theory [48], the supersymmetric QED presented in Section 10.6.9. It is instructive to start from its nonsupersymmetric version, scalar QED with Lagrangian

$$\mathcal{L} = -\frac{1}{4e^2} F_{\mu\nu} F^{\mu\nu} + \left|\left(\partial_\mu - iA_\mu\right)\varphi\right|^2 - h\left(\left|\varphi^2\right| - v^2\right)^2, \qquad (10.228)$$

where φ is a complex field, v is a real parameter, and h is a coupling constant (which at the very end will be assumed to be small, $h \to 0$). One can parametrize φ by its modulus and phase,

$$\varphi \equiv \rho \, \exp(i\alpha). \qquad (10.229)$$

Then the potential term in (10.228) forces ρ to develop a vacuum expectation value,

$$\rho_{\text{vac}} = v. \qquad (10.230)$$

The phase α can be gauged away if one imposes the (unitary) gauge condition $\varphi \equiv \rho$. We are left with a real scalar field (the physical Higgs particle), described by the fluctuations in ρ near its VEV, plus a massive vector boson, i.e., a "W boson," with mass

$$m_W^2 = 2e^2 \rho_{\text{vac}}^2 = 2e^2 v^2. \qquad (10.231)$$

The mass of the Higgs particle is

$$m_H^2 = 4hv^2. \qquad (10.232)$$

It tends to zero at $h \to 0$. The balance of degrees of freedom is as follows: before the launch of the Higgs mechanism we have $2 + 2$ and after its launch we have $3 + 1$, where "3" represents the degrees of freedom of the massive vector field (with three polarizations) while "1" represents the single degree of freedom residing in the real scalar field.

Now we turn to supersymmetric QED. Let us have a closer look at the Lagrangian (10.191) with scalar potential (10.190) at $m = 0$. When $m = 0$ the theory has a flat direction; the selectron fields acquire VEVs that can be parametrized as in Eq. (10.198). In the latter equation a possible phase difference between q and $\tilde{q}$ has already been gauged away. Thus, it presents an analog of Eq. (10.229) with the phase set to zero (i.e., $\varphi \to \rho$).

Substituting the selectron VEVs into Eq. (10.191) we get for the W-boson mass

$$m_W^2 = 2e^2 \xi \cosh 2\rho = 2\, e^2 \xi \sqrt{1 + \bar{\varphi}\varphi}, \qquad (10.233)$$

where the moduli field φ was defined in Eq. (10.199). The same mass is acquired by a real scalar field and a Dirac spinor (two Weyl spinors). Before the onset of the Higgs

Super-Higgs
mechanism
in the Wess–
Zumino
gauge
regime we have three chiral superfields, $W_\alpha, Q,$ and $\tilde{Q}$ ($3\times(2+2)$ degrees of freedom). After the onset of the Higgs regime we have one massive vector supermultiplet (4 + 4 degrees of freedom) and one massless chiral superfield Φ (2 + 2 degrees of freedom), which has a VEV on the flat direction. All degrees of freedom are balanced.

The vacuum energy density vanishes and supersymmetry remains unbroken. At the same time, the U(1) gauge symmetry is realized in the Higgs regime in any vacuum on the flat direction. This explains the origin of the term "Higgs branch."

Unitary
gauge
The above consideration was carried out in the Wess–Zumino gauge. Needless to say, one could choose another gauge. A supergeneralization of the unitary gauge is singled out. Using the supergauge transformation for $\tilde{Q}$ one can always reduce $\tilde{Q}$ to an arbitrary c-numerical constant. We will impose the following gauge condition:

$$\tilde{Q} = \sqrt{\xi}. \tag{10.234}$$

Then the chiral invariant $Q\tilde{Q}$ is given by

$$Q\tilde{Q} = \sqrt{\xi}\,Q \equiv \frac{\xi}{2}\Phi. \tag{10.235}$$

In other words, the moduli superfield Φ becomes a linear function of the original chiral matter superfield Q:

$$\Phi = \frac{2}{\sqrt{\xi}}Q. \tag{10.236}$$

Thus the physical Higgs particle and its fermion superpartner – the component fields of Φ – coincide up to normalization with the component fields of Q.

In this gauge the Lagrangian (10.181) (at $m = 0$ and with the Fayet–Iliopoulos term switched on) takes the form

$$\mathcal{L} = \left(\frac{1}{4e^2}\int d^2\theta\, W^2 + \text{H.c.}\right) + \int d^4\theta \left(\frac{\xi}{4}\bar{\Phi}e^V\Phi + \xi e^{-V} - \xi V\right). \tag{10.237}$$

To study the vacuum structure one should discard the massive degrees of freedom, which amounts to crossing out the kinetic term $\int d^2\theta\, W^2$. Then the superfield V becomes nondynamical and can be determined in terms of $\bar{\Phi}\Phi$ by virtue of the equation of motion. The latter is obtained by differentiating the second term in Eq. (10.237) over V and setting the result equal to 0,

$$\frac{\xi}{4}\bar{\Phi}\Phi e^V = \xi\left(e^{-V} + 1\right), \tag{10.238}$$

implying that

$$e^{-V_0} = -\frac{1}{2} + \frac{1}{2}\sqrt{1 + \bar{\Phi}\Phi}, \qquad V_0 = -\ln\frac{\sqrt{1 + \bar{\Phi}\Phi} - 1}{2}. \tag{10.239}$$

This in turn allows one immediately to obtain the Kähler potential for the moduli superfield (10.236). Indeed, let us substitute Eqs. (10.239) into the second term in (10.237) remembering that the kinetic term for the vector field is omitted. Then we get

$$\mathcal{L}_\Phi = \xi \int d^4\theta \left(\sqrt{1 + \bar{\Phi}\Phi} + \ln \frac{\sqrt{1 + \bar{\Phi}\Phi} - 1}{2} \right). \qquad (10.240)$$

Observe that the Kähler potential is defined modulo an arbitrary function $f(\Phi)$+H.c., which drops out upon integration over $d^4\theta$. Adding $-\frac{1}{2}\ln\bar{\Phi}\Phi$, we derive from Eq. (10.240) precisely the Kähler potential obtained in Section 10.6.10; see Eq. (10.205).

To find the spectrum of massive excitations residing in the superfield V and their scattering amplitudes, we split the vector superfield V in two parts, the vacuum field and the quantum fluctuations, writing

$$V = V_0 + \delta V,$$

$$\delta V \equiv c + i\theta\chi - i\bar{\theta}\bar{\chi} + \frac{i}{\sqrt{2}}\theta^2 M - \frac{i}{\sqrt{2}}\bar{\theta}^2 \bar{M}$$

$$- 2\theta^\alpha \bar{\theta}^{\dot\alpha} v_{\alpha\dot\alpha} + \left[2i\theta^2 \bar{\theta}_{\dot\alpha} \left(\bar{\lambda}^{\dot\alpha} - \frac{i}{4}\partial^{\alpha\dot\alpha}\chi_\alpha \right) + \text{H.c.} \right]$$

$$+ \theta^2 \bar{\theta}^2 \left(D - \frac{1}{4}\partial^2 c \right). \qquad (10.241)$$

We then substitute the expression for δV into the Lagrangian:

$$\mathcal{L} = \left(\frac{1}{4e^2} \int d^2\theta\, W^2 + \text{H.c.} \right)$$

$$+ \int d^4\theta \left[\frac{\xi}{4}\bar{\Phi}\Phi \left(e^{V_0} e^{\delta V} \right) + \xi e^{-V_0} e^{-\delta V} - \xi(V_0 + \delta V) \right]. \qquad (10.242)$$

Expanding in δV and using Eq. (10.239) confirms the expression for the W-boson mass quoted in (10.233).

It is instructive to trace the fate of the various component fields in the vector superfield. The field χ becomes dynamical and pairs up with λ to form a Dirac spinor. The real field c becomes dynamical too; together with the three polarizations of $v_{\alpha\dot\alpha}$ it forms the bosonic sector of the massive vector supermultiplet. The field M enters with no derivatives and is nondynamical, and so is D.

The "superunitary" gauge has its advantages and disadvantages. It makes explicit the bookkeeping of the degrees of freedom in two distinct supermultiplets: the massive vector field and its superpartners in one superfield plus the moduli fields in the other. The moduli superfield is just Q. However, this gauge is inconvenient for practical calculations of the scattering amplitudes since the dependence on c in the Lagrangian is nonpolynomial.

Exercise

10.9.1 Write down the mass matrix for the fermion fields in the Lagrangian (10.191), with scalar potential (10.196), at the following point on the vacuum manifold: $q\tilde{q} = (\xi/2)\varphi$ (see Eq. (10.199)). Determine the masses and eigenstates in the fermion sector of the theory by diagonalization of this mass matrix.

10.10 Spontaneous Breaking of Supersymmetry

From Section 10.4.3 we know that in theories with unbroken Lorentz symmetry, supersymmetry is *spontaneously broken* if any supercharge does *not* annihilate the vacuum state. The inverse is also true: if the vacuum state is annihilated by all supercharges then supersymmetry is unbroken and the vacuum energy vanishes. Let us ask ourselves what this implies in terms of the order parameters signaling supersymmetry breaking.

To answer this question we must examine the supertransformations (10.120) and (10.185), namely,

$$\left[\left(Q\epsilon + \bar{Q}\bar{\epsilon}\right), \psi_\alpha\right] \sim \delta\psi_\alpha = -\sqrt{2}i\,\partial_{\alpha\dot\alpha}\phi\bar{\epsilon}^{\dot\alpha} + \sqrt{2}\epsilon_\alpha F,$$

$$[Q\epsilon, \lambda_\alpha] \sim \delta\lambda_\alpha = iD\epsilon_\alpha - F_{\alpha\beta}\epsilon^\beta. \tag{10.243}$$

Averaging the left- and right-hand sides of these relations over the ground state we conclude that supersymmetry is spontaneously broken if either the F or the D component has a nonvanishing VEV.[34] If so then the supercharge, acting on the vacuum, instead of annihilating it creates the corresponding fermion: either ψ or λ (see Section 10.11 below). Note that in the Lorentz-invariant vacuum neither $\partial_{\alpha\dot\alpha}\phi$ nor $F_{\alpha\beta}$ can have expectation values. An additional lesson one should remember is that an x-independent vacuum expectation value of the lowest component of the chiral superfield does not lead to supersymmetry breaking, generally speaking. This fact was used in Section 10.8.

Out of a variety of models exhibiting spontaneous supersymmetry breaking, the majority reduce – either directly or in the low-energy limit – to one of two patterns: F-term breaking or D-term breaking.

10.10.1 The O'Raifeartaigh Mechanism

The F-term-based mechanism, also known as the O'Raifeartaigh mechanism [49] (it was devised by O'Raifeartaigh), works by a "conflict of interests" between the F terms of the various fields belonging to the matter sector. The necessary and sufficient condition for the existence of supersymmetric vacua is the vanishing of *all* F terms. For generic superpotentials this is possible to achieve.

In the O'Raifeartaigh construction, the superpotentials are arranged in such a way that it is impossible to make all F terms vanish simultaneously.

One needs at least three matter fields to realize the phenomenon in renormalizable models with polynomial superpotentials. With one or two matter fields and a polynomial superpotential a supersymmetric vacuum solution always exists. With three superfields and a generic superpotential, a supersymmetric solution exists too but it ceases to exist for some degenerate superpotentials.

Consider the superpotential

$$\mathcal{W}(\Phi_1, \Phi_2, \Phi_3) = \lambda_1\Phi_1(\Phi_3^2 - M^2) + \mu\Phi_2\Phi_3. \tag{10.244}$$

---

[34] This statement assumes that the vacuum does not break the Lorentz symmetry.

Then

$$\bar{F}_i = -\frac{\partial \mathcal{W}}{\partial \Phi_i} = \begin{cases} \lambda_1(\phi_3^2 - M^2), & i = 1, \\ \mu \phi_3, & i = 2, \\ 2\lambda_1 \phi_1 \phi_3 + \mu \phi_2, & i = 3. \end{cases} \tag{10.245}$$

The vanishing of the second line implies that $\phi_3 = 0$; then the first line cannot vanish. There is no solution for which $F_1 = F_2 = F_3 = 0$; therefore supersymmetry is spontaneously broken.

What is the minimal energy configuration? It depends on the ratio $\lambda_1 M/\mu$. For instance, at $M^2 < \mu^2/(2\lambda_1^2)$ the minimum of the scalar potential occurs at $\phi_2 = \phi_3 = 0$. The value of ϕ_1 can be arbitrary: an indefinite equilibrium takes place at the tree level. (The loop corrections to the Kähler potential lift this degeneracy and lock the vacuum at $\phi_1 = 0$.) Then $F_2 = F_3 = 0$ and the vacuum energy density is obviously $\mathcal{E} = |F_1|^2 = \lambda_1^2 M^4$.

Since $F_1 \neq 0$ the fermion from the same superfield, ψ_1, is the massless Goldstino (see Section 10.11):

$$m_{\psi_1} = 0.$$

It is not difficult to calculate the masses of other particles. Assume that the vacuum expectation value of the field ϕ_1 vanishes. Then the fluctuations of ϕ_1 remain massless (and degenerate with ψ_1). The Weyl field ψ_2 and the quanta of ϕ_2 are also degenerate; their common mass is μ. At the same time, the fields from Φ_3 split: the Weyl spinor ψ_3 has mass μ, while

$$m_a^2 = \mu^2 - 2\lambda_1^2 M^2, \qquad m_b^2 = \mu^2 + 2\lambda_1^2 M^2, \tag{10.246}$$

where the real fields a and b are defined by

$$\phi_3 \equiv \frac{1}{\sqrt{2}}(a + ib).$$

Note that, despite the mass splitting,

$$m_a^2 + m_b^2 - 2m_{\psi_3}^2 = 0, \tag{10.247}$$

as if there were no supersymmetry breaking. Equation (10.247) is a particular example of the general supertrace relation [50]

$$\text{Str } \mathcal{M}^2 \equiv \sum_J (-1)^{2J}(2J + 1)m_J^2 = 0, \tag{10.248}$$

Supertrace mass formula

where Str stands for the supertrace, $\mathcal{M}^2$ is the squared mass matrix of the real fields in the supermultiplet; the subscript J indicates the spin of the particle. Equation (10.248) is valid at tree level for spontaneous supersymmetry breaking through F terms. Quantum (loop) corrections, generally speaking, modify it; it also becomes modified in theories where (part of the) supersymmetry breaking occurs by the Fayet–Iliopoulos (D-term-based) mechanism.

A combined conclusion to the first part of Section 10.8 and to Section 10.10.1 is in order here. *Theorem: If supersymmetry is unbroken at tree level in a given model, i.e., all F terms vanish for a certain field configuration, then supersymmetry is not broken to any order in perturbation theory. Reservation: This assertion refers to models without a U(1) gauge subsector. For such models a Fayet–Iliopoulos term is possible.*

10.10.2 The Fayet–Iliopoulos Mechanism

The Fayet–Iliopoulos mechanism [41], also called the D-term mechanism, applies in models where the gauge sector includes a U(1) subgroup. The simplest and most transparent example is supersymmetric QED (Section 10.6.10). The D component of the vector superfield develops a nonvanishing VEV, implying spontaneous supersymmetry breaking provided that neither the Fayet–Iliopoulos term nor the mass term $\int d^2\theta \, m Q\tilde{Q}$ vanish.

Equation (10.190) shows that for massive matter ($m \neq 0$) a zero vacuum energy is not attainable. The mass terms in the scalar potential require q and $\tilde{q}$ to vanish in the vacuum, while the D term in the scalar potential requires $|q|^2 = |\tilde{q}|^2 + \xi$. If $\xi \neq 0$, both conditions cannot be met simultaneously. Where then does the vacuum state lie?

Qualitatively it is clear that, on the one hand, when $|m|$ is very large, the F terms prevail over the D term, pushing the vacuum field configuration towards the origin. On the other hand, when ξ is very large the D term prevails over the F terms. Quantitatively, if $\xi > m^2/e^2$ then the minimal energy is achieved at

$$\bar{\tilde{q}}\tilde{q} = 0, \qquad \bar{q}q = \xi - \frac{m^2}{e^2}. \tag{10.249}$$

The vacuum energy density is

$$\mathcal{E} = m^2\left(\xi - \frac{m^2}{2e^2}\right) \neq 0. \tag{10.250}$$

The gauge U(1) symmetry is broken too. The phase of q is eaten up in the super-Higgs mechanism and the photon becomes massive:

$$m_W = e\sqrt{2}\sqrt{\xi - (m^2/e^2)}. \tag{10.251}$$

It is instructive to compare these results and expressions with those obtained in Section 10.9 for $m = 0$.

One linear combination of the photino λ and $\tilde{\psi}$ is the Goldstino; it is massless. Other linear combinations, and the scalar and spinor fields from $\tilde{Q}$ are massive.

If $\xi < m^2/e^2$ the selectron fields develop no VEVs and the vacuum configuration corresponds to

$$\bar{\tilde{q}}\tilde{q} = \bar{q}q = 0. \tag{10.252}$$

The D term becomes equal to $e^2\xi$, while the vacuum energy is $\mathcal{E} = e^2\xi^2/2$. The gauge U(1) symmetry remains unbroken: the photon is massless, while the photino assumes the role of the Goldstino. The fermion part of the matter sector does not feel the broken supersymmetry (at the tree level),

$$m_\psi = m_{\tilde{\psi}} = m, \tag{10.253}$$

while the boson part does,

$$m_{\tilde{q}}^2 = m^2 + e^2\xi, \qquad m_q^2 = m^2 - e^2\xi. \tag{10.254}$$

Exercise

10.10.1 Write down the mass matrix for the fermion fields in the Lagrangian (10.191) with scalar potential (10.190) in the vacuum (10.249). Determine the masses in the fermion sector of the theory by diagonalizing this mass matrix for the small-ξ case; see Eq. (10.252). Do the same for the bosons. Repeat the same exercise for large ξ.

10.11 Goldstinos

As mentioned earlier, if the F or D terms develop a nonvanishing expectation value then the fermion fields from the corresponding superfields produce massless Goldstinos. In this section I will give a formal proof of this statement that is valid irrespective of whether the theory under consideration is at weak or strong coupling [51]. Moreover, we will determine the Goldstino's coupling to the (spontaneously broken) supercharges and prove that this coupling is proportional to the square root of the vacuum energy density.

First, we consider the two-point vacuum correlation function

$$G_\mu^{\dot\alpha\beta}(p) = -i \int d^4x\, e^{ipx} \left\langle \mathrm{vac} \left| T\left(\bar{J}_\mu^{\dot\alpha}(x), \bar\psi^\beta(0) \right) \right| \mathrm{vac} \right\rangle, \qquad (10.255)$$

where T stands for the T product and J_μ^α is the supercurrent, cf. Eq. (10.70). In what follows we will omit explicit mention of "vac" since the only correlators considered here are those averaged over the vacuum state. We will calculate the limiting value

$$\lim_{p\to 0} \left[p^\mu\, G_\mu^{\dot\alpha\beta}(p) \right] = \lim_{p\to 0} \int d^4x\, e^{ipx} \left\langle \partial^\mu T\left(\bar{J}_\mu^{\dot\alpha}(x), \bar\psi^\beta(0) \right) \right\rangle. \qquad (10.256)$$

Since $\partial^\mu \bar{J}_\mu^{\dot\alpha} = 0$, the derivative acts only on T, producing

$$\lim_{p\to 0} \left[p^\mu\, G_\mu^{\dot\alpha\beta}(p) \right] = \lim_{p\to 0} \int d^4x\, e^{ipx} \left\langle \left\{ \bar{J}_0^{\dot\alpha}(x), \bar\psi^\beta(0) \right\} \right\rangle \delta(t)$$
$$= \left\langle \bar{Q}^{\dot\alpha} \bar\psi^\beta + \bar\psi^\beta \bar{Q}^{\dot\alpha} \right\rangle = i\sqrt{2}\, \bar{F}_{\mathrm{vac}}\, \varepsilon^{\dot\alpha\beta}. \qquad (10.257)$$

By assumption $\bar{F}_{\mathrm{vac}} \neq 0$. Let us ask how the left-hand side, which contains an explicit factor of p, can remain nonvanishing in the limit $p \to 0$. The only solution is a $1/p$ pole in $G_\mu^{\dot\alpha\beta}(p)$. More exactly,

$$\lim_{p\to 0} G_\mu^{\dot\alpha\beta} = \left(i\sqrt{2}\, \bar{F}_{\mathrm{vac}}\, \varepsilon^{\dot\alpha\beta} \right) \times \frac{p_\mu}{p^2}. \qquad (10.258)$$

Then Eq. (10.257) is satisfied. The pole in $G_\mu^{\dot\alpha\beta}$ proves the existence of a massless fermion – the Goldstino – produced from the vacuum by the operator $\bar\psi^\beta$ and annihilated by the supercurrent, with the following constants:

$$\left\langle G \left| \bar\psi^\beta \right| \mathrm{vac} \right\rangle \sim \bar{u}^\beta, \qquad \left\langle \mathrm{vac} \left| \bar{J}_\mu^{\dot\alpha} \right| G \right\rangle \sim i\, \bar{F}_{\mathrm{vac}} \left(\bar\sigma_\mu \right)^{\dot\alpha\beta} u_\beta, \qquad (10.259)$$

where G stands for the Goldstino and u_β is its polarization spinor. One should take into account that

$$u_\beta \bar{u}_{\dot{\beta}} = p_{\beta\dot{\beta}}.$$

(10.260)

Now, let us perform a more general calculation of the Goldstino's coupling to the supercurrent, which in the general case, is defined as

$$\left\langle G \left| J_\mu^\beta \right| \text{vac} \right\rangle = -i f_G\, \bar{u}_{\dot{\alpha}} \left(\bar{\sigma}_\mu \right)^{\dot{\alpha}\beta}, \qquad \left\langle \text{vac} \left| \bar{J}_\nu^{\dot{\alpha}} \right| G \right\rangle = i f_G (\bar{\sigma}_\nu)^{\dot{\alpha}\beta} u_\beta,$$

(10.261)

where f_G can be chosen to be real. To this end, following the same line of reasoning as above, we will consider the correlation function

$$G_{\nu\mu}^{\dot{\alpha}\beta}(p) = -i \int d^4x\, e^{ipx} \left\langle \text{vac} \left| T \left(\bar{J}_\nu^{\dot{\alpha}}(x),\, J_\mu^\beta(0) \right) \right| \text{vac} \right\rangle,$$

(10.262)

multiply it by p^ν, and then let p tend to 0. We then obtain

$$\lim_{p\to 0} \left[p^\nu G_{\nu\mu}^{\dot{\alpha}\beta}(p) \right] = \lim_{p\to 0} \int d^4x\, e^{ipx} \left\langle \left\{ \bar{J}_0^{\dot{\alpha}}(x),\, J_\mu^\beta(0) \right\} \right\rangle \delta(t).$$

(10.263)

Next we use the fact that the anticommutator of the supercharge with the supercurrent is proportional to the energy–momentum tensor $T_{\mu\nu}$,

$$\left\{ \bar{Q}^{\dot{\alpha}},\, J_\mu^\beta \right\} = (\bar{\sigma}^\nu)^{\dot{\alpha}\beta} \left(2T_{\mu\nu} + 2\Sigma_{[\mu\nu]} \right),$$

(10.264)

Semilocal form of superalgebra

where $\Sigma_{[\mu\nu]}$ is an antisymmetric operator whose $0i$ components are full spatial derivatives. We will derive this relation in Section 10.16. If we set $\mu = 0$ and integrate over 3-space, we arrive at the superalgebra relation (10.74). Moreover, owing to the Lorentz invariance of the vacuum state, for the vacuum expectation value of $T_{\mu\nu}$ we have

$$\left\langle T_{\mu\nu} \right\rangle = \mathcal{E}_{\text{vac}}\, g_{\mu\nu}.$$

(10.265)

Combining Eqs. (10.263)–(10.265) we get

$$\lim_{p\to 0} \left[p^\nu G_{\nu\mu}^{\dot{\alpha}\beta}(p) \right] = 2\mathcal{E}_{\text{vac}} \left(\bar{\sigma}_\mu \right)^{\dot{\alpha}\beta}.$$

(10.266)

The Goldstino contribution to (10.262) produces a pole at small p whose residue is determined by (10.261),

$$G_{\nu\mu}^{\dot{\alpha}\beta}(p) = f_G^2 \frac{p^\rho}{p^2} \left(\bar{\sigma}_\nu \sigma_\rho \bar{\sigma}_\mu \right)^{\dot{\alpha}\beta}.$$

(10.267)

Multiplying by p^ν and comparing with (10.266) we obtain

$$f_G^2 = 2\,\mathcal{E}_{\text{vac}},$$

(10.268)

as required.

An immediate consequence of the above consideration is the following theorem.

Theorem: If a given theory has no fermion(s) that could play the role of the massless Goldstino, supersymmetry cannot be spontaneously broken. This is the case for instance in weakly coupled theories that are supersymmetric at tree level, in which all fermions are massive. Another obstacle to the occurrence of a Goldstino even in the presence of massless fermions is a mismatch in the global quantum numbers. Assume that the theory under consideration has an unbroken global symmetry. If the charge of the massless fermion with respect to this symmetry does not coincide with that of the supercurrent, the fermion cannot assume the Goldstino role.

A concluding remark is in order here. The supercurrent may create a massless fermion from the vacuum state by virtue of a derivative coupling,

$$\left\langle \text{ferm} \left| J_\mu^\beta \right| \text{vac} \right\rangle = g p_\mu u^\beta, \tag{10.269}$$

Derivative coupling does not give rise to super-symmetry breaking.

where p_μ is the fermion's momentum and g is a coupling constant. Such a derivative coupling gives a matrix element that vanishes at small p, which implies, in turn, that the derivatively coupled fermion cannot produce a pole in (10.262) and hence is not the Goldstino. When one says "the supercurrent creates a massless fermion from the vacuum," one usually means a nonderivative coupling as in Eq. (10.261).

10.12　Digression: Two-Dimensional Supersymmetry

There is no genuine spin in two dimensions because there are no spatial rotations. Nevertheless spinors can be introduced. Moreover, in two dimensions one can require spinors to be both chiral and Majorana simultaneously. Therefore there exist a number of "exotic" supersymmetries. They will not be considered here. We will focus on the simplest cases, $\mathcal{N} = 1$ (two real supercharges, one left-handed, one right-handed; this supersymmetry is also referred to as $\mathcal{N} = (1, 1)$), and $\mathcal{N} = 2$ (four real supercharges, or two complex, half left-handed, another half right-handed; also known as the $\mathcal{N} = (2, 2)$ supersymmetry).

Cf. Sections 2.4, 2.5, 6.1, 8.1, 9.4, 9.5.

10.12.1　Superspace for $\mathcal{N} = 1$ in Two Dimensions

The two-dimensional space $x^\mu = \{t, z\}$ can be promoted to a superspace by adding a two-component real Grassmann variable $\theta_\alpha = \{\theta_1, \theta_2\}$. The coordinate transformations

$$\theta_\alpha \to \theta_\alpha + \epsilon_\alpha, \qquad x^\mu \to x^\mu - i\bar\theta\gamma^\mu\epsilon \tag{10.270}$$

supplement the translations and Lorentz boosts. A convenient representation for the two-dimensional Majorana γ matrices was given in Section 10.2.2, i.e., $\gamma^0 = \sigma_y$, $\gamma^1 = -i\sigma_x$. Chiral subspaces are not introduced, and there is no need for spinors with both upper and lower indices; all spinorial indices are taken to be lower indices. Moreover, $\int d^2x = \int dt\, dz$ and the spinorial derivatives are defined as follows:

$$D_\alpha = \frac{\partial}{\partial\bar\theta_\alpha} - i(\gamma^\mu\theta)_\alpha\partial_\mu, \qquad \bar D_\alpha = -\frac{\partial}{\partial\theta_\alpha} + i(\bar\theta\gamma^\mu)_\alpha\partial_\mu, \tag{10.271}$$

so that

$$\{D_\alpha, \bar D_\beta\} = 2i(\gamma^\mu)_{\alpha\beta}\partial_\mu; \tag{10.272}$$

in (10.271)

$$\bar\theta = \theta\gamma^0. \tag{10.273}$$

We will define the two-dimensional Levi–Civita tensor and the norm of Grassmann integration as follows:

$$\varepsilon_{12} = 1, \qquad \int d^2\theta \, \bar{\theta}\theta = 1. \tag{10.274}$$

With this notation $\left(\gamma^0\right)_{\alpha\beta} = -i\varepsilon_{\alpha\beta}$ and $\bar{\theta}\theta' = \bar{\theta}'\theta$. Moreover, the superalgebra takes the form

$$\{Q_\alpha, \bar{Q}_\beta\} = 2P_\mu(\gamma^\mu)_{\alpha\beta}, \tag{10.275}$$

which implies, in turn, that

$$\boxed{\mathcal{N} = (1,1)\ \text{superalgebra}} \qquad \{Q_\alpha, Q_\beta\} = 2\begin{pmatrix} P_0 - P_z & 0 \\ 0 & P_0 + P_z \end{pmatrix}_{\alpha\beta}. \tag{10.276}$$

10.12.2 $\mathcal{N} = 1$ Superfields and Supersymmetric Kinetic Term

We will deal with a real superfield $\Phi(x, \theta)$ that has the form

$$\Phi(x, \theta) = \phi(x) + \bar{\theta}\psi(x) + \tfrac{1}{2}\bar{\theta}\theta F(x), \qquad \Phi^\dagger = \Phi, \tag{10.277}$$

where θ, ψ are real two-component spinors and ϕ is a real scalar field. The superspace transformations (10.270) generate the following supersymmetry transformations of the component fields:

$$\delta\phi = \bar{\epsilon}\psi, \qquad \delta\psi = -i\partial_\mu\phi\,\gamma^\mu\epsilon + F\epsilon, \qquad \delta F = -i\bar{\epsilon}\gamma^\mu\partial_\mu\psi. \tag{10.278}$$

As usual, the F component is nondynamical (see Eq. (10.282) below). The physical degrees of freedom in Φ are one bosonic (the real scalar field ϕ) and one fermionic (the Majorana spinor ψ). This is in accord with the supermultiplet structure in $\mathcal{N} = 1$ theories in two dimensions. Indeed, following the line of reasoning presented in Section 10.4.6, we will rewrite the two real supercharges in terms of two complex supercharges:

$$Q = \tfrac{1}{\sqrt{2}}(Q_1 + iQ_2), \qquad Q^\dagger = \tfrac{1}{\sqrt{2}}(Q_1 - iQ_2), \tag{10.279}$$

with algebra

$$\left\{Q^\dagger, Q\right\} = 2P_0, \qquad \{Q, Q\} = -2P_z. \tag{10.280}$$

For massive particles we can choose a reference frame in which $P_z = 0$; then only the first anticommutation relation remains informative. If Q annihilates a state $|a\rangle$, its only superpartner is $Q^\dagger|a\rangle$. If the first state is bosonic then the second is fermionic and vice versa.

The supertransformation of the F term reduces to a full derivative; therefore, projecting it out by virtue of $\int d^2\theta$ produces a supersymmetric action. Here it is in order to derive the kinetic term. To this end we first perform spinorial differentiation of the superfield Φ,

$$D_\alpha\Phi = \psi_\alpha + \theta_\alpha F - i\left(\gamma^\mu\theta\right)_\alpha \partial_\mu\phi + \tfrac{1}{2}i\,\bar{\theta}\theta\left(\gamma^\mu\partial_\mu\psi\right)_\alpha,$$

$$\bar{D}_\alpha\Phi = \bar{\psi}_\alpha + \bar{\theta}_\alpha F + i\left(\bar{\theta}\gamma^\mu\right)_\alpha \partial_\mu\phi - \tfrac{1}{2}i\,\bar{\theta}\theta\left(\bar{\psi}\gamma^\mu\overleftarrow{\partial}_\mu\right)_\alpha. \tag{10.281}$$

A simple inspection of the above expressions suggests that the product $\bar{D}_\alpha \Phi D_\alpha \Phi$ gives rise to the desired structure. Indeed,

$$S_{\text{kin}} = \int d^2\theta \, d^2x \left(\tfrac{1}{2} \bar{D}_\alpha \Phi D_\alpha \Phi \right)$$

$$= \tfrac{1}{2} \int d^2x \left(\partial^\mu \phi \, \partial_\mu \phi + \tfrac{1}{2} i \bar\psi \gamma^\mu \overleftrightarrow{\partial}_\mu \psi + F^2 \right). \tag{10.282}$$

10.12.3 Models

Below we will consider the two most popular models, which appear in numerous applications.

10.12.3.1 Minimal Wess–Zumino Model in Two Dimensions

The action in the minimal two-dimensional Wess–Zumino model is

$$S = \int d^2\theta \, d^2x \left(\tfrac{1}{2} \bar{D}_\alpha \Phi D_\alpha \Phi + 2\mathcal{W}(\Phi) \right), \tag{10.283}$$

where $\mathcal{W}(\Phi)$ will be referred to as the superpotential, keeping in mind a parallel with the four-dimensional Wess–Zumino model although in the two-dimensional case the superpotential term is the integral over the full superspace, and is not chiral.

Lack of the holomorphy in 2D $\mathcal{N}(1,1)$ superpotential

The standard mass term is obtained from $\mathcal{W} = \tfrac{1}{2} m \Phi^2$ while the interaction terms are generated by Φ^3 and higher orders. Note that, while in four dimensions $\mathcal{W}$ is an analytic function of a complex argument, in the minimal two-dimensional Wess–Zumino model $\mathcal{W}$ is just a function of a real argument. In two dimensions any such function leads to a renormalizable field theory and is thus allowed.

In components the Lagrangian takes the form

$$\mathcal{L} = \tfrac{1}{2} \left(\partial_\mu \phi \partial^\mu \phi + \bar\psi i \not\partial \psi + F^2 \right) + \mathcal{W}'(\phi) F - \tfrac{1}{2} \mathcal{W}''(\phi) \bar\psi \psi. \tag{10.284}$$

Superficially this Lagrangian looks similar to that considered in Section 10.6.4; there is a deep difference, however. In four dimensions the field Φ is complex, and, as a result, we have four conserved supercharges (i.e., an $\mathcal{N} = (2,2)$ superalgebra), while the fields in (10.284) are real and the number of conserved supercharges is two, i.e., the supersymmetry with which we are dealing is $\mathcal{N} = (1,1)$.

What is so special about the model (10.284)? The answer is that it gives an example of a "global anomaly" [52]. Let me explain this in more detail. The model (10.284) has no fermion current. Indeed, for the Majorana spinors both $\bar\psi \gamma^\mu \psi$ and $\bar\psi \gamma^\mu \gamma^5 \psi$ vanish identically. However, $(-1)^F$ (i.e., the fermion number modulo 2) is defined. There is no genuine spin in two dimensions. What distinguishes the boson fields from the fermion fields in the Lagrangian (10.284) is the way in which quantization is achieved (i.e., the statistics). The boson fields are quantized by imposing a quantization condition on the canonical commutators, while for the fermions a quantization condition is imposed on the anticommutators. This allows one to introduce $(-1)^F$.

It turns out that beyond perturbation theory $(-1)^F$ is lost [52]; see Section 11.2.8. In the soliton sector $(-1)^F$ ceases to exist. This implies the disappearance of the boson–fermion classification, resulting in abnormal statistics. The fact of the abnormal statistics in the model (10.284) is well established.

10.12.3.2 Supersymmetric O(3) Sigma Model

Look through Chapter 6.

This is a supergeneralization [53] of its famous nonsupersymmetric parent. The model is built on a triplet of real superfields $\sigma^a(x, \theta)$ where the vectorial index $a = 1, 2, 3$ refers to the target space. In components,

$$\sigma^a(x, \theta) = S^a + \bar{\theta}\chi^a + \tfrac{1}{2}\bar{\theta}\theta\, F^a \tag{10.285}$$

where $\vec{S}$ and $\vec{F}$ are bosonic fields while $\vec{\chi}$ denotes a two-component Majorana field.

Formally the Lagrangian has the form of a free kinetic term:

$$\begin{aligned}
\mathcal{L} &= \frac{1}{g^2} \int d^2\theta \left(\tfrac{1}{2}\bar{D}_\alpha \sigma^a D_\alpha \sigma^a \right) \\
&= \frac{1}{2g^2} \left(\partial^\mu S^a \partial_\mu S^a + \tfrac{1}{2} i \bar{\chi}^a \gamma^\mu \overleftrightarrow{\partial}_\mu \chi^a + \vec{F}^2 \right).
\end{aligned} \tag{10.286}$$

However, in fact an interaction is there, hidden in the constraint on the superfields,

$$\sigma^a(x, \theta)\sigma^a(x, \theta) = 1, \tag{10.287}$$

which replaces the nonsupersymmetric version of the constraint, $\vec{S}^2 = 1$.

Decomposing (10.287) into components we get

$$\vec{S}^2 = 1, \qquad \vec{S}\vec{\chi} = 0, \qquad \vec{F}\vec{S} = \tfrac{1}{2}\left(\bar{\chi}^a \chi^a\right). \tag{10.288}$$

As usual, the F term enters with no derivatives. In eliminating F by using the equations of motion one must proceed with care, combining the information encoded in (10.286) and (10.288). The last equation in (10.288) unambiguously determines the longitudinal part of F, while its transverse part must be determined from the equations of motion.

In more detail, let us split F^a as follows:

$$F^a = F^a_\| + F^a_\perp \equiv F_0 S^a + F_1 n^a_1 + F_2 n^a_2, \tag{10.289}$$

where

$$S^a n^a_{1,2} = 0, \qquad n^a_1 n^a_2 = 0, \tag{10.290}$$

and $F_{0,1,2}$ are scalars on the target space. For instance, we can choose

$$n^a_1 = S^3 S^a - \delta^{3a}, \qquad n^a_2 = \left(\vec{v}\vec{S}\right) S^a - v^a,$$

$$\vec{v} = \left\{ 1, 0, \frac{S^1 S^3}{1 - S^3 S^3} \right\}. \tag{10.291}$$

The last equation in (10.288) implies that

$$F_0 = \tfrac{1}{2}\left(\bar{\chi}^a \chi^a\right), \qquad F^a_\| = \tfrac{1}{2} S^a \left(\bar{\chi}^a \chi^a\right). \tag{10.292}$$

Furthermore, $F_{1,2}$ must be determined through minimization of the F^2 term in the Lagrangian, which obviously leads to $F_1 = F_2 = 0$. As a result, the component Lagrangian of the supersymmetric O(3) sigma model takes the form

$$\mathcal{L} = \frac{1}{2g^2} \left\{ \partial^\mu S^a \, \partial_\mu S^a + \tfrac{1}{2} i \bar{\chi}^a \gamma^\mu \overset{\leftrightarrow}{\partial}_\mu \chi^a + \tfrac{1}{4} (\bar{\chi}^a \chi^a)^2 \right\}, \tag{10.293}$$

plus the first two constraints in Eq. (10.288).

The global O(3) symmetry is explicit in this Lagrangian. Moreover, $\mathcal{N} = 1$ supersymmetry is built in. The conserved supercurrent corresponding to this symmetry is

$$J^\mu = \frac{1}{g^2} (\partial_\lambda S^a) \, \gamma^\lambda \gamma^\mu \chi^a. \tag{10.294}$$

The reader may be surprised to know that there is another, "extra," supercurrent whose conservation is not obvious in the $\mathcal{N} = 1$ formalism. Indeed, following the same line of reasoning as in the problem above one can show (after some algebra) that the supercurrent

$$J'^\mu = \frac{1}{g^2} \varepsilon^{abc} S^a \left(\partial_\lambda S^b \right) \gamma^\lambda \gamma^\mu \chi^c \tag{10.295}$$

Extra supercurrent is conserved too. Thus, the $\mathcal{N} = (1, 1)$ superextension of the O(3) sigma model automatically has an extended $\mathcal{N} = (2, 2)$ supersymmetry, i.e., four rather than two conserved supercharges. The reason for the "unexpected" emergence of this $\mathcal{N} = 2$ superalgebra is the Kählerian nature of the target space manifold, the two-dimensional sphere S_2.[35] As elucidated by Zumino [54], any Kähler sigma model with $\mathcal{N} = (1, 1)$ supersymmetry is, in fact, endowed with $\mathcal{N} = (2, 2)$ supersymmetry also. The easiest way to make this extended supersymmetry explicit is the use of $\mathcal{N} = 2$ superfields in two dimensions rather than $\mathcal{N} = 1$ superfields (10.285). In Section 10.12.3.3 we will construct the $\mathcal{N} = 2$ superspace, develop the corresponding $\mathcal{N} = 2$ superfield formalism, and rederive the supersymmetric O(3) sigma model, which in this formalism is more often referred to as the CP(1) model.

One last remark before concluding this section. The model under consideration has two (classically) conserved bifermion currents, vector and axial,

$$V^\mu = -\frac{i}{2g^2} \varepsilon^{abc} S^a \bar{\chi}^b \gamma^\mu \chi^c \quad \text{and} \quad A^\mu = -\frac{i}{2g^2} \varepsilon^{abc} S^a \bar{\chi}^b \gamma^\mu \gamma^5 \chi^c. \tag{10.296}$$

The vector current V^μ is strictly conserved, while A^μ acquires a quantum anomaly upon regularization [55],

$$\partial_\mu A^\mu = \frac{1}{2\pi} \varepsilon^{\mu\nu} \varepsilon^{abc} S^a \partial_\mu S^b \partial_\nu S^c. \tag{10.297}$$

The θ term in O(3) sigma model In such theories, typically a θ term exists; the O(3) sigma model is no exception. Here θ is a vacuum angle. The physics is periodic in θ with periodicity 2π. The θ term $\mathcal{L}_\theta$, to be added to the Lagrangian (10.293), is proportional to the right-hand side of Eq. (10.297),

$$\mathcal{L}_\theta = -\frac{\theta}{8\pi} \varepsilon^{\mu\nu} \varepsilon^{abc} S^a \, \partial_\mu S^b \, \partial_\nu S^c. \tag{10.298}$$

---

[35] At this point the reader is advised to return to Section 10.6.7 and study it carefully.

10.12.3.3 The $\mathcal{N} = 2$ Superspace in Two Dimensions

Now we will start our journey into extended supersymmetries. Our first encounter is with the $\mathcal{N} = 2$ supersymmetry, which (in two dimensions) has four supercharges. Our first step is to build the corresponding superspace.

The $\mathcal{N} = 2$ superspace in two dimensions can be obtained by the dimensional reduction of the $\mathcal{N} = 1$ superspace in four dimensions; see Section 10.5.1. By such a reduction I mean that all objects of interest are assumed to depend only on t and z and to have no dependence on x and y. Thus, we completely ignore x and y but keep all four Grassmann coordinates, i.e., the two complex components of the spinor θ_α. This corresponds to four supercharges in the $\mathcal{N} = 2$ superalgebra in two dimensions. In two dimensions there is no difference between dotted and undotted spinorial indices; thus, we will omit the dots over spinorial indices in complex-conjugated spinors such as $\theta_\alpha^\dagger$. All spinorial quantities carry *lower* indices, for instance, we have ψ_α or $\bar\psi_\beta$ where $\bar\psi \equiv \psi^\dagger \gamma^0$. Adapting Eq. (10.5) to two dimensions we find that the following quantities are Lorentz invariant:

$$\psi_1\psi_2 + \text{H.c.}, \quad \psi_1^\dagger\psi_2, \quad \psi_2^\dagger\psi_1. \tag{10.299}$$

In more conventional notation these Lorentz invariants can be written as

$$\bar\psi \left(1 \pm \gamma^5\right)\psi, \quad \varepsilon^{\alpha\beta}\psi_\alpha\psi_\beta. \tag{10.300}$$

Next, we will rename the two space–time coordinates as

$$x^\mu = \{t, z\}, \quad \mu = 1, 2. \tag{10.301}$$

Thus, the superspace is spanned by

$$\{x^\mu, \theta_\alpha, \bar\theta_\beta\}, \quad \mu = 0, 1, \quad \alpha, \beta = 1, 2. \tag{10.302}$$

Warning: in contradistinction with four dimensions, in two dimensions we have

$$\bar\theta \equiv \theta^\dagger\gamma^0, \tag{10.303}$$

where the two-dimensional γ matrices were defined in Section 10.2.2. The same definition applies to all other spinors.

The supertransformations of the superspace coordinates take the form

$$\delta\theta_\alpha = \epsilon_\alpha, \quad \delta\bar\theta_\alpha = \bar\epsilon_\alpha, \quad \delta x^\mu = i\bar\epsilon\gamma^\mu\theta - i\bar\theta\gamma^\mu\epsilon, \tag{10.304}$$

i.e., they are exactly the same as in (10.105) except that here μ runs over 0 and 1. Moreover, we can introduce the same invariant subspaces as in four dimensions, $\{x_L^\mu, \theta_\alpha\}$ and $\{x_R^\mu, \bar\theta_\alpha\}$, which are relevant for chiral superfields (see below):

$$\begin{aligned}
\text{for } \{x_L^\mu, \theta_\alpha\}, \quad &\delta\theta_\alpha = \epsilon_\alpha, \quad \delta x_L^\mu = 2i\bar\epsilon\gamma^\mu\theta, \\
\text{for } \{x_R^\mu, \bar\theta_\alpha\}, \quad &\delta\bar\theta_\alpha = \bar\epsilon_\alpha, \quad \delta x_R^\mu = -2i\bar\theta\gamma^\mu\epsilon,
\end{aligned} \tag{10.305}$$

where

$$x_L{}^\mu = x^\mu + i\bar\theta\gamma^\mu\theta, \quad x_R{}^\mu = x^\mu - i\bar\theta\gamma^\mu\theta. \tag{10.306}$$

The spinorial derivatives are defined as

$$D_\beta = -\frac{\partial}{\partial\theta_\beta} + i\left(\bar{\theta}\gamma^\mu\right)_\beta \partial_\mu, \qquad \bar{D}_\alpha = \frac{\partial}{\partial\bar{\theta}_\alpha} - i\left[\,_\alpha(\gamma^\mu\theta)\right]\partial_\mu. \qquad (10.307)$$

Then

$$D_\beta x_R^\mu = 0, \qquad \bar{D}_\alpha x_L^\mu = 0. \qquad (10.308)$$

10.12.3.4 Supersymmetric *CP*(1) Model

In Chapter 6 we learned that the O(3) sigma model can be rewritten in the form of the CP(1) model, in which the Kähler geometry of the target space is explicit. This can also be done for supersymmetric versions. Here we will study a geometrical formulation of the supersymmetric CP(1) model.

Before starting this section, however, it is suggested that the reader might like to return to Section 10.6.7, setting $\mathcal{W} = 0$ there. With the superpotential switched off, every point in the target space becomes a valid vacuum of the theory (at the classical level), i.e., the vacuum energy density vanishes for all coordinate-independent field configurations. Thus the model has a vacuum manifold, which is characteristic of sigma models.

As in four dimensions, we can introduce chiral and antichiral superfields:

$$\Phi(x^\mu + i\bar{\theta}\gamma^\mu\theta, \theta), \qquad \Phi^\dagger(x^\mu - i\bar{\theta}\gamma^\mu\theta, \bar{\theta}). \qquad (10.309)$$

In Section 10.6.7 I discuss general Kähler manifolds as the target space.

The component decomposition of, say, the chiral superfield is

$$\Phi(x_L, \theta) = \phi(x_L) + \sqrt{2}\,\varepsilon^{\alpha\beta}\theta_\alpha\psi_\beta(x_L) + \varepsilon^{\alpha\beta}\theta_\alpha\theta_\beta F(x_L). \qquad (10.310)$$

The target space S_2 is the Kähler manifold of complex dimension 1 (real dimension 2) parametrized by the fields $\phi, \phi^\dagger$, which are the lowest components of the chiral and antichiral superfields. As we already know from Section 10.6.7, the superinvariant Lagrangian has the form

Kähler potential for CP(1) in the Fubini–Study form

$$\mathcal{L} = \int d^4\theta\, \mathcal{K}(\Phi, \Phi^\dagger), \qquad (10.311)$$

where $\mathcal{K}(\Phi, \Phi^\dagger)$ is the Kähler potential corresponding to the two-dimensional sphere. The standard choice of this Kähler potential is

$$\mathcal{K}(\Phi, \Phi^\dagger) = \frac{2}{g^2}\ln\left(1 + \Phi^\dagger\Phi\right), \qquad (10.312)$$

where g^2 is the same coupling constant as in Eq. (10.313). Let us examine the metric following from (10.312),

$$G \equiv G_{1\bar{1}} = \partial_\phi\partial_{\phi^\dagger}\mathcal{K}\Big|_{\theta=\bar{\theta}=0} = \frac{2}{g^2\chi^2}, \qquad (10.313)$$

where

$$\chi \equiv 1 + \phi\phi^\dagger. \qquad (10.314)$$

This is the Kähler metric of the two-dimensional sphere of radius $1/g$ (see below) in Fubini–Study form.[36] Note that in the case at hand the metric tensor, the Riemann curvature tensor, and the Ricci tensor all have just one component, while there are two independent Christoffel symbols Γ and $\bar{\Gamma}$. More exactly,

$$\Gamma = \Gamma^1_{11} = -2\frac{\phi^\dagger}{\chi}, \qquad \bar{\Gamma} = \Gamma^{\bar{1}}_{\bar{1}\bar{1}} = -2\frac{\phi}{\chi}. \tag{10.315}$$

We then calculate the Riemann and Ricci tensors according to the standard rules of the Kähler geometry (specified for CP(1)),

$$R^{\bar{1}}_{\bar{1}\bar{1}1} = R_{\bar{1}1} = -\frac{\partial}{\partial\phi}\Gamma^{\bar{1}}_{\bar{1}\bar{1}} = \frac{2}{(1+\phi\bar{\phi})^2}, \tag{10.316}$$

$$R^1_{11\bar{1}} = R_{1\bar{1}} = -\frac{\partial}{\partial\bar{\phi}}\Gamma^1_{11} = \frac{2}{(1+\phi\bar{\phi})^2}. \tag{10.317}$$

Finally, using the metric with the upper components

$$G^{-1} \equiv G^{1\bar{1}} = \frac{g^2}{2}(1+\phi\bar{\phi})^2$$

we arrive at the corresponding scalar curvature

$$\mathcal{R} = G^{-1}\left(R_{\bar{1}1} + R_{1\bar{1}}\right) = 2g^2. \tag{10.318}$$

For two-dimensional surfaces, such as the one we deal with here, the scalar curvature $\mathcal{R}$ coincides, up to a normalization constant, with the *Gaussian curvature* K of the surface [56],

$$\mathcal{R} = 2K = \frac{2}{\rho_1\rho_2}, \tag{10.319}$$

where ρ_1 and ρ_2 are the principal radii of curvature of the surface at the given point of the surface. For S_2,

$$\rho_1 = \rho_2 = \frac{1}{g}. \tag{10.320}$$

At weak coupling the radius of the target space sphere is very large.

Next, we can use either the general expression (10.165) or directly calculate the integral $\int d^4\theta \ln\left(1 + \Phi^\dagger\Phi\right)$ to obtain the Lagrangian of the supersymmetric CP(1) model [9, 54, 55] in components,

$$\mathcal{L} = G\left[\partial_\mu\phi^\dagger\partial^\mu\phi + i\bar{\psi}\gamma^\mu\partial_\mu\psi - \frac{2i}{\chi}\phi^\dagger\partial_\mu\phi\bar{\psi}\gamma^\mu\psi + \frac{1}{\chi^2}(\bar{\psi}\psi)^2\right]. \tag{10.321}$$

Needless to say, $\mathcal{N} = 2$ supersymmetry is built in by the construction based on $\mathcal{N} = 2$ superfields. What about the target space symmetry? The U(1) symmetry corresponding to rotation around the third axis in the target space is realized linearly in Eqs. (10.311) and (10.312),

$$\Phi \to \Phi + i\alpha\Phi, \qquad \Phi^\dagger \to \Phi^\dagger - i\alpha\Phi^\dagger, \tag{10.322}$$

---

[36] This metric was originally described in 1904 and 1905 by Guido Fubini and Eduard Study.

where α is a real parameter. At the same time, two other symmetry rotations are realized nonlinearly,

$$\Phi \to \Phi + \beta + \beta^* \Phi^2, \quad \Phi^\dagger \to \Phi^\dagger + \beta^* + \beta \left(\Phi^\dagger\right)^2, \tag{10.323}$$

with complex parameter β.

As noted in Section 10.12.3.2 one can introduce the θ term, which in this formalism is

$$\mathcal{L}_\theta = \frac{i\theta}{2\pi} \frac{\varepsilon^{\mu\nu} \partial_\mu \phi^\dagger \partial_\nu \phi}{\chi^2}. \tag{10.324}$$

| *Connecting CP(1) to O(3)* |

It is instructive to check that the Lagrangian (10.321) and the Lagrangian (10.293) discussed in Section 10.12.3.2 in fact describe the same model, i.e., CP(1) = O(3). The fields ϕ and ψ are related to the real fields S^a and χ^a introduced in Section 10.12.3.2 through the stereographic projection

$$\phi = \frac{S^1 + iS^2}{1 + S^3}. \tag{10.325}$$

The complex field ϕ replaces the two independent components of S^a. The unconstrained two-component complex fermion field ψ is related to χ^a as follows:

$$\psi = \frac{\chi^1 + i\chi^2}{1 + S^3} - \frac{S^1 + iS^2}{(1 + S^3)^2} \chi^3. \tag{10.326}$$

The inverse transformations have the form

$$S^1 = \frac{2\,\mathrm{Re}\,\phi}{1 + |\phi|^2}, \quad S^2 = \frac{2\,\mathrm{Im}\,\phi}{1 + |\phi|^2}, \quad S^3 = \frac{1 - |\phi|^2}{1 + |\phi|^2}, \tag{10.327}$$

and

$$\chi^1 = \frac{2\,\mathrm{Re}\,\psi}{1 + |\phi|^2} - \frac{2\,\mathrm{Re}\,\phi(\phi^\dagger \psi + \mathrm{H.c.})}{(1 + |\phi|^2)^2},$$

$$\chi^2 = \frac{2\,\mathrm{Im}\,\psi}{1 + |\phi|^2} - \frac{2\,\mathrm{Im}\,\phi(\phi^\dagger \psi + \mathrm{H.c.})}{(1 + |\phi|^2)^2}, \tag{10.328}$$

$$\chi^3 = -2\frac{\phi^\dagger \psi + \mathrm{H.c.}}{(1 + |\phi|^2)^2}.$$

The reader is invited to carry out explicit and direct verification of the equivalence of the two Lagrangians; for some hints, see Appendix section 10.27.3.

10.12.3.5 CP(1) Generalities

| *N = 1 super-Yang–Mills is found in Section 10.14.* |

The CP(1) model we consider here (with four conserved supercharges, i.e., $\mathcal{N} = (2, 2)$ supersymmetry) is recognized [55] to be an excellent theoretical laboratory for studying, in a simplified setting, the highly nontrivial nonperturbative effects inherent to $\mathcal{N} = 1$ Yang–Mills theory in four dimensions. Namely, it is asymptotically free [57], strongly coupled in the infrared, exhibits dynamical scale generation and mass gap generation, possesses instantons, and has a discrete global Z_4 symmetry spontaneously broken down to Z_2 by a bifermion vacuum condensate [55] – all features we also find in four-dimensional supersymmetric Yang–Mills theory. However, the CP(1) model is very much simpler, which makes it an ideal testing ground for the various new methods that theorists design to deal with strongly coupled gauge theories.

10.12.3.6 Mass Deformation

Concluding our consideration of the O(3) or CP(1) sigma model, we will discuss a mass deformation of the model that eliminates the vacuum degeneracy on the target space and breaks O(3) symmetry down to U(1) but preserves full $\mathcal{N} = 2$ supersymmetry, i.e., all four supercharges remain conserved. This mass deformation is unique [58]. It makes the model under consideration weakly coupled.

In terms of the O(3) sigma model the mass-deformed action that preserves $\mathcal{N} = 2$ is

$$S = \frac{1}{2g^2} \int d^2x \, d^2\theta \left[(\bar{D}_\alpha \sigma^a)(D_\alpha \sigma^a) + 4m\sigma^3 \right], \tag{10.329}$$

where the σ superfield is defined in (10.285), σ^3 is the third component of σ^a, and m is a mass parameter. Note that the $\mathcal{N} = 2$ symmetry is preserved only because the added term is a special case – it is linear in σ^a. The fact of the explicit breaking of O(3) down to O(2), corresponding to rotations in the 12 plane, is obvious. The fact that the four supercharges are conserved is less obvious in this formulation. The conserved supercurrents are

$$J^\mu = \frac{1}{g^2} \left[(\partial_\lambda S^a) \, \gamma^\lambda \gamma^\mu \chi^a - im\gamma^\mu \chi^3 \right],$$
$$\tilde{J}^\mu = \frac{1}{g^2} \left[\varepsilon^{abc} S^a \left(\partial_\lambda S^b \right) \gamma^\lambda \gamma^\mu \chi^c + im \, \varepsilon^{3ab} S^a \gamma^\mu \chi^b \right]. \tag{10.330}$$

In components the Lagrangian in (10.329) has the form[37]

$$\mathcal{L} = \frac{1}{2g^2} \left[\left(\partial_\mu S^a \right)^2 + \bar{\chi}^a i \, \partial\!\!\!/ \chi^a + \tfrac{1}{4}(\bar{\chi}\chi)^2 \right.$$
$$\left. -m^2 \left(1 - S^3 S^3 \right) + mS^3 \, \bar{\chi}\chi \right]. \tag{10.331}$$

To find the F term one must use the decomposition (10.289), which implies that $F_{0,2}$ remain the same, $F_0 = \frac{1}{2}\bar{\chi}^a \chi^a$, $F_2 = 0$, while F_1 changes, i.e.,

$$F_1 = m, \tag{10.332}$$

which results in

$$F^a = \tfrac{1}{2}(\bar{\chi}\chi) \, S^a + mS^3 S^a - m\delta^{3a}. \tag{10.333}$$

It is obvious that the mass-deformed model (10.331) has two discrete degenerate vacua, at the north and south poles of the sphere, i.e., at $S^3 = \pm 1$. Both vacua are supersymmetric; the corresponding energy density vanishes. Later we will use this fact in calculating Witten's index for $\mathcal{N} = 2$ sigma models in two dimensions.

---

[37] In what follows the mass parameter of the fermion term is real. One can introduce a phase into the fermion term, e.g., through the θ term, which is omitted in Eq. (10.331).

Since we already know that the O(3) and CP(1) formulations of the sigma model are equivalent, let us ask ourselves how the above mass deformation will look in the language of CP(1). The answer is as follows:

$$\mathcal{L} = G \left[\partial_\mu \phi^\dagger \partial^\mu \phi - |m|^2 \phi^\dagger \phi + i \bar{\psi} \gamma^\mu \partial_\mu \psi - \frac{1 - \phi^\dagger \phi}{\chi} \bar{\psi} \mu \psi \right.$$
$$\left. - \frac{2i}{\chi} \phi^\dagger \partial_\mu \phi \bar{\psi} \gamma^\mu \psi + \frac{1}{\chi^2} (\bar{\psi}\psi)^2 \right],$$
(10.334)

where

$$\mu = m \frac{1 + \gamma_5}{2} + m^* \frac{1 - \gamma_5}{2}.$$
(10.335)

First appearance of "twisted mass"

This mass parameter is usually referred to as the *twisted mass*. The phase of the mass parameter m appears in physical quantities only in combination with the vacuum angle θ, namely, as $\theta + 2 \arg m$. Therefore, one can always include the phase of m in θ, thus transforming m into a real parameter. The conserved (complex) supercurrent is

$$J_\alpha^\mu = \sqrt{2} G \left[\left(\partial_\nu \phi^\dagger \right) \gamma^\nu \gamma^\mu \psi + i \phi^\dagger \gamma^\mu \mu \psi \right]_\alpha.$$
(10.336)

It should be emphasized that, in $\mathcal{N} = 2$ superfield language, the twisted mass does *not* come from a superpotential. Indeed, there are no nontrivial *holomorphic* nonsingular functions on the sphere[38] that could play the role of a conventional superpotential. I will not explain here how the ($\mathcal{N} = 2$)-preserving mass deformation of the CP(1) model emerged in theoretical constructions [58] or the origin of the term "twisted mass;" this would lead us too far astray. I will say only that the possibility of this mass deformation strongly enhances the potential of the O(3)/CP(1) model as a theoretical laboratory and testing ground for strongly coupled gauge theories in four dimensions.

10.12.4 CP(N − 1)

From Chapter 6 we know that the O(3) or CP(1) models allow for generalizations to arbitrary N in two distinct ways:

$$O(3) \to O(N), \ N \geq 4 \quad \text{and} \quad CP(1) \to CP(N-1), \ N \geq 3.$$
(10.337)

Look through Section 6.2. Gauged formulation is in Appendix section 10.27.1.

The same is valid for the supersymmetric versions. The first case deals with the $\mathcal{N} = (1, 1)$ supersymmetry; in the second, the supersymmetry is extended to $\mathcal{N} = (2, 2)$. In this section we will build the supersymmetric CP($N - 1$) model in a geometric formulation generalizing (10.321). In fact, all the general expressions we need are collected in Section 10.6.7, devoted to the generalized Wess–Zumino

---

[38] In discussing the O(3) sigma model we have used $\mathcal{N} = 1$ superfield language. It is obvious that the $\mathcal{N} = 1$ superpotential does not have the property of holomorphy. The fact of the absence of appropriate $\mathcal{N} = 2$ superpotentials is transparent in the O(3) formulation. For instance, the seemingly innocuous superpotential $\mathcal{W} = m\Phi^2$ leads to the "south pole" singularity $\sim (1 + S^3)^{-3}$. Such a singularity effectively destroys the topology of the target space sphere, transforming the compact manifold into a noncompact manifold.

model. We need to reduce the number of dimensions to 2, discard the superpotential part, and specify the Kähler metric,

$$\mathcal{K} = \frac{2}{g^2} \ln \left(1 + \sum_{i,\bar{j}=1}^{N-1} \Phi^{\dagger \bar{j}} \delta_{\bar{j}i} \Phi^i \right) \tag{10.338}$$

(the above expression corresponds to the round Fubini–Study metric). For $CP(N-1)$ the Riemann tensor is locally related to the metric,

$$R_{i\bar{j}k\bar{m}} = -\frac{g^2}{2} \left(G_{i\bar{j}} G_{k\bar{m}} + G_{i\bar{m}} G_{k\bar{j}} \right), \tag{10.339}$$

while the Ricci tensor $R_{i\bar{j}}$ is simply proportional to the metric,

$$R_{i\bar{j}} = \frac{g^2}{2} N G_{i\bar{j}}. \tag{10.340}$$

The Lagrangian is [54]

$$\mathcal{L} = \int d^2\theta \, d^2\bar{\theta} \mathcal{K}(\Phi, \Phi^{\dagger}) = G_{i\bar{j}} \left(\partial_\mu \phi^{\dagger \bar{j}} \partial_\mu \phi^i + i \bar{\psi}^{\bar{j}} \gamma^\mu D_\mu \psi^i \right) - \tfrac{1}{4} R_{i\bar{j}k\bar{l}} (\bar{\psi}^{\bar{j}} \psi^i)(\bar{\psi}^{\bar{l}} \psi^k), \tag{10.341}$$

where the covariant derivative D_μ acting on ψ was defined in Eq. (10.168).

Equation (10.339) implies that $R_{i\bar{j}k\bar{m}}$ is symmetric under the interchange $i \leftrightarrow k$ or $\bar{j} \leftrightarrow \bar{m}$. Therefore the last (four-fermion) term in (10.341) can be rewritten as

$$+ R_{i\bar{j}k\bar{l}} \, \bar{\psi}_2^{\bar{j}} \psi_2^i \bar{\psi}_1^{\bar{l}} \psi_1^k,$$

where the subscripts $1, 2$ label the upper and lower components of the spinor ψ. Sometimes, the subscripts L, R respectively are used to this end. Substituting (10.316) in the preceding expression we reproduce the four-fermion term in $CP(1)$ model; see Eq. (10.321).

Exercises

10.12.1 Prove the equivalence of the Lagrangians (10.334) and (10.331) plus the constraints $S^a S^a = 1$, $S^a \chi^a = 0$. Prove the equivalence of the supercurrents (10.336) and (10.330).

10.12.2 Derive the equations of motion following from (10.293) and use them to prove that $\partial_\mu J^\mu = 0$. The supercurrent is defined in (10.294).

10.12.3 Prove that $\varepsilon^{\mu\nu} \varepsilon^{abc} S^a \partial_\mu S^b \partial_\nu S^c$ is a full derivative,

$$\varepsilon^{\mu\nu} \varepsilon^{abc} S^a \partial_\mu S^b \partial_\nu S^c \equiv \partial_\mu K^\mu, \tag{10.342}$$

where K^μ is a local function of S^a. Calculate K^μ. *Hint:* One should not assume that K^μ is O(3) invariant; in fact, it is not. *Another hint:* The solution of this problem could be deferred until the reader is acquainted with the contents of Section 10.12.3.4.

10.12.4 Derive Eqs. (10.315)–(10.318), starting from the Kähler metric (10.313).

10.12.5 Prove that $\chi^{-2} \varepsilon^{\mu\nu} \partial_\mu \phi^{\dagger} \partial_\nu \phi \equiv \partial_\mu K^\mu$.

10.12.6 Verify that the two expressions for $\mathcal{L}_\theta$ in Eqs. (10.324) and (10.298) are identically equal.

10.12.7 Prove that the one-loop β function in the supersymmetric CP(1) model is the same as in its nonsupersymmetric version; see Section 6.3.5.

10.13 Supersymmetric Yang–Mills Theories

We already know how to construct supersymmetric Abelian gauge theories (see Sections 10.6.8 and 10.6.9). Now it is time to proceed to non-Abelian theories.

10.13.1 Gauge Sector

It is convenient to start with the matter fields. For the time being we will consider nonchiral theories and the gauge (color) group SU(N). The matter fields are replaced by chiral superfields that belong to certain representations R of SU(N) and are endowed with color indices. If representation R is complex, for instance fundamental, then the corresponding superfield should be supplemented by another belonging to the complex-conjugate representation. For example, each "quark" flavor is represented by two superfields, Q^i and $\tilde{Q}_j$, belonging to the fundamental and antifundamental representations, respectively (for SU(N) the color indices are $i, j = 1, 2, \ldots, N$). The two-index representations $Q^{\{ij\}}, \tilde{Q}_{\{ij\}}$ and $Q^{[ij]}, \tilde{Q}_{[ij]}$ are also sometimes employed ($\{\ldots\}$ and $[\ldots]$ stand for symmetrization and antisymmetrization). Another matter superfield with which we will deal below is that in the adjoint representation, Φ^a where $a = 1, 2, \ldots, N^2 - 1$, or, equivalently, Φ^i_j. This representation is real; therefore, one can introduce just one adjoint chiral superfield.

Let T^a denote the (Hermitian) generators of the gauge group in the representation R. The supergauge transformations (10.176) are now generalized as follows:

$$Q(x_L, \theta) \to e^{i\Lambda(x_L, \theta)} Q(x_L, \theta), \qquad \bar{Q}(x_R, \bar{\theta}) \to \bar{Q}(x_R, \bar{\theta}) e^{-i\bar{\Lambda}(x_R, \bar{\theta})}, \quad (10.343)$$

where Λ and $\bar{\Lambda}$ are matrices representing two sets of chiral superfields, each set containing $N^2 - 1$ superfields

$$\Lambda(x_L, \theta) \equiv \Lambda^a T^a, \qquad \bar{\Lambda}(x_R, \theta) \equiv \bar{\Lambda}^a T^a. \quad (10.344)$$

The generators obey the standard commutation relations

$$[T^a, T^b] = i f^{abc} T^c, \quad (10.345)$$

f^{abc} being the structure constants of the gauge group, and are normalized in a conventional manner,

$$T^a T^a = C_2(R), \qquad \text{Tr}\, T^a T^b = T(R) \delta^{ab},$$

$$T(R) = C_2(R) \frac{\dim(R)}{\dim(\text{adj})}, \quad (10.346)$$

Definition of quadratic Casimir operators

where $C_2(R)$ is the quadratic Casimir operator and $2T(R)$ is known as the Dynkin index in the mathematical literature (see Table 10.3). Sometimes $T(\text{adj}) \equiv T_G$ is referred to as the dual Coxeter number. For the fundamental representation we have

Table 10.3. The group coefficients for the fundamental, adjoint, and two-index antisymmetric and symmetric representations of SU(N)

	Fundamental	Adjoint	Two-index A	Two-index S
$T(R)$	$\dfrac{1}{2}$	N	$\dfrac{N-2}{2}$	$\dfrac{N+2}{2}$
$C_2(R)$	$\dfrac{N^2-1}{2N}$	N	$\dfrac{(N-2)(N+1)}{N}$	$\dfrac{(N+2)(N-1)}{N}$

$T(\text{fund}) = \frac{1}{2}$. Note that the generators of a given complex representation R are related to those of the complex-conjugate representation $\bar{R}$ by the formula

$$\bar{T}^a = -\tilde{T}^a = -T^{a*}, \tag{10.347}$$

where the tilde denotes the transposed matrix.

The vector superfield V in which all gauge bosons and gauginos reside is now a matrix too,

$$V(x, \theta, \bar{\theta}) \equiv V^a T^a. \tag{10.348}$$

The kinetic term $\bar{Q}e^V Q$ is gauge invariant provided that we supplement the supergauge transformation (10.343) by the following transformation of the vector superfield:

$$e^{V(x,\theta,\bar{\theta})} \to e^{i\bar{\Lambda}(x_R,\bar{\theta})} e^{V(x,\theta,\bar{\theta})} e^{-i\Lambda(x_L,\theta)} \tag{10.349}$$

If we assume Λ, $\bar{\Lambda}$ to be small, neglect all fermion components in Λ, $\bar{\Lambda}$, and expand (10.349) in powers of Λ and V keeping the leading and the next-to-leading terms, we get

$$\delta A_\mu^a = \mathcal{D}_\mu \omega^a,$$
$$\omega^a = 2\,\mathrm{Re}\,\varphi^a,$$
$$\mathcal{D}_\mu \omega^a \equiv \partial_\mu \omega^a + f^{abc} A_\mu^b \omega^c, \tag{10.350}$$

i.e., the standard gauge transformation law for the gauge 4-potential. Here φ is defined in (10.178).

One can use the supergauge transformation to impose the Wess–Zumino gauge, in just the same way as in supersymmetric QED. In this gauge the C^a, χ^a, and M^a components of the vector superfield are eliminated, leaving us with the following expression:

| Wess–
| Zumino
| gauge

$$V^a = -2\theta^\alpha \bar{\theta}^{\dot{\alpha}} A_{\alpha\dot{\alpha}}^a - 2i\bar{\theta}^2 (\theta\lambda^a) + 2i\theta^2 (\bar{\theta}\bar{\lambda}^a) + \theta^2 \bar{\theta}^2 D^a. \tag{10.351}$$

As in supersymmetric QED, V^3 and all higher powers of V vanish; therefore in the action we can expand e^V keeping only terms up to quadratic.

To construct the non-Abelian field strength tensor superfield analogous to (10.182) it is necessary to generalize the supersymmetric covariant derivatives to make them both supersymmetric and *gauge* covariant.

Let us indicate supergauge-transformed quantities by primes, while supersymmetric and gauge-covariant derivatives will be denoted as ∇_A, where $A = \mu$, α, or $\dot{\alpha}$. As usual, their definition will depend on which particular field they act. As an instructive example let us consider a chiral superfield Q in a nontrivial representation of the gauge group. Then $Q' = e^{i\Lambda}Q$, and therefore from the covariant derivative we require

$$(\nabla_A Q)' = e^{i\Lambda}\nabla_A Q, \tag{10.352}$$

which implies in turn that

$$(\nabla_A)' = e^{i\Lambda}\nabla_A e^{-i\Lambda}. \tag{10.353}$$

> ∇_A covariantizes $\mathcal{D}_\mu$, D_α, and $\bar{D}_{\dot\alpha}$

Since Λ is a chiral superfield and hence $\bar{D}_{\dot\alpha}\Lambda = 0$, we can choose

$$\nabla_{\dot\alpha} \equiv \bar{D}_{\dot\alpha} \tag{10.354}$$

and, correspondingly,

$$\nabla'_{\dot\alpha} = \nabla_{\dot\alpha}. \tag{10.355}$$

As for the left-handed covariant derivative we define

$$\nabla_\alpha \equiv e^{-V}D_\alpha e^V. \tag{10.356}$$

Then

$$\nabla_\alpha' = e^{-V'}D_\alpha e^{V'} = e^{i\Lambda}e^{-V}e^{-i\bar\Lambda}D_\alpha e^{i\bar\Lambda}e^V e^{-i\Lambda}$$

$$= e^{i\Lambda}e^{-V}D_\alpha e^V e^{-i\Lambda} = e^{i\Lambda}\nabla_\alpha e^{-i\Lambda}, \tag{10.357}$$

as required according to (10.353). Finally, the vectorial covariant derivative must be defined as

$$\left\{\nabla_\alpha, \bar\nabla_{\dot\alpha}\right\} = 2i\nabla_{\alpha\dot\alpha}, \tag{10.358}$$

cf. Eq. (10.114). Here we have used the fact that $D_\alpha\bar\Lambda = 0$. It is useful to rewrite the left-handed covariant derivative as[39]

$$\nabla_\alpha = D_\alpha + e^{-V}\left(D_\alpha e^V\right). \tag{10.359}$$

By analogy with the gauge-covariant derivative we can call the second term on the right-hand side a *supersymmetric gauge connection*,

$$\Gamma_\alpha \equiv ie^{-V}\left(D_\alpha e^V\right). \tag{10.360}$$

Making use of Eq. (10.349) we get

$$\Gamma_\alpha' = ie^{-V'}\left(D_\alpha e^{V'}\right) = e^{i\Lambda}\Gamma_\alpha e^{-i\Lambda} + ie^{i\Lambda}\left(D_\alpha e^{-i\Lambda}\right), \tag{10.361}$$

a transformation law typical of those for gauge connections.

Finally we are ready to construct a non-Abelian field strength tensor superfield analogous to (10.182) in terms of the gauge connection defined above, namely,

$$W_\alpha = -\tfrac{1}{8}i\bar{D}^2\Gamma_\alpha = \tfrac{1}{8}\bar{D}^2 e^{-V}\left(D_\alpha e^V\right)$$

$$= i\left(\lambda_\alpha + i\theta_\alpha D - \theta^\beta G_{\alpha\beta} - i\theta^2\mathcal{D}_{\alpha\dot\alpha}\bar\lambda^{\dot\alpha}\right), \tag{10.362}$$

---

[39] In (10.356) the spinorial derivative D_α acts on everything to its right, i.e., $\nabla_\alpha X = e^{-V}(D_\alpha e^V X)$, while in the second term in (10.359) D_α acts only on e^V.

where the gluon field strength tensor is denoted by $G_{\alpha\beta}$ (in the Abelian case it is denoted by $F_{\alpha\beta}$). This serves as a reminder that here the gluon field strength tensor includes terms linear and quadratic in the gauge 4-potential:

$$G_{\mu\nu}^a = \partial_\mu A_\nu^a - \partial_\nu A_\mu^a + f^{abc} A_\mu^a A_\nu^c. \tag{10.363}$$

The component decomposition in (10.362) refers to the Wess–Zumino gauge. Each component field in Eq. (10.362) is a matrix in color space; for instance $G_{\alpha\beta} = G_{\alpha\beta}^a T^a$ and $D = D^a T^a$. For simplicity we will assume below that the generators T^a in Eq. (10.362) are taken to be in the fundamental representation.

The emergence of the quadratic term above can be seen by expanding the expression for Γ_α up to terms quadratic in V (in the Wess–Zumino gauge),

$$\Gamma_\alpha^a = i \left[(D_\alpha V^a) + \tfrac{1}{2} i f^{abc} (D_\alpha V^b) V^c \right], \tag{10.364}$$

where we can drop all terms in V except $V^a = -2\theta^\alpha \bar{\theta}^{\dot{\alpha}} (\sigma^\mu)_{\alpha\dot{\alpha}} A_\mu^a$. The spinorial derivatives were defined in Section 10.5.2. Two helpful relations used in the derivation are

$$\bar{D}^2 \bar{\theta}^2 = -4 \quad \text{and} \quad G_{\mu\nu}^a (\sigma^\mu)_{\alpha\dot{\alpha}} (\sigma^\nu)_\beta^{\dot{\alpha}} = -2 G_{\alpha\beta}^a, \tag{10.365}$$

cf. Eq. (10.26). Using these relations and calculating $-\tfrac{1}{8} i \bar{D}^2 \Gamma_\alpha^a$, after some straight-forward but rather tedious algebra we arrive at

$$W_\alpha^a \to -i\, G_{\alpha\beta}^a \theta^\beta$$

with the standard non-Abelian expression for $G_{\mu\nu}^a$ (see Eq. (10.363)). Moreover, the second term in (10.364) converts the regular derivative $\partial_{\alpha\dot{\alpha}} \bar{\lambda}^{\dot{\alpha}}$ into the covariant derivative $\mathcal{D}_{\alpha\dot{\alpha}} \bar{\lambda}^{\dot{\alpha}}$.

Unlike in supersymmetric QED (Section 10.6.9), $G_{\alpha\beta}$ and the superfield W_α in its entirety are not invariant under gauge transformations. Equation (10.361) implies that

$$W_\alpha' = e^{i\Lambda} W_\alpha e^{-i\Lambda}. \tag{10.366}$$

At the same time $\mathrm{Tr}\, W^2 \sim W^a W^a$ is supergauge invariant. For convenience I will reproduce here the component decomposition of $\mathrm{Tr}\, W^2$, which is very similar to that in supersymmetric QED (Section 10.6.9),

$$\begin{aligned} W^2(x_L, \theta) = &-\lambda^2 - 2i(\lambda\theta) D + 2\lambda^\alpha G_{\alpha\beta} \theta^\beta \\ &+ \theta^2 \left(D^2 - \tfrac{1}{2} G^{\alpha\beta} G_{\alpha\beta} \right) + 2i\theta^2 \bar{\lambda}_{\dot{\alpha}} \mathcal{D}^{\dot{\alpha}\alpha} \lambda_\alpha. \end{aligned} \tag{10.367}$$

10.13.2 Matter Sector

Now we are ready to construct a generic supersymmetric $\mathcal{N} = 1$ gauge theory, with matter sector $Q = \{Q_i\}$ in the representations R of G. The most general form of the Lagrangian is

$$\begin{aligned} \mathcal{L} = &\left(\frac{1}{4g^2} \int d^2\theta\, W^{a\alpha} W_\alpha^a + \mathrm{H.c.} \right) + \sum_{\text{all flavors}} \int d^2\theta\, d^2\bar{\theta}\, \bar{Q}^f e^V Q_f \\ &+ \left(\int d^2\theta\, \mathcal{W}(Q_f) + \mathrm{H.c.} \right), \end{aligned} \tag{10.368}$$

where $\mathcal{W}$ is a superpotential that depends on the chiral superfields Q_f of all flavors, generally speaking. It must be (super)gauge invariant. For instance, $Q^{\{ij\}}\tilde{Q}_i\tilde{Q}_j$ is allowed while $Q^{\{ij\}}Q^iQ^j$ is not. The gauge coupling constant is complexified,

$$\frac{1}{g^2} \to \frac{1}{g^2} - i\frac{\theta}{8\pi^2}, \tag{10.369}$$

where θ is the vacuum angle.

Following the standard procedure it is easy to derive from Eq. (10.368) the F terms:

$$\bar{F}_f = -\frac{\partial \mathcal{W}(Q)}{\partial Q_f}\bigg|_{\theta=0}, \qquad \text{for all flavors.} \tag{10.370}$$

The D term has the form

$$D^a = -g^2 \sum_f \overline{q_f} T^a q_f. \tag{10.371}$$

The scalar potential is the sum of F and D terms,

$$V = \frac{1}{2g^2}D^a D^a + \sum_f (\bar{F}F)_f. \tag{10.372}$$

The generic $\mathcal{N} = 1$ non-Abelian theory presented above was first worked out in [59].

The gauge group G can be a direct product of several factors: $G = G_1 \times G_2 \times \ldots$ Then the gauge kinetic term in the first line of (10.368) must be replaced by a sum of such terms, each with its own complexified gauge coupling. If G contains a U(1) factor (or factors), one can add the Fayet–Iliopoulos term

$$\Delta\mathcal{L}_\xi = -\xi \int d^2\theta\, d^2\bar{\theta}\, V(x, \theta, \bar{\theta}) \equiv \xi D \tag{10.373}$$

for each U(1) factor. If not stated otherwise, in what follows we will consider only theories that have no Fayet–Iliopoulos term.

10.14 Supersymmetric Gluodynamics

Polyakov coined the term *supersymmetric gluodynamics* for a super-Yang–Mills theory without matter superfields. Let us discuss this theory of gluons and gluinos in some detail.

The Lagrangian of the theory is

$$\mathcal{L} = \left(\frac{1}{4g^2}\int d^2\theta\, W^{a\alpha} W^a_\alpha + \text{H.c.}\right)$$

$$= -\frac{1}{4g^2}G^a_{\mu\nu}G^a_{\mu\nu} + \frac{i}{g^2}\lambda^{a\alpha}\mathcal{D}_{\alpha\dot{\beta}}\bar{\lambda}^{a\dot{\beta}} + \frac{\theta}{32\pi^2}G^a_{\mu\nu}\tilde{G}^a_{\mu\nu}. \tag{10.374}$$

Supersymmetric gluodynamics is a close relative of one-flavor QCD. The distinction is that in the former the fermion sector consists of one Weyl (or Majorana) spinor in the adjoint representation while in one-flavor QCD the quark is the Dirac fermion in the fundamental representation of the gauge group.

The theory (10.374) is supersymmetric. The conserved spin-$\frac{3}{2}$ current J_β^μ has the form (in spinorial notation)

$$J_{\beta\alpha\dot{\alpha}} \equiv (\sigma_\mu)_{\alpha\dot{\alpha}} J_\beta^\mu = \frac{2i}{g^2} G_{\alpha\beta}^a \bar{\lambda}_{\dot{\alpha}}^a. \tag{10.375}$$

In the Majorana representation

$$J_\alpha^\mu = \frac{2i}{g^2} \text{Tr}\left[G_{\nu\omega} \left(\sigma^{\nu\omega} \gamma^\mu \lambda \right)_\alpha \right].$$

At the classical level this supercurrent is conserved, $\partial_\mu J_\alpha^\mu = 0$, and, quite obviously, $\gamma_\mu J_\alpha^\mu = 0$. The former relation remains true at the quantum level too, while the latter acquires an anomaly, $\gamma_\mu J_\alpha^\mu \neq 0$; see Section 10.16.4.

What other global symmetries (besides supersymmetry) are intrinsic to this theory? Needless to say, the energy–momentum tensor $T_{\mu\nu}$ is conserved. Moreover classically the trace of the energy–momentum tensor T_μ^μ vanishes. Equivalently, one can say that the classical action is scale invariant. As explained in Appendix section 1.4, scale invariance when combined with Poincaré invariance of the action implies full conformal symmetry. Then supersymmetry promotes it to the super-conformal symmetry of the Lagrangian (10.374) at the classical level. In addition, (10.374) is invariant under the U(1) chiral rotation

$$\lambda \to e^{i\alpha} \lambda, \qquad \bar{\lambda} \to e^{-i\alpha} \bar{\lambda} \tag{10.376}$$

generated by the chiral charge

$$Q = \int d^3x \, R_0, \qquad R^\mu = \frac{1}{g^2} \bar{\lambda}^a \bar{\sigma}^\mu \lambda^a. \tag{10.377}$$

The chiral transformation (10.376) is nothing other than the R symmetry of supersymmetric gluodynamics. The R charges are as follows:

$$r(\lambda) = 1, \qquad r(G_{\alpha\beta}) = r(\bar{G}_{\dot{\alpha}\dot{\beta}}) = 0, \qquad r(\bar{\lambda}) = -1, \tag{10.378}$$

cf. Eqs. (10.216).

The R current in the theory at hand is the only current that could play the role of the fermion current. However, the R symmetry of the classical Lagrangian (10.374) is broken by a chiral anomaly,[40] namely,

$$\partial_\mu R^\mu = \frac{T_G}{16\pi^2} G_{\mu\nu}^a \tilde{G}^{a\mu\nu}, \tag{10.379}$$

> **Chiral anomaly in supersymmetric gluodynamics**

where R^μ is defined in (10.377) and $T_G \equiv T$ (adj). For SU(N), as can be readily deduced from Eq. (10.346), we have

$$T_{\text{SU}(N)} = N.$$

---

[40] Simultaneously, owing to supersymmetry anomalous terms in T_μ^μ and $\varepsilon^{\beta\alpha} J_{\beta\alpha\dot{\alpha}}$ are generated, see Section 10.16, destroying the conformal and superconformal invariance of the theory. For instance, $T_\mu^\mu = -(3T_G/32\pi^2) G_{\mu\nu}^a G^{a\mu\nu}$.

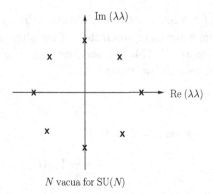

N vacua for SU(N)

Fig. 10.2 The gluino condensate $\langle \lambda \lambda \rangle$ is the order parameter labeling the distinct vacua in supersymmetric gluodynamics. For the SU(N) gauge group there are N discrete degenerate vacua.

For other groups see Table 10.10 in Section 10.22. A discrete Z_{2N} subgroup, for which

$$\lambda \to e^{\pi i k/N} \lambda,$$

is nonanomalous, however.

The Z_{2N} symmetry, a remnant of the R symmetry, is known to be dynamically broken down to Z_2. The order parameter, the gluino condensate $\langle \lambda \lambda \rangle$,[41] can take N distinct values,

$$\langle \lambda_\alpha^a \lambda^{a,\alpha} \rangle = -12N\Lambda^3 \exp\left(\frac{2\pi i k}{N}\right), \qquad k = 0, 1, \ldots, N-1, \qquad (10.380)$$

Here N corresponds to Witten's index, Section 10.22.

labeling the N distinct vacua of the theory (10.374), see Fig. 10.2. Here Λ is a dynamical scale, defined in the standard manner in terms of the ultraviolet parameters:

$$\Lambda^3 = \tfrac{2}{3} M_{\text{uv}}^3 \frac{8\pi^2}{Ng_0^2} \exp\left(-\frac{8\pi^2}{Ng_0^2}\right), \qquad (10.381)$$

where M_{uv} is the ultraviolet (UV) regulator mass while g_0^2 is the bare coupling constant. For the time being we will set $\theta = 0$.

If the reader has enough patience to go through Section 10.19 in which supersymmetric instanton calculus is studied, it will be seen that Eq. (10.381) is exact in supersymmetric gluodynamics. If $\theta \neq 0$, the exponent in Eq. (10.380) is replaced by

$$\exp\left(\frac{2\pi i k}{N} + \frac{i\theta}{N}\right).$$

---

[41] The gluino condensate in supersymmetric gluodynamics was first conjectured, on the basis of the value of Witten's index [60]; see Section 10.22. It was calculated exactly (using holomorphy and analytic continuation in mass parameters) by Shifman and Vainshtein [61]. The exact value of the factor $12N$ in Eq. (10.380) can be extracted from several sources. All numerical factors are carefully collected for SU(2) in the review paper [62]. A weak-coupling calculation for SU(N) with arbitrary N was carried out in [63]. Note, however, that an unconventional definition of the scale parameter Λ was used in [63]. One can pass to the conventional definition of Λ either by normalizing the result to the SU(2) case [62] or by analyzing the context of [63]. Both methods give the same result.

Since supersymmetric gluodynamics has no conserved fermion current, the fermion number F is not defined. However, $(-1)^F$ is well defined. In other words, owing to the surviving Z_2 symmetry one can determine whether F is even or odd.

The theory is believed to be confining, with a mass gap. Although there is no proof of this statement, there are solid arguments, partly theoretical and partly empirical, that substantiate this point of view (see, e.g., [64], Section 6.3).

The spectrum of supersymmetric gluodynamics comprises composite (color-singlet) hadrons, which enter in degenerate supermultiplets. The simplest of these is the chiral supermultiplet, which includes two (massive) spin-zero mesons, with opposite parities, and a Majorana fermion with the Majorana mass (alternatively one can treat it as a Weyl fermion). The interpolating operators producing the corresponding hadrons from the vacuum are G^2, $G\tilde{G}$, and $G\lambda$. The vector supermultiplet consists of a spin-1 massive vector particle, a 0^+ scalar, and a Dirac fermion. All particles from a particular supermultiplet have degenerate masses. The two-point functions are degenerate also (modulo obvious kinematical spin factors). For instance,

$$\langle G^2(x), G^2(0)\rangle = \langle G\tilde{G}(x), G\tilde{G}(0)\rangle = \langle G\lambda(x), G\lambda(0)\rangle. \tag{10.382}$$

Unlike in conventional QCD, both the meson and "baryon" masses are expected to scale as N^0 at large N.

A remarkable feature of supersymmetric gluodynamics is that in the limit $N \to \infty$ it is equivalent (in the bosonic sector) to two *nonsupersymmetric* theories [64, 65], namely, SU(N) Yang–Mills theory with one Dirac field either in the symmetric ($Q^{\{ij\}}$) or the antisymmetric ($Q^{[ij]}$) two-index representation. At $N = 3$ the antisymmetric field $Q^{[ij]}$ coincides with the conventional fundamental quark field (i.e., Q_i); see Section 9.2.6.

| Planar equivalence |

10.15 One-Flavor Supersymmetric QCD

Here we will limit ourselves to the gauge group SU(2) with a matter sector consisting of one flavor. The gauge sector consists of three gluons and their superpartners, gluinos.

As in supersymmetric QED, the matter sector is built from two superfields. Instead of the electric charges now we must choose certain representations of SU(2). In supersymmetric QED the fields Q and $\tilde{Q}$ have opposite electric charge. Analogously, in supersymmetric QCD one superfield must be in the fundamental representation and the other in the antifundamental representation. The specific feature of SU(2) is the equivalence of the doublets and antidoublets. Thus, the matter is described by a set of superfields Q_f^i, where $i = 1, 2$ is the color index and $f = 1, 2$ is a "subflavor" index; two subflavors comprise one flavor. In components,

$$Q_f^i = q_f^i + \sqrt{2}\,\theta\psi_f^i + \theta^2 F_f^i, \qquad i = 1, 2, \quad f = 1, 2, \tag{10.383}$$

where q_f^i and ψ_f^i are the squark and quark fields, respectively.

The Lagrangian of the model is given by Eq. (10.368) with superpotential

| Mass term |

$$\mathcal{W} = \frac{m}{2} Q_i^f Q_f^i. \tag{10.384}$$

Note that the SU(2) model under consideration, with one flavor, possesses a global SU(2) (subflavor) invariance allowing one freely to rotate the superfields Q_f. All indices corresponding to the SU(2) groups (gauge, Lorentz, and subflavor) can be lowered and raised by means of the Levi–Civita $\varepsilon^{\alpha\beta}$ symbol, according to the general rules.

The superpotential presented in Eq. (10.384) is unique if the requirement of renormalizability is imposed. Without this requirement this superpotential could be supplemented, e.g., by the quartic color invariant $\left(Q^{if}Q_{if}\right)^2$. The cubic term is not allowed in SU(2). In general, renormalizable models with a richer matter sector may allow terms cubic in Q in the superpotential.

It is instructive to pass from the superfield notation to components. We will do this exercise for W^2. The F component of W^2 includes the kinetic term of the gluons and gluinos, as well as the square of the D term,

$$\frac{1}{4g^2}\int d^2\theta\, W^{a\alpha}W_\alpha^a = -\frac{1}{8g^2}\left(G_{\mu\nu}^a G^{a\mu\nu} - iG_{\mu\nu}^a \tilde{G}^{a\mu\nu}\right) + \frac{1}{4g^2}D^a D^a + \frac{i}{2g^2}\lambda^a\sigma^\mu \mathcal{D}_\mu\bar\lambda^a.$$
(10.385)

Gauge sector in components

The next term to be considered is $\int d^2\theta\, d^2\bar\theta\, \bar Q e^V Q$. Calculation of the D component of $\bar Q e^V Q$ is a more time-consuming exercise, since we must take into account the fact that Q depends on x_L while $\bar Q$ depends on x_R: both arguments differ from x. Therefore, one has to expand in this difference. The factor e^V sandwiched between $\bar Q$ and Q "covariantizes" all derivatives. Taking the field V in the Wess–Zumino gauge one gets

$$\int d^2\theta\, d^2\bar\theta\, \bar Q^f e^V Q_f = \mathcal{D}^\mu\bar q^f \mathcal{D}_\mu q_f + \bar F^f F_f + D^a\bar q^f T^a q_f$$
$$+ i\psi_f\sigma^\mu \mathcal{D}_\mu\bar\psi^f + \left[i\sqrt{2}(\psi_f\lambda)\bar q^f + \text{H.c.}\right],\quad (10.386)$$

Matter sector in components

where $T^a = \frac{1}{2}\sigma^a$. Finally, we present the superpotential term,

$$\frac{m}{2}\int d^2\theta\, Q_i^f Q_f^i = mq_i^f F_f^i - \frac{m}{2}\psi_i^f\psi_f^i.$$
(10.387)

The fields D and F are auxiliary and can be eliminated by virtue of the equations of motion. In this way we arrive at the scalar potential in the form

$$V = V_D + V_F,\qquad V_D = \frac{1}{2g^2}D^a D^a,\qquad V_F = \bar F_i^f F_f^i,$$
(10.388)

where

$$D^a = -g^2\bar q^f T^a q_f,\qquad F_f^i = -\bar m\, \bar q_f^i.$$
(10.389)

Assembling (10.386), (10.387), and (10.389) and eliminating the auxiliary fields we arrive at

$$\mathcal{L} = -\frac{1}{4g^2}G_{\mu\nu}^a G_{\mu\nu}^a + \frac{\theta}{32\pi^2}G_{\mu\nu}^a \tilde{G}_{\mu\nu}^a + \frac{i}{g^2}\bar\lambda^a\bar\sigma^\mu \mathcal{D}_\mu\lambda^a$$
$$+ \sum_f\left(\mathcal{D}^\mu\bar q_f\mathcal{D}_\mu q_f + i\bar\psi_f\bar\sigma^\mu \mathcal{D}_\mu\psi_f\right)$$
$$+ \left[-\frac{m}{2}\psi_i^f\psi_f^i + i\sqrt{2}\left(\psi_f\lambda^a T^a\right)\bar q_f + \text{H.c.}\right] - V(q_f),$$
(10.390)

where

$$V(q_f) = \frac{g^2}{2} \left(\sum_f \bar{q}_f T^a q_f \right)^2 + \sum_f |m|^2 \left| q_f \right|^2. \tag{10.391}$$

The D part of the scalar potential (the first term in (10.391)) represents a quartic self-interaction of the scalar fields, of a peculiar form. There is a continuous vacuum degeneracy: the minimal (zero) energy is achieved on an infinite set of field configurations that are *not* physically equivalent.

To examine the vacuum manifold let us start from the case of vanishing superpotential, i.e., $m = 0$. From Eq. (10.389) it is clear that the classical space of vacua is defined by the D-flatness condition

$$D^a = -g^2 \sum_f \bar{q}_f T^a q_f = 0, \qquad a = 1, 2, 3. \tag{10.392}$$

It is not difficult to find the D-flat direction explicitly. Indeed, consider squark fields of the form

$$q_f^i = v \begin{pmatrix} 1 & 0 \\ 0 & 1 \end{pmatrix}, \tag{10.393}$$

where v is an arbitrary complex constant. It is obvious that for any value of v all Ds vanish, D^1 and D^2 because $\sigma^{1,2}$ are off-diagonal matrices and D^3 because there is summation over the two subflavors.

It is quite obvious that if $v \neq 0$ then the original gauge symmetry SU(2) is totally Higgsed. Indeed, in the vacuum field (10.393) all three gauge bosons acquire a mass $M_W = g|v|$. Needless to say, supersymmetry is not broken. It is instructive to trace the reshuffling of degrees of freedom by the Higgs phenomenon. In the unbroken phase, corresponding to $v = 0$, we have three massless gauge bosons (six degrees of freedom), three massless gauginos (six degrees of freedom), four matter Weyl fermions (eight degrees of freedom), and four complex matter scalars (eight degrees of freedom). In the broken phase, three matter fermions combine with the gauginos to form three massive Dirac fermions (twelve degrees of freedom). Moreover, three matter scalars combine with the gauge fields to form three *massive* vector fields (nine degrees of freedom) plus three massive (real) scalars. What remains massless? One complex scalar field corresponding to the motion along the bottom of the valley, v, and its fermion superpartner, a Weyl fermion. The balance between the fermion and boson degrees of freedom is explicit.

Thus, we see that in the effective low-energy theory only one chiral superfield Φ survives. This chiral superfield can be introduced as a supergeneralization of Eq. (10.393),

$$Q_f^i = \Phi \begin{pmatrix} 1 & 0 \\ 0 & 1 \end{pmatrix}. \tag{10.394}$$

Substituting this expression into the original Lagrangian (10.368) we get

$$\mathcal{L}_{\text{eff}} = 2 \int d^2\theta \, d^2\bar{\theta} \, \bar{\Phi}\Phi + \left(m \int d^2\theta \, \Phi^2 + \text{H.c.} \right). \tag{10.395}$$

Here I have also included the superpotential term, assuming that $|m| \ll g|v|$. Thus, the low-energy theory is that of the free chiral superfield with mass m. The mass term obviously lifts the flat direction; the solution for the vacuum field is unique, $\phi_{\text{vac}} = 0$. As we will see later, in fact there are two isolated vacua in the model at hand. In the tree approximation, which we have so far used, these vacua coalesce into a single point.

The point $\phi_{\text{vac}} = 0$ lies in the middle of the domain $|\phi| < \Lambda$, where Λ is the dynamical scale parameter of supersymmetric QCD. This is the domain of strong coupling, where the tree-level discussion presented above is invalid. In particular, the Kähler potential (which is flat in Eq. (10.395)) receives quantum corrections even in perturbation theory. The expansion parameter is $(\ln|\phi|/\Lambda)^{-1}$; it is small if $|\phi|/\Lambda \gg 1$. However, it explodes in the domain $|\phi|/\Lambda \lesssim 1$.

Quantum corrections to the superpotential vanish in perturbation theory (Section 10.8). One-flavor supersymmetric QCD is an example of a theory in which the superpotential gets modified *nonperturbatively* [66], as we will see later. This modification drastically changes the vacuum structure of the theory, pushing it out of the strong-coupling domain $|\phi| < \Lambda$.

Before discussing a possible form of nonperturbative correction to the superpotential, I will pause to make a remark. The chiral (supergauge) invariant[42] describing the moduli fields is $X \equiv Q_\alpha^f Q_f^\alpha = 2\Phi^2$. Taking the square root introduces a "double-valuedness" that is an artifact of this coordinate choice. From this point of view it would be more transparent to use the superfield X directly to describe the moduli fields. A disadvantage of X compared with Φ is the more complicated form of Kähler term. In terms of X,

| Low-energy limit in one-flavor SU(2) SQCD |

$$\mathcal{L}_{\text{eff}} = \int d^2\theta\, d^2\bar{\theta} \sqrt{\bar{X}X} + \left(\frac{m}{2} \int d^2\theta\, X + \text{H.c.}\right). \tag{10.396}$$

Needless to say, the Kähler metric remains flat.

Now let us examine the global symmetries of the theory. We have already mentioned the global subflavor SU(2) symmetry. It is contained in Eq. (10.396) already since the chiral invariant X is obviously also invariant under the subflavor SU(2) transformations.

At $m = 0$ the theory (10.368) has two U(1) symmetries: one is the R symmetry, the other is the global symmetry

$$Q \to e^{i\alpha} Q, \qquad \tilde{Q} \to e^{i\alpha} \tilde{Q}. \tag{10.397}$$

Both symmetries are anomalous at the quantum level. The currents generating the R transformation and the U(1) transformation (10.397) are

$$R^\mu = \frac{1}{g^2} \bar{\lambda}^a \bar{\sigma}^\mu \lambda^a + \frac{1}{3} \sum_f \left(2i\bar{q}_f \overleftrightarrow{\mathcal{D}}_\mu q_f - \bar{\psi}_f \bar{\sigma}^\mu \psi_f\right), \tag{10.398}$$

$$j^\mu = \sum_f \left(\bar{\psi}_f \bar{\sigma}^\mu \psi_f + i\bar{q}_f \overleftrightarrow{\mathcal{D}}_\mu q_f\right). \tag{10.399}$$

| Cf. Section 10.16. |

Their anomalies are well known, namely,

$$\partial_\mu R^\mu = \frac{1}{16\pi^2} \frac{5}{3} G_{\mu\nu}^a \tilde{G}^{a\mu\nu}, \qquad \partial_\mu j^\mu = \frac{1}{16\pi^2} G_{\mu\nu}^a \tilde{G}^{a\mu\nu}. \tag{10.400}$$

---

[42] This is the only chiral invariant that one can construct in the model under consideration.

Table 10.4. The R and $\tilde{R}$ charges

Fields or parameters	R charge	$\tilde{R}$ charge
Q_f^i	$\frac{2}{3}$	-1
ψ_f^i	$-\frac{1}{3}$	-2
λ	1	1
θ	1	1
X	$\frac{4}{3}$	-2

Therefore, the current

$$\tilde{R}^\mu = R^\mu - \frac{5}{3}\, j^\mu \tag{10.401}$$

is anomaly-free: it is strictly conserved. The corresponding $\tilde{R}$ charges are shown in Table 10.4. Soon we shall omit the tildes and will refer to conserved R currents and charges where there is no danger of confusion.

From this table it is clear that the $\tilde{R}$ symmetry of one-flavor supersymmetric QCD (which is exact at $m = 0$) does not forbid the emergence of a nonperturbative superpotential term,

$$\mathcal{W}_{\text{np}} = \frac{\Lambda^5}{X}, \tag{10.402}$$

in the effective low-energy Lagrangian (10.396). The fifth order of the dynamical scale parameter Λ in the numerator appears on dimensional grounds, since the superfield X has dimension 2 while the dimension of the superpotential must be 3. Those who will follow the author into supersymmetric instanton calculus in Section 10.19 will learn how Eq. (10.402) is actually derived. For the time being let us take it as given [66]. With this superpotential the vacuum energy vanishes only at $|X| = \infty$. The theory is said to have a run-away vacuum. Such theories can only be considered in a cosmological context. From the point of view of field theory, there is no stable vacuum in the case at hand.

Affleck–Dine–Seiberg superpotential, Section 10.20

However, we should not come to hasty conclusions and should not forget about the small mass term present in the Lagrangian (10.396) at tree level. If both terms are assembled, the total effective superpotential takes the form[43]

$$\mathcal{W}_{\text{eff}} = \frac{m}{2}X + \frac{\Lambda^5}{X}. \tag{10.403}$$

Hence, we have for the corresponding F term

$$\bar{F} \propto \frac{\partial \mathcal{W}_{\text{eff}}}{\partial X} = \frac{m}{2} - \frac{\Lambda^5}{X^2}, \tag{10.404}$$

---

[43] One should not forget that $|m| \ll \Lambda$ by assumption; only in this case are the moduli fields much lighter than the Higgsed gauge bosons, so that their dynamics can be considered separately. In the following we will assume that both m and Λ are real and positive. This can be always achieved by an appropriate choice of parameters.

which vanishes at

$$X_{\mathrm{vac}} = \pm\Lambda^2\sqrt{\frac{2\Lambda}{m}}. \tag{10.405}$$

We have two well-defined vacua. The mass term stabilizes the run-away direction. Note that at small m both vacua lie well beyond the dangerous strong-coupling domain $|X| < \Lambda^2$. This confirms the statement made at the beginning of this section: one-flavor supersymmetric QCD with gauge group $\mathrm{SU}(2)$ has two discrete vacua.

Warning: In concluding this section I need to make a comment regarding the determination of the D-flat directions. In the one-flavor case, when there is only a single chiral invariant, it is easy to identify and parametrize the flat direction. If, instead, we consider an arbitrary gauge group and a generic matter sector (see Eq. (10.368)), the analysis of the D-flat direction is a difficult (and not always analytically solvable) technical problem, generally speaking. We will not dwell on this issue. The interested reader can acquaint himself or herself with the elements of the general theory of D-flat directions in more specialized works, e.g., [22] or Sections 2.4–2.7 in [62].

10.16 Hypercurrent and Anomalies

In Section 10.7 we learned that some supersymmetric theories have an exact R symmetry and that the latter can play an important role in dynamical analyses. The R symmetry is, in a sense, inherent to supersymmetric theories because of its geometric nature.[44] In superspace an R transformation is expressed by phase rotations of the Grassmann coordinates θ and $\bar{\theta}$,

$$\theta \to e^{i\alpha}\theta, \qquad \bar{\theta} \to e^{-i\alpha}\bar{\theta}, \qquad x_\mu \to x_\mu. \tag{10.406}$$

The R transformation of a generic superfield is

$$\Phi(x, \theta, \bar{\theta}) \to e^{ir\alpha}\Phi(x, e^{-i\alpha}\theta, e^{i\alpha}\bar{\theta}), \tag{10.407}$$

where r is the R charge of the field Φ.[45] It is perfectly natural that the chiral symmetry of the geometric origin is combined with the supersymmetry and energy–momentum conservation in one common superfield.

Indeed, the commutators of the R charge with the supercharges are proportional to the supercharges; see Eq. (10.209). Hence, the supercurrent can be obtained from the R current by the action of appropriate supercharges. Moreover, as we already know from Section 10.11, by anticommuting the supercurrent with the supercharges we get the energy–momentum tensor, see (10.264). Then all three conserved operators – the R current, supercurrent, and energy–momentum tensor – must belong to the same supermultiplet, as was first pointed out by Ferrara and Zumino [40]. In Section 10.6.6 we coined the term *hypercurrent* for this supermultiplet and denoted it as $\mathcal{J}_{\alpha\dot{\alpha}}$. In the

---

[44] These introductory remarks are imprecise. Gradually, we will make them more precise; just be patient!
[45] If the R charges are canonical, (10.217), we will call this R symmetry geometric. For instance, this is the case in the Wess–Zumino model, with a purely cubic superpotential. Generally speaking, the set of r values need not be canonical.

present section the construction of the hypercurrent will be considered in detail for super-Yang–Mills theories. The Konishi anomaly will be included in our narrative as a necessary element. An exceptional case – Yang–Mills theory with the Fayet–Iliopoulos term – will be briefly discussed. First, however, we will study some general aspects of this topic, which are applicable to all supersymmetric theories in four dimensions [38, 40].

10.16.1 Generalities

All supersymmetric theories can be naturally divided into two classes – those with an exact R symmetry[46] and those with a broken R symmetry. The first class is quite narrow (see Eq. (10.411) below), such theories being quite rare,[47] while the majority of (four-dimensional) supersymmetric theories belong to the second class. In these theories the hypecurrent $\mathcal{J}_{\alpha\dot\alpha}$ is defined (following Ferrara and Zumino) by the condition

Defining relation for hypercurrent in majority of theories

$$D^\alpha \mathcal{J}_{\alpha\dot\alpha} = \bar{D}_{\dot\alpha} \bar{X} \tag{10.408}$$

where $\bar{X}$ is an (anti)chiral superfield. I will continue to refer to the lowest component of this superfield as R current,

$$\mathcal{J}_{\alpha\dot\alpha}(\theta = \bar\theta = 0) = R_{\alpha\dot\alpha}, \quad R_\mu = \frac{1}{2}(\bar\sigma^\mu)^{\dot\alpha\alpha} R_{\alpha\dot\alpha}, \tag{10.409}$$

although generally speaking it is not conserved. (The conserved R current will be denoted by $\tilde{R}$). Higher components in $\mathcal{J}_{\alpha\dot\alpha}$ will include the supercurrent and the energy-momentum tensor.

As we know from Section 10.5.2, the anticommutator $\{D^\alpha, \bar{D}^{\dot\alpha}\} = 2i\partial^{\dot\alpha\alpha}$. Differentiating $\mathcal{J}_{\alpha\dot\alpha}$ in the defining relation (10.408) in two different ways and adding the results we arrive at

$$\partial^{\dot\alpha\alpha} \mathcal{J}_{\alpha\dot\alpha} = -\frac{i}{2}\left(D^2 X - \bar{D}^2 \bar{X}\right). \tag{10.410}$$

Combining the two equations above we see that if $X = 0$ the R current is conserved. Moreover, as a consequence of the hypercurrent conservation, $D^\alpha \mathcal{J}_{\alpha\dot\alpha} = 0$, the corresponding theory turns out to be superconformal.

The superconformality condition $X = 0$ can be somewhat relaxed. (This is described in Section 7.10 of the textbook [10]). Indeed, assume that instead of (10.408) we have

$$D^\alpha \mathcal{J}_{\alpha\dot\alpha}^R = \bar{\chi}_{\dot\alpha} \tag{10.411}$$

where $\bar{\chi}$ is an antichiral superfield satisfying an analog of the Bianchi identity (cf. Eq. (10.183)),

$$D^\alpha \chi_\alpha = \bar{D}_{\dot\alpha} \bar{\chi}^{\dot\alpha}. \tag{10.412}$$

---

[46] The corresponding R current can be a combination of the geometric R current and the flavor currents, see Sections 10.7 and 10.16.7.1.

[47] Not only are such theories hard to find, they carry an intrinsic problem associated with the conserved R charge. It is believed that no global symmetries of this type can survive after gravity is switched on (e.g., [67] and references therein). This is a separate question, however, which will not be treated in this text.

The superscript R in (10.411) shows that the lowest component of $\mathcal{J}^R_{\alpha\dot\alpha}$, the R current, is conserved. This can be readily proved,

$$\bar{D}^{\dot\alpha} D^\alpha \mathcal{J}_{\alpha\dot\alpha} = -\bar{D}\bar\chi = -D\chi = -D^\alpha \bar{D}^{\dot\alpha} \mathcal{J}^R_{\alpha\dot\alpha} \tag{10.413}$$

implying

$$2i\partial^{\dot\alpha\alpha} \mathcal{J}^R_{\alpha\dot\alpha} = 0. \tag{10.414}$$

Note that Eq. (10.408) with $X \neq 0$ reduces to (10.411) provided that X satisfies the following condition,

$$\bar{X} = D^2 Y, \tag{10.415}$$

where Y is a chiral superfield. Indeed, if we introduce

$$\mathcal{J}^R_{\alpha\dot\alpha} = \mathcal{J}_{\alpha\dot\alpha} + 2i\partial_{\alpha\dot\alpha}(Y - \bar{Y}), \quad \text{and} \quad \bar\chi_{\dot\alpha} = D^2 \bar{D}_{\dot\alpha} Y, \tag{10.416}$$

the transiton from (10.408) to (10.411) becomes trivial. An example of an $\mathcal{J}^R_{\alpha\dot\alpha}$ hypercurrent is provided by Yang-Mills theories without matter at the classical level. We will discuss this in Section 10.16.3. After quantum effects are taken into account a nontrivial X is generated through anomalies (Section 10.16.4).

As I have already mentioned, there are a few exceptional theories in which the Ferrara-Zumino hypercurrent (10.408) does not exist. The most notable example of this type is provided by Yang-Mills theories with the gauge group including, among others, a U(1) factor. If we add the Fayet-Iliopoulos term in this theory, an *extended* hypercurrent should be introduced (I suggest to call it Komargodski-Seiberg hypercurrent) interpolating between (10.408) and (10.411). The equation defining the Komargidski-Seiberg hypercurrent is

Ferrara–Zumino hypercurrent vs. Komargodski-Seiberg hypercurrent

$$D^\alpha \mathcal{J}_{\alpha\dot\alpha} = \bar{D}_{\dot\alpha} X + \bar\chi_{\dot\alpha} \tag{10.417}$$

with the constraint (10.412). Below I will not discuss this case further referring the reader to [38] for details.

Returning to the the Ferrara-Zumino hypercurrent, I will present here its lowest components,

$$\mathcal{J}_{\alpha\dot\alpha} = R_{\alpha\dot\alpha} - \left(i\theta^\beta J_{\beta\dot\alpha} + \sqrt{2}\theta_\alpha \bar\eta_{\dot\alpha} + \text{H.c.}\right) - 2\theta^\beta \bar\theta^{\dot\beta} T_{\alpha\dot\alpha\beta\dot\beta}$$
$$- 4\theta_\alpha \bar\theta_{\dot\alpha} (\text{Re } Z) - 2i\theta_\alpha \bar\theta_{\dot\beta} \left(\partial_\rho R_\gamma\right) (\bar\sigma^{\rho\gamma})^{\dot\beta}_{\dot\alpha} + \cdots. \tag{10.418}$$

where J and T are the *conserved* supersurrent and the energy-momentum tensor (symmetric),

$$J_{\beta\dot\alpha} = (\sigma^\mu)_{\alpha\dot\alpha} J_{\mu,\beta}, \qquad T_{\alpha\dot\alpha\beta\dot\beta} = (\sigma^\mu)_{\alpha\dot\alpha} (\sigma^\nu)_{\beta\dot\beta} T_{\mu\nu}. \tag{10.419}$$

The fields η and Z are defined through the components of the chiral X superfield appearing in Eq. (10.408),

$$X = \sigma + \sqrt{2}\theta^\alpha \eta_\alpha + \theta^2 Z, \qquad \bar{X} = \bar\sigma + \sqrt{2}\bar\eta_{\dot\alpha}\bar\theta^{\dot\alpha} + \bar\theta^2 \bar{Z}. \tag{10.420}$$

10.16.2 Supercurrent and Energy-Momentum Tensor Conservation: Confirmation

Let us explicitly check that the supercurrent and the energy-momentum tensor from (10.418) are conserved even though the current $R_{\alpha\dot\alpha}$ is not. I will start from the energy-momentum tensor,

$$\partial^{\dot\alpha\alpha} \mathcal{J}_{\alpha\dot\alpha}\big|_{\theta\bar\theta} = \partial^{\dot\alpha\alpha}\left(-2\theta^\beta\bar\theta^{\dot\beta} T_{\alpha\dot\alpha\beta\dot\beta} - 4\theta_\alpha\bar\theta_{\dot\alpha}\,(\mathrm{Re}\,Z)\right)$$

$$= 4\left(\partial^\mu T_{\mu\nu}\right)v^\nu + 4v^\mu\partial_\mu(\mathrm{Re}\,Z),$$

$$v^\nu = \bar\theta_{\dot\alpha}\,(\bar\sigma^\nu)^{\dot\alpha\alpha}\,\theta_\alpha, \tag{10.421}$$

to be compared with the right-hand side of Eq. (10.410),

$$-\frac{i}{2}\left(D^2 X - \bar{D}^2\bar{X}\right)\Big|_{\theta\bar\theta} = 4v^\mu\partial_\mu(\mathrm{Re}\,Z). \tag{10.422}$$

Combining (10.421) and (10.422), we see that see that $\partial^\mu T_{\mu\nu} = 0$ indeed.

Next, we will examine the θ component of $\partial^{\dot\alpha\alpha} \mathcal{J}_{\alpha\dot\alpha}$,

$$\partial^{\dot\alpha\alpha} \mathcal{J}_{\alpha\dot\alpha}\big|_{\theta} = -\partial^{\dot\alpha\alpha}\left(i\theta^\beta J_{\beta\alpha\dot\alpha} + \sqrt{2}\theta_\alpha\bar\eta_{\dot\alpha}\right)$$

$$= -2i\theta^\beta\left(\partial_\mu J^\mu_\beta\right) - \sqrt{2}\theta^\alpha\partial_{\alpha\dot\alpha}\bar\eta^{\dot\alpha}, \tag{10.423}$$

to be compared with

$$\frac{i}{2}\bar{D}^2\bar{X} = -\sqrt{2}\theta^\alpha\partial_{\alpha\dot\alpha}\bar\eta^{\dot\alpha}, \tag{10.424}$$

which, in turn, implies that $\partial_\mu J^\mu_\beta = 0$.

The remainder of Section 10.16 will be devoted to super-Yang–Mills theories with or without matter.

10.16.3 Supersymmetric Gluodynamics at the Classical Level

To become further acquainted with the Ferrara–Zumino construction let us consider first supersymmetric gluodynamics, the simplest non-Abelian gauge theory, discussed in Section 10.14. The Lagrangian of this theory is given in Eq. (10.374). Since the gluino field is massless, the Lagrangian (10.374) is obviously invariant under the chiral rotation $\lambda \to \lambda e^{i\alpha}$ at the classical level. This corresponds to the chiral transformation of the vector superfield with R charge 0 and that of W with R charge 1. The classically conserved R current that exists in this theory [48, 68] was defined in (10.377). The R charge is given by

$$R = \int d^3x\, R_0. \tag{10.425}$$

Classically, the theory at hand is superconformal. The conserved hypercurrent is defined through Eq. (10.411) with $\chi = 0$. Namely,

$$\mathcal{J}_{\alpha\dot\alpha} = -\frac{4}{g^2}\,\mathrm{Tr}\left(e^V W_\alpha e^{-V}\bar{W}_{\dot\alpha}\right) = R_{\alpha\dot\alpha} - \left(i\theta^\beta J_{\beta\alpha\dot\alpha} + \mathrm{H.c.}\right)$$

$$- 2\theta^\beta\bar\theta^{\dot\beta} T_{\alpha\dot\alpha\beta\dot\beta} - \tfrac{1}{2}\left(\theta_\alpha\bar\theta_{\dot\beta}i\partial^{\dot\beta\gamma} R_{\gamma\dot\alpha}\right) + \dots, \tag{10.426}$$

where $J_{\beta\alpha\dot\alpha}$ is the supercurrent and $T_{\alpha\dot\alpha\beta\dot\beta}$ is the energy–momentum tensor:

$$J_{\beta\alpha\dot\alpha} = (\sigma_\mu)_{\alpha\dot\alpha} J_\beta^\mu = \frac{4i}{g^2} \, \mathrm{Tr}(G_{\alpha\beta} \bar\lambda_{\dot\alpha}),$$

$$T_{\alpha\dot\alpha\beta\dot\beta} = (\sigma^\mu)_{\alpha\dot\alpha} (\sigma^\nu)_{\beta\dot\beta} T_{\mu\nu}$$

$$= \frac{2}{g^2} \, \mathrm{Tr} \left(i\lambda_{\{\alpha} \mathcal{D}_{\beta\}\dot\beta} \bar\lambda_{\dot\alpha} - i\mathcal{D}_{\beta\{\dot\beta} \lambda_\alpha \bar\lambda_{\dot\alpha\}} + G_{\alpha\beta} \bar G_{\dot\alpha\dot\beta} \right). \qquad (10.427)$$

Symmetrization over α, β or $\dot\alpha, \dot\beta$ is indicated by braces.[48] The above expressions should be compared with the general formula (10.418) in which we must put $\eta = Z = 0$.

The classical equation for $\mathcal{J}_{\alpha\dot\alpha}$ is

$$\bar D^{\dot\alpha} \mathcal{J}_{\alpha\dot\alpha} = 0. \qquad (10.428)$$

In addition to the conservation of all three operators, R^μ, J_α^μ, and $T_{\mu\nu}$, Eq. (10.428) contains the following relations also:

$$T_\mu^\mu = 0, \qquad (\bar\sigma_\mu)^{\dot\alpha\alpha} J_\alpha^\mu = 0. \qquad (10.429)$$

In conjunction with the conservation of $T_{\mu\nu}$ and J_α^μ these relations express the *classical* conformal and superconformal symmetries.

10.16.4 Supersymmetric Gluodynamics at the Quantum Level

As we know from Section 10.14, conservation of the R_μ current is lost at the quantum level owing to the chiral anomaly; see Eq. (10.379). The superfield generalization of Eq. (10.379) is now based on Eq. (10.408) with

$$X = -\frac{T_G}{8\pi^2} \, \mathrm{Tr} \, W^2. \qquad (10.430)$$

Equation (10.426) is no longer valid. Additional terms must be added. Simultaneously, in (10.427) the tracelessness of $T_{\mu\nu}$ is lost, as well as $\gamma_\mu J_\alpha^\mu = 0$ is replaced by $\gamma_\mu J_\alpha^\mu \neq 0$. One must use Eqs. (10.156) and (10.158).

In components, identification of the X field is as follows,

$$\sigma = \frac{T_G}{8\pi^2} \, \mathrm{Tr} \, \lambda^2, \qquad \eta_\alpha = \frac{T_G}{8\pi^2} \, \sqrt{2} \, \mathrm{Tr} \, G_{\alpha\beta} \lambda^\beta,$$

$$Z = -\frac{T_G}{8\pi^2} \, \mathrm{Tr} \left(-\frac{1}{2} G^{\alpha\beta} G_{\alpha\beta} + 2\bar\lambda_{\dot\alpha} \left(i \mathcal{D}^{\dot\alpha\alpha} \right) \lambda_\alpha \right). \qquad (10.431)$$

Equation (10.379) reduces to

$$\frac{1}{2} \partial^{\alpha\dot\alpha} \left(R_{\alpha\dot\alpha} \right)_{\theta=\bar\theta=0} = -2 \, \mathrm{Im} \, Z. \qquad (10.432)$$

The anomalies in T_μ^μ and $\gamma_\mu J_\alpha^\mu$ can be derived from Eq. (10.408). Its $\bar\theta$ component results in

$$T_\mu^\mu = \frac{-3T_G}{16\pi^2} \, \mathrm{Tr} \left(G_{\mu\nu} G^{\mu\nu} - 4i\bar\lambda_{\dot\alpha} \mathcal{D}^{\dot\alpha\alpha} \lambda_\alpha \right) \qquad (10.433)$$

---

[48] The component decompositions in the present section predominantly refer to the Wess–Zumino gauge, although some are more general.

while the lowest component yields $\varepsilon^{\beta\alpha} J_{\beta\alpha\dot\alpha}$. Let us demonstrate the latter statement. Indeed,

$$D^\alpha \mathcal{J}_{\alpha\dot\alpha}\big|_{\theta=\bar\theta=0} = i\varepsilon^{\beta\alpha} J_{\beta\alpha\dot\alpha} - 2\sqrt{2}\,\bar\eta_{\dot\alpha}$$

and

$$\bar D_{\dot\alpha} \bar X\big|_{\theta=\bar\theta=0} = \sqrt{2}\,\bar\eta_{\dot\alpha}.$$

Equating the preceding expressions as prescribed by (10.408) we arrive at

$$\varepsilon^{\beta\alpha} J_{\beta\alpha\dot\alpha} = -3i\sqrt{2}\bar\eta_{\dot\alpha} = -3i\frac{T_G}{4\pi^2}\mathrm{Tr}\left(\bar G_{\dot\alpha\dot\beta}\bar\lambda^{\dot\beta}\right) \tag{10.434}$$

For convenience I will rewrite here Eq. (10.432) in the vectorial form,

$$\partial_\mu R^\mu = \frac{T_G}{8\pi^2}\mathrm{Tr}\left(G_{\mu\nu}\tilde G^{\mu\nu}\right). \tag{10.435}$$

The supermultiplet structure of the anomalies in $\partial^\mu R_\mu$, in the trace of the energy–momentum tensor T^μ_μ, and in $J^\alpha_{\alpha\dot\alpha}$ (the three "geometric" anomalies) was discovered and discussed by Grisaru [69].

10.16.5 Including Matter

The inclusion of matter fields typically results in additional global symmetries, and, in particular, in additional U(1) symmetries. Some of them act exclusively in the matter sector. These are usually quite evident and are immediately detectable. Here I will present a classification of the anomalous and nonanomalous U(1) symmetries. At this first stage it is convenient to assume that there is no superpotential, i.e.,

$$\mathcal{W} = 0$$

to any finite order of perturbation theory.

The general Lagrangian of the gauge theory with matter is given in Eq. (10.368), where, in the absence of a superpotential, we set $\mathcal{W}(Q_f) = 0$. The matter sector consists of a number of irreducible representations of the gauge group. Every irreducible representation will be referred to as a "flavor." It is clear that, additionally to the U(1)$_R$ symmetry discussed above, one can make phase rotations of each of the N_f matter fields independently. Thus altogether we have $N_f + 1$ chiral rotations. It would be in order here to summarize these chiral rotations.

1. The R transformation. The action is invariant under the following transformation:

$$V(x, \theta, \bar\theta) \to V(x, e^{-i\alpha}\theta, e^{i\alpha}\bar\theta), \qquad Q(x_L, \theta) \to e^{2i\alpha/3}Q(x_L, e^{-i\alpha}\theta). \tag{10.436}$$

In components the same transformations are given as

$$A_\mu \to A_\mu, \qquad \lambda_\alpha \to e^{i\alpha}\lambda_\alpha, \qquad \psi^f_\alpha \to e^{-i\alpha/3}\psi^f_\alpha, \qquad q^f \to e^{2i\alpha/3}q^f. \tag{10.437}$$

The corresponding chiral current, the "geometric" R current, which can be viewed as a generalization of the current (10.377), has the form

$$R_\mu = -\frac{1}{g^2}\lambda^a\sigma_\mu\bar\lambda^a + \frac{1}{3}\sum_f\left(\psi_f\sigma_\mu\bar\psi_f - 2i\phi_f\overleftrightarrow{D}_\mu\bar\phi_f\right). \tag{10.438}$$

This current is the lowest component of the "geometric" hypercurrent $\mathcal{J}_{\alpha\dot\alpha}$,

$$\mathcal{J}_{\alpha\dot\alpha} = \frac{4}{g^2} \operatorname{Tr}\left(\bar{W}_{\dot\alpha} e^V W_\alpha e^{-V}\right)$$

$$- \frac{1}{3} \sum_f \bar{Q}_f \left(\overleftarrow{\bar{\nabla}}_{\dot\alpha} e^V \nabla_\alpha - e^V \bar{D}_{\dot\alpha} \nabla_\alpha + \overleftarrow{\bar{\nabla}}_{\dot\alpha} \overleftarrow{D}_\alpha e^V\right) Q_f, \qquad (10.439)$$

where the spinorial gauge-covariant derivatives were introduced in Section 10.13. For the reader's convenience I reproduce the relevant definitions:

$$\nabla_\alpha Q = e^{-V} D_\alpha \left(e^V Q\right), \qquad \bar{\nabla}_{\dot\alpha} \bar{Q} = e^V \bar{D}_{\dot\alpha} \left(e^{-V} \bar{Q}\right). \qquad (10.440)$$

Equation (10.439) extends the first formula in (10.426) in a natural way to include matter. In particular, the $\theta\bar\theta$ component now contains the energy–momentum tensor with inclusion of the matter contribution.

2. *The flavor* U(1) *transformations.* The remaining N_f currents are due to phase rotations of each flavor superfield independently,

$$Q_f(x_L, \theta) \to e^{i\alpha_f} Q_f(x_L, \theta). \qquad (10.441)$$

Note that θ is not affected by these transformations. The corresponding chiral currents are

$$R_\mu^f = -\psi_f \sigma_\mu \bar\psi_f - \phi_f i\overleftrightarrow{\mathcal{D}}_\mu \bar\phi_f, \qquad (10.442)$$

also known as the Konishi currents in the context of super-Yang–Mills theories. In superfield language R_μ^f is the $\theta\bar\theta$ component of the Konishi operator [70],

$$\mathcal{J}^f = \bar{Q}_f e^V Q_f. \qquad (10.443)$$

Konishi operator

In order to derive from the Konishi operator an object similar to $\mathcal{J}_{\alpha\dot\alpha}$ (i.e., belonging to the representation $\left(\frac{1}{2}, \frac{1}{2}\right)$ of the Lorentz group) we can form a flavor superfield $\mathcal{J}_{\alpha\dot\alpha}^f$, defined as

$$\mathcal{J}_{\alpha\dot\alpha}^f = -\frac{1}{2}[D_\alpha, \bar{D}_{\dot\alpha}]\mathcal{J}^f = -\frac{1}{2}[D_\alpha, \bar{D}_{\dot\alpha}]\bar{Q}_f e^V Q_f, \qquad (10.444)$$

of which R_μ^f is the lowest component. There is a deep difference between the Konishi current $\mathcal{J}_{\alpha\dot\alpha}^f$ and the geometric hypercurrent $\mathcal{J}_{\alpha\dot\alpha}$: the latter contains (in its higher components) the supercurrent and the energy–momentum tensor while the higher components of the Konishi currents $\mathcal{J}_{\alpha\dot\alpha}^f$ are conserved trivially (nondynamically).

10.16.6 Anomalies in Theories with Matter

In this subsection we will consider all the U(1) currents discussed above and derive their anomalies. For the time being we will set $\mathcal{W} = 0$. The latter condition will be lifted shortly.

Let us start from the hypercurrent (10.439). Our task is to generalize the gluodynamics formula (10.430) to include matter. Then, instead of (10.430) we obtain

$$X = -\frac{2}{3}\left[\frac{3T_G - \sum_f T(R_f)}{16\pi^2} \operatorname{Tr} W^2 + \frac{1}{8} \sum_f \gamma_f \bar{D}^2 \left(\bar{Q}_f e^V Q_f\right)\right], \qquad (10.445)$$

where the γ_f are the anomalous dimensions of the matter fields,

$$\gamma_f \equiv -\frac{d \ln Z_f}{d \ln M_{\mathrm{uv}}}, \tag{10.446}$$

Z_f is the Z factor of the matter field f (Z is defined as the coefficient in front of the corresponding kinetic term in the effective action; see Eq. (10.448) below), and M_{uv} is the ultraviolet cutoff. When understood in operator form, Eq. (10.445) is exact [46]. We will derive it in two steps.

If we compare Eq. (10.445) with its counterpart (10.430) in supersymmetric gluodynamics, two distinctions are apparent. First, the coefficient in front of the gauge term $\mathrm{Tr}\, W^2$ is different,

$$T_G \to T_G - \tfrac{1}{3} \sum_f T(R_f). \tag{10.447}$$

Second, the term proportional to $\bar{D}^2(\bar{Q}_f e^V Q_f)$ has appeared on the right-hand side, with a coefficient proportional to the anomalous dimension of the given matter field. Formally $\bar{D}^2(\bar{Q}_f e^V Q_f)$ vanishes by virtue of the equations of motion. In fact it has its own anomaly, as we will see shortly, and therefore should be kept in the formula.

Look back through Section 8.2.1.

The easiest way to derive (10.447) is through the lowest component of $\mathcal{J}_{\alpha\dot{\alpha}}$, given in (10.438). The anomaly in the divergence of the axial current comes exclusively from fermion triangles. The R current, R^μ, has the gaugino component and that of the matter fermions; by definition the latter carries the relative coefficient $-\tfrac{1}{3}$. Taking into account that the anomalous triangle for gauginos is proportional to T_G and that for the matter fermions to $T(R_f)$, while everything else is the same,[49] and including the above factor $1/3$ we immediately confirm (10.447).

Now let us deal with $\bar{D}^2(\bar{Q}_f e^V Q_f)$. This term is best traced as a response of the theory to the scale transformation.

If $\mathcal{W} = 0$ then the theory under consideration is classically scale invariant. However, already at one loop the scale invariance is lost owing to ultraviolet divergences. In calculating the relevant effective Lagrangian one must introduce an ultraviolet cutoff M_{uv} (e.g., the Pauli–Villars mass) and a renormalization point μ regularizing logarithmically divergent integrals at small momenta (in the infrared). To find the scale noninvariance of the effective Lagrangian we must scale μ keeping M_{uv} fixed or, alternatively, scale M_{uv} keeping μ fixed. Both procedures give the same result since ultraviolet logarithms depend only on the ratio M_{uv}/μ.

Look back through Section 8.4.

The original Lagrangian (10.368) was formulated in the ultraviolet; for the present we denote the gauge coupling at the ultraviolet cutoff in this Lagrangian as g_0^2 (the subscript 0 indicates the bare coupling constant as opposed to g^2). Next, we evolve the Lagrangian from M_{uv} to μ and obtain[50]

$$\mathcal{L}_{\mathrm{eff}} = \left(\frac{1}{2g_0^2} - \frac{\beta_0}{16\pi^2} \ln \frac{M_{\mathrm{uv}}}{\mu} \right) \int d^2\theta \, \mathrm{Tr}\,(W^\alpha W_\alpha)$$

$$+ \sum_{\mathrm{all\ flavors}} \frac{1}{8} Z_f(\mu) \int d^2\theta \, \bar{D}^2 \bar{Q}^f e^V Q_f + \mathrm{H.c.}, \tag{10.448}$$

---

[49] Both gaugino and matter fermions are counted in terms of Weyl fields.
[50] The coefficient $\tfrac{1}{8}$ in front of $Z_f(\mu)$ is not a mistake. Question: Where does it come from?

where

$$\beta_0 = 3T_G - \sum_f T(R_f) \tag{10.449}$$

is the first coefficient in the β function. The above answer for the effective Lagrangian is exact if the latter is treated as Wilsonian.

The noninvariance of the effective action with regard to the scale transformation is represented by the factor $\ln M_{\mathrm{uv}}$ in the first line of Eq. (10.448), and similar logarithms reside in Z_f in the second line. Differentiating with respect to $\ln M_{\mathrm{uv}}$, we can verify the presence of $\bar{D}^2 \bar{Q}^f e^V Q_f$ in the anomaly equation (10.445), with its coefficient $\frac{1}{8}\gamma_f$.

10.16.6.1 Konishi Anomaly

In Section 10.16.5 we started discussing the U(1) currents of the matter fields; see Eq. (10.441). The Konishi operator $\bar{Q}_f e^V Q_f$ contains the corresponding current (10.442) in the $\bar{\theta}^{\dot\alpha}\theta^\alpha$ component. The statement that $\bar{D}^2(\bar{Q}_f e^V Q_f) = 0$ is nothing other than the classical equation of motion. Indeed, its lowest component is $\bar{F}_f q_f = 0$, implying the vanishing of the F_f term in the absence of a superpotential.

Explorations of super-Yang–Mills theories with matter revealed that these classical equations of motion have anomalies at the quantum level [70]. In the supersymmetric formulation this fact is known as the Konishi anomaly and can be expressed as follows:

$$\bar{D}^2 \mathcal{J}^f = \bar{D}^2(\bar{Q}_f e^V Q_f) = \frac{T(R_f)}{2\pi^2}\,\mathrm{Tr}\,W^2. \tag{10.450}$$

| The Konishi formula |

This operator result is exact. The easiest way to confirm the coefficient $T/(2\pi^2)$ on the right-hand side is through consideration of the θ^2 components on the left- and right-hand sides of Eq. (10.450). More exactly, we will focus on the imaginary part of the coefficient of θ^2. To this end it is sufficient to make the replacements

$$\bar{D}^2 \to -2i\frac{\partial}{\partial\bar{\theta}^{\dot\alpha}}\theta^\gamma\partial_\gamma^{\dot\alpha}$$
$$\bar{Q}_f e^V Q_f \to 2\theta^\beta\bar{\theta}^{\dot\beta}\left[\left(\psi_\beta^f\bar{\psi}_{\dot\beta}^f\right) + \frac{1}{2}\phi_f\overleftrightarrow{D}_{\beta\dot\beta}\bar{\phi}_f\right],$$
$$\mathrm{Tr}\,W^2 \to \frac{1}{2}i\theta^2\,\mathrm{Tr}\,G^{\mu\nu}\tilde{G}_{\mu\nu}. \tag{10.451}$$

Then

$$\bar{D}^2(\bar{Q}_f e^V Q_f) \to 2i\theta^2\left(\partial^\mu R_\mu^f\right) \to 2i\theta^2\frac{T(R_f)}{8\pi^2}\,\mathrm{Tr}\,G^{\mu\nu}\tilde{G}_{\mu\nu}, \tag{10.452}$$

where R_μ^f was defined in Eq. (10.442). Combining (10.452) with the last line in (10.451) we arrive at (10.450).

In terms of the $(\frac{1}{2},\frac{1}{2})$ operator $\mathcal{J}_{\alpha\dot\alpha}^f$, defined in (10.444), the Konishi anomaly takes the form[51]

---

[51] Here we use the algebraic relation

$$\partial^{\alpha\dot\alpha}[D_\alpha, \bar{D}_{\dot\alpha}] = -\frac{1}{4}i\left(D^2\bar{D}^2 - \bar{D}^2 D^2\right).$$

$$\partial^{\alpha\dot\alpha} \mathcal{J}^f_{\alpha\dot\alpha} = iD^2 \left[\frac{T(R_f)}{16\pi^2} \operatorname{Tr} W^2 \right] + \text{H.c.} \tag{10.453}$$

Note that in this operator relation there are no higher-order corrections, in contrast with the situation for the geometric anomalies (10.445), where higher-order corrections enter through the anomalous dimensions γ_f.

10.16.6.2 Combining the Anomalies

Let us return to the analysis of the geometric anomalies in Eq. (10.445), adding information from the Konishi anomaly (10.450). Both are exact operator equalities. Hence, one can substitute (10.450) on the right-hand side of (10.445) to get

General formula

$$X = -\frac{2}{3} \left\{ \frac{1}{16\pi^2} \left[3T_G - \sum_f \left(1 - \gamma_f \right) T(R_f) \right] \operatorname{Tr} W^2 \right\}. \tag{10.454}$$

Alternatively,

$$\partial^{\alpha\dot\alpha} \mathcal{J}_{\alpha\dot\alpha} = \frac{i}{48\pi^2} D^2 \left[3T_G - \sum_f \left(1 - \gamma_f \right) T(R_f) \right] \operatorname{Tr} W^2 + \text{H.c.} \tag{10.455}$$

Among its other components, $\bar{D}^{\dot\alpha} \mathcal{J}_{\alpha\dot\alpha}$ contains (in its θ component) the anomaly in the trace of the energy–momentum tensor T^μ_μ. The trace of the energy–momentum tensor T^μ_μ describes the response of the theory to scale transformations, i.e., $T^\mu_\mu \propto G^a_{\mu\nu} G^{a,\mu\nu}$ (Section 8.4). The proportionality coefficient is related to the β function governing the running of the gauge coupling constant. Equation (10.454) implies this

$$\beta \propto \left(3T_G - \sum_f (1 - \gamma_f) T(R_f) \right). \tag{10.456}$$

The first appearance of the NSVZ beta function

Equation (10.456) should be committed to memory; we will use it in Section 10.21 in deriving the exact Novikov–Shifman–Vainshtein–Zakharov (NSVZ) beta function.

From Eqs. (10.453) and (10.455) it is clear that there exists a linear combination of the chiral currents that is free from the gauge anomaly:

$$\tilde{\mathcal{J}}_{\alpha\dot\alpha} = \mathcal{J}_{\alpha\dot\alpha} - \frac{3T_G - \sum_f \left(1 - \gamma_f \right) T(R_f)}{3 \sum_f T(R_f)} \sum_f \mathcal{J}^f_{\alpha\dot\alpha}. \tag{10.457}$$

The hypercurrent $\tilde{\mathcal{J}}_{\alpha\dot\alpha}$ defined in this way is exactly conserved: $\partial^{\alpha\dot\alpha} \tilde{\mathcal{J}}_{\alpha\dot\alpha} = 0$ in the absence of a superpotential.[52] In other words, its lowest component, the R current, is conserved and so are the components $O(\theta)$, $O(\bar\theta)$, and $O(\theta\bar\theta)$. The former is

---

[52] There exists an interesting class of super-Yang–Mills theories which flow to the conformal limit in the infrared. In particular, $\mathcal{N} = 1$ SQCD with the gauge group SU(N) and N_f flavors in the fundamental representation belongs to this class [71] provided that $3N/2 < N_f < 3N$. Conformality in the infrared implies that the β function vanishes, see the remark leading to Eq. (10.456). Technically this means that the anomalous dimensions γ_f flow, in the infrared, to a set of values that guarantee the vanishing of the right-hand side of (10.456). Then $\tilde{\mathcal{J}}_{\alpha\dot\alpha} = \mathcal{J}_{\alpha\dot\alpha}$, i.e., the conserved hypercurrent and the geometric hypercurrent coincide. In this limit the conserved R charge of the gluino is 1, while its scale dimension is 3/2. The ratio 3/2 of the scale dimension and the R charge is characteristic of superconformal theories.

an improved supercurrent while the latter is an improved energy–momentum tensor. Moreover, it is not difficult to prove that $\tilde{\mathcal{J}}$ is, in fact, the R hypercurrent of the type (10.411), $\tilde{\mathcal{J}}_{\alpha\dot\alpha} = \mathcal{J}^R_{\alpha\dot\alpha}$.

Indeed,

$$D^\alpha \left(D_\alpha \bar D_{\dot\alpha} - \bar D_{\dot\alpha} D_\alpha \right) \equiv \tfrac{3}{2} D^2 \bar D_{\dot\alpha} + \tfrac{1}{2} \bar D_{\dot\alpha} D^2, \tag{10.458}$$

where the differential operators on the left- and right-hand sides are assumed to be acting on a real superfield. Using this identity in conjunction with (10.444), (10.454), and (10.457) we arrive at

$$D^\alpha \tilde{\mathcal{J}}_{\alpha\dot\alpha} = \tfrac{3}{4} \frac{[T_G - \sum_f \tfrac{1}{3}\left(1 - \gamma_f\right) T(R_f)]}{\sum_f T(R_f)} D^2 \bar D_{\dot\alpha} \sum_f \overline{Q_f}\, e^V Q_f. \tag{10.459}$$

The operator on the right-hand side is obviously an antichiral superfield with spinorial index $\dot\alpha$. Denoting it by $\bar\chi_{\dot\alpha}$ and comparing with (10.417) we observe, with satisfaction, full agreement. It is simple to check that this operator satisfies the additional constraint (10.412), which is also required. Moreover, if

$$T_G - \sum_f \tfrac{1}{3}\left(1 - \gamma_f\right) T(R_f)$$

vanishes then so does $\bar\chi_{\dot\alpha}$, and the theory must be superconformal. This is indeed the case since the above combination constitutes the numerator of the NSVZ β function, and the vanishing of the β function in the case at hand is the necessary and sufficient condition for conformality.

The remaining $N_f - 1$ anomaly-free currents can be chosen as

$$\mathcal{J}^{fg}_{\alpha\dot\alpha} = T(R_g)\mathcal{J}^f_{\alpha\dot\alpha} - T(R_f)\mathcal{J}^g_{\alpha\dot\alpha}, \tag{10.460}$$

where one can fix g and consider all $f \neq g$.

I pause here to make a remark. Equation (10.456) is valid even in those theories in which $\mathcal{W} \neq 0$ (Section 10.16.7). A nonvanishing superpotential introduces, generally speaking, super-Yukawa constants to the theory, to be referred to as h_i. These super-Yukawa interactions manifest themselves in (10.456) only implicitly, through the anomalous dimensions γ_f, which depend, generally speaking, on all the gauge constants and super-Yukawa constants.

10.16.6.3 Digression: Gaugino Condensate as the Order Parameter

The Konishi anomaly has a practically important implication. Assume that we have a super-Yang–Mills theory with a matter sector in which (at least) one flavor is *absent from the superpotential*. Then, for this particular flavor Eq. (10.450) is valid. The left-hand side is a full superderivative. Consequently, unbroken supersymmetry implies that the vacuum expectation value $\langle \bar D^2 \mathcal{J}^f \rangle$ vanishes identically. This means that the vacuum expectation value of the right-hand side must vanish too, $\langle \mathrm{Tr}\, W^2 \rangle = 0$, implying in turn that

$$\left\langle \mathrm{Tr}\, \lambda^2 \right\rangle = 0 \tag{10.461}$$

since the lowest component of W^2 is λ^2. The converse is also true. If $\langle \text{Tr } \lambda^2 \rangle \neq 0$ then $\langle \text{Tr } W^2 \rangle \neq 0$, which means that $\langle \bar{D}^2 \mathcal{J}^f \rangle$ cannot vanish. Supersymmetry is spontaneously broken.

Hence, in such theories the gaugino condensate $\langle \text{Tr } \lambda^2 \rangle$ is the order parameter signaling the presence or absence of spontaneous supersymmetry breaking.

10.16.7 $\mathcal{W} \neq 0$

Switching on a nonvanishing superpotential

Now we are ready to discuss a generic super-Yang–Mills theory with matter *and* a superpotential. To avoid cumbersome expressions we will assume that all coupling constants are asymptotically free and that all operators presented in this section are normalized at a high ultraviolet point $\mu = M_{\text{uv}}$. At this point all anomalous dimensions vanish since they are proportional to powers of the coupling constants: $\gamma_f \rightarrow 0$. Setting $\gamma_f = 0$ simplifies the superanomaly formulas. (The complete expressions with $\gamma_f \neq 0$ can be found, e.g., in [62] or in Appendix section 10.27.4.)

The impact of a superpotential on the U(1) currents considered above is fairly clear: it appears at tree level and can be obtained readily from the classical equations of motion. Omitting the details of this quite straightforward calculation, I present here the final results for current nonconservation due both to the classical superpotential and to the quantum anomalies. For the geometric current $\mathcal{J}_{\alpha\dot\alpha}$ one has

$$\bar{D}^{\dot\alpha} \mathcal{J}_{\alpha\dot\alpha} = \frac{2}{3} D_\alpha \left[\left(3\mathcal{W} - \sum_f Q_f \frac{\partial W}{\partial Q_f} \right) - \frac{3T_G - \sum_f T(R_f)}{16\pi^2} \, \text{Tr } W^2 \right] \tag{10.462}$$

and

$$\partial^{\alpha\dot\alpha} \mathcal{J}_{\alpha\dot\alpha} = -\frac{1}{3} i D^2 \left[\left(3\mathcal{W} - \sum_f Q_f \frac{\partial W}{\partial Q_f} \right) - \frac{3T_G - \sum_f T(R_f)}{16\pi^2} \text{Tr } W^2 \right] + \text{H.c.} \tag{10.463}$$

The first terms in (10.462) and (10.463) are purely classical; the remainder is due to the anomaly. It can be seen that the classical part vanishes for a superpotential that is cubic in Q when the theory is classically conformally invariant.

The Konishi equations take the form

$$\bar{D}^2 \mathcal{J}^f = \bar{D}^2 (\bar{Q}_f e^V Q_f) = 4Q_f \frac{\partial W}{\partial Q_f} + \frac{T(R_f)}{2\pi^2} \text{Tr } W^2 \tag{10.464}$$

and

$$\partial^{\alpha\dot\alpha} \mathcal{J}^f_{\alpha\dot\alpha} = i D^2 \left[\frac{1}{2} Q_f \frac{\partial W}{\partial Q_f} + \frac{T(R_f)}{16\pi^2} \text{Tr } W^2 \right] + \text{H.c.} \tag{10.465}$$

Again the first terms on the right-hand sides are classical and the remainder is due to the anomaly.

10.16.7.1 Anomalies, Nonvanishing Superpotential, and Exact R Symmetry

Anomalies plus a nonvanishing superpotential leave the theory with no exact R symmetry, generally speaking. Indeed, all the anomaly-free global U(1) currents were given in Section 10.16.6.2. Adding a generic superpotential $\mathcal{W} \neq 0$ breaks all these U(1) symmetries at the classical level. For nonvanishing superpotentials the classical parts in the divergences of the currents (10.457) and (10.460) can be simply read off from Eqs. (10.463) and (10.465).

For "exceptional" superpotentials it may happen, however, that one *can* find the desired exactly conserved combination of currents. Such situations arise in some problems discussed in the literature. An example of particular importance – the SU(5) model with two matter generations,[53] each consisting of one quintet V and one antidecuplet X – will be considered now.

Let us denote the two quintets present in this model as V_f^i, $f = 1, 2$, and the two antidecuplets as $(X_{\bar{g}})_{ij}$, $\bar{g} = 1, 2$; the matrices $X_{\bar{g}}$ are antisymmetric in the color indices i, j. The requirement of renormalizability tells us that the superpotential must be chosen as a linear combination of two terms:

$$\mathcal{W} = \sum_{\bar{g}} = c_{\bar{g}} V_k X_{\bar{g}} V_l \varepsilon^{kl}, \qquad (10.466)$$

where the gauge indices in $V^i X_{ij} V^j$ are convoluted in a straightforward manner. There are no other gauge-invariant cubic combinations of the matter superfields.

One can always redefine the antidecuplet fields $X_{\bar{g}}$, $\bar{g} = 1, 2$, so that $\sum_{\bar{g}} c_{\bar{g}} X_{\bar{g}}$ becomes $X_{\bar{1}}$ while the orthogonal combination is $X_{\bar{2}}$. Then the superpotential of the model takes the form

$$\mathcal{W} = V_k X_{\bar{1}} V_l \varepsilon^{kl}, \qquad (10.467)$$

$X_{\bar{2}}$ being absent.

To derive an anomaly-free and conserved R current we need to know the relevant group-theoretic factors. The first coefficient of the β function is

$$\beta_0 = 3T_G - \sum_i T(R_i) = 11.$$

We recall that[54] $T(V) = \frac{1}{2}$ and $T(X) = \frac{3}{2}$ while $T_{SU(5)} = 5$.

Let us assign the R charges $(0, 1)$ to the superfields X_1 and $V_{1,2}$, respectively; see Table 10.5. (Self-consistency will be checked later.) It is obvious that under this assignment the superpotential has $\tilde{r} = 2$, implying the invariance of the superpotential contribution to the action. Since X_2 is absent from the superpotential, at this stage its R charge is arbitrary; we can determine it from the anomaly cancelation condition in the R current. The result is quoted in Table 10.5. Indeed,

Constructing conserved anomaly-free R current

---

[53] Historically this model was the first to exhibit dynamical supersymmetry breaking at weak coupling [72, 73].

[54] Incidentally, instanton calculus provides the easiest and fastest way of calculating the Dynkin indices, if you do not have handy an appropriate text book where they are tabulated. The procedure is as follows. Assume that a group G and a representation R of this group are given. Then choose an SU(2) subgroup of G and decompose R with respect to this SU(2) subgroup. For each irreducible SU(2) multiplet of spin j the index T is given by $\frac{1}{3}j(j+1)(2j+1)$. Hence the number of zero modes in the SU(2) instanton background is $\frac{2}{3}j(j+1)(2j+1)$. In this way one readily establishes the total number of zero modes for the given representation R. This is nothing other than the Dynkin index. The value of $T(R)$ is one-half this number. For instance, in SU(5) a good choice of SU(2) subgroup to be used for decomposition is the weak isospin SU(2) group. Each quintet has one weak isospin doublet; the remaining elements are singlets. Each doublet has one zero mode. As a result, $T(V) = \frac{1}{2}$. Moreover, each decuplet has three weak isospin doublets while the remaining elements are singlets. Hence, $T(X) = \frac{3}{2}$.

Table 10.5. The $\tilde{R}$ charge in the SU(5) model for two generations

Field	X_1	X_2	$V_{1,2}$	ψ_{X_1}	ψ_{X_2}	$\psi_{V_{1,2}}$
$\tilde{R}$ charge	0	$-\frac{4}{3}$	1	-1	$-\frac{7}{3}$	0

using Eqs. (10.463) and (10.465) and the R charges from this table we find that the coefficient of the anomaly term in $\partial_\mu R^\mu$ is

$$T_G - T(X) - \tfrac{7}{3} T(X) = 5 - \tfrac{3}{2} - \tfrac{7}{2} = 0. \tag{10.468}$$

In terms of the "geometric" R current (10.438) and the flavor U(1) currents (10.442) the anomaly-free and conserved $\tilde{R}$ current then takes the form

$$\tilde{R}_\mu = R_\mu + \tfrac{1}{3}\left(R_\mu^{V_1} + R_\mu^{V_2}\right) - \tfrac{2}{3} R_\mu^{X_1} - 2 R_\mu^{X_2}. \tag{10.469}$$

The conservation of $\tilde{R}_\mu$ both at the classical and quantum levels follows directly from Eqs. (10.462) and (10.463).

The $\tilde{R}$ current (10.469) and the assignment in Table 10.5 are not unique in the model at hand. This is due to the fact that in addition to (10.469) there exists a strictly conserved flavor U(1) current, which can be added to (10.469) with an arbitrary coefficient.

Two general lessons that one can draw from the above example are as follows.

Theorem 1 *In the class of theories with a purely cubic superpotential the R hypercurrent is guaranteed to exist if one of the flavors Q_{f_0} does not appear in the superpotential $\mathcal{W}(Q_f)$.*

Theorem 2 *If there is more than one conserved R symmetry, say, R and R', then the difference between them is due to a flavor symmetry, and,*

$$\mathcal{J}_{\alpha\dot\alpha}^R - \left(\mathcal{J}_{\alpha\dot\alpha}^R\right)' = \left[D_\alpha, \bar{D}_{\dot\alpha}\right] J,$$

where J is a combination of the Konishi operators.

10.16.8 Supercurrent

Up to now the focus of our considerations has been the lowest component of the hypercurrent (10.439). Now a few words are in order regarding its θ component, the supercurrent. For a generic matter sector,

$$J_{\alpha\beta\dot\beta} = 2\left\{ \frac{1}{g^2}\left(iG_{\beta\alpha}^a \bar\lambda_{\dot\beta}^a + \varepsilon_{\beta\alpha} D^a \bar\lambda_{\dot\beta}^a\right) + \sqrt{2}\sum_f \left[(\mathcal{D}_{\alpha\beta}\phi^\dagger)\psi_\beta - i\varepsilon_{\beta\alpha} F\bar\psi_{\dot\beta}\right] \right.$$

$$\left. - \sum_f \frac{\sqrt{2}}{6}\left[\partial_{\alpha\beta}(\psi_\beta\phi^\dagger) + \partial_{\beta\dot\beta}(\psi_\alpha\phi^\dagger) - 3\varepsilon_{\beta\alpha}\partial_{\dot\beta}^\gamma(\psi_\gamma\phi^\dagger)\right]\right\}, \tag{10.470}$$

Cf. Section 10.6.6.

where the sum runs over all matter flavors. The expressions for the F and D terms are those quoted in Eqs. (10.370) and (10.371), up to field renaming.

The third line in Eq. (10.470), being a full derivative, does not change supercharges defined as

$$Q_\alpha = \int d^3x \frac{1}{2} \left(\bar{\sigma}^0\right)^{\dot{\beta}\beta} J_{\alpha\beta\dot{\beta}}.$$

(10.471)

This is due to the fact that only spatial derivatives survive in the time component, given by the following expression

$$\left(\bar{\sigma}^0\right)^{\dot{\beta}\beta} \left[\partial_{\alpha\dot{\beta}}(\psi_\beta \phi^\dagger) + \partial_{\beta\dot{\beta}}(\psi_\alpha \phi^\dagger) - 3\varepsilon_{\beta\alpha} \partial_{\dot{\beta}}^\gamma (\psi_\gamma \phi^\dagger) \right].$$

Thus Eq. (10.434) is replaced by

$$(\bar{\sigma}_\mu)^{\dot{\alpha}\alpha} J_\alpha^\mu = J_{\alpha\dot{\alpha}}^\alpha = -3i \frac{T_G - \sum_f \frac{1}{3}(1 - \gamma_f)T(R_f)}{4\pi^2} \mathrm{Tr}\left(\bar{G}_{\dot{\alpha}\dot{\beta}} \bar{\lambda}^{\dot{\beta}}\right).$$

(10.472)

10.16.9 U(1) Gauge Factors with the Fayet–Iliopoulos Term

While the above construction of the hypercurrent in super-Yang–Mills theories was quite general, the reader should be aware of a crucial peculiarity arising in one particular case: if the gauge group contains U(1) factors then, with the corresponding Fayet–Iliopoulos terms added (Section 10.6.9), the Ferrara–Zumino hypercurrent ceases to be (super)gauge invariant [38]. For instance, in pure super-Maxwell theory (i.e., supersymmetric QED with no matter) $\mathcal{J}_{\alpha\dot{\alpha}}$ takes the form [38]

Cf. Eq. (10.426).

$$\mathcal{J}_{\alpha\dot{\alpha}} = -\frac{2}{e^2} W_\alpha \bar{W}_{\dot{\alpha}} + \frac{\xi}{3}[D_\alpha, \bar{D}_{\dot{\alpha}}]V,$$

(10.473)

where V is the Abelian vector superfield. It is not difficult to show that the operator X in the general formula (10.408) reduces to

$$X = \frac{\xi}{6} \bar{D}^2 V,$$

(10.474)

where we have used the equation of motion for the super-Maxwell theory with Fayet–Iliopoulos (FI) term,

$$D_\alpha W^\alpha = -2e^2 \xi,$$

(10.475)

Equation of motion for super-Maxwell theory with FI

and Eq. (10.458). At the same time the supercharges $Q_\alpha, \bar{Q}_{\dot{\alpha}}$ and the energy–momentum operator P_μ are gauge invariant, so that the theory *per se* is consistent. What requires a change, a deviation from the standard route, is the construction of a new Komargodski–Seiberg hypercurrent that has extra components in comparison with the Ferrara–Zumino hypercurrent; these are needed for the embedding of this theory in supergravity. Since it will not be discussed in this course, the interested reader is referred to [38].[55]

The fact that the hypercurrent (10.473) is not supergauge invariant implies that the Fayet–Iliopoulos term cannot be generated in the low-energy effective

---

[55] The second of these papers points out another "exceptional" situation, which may arise in the generalized Wess-Zumino models of Section 10.6.7. For the Kähler manifolds of nonzero Kähler classes, i.e., those with nontrivial homology, in particular all compact Kähler manifolds, the standard R current is not invariant under Kähler transformations, i.e., it is not a good operator. The Komargodski–Seiberg hypercurrent still exists. Particularly curious readers, with hungry minds, are advised to look through [39].

Lagrangian unless it is introduced "by hand" from the very beginning, in the original Lagrangian given at the ultraviolet cutoff. For the same reason, the value of ξ is not renormalized upon evolution from M_{uv} down to μ. This fact was discussed in Section 10.8 from a different point of view.

Exercises

10.16.1 Derive the $\theta\bar{\theta}$ components in Eq. (10.426). Find the component expansion of $\mathcal{J}_{\alpha\dot{\alpha}}$ in vector/tensor notation (the θ, $\bar{\theta}$ and $\theta\bar{\theta}$ components).

10.16.2 Assume that you have constructed a hypercurrent such that

$$D^\alpha \mathcal{J}_{\alpha\dot{\alpha}} = \bar{D}_{\dot{\alpha}} D^2 Y,$$

where Y is a chiral superfield. Prove that

$$D^\alpha \left[\mathcal{J}_{\alpha\dot{\alpha}} + 4i\partial_{\alpha\dot{\alpha}}(Y - \bar{Y}) \right] = 0.$$

10.16.3 Find the higher components of $\mathcal{J}^R_{\alpha\dot{\alpha}}$ in (10.418) and of $\mathcal{J}^f_{\alpha\dot{\alpha}}$ in (10.443) and demonstrate that to prove their conservation one does not need to use the equations of motion.

10.16.4 Prove the identity (10.458).

10.16.5 Find the R hypercurrent in the model of Section 10.16.7 with superpotential (10.467). With this hypercurrent determine $\bar{X}$ in the formula (10.408).

10.16.6 Assuming that the Fayet–Iliopoulos term $\xi \neq 0$ and that the hypercurrent is given by (10.473) find $\partial^{\dot{\alpha}\alpha} \mathcal{J}_{\alpha\dot{\alpha}}$ in terms of the V components using Eqs. (10.474) and (10.410). Apply equations of motion. Demonstrate that $\partial^{\dot{\beta}\beta} J_{\alpha\beta\dot{\beta}} = 0$.

10.17 R Parity

In many theories without an (exactly) conserved R current one can still introduce a *discrete* symmetry of the R type, referred to as R parity. By definition, the R parity transformations are given by

$$\theta \to -\theta, \qquad \bar{\theta} \to -\bar{\theta}, \tag{10.476}$$
$$d^2\theta \to d^2\theta, \qquad d^2\bar{\theta} \to d^2\bar{\theta},$$

while the superfields transform as follows:

$$V(x, \theta, \bar{\theta}) \to V(x, -\theta, -\bar{\theta}),$$
$$W(x_L, \theta) \to -W(x_L, -\theta), \tag{10.477}$$
$$Q_f(x_L, \theta) \to (-1)^{\kappa_f} Q_f(x_L, -\theta),$$

where κ_f takes two values, 0 or 1, assigned to each flavor on an individual basis. In components,

$$\lambda \to -\lambda, \qquad A_\mu \to A_\mu, \qquad G^a_{\mu\nu} \to G^a_{\mu\nu},$$
$$q_f \to (-1)^{\kappa_f} q_f, \qquad \psi_f \to -(-1)^{\kappa_f} \psi_f. \tag{10.478}$$

The κ_f assignment must be performed in such a way that $\mathcal{W} \to \mathcal{W}$. This constrains the form of superpotential.

The conservation of R parity implies that the particle spectrum of such theories can be divided into two classes, having positive and negative R parities, respectively. The lightest particle in the negative R parity class is stable. It bears a special name, LSP (lightest superpartner).

10.18 Extended Supersymmetries in Four Dimensions

In four dimensions one can have at most 16 conserved supercharges. With more supercharges, supermultiplets will necessarily include states with spins higher than 1. The only consistent field theory with spins higher than 1 is supergravity or local supersymmetry: it has spin-2 fields (gravitons) and spin-$\frac{3}{2}$ fields (gravitinos). In this text we are limiting ourselves to global supersymmetry; hence, the maximal number of supercharges is 16.

At the same time, supersymmetric field theories based on minimal supersymmetry, i.e., $\mathcal{N} = 1$ theories, have four conserved supercharges. This opens the possibility of extensions to $\mathcal{N} = 2$ and $\mathcal{N} = 4$.

Gauge theories of this type are known: these are the $\mathcal{N} = 2$ and $\mathcal{N} = 4$ super-Yang–Mills theories. They obtained by dimensional reduction from minimal super-Yang–Mills theories in six and 10 dimensions, respectively. Although they are unsuitable for phenomenology, because the fermion fields they contain are all nonchiral, they have rich dynamics, the study of which provides deep insights into a large number of problems in mathematical physics that defied solution for decades. Extended supersymmetry produces powerful tools.

| $\mathcal{N} = 4$ is the maximal global SUSY in 4D. |

10.18.1 Algebraic Aspects

Before starting this section the reader is directed to Section 10.4.2 dealing with algebraic aspects of minimal supersymmetry. Extended superalgebras have the form

$$\{Q^I_\alpha, \bar{Q}^J_{\dot\beta}\} = 2P_\mu(\sigma^\mu)_{\alpha\dot\beta} = 2P_{\alpha\dot\beta}\,\delta^{IJ},$$
$$\{Q^I_\alpha, Q^J_\beta\} = \{Q^I_{\dot\alpha}, Q^J_{\dot\beta}\} = 0, \tag{10.479}$$

where I and J are "extension indices" with the following ranges:

$$I, J = 1, 2, \qquad\qquad \mathcal{N} = 2;$$
$$I, J = 1, 2, 3, 4, \qquad \mathcal{N} = 4. \tag{10.480}$$

Equation (10.479) does not include possible central charges, which we will discuss in Section 10.25. Needless to say, such properties as the vanishing vacuum energy and the spectral degeneracy between boson and fermion states remain intact. Now we will discuss the irreducible representations of extended supersymmetries both for massive and massless states. The reader is recommended to start by reading Section 10.4.6.

10.18.1.1 $\mathcal{N} = 2$

For massive particles, we can boost to a frame in which the particle is at rest, $P_\mu = (m, 0, 0, 0)$. The massive particles belong to representations of SO(3) labeled by the spin j, which can be either integer (for bosons) or half-integer (for fermions). Any given spin-j representation is $(2j + 1)$-dimensional with states labeled by j_z:

$$|j, j_z\rangle, \qquad j_z = -j, -j + 1, \ldots, j - 1, j. \qquad (10.481)$$

Let us start with the supermultiplets of $\mathcal{N} = 2$. For a state $|a\rangle$ at rest the supersymmetry algebra (10.479) takes the form

$$\{Q_\alpha^I, (Q_\beta^J)^\dagger\} = 2m\delta_{\alpha\beta}\,\delta^{IJ}, \qquad \{Q_\alpha^I, Q_\beta^J\} = 0, \qquad I, J = 1, 2. \qquad (10.482)$$

To construct representations of this algebra we note that this is an algebra of four creation and four annihilation operators (up to a rescaling of Q by $\sqrt{2m}$). If we assume Q_α^I to annihilate the state $|a\rangle$, i.e., $Q_\alpha^I|a\rangle = 0$, then we find the following representation:

$$|a\rangle, \quad (Q_{[\alpha}^1)^\dagger (Q_{\beta]}^1)^\dagger |a\rangle, \quad (Q_{[\alpha}^2)^\dagger (Q_{\beta]}^2)^\dagger |a\rangle, \quad (Q_{[\alpha}^1)^\dagger (Q_{\beta]}^2)^\dagger |a\rangle,$$

$$(Q_{[\alpha}^1)^\dagger (Q_{\beta]}^1)^\dagger (Q_{[\alpha}^2)^\dagger (Q_{\beta]}^2)^\dagger |a\rangle,$$

$$(Q_\beta^1)^\dagger |a\rangle, \quad (Q_\beta^2)^\dagger |a\rangle, \qquad\qquad\qquad\qquad\qquad (10.483)$$

$$(Q_{[\alpha}^2)^\dagger (Q_{\beta]}^2)^\dagger (Q_\beta^1)^\dagger |a\rangle, \quad (Q_{[\alpha}^1)^\dagger (Q_{\beta]}^1)^\dagger (Q_\beta^2)^\dagger |a\rangle,$$

$$(Q_{\{\alpha}^1)^\dagger (Q_{\beta\}}^2)^\dagger |a\rangle,$$

where $[\ldots]$ means antisymmetrization, and $\{\ldots\}$ symmetrization of the spinorial indices. Suppose for simplicity that $|a\rangle$ is a spin-0 particle. Then the states listed in the first two lines in Eq. (10.483) have spin 0, the states in the third and the fourth lines have spin $\frac{1}{2}$, and, finally, the states in the fifth line have spin 1. Altogether we have eight bosonic states (three spin-1 and five spin-0) and eight fermionic states, of spin $\frac{1}{2}$. The overall number of states in the supermultiplet (counting spin) is

$$\nu = 2^{2\mathcal{N}}. \qquad (10.484)$$

If, instead of a spin-0 state, we started from spin $j \neq 0$ we would obtain supermultiplets with multiplicity

$$\nu_j = 2^{2\mathcal{N}}(2j + 1). \qquad (10.485)$$

Multiplicity of states in $\mathcal{N} = 2$

In the practical applications below we will limit ourselves to $j = 0$.

Now let us pass to massless states. For such states we choose a reference frame in which $P_\mu = (E, 0, 0, E)$. Then the superalgebra (10.479) takes the form

$$\{Q_\alpha^I, (Q_\beta^J)^\dagger\} = 4E\delta^{IJ} \begin{pmatrix} 1 & 0 \\ 0 & 0 \end{pmatrix}, \qquad I, J = 1, 2; \qquad (10.486)$$

all other anticommutators vanish. In constructing supermultiplets we are left with two nontrivial creation and two nontrivial annihilation operators, namely, Q_1^I and $(Q_1^I)^\dagger$, where $I = 1, 2$.

As in Section 10.4.6, we start from a state $|b\rangle$ with helicity λ. Then the two states $(Q_1^I)^\dagger|b\rangle$ have helicity $\lambda + \frac{1}{2}$. In addition, the state $(Q_1^1)^\dagger(Q_1^2)^\dagger|b\rangle$ has helicity $\lambda + 1$. This is a dimension-4 representation

$$\nu = 2^N. \tag{10.487}$$

However, CPT transformation, generally speaking, does not map the above representation onto itself, as required in field theory, unless we start from $\lambda = -\frac{1}{2}$. Thus, keeping in mind the field-theoretic implementation of $N = 2$ supersymmetry, we should consider two options:

(i) a massless *hypermultiplet*

$$\lambda = \{-\tfrac{1}{2}, 0, 0, \tfrac{1}{2}\} \text{ or} \tag{10.488}$$

(ii) two supermultiplets

$$\lambda = \{0, \tfrac{1}{2}, \tfrac{1}{2}, 1\} \quad \text{and} \quad \lambda = \{-1, -\tfrac{1}{2}, -\tfrac{1}{2}, 0\}, \tag{10.489}$$

comprising a massless *vector* $N = 2$ supermultiplet. The overall dimension of the representation in the second case is 8.

| *Centrally extended superalge-bras* |

For centrally extended superalgebras (see Section 10.25 below) the construction of saturated (critical) supermultiplets is similar to that of massless supermultiplets. Here I will briefly mention just one case of the monopole central charge, for which

$$\{Q_\alpha^I, Q_\beta^J\} = 2\varepsilon_{\alpha\beta}\varepsilon^{IJ}Z \tag{10.490}$$

plus a corresponding expression for the conjugated supercharges. The particles are massive, hence we choose a reference frame in which $P_\mu = (m, 0, 0, 0)$. However, if $m = |Z|$ then only two linear combinations of supercharges act nontrivially; the other two act trivially. For instance, if Z is real and positive then $Q_\alpha^2 \equiv \frac{1}{\sqrt{2}}(Q_\alpha^1 - \varepsilon_{\alpha\beta}\bar{Q}_{\dot\beta}^2)$

| *Cf. Sections 10.25.4 and 10.26.* |

and its complex conjugate act trivially while $Q_\alpha^1 \equiv \frac{1}{\sqrt{2}}(Q_\alpha^1 + \varepsilon_{\alpha\beta}\bar{Q}_{\dot\beta}^2)$ and its complex conjugate act nontrivially. This is similar to what happens for the massless supermultiplet. In constructing the "short" (saturated) supermultiplet one needs to take into account only Q^1 and $(Q^1)^\dagger$. Hence, the multiplet is four dimensional and consists of four states: $|a\rangle$, the two states $(Q_\alpha^1)^\dagger|a\rangle$, and $(Q_1^1)^\dagger(Q_2^1)^\dagger|a\rangle$. If $|a\rangle$ has spin 0, so does $(Q_1^1)^\dagger(Q_2^1)^\dagger|a\rangle$. The two states $(Q_\alpha^1)^\dagger|a\rangle$ form a spin-$\frac{1}{2}$ spin representation. Thus, in this case the short massive supermultiplet coincides with the massless hypermultiplet (10.488) of the unextended algebra (10.482). This is a common occurrence.

10.18.1.2 $N = 4$

For massive supermultiplets Eqs. (10.484) and (10.485) remain valid. We will consider in some detail one massless example, the massless vector supermultiplet. We start from a state $|b\rangle$ with helicity $\lambda = -1$. Then the four states $(Q_1^I)^\dagger|b\rangle$ (with $I = 1, 2, 3, 4$) have helicity $-\frac{1}{2}$. The six states $(Q_1^{[I})^\dagger(Q_1^{J]})^\dagger|b\rangle$ have helicity 0. The four

Table 10.6. The massless $\mathcal{N} = 4$ vector supermultiplet

Helicity	-1	$-\frac{1}{2}$	0	$\frac{1}{2}$	1
Number of states	1	4	6	4	1

states $(Q_1^{[I]})^{\dagger}(Q_1^{J})^{\dagger}(Q_1^{F]})^{\dagger}|b\rangle$ (with antisymmetrized indices I, J, F) have helicity $\frac{1}{2}$. Finally, one state, $(Q_1^{[I]})^{\dagger}(Q_1^{J})^{\dagger}(Q_1^{F})^{\dagger}(Q_1^{G]})^{\dagger}|b\rangle$ with fully antisymmetrized indices I, J, F, G, has helicity 1. Altogether we have eight bosonic and eight fermionic states. This is summarized in Table 10.6.

10.18.2 Field-Theoretic Implementation for $\mathcal{N} = 2$

To construct $\mathcal{N} = 2$ supersymmetric theories (eight supercharges in four dimensions) it seems most natural to use an $\mathcal{N} = 2$ superfield formalism based on two θs. Such a formalism exists, but it is rather cumbersome because the number of auxiliary components proliferates. In this introductory presentation we will continue to use $\mathcal{N} = 1$ superfields to construct $\mathcal{N} = 2$ theories.

Constructing $\mathcal{N} = 2$ SYM

To begin with, let us consider the $\mathcal{N} = 2$ generalization of pure super-Yang–Mills theory. This means that we have no $\mathcal{N} = 2$ matter fields. However, in terms of $\mathcal{N} = 1$ superfields, the introduction of matter superfields is inevitable. Indeed, the massless $\mathcal{N} = 2$ vector multiplet has four bosonic and four fermionic degrees of freedom (Section 10.18.1.1). The $\mathcal{N} = 1$ gauge superfield has two physical bosonic and two fermionic degrees of freedom. Hence, we must add a chiral superfield ($2 + 2$ physical degrees of freedom) belonging to the adjoint representation of the gauge group:

$$\mathcal{A}^a(x_L, \theta) = a^a(x_L) + \sqrt{2}\theta^a \chi^a(x_L) + \theta^2 F_{\mathcal{A}}. \tag{10.491}$$

The Lagrangian of $\mathcal{N} = 2$ super-Yang–Mills theory (without $\mathcal{N} = 2$ matter) is

$$\mathcal{L} = \left(\frac{1}{4g^2} \int d^2\theta\, W^{a\alpha}W_{\alpha}^a + \text{H.c.} \right) + \frac{1}{g^2}\int d^2\theta\, d^2\bar{\theta}\, \bar{\mathcal{A}}e^V \mathcal{A}. \tag{10.492}$$

Let us present it in components. I recall that, for the adjoint representation of SU(N),

$$(T^a)_{bd} = if_{bad}. \tag{10.493}$$

Then we obtain

$$\begin{aligned} \mathcal{L} = \frac{1}{g^2}\Big[&-\tfrac{1}{4}F^{a\mu\nu}F_{\mu\nu}^a + \lambda^{\alpha,a}i\mathcal{D}_{\alpha\dot\alpha}\bar\lambda^{\dot\alpha,a} + \tfrac{1}{2}D^a D^a \\ &+ (\mathcal{D}^{\mu}\bar{a})(\mathcal{D}_{\mu}a) + \chi^{\alpha,a}i\mathcal{D}_{\alpha\dot\alpha}\bar\chi^{\dot\alpha,a} - if_{abc}D^a\bar{a}^b a^c \\ &-\sqrt{2}f_{abc}\left(\bar{a}^a\lambda^{\alpha,b}\chi_{\alpha}^c + a^a\bar\lambda_{\dot\alpha}^b\bar\chi^{\dot\alpha,c} \right)\Big]. \end{aligned} \tag{10.494}$$

$\mathcal{N} = 2$ SYM

As usual, the D field is auxiliary and can be eliminated via the equation of motion

$$D^a = if_{abc}\bar{a}^b a^c. \tag{10.495}$$

In $\mathcal{N} = 2$ super-Yang–Mills theory there are flat directions: for instance, if the field a is purely real or purely imaginary then all D terms vanish. More generally, the D terms vanish if a and $\bar{a}$ can be aligned, e.g., for SU(2) $a^1 = a^2 = \bar{a}^1 = \bar{a}^2 = 0$, with arbitrary a^3. If a is purely real or purely imaginary then one can always perform such an alignment.

The theory (10.492) is explicitly supergauge invariant and $\mathcal{N} = 1$ supersymmetric. The $\mathcal{N} = 2$ supersymmetry is implicit. Its manifestation is a global SU(2) symmetry (referred to as SU(2)$_R$, see below), which becomes obvious in (10.494) if we introduce an SU(2)$_R$ doublet,

$$\left(\lambda^f\right)^{\alpha,a} = \begin{pmatrix} \lambda^{\alpha,a} \\ \chi^{\alpha,a} \end{pmatrix}, \tag{10.496}$$

where $f = 1, 2$ is the index of the fundamental representation of SU(2)$_R$. Rewritten in terms of λ^f, the Lagrangian (10.494) explicitly exhibits symmetry under global unitary SU(2) rotations of λ^f:

$$\mathcal{L} = \frac{1}{g^2} \left\{ -\tfrac{1}{4} F^{a\mu\nu} F^a_{\mu\nu} + \left(\lambda^f\right)^{\alpha,a} i \mathcal{D}_{\alpha\dot\alpha} \left(\bar\lambda_f\right)^{\dot\alpha,a} - \tfrac{1}{2} \left(i f_{abc} \bar a^b a^c\right)^2 \right.$$

$$\left. + (\mathcal{D}^\mu \bar a)(\mathcal{D}_\mu a) + \tfrac{1}{\sqrt2} \varepsilon_{fg} f_{abc} \left[\bar a^a \left(\lambda^f\right)^{\alpha,b} (\lambda^g)^c_\alpha + \text{H.c.} \right] \right\}, \tag{10.497}$$

where the Levi–Civita tensor ε_{fg} is defined in the same way as in Eq. (10.10). In particular, (10.497) is symmetric under the interchanges $\lambda^1 \to -\lambda^2$, $\lambda^2 \to \lambda^1$. This implies in turn that, in addition to the standard $\mathcal{N} = 1$ supercurrent which exists in all $\mathcal{N} = 1$ theories (Section 10.16.8), there is another conserved supercurrent. The two can be written in the following unified form:

Supercurrent in $\mathcal{N} = 2$ SYM. Improvement terms are omitted, cf. (11.212) below.

$$(J_f)_{\alpha\beta\dot\beta} = \frac{2}{g^2} \left[i G^a_{\beta\alpha} \left(\bar\lambda_f\right)^a_{\dot\beta} + \varepsilon_{\beta\alpha} D^a \left(\bar\lambda_f\right)^a_{\dot\beta} \right] - \frac{2\sqrt2}{g^2} \varepsilon_{fg} (\mathcal{D}_{\alpha\dot\beta} \bar a^a) (\lambda^g)^a_\beta, \tag{10.498}$$

where $f = 1, 2$. In this regard we encounter the same situation as in the O(3) sigma model (Section 10.12.3.2).

The origin of the full $\mathcal{N} = 2$ supersymmetry seen in the Lagrangian (10.494) becomes explicit if we look at it from a different standpoint. Assume that we are starting from $\mathcal{N} = 1$ super-Yang–Mills theory in six rather than four dimensions. In six dimensions the minimal number of supercharges is eight. In six dimensions the gauge field contains four physical (bosonic) degrees of freedom and so does the six-dimensional Weyl spinor, which has four fermionic degrees of freedom. Now, we take this six-dimensional $\mathcal{N} = 1$ super-Yang–Mills theory and reduce it to four dimensions. This means that we ignore the dependence of all fields on x_4 and x_5. The fourth and fifth components of the gauge potential now become scalar fields, and we combine them as follows: $a = A_4 + iA_5$. The six-dimensional Weyl spinor can be decomposed into two four-dimensional Weyl spinors. In this way, we arrive directly at the Lagrangian (10.494). This procedure makes explicit the origin of the above-mentioned global SU(2)$_R$ symmetry.[56] It is a manifestation of the part of the Lorentz invariance of the six-dimensional theory which became an internal symmetry upon the reduction to four dimensions.

As mentioned above, $\mathcal{N} = 2$ super-Yang–Mills theories have flat directions. For instance, for SU(2)$_{\text{gauge}}$ the flat direction can be parametrized by Tr a^2. If $a^3 \neq 0$ then the gauge group SU(2) is broken down to U(1). The theory is Higgsed and the

---

[56] The fact that this is the R symmetry is seen in the $\mathcal{N} = 2$ formalism. A clear-cut indication is that distinct θ components of superfields transform differently.

spectrum is rearranged. Instead of all massless supermultiplets we now have two massive vector supermultiplets ("W" bosons) and one massless (a "photon"). Since the massive supermultiplets have the same number of components as the massless one, they must be short (Section 10.26).

In concluding this section we will discuss how to add $\mathcal{N} = 2$ matter fields. To this end we will use short supermultiplets similar to (10.488). For simplicity we will limit ourselves to one flavor in the fundamental representation. Generalization to more than one flavor and other representations is straightforward.

We introduce a chiral $\mathcal{N} = 1$ superfield Q in the fundamental representation and a partner superfield $\tilde{Q}$ in the antifundamental representation,

$$Q^k(x_L, \theta) = q^k(x_L) + \sqrt{2}\theta^\alpha \psi_\alpha^k(x_L) + \theta^2 F_q^k,$$
$$\tilde{Q}_k(x_L, \theta) = \tilde{q}_k(x_L) + \sqrt{2}\theta^\alpha \tilde{\psi}_{k,\alpha}(x_L) + \theta^2 \left(\tilde{F}_{\tilde{q}}\right)_k, \tag{10.499}$$

where, for $SU(N)_{\text{gauge}}$ the index k runs over $k = 1, 2, \ldots, N$. Each expression describes two bosonic and two fermionic degrees of freedom (per each value of k). The superfields Q and $\tilde{Q}$ together comprise one $\mathcal{N} = 2$ *hypermultiplet* with four bosonic and four fermionic degrees of freedom (this is a short massive supermultiplet). The gauge sector of the theory is given by the Lagrangian (10.492). The matter sector is

$$\mathcal{L}_{\text{matter}} = \int d^2\theta\, d^2\bar{\theta} \left(\bar{Q}e^V Q + \tilde{Q}e^V \bar{\tilde{Q}}\right) + \left[\int d^2\theta\, \mathcal{W}(Q, \tilde{Q}, \mathcal{A}) + \text{H.c.}\right], \tag{10.500}$$

$\boxed{\begin{array}{c}\mathcal{N} = 2 \; SYM \\ \text{with matter}\end{array}}$ where the superpotential $\mathcal{W}$ has the form

$$\mathcal{W} = m\tilde{Q}Q + \sqrt{2}\tilde{Q}\mathcal{A}Q. \tag{10.501}$$

Here m is the mass parameter, and the convolution of the color indices is self-evident.

This expression appears quite concise but becomes rather bulky when written in components. Then the bosonic part of the Lagrangian takes the form

$$\mathcal{L}_{\text{bos}} = -\frac{1}{4g^2}\left(F_{\mu\nu}^a\right)^2 + \frac{1}{g^2}\left|\mathcal{D}_\mu a^a\right|^2$$
$$+ \left|\mathcal{D}_\mu q\right|^2 + \left|\mathcal{D}_\mu \bar{\tilde{q}}\right|^2 - V(q, \tilde{q}, a^a). \tag{10.502}$$

Here $\mathcal{D}_\mu$ is the covariant derivative acting in the appropriate representation of $SU(N)$. The scalar potential $V(q, \tilde{q}, a^a)$ in the Lagrangian (10.502) is a sum of D and F terms,

$$V(q, \tilde{q}, a^a) = \tfrac{1}{2}g^2\left(\frac{i}{g^2}f^{abc}\bar{a}^b a^c - \bar{q}T^a q + \tilde{q}T^a\bar{\tilde{q}}\right)^2 + 2g^2\,|\tilde{q}T^a q|^2$$
$$+ \tfrac{1}{2}\left\{\left|(\sqrt{2}m + 2T^a a^a)q\right|^2 + \left|(\sqrt{2}\bar{m} + 2T^a\bar{a}^a)\bar{\tilde{q}}\right|^2\right\}. \tag{10.503}$$

The first term in the first line represents the D term, the second term in the first line represents the $F_{\mathcal{A}}$ term, while the second line represents the F_q and $F_{\tilde{q}}$ terms.

Before passing to the fermion part of the Lagrangian I want to introduce a convenient notation, which will make the $SU(2)_R$ symmetry of the matter sector explicit. For the matter fields the two relevant $SU(2)_R$ doublets are

$$q^f \equiv \{q, \bar{\tilde{q}}\}, \qquad \bar{q}_f \equiv \{\bar{q}, \tilde{q}\}, \qquad f = 1, 2. \tag{10.504}$$

In the first case we are dealing with the $SU(2)_R$ doublet of fundamentals and in the second case that of antifundamentals.

In this notation the expression in the second line of Eq. (10.502) takes the form

$$
\mathcal{D}_\mu \bar{q}_f \mathcal{D}^\mu q^f - \left\{ \frac{1}{2g^2} \left| f^{abc} \bar{a}^b a^c \right|^2 + \bar{q}_f \left(|m|^2 + \bar{a}a + a\bar{a} \right) q^f \right.
$$

$$
+ \sqrt{2} \bar{q}_f (\bar{m}a + m\bar{a}) q^f - g^2 \bar{q}_f T^a q^{f'} \bar{q}_g T^a q^{g'} \varepsilon^{fg} \varepsilon_{f'g'}
$$

$$
\left. + \frac{g^2}{2} \bar{q}_f T^a q^f \bar{q}_g T^a q^g \right\},
\tag{10.505}
$$

where summation over the repeated $SU(2)_R$ indices is implied.

Now we are ready for the fermion part of the Lagrangian. In the same notation it has the form

$$
\mathcal{L}_{\text{ferm}} = \frac{i}{g^2} \bar{\lambda}_f^a \bar{\mathcal{D}} \lambda^{af} + \bar{\psi} i \bar{\mathcal{D}} \psi + \tilde{\psi} i \mathcal{D} \bar{\tilde{\psi}} + \frac{1}{\sqrt{2}} f^{abc} \bar{a}^a (\lambda_f^b \lambda^{cf})
$$

$$
+ \frac{1}{\sqrt{2}} f^{abc} (\bar{\lambda}^{bf} \bar{\lambda}_f^c) a^c + i\sqrt{2} \left[\bar{q}_f (\lambda^f \psi) + (\tilde{\psi} \lambda_f) q^f + (\bar{\psi} \bar{\lambda}_f) q^f + \bar{q}^f (\bar{\lambda}_f \bar{\tilde{\psi}}) \right]
$$

$$
+ \tilde{\psi} \left(m + \sqrt{2} a \right) \psi + \bar{\psi} \left(\bar{m} + \sqrt{2} \bar{a} \right) \bar{\tilde{\psi}},
\tag{10.506}
$$

where λ_f was defined in Eq. (10.496) and the contraction of spinor indices is assumed inside parentheses; for example, $(\lambda\psi) \equiv \lambda_\alpha \psi^\alpha$.

10.18.3 Field-Theoretic Implementation for $\mathcal{N} = 4$

Here we will discuss $\mathcal{N} = 4$ super-Yang–Mills theory (with 16 conserved supercharges). The $\mathcal{N} = 4$ superspace formalism is too complicated for this textbook.[57] We will base our considerations on the idea mentioned in Section 10.18.1.2. One can obtain $\mathcal{N} = 4$ super-Yang–Mills theory in four dimensions by dimensionally reducing $\mathcal{N} = 1$ super-Yang–Mills theory from 10 dimensions. In 10 dimensions the gauge potential has eight physical degrees of freedom, and so does the Majorana–Weyl spinor field. The balance between the numbers of bosonic and fermionic degrees of freedom is evident.

To reduce the theory we assume that none of the fields depends on $x_4, x_5, \ldots, x_9$. The six components of the gauge potential $A_4, A_5, \ldots, A_9$ become six real scalar fields, or, equivalently, three complex fields. In addition, we must decompose the 10-dimensional Majorana–Weyl spinor into four-dimensional spinors. This decomposition leaves us with four four-dimensional Weyl spinors.

In terms of $\mathcal{N} = 1$ supermultiplets, the $\mathcal{N} = 4$ supersymmetric gauge theory contains a vector superfield consisting of the gauge field A_μ^a and gaugino λ^{aa}, and three chiral superfields $\Phi_{1,2,3}^a$. The superpotential of the $\mathcal{N} = 4$ gauge theory is

Super-potential in $\mathcal{N} = 4$ SYM

$$
\mathcal{W}_{\mathcal{N}=4} = \frac{\sqrt{2}}{g^2} f_{abc} \Phi_1^a \Phi_2^b \Phi_3^c.
\tag{10.507}
$$

---

[57] The interested reader is referred to [74].

In components the Lagrangian of $\mathcal{N} = 4$ Yang–Mills theory can be cast in the form

$$\mathcal{L} = \frac{1}{g^2} \, \text{Tr} \left\{ -\frac{1}{2} F_{\mu\nu} F^{\mu\nu} + 2i \bar{\lambda}_{\dot{\alpha}A} \mathcal{D}^{\dot{\alpha}\beta} \lambda_{\beta}^A + \frac{1}{2} \left(\mathcal{D}_\mu \phi^{AB} \right) \left(\mathcal{D}^\mu \bar{\phi}_{AB} \right) \right.$$
$$- h_3 \left(\lambda^{\alpha A} \left[\bar{\phi}_{AB}, \lambda_\alpha^B \right] - \bar{\lambda}_{\dot{\alpha}A} \left[\phi^{AB}, \bar{\lambda}_B^{\dot{\alpha}} \right] \right)$$
$$\left. - h_4 \left[\phi^{CD}, \bar{\phi}_{AB} \right] \left[\phi^{AB}, \bar{\phi}_{CD} \right] \right\}, \tag{10.508}$$

with gauge fields, gauginos, and scalars

$$X = \{ A_\mu, \lambda^A, \phi^{AB} \}$$

in the adjoint representation of the gauge group $X = X^a T^a$, where the T^a are generators in the fundamental representation; hence,

$$\mathcal{D}_\mu = \partial_\mu - i[A_\mu,].$$

Moreover,

$$h_3 = \sqrt{2}, \qquad h_4 = \frac{1}{8}. \tag{10.509}$$

The indices A, B run over

$$A, B = 1, \dots, 4. \tag{10.510}$$

The gauginos are described by the Weyl fermion λ^A that belongs to the fundamental representation of the global $SU(4)_R$ symmetry group, which extends $SU(2)_R$ of $\mathcal{N} = 2$.[58] The three complex scalar fields are assembled into an antisymmetric tensor

$$\phi^{AB} = -\phi^{BA}, \tag{10.511}$$

with the additional condition

$$\phi^{AB} = \frac{1}{2} \varepsilon^{ABCD} \bar{\phi}_{CD}, \qquad \bar{\phi}_{CD} = \left(\phi^{CD} \right)^*. \tag{10.512}$$

The Lagrangian $\mathcal{L}_{\mathcal{N}=4}$ is invariant under the following supertransformation rules

$$\delta A^\mu = -i \epsilon^{\alpha A} \bar{\sigma}^\mu_{\ \alpha\dot{\beta}} \bar{\lambda}_A^{\dot{\beta}} - i \bar{\epsilon}_{\dot{\alpha}A} \sigma^{\mu\dot{\alpha}\beta} \lambda_\beta^A,$$

$$\delta \phi^{AB} = -i\sqrt{2} \left(\epsilon^{\alpha A} \lambda_\alpha^B - \epsilon^{\alpha B} \lambda_\alpha^A - \varepsilon^{ABCD} \bar{\epsilon}_{\dot{\alpha}C} \bar{\lambda}_D^{\dot{\alpha}} \right),$$

$$\delta \lambda_\alpha^A = \frac{1}{2} i F_{\mu\nu} (\sigma^{\mu\nu})_\alpha^{\ \beta} \epsilon_\beta^A - \sqrt{2} \left(\mathcal{D}_\mu \phi^{AB} \right) \bar{\sigma}^\mu_{\alpha\dot{\beta}} \bar{\epsilon}_B^{\dot{\beta}} + ig \left[\phi^{AB}, \bar{\phi}_{BC} \right] \epsilon_\alpha^C,$$

$$\delta \bar{\lambda}_A^{\dot{\alpha}} = \frac{1}{2} i F_{\mu\nu} (\bar{\sigma}^{\mu\nu})_{\dot{\beta}}^{\dot{\alpha}} \bar{\epsilon}_A^{\dot{\beta}} + \sqrt{2} \left(\mathcal{D}_\mu \bar{\phi}_{AB} \right) (\sigma^\mu)^{\dot{\alpha}\beta} \epsilon_\alpha^B + ig \left[\bar{\phi}_{AB}, \phi^{BC} \right] \bar{\epsilon}_C^{\dot{\alpha}}, \tag{10.513}$$

where the ϵ^A are the supertransformation parameters ($A = 1, \dots, 4$).

The Lagrangian (10.508) is presented in a form that also covers the case $\mathcal{N} = 2$. To obtain $\mathcal{L}_{\mathcal{N}=2}$ we must substitute the following into (10.508):

$$A, B = 1, 2, \qquad \phi^{AB} = \sqrt{2} \, \varepsilon^{AB} \phi,$$
$$h_3 = 1, \qquad h_4 = \frac{1}{16}. \tag{10.514}$$

---

[58] Much as in the $\mathcal{N} = 2$ case, the $\mathcal{N} = 4$ theory has an extended R symmetry. In the $\mathcal{N} = 1$ superfield formulation the manifest global symmetry is $SU(3) \times U(1)$. However, the action written in terms of the component fields exhibits the full $SU(4)_R$ symmetry. The complex scalar fields, which are equivalent to six real scalar fields, can be assigned to the real representation **6** of $O(6) = SU(4)$.

10.19 Instantons in Supersymmetric Yang–Mills Theories

The reader is advised to look through Chapter 5.

Instantons are related to the tunneling amplitudes connecting the vacuum state to itself. In gauge theories at weak coupling this is the main source of the nonperturbative physics shaping the vacuum structure.

In the semiclassical treatment of tunneling transitions, instantons present the extremal trajectories (classical solutions) in *imaginary* time. Thus, the analytical continuation to imaginary time becomes a necessity. In imaginary time the theory can often be formulated as a field theory in Euclidean space.

However, a Euclidean formulation does not exist in minimal supersymmetric theories in four dimensions, because they contain the Weyl (or, equivalently, Majorana) fermions. The easy way to see this is to observe that it is impossible to find four purely imaginary 4×4 matrices with the algebra $\{\gamma_\mu, \gamma_\nu\} = \delta_{\mu\nu}$ necessary for constructing a Euclidean version of the theory with Majorana spinors. The fermionic integration in the functional integral runs over the holomorphic variables, and the operation of involution (i.e., complex conjugation) that relates ψ_α and $\bar{\psi}_{\dot\alpha}$ has no Euclidean analog. In Euclidean space ψ_α and $\bar{\psi}_{\dot\alpha}$ must be considered as independent variables. Only theories with extended superalgebras, $\mathcal{N} = 2$ or 4, where all spinor fields can be written in Dirac form, admit a Euclidean formulation [75].

Passing to Euclidean time in the case of $\mathcal{N} = 1$ instantons

A Euclidean formulation of the theory is by no means necessary for imaginary time analysis [76]. All we need to do is to replace the time t by the Euclidean time τ in all fields and in the definition of the action,

$$t = -i\tau, \quad \phi(t, \vec{x}) \to \phi(-i\tau, \vec{x}),$$

$$\int d^4x \to -i \int d\tau\, d^3\vec{x},$$

$$i \int d^4x\, \mathcal{L}_{\text{Mink}} \to - \int d\tau\, d^3\vec{x}\, \mathcal{L}_{\text{Eucl}},$$

$$\mathcal{L}_{\text{Eucl}} = -\mathcal{L}_{\text{Mink}}\big|_{t=-i\tau}. \tag{10.515}$$

The weight factor in the functional integral is given by

$$\int \prod_\phi D\phi \exp\left(-S_{\text{Eucl}}\right), \quad S_{\text{Eucl}} = \int d^4x_E\, \mathcal{L}_{\text{Eucl}}. \tag{10.516}$$

I stress again that no redefinition of fields is made; the integration in the path integral is over the same variables as in Minkowski space. In particular, the gauge 4-potential remains as $\{A_0, \vec{A}\}$. The fermion part remains as it is, too. Then we can find the extremal trajectories (both the bosonic and fermionic parts) by solving the classical equations of motion. In this formalism some components of the fields involved in the instanton solution will be purely imaginary. We have to accept this. Quantities that must be real, such as the action, remain real, of course.

To illustrate the procedure we will consider first the Belavin–Polyakov–Schwarz–Tyupkin (BPST) instanton [77] and the gluino zero modes in supersymmetric Yang–Mills theory. The gauge group is SU(2).[59]

---

[59] Section 10.19 is based on [62, 76].

10.19.1 Instanton Solution in Spinor Notation

Here we develop the spinorial formalism as applied to the instantons; this is especially convenient in supersymmetric theories, where the bosons and fermions are related. An additional bonus is that there is no need to introduce the 't Hooft symbols.

The spinor notation introduced in Section 10.2.1 is based on the $SU(2)_L \times SU(2)_R$ algebra of the Lorentz group (the undotted and dotted indices corresponding to the two subalgebras). In Minkowski space these two $SU(2)$ subalgebras are related by complex conjugation (involution). In particular, this allows one to define the notion of a real vector as $(A_{\alpha\dot\beta})^* = A_{\beta\dot\alpha}$. As mentioned above, the property of involution is lost after the continuation to imaginary time.

> *Instanton:*
> $G_{\alpha\beta} = 0$,
> $\bar G_{\dot\alpha\dot\beta} \neq 0$;
> *Anti-inst.:*
> $G_{\alpha\beta} \neq 0$,
> $\bar G_{\dot\alpha\dot\beta} = 0$

Consider the simplest non-Abelian gauge theory – supersymmetric $SU(2)$ gluodynamics. The Lagrangian was given in Eq. (10.374). As explained above, the classical equations are the same as for Minkowski space, with the substitution $t = -i\tau$, while no substitution is made for the fields. In particular, the duality equation has the form

$$\bar G_{\dot\alpha\dot\beta} = \left(E^j + iB^j\right)\left(\tau^j\right)_{\dot\alpha\dot\beta} = 0, \tag{10.517}$$

where the matrices $(\vec\tau)_{\dot\alpha\dot\beta}$ were defined in Eq. (10.27). The antiduality relation is similar, namely,

$$G_{\alpha\beta} = \left(E^j - iB^j\right)\left(\tau^j\right)_{\alpha\beta} = 0. \tag{10.518}$$

The (anti)instanton 4-potential – the solution of Eq. (10.517) – is

$$A^{\{\alpha\gamma\}}_{\beta\dot\beta} = -2i\frac{1}{x^2 + \rho^2}\left(\delta^\alpha_\beta x^\gamma_{\dot\beta} + \delta^\gamma_\beta x^\alpha_{\dot\beta}\right). \tag{10.519}$$

Here ρ is a collective coordinate (or modulus) of the instanton solution, known as the *instanton size*.

Where is the familiar color index $a = 1, 2, 3$? It has been exchanged for two spinorial indices $\{\alpha\gamma\}$,

$$A^{\{\alpha\gamma\}} \equiv A^a(\tau^a)_{\alpha\gamma}. \tag{10.520}$$

The tensor $A^{\{\alpha\gamma\}}$, which is symmetric in α and γ, presents the adjoint representation of the color $SU(2)$. The instanton is a"hedgehog" configuration, with entangled color and Lorentz indices. It is invariant under simultaneous rotations in the $SU(2)_{color}$ and $SU(2)_L$ spaces (see Eq. (10.523) below). This invariance is explicit in Eq. (10.519). The superscript braces remind us that this symmetric pair of spinorial indices is connected with the color index a.

All the definitions above are obviously taken from Minkowski space. The Euclidean aspect of the problem reveals itself only in the fact that x_0 (the time component of x_μ) is purely imaginary. As a concession to the Euclidean nature of the instantons we will define and consistently use[60]

$$x^2_E \equiv -x_\mu x^\mu = \vec x^2 - x^2_0 = \vec x^2 + \tau^2. \tag{10.521}$$

---

[60] The subscript E will be omitted hereafter.

The minus sign in Eq. (10.521) is by no means necessary; it turns out to be rather convenient, though.

It is instructive to check that the field configuration (10.519) reduces to the standard BPST anti-instanton [77]. Indeed,

$$A^a_\mu = \tfrac{1}{4} A^{\{\alpha\gamma\}}_{\beta\dot\beta} (-\tau^a)_{\gamma\alpha} (\bar\sigma_\mu)^{\dot\beta\beta} \tag{10.522}$$

$$= \begin{cases} 2i\, x^a (x^2 + \rho^2)^{-1} & \text{at } \mu = 0, \\ 2\, (\varepsilon^{amj} x^j - \delta^{am} x_4)\, (x^2 + \rho^2)^{-1} & \text{at } \mu = m. \end{cases}$$

This can be seen to be the standard anti-instanton solution (in the nonsingular gauge), provided that one takes into account that

$$A^a_0 = i A^a_4.$$

Let us stress that it is A_μ, with the *lower vectorial index*, which is related to the standard Euclidean solution; for further details see Section 5.3. The time component of A^a_μ in Eq. (10.522) is purely imaginary. This is all right – in fact, A_0 is not the integration variable in the canonical representation of the functional integral. The spatial components A^a_m are real.

Anti-
instanon in
spinorial
notation

From Eq. (10.519) it is not difficult to get the anti-instanton gluon field strength tensor,

$$G^{\{\gamma\delta\}}_{\alpha\beta} = \left(E^j - i\, B^j \right)^{\{\gamma\delta\}} \left(\tau^j \right)_{\alpha\beta}$$

$$= 8i \left(\delta^\gamma_\alpha \delta^\delta_\beta + \delta^\delta_\alpha \delta^\gamma_\beta \right) \frac{\rho^2}{(x^2 + \rho^2)^2}. \tag{10.523}$$

The last expression implies that

$$E^a_n = 4i\delta^a_n \frac{\rho^2}{(x^2 + \rho^2)^2}, \qquad B^a_n = -4\delta^a_n \frac{\rho^2}{(x^2 + \rho^2)^2}. \tag{10.524}$$

This completes the construction of the anti-instanton. As for the instanton, it is the solution for the constraint $G_{\alpha\beta} = 0$ that can be obtained by the replacement of all dotted indices by undotted, and vice versa.

The advantages of the approach presented here become fully apparent when the fermion fields are included. Below we briefly discuss the impact of the fermion fields in SU(2) supersymmetric gluodynamics.

The supersymmetry transformations in supersymmetric gluodynamics take the form

$$\delta\lambda^a_\alpha = G^a_{\alpha\beta} \epsilon^\beta, \qquad \delta\bar\lambda^a_{\dot\alpha} = \bar G^a_{\dot\alpha\dot\beta} \bar\epsilon^{\dot\beta}. \tag{10.525}$$

Since in the anti-instanton background $\bar G_{\dot\alpha\dot\beta} = 0$, supertransformations with dotted index parameter, $\bar\epsilon^{\dot\beta}$, do not act on the background field. Thus, the half supersymmetry is preserved.

Super-
symmetric
zero modes

However, supertransformations with parameter with undotted index, ϵ^β, do act nontrivially. When applied to the gluon background field, they create two fermion zero modes,

$$\lambda^{\{\gamma\delta\}}_{\alpha(\beta)} \propto G^{\{\gamma\delta\}}_{\alpha\beta}$$

$$\propto \left(\delta^{\gamma}_{\alpha}\delta^{\delta}_{\beta} + \delta^{\delta}_{\alpha}\delta^{\gamma}_{\beta}\right) \frac{\rho^2}{(x^2 + \rho^2)^2}; \tag{10.526}$$

the subscript $\beta = 1, 2$ performs the numeration of the zero modes. These two zero modes are built on the basis of supersymmetry, hence they are called *supersymmetric*. Somewhat less obvious is the existence of two extra zero modes. They are related to superconformal transformations (see Section 10.14) and thus are called *superconformal*. Superconformal transformations have the same form as in Eq. (10.525) but with the parameter ϵ substituted by a linear function of the coordinates x_μ:

$$\epsilon^\alpha \to x^\alpha_{\dot\gamma}\bar\beta^{\dot\gamma}. \tag{10.527}$$

| Super-
| conformal   In this way we get
| zero modes

$$\lambda^{\{\gamma\delta\}}_{\alpha(\dot\gamma)} \propto G^{\{\gamma\delta\}}_{\alpha\beta} x^{\beta}_{\dot\gamma}$$

$$\propto \left(\delta^{\gamma}_{\dot\alpha}x^{\delta}_{\dot\gamma} + \delta^{\delta}_{\dot\alpha}x^{\gamma}_{\dot\gamma}\right) \frac{\rho^2}{(x^2 + \rho^2)^2}, \tag{10.528}$$

where the subscript $\dot\gamma = 1, 2$ enumerates two modes.

Thus we have constructed four zero modes, in full accord with the index theorem following from the chiral anomaly (10.379). It is instructive to verify that they satisfy the Dirac equation $\mathcal{D}_{\alpha\dot\alpha}\lambda^\alpha = 0$. For the supersymmetric zero modes (10.526) this equation reduces to the equation $\mathcal{D}^\mu G_{\mu\nu} = 0$ for the instanton field. As far as the superconformal modes (10.528) are concerned, the additional term containing $\partial^\alpha_{\dot\alpha}x^\beta_{\dot\gamma} \propto \varepsilon^{\alpha\beta}\varepsilon_{\dot\alpha\dot\gamma}$ vanishes upon contraction with $G_{\alpha\beta}$.

All four zero modes are chiral (left-handed). There are no right-handed zero modes for the anti-instanton, i.e., the equation $\mathcal{D}_{\alpha\dot\alpha}\bar\lambda^{\dot\alpha} = 0$ has no normalizable solutions. This is another manifestation of the loss of involution; the operator $\mathcal{D}_{\alpha\dot\alpha}$ ceases to be Hermitian.

We will use the anti-instanton field as a reference point in what follows. In the instanton field the roles of λ and $\bar\lambda$ interchange, together with the dotted and undotted indices.

This concludes our explanatory remarks regarding the analytic continuation necessary in developing instanton calculus in $\mathcal{N} = 1$ supersymmetric Yang–Mills theories.

In the subsequent sections which can be viewed as an "ABC of superinstantons," we will discuss the basic elements of instanton calculus in supersymmetric gauge theories. These elements are: collective coordinates (instanton moduli) both for the gauge and matter fields, the instanton measure in the moduli space, and the cancelation of the quantum corrections.

10.19.2 Collective Coordinates

| Collective   The instanton solution (10.519) has only one collective coordinate, the instanton size
| coordinates ρ. In fact, the classical BPST instanton depends on eight collective coordinates; the
| $\equiv$ instanton   instanton size ρ, its center $(x_0)_\mu$, and three angles that describe the orientation of the
| moduli   instanton in one of the SU(2) subgroups of the Lorentz group (or, equivalently, in
                 SU(2) color space). If the gauge group is larger than SU(2), additional coordinates

are needed to describe the embedding of the instantonic SU(2) "corner" in the full gauge group G.

The procedure allowing one to introduce these eight coordinates is already known to us (see Chapter 5); here our focus is mainly on the Grassmann collective coordinates and on the way that supersymmetry acts in the space of the collective coordinates.

The general strategy is as follows. One starts by finding the symmetries of the classical field equations. These symmetries form some group G. The next step is to consider a particular classical solution (an instanton). This solution defines a *stationary group* $\mathcal{H}$ of transformations – i.e., those that act trivially, leaving unchanged the original solution. It is evident that $\mathcal{H}$ is a subgroup of G. The space of the collective coordinates is determined by the quotient $G/\mathcal{H}$. The construction of this quotient is a convenient way of introducing the collective coordinates.

An example of a transformation belonging to the stationary subgroup $\mathcal{H}$ for the anti-instanton (10.519) is the SU(2)$_R$ subgroup of the Lorentz group. An example of transformations that act nontrivially is given by the four-dimensional translations. The latter are part of the group G.

An important comment is in order here. In supersymmetric gluodynamics the construction sketched above generates the full one-instanton moduli space. However, in the multi-instanton problem, or in the presence of matter, some extra moduli appear that are not tractable via the classical symmetries. An example is the 't Hooft zero mode for matter fermions. Even in such situations supersymmetry acts on these extra moduli in a certain way, and we will study this issue below.

10.19.3 Superconformal Symmetry

Following the program outlined above let us start by identifying the symmetry group G of the classical equations in supersymmetric gluodynamics. An obvious symmetry is Poincaré invariance, extended to include the supercharges $Q_\alpha, \bar{Q}_{\dot\alpha}$. The Poincaré group includes the translations $P_{\alpha\dot\alpha}$ and the Lorentz rotations $M_{\alpha\beta}, \bar{M}_{\dot\alpha\dot\beta}$. Additionally the fermions bring in the chiral rotation R.

In fact, the classical Lagrangian (10.374) has a wider symmetry – the super-conformal group (a pedagogical introduction to the superconformal group can be found in [78]; see also Appendix section 1.4 at the end of Chapter 1). The additional generators are the dilatation D, the special conformal transformations $K_{\alpha\dot\alpha}$, and the superconformal transformations S_α and $\bar{S}_{\dot\alpha}$.

Thus, the superconformal algebra in four dimensions includes 16 bosonic and eight fermionic generators. They all are of a geometric nature – they can be realized as coordinate transformations in superspace. Correspondingly, the 24 generators can be presented as differential operators acting in the superspace, in particular,

$$P_{\alpha\dot\alpha} = i\partial_{\alpha\dot\alpha}, \qquad \bar{M}_{\dot\alpha\dot\beta} = -\tfrac{1}{2}x^\gamma_{\{\dot\alpha}\partial_{\gamma\dot\beta\}} - \bar{\theta}_{\{\dot\alpha}\frac{\partial}{\partial\bar{\theta}^{\dot\beta\}}},$$

$$D = \frac{i}{2}\left(x^{\alpha\dot\alpha}\partial_{\alpha\dot\alpha} + \theta^\alpha\frac{\partial}{\partial\theta^\alpha} + \bar{\theta}^{\dot\alpha}\frac{\partial}{\partial\bar{\theta}^{\dot\alpha}}\right), \qquad R = \theta^\alpha\frac{\partial}{\partial\theta^\alpha} - \bar{\theta}^{\dot\alpha}\frac{\partial}{\partial\bar{\theta}^{\dot\alpha}}, \qquad (10.529)$$

$$Q_\alpha = -i\frac{\partial}{\partial\theta^\alpha} + \bar{\theta}^{\dot\alpha}\partial_{\alpha\dot\alpha}, \qquad \bar{Q}_{\dot\alpha} = i\frac{\partial}{\partial\bar{\theta}^{\dot\alpha}} - \theta^\alpha\partial_{\alpha\dot\alpha},$$

$$S_\alpha = -(x_R)_{\alpha\dot\alpha}\bar{Q}^{\dot\alpha} - 2\theta^2 D_\alpha, \qquad \bar{S}_{\dot\alpha} = -(x_L)_{\alpha\dot\alpha}Q^\alpha + 2\bar{\theta}^2\bar{D}_{\dot\alpha}.$$

Here, symmetrization in $\dot{\alpha}$, $\dot{\beta}$ is indicated as before by braces. The generators as given above act on the superspace coordinates. In applications to fields, the generators must be supplemented by extra terms (the spin term in $\bar{M}$, the conformal weight in D, etc.).

The differential realization (10.529) allows one to establish a full set of (anti)commutation relations in the superconformal group. This set can be found in [78].[61] What we will need for the supersymmetry transformations of the collective coordinates is the commutators of the supercharges with all generators:

$$\{Q_\alpha, \bar{Q}_{\dot{\beta}}\} = 2P_{\alpha\dot{\beta}}, \quad \{Q_\alpha, \bar{S}_{\dot{\beta}}\} = 0, \quad \{\bar{Q}_{\dot{\beta}}, \bar{S}_{\dot{\alpha}}\} = -4i\bar{M}_{\dot{\alpha}\dot{\beta}} + 2D\varepsilon_{\dot{\alpha}\dot{\beta}} + 3iR\varepsilon_{\dot{\alpha}\dot{\beta}},$$

$$[Q_\alpha, D] = \tfrac{1}{2}i\, Q_\alpha, \quad [\bar{Q}_{\dot{\alpha}}, D] = \tfrac{1}{2}i\, \bar{Q}_{\dot{\alpha}}, \quad [Q_\alpha, R] = Q_\alpha, \quad [\bar{Q}_{\dot{\alpha}}, R] = -\bar{Q}_{\dot{\alpha}},$$

Super-
conformal
algebra

$$[Q_\alpha, M_{\beta\gamma}] = -\tfrac{1}{2}(Q_\beta\varepsilon_{\alpha\gamma} + Q_\gamma\varepsilon_{\alpha\beta}), \quad [\bar{Q}_{\dot{\alpha}}, M_{\beta\gamma}] = 0,$$

$$[Q_\alpha, K_{\beta\dot{\beta}}] = 2i\varepsilon_{\alpha\beta}\bar{S}_{\dot{\beta}}. \tag{10.530}$$

10.19.4 Collective Coordinates: Continuation

Now, what is the stationary group $\mathcal{H}$ for the anti-instanton solution (10.519)? This bosonic solution is obviously invariant under the chiral transformation R, which acts only on fermions. Furthermore, the transformation $K_{\alpha\dot{\alpha}} + \tfrac{1}{2}\rho^2 P_{\alpha\dot{\alpha}}$ does not act on this solution. (The subtlety to be taken into account is that this and other similar statements are valid modulo a gauge transformation.) A simple way to verify that $K_{\alpha\dot{\alpha}} + \tfrac{1}{2}\rho^2 P_{\alpha\dot{\alpha}}$ does not act is to apply it to a gauge-invariant object such as Tr $G_{\alpha\beta}G_{\gamma\delta}$. Another possibility is to observe that a conformal transformation is a combination of a translation and inversion. Under inversion an instanton in the regular gauge becomes the very same instanton in the singular gauge.

Unraveling the gauge transformations is particularly important for the instanton orientations. At first glance, it would seem that neither $SU(2)_R$ nor $SU(2)_L$ Lorentz rotations act on the instanton solution: the expression (10.523) for the gluon field strength tensor contains no dotted indices, which explains the first part of the statement, while the $SU(2)_L$ rotations of $G_{\alpha\beta}$ can be compensated by those in the gauge group. This conclusion would be misleading, however. In Section 10.19.8 we will show that the instanton orientations are coupled to the $SU(2)_R$ Lorentz rotations, i.e., to the $\bar{M}_{\dot{\alpha}\dot{\beta}}$ generators, while the $SU(2)_L$ rotations are compensated by gauge transformations.

Thus, we can count eight bosonic generators of the stationary group $\mathcal{H}$. It also contains four fermionic, $\bar{Q}_{\dot{\alpha}}$ and S_α. It is easy to check that these 12 generators do indeed form a graded algebra. To guide the reader, the generators of $\mathcal{G}$ and $\mathcal{H}$ are collected in Table 10.7.

Now we are ready to introduce the set of collective coordinates (the instanton moduli) parametrizing the quotient $\mathcal{G}/\mathcal{H}$. To this end let us start from the purely bosonic anti-instanton solution (10.519) of size $\rho = 1$ and centered at the origin and apply to it a generalized shift operator [79]:

$$\Phi(x, \theta, \bar{\theta}; x_0, \rho, \bar{\omega}, \theta_0, \bar{\beta}) = \mathcal{V}(x_0, \rho, \bar{\omega}, \theta_0, \bar{\beta})\Phi_0(x, \theta, \bar{\theta}),$$

$$\mathcal{V}(x_0, \rho, \bar{\omega}, \theta_0, \bar{\beta}) = e^{iPx_0}e^{-iQ\theta_0}e^{-i\bar{S}\bar{\beta}}e^{i\bar{M}\bar{\omega}}e^{iD\ln\rho}, \tag{10.531}$$

---

[61] Warning: my normalization of some generators differs from that in Ref. [78].

Table 10.7. The generators of the classical symmetry group $\mathcal{G}$ and the stationary subgroup $\mathcal{H}$		
Group	Bosonic generators	Fermionic generators
$\mathcal{G}$	$P_{\alpha\dot\alpha}, M_{\alpha\beta}, \bar{M}_{\dot\alpha\dot\beta}, D, R, K_{\alpha\dot\alpha}$	$Q_\alpha, \bar{Q}_{\dot\alpha}, S_\alpha, \bar{S}_{\dot\alpha}$
$\mathcal{H}$	$R, K_{\alpha\dot\alpha} + \frac{1}{2}\rho^2 P_{\alpha\dot\alpha}, M_{\alpha\beta}$	$\bar{Q}_{\dot\alpha}, S_\alpha$

where $\Phi_0(x, \theta, \bar\theta)$ is a superfield constructed from the original bosonic solution (10.519). Moreover, $P_{\alpha\dot\alpha}, Q_\alpha, \bar{S}_{\dot\alpha}, \bar{M}_{\alpha\beta}, D$ are the generators in differential form (10.529) (plus non-derivative terms relating to the conformal weights and spins of the fields). The relevant representation is differential because we are dealing with classical fields. In operator language the action of the operators at hand would correspond to standard commutators, e.g., $[P_{\alpha\dot\alpha}, \Phi] = i\partial_{\alpha\dot\alpha}\Phi$.

To illustrate how the generalized shift operator $\mathcal{V}$ acts, we will apply it to the superfield Tr W^2:

$$\text{Tr}(W^\alpha W_\alpha)_0 = \theta^2 \frac{96}{(x^2 + 1)^4} = \theta^2 \frac{96}{(x_L^2 + 1)^4}. \tag{10.532}$$

Applying $\mathcal{V}$ to this expression one obtains

$$\text{Tr}\, W^\alpha W_\alpha = \mathcal{V}(x_0, \rho, \bar\omega, \theta_0, \bar\beta) \frac{96\theta^2}{(x_L^2 + 1)^4} = \frac{96\tilde\theta^2 \rho^4}{[(x_L - x_0)^2 + \rho^2]^4}, \tag{10.533}$$

W^2 in the instanton field

where

$$\tilde\theta_\alpha = (\theta - \theta_0)_\alpha + (x_L - x_0)_{\alpha\dot\alpha}\bar\beta^{\dot\alpha}. \tag{10.534}$$

In deriving this expression we used the representation (10.529) for the generators. Note that the generators $\bar{M}$ act trivially on the Lorentz scalar W^2. Regarding the dilatation D, a nonderivative term should be added to account for the nonvanishing dimension of W^2, equal to 3.

The value of Tr W^2 depends on the variables x_L and θ and on the moduli x_0, ρ, θ_0, and $\bar\beta$. It does not depend on $\bar\omega$ because Tr W^2 is the Lorentz and color singlet.

Of course, the most detailed information is contained in the superfield V. Applying the generalized shift operator $\mathcal{V}$ to V_0, where

$$V_0^{\{\alpha\gamma\}} = 4i\frac{1}{x^2 + 1}\left(\theta^\alpha x_{\dot\beta}^\gamma \bar\theta^{\dot\beta} + \theta^\gamma x_{\dot\beta}^\alpha \bar\theta^{\dot\beta}\right), \tag{10.535}$$

we obtain a generic instanton configuration that depends on all the collective coordinates. One should keep in mind, however, that, in contradistinction to Tr W^2, the superfield $V^{\{\alpha\gamma\}}$ is not a gauge-invariant object. Therefore the action of $\mathcal{V}$ should be supplemented by a subsequent gauge transformation,

$$e^V \to e^{i\bar\Lambda}e^V e^{-i\Lambda}, \tag{10.536}$$

where the chiral superfield Λ must be chosen in such a way that the original gauge is maintained.

10.19.5 The Symmetry Transformations of the Moduli

Once all the relevant collective coordinates have been introduced, it is natural to pose the question: how does the classical symmetry group act on them? Although a complete set of superconformal transformations of the instanton moduli could be readily found, we will focus on the exact symmetries – the Poincaré group plus supersymmetry. Only exact symmetries are preserved by the instanton measure, and we will use them for its reconstruction.

The following discussion will show how to find the transformation laws for the collective coordinates. Assume that we are interested in translations $x \to x + a$. The operator generating the translation is $\exp(iPa)$. Let us apply it to the configuration $\Phi(x, \theta, \bar{\theta}; x_0, \rho, \bar{\omega}, \theta_0, \bar{\beta})$; see Eq. (10.531):

$$e^{iPa}\Phi(x, \theta, \bar{\theta}; x_0, \rho, \bar{\omega}, \theta_0, \bar{\beta}) = e^{iPa} e^{iPx_0} e^{-iQ\theta_0} e^{-i\bar{S}\bar{\beta}} e^{i\bar{M}\bar{\omega}} e^{iD\ln\rho}\Phi_0(x, \theta, \bar{\theta})$$

$$= \Phi(x, \theta, \bar{\theta}; x_0 + a, \rho, \bar{\omega}, \theta_0, \bar{\beta}). \tag{10.537}$$

Thus, we obviously get the original configuration with x_0 replaced by $x_0 + a$ and no change in the other collective coordinates. Alternatively, one can say that the interval $x - x_0$ is an invariant of the translations; the instanton field configuration does not depend on x and x_0 separately, but on invariant combinations.

Passing to supersymmetry, the transformation generated by $\exp(-iQ\epsilon)$ is the simplest to deal with, i.e.,

$$\theta_0 \to \theta_0 + \epsilon. \tag{10.538}$$

The other moduli stay intact.

For supertranslations with the parameter $\bar{\epsilon}$, we act with $\exp(-i\bar{Q}\bar{\epsilon})$ on the configuration Φ,

$$e^{-i\bar{Q}\bar{\epsilon}}\Phi(x, \theta, \bar{\theta}; x_0, \rho, \bar{\omega}, \theta_0, \bar{\beta}) = e^{-i\bar{Q}\bar{\epsilon}} e^{iPx_0} e^{-iQ\theta_0} e^{-i\bar{S}\bar{\beta}} e^{i\bar{M}\bar{\omega}} e^{iD\ln\rho}\Phi_0(x, \theta, \bar{\theta}). \tag{10.539}$$

Our goal is to move $\exp(-i\bar{Q}\bar{\epsilon})$ to the rightmost position, since when $\exp(-i\bar{Q}\bar{\epsilon})$ acts on the original anti-instanton solution $\Phi_0(x, \theta, \bar{\theta})$ it produces unity. On the way we get the various commutators listed in Eq. (10.530). For instance, the first nontrivial commutator that we encounter is $[\bar{Q}\bar{\epsilon}, Q\theta_0]$. This commutator produces P, which effectively shifts x_0 by $-4i\theta_0\bar{\epsilon}$. Proceeding further in this way we arrive at the following results [79] for the supersymmetric transformations of the moduli:

$$\delta(x_0)_{\alpha\dot{\alpha}} = -4i(\theta_0)_\alpha \bar{\epsilon}_{\dot{\alpha}}, \qquad \delta\rho^2 = -4i(\bar{\epsilon}\bar{\beta})\rho^2,$$

$$\delta(\theta_0)_\alpha = \epsilon_\alpha, \qquad \delta\bar{\beta}_{\dot{\alpha}} = -4i\bar{\beta}_{\dot{\alpha}}(\bar{\epsilon}\bar{\beta}), \tag{10.540}$$

$$\delta\Omega_{\hat{\beta}}^{\hat{\alpha}} = 4i\left[\bar{\epsilon}^{\hat{\alpha}}\bar{\beta}_{\hat{\gamma}} + \tfrac{1}{2}\delta_{\hat{\gamma}}^{\hat{\alpha}}(\bar{\epsilon}\bar{\beta})\right]\Omega_{\hat{\beta}}^{\hat{\gamma}},$$

where we have introduced the rotation matrix Ω, defined as

$$\Omega_{\hat{\beta}}^{\hat{\alpha}} = \exp\left(-i\bar{\omega}_{\hat{\beta}}^{\hat{\alpha}}\right). \tag{10.541}$$

This definition of the rotation matrix Ω corresponds to the rotation of spin-1/2 objects.

Once the transformation laws for the instanton moduli are established, one can construct invariant combinations of these moduli. It is easy to verify that such invariants are

$$\frac{\bar{\beta}}{\rho^2}, \qquad \bar{\beta}^2 F(\rho), \tag{10.542}$$

where $F(\rho)$ is an arbitrary function of ρ.

A priori, one might have expected that the above invariants would appear in the quantum corrections to the instanton measure. In fact, the transformation properties of the collective coordinates under the chiral U(1) symmetry preclude this possibility. The chiral charges of all fields are given in Section 10.14. In terms of the collective coordinates, the chiral charges of θ_0 and $\bar{\beta}$ are unity while those of x_0 and ρ are zero. This means that the invariants (10.542) are chiral nonsinglets and cannot appear in the corrections to the measure.

The chiral U(1) symmetry is anomalous. For $SU(2)_{\mathrm{gauge}}$ it has a nonanomalous discrete subgroup Z_4, however (see Section 10.14). This subgroup is sufficient to disallow the invariants (10.542) nonperturbatively.

A different type of invariants is built from the superspace coordinates and the instanton moduli. An example from nonsupersymmetric instanton calculus is the interval $x - x_0$, which is invariant under translations. Now it is time to elevate this notion to superspace.

The first invariant of this type is evidently

$$(\theta - \theta_0)_\alpha. \tag{10.543}$$

Furthermore, $x_L - x_0$ does not change under translations or under the part of the supertransformations generated by Q_α. It does change, however, under $\bar{Q}_{\dot\alpha}$ transformations. Using Eqs. (10.106) and (10.540) one can built a combination of $\theta - \theta_0$ and $x_L - x_0$ that is invariant,

$$\frac{\tilde{\theta}_\alpha}{\rho^2} = \frac{1}{\rho^2}\left[(\theta - \theta_0)_\alpha + (x_L - x_0)_{\alpha\dot\alpha}\bar{\beta}^{\dot\alpha}\right]. \tag{10.544}$$

The superfield $\mathrm{Tr}\, W^2$ given in Eq. (10.533) can be used as a check. It can be presented as follows:

$$\mathrm{Tr}\, W^2 = \frac{\tilde{\theta}^2}{\rho^4} F\left[\frac{(x_L - x_0)^2}{\rho^2}\right]. \tag{10.545}$$

Although the first factor is invariant, the ratio $(x_L - x_0)^2/\rho^2$ is not. Its variation is proportional to $\tilde{\theta}$, however; therefore the product (10.545) is invariant (the factor $\tilde{\theta}^2$ acts as $\delta(\tilde{\theta})$). The instanton collective coordinate θ_0 is a companion to θ while $\bar{\theta}$ has no companions in supersymmetric gluodynamics. It will appear in Section 10.19.7.

10.19.6 The Measure in Moduli Space

Now that the appropriate collective coordinates have been introduced, we come to an important ingredient of superinstanton calculus – the *instanton measure*, or the formula for integration in the space of the collective coordinates. The general procedure for obtaining the measure is well known; it is based on a path integral representation. In terms of a mode expansion this representation reduces to an integral over the coefficients of the mode expansion. The integration measure splits into two factors; integrals over the zero and nonzero mode coefficients. Only the zero mode coefficients are related to the moduli.

We will follow the route pioneered by 't Hooft [80]. In the one-loop approximation the functional integral, say, over the scalar field can be written as

$$\left[\frac{\det(L_2 + M_{\mathrm{PV}}^2)}{\det L_2}\right]^{1/2}, \tag{10.546}$$

where L_2 is a differential operator appearing in the expansion of the Lagrangian near the given background in the quadratic approximation, $L_2 = -\mathcal{D}^2$. The numerator is due to ultraviolet regularization. We will use the Pauli–Villars regularization – there is no alternative in instanton calculations. The mass term of the regulator fields is M_{PV}. Each given eigenmode of L_2 with eigenvalue ϵ^2 contributes M_{PV}/ϵ. For a scalar field there are no zero eigenvalues. However, for vector and spinor fields zero modes do exist: the set of zero modes corresponds to the set of moduli, generically denoted in this section as η_i. For the bosonic zero modes the factor $1/\epsilon$ (which, of course, explodes at $\epsilon \to 0$) is replaced by an integral over the corresponding collective coordinate $d\eta^b$, up to a normalization factor. Similarly, for the fermion zero mode $\sqrt{\epsilon} \to d\eta^f$; see the discussion below.

The zero modes can be obtained by differentiating the field $\Phi(x, \theta, \bar{\theta}; \eta)$ over the collective coordinates η_i at a generic point in the space of the instanton moduli. In the instanton problem, $\{\eta_i\} = \{x_0, \rho, \bar{\omega}, \theta_0, \bar{\beta}\}$. The derivatives $\partial\Phi/\partial\eta_i$ differ from the corresponding zero modes by a normalization factor. It is these normalization factors that determine the measure:

$$d\mu = e^{-8\pi^2/g^2} \prod_i d\eta_i^b \frac{M_{\mathrm{PV}}}{\sqrt{2\pi}} \left\|\frac{\partial\Phi(\eta)}{\partial\eta_i^b}\right\| \prod_k d\eta_k^f \frac{1}{\sqrt{M_{\mathrm{PV}}}} \left\|\frac{\partial\Phi(\eta)}{\partial\eta_i^f}\right\|^{-1}, \tag{10.547}$$

where the norm $\|\Phi\|$ is defined as the square root of the integral over $|\Phi|^2$. The superscripts b and f indicate the bosonic and fermionic collective coordinates, respectively. Note that we have also included $\exp(-\mathcal{S})$ in the measure (the instanton action $\mathcal{S} = 8\pi^2/g^2$). In the expression above it is implied that the zero modes are orthogonal. If this is not the case, which often happens in practice, the measure is given by a more general formula:

$$d\mu = e^{-8\pi^2/g^2} (M_{\mathrm{PV}})^{n_b - n_f/2} (2\pi)^{-n_b/2} \prod_i d\eta_i \left[\mathrm{Ber}\left\langle\frac{\partial\Phi(\eta)}{\partial\eta_j}\left|\frac{\partial\Phi(\eta)}{\partial\eta_k}\right.\right\rangle\right]^{1/2}, \tag{10.548}$$

| General formula |

where Ber stands for the Berezinian (superdeterminant). The normalization of the fields is fixed by the requirement that their kinetic terms are canonical.

I pause here to make a remark regarding the fermion part of the measure. The fermion part of the Lagrangian is $i\bar{\lambda}_{\dot{\beta}}\bar{\mathcal{D}}^{\dot{\beta}\alpha}\lambda_\alpha$ or $i\lambda^\alpha \mathcal{D}_{\alpha\dot{\alpha}}\bar{\lambda}^{\dot{\alpha}}$, where $\bar{\mathcal{D}}^{\dot{\beta}\alpha} \equiv (\bar{\sigma}^\mu)^{\dot{\beta}\alpha} \mathcal{D}_\mu$. For the mode expansion of the field λ^α it is convenient to use the Hermitian operator

$$(L_2)_\beta{}^\alpha = -\mathcal{D}_{\beta\dot{\beta}} \bar{\mathcal{D}}^{\dot{\beta}\alpha} = -\left(\mathcal{D}_\mu\right)^2 \delta_\beta^\alpha - i\, G_\beta{}^\alpha\,;$$

$$(L_2)_\beta{}^\alpha \lambda_\alpha = \epsilon^2 \lambda_\beta. \tag{10.549}$$

The operator determining the $\bar{\lambda}$ modes is

$$(\tilde{L}_2)^{\dot{\beta}}_{\ \dot{\alpha}} = -\bar{\mathcal{D}}^{\dot{\beta}\beta}\mathcal{D}_{\beta\dot{\alpha}} = -\left(\mathcal{D}_\mu\right)^2 \delta^{\dot{\beta}}_{\ \dot{\alpha}} + i\,\bar{G}^{\dot{\beta}}_{\ \dot{\alpha}}\,;$$

$$(\tilde{L}_2)^{\dot{\beta}}_{\ \dot{\alpha}}\bar{\lambda}^{\dot{\alpha}} = \epsilon^2\,\bar{\lambda}^{\dot{\beta}}. \tag{10.550}$$

The operators $(L_2)^\alpha_\beta$ and $(\tilde{L}_2)^{\dot{\alpha}}_{\dot{\beta}}$ are not identical.

In the anti-instanton background the operator L_2 has four zero modes, discussed above, while $\tilde{L}_2$ has none. As far as the nonzero modes are concerned, they are degenerate and are related as follows:

$$\bar{\lambda}^{\dot{\alpha}} = \frac{i}{\epsilon}\mathcal{D}^{\dot{\alpha}}_\alpha\lambda^\alpha. \tag{10.551}$$

Taking into account the relations above, we find that the modes with a given ϵ appear in the mode decomposition of the fermion part of the action, in the form $\epsilon \int d^4x\lambda^2$. For a given mode $\lambda^2 = \epsilon^{\alpha\beta}\lambda_\beta\lambda_\alpha$ vanishes, literally speaking. However, there are two modes, $\lambda_{(1)}$ and $\lambda_{(2)}$, for each ϵ and in fact it is the product $\lambda_{(1)}\lambda_{(2)}$ that enters. This consideration provides us with a definition of the norm matrix for the fermion zero modes, namely

$$\int d^4x\lambda_{(i)}\lambda_{(j)}, \tag{10.552}$$

which should be used in calculating the Berezinian.

The norm factors depend on η_i, generally speaking. Equation (10.548) gives the measure at any point in the instanton moduli space. Thus, (10.548) conceptually solves the problem of constructing the measure.

In practice, the measure turns out to be simple at certain points on the moduli space. For instance, instanton calculus always starts from a purely bosonic instanton. Then, to reconstruct the measure everywhere on the instanton moduli space one can apply the exact symmetries of the theory; by exact, I mean those symmetries that are preserved at the quantum level, i.e. the Poincaré symmetries plus supersymmetry, in the case at hand, rather than the full superconformal group. As we will see, this is sufficient to obtain the full measure in supersymmetric gluodynamics but not in theories with matter. For nonsupersymmetric Yang–Mills theories, the instanton measure was found in [80]. After a brief summary of 't Hooft's construction, we will add the fermion part specific to supersymmetric gluodynamics.

Transition to canonically normalized field

Translations: The translational zero modes are obtained by differentiating the instanton field A_ν/g over $(x_0)_\mu$, where μ performs the numeration of the modes: there are four of them. The factor $1/g$ reflects the transition to the canonically normalized field, a requirement mentioned after Eq. (10.548). Up to a sign, differentiation over $(x_0)_\mu$ is the same as differentiation over x_μ. The field $a^{(\mu)}_\nu = g^{-1}\partial_\mu A_\nu$ obtained in this way does not satisfy the gauge condition $\mathcal{D}^\nu a_\nu = 0$. Therefore, it must be supplemented by a gauge transformation, $\delta a_\nu = g^{-1}\mathcal{D}_\nu\varphi$. In the case at hand the gauge function $\varphi^{(\mu)} = -A_\mu$. As a result, the translational zero modes take the form

$$a^{(\mu)}_\nu = g^{-1}\left(\partial_\mu A_\nu - \mathcal{D}_\nu A_\mu\right) = g^{-1}G_{\mu\nu}. \tag{10.553}$$

Note that now the gauge condition is satisfied. The norm of each translational mode is obviously $\sqrt{8\pi^2/g^2}$.

Dilatation: The dilatational zero mode is

$$a_\nu = \frac{1}{g}\frac{\partial A_\mu}{\partial \rho} = \frac{1}{g\rho}G_{\nu\mu}x^\mu, \qquad \|a_\nu\| = \frac{4\pi}{g}. \tag{10.554}$$

The gauge condition is not broken by the differentiation over ρ.

Orientations: The orientation zero modes look like a particular gauge transformation of A_ν [80],

$$(a_\nu)^\alpha_\beta = g^{-1}(\mathcal{D}_\nu\Lambda)^\alpha_\beta. \tag{10.555}$$

Here the spinor notation for color is used and the gauge function Λ has the form

$$\Lambda^\alpha_\beta = \left(U\bar\omega U^T\right)^\alpha_\beta = U^\alpha_{\dot\alpha} U^{\dot\beta}_\beta \bar\omega^{\dot\alpha}_{\dot\beta}, \tag{10.556}$$

where

$$U^\alpha_{\dot\alpha} = \frac{x^\alpha_{\dot\alpha}}{\sqrt{x^2 + \rho^2}} \tag{10.557}$$

and the $\bar\omega^{\dot\alpha}_{\dot\beta}$ are three orientation parameters. It is easy to check that Eqs. (10.555), (10.556) do indeed produce the normalized zero modes, satisfying the condition $\mathcal{D}^\nu a_\nu = 0$. The gauge function (10.556) presents special gauge transformations that are absent in the topologically trivial sector.

This description of the procedure that leads to the occurrence of the $\bar\omega^{\dot\alpha}_{\dot\beta}$ as the orientation collective coordinates is rather sketchy. We will return to the geometrical meaning of these coordinates in Section 10.19.8, after we have introduced the matter fields in the fundamental representation.

Note that the matrix U in (10.557) satisfies the equation

$$\mathcal{D}^2 U_{\dot\alpha} = 0, \tag{10.558}$$

where the undotted index of U is understood as the color index. Correspondingly, the operator $\mathcal{D}$ in Eq. (10.558) acts as the covariant derivative in the fundamental representation. Equation (10.558) will be exploited below when we are considering matter fields in the fundamental representation. Note also that

$$\mathcal{D}^2 \Lambda = 0. \tag{10.559}$$

This construction – building a "string" from several matrices U – can be extended to arbitrary representations of SU(2). The representation with spin j is obtained by multiplying $2j$ matrices U in a manner analogous to that exhibited in Eq. (10.556).

Calculating $\mathcal{D}_\nu\Lambda$ explicitly, we arrive at the following expression for the orientation modes and their norms:

$$a^{\{\alpha\gamma\}}_{\beta\dot\beta} = \frac{1}{4g}G^{\{\alpha\gamma\}}_{\beta\sigma}x^{\sigma\dot\sigma}\bar\omega_{\dot\sigma\dot\beta}, \qquad \left\|\frac{\partial a^a_\nu}{\partial\bar\omega^b}\right\| = \frac{2\pi\rho}{g}. \tag{10.560}$$

Supersymmetric modes: We started discussing these modes in Section 10.19.1:

$$\lambda^{\{\gamma\delta\}}_{\alpha(\beta)} = \frac{1}{g}G^{\{\gamma\delta\}}_{\alpha\beta}, \qquad \langle\lambda_{(1)}|\lambda_{(2)}\rangle = \frac{32\pi^2}{g^2}. \tag{10.561}$$

Up to a numerical matrix, the supersymmetric modes coincide with the translational modes. There are four translational modes and two supersymmetric modes.

Table 10.8. The contribution of the zero modes to the instanton measure. The notation is as follows: 4 T stands for the four translational modes, 1 D for the one dilatational mode, 3 GCR for the three modes associated with the orientations (the global color rotations; the group volume is included), 2 SS for the two supersymmetric gluino modes, 2 SC for the two superconformal gluino modes, and 2 MF for the two matter fermion zero modes; $S \equiv 8\pi^2/g^2$

Boson modes	Fermion modes
$4\,\text{T} \to S^2(2\pi)^{-2}M_{\text{PV}}^4 d^4x_0$	$2\,\text{SS} \to S^{-1}(4M_{\text{PV}})^{-1}d^2\theta_0$
$1\,\text{D} \to S^{1/2}(\pi)^{-1/2}M_{\text{PV}}d\rho$	$2\,\text{SC} \to S^{-1}(8M_{\text{PV}})^{-1}\rho^{-2}d^2\bar\beta$
$3\,\text{GCR} \to S^{3/2}(\pi)^{1/2}M_{\text{PV}}^3\rho^3$	$2\,\text{MF} \to (M_{\text{PV}})^{-1}(8\pi^2|v|^2\rho^2)^{-1}d^2\bar\theta_0$

The factor 2, the ratio of the numbers of the bosonic and fermionic modes, reflects the difference in the numbers of spin components. This is, of course, a natural consequence of supersymmetry.

Superconformal modes: These modes were also briefly discussed in Section 10.19.1:

$$\lambda_{\alpha(\dot\beta)}^{\{\gamma\delta\}} = \frac{1}{g}x_{\dot\beta}^{\beta}G_{\alpha\beta}^{\{\gamma\delta\}}, \qquad \langle\lambda_{(1)}|\lambda_{(2)}\rangle = \frac{64\pi^2\rho^2}{g^2}. \tag{10.562}$$

The superconformal modes have the form $x\,G$, the same as that for the orientational and dilatational modes. Again we have four bosonic and two fermionic modes.

The relevant normalization factors, as well as the accompanying factors from the regulator fields, are collected for all modes in Table 10.8. Assembling all factors together we get the measure for a specific point in moduli space: near the original bosonic anti-instanton solution (10.519) we have

$$d\mu_0 = \frac{1}{256\pi^2}e^{-8\pi^2/g^2}(M_{\text{PV}})^6\left(\frac{8\pi^2}{g^2}\right)^2\frac{d^3\bar\omega}{8\pi^2}d^4x_0 d^2\theta_0 d\rho^2 d^2\bar\beta. \tag{10.563}$$

How does this measure transform under the exact symmetries of the theory? First, let us check the supersymmetry transformations (10.540). They imply that d^4x_0 and $d^2\theta_0$ are invariant. For the last two differentials,

$$d\rho^2 \to d\rho^2[1-4i(\bar\epsilon\bar\beta)], \qquad d^2\bar\beta \to d^2\bar\beta[1+4i(\bar\epsilon\bar\beta)], \tag{10.564}$$

so that their product is invariant too.

The only noninvariance in the measure (10.563) is that of $d^3\bar\omega$ under the SU(2)$_R$ Lorentz rotation generated by $\bar M_{\dot\alpha\dot\beta}$. It is clear that, for a generic instanton orientation $\bar\omega$, the differential $d^3\bar\omega$ is replaced by the SU(2) group measure $d^3\Omega_{\text{SU(2)}} = d^3\bar\omega\sqrt{G}$, where G is the determinant of the Killing metric on the group SU(2) and the matrix Ω defined in Eq. (10.541) is a general element of the group. In fact, this determinant is a part of the Berezinian in the general expression (10.548). The SU(2) group is compact: an integral over all orientations yields the volume of the group,[62] which is equal to $8\pi^2$. Performing this integration we arrive at the final result for the instanton measure in supersymmetric gluodynamics with SU(2)$_{\text{gauge}}$ symmetry:

Instanton measure in SYM

---

[62] Actually, the group of instanton orientations is O(3) = SU(2)/Z_2 rather than SU(2). This distinction is unimportant for the algebra but it is important for the group volume.

$$d\mu_{\mathrm{SU}(2)} = \frac{1}{256\pi^2} e^{-8\pi^2/g^2} (M_{\mathrm{PV}})^6 \left(\frac{8\pi^2}{g^2}\right)^2 d^4x_0 \, d^2\theta_0 \, d\rho^2 \, d^2\bar{\beta}. \tag{10.565}$$

Note that the regulator mass M_{PV} can be viewed as a complex parameter. It arose from the regularization of the operator (10.549), which has a certain chirality.

10.19.7 Including Matter: Supersymmetric QCD with One Flavor

Now we will extend the analysis of the previous sections to include matter. A particular model to be considered is SU(2) SQCD with one flavor (two subflavors); see Section 10.15.

In the Higgs phase the instanton configuration is an approximate solution. A manifestation of this fact is the ρ dependence of the classical action [80]. The solution becomes exact in the limit $\rho \to 0$. For future applications only this limit is of importance, as we will see later. A new feature of theories with matter is the occurrence of extra fermionic zero modes in the matter sector, which gives rise to additional collective coordinates. Supersymmetry provides a geometrical meaning for these collective coordinates.

As above, we start from a bosonic field configuration and apply supersymmetry to build the full instanton orbit. In this way we find a realization of supersymmetry in the instanton moduli space.

We already know that classically SQCD with one flavor has a one-dimensional D-flat direction,

$$(\phi_f^i)_{\mathrm{vac}} = v\delta_f^i, \qquad (\bar{\phi}_f^i)_{\mathrm{vac}} = \bar{v}\delta_f^i, \tag{10.566}$$

where v is an arbitrary complex parameter, the vacuum expectation value of the squark fields. Here i is the color index while f is the subflavor index; $i, f = 1, 2$. The color and flavor indices get entangled, even in the topologically trivial sector, although in a rather trivial manner.

What changes occur in the instanton background? The equation for the scalar field ϕ_f^i becomes

$$\mathcal{D}_\mu^2 \phi_f = 0, \qquad \mathcal{D}_\mu = \partial_\mu - \tfrac{1}{2} i A_\mu^a \tau^a. \tag{10.567}$$

Its solution in the anti-instanton background (10.519) has the form

$$\phi_f^i = v U_f^i = v \frac{x_f^i}{\sqrt{x^2 + \rho^2}}. \tag{10.568}$$

Asymptotically, at $x \to \infty$,

$$\phi_f^i \to \tilde{U}_f^i v, \qquad A_\mu \to i\tilde{U}\partial_\mu \tilde{U}^\dagger, \qquad \tilde{U}_f^i = \frac{x_f^i}{\sqrt{x^2}}, \tag{10.569}$$

i.e., the configuration is gauge equivalent to the flat vacuum (10.566). Note that the equation for the field $\bar{\phi}$ is the same. With the boundary conditions (10.566) the solution is

$$\bar{\phi}_f^i = \bar{v} U_f^i = \bar{v} \frac{x_f^i}{\sqrt{x^2 + \rho^2}}. \tag{10.570}$$

To generate the full instanton orbit, with all collective coordinates switched on, we again apply the generators of the superconformal group to the field configuration Φ_0, which now presents a set of superfields, V_0, Q_0, and $\bar{Q}_0$. The bosonic components are given in Eqs. (10.519), (10.568), and (10.570); the fermionic components vanish. The superconformal group is still the symmetry group of the classical equations. Unlike SUSY gluodynamics, now, at $v \neq 0$, all generators act nontrivially. At first glance we might suspect that we need to introduce $16 + 8$ collective coordinates.

In fact, some of the generators act nontrivially even in a flat (i.e., "instantonless") vacuum with $v \neq 0$. For example, the action of $\exp(iR\alpha)$ changes the phase of v. Since we want to consider a theory with the given vacuum state such a transformation should be excluded from the set generating the instanton collective coordinates. This situation is rather general [81]; see Section 10.19.8.

As a result, the only new collective coordinates to be added are conjugate to $\bar{Q}_{\dot\alpha}$. The differential operators $\bar{Q}_{\dot\alpha}$, defined in Eq. (10.113),[63] annihilate V_0 (modulo a supergauge transformation) and $\bar{Q}_0$. They act nontrivially on Q_0, producing the 't Hooft zero modes of the matter fermions,

$$\bar{Q}^{\dot\alpha}(Q_0)_{\hat{f}}^\alpha = -2\theta^\beta \left(\frac{\partial}{\partial x_L} - iA\right)_\beta^{\dot\alpha} \left[v U_{\hat{f}}^\alpha(x_L)\right] = 4\delta_{\hat{f}}^{\dot\alpha}\theta^\alpha v \frac{\rho^2}{(x_L^2 + \rho^2)^{3/2}}. \quad (10.571)$$

The 't Hooft zero mode explained, (10.571)

I recall that the superscript of Q_0 is the color index while the subscript stands for the subflavor, and they are entangled with the Lorentz spinor index of the supercharge. Note that only the left-handed matter fermion fields have zero modes, as in the case of the gluino. We see how the 't Hooft zero modes get a geometrical interpretation *through supersymmetry*. It is natural to call the corresponding fermionic coordinates $(\bar{\theta}_0)_{\dot\alpha}$. The supersymmetry transformations shift them by $\bar\epsilon$.

In order to determine the action of supersymmetry in the expanded moduli space let us write down the generalized shift operator,

$$\mathcal{V}(x_0, \theta_0, \bar\beta, \bar\zeta, \bar\omega, \rho) = e^{iPx_0} e^{-iQ\theta_0} e^{-i\bar{S}\bar\beta} e^{-i\bar{Q}\bar\zeta} e^{i\bar{M}\bar\omega} e^{iD\ln\rho}. \quad (10.572)$$

Here new Grassmann coordinates $\bar\zeta^{\dot\alpha}$ conjugate to $\bar{Q}_{\dot\alpha}$ are introduced. Repeating the procedure described in Section 10.19.5 but now including $\bar\zeta$ we obtain the supersymmetry transformations of the moduli. They are the same as in Eq. (10.540) but with the addition of the transformations of $\bar\zeta$, i.e.,

$$\delta\bar\zeta_{\dot\alpha} = \bar\epsilon_{\dot\alpha} - 4i\bar\beta_{\dot\alpha}(\bar\zeta\bar\epsilon). \quad (10.573)$$

At linear order in the fermionic coordinates the SUSY transformation of $\bar\zeta$ is the same as that of $\bar\theta$, but the former contains nonlinear terms. A combination that transforms linearly, exactly as $\bar\theta$, is

---

[63] The supercharges and the matter superfields are denoted by the same letter Q. It is hoped that this unfortunate coincidence will cause no confusion. The indices help us to work out what is meant in a given context. For supercharges we usually indicate the spinorial indices, using Greek letters from the beginning of the alphabet. The matter superfields carry the flavor indices (the Latin letters) typically f or g. However, Q_0 and $\bar{Q}_0$, with subscript 0, represent the starting purely bosonic configuration of the matter superfields.

$$(\bar{\theta}_0)^{\dot{\alpha}} = \bar{\zeta}^{\dot{\alpha}}[1 - 4i(\bar{\beta}\bar{\zeta})], \qquad \delta(\bar{\theta}_0)^{\dot{\alpha}} = \bar{\epsilon}^{\dot{\alpha}}. \tag{10.574}$$

The variable $\bar{\theta}_0$ joins the set $\{x_0, \theta_0\}$ describing the superinstanton center, $\{x_0, \theta_0\} \rightarrow \{x_0, \theta_0, \bar{\theta}_0\}$.

The supervinvariant combinations of the (anti)instanton moduli are

$$
\begin{aligned}
\bar{\beta}_{\text{inv}} &= \bar{\beta}\left[1 + 4i\left(\bar{\beta}\bar{\zeta}\right)\right] = \frac{\bar{\beta}}{\left[1 - 4i\left(\bar{\beta}\bar{\theta}_0\right)\right]}, \\
\rho_{\text{inv}}^2 &= \rho^2\left[1 + 4i\left(\bar{\beta}\bar{\zeta}\right)\right] = \frac{\rho^2}{\left[1 - 4i\left(\bar{\beta}\bar{\theta}_0\right)\right]},
\end{aligned}
\tag{10.575}
$$

and

$$[\Omega_{\text{inv}}]_{\dot{\beta}}^{\dot{\alpha}} \equiv \left[e^{-i\bar{\omega}_{\text{inv}}}\right]_{\dot{\beta}}^{\dot{\alpha}} = \exp\left\{-4i\left[\bar{\zeta}^{\dot{\alpha}}\bar{\beta}_{\dot{\gamma}} + \frac{1}{2}\delta_{\dot{\gamma}}^{\dot{\alpha}}\left(\bar{\zeta}\bar{\beta}\right)\right]\right\}\Omega_{\dot{\beta}}^{\dot{\gamma}}. \tag{10.576}$$

Let us emphasize that all these superinvariants became possible only due to the emergence of the collective coordinate $\bar{\zeta}$ conjugated to $\bar{Q}$ and, accordingly, $\bar{\theta}_0$ defined in Eq. (10.574).

We recall that in the theory with matter there is a nonanomalous R symmetry; see Section 10.15. We did not introduce the corresponding collective coordinate because it is not new in relation to the moduli of the flat vacua. Nevertheless, it is instructive to consider the R charges of the collective coordinates. We have collected these charges in Table 10.9.

From this table it can be seen that the only invariant with a vanishing R charge is ρ_{inv}^2. This fact has a drastic impact. In supersymmetric gluodynamics no combination of moduli was invariant under both supersymmetry and $U(1)_R$. This fact was used, in particular, in constructing the instanton measure; the expression for the measure comes out unambiguously. In a theory with matter, generally speaking, corrections to the instanton measure proportional to powers of $|v|^2 \rho_{\text{inv}}^2$ can emerge. And they do emerge, although all terms beyond the leading $|v|^2 \rho_{\text{inv}}^2$ term are accompanied by powers of the coupling constant g^2.

$\tilde{\theta}_{\dot{\alpha}}$ is defined in (10.534).

Let us now pass to the invariants constructed from the coordinates in the superspace and the moduli – analogs of (10.544). First, as a partner to $\tilde{\theta}_{\dot{\alpha}}$, we introduce

$$\tilde{x}_{\alpha\dot{\alpha}} = (x_L - x_0)_{\alpha\dot{\alpha}} + 4i\tilde{\theta}_\alpha\left(\bar{\zeta}\right)_{\dot{\alpha}} \tag{10.577}$$

with the following supertransformations,

Table 10.9. The R charges of the instanton collective coordinates					
Coordinates	θ_0	$\bar{\beta}$	η	$\bar{\theta}_0$	x_0	ρ
R charges	1	1	1	−1	0	0

$$\delta \tilde{\theta}_{\dot{\alpha}} = -4i \left(\bar{\varepsilon} \bar{\beta} \right) \tilde{\theta}_{\dot{\alpha}}, \quad \delta \tilde{x}_{\alpha \dot{\alpha}} = 4i \tilde{x}_{\alpha \dot{\gamma}} \bar{\beta}^{\dot{\gamma}} \bar{\varepsilon}_{\dot{\alpha}},$$

$$\delta \tilde{x}^2 = -4i \left(\bar{\varepsilon} \bar{\beta} \right) \tilde{x}^2. \tag{10.578}$$

It is easy to see that the superinvariants analogous to $\theta - \theta_0$ are

$$\rho^{-2} \tilde{\theta} \quad \text{and} \quad \rho^{-2} \tilde{x}^2. \tag{10.579}$$

One can exploit these invariants to generate in a straightforward manner various superfields with the collective coordinates switched on starting from the bosonic anti-instanton contribution. For example [76],

$$\text{Tr}\, W^\alpha W_\alpha \;\longrightarrow\; 96\, \tilde{\theta}^2 \frac{\rho^4}{\left(\tilde{x}^2 + \rho^2 \right)^4},$$

$$Q^2 \;\longrightarrow\; 2v^2 \frac{\tilde{x}^2}{\tilde{x}^2 + \rho^2}, \quad \bar{Q}^2 \longrightarrow 2\bar{v}^2 \frac{\tilde{x}^2}{\tilde{x}^2 + \rho^2} \tag{10.580}$$

<table>
<tr><td>Q^2 in the instanton field</td><td>The difference between $\tilde{x}$ and $x_L - x_0$ is unimportant in $\text{Tr}\, W^2$ because of the factor $\tilde{\theta}^2$. Thus, the superfield $\text{Tr}\, W^2$ remains intact: the matter fields do not alter the result for $\text{Tr}\, W^2$ obtained in SUSY gluodynamics. The difference between $\tilde{x}$ and</td></tr>
</table>

$x_L - x_0$ is very important, however, in the superfield Q^2. Indeed, putting $\theta_0 = \bar{\beta} = 0$ and expanding Eq. (10.580) in $\bar{\theta}_0$ we recover, in the linear approximation, the same 't Hooft zero modes as in Eq. (10.571):

$$\psi_\gamma^{\alpha \dot{f}} = 2\sqrt{2} i v (\bar{\theta}_0)^{\dot{f}} \delta_\gamma^\alpha \frac{\rho^2}{[(x - x_0)^2 + \rho^2]^{3/2}}. \tag{10.581}$$

Note that the superfield $\bar{Q}^{\dot{\alpha} f} \bar{Q}_{\dot{\alpha} f}$ contains a fermion component if $\theta_0 \neq 0$. What is the meaning of this fermion field? (We keep in mind that the Dirac equation for $\bar{\psi}$ has no zero modes.) The origin of this fermion field is the Yukawa interaction $(\psi \lambda) \bar{\phi}$ generating a source term in the classical equation for $\bar{\psi}$, namely, $\mathcal{D}_{\alpha \dot{\alpha}} \bar{\psi}^{\dot{\alpha}} \propto \lambda_\alpha \bar{\phi}$.

10.19.8 Orientation Collective Coordinates as Lorentz SU(2)$_R$ Rotations

<table>
<tr><td>Cf. Section 5.4.5.</td><td>In this section we focus on the orientation collective coordinates $\tilde{\omega}_{\dot{\beta}}^{\dot{\alpha}}$, in an attempt to explain their origin in the most transparent manner. The presentation below is adapted from [81]. The main technical problem with the introduction of orientations</td></tr>
</table>

is the necessity of untangling them from the nonphysical gauge degrees of freedom. The introduction of matter is the most straightforward way to make this untangling transparent.

First, we define a gauge invariant vector field W_μ

$$(W_\mu)^{\dot{f} \dot{g}} = \frac{i}{|v|^2} \left[\bar{\phi}^{\dot{f}} \mathcal{D}_\mu \phi^{\dot{g}} - (\mathcal{D}_\mu \bar{\phi}^{\dot{f}}) \phi^{\dot{g}} \right], \tag{10.582}$$

where $\dot{f}, \dot{g}$ are the SU(2) (sub)flavor indices, $\phi_{\dot{g}}$ is the lowest component of the superfield $Q_{\dot{g}}$, and the color indices are suppressed. In the flat vacuum (10.393) the field W_μ coincides with the gauge field A_μ (in the unitary gauge).

What are the symmetries of the flat vacuum? They obviously include the Lorentz $SU(2)_L \times SU(2)_R$ group. In addition, the vacuum is invariant under flavor $SU(2)$ rotations. Indeed, although $\phi_f^\alpha \propto \delta_f^\alpha$ is not invariant under the multiplication by the unitary matrix S_g^f, this noninvariance is compensated by a rotation in the gauge $SU(2)$ group. Another way to see this is to observe that the only modulus field $\phi_f^\alpha \phi_\alpha^f$ in the model at hand is a flavor singlet.

For the instanton configuration, see (10.519) for A_μ and (10.568) for ϕ, the field $W_{\alpha\dot\alpha}$ reduces to

> *The field $W_{\alpha\dot\alpha}$ is not to be confused with supergauge strength tensor W_α.*

$$(W_{\alpha\dot\alpha}^{\text{inst}})^{\dot f \dot g} = 2i\,\frac{\rho^2}{(x^2+\rho^2)^2}\left(x_\alpha^{\dot g}\delta_{\dot\alpha}^{\dot f} + x_\alpha^{\dot f}\delta_{\dot\alpha}^{\dot g}\right). \tag{10.583}$$

The next task is to examine the impact of $SU(2)_L \times SU(2)_R \times SU(2)_{\text{flavor}}$ rotations on W_μ^{inst}. It can be seen immediately that Eq. (10.583) is invariant under the action of $SU(2)_L$. It is also invariant under simultaneous rotations from $SU(2)_R$ and $SU(2)_{\text{flavor}}$. Thus, only one $SU(2)$ acts on W_μ^{inst} nontrivially. We can choose it to be the $SU(2)_R$ subgroup of the Lorentz group. This explains why we introduced the orientation coordinates through $\bar{M}\bar\omega$.

Note that the scalar fields play an auxiliary role in the construction presented; they allow one to introduce a relative orientation. At the end one can take the limit $v \to 0$ (the unbroken phase).

Another comment relates to higher groups. Extra orientation coordinates describe the orientation of the instanton $SU(2)$ within the given gauge group. Considering the theory in the Higgs regime allows one again to perform the analysis in a gauge-invariant manner. The crucial difference, however, is that the extra orientations, unlike the three $SU(2)$ orientations, are not related to exact symmetries of the theory in the Higgs phase. Generally speaking, the classical action becomes dependent on the extra orientations [81].

10.19.9 The Instanton Measure in the One-Flavor Model

The approximate nature of the instanton configuration at $\rho v \neq 0$ implies that the classical action is ρ-dependent. From 't Hooft's calculation [80] it is well known that in the limit $\rho v \to 0$ we have for the action (Section 5.4.12.1)

$$\frac{8\pi^2}{g^2} \longrightarrow \frac{8\pi^2}{g^2} + 4\pi^2 |v|^2 \rho^2. \tag{10.584}$$

> *Derivation of the 't Hooft term; cf. Section 5.4.12.*

The coefficient of $|v|^2 \rho^2$ is twice as large as in the 't Hooft case because there are two scalar (squark) fields in the model at hand, as compared with the one scalar doublet in 't Hooft's calculation. Let us recall that the $|v|^2 \rho^2$ term (which is often referred to in the literature as the 't Hooft term) is entirely due to a surface contribution in the action,

$$\int \mathcal{D}_\mu \bar\phi \mathcal{D}_\mu \phi \, d^4x = -\int \bar\phi \mathcal{D}^2 \phi \, d^4x + \int d\Omega_\mu\, \partial^\mu\left(\bar\phi \mathcal{D}_\mu \phi\right) d^4x$$

$$= \int d\Omega_\mu\, \partial^\mu\left(\bar\phi \mathcal{D}_\mu \phi\right) d^4x. \tag{10.585}$$

Since the 't Hooft term is saturated on the large sphere, the question of a possible ambiguity in its calculation immediately comes to mind. Indeed, what would happen if from the very beginning one used in the bosonic Lagrangian a kinetic term $-\bar{\phi}\mathcal{D}^2\phi$ rather than $\mathcal{D}_\mu\bar{\phi}\mathcal{D}_\mu\phi$? Alternatively, perhaps one could start from an arbitrary linear combination of these two kinetic terms; in fact, such a linear combination appears naturally in supersymmetric theories deriving from $\int d^4\theta \bar{Q} e^V Q$. These questions are fully legitimate. In Section 10.19.10 we demonstrate that the result quoted in Eq. (10.584) is unambiguous and correct: it is substantiated by a dedicated analysis.

The term $4\pi^2|v|^2\rho^2$ is obtained for the purely bosonic field configuration. For nonvanishing fermion fields an additional contribution to the action comes from the Yukawa term $(\psi\lambda)\bar{\phi}$. We could have calculated this term by substituting the classical field ϕ and the zero modes for ψ and λ. However, it is much easier to find the answer indirectly, by using the superinvariance of the action. Since ρ_{inv}^2 (see Eq. (10.575)) is the only appropriate invariant that can be constructed from the moduli, the action at $\bar{\theta}_0 \neq 0$ and $\bar{\beta} \neq 0$ becomes

$$\frac{8\pi^2}{g^2} + 4\pi^2|v|^2\rho_{\mathrm{inv}}^2. \tag{10.586}$$

To obtain the full instanton measure we proceed in the same way as in Section 10.19.6. In addition to the term $|v|^2\rho_{\mathrm{inv}}^2$ in the classical action, the change is due to the extra integration over $d^2\bar{\theta}_0$. From the general formula (10.548) we infer that this brings in an extra power of M_{PV}^{-1} and a normalization factor that can be read off from the expression (10.581). Overall, the extra integration takes the form (see Table 10.8),

$$\frac{1}{M_{\mathrm{PV}}}\frac{1}{8\pi^2 v^2 \rho^2}d^2\zeta = \frac{1}{M_{\mathrm{PV}}}\frac{1}{8\pi^2 v^2 \rho_{\mathrm{inv}}^2}d^2\bar{\theta}_0. \tag{10.587}$$

Note that the supertransformations (10.540) and (10.573) leave this combination invariant. Note also that the 't Hooft zero modes are chiral: it is $1/v^2$ that appears rather than $1/|v|^2$. The instanton measure "remembers" the phase of the vacuum expectation value of the scalar field. As we will see shortly, this is extremely important for recovering correct chiral properties for the instanton-induced superpotentials.

Combining the $d^2\bar{\theta}_0$ integration with the previous result (10.563) one arrives at

$$d\mu_{\text{one-flavor}} = \frac{1}{2^{11}\pi^4 v^2}M_{\mathrm{PV}}^5\left(\frac{8\pi^2}{g^2}\right)^2 \exp\left(-\frac{8\pi^2}{g^2} - 4\pi^2|v|^2\rho_{\mathrm{inv}}^2\right)$$

$$\times \frac{d\rho^2}{\rho^2}d^4 x_0\, d^2\theta_0\, d^2\bar{\beta}_{\mathrm{inv}}\, d^2\bar{\theta}_0. \tag{10.588}$$

This measure is explicitly invariant under supertransformations. Indeed, $d\rho^2/\rho^2$ reduces to $d\rho_{\mathrm{inv}}^2/\rho_{\mathrm{inv}}^2$, up to a subtlety at the singular point $\rho^2 = 0$, to be discussed later.

Let us recall that the expression (10.588) is obtained under the assumption that the parameter $\rho^2|v|^2 \ll 1$ and so accounts for the zero- and first-order terms in the expansion of the action in this parameter. Summing up the higher orders leads to some function of $\rho_{\mathrm{inv}}^2|v|^2$ in the exponent.

10.19.10 Verification of the 't Hooft Term

In the previous section we mentioned the ambiguity in the 't Hooft term due to its surface nature. Discussion of the surface terms calls for careful consideration of the boundary conditions. However, there is an alternative route via the scattering amplitude technique [82]; calculation of the scattering amplitude takes care of the correct boundary conditions automatically.

As a simple example let us consider the nonsupersymmetric SU(2) model with one Higgs doublet ϕ^α. Our task is to demonstrate that the instanton-induced effective interaction of the ϕ field is

$$\Delta \mathcal{L} = \int d\mu \exp \left\{ -2\pi^2 \rho^2 \left[\bar{\phi}(x)\phi(x) - |v|^2 \right] \right\}, \tag{10.589}$$

where $d\mu$ is the instanton measure of the model. Note that this includes, in particular, the factor $\exp(-2\pi^2 \rho^2 |v|^2)$.

We want to compare two alternative calculations of a particular amplitude – one based on the instanton calculus and the other based on the effective Lagrangian (10.589). Let us start from the emission of one physical Higgs particle by a given instanton with collective coordinates fixed. The interpolating field σ for the physical Higgs can be defined as

$$\sigma(x) = \frac{1}{\sqrt{2}|v|} \left[\bar{\phi}(x)\phi(x) - |v|^2 \right]. \tag{10.590}$$

The Lagrangian (10.589) implies that the emission amplitude A is equal to

$$A = -2\sqrt{2}\pi^2 \rho^2 |v|. \tag{10.591}$$

Let us now calculate the expectation value of $\sigma(x)$ in the instanton background. In the leading (classical) approximation,

$$\langle \sigma(x) \rangle_{\text{inst}} = \frac{1}{\sqrt{2}|v|} \left[\bar{\phi}_{\text{inst}}(x)\phi_{\text{inst}}(x) - |v|^2 \right] = -\frac{|v|}{\sqrt{2}} \frac{\rho^2}{x^2 + \rho^2}. \tag{10.592}$$

Taking $x \gg \rho$ we find that

$$\langle \sigma(x) \rangle_{\text{inst}} \to -2\sqrt{2}\pi^2 \rho^2 |v| \frac{1}{4\pi^2 x^2}. \tag{10.593}$$

The first factor is the emission amplitude A and the second factor is the free particle propagator.

Thus, the effective Lagrangian (10.589) is verified in the order linear in σ. To verify the exponentiation it is sufficient to show the factorization of the amplitude for the emission of an arbitrary number of σ particles. In the classical approximation this factorization is obvious.

10.19.11 Cancelation of the Quantum Corrections to the Measure

So far, our analysis of the instanton measure has been in essence classical. Strictly speaking, though, it would be better to call it semiclassical. Indeed, let us not forget that calculation of the pre-exponent is related to the one-loop corrections.

In our case the pre-exponent is given by an integral over the collective coordinates. In nonsupersymmetric theories the pre-exponent is not exhausted by this integration – the nonzero modes contribute as well. Here we will show that the nonzero modes cancel out in SUSY theories. Moreover, in the unbroken phase the cancelation of the nonzero modes persists to any order in perturbation theory and even beyond, i.e., nonperturbatively. Thus, we will obtain the extension of the F term nonrenormalization theorem [44] to the instanton background. The specific feature of this background, responsible for the extension, is the preservation of half the supersymmetry. Note that in the Higgs phase the statement of cancelation is also valid in terms of zero order and of first order in the parameter $\rho^2 |v|^2$.

In the first loop the cancelation is fairly obvious. Indeed, in supersymmetric gluodynamics the differential operator L_2 defining the mode expansion has the same form, see Eq. (10.549), for both the gluon and gluino fields,

This is why the nonzero modes cancel.

$$-\mathcal{D}^{\alpha\dot\alpha}\mathcal{D}_{\beta\dot\alpha}a_n^{\beta\dot\gamma} = \omega_n^2 a_n^{\alpha\dot\gamma},$$
$$-\mathcal{D}^{\alpha\dot\alpha}\mathcal{D}_{\beta\dot\alpha}\lambda_n^\beta = \omega_n^2 \lambda_n^\alpha. \tag{10.594}$$

The residual supersymmetry (generated by $\bar{Q}_{\dot\alpha}$) is reflected in L_2 in the absence of free dotted indices. Therefore, if the boundary conditions respect the residual supersymmetry – which we assume to be the case – the eigenvalues and eigenfunctions are the same for $a^{\alpha 1}$, $a^{\alpha 2}$, and λ^α. For the field $\bar{\lambda}^{\dot\alpha}$ the relevant operator is $-\mathcal{D}^{\alpha\dot\alpha}\mathcal{D}_{\alpha\dot\beta} = -\frac{1}{2}\delta_{\dot\beta}^{\dot\alpha}\mathcal{D}^{\alpha\dot\gamma}\mathcal{D}_{\alpha\dot\gamma}$, where

$$-\mathcal{D}^{\alpha\dot\gamma}\mathcal{D}_{\alpha\dot\gamma}\bar{\lambda}_n^{\dot\alpha} = \omega_n^2 \bar{\lambda}_n^{\dot\alpha}. \tag{10.595}$$

This equation shows[64] that the modes of $\bar{\lambda}$ coincide with those of the scalar field ϕ in the same representation of the gauge group,

$$-\mathcal{D}^{\alpha\dot\gamma}\mathcal{D}_{\alpha\dot\gamma}\phi_n = \omega_n^2 \phi_n. \tag{10.596}$$

Moreover, *all* nonzero modes are expressible in terms of ϕ_n (this nice feature was noted in [83]). This is evident for $\bar{\lambda}^1$ and $\bar{\lambda}^2$. The nonzero modes of a and λ are given by

$$a_n^{\alpha\dot1(\dot\beta)} = a_n^{\alpha\dot2(\dot\beta)} = \frac{1}{\omega_n}\mathcal{D}^{\alpha\dot\beta}\phi_n, \qquad \lambda_n^{\alpha(\dot\beta)} = \frac{1}{\omega_n}\mathcal{D}^{\alpha\dot\beta}\phi_n. \tag{10.597}$$

Thus, the integration over a produces $1/\omega_n^4$ for each given eigenvalue. The integration over λ and $\bar{\lambda}$ produces ω_n^2. The balance is restored by the contribution of the scalar ghosts, which provides the remaining ω_n^2.

The same cancelation is extended to the matter sector. In every supermultiplet each mode of the scalar field ϕ is accompanied by two modes in ψ^α and $\bar\psi^{\dot\alpha}$; see Eq. (10.597). Correspondingly, one obtains ω_n^2/ω_n^2 for each eigenvalue.

From the above one-loop discussion it is clear that the cancelation is due to boson–fermion pairing enforced by the residual supersymmetry of the instanton background. This same supersymmetry guarantees cancelation in higher loops. On general symmetry grounds corrections, if present, could not be functions of the collective coordinates: it has been shown previously that no appropriate invariants exist. Therefore, the only possibility left is a purely numerical series in powers of g^2.

---

[64] The equality $\mathcal{D}^{\alpha\dot\alpha}\mathcal{D}_{\alpha\dot\beta} = (\frac{1}{2})\delta_{\dot\beta}^{\dot\alpha}\mathcal{D}^{\alpha\dot\gamma}\mathcal{D}_{\alpha\dot\gamma}$ exploits the fact that $\bar{G}_{\dot\alpha\dot\beta} = 0$ for the anti-instanton.

In fact, not even this type of series appears. Indeed, let us consider the two-loop supergraph in the instanton background. It was presented in Fig. 10.1 in Section 10.8, where each line is to be understood as the gluon or gluino Green's function in the instanton background field. This graph has two vertices. Its contribution is equal to the integral over the supercoordinates of both vertices, i.e., $\{x, \theta, \bar{\theta}\}$ and $\{x', \theta', \bar{\theta}'\}$, respectively. If we integrate over the supercoordinates of the second vertex and over the coordinates x and θ (but not $\bar{\theta}$!) of the first vertex then the graph can be presented as the integral $\int d^2\bar{\theta}\, F(\bar{\theta})$. The function F is invariant under simultaneous supertransformations of $\bar{\theta}$ and the instanton collective coordinates. As was shown in Section 10.19.5, in supersymmetric gluodynamics there are no invariants containing $\bar{\theta}$. Therefore, the function $F(\bar{\theta})$ can only be a constant; thus the integration over $\bar{\theta}$ yields zero [84].

The proof above is a version of arguments based on the residual supersymmetry. Indeed, no invariant can be built from $\bar{\theta}$ because there is no collective coordinate $\bar{\theta}_0$. The absence of $\bar{\theta}_0$ is, in turn, a consequence of the residual supersymmetry. The introduction of matter in the Higgs phase changes the situation. At $v \neq 0$ no residual supersymmetry survives. In terms of the collective coordinates this is reflected in the emergence of $\bar{\theta}_0$. Correspondingly, the function $F(\bar{\theta})$ becomes a function of the invariant $\bar{\theta} - \bar{\theta}_0$ (see Eq. (10.577)), and the integral does not vanish.

Therefore, in theories with matter, in the Higgs phase the instanton does acquire corrections. However, these corrections vanish [85] in the limit $|v|^2\rho^2 \to 0$. Technically, the invariant above containing $\bar{\theta}$ disappears at small v because $\bar{\theta}_0$ is proportional to $1/v$.

Summarizing, the instanton measure acquires no quantum corrections in SUSY gluodynamics or in the unbroken phase, in the presence of matter. In the Higgs phase, corrections start with the terms $g^2|v|^2\rho^2$.

An important comment is in order here regarding the discussion above. Our proof assumes that there exists a supersymmetric ultraviolet regularization of the theory. At one-loop level the Pauli–Villars regulators do the job. In higher loops the regularization is achieved by a combination of the Pauli–Villars regulators and higher-derivative terms. We do not use this regularization explicitly; rather, we rely on the theorem that it exists.[*] This is all we need. As for infrared regularization, it is provided by the instanton field itself. Indeed, at fixed collective coordinates all eigenvalues are nonvanishing. The zero modes should not be included in the set when the collective coordinates are fixed.

Exercise

10.19.1 Find the overall number of the fermion zero modes in the instanton background in the model considered in Section 10.16.7.1 (SU(5) with two generation and $\mathcal{W} = 0$).

---

[*] This regularization was developed and explicitly implemented in [100].

10.20 Affleck–Dine–Seiberg Superpotential

The stage is set, and we are ready to apply the formalism outlined above in concrete problems that arise in super-Yang–Mills theories. In this section we start by discussing applications of instanton calculus that are of practical interest. Our first problem is a calculation of the Affleck–Dine–Seiberg superpotential in one-flavor SQCD.

The ADS superpotential is a crucial element in many problems in $\mathcal{N} = 1$.

The classical structure of SQCD, with gauge group SU(2) and one flavor, was discussed in Section 10.15. The model has one modulus,

$$\Phi = \sqrt{\tfrac{1}{2} Q_\alpha^f Q_f^\alpha}. \tag{10.598}$$

In the absence of a superpotential all vacua with different Φ are degenerate. The degeneracy is not lifted to any finite order of perturbation theory. As shown below it is lifted nonperturbatively [66] by an instanton-generated superpotential $\mathcal{W}(\Phi)$.

Far from the origin of the moduli space, where $|\Phi| \gg \Lambda$, the gauge SU(2) is spontaneously broken, the theory is in the Higgs regime, and the gauge bosons are heavy. In addition the gauge coupling is small, so that a quasiclassical treatment is reliable. At weak coupling the leading nonperturbative contribution is due to instantons. Thus, our task is to find the instanton-induced effects.

The exact R invariance of the model (Section 10.15) is sufficient to establish the functional form of the effective superpotential $\mathcal{W}(\Phi)$:

$$\mathcal{W}(\Phi) \propto \frac{\Lambda_{\text{one-flavor}}^5}{\Phi^2}, \tag{10.599}$$

where the power of Φ is determined by its R charge ($R_\Phi = -1$; see the $\tilde{R}$ charge of Q_f^α in Table 10.4, Section 10.15) and the power of Λ is fixed by dimensional considerations. Here we have introduced the notation[65]

$$\Lambda_{\text{one–flavor}}^5 = \frac{e^{-8\pi^2/g^2}}{Z g^4} (M_{\text{PV}})^5. \tag{10.600}$$

To see that one instanton induces this superpotential, we consider an instanton transition in a background field $\Phi(x_L, \theta)$ weakly depending on the superspace

---

[65] The Z factor of the matter fields is introduced in Eq. (10.600). To avoid confusion I have to stress that there are two different definitions of the Z factors in the Wilsonean procedure. The first and more common definition is that in the UV one starts from the Lagrangian (symbolically) $\mathcal{L}_{\text{UV}} = \int d^4\theta \bar{Q}Q + \int d^2\theta\, m_0 Q\tilde{Q}$, and then evolves down to normalization point μ to obtain

$$\mathcal{L}_\mu = \int d^4\theta\, Z(\mu) \bar{Q}Q + \int d^2\theta\, m_0 Q\tilde{Q}.$$

Then $m_{\text{bare}} = m_0$ and $m_{\text{phys}} = m_0/Z$. Another definition that I consistently use here is the same as in Ref. [62] and our previous works. Namely, in the ultraviolet (i.e., at $\mu = M_{\text{PV}}$) we define

$$\mathcal{L}_{\text{UV}} = \int d^4\theta\, Z_0\, \bar{Q}Q + \int d^2\theta\, m_0\, Q\tilde{Q}.$$

The Z factor above, which I denoted here as Z_0, is chosen in such a way, that when we evolve the above UV Lagrangian down to a low renormalization point μ the Z factor reduces to unity in the infrared. With this definition $m_{\text{bare}} = m_0/Z_0$ and $m_{\text{phys}} = m_0$. In Eqs. (10.608), (10.609), and (10.613) the second definition is used. The Z factors appearing in these equations should be understood as Z_0, but I omit this subscript, since the first definition is never mentioned.

coordinates. To this end one generalizes the result (10.588), which assumes that $\Phi = v$ at distances much larger than ρ, to a variable superfield Φ:

$$d\mu = \frac{1}{2^5} \frac{\Lambda^5_{\text{one-flavor}}}{\Phi^2(x_0, \theta_0)} \exp\left(-4\pi^2 \bar{\Phi}\Phi\rho^2_{\text{inv}}\right) \frac{d\rho^2}{\rho^2} d^4 x_0 d^2\theta_0 d^2\bar{\beta} d^2\bar{\theta}_0. \tag{10.601}$$

There exist many alternative ways to verify that this generalization is correct. For instance, one could calculate the propagator of the quantum part of $\Phi = v + \Phi_{\text{qu}}$ using a constant background $\Phi = v$ in the measure; see Section 10.19.10 for more details.

The effective superpotential is obtained by integrating over $\rho, \bar{\beta}$, and $\bar{\theta}_0$. Since these variables enter the measure only through ρ^2_{inv}, at first glance the integral would seem to be zero; indeed, changing the variable ρ^2 to ρ^2_{inv} makes the integrand independent of $\bar{\beta}$ and $\bar{\theta}_0$. The integral does not vanish, however. The loophole is due to the singularity at $\rho^2_{\text{inv}} = 0$. To resolve the singularity let us integrate first over the fermionic variables. For an arbitrary function $F(\rho^2_{\text{inv}})$ the integral takes the form

$$\int \frac{d\rho^2}{\rho^2} d^2\bar{\beta} d^2\bar{\theta}_0 F\left(\rho^2(1 + 4i\bar{\beta}\bar{\theta}_0)\right) = \int \frac{d\rho^2}{\rho^2} 16\rho^4 F''(\rho^2) = 16 F(\rho^2 = 0). \tag{10.602}$$

The integration over ρ^2 was performed by integrating by parts twice. It can be assumed that $F(\rho^2 \to \infty) = 0$. It can be seen that the result depends only on the zero-size instantons. In other words,

$$\frac{d\rho^2}{\rho^2} d^2\bar{\beta} d^2\bar{\theta}_0 F(\rho^2_{\text{inv}}) = 16 d\rho^2_{\text{inv}} \delta(\rho^2_{\text{inv}}) F(\rho^2_{\text{inv}}). \tag{10.603}$$

This expression presents historically the first example [76] of what is currently called instanton (or Nekrasov) localization [101]

The instanton-generated superpotential is

$$\mathcal{W}_{\text{inst}}(\Phi) = \frac{\Lambda^5_{\text{one-flavor}}}{\Phi^2}. \tag{10.604}$$

The result presented in Eq. (10.604) bears a topological nature: it does not depend on the particular form of the integrand $F(\rho^2_{\text{inv}})$ since the integral is determined by the value of the integrand at $\rho^2 = 0$; the integrand is given by the exponent only at small ρ^2. No matter how it behaves as a function of ρ^2, the formula for the superpotential is the same provided that the integration over ρ^2 is convergent at large ρ^2.

Technically, the saturation at $\rho^2 = 0$ makes the calculation self-consistent (remember, at $\rho^2 = 0$ the instanton solution becomes exact in the Higgs phase) and explains why the result (10.604) acquires no perturbative corrections in higher orders.

We see that in the model at hand the instanton does indeed generate a superpotential that lifts the vacuum degeneracy. (This superpotential bears the name of Affleck, Dine, and Seiberg, ADS for short.) This result is *exact both perturbatively and nonperturbatively*.

In the absence of a tree-level superpotential the induced superpotential leads to a runaway vacuum – the lowest energy state is achieved at an infinite value of Φ. One can stabilize the theory by adding the mass term $m\Phi^2$ to the classical superpotential. The total superpotential then takes the form

$$\mathcal{W}(\Phi) = m\Phi^2 + \mathcal{W}_{\text{inst}}(\Phi). \tag{10.605}$$

One can trace the origin of the second term to the anomaly in (10.462) in the original full theory (i.e., the theory before the gauge fields are integrated out).

Determining the critical points of the ADS superpotential we find two supersymmetric vacua at

$$\langle \Phi^2 \rangle = \pm \left(\frac{\Lambda^5_{\text{one–flavor}}}{m} \right)^{1/2}. \tag{10.606}$$

Now, with the ADS superpotential in hand, we are able to calculate the gluino condensate using the Konishi relation (10.450) (see Section 10.16.6.1), which, in the present case, implies that

$$\langle \text{Tr}\, \lambda^2 \rangle = 16\pi^2 m \langle \Phi^2 \rangle$$

$$= \pm 16\pi^2 \left[m\Lambda^5_{\text{one–flavor}} \right]^{1/2} = \pm 16\pi^2 \left[\frac{me^{-8\pi^2/g^2}}{Zg^4} (M_{\text{PV}})^5 \right]^{1/2}. \tag{10.607}$$

Our convention for the Z factors of the matter fields is as follows:

<table>
<tr><td>

The dependence of the gluino condensate on m_{bare} is holomorphic.

</td><td>

$$\mathcal{L}_{\text{matter}} = \sum_i Z_i \int d^2\theta\, d^2\bar{\theta}\, \bar{Q}_i e^V Q_i + \left[\int d^2\theta\, \mathcal{W}(Q_i) + \text{H.c.} \right]. \tag{10.608}$$

Then the bare quark mass m_{bare} is given by

$$m_{\text{bare}} = \frac{m}{Z}. \tag{10.609}$$

</td></tr>
</table>

Therefore, the gluino condensate dependence on m_{bare} is holomorphic. In fact its square root dependence on m_{bare} bare is an exact statement [61]. It follows from an extended R symmetry that requires m_{bare} to rotate with R charge $+4$ (see the last column in Table 10.4, Section 10.15). Given that the R charge of λ^2 is $+2$, the exact law $\lambda^2 \propto \sqrt{m_{\text{bare}}}$ ensues immediately.

This allows one to pass to large m_{bare}, where the matter field can be viewed as one of the regulators. Setting $m_{\text{bare}} = M_{\text{PV}}$ we return to supersymmetric gluodynamics, recovering Eq. (10.380) considered in Section 10.14.[66] There we passed from SU(2) to SU(N) with arbitrary N.

In addition to its holomorphic dependence on m_{bare}, the gluino condensate depends holomorphically on the regulator mass M. Regarding the gauge coupling, the factor $1/g^2$ in the exponent can and must be complexified according to Eq. (10.369), but in the pre-exponential factor it is $\text{Re}\, g^{-2}$ that enters. This is the so-called holomorphic anomaly [86].

10.21 Novikov–Shifman–Vainshtein–Zakharov β Function

The exact results obtained above, in conjunction with renormalizability, can be converted into exact relations for the β functions, usually referred to as the Novikov–Shifman–Vainshtein–Zakharov (NSVZ) β functions.

---

[66] We also learn that the ultraviolet cutoff M_{uv} appearing in Section 10.14 must be identified with M_{PV}.

10.21.1 Exact β Function in Supersymmetric Gluodynamics

Consider first supersymmetric gluodynamics. The gauge group G can be arbitrary. The gluino condensate (10.380) is a physically measurable quantity. As such, it must be expressible through a combination of parameters – the bare coupling constant and the ultraviolet cutoff – that is cutoff independent. The renormalizability of the theory implies that the ultraviolet cutoff M_{PV} must conspire with the bare coupling g to make the gluino condensate expression independent of M_{PV}. In other words, g should be understood as a function $g(M_{PV})$ such that the combination entering the gluino condensate (10.381) does not depend on M_{PV}. Let us write it as follows:

$$(M_{PV})^{3T_G} \left[\frac{1}{g^2(M_{PV})} \right]^{T_G} \exp\left[-\frac{8\pi^2}{g^2(M_{PV})} \right] = \text{const}, \qquad (10.610)$$

where I have replaced the parameter N in (10.380), relevant for $SU(N)_{\text{gauge}}$, by T_G, making this expression valid for arbitrary gauge group G.

That the left-hand side of (10.610) must be independent of M_{PV} gives the law for the running of the gauge coupling, $\alpha(\mu) = g^2(\mu)/(4\pi)$ (in the Pauli–Villars scheme). The result can be formulated, of course, in terms of the exact β function. Taking the logarithm and differentiating with respect to $\ln M_{PV}$, we arrive at

NSVZ for supersymmetric gluodynamics

$$\beta(\alpha) \equiv \frac{d\,\alpha(M_{PV})}{d \ln M_{PV}} = -\frac{3T_G\alpha^2}{2\pi}\left(1 - \frac{T_G\alpha}{2\pi}\right)^{-1}. \qquad (10.611)$$

In the derivation above we have assumed that both the gauge coupling g and the Pauli–Villars regulator mass M_{PV} are real.

10.21.2 Theories with Matter

First, let us return to Eq. (10.607), which is valid in $SU(2)$ SQCD with one flavor (two subflavors). This expression implies that

$$\frac{m\, e^{-8\pi^2/g^2}}{Z g^4}(M_{PV})^5 = \text{const}. \qquad (10.612)$$

Here m is the physical (s)quark mass, and as such is M_{PV}-independent. At the same time, the bare coupling g and the Z factor do depend on M_{PV}. Taking the logarithm of the left-hand side, differentiating with respect to $\ln M_{PV}$, and using the fact that the anomalous dimension of the ith flavor can be defined as

$$\gamma_i \equiv -\frac{d \ln Z_i}{d \ln M_{PV}}, \qquad (10.613)$$

NSVZ in SQCD with arbitrary super-Yukawa terms, cf. Eq. (10.456)

we arrive at

$$\beta(\alpha) = -\frac{\alpha^2}{2\pi}\frac{5+\gamma}{1-\alpha/\pi} = -\frac{\alpha^2}{2\pi}\frac{3\times 2 - (1-\gamma)}{1-2\alpha/2\pi}. \qquad (10.614)$$

The second equality here is arranged to reveal the nature of the various coefficients, making possible an easy transition from $SU(2)_{\text{gauge}}$ and one flavor to an arbitrary gauge group G and an arbitrary set of flavors. To this end we note that $T_{SU(2)} = 2$

and $T_{\text{fund}} = 1$ and compare Eq. (10.614) with the general expressions (10.456) and (10.611). The following NSVZ formula ensues:[67]

$$\beta(\alpha) \equiv \frac{d\,\alpha(M_{\text{PV}})}{d\ln M_{\text{PV}}} = -\frac{\alpha^2}{2\pi}\left[3T_G - \sum_i T(R_i)(1 - \gamma_i)\right]\left(1 - \frac{T_G\alpha}{2\pi}\right)^{-1}. \quad (10.615)$$

A few explanatory remarks are in order with regard to this formula. The matter fields are in an arbitrary representation R. This representation can be reducible, so that $R = \sum R_i$. The sum in (10.615) runs over all irreducible representations, or, equivalently, over all flavors. Besides the gauge interaction, the matter fields can have arbitrary (self-)interactions through super-Yukawa terms, i.e., an arbitrary renormalizable superpotential is allowed. Such a superpotential would not show up explicitly in the NSVZ formula (10.615). It would be hidden in the anomalous dimensions, which certainly do depend on the presence or absence of a superpotential. In contradistinction to the pure gauge case, Eq. (10.615) does not *per se* fix the running of the gauge coupling; rather, it expresses the running of the gauge coupling via the anomalous dimensions of the matter fields (10.613). The denominator in Eq. (10.615) is due to the holomorphic anomaly [86] mentioned in passing in Section 10.20.

It is instructive to examine how the general formula (10.615) works in some particular cases. Let us start from theories with extended supersymmetry, $\mathcal{N} = 2$. The simplest such theory can be presented as an $\mathcal{N} = 1$ theory containing one matter field in the adjoint representation (which enters the same extended $\mathcal{N} = 2$ supermultiplet as the gluon field; see Section 10.18). Therefore, its Z factor equals $1/g^2$ and γ equals β/α. In addition, we can allow for some number of matter hypermultiplets in arbitrary color representations (remembering that every hypermultiplet consists of two $\mathcal{N} = 1$ chiral superfields). The $\mathcal{N} = 2$ supersymmetry leads to $Z = 1$ for all hypermultiplets. Indeed, for $\mathcal{N} = 2$ the Kähler potential and, hence, the kinetic term of the matter fields are in one-to-one correspondence with the superpotential. The latter is not renormalized perturbatively owing to the $\mathcal{N} = 1$ supersymmetry. Hence, the Kähler potential for the hypermultiplets is not renormalized either, implying that $Z = 1$.

Taking into account these facts, we can derive from Eq. (10.615) the following gauge coupling β function:

$$\beta_{\mathcal{N}=2}(\alpha) = -\frac{\alpha^2}{2\pi}\left[2T_G - \sum_i T(R_i)\right]. \quad (10.616)$$

Here the summation runs over the $\mathcal{N} = 2$ matter hypermultiplets. This result proves that the β function is one-loop in $\mathcal{N} = 2$ theories.

We can now make one step further, passing to $\mathcal{N} = 4$. In terms of $\mathcal{N} = 2$ this theory corresponds to one matter hypermultiplet in the adjoint representation. Substituting $\sum T(R_i) = 2T_G$ into Eq. (10.616) produces a vanishing β function. Thus the $\mathcal{N} = 4$ theory is finite.

In fact Eq. (10.616) shows that the class of finite theories is much wider. Any $\mathcal{N} = 2$ theory whose matter hypermultiplets satisfy the condition $2T_G - \sum_i T(R_i) = 0$

---

[67] The relation between the NSVZ β function and standard perturbative calculations based on dimensional reduction is discussed in, e.g., [87] and [100].

is finite. An example is provided by the T_G hypermultiplets in the fundamental representation.

10.22 The Witten Index

The spontaneous breaking of supersymmetry is a rather subtle issue. As we already know, the order parameter is the vacuum energy. Supersymmetry is spontaneously broken if and only if the vacuum energy is strictly higher than zero. The presence of a Goldstino is a clear-cut signature of this spontaneous breaking. Though weakly coupled theories are usually amenable to solution this is not the case for strongly coupled theories, in which it is typically very hard (if possible at all) to establish directly the positivity of the vacuum energy or the Goldstino existence and its coupling to the supercurrent. Even in weakly coupled theories it may happen that the supersymmetry is unbroken to any finite order in perturbation theory but an exponentially small shift of the vacuum energy is induced by nonperturbative effects (e.g., instantons).

Therefore, it is highly desirable to develop a method which could tell us beforehand that this or that given theory has an *exactly vanishing* ground state energy and, therefore, under no circumstances can be considered as a candidate for spontaneous supersymmetry breaking. Such a method was devised by Witten [60], who suggested that one should define an index (now known as Witten's index) that, for each supersymmetric theory, counts the number of supersymmetric vacuum states.

When mathematicians and physicists speak of an index they mean a quantity (usually integer-valued) that does not change under any *continuous* deformation of the parameters defining the object under consideration. Thus, we are dealing with a topological characteristic. An index well-known to theoretical physicists for many years is the Dirac operator index. Supersymmetry allows one to introduce an index technically defined as

$$I_W = \mathrm{Tr}(-1)^F \equiv \sum_a \langle a | (-1)^F | a \rangle, \qquad (10.617)$$

where the sum runs over all physical states of the theory under consideration and F is the fermion number operator. To discretize the spectrum one can think of the theory as being formulated in a large box; this is a routine procedure in many texts on quantum field theory.

Why is (10.617) an index?

In any supersymmetric theory there are several conserved supercharges. One can always define a linear combination Q such that $Q^\dagger = Q$ and $H = 2Q^2$, where H is the Hamiltonian of the system. We will restrict ourselves to the sector of Hilbert space with vanishing total spatial momentum, $\vec{P} = 0$. This can be done without loss of generality.

Since $Q^2 = \frac{1}{2}H$, any state with vanishing energy must nullify upon the action of Q, i.e., $Q|a_{E=0}\rangle = 0$. If $E > 0$, however, then the action of Q on a bosonic state $|b\rangle$ produces a fermionic state $|f\rangle$ with the same energy and vice versa,[68]

---

[68] The set $|b\rangle$ and $|f\rangle$ is by no means restricted to one-particle states. It includes all states of the theory. The fermion number of the $|b\rangle$ states is even, while that of the $|f\rangle$ states is odd.

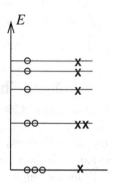

Fig. 10.3 A possible pattern for the spectrum of a supersymmetric theory. The closed circles indicate bosonic states, with even fermion number, while the crosses indicate fermion states, with odd fermion number.

$$Q|b\rangle = \sqrt{\tfrac{1}{2}E}|f\rangle, \qquad Q|f\rangle = \sqrt{\tfrac{1}{2}E}|b\rangle, \tag{10.618}$$

where both states are normalized to unity, $\langle b|b \rangle = \langle f|f \rangle = 1$. Thus, all positive energy states are subject to this boson–fermion degeneracy, a fact that we have already discussed more than once. Owing to this degeneracy the Witten index actually reduces to

$$I_W = n^b_{E=0} - n^f_{E=0}, \tag{10.619}$$

where $n^b_{E=0}$ and $n^f_{E=0}$ are the numbers of bosonic and fermionic zero-energy states, respectively; the zero-energy states (vacua) need not come in pairs. (Moreover, in more than two dimensions in the infinite-volume limit all vacua are bosonic in theories with a mass gap.)

We still have to answer the questions why the Witten index is independ of continuous deformations of the parameters of the theory and which particular deformations can be considered as continuous.

The (discretized) spectrum of a supersymmetric theory is symbolically depicted in Fig. 10.3. In this figure there are four zero-energy states, three bosonic and one fermionic, implying that $I_W = 2$. What happens when we vary the parameters of the theory, such as the box volume, the mass terms in the Lagrangian, the coupling constants, etc.? Under such deformations the states of the system breathe; they can come to or leave zero. As long as the Hamiltonian is supersymmetric, however, once a bosonic state, say, descends to zero it must be accompanied by its fermionic counterpartner, so that I_W does not change. And vice versa, the lifting of states from zero can occur only in boson–fermion pairs (Fig. 10.4). Thus, as was realized by Witten [60], I_W is indeed invariant under any continuous deformation of the theory.

A continuous deformation, what does that mean? Gradually changing the volume of a "large" box (i.e., making it smaller) is a continuous procedure. Changing the values of parameters in front of various terms in the Lagrangian is a continuous procedure too. Adding mass terms to those theories where they are allowed is a continuous deformation of the theory. Indeed, the mass terms are quadratic in the fields – and are thus of the same order as the kinetic terms. However, adding terms of higher orders than those already present in the Lagrangian is potentially

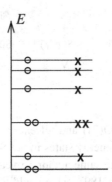

Fig. 10.4 The spectrum of Fig. 10.3 "breathes" as a result of parameter deformations. Depicted is the uplift of two states from zero. Once a state leaves zero, so – of necessity – does its degenerate superpartner.

a discontinuous deformation: "extra" vacua can come in from infinity. If the superpotential is, say, quadratic in the fields then adding a cubic term will change I_W.

If $I_W \neq 0$ then the theory has at least I_W zero-energy states. The existence of a zero-energy vacuum state is the necessary and sufficient condition for a supersymmetry to be realized linearly, i.e., to stay unbroken. Thus, in search of dynamical supersymmetry breaking one should focus on $I_W = 0$ theories.

Now when we know that I_W is invariant under continuous deformations, we can take advantage of this and deform supersymmetric theories as we see fit (without losing the supersymmetry) in order to simplify them to an extent such that a reliable calculation of the zero-energy states becomes possible.

10.22.1 Witten's Index in Super-Yang–Mills Theories

Witten's index in super-Yang–Mills theories with arbitrary nonchiral matter

Witten's index was first calculated for super-Yang–Mills theories with arbitrary Lie groups, without matter. Its value is

$$I_W = T_G. \tag{10.620}$$

The values of T_G for various semi-simple Lie groups are collected in Table 10.10. In theories where the gauge group is a product of semi-simple groups, $G = G_1 \times G_2 \times \ldots$, Witten's index is given by

$$I_W = T_{G_1} \times T_{G_2} \times \ldots \tag{10.621}$$

Two alternative calculations of I_W are known in the literature. The first is the original calculation of Witten, who deformed the theory by putting it into a finite three-dimensional volume $V = L^3$. The length L is such that the coupling $\alpha(L)$ is weak, $\alpha(L) \ll 1$. The field-theoretical problem of counting the number of zero-energy states becomes, in the limit $L \to 0$, a quantum-mechanical problem of counting the gluon and gluino zero modes. In practice, the problem is still quite tricky because of the subtleties associated with quantum mechanics on group spaces.

The story has a dramatic development. The result obtained in the original paper in [60] was $I_W = r + 1$, where r stands for the rank of the group. For the unitary and simplectic groups $r + 1$ coincides with T_G. However, for the orthogonal groups

Table 10.10. The dual Coxeter number (equal to one-half the Dynkin index) for various groups

Group	SU(N)	SO(N)	Sp($2N$)	G_2	F_4	E_6	E_7	E_8
T_G	N	$N-2$	$N+1$	4	9	12	18	30

(starting from SO(7)) and all exceptional groups, $r + 1$ is smaller than T_G. The overlooked zero-energy states in the SO(N) quantum mechanics of the zero modes were found by the same author 15 years later! (See [88]). Further useful comments can be found in [89], where additional states in the exceptional groups were exhibited.

An alternative calculation of I_W [61, 90] resorts to another deformation, which, in a sense, is an opposite extreme. Adding heavy matter fields, in the fundamental representation (with quadratic superpotential), to super-Yang–Mills theories obviously does not change the Witten index of the latter, since heavy matter has no impact on the zero-energy states. In the limit of a very large mass parameter one can integrate out all heavy matter fields, thus returning to the original super-Yang–Mills theory. On the other hand, I_W stays intact under variations of the mass parameters. Therefore, without changing I_W one can make the mass parameters small (but nonvanishing) in such a way that the theory becomes completely Higgsed and weakly coupled. Moreover, for a certain ratio of the mass parameters the pattern of the gauge symmetry breaking is hierarchical, e.g.,

$$\text{SU}(N) \to \text{SU}(N-1) \to \cdots \to \text{SU}(2) \to \text{nothing.} \tag{10.622}$$

In this weakly coupled theory everything is calculable. In particular, one can find the vacuum states and count them. This was done in [61, 90]. As mentioned, the gluino condensate is a convenient indicator of the vacua – it takes distinct values in the various vacua.[69] The gluino condensate $\langle \lambda\lambda \rangle$ was calculated *exactly* in [61, 90]; the result is multiple-valued,

| *Cf.* |
| Section 10.14. |

$$\langle \lambda\lambda \rangle \propto e^{2\pi i k/T_G}, \qquad k = 0, 1, \ldots, T_G - 1. \tag{10.623}$$

All vacuum states are, of course, bosonic, implying that $I_W = T_G$.

The crucial element of the index analysis is the assumption that no vacuum state runs away to infinity, in the space of fields, in the process of parameter deformation. For instance, in Witten's analysis [60] it was tacitly assumed that at $L \to \infty$ no fields develop infinitely large expectation values. An analysis based on Higgsing [61, 90] confirms this assumption, at least in theories with fundamental matter. Generally speaking, if the theory under consideration has flat (or nearly flat) directions then it can develop asymptotically large expectation values of certain operators in the process of parameter deformation, so that the calculation of I_W will be contaminated. If vacua characterized by infinitely large expectation values exist, they are referred to as *run-away vacua*.

---

[69] Actually, using the gluino condensate as an order parameter was suggested by Witten [60]; he realized that there was a mismatch for orthogonal groups.

10.22.2 Nongauge Theories

In Wess–Zumino models with polynomial superpotentials, Witten's index is determined by the number of solutions of the equation

$$\frac{\partial \mathcal{W}}{\partial \phi} = 0. \tag{10.624}$$

Each solution corresponds to a vacuum. Supersymmetry cannot be spontaneously
broken since, in the general case, there are no massless fermions in the Wess–
Zumino models that could become Goldstinos. If $\mathcal{W}$ is a polynomial of nth order, it
is clear that

$$I_W = n - 1. \tag{10.625}$$

In particular, in the renormalizable case, $\mathcal{W}$ is cubic and $I_W = 2$. That we have two
vacua in this case was discussed in Section 10.6.4.

The Wess–Zumino model, being very simple, presents a good pedagogical
example in which one can trace the property of the volume independence of the
Witten index, as well as its independence of the mass parameter in the superpotential.
In Appendix section 10.27.5 at the end of Chapter 10 I calculate, as an exercise, the
Witten index for cubic superpotentials in the limits $L \to 0$ and $m \to 0$. At $L \to 0$ the
problem reduces to a quantum-mechanical one, since we can completely ignore the
$\vec{x}$-dependence of all fields, keeping only the time dependence. We recover $I_W = 2$ in
this limit.

The Witten index for supersymmetric CP$(N-1)$ models is N. In particular, in the
CP(1) model $I_W = 2$. In Section 10.12.3.6, where a mass deformation was studied,
we saw that this model has two vacua, at $S^3 = \pm 1$. Similar mass deformations can be
constructed for CP$(N-1)$. Witten's original derivation [60] was carried out in the
$L \to 0$ limit.

10.23 Q-Closed and Q-Exact Operators

In this section we will establish a very general result in a class of supersymmetric
theories in which supersymmetry is *not* spontaneously broken. Assume that we can
find a supercharge Q and an operator φ obeying the equality

$$[\epsilon Q, \varphi] = 0. \tag{10.626}$$

Then the operator φ is called *Q-closed*. If, in addition, the operator $\partial_\mu \varphi$ can be
obtained as the commutator of the above supercharge with another operator,

$$[\epsilon Q, \psi] \sim \partial_\mu \varphi, \tag{10.627}$$

it is called *Q-exact*. Then the two-point function

$$\langle \text{vac}|T\{\varphi(x)\,\varphi(0)\}|\text{vac}\rangle = x\text{-independentc constant } C. \tag{10.628}$$

The same theorem is valid also for the n-point functions of the type

$$\langle \text{vac}|T\{\varphi(x_1)\,\varphi(x_2)\,\ldots\,\varphi(x_{n-1})\,\varphi(0)\}|\text{vac}\rangle. \tag{10.629}$$

Next, I will outline the proof of this theorem and give some examples. Let us start with the correlator

$$\langle \text{vac}|T\{\psi(x)\,(\epsilon\,Q)\,\varphi(0)\}|\text{vac}\rangle = \langle \text{vac}|T\{\psi(x)\,[(\epsilon\,Q),\varphi(0)]\}|\text{vac}\rangle = 0, \quad (10.630)$$

where I used the fact that Q acting on the vacuum produces zero and that φ is Q-closed; see Eq. (10.626). On the other hand, the first line in (10.630) can be rewritten as

$$\langle \text{vac}|T\{[\psi(x),(\epsilon\,Q)]\,\varphi(0)\}|\text{vac}\rangle \qquad\qquad (10.631)$$

$$\sim \langle \text{vac}|T\{\partial_\mu\varphi(x)\,\varphi(0)\}|\text{vac}\rangle = \partial_\mu\langle \text{vac}|T\{\varphi(x)\,\varphi(0)\}|\text{vac}\rangle.$$

Comparing (10.630) and (10.631) we arrive at (10.628). Since the two-point function is x independent we can take the limit $x \to \infty$ and conclude that

$$\langle \text{vac}|\varphi|\text{vac}\rangle\,\langle \text{vac}|\varphi|\text{vac}\rangle = C. \qquad\qquad (10.632)$$

This result is exact.

Example a

Let us start from the simplest example, the Wess–Zumino model (Section 49.4). In this model we have

$$[\bar{Q}_{\dot\beta}\bar\epsilon^{\dot\beta},\,\phi] = 0, \quad [\bar{Q}_{\dot\beta}\bar\epsilon^{\dot\beta},\,\psi_\alpha] = \sqrt{2}\,i\,(\partial_\mu\phi)\,(\sigma^\mu)_{\alpha\dot\beta}\,\bar\epsilon^{\dot\beta}. \qquad (10.633)$$

The first equality tells us that ϕ, the lowest component of a chiral superfield, is Q-closed, while the second equality tells that $\partial\phi$ is Q-exact. Therefore,

$$\langle \text{vac}|T\{\phi(x)\,\phi(0)\}|\text{vac}\rangle = \langle\phi\rangle^2 = \frac{m^2}{4\lambda^2}. \qquad (10.634)$$

Equations (10.633) and (10.634) are based on (10.120) and (10.144).

This exercise is, of course, trivial since we could have obtained the same result by standard methods.

Example b: super-Yang–Mills theory

This exercise is much less trivial because supersymmetric gluodynamics discussed in Section 10.14 is a strongly coupled theory and standard perturbative methods do not apply.

Let us start with the commutation relations for the following gauge invariant operators,

$$\lambda^2 \equiv \lambda^{\alpha\,a}\lambda_\alpha^a, \qquad \lambda G \equiv \lambda^{\alpha\,a}G_{\alpha\beta}^a. \qquad (10.635)$$

They are the lowest and next-to-lowest components of the superfield W^2; see Eq. (10.367). We should not forget that in super-Yang–Mills theory without matter the D component vanishes. Then we conclude

$$[\bar{Q}_{\dot\beta}\bar\epsilon^{\dot\beta},\,\lambda^2] = 0, \quad [\bar{Q}_{\dot\beta}\bar\epsilon^{\dot\beta},\,(\lambda G)_\beta] = -2i\left(\partial_{\beta\dot\beta}\lambda^2\right)\bar\epsilon^{\dot\beta}. \qquad (10.636)$$

Thus, the operator λ^2 is Q-closed and $\left(\partial_{\beta\dot\beta}\lambda^2\right)$ is Q-exact. As a result, the two-point function

$$\langle \text{vac}|T\{\lambda^2(x)\,\lambda^2(0)\}|\text{vac}\rangle = x\text{-independentc constant } C. \qquad (10.637)$$

In SU(2) Yang–Mills theory the small-size instanton contribution to the constant on the right-hand side is nonvanishing. Hence, taking the large-x limit we conclude that $\langle \lambda^2 \rangle \neq 0$ and, moreover, it is double-valued. For SU(N) one must consider the N-point function of the type (10.629) to arrive at a nonvanishing N-valued gluino condensate (see Fig. 10.2 on page 490).

This theorem was first formulated for the two-point function (10.629) in [91].

10.24 Soft versus Hard Explicit Violations of Supersymmetry

Many models of supersymmetry breaking reduce, at low energies, to explicit supersymmetry-breaking terms in a generic renormalizable super-Yang–Mills theory (10.368). Renormalizability of the theory implies that the superpotential is a polynomial which is at most cubic in the chiral superfields. In such theories there are no quadratic ultraviolet divergences in loops, in spite of the presence of spin-0 and spin-1 fields.[70] Only logarithmic divergences occur. In introducing explicit supersymmetry-breaking terms, we want to preserve this property.

Terms that keep all ultraviolet loop divergences purely logarithmic are referred to as *soft* supersymmetry-breaking terms, as opposed to hard breaking, which does induce quadratic divergences. By loops I mean radiative corrections to the various terms in the effective Lagrangian. We will not discuss the impact of explicit supersymmetry breaking on radiative corrections to the vacuum energy density. Needless to say, the latter no longer vanish when explicit supersymmetry breaking is switched on.

The problem of cataloging the soft terms was solved by Girardello and Grisaru [92]. Out of a large set of terms explicitly breaking supersymmetry, very few are soft. Below I present a full list of such terms and briefly outline the logic of the analysis of Girardello and Grisaru.

Before starting our discussion the reader is advised to revisit Section 10.6.11. In that subsection we introduced auxiliary nondynamical "spurion" superfields, whose lowest components coincided with various couplings, to prove complexification and holomorphy. Now we will use a similar device to derive the possible soft terms. Unlike in Section 10.6.11, the spurion superfields will be endowed with nonvanishing *last* components; D terms for general superfields and F terms for chiral superfields. This will make explicit supersymmetry breaking look spontaneous. The advantage of this construction is obvious – as far as ultraviolet divergences are concerned, it allows one to carry out all calculations as if the theory were supersymmetric. Only at the very end, when an effective Lagrangian is obtained at the desired loop order, can one substitute D, $F \neq 0$ in the spurion fields.

No quadratic divergences arise from these terms. The limitations imposed on the generic super-Yang–Mills Lagrangian mentioned above imply that only four classes of spurion-containing terms are possible,

---

[70] We will not consider here theories with the Fayet–Iliopoulos term, in which there may be subtleties.

$$\mathcal{L}_2 = \int d^2\theta\, \eta_1\, \mathrm{Tr}\, W^2 + \mathrm{H.c.,} \tag{10.638}$$

$$\mathcal{L}_2 = \int d^4\theta\, Z\bar{Q}e^V Q, \tag{10.639}$$

$$\mathcal{L}_3 = \int d^2\theta\, \eta_2 Q^2 + \mathrm{H.c.,} \tag{10.640}$$

$$\mathcal{L}_4 = \int d^2\theta\, \eta_3 Q^3 + \mathrm{H.c.,} \tag{10.641}$$

where $\eta_{1,2,3}$ are chiral superfields while Z is a general superfield. At the end we must set

$$\begin{aligned} \eta_i &= F_i\theta^2, & i &= 1,2,3, & F_i &\neq 0, \\ Z &= D\theta^2\bar{\theta}^2, & D &\neq 0. \end{aligned} \tag{10.642}$$

All spurion fields are dimensionless and gauge invariant. They can carry flavor indices, however. In particular, if the gauge group in the theory under consideration is actually a product of gauge groups, we will have several gauge kinetic terms and $\mathcal{L}_1$ can take the form

$$\sum_g \int d^2\theta(\eta_1)_g\, \mathrm{Tr}\, W_g^2.$$

By the same token the symbolic notation used in (10.640) must be understood as

$$\sum_{f,g} \int d^2\theta(\eta_2)_{fg}Q^f Q^g,$$

and so on. At first sight it might seem that other relevant operators exist that do not belong to the list above, for instance $\int d^4\theta\, \bar{\eta}Q^2$ However, this is not the case. In particular, the operator just mentioned reduces to $\int d^2\theta(\bar{D}^2\bar{\eta})Q^2 \rightarrow \mathrm{const} \times \int d^2\theta\, Q^2$, which is superinvariant. It does not introduce supersymmetry breaking.

Substituting (10.642), we see that $\mathcal{L}_1$ becomes the gaugino mass term, $\mathcal{L}_2$ and $\mathcal{L}_3$ become the mass terms for the scalar components of the chiral superfields (the elements of the mass matrix), of type $m\bar{q}q$ and mq^2, respectively, while $\mathcal{L}_4$ generates cubic interactions, between the scalar components, of a special form.

Now it is time to explain why other possible supersymmetry breaking terms are in fact unsuitable. Let us first add (10.638)–(10.641) to the original Lagrangian at a high normalization point (i.e., at an ultraviolet cutoff) and then let the theory evolve down, calculating the effective Lagrangian at a current normalization point $\mu \ll M_{\mathrm{uv}}$. The additional $\eta_{1,2,3}$ terms as well as the Z term in (10.639) generate polynomial terms in the effective Lagrangian. Since the initial supersymmetric theory *per se* does not have quadratic divergences, we must focus only on terms in the effective Lagrangian that are proportional to powers of $\eta_{1,2,3}$ and/or Z in addition to powers of other fields present in theory. (Terms that contain only $\eta_{1,2,3}$ and/or Z, without other fields, are relevant only to a vacuum energy calculation and will not be considered here.)

Let us examine the possible impact of (10.638)–(10.641) on the effective Lagrangian. In search of quadratic divergences we can limit ourselves to induced terms of dimension less than those in (10.638)–(10.641); other terms are either convergent or diverge only logarithmically. It is obvious that the induced terms to be analyzed must be gauge and Lorentz invariant.

In the case of $\mathcal{L}_2$ only one such term exists, namely

$$\int d^4\theta\, ZQ_0, \tag{10.643}$$

where Q_0 is a chiral superfield.

The integrand has mass dimension 1. It can emerge in loops only if the theory at hand contains, among other fields, a gauge-invariant chiral superfield Q_0. Then dimensional counting tells us that the coefficient in front of (10.643) must be linearly divergent. But there are no linear divergences in four-dimensional field theory. This means that the term (10.643) can appear only multiplied by some mass parameter of the theory having a logarithmic divergence.

Now let us pass to $\mathcal{L}_3$. Integrals over a chiral subspace of the type $\int d^2\theta\, \eta_2(\ldots)$ do not appear in perturbation theory (see Section 10.8). A new divergence could be introduced through

$$\int d^4\theta\, \bar{\eta}_2 Q_0 \qquad \text{or} \qquad \int d^4\theta\, \bar{\eta}_2 \eta_2 Q_0 \tag{10.644}$$

but, if so, it is a logarithmic divergence for the same reason as above.

The term (10.638) generates a number of induced terms in the effective Lagrangian, for instance those similar to (10.644), namely,

$$\int d^4\theta\, \bar{\eta}_1 Q_0 \qquad \text{or} \qquad \int d^4\theta\, \bar{\eta}_1 \eta_1 Q_0, \tag{10.645}$$

Formally speaking, on purely dimensional grounds there is a linear divergence in (10.645) but in fact the corresponding coefficients are at most logarithmically divergent. All other induced terms are either clearly logarithmically divergent or convergent.

The same assertions are valid with regard to $\mathcal{L}_4$.

It is instructive to consider examples of supersymmetry-breaking terms that do not preserve the logarithmic nature of loop divergences, i.e., they are hard. For instance, what would happen if supersymmetry were broken explicitly through a mass term of the matter fermion field of the type $m\psi^2 + $ H.c.? If the field ψ belongs to a gauge-invariant chiral superfield Q_0, in superfield language the operator from which $m\psi^2$ is generated is

$$\mu^{-1} \int d^4\theta\, Z(D_\alpha Q_0)\,(D^\alpha Q_0)\,, \tag{10.646}$$

where the background factor Z was defined in (10.642) and μ is a constant with the dimension of mass. The mass dimension of the integrand here is 3; thus it is higher than the normal dimension of D terms, 2. This means that the operator (10.646) will mix with $\int d^4\theta\, ZQ_0$, with a quadratically divergent coefficient. The same is true with regard to, say,

$$\mu^{-1} \int d^4\theta\, ZQ_0^2\bar{Q}_0,$$

an operator which gives rise to supersymmetry-breaking q^3 terms in the Lagrangian under consideration. Their structure is different from that in (10.641).

Note, however, that the operator similar to (10.646) in supersymmetric QCD,

$$\mu^{-1} \int d^4\theta \, Z \, \nabla_\alpha Q \, \nabla^\alpha \tilde{Q}, \qquad (10.647)$$

will not lead to quadratic divergences since there are no gauge-invariant matter superfields in the Lagrangian of supersymmetric QCD, and, correspondingly, no mixing with $\int d^4\theta \, Q$. However, the term (10.647) can mix with

$$\int d^4\theta \, Z \left(\bar{Q} e^V Q + \bar{\tilde{Q}} e^V \tilde{Q} \right).$$

The formal degree of divergence is linear. In fact, it will mix with a logarithmic divergence in the coefficient.

In summary, gaugino masses and those of scalar matter fields break supersymmetry in a soft way. The quadratic and cubic holomorphic operators $\mu^2 qq$ and μqqq (and their complex conjugates), whose structure repeats that of the superpotential, are soft too.

10.25 Central Charges

For a more detailed discussion of centrally extended algebras and their implications see Chapter 11.

In Section 10.6.4 we discussed the Wess–Zumino model. It must be admitted that the whole truth was not told there. Since the model was obtained in a superfield formalism, the reader might have tacitly assumed that supersymmetry of this model is expressed through the standard superalgebra (10.74), (10.75). Well . . . this is not the case. In fact, the superalgebra in the Wess–Zumino model is centrally extended. This present section is devoted to central charges. We will become acquainted with them by focusing on the simplest model, a two-dimensional reduction of the Wess–Zumino model. Reducing from four to two dimensions will allow us to get rid of inessential technicalities, which, at this stage, would only blur our picture of the given phenomenon. Reducing the model to two dimensions amounts to saying that nothing depends on the two spatial coordinates x and y. In addition, instead of four matrices $(\sigma^\mu)_{\alpha\dot{\beta}}$, we will use the two-dimensional gamma matrices defined in Eqs. (10.55) and (10.56). In two dimensions there is no distinction between dotted and undotted indices, since the Lorentz group includes only one transformation – the Lorentz boost – which acts in the same way on dotted and undotted spinors.

Two-dimensional reduction from (10.140) to $\mathcal{L}_{\text{WZ2D}}$ is a straightforwards exercise. For instance, the fermion kinetic term takes the form $i\bar{\psi} \left(\gamma^0 \partial_t + \gamma^1 \partial_z \right) \psi$ where ψ is a two-dimensional Dirac spinor.

Needless to say, the dimensionally reduced Wess–Zumino model has four supercharges, just as in four dimensions. From the standpoint of two dimensions it is an $\mathcal{N} = 2$ supersymmetry.

We will approach the issue gradually, in two steps.

10.25.1 Bogomol'nyi Completion

Look back through Section 2.1.5.

The Hamiltonian of the Wess–Zumino model can be derived immediately from Eq. (10.140). If we limit ourselves to time-independent field configurations and ignore, for the time being, the fermion degrees of freedom, we obtain an energy functional in the form

$$\mathcal{E} = \int dz \left(\left| \frac{\partial \phi}{\partial z} \right|^2 + \left| \frac{\partial \mathcal{W}}{\partial \phi} \right|^2 \right), \tag{10.648}$$

where the superpotential $\mathcal{W}$ was given in Eq. (10.144) and we will assume for simplicity that both the parameters, m and λ, are real and positive.[71]

To perform the Bogomol'nyi completion [93] we add and subtract a term that can be expressed as a full derivative:

$$\mathcal{E} = \int dz \left\{ \left[\frac{\partial \phi}{\partial z} - \left(\frac{\partial \mathcal{W}}{\partial \phi} \right)^* \right] \left[\frac{\partial \phi}{\partial z} - \left(\frac{\partial \mathcal{W}}{\partial \phi} \right)^* \right]^* + 2 \operatorname{Re} \frac{\partial \phi}{\partial z} \frac{\partial \mathcal{W}(\phi)}{\partial \phi} \right\}. \tag{10.649}$$

The last term clearly reduces to

$$2 \int dz \operatorname{Re} \frac{\partial \mathcal{W}}{\partial z} = 2 \operatorname{Re} \left[\mathcal{W}(z = \infty) - \mathcal{W}(z = -\infty) \right]. \tag{10.650}$$

Deriving the Bogomol'nyi bound in the WZ model

We see that it depends only on the boundary conditions and in this sense is topological.

Let us consider topologically nontrivial boundary conditions, i.e., at $z = -\infty$ the field ϕ resides in one vacuum ($\phi = -m/2\lambda$), and at $z = \infty$ in the other ($\phi = m/2\lambda$), see Eq. (10.145). This is a topologically stable field configuration which, in two dimensions, presents a kink, a localized object that must be treated as a particle.

Combining (10.649) and (10.650) we conclude that

$$\mathcal{E}_{\text{kink}} \geq 2 \operatorname{Re} \left[\mathcal{W}(z = \infty) - \mathcal{W}(z = -\infty) \right]. \tag{10.651}$$

Equality is achieved if and only if

$$\frac{\partial \phi}{\partial z} = \left(\frac{\partial \mathcal{W}}{\partial \phi} \right)^*. \tag{10.652}$$

Anticipating that, with positive m and λ, the solution will be real as well as the values of the superpotential at the infinities, we can write, instead,

$$\frac{\partial \phi}{\partial z} = \frac{\partial \mathcal{W}}{\partial \phi}, \tag{10.653}$$

$$\mathcal{E}_{\text{kink}} = 2 \left[\mathcal{W}(z = \infty) - \mathcal{W}(z = -\infty) \right] = \frac{m^3}{3\lambda^2}. \tag{10.654}$$

The first-order equation (10.653), known as the Bogomol'nyi–Prasad–Sommerfield (BPS) equation [93, 94], replaces the classical equation of motion. The latter follows from the BPS equation but the converse is not true. In this sense the BPS equation is stronger than the equation of motion. If the solution of (10.653) with appropriate boundary conditions exists, the kink is referred to as BPS-saturated (or, sometimes, critical). Then its mass is given by (10.654). In the case at hand the boundary conditions are

$$\phi(z = -\infty) = -\frac{m}{2\lambda}, \qquad \phi(z = \infty) = \frac{m}{2\lambda}, \tag{10.655}$$

---

[71] In fact, they can be arbitrary complex numbers; generalization to this case is straightforward. All expressions given in this section depend crucially on the fact that $\mathcal{W}(\phi_{\text{vac}})$ is real. Passing to the complex plane changes the particular form of these expressions but not the general idea.

Cf.
Eq. (2.11).

and the solution of (10.653) with these boundary conditions is

$$\phi(z) = \frac{m}{2\lambda} \tanh \, mz. \tag{10.656}$$

To understand better the physical meaning of Bogomol'nyi completion in the context of supersymmetry, let us examine the field supertransformations (10.120). More exactly, we will focus on the second, which in two dimensions takes the form

$$\delta\psi = \sqrt{2}\left(\frac{\partial}{\partial t}\gamma^0 + \frac{\partial}{\partial z}\gamma^1\right)\left(\epsilon^*\phi\right) - \sqrt{2}\left(\frac{\partial W}{\partial \phi}\right)^* \epsilon. \tag{10.657}$$

If $\gamma^1\epsilon^* = \epsilon$ then the condition $\delta\psi = 0$ is equivalent to Eq. (10.652).[72] This means that the solution presented above preserves a part of supersymmetry. Namely, there are two linear combination of supercharges that annihilate the kink,

$$Q_1 + i\,Q_2^* \qquad \text{and} \qquad Q_1^* - i\,Q_2. \tag{10.658}$$

On general grounds this should not happen. Indeed, the kink solution breaks translational invariance. Generally speaking, then, one should expect that all four supercharges are broken in the case of this solution. In fact, the BPS saturated kink breaks only two out of the four supercharges,[73] i.e.,

$$Q_1 - i\,Q_2^* \qquad \text{and} \qquad Q_1^* + i\,Q_2. $$

This is possible only if the superalgebra is centrally extended.

10.25.2 Central Extensions

In the Golfand–Likhtman superalgebra we have $\{Q_\alpha, Q_\beta\} = 0$; see Eq. (10.75). Let us calculate this anticommutator again, taking into account that, for topologically nontrivial field configurations, $\Delta W \equiv W(z = \infty) - W(z = -\infty) \neq 0$.

To this end we will need the time component of the supercurrent, which can be extracted easily from the general expression (10.470) if we discard irrelevant terms, i.e., the first and third terms, replace the covariant derivative in the second term by an ordinary partial derivative (there are no gauge fields in the Wess–Zumino model), and take into account the fact that in two dimensions the only derivatives to be retained are ∂_t and ∂_z. In this way we get

$$J_\alpha^0 = \sqrt{2}\left[\dot{\phi}^\dagger\psi_\alpha - \frac{\partial\phi^\dagger}{\partial z}\left(\gamma^5\psi\right)_\alpha + F\left(\gamma^0\psi^\dagger\right)_\alpha\right]. \tag{10.659}$$

*Derivation
of the central
charge in the
WZ model*

Next we calculate the anticommutator $\{Q_\alpha, Q_\beta\}$ using the canonical commutation relations. It is easy to see that the anticommutators of $\dot{\phi}^\dagger\psi_\alpha$ with two other terms vanish. A contribution due to $\{\gamma^5\psi, \gamma^0\psi^\dagger\}$ remains namely,

$$\{Q_\alpha, Q_\beta\} = 4\left(\gamma^1\right)_{\alpha\beta}\int_{-\infty}^{\infty} dz\frac{\partial\bar{W}}{\partial z}$$
$$\equiv 4\left(\gamma^1\right)_{\alpha\beta}\Delta\bar{W}, \tag{10.660}$$

---

[72] Remember that we are considering static, i.e., time-independent, solutions. Moreover ϵ, ϵ^* in Eq. (10.657) are two-component spinors with *lower* indices, in contradistinction with Eq. (10.120), which contains $\bar{\epsilon}^{\dot{\alpha}}$.

[73] Field configurations preserving two out of four supercharges are referred to as 1/2 BPS saturated. If two out of eight supercharges were preserved, this would be called 1/4 BPS saturation, and so on.

where we have used the fact that $F = -\partial\bar{W}/\partial\phi^\dagger$. It is obvious that, for topologically nontrivial field configurations, $\{Q_\alpha, Q_\beta\} \neq 0$. Note that the right-hand side is symmetric in α, β as it should be, given the symmetry property of $\{Q_\alpha, Q_\beta\}$.

As a result, in the 2D reduction of the Wess-Zumino the superalgebra takes the following (covariant) form:

$$\left\{Q_\alpha, \left(Q^\dagger\gamma^0\right)_\beta\right\} = 2P_\mu\left(\gamma^\mu\right)_{\alpha\beta},$$
$$\left\{Q_\alpha, \left(Q\gamma^0\right)_\beta\right\} = -2Z\left(\gamma^5\right)_{\alpha\beta},$$

(10.661)

where Z is the central charge,

$$Z = 2\Delta\bar{W}.$$

(10.662)

The result for $\{Q_\alpha^\dagger, Q_\beta^\dagger\}$ is the complex conjugate of that in Eq. (10.661).

10.25.3 Central Extensions: Generalities

More details
are given in
Chapter 11.

If we consider superalgebras with $\mathcal{N} > 1$ and limit ourselves to Lorentz-scalar central charges then the centrally extended anticommutators take the form

$$\{Q_\alpha^I, Q_\beta^J\} = \varepsilon_{\alpha\beta} Z^{IJ},$$

(10.663)

where $I, J = 1, \ldots, \mathcal{N}$. The matrix is obviously antisymmetric in I, J.

10.25.4 Implications of the Central Extension

The Golfand–Likhtman superalgebra (10.75) implied that the vacuum energy density vanishes; the centrally extended superalgebra (10.661), in addition to this, provides us with a new prediction. The masses of those states on which part of the supercharges is conserved are "equal" to the central charge.[74] To see that this is indeed the case, and (simultaneously) to outline a general strategy, let us consider a 4×4 matrix κ_{ij} of all possible anticommutators $\{Q_i, Q_j^\dagger\}$, where

$$Q_i = \begin{pmatrix} Q_1 \\ Q_2 \\ Q_1^\dagger \\ Q_2^\dagger \end{pmatrix}, \qquad Q_i^\dagger = \begin{pmatrix} Q_1^\dagger & Q_2^\dagger & Q_1 & Q_2 \end{pmatrix}.$$

(10.664)

If we limit ourselves to the rest frame (in which $P^0 = M$ and $P_z = 0$), this matrix takes the form

$$\kappa_{ij} = 2\left(\begin{array}{cc|cc} M & 0 & 0 & -iZ \\ 0 & M & -iZ & 0 \\ \hline 0 & iZ^* & M & 0 \\ iZ^* & 0 & 0 & M \end{array}\right).$$

(10.665)

---

[74] The word equal is in quotation marks because this statement requires clarification, to be provided shortly. See Exercise 10.25.1 on page 561 and its solution on page 710

To make transparent the consequences ensuing from the central extension it is instructive to cast the matrix κ_{ij} into diagonal form. To this end we introduce four linear combinations of the original supercharges,

$$\tilde{Q}_1 = \frac{Q_1 + ie^{-i\alpha}Q_2^\dagger}{\sqrt{2}}, \qquad \tilde{Q}_1^\dagger = \frac{Q_1^\dagger - ie^{i\alpha}Q_2}{\sqrt{2}},$$

$$\tilde{Q}_2 = \frac{Q_1 - ie^{-i\alpha}Q_2^\dagger}{\sqrt{2}}, \qquad \tilde{Q}_2^\dagger = \frac{Q_1^\dagger + ie^{i\alpha}Q_2}{\sqrt{2}}, \tag{10.666}$$

where the phase α coincides with that of Z^*:

$$\alpha \equiv \arg Z^* = -\arg Z. \tag{10.667}$$

In the new basis the matrix κ takes the form

$$\tilde{\kappa} \equiv \{\tilde{Q}_i, \tilde{Q}_j^\dagger\} = 2 \left(\begin{array}{cc|cc} M - |Z| & 0 & 0 & 0 \\ 0 & M + |Z| & 0 & 0 \\ \hline 0 & 0 & M - |Z| & 0 \\ 0 & 0 & 0 & M + |Z| \end{array} \right). \tag{10.668}$$

If there exist states $|a\rangle$ that are annihilated by $\tilde{Q}_1$ and $\tilde{Q}_1^\dagger$ then the mass of such states, M_a, is fixed:

$$M_a = |Z|. \tag{10.669}$$

In the general case, for arbitrary states,

$$M_a \geq |Z|. \tag{10.670}$$

Bogomol'nyi bound in the context of supersymmetry

The latter inequality is referred to as the Bogomol'nyi bound, while Eq. (10.669) holds if BPS saturation takes place (more exactly, in the case at hand we are dealing with 1/2 BPS saturation).

As a matter of fact, there is a fast way of finding the masses of the BPS states, without carrying out explicit diagonalization of the matrix κ_{ij}. To this end it suffices to calculate the determinant of κ_{ij} and solve the equation

$$\det(\kappa_{ij}) = 0. \tag{10.671}$$

For instance, in our problem, calculating the determinant of the matrix (10.665) is trivial, yielding

$$\det(\kappa_{ij}) = 2^4 (M - |Z|)^2 (M + |Z|)^2. \tag{10.672}$$

Equation (10.669) ensues immediately.

In concluding this subsection it is worth noting that if Z is real and positive (as assumed in Section 10.25.1) then $\alpha = 0$ and the unbroken supercharges $\tilde{Q}_1, \tilde{Q}_1^\dagger$ coincide with those found in Eq. (10.658), by virtue of the Bogomol'nyi completion. This remark prompts us as to how to carry out the Bogomol'nyi completion for complex values of the central charge.

Exercise

10.25.1 Derive the Bogomol'nyi bound using a representation similar to (10.649) and assuming that the central charge $Z = 2\Delta \bar{W}$ is an arbitrary complex number.

10.26 Long versus Short Supermultiplets

In this section we will discuss the multiplicity of representations for centrally extended superalgebras. Rather than performing a general analysis, I will outline the basic idea using the example of Section 10.25. Before delving into the topic of long versus short supermultiplets the reader is recommended to return to Section 10.4.6.

The centrally extended superalgebra (10.661), built on four supercharges, can be cast in all cases into diagonal form, as in Eq. (10.668). The representation multiplicity crucially depends on whether we are dealing with BPS saturated or nonsaturated (noncritical) states. Indeed, for noncritical states (i.e., $M > Z$), normalizing appropriately the supercharges with tildes, $\tilde{Q}_{1,2}$, one can write the superalgebra as

$$\left\{ \tilde{Q}_\alpha, \left(\tilde{Q}_\beta \right)^\dagger \right\} = \delta_{\alpha\beta}, \qquad \{ \tilde{Q}_\alpha, \tilde{Q}_\beta \} = 0, \qquad \left\{ \left(\tilde{Q}_\alpha \right)^\dagger, \left(\tilde{Q}_\beta \right)^\dagger \right\} = 0,$$
$$\alpha, \beta = 1, 2. \tag{10.673}$$

Repeating the arguments after Eq. (10.93) we conclude that the noncritical supermultiplet consists of four states, two bosonic and two fermionic.

However, if BPS saturation is achieved (i.e., $M = |Z|$), the corresponding superalgebra takes a form similar to (10.96),

$$\left\{ \tilde{Q}_\alpha, \left(\tilde{Q}_\beta \right)^\dagger \right\} = \begin{pmatrix} 0 & 0 \\ 0 & 1 \end{pmatrix}; \tag{10.674}$$

all other anticommutators vanish. As a result, the supermultiplet is two dimensional: it includes just one bosonic state and one fermionic. This phenomenon is referred to as *multiplet shortening* for BPS states. In supersymmetric theories with central charges, two types of massive supermultiplets coexist: long multiplets for noncritical states and short multiplets for BPS saturated states.

Sometimes, the class of short multiplets is further divided into subclasses. An example in which distinct short multiplets can appear is $N = 2$ theory in four dimensions. There are eight supercharges in such theory. The simplest long representation is 16 dimensional, with eight bosonic and eight fermionic states. Half-BPS-saturated massive solitons form a four-dimensional representation $(2 + 2)$. If quarter-BPS-saturated states exist, they will form a two-dimensional representation $(1 + 1)$.

If $N > 2$ then we have a spectrum of possibilities (even if we limit ourselves to Lorentz-scalar central charges). A generic massive $N = 4$ multiplet contains $2^{2N} = 256$ states, including the helicities ± 2. Thus, such a theory must include a massive spin-2 particle, which is impossible in globally supersymmetric field theories. Short multiplets can contain $2^{2(N-k)}$ states, where $k = 1$ or 2 (generically, k runs from 1 to

More details are given in [95].

$\frac{1}{2}\mathcal{N}$ for even $\mathcal{N}$). If $k = \frac{1}{2}\mathcal{N}$ then we get the shortest multiplets, with only $2^{\mathcal{N}} = 16$ states. This is exactly the number of states in the massless representation. Such BPS multiplets are called *ultrashort*. They are analogs of the massless supermultiplets.

10.27 Appendices

10.27.1 Supersymmetric $CP(N-1)$ in Gauged Formulation

Let us outline the construction of the $N = 2$ $CP(N-1)$ model with twisted masses in the so-called gauged formulation [96]. This formulation is built on an N-plet of complex scalar fields n^i, where $i = 1, 2, \ldots, N$. We will impose the constraint

Section 10.12.4 presented a geometric formulation of $N = 2$ $CP(N-1)$; see also Section 9.4.

$$n_i^\dagger n^i = 1. \tag{10.675}$$

This leaves us with $2N - 1$ real bosonic degrees of freedom. To eliminate the extra degree of freedom we impose a local U(1) invariance, $n^i(x) \to e^{i\alpha(x)} n^i(x)$. To this end we introduce a gauge field A_μ, which converts the partial derivative into a covariant derivative,

$$\partial_\mu \to \mathcal{D}_\mu \equiv \partial_\mu - iA_\mu. \tag{10.676}$$

The field A_μ is auxiliary; it enters the Lagrangian without derivatives. The kinetic term of the n fields is

$$\mathcal{L} = \frac{2}{g_0^2} \left| \mathcal{D}_\mu n^i \right|^2. \tag{10.677}$$

The superpartner to the field n^i is an N-plet of complex two-component spinor fields ξ^i,

$$\xi^i = \begin{cases} \xi_R^i, \\ \xi_L^i. \end{cases} \tag{10.678}$$

The auxiliary field A_μ has a complex scalar superpartner σ and a two-component complex spinor superpartner λ; both enter without derivatives. The full $\mathcal{N} = 2$ symmetric Lagrangian is

$$\mathcal{L} = \frac{2}{g^2} \left\{ \left| \mathcal{D}_\mu n^i \right|^2 + \xi_i^\dagger i\gamma^\mu \mathcal{D}_\mu \xi^i + 2 \sum_i \left| \sigma - \frac{m_i}{\sqrt{2}} \right|^2 |n^i|^2 + D \left(|n^i|^2 - 1 \right) \right.$$
$$\left. + \left[i\sqrt{2} \sum_i \left(\sigma - \frac{m_i}{\sqrt{2}} \right) \xi_{iR}^\dagger \xi_L^i + i\sqrt{2} n_i^\dagger \left(\lambda_R \xi_L^i - \lambda_L \xi_R^i \right) + \text{H.c.} \right] \right\}, \tag{10.679}$$

where the m_i are twisted mass parameters. Equation (10.679) is valid in the special case for which

$$\sum_{i=1}^N m_i = 0. \tag{10.680}$$

Of particular elegance is a special (Z_N-symmetric) choice of the parameters m_i, namely,

$$m_i = m \left\{ e^{2\pi i/N}, e^{4\pi i/N}, \ldots, e^{2(N-1)\pi i/N}, 1 \right\}, \tag{10.681}$$

where m is a single complex parameter. If desired, m can be chosen to be real since its phase can be hidden in the θ term. The constraint (10.680) is automatically satisfied. Without loss of generality m can be assumed to be real and *positive*. The U(1) gauge symmetry is built in. This symmetry eliminates one bosonic degree of freedom, leaving us with $2N - 2$ dynamical bosonic degrees of freedom intrinsic to the CP($N - 1$) model [96].

For CP(1) we have $N = 2$, and the two mass parameters must be chosen as follows:

$$m_1 = -m_2 \equiv m. \tag{10.682}$$

In this case the relations between the fields of the gauge formulation of the model and those of the O(3) formulation are given by

$$S^a = n^\dagger \sigma^a n. \tag{10.683}$$

In Section 9.4 we discussed the large-N solution of the nonsupersymmetric CP($N - 1$) model. It is not difficult to generalize it to include supersymmetry. This can be done both with or without twisted masses [97, 98]. I will briefly outline the solution for vanishing twisted masses, referring the reader interested in the effect of nonzero twisted mass to [98].

When we switch on $\mathcal{N} = 2$ supersymmetry, the auxiliary field A_μ acquires superpartners. Its bosonic superpartners are the complex field σ and the real Lagrange multiplier D implementing the constraint (10.675).[75] The relevant Lagrangian is obtained from (10.679) by setting $m_i = 0$. Assuming σ and D to be constant background fields and integrating out the boson and fermion fields at one loop, n^i and ξ^i, respectively, we get the following effective Lagrangian $\mathcal{L}_{\text{eff}}(\sigma, D)$:

$$-\mathcal{L}_{\text{eff}} = \frac{N}{4\pi} \left\{ -\left(D + 2|\sigma|^2\right) \ln \frac{D + 2|\sigma|^2}{\Lambda^2} + D + 2|\sigma|^2 \ln \frac{2|\sigma|^2}{\Lambda^2} \right\}, \tag{10.684}$$

where

$$\Lambda^2 = M_{\text{uv}}^2 \exp\left(-\frac{8\pi}{Ng^2}\right). \tag{10.685}$$

Minimizing the above expression with respect to D and σ we arrive at an analog of Eq. (9.61),

$$\frac{N}{4\pi} \ln \frac{D + 2|\sigma|^2}{\Lambda^2} = 0, \qquad \ln \frac{D + 2|\sigma|^2}{2|\sigma|^2} = 0, \tag{10.686}$$

implying that in the vacuum

$$D = 0, \qquad 2|\sigma|^2 = \Lambda^2. \tag{10.687}$$

---

[75] In comparing this section with Section 9.4, the reader is warned not to be confused about the change in notation. In Section 9.4 the real Lagrange multiplier is σ; it parallels D of this section. There is no analog of the complex field σ with which we are dealing here. However, the general strategy is the same.

The phase factor of σ cannot be determined from (10.687). We can find it by taking into account the spontaneous breaking of the discrete chiral Z_{2N} down to Z_2, inherent to the model at hand;[76] we conclude that the theory has N vacua at

$$\sigma = \frac{1}{\sqrt{2}}\Lambda \exp\left(\frac{2\pi i k}{N}\right), \quad k = 0,\ldots,N-1, \tag{10.688}$$

Witten's index in $CP(N-1)$ is N. See Section 10.22.

in full accord with Witten's index. All these vacua are supersymmetric (i.e., the vacuum energy vanishes). The vacuum degeneracy we observe here is in contradistinction with the nonsupersymmetric version of the model; see Section 9.4.4. This has crucial consequences. Namely, the charged fields, such as n^i, are confined in the nonsupersymmetric model, while supersymmetry liberates them. This is easy to understand if you look at Fig. 9.32: the energy densities in vacuum 1 "outside" and vacuum 2 "inside" are now the same. Technically, deconfinement occurs because the formerly massless photon acquires a nonvanishing mass from the mixing of Im σ and F^* [97]. The mass of the n quantum is Λ and that of the photon is 2Λ. The mixing is related to the chiral anomaly in two dimensions; see Chapter 8. Therefore, at distances $\gg 1/\Lambda$ the attraction between n and $\bar{n}$ (or their superpartners) is screened; their interaction falls off exponentially at large distances.

10.27.2 Moduli Space of Vacua in Supersymmetric QED

Here, I give a solution of the problem discussed in Section 10.6.10. This problem obviously has U(1) axial symmetry. The two-dimensional hyperboloid is a surface described by the equation

$$z^2 - a(x^2 + y^2) = b, \tag{10.689}$$

where a and b are positive constants; see Fig. 10.5. It is convenient to pass to polar coordinates in the xy plane. We will introduce $r = \sqrt{x^2 + y^2}$ and the polar angle α. Then for the hyperboloid (10.689), we have (Fig. 10.6)

$$z = \begin{cases} \sqrt{b} + \frac{a}{2\sqrt{b}}r^2 + O(r^4), & r \to 0, \\ \sqrt{a}\,r + O(1/r), & r \to \infty. \end{cases} \tag{10.690}$$

We will assume that the surface corresponding to the metric (10.204) is described by a function $z(r)$, to be determined below. For simplicity we will set $\xi = 4$, although this is inessential to the argument. If φ is parametrized as

$$\varphi = \rho e^{i\alpha} \tag{10.691}$$

the metric (10.204) implies, on the one hand, the following expression for the interval:

$$ds^2 = \frac{1}{\sqrt{1+\rho^2}}d\rho^2 + \frac{\rho^2}{\sqrt{1+\rho^2}}d\alpha^2. \tag{10.692}$$

---

[76] This is explained in great detail in [97]. *Hint: the remnant of the axial symmetry broken down to Z_{2N} by anomaly/instantons.*

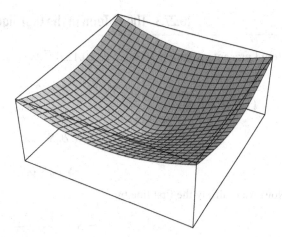

Fig. 10.5   Two-dimensional hyperboloid.

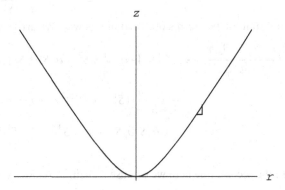

Fig. 10.6   The surface in Fig. 10.5 is described by a function $z(r)$, with no α dependence.

On the other hand, an interval on a surface $z(r)$ is given in general by

$$ds^2 = dr^2 \left[1 + \left(\frac{dz}{dr} \right)^2 \right] + r^2 \, d\alpha^2. \tag{10.693}$$

Comparing the last terms in Eqs. (10.692) and (10.693) we can deduce that

$$r^2 = \frac{\rho^2}{\sqrt{1 + \rho^2}}, \quad r = \begin{cases} \rho(1 - \frac{1}{4}\rho^2), & r \to 0, \\ \sqrt{\rho}, & r \to \infty. \end{cases} \tag{10.694}$$

Calculating dr in terms of $d\rho$ and comparing the result with the first term in (10.693) we find dz/dr and, hence, $z(r)$ up to a constant,

$$z(r) = \begin{cases} c + \frac{1}{2}r^2 + O(r^4), & r \to 0, \\ \sqrt{3}r + O(1/r), & r \to \infty, \end{cases} \tag{10.695}$$

where c is an integration constant. We see that if the subleading corrections are neglected then Eq. (10.695) is compatible with (10.690) when $a = 3$ and $b = 9$; then $c = 3$.

10.27.3 The θ Term in the O(3) Sigma Model

Here I present a solution of Exercise 10.12.6.[77] We start from the observation that

$$
i\frac{\varepsilon^{\mu\nu}\partial_\mu\phi^\dagger\partial_\nu\phi}{\left(1+\phi^\dagger\phi\right)^2} = \tfrac{1}{4}\varepsilon^{\mu\nu}\Big[-\partial_\mu S^1\partial_\nu S^2 + \partial_\mu S^2\partial_\nu S^1
$$
$$
+ \frac{1}{1+S^3}\left(S^1\partial_\mu S^3\partial_\nu S^2 - S^2\partial_\mu S^3\partial_\nu S^1\right.
$$
$$
\left.-S^1\partial_\mu S^2\partial_\nu S^3 + S^2\partial_\mu S^1\partial_\nu S^3\right)\Big]. \quad (10.696)
$$

Now, we multiply the first line by

$$
S^3 + 1 - S^3 = S^3 + \frac{1-(S^3)^2}{1+S^3}
$$
$$
= S^3 + \frac{(S^1)^2 + (S^2)^2}{1+S^3}, \quad (10.697)
$$

and split the two terms obtained in this way. We arrive at

$$
i\frac{\varepsilon^{\mu\nu}\partial_\mu\phi^\dagger\partial_\nu\phi}{(1+\phi^\dagger\phi)^2} = \tfrac{1}{4}\varepsilon^{\mu\nu}\Big\{S^3\left(-\partial_\mu S^1\partial_\nu S^2 + \partial_\mu S^2\partial_\nu S^1\right)
$$
$$
+ \frac{1}{1+S^3}\Big[\left((S^1)^2 + (S^2)^2\right)\left(-\partial_\mu S^1\partial_\nu S^2 + \partial_\mu S^2\partial_\nu S^1\right)
$$
$$
+ \left(S^1\partial_\mu S^3\partial_\nu S^2 - S^2\partial_\mu S^3\partial_\nu S^1 - S^1\partial_\mu S^2\partial_\nu S^3 + S^2\partial_\mu S^1\partial_\nu S^3\right)\Big]\Big\}. \quad (10.698)
$$

In the second line here we can use a chain of relations, e.g.,

$$
(S^1)^2\partial_\mu S^1\partial_\nu S^2 = \tfrac{1}{2}S^1\partial_\mu(S^1)^2\partial_\nu S^2 = \frac{1}{2}S^1\partial_\mu\left[1-(S^2)^2 - (S_3)^2\right]\partial_\nu S^2
$$
$$
\to -S^1 S^3\partial_\mu S^3\partial_\nu S^2, \quad (10.699)
$$

and others of this type. In the last transition (in 10.699) we took into account convolution with $\varepsilon^{\mu\nu}$.

Assembling everything we obtain

$$
i\frac{\varepsilon^{\mu\nu}\partial_\mu\phi^\dagger\partial_\nu\phi}{(1+\phi^\dagger\phi)^2} = -\tfrac{1}{4}\varepsilon^{\mu\nu}\varepsilon^{abc}\,S^a\partial_\mu S^b\partial_\nu S^c. \quad (10.700)
$$

10.27.4 Hypercurrents at $\gamma_f \neq 0$

For the geometric current $\mathcal{J}_{a\dot\alpha}$ one has (in theories with an arbitrary matter sector)

$$
\bar{D}^{\dot\alpha}\mathcal{J}_{\alpha\dot\alpha} = \frac{2}{3}D_\alpha\Bigg\{\left[3\mathcal{W} - \sum_f Q_f\frac{\partial\mathcal{W}}{\partial Q_f}\right]
$$
$$
-\left[\frac{3T_G - \sum_f T(R_f)}{16\pi^2}\operatorname{Tr}W^2 + \frac{1}{8}\sum_f \gamma_f\bar{D}^2(\bar{Q}_f e^V Q_f)\right]\Bigg\}, \quad (10.701)
$$

<hr/>

[77] A different solutions is presented on pages 703, 704.

and

$$\partial^{\alpha\dot\alpha} \mathcal{J}_{\alpha\dot\alpha} = -\frac{i}{3} D^2 \left\{ \left[3\mathcal{W} - \sum_f \left(1 + \frac{\gamma_f}{2} \right) Q_f \frac{\partial W}{\partial Q_f} \right] \right.$$
$$\left. - \frac{1}{16\pi^2} \left[3T_G - \sum_f \left(1 - \gamma_f \right) T(R_f) \right] \mathrm{Tr}\, W^2 \right\} + \mathrm{H.c.},$$

$$(10.702)$$

where the γ_f are the anomalous dimensions of the matter fields Q_f. These expressions are more exact than those presented in Eqs. (10.462) and (10.463), in which the γ_f terms were (deliberately) omitted.

The Konishi anomaly stays intact; see Eqs. (10.450) and (10.465). The γ_f terms have no effect on the Konishi anomaly.

10.27.5 The Witten Index in the Wess–Zumino Model, in the Quantum-Mechanical Limit $L \to 0$

The problem that I will address here is to find Witten's index for a system described by the Lagrangian

$$\mathcal{L} = \frac{d\phi^\dagger}{dt} \frac{d\phi}{dt} + i\psi^\dagger \frac{d\psi}{dt} + F^\dagger F + \left\{ m[\phi F - \tfrac{1}{2}(\psi)^2] + g[\phi^2 F - \phi(\psi)^2] + \mathrm{H.c.} \right\},$$

$$(10.703)$$

where ϕ and F are complex variables, ψ is a *two-component* Grassmann variable, $\psi = (\psi^1, \psi^2)$, and $(\psi)^2 \equiv \psi^2\psi^1 - \psi^1\psi^2$. This Lagrangian occurs under the reduction of the Wess–Zumino model from four to one dimension. The auxiliary variable F enters without the kinetic term; thus, it can be eliminated via the equations of motion. The solution of the above problem is as follows.

First it is instructive to check that the model (10.703) is indeed supersymmetric and to write down the corresponding supercharges. We have four supercharges,

$$Q^1 = \sqrt{2} \left(\frac{d\phi^\dagger}{dt} \psi^1 - i\psi^{2\dagger} F \right), \qquad Q^2 = \sqrt{2} \left(\frac{d\phi^\dagger}{dt} \psi^2 + i\psi^{1\dagger} F \right), \qquad (10.704)$$

plus the Hermitian conjugates. Next, using the equations of motion

$$\frac{d^2}{dt^2} \phi^\dagger = mF + 2g\phi F + 2g\psi^1\psi^2,$$

$$i\frac{d}{dt}\psi^{1\dagger} = -m\psi^2 - 2g\phi\psi^2, \qquad i\frac{d}{dt}\psi^{2\dagger} = m\psi^1 + 2g\phi\psi^1,$$

$$F^\dagger = -(m\phi + g\phi^2), \qquad (10.705)$$

it is not difficult to check that

$$\frac{d}{dt}Q^1 = \frac{d}{dt}Q^2 = 0. \qquad (10.706)$$

Clearly, the complex-conjugate supercharges are also conserved.

The algebra of the supercharges takes the form

$$\{Q^1, Q^{1\dagger}\} = \{Q^2, Q^{2\dagger}\} = 2H \qquad (10.707)$$

(all other commutators vanish). Here H is the Hamiltonian of the system,

$$H = \pi_\phi \pi_{\phi^\dagger} + |F|^2 + \left[\tfrac{1}{2}(m + 2g\phi)(\psi)^2 + \text{H.c.} \right],$$

where

$$\pi_\phi = -i\frac{\partial}{\partial \phi}, \qquad \pi_{\phi^\dagger} = -i\frac{\partial}{\partial \phi^\dagger}, \qquad (\psi)^2 \equiv \psi^2 \psi^1 - \psi^1 \psi^2.$$

At the next stage we must realize the fermion variables ψ^α in a matrix representation. In the problem at hand there are two fermion variables (plus their complex conjugates). The procedure of constructing the matrix representation ensuring the canonial commutation relations

$$\{\psi^\alpha,\ \psi^\beta\} = \{\psi^{\alpha\dagger},\ \psi^{\beta\dagger}\} = 0, \quad \{\psi^\alpha,\ \psi^{\beta\dagger}\} = \delta^{\alpha\beta}, \tag{10.708}$$

is well known; see, e.g., [99]. The minimal dimension of matrices implementing (10.708) is 4×4.

Let us build ψ^α in the form of a direct product of two 2×2 matrices,

$$\psi^1 = \sigma^- \otimes 1, \qquad \psi^2 = \sigma^3 \otimes \sigma^-;$$
$$\psi^{1\dagger} = \sigma^+ \otimes 1, \qquad \psi^{2\dagger} = \sigma^3 \otimes \sigma^+, \tag{10.709}$$

where $\sigma^\pm = \tfrac{1}{2}(\sigma^1 \pm i\sigma^2)$. Then in the matrix representation the expression for the Hamiltonian reduces to

$$H = \pi_\phi \pi_{\phi^\dagger} + |F|^2 - \left[(m + 2g\phi)\sigma^- \otimes \sigma^- + \text{H.c.} \right]. \tag{10.710}$$

The Hamiltonian acts on wave functions with two "spins."

In the classical approximation (which applies for $m/g \gg 1$) the spin interaction can be neglected while the potential of the system,

$$V_{\text{pot}} = |F|^2 = \left| m\phi + g\phi^2 \right|^2,$$

Verifying that two degenerate (supersymmetric) ground states survive at $m = 0$, when the classical approximation is no longer valid

has two minima, at $\phi = 0$ and at $\phi = -m/g$, corresponding to zero energy. Furthermore, as will be seen below, both ground states are of the "boson" type and have no "fermion" partners. We conclude that Witten's index for the system is

$$I_\text{W} = 2. \tag{10.711}$$

Supersymmetry cannot be broken. Because of the supersymmetry, continuous variations in m should not affect I_W. It is instructive to examine how the ground state of the quantum problem $H\Psi = E\Psi$ continues to be doubly degenerate (at $E = 0$) in the limit $m = 0$. In this limit $V_{\text{pot}} = g\phi^4$, so that classically there exists only one zero-energy state.

Now we will examine this massless case. Let us choose first the spin state, $| \uparrow\downarrow\rangle$ or $| \downarrow\uparrow\rangle$. Then the spin part of the Hamiltonian (10.710) acting on these states vanishes. It is obvious that the wave functions corresponding to these spin states are characterized by $E > 0$. Indeed, if the coordinate part of the wave function is denoted by Φ then, with the spin term switched off, we have

$$\langle\Phi|(\pi_\phi \pi_{\phi^\dagger} + |F|^2)|\Phi\rangle > 0.$$

Thus, the wave function of the ground state should have the form

$$\Psi = \Phi_1| \uparrow\uparrow\rangle + \Phi_2| \downarrow\downarrow\rangle. \tag{10.712}$$

Now we take this *ansatz*, act on it with the supercharges, and require the result to be zero, $Q\Psi = 0$.

The Lagrangian (10.703) is invariant under the following transformations:

$$\phi \leftrightarrow \phi^\dagger, \qquad F \leftrightarrow F^\dagger, \qquad \psi^1 \leftrightarrow \psi^{2\dagger}, \qquad \psi^2 \leftrightarrow \psi^{1\dagger}. \qquad (10.713)$$

(In field theory these transformations would correspond to C-parity.) Under the transformations (10.713),

$$Q^1 \leftrightarrow Q^{2\dagger}, \qquad Q^2 \leftrightarrow Q^{1\dagger}.$$

Therefore, instead of considering four supercharges, Q^α and $Q^{\alpha\dagger}$, which must annihilate the vacuum state it is quite sufficient to keep two:

$$\frac{1}{\sqrt{2}}Q^1\Psi = \pi_\phi \Phi_1 | \downarrow\uparrow\rangle + iF\Phi_2 | \downarrow\uparrow\rangle = 0,$$
$$\frac{1}{\sqrt{2}}Q^{1\dagger}\Psi = \pi_{\phi^\dagger} \Phi_2 | \uparrow\downarrow\rangle + iF^\dagger\Phi_1 | \uparrow\downarrow\rangle = 0.$$

These equations, written down in an explicit form, imply that

$$\frac{\partial\Phi_1(\phi,\phi^\dagger)}{\partial\phi} = -g(\phi^\dagger)^2\Phi_2(\phi,\phi^\dagger), \qquad \frac{\partial\Phi_2(\phi,\phi^\dagger)}{\partial\phi^\dagger} = -g\phi^2\Phi_1(\phi,\phi^\dagger).$$
$$(10.714)$$

After some reflection it is not difficult to see that the solutions of the system (10.714) are

$$\Phi_1 = X(r), \qquad \Phi_2 = Y(r)e^{i\alpha} \qquad (10.715)$$

and

$$\Phi_1 = Y(r)e^{-i\alpha}, \qquad \Phi_2 = X(r), \qquad (10.716)$$

where $r \equiv |\phi|$ and $\alpha \equiv \arg\phi$, while the functions X, Y satisfy the following system of first-order linear differential equations:

$$X' = -2gr^2Y, \qquad Y' - \frac{Y}{r} = -2gr^2X. \qquad (10.717)$$

The solution is expressible in terms of the McDonald functions,

$$X = -r^2 K_{2/3}\left(\frac{2gr^3}{3}\right), \qquad Y = r^2 K_{1/3}\left(\frac{2gr^3}{3}\right), \qquad (10.718)$$

which fall off exponentially at large r. Thus, we see that there are indeed two ground states,

$$\Psi_{(1)} = -r^2 K_{2/3} | \uparrow\uparrow\rangle + r^2 K_{1/3}e^{i\alpha} | \downarrow\downarrow\rangle,$$
$$\Psi_{(2)} = r^2 K_{1/3}e^{-i\alpha} | \uparrow\uparrow\rangle - r^2 K_{2/3} | \downarrow\downarrow\rangle, \qquad (10.719)$$

where the argument of the McDonald function is $2gr^3/3$. The orthogonality of $\Psi_{(1)}$ and $\Psi_{(2)}$ is trivially ensured by the angular factor $\exp(i\alpha)$.

Finally, we note that both states (10.719) are of the boson type. The states of fermion type are obtained from these if one acts with the supercharge operators, and they obviously have the structure $| \uparrow\downarrow\rangle$ or $| \downarrow\uparrow\rangle$.

References for Chapter 10

[1] E. Witten, in G. Kane, *Supersymmetry: Unveiling the Ultimate Laws of Nature* (Perseus Publishing, 2000).

[2] N. Seiberg and E. Witten, *Nucl. Phys. B* **426**, 19 (1994). Erratum: *ibid.* **430**, 485 (1994) [hep-th/9407087].

[3] J. Wess, From symmetry to supersymmetry, in G. Kane and M. Shifman (eds.), *The Supersymmetric World* (World Scientific, Singapore, 2000), pp. 67–86.

[4] Yu. A. Golfand and E. P. Likhtman, *JETP Lett.* **13**, 323 (1971) [reprinted in S. Ferrara (ed.), *Supersymmetry* (North Holland/World Scientific, 1987), Vol. 1, pp. 7–10].

[5] D. V. Volkov and V. P. Akulov, *JETP Lett.* **16**, 438 (1972).

[6] A. Neveu and J. H. Schwarz, *Nucl. Phys. B* **31**, 86 (1971).

[7] J. L. Gervais and B. Sakita, *Nucl. Phys. B* **34**, 632 (1971).

[8] J. Wess and B. Zumino, *Nucl. Phys. B* **70**, 39 (1974).

[9] J. Wess and J. Bagger, *Supersymmetry and Supergravity*, Second Edition (Princeton University Press, 1992).

[10] S. J. Gates, Jr., M.T. Grisaru, M. Roček, and W. Siegel, *Superspace, or One Thousand and One Lessons in Supersymmetry* (Benjamin/Cummings Publishing, 1983), [arXiv:hep-th/0108200].

[11] D. Bailin and A. Love, *Supersymmetric Gauge Field Theory and String Theory* (IOP Publishing, 1994).

[12] H. J. W. Mmüller-Kirsten and A. Wiedemann, *Introduction to Supersymmetry*, Second Edition (World Scientific, Singapore, 2010).

[13] P. Srivastava, *Supersymmetry, Superfields and Supergravity: An Introduction*, (IOP Publishing, Bristol, 1986).

[14] J. Terning, *Modern Supersymmetry: Dynamics and Duality* (Clarendon Press, Oxford, 2006).

[15] M. Dine, *Supersymmetry and String Theory: Beyond the Standard Model* (Cambridge University Press, 2007).

[16] D. Olive and P. West (eds.), *Duality and Supersymmetric Theories* (Cambridge University Press, 1999).

[17] H. Baer and X. Tata, *Weak Scale Supersymmetry: From Superfields to Scattering Events* (Cambridge University Press, 2006).

[18] P. M. R. Binétruy, *Supersymmetry: Theory, Experiment, and Cosmology* (Oxford University Press, 2006).

[19] I. Aitchison, *Supersymmetry in Particle Physics: An Elementary Introduction* (Cambridge University Press, 2007).

[20] P. West, *Introduction to Supersymmetry and Supergravity*, Second Edition (World Scientific, Singapore, 1990).

[21] S. Weinberg, *The Quantum Theory of Fields* (Cambridge University Press, 2000), Vol. 3.

[22] P. Deligne and J. Morgan, Notes on supersymmetry, in P. Deligne *et al.* (eds.), *Quantum Fields and Strings: A Course for Mathematicians* (American Mathematical Society, 1999), Vol. 1, p. 41.

[23] S. Ferrara (ed.), *Supersymmetry* (North Holland/World Scientific, 1987), Vol. 1.

[24] V. Berestetskii, E. Lifshitz, and L. Pitaevskii, *Quantum Electrodynamics* (Pergamon, 1980), Section 17.

[25] M. Peskin and D. Schroeder, *An Introduction to Quantum Field theory*, (CRC Press, Boca Raton, Florida, 2018)

[26] S. R. Coleman and J. Mandula, *Phys. Rev.* **159**, 1251 (1967).

[27] E. Witten, Introduction to supersymmetry, in A. Zichichi (ed.), *The Unity of the Fundamental Interactions* (Plenum Press, New York, 1983), pp. 305–355.

[28] R. Haag, J. T. Łopuszański, and M. Sohnius, *Nucl. Phys. B* **88**, 257 (1975) [reprinted in S. Ferrara (ed.), *Supersymmetry* (North Holland/World Scientific, 1987) Vol. 1, pp. 51–68].

[29] C. M. Hull and E. Witten, *Phys. Lett. B* **160**, 398 (1985).

[30] B. Zumino, Supersymmetric sigma models in 2 dimensions, in D. Olive and P. West (eds.), *Duality and Supersymmetric Theories* (Cambridge University Press, 1999), pp. 49–61.

[31] O. Aharony, O. Bergman, D. L. Jafferis, and J. Maldacena, *JHEP* **0810**, 091 (2008) [arXiv:0806.1218 [hep-th]].

[32] E. Witten and D. I. Olive, *Phys. Lett. B* **78**, 97 (1978).

[33] W. Pauli, *Pauli Lectures on Physics, Selected Topics in Field Quantization* (MIT Press, Cambridge, 1973), Vol. 6, p. 33.

[34] A. Salam and J. A. Strathdee, *Nucl. Phys. B* **76**, 477 (1974); *Nucl. Phys. B* **86**, 142 (1975) [reprinted in A. Ali *et al.* (eds.), *Selected Papers of Abdus Salam* (World Scientific, Singapore, 1994) pp. 438–448].

[35] S. Ferrara, J. Wess, and B. Zumino, *Phys. Lett. B* **51**, 239 (1974).

[36] J. Wess and B. Zumino, *Phys. Lett. B* **49**, 52 (1974) [reprinted in S. Ferrara (ed.), *Supersymmetry*, (North-Holland/World Scientific, Amsterdam–Singapore, 1987), Vol. 1, p. 77].

[37] F. A. Berezin, *Method of Second Quantization* (Academic Press, New York, 1966); *Introduction to Superanalysis* (Springer-Verlag, Berlin, 2001).

[38] Z. Komargodski and N. Seiberg, *JHEP* **0906**, 007 (2009) [arXiv:0904.1159 [hep-th]]; *JHEP* **1007**, 017 (2010) [arXiv:1002.2228 [hep-th]].

[39] T. Dumitrescu and N. Seiberg, *JHEP* **1107**, 095 (2011) [arXiv:1106.0031].

[40] S. Ferrara and B. Zumino, *Nucl. Phys. B* **87**, 207 (1975).

[41] P. Fayet and J. Iliopoulos, *Phys. Lett. B* **51**, 461 (1974).

[42] K. Evlampiev and A. Yung, *Nucl. Phys. B* **662**, 120 (2003) [arXiv:hep-th/0303047].

[43] M. Shifman and A. Yung, *Supersymmetric Solitons* (Cambridge University Press, 2009).

[44] J. Wess and B. Zumino, *Phys. Lett. B* **49**, 52 (1974); J. Iliopoulos and B. Zumino, *Nucl. Phys. B* **76**, 310 (1974); P. West, *Nucl. Phys. B* **106**, 219 (1976); M. Grisaru, M. Roček, and W. Siegel, *Nucl. Phys. B* **159** 429 (1979).

[45] N. Seiberg, *Phys. Lett. B* **318**, 469 (1993) [arXiv:hep-ph/9309335].

[46] M. A. Shifman and A. I. Vainshtein, *Nucl. Phys. B* **277**, 456 (1986).

[47] W. Fischler, H. P. Nilles, J. Polchinski, S. Raby, and L. Susskind, *Phys. Rev. Lett.* **47**, 757 (1981).

[48] P. Fayet, *Nucl. Phys. B* **90**, 104 (1975).

[49] L. O'Raifeartaigh, *Nucl. Phys. B* **96**, 331 (1975).

[50] S. Ferrara, L. Girardello, and F. Palumbo, *Phys. Rev. D* **20**, 403 (1979).

[51] A. Salam and J. A. Strathdee, *Phys. Lett. B* **49**, 465 (1974) [reprinted in A. Ali *et al.* (eds.), *Selected Papers of Abdus Salam*, (World Scientific, Singapore, 1994) pp. 423–437]; J. Iliopoulos and B. Zumino, *Nucl. Phys. B* **76**, 310 (1974).

[52] A. Losev, M. A. Shifman, and A. I. Vainshtein, *New J. Phys.* **4**, 21 (2002) [arXiv:hep-th/0011027]; *Phys. Lett. B* **522**, 327 (2001) [arXiv:hep-th/0108153].

[53] E. Witten, *Phys. Rev. D* **16**, 2991 (1977); P. Di Vecchia and S. Ferrara, *Nucl. Phys. B* **130**, 93 (1977).

[54] B. Zumino, *Phys. Lett. B* **87**, 203 (1979).

[55] V. A. Novikov, M. A. Shifman, A. I. Vainshtein, and V. I. Zakharov, *Phys. Rept.* **116**, 103 (1984).

[56] L. D. Landau and E. M Lifshitz, *The Classical Theory of Fields* (Pergamon Press, 1987), Section 92.

[57] A. M. Polyakov, *Phys. Lett. B* **59**, 79 (1975).

[58] L. Alvarez-Gaumé and D. Z. Freedman, *Commun. Math. Phys.* **91**, 87 (1983); S. J. Gates, *Nucl. Phys. B* **238**, 349 (1984); S. J. Gates, C. M. Hull, and M. Roček, *Nucl. Phys. B* **248**, 157 (1984).

[59] S. Ferrara and B. Zumino, *Nucl. Phys. B* **79**, 413 (1974) [reprinted in S. Ferrara (ed.), *Supersymmetry* (North Holland/World Scientific, Amsterdam–Singapore, 1987), Vol. 1, p. 93]; A. Salam and J. A. Strathdee, *Phys. Lett. B* **51**, 353 (1974) [reprinted in S. Ferrara (ed.), *Supersymmetry* (North Holland/World Scientific, Amsterdam–Singapore, 1987), Vol. 1, p. 102].

[60] E. Witten, *Nucl. Phys. B* **202**, 253 (1982) [reprinted in S. Ferrara (ed.), *Supersymmetry* (North Holland/World Scientific, Amsterdam-Singapore, 1987), Vol. 1, p. 490].

[61] M. A. Shifman and A. I. Vainshtein, *Nucl. Phys. B* **296**, 445 (1988).

[62] M. A. Shifman and A. I. Vainshtein, Instantons versus supersymmetry: fifteen years later, in M. Shifman (ed.), *ITEP Lectures on Particle Physics and Field Theory* (World Scientific, Singapore, 1999) Vol. 2, pp. 485–647 [hep-th/9902018].

[63] N. M. Davies, T. J. Hollowood, V. V. Khoze, and M. P. Mattis, *Nucl. Phys. B* **559**, 123 (1999) [hep-th/9905015].

[64] A. Armoni, M. Shifman, and G. Veneziano, From super-Yang–Mills theory to QCD: Planar equivalence and its implications, in M. Shifman, A. Vainshtein, and J. Wheater (eds.), *From Fields to Strings: Circumnavigating Theoretical Physics* (World Scientific, Singapore, 2004), Vol. 1, pp. 353–444 [arXiv:hep-th/0403071].

[65] A. Armoni, M. Shifman, and G. Veneziano, *Nucl. Phys. B* **667**, 170 (2003) [arXiv:hep-th/0302163]; *Phys. Rev. D* **71**, 045 015 (2005) [arXiv:hep-th/0412203].

[66] I. Affleck, M. Dine, and N. Seiberg, *Nucl. Phys. B* **241**, 493 (1984).

[67] T. Banks and N. Seiberg, Symmetries and strings in field theory and gravity, *Phys. Rev. D* **83**, 084019 (2011) [arXiv:1011.5120].

[68] A. Salam and J. A. Strathdee, *Nucl. Phys. B* **87**, 85 (1975).

[69] M. Grisaru, Anomalies in supersymmetric theories, in M. Lévy and S. Deser (eds.), *Recent Developments in Gravitation* (Plenum Press, New York, 1979), p. 577; an updated version of this paper is published in M. Shifman (ed.), *The Many Faces of the Superworld* (World Scientific, Singapore, 2000), p. 370.

[70] T. E. Clark, O. Piguet, and K. Sibold, *Nucl. Phys. B* **159**, 1 (1979); K. Konishi, *Phys. Lett. B* **135**, 439 (1984); K. Konishi and K. Shizuya, *Nuov. Cim. A* **90**, 111 (1985).

[71] N. Seiberg, *Nucl. Phys. B* **435**, 129 (1995) [arXiv:hep-th/9411149].

[72] Y. Meurice and G. Veneziano, *Phys. Lett. B* **141**, 69 (1984).

[73] I. Affleck, M. Dine, and N. Seiberg, *Phys. Rev. Lett.* **52**, 1677 (1984).

[74] A. S. Galperin, E. A. Ivanov, V. I. Ogievetsky, and E. S. Sokatchev, *Harmonic Superspace* (Cambridge University Press, 2001).

[75] B. Zumino, *Phys. Lett. B* **69**, 369 (1977).

[76] V. A. Novikov, M. A. Shifman, A. I. Vainshtein, and V. I. Zakharov *Nucl. Phys. B* **260**, 157 (1985) [reprinted in M. Shifman (ed.), *Instantons in Gauge Theories* (World Scientific, Singapore, 1994), p. 311].

[77] A. A. Belavin, A. M. Polyakov, A. S. Schwarz, and Yu. S. Tyupkin, *Phys. Lett. B* **59** 85 (1975) [reprinted in M. Shifman (ed.), *Instantons in Gauge Theories* (World Scientific, Singapore, 1994), p. 22].

[78] P. Fayet and S. Ferrara, *Phys. Rept.* **32**, 249 (1977).

[79] V. A. Novikov, M. A. Shifman, A. I. Vainshtein, M. B. Voloshin, and V. I. Zakharov, *Nucl. Phys. B* **229**, 394 (1983) [reprinted in M. Shifman (ed.), *Instantons in Gauge Theories* (World Scientific, Singapore, 1994), p. 298].

[80] G 't Hooft, *Phys. Rev. D* **14**, 3432 (1976). Erratum: *ibid.* **18**, 2199 (1978) [reprinted in M. Shifman (ed.), *Instantons in Gauge Theories* (World Scientific, Singapore, 1994), p. 70; note that in the reprinted version the numerical errors summarized in the Erratum above are corrected].

[81] M. Shifman and A. Vainshtein, *Nucl. Phys. B* **362**, 21 (1991) [reprinted in M. Shifman (ed.), *Instantons in Gauge Theories* (World Scientific, Singapore, 1994), p. 97].

[82] M. Shifman, A. Vainshtein, and V. Zakharov, *Nucl. Phys. B* **163**, 46 (1980); **165**, 45 (1980).

[83] L. Brown, R. Carlitz, D. Creamer, and C. Lee, *Phys. Rev. D* **17**, 1583 (1978) [reprinted in M. Shifman (ed.), *Instantons in Gauge Theories*, (World Scientific, Singapore, 1994), p. 168].

[84] V. A. Novikov, M. A. Shifman, A. I. Vainshtein, and V. I. Zakharov, *Nucl. Phys. B* **229**, 381 (1983); *Phys. Lett. B* **166**, 329 (1986).

[85] V. Novikov, M. Shifman, A. Vainshtein, and V. Zakharov, *Phys. Lett. B* **217**, 103 (1989).

[86] M. Shifman and A. Vainshtein, *Nucl. Phys. B* **359**, 571 (1991).

[87] I. Jack, D.R.T. Jones, and A. Pickering, *Phys. Lett. B* **435**, 61 (1998) and references therein.

[88] E. Witten, *JHEP* **9802**, 006 (1998) [arXiv:hep-th/9712028] (see Appendix).

[89] A. Keurentjes, A. Rosly, and A. Smilga, *Phys. Rev. D* **58**, 081 701 (1998); V. Kǎc and A. Smilga [hep-th/9902029], in M. Shifman (ed.), *The Many Faces of the Superworld*, (World Scientific, Singapore, 1999) pp. 185–234.

[90] A. Morozov, M. Olshanetsky, and M. Shifman, *Nucl. Phys. B* **304**, 291 (1988).

[91] V. A. Novikov, M. A. Shifman, A. I. Vainshtein, and V. I. Zakharov, *Nucl. Phys. B* **229**, 407 (1983) [reprinted in S. Ferrara, *Supersymmetry* (World Scientific, 1987), Vol. 1, p. 606].

[92] L. Girardello and M. T. Grisaru, *Nucl. Phys. B* **194**, 65 (1982).

[93] E. B. Bogomol'nyi, *Sov. J. Nucl. Phys.* **24**, 449 (1976) [reprinted in C. Rebbi and G. Soliani (eds.), *Solitons and Particles* (World Scientific, Singapore, 1984) p. 389].

[94] M. K. Prasad and C. M. Sommerfield, *Phys. Rev. Lett.* **35**, 760 (1975) [reprinted in C. Rebbi and G. Soliani (eds.), *Solitons and 94Particles* (World Scientific, Singapore, 1984) p. 530].

[95] A. Bilal, *Introduction to Supersymmetry*, lecture at *Ecole de Gif 2000, Supercordes et Dimensions Supplémentaires*, September, 2000 [arXiv:hep-th/0101055].

[96] H. Eichenherr, *Nucl. Phys. B* **146**, 215 (1978). Erratum: *ibid*. **155**, 544 (1979); V. L. Golo and A. M. Perelomov, *Lett. Math. Phys.* **2**, 477 (1978); E. Cremmer and J. Scherk, *Phys. Lett. B* **74**, 341 (1978).

[97] E. Witten, *Nucl. Phys. B* **149**, 285 (1979).

[98] M. Shifman and A. Yung, *Phys. Rev. D* **77**, 125 017 (2008). Erratum: *ibid*. **81**, 089 906 (2010) [arXiv:0803.0698 [hep-th]]; P. A. Bolokhov, M. Shifman, and A. Yung, *Phys. Rev. D* **82**, 025 011 (2010) [arXiv:1001.1757 [hep-th]].

[99] J. D. Bjorken and S. D. Drell, *Relativistic Quantum Fields* (McGraw-Hill, 1965).

[100] K. V. Stepanyantz, Nucl. Phys. B **852**, 71-107 (2011); A. L. Kataev and K. V. Stepanyantz, Nucl. Phys. B **875**, 459-482 (2013); I. L. Buchbinder and K. V. Stepanyantz, Nucl. Phys. B **883**, 20-44 (2014); A. L. Kataev, A. E. Kazantsev and K. V. Stepanyantz, Phys. Part. Nucl. **49**, no.5, 911-913 (2018); K. V. Stepanyantz, JHEP **01**, 192 (2020).

[101] N. Nekrasov, Adv. Theor. Math. Phys. **7** 831 (2004).

11 Supersymmetric Solitons

Classifying centrally extended superalgebras. — Meet Bogomol'nyi, Prasad, and Sommerfield. — BPS-saturated (or critical or supersymmetric) solitons. — Bogomol'nyi completion, topological and central charges. — Kinks and domain walls. — Vortices and strings. — Monopoles. — Semiclassical quantization of moduli.

11.1 Central Charges in Superalgebras

In this section we will briefly review general issues related to central charges (CCs) in superalgebras.

11.1.1 History

The first superalgebra in four-dimensional field theory was derived by Golfand and Likhtman [1] in the form

$$\{\bar{Q}_{\dot\alpha}, Q_\beta\} = 2P_\mu\,(\sigma^\mu)_{\alpha\beta}\,, \qquad \{\bar{Q}_\alpha, \bar{Q}_\beta\} = \{Q_\alpha, Q_\beta\} = 0; \qquad (11.1)$$

thus it has no central charges. The possible occurrence of CCs (the elements of the superalgebra that commute with all other generators) was first mentioned in an unpublished paper of Łopuszański and Sohnius [2] where the last two anticommutators were modified to

$$\{Q_\alpha^I, Q_\beta^G\} = Z_{\alpha\beta}^{IG}. \qquad (11.2)$$

Look through Section 10.25.

The superscripts I, G indicate extended supersymmetry. A more complete description of superalgebras with CCs in quantum field theory was worked out in [3]. The central charge derived in this paper was for $\mathcal{N} = 2$ superalgebra in four dimensions, $Z_{\alpha\beta}^{IG} \sim \varepsilon_{\alpha\beta}\varepsilon^{IG}$. It is Lorentz scalar.

A few years later, Witten and Olive [4] showed that, in supersymmetric theories with solitons, the central extension of superalgebras is typical; topological quantum numbers play the role of central charges.

It was generally understood that superalgebras with (Lorentz-scalar) central charges can be obtained from superalgebras without central charges in higher-dimensional space–time by interpreting some of the extra components of the momentum as CCs (see, e.g., [5]). When one compactifies the extra dimensions one obtains an extended supersymmetry; the extra components of the momentum act as scalar central charges.

Algebraic analysis extending that of [3], carried out in the early 1980s (see, e.g., [6]), indicated that the super-Poincaré algebra admits "central charges" of a more general form, but the dynamical role of the additional tensorial charges was not recognized until much later, when it was finally realized that extensions with Lorentz-noninvariant "central charges" (such as $(1,0) + (0,1)\, Z_{\{\alpha\beta\}}$ or $(1/2, 1/2)\, Z_\mu$) not only exist but play a very important role in the theory of supersymmetric solitons. Above, I have put central charges in quotation marks because $Z_{\{\alpha\beta\}}$ or Z_μ or other Lorentz-noninvariant elements of superalgebras in various dimensions are not central in the strict sense: they only commute with $Q_\alpha, \bar{Q}_{\dot{\alpha}}$, and P_μ, not with Lorentz rotations since they carry Lorentz indices. They are associated with extended topological defects – such as domain walls or strings – and could be called *brane charges*. Leaving this subtlety aside, I will continue to refer to these elements as central charges, or, sometimes, tensorial central charges. I want to stress again that the latter originate from operators other than the energy–momentum operator in higher dimensions.

Brane charges

Central charges that are antisymmetric tensors in various dimensions were introduced (in the supergravity context, in the presence of *p*-branes) in [7] (see also [8, 9]). These CCs are relevant to extended objects of domain-wall type (i.e., branes). Their occurrence in four-dimensional super-Yang–Mills theory, as a quantum anomaly, was first observed in [10]. A general theory of central extensions of superalgebras in three and four dimensions was discussed in [11]. It is worth noting that central charges that have the Lorentz structure of Lorentz vectors were not considered in [11]. This gap was closed in [12].

11.1.2 Minimal Supersymmetry

The minimal number of supercharges ν_Q in various dimensions is given in Table 11.1. Two-dimensional theories with a single supercharge, although algebraically possible, are quite exotic. In "conventional" models in $D = 2$ with local interactions the minimal number of supercharges is 2.

The minimal number of supercharges in Table 11.1 is given for a real representation. It is clear that, generally speaking, the maximal possible number of CCs is determined by the dimension of the symmetric matrix $\{Q_i, Q_j\}$ of size $\nu_Q \times \nu_Q$, namely,

$$\nu_{CC} = \frac{\nu_Q(\nu_Q + 1)}{2}. \tag{11.3}$$

In fact, the D anticommutators have the Lorentz structure of the energy–momentum operator P_μ. Therefore, up to D central charges could be absorbed in P_μ, generally speaking. In particular situations this number can be smaller, since although algebraically the corresponding CCs have the same structure as P_μ, they are dynamically distinguishable. The point is that P_μ is uniquely defined through the conserved and symmetric energy–momentum tensor of the theory.

Additional dynamical and symmetry constraints can diminish further the number of independent central charges; see Section 11.1.2.1 below.

Table 11.1. For varying dimension D, the minimal number of supercharges, the complex dimension of the spinorial representation, and the number of additional conditions (i.e., the Majorana and/or Weyl conditions)

D	2	3	4	5	6	7	8	9	10
ν_Q	$(1^*)2$	2	4	8	8	8	16	16	16
$\mathrm{Dim}(\psi)_C$	2	2	4	4	8	8	16	16	32
No. cond.	2	1	1	0	1	1	1	1	2

The total set of CCs can be arranged by classifying the CCs with respect to their Lorentz structure. Below I will present this classification for $D = 2, 3$, and 4, with special emphasis on the four-dimensional case. In Section 11.1.3 we will deal with $\mathcal{N} = 2$ superalgebras.

11.1.2.1 $D = 2$

Consider two-dimensional theories with two supercharges. From the discussion above, on purely algebraic grounds three CCs are possible: one Lorentz scalar, and a two-component vector

$$\{Q_\alpha, Q_\beta\} = 2(\gamma^\mu \gamma^0)_{\alpha\beta}(P_\mu + Z_\mu) + i(\gamma^5 \gamma_0)_{\alpha\beta} Z. \tag{11.4}$$

The condition $Z^\mu \neq 0$ would require the existence of a vector order parameter taking distinct values in different vacua. Indeed, if this CC existed, its current would have the form

$$\zeta_\nu^\mu = \varepsilon_{\nu\rho} \partial^\rho A^\mu, \qquad Z^\mu = \int dz \zeta_0^\mu,$$

where A^μ is the above-mentioned order parameter. However, $\langle A^\mu \rangle \neq 0$ would break the Lorentz invariance and supersymmetry of the vacuum state. This option will not be considered. Limiting ourselves to supersymmetric vacua we conclude that a single (real) Lorentz-scalar central charge Z is possible in $\mathcal{N} = 1$ theories. This central charge is saturated by kinks.

11.1.2.2 $D = 3$

The CC allowed in this case is a Lorentz vector Z_μ, i.e.,

$$\{Q_\alpha, Q_\beta\} = 2(\gamma^\mu \gamma^0)_{\alpha\beta}(P_\mu + Z_\mu). \tag{11.5}$$

One should arrange Z_μ to be orthogonal to P_μ. In fact, this is the scalar central charge of Section 11.1.2.1 elevated by one dimension. Its topological current can be written as

$$\zeta_{\mu\nu} = \varepsilon_{\mu\nu\rho} \partial^\rho A, \qquad Z_\mu = \int d^2x \, \zeta_{\mu 0}. \tag{11.6}$$

By an appropriate choice of reference frame, Z_μ can always be reduced to a real number times $(0, 0, 1)$. This CC is associated with a domain line oriented along the second axis.

Although from the general relation (11.5) it is fairly clear why BPS vortices cannot appear in theories with two supercharges, it is instructive to discuss this question from a slightly different standpoint. Vortices in three-dimensional theories are localized objects, i.e., particles (BPS vortices in 2+1 dimensions were considered in [13]). The number of broken translational generators is d, where d is the soliton's codimension; $d = 2$ in the case at hand. Then *at least* d supercharges are broken. Since we have only two supercharges in the present case, both must be broken. This simple argument tells us that for a 1/2-BPS vortex the minimal matching between the bosonic and fermionic zero modes in the (super)translational sector is one-to-one.

Consider now a putative BPS vortex in a theory with minimal $\mathcal{N} = 1$ supersymmetry in $2 + 1$ dimensions. Such a configuration would require a world volume description with two bosonic zero modes but only one fermionic mode. This is not permitted, by the argument above, and indeed no configuration of this type is known. Vortices always exhibit at least two fermionic zero modes and can be BPS-saturated only in $\mathcal{N} = 2$ theories.

11.1.2.3 $D = 4$

Maximally one can have 10 CCs, which are decomposed into Lorentz representations as $(0, 1) + (1, 0) + (\frac{1}{2}, \frac{1}{2})$:

$$\{Q_\alpha, \bar{Q}_{\dot\alpha}\} = 2(\gamma^\mu)_{\alpha\dot\alpha}(P_\mu + Z_\mu), \tag{11.7}$$

$$\{Q_\alpha, Q_\beta\} = (\Sigma^{\mu\nu})_{\alpha\beta} Z_{[\mu\nu]}, \tag{11.8}$$

$$\{\bar{Q}_{\dot\alpha}, \bar{Q}_{\dot\beta}\} = (\bar\Sigma^{\mu\nu})_{\dot\alpha\dot\beta} \bar{Z}_{[\mu\nu]}, \tag{11.9}$$

where $(\Sigma^{\mu\nu})_{\alpha\beta} = (\sigma^\mu)_{\alpha\dot\alpha}(\bar\sigma^\nu)^{\dot\alpha}_\beta$ is a chiral version of $\sigma^{\mu\nu}$ (see Section 10.2, Eq. (10.34)). The antisymmetric tensors $Z_{[\mu\nu]}$ and $\bar{Z}_{[\mu\nu]}$ are associated with the domain walls and reduce to a complex number and a spatial vector orthogonal to a domain wall. The $(\frac{1}{2}, \frac{1}{2})$ CC Z_μ is a Lorentz vector orthogonal to P_μ. It is associated with strings (flux tubes) and reduces to one real number and a three-dimensional unit spatial vector parallel to the string.

11.1.3 Extended SUSY

In four dimensions one can extend the superalgebra up to $\mathcal{N} = 4$, which corresponds to 16 supercharges. Reducing this to lower dimensions, we obtain a rich variety of extended superalgebras in $D = 3$ and 2. In fact, in two dimensions Lorentz invariance provides a much weaker constraint than in higher dimensions, and one can consider a wider set of (p, q) superalgebras comprising $p + q = 4, 8$, or 16 supercharges. We will not pursue a general solution; instead, we will limit our task to: (i) analysis of the CCs in $\mathcal{N} = 2$ in four dimensions; (ii) reduction of the minimal SUSY algebra in $D = 4$ to $D = 2$ and 3, i.e., to the $\mathcal{N} = 2$ SUSY algebra in those dimensions. Thus, in two dimensions we will consider only the nonchiral $\mathcal{N} = (2, 2)$ case. As should be clear from the discussion above, in the dimensional reduction the maximal number of CCs in a sense stays intact. What changes is the decomposition into Lorentz and R symmetry irreducible representations.

Two-dimensional chiral superalgebras $\mathcal{N} = (0, 2)$ or $\mathcal{N} = (2, 0)$ are also of interest. In a sense, they are close relatives of $4D$ super-Yang-Mills. I will not consider them in this textbook.

11.1.3.1 The Case $\mathcal{N} = 2$ in $D = 2$

Let us focus on the nonchiral $\mathcal{N} = (2, 2)$ case corresponding to the dimensional reduction of the $\mathcal{N} = 1$, $D = 4$ algebra. The tensorial decomposition is as follows:

$$\{Q^I_\alpha, Q^J_\beta\} = 2(\gamma^\mu\gamma^0)_{\alpha\beta}\left[(P_\mu + Z_\mu)\delta^{IJ} + Z^{(IJ)}_\mu\right] + 2i(\gamma^5\gamma^0)_{\alpha\beta}Z^{\{IJ\}}$$
$$+ 2i\gamma^0_{\alpha\beta}Z^{[IJ]}, \qquad I, J = 1, 2. \tag{11.10}$$

Here $Z^{[IJ]}$ is antisymmetric in I, J; $Z^{\{IJ\}}$ is symmetric; while $Z^{(IJ)}$ is symmetric and traceless. We can discard both vectorial CCs $Z_\mu^{(IJ)}$ and Z_μ for the same reasons as in Section 11.1.2.1. Then we are left with three Lorentz singlets $Z^{\{IJ\}}$, which represent the reduction of the domain-wall charges in $D = 4$ and one Lorentz singlet $Z^{[IJ]}$, arising from P_2 and the vortex charge in $D = 3$ (see Section 11.1.3.2). These CCs are saturated by kinks.

Summarizing, the $(2,2)$ superalgebra in $D = 2$ is

$$\{Q_\alpha^I, Q_\beta^J\} = 2(\gamma^\mu \gamma^0)_{\alpha\beta} P_\mu \delta^{IJ} + 2i(\gamma^5 \gamma^0)_{\alpha\beta} Z^{\{IJ\}} + 2i\gamma_{\alpha\beta}^0 Z^{[IJ]}. \tag{11.11}$$

Altogether we deal with four real components of $Z^{\{IJ\}}$ and $Z^{[IJ]}$ which can be rearranged into two complex Lorentz-invariant charges, Z and Z', see (11.14).

It is instructive to rewrite Eq. (11.11) in terms of the complex supercharges Q_α and $Q_\beta^\dagger$ corresponding to the four-dimensional Q_α, $\bar{Q}_{\dot\alpha}$; see Section 11.1.2.3. Then

$$\{Q_\alpha, Q_\beta^\dagger\}(\gamma^0)_{\beta\gamma} = 2\left(P_\mu \gamma^\mu + Z\frac{1-\gamma_5}{2} + Z^\dagger\frac{1+\gamma_5}{2}\right)_{\alpha\gamma},$$

$$\{Q_\alpha, Q_\beta\}(\gamma^0)_{\beta\gamma} = -2Z'\,(\gamma_5)_{\alpha\gamma}, \tag{11.12}$$

$$\{Q_\alpha^\dagger, Q_\beta^\dagger\}(\gamma^0)_{\beta\gamma} = 2Z'^\dagger\,(\gamma_5)_{\alpha\gamma}.$$

The algebra contains two complex CCs, Z and Z'. In terms of components $Q_\alpha = (Q_R, Q_L)$, the nonvanishing anticommutators are

$$\{Q_L, Q_L^\dagger\} = 2(H + P), \qquad \{Q_R, Q_R^\dagger\} = 2(H - P),$$

$$\{Q_L, Q_R^\dagger\} = 2iZ, \qquad \{Q_R, Q_L^\dagger\} = -2iZ^\dagger,$$

$$\{Q_L, Q_R\} = 2iZ', \qquad \{Q_R^\dagger, Q_L^\dagger\} = -2iZ'^\dagger. \tag{11.13}$$

These anticommutators exhibit the automorphism $Q_R \leftrightarrow Q_R^\dagger$, $Z \leftrightarrow Z'$ (see [14]). The complex CCs Z and Z' can be readily expressed in terms of real CCs $Z^{\{IJ\}}$ and $Z^{[IJ]}$:

$$Z = Z^{[12]} + \frac{i}{2}\left(Z^{\{11\}} + Z^{\{22\}}\right), \qquad Z' = \frac{Z^{\{12\}} + Z^{\{21\}}}{2} - i\frac{Z^{\{11\}} - Z^{\{22\}}}{2}. \tag{11.14}$$

Typically, in a given model either Z or Z' vanish. A practically important example to which we will repeatedly turn below is provided by the twisted-mass deformed $CP(N-1)$ model [15] (Section 10.12.3.6). The CC Z emerges in this model at the classical level. At the quantum level it acquires additional anomalous terms [16, 17].

11.1.3.2 The Case $\mathcal{N} = 2$ in $D = 3$

The superalgebra in this case can be decomposed into Lorentz and R symmetry tensorial structures as follows:

$$\{Q_\alpha^I, Q_\beta^J\} = 2(\gamma^\mu \gamma^0)_{\alpha\beta}[(P_\mu + Z_\mu)\delta^{IJ} + Z_\mu^{(IJ)}] + 2i\gamma_{\alpha\beta}^0 Z^{[IJ]}, \tag{11.15}$$

where all the CCs above are real. The maximal set of 10 CCs enters as a triplet of space–time vectors Z_μ^{IJ} and a singlet $Z^{[IJ]}$. The singlet CC is associated with vortices (or lumps) and corresponds to the reduction of the $(\frac{1}{2}, \frac{1}{2})$ charge or the fourth component of the momentum vector in $D = 4$. The triplet Z_μ^{IJ} is decomposed into an R symmetry singlet Z_μ, algebraically indistinguishable from the momentum, and a traceless symmetric combination $Z_\mu^{(IJ)}$. The former is equivalent to the vectorial charge in the $\mathcal{N} = 1$ algebra, while $Z_\mu^{(IJ)}$ can be reduced to a complex number and vectors specifying the orientation. We see that these are the direct reduction of the $(0, 1)$ and $(1, 0)$ wall charges in $D = 4$. They are saturated by domain lines.

11.1.3.3 Extended Supersymmetry (Eight Supercharges) in $D = 4$

The complete algebraic analysis of all tensorial central charges in this problem is analogous to the previous cases and is rather straightforward. With eight supercharges the maximal number of CCs is 36. The dynamical aspect is less developed – only a modest fraction of the above 36 CCs are known to be nontrivially realized in the models studied in the literature. We will limit ourselves to a few remarks regarding the well-established CCs. We use a complex (holomorphic) representation of the supercharges. Then the supercharges are labeled as follows:

$$Q_\alpha^F, \quad \bar{Q}_{\dot{\alpha}G}, \quad \alpha, \dot{\alpha} = 1, 2, \quad F, G = 1, 2. \tag{11.16}$$

On general grounds one can write

$$\{Q_\alpha^F, \bar{Q}_{\dot{\alpha}G}\} = 2\delta_G^F P_{\alpha\dot{\alpha}} + 2(Z_G^F)_{\alpha\dot{\alpha}},$$

$$\{Q_\alpha^F, Q_\beta^G\} = 2Z_{\{\alpha\beta\}}^{\{FG\}} + 2\varepsilon_{\alpha\beta}\varepsilon^{FG}Z, \tag{11.17}$$

$$\{\bar{Q}_{\dot{\alpha}F}, \bar{Q}_{\dot{\beta}G}\} = 2(\bar{Z}_{\{FG\}})_{\{\dot{\alpha}\dot{\beta}\}} + 2\varepsilon_{\dot{\alpha}\dot{\beta}}\varepsilon_{FG}\bar{Z}.$$

Here the $(Z_G^F)_{\alpha\dot{\alpha}}$ are four vectorial CCs $(\frac{1}{2}, \frac{1}{2})$, (16 components altogether) while $Z_{\{\alpha\beta\}}^{\{FG\}}$ and its complex conjugate are $(1, 0)$ and $(0, 1)$ CCs. Since the matrix $Z_{\{\alpha\beta\}}^{\{FG\}}$ is symmetric with respect to F and G there are three flavor components, while the total number of components residing in $(1, 0)$ and $(0, 1)$ CCs is 18. Finally, there are two scalar CCs, Z and $\bar{Z}$.

Dynamically the above CCs can be described as follows. The scalar CCs Z and $\bar{Z}$ are saturated by monopoles or dyons. One vectorial CC Z_μ (with the additional condition $P^\mu Z_\mu = 0$) is saturated [18] by an Abrikosov–Nielsen–Olesen string (ANO) [19]. A $(1, 0)$ CC with $F = G$ is saturated by domain walls [20].

Let us briefly discuss the Lorentz-scalar CCs in Eq. (11.17), which are saturated by monopoles or dyons. They will be referred to as monopole CCs. A rather dramatic story is associated with them. Historically they were the first to be introduced within the framework of an extended four-dimensional superalgebra [2, 3]. On the dynamical side, they appeared as the first example of the "topological charge ↔ central charge" relation revealed by Witten and Olive in their pioneering paper [4]. Twenty years later, the $\mathcal{N} = 2$ model, where these CCs first appeared, was solved by Seiberg and Witten [21, 22] and the exact masses of the BPS-saturated monopoles or dyons were found. No direct comparison with the operator expression for the CCs was carried out, however. In [23] it was noted that for the Seiberg–Witten formula to be valid, a boson-term anomaly should exist in the monopole CCs. Even before [23] a fermion-term anomaly was identified [20], which plays a crucial role [24] for monopoles in the Higgs regime (i.e., confined monopoles).

11.1.4 Which Supersymmetric Solitons Will Be Considered

Scott-
Russell's
discovery of
a solitary
wave
The term "soliton" was introduced in the 1960s but scientific research on solitons
had started much earlier, in the nineteenth century, when a Scottish engineer, John
Scott-Russell, observed a large solitary wave in a canal near Edinburgh.

We are already familiar with a few topologically stable (topological for short)
solitons, such as:

(i) kinks in $D = 1 + 1$ (when elevated to $D = 1 + 3$ they represent domain walls);
(ii) vortices in $D = 1 + 2$ (when elevated to $D = 1 + 3$ they represent strings or flux
     tubes);
(iii) magnetic monopoles in $D = 1 + 3$.

In the three cases above the topologically stable solutions have been known since the
1930s, 1950s, and 1970s, respectively. Then it was shown that all these solitons can
be embedded in supersymmetric theories [25]. To this end one adds an appropriate
fermion sector and, if necessary, expands the boson sector.

The presence of fermions leads to a variety of novel physical phenomena inherent
to BPS-saturated solitons.

Now we will explain why supersymmetric solitons are especially interesting. We
will start with the simplest model: one (real) scalar field in two dimensions plus the
minimal set of superpartners.

11.2 $\mathcal{N} = 1$: Supersymmetric Kinks

Look
through
Chapter 2,
especially
Section 2.1
and 2.5.
The embedding of bosonic models supporting kinks in $\mathcal{N} = 1$ supersymmetric models
in two dimensions was first discussed in [4, 26]. Occasional remarks about kinks
in models with four supercharges of the type found in Wess–Zumino models [27]
appeared in the literature of the 1980s but they went unnoticed. The question which
caused much interest and debate was that of quantum corrections to the BPS kink
mass in two-dimensional models with $\mathcal{N} = 1$ supersymmetry. By now this question
has been completely solved [28]. We will go through its solution in this section.

11.2.1 The Case $D = 1 + 1$ and $\mathcal{N} = 1$

The simplest BPS-saturated soliton in two dimensions is a kink of a special type. In
this subsection we will consider the simplest supersymmetric model in $D = 1 + 1$
that admits solitons. We met this model in Section 10.12.3.1. Its Lagrangian is

$$\mathcal{L} = \frac{1}{2} \left[\partial_\mu \phi \partial^\mu \phi + \bar{\psi} i \, \partial\!\!\!/ \, \psi - \left(\frac{\partial W}{\partial \phi} \right)^2 - \frac{\partial^2 W}{\partial \phi^2} \bar{\psi} \psi \right], \qquad (11.18)$$

where ϕ is a real scalar field, ψ is a Majorana spinor, and

$$\psi = \begin{pmatrix} \psi_1 \\ \psi_2 \end{pmatrix}, \qquad (11.19)$$

with $\psi_{1,2}$ real. Needless to say, the gamma matrices for the model must be chosen to be in the Majorana representation. A convenient choice is

$$\gamma^0 = \sigma_2, \qquad \gamma^1 = i\sigma_3, \tag{11.20}$$

where $\sigma_{2,3}$ are the Pauli matrices. (*Warning: this is in contradistinction with Section 10.2.2, in which we defined the γ matrices in two dimensions in the chiral representation.*) For future reference we will introduce a "γ_5" matrix, $\gamma^5 = \gamma^0\gamma^1 = -\sigma_1$. Moreover,

The scalar potential is related to the superpotential by
$U(\phi) = \frac{1}{2}(\partial \mathcal{W}/\partial\phi)^2.$

$$\bar{\psi} = \psi\gamma^0.$$

The *superpotential function* $\mathcal{W}(\phi)$ is, in principle, arbitrary. The model (11.18) with any $\mathcal{W}(\phi)$ is supersymmetric provided that $\mathcal{W}' \equiv \partial\mathcal{W}/\partial\phi$ vanishes at some value of ϕ. The points ϕ_i where

$$\frac{\partial \mathcal{W}}{\partial \phi} = 0$$

are called critical. As is seen from Eq. (11.18), they correspond to vanishing energy density,

$$U(\phi) = \frac{1}{2}\left(\frac{\partial\mathcal{W}}{\partial\phi}\right)^2_{\phi=\phi_i} = 0. \tag{11.21}$$

The critical points are the classical minima of the potential energy – the classical vacua. For our purposes, soliton studies, we require the existence of at least two distinct critical points in the problem at hand. The kink will interpolate between the two distinct vacua.

Superpolynomial and super-sine-Gordon

Two popular choices of superpotential function are:

$$\mathcal{W}(\phi) = \frac{m^2}{4\lambda}\phi - \frac{\lambda}{3}\phi^3, \tag{11.22}$$

and

$$\mathcal{W}(\phi) = mv^2 \sin\frac{\phi}{v}. \tag{11.23}$$

Here m, λ, and v are real (positive) parameters. The first model is referred to as superpolynomial (SPM), the second as super-sine-Gordon (SSG). The classical vacua in SPM are at $\phi = \pm m(2\lambda)^{-1}$. We will assume that $\lambda/m \ll 1$ to ensure the applicability of a quasiclassical treatment. This is the weak coupling regime for SPM. A kink solution interpolates between $\phi_*^- = -m/(2\lambda)$ at $z = -\infty$ and $\phi_*^+ = m/(2\lambda)$ at $z = \infty$, while an antikink interpolates between $\phi_*^+ = m/(2\lambda)$ and $\phi_*^- = -m/(2\lambda)$. The classical kink solution has the form

$$\phi_0 = \frac{m}{2\lambda}\tanh\frac{mz}{2}. \tag{11.24}$$

The weak coupling regime in the SSG case is attained for $v \gg 1$. In the sine-Gordon model there are infinitely many vacua; they lie at

$$\phi_*^k = v\left(\frac{\pi}{2} + k\pi\right), \tag{11.25}$$

Cf. Eq.
(9.195).

where k is an integer, either positive or negative. Correspondingly, there exist solitons connecting any pair of vacua. In this case we will limit ourselves to consideration of the "elementary" solitons connecting adjacent vacua, e.g., $\phi_*^{0,-1} = \pm\pi v/2$,

$$\phi_0 = v \arcsin[\tanh(mz)]. \tag{11.26}$$

In $D = 1 + 1$ the real scalar field represents one degree of freedom (bosonic) and so does the two-component Majorana spinor (fermionic). Thus, the number of bosonic and fermionic degrees of freedom matches, a necessary condition for supersymmetry. One can show in many different ways that the Lagrangian (11.18) possesses supersymmetry. For instance, let us consider the *supercurrent*,

Supercurrent for N = 1 in 2D

$$J^\mu = (\partial\phi)\gamma^\mu\psi + i\frac{\partial W}{\partial\phi}\gamma^\mu\psi. \tag{11.27}$$

On the one hand, this object is linear in the fermion field; therefore, it is obviously fermionic. On the other hand, it is conserved. Indeed,

$$\partial_\mu J^\mu = (\partial^2\phi)\psi + (\phi)(\partial\psi) + i\frac{\partial^2 W}{\partial\phi^2}(\partial\phi)\psi + i\frac{\partial W}{\partial\phi}\partial\psi. \tag{11.28}$$

The first, second, and third terms can be reexpressed by virtue of the equations of motion; this immediately results in various cancelations. After these cancelations the only term left in the divergence of the supercurrent is

$$\partial_\mu J^\mu = -\frac{1}{2}\frac{\partial^3 W}{\partial\phi^3}(\bar\psi\psi)\psi. \tag{11.29}$$

If one takes into account (i) the fact that the spinor ψ is real and two-component, and (ii) the Grassmannian nature of $\psi_{1,2}$, one can immediately conclude that the right-hand side in Eq. (11.29) vanishes.

The supercurrent conservation implies the existence of *two* conserved charges,[1]

$$Q_\alpha = \int dz\, J_\alpha^0 = \int dz\left[\left(\partial\phi + i\frac{\partial W}{\partial\phi}\right)\left(\gamma^0\psi\right)\right]_\alpha, \qquad \alpha = 1, 2. \tag{11.30}$$

These supercharges form a doublet with respect to the Lorentz group in $D = 1 + 1$. They generate supertransformations of the fields, for instance,

$$[Q_\alpha, \phi] = -i\psi_\alpha, \qquad \{Q_\alpha, \bar\psi_\beta\} = (\partial)_{\alpha\beta}\phi + i\frac{\partial W}{\partial\phi}\delta_{\alpha\beta}, \tag{11.31}$$

and so on. In deriving Eqs. (11.31) we have used the canonical commutation relations

$$[\phi(t, z), \dot\phi(t, z')] = i\delta(z - z'), \qquad \left\{\psi_\alpha(t, z), \bar\psi_\beta(t, z')\right\} = \left(\gamma^0\right)_{\alpha\beta}\delta(z - z'). \tag{11.32}$$

Note that by acting with Q on the bosonic field we get a fermionic field and vice versa. This demonstrates, once again, that the supercharges are symmetry generators with a fermion nature.

Given the expression (11.30) for the supercharges and the canonical commutation relations (11.32) it is not difficult to find the superalgebra:

$$\{Q_\alpha, \bar Q_\beta\} = 2(\gamma^\mu)_{\alpha\beta}P_\mu + 2i(\gamma^5)_{\alpha\beta}\mathcal{Z}. \tag{11.33}$$

---

[1] Remember, two-dimensional theories with two conserved supercharges are referred to as $N = 1$.

Here P_μ is the operator of the total energy and momentum,

$$P^\mu = \int dz\, T^{\mu 0},$$ (11.34)

Energy–momentum tensor
---

where $T^{\mu\nu}$ is the energy–momentum tensor,

$$T^{\mu\nu} = \partial^\mu \phi\, \partial^\nu \phi + \tfrac{1}{2}\bar\psi \gamma^\mu i \partial^\nu \psi - \tfrac{1}{2}g^{\mu\nu}\left[\partial_\gamma \phi \partial^\gamma \phi - (W')^2\right],$$ (11.35)

and $\mathcal{Z}$ is the central charge,

$$\mathcal{Z} = \int dz\, \partial_z W(\phi) = W[\phi(z = \infty)] - W[\phi(z = -\infty)].$$ (11.36)

The local form of the superalgebra (11.33) is

Local form of the superalgebra

$$\left\{J_\alpha^\mu, \bar Q_\beta\right\} = 2(\gamma_\nu)_{\alpha\beta}T^{\mu\nu} + 2i(\gamma^5)_{\alpha\beta}\zeta^\mu,$$ (11.37)

where ζ^μ is the conserved topological current,

$$\zeta^\mu = \varepsilon^{\mu\nu}\partial_\nu W.$$ (11.38)

Symmetrization (antisymmetrization) over the bosonic (fermionic) operators in the products is implied in the above expressions.

The CC $\mathcal{Z}$ replaces the topological charges of nonsupersymmetric theories.

I will pause here to make a comment. Since the CC is the integral of the full derivative, it is independent of the details of the soliton solution and is determined only by the boundary conditions. To ensure that $\mathcal{Z} \neq 0$ the field ϕ must tend to distinct limits at $z \to \pm\infty$.

11.2.2 Critical (BPS-Saturated) Kinks

A kink in $D = 1 + 1$ is a particle. Any given soliton solution obviously breaks translational invariance. Since $\{Q, \bar Q\} \propto P$, typically both supercharges are broken on the soliton solutions,

$$Q_\alpha|\text{sol}\rangle \neq 0, \qquad \alpha = 1, 2.$$ (11.39)

However, for certain special kinks one can preserve half the supersymmetry, say,

$$Q_1|\text{sol}\rangle \neq 0 \quad \text{but} \quad Q_2|\text{sol}\rangle = 0,$$ (11.40)

or vice versa. Therefore, here we will deal with critical, or BPS-saturated kinks.[2]

A critical kink must satisfy a first-order differential equation; this fact, as well as the particular form of the equation, follows from the inspection of Eq. (11.30) or the second equation in (11.31). Indeed, for static fields $\phi = \phi(z)$ the supercharge Q_α is proportional to a matrix:

$$Q_\alpha \propto \begin{pmatrix} \partial_z \phi + W' & 0 \\ 0 & -\partial_z \phi + W' \end{pmatrix}.$$ (11.41)

---

[2] More exactly, in the case at hand we are dealing with 1/2-BPS-saturated kinks. As already mentioned, BPS stands for Bogomol'nyi, Prasad, and Sommerfield [29, 30]. In fact, these authors considered solitons in a nonsupersymmetric setting. They found, however, that under certain conditions solitons can be described by first-order differential equations rather than the second-order equations of motion. Moreover, under these conditions the soliton mass was shown to be proportional to the topological charge. We understand now that the limiting models considered in [29] correspond to the bosonic sectors of supersymmetric models [25].

One of the supercharges vanishes provided that

$$\frac{\partial \phi(z)}{\partial z} = \pm \frac{\partial \mathcal{W}(\phi)}{\partial \phi}, \qquad (11.42)$$

which can be abbreviated to

$$\partial_z \phi = \pm \mathcal{W}'. \qquad (11.43)$$

The plus and minus signs correspond to a kink and an antikink, respectively. Generically, equations expressing conditions for the vanishing of certain supercharges are called the *BPS equations*.

The first-order BPS equation (11.43) implies that the kink automatically satisfies the general second-order equation of motion. Indeed, let us differentiate both sides of Eq. (11.43) with respect to z. Then we get

$$\partial_z^2 \phi = \pm \partial_z \mathcal{W}' = \pm \mathcal{W}'' \partial_z \phi$$

$$= \mathcal{W}'' \mathcal{W}' = \frac{\partial U}{\partial \phi}. \qquad (11.44)$$

The latter presents the equation of motion for static (time-independent) field configurations. This is a general feature of supersymmetric theories: in any theory, compliance with the BPS equations entails compliance with the equations of motion.

> *Not all solitons are critical.*

The inverse statement is generally speaking wrong – not all solitons that are static solutions of the second-order equations of motion satisfy the BPS equations. However, in the model at hand, with a single scalar field, the converse *is* true: in this model, any static solution of the equation of motion satisfies the BPS equation. This is due to the fact that there exists an "integral of motion." Indeed, let us reinterpret z as a "time," for a short while. Then the equation $\partial_z^2 \phi - U' = 0$ can be reinterpreted as $\ddot{\phi} - U' = 0$, i.e., the one-dimensional motion of a particle of mass 1 in a potential $-U(\phi)$. The conserved "energy" is $\frac{1}{2}\dot{\phi}^2 - U$. At $-\infty$ both the "kinetic" and "potential" terms tend to zero. This boundary condition emerges because the kink solution interpolates between two critical points, the vacua of the model, while supersymmetry ensures that $U(\phi_*) = 0$. Thus, for the kink configuration we have $\frac{1}{2}\dot{\phi}^2 = U$, implying that $\dot{\phi} = \pm \mathcal{W}'$.

We have already learned that the BPS saturation in a supersymmetric setting means the preservation of a part of supersymmetry. Now, let us ask why this feature is so precious.

To answer this question we will have a closer look at the superalgebra (11.33). In the kink's rest frame it reduces to

$$(Q_1)^2 = M + \mathcal{Z}, \qquad (Q_2)^2 = M - \mathcal{Z}, \qquad (11.45)$$

$$\{Q_1, Q_2\} = 0,$$

where M is the kink mass. Since Q_2 vanishes for the critical kink, we see that

$$M = \mathcal{Z}. \qquad (11.46)$$

Thus, the kink mass is equal to the central charge, a nondynamical quantity that is determined only by the boundary conditions on the field ϕ (more exactly, by the values of the superpotential in the vacua between which the kink under consideration interpolates).

11.2.3 The Kink Mass (Classical)

The classical expression for the central charge is given in Eq. (11.36). (Anticipating a turn of events, I hasten to add that a quantum anomaly will modify it; see Section 11.2.7 below.) Now we will discuss the critical kink mass.

In SPM we have

$$\phi_* = \frac{m}{2\lambda}, \qquad \mathcal{W}_0 \equiv \mathcal{W}[\phi_*] = \frac{m^3}{12\lambda^2} \qquad (11.47)$$

and, hence,

$$M_{\text{SPM}} = \frac{m^3}{6\lambda^2}. \qquad (11.48)$$

In the SSG model,

$$\phi_* = v\frac{\pi}{2}, \qquad \mathcal{W}_0 \equiv \mathcal{W}[\phi_*] = mv^2. \qquad (11.49)$$

Therefore

$$M_{\text{SSG}} = 2mv^2. \qquad (11.50)$$

| Kink masses | Applicability of the quasiclassical approximation demands that $m/\lambda \gg 1$ and $v \gg 1$.

11.2.4 Interpretation of the BPS Equations. Morse Theory

In the model described above we are dealing with a single scalar field. Since the BPS equation is of first order, it can always be integrated by quadratures. Examples of the solution for two popular choices of superpotential are given in Eqs. (11.24) and (11.26).

| Multifield generalization of (11.18) | The one-field model is the simplest but certainly not the only model with interesting applications. The generic multifield $\mathcal{N}$ = 1 SUSY model of Landau–Ginzburg type has a Lagrangian of the form

$$\mathcal{L} = \frac{1}{2}\left(\partial_\mu\phi^a\partial^\mu\phi^a + i\bar{\psi}^a\gamma^\mu\partial_\mu\psi^a - \frac{\partial\mathcal{W}}{\partial\phi^a}\frac{\partial\mathcal{W}}{\partial\phi^a} - \frac{\partial^2\mathcal{W}}{\partial\phi^a\partial\phi^b}\bar{\psi}^a\psi^b\right), \qquad (11.51)$$

where the superpotential $\mathcal{W}$ now depends on n variables, $\mathcal{W} = \mathcal{W}(\phi^a)$; in what follows a, b will be referred to as "flavor" indices, $a, b = 1, \ldots, n$. Sums over a and b are implied in Eq. (11.51). The vacua (critical points) of the generic model are determined by the set of equations

$$\frac{\partial\mathcal{W}}{\partial\phi^a} = 0, \qquad a = 1, \ldots, n, \qquad (11.52)$$

If one views $\mathcal{W}(\phi^a)$ as a "mountain profile," the critical points are the extremal points of this profile – the minima, maxima, and saddle points. At the critical points the potential energy,

$$U(\phi^a) = \frac{1}{2}\left(\frac{\partial\mathcal{W}}{\partial\phi^a}\right)^2, \qquad (11.53)$$

is minimal $-U(\phi_*^a)$ vanishes. The kink solution is a trajectory $\phi^a(z)$ interpolating between a selected pair of critical points.

The BPS equations take the form

$$\frac{\partial \phi^a}{\partial z} = \pm \frac{\partial \mathcal{W}}{\partial \phi^a}, \qquad a = 1,\dots,n. \tag{11.54}$$

For $n > 1$ not all solutions of the equations of motion are solutions of the BPS equations, generally speaking. In this case the critical kinks represent a subclass of all possible kinks. Needless to say, as a general rule the set of equations (11.54) cannot be integrated analytically.

A mechanical analogy exists allowing one to use rich intuition that one has from mechanical motion to answer the question whether a solution interpolating between two given critical points exists. Indeed, let us again interpret z as a "time." Then Eq. (11.54) can be read as follows: the velocity vector is equal to the force (the gradient of the superpotential profile). This is the equation describing the flow of a very viscous fluid, such as honey. One places a droplet of honey at a given extremum of the profile $\mathcal{W}$ and then one asks oneself whether this droplet will flow into another given extremum of this profile. If there is no obstruction in the form of an abyss or an intermediate extremum, the answer is yes. Otherwise it is no.

Mathematicians have developed an advanced theory regarding gradient flows, called Morse theory. Here I will not go into further details, referring the interested reader to Milnor's well-known textbook [31].

11.2.5 Quantization. Zero Modes: Bosonic and Fermionic

So far we have been discussing classical kink solutions. Now we will proceed to quantize the theory; this will be carried out in the quasiclassical approximation (i.e., at weak coupling).

The quasiclassical quantization procedure is quite straightforward. If the classical solution is denoted by ϕ_0 then one represents the field ϕ as a sum of the classical solution plus small deviations,

$$\phi = \phi_0 + \chi. \tag{11.55}$$

One then expands χ, and the fermion field ψ, in modes of appropriately chosen differential operators, in such a way as to diagonalize the Hamiltonian. The coefficients in the mode expansion are canonical coordinates, to be quantized. The zero modes in the mode expansion – they are associated with the collective coordinates of the kink – must be treated separately. As we will see, for critical solitons in the ground state all nonzero modes cancel (this is a manifestation of the Bose–Fermi cancelation instrinsic to supersymmetric theories).[3] In this sense, the quantization of supersymmetric solitons is simpler than that of their nonsupersymmetric brethren. We have to deal exclusively with the zero modes. The cancelation of the nonzero modes will be discussed in the next subsection.

---

[3] Statements contradicting this assertion can be found in the literature quite often. People say that "continuum contributions to the spectral density are asymmetric" or "the densities of the bosonic and fermionic excitations in the continuum are unequal." This is due to the fact that the boundary conditions they impose on the modes do not respect the residual supersymmetry. If supersymmetry is maintained by the boundary conditions then the Bose–Fermi cancelation takes place for each level separately, as we will see shortly.

To define the mode expansion properly we have to discretize the spectrum, i.e., introduce infrared regularization. To this end we place the system in a large spatial box, i.e., impose the boundary conditions at $z = \pm L/2$, where L is a large auxiliary size (at the very end, $L \to \infty$). The conditions we will choose are as follows:

$$\left[\partial_z \phi - \mathcal{W}'(\phi)\right]_{z=\pm L/2} = 0, \qquad \psi_1|_{z=\pm L/2} = 0,$$
$$\left[\partial_z - \mathcal{W}''(\phi)\right] \psi_2|_{z=\pm L/2} = 0, \tag{11.56}$$

where $\psi_{1,2}$ denote the components of the spinor ψ_α. The first line is simply a supergeneralization of the BPS equation for the classical kink solution. The second line is the consequence of the Dirac equation of motion; if ψ satisfies the Dirac equation then there are essentially no boundary conditions for ψ_2. Therefore, the second line is not an independent boundary condition – it follows from the first line. We will use these boundary conditions for the construction of modes in the differential operators of second order.

The above choice of boundary conditions is not unique, but it is particularly convenient because it is compatible with the residual supersymmetry in the presence of the BPS soliton. The boundary conditions (11.56) are consistent with the classical solutions, both for the spatially constant vacuum configurations and for the kink. In particular, the soliton solution ϕ_0 of (11.24) (for the superpolynomial case) or (11.26) (for the super-sine-Gordon model) satisfies $\partial_z \phi - \mathcal{W}' = 0$ everywhere. Note that the conditions (11.56) are not periodic.

Associated pairs $(L_2, \tilde{L}_2)$ and $P, P^\dagger$

Now, for the mode expansion we will use the second-order Hermitian differential operators L_2 and $\tilde{L}_2$,

$$L_2 = P^\dagger P, \qquad \tilde{L}_2 = PP^\dagger, \tag{11.57}$$

where

$$P = \partial_z - \mathcal{W}''|_{\phi=\phi_0(z)}, \qquad P^\dagger = -\partial_z - \mathcal{W}''|_{\phi=\phi_0(z)}. \tag{11.58}$$

The operator L_2 defines the modes of $\chi \equiv \phi - \phi_0$ and those of the fermion field ψ_2, while $\tilde{L}_2$ does this job for ψ_1. The boundary conditions for $\psi_{1,2}$ are given in Eq. (11.56); for χ they follow from the expansion of the first condition in Eq. (11.56),

$$\left[\partial_z - \mathcal{W}''(\phi_0(z))\right] \chi\big|_{z=\pm L/2} = 0. \tag{11.59}$$

It would be natural at this point to ask why it is the differential operators L_2 and $\tilde{L}_2$ that are chosen for the mode expansion. In principle, any Hermitian operator has an orthonormal set of eigenfunctions. The choice above is singled out because it ensures diagonalization. Indeed, the quadratic form following from the Lagrangian (10.284) for small deviations from the classical kink solution is

$$S^{(2)} \to \tfrac{1}{2} \int d^2x \left(-\chi L_2 \chi - i\psi_1 P\psi_2 + i\psi_2 P^\dagger \psi_1\right), \tag{11.60}$$

Zero mode in L_2

where we have neglected time derivatives and used the fact that $d\phi_0/dz = \mathcal{W}'(\phi_0)$ for the kink under consideration. If the diagonalization is not yet transparent, wait for the explanatory comment in the next subsection.

It is easy to verify that there is only one zero mode $\chi_0(z)$ for the operator L_2. It has the form

$$\chi_0 \propto \frac{d\phi_0}{dz} \propto \mathcal{W}''\big|_{\phi=\phi_0(z)} \propto \begin{cases} \dfrac{1}{\cosh^2(mz/2)} & \text{(SPM)}, \\[2mm] \dfrac{1}{\cosh(mz)} & \text{(SSG)}. \end{cases} \tag{11.61}$$

It is obvious that this zero mode is due to translations. The corresponding collective coordinate z_0 can be introduced through the substitution $z \longrightarrow z - z_0$ in the classical kink solution. Then

$$\chi_0 \propto \frac{\partial \phi_0(z - z_0)}{\partial z_0}. \tag{11.62}$$

The existence of a zero mode for the fermion component ψ_2, which is functionally the same as that in χ (in fact, this is the zero mode in P), is due to supersymmetry. The translational bosonic zero mode entails a fermionic one – it is usually referred to as the "supersymmetric (or supertranslational) mode."

The operator $\tilde{L}_2$ has no zero modes at all.

The translational and supertranslational zero modes discussed above imply that the kink is described by two collective coordinates, its center z_0 and a fermionic "center" η, where

$$\phi = \phi_0(z - z_0) + \text{nonzero modes}, \qquad \psi_2 = \eta\chi_0 + \text{nonzero modes}, \tag{11.63}$$

where χ_0 is the normalized mode obtained from Eq. (11.61) after normalization. The nonzero modes in Eq. (11.63) are those of the operator L_2. Regarding ψ_1, it is given by the sum over nonzero modes of the operator $\tilde{L}_2$.

> η is a Grassmann parameter.

Now we are ready to derive a Lagrangian describing the moduli dynamics. To this end we substitute Eqs. (11.63) into the original Lagrangian (10.284), ignoring the nonzero modes and assuming that the time dependence enters only through (an adiabatically slow) time dependence of the moduli z_0 and η:

$$\begin{aligned} \mathcal{L}_{\text{QM}} &= -M + \tfrac{1}{2}\dot{z}_0^2 \int dz \left[\frac{d\phi_0(z)}{dz}\right]^2 + \tfrac{1}{2}i\eta\dot{\eta} \int dz[\chi_0(z)]^2 \\ &= -M + \tfrac{1}{2}M\dot{z}_0^2 + \tfrac{1}{2}i\eta\dot{\eta}, \end{aligned} \tag{11.64}$$

where M is the kink mass and the subscript QM emphasizes the fact that the original field theory is now reduced to the quantum mechanics of the kink moduli. The bosonic part of this Lagrangian is evident: it corresponds to the free nonrelativistic motion of a particle with mass M.

A priori one might expect the fermionic part of $\mathcal{L}_{\text{QM}}$ to give rise to a Fermi–Bose doubling. While generally speaking this is the case, in the simple example at hand there is no doubling and the "fermion center" modulus does not manifest itself.

Indeed, the (quasiclassical) quantization of the system amounts to imposing the commutation and anticommutation relations

$$[p, z_0] = -i, \qquad \eta^2 = \tfrac{1}{2}, \tag{11.65}$$

where $p = M\dot{z}_0$ is the canonical momentum conjugate to z_0. These relations mean that in the quantum dynamics of the soliton moduli z_0 and η, the operators p and η can be realized as

$$p = M\dot{z}_0 = -i\frac{d}{dz_0}, \qquad \eta = \tfrac{1}{\sqrt{2}}. \tag{11.66}$$

(It is clear that we could have chosen $\eta = -\tfrac{1}{\sqrt{2}}$. The two choices are physically equivalent.)

Thus, η reduces to a constant; the Hamiltonian of the system is then

$$H_{\text{QM}} = M - \frac{1}{2M}\frac{d^2}{dz_0^2}. \tag{11.67}$$

The wave function on which this Hamiltonian acts is *single-component*.

One can obtain the same Hamiltonian by calculating the supercharges. Substituting the mode expansion into the supercharges (11.30) we arrive at

$$Q_1 = 2\sqrt{\mathcal{Z}}\eta + \dots, \qquad Q_2 = \sqrt{\mathcal{Z}}\dot{z}_0\eta + \dots, \tag{11.68}$$

where $\mathcal{Z}$ is the central change and $Q_2^2 = H_{\text{QM}} - M$. (Here the ellipses stand for the omitted nonzero modes.) The supercharges depend only on the canonical momentum p:

$$Q_1 = \sqrt{2\mathcal{Z}}, \qquad Q_2 = \frac{p}{\sqrt{2\mathcal{Z}}}. \tag{11.69}$$

In the rest frame in which we are working, $\{Q_1, Q_2\} = 0$; the only value of p consistent with this is $p = 0$. Thus, for a kink at rest we have $Q_1 = \sqrt{2\mathcal{Z}}, Q_2 = 0$, in full agreement with the general construction. The representation (11.69) can be used at nonzero p as well. It reproduces the superalgebra (11.33) in the nonrelativistic limit; p has the meaning of the total spatial momentum P_1.

The conclusion that there is no Fermi–Bose doubling for the supersymmetric kink rests on the fact that there is only *one* (real) fermion zero mode in the kink background and, consequently, a single fermionic modulus. This is totally counterintuitive and is, in fact, a manifestation of an *anomaly*. We will discuss this issue in more detail later (see Section 11.2.8).

11.2.6 Cancelation of the Nonzero Modes

Above we have omitted the nonzero modes altogether. Now I want to show that for a kink in the ground state the effect of the bosonic nonzero modes is canceled by that of the fermionic nonzero modes.

For each given nonzero eigenvalue there is one bosonic eigenfunction (in the operator L_2), the same eigenfunction in ψ_2, and one eigenfunction in ψ_1 (that of the operator $\tilde{L}_2$) with the same eigenvalue. The operators L_2 and $\tilde{L}_2$ have the same spectrum, except for the zero modes, and their eigenfunctions are related. They can be called *associated operators*.

Indeed, let χ_n be a normalized eigenfunction of L_2,

$$L_2\chi_n(z) = \omega_n^2\chi_n(z). \tag{11.70}$$

Introduce

$$\tilde{\chi}_n(z) = \frac{1}{\omega_n}P\chi_n(z). \tag{11.71}$$

Then, $\tilde{\chi}_n(z)$ is a normalized eigenfunction of $\tilde{L}_2$ with the same eigenvalue,

$$\tilde{L}_2\tilde{\chi}_n(z) = PP^\dagger \frac{1}{\omega_n}P\chi_n(z) = \frac{1}{\omega_n}P\,\omega_n^2\chi_n(z) = \omega_n^2\tilde{\chi}_n(z). \tag{11.72}$$

In turn,

$$\chi_n(z) = \frac{1}{\omega_n}P^\dagger\tilde{\chi}_n(z). \tag{11.73}$$

The quantization of the nonzero modes is quite standard. Let us denote the Hamiltonian *density* by $\mathcal{H}$,

$$H = \int dz\,\mathcal{H}.$$

Then, in the approximation that is quadratic in the quantum fields χ the Hamiltonian density takes the following form:

$$\mathcal{H} - \partial_z\mathcal{W} = \tfrac{1}{2}\left\{\dot{\chi}^2 + [(\partial_z - \mathcal{W}'')\chi]^2 + i\psi_2(\partial_z + \mathcal{W}'')\psi_1 + i\psi_1(\partial_z - \mathcal{W}'')\psi_2\right\}, \tag{11.74}$$

where $\mathcal{W}''$ is evaluated at $\phi = \phi_0$. We recall that the prime denotes differentiation over ϕ,

$$\mathcal{W}'' = \frac{d^2\mathcal{W}}{d\phi^2}.$$

The expansions in eigenmodes have the forms

$$\chi(x) = \sum_{n\neq 0} b_n(t)\chi_n(z), \qquad \psi_2(x) = \sum_{n\neq 0}\eta_n(t)\chi_n(z),$$
$$\psi_1(x) = \sum_{n\neq 0}\xi_n(t)\tilde{\chi}_n(z). \tag{11.75}$$

Note that the summations do not include the zero mode $\chi_0(z)$. This mode is not present in ψ_1 at all. As for the expansions of χ and ψ_2, the inclusion of the zero mode would correspond to a shift in the collective coordinates z_0 and η. Their quantization has been already considered in the previous section. Here we set $z_0 = 0$.

The coefficients b_n, η_n, and ξ_n are time-dependent operators. Their equal-time commutation relations are determined by the canonical commutators (11.32),

$$[b_m, \dot{b}_n] = i\delta_{mn}, \qquad \{\eta_m, \eta_n\} = \delta_{mn}, \qquad \{\xi_m, \xi_n\} = \delta_{mn}. \tag{11.76}$$

Thus, the mode decomposition reduces the dynamics of the system under consideration to the quantum mechanics of an infinite set of supersymmetric harmonic oscillators (in higher orders the oscillators become anharmonic). The *ground state* of the quantum kink corresponds to each oscillator in the set being in the ground state.

Constructing the creation and annihilation operators in the standard way, we find the following nonvanishing expectation values of the bilinears built from the operators b_n, η_n, and ξ_n in the ground state:

$$\langle \dot{b}_n^2\rangle_{\text{sol}} = \frac{\omega_n}{2}, \qquad \langle b_n^2\rangle_{\text{sol}} = \frac{1}{2\omega_n}, \qquad \langle \eta_n\xi_n\rangle_{\text{sol}} = \frac{i}{2}. \tag{11.77}$$

The expectation values of other bilinears obviously vanish. Combining Eqs. (11.74), (11.75), and (11.77) we get

$$\langle \text{sol}|\mathcal{H}(z) - \partial_z \mathcal{W}|\text{sol}\rangle = \frac{1}{2} \sum_{n \neq 0} \left\{ \frac{\omega_n}{2} \chi_n^2 + \frac{1}{2\omega_n} [(\partial_z - \mathcal{W}'')\chi_n]^2 - \frac{\omega_n}{2} \chi_n^2 \right.$$
$$\left. - \frac{1}{2\omega_n} [(\partial_z - \mathcal{W}'')\chi_n]^2 \right\} \equiv 0. \qquad (11.78)$$

Mode decomposition of the Hamiltonian density

For critical solitions, quantum corrections cancel altogether; $M = \mathcal{Z}$ is exact.

In other words, for the critical kink in the ground state the Hamiltonian density is *locally* equal to $\partial_z \mathcal{W}$ – this statement is valid at the level of quantum corrections!

The four terms in the braces in Eq. (11.78) are in one-to-one correspondence with the four terms in Eq. (11.74). Note that in proving the vanishing of the right-hand side of (11.78) we did not perform integration by parts. The vanishing of the right-hand side of (11.74) demonstrates explicitly the residual supersymmetry – i.e., the conservation of Q_2 and the fact that $M = \mathcal{Z}$. Equation (11.78) must be considered as a local version of BPS saturation (i.e., the conservation of a residual supersymmetry).

Multiplet shortening guarantees that the equality $M = \mathcal{Z}$ is not corrected in higher orders.

What lessons can one draw from the discussion in the subsection? In the case of the polynomial model the target space is noncompact, while in the sine-Gordon case it can be viewed as a compact target manifold S^1. In both cases we get the same result: a short (one-dimensional) soliton multiplet defying fermion parity (further details will be given in Section 11.2.8).

11.2.7 Anomaly I

We have demonstrated explicitly that the equality between the kink mass M and the central charge $\mathcal{Z}$ survives at the quantum level. The classical expression for the central charge is given in Eq. (11.36). If one takes proper care of the ultraviolet regularization one can show [28] that quantum corrections modify Eq. (11.36). Here I will present a simple argument demonstrating the emergence of an anomalous term in the central charge and discuss its physical meaning.

To begin with, let us consider $\gamma^\mu J_\mu$, where J_μ is the supercurrent defined in Eq. (11.27). This quantity is related to the superconformal properties of the model under consideration. At the classical level,

$$\left(\gamma^\mu J_\mu \right)_{\text{class}} = 2i\mathcal{W}'\psi. \qquad (11.79)$$

Note that the first term in the supercurrent (11.27) gives no contribution in Eq. (11.79) due to the fact that in two dimensions $\gamma_\mu \gamma^\nu \gamma^\mu = 0$.

The local form of the superalgebra is given in Eq. (11.37). Multiplying Eq. (11.37) by γ_μ from the left we get the supertransformation of $\gamma_\mu J^\mu$,

$$\tfrac{1}{2} \left\{ \gamma^\mu J_\mu, \bar{Q} \right\} = T_\mu^\mu + i\gamma_\mu \gamma^5 \zeta^\mu, \qquad \gamma^5 = \gamma^0 \gamma^1 = -\sigma_1. \qquad (11.80)$$

This equation establishes a supersymmetric relation between $\gamma^\mu J_\mu$, T_μ^μ, and ζ^μ and, as mentioned above, remains valid then quantum corrections are included. But the expression for these operators can (and will) change. Classically the trace of the energy–momentum tensor is

$$\left(T_\mu^\mu\right)_{\text{class}} = (W')^2 + \tfrac{1}{2}W''\bar\psi\psi, \tag{11.81}$$

as follows from Eq. (11.35). The zero component of ζ^μ in the second term in Eq. (11.80) classically coincides with the density of the central charge, $\partial_z W$; see Eq. (11.38). It can be seen that the trace of the energy–momentum tensor and the density of the central charge appear in this relation together.

It is well known that, in renormalizable theories with ultraviolet logarithmic divergences, both the trace of the energy–momentum tensor and $\gamma^\mu J_\mu$ have anomalies. We will use this fact, in conjunction with Eq. (11.80), to establish the general form of the anomaly in the density of the central charge.

To get an idea of this anomaly, it is convenient to use dimensional regularization. If we assume that the number of dimensions D is $2 - \varepsilon$ rather than 2, then the first term in Eq. (11.27) generates a nonvanishing contribution to $\gamma^\mu J_\mu$ that is proportional to $(D - 2)(\partial_\nu\phi)\gamma^\nu\psi$. At the quantum level this operator acquires an ultraviolet logarithm (i.e., a factor $(D - 2)^{-1}$ in the dimensional regularization), so that the factor $D - 2$ cancels and we are left with an anomalous term in $\gamma^\mu J_\mu$.

To do the one-loop calculation, here, as well as in some other instances in this textbook, we will use the background field technique: we split the field ϕ into its background and quantum parts, ϕ and χ, respectively,

$$\phi \to \phi + \chi. \tag{11.82}$$

Specifically, for the anomalous term in $\gamma^\mu J_\mu$, we obtain

$$\left(\gamma^\mu J_\mu\right)_{\text{anom}} = (D - 2)(\partial_\nu\phi)\gamma^\nu\psi = -(D - 2)\chi\gamma^\nu\partial_\nu\psi$$
$$= i(D - 2)\chi W''(\phi + \chi)\psi, \tag{11.83}$$

Anomaly in the supercurrent

where integration by parts has been carried out, and a total derivative term is omitted (on dimensional grounds it vanishes in the limit $D = 2$). We have also used the equation of motion for the ψ field. The quantum field χ then forms a loop and we get, for the anomaly,

$$\left(\gamma^\mu J_\mu\right)_{\text{anom}} = i(D - 2)\langle 0|\chi^2|0\rangle W'''(\phi)\psi$$
$$= -(D - 2)\int\frac{d^D p}{(2\pi)^D}\frac{1}{p^2 - m^2}W'''(\phi)\psi$$
$$= \frac{i}{2\pi}W'''(\phi)\psi. \tag{11.84}$$

The supertransformation of the anomalous term in $\gamma^\mu J_\mu$ is

$$\tfrac{1}{2}\left\{\left(\gamma^\mu J_\mu\right)_{\text{anom}}, \bar{Q}\right\} = \left(\frac{1}{8\pi}W''''\bar\psi\psi + \frac{1}{4\pi}W'''W'\right)$$
$$+ i\gamma_\mu\gamma^5\,\varepsilon^{\mu\nu}\partial_\nu\left(\frac{1}{4\pi}W''\right). \tag{11.85}$$

Anomaly in the topological current

The first line on the right-hand side is the anomaly in the trace of the energy–momentum tensor and the second line represents the anomaly in the topological current; the *corrected current* has the form

$$\zeta^\mu = \varepsilon^{\mu\nu}\partial_\nu\left(W + \frac{1}{4\pi}W''\right). \tag{11.86}$$

Consequently, at the quantum level, after inclusion of the anomaly the central charge becomes

$$\mathcal{Z} = \left(\mathcal{W} + \frac{1}{4\pi}\mathcal{W}''\right)_{z=+\infty} - \left(\mathcal{W} + \frac{1}{4\pi}\mathcal{W}''\right)_{z=-\infty}. \tag{11.87}$$

11.2.8 Anomaly II (Shortening the Supermultiplet down to One State)

In the model under consideration, see Eq. (10.284), the fermion field is real, which implies that the fermion number is not defined. What is defined, however, is the fermion parity, $G = (-1)^F$. The action of G reduces to that of changing the sign for the fermion operators but leaving the boson operators intact, for instance,

$$G Q_\alpha G^{-1} = -Q_\alpha, \qquad G P_\mu G^{-1} = P_\mu. \tag{11.88}$$

The fermion parity G realizes the Z_2 symmetry associated with changing the sign of the fermion fields. This symmetry is obvious at the classical level (and, in fact, in any finite order of perturbation theory). It is intuitive – it is the symmetry that distinguishes fermion states from boson states in the model at hand, with Majorana fermions.

Here I will demonstrate (without delving too deep into technicalities) that in the soliton sector the very classification of states as either bosonic or fermionic is broken. The disappearance of the fermion parity in the BPS soliton sector is a global anomaly [32].

Let us consider the algebra (11.45) in the special case $M^2 = \mathcal{Z}^2$. Assuming $\mathcal{Z}$ to be positive we consider the BPS soliton, $M = \mathcal{Z}$, for which the supercharge Q_2 is trivial, $Q_2 = 0$. Thus we are left with a single supercharge Q_1 realized nontrivially. The algebra reduces to a single relation,

$$(Q_1)^2 = 2\mathcal{Z}. \tag{11.89}$$

The irreducible representations of this algebra are one-dimensional. There are two such representations,

$$Q_1 = \pm\sqrt{2\mathcal{Z}}, \tag{11.90}$$

Fermion parity has gone in the soliton sector, in the minimal model (10.284).

i.e., two types of soliton,

$$Q_1|\text{sol}_+\rangle = \sqrt{2\mathcal{Z}}|\text{sol}_+\rangle, \qquad Q_1|\text{sol}_-\rangle = -\sqrt{2\mathcal{Z}}|\text{sol}_-\rangle. \tag{11.91}$$

It is clear that these two representations are unitarily nonequivalent.

The one-dimensional irreducible representation of supersymmetry implies multiplet shortening: the short BPS supermultiplet contains only one state while non-BPS supermultiplets contain two. The possibility of such supershort one-dimensional multiplets was discounted in the literature for many years. This was for a good reason: while the fermion parity $(-1)^F$ is valid in any local field theory based on fermionic and bosonic fields, it is *not defined* in the one-dimensional irreducible representation. Indeed, if it were defined then it would be -1 for Q_1, which would be incompatible with the equations (11.91). The only way to recover $(-1)^F$ is to have a reducible representation containing both $|\text{sol}_+\rangle$ and $|\text{sol}_-\rangle$. Then

$$Q_1 = \sigma_3\sqrt{2\mathcal{Z}}, \qquad (-1)^F = \sigma_1. \tag{11.92}$$

Does this mean that a single-state supermultiplet is not a possibility in the local field theory? As I argued above, in the simplest two-dimensional supersymmetric model (11.18) BPS solitons do exist and do realize such supershort multiplets that defy $(-1)^F$. These BPS solitons are neither bosons nor fermions [32]. Needless to say, this is possible only in two dimensions.

The important point is that short multiplets of BPS states are protected against becoming non-BPS under small perturbations. Although the overall sign of Q_1 in the irreducible representation is not observable, the relative sign is observable. For instance, there are two types of reducible representations of dimension 2: one is $\{+, -\}$ (see Eq. (11.92)) and the other is $\{+, +\}$ (which is equivalent to $\{-, -\}$). In the first case two states can pair up and leave the BPS bound as soon as appropriate perturbations are introduced. In the second case the BPS relation $M = \mathcal{Z}$ is "bullet-proof."

To reiterate, the discrete Z_2 symmetry $G = (-1)^F$ discussed above is nothing other than the change in sign of all fermion fields, $\psi \to -\psi$. This symmetry is seemingly present in any theory with fermions. How on earth can this symmetry be lost in the soliton sector?

Technically the loss of $G = (-1)^F$ is due to the fact that there is only *one* (real) fermion zero mode for the soliton in the model at hand. Normally, the fermion degrees of freedom enter in holomorphic pairs $\{\bar{\psi}, \psi\}$. In our case, that of a single fermion zero mode, we have "half" such a pair. The second fermion zero mode, which would produce the missing half, turns out to be delocalized. More exactly, it is not localized on the soliton but, rather, on the boundary of the "large box" one introduces for quantization (see Section 11.2.6 above). For physical measurements made far from the auxiliary box boundary the fermion parity G is lost, and a supermultiplet consisting of a single state becomes a physical reality. In a sense, the phenomenon is akin to that of charge fractionalization [33] (Section 2.5): the total charge, which includes that concentrated on the box boundaries, is always integer but local measurements on a Jackiw–Rebbi soliton will yield a fractional charge.

11.3 $\mathcal{N} = 2$: Kinks in Two-Dimensional Supersymmetric CP(1) Model

See also the two subsections following Section 10.12.3.4, where "twisted" mass was introduced.

We are already familiar with the two-dimensional supersymmetric CP(1) model from Section 10.12.3.4. The supersymmetry of this model is extended (it is more than minimal).

The model has four conserved supercharges rather than two, as was the case in Section 11.2. Solitons in the $\mathcal{N} = 2$ sigma model present a showcase for a variety of intriguing dynamical phenomena. One is charge "irrationalization:" in the presence of the θ term (the topological term) the U(1) charge of the soliton acquires an extra $\theta/(2\pi)$. This phenomenon was first discovered by Witten [34] in 't Hooft–Polyakov monopoles [35, 36] (see Section 4.1.10).

The Lagrangian of the CP(1) model with twisted mass [37] was presented in Eqs. (10.334) and (10.324) in Section 10.12.3. The chiral components of the supercurrent are [17]

$$J_R^+ = \sqrt{2}G(\partial_R\bar{\phi})\psi_R, \qquad J_R^- = -\sqrt{2}iG\bar{m}\bar{\phi}\psi_L;$$
$$J_L^- = \sqrt{2}G(\partial_L\bar{\phi})\psi_L, \qquad J_L^+ = \sqrt{2}iGm\bar{\phi}\psi_R, \qquad (11.93)$$

where the metric G is given in (10.313). The superalgebra generated by the four supercharges is as follows:

$$\{\bar{Q}_L, Q_L\} = 2(H+P), \qquad \{\bar{Q}_R\,Q_R\} = 2(H-P); \qquad (11.94)$$

$$\left.\begin{array}{l} \{Q_L,\ Q_R\} = 0 \\ \{Q_R,\ Q_R\} = 0 \\ \{Q_L,\ Q_L\} = 0 \end{array}\right\} \text{ and H.c.}, \qquad (11.95)$$

$$\{\bar{Q}_R,\ Q_L\} = 2i\mathcal{Z}, \qquad \{\bar{Q}_L\,Q_R\} = -2i\mathcal{Z}^\dagger, \qquad (11.96)$$

where (H, P) is the energy–momentum operator,

$$(H, P) = \int dz\, T^{0i}, \qquad i = 0, 1,$$

and $T^{\mu\nu}$ is the energy–momentum tensor. Moreover, the central charge $\mathcal{Z}$ consists of two terms – the Noether and topological parts, respectively:

$$\mathcal{Z} = mq_{U(1)} - i\int dz\,\partial_z O, \qquad (11.97)$$

where

$$q_{U(1)} \equiv \int dz\,\mathcal{J}_{U(1)}^0,$$
$$\mathcal{J}_{U(1)}^\mu = G\left(\bar{\phi}i\,\overleftrightarrow{\partial}^\mu\phi + \bar{\psi}\gamma^\mu\psi - 2\frac{\phi\bar{\phi}}{\chi}\bar{\psi}\gamma^\mu\psi\right), \qquad (11.98)$$

and O in turn is composed of two parts: the first is canonical while the second is an anomaly [16, 17],

$$O = mh - \frac{g^2}{2\pi}\left(mh + G\bar{\psi}_R\psi_L\right), \qquad (11.99)$$

$$h = \frac{2}{g^2}\frac{\bar{\phi}\phi}{\chi}. \qquad (11.100)$$

| Recall that χ was defined in (10.314). | The second term on the right-hand side in (11.99) vanishes at the classical level. These anomalies will not be used in what follows. I will quote them here only for the sake of completeness. Equations (11.96) and (11.97) clearly demonstrate that the very possibility of introducing twisted masses is due to U(1) symmetry. The model (10.334) is asymptotically free [38] (see Section 6.3). The scale parameter of the model is |

$$\Lambda^2 = M_{uv}^2 \exp\left(-\frac{4\pi}{g_0^2}\right). \qquad (11.101)$$

Our task is to study kinks in this model in a pedagogical setting, which means by default that the theory must be weakly coupled. The model (10.334) is indeed weakly coupled, still preserving $\mathcal{N} = 2$ supersymmetry, provided that $m \gg \Lambda$, which will be assumed. Then the solitons emerging in this model can and will be treated quasiclassically.

11.3.1 Symmetry

One can always eliminate the phase of m by a chiral rotation of the fermion fields. Owing to the chiral anomaly this will lead to a shift in the vacuum angle θ. In fact, it is the combination $\theta_{\text{eff}} = \theta + 2 \arg m$ on which the physics depends. We will choose m to be real.

With the mass term included the symmetry of the model, i.e., of the target space, is reduced to a global U(1),

$$\phi \to e^{i\alpha}\phi, \qquad \bar{\phi} \to e^{-i\alpha}\bar{\phi},$$
$$\psi \to e^{i\alpha}\psi, \qquad \bar{\psi} \to e^{-i\alpha}\bar{\psi}. \tag{11.102}$$

11.3.2 BPS Solitons at the Classical Level

The target space of the model is S_2. The U(1)-invariant scalar potential term

$$V = |m|^2 G\bar{\phi}\phi \tag{11.103}$$

lifts the vacuum degeneracy, leaving us with discrete vacua at the south and north poles of the sphere (Fig. 11.1), i.e., at $\phi = 0$ and $\phi = \infty$.

The kink solutions interpolate between these two vacua. Let us focus for definiteness, on the kink with boundary conditions

$$\phi \to 0 \quad \text{as} \quad z \to -\infty, \qquad \phi \to \infty \quad \text{as} \quad z \to \infty. \tag{11.104}$$

Consider the following linear combinations of supercharges:

$$q = \tfrac{1}{\sqrt{2}}\left(Q_R - e^{-i\beta}Q_L\right), \qquad \bar{q} = \tfrac{1}{\sqrt{2}}\left(\bar{Q}_R - e^{i\beta}\bar{Q}_L\right), \tag{11.105}$$

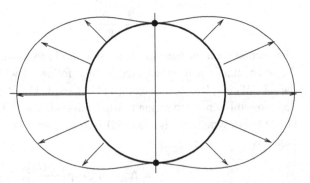

Fig. 11.1 A meridian slice of the target space sphere (thick solid line). The arrows present the scalar potential (11.103), their length corresponding to the strength of the potential. The two vacua of the model are shown by the solid circles.

where β is the argument of the mass parameter

$$m = |m|e^{i\beta}.$$ (11.106)

Then

$$\{q, \bar{q}\} = 2H - 2|m| \int dz\, \partial_z h, \qquad \{q, q\} = \{\bar{q}, \bar{q}\} = 0.$$ (11.107)

Now, we require q and $\bar{q}$ to vanish on the classical solution. Since, for static field configurations,

$$q = -\left(\partial_z \bar{\phi} - |m|\bar{\phi}\right)\left(\psi_R + ie^{-i\beta}\psi_L\right),$$

the vanishing of these two supercharges implies that

$$\partial_z \bar{\phi} = |m|\bar{\phi} \qquad \text{or} \qquad \partial_z \phi = |m|\phi.$$ (11.108)

This is the BPS equation for the sigma model with twisted mass.

The BPS equation (11.108) has a number of peculiarities compared to those in the more familiar Wess–Zumino $\mathcal{N} = 2$ models. The most important feature is its complexification, i.e., the fact that Eq. (11.108) is holomorphic in ϕ. The solution of this equation is, of course, trivial, and can be written as

$$\phi(z) = e^{|m|(z-z_0)-i\alpha}.$$ (11.109)

Here z_0 is the kink center while α is an arbitrary phase. In fact, these two parameters enter only in the combination $|m|z_0 + i\alpha$. We see that the notion of the kink center also gets complexified.

The physical meaning of the modulus α is obvious: there is a continuous family of solitons interpolating between the north and south poles of the target space sphere. This is due to the U(1) symmetry. The soliton trajectory can follow any meridian (Fig. 11.2).

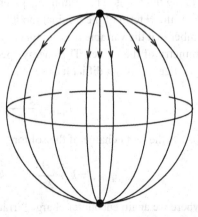

Fig. 11.2 The soliton solution family. The collective coordinate α in Eq. (11.109) spans the interval $0 \leq \alpha \leq 2\pi$. For given α the soliton trajectory on the target space sphere follows a meridian, so that when α varies from 0 to 2π all meridians are covered.

It is instructive to derive the BPS equation directly from the (bosonic part of the) Lagrangian, performing Bogomol'nyi completion:

$$\int d^2x \mathcal{L} = \int d^2x \, G \left(\partial_\mu \bar{\phi} \, \partial^\mu \phi - |m|^2 \bar{\phi} \phi \right)$$

$$\rightarrow - \left[\int dz \, G \left(\partial_z \bar{\phi} - |m| \bar{\phi} \right) \left(\partial_z \phi - |m| \phi \right) \right.$$

$$\left. + |m| \int dz \, \partial_z h \right], \tag{11.110}$$

Bogomol'nyi completion | where we have assumed ϕ to be time independent and used the following identity:

$$\partial_z h \equiv G(\phi \, \partial_z \bar{\phi} + \bar{\phi} \, \partial_z \phi).$$

Equation (11.108) ensues immediately. In addition, Eq. (11.110) implies that classically the kink mass is

$$M_0 = |m| \left[h(\infty) - h(0) \right] = \frac{2|m|}{g^2}. \tag{11.111}$$

The subscript 0 emphasizes that this result is obtained at the classical level. Quantum corrections will be considered shortly.

11.3.3 Quantization of the Bosonic Moduli

To carry out conventional quasiclassical quantization we assume, as usual, that the moduli z_0 and α in Eq. (11.109) are weakly time dependent, substitute (11.109) into the bosonic Lagrangian (11.110), integrate over z, and thus derive a quantum-mechanical Lagrangian describing the moduli dynamics,

$$\mathcal{L}_{QM} = -M_0 + \frac{M_0}{2} \dot{z}_0^2 + \left(\frac{1}{g^2 |m|} \dot{\alpha}^2 - \frac{\theta}{2\pi} \dot{\alpha} \right). \tag{11.112}$$

The first term is the classical kink mass and the second describes the free motion of the kink along the z axis. The term in the parentheses is the most interesting, being a reflection of the θ term of the original model.

Remember that the variable α is compact. Its very existence is related to the exact U(1) symmetry of the model. The energy spectrum corresponding to the dynamics of α is quantized. It is not difficult to see that

$$E_{[\alpha]} = \frac{g^2 |m|}{4} \, q_{U(1)}^2, \tag{11.113}$$

where $q_{U(1)}$ is the U(1) charge of the soliton,

$$q_{U(1)} = k + \frac{\theta}{2\pi}, \qquad k \text{ is an integer.} \tag{11.114}$$

The QM Hamiltonian and Witten's effect | This is where we again encounter charge "irrationalization" (the Witten effect) – the soliton's U(1) charge is no longer integer in the presence of the θ term since it is shifted by $\theta/(2\pi)$. This is the same effect as the shift of the dyon's electric charge by $\theta/(2\pi)$ discussed in Section 4.1.10.

A brief comment regarding Eqs. (11.113) and (11.114) is in order here. The dynamics of the compact modulus α is described by the Hamiltonian

$$H_{QM} = \frac{1}{g^2|m|}\dot{\alpha}^2, \tag{11.115}$$

while the canonical momentum conjugate to α is

$$p_{[\alpha]} = \frac{\delta \mathcal{L}_{QM}}{\delta \dot{\alpha}} = \frac{2}{g^2|m|}\dot{\alpha} - \frac{\theta}{2\pi}. \tag{11.116}$$

In terms of the canonical momentum the Hamiltonian takes the form

$$H_{QM} = \frac{g^2|m|}{4}\left(p_{[\alpha]} + \frac{\theta}{2\pi}\right)^2. \tag{11.117}$$

The eigenfunctions are obviously

$$\Psi_k(\alpha) = e^{ik\alpha}, \qquad k \text{ is an integer}, \tag{11.118}$$

which immediately leads to $E_{[\alpha]} = (g^2|m|/4)[k + \theta(2\pi)^{-1}]^2$.

Let us now calculate the U(1) charge of the kth state. Starting from Eq. (11.98) we arrive at

$$q_{U(1)} = \frac{2}{g^2|m|}\dot{\alpha} = p_{[\alpha]} + \frac{\theta}{2\pi} \rightarrow k + \frac{\theta}{2\pi}, \tag{11.119}$$

as required; cf. Eq. (11.114).

11.3.4 The Soliton Mass and Holomorphy

Taking account of $E_{[\alpha]}$ – the energy of the "internal motion" – the kink mass can be written as

$$\begin{aligned}
M &= \frac{2|m|}{g^2} + \frac{g^2|m|}{4}\left(k + \frac{\theta}{2\pi}\right)^2 \\
&= \frac{2|m|}{g^2}\left[1 + \frac{g^4}{4}\left(k + \frac{\theta}{2\pi}\right)^2\right]^{1/2} \\
&= |m|\left|\frac{2}{g^2} + i\frac{\theta + 2\pi k}{2\pi}\right|.
\end{aligned} \tag{11.120}$$

Formally, the second equality is approximate, valid only to leading order in the coupling constant. In fact, though, it is exact! We will return to this point later.

The important circumstance to be stressed is that the kink mass depends on a special combination of the coupling constant and θ, namely,

$$\tau = \frac{1}{g^2} + i\frac{\theta}{4\pi}. \tag{11.121}$$

Complexified coupling constant

In other words, it is a complexified coupling constant that enters.

It is instructive to pause here and examine the issue of the kink mass from a slightly different angle. Equation (11.96) tells us that there is a central charge $\mathcal{Z}$ in the anticommutator $\{Q_L, \bar{Q}_R\}$, which, after omitting the anomaly term in (11.97),[4] takes the form

---

[4] Omitting the anomaly term is fully justified at weak coupling.

$$\mathcal{Z} = m\left(q_{U(1)} - i \int dz\, \partial_z h\right).\tag{11.122}$$

If the soliton under consideration is critical – and it is – its mass *must* be equal to the absolute value of $\mathcal{Z}$. This leads us directly to Eq. (11.120). One can say more, however.

Indeed, the factor $1/g^2$ in Eq. (11.120) is the bare coupling constant. It is quite clear that the kink mass, being a physical parameter, should contain the renormalized constant $1/g^2(m)$, after account has been taken of radiative corrections. In other words, switching on the radiative corrections in $\mathcal{Z}$ replaces the bare $1/g^2$ by the renormalized $1/g^2(m)$. We will now derive this result, verifying en route a very important assertion – that the dependence of $\mathcal{Z}$ on the relevant parameters, τ and m, is holomorphic.

We will perform a one-loop calculation in two steps. First, we rotate the mass parameter m in such a way as to make it real, $m \leftrightarrow |m|$. Simultaneously, the θ angle is replaced by θ_{eff}, where

$$\theta_{\text{eff}} = \theta + 2\beta\tag{11.123}$$

| One-loop calculation of $\mathcal{Z}$ |

and the phase β was defined in Eqs. (11.105). Next we decompose the field ϕ into a classical plus a quantum part:

$$\phi \to \phi + \delta\phi.$$

Then the h part of the central charge $\mathcal{Z}$ takes the form

$$h \to h + \frac{2}{g^2}\frac{1 - \bar{\phi}\phi}{(1 + \bar{\phi}\phi)^3}\delta\bar{\phi}\,\delta\phi.\tag{11.124}$$

Contracting $\delta\bar{\phi}\delta\phi$ into a loop (Fig. 11.3) and calculating this loop – an easy exercise – we find that

$$h \to \left(\frac{2}{g_0^2} - \frac{1}{2\pi}\ln\frac{M_{uv}^2}{|m|^2}\right)\frac{\bar{\phi}\phi}{\chi}.\tag{11.125}$$

| Holomorphy! |

Combining this result with Eqs. (11.121) and (11.123) we arrive at

$$\mathcal{Z} = 2m\left(\tau - \frac{1}{4\pi}\ln\frac{M_{uv}^2}{m^2} + i\frac{k}{2}\right)\tag{11.126}$$

(remember that the kink mass $M = |\mathcal{Z}|$). A salient feature of this formula, to be noted, is the holomorphic dependence of $\mathcal{Z}$ on m and τ. Such a holomorphic dependence would be impossible if two or more loops contributed to the renormalization

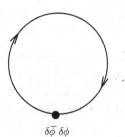

$\delta\bar{\phi}\,\delta\phi$

Fig. 11.3   Renormalization of h.

of h. Thus, h-renormalization beyond one loop must cancel, and it does.[5] Note also that the bare coupling in Eq. (11.126) conspires with the logarithm to replace the bare coupling by that renormalized at $|m|$, as expected.

The analysis carried out above is quasiclassical. It tells us nothing about the possible occurrence of nonperturbative terms in $\mathcal{Z}$. In fact, all terms of the type

$$\left[\frac{M_{\text{uv}}^2}{m^2}\exp(-4\pi\tau)\right]^{\ell}, \qquad \ell \text{ is an integer,}$$

are fully compatible with holomorphy; they can and do emerge from instantons [14].

11.3.5 Switching on Fermions

The nonzero modes are irrelevant for our discussion since, when combined with the boson nonzero modes, they cancel for critical solitons; the usual story. Thus, for our purposes it is sufficient to focus on the (static) zero modes in the kink background (11.109). The coefficients in front of the fermion zero modes will become (time-dependent) fermion moduli, for which we are going to build the corresponding quantum mechanics. There are two such moduli, $\bar{\eta}$ and η.

The equations for the fermion zero modes are

$$\partial_z \psi_L - \frac{2}{\chi}(\bar{\phi}\,\partial_z\phi)\psi_L - i\frac{1-\bar{\phi}\phi}{\chi}|m|e^{i\beta}\psi_R = 0,$$

$$\partial_z \psi_R - \frac{2}{\chi}(\bar{\phi}\,\partial_z\phi)\psi_R + i\frac{1-\bar{\phi}\phi}{\chi}|m|e^{-i\beta}\psi_L = 0 \tag{11.127}$$

(plus similar equations for $\bar{\psi}$; since our operator is Hermitian we do not need to consider them separately).

It is not difficult to find solution to these equations, either directly or using supersymmetry. Indeed, since we know the bosonic solution (11.109), its fermionic superpartner – and the fermion zero modes are such superpartners – is obtained from (11.109) by two supertransformations which act nontrivially on $\bar{\phi}, \phi$. In this way we conclude that the functional form of the fermion zero mode must coincide with the functional form of the boson solution (11.109). Concretely,

$$\begin{pmatrix} \psi_R \\ \psi_L \end{pmatrix} = \eta \left(\frac{g^2|m|}{2}\right)^{1/2} \begin{pmatrix} -ie^{-i\beta} \\ 1 \end{pmatrix} e^{|m|(z-z_0)} \tag{11.128}$$

and

$$\begin{pmatrix} \bar{\psi}_R \\ \bar{\psi}_L \end{pmatrix} = \bar{\eta} \left(\frac{g^2|m|}{2}\right)^{1/2} \begin{pmatrix} ie^{i\beta} \\ 1 \end{pmatrix} e^{|m|(z-z_0)}, \tag{11.129}$$

Fermion zero modes

where the numerical factor is introduced to ensure the proper normalization of the quantum-mechanical Lagrangian. Another solution, which asymptotically, at large z, behaves as $e^{3|m|(z-z_0)}$, must be discarded as non-normalizable.

Now, to perform quasiclassical quantization we follow the standard route: the moduli are assumed to be time dependent and we derive the quantum mechanics of

---

[5] Fermions are important for this cancelation.

moduli starting from the original Lagrangian (10.334). Substituting the kink solution and the fermion zero modes for ψ, one obtains

$$\mathcal{L}'_{\mathrm{QM}} = i\,\bar{\eta}\dot{\eta}. \tag{11.130}$$

In the Hamiltonian approach the only remnants of the fermion moduli are the anticommutation relations

$$\{\bar{\eta}, \eta\} = 1, \qquad \{\bar{\eta}, \bar{\eta}\} = 0, \qquad \{\eta, \eta\} = 0, \tag{11.131}$$

Short super-multiplet

which tell us that the wave function is *two-component* (i.e., the kink supermultiplet is two-dimensional). One can implement Eq. (11.131) by choosing, e.g., $\bar{\eta} = \sigma^+$, $\eta = \sigma^-$.

The fact that there are two critical kink states in the supermultiplet is consistent with the multiplet shortening in $\mathcal{N} = 2$. Indeed, in two dimensions the full $\mathcal{N} = 2$ supermultiplet must consist of four states; two bosonic and two fermionic. Half-BPS multiplets are shortened – they contain twice fewer states than the full supermultiplets: one bosonic and one fermionic. This is to be contrasted with the single-state kink supermultiplet in the minimal supersymmetric model of Section 11.2. The notion of fermion parity remains well defined in the kink sector of the CP(1) model.

11.3.6 Combining the Bosonic and Fermionic Moduli

The quantum dynamics of the kink under discussion is summarized by the Hamiltonian

$$H_{\mathrm{QM}} = \frac{M_0}{2}\dot{\bar{\zeta}}\dot{\zeta} \tag{11.132}$$

acting in the space of *two-component wave functions*. The variable ζ here is a complexified kink center,

$$\zeta = z_0 + \frac{i}{|m|}\alpha. \tag{11.133}$$

For simplicity, we will set the vacuum angle θ to 0 for the time being (it will be reinstated later).

The original field theory with which we are dealing has four conserved super-charges. Two of them, q and $\bar{q}$, see Eq. (11.105), act trivially in the critical kink sector. In the moduli quantum mechanics they take the form

$$q = \sqrt{M_0}\dot{\zeta}\eta, \qquad \bar{q} = \sqrt{M_0}\dot{\bar{\zeta}}\bar{\eta}, \tag{11.134}$$

explicitly demonstrating their vanishing provided that the kink is at rest. The superalgebra describing the kink quantum mechanics is $\{\bar{q}, q\} = 2\,H_{\mathrm{QM}}$. This is simply Witten's $\mathcal{N} = 1$ supersymmetric quantum mechanics [39] (two supercharges). The realization that we are dealing with is peculiar and distinct from that of Witten. Indeed, the standard quantum mechanics of Witten includes one (real) bosonic degree of freedom and two fermionic, while we have two bosonic degrees of freedom, x_0 and α. Nevertheless, the superalgebra remains the same due to the fact that the bosonic coordinate is complexified.

Finally, to conclude this section, let us calculate the U(1) charge of the kink states. We start from Eq. (11.98), substitute the fermion zero modes, and obtain[6]

$$\Delta q_{U(1)} = \frac{1}{2}[\bar{\eta}\eta] \qquad (11.135)$$

(this is to be added to the bosonic part, Eq. (11.119)). Given that $\bar{\eta} = \sigma^+$ and $\eta = \sigma^-$ we arrive at $\Delta q_{U(1)} = \frac{1}{2}\sigma_3$. This means that the U(1) charges of the two kink states in the supermultiplet split from the value given in Eq. (11.119) and become

$$k + \frac{1}{2} + \frac{\theta}{2\pi} \qquad \text{and} \qquad k - \frac{1}{2} + \frac{\theta}{2\pi}, \text{respectively.}$$

11.3.7 What Happens When One Moves to Small m?

Let us ask, in passing, what happens at small m. Needless to say the small-m domain is that of strong coupling. Quasiclassical methods are inapplicable. However, the soliton mass spectrum was found exactly by Dorey [14]. If $m = 0$ then there are two degenerate two-dimensional kink supermultiplets, corresponding to a nonvanishing Cecotti–Fendley–Intriligator–Vafa (CFIV) index [40]. (This index will be discussed in Section 11.3.8.) These kink supermultiplets have quantum numbers $\{q, T\} = (0, 1)$ and $(1, 1)$, respectively. Away from the point $m = 0$ the masses of these states are no longer equal. There is one singular point, where one of the two kink supermultiplets becomes massless [17]. The region containing the point $m = 0$ is separated from the quasiclassical region of large $|m|$ by the curve of marginal stability (CMS), on which an infinite number of other BPS states, visible quasiclassically, decay; see Fig. 11.4.

The CMS is also referred to in the current literature as the "wall." Correspondingly, people speak of the wall crossing.

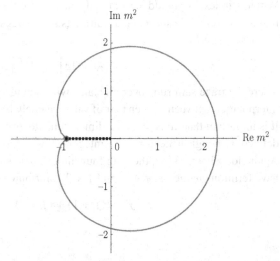

Fig. 11.4   The curve of marginal stability in CP(1) with twisted mass. We set $4\Lambda^2$ equal to 1. From [17]. The point $m^2 = -1$ is the so-called Argyres–Douglas point, at which one of the two kink supermultiplets becomes massless.

---

[6] To set the scale properly, so that the U(1) charge of the vacuum state vanishes, one must antisymmetrize the fermion current, $\bar{\psi}\gamma^\mu\psi \to \frac{1}{2}(\bar{\psi}\gamma^\mu\psi - \bar{\psi}^c\gamma^\mu\psi^c)$ where the superscript c denotes C conjugation. See Section 4.1.10.

Thus, the infinite tower of $\{q, T\}$ BPS states existing in the quasiclassical domain degenerates into just two stable BPS states in the vicinity of $m = 0$.[7]

11.3.8 The Cecotti–Fendley–Intriligator–Vafa Index

To put things into the proper perspective and refresh the reader's memory, I will start with Witten's index, $I_W = \text{Tr}(-1)^F \equiv \sum_a \langle a | (-1)^F | a \rangle$ (Section 10.22). This index is defined for $\mathcal{N} \geq 1$ theories in any number of dimensions. Nonvanishing-energy states always come in boson–fermion pairs and, thus, do not contribute to I_W. Vanishing-energy states, i.e., vacua, may or may not be paired. If they are not paired, $I_W \neq 0$, then a supersymmetric vacuum (or vacua) exists, and supersymmetry is not spontaneously broken.

Unfortunately, Witten's index says nothing about massive states which are always paired. Is there an analog of Witten's index which might tell us whether BPS-saturated solitons exist in the given theory?

An "index" I_{CFIV} acting as a litmus test for the presence of short multiplets was devised by Cecotti, Fendley, Intriligator, and Vafa [40]. I have put the word index in quotation marks because I_{CFIV} is independent of the D terms but it may depend on the F terms. Moreover, it is applicable only to $\mathcal{N} = 2$ theories in two dimensions. If $I_{\text{CFIV}} \neq 0$ then short (i.e., BPS-saturated) supermultiplets of kinks are guaranteed to exist. The converse is also true. We will see this shortly.

If the given two-dimensional theory has two or more supersymmetric vacua, it supports kinks – solitons interpolating between distinct vacua a and b. These kinks may or may not be BPS-saturated. In the former case they belong to short supermultiplets, in the latter to long supermultiplets. To reveal the parallel with Witten's index, it should be pointed out that long supermultiplets are analogs of massive states while short supermultiplets are analogs of the vacua.

Loosely speaking,[8]

$$I_{\text{CFIV}} = \text{Tr}\left[F(-1)^F \right], \tag{11.136}$$

where the trace sum runs over all states with boundary conditions corresponding to interpolation between a given pair of vacua, namely $|a\rangle$ at $z \to -\infty$ and $|b\rangle$ at $z \to \infty$. It is important that, in $\mathcal{N} = 2$ two-dimensional theories, the fermion charge F is well defined, although it need not be integer, as we learned from, e.g., Section 11.3.6. Again, loosely speaking, the long four-dimensional supermultiplets whose members have fermion charges f, $f + 1$, and $f + 2$ contribute (up to an overall phase)

| Look |
| through |
| Section 2.5. |

$$f - 2(f + 1) + (f + 2) = 0.$$

---

[7] CMS in $CP(N - 1)$ models with $N > 2$ are discussed in [41].

[8] In fact, this sum should be made convergent and well defined through an appropriate regularization. (The same is true, though, with regards to Witten's index.) In particular, IR regularization implies discretizing the spectrum of excitations in the soliton sector. The boundary conditions should be carefully chosen so as not to break the residual supersymmetry; cf. Section 11.2.5. "Residual" means the half of supersymmetry unbroken on the BPS-saturated kink.

At the same time, the short two-dimensional multiplets $\{f, f + 1\}$ contribute

$$f - (f + 1) = -1 \neq 0.$$

In particular, in the problem considered in Section 11.3.7, $I_{\text{CFIV}} = -2$.

If there are more than two vacua then the CFIV index is a matrix I_{ab}, with entries depending on the choice of the vacua at $z \to \pm\infty$. Taking into account the condition of CPT invariance[9] one can show [40] that the matrix I_{ab} is purely real and antisymmetric.

11.4 Domain Walls

The reader is advised to return to Section 2.1.

In four dimensions, domain walls are extended two-dimensional objects. In three dimensions they become domain lines, while in two dimensions they reduce to kinks, considered in Sections 11.2 and 11.3. Alternatively, one can say that the domain walls are obtained by elevating kinks from two to four dimensions. As in the kink case the domain wall is a field configuration of codimension 1 interpolating between vacuum i and vacuum f with some transitional domain in the middle (Fig. 11.5).

Critical domain walls in $\mathcal{N} = 1$ four-dimensional theories (i.e., theories with four supercharges) started attracting attention in the 1990s. The very existence of BPS-saturated domain walls (also known as *branes*) is due to nonvanishing $(1, 0)$ and $(0, 1)$ central charges; see Eqs. (11.8) and (11.9).[10]

Early on, domain-wall studies were limited to the generalized Wess–Zumino model (Section 10.6.7) with Lagrangian

$$\mathcal{L} = \int d^2\theta \, d^2\bar{\theta} \, \mathcal{K}(\bar{\Phi}_a, \Phi_a) + \left(\int d^2\theta \, \mathcal{W}(\Phi) + \text{H.c.} \right), \tag{11.137}$$

where $\mathcal{K}$ is the Kähler potential and Φ_a stands for a set of chiral superfields. The number of chiral superfields is arbitrary, but the superpotential $\mathcal{W}$ must have at least two critical points, i.e., two vacua. One can achieve BPS saturation provided that the following first-order differential equations [8, 43–46] are satisfied:

$$g_{\bar{a}b} \partial_z \Phi^b = e^{i\eta} \partial_{\bar{a}} \bar{\mathcal{W}}, \tag{11.138}$$

where the Kähler metric is given by

$$g_{\bar{a}b} = \frac{\partial^2 K}{\partial \bar{\Phi}^{\bar{a}} \partial \Phi^b} \equiv \partial_{\bar{a}} \partial_b \mathcal{K} \tag{11.139}$$

---

[9] Under CPT the initial and final vacua interchange, $|a\rangle \leftrightarrow |b\rangle$, and, simultaneously, $f \to -f$.

[10] Townsend was the first to note [42] that starting from solitons in D dimensions one can obtain "extended solitons" of supersymmetric field theories in $D + 1, D + 2$, etc. dimensions for which a Bogomol'nyi-type bound of the form (mass/unit p-volume)$\geq$ topological charge is saturated. This would require the existence of tensorial central charges with Lorentz indices. That the anticommutator $\{Q_\alpha, Q_\beta\}$ in the four-dimensional Wess–Zumino model contains the $(1, 0)$ central charge is obvious. This anticommutator vanishes, however, in super-Yang–Mills theory at the classical level (Section 11.4.2). It appears as a result of the quantum anomaly [10].

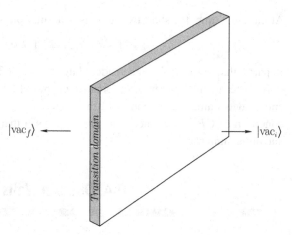

Fig. 11.5 A field configuration interpolating between two distinct degenerate vacua.

and η is the phase of the $(1,0)$ central charge Z defined in (11.8). The phase η depends on the choice of the vacua between which the given domain wall interpolates,

$$\eta = \arg Z = \arg \left[2 \left(\mathcal{W}_{\text{vac}_f} - \mathcal{W}_{\text{vac}_i} \right) \right]. \qquad (11.140)$$

A useful consequence of the BPS equations is

$$\partial_z \mathcal{W} = e^{i\eta} \| \partial_a \mathcal{W} \|^2, \qquad (11.141)$$

which entails, in turn, that the domain wall describes a straight line in the $\mathcal{W}$-plane connecting the two vacua (see Eq. (11.164) below and the subsequent comment). Needless to say, the first-order BPS equation (11.138) guarantees the validity of the second-order equation of motion. The converse is not true.[11]

11.4.1 Domain Wall in the Minimal Wess–Zumino Model

Here we will consider the minimal Wess–Zumino model [48] (Section 10.6.4), with one chiral superfield. In components the Lagrangian has the form

$$\mathcal{L} = (\partial^\mu \bar\phi)(\partial_\mu \phi) + \psi^\alpha i \partial_{\alpha\dot\alpha} \bar\psi^{\dot\alpha} + \bar{F} F + \left[F W'(\phi) - \frac{1}{2} W''(\phi) \psi^2 + \text{H.c.} \right]. \qquad (11.142)$$

As usual, F can be eliminated by virtue of the classical equation of motion,

$$\bar{F} = -\frac{\partial W(\phi)}{\partial \phi}, \qquad (11.143)$$

Scalar potential

so that the scalar potential describing the self-interaction of the field ϕ is

$$V(\phi, \bar\phi) = \left| \frac{\partial W(\phi)}{\partial \phi} \right|^2. \qquad (11.144)$$

---

[11] However, if one is dealing with a *single* chiral field Φ, then one can prove [47] that the BPS equation does follow from the second-order equation of motion. The proof of this assertion is presented in Exercise 2.1.5.

If one limits oneself to renormalizable theories, the superpotential $\mathcal{W}$ must be a polynomial function of Φ of power not higher than 3. In the model at hand, with one chiral superfield, the generic superpotential can be reduced to the following "standard" form:

$$\mathcal{W}(\Phi) = \frac{m^2}{\lambda}\Phi - \frac{\lambda}{3}\Phi^3. \tag{11.145}$$

The quadratic term can be eliminated by a redefinition of the field Φ. Moreover, by using symmetries of the model one can choose the phases of the constants m and λ at will.

The superpotential (11.145) implies two degenerate classical vacua,

$$\phi_{\text{vac}} = \pm\frac{m}{\lambda}. \tag{11.146}$$

These vacua are physically equivalent. This equivalence can be explained by the spontaneous breaking of Z_2 symmetry, $\Phi \to -\Phi$, present in the action.

The field configurations interpolating between two degenerate vacua are the domain walls. They have the following properties: (i) the corresponding solutions are static and depend only on one spatial coordinate; (ii) they are topologically stable and indestructible – once a wall is created it cannot disappear. Assume for definiteness that the wall lies in the xy plane. This is the geometry that we will keep in mind throughout our discussion. Then the wall solution ϕ_{w} depends only on z. Since the wall extends indefinitely in the xy plane, its energy E_{w} is infinite. However, the wall tension T_{w}, the energy per unit area $T_{\text{w}} = E_{\text{w}}/A$, is finite, in principle measurable, and has a clear-cut physical meaning.

The wall solution of the classical equations of motion superficially looks very similar to that of the kink,

$$\phi_{\text{w}} = \frac{m}{\lambda}\tanh(|m|z). \tag{11.147}$$

Note, however, that the parameters m and λ are not assumed to be real; the field ϕ is complex in the four-dimensional Wess–Zumino model. A remarkable feature of this solution is that it preserves half the supersymmetry, in much the same way as the kink considered in Section 11.2. The difference is that 1/2 BPS in the two-dimensional model meant one supercharge; now it means two supercharges.

The supertransformations of the fields are

$$\delta\phi = \sqrt{2}\epsilon\psi, \qquad \delta\psi^\alpha = \sqrt{2}\left[\epsilon^\alpha F + i\partial_\mu\phi(\sigma^\mu)^{\alpha\dot\alpha}\bar\epsilon_{\dot\alpha}\right]. \tag{11.148}$$

The domain wall we are considering is purely bosonic, $\psi = 0$. Moreover, the BPS equation is

$$F|_{\bar\phi=\phi_{\text{w}}^*} = -e^{-i\eta}\partial_z\phi_{\text{w}}(z), \tag{11.149}$$

where, in the case at hand,

$$\eta = \arg\frac{m^3}{\lambda^2} \tag{11.150}$$

and $F = -\partial\bar{\mathcal{W}}/\partial\bar\phi$. Equation (11.149) is a first-order differential equation. The solution quoted in (11.147) satisfies both (11.149) and the boundary conditions.

The reason for the occurrence of the phase factor $\exp(-i\eta)$ on the right-hand side of Eq. (11.149) will become clear shortly. Note that no analog of this phase factor exists in the two-dimensional $\mathcal{N} = 1$ problem with which we dealt in Section 11.2. There was only a sign ambiguity: two choices of sign were possible, corresponding to a kink or an antikink.

If the BPS equation is satisfied then the second supertransformation in Eq. (11.148) reduces to

$$\delta\psi_\alpha \propto \epsilon_\alpha + ie^{i\eta}(\sigma^z)_{\alpha\dot\alpha}\bar\epsilon^{\dot\alpha}. \tag{11.151}$$

The right-hand side of (11.151) vanishes provided that

$$\epsilon_\alpha = -ie^{i\eta}(\sigma^z)_{\alpha\dot\alpha}\bar\epsilon^{\dot\alpha}. \tag{11.152}$$

This leaves up to two supertransformations (out of four) that do not act on the domain wall (alternatively it is often said that they act trivially), as we set out to show.

Now let us calculate the wall tension. To this end, we perform Bogomol'nyi completion for the energy functional,

$$\begin{aligned}
\mathcal{E} &= \int_{-\infty}^{+\infty} dz \left(\partial_z\bar\phi\,\partial_z\phi + \bar F F\right) \\
&\equiv \int_{-\infty}^{+\infty} dz \left[\left(e^{-i\eta}\partial_z\mathcal{W} + \text{H.c.}\right) + \left|\partial_z\phi + e^{i\eta}F\right|^2\right],
\end{aligned} \tag{11.153}$$

where ϕ is assumed to depend only on z. The second term on the right-hand side is non-negative – its minimal value is zero. The first term, being a full derivative, depends only on the boundary conditions for ϕ at $z = \pm\infty$.

Equation (11.153) implies that $\mathcal{E} \geq 2\operatorname{Re}(e^{-i\eta}\Delta\mathcal{W})$. Bogomol'nyi completion can be performed with any η; however, the strongest bound is achieved when $e^{-i\eta}\Delta\mathcal{W}$ is real. This explains the emergence of the phase factor (11.140) in the BPS equations. In the model at hand, to make $e^{-i\eta}\Delta\mathcal{W}$ real we have to choose η according to Eq. (11.150).

When the energy functional is written in the form (11.153), it is perfectly obvious that the absolute minimum is achieved provided that the BPS equation (11.149) is satisfied. In fact, Bogomol'nyi completion provides us with an alternative way of deriving the BPS equations. Then the result for the minimum of the energy functional, i.e., the wall tension T_w, is

$$T_w = |\mathcal{Z}|, \tag{11.154}$$

where the topological charge $\mathcal{Z}$ is defined as

$$\mathcal{Z} = 2[\mathcal{W}(\phi(z=\infty)) - \mathcal{W}(\phi(z=-\infty))] = \frac{8m^3}{3\lambda^2}. \tag{11.155}$$

An explanatory comment is in order here. In the present problem the extension of the superalgebra is tensorial, with Lorentz structure $(1,0) + (0,1)$:

$$\left\{Q_\alpha, Q_\beta\right\} = -4\Sigma_{\alpha\beta}\bar{\mathcal{Z}}, \qquad \left\{\bar Q_{\dot\alpha}, \bar Q_{\dot\beta}\right\} = -4\bar\Sigma_{\dot\alpha\dot\beta}\mathcal{Z}, \tag{11.156}$$

where

$$\Sigma_{\alpha\beta} = -\tfrac{1}{2}\int dx_{[\mu}dx_{\nu]}(\sigma^\mu)_{\alpha\dot\alpha}(\bar\sigma^\nu)^{\dot\alpha}_\beta \tag{11.157}$$

is the wall area tensor. Equation (11.156) is primary, while Eq. (11.155) is a reduction of (11.156) in which the tensorial structure is separated and discarded.

The expressions for the two supercharges $\tilde{Q}_\alpha$ that annihilate the wall are

$$\tilde{Q}_\alpha = e^{i\eta/2} Q_\alpha - \frac{2}{A} e^{-i\eta/2} \Sigma_{\alpha\beta} n^\beta_{\dot\alpha} \bar{Q}^{\dot\alpha}, \tag{11.158}$$

where

$$n_{\alpha\dot\alpha} = \frac{P_{\alpha\dot\alpha}}{T_{\rm w} A} \tag{11.159}$$

is the unit vector proportional to the wall's 4-momentum $P_{\alpha\dot\alpha}$; only its time component is nonvanishing in the wall's rest frame. The subalgebra of these "residual" (unbroken) supercharges in the rest frame is

$$\left\{ \tilde{Q}_\alpha, \tilde{Q}_\beta \right\} = 8\Sigma_{\alpha\beta}(T_{\rm w} - |\mathcal{Z}|). \tag{11.160}$$

The existence of the subalgebra (11.160) immediately proves that the wall tension $T_{\rm w}$ is equal to the central charge $\mathcal{Z}$. Indeed, $\tilde{Q}|{\rm wall}\rangle = 0$ implies that $T_{\rm w} - |\mathcal{Z}| = 0$. This equality is valid both to any order in perturbation theory and nonperturbatively.

From the nonrenormalization theorem for the superpotential [48, 49] (Section 10.8) we can infer in addition that the central charge $\mathcal{Z}$ is not renormalized. This is in contradistinction with the situation in the two-dimensional model of Section 11.2. The fact that in four dimensions there are more conserved supercharges than in two turns out to be crucial. As a consequence, the result

$$T_{\rm w} = \frac{8}{3} \left| \frac{m^3}{\lambda^2} \right| \tag{11.161}$$

Nonrenor-
malization of
$T_{\rm w} \leftrightarrow$
nonrenor-
malization of
superpoten-
tial

for the wall tension is exact [46].

The wall tension $T_{\rm w}$ is a physical parameter and, as such, should be expressible in terms of the physical (renormalized) parameters $m_{\rm ren}$ and $\lambda_{\rm ren}$. One can easily verify that this is compatable with the nonrenormalization of $T_{\rm w}$. Indeed,

$$m = Z m_{\rm ren}, \qquad \lambda = Z^{3/2} \lambda_{\rm ren},$$

where the Z factor comes from the kinetic term. Consequently,

$$\frac{m^3}{\lambda^2} = \frac{m^3_{\rm ren}}{\lambda^2_{\rm ren}}.$$

Thus, the absence of quantum corrections to Eq. (11.161), the renormalizability of the theory, and the nonrenormalization theorem for superpotentials are all intertwined with each other. In fact, any two of these features imply the third.

What lessons can we draw from the domain-wall example? In centrally extended superalgebras the exact relation $E_{\rm vac} = 0$ is replaced by the exact relation $T_{\rm w} - |\mathcal{Z}| = 0$. Although this statement is valid both perturbatively and nonperturbatively, it is very instructive to visualize it as an explicit cancelation between the bosonic and fermionic modes in perturbation theory. The nonrenormalization of $\mathcal{Z}$ is a specific feature of the four-dimensional Wess–Zumino model. We have seen previously that it does not take place in minimally supersymmetric models in two dimensions.

11.4.1.1 Finding the Solution to the BPS Equation

In the two-dimensional theory considered in Section 11.2, integrating the first-order BPS equation (11.42) was trivial. The BPS equation (11.149) presents two equations, one for the real part and one for the imaginary part. Nevertheless finding the solution is still trivial; this is due to the existence of an "integral of motion,"

$$\frac{\partial}{\partial z}\left(\text{Im}\, e^{-i\eta}\mathcal{W}\right) = 0. \tag{11.162}$$

The proof of the formula is straightforward and is valid in the generic Wess–Zumino model with arbitrary number of fields. Indeed, differentiating $\mathcal{W}$ and using the BPS equation we get

$$\frac{\partial}{\partial z}\left(e^{-i\eta}\mathcal{W}\right) = \left|\frac{\partial \mathcal{W}}{\partial \phi}\right|^2, \tag{11.163}$$

which immediately entails Eq. (11.162). The constraint

$$\text{Im}\, e^{-i\eta}\mathcal{W} = \text{const} \tag{11.164}$$

can be interpreted as follows: in the complex $\mathcal{W}$ plane the domain-wall trajectory is a straight line.

11.4.1.2 Living on a Wall

What is the fate of two broken supercharges? As we already know, two out of the four supercharges annihilate the wall – these supersymmetries are preserved in the given wall background. Two other supercharges are broken: when applied to the wall solution they create two fermion zero modes. These zero modes correspond to a $(2+1)$-dimensional Majorana (massless) spinor field $\psi(t, x, y)$ localized on the wall.

To elucidate the above assertion it is convenient to turn first to the fate of another symmetry of the original theory, which is spontaneously broken for each given wall, namely, translational invariance in the z direction.

Cf.
Section 2.1.8.

Indeed, each wall solution, e.g., Eq. (11.147), breaks this invariance. This means that in fact we must deal with a family of solutions: if $\phi(z)$ is a solution, then so is $\phi(z - z_0)$. The parameter z_0 is a collective coordinate, the wall center. People also refer to it as a *modulus* (plural *moduli*). For a static wall z_0 is a fixed constant.

Assume, however, that the wall bends slightly. The bending should be negligible compared to the wall thickness (which is of order m^{-1}). It can be described as an adiabatically slow dependence of the wall center z_0 on t, x, and y. We will write a slightly bent wall field configuration as

$$\phi(t, x, y, z) = \phi_w(z - \zeta(t, x, y)). \tag{11.165}$$

Substituting this field into the original action we arrive at the following effective $(2+1)$-dimensional action for the field $\zeta(t, x, y)$:

$$S_{2+1}^{\zeta} = \frac{T_w}{2}\int d^3x\,(\partial^m\zeta)\,(\partial_m\zeta), \qquad m = 0, 1, 2. \tag{11.166}$$

It is clear that $\zeta(t, x, y)$ can be viewed as a massless scalar field (called the translational modulus) that lives on the wall. It is simply a Goldstone field corresponding to the spontaneous breaking of the translational invariance.

Returning to the two broken supercharges, they generate a Majorana $(2+1)$-dimensional Goldstino field $\psi_\alpha(t, x, y)$, $\alpha = 1, 2$, localized on the wall. The total $(2+1)$-dimensional effective action on the wall world volume takes the form

$$S_{2+1} = \frac{T_w}{2} \int d^3 x \left[(\partial^m \zeta)(\partial_m \zeta) + i \bar{\psi} \partial_m \gamma^m \psi \right], \tag{11.167}$$

| World-sheet |
| theory on the |
| wall |

where the γ^m are the three-dimensional gamma matrices.

The effective theory of the moduli fields on the wall's world volume is supersymmetric, with two conserved supercharges. This is the minimal supersymmetry in $2+1$ dimensions. It corresponds to the fact that two out of the four supercharges are conserved.

11.4.2 D-Branes in Gauge Field Theory

The $(1,0)$ central extension in $\mathcal{N} = 1$ superalgebra is not seen at the classical level in supersymmetric gluodynamics. Nevertheless, it exists [10] as a quantum anomaly.[12] The above central charge is saturated on domain walls that interpolate between vacua with distinct values of the order parameter, the gluino condensate $\langle \lambda\lambda \rangle$, labeling N distinct vacua of super-Yang–Mills theory (see Section 10.14) with gauge group $SU(N)$.

Supersymmetric gluodynamics is described by the Lagrangian (10.374). There is a large variety of domain walls in supersymmetric gluodynamics, as shown in Fig. 11.6. Minimal, or elementary, walls interpolate between vacua n and $n+1$, while k-walls interpolate between n and $n+k$.

In $\mathcal{N} = 1$ gauge theories with arbitrary matter content and superpotentials the general relation (11.8) takes the form

$$\{ Q_\alpha, Q_\beta \} = -4 \Sigma_{\alpha\beta} \bar{Z}, \tag{11.168}$$

where

$$\Sigma_{\alpha\beta} = -\frac{1}{2} \int dx_{[\mu} dx_{\nu]} (\sigma^\mu)_{\alpha\dot\alpha} (\bar\sigma^\nu)^{\dot\alpha}_\beta \tag{11.169}$$

is the wall area tensor and [46, 52]

$$Z = \frac{2}{3} \Delta \left\{ \left(3\mathcal{W} - \sum_f Q_f \frac{\partial \mathcal{W}}{\partial Q_f} \right) \right.$$
$$\left. - \left[\frac{3N - \sum_f T(R_f)}{16\pi^2} \mathrm{Tr}\, W^2 + \frac{1}{8} \sum_f \gamma_f \bar{D}^2 (\bar{Q}_f e^V Q_f) \right] \right\}_{\theta=0} ; \tag{11.170}$$

cf. Eq. (10.445). In (11.170), the action of the symbol Δ is to take the difference at two spatial infinities in a direction perpendicular to the surface of the wall. The

---

[12] A remark in passing: Witten interpreted BPS walls in supersymmetric gluodynamics as analogs of D-branes [50]. The reason was that their tension scales as $N \sim 1/g_s$ rather than $1/g_s^2$, the later scaling being typical of solitonic objects (g_s is the string constant). Many promising consequences ensued. One was the Acharya–Vafa derivation of the wall world-volume theory [51]. Using a wrapped D-brane picture and certain dualities they identified the k-wall world-volume theory as a $(1+2)$-dimensional $U(k)$ gauge theory with the field content of $\mathcal{N} = 2$ and the Chern–Simons term at level N breaking $\mathcal{N} = 2$ down to $\mathcal{N} = 1$. This allowed them to calculate the wall multiplicity; see the end of this subsection.

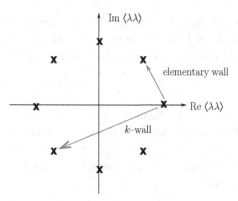

Fig. 11.6   The N vacua for SU(N). The vacua are labeled by the vacuum expectation value $\langle \lambda \lambda \rangle = -6N\Lambda^3 \exp(2\pi i k / N)$, where $k = 0, 1, \ldots, N-1$. The elementary walls interpolate between two neighboring vacua.

first term in the second line presents the gauge anomaly in the central charge. The second term is a total superderivative; therefore, it vanishes after averaging over any supersymmetric vacuum state and hence, can safely be omitted. The first line presents the classical result; see Section 10.16.7. At the classical level $Q_f(\partial \mathcal{W}/\partial Q_f)$ is a total superderivative; this can be seen from the Konishi anomaly (10.464). If we discard all anomalies and total superderivatives (just for a short while), we return to $Z = 2\Delta(\mathcal{W})$, the formula obtained in the Wess–Zumino model; see Eq. (11.155). At the quantum level, with anomalies included, $Q_f(\partial \mathcal{W}/\partial Q_f)$ ceases to be a total superderivative because of the Konishi anomaly. It is still convenient to eliminate $Q_f(\partial \mathcal{W}/\partial Q_f)$ in favor of $\operatorname{Tr} W^2$ by virtue of the Konishi relation (10.464). In this way one arrives at

$$Z = 2\Delta \left[\mathcal{W} - \frac{N - \sum_f T(R_f)}{16\pi^2} \operatorname{Tr} W^2 \right]_{\theta=0}. \tag{11.171}$$

We see that the superpotential $\mathcal{W}$ is amended by the anomaly; in operator form we have

$$\mathcal{W} \to \mathcal{W} - \frac{N - \sum_f T(R_f)}{16\pi^2} \operatorname{Tr} W^2. \tag{11.172}$$

Of course, in pure super-Yang–Mills theory only the anomaly term survives.

Equation (11.170) implies that in pure gluodynamics (super-Yang–Mills theory without matter) the domain-wall tension is

$$T = \frac{N}{8\pi^2} \left| \langle \operatorname{Tr}\lambda^2 \rangle_{\text{vac}_f} - \langle \operatorname{Tr}\lambda^2 \rangle_{\text{vac}_i} \right| \tag{11.173}$$

where $\text{vac}_{i,f}$ stands for the initial or final vacuum between which the given wall interpolates. Furthermore, the gluino condensate $\langle \operatorname{Tr} \lambda^2 \rangle_{\text{vac}}$ was calculated – exactly – long ago [53], using the same methods as those which were later advanced and perfected by Seiberg and Witten in their quest for the dual Meissner effect in $\mathcal{N} = 2$ (see [21, 22]):

Cf.
Section 10.14.

$$2\langle \operatorname{Tr}\lambda^2 \rangle = \langle \lambda_\alpha^a \lambda^{a,\alpha} \rangle = -6N\Lambda^3 \exp\left(\frac{2\pi i k}{N} \right), \qquad k = 0, 1, \ldots, N-1. \tag{11.174}$$

Here k labels the N distinct vacua of the theory; see Fig. 11.6. The dynamical scale Λ is defined in the standard manner, i.e., in accordance with [54], in terms of the ultraviolet parameters M_{uv} (the ultraviolet regulator mass) and g_0^2 (the bare coupling constant):

$$\Lambda^3 = \frac{2}{3} M_{uv}^3 \left(\frac{8\pi^2}{Ng_0^2}\right) \exp\left(-\frac{8\pi^2}{Ng_0^2}\right). \tag{11.175}$$

In each given vacuum the gluino condensate scales with the number of colors as N. However, the difference in the values of the gluino condensates in two vacua that lie not too far from each other scales as N^0. From Eq. (11.173) we can conclude that the wall tension in supersymmetric gluodynamics satisfies

$$T \sim N.$$

Since the string coupling constant $g_s \sim 1/N$, see Section 9.2.3, $T \sim 1/g_s$ rather than $1/g_s^2$. Therefore, this is not a "normal" soliton but, rather, a D-brane. (This is the essence of Witten's argument regarding why the above walls should be considered as analogs of D-branes.)

As mentioned, there is a large variety of walls in supersymmetric gluodynamics as they can interpolate between vacua with arbitrary values of k. Even if $k_f = k_i + 1$, i.e., the wall is elementary, in fact we are dealing with several walls, all having the same tension – let us call them *degenerate walls*.[13] The fact that distinct walls can have the same tension is specific to supersymmetry. It was discovered in studies of BPS-saturated walls – in such walls, even if their internal structures are different, tension degeneracy is a consequence of the general law $T = |Z|$.

Multiplicity of walls interpolating between the given initial and final vacua

The k-wall multiplicity is

$$\nu_k = C_N^k = \frac{N!}{k!\,(N-k)!}. \tag{11.176}$$

For $N = 2$, only elementary walls exist and $\nu = 2$. In a field-theoretic setting, Eq. (11.176) was derived in [56]. This derivation was based on the fact that the index ν is topologically stable – continuous deformations of the theory do not change ν. Thus one can add an appropriate set of matter fields sufficient for the complete Higgsing of supersymmetric QCD. The domain wall multiplicity in the effective low-energy theory obtained in this way is the same as in supersymmetric gluodynamics, although the effective low-energy theory, a Wess–Zumino-type model, is much simpler. Varying the matter mass term m from small ($m \ll \Lambda$) to large ($m \gg \Lambda$) we pass (at $m \sim \Lambda$) from ν_k classically distinct walls in Wess-Zumino to an appropriate Chern-Simons theory on the world-sheet of the super-Yang-Mills domain wall without changing ν_k (Z. Komagodski, M. Shifman, unpublished).

11.4.3 1/4-BPS Saturated Domain-Wall Junctions

If distinct walls can have the same tension, two degenerate domain walls can coexist in one plane – a new phenomenon discovered in [57]. It is illustrated in Fig. 11.7.

---

[13] The first indication on wall degeneracy was obtained in [55], where two degenerate walls were observed in SU(2) theory. Later, Acharya and Vafa calculated the k-wall multiplicity [51] within the framework of D-brane and string formalism.

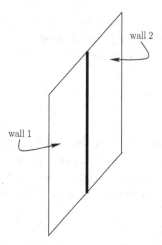

wall 2

wall 1

Two distinct degenerate domain walls separated by a wall junction.

Two distinct degenerate domain walls lie in a plane; the transition domain between wall 1 and wall 2 is a domain wall junction (domain line).

Each individual domain wall is 1/2-BPS-saturated. A wall configuration with a junction line (Fig. 11.7) is 1/4-BPS-saturated.

11.5 Vortices in $D = 3$ and Flux Tubes in $D = 4$

For in-depth study delve into [58].

Vortices were among the first examples of topological defects treated in the Bogomol'nyi limit (see, e.g., [4, 25, 29]). The explicit embedding of the bosonic sector in supersymmetric models dates back to the 1980s. The three-dimensional Abelian Higgs model is the simplest model supporting vortices [59]. This model has $\mathcal{N} = 1$ supersymmetry (two supercharges) and thus, according to Section 11.1.2.2, contains no central charge that could be saturated by vortices. Hence the vortices discussed in [59] were noncritical. However, BPS-saturated vortices can and do occur in $\mathcal{N} = 2$ three-dimensional models (four supercharges) with a nonvanishing Fayet–Iliopoulos term [60, 61]. Such a model can be obtained by dimensional reduction from four-dimensional $\mathcal{N} = 1$ SQED, a model that we discussed in detail in Section 10.6.9. We will start by performing such a reduction. The bosonic sector of the model, as well as the bosonic solutions, were considered in Chapter 3.

11.5.1 $\mathcal{N} = 2$ SQED in Three Dimensions

The starting point is SQED, with the Fayet–Iliopoulos term ξ, in four dimensions. The SQED Lagrangian is

$$
\begin{aligned}
\mathcal{L} = {} & \left(\frac{1}{4e^2} \int d^2\theta \, W^2 + \text{H.c.} \right) + \int d^4\theta \, \bar{Q} \, e^{n_e V} Q \\
& + \int d^4\theta \, \bar{\tilde{Q}} e^{-n_e V} \tilde{Q} - n_e \xi \int d^2\theta \, d^2\bar{\theta} \, V(x, \theta, \bar{\theta}),
\end{aligned}
\tag{11.177}
$$

SQED
Lagrangian

where e is the electric coupling constant and Q and $\tilde{Q}$ are chiral matter superfields (with charges n_e and $-n_e$, respectively). This expression differs from (10.181) in two aspects. In (11.177) we do not assume the electric charge of matter to be 1 (in units of e), and we set the matter mass term m equal to 0.

In four dimensions the absence of the chiral anomaly in SQED requires the matter superfields to enter in pairs of opposite charges, e.g.,

$$i\mathcal{D}_\mu \psi = (i\partial_\mu + n_e A_\mu)\psi, \qquad i\mathcal{D}_\mu \tilde{\psi} = \left(i\partial_\mu - n_e A_\mu\right)\tilde{\psi}. \qquad (11.178)$$

Otherwise the theory would be anomalous; the chiral anomaly would render it noninvariant under gauge transformations. Thus, the minimal matter sector includes two chiral superfields Q and $\tilde{Q}$, with charges n_e and $-n_e$, respectively.

In three dimensions there is no chirality. Therefore, one can consider three-dimensional SQED with a single matter superfield Q, with charge n_e. Surprising though it is, this theory is more complicated than that with two chiral superfields, Q and $\tilde{Q}$, because of a quantum anomaly on which we will not dwell here. We will limit ourselves to a nonminimal matter sector, in which both Q and $\tilde{Q}$ are present.

The (integer) charge of Q is n_e.

Now we keep the three coordinates, t, x, and z, uncompactified while $y \equiv x_2$ is reduced. After reduction to three dimensions and passing to components (in the Wess–Zumino gauge) we arrive at the action in the following form, in three-dimensional notation:

$$S = \int d^3x \left\{ -\frac{1}{4e^2} F_{\mu\nu}F^{\mu\nu} + \frac{1}{2e^2}\left(\partial_\mu a\right)^2 + \frac{1}{e^2}\bar{\lambda}i\,\slashed{\partial}\lambda \right.$$
$$+ \frac{1}{2e^2}D^2 - n_e\xi D + n_e D\left(\bar{q}q - \bar{\tilde{q}}\tilde{q}\right)$$
$$+ \left[\mathcal{D}^\mu\bar{q}\mathcal{D}_\mu q + \bar{\psi}\,i\,\slashed{\mathcal{D}}\psi\right] + \left[\mathcal{D}^\mu\bar{\tilde{q}}\mathcal{D}_\mu\tilde{q} + \bar{\tilde{\psi}}\,i\,\slashed{\mathcal{D}}\tilde{\psi}\right]$$
$$- a^2\bar{q}q - a^2\bar{\tilde{q}}\tilde{q} + a\bar{\psi}\psi - a\bar{\tilde{\psi}}\tilde{\psi}$$
$$\left. + n_e\left[\sqrt{2}(\bar{\lambda}\psi)\bar{q} + \text{H.c.}\right] - n_e\left[\sqrt{2}\left(\bar{\lambda}\tilde{\psi}\right)\bar{\tilde{q}} + \text{H.c.}\right]\right\}. \qquad (11.179)$$

Here a is a real scalar field,

$$a = -n_e A_2,$$

λ is the photino field, and $q, \tilde{q}$ and $\psi, \tilde{\psi}$ are matter fields belonging to Q and $\tilde{Q}$, respectively. The covariant derivatives were defined in Eq. (11.178). Finally, D is an auxiliary field, the last component of the superfield V. Eliminating D via the equation of motion we get the scalar potential

$$V = \frac{e^2}{2}n_e^2\left[\xi - \left(\bar{q}q - \bar{\tilde{q}}\tilde{q}\right)\right]^2 + a^2\bar{q}q + a^2\bar{\tilde{q}}\tilde{q}. \qquad (11.180)$$

We will assume that

$$\xi > 0. \qquad (11.181)$$

For our purposes – the consideration of BPS-saturated vortices – only the Higgs branch is of importance. Hence we will set $a = 0$; the field a will play no role in what follows. Then the bosonic sector is essentially the same as considered in Chapter 3, with one exception: we have two scalar fields q and $\tilde{q}$. In the vacuum they are subject to the constraint $\xi = \bar{q}q - \bar{\tilde{q}}\tilde{q}$. This demonstrates the existence of a flat direction with complex dimension 1 (see Section 10.6.10). Correspondingly, there are gapless modes – a massless modulus and its superpartner – which render the theory ill defined in the infrared. We will discuss this issue in more detail in Section 11.5.2. If we choose a generic vacuum belonging to the flat direction then infinite-length flux tubes with finite tension do not exist [63]. A classical solution to the BPS equations can be found only at the base of the Higgs branch, i.e., at $\tilde{q} = 0$ (then $q_{\text{vac}} = \sqrt{\xi}$). To be in the weak coupling regime requires $e^2/\sqrt{\xi} \ll 1$. Up to gauge transformations, the vacuum $q_{\text{vac}} = \sqrt{\xi}$ is unique. The fields $\tilde{q}, \tilde{\psi}$ play a role only at the level of quantum corrections, in loops.

11.5.1.1 Central Charge

The general form of the centrally extended $\mathcal{N} = 2$ superalgebra in $D = 3$ was discussed in Section 11.1.3.2. The central charge relevant in the problem at hand – vortices – is given by the last term in Eq. (11.15). It can be conveniently derived using the complex representation for supercharges and reducing from $D = 4$ to $D = 3$. In four dimensions [12],

$$\{Q_\alpha, \bar{Q}_{\dot\alpha}\} = 2P_{\alpha\dot\alpha} + 2Z_{\alpha\dot\alpha} \equiv 2\left(P_\mu + Z_\mu\right)(\sigma^\mu)_{\alpha\dot\alpha}, \tag{11.182}$$

where P_μ is the momentum operator and

$$Z_\mu = n_e \xi \int d^3x \varepsilon_{0\mu\nu\rho}(\partial^\nu A^\rho) + \ldots \tag{11.183}$$

Here ellipses denote full spatial derivatives of currents that fall off exponentially fast at infinity. Such terms are clearly inessential for the soliton vortices. They are essential, however, for wall junctions; see Section 2.2.

In three dimensions the central charge of interest reduces to $P_2 + Z_2$. Thus, in terms of complex supercharges the appropriate centrally extended algebra takes the form[14]

$$\{Q, (Q^\dagger)\gamma^0\} = 2\left(P_0\gamma^0 + P_1\gamma^x + P_3\gamma^z\right)$$
$$+ 2\left[\frac{1}{e^2}\int d^2x\, \vec{\nabla}\left(\vec{E}a\right) - n_e\,\xi\int d^2x\, B\right], \tag{11.184}$$

| Extended |
| superalgebra |
| in 3D |

where $\vec{E}$ is the electric field and B is the magnetic field,

$$B = \frac{\partial A_z}{\partial x} - \frac{\partial A_x}{\partial z}. \tag{11.185}$$

The second line in Eq. (11.184) gives the vortex-related central charge. In the problem at hand the ξ term in the central charge is not renormalized in loops (see the remark after Eq. (10.222)) and neither is the vortex mass.

11.5.1.2 BPS Equations for a Vortex

The BPS equations for a vortex were considered in detail in Section 3.1.3. For completeness, I will reproduce them here in the notation that we are using for

---

[14] In the following expression terms containing equations of motion of the type $a\left(\vec{\nabla}\vec{E} - J_0\right)$ are omitted.

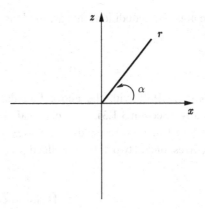

Fig. 11.8   Polar coordinates in the xz plane.

supersymmetric theories. The first-order equations describing the ANO vortex in the Bogomol'nyi limit [4, 25, 29] take the form

$$B - n_e e^2 \left(|q|^2 - \xi \right) = 0, \tag{11.186}$$

$$(\mathcal{D}_x + i\mathcal{D}_z)q = 0,$$

with boundary conditions

$$q \rightarrow \sqrt{\xi} e^{ik\alpha} \quad \text{as} \quad r \rightarrow \infty,$$

$$q \rightarrow 0 \qquad \text{as} \quad r \rightarrow 0. \tag{11.187}$$

Cf.
Section 3.1.3.

Here α is the polar angle in the xz plane, while r is the distance from the origin in the same plane (Fig. 11.8). Moreover k is an integer counting the number of windings.

If Eqs. (11.186) are satisfied, the flux of the magnetic field is $2\pi k$ (the winding number k determines the quantized magnetic flux), and the k-vortex mass (the string tension) is

$$M_{\text{vortex}} = 2\pi\xi k. \tag{11.188}$$

Vortex mass

The linear dependence of the k-vortex mass on k implies the absence of a potential between the vortices. In the model at hand – with four supercharges – a nonrenormalization theorem protects the central charge (i.e., ξ) and M_{vortex} from renormalization. Equation (11.188) is *exact*. For the curious reader, I would like to add that breaking $\mathcal{N} = 2$ down to $\mathcal{N} = 1$ in three-dimensional SQED leads to subtle and intriguing effects [63], which cannot be discussed here.

For the elementary $k = 1$ vortex it is convenient to introduce two profile functions $\phi(r)$ and $f(r)$ as follows:

$$q(x) = \phi(r)e^{i\alpha}, \qquad A_n(x) = -\frac{1}{n_e}\varepsilon_{nm}\frac{x_m}{r^2}[1 - f(r)]. \tag{11.189}$$

Bogomol'nyi
ansatz and
equation

The *ansatz* (11.189) can be substituted into the set of equations (11.186). It is consistent with this set, and we get the following two equations for the profile functions:

$$-\frac{1}{r}\frac{df}{dr} + n_e^2 e^2 \left(\phi^2 - \xi \right) = 0, \qquad r\frac{d\phi}{dr} - f\phi = 0, \tag{11.190}$$

with the boundary conditions that are obvious from the form of the *ansatz* (11.189):

$$\phi(\infty) = \sqrt{\xi}, \qquad f(\infty) = 0, \qquad (11.191)$$

$$\phi(0) = 0, \qquad f(0) = 1. \qquad (11.192)$$

Equations (11.190) with the above boundary conditions can readily be solved numerically (Section 3.1.3). The classical solution is BPS-saturated. It has two bosonic zero modes corresponding to vortex shifts in two spatial dimensions. These modes correspond to two bosonic collective coordinates describing the vortex center.

11.5.1.3 Fermion Zero Modes

To complete the quantization procedure we must know the fermion zero modes for the given classical solution. More precisely, since the solution under consideration is static, we are interested in the zero-eigenvalue solutions of the static fermion equations, which, thus, effectively become two rather than three dimensional:

$$i \left(\gamma^x \mathcal{D}_x + \gamma^z \mathcal{D}_z \right) \psi + n_e \sqrt{2} \lambda \, q = 0. \qquad (11.193)$$

This equation is obtained from (11.179) where we have dropped the terms involving a tilde (since $\tilde{q} = 0$). The fermion operator is Hermitian implying that every solution for $\{\psi, \lambda\}$ is accompanied by one for $\{\bar{\psi}, \bar{\lambda}\}$.

Since the solution to Eqs. (11.186) discussed above is 1/2-BPS, two of the four supercharges annihilate it while the other two generate the fermion zero modes – the superpartners of the translational modes. These are the only normalizable fermion zero modes in the problem at hand [64]. There are two extra modes, whose normalization diverges logarithmically.

Side remark: This situation – the logarithmic divergence of the norm – is subtle. Those modes whose normalization diverges as powers of the distance obviously belong to the bulk and should not be included in the soliton analysis. The normalizable modes obviously belong to the soliton and should be included. The logarithmically divergent modes are in the middle; they require special analysis through an appropriate infrared regularization.

11.5.1.4 Short versus Long Representations

The $(1 + 2)$-dimensional model under consideration has four supercharges. The corresponding regular superrepresentation is four dimensional (i.e., it contains two bosonic and two fermionic states).

The vortex we are discussing has two fermion zero modes. Hence, viewed as a particle in $1 + 2$ dimensions, it forms a superdoublet (one bosonic state plus one fermionic). This is a short multiplet.

Look through Section 10.6.9.

11.5.2 Four-Dimensional SQED and the ANO String

In this subsection we will discuss $\mathcal{N} = 1$ SQED (four supercharges) in four dimensions.

The Lagrangian is the same as that in Eq. (10.181). We will consider the simplest case: one chiral superfield Q with charge n_e and one chiral superfield $\tilde{Q}$ with charge $-n_e$. The scalar potential can be obtained from Eq. (11.180) by setting $a = 0$,

$$V = \frac{e^2}{2} n_e^2 \left[\xi - \left(\bar{q}q - \bar{\tilde{q}}\tilde{q} \right) \right]^2. \tag{11.194}$$

Just as in three dimensions, we are dealing here with the Higgs branch of real dimension 2. In fact, the vacuum manifold can be parametrized by a complex modulus $\tilde{q}q$. On this Higgs branch the photon field and superpartners form a massive supermultiplet, while $\tilde{q}q$ and its superpartners form a massless supermultiplet.

As shown in [63], no finite-thickness vortices exist at a generic point on the vacuum manifold owing to the absence of a mass gap (i.e., the presence of massless Higgs excitations). The moduli fields are involved in the solution at the classical level, generating a logarithmically divergent tail. An infrared regularization must be applied to remove this logarithmic divergence. To this end one can embed SQED in a slightly more complicated model, which bears the name of the M model [65].

Infrared regularization through the M model

We now introduce an extra *neutral* chiral superfield M, which interacts with Q and $\tilde{Q}$ through the super-Yukawa coupling,

$$\mathcal{L}_M = \int d^2\theta \, d^2\bar{\theta} \frac{1}{h} \bar{M}M + \left(\int d^2\theta \, QM\tilde{Q} + \text{H.c.} \right). \tag{11.195}$$

Here h is a coupling constant. As we will see shortly the Higgs branch is lifted. This is probably the simplest $\mathcal{N} = 1$ model that supports BPS-saturated ANO strings without any infrared problem.

The scalar potential (11.194) is now replaced by

$$V_M = \frac{e^2}{2} n_e^2 \left[\xi - \left(\bar{q}q - \bar{\tilde{q}}\tilde{q} \right) \right]^2 + h|q\tilde{q}|^2 + |qM|^2 + |M\tilde{q}|^2. \tag{11.196}$$

The vacuum is unique modulo a gauge transformation:

$$q = \bar{q} = \sqrt{\xi}, \quad \tilde{q} = 0, \quad M = 0. \tag{11.197}$$

The classical ANO flux-tube solution considered above remains valid as long as we set, additionally, $\tilde{q} = M = 0$. The string tension is the same, $T_{\text{string}} = 2\pi\xi$. (Note that in Eq. (11.196) the parameter ξ is defined with n_e^2 factored out.) The quantization procedure is straightforward, since one encounters no infrared problems whatsoever –

The first occurrence of the chiral $\mathcal{N} = (0, 2)$ SUSY.

all particles in the bulk are massive. In particular, there are four normalizable fermion zero modes (more details can be found in [18]). The string world-sheet theory has two supercharges, although – remarkably – we are not dealing here with the conventional $\mathcal{N} = 1$ supersymmetry in two dimensions but, rather, with the so-called chiral supersymmetry $\mathcal{N} = (0, 2)$ [66]. This will not be discussed further here.

11.5.3 Supersymmetric Non-Abelian Vortex Strings

The simplest nonsupersymmetric example of such strings is presented in Section 3.2. Now we will discuss its generalization. Our starting point is U(2) super-Yang–Mills theory with $\mathcal{N} = 2$ supersymmetry and two matter hypermultiplets in the *fundamental* representation of SU(2). We will denote the matter supermultiplets as

$$\{q^A, \tilde{q}^A\},$$

where $A = 1, 2$ is the flavor index, and the color indices are suppressed for a while. The superfields q^A and $\tilde{q}^A$ contain squarks and quarks. The electric charges of the tilded and untilded fields (with respect to U(1)) are opposite. Also, untilded fields are color doublets while the tilded ones are antidoublets.

Compared to the basic $\mathcal{N} = 2$ SQCD considered in Section 10.18.2 the following changes should be made to make this theory support 1/2-BPS vortex strings:

(i) The gauge theory must be extended from SU(2) to U(2). In other words, an extra (Abelian) field will be added, a "photon" and its superpartners;

(ii) To lift the vacuum degeneracy we will add the Fayet–Iliopoulos term (10.373). This term will trigger the super-Higgs mechanism.

In terms of $\mathcal{N} = 1$ superfields the action of the model has the form

$$
S = \left(\frac{1}{4g_2^2} \int d^2\theta\, W^{a\,\alpha} W_\alpha^a + \frac{1}{4g_1^2} \int d^2\theta\, W^\alpha W_\alpha + \text{H.c} \right)
$$
$$
+ \int d^2\theta\, d^2\bar{\theta} \left(\frac{1}{g_1^2} \bar{\mathcal{A}}\mathcal{A} + \frac{1}{g_2^2} \bar{\mathcal{A}}^a e^V \mathcal{A}^a + \bar{q}_A e^V q^A + \tilde{q}_A e^V \bar{\tilde{q}}^A - \xi V \right)
$$
$$
+ \int d^2\theta\, \frac{1}{\sqrt{2}} \left(\tilde{q}_A \mathcal{A} q^A + \tilde{q}_A \mathcal{A}^a \tau^a q^A + \text{H.c.} \right), \qquad (11.198)
$$

where $\mathcal{A}$ and $\mathcal{A}^a$ are the chiral superfields – the former is the superpartner to the U(1) "photon" and contains the "second" photino, while the latter is in the adjoint representation and contains the "second" gluino. The last term in the second line is the Fayet–Iliopoulos term with the coefficient ξ assumed to be positive. All matter superfields are massless. It is is convenient to write the superfield q^{iA} as a two-by-two matrix:

$$
\{ q^{iA} \} = \begin{pmatrix} q^{11} & q^{12} \\ q^{21} & q^{22} \end{pmatrix} \qquad (11.199)
$$

(and the same for $\tilde{q}$) where the first index (small Latin letter from the middle of the alphabet) stands for color and the second (capital Latin) for flavor.

In components, after integration over the auxiliary F and D fields we obtain

$$
\mathcal{L} = -\frac{1}{4g_2^2} F_{\mu\nu}^a F^{\mu\nu\,a} - \frac{1}{4g_1^2} F_{\mu\nu} F^{\mu\nu} + \frac{1}{g_2^2} \left| \mathcal{D}_\mu a^a \right|^2 + \frac{1}{g_1^2} \left| \mathcal{D}_\mu a \right|^2
$$
$$
+ \left| \mathcal{D}_\mu q^A \right|^2 + \left| \mathcal{D}_\mu \bar{\tilde{q}}^A \right|^2 - V(q^A, \tilde{q}_A, a_\mu^a, a_\mu) + \text{fermions}, \qquad (11.200)
$$

where the covariant derivative acting on matter fields are defined as

$$
\mathcal{D}_\mu = \partial_\mu - \frac{i}{2} A_\mu - A_\mu^a T^a = \partial_\mu - \frac{i}{2} A_\mu - \frac{i}{2} A_\mu^a \tau^a \qquad (11.201)
$$

and $T^a = \frac{1}{2}\tau^a$ (here τ^a stand for the Pauli matrices). Moreover, a^a and a are the lowest components of the superfields $\mathcal{A}^a$ and $\mathcal{A}$. Finally, the scalar potential has the form

$$
V(q, \tilde{q}) = \frac{g_2^2}{2} \left(\bar{q}_A T^a q^A - \tilde{q}_A T^a \bar{\tilde{q}}^A \right)^2
$$
$$
+ \frac{g_1^2}{8} \left(\bar{q}_A q^A - \tilde{q}_A \bar{\tilde{q}}^A - 2\xi \right)^2 + 2g_2^2 \left| \tilde{q}_A T^a q^A \right|^2 + \frac{g_1^2}{2} \left| \tilde{q}_A q^A \right|^2
$$
$$
+ \cdots \qquad (11.202)
$$

The ellipses in the above expression denote a number of terms containing a and a^a which will play no role in the vortex string solution because a and a^a vanish by virtue of the equations of motion. Reconstruction of the full scalar potential as well as the fermion part in (11.200) will be suggested for the reader as an exercise. See Eqs. (PS.488)-(PS.490).

The theory described by the action in Eq. (11.198) has eight conserved supercharges. 1/2-BPS saturated vortex string will preserve 4 supercharges. This means that the two-dimensional world sheet theory on the vortex string will have $\mathcal{N} = 2$ supersymmetry, four conserved supercharges. But first, we will analyze the vacuum structure.

The solution of the equation $V = 0$ is rather obvious from the structure of (11.202). Up to gauge transformations the solution is

| color-flavor locking |

$$\{q^{iA}\} = \sqrt{\xi} \begin{pmatrix} 1 & 0 \\ 0 & 1 \end{pmatrix}, \qquad \{\tilde{q}_{iA}\} = 0. \tag{11.203}$$

The theory is fully Higgsed and the flavor SU(2) obvious in (11.198) is spontaneously broken. However, the global diagonal SU(2) symmetry is preserved, see Eqs. (3.44), (3.45) and (3.47). This phenomenon is called *color-flavor locking*.

Passing to the vortex string construction we note that at the classical level the subsequent consideration is very similar to that in Section 3.2. Topologically stable vortex strings are obtained if we wind the matrix (11.203) around the string axis; say, at large r there are two options:

$$\sqrt{\xi} \begin{pmatrix} 1 & 0 \\ 0 & 1 \end{pmatrix} e^{i\alpha(x,y)} \text{ or } \sqrt{\xi} \begin{pmatrix} 1 & 0 \\ 0 & e^{i\alpha(x,y)} \end{pmatrix} \text{ at } r \to \infty, \tag{11.204}$$

see Fig. 3.5, page 102, for our notation, $r = \sqrt{x^2 + y^2}$. Both string constructions are topologically stable. The first one, the ANO string, is stable due to the fact that

$$\pi_1(U(1)) = Z.$$

The number of windings is arbitrary in the Abrikosov string. The second *ansatz* in (11.204) is topologically stable because

$$\pi_1 \{(SU(2) \times U(1))/Z_2\} = Z_2.$$

The Z_2 denominator in the braces reflects a common center in the SU(2) and U(1), namely, -1 belongs to both. The string based on the second *ansatz* in (11.204) is stable for a single winding and unstable for double winding.

Of crucial importance in this construction is the fact that the second *ansatz* in (11.204) spontaneously breaks the vacuum symmetry of color-flavor locked SU(2) down to U(1). Therefore, the existence of a single vortex string solution of the type (11.204) immediately implies the existence of a family of solutions, with two "rotational" moduli forming the CP(1) model on the world sheet. Remember, the target space of CP(1) is SU(2)/U(1).

The above statement is in correspondence with Eq. (3.76) which presents the O(3) model on the world sheet of a similar (nonsupersymmetric) vortex string. The O(3) sigma model is equivalent to CP(1). The bosonic solution for the string is 1/2-BPS saturated – four out of eight supercharges are spontaneously broken on the given solution. They create four supertranslational modes; therefore we have $8 - 4 = 4$

fermion zero modes. They pair up with the standard translational zero modes. The other four supercharges that do not act on the straight vortex string create four superorientational fermion zero modes when we make the $\vec{S}$ fields in Eq. (3.76) t and z dependent. After we elevate them to the world-sheet fields, they comprise the two-dimensional Dirac fermion ψ appearing in $\mathcal{N} = 2$ CP(1) model or, alternatively, the Majorana fermion field of the O(3) sigma model. See Sections 3.2.6, 10.12.3.5, and 10.12.4.

What is the difference between nonsupersymmetric and supersymmetric non-Abelian vortex strings? The nonsupersymmetric CP(1) model upon quantization at strong coupling possesses a unique vacuum state.[15] On the other hand, in the $\mathcal{N} = 2$ CP(1) model there are two distinct vacua at strong coupling. This follows, for instance, from the value of the Witten index.[16] These two vacua are labeled by the *bifermion* order parameter,

$$\langle G\bar{\psi}\psi \rangle = \pm\Lambda \tag{11.205}$$

at $\theta = 0$. The fact that the vacua of $\mathcal{N} = 2$ CP(1) are discrete and there are two of them is not seen at the classical level because of the fermion nature of the order parameter.[17] A transition domain interpolating by the two vacua in (11.205), a kink, presents in fact a confined monopole. I will not explain this statement, and instead refer the reader to the original papers [67].

11.5.4 Boojums

There exist a number of gauge theories, weakly coupled in the four-dimensional bulk (and, thus, fully controllable), which support both BPS walls and BPS flux tubes. A particular example is $\mathcal{N} = 2$ SQED with several flavors, and some non-Abelian generalizations. In such theories a U(1) gauge field can be localized on the minimal wall; in addition, they support a BPS wall–string junction. A field-theoretical string does end on a BPS wall, after all! The endpoint of the string on the wall, after Polyakov's dualization, becomes an electric field source localized on the wall. Norisuke Sakai and David Tong analyzed [68] generic wall–string configurations. Following condensed matter physicists they called them boojums. The word "boojum" comes from Lewis Carroll's children's book, the *Hunting of the Snark*. Apparently, it is fun to hunt a snark, but if the snark turns out to be a boojum, you are in trouble! Condensed matter physicists adopted the name to describe solitonic objects of the wall-string-junction type in helium-3. Furthermore, the boojum tree (Mexico) is the strangest plant imaginable. For most of the year it is leafless and looks like a giant upturned turnip. G. Sykes found it in 1922 and said, referring to Carroll, "It must be a boojum!" The Spanish common name for this tree is Cirio, referring to its candle-like appearance.

Section 9.7 treats Polyakov's dualization.

---

[15] In order to make two distinct strings and the ensuing two-string junctions we had to introduce a perturbation $\Delta\mathcal{L}_m = -m^2 \left\{ 1 - (S_3)^2 \right\}$ at the classical level. See the second Appendix in Chapter 3.

[16] See Section 10.22.2, presenting Witten's result $I_W\,(\mathrm{CP}(N-1)) = N$.

[17] Large-N solution of $\mathcal{N} = 2$ CP($N-1$) model analogous to that presented in Section 9.4 demonstrates N discrete degenerate vacua in this case.

Exercise

11.5.1 Calculate the full scalar potential and the fermion part of the action (11.198).

11.6 Critical Monopoles

11.6.1 Monopoles and Fermions

Critical 't Hooft–Polyakov monopoles emerge in $\mathcal{N} = 2$ super-Yang–Mills theories. There are no $\mathcal{N} = 1$ models with BPS-saturated monopoles since the $\mathcal{N} = 1$ theories have no monopole central charge. The minimal model with a BPS-saturated monopole is the $\mathcal{N} = 2$ generalization of supersymmetric gluodynamics, with gauge group SU(2). In terms of $\mathcal{N} = 1$ superfields it contains one vector superfield in the adjoint describing the gluon and gluino plus one chiral superfield in the adjoint describing a scalar $\mathcal{N} = 2$ superpartner for the gluon and a Weyl spinor, an $\mathcal{N} = 2$ superpartner for the gluino (Section 10.18).

The couplings of the fermion fields to the boson fields are of a special form; they are fixed by $\mathcal{N} = 2$ supersymmetry. In this section we will focus mostly on effects due to the adjoint fermions.

11.6.1.1 $\mathcal{N} = 2$ Super-Yang–Mills (without Matter)

The Lagrangian of the model can be obtained from Eq. (10.494) by specifying the gauge group to be SU(2),

$$\mathcal{L} = \frac{1}{g^2}\left[-\frac{1}{4}F^{a\mu\nu}F^a_{\mu\nu} + \lambda^{\alpha,a}i\mathcal{D}_{\alpha\dot\alpha}\bar\lambda^{\dot\alpha,a} + \frac{1}{2}D^a D^a \right.$$
$$+ \psi^{\alpha,a}i\mathcal{D}_{\alpha\dot\alpha}\bar\psi^{\dot\alpha,a} + (\mathcal{D}^\mu\bar a)(\mathcal{D}_\mu a)$$
$$\left. -\sqrt{2}\varepsilon_{abc}\left(\bar a^a\lambda^{\alpha,b}\psi^c_\alpha + a^a\bar\lambda^b_{\dot\alpha}\bar\psi^{\dot\alpha,c}\right) - i\varepsilon_{abc}D^a\bar a^b a^c \right]. \qquad (11.206)$$

$\boxed{\mathcal{N} = 2\ SYM}$ As usual, the D field is auxiliary,

$$D^a = i\varepsilon_{abc}\bar a^b a^c, \qquad (11.207)$$

and can be eliminated via the equation of motion. There is a flat direction: if the field a is real then all D terms vanish. If a is chosen to be purely real or purely imaginary and the fermion fields are ignored then we return to the Georgi–Glashow model.

Let us perform a Bogomol'nyi completion of the bosonic part of the Lagrangian (11.206) for static field configurations. Neglecting all time derivatives and, as usual, setting $A_0 = 0$, one can write the energy functional as follows:

$$\mathcal{E} = \sum_{i=1,2,3;\,a=1,2,3}\left[\int d^3x\left(\frac{1}{\sqrt{2}g}F_i^{*a} \pm \frac{1}{g}D_i a^a\right)^2 \mp \frac{\sqrt{2}}{g^2}\int d^3x\,\partial_i\left(F_i^{*a}a^a\right)\right],$$
$$(11.208)$$

where

$$F_m^* = \frac{1}{2} \varepsilon_{mnk} F_{nk},$$

and the square of the D term (11.207) is omitted – the D term vanishes provided a is real, which we will assume. This assumption also allows us to replace the absolute value in the first line of (11.208) by the contents of the parentheses. The term in the second line can be written as an integral over a large sphere,

$$\frac{\sqrt{2}}{g^2} \int d^3x \, \partial_i \left(F_i^{*a} a^a\right) = \frac{\sqrt{2}}{g^2} \int dS_i \left(a^a F_i^{*a}\right). \tag{11.209}$$

Bogomol'nyi equations

The Bogomol'nyi equations for the monopole are

$$F_i^{*a} \pm \sqrt{2} D_i a^a = 0. \tag{11.210}$$

See Section 4.1.1

This coincides with parallel expressions in the Georgi–Glashow model, up to normalization. (The field a is complex, generally speaking, and its kinetic term is normalized differently.) If the Bogomol'nyi equations are satisfied then the monopole mass M_M is determined by the surface term (classically). Assuming that in the "flat" vacuum a^a is aligned along the third direction and taking into account that in our normalization the magnetic flux is 4π we obtain

$$M_M = \frac{\sqrt{2} \, a_{\text{vac}}^3}{g^2} 4\pi, \tag{11.211}$$

where a_{vac}^3 is assumed to be real and positive.

11.6.1.2 Supercurrents and the Monopole Central Charge

A general classification of central charges in $\mathcal{N} = 2$ theories in four dimensions was presented in Section 11.1.3.3. Here we will briefly discuss the Lorentz-scalar central charge Z in the theory (11.206). It is this central charge that is saturated by critical monopoles.

The model, being $\mathcal{N} = 2$, possesses two conserved supercurrents; see Eq. (10.498). For convenience, I will quote these supercurrents, including the full derivative terms omitted in (10.498):

$$J_{\alpha\beta\dot{\beta},f} = \frac{2}{g^2} \left\{ iF_{\beta\alpha}^a \bar{\lambda}_{\dot{\beta},f}^a + \varepsilon_{\beta\alpha} D^a \bar{\lambda}_{\dot{\beta},f}^a - \sqrt{2} \left(D_{\alpha\dot{\beta}} \bar{a}^a\right) \lambda_{\beta,f}^a \right.$$

$$\left. + \frac{\sqrt{2}}{6} \left[\partial_{\alpha\dot{\beta}}(\lambda_{\beta,f} \bar{a}) + \partial_{\beta\dot{\beta}}(\lambda_{\alpha,f} \bar{a}) - 3\varepsilon_{\beta\alpha} \partial_{\dot{\beta}}^{\gamma}(\lambda_{\gamma,f} \bar{a})\right] \right\}; \tag{11.212}$$

see (10.498) for the notation. Classically the commutator of the corresponding supercharges is

$$\{Q_\alpha^I, Q_\beta^{II}\} = 2Z\varepsilon_{\alpha\beta} = -\frac{2\sqrt{2}}{g^2} \varepsilon_{\alpha\beta} \int d^3x \, \text{div} \left[\bar{a}^a \left(\vec{E}^a - i\vec{B}^a\right)\right]$$

Classical monopole central charge

$$= -\frac{2\sqrt{2}}{g^2} \varepsilon_{\alpha\beta} \int dS_j \left[\bar{a}^a \left(E_j^a - iB_j^a\right)\right]. \tag{11.213}$$

The central charge Z in Eq. (11.213) is referred to as the monopole central charge. For BPS-saturated monopoles $M_M = Z$.

The quantum corrections in the monopole central charge and, hence, in the mass of BPS-saturated monopoles do not vanish. They were first discussed in [69–71] in the late 1970s and 1980s. The monopole central charge is renormalized at the one-loop level. This is obviously due to the fact that the corresponding quantum correction must convert the bare coupling constant in Eq. (11.213) into a renormalized one. The logarithmic renormalizations of the monopole mass and the gauge coupling constant match. One can readily verify this. However, there is a residual nonlogarithmic effect, which cannot be obtained from Eq. (11.213). It was not until 2004 that people realized that the monopole central charge (11.213) must be supplemented by an anomalous term [24].

To elucidate the point, let us consider [23] the formula for the monopole or dyon mass obtained in the Seiberg–Witten exact solution [21],

$$M_{n_e,n_m} = \sqrt{2}\left|a\left(n_e - \frac{a_D}{a}n_m\right)\right|, \tag{11.214}$$

where $n_{e,m}$ are integer electric and magnetic numbers (we will consider here only the particular cases when either $n_e = 0, 1$ or $n_m = 0, 1$) and

$$a_D = i\,a\left(\frac{4\pi}{g_0^2} - \frac{2}{\pi}\ln\frac{M_0}{a}\right). \tag{11.215}$$

The quasiclassical limit $|a| \gg \Lambda$ is implied. The subscript 0 is introduced for clarity to indicate the bare charge. The renormalized coupling constant is defined in terms of the ultraviolet parameters as follows:

$$\frac{\partial a_D}{\partial a} \equiv \frac{4\pi i}{g^2}. \tag{11.216}$$

Because of the $a \ln a$ dependence in (11.215), $\partial a_D/\partial a$ differs from a_D/a by a constant (nonlogarithmic) term, namely,

$$\frac{a_D}{a} = i\left(\frac{4\pi}{g^2} - \frac{2}{\pi}\right). \tag{11.217}$$

Combining Eqs. (11.214) and (11.217) we get

$$M_{n_e,n_m} = \sqrt{2}\left|a\left[n_e - i\left(\frac{4\pi}{g^2} - \frac{2}{\pi}\right)n_m\right]\right|. \tag{11.218}$$

This equation does not match the renormalization of Eq. (11.213) in the nonlogarithmic part (i.e., the term $2\sqrt{2}a\,n_m/\pi$). Since the relative weight of the electric and magnetic parts in Eq. (11.213) is unambiguously determined by g^2, the presence of the above nonlogarithmic term implies that in fact the chiral structure $E_j^a - iB_j^a$ obtained at the canonical commutator level cannot be maintained once the quantum corrections are switched on. This is a quantum anomaly.

No direct calculation of the anomalous contribution in $\{Q_\alpha^I, Q_\beta^{II}\}$ in operator form has been carried out. However, it is not difficult to construct it indirectly, using Eq. (11.218) and the close parallel between $\mathcal{N} = 2$ super-Yang–Mills theory and the $\mathcal{N} = 2\,CP(N-1)$ model with twisted mass in two dimensions, in which, in essence, the same puzzle is solved [17]. (In fact this is more than a close parallel: it is a manifestation of a 4D–2D correspondence.) The anomalous contribution takes the form

$$\left\{ Q_\alpha^I, Q_\beta^{II} \right\}_{\text{anom}} = 2\varepsilon_{\alpha\beta}\delta Z_{\text{anom}} = -\left(\varepsilon_{\alpha\beta}\right) 2\sqrt{2}\frac{1}{4\pi^2}\int dS_j \Sigma^j, \qquad (11.219)$$

where

$$\Sigma^j = \frac{i}{2}\frac{\partial}{\partial\bar\theta^{\dot\beta}}\left(\bar{\mathcal{A}}^a \bar{W}_{\dot\alpha}^a\right)\left(\sigma^j\right)^{\dot\alpha\dot\beta}\Big|_{\bar\theta=0} = \bar{a}^a\left(\vec{E}^a + i\vec{B}^a\right)^j - \frac{\sqrt{2}}{2}\bar\lambda_{\dot\alpha}^a\left(\sigma^j\right)^{\dot\alpha\dot\beta}\bar\chi_{\dot\beta}^a. \quad (11.220)$$

| Anomaly in the monopole central charge |

It should be added to Eq. (11.213). The $(1,0)$ conversion matrix $(\sigma^j)^{\dot\alpha\dot\beta}$ was defined in Section 10.2.1, in which all the notation pertinent to spinors is collected.[18] In SU(N) theory we would have $N/(8\pi^2)$ instead of $1/(4\pi^2)$ in Eq. (11.219).

Adding the canonical and anomalous terms in $\left\{ Q_\alpha^I, Q_\beta^{II} \right\}$ we see that the fluxes generated by the color-electric and color-magnetic terms are now shifted, untied from each other, by the nonlogarithmic term in the magnetic part. Normalizing to the electric term, $M_W = \sqrt{2}a$, we get for the magnetic term

$$M_M = \sqrt{2}a\left(\frac{4\pi}{g^2} - \frac{2}{\pi}\right), \qquad (11.221)$$

as is necessary for consistency with the exact Seiberg–Witten solution.

11.6.1.3 Zero Modes for Adjoint Fermions

Equations for the fermion zero modes can be readily derived from the Lagrangian (11.206):

$$i\mathcal{D}_{\alpha\dot\alpha}\lambda^{a,c} - \sqrt{2}\varepsilon_{abc}\,a^a\bar\psi_{\dot\alpha}^b = 0,$$
$$i\mathcal{D}_{\alpha\dot\alpha}\psi^{a,c} + \sqrt{2}\varepsilon_{abc}\,a^a\bar\lambda_{\dot\alpha}^b = 0, \qquad (11.222)$$

plus the Hermitian conjugates. After a brief reflection we see that there are two complex or four real zero modes.[19] Two solutions are obtained if we substitute

$$\lambda^\alpha = F^{\alpha\beta}, \qquad \bar\psi_{\dot\alpha} = \sqrt{2}\mathcal{D}_{\alpha\dot\alpha}\,\bar{a}. \qquad (11.223)$$

The other two solutions correspond to the substitution

$$\psi^\alpha = F^{\alpha\beta}, \qquad \bar\lambda_{\dot\alpha} = \sqrt{2}\mathcal{D}_{\alpha\dot\alpha}\,\bar{a}. \qquad (11.224)$$

This result is easy to understand. Our starting theory has eight supercharges. The classical monopole solution is BPS-saturated, implying that four of the eight supercharges annihilate the solution (these correspond to the Bogomol'nyi equations) while the action of the other four supercharges produces the fermion zero modes.

Having four real fermion collective coordinates, the monopole supermultiplet is four dimensional: it includes two bosonic states and two fermionic. (The above

---

[18] In fact, the bifermion term $\bar\lambda\bar\chi$ in δZ_{anom} (see the second line in (11.220) *was* calculated in [24]. Invoking the fact that $\bar{\mathcal{A}}^a\,\bar{W}_{\dot\alpha}^a$ is the only color-singlet operator with the appropriate dimension and quantum numbers, one can unambiguously obtain the coefficient in front of $\int dS_j \Sigma^j$ without reference to the Seiberg–Witten solution.

[19] This means that the monopole is described by two complex fermion collective coordinates, or four real ones.

counting refers just to the monopole, without its antimonopole partner. The anti-monopole supermultiplet also includes two bosonic and two fermionic states.) From the standpoint of $\mathcal{N} = 2$ supersymmetry in four dimensions this is a short multiplet. Hence, the monopole states remain BPS-saturated to all orders in perturbation theory (in fact, the criticality of the monopole supermultiplet is valid beyond perturbation theory [21, 22]).

11.6.1.4 The Monopole Supermultiplet: Dimension of the BPS Representations

As was first noted by Montonen and Olive [72], the states in the $\mathcal{N} = 2$ model with a small enough magnetic charge – W bosons and monopoles alike – are BPS-saturated.[20] As a result the supermultiplets of this model are short. Regular (long) supermultiplets would contain $2^{2\mathcal{N}} = 16$ helicity states while the short ones contain $2^{\mathcal{N}} = 4$ helicity states, two bosonic and two fermionic. This is in full accord with the fact that the number of fermion zero modes in the given monopole solution is four, resulting in a four-dimensional representation of the supersymmetry algebra. If we combine the particles and antiparticles, as is customary in field theory, we will have one Dirac spinor on the fermion side of the supermultiplet. This statement is valid in both cases, that of the monopole supermultiplet and that of W bosons.

References for Chapter 11

[1] Y. A. Golfand and E. P. Likhtman, *Pisma Zh. Eksp. Teor. Fiz.* **13**, 452 (1971) [*JETP Lett.* **13**, 323 (1971)] [reprinted in S. Ferrara (ed.), *Supersymmetry* (North-Holland/World Scientific, 1987) Vol. 1, p. 7].

[2] J. T. Łopuszański, and M. Sohnius, Karlsruhe Report Print-74-1269 (unpublished).

[3] R. Haag, J. T. Łopuszański, and M. Sohnius, *Nucl. Phys.* B **88**, 257 (1975) [reprinted in S. Ferrara (ed.), *Supersymmetry* (North-Holland/World Scientific, 1987) Vol. 1, p. 51].

[4] E. Witten and D. I. Olive, *Phys. Lett.* B **78**, 97 (1978).

[5] S. Gates, Jr., M. Grisaru, M. Roček, and W. Siegel, *Superspace, or One Thousand and One Lessons in Supersymmetry* (Benjamin/Cummings, 1983) [hep-th/0108200].

[6] J. W. van Holten and A. Van Proeyen, *J. Phys.* A **15**, 3763 (1982).

[7] J. A. de Azcarraga, J. P. Gauntlett, J. M. Izquierdo, and P. K. Townsend, *Phys. Rev. Lett.* **63**, 2443 (1989).

[8] E. R. Abraham and P. K. Townsend, *Nucl. Phys.* B **351**, 313 (1991).

[9] P. K. Townsend, *P*-brane democracy, in M. Duff (ed.), *The World in Eleven Dimensions: Supergravity, Supermembranes and M-theory* (IOP, 1999) pp. 375–389 [hep-th/9507048].

---

[20] For instance, in the minimal pure $\mathcal{N} = 2$ theory with SU(2) gauge group, those states that carry a magnetic charge greater than 1 are non-BPS.

[10] G. R. Dvali and M. A. Shifman, *Phys. Lett. B* **396**, 64 (1997). Erratum: *ibid.* **407**, 452 (1997) [hep-th/9612128].

[11] S. Ferrara and M. Porrati, *Phys. Lett. B* **423**, 255 (1998) [hep-th/9711116].

[12] A. Gorsky and M. Shifman, *Phys. Rev. D* **61**, 085 001 (2000) [hep-th/9909015].

[13] Z. Hloušek and D. Spector, *Nucl. Phys. B* **370**, 143 (1992); J. D. Edelstein, C. Nuñez, and F. Schaposnik, *Phys. Lett. B* **329**, 39 (1994) [hep-th/9311055]; S. C. Davis, A. C. Davis, and M. Trodden, *Phys. Lett. B* **405**, 257 (1997) [hep-ph/9702360].

[14] N. Dorey, *JHEP* **9811**, 005 (1998) [hep-th/9806056].

[15] L. Alvarez-Gaumé and D. Z. Freedman, *Commun. Math. Phys.* **91**, 87 (1983); S. J. Gates, *Nucl. Phys. B* **238**, 349 (1984); S. J. Gates, C. M. Hull, and M. Roček, *Nucl. Phys. B* **248**, 157 (1984).

[16] A. Losev and M. Shifman, *Phys. Rev. D* **68**, 045 006 (2003) [hep-th/0304003].

[17] M. Shifman, A. Vainshtein, and R. Zwicky, *J. Phys. A* **39**, 13005 (2006) [hep-th/0602004].

[18] A. I. Vainshtein and A. Yung, *Nucl. Phys. B* **614**, 3 (2001) [hep-th/0012250].

[19] A. Abrikosov, *Sov. Phys. JETP* **32**, 1442 (1957) [reprinted in C. Rebbi and G. Soliani (eds.), *Solitons and Particles* (World Scientific, Singapore, 1984), p. 356]; H. Nielsen and P. Olesen, *Nucl. Phys. B* **61**, 45 (1973) [reprinted in C. Rebbi and G. Soliani (eds.), *Solitons and Particles* (World Scientific, Singapore, 1984), p. 365].

[20] M. Shifman and A. Yung, *Phys. Rev. D* **70**, 025 013 (2004) [hep-th/0312257].

[21] N. Seiberg and E. Witten, *Nucl. Phys. B* **426**, 19 (1994). Erratum: *ibid.* **430**, 485 (1994) [hep-th/9407087].

[22] N. Seiberg and E. Witten, *Nucl. Phys. B* **431**, 484 (1994) [hep-th/9408099].

[23] A. Rebhan, P. van Nieuwenhuizen, and R. Wimmer, *Phys. Lett. B* **594**, 234 (2004) [hep-th/0401116].

[24] M. Shifman and A. Yung, *Phys. Rev. D* **70**, 045 004 (2004) [hep-th/0403149].

[25] H. J. de Vega and F. A. Schaposnik, Phys. Rev. D **14**, 1100 (1976), reprinted in C. Rebbi and G. Soliani (eds.), *Solitons and Particles* (World Scientific, Singapore, 1984) p. 382.

[26] P. Di Vecchia and S. Ferrara, *Nucl. Phys. B* **130**, 93 (1977).

[27] J. Bagger and J. Wess, *Supersymmetry and Supergravity*, Second Edition (Princeton University Press, 1992).

[28] M. Shifman, A. Vainshtein, and M. Voloshin, *Phys. Rev. D* **59**, 045016 (1999) [hep-th/9810068].

[29] E. B. Bogomol'nyi, *Yad. Fiz.* **24**, 861 (1976) [*Sov. J. Nucl. Phys.* **24**, 449 (1976)] [reprinted in C. Rebbi and G. Soliani (eds.), *Solitons and Particles* (World Scientific, Singapore, 1984) p. 389].

[30] M. K. Prasad and C. M. Sommerfield, Phys. Rev. Lett. **35**, 760 (1975), reprinted in C. Rebbi and G. Soliani (eds.), *Solitons and Particles* (World Scientific, Singapore, 1984) p. 530.

[31] J. Milnor, *Morse Theory* (Princeton University Press, 1973).

[32] A. Losev, M. A. Shifman, and A. I. Vainshtein, *Phys. Lett. B* **522**, 327 (2001) [hep-th/0108153]; *New J. Phys.* **4**, 21 (2002) [hep-th/0011027] [reprinted in M. Olshanetsky and A. Vainshtein (eds.), *Multiple Facets of Quantization and*

Supersymmetry, the Michael Marinov Memorial Volume (World Scientific, Singapore, 2002), pp. 585–625].

[33] R. Jackiw and C. Rebbi, *Phys. Rev. D* **13**, 3398 (1976), reprinted in C. Rebbi and G. Soliani (eds.), *Solitons and Particles* (World Scientific, Singapore, 1984), p. 331.

[34] E. Witten, *Phys. Lett. B* **86**, 283 (1979) [reprinted in C. Rebbi and G. Soliani (eds.), *Solitons and Particles* (World Scientific, Singapore, 1984) p. 777].

[35] G. 't Hooft, *Nucl. Phys. B* **79**, 276 (1974).

[36] A. M. Polyakov, *Pisma Zh. Eksp. Teor. Fiz.* **20**, 430 (1974) [Engl. transl. *JETP Lett.* **20**, 194 (1974), reprinted in C. Rebbi and G. Soliani (eds.), *Solitons and Particles* (World Scientific, Singapore, 1984), p. 522].

[37] L. Alvarez-Gaumé and D. Z. Freedman, *Commun. Math. Phys.* **91**, 87 (1983).

[38] A. M. Polyakov, *Phys. Lett. B* **59**, 79 (1975).

[39] E. Witten, *Nucl. Phys. B* **188**, 513 (1981).

[40] S. Cecotti and C. Vafa, *Nucl. Phys. B* **367**, 359 (1991); S. Cecotti, P. Fendley, K. A. Intriligator, and C. Vafa, *Nucl. Phys. B* **386**, 405 (1992) [hep-th/9204102]; P. Fendley and K. A. Intriligator, *Nucl. Phys. B* **372**, 533 (1992) [hep-th/9111014]; S. Cecotti and C. Vafa, *Commun. Math. Phys.* **158**, 569 (1993) [hep-th/9211097].

[41] P. A. Bolokhov, M. Shifman and A. Yung, *Phys. Rev. D* **85**, 085028 (2012).

[42] P. K. Townsend, *Phys. Lett. B* **202**, 53 (1988).

[43] P. Fendley, S. D. Mathur, C. Vafa, and N. P. Warner, *Phys. Lett. B* **243**, 257 (1990).

[44] M. Cvetič, F. Quevedo, and S. J. Rey, *Phys. Rev. Lett.* **67**, 1836 (1991).

[45] S. Cecotti and C. Vafa, *Commun. Math. Phys.* **158**, 569 (1993) [hep-th/9211097].

[46] B. Chibisov and M. A. Shifman, *Phys. Rev. D* **56**, 7990 (1997). Erratum: *ibid* **58**, 109 901 (1998) [hep-th/9706141].

[47] D. Bazeia, J. Menezes, and M. M. Santos, *Nucl. Phys. B* **636**, 132 (2002) [hep-th/0103041]; *Phys. Lett. B* **521**, 418 (2001) [hep-th/0110111].

[48] J. Wess and B. Zumino, *Phys. Lett. B* **49**, 52 (1974) [reprinted in S. Ferrara (ed.), *Supersymmetry*, (North-Holland/World Scientific, Amsterdam–Singapore, 1987), Vol. 1, p. 77].

[49] J. Iliopoulos and B. Zumino, *Nucl. Phys. B* **76**, 310 (1974); P. West, *Nucl. Phys. B* **106**, 219 (1976); M. Grisaru, M. Roček, and W. Siegel, *Nucl. Phys. B* **159**, 429 (1979).

[50] E. Witten, *Nucl. Phys. B* **507**, 658 (1997) [hep-th/9706109].

[51] B. S. Acharya and C. Vafa, On domain walls of $\mathcal{N} = 1$ supersymmetric Yang–Mills in four dimensions [hep-th/0103011].

[52] I. I. Kogan, M. A. Shifman, and A. I. Vainshtein, *Phys. Rev. D* **53**, 4526 (1996). Erratum: *ibid.* **59**, 109 903 (1999) [arXiv:hep-th/9507170].

[53] M. A. Shifman and A. I. Vainshtein, *Nucl. Phys. B* **296**, 445 (1988).

[54] G. Dissertori and G. P. Salam, Review on quantum chromodynamics, in K. Nakamura *et al.* (Particle Data Group), *J. Phys. G* **37**, 075 021 (2010).

[55] A. Kovner, M. A. Shifman, and A. Smilga, *Phys. Rev. D* **56**, 7978 (1997) [hep-th/9706089].

[56] A. Ritz, M. Shifman, and A. Vainshtein, *Phys. Rev. D* **66**, 065 015 (2002) [hep-th/0205083].

[57] A. Ritz, M. Shifman, and A. Vainshtein, *Phys. Rev. D* **70**, 095 003 (2004) [hep-th/0405175].

[58] M. Shifman and A. Yung, *Supersymmetric Solitons* (Cambridge University Press, 2009).

[59] E. R. Bezerra de Mello, *Mod. Phys. Lett. A* **5**, 581 (1990).

[60] J. R. Schmidt, *Phys. Rev. D* **46**, 1839 (1992).

[61] J. D. Edelstein, C. Nuñez, and F. Schaposnik, *Phys. Lett. B* **329**, 39 (1994) [hep-th/9311055].

[62] S. Ölmez and M. Shifman, *Phys. Rev. D* **78**, 125 021 (2008) [arXiv:0808.1859 [hep-th]].

[63] A. A. Penin, V. A. Rubakov, P. G. Tinyakov, and S. V. Troitsky, *Phys. Lett. B* **389**, 13 (1996) [hep-ph/9609257].

[64] A. Rebhan, P. van Nieuwenhuizen, and R. Wimmer, *Nucl. Phys. B* **679**, 382 (2004) [hep-th/0307282].

[65] A. Gorsky, M. Shifman, and A. Yung, *Phys. Rev. D* **75**, 065 032 (2007) [hep-th/0701040].

[66] P. A. Bolokhov, M. Shifman, and A. Yung, *Phys. Rev. D* **79**, 106 001 (2009) [arXiv:0903.1089 [hep-th]].

[67] R. Auzzi, S. Bolognesi, J. Evslin, K. Konishi and A. Yung, *Nucl. Phys. B* **673**, 187 (2003) [hep-th/0307287]; M. Shifman and A. Yung, *Phys. Rev. D* **70**, 045004 (2004) [arXiv:hep-th/0403149 [hep-th]]; A. Hanany and D. Tong, *JHEP* **0404**, 066 (2004) [hep-th/0403158].

[68] N. Sakai and D. Tong, *JHEP* **0503**, 019 (2005) [arXiv:hep-th/0501207].

[69] A. D'Adda, R. Horsley, and P. Di Vecchia, *Phys. Lett. B* **76**, 298 (1978).

[70] R. K. Kaul, *Phys. Lett. B* **143**, 427 (1984).

[71] C. Imbimbo and S. Mukhi, *Nucl. Phys. B* **247**, 471 (1984).

[72] C. Montonen and D. I. Olive, *Phys. Lett. B* **72**, 117 (1977).

PART III

SOLUTIONS TO EXERCISES

Prepared in Collabaration with G. Tallarita

Chapter 1

Exercise 1.1.1 , page 19

The minimal-length design is obtained by minimization over the angle $\{Q\mathcal{M}\mathcal{L}\}$, which turns out $\pi/6$ at the minimum. The total length is $(1 + \sqrt{3})L$, where L is the distance between, say, $\mathcal{A}$ and $\mathcal{B}$. The spontaneous symmetry, breaking pattern is $Z_4 \to Z_2$. Therefore, two equivalent solutions must exist. It is instructive to compare the minimal-length solutions with the one-connection railroad consisting of two diagonals: $\mathcal{A}\mathcal{L}$ and $\mathcal{B}\mathcal{M}$. The length of the latter railroad (which does *not* break Z_4) is $2\sqrt{2}\, L$, i.e., it is approximately 4% larger.

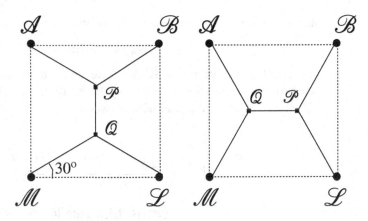

Fig. S.1 Two distinct solutions depicting the minimal-length railroad (it has two connections, in $\mathcal{P}$ and $\mathcal{Q}$). In both cases the length is $(1 + \sqrt{3})L$. The symmetry of each map is Z_2 (rotation by π). The original symmetry Z_4 (see Fig. 1.6 on page 19) is spontaneously broken down to Z_2 by the choice of one of the two equivalent minimal-length solutions.

Exercise 1.2.1, page 26

The easiest way to solve this problem is to apply the general theory presented in Section 4.3, although this is not necessary, of course.

In SU(3), the adjoint field ϕ^a ($a = 1, 2, \ldots, 8$) can be written as a 3×3 Hermitean traceless matrix,

$$\phi = \phi^a T^a = \frac{1}{2}\phi^a \lambda^a, \tag{PS.1}$$

where λ^a are the Gell–Mann matrices.

Using the gauge freedom $\phi \to U\phi U^\dagger$ can always reduce $\langle\phi\rangle_{\text{vac}}$ to a diagonal and traceless form, parametrizing it as

$$\langle\phi\rangle_{\text{vac}} = \frac{1}{2}\begin{pmatrix} v_1 + \frac{v_2}{\sqrt{3}} & 0 & 0 \\ 0 & -v_1 + \frac{v_2}{\sqrt{3}} & 0 \\ 0 & 0 & -\frac{2v_2}{\sqrt{3}} \end{pmatrix}. \tag{PS.2}$$

This, in turn, implies that

$$(\phi^3)_{\text{vac}} = v_1 \quad (\phi^8)_{\text{vac}} = v_2, \quad \phi^{1,2,4,5,6,7} = 0. \tag{PS.3}$$

Here v_1 and v_2 are two independent real parameters.

Since the covariant derivative acts on the adjoint field ϕ defined in (PS.1) as $\mathcal{D}_\mu \phi = \partial_\mu \phi - i[A_\mu \, \phi]$, it is obvious that the fields A_μ^3 and A_μ^8 remain *massless* and can be called "photons." The gauge symmetry-breaking pattern is SU(3) $\rightarrow$ U(1)^2 (in the general case, SU(N) $\rightarrow$ U(1)$^{N-1}$).

The masses of six Higgsed mesons can be obtained directly from the expression

$$\frac{1}{2}(\mathcal{D}_\mu \phi^a)(\mathcal{D}^\mu \phi^a) \rightarrow \frac{1}{2}g^2 \left[v_1^2 f^{ab'3} f^{ab3} + v_2^2 f^{ab'8} f^{ab8} \right.$$

$$\left. + \left(v_1 v_2 \, f^{ab'3} f^{ab8} + 3 \leftrightarrow 8 \right) \right] A_\mu^b \, A^{\mu \, b'}. \tag{PS.4}$$

Alternatively, one can use the general theory briefly reviewed in Section 4.3,

$$(m_W)_{\boldsymbol{\alpha}} = g \, \boldsymbol{h} \, \boldsymbol{\alpha}, \tag{PS.5}$$

where $\boldsymbol{\alpha}$ is the positive root of SU(N) (there are three of them for SU(3)) while $\boldsymbol{h}$ is the two-component vector $\boldsymbol{h} = \{v_1, v_2\}$. Both calculations give one and the same result, namely, the set of nonvanishing masses is

$$m_W = \left\{ \underbrace{\frac{g}{2}v_1}_{W^{1,2}}, \; \underbrace{\frac{g}{4}\left(v_1 + \sqrt{3}v_2\right)}_{W^{4,5}}, \; \underbrace{\frac{g}{4}\left|-v_1 + \sqrt{3}v_2\right|}_{W^{6,7}} \right\}. \tag{PS.6}$$

Exercise 1.3.1, page 34

In this problem it is assumed that the electron mass m_e is negligible, i.e., $m_e \ll \mu$.

To calculate renormalization of the electric charge in QED it is most convenient to use the effective Lagrangian (background field) method. We start from the bare Lagrangian

$$\mathcal{L}_{\text{QED}} = -\frac{1}{4e_0^2} F_{\mu\nu} F^{\mu\nu} + \bar{\psi} \, (i\mathcal{D}) \, \psi \tag{PS.7}$$

and evolve it down, integrating out virtual momenta in the interval $[\mu, M_0]$, where μ is the normalization point and M_0 is the ultraviolet cut-off. The bare coupling is $e_0^2 \equiv e_0^2(M_0^2)$, while the running coupling $e^2(\mu)$ is defined through the effective Lagrangian

$$\mathcal{L}_{\text{eff}} = -\frac{1}{4e^2(\mu^2)} F_{\mu\nu} F^{\mu\nu} + Z\bar{\psi} \, (i\mathcal{D}) \, \psi. \tag{PS.8}$$

Calculation of $\mathcal{L}_{\text{eff}}$ from (PS.7) is straightforward (especially since only logarithmic accuracy is required),

$$i\mathcal{L}_{\text{eff}}^{\text{one-loop}} = \frac{i^2}{2} A_\mu \, \Pi^{\mu\nu} \, A_\nu. \tag{PS.9}$$

The factor 1/2 appears because $\mathcal{L}_{\text{eff}}$ is the second-order iteration of the fermion term in the original Lagrangian. Moreover, $\Pi^{\mu\nu}$ is determined from the diagram depicted in Fig. S.2.

Fermion loop determining the running of e^2 in QED.

Equation (PS.9) implies

$$\Pi^{\mu\nu} = (-1) \int \frac{d^4k}{(2\pi)^4} \frac{i}{(k-p)^2} \frac{i}{k^2} \mathrm{Tr}\left[\gamma^\mu (\not{k} - \not{p})\gamma^\nu \not{k}\right], \tag{PS.10}$$

where the factor -1 is due to the fermion loop.

Formally, the integral in (PS.10) is quadratically divergent. However, any gauge-invariant regularization will automatically result in a transversal expresson

$$\Pi^{\mu\nu} = \Pi(p^2)\left(g^{\mu\nu}p^2 - p^\mu p^\nu\right) \tag{PS.11}$$

where $\Pi(p^2)$ is only logarithmically divergent. Keeping this in mind and pursuing the logarithmic accuracy, as was mentioned above, we can just drop quadratically divergent terms should they appear. Moreover, from (PS.11) it is obvious that

$$\Pi = \frac{1}{3p^2} \Pi^\alpha_{\ \alpha}. \tag{PS.12}$$

The trace in (PS.10), upon convolution of μ and ν, becomes

$$\mathrm{Tr}\left[\gamma^\alpha (\not{p} - \not{k})\gamma_\alpha \not{k}\right] = 4\left[(p-k)^2 + k^2 - p^2\right]. \tag{PS.13}$$

The first two terms on the right-hand side can be ignored since they cancel one of the denominators in (PS.10) and, hence, produce only quadratic divergences in Π. Thus, we conclude that for our purposes we can rewrite

$$\Pi = \frac{4p^2}{3p^2} \int \frac{d^4k}{(2\pi)^4} \frac{1}{(p-k)^2} \frac{1}{k^2} \rightarrow \frac{i}{12\pi^2} \int_{\mu^2}^{M_0^2} \frac{dk^2}{k^2}, \tag{PS.14}$$

where we performed the Euclidean rotation in the integrand and approximated the integral in an appropriate way, keeping in mind the desired logarithmic accuracy.

Now, assembling (PS.9), (PS.11) and (PS.14) we arrive at

$$\Pi^{\mu\nu} = \left(g^{\mu\nu}p^2 - p^\mu p^\nu\right) \times \frac{i}{12\pi^2} \log \frac{M_0^2}{\mu^2},$$

$$\mathcal{L}_{\mathrm{eff}}^{\mathrm{one-loop}} = \frac{i}{2} A_\mu \left(g^{\mu\nu}p^2 - p^\mu p^\nu\right) A_\nu \times \left(\frac{i}{12\pi^2} \log \frac{M_0^2}{\mu^2}\right)$$

$$\rightarrow -\frac{1}{48\pi^2} \log \frac{M_0^2}{\mu^2} F_{\mu\nu}F^{\mu\nu}. \tag{PS.15}$$

Let us compare this expression to (PS.7). Inspecting the sum of the bare Lagrangian and $\mathcal{L}_{\mathrm{eff}}^{\mathrm{one-loop}}$ we observe that

$$\frac{1}{e^2(\mu^2)} = \frac{1}{e_0^2} + \frac{1}{12\pi^2} \log \frac{M_0^2}{\mu^2}, \tag{PS.16}$$

implying that $e^2(\mu)$ decreases with decreasing μ. In a slightly different notation Eq. (PS.16) coincides with the formula presented on page 34 (namely, one should replace $M_0 \leftrightarrow p$ and $e_0 \leftrightarrow e(p)$).

Chapter 2

Exercise 2.1.1, page 55

In classical theory, the vacuum (ground) states are classical solutions with minimal energy. No fluctuations or tunneling probabilities are included. In quantum systems vacuum fluctuations and tunneling should be taken into account. In quantum mechanics, tunneling between two classical minima of the double well potential is possible; moreover, it leads to a "smearing" of the wave function. Therefore, in quantum mechanics the true vacuum is a superposition of the wave functions peaked at the left and the right classical minima. It is unique. Therefore, the Z_2 invariance of the Hamiltonian is unbroken in the ground state.

In field theory, if it is considered in the infinite (or very large) volume V, the probability for the system to tunnel from the left to the right classical minimum (and vice versa) vanishes as $\exp\left(-CVm^3\right)$ where C is a numerical constant and m is a mass parameter characterizing the field theory under consideration,

$$P = \lim_{V \to \infty} \exp\left(-CVm^3\right) \to 0. \tag{PS.17}$$

As a results, there are two distinct vacua. The Z_2 invariance is spontaneously broken.

Exercise 2.1.2, page 56

The Bogomol'nyi completion in the case at hand has a sign ambiguity,

$$T_w = \int dz \frac{1}{2}\left[\left(\frac{d\phi}{dz}\right)^2 + \left(W'(\phi)\right)^2\right]$$

$$= \int dz \left[\left(\frac{1}{2}\frac{d\phi}{dz} \pm W'(\phi)\right)^2 \mp \frac{d\phi}{dz}W'(\phi)\right]. \tag{PS.18}$$

For the antikink one must pick the minus sign,

$$T_w = \frac{1}{2}\int dz \left(\frac{d\phi}{dz} - W'(\phi)\right)^2 + \Delta W, \tag{PS.19}$$

where

$$\Delta W = W(-v) - W(+v) = \frac{4}{3\sqrt{2}}\frac{\mu^3}{g^2} = \frac{m^3}{3g^2}. \tag{PS.20}$$

showing that the kink and the antikink have the same lower BPS bound.

Exercise 2.1.3, page 56

Without loss of generality we can put $z_0 = 0$. The energy functional has the form

$$\mathcal{E}(z) = \frac{1}{2}\left(\frac{d\phi_w}{dz}\right)^2 + \frac{g^2}{4}\left(\phi_w^2 - v^2\right)^2. \tag{PS.21}$$

Substituting the classical solution

$$\phi_w(z) = v\tanh\left(\frac{\mu}{\sqrt{2}}z\right), \tag{PS.22}$$

we obtain

$$\mathcal{E}(z) = \frac{1}{8}\frac{m^4}{g^2}\left[\operatorname{sech}\left(\frac{m}{2}z\right)\right]^4. \tag{PS.23}$$

Note that

$$\int_{-\infty}^{\infty} dz\,\mathcal{E}(z) = \frac{1}{3}\frac{m^3}{g^2} = T_w, \tag{PS.24}$$

which coincides with Eq. (2.19) in the textbook. To find the variance of the above distribution we must calculate

$$\sigma^2 = T_w^{-1}\int_{-\infty}^{\infty} dz\,z^2\,\mathcal{E}(z) = \frac{1}{9}\left(\pi^2 - 6\right) \times 3m^{-2}. \tag{PS.25}$$

Thus, the thickness of the wall ℓ is

$$\ell = \sqrt{\frac{\pi^2 - 6}{3}}\,m^{-1} \sim m^{-1}. \tag{PS.26}$$

Far from the wall center (i.e., at $z \to \infty$),

$$\phi_w(z) \to v\left(1 - 2e^{-mz}\right), \qquad \mathcal{E}(z) \to 2\frac{m^4}{g^2}e^{-2mz}. \tag{PS.27}$$

Exercise 2.1.4, page 56

We must check that the moving kink solution

$$\phi(t, z|V) = v\tanh\left(\frac{gv}{\sqrt{2}}\frac{z - z_0 - Vt}{\sqrt{1 - V^2}}\right) \tag{PS.28}$$

satisfies the equation

$$\left(\frac{\partial^2}{\partial t^2} - \frac{\partial^2}{\partial z^2}\right)\phi(t, z|V) + \frac{\partial U(\phi(t, z|V))}{\partial \phi} = 0, \tag{PS.29}$$

where

$$U(\phi(t, z|V)) = \frac{g^2}{4}\left[\phi(t, z|V)^2 - v^2\right]^2. \tag{PS.30}$$

We obtain

$$\frac{\partial\phi}{\partial t} = -\frac{gv^2V}{\sqrt{2(1 - V^2)}}\left[\operatorname{sech}(x)\right]^2, \tag{PS.31}$$

where

$$x = \frac{gv}{\sqrt{2}}\xi, \qquad \xi \equiv \frac{z - z_0 - Vt}{\sqrt{1 - V^2}}. \tag{PS.32}$$

Furthermore,

$$\frac{\partial^2 \phi}{\partial t^2} = -\frac{g^2 v^3 V^2}{(1 - V^2)} \left[\mathrm{sech}(x) \right]^2 \tanh(x). \tag{PS.33}$$

Similarly

$$\frac{\partial \phi}{\partial z} = \frac{g v^2 V}{\sqrt{2(1 - V^2)}} \left[\mathrm{sech}(x) \right]^2 \tag{PS.34}$$

and

$$\frac{\partial^2 \phi}{\partial z^2} = -\frac{g^2 v^3}{(1 - V^2)} \left[\mathrm{sech}(x) \right]^2 \tanh(x). \tag{PS.35}$$

As a result,

$$\frac{\partial^2 \phi}{\partial t^2} - \frac{\partial^2 \phi}{\partial z^2} = g^2 v^3 \left[\mathrm{sech}(x) \right]^2 \tanh(x). \tag{PS.36}$$

It is easy to check that

$$\frac{\partial U}{\partial \phi} = -g^2 v^3 \left[\mathrm{sech}(x) \right]^2 \tanh(x). \tag{PS.37}$$

Assembling everything together,

$$\frac{\partial^2 \phi}{\partial t^2} - \frac{\partial^2 \phi}{\partial z^2} + \frac{\partial U}{\partial \phi} = (g^2 v^3 - g^2 v^3) \left[\mathrm{sech}(x) \right]^2 \tanh(x) = 0, \tag{PS.38}$$

which shows that the moving kink solution is indeed a solution of the equations of motion.

The relevant components of the energy–momentum tensor of the model (see Eq. (2.84) in the book) are

$$T^{00} = \frac{1}{2} \left(\dot{\phi} \right)^2 + \frac{1}{2} \left(\frac{\partial \phi}{\partial z} \right)^2 + U(\phi), \qquad T^{0j} = \dot{\phi} \, \partial_z \phi. \tag{PS.39}$$

Let us first calculate the energy of the moving kink (per unit area). Using the definition of ξ in (PS.32) we can write

$$E = \int dz \, T^{00} = \int \sqrt{1 - V^2} \, d\xi \left[\frac{1}{2} \left(\frac{\partial \phi}{\partial \xi} \right)^2 \frac{1 + V^2}{1 - V^2} + U(\phi) \right]$$

$$= \frac{1}{\sqrt{1 - V^2}} \int d\xi \left[\frac{1}{2} \left(\frac{\partial \phi}{\partial \xi} \right)^2 + U(\phi) \right]$$

$$+ \frac{V^2}{\sqrt{1 - V^2}} \int d\xi \left[\frac{1}{2} \left(\frac{\partial \phi}{\partial \xi} \right)^2 - U(\phi) \right]. \tag{PS.40}$$

For the static (i.e., t independent solution at $V \to 0$) there exists an "integral of motion," namely

$$\frac{1}{2}\left(\frac{\partial \phi}{\partial z}\right)^2 - U(\phi) = 0. \tag{PS.41}$$

This can be proven by using the equation of motion. Therefore, the third line in (PS.40) vanishes and we arrive at

$$E_w = \frac{T_w}{\sqrt{1 - V^2}}. \tag{PS.42}$$

Next, the momentum of the wall (per unit area) in the z direction is

$$p^z = \int dz\, T^{0z} = \frac{V}{\sqrt{1 - V^2}} \int d\xi \left(\frac{\partial \phi}{\partial \xi}\right)^2 = \frac{T_w V}{\sqrt{1 - V^2}}, \tag{PS.43}$$

where we used the definition of T^{0z} in Eq. (PS.39) in conjunction with a consequence of Eq. (PS.41). It is obvious that

$$E^2 - (p^z)^2 = T_w^2,$$

q.e.d. Moreover, if the static domain wall satisfies the equation

$$\frac{\partial \phi}{\partial z} = -\mathcal{W}'(\phi) \tag{PS.44}$$

(see Eq. (2.18), on page 49) then for the moving domain wall one has

$$\frac{\partial}{\partial z}\, \phi(t, z|V) = -\frac{1}{\sqrt{1 - V^2}} \mathcal{W}'(\phi),$$

$$\frac{\partial}{\partial t}\, \phi(t, z|V) = \frac{V}{\sqrt{1 - V^2}} \mathcal{W}'(\phi). \tag{PS.45}$$

Exercise 2.1.5, page 56

The minima of the potential, the so-called critical points, are determined by the condition $\mathcal{W}' = 0$. The kink solution interpolates between two distinct critical points. It is obvious that at $z \to \mp\infty$ the solution must approach the initial (final) critical point, while $\partial\phi/\partial z \to 0$. The second-order equation of motion

$$\frac{\partial^2 \phi}{\partial z^2} = \mathcal{W}'(\phi)\overline{\mathcal{W}''(\phi)} \tag{PS.46}$$

implies that

$$\frac{\partial}{\partial z} \left|\mathcal{W}'(\phi)\right|^2 = \frac{\partial}{\partial z} \left|\frac{\partial \phi}{\partial z}\right|^2, \tag{PS.47}$$

from which we conclude that

$$\left|\mathcal{W}'(\phi)\right|^2 - \left|\frac{\partial \phi}{\partial z}\right|^2 = \text{const} = 0. \tag{PS.48}$$

That the constant vanishes follows from the boundary conditions near either of the two critical points. To derive Eq. (PS.48) we started from (PS.47) and then used (PS.46).

Now, following Bazeia *et al.*,[1] let us consider the ratio

$$R(\phi) = \left(\overline{W'(\phi)}\right)^{-1} \frac{\partial \phi}{\partial z}. \tag{PS.49}$$

Differentiating this ratio with respect to z we arrive at

$$\frac{\partial R}{\partial z} = \left(\overline{W'(\phi)}\right)^{-2} \left(|W'(\phi)|^2 - \left|\frac{\partial \phi}{\partial z}\right|^2\right) \overline{W''(\phi)} = 0, \tag{PS.50}$$

by virtue of Eq. (PS.48). This implies that R is a z-independent constant, while Eq. (PS.48) tells us that the absolute value of this constant is 1. Hence

$$\frac{\partial \phi}{\partial z} = e^{i\alpha} \overline{W'(\phi)}, \tag{PS.51}$$

where α is a constant phase that is to be determined from the boundary conditions. It is not difficult to see that $\alpha = \arg \Delta W$, where ΔW is the difference between the superpotentials at the final and initial critical points.

Exercise 2.2.1, page 67

In the model (2.1) the domain wall separates vacuum I from vacuum II; these vacua are degenerate. Let us try to arrange a junction as it is shown in Fig. S.3a. It is quite obvious that it is unstable under the deformation depicted in Fig. S.3b. Thus, topology does not forbid decay of the junction on the left into a two-wall configuration on the right with energy release.

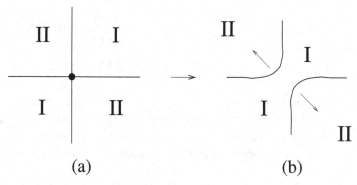

(a)   (b)

Fig. S.3   A domain junction that is not topologically stable.

Exercise 2.2.2, page 67

In the left-hand side of Eq. (2.55) we differentiate over ξ, where ξ is a *complex* variable, unlike x in (2.54). One can always replace ξ by a new complex variable

$$\xi' = e^{i\alpha}\xi.$$

Rewriting (2.55) in terms of ξ' we eliminate α.

[1] D. Bazeia, J. Menezes, and M. M. Santos, *Phys. Lett. B* **521**, 418 (2001) [arXiv:hep-th/0110111].

Exercise 2.2.3, page 67

The superpotential for the model (2.47) has the form

$$\mathcal{W}(\phi) = \mu\left(\phi - \frac{\nu}{n+1}\phi^{n+1}\right),$$

$$\frac{\partial \mathcal{W}(\phi)}{\partial \phi} = \mu\left(1 - \nu\phi^n\right), \tag{PS.52}$$

while the BPS equation is

$$\frac{\partial \phi}{\partial x} = e^{i\alpha}\mu\left(1 - \nu\bar{\phi}^n\right). \tag{PS.53}$$

The tension of the wall is given by

$$T_{\rm w} = \int dz\left(\left|\frac{\partial \phi}{\partial z}\right|^2 + |\mathcal{W}'(\phi)|^2\right). \tag{PS.54}$$

Applying the Bogomol'nyi completion, we rewrite it as follows:

$$T_{\rm w} = \int dz\left|\frac{\partial \phi}{\partial z} - e^{i\alpha}\overline{\mathcal{W}}'\right|^2 + 2{\rm Re}\left(e^{-i\alpha}\int dz\frac{d\mathcal{W}}{dz}\right), \tag{PS.55}$$

where the extra factor of 2 comes from the summation over the contributions of ϕ and $\bar{\phi}$. The first term above vanishes for the BPS-saturated solution while the second reduces to

$$T_{\rm w} = 2{\rm Re}\left[e^{-i\alpha}\left(\mathcal{W}_{\rm final} - \mathcal{W}_{\rm initial}\right)\right] = 4\frac{n}{n+1}\mu\nu^{-\frac{1}{n}}\sin\frac{\pi}{n},$$

$$\rightarrow \frac{4\pi}{n}\mu\nu^{-\frac{1}{n}} \tag{PS.56}$$

where we used Eqs. (2.60) and (2.62). In particular, Eq. (2.62) determines the phase α.

It would be useful to check mass dimensions of the parameters above,

$$\mu = [m^2], \quad \nu^{\frac{-1}{n}} = [m], \tag{PS.57}$$

implying cubic mass dimension for $T_{\rm w}$. At the same time, m_φ in the expression below has dimension of mass.

The masses of the elementary excitations are

$$m_\varphi = n\mu\nu^{\frac{1}{n}}. \tag{PS.58}$$

The validity of the quasiclassical approximation is determined by the condition $T_{\rm w}/m_\varphi^3 \gg 1$, implying

$$\frac{\nu^{-\frac{4}{n}}}{\mu^2} \gg \frac{n^2(n+1)}{4\sin\frac{\pi}{n}}. \tag{PS.59}$$

If n is large the left-hand side above must be much larger than $n^4/(4\pi)$.

Exercise 2.2.4*, page 67

The wall-junction tension T_j in the Z_n models is defined in Eq. (2.45). As we will learn from Part II of this textbook, the walls per se are 1/2-BPS saturated. This means that the given wall solutions conserves two out of four supercharges. The wall junctions are 1/4-BPS saturated – they conserve one supercharge out of four. In both cases instead of the second-order equations of motion one can apply the BPS equations given in Eqs. (2.54) and (2.55). Now we will focus on the latter equation. Given a specific form of the superpotential (see (PS.52)) the corresponding BPS equation simplifies in the limit $n \to \infty$. Indeed, it is intuitively clear that the field $\phi \to 0$ in the junction center. If so, the second term in $\mathcal{W}'$ is much smaller than the first one and can be negected,

$$\mathcal{W}' = \mu \text{ if } |\phi| < v^{\frac{-1}{n}}. \tag{PS.60}$$

Then Eq. (2.55) implies

$$\phi = \frac{\mu}{2}\xi \equiv \frac{\mu}{2} r e^{i\alpha} \text{ if } r < r_* = \frac{2}{\mu} v^{\frac{-1}{n}}. \tag{PS.61}$$

Here $\vec{r}$ is the radius in the plane perpendicular to the junction axis,

$$r = \sqrt{x^2 + y^2}, \quad \tan\alpha = \frac{y}{x}. \tag{PS.62}$$

If $|\phi| > v^{\frac{-1}{n}}$ the second term in $\mathcal{W}'$ tends to infinity and Eq. (2.55) becomes meaningless; it does not define a solution. But the solution at $r > r_*$ can be inferred indirectly.

Let us look at Fig. S.4 presenting a sketch of a wall junction. A shaded circle of radius r_* is a hub. There are n spokes – domains walls – that are attached to the hub in the maximal junction. Inside the hub they overlap and their individual structure is not resolved in the large-n limit. Indeed, according to (PS.58), the individual wall thickness can be estimated as

$$\ell \sim m_\varphi^{-1} \sim \frac{1}{n\mu} v^{\frac{-1}{n}}, \tag{PS.63}$$

implying that

$$n\ell \sim \frac{1}{\mu} v^{\frac{-1}{n}} \sim r_*, \tag{PS.64}$$

see Eq. (PS.61). The above solution $\phi = \frac{\mu}{2} r e^{i\alpha}$ is homogeneous in α.

Outside the circle $r = r_*$ the field ϕ between the spokes takes the vacuum values

$$(\phi_{\text{vac}})_k = v^{\frac{-1}{n}} \exp\left[i\frac{2\pi k}{n}\right]. \tag{PS.65}$$

The vacua to the right and to the left of a given spoke differ only in a subleading order $O(1/n)$. At $r \gg r_*$ the spokes can be be viewed as well-separated, "undeformed" domain walls. The matching at $r = r_*$ and along the spokes is not smooth. One should take into account $O(1/n)$ corrections to the leading order result to make it smooth.

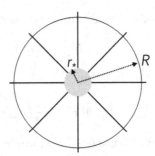

Hub and spokes of the wall junction.

The energy of the configuration shown in Fig. S.4 (per unit length in the z direction) is

$$T = nT_{\rm w}\,(R - r_*) + 2\pi \int_{r<r_*} r\,dr \left\{ 2\left(\partial_\xi \phi \partial_\xi \bar{\phi} + \phi \leftrightarrow \bar{\phi}\right) + \left|\frac{\partial W}{\partial \phi}\right|^2 \right\}. \qquad \text{(PS.66)}$$

Assembling Eqs. (PS.56) and (PS.61) and comparing with the general definition (2.45) we obtain the junction tension, namely,

$$T_{\rm j} = -2\pi v^{\frac{-2}{n}}. \qquad \text{(PS.67)}$$

Note that the junction tension is negative due to the fact that the r_* term in the beginning of Eq. (PS.66) is negative. This does not make the junction unstable since it is always attached to the walls.

Further details and relevant references can be found in G. Gabadadze and M. Shifman, *Phys. Rev.* D **61**, 075014 (2000) [hep-th/9910050].

Exercise 2.3.1, page 72

As was mentioned with regard to Eq. (2.82) the easiest way to obtain the energy–momentum tensor is to insert in the Lagrangian at hand the metric tensor $g^{\mu\nu}$, and then differentiate with regard to $g^{\mu\nu}$. Since the photon Lagrangian does not contain derivatives of the metric tensor, the expression for $T_{\mu\nu}$ for the free photon fields takes the form

$$T_{\mu\nu} = \left[2\frac{\delta\mathcal{L}}{\delta g^{\mu\nu}} - g_{\mu\nu}\mathcal{L} \right]_{g_{\mu\nu}=\eta_{\mu\nu}}$$

$$= -F_{\mu\alpha}F_{\nu\beta}\eta^{\alpha\beta} + \frac{1}{4}\eta_{\mu\nu}F_{\alpha\beta}F^{\alpha\beta}. \qquad \text{(PS.68)}$$

The energy–momentum tensor we obtain is automatically symmetric under $\mu \leftrightarrow \nu$. Convoluting μ and ν we immediately see the tracelessness of (PS.68),

$$T_\mu{}^\mu = 0. \qquad \text{(PS.69)}$$

As for conservation, let us consider

$$\partial^\mu T_\mu{}^\nu = \frac{1}{2}F_{\alpha\beta}\,\partial^\nu F^{\alpha\beta} - F_{\mu\alpha}\,\partial^\mu F^{\nu\alpha} = 0. \qquad \text{(PS.70)}$$

In calculating the second term above one needs to use the Bianchi identity

$$\partial^\mu F^{\nu\alpha} + \partial^\alpha F^{\mu\nu} + \partial^\nu F^{\alpha\mu} = 0. \tag{PS.71}$$

Then it is easy to show that

$$F_{\mu\alpha}\,\partial^\mu F^{\nu\alpha} = \frac{1}{2}\left(F_{\mu\alpha}\,\partial^\mu F^{\nu\alpha} + F_{\alpha\mu}\partial^\alpha F^{\nu\mu}\right). \tag{PS.72}$$

Combining (PS.71) and (PS.72) we arrive at (PS.70).

In three dimensions, the Maxwell Lagrangian is the same, and, hence, the energy-momentum tensor remains intact. However, the convolution $\eta_{\mu\nu}\eta^{\mu\nu} = 3$, unlike in the four-dimensional case. This means that the trace of the energy–momentum no longer vanishes,

$$T_\mu^{\ \mu} = -\frac{1}{4}\,F_{\alpha\beta}F^{\alpha\beta} \quad (3D). \tag{PS.73}$$

Nevertheless, a conserved dilatation current can by constructed by virtue of an "improvement,"

$$J_{\text{scale}}^\mu = T^\mu_{\ \nu}\,x^\nu + \frac{1}{2}F^{\mu\nu}\,A_\nu, \tag{PS.74}$$

see R. Jackiw and S.-Y. Pi, *Tutorial on Scale and Conformal Symmetries in Diverse Dimensions,* J. Phys. A **44**, 223001 (2011) [arXiv:1101.4886] and S. El-Showk, Y. Nakayama and S. Rychkov, What Maxwell Theory in $D \neq 4$ Teaches Us about Scale and Conformal Invariance, *Nucl. Phys.* B **848**, 578 (2011) [arXiv:1101.5385]. Note that J_{scale}^μ is not gauge invariant because of the improvement term. However, $\partial_\mu J_{\text{scale}}^\mu = 0$ and is gauge invariant.

The Maxwell theory in three dimensions is an example of a rare exceptional theories mentioned on page 38: it has scale and Lorentz invariances but is not conformally invariant.

Exercise 2.3.2*, page 73

In the case where $n = 1$ and $\delta = 2$, we see that Eq. (2.100) becomes infinite because of the gamma function in the numerator. However, as we will see momentarily, the infinite part is independent of $|\vec{x}|$ and therefore is irrelevant in interactions. The $|\vec{x}|$ dependent part is finite.

To see that this is the case, let us put $n = 1$ and $\delta = 2 + 2\varepsilon$ with the intension to tend $\varepsilon \to 0$ at the very end of our calculations. In this limit we have

$$\Gamma\left(\frac{\delta}{2} - n\right) = \Gamma(\varepsilon) = \frac{1}{\varepsilon} + O(\varepsilon^0),$$

$$x^{2n-\delta} = \exp\left(-2\varepsilon \log x\right) = 1 - 2\varepsilon \log x, \tag{PS.75}$$

and all other factors being assembled combine into

$$\pi\left(1 + O(\varepsilon^1)\right). \tag{PS.76}$$

Equation (2.101) takes the form

$$-\left[\frac{\pi}{\varepsilon} + O(\varepsilon^0)\right] + 2\pi \log x. \tag{PS.77}$$

The expression in the square brackets is singular in the limit $\varepsilon \to 0$ but $|\vec{x}|$ independent while the x dependent part is logarithmic. This case corresponds to interaction of two infinitely long straight strings aligned parallel to each other or a string and a ball. Since the two-dimensional Laplacian of $\log x$ vanishes everywhere except $x = 0$, there is no gravitational interaction at any distance induced by an infinitely long straight string. However, there is a global effect that generates a singularity in the metric at the origin; see Section 3.4.

Exercise 2.4.1, page 81

The scalar potential is

$$V(\phi) = \frac{g^2}{4} \left(\phi^2 - v^2\right)^2. \tag{PS.78}$$

In one of the vacua (it is not important which one), we shift the field ϕ, say, $\phi = v + \chi$, where χ is a small deviation,

$$V(\chi) = g^2 \left(v^2 \chi^2 + v\chi^3 + \frac{1}{4}\chi^4\right). \tag{PS.79}$$

The shift is performed to place the classical vacuum value of the field at the origin. Indeed, we can choose the solution of the equation

$$V(\chi)' \equiv g^2 \left(2v^2\chi + 3v\chi^2 + \chi^3\right) = 0 \tag{PS.80}$$

at $\chi = 0$, thus choosing one of two possible vacua.

Then

$$m^2 = V(\chi)''\big|_{\chi=0} = g^2 \left(2v^2 + 6v\chi + 3\chi^2\right)\big|_{\chi=0} = 2g^2v^2, \tag{PS.81}$$

which coincides with the mass of elementary excitation given *after* Eq. (2.104).

Now, let us analyze one-loop quantum correction to m^2 in the logarithmic approximation. A direct contribution to the mass term $g^2v^2\chi^2$ in (PS.79) comes from the χ^4 term through convolution of two of the χ fields in a loop, depicted in Fig. S.5. This is not the end of the story, however, since the same diagram transforms the χ^3 term in (PS.79) into a linear in χ correction in the potential which, in turn, slightly shifts the vacuum expectation value of χ from zero. This shift also contributes to the one-loop renormalization of mass.

The diagram in Fig. S.5 is trivially calculable by virtue of the Wick rotation $p_0 \to ip_0$,

$$\langle\chi^2\rangle = \int \frac{d^2p}{(2\pi)^2} \frac{i}{p^2 - m^2} = \frac{1}{4\pi} \log \frac{M_{uv}^2}{m^2} \equiv \kappa, \tag{PS.82}$$

where M_{uv} is the ultraviolet cutoff and we introduced κ, a dimensionless loop expansion parameter small compared to v^2,

$$\kappa/v^2 \ll 1.$$

χ

χ^2

Fig. S.5   A loop Feynman graph for $\langle\chi^2\rangle$ in the theory of Section 2.4.1.

Now, using (PS.82) we will find the potential (PS.79) in the one-loop approximation keeping only those terms that affect the mass renormalization. Taking into account combinatorial factors, we arrive at

$$V(\chi) \rightarrow g^2 \left(v^2\chi^2 + 3v\kappa\chi + v\chi^3 + \frac{3\kappa}{2}\chi^2 + \frac{1}{4}\chi^4 + \cdots \right), \tag{PS.83}$$

where the ellipses stand for irrelevant higher-order terms. We find a shifted vacuum χ_* from the equation $V(\chi)' = 0$ implying

$$\chi_* = -\frac{3}{2}\frac{\kappa}{v} + O(\kappa^2). \tag{PS.84}$$

Including the one-loop correction in Eq. (PS.81) we obtain

$$m^2 = V(\chi)''\big|_{\chi=\chi_*} = g^2 \left(2v^2 + 6v\chi_* + 3\kappa + O(\kappa^2) \right). \tag{PS.85}$$

Combining Eqs. (PS.84) and (PS.85) we conclude that

$$m_R^2 = m^2 - \frac{3g^2}{2\pi}\log\frac{M_{uv}^2}{m^2}, \tag{PS.86}$$

q.e.d.

Exercise 2.4.2, page 81

The simplest way to check the equality $L_2\,\chi_0(z) = 0$ is as follows. Let us represent L_2 in the factorized form

$$L_2 = P^\dagger P, \tag{PS.87}$$

where

$$\begin{aligned} P &= \partial_z + \mathcal{W}''(\phi_k) = \partial_z + m\tanh(mz/2), \\ P^\dagger &= -\partial_z + \mathcal{W}''(\phi_k) = -\partial_z + m\tanh(mz/2). \end{aligned} \tag{PS.88}$$

This decomposition reduces the second-order equation (2.148) to the first-order equation

$$P \chi_0 \equiv [\partial_z + m \tanh(mz/2)] \frac{1}{[\cosh(mz/2)]^2} = 0, \qquad \text{(PS.89)}$$

which is obviously satisfied.

Exercise 2.4.3, page 81

The operator L_2 is presented in Eq. (2.112). Given its definition we see that Eq. (2.148) has the form of the standard one-dimensional Schrödinger equation).[2] The famous *oscillation theorem* in one-dimensional quantum mechanics states that the eigenfunction χ_n has exactly n zeroes. In particular, the ground state χ_0 has no zeroes, the first excitation has one zero and so on (see e.g. section 21 in Landau and Lifshitz[3]). Now, look at the expression (2.120) for $\chi_0(z)$. It is obvious that the right-hand side of (2.120) has no zeroes in the interval $-L/2 \leq z \leq L/2$ (remember, at the very end $L \to \infty$). This implies that $\omega_0^2 = 2$ is the lowest eigenvalue. Therefore, there are no eigenfunctions with negative eigenvalues.

One can derive the same result in a more general way. From Eq. (2.148) we know that $L_2 = P^\dagger P$. This means that for any state $|a\rangle$ – not necessarily the eigenstate – we have

$$\langle a|L_2|a\rangle = \langle a|P^\dagger P|a\rangle = \langle b|b\rangle \geq 0. \qquad \text{(PS.90)}$$

Here

$$|b\rangle = P|a\rangle.$$

Equation (PS.90) implies that the eigenvalues are non-negative for all eigenfunctions of L_2.

Hamiltonians of the Schrödinger type which can be represented as

$$L_2 = P^\dagger P \text{ or } PP^\dagger \qquad \text{(PS.91)}$$

are called *factorizable Schrödinger operators*. Their discovery can be traced back to old works of Schrödinger, Dirac, Infeld, and Hull (for references, see, e.g., J. O. Rosas-Ortiz, *On the Factorization Method in Quantum Mechanics,* quant-ph/9812003). The exact spectrum in factorizable Schrödinger Hamiltonians can be found purely algebraically, through a recursion. A good pedagogical review of this topic is presented in the textbook by F. Schwabl, *Quantum Mechanics,* Springer, 1990, chapter 19. The factorization method is closely related to Witten's supersymmetric quantum mechanics (E. Witten, Dynamical Breaking of Supersymmetry, *Nucl. Phys.* B188 (1981) 513; for instance, L_2 and its generalizations were considered in L. E. Gendenshtein, Derivation of Exact Spectra of the Schrödinger Equation by Means of Supersymmetry, *JETP Lett.* 38, 356 (1983). For a review, see M. Shifman, *ITEP Lectures on Particle Physics and Field Theory,* World Scientific, 1999, Vol. 1, chapter 4).

---

[2] One should not forget about the boundary conditions – two infinite walls at $z = \pm L/2$. They discretize the spectrum for positive eigenvalues.

[3] L. D. Landau and E. M. Lifshitz, *Quantum Mechanics: Non-Relativistic Theory,* Elsevier, 2013.

Chapter 3

Exercise 3.1.1, page 99

Prove that the gauge potential

$$A_i = -\varepsilon_{ij}\frac{x_j}{r^2}, \qquad r^2 = x_1^2 + x_2^2 \tag{PS.92}$$

is pure gauge everywhere except the origin. Find the value of F_{12} at the origin.

To prove the statement, it is sufficient to calculate F_{12} using Eq. (PS.92) for $A_{1,2}$,

$$F_{12} = \partial_1 A_2 - \partial_2 A_1$$

$$= \frac{2}{r^2}\left(1 - \frac{x_1^2 + x_2^2}{r^2}\right) = 0 \tag{PS.93}$$

everywhere except a singular point at $r = 0$.

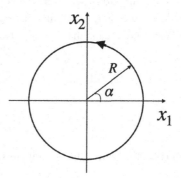

Large circular contour C of radius R.

The singularity of F_{12} near the origin must be of the δ-function type. Indeed, let us integrate F_{12} over the area of the disk with $r \leq R$ (see Fig. S.6) and use Green's theorem to express this integral in terms of the contour integral over a large circle (with radius $R \to \infty$),

$$\int_{r<R} F_{12}\, d^2x = \int_{r<R} F_{12}\, r\, dr\, d\alpha$$

$$= \oint_C A_i\, dx_i = \int_C |\vec{A}|R\, d\alpha, \tag{PS.94}$$

where we pass from Cartesian to polar coordinates; see Figs. 3.2 and S.6. Moreover, on the contour C

$$|\vec{A}| = \frac{1}{R} \quad \text{and hence} \quad \int_C |\vec{A}|R\, d\alpha = 2\pi. \tag{PS.95}$$

We conclude that the flux

$$\int_{r<R} F_{12}\, d^2x = 2\pi \tag{PS.96}$$

and the delta function is as follows:

$$\delta F_{12} = 2\delta(r^2). \tag{PS.97}$$

Exercise 3.2.1, page 109

The matrix Φ in (3.41) is composed of four complex fields or eight real fields. It is most convenient to choose a parametrization in which the imaginary parts of the fields (i.e., four components) are eaten up by the four gauge bosons, providing masses to these bosons, while four real parts become massive "physical" Higgs particles,

$$\Phi = (v + \chi_0)\, \mathbb{1} + \chi_i \tau_i \tag{PS.98}$$

where τ_i are the Pauli matrices and χ_j $(j = 0, 1, 2, 3)$ are complex fields,

$$\chi_j = \mathrm{Re}\, \chi_j + i\, \mathrm{Im}\, \chi_j.$$

Substituting (PS.98) in the potential (3.43), expanding the result in χ_j up to the second order, and using the fact that the parameter v in (3.46) is real, we arrive at the mass terms of the form

$$U_{\mathrm{mass}} = \sum_{j=1}^{3} 2g_2^2\, v^2 \left(\mathrm{Re}\chi_j\right)^2 + 2g_1^2\, v^2 \left(\mathrm{Re}\chi_0\right)^2. \tag{PS.99}$$

As was expected, no mass terms for $\mathrm{Im}\, \chi_j$ are generated.

To derive the masses of the physical Higgs particles, we have to compare (PS.99) with the normalization of the kinetic terms. To this end we substitute (PS.99) in (3.42) and obtain

$$2 \sum_{j=0}^{3} \left(\partial_\mu \mathrm{Re}\chi_j\right)^2. \tag{PS.100}$$

Comparing (PS.99) and (PS.100) we conclude that the mass terms are

$$m_j^2 = v^2 g_2^2, \quad (j = 1, 2, 3), \quad m_0^2 = v^2 g_1^2. \tag{PS.101}$$

This result is to be compared with Eq. (3.48). The fact that the masses of the physical Higgs particles and the vector bosons coincide can be traced to the origin of the potential (3.39). This potential emerges as a truncation of a supersymmetric model; see Part II. The above equality does not extend beyond the leading order because of the truncation.

Exercise 3.2.2, page 109

If we start from the *ansatz* (3.52) the generator $\exp(i\omega_3 \tau_3/2)$ acts trivially (i.e., does not change the form of the *ansatz*). The action of

$$\exp(i\omega_1 \tau_1/2 + i\omega_2 \tau_2/2)$$

is nontrivial. For each value of $\vec{S}$ in (3.65) only one particular combination acts trivially, namely, $\vec{\omega} = \vec{S}$. In such cases, mathematicians say that the solution family spans the manifold SU(2)/U(1). This is equivalent to CP(1). See Sections 6.1.1 and 6.1.2.

Exercise 3.4.1, page 118

From Eq. (3.122) we know that the only nonvanishing components of $h_{\mu\nu}$ are h_{11} and h_{22}. From the geometry of the problem it is clear that both depend only on x_1 and x_2 (i.e., perpendicular coordinates). Then

$$
g_{\mu\nu} = \begin{pmatrix} 1 & 0 & 0 & 0 \\ 0 & h(x_1, x_2) - 1 & 0 & 0 \\ 0 & 0 & h(x_1, x_2) - 1 & 0 \\ 0 & 0 & 0 & -1 \end{pmatrix}. \tag{PS.102}
$$

From the above metric it is easy to calculate the Christoffel symbols in the linear in h approximation. The only nonvanishing symbols are

$$
-\Gamma^1_{11} = \Gamma^1_{22} = -\Gamma^2_{12} = -\Gamma^2_{21} = \frac{1}{2}\partial_1 h \,;
$$

$$
-\Gamma^1_{12} = -\Gamma^1_{21} = -\Gamma^2_{22} = \Gamma^2_{11} = \frac{1}{2}\partial_2 h. \tag{PS.103}
$$

In the linear in h approximation the nonvanishing components of the Riemann tensor following from the preceding expressions are

$$
-R^1_{221} = R^1_{212} = \partial_1 \Gamma^1_{22} - \partial_2 \Gamma^1_{21} = \frac{1}{2}\square h,
$$

$$
-R^2_{121} = R^2_{112} = \partial_1 \Gamma^2_{12} - \partial_2 \Gamma^2_{11} = -\frac{1}{2}\square h, \tag{PS.104}
$$

where

$$
\square = \partial_1^2 + \partial_2^2.
$$

Then we easily obtain the Ricci tensor and scalar curvature in the form

$$
R_{11} = R_{22} = \frac{1}{2}\square h, \quad R = -\square h. \tag{PS.105}
$$

This is fully consistent with the Einstein equation (3.116) and $T^{\mu\nu}$ in Eq. (3.118). Both sides of the Einstein equation for the components $\mu, \nu = 11$ and 22 vanish identically thanks to Eq. (PS.105). The Ricci tensor drops out from consideration of the 00 and 33 components of (3.116), leaving only $-\frac{1}{2}g_{00}$ or $-\frac{1}{2}g_{33}$ in the left-hand side, which matches the non-vanishing components of $T_{\mu\nu}$. Both equations lead to one and the same condition for h, namely

$$
\frac{1}{2}\square h = 8\pi G T_{\text{str}} \, \delta^{(2)}(x_\perp). \tag{PS.106}
$$

This equation implies that $\square h = 0$ for all $r^2 > 0$ where $r^2 = x_1^2 + x_2^2$. Since the left-hand side in (PS.106) is proportional to $\square h$, we guess that $h \sim \log r$. Now we can easily complete the solution of this problem, for instance, by virtue of the Fourier transform of (PS.106), leading to

$$
h(x) = 4G T_{\text{str}} \log\left(x_1^2 + x_2^2\right), \tag{PS.107}
$$

to be compared with (3.122).

Chapter 4

Exercise 4.1.1, page 150

Starting from

$$a_{i(k)}^{a,zm} = -\frac{1}{g}\left(\partial_k A_i^{a(0)} - D_i A_k^{a(0)}\right) = -\frac{1}{g}G_{ki}^{a(0)} = \frac{1}{g}G_{ik}^{a(0)},$$

$$\delta\phi_{(k)}^{a,zm} = -D_k\phi^{a(0)} \tag{PS.108}$$

we have to check that the gauge condition (4.39) is satisfied, namely,

$$\frac{1}{g}D_i a_i^a + \varepsilon^{abc}\phi^{b(0)}\delta\phi^c = 0. \tag{PS.109}$$

To this end we first derive the classical equations of motion for the background fields starting from (4.1) and keeping only the spatial components,

$$\mathcal{L} = -\frac{1}{4g^2}G_{ik}^a G_{ik}^a - \frac{1}{2}(D_i\phi^a)(D_i\phi)^a + \cdots \tag{PS.110}$$

Upon variation with respect to A_k^a we arrive at the following equation:

$$\frac{1}{g^2}D_i\, G_{ik}^{a(0)} - \varepsilon^{abc}\phi^{b(0)}D_k\phi^{c(0)} = 0. \tag{PS.111}$$

Now, substituting the expressions for the zero modes (PS.108) in the gauge condition (PS.109) we obtain (PS.111), q.e.d.

Exercise 4.1.2, page 150

(a)

We know that

$$\{a_0^i, a_0^j\} = \{a_0^{i\dagger}, a_0^{j\dagger}\} = 0,$$

$$\{a_0^i, a_0^{j\dagger}\} = \delta_{ij} \tag{PS.112}$$

and want to construct operators obeying the SU(N_f) Lie algebra in terms of (PS.112). Consider the set

$$O^a = (T^a)_{ij}\, a_i^\dagger a_j, \quad a = 1, 2, \ldots, N^2 - 1, \tag{PS.113}$$

where T^a are the generator matrices in the fundamental representation and for brevity all subscripts 0 are omitted. Then the commutators of the above operators are

$$[O^a O^b] = (T^a)_{ij}\left(T^b\right)_{i'j'}\left[a_i^\dagger a_j, a_{i'}^\dagger a_{j'}\right]$$

$$= (T^a)_{ij}\left(T^b\right)_{i'j'}\left(a_i^\dagger a_{j'}\,\delta_{i'j} - a_{i'}^\dagger a_j\delta_{ij'}\right)$$

$$= \left\{\left(T^a T^b - T^b T^a\right)_{ij} a_i^\dagger a_j\right\} = if^{abc}\,(T^c)_{ij}\, a_i^\dagger a_j$$

$$= if^{abc}O^c. \tag{PS.114}$$

(b)

Each of the zero levels can be filled or empty. Hence, altogether we will have 2^{N_f} degenerate zero-energy states. These degenerate states form a number of irreducible representations with regard to the flavor $\text{SU}(N_f)$ symmetry. The singlet state $|0\rangle$ (i.e., the state with all zero levels empty) is defined by the condition

$$a_i|0\rangle = 0, \quad \text{for all } i. \tag{PS.115}$$

Then the states in the fundamental representation are

$$a_i^\dagger|0\rangle, \quad \text{for all } i, \tag{PS.116}$$

$$\text{multiplicity of the representation} = N.$$

A generic representation has k indices and they are *automatically antisymmetric*,

$$a_{i_1}^\dagger a_{i_2}^\dagger \dots a_{i_k}^\dagger |0\rangle, \quad k = 0, 1, 2, \dots N. \tag{PS.117}$$

$$\text{Multiplicity of the representation} = C_k^N,$$

where

$$C_k^N = \frac{N!}{k!\,(N-k)!}. \tag{PS.118}$$

Adding the multiplicities of all k-index *antisymmetric* representations we get

$$\sum_{k=0}^{N} C_k^N = 2^N. \tag{PS.119}$$

Exercise 4.1.3, page 150

We must analyze how Eqs. (4.30) and (4.33) transform under simultaneous and identical rotations in two O(3) spaces: configurational and "isospace." In other words, we apply $(L^a + T^a)\,\omega^a$ to both sides of Eqs. (4.27) and (4.29) observing that these equations are satisfied in a certain reference frame in both O(3) spaces (which are identified). The latter fact explains why above we use only three independent rotation parameters ω^a, rather than six. Note that $r = \sqrt{\vec{x}^2}$ is invariant and so are the Kronecker and Levi-Civita tensors – they are the same in any reference frame. For instance, δ_{ij} is transformed as $M_{ii'}M_{jj'}\delta_{i'j'} \to M_{ij'}M_{jj'} \to \delta_{ij}$ because for rotations $MM^T = 1$. Here M_{ij} is a generic O(3) rotation matrix depending on ω^a.

Then, the overall structure in all six terms in (4.30) and (4.33)

$$\phi^a \sim n^a, \quad A_i^a \sim \varepsilon^{aij} n^j, \quad B_i^a \sim \delta^{ai}, \quad B_i^a \sim n^a n^i$$
$$\mathcal{D}_i \phi^a \sim \delta^{ai}, \quad \mathcal{D}_i \phi^a \sim n^a n^i \tag{PS.120}$$

remain intact after any rotation described above, and so are the invariant functions $H(r)$ and $F(r)$. Simultaneous application of $(L^a + T^a)\,\omega^a$ does not change the *ansatz*.

Exercise 4.2.1, page 169

In calculating the divergence of the baryon Skyrme current

$$J^\mu = -\frac{\epsilon^{\mu\nu\alpha\beta}}{24\pi^2}\text{Tr}\left(U^\dagger\partial_\nu U\right)\left(U^\dagger\partial_\alpha U\right)\left(U^\dagger\partial_\beta U\right) \tag{PS.121}$$

we must take into account the fact that we must differentiate only the $U^\dagger$ matrices – differentiating any of the U matrices we automatically obtain zero because of the antisymmetric Levi-Civita tensor in the definition (PS.121).

Differentiating one of the $U^\dagger$ matrices we arrive at three terms that reduce to each other by virtue of the cyclic permutations,

$$\partial_\mu J^\mu = -\frac{\epsilon^{\mu\nu\alpha\beta}}{8\pi^2}\text{Tr}\left(\partial_\mu U^\dagger\partial_\nu UU^\dagger\partial_\alpha UU^\dagger\partial_\beta U\right). \tag{PS.122}$$

Note that

$$U\partial_\mu U^\dagger = -\partial_\mu UU^\dagger \text{ and hence } \partial_\mu U^\dagger = -U^\dagger\partial_\mu UU^\dagger, \tag{PS.123}$$

implying that the right-hand side in (PS.122) can be written as

$$\partial_\mu J^\mu = \frac{\epsilon^{\mu\nu\alpha\beta}}{8\pi^2}\text{Tr}\left[U^\dagger\partial_\mu UU^\dagger\partial_\nu UU^\dagger\partial_\alpha UU^\dagger\partial_\beta U\right]$$

$$= \frac{\epsilon^{\mu\nu\alpha\beta}}{8\pi^2}\text{Tr}\left[\mathcal{J}_\mu\mathcal{J}_\nu\mathcal{J}_\alpha\mathcal{J}_\beta\right] = 0 \tag{PS.124}$$

where

$$\mathcal{J}_\mu = U^\dagger\partial_\mu U. \tag{PS.125}$$

The vanishing in (PS.124) is immediately seen by virtue of the cyclic permutation of $\mathcal{J}_\mu$ in $\text{Tr}(\dots)$.

Perturbative expansion means the expansion of the exponent in (4.128) in powers of $\pi^a(x)$ up to any finite order in π^a assuming, as usual that $\pi^a(x) \to 0$ at $|x| \to \infty$. For instance, to the leading order

$$B = -\frac{1}{24\pi^2}\varepsilon_{ijk}\int d^3x\ \text{Tr}\left(U^\dagger\partial_i UU^\dagger\partial_j UU^\dagger\partial_k U\right)$$

$$= -\frac{1}{12\pi^2}\varepsilon_{ijk}\varepsilon^{abc}\int d^3x\ F_\pi^{-3}\ \partial_i\pi^a\partial_j\pi^b\partial_k\pi^c$$

$$= -\frac{1}{12\pi^2}\varepsilon_{ijk}\varepsilon^{abc}F_\pi^{-3}\int d^3x\ \partial_i\left(\pi^a\partial_j\pi^b\partial_k\pi^c\right) = 0. \tag{PS.126}$$

If the matrix $U(x)$ is not close to unity, in the vicinity of an arbitrary point x_0 one can always split it into two parts:

$$U(x) = \tilde{U}(x)U_0, \quad U_0 \equiv U(x_0), \tag{PS.127}$$

where all dependence on x resides in $\tilde{U}(x)$ and $\tilde{U}$ is close to 1, so that one can expand it,

$$\tilde{U}(x) = \exp\left(i\tau^a \vartheta^a(x)\right) = 1 + i\tau^a \vartheta^a(x) + \dots,$$

$$\partial_i \tilde{U}(x) = i\tau^a \partial_i \vartheta^a(x) + \dots, \tag{PS.128}$$

cf. Eq. (4.14). Then the baryon charge takes the form

$$B = -\frac{1}{24\pi^2} \varepsilon_{ijk}$$

$$\times \int d^3x \, \mathrm{Tr}\left(U_0^\dagger \tilde{U}^\dagger(x) \partial_i \tilde{U}(x) U_0 U_0^\dagger \tilde{U}^\dagger(x) \partial_j \tilde{U}(x) U_0 U_0^\dagger \tilde{U}^\dagger(x) \partial_k \tilde{U}(x) U_0\right)$$

$$= -\frac{1}{12\pi^2} \varepsilon_{ijk} \varepsilon^{abc} \int d^3x \, \partial_i \vartheta^a \partial_j \vartheta^b \partial_k \vartheta^c. \tag{PS.129}$$

It is easy to see that the integrand is a total derivative as in Eq. (PS.126). Terms of higher (but finite) orders in the expansion in ϑ can be treated in a similar manner although it is time and labor-consuming.

The above argument is not valid for topologically nontrivial matrices U, which are not defined perturbatively, as an expansion in powers of ϑ. This was demonstrated in Section 4.2.4, see Eqs. (4.146), (4.147).

Exercise 4.2.2, page 169

First, let us note that the general expression for the SO(3) group element in the case at hand can be written as

$$G = \exp\left\{-i\theta \vec{v}\vec{T}\right\}, \tag{PS.130}$$

$$G_{ij} = \delta_{ij}\cos\theta + v_i v_j(1 - \cos\theta) + v^a \epsilon_{iaj}\sin\theta, \tag{PS.131}$$

$$\vec{v}^2 = \sum_{i=1,2,3} v_i^2 = 1, \qquad i,j = 1,2,3,$$

where the generators $\vec{T}$ in (PS.130) are in the vector (adjoint) representation,

$$(T^a)_{ij} = i\varepsilon_{iaj}, \qquad a = 1,2,3, \tag{PS.132}$$

and the unit vector $\vec{v}$ indicates the axis in the three-dimensional space around which one rotates by the angle θ in the anti-clockwise direction. Equation (PS.131) is sometimes referred to as the *Rodrigues formula*.

We remind that the $\vec{T}$ generators in the fundamental representation of SU(2) are

$$T^a = \frac{1}{2}\tau^a, \tag{PS.133}$$

and therefore in (4.155) we have

$$\mathcal{B} = \exp\left(\frac{i\theta}{2}\vec{v}\vec{\tau}\right). \tag{PS.134}$$

Let us substitute the above expression (PS.134) in our definition (4.155),

$$O_{ij} = \frac{1}{2}\mathrm{Tr}\left(\tau_i \mathcal{B}^\dagger \tau_j \mathcal{B}\right)$$

$$= \frac{1}{2}\mathrm{Tr}\left[\tau_i\left(\cos\frac{\theta}{2} - i\vec{v}\vec{\tau}\sin\frac{\theta}{2}\right)\tau_j\left(\cos\frac{\theta}{2} + i\vec{v}\vec{\tau}\sin\frac{\theta}{2}\right)\right]. \qquad (PS.135)$$

After taking the trace, Eq. (PS.135) reduces to the Rodrigues formula (PS.131).

Now, we want to rotate simultaneously – by one and the same angle and around the same axis – the vector x_i in the coordinate space and the vector τ^i in the "isotopic" SU(2) space. The first rotation is performed by the matrix O_{ij}, namely, $x_i' = O_{ij}x_j$ while the second rotation is performed by $\tau^i \to U\tau^i U^\dagger$ in the isospace. Here O_{ij} is a matrix belonging to the orthogonal O(3) group, $OO^T = 1$.

Let us check that

$$O_{ij} = \frac{1}{2}\mathrm{Tr}(\tau^i U^\dagger \tau^j U) \qquad (PS.136)$$

does exactly this. Indeed,

$$\tau^i x_i \to U\tau^i U^\dagger x_i' = U\tau^i U^\dagger O_{ij} x_j =$$

$$= U\tau_i\left[\frac{1}{2}\mathrm{Tr}(\tau^i U^\dagger \tau^j U)\right]U^\dagger x_j = UU^\dagger \tau^j UU^\dagger x_j = \tau^i x_i. \qquad (PS.137)$$

Here I applied completeness formula from page xvi. We see that the product $\tau^i x_i$ is indeed invariant under the above simultaneous rotations. This proves Eqs. (4.40) and (4.41).

Exercise 4.2.3, page 169

At first, it is instructive to see how Γ (see Eqs. (4.172) and (4.173)) changes under the gauge transformation (4.185). To this end, we observe that

$$U^\dagger \partial_j U \to e^{i\epsilon Q}U^\dagger\left\{\partial_j U + i(\partial_j\epsilon)[Q, U]\right\}e^{-i\epsilon Q},$$

$$\partial_j UU^\dagger \to e^{i\epsilon Q}\left\{\partial_j U + i(\partial_j\epsilon)[Q, U]\right\}U^\dagger e^{-i\epsilon Q}, \qquad (PS.138)$$

and, therefore,

$$\delta\Gamma = \int_Q d\Sigma^{ijklm}\,\frac{5}{240\pi^2}(\partial_i\epsilon)\mathrm{Tr}\left\{U^\dagger[Q, U]U^\dagger\partial_j UU^\dagger\partial_k UU^\dagger\partial_l UU^\dagger\partial_m U\right\}$$

$$= \int_Q d\Sigma^{ijklm}\,\frac{\partial_i\epsilon}{48\pi^2}\,\mathrm{Tr}\left\{Q\,\partial_j U\partial_k U^\dagger\partial_l U\partial_m U^\dagger - Q\,\partial_j U^\dagger\partial_k U\partial_l U^\dagger\partial_m U\right\}$$

$$= -\int_Q d\Sigma^{jiklm}\,\frac{\partial_i\epsilon}{48\pi^2}\partial_j\left\{\mathrm{Tr}Q(\partial_k UU^\dagger\,\partial_l UU^\dagger\,\partial_m UU^\dagger + U^\dagger\partial_k U\,U^\dagger\partial_l U\,U^\dagger\partial_m U)\right\}$$

$$= -\int d^4x\,\varepsilon^{\mu\nu\alpha\beta}\,\frac{\partial_\mu\epsilon}{48\pi^2}\left\{\mathrm{Tr}Q(\partial_\nu UU^\dagger\,\partial_\alpha UU^\dagger\,\partial_\beta UU^\dagger\right.$$

$$\left. + U^\dagger\partial_\nu U\,U^\dagger\partial_\alpha U\,U^\dagger\partial_\beta U)\right\} = -\int d^4x\,(\partial_\mu\epsilon)\,J^\mu, \qquad (PS.139)$$

cf. Eq. (4.189). In passing from the third to the fifth line we integrated over full derivative, thus reducing the result to an integral over the four-dimensional space. We inserted $UU^\dagger = U^\dagger U \equiv 1$ where necessary in the integrand in the third and fourth lines. In passing from the second to third line, we changed the ordering of upper indices in $d\Sigma$ and, correspondingly changed the overall sign. Equation (PS.138) is valid in the linear order in $\partial\epsilon$; quadratic and higher-order terms in $\partial\epsilon$ are neglected.

The U(1) gauge invariant generalization of the WZNW term is defined as (cf. Eq. (4.189))

$$\tilde{\Gamma}(U, A_\mu) = \Gamma(U) + e \int d^4x A_\mu J^\mu$$

$$+ \frac{ie^2}{24\pi^2} \int d^4x \epsilon^{\mu\nu\alpha\beta} \partial_\mu A_\nu \, A_\alpha \mathrm{Tr}\Big[Q^2(\partial_\beta U)U^\dagger + Q^2 U^\dagger(\partial_\beta U)$$

$$+ QUQU^\dagger(\partial_\beta U)U^\dagger\Big]. \tag{PS.140}$$

First, note that the gauge transformation

$$\delta A_\mu = \frac{1}{e}(\partial_\mu \epsilon) \tag{PS.141}$$

upon substitution in the second term in the first line in (PS.140) cancels $\delta\Gamma$ in (PS.139). What remains to be done is to calculate δJ^μ and δY, where

$$Y = \epsilon^{\mu\nu\alpha\beta} \partial_\mu A_\nu \, A_\alpha \mathrm{Tr}\Big[Q^2(\partial_\beta U)U^\dagger + Q^2 U^\dagger(\partial_\beta U) + QUQU^\dagger(\partial_\beta U)U^\dagger\Big]. \tag{PS.142}$$

We will carry out these calculations one by one.

δJ^μ :

$$\delta J^\mu = \frac{1}{48\pi^2}\varepsilon^{\mu\nu\alpha\beta} \, \delta \, \mathrm{Tr}\Big[Q\partial_\nu UU^\dagger \, \partial_\alpha UU^\dagger \, \partial_\beta UU^\dagger - (U \leftrightarrow U^\dagger)\Big], \tag{PS.143}$$

cf. Eq. (4.188). For a short while let us suppress the factor

$$\varepsilon^{\mu\nu\alpha\beta}/(48\pi^2), \tag{PS.144}$$

keeping it in mind. We will also suppress the second term in (PS.150) Then

$$\mathrm{Tr}\,\delta\Big[Q\partial_\nu UU^\dagger \, \partial_\alpha UU^\dagger \, \partial_\beta UU^\dagger\Big] = i\mathrm{Tr}\,\Big[Q(\partial_\nu \epsilon)[Q, U]U^\dagger \, \partial_\alpha UU^\dagger \, \partial_\beta UU^\dagger$$

$$+ Q\partial_\nu UU^\dagger \, (\partial_\alpha \epsilon)[Q, U]U^\dagger \, \partial_\beta UU^\dagger + Q\partial_\nu UU^\dagger \, \partial_\alpha UU^\dagger \, (\partial_\beta \epsilon)[Q, U]U^\dagger\Big]$$

$$= i\mathrm{Tr}\Big[-(\partial_\nu \epsilon)Q^2 \partial_\alpha U\partial_\beta U^\dagger + (\partial_\nu \epsilon)QUQU^\dagger \partial_\alpha U\partial_\beta U^\dagger \tag{PS.145}$$

$$+ (\partial_\alpha \epsilon)Q\partial_\nu UU^\dagger Q\partial_\beta UU^\dagger + (\partial_\alpha \epsilon)Q\partial_\nu UQ\partial_\beta U^\dagger \tag{PS.146}$$

$$- (\partial_\beta \epsilon)Q^2 \partial_\nu U\partial_\alpha U^\dagger + (\partial_\beta \epsilon)QU^\dagger QU\partial_\nu U^\dagger \partial_\alpha U\Big]. \tag{PS.147}$$

The first term in (PS.146) vanishes since it is symmetric under $\nu \leftrightarrow \beta$; hence, convolution with the Levi-Civita tensor in (PS.144) eliminates it. The first terms in (PS.145) and (PS.147) proportional to Q^2 as well as the second term in (PS.146), given their convolution with the Levi-Civita tensor, reduce to full derivatives in an obvious manner. What remains to examine is the sum of the last terms in (PS.145) and (PS.147),

$$i\mathrm{Tr}\Big[(\partial_\nu \epsilon)QUQU^\dagger \partial_\alpha U\partial_\beta U^\dagger - (\partial_\nu \epsilon)QU^\dagger QU\partial_\beta U^\dagger \partial_\alpha U\Big]. \tag{PS.148}$$

They are canceled by their counterparts from the $(U \leftrightarrow U^\dagger)$ term in (PS.143),

$$i\mathrm{Tr}\left[(\partial_\nu\epsilon)QUQU^\dagger\partial_\beta U\partial_\alpha U^\dagger - (\partial_\nu\epsilon)QU^\dagger QU\partial_\alpha U^\dagger\partial_\beta U\right]. \qquad (\mathrm{PS.}149)$$

Assembling Eqs. (PS.148) and (PS.149) we get an expression symmetric under $\alpha \leftrightarrow \beta$, which vanishes under convolution with the Levi-Civita tensor.

Three surviving terms which are total derivatives reduce to (after using antisymmetry due to the Levi-Civita tensor):

$$\delta J^\mu = \frac{i}{48\pi^2}\epsilon^{\mu\nu\alpha\beta}\left\{\mathrm{Tr}\,\partial_\beta\left[-2(\partial_\nu\epsilon)Q^2\partial_\alpha UU^\dagger + (\partial_\nu\epsilon)QUQ\partial_\alpha U^\dagger\right]\right.$$

$$\left. + \mathrm{Tr}\,\partial_\beta\left[-2(\partial_\nu\epsilon)Q^2 U^\dagger\partial_\alpha U + (QUQ\partial_\alpha U^\dagger)\right]\right\} \qquad (\mathrm{PS.}150)$$

where the second line here comes from the $(U \leftrightarrow U^\dagger)$ term in (PS.143).

The following formula summarizes the above calculations:

$$\delta\left\{\Gamma(U) + e\int d^4x A_\mu J^\mu\right\}$$

$$= \frac{-ie}{24\pi^2}\epsilon^{\mu\nu\alpha\beta}\int d^4x\, A_\mu\,(\partial_\nu\epsilon)\,\partial_\beta\,\mathrm{Tr}\left[Q^2\partial_\alpha UU^\dagger - (\partial_\nu\epsilon)QUQ\partial_\alpha U^\dagger + Q^2 U^\dagger\partial_\alpha U\right].$$
$$(\mathrm{PS.}151)$$

Next, we have to calculate δY defined in Eq. (PS.142).

Truncated δY: For a short while we will truncate the A fields and the Levi-Civita tensor keeping it in mind. Then

$$\delta\mathrm{Tr}\left[Q^2(\partial_\beta U)U^\dagger + \ Q^2 U^\dagger(\partial_\beta U) - QUQ\partial_\beta U^\dagger\right]$$

$$i\partial_\beta\epsilon\,\mathrm{Tr}\left[Q^2[Q,U]U^\dagger + Q^2 U^\dagger[Q,U] - QUQ[Q,U^\dagger]\right]$$

$$i\partial_\beta\epsilon\,\mathrm{Tr}\left[Q^3 - Q^2 UQU^\dagger + Q^2 U^\dagger QU - Q^3 - QUQ^2 U^\dagger + Q^2 UQU^\dagger\right] = 0.$$
$$(\mathrm{PS.}152)$$

The latter result implies that

$$\delta Y = \epsilon^{\mu\nu\alpha\beta}\partial_\mu A_\nu \frac{1}{e}\partial_\alpha\epsilon\mathrm{Tr}\left[Q^2(\partial_\beta U)U^\dagger + \ Q^2 U^\dagger(\partial_\beta U) - QUQ\partial_\beta U^\dagger\right]$$

$$= -\frac{1}{e}\epsilon^{\mu\nu\alpha\beta}\,\partial_\beta A_\mu\,(\partial_\nu\epsilon)\,\mathrm{Tr}\left[Q^2(\partial_\alpha U)U^\dagger + \ Q^2 U^\dagger(\partial_\alpha U) - QUQ\partial_\alpha U^\dagger\right].$$
$$(\mathrm{PS.}153)$$

Finally, we can assemble Eqs. (PS.140), (PS.142), (PS.151), and (PS.153) to confirm that

$$\delta\tilde{\Gamma}(U, A_\mu) = 0.$$

To this end we must integrate by parts the expression in (PS.153) after substituting it to (PS.140).

Exercise 4.2.4, page 169

We start from the *ansatz*

$$U_0(x) = \exp\left(iF(r)\frac{x_i \tau_i}{r}\right) = \cos F + i\vec{\tau}\vec{n}\,\sin F, \qquad (\text{PS.154})$$

$$\vec{n} = \frac{\vec{x}}{r}.$$

Note that $U^\dagger \partial_i U$ is an element of the algebra and, therefore, can be written as

$$U^\dagger \partial_i U = \mathcal{M}_{ij}\tau_j, \qquad (\text{PS.155})$$

where $\mathcal{M}_{ij}$ is a 3×3 matrix and τ_j (with $j = 1, 2, 3$) is the set of the Pauli matrices.

After some algebra – rather tedious but straightforward – we obtain $\mathcal{M}_{ij}$. In fact, all we need is $\mathrm{Det}\mathcal{M}$.

Indeed, the *ansatz* (PS.154) is SO(3) invariant. The expression for the baryon charge (4.135) is SO(3) scalar – all indices are convoluted. Therefore, in the calculation that will follow, at the end [4] we can make any choice of $\vec{n}$. It is convenient to choose

$$\vec{n} = \{0, 0, 1\}. \qquad (\text{PS.156})$$

Then, $\mathcal{M}_{ij}$ takes the form

$$\mathcal{M}_{ij} = i \begin{pmatrix} \frac{\sin F \cos F}{r} & \frac{-\sin^2 F}{r} & 0 \\ \frac{\sin^2 F}{r} & \frac{\sin F \cos F}{r} & 0 \\ 0 & 0 & F' \end{pmatrix}, \qquad (\text{PS.157})$$

where prime means differentiation over r. Its determinant is

$$\mathrm{Det}\mathcal{M} = -i\, F'\, \frac{1 - \cos 2F}{2\, r^2}. \qquad (\text{PS.158})$$

Taking the trace in (4.135) results in

$$B = -i\, \frac{\varepsilon_{ijk}\, \varepsilon_{abc}}{12\pi^2} \int d^3x\, \mathcal{M}_{ia}\mathcal{M}_{jb}\mathcal{M}_{kc} = \frac{-i}{2\pi^2} \int d^3x\, \mathrm{Det}\mathcal{M}$$

$$= -\frac{1}{4\pi^2} \int d^3x \left\{F'(1 - \cos 2F)\right\} \frac{1}{r^2}$$

$$= -\frac{1}{\pi} \int_0^\infty dr \left\{F' - \frac{1}{2}(\sin 2F)'\right\}. \qquad (\text{PS.159})$$

The integrand is a full derivative, hence,

$$B = -\frac{1}{\pi}\left[F(r) - \frac{1}{2}\sin 2F(r)\right]\Big|_0^\infty \qquad (\text{PS.160})$$

in full agreement with (4.146).

---

[4] Differentiation in ∂U must be done first for arbitrary $\vec{n}$. After the diffeentiation we can use (PS.156).

Chapter 5

Exercise 5.1.1, page 182

Let us rewrite Eq. (5.11) as

$$U(x) = \exp\left[-i\pi\,\tau\,n\,\frac{|x|}{\left(x^2 + \rho^2\right)^{1/2}}\right],$$

where the unit vector n is defined as

$$n = \frac{x}{|x|}.$$

Using (5.12) and (5.13) we can write

$$\mathcal{K} = \int d^3x\,\frac{1}{8\pi^2}\,\varepsilon^{ijk}\,\mathrm{Tr}\left(A_i\partial_j A_k - \frac{2i}{3}\,A_i A_j A_k\right),\quad A_i \equiv g\,\frac{\tau^a}{2}\,A_i^a.$$

If we use (5.5) and the identity for the unitary matrices $\partial_j U = -U\left(\partial_j U^\dagger\right)U$ we arrive at

$$\mathcal{K} = \int d^3x\,\frac{1}{24\pi^2}\,\varepsilon^{ijk}\,\mathrm{Tr}\left[\left(U\partial_i U^\dagger\right)\left(U\partial_j U^\dagger\right)\left(U\partial_k U^\dagger\right)\right],$$

where in what follows we will abbreviate $\left(U\partial_i U^\dagger\right) \longrightarrow U\partial_i U^\dagger$.

In order to complete the exercise it is convenient to deal with more general matrices

$$U(x) = \exp\left[-i\,\tau\,n\,F(|x|)\right] \equiv \cos F - i\,\tau\,n\,\sin F,$$

where at the very end we will specify

$$F = \pi\,\frac{|x|}{\left(x^2 + \rho^2\right)^{1/2}}. \tag{PS.161}$$

A few useful relations and notation follow:

$$r \equiv |x|,\qquad n_k \equiv x_k/r,\qquad \gamma_{kl} = \delta_{kl} - n_k n_l,$$

$$U^\dagger = \cos F + i\,\tau\,n\,\sin F,$$

$$\partial_k F = F'\,n_k,\qquad F' \equiv \frac{\partial F}{\partial r}.$$

Moreover,

$$\partial_k n_l = \frac{1}{r}\,\gamma_{kl}$$

and

$$U\partial_k U^\dagger = i\,\tau n\,F'\,n_k + \sin F\cos F\,\frac{i}{r}\,\gamma_{k\ell}\,\tau_\ell + n_p\,\frac{i}{r}\,\varepsilon^{p\ell m}\,\tau_m\,(\sin F)^2\,\gamma_{kl}.$$

Now we have to take the product $U\partial U^\dagger\, U\partial U^\dagger\, U\partial U^\dagger$ and then the trace. After all differentiations are done we are free to choose the reference frame at any given point in the most convenient way. We will choose

$$n = (0, 0, 1).$$

Then $\gamma_{k\ell} = 0$ if either k or $\ell = 3$, and

$$\gamma_{\tilde{k}\tilde{\ell}} = \delta_{\tilde{k}\tilde{\ell}}, \qquad \tilde{k}, \tilde{\ell} = 1, 2.$$

Correspondingly,

$$-iA_k = U\partial_k U^\dagger = i\tau_3 F' \delta_{k3} + \frac{i}{r}(\sin F)(\cos F)\tau_{\tilde{k}} + \frac{i}{r}\varepsilon^{\tilde{k}\tilde{m}}\tau_{\tilde{m}}(\sin F)^2.$$

This implies in turn that in the product $\mathrm{Tr}\,\varepsilon^{ijk} A_i A_j A_k$ the first term can (and must) enter only once being multiplied either by the second term times the second term or by the third term times the third term. As a result we obtain

$$\int d^3x\, \varepsilon^{ijk}\, \mathrm{Tr}\left[U\partial_i U^\dagger\, U\partial_j U^\dagger\, U\partial_k U^\dagger\right] = \int d^3x\, 6\frac{F'(1 - \cos 2F)}{r^2}.$$

Now, we can demonstrate that $\mathcal{K}$ reduces to the surface term, namely after performing the angular integration

$$\mathcal{K} = \int_0^\infty dr\, \frac{1}{\pi}\, F'\,(1 - \cos 2F) = \frac{1}{\pi}\,\left(F - \frac{1}{2}\sin 2F\right)\Big|_0^\infty = 1,$$

q.e.d.

The derivation presented above can be trivially generalized to $U_n(x) = \{U(x)\}^n$. To this end, F in Eq. (PS.161) must be replaced by nF, which obviously implies $\mathcal{K} = n$.

Exercise 5.2.1, page 185

It will be instructive to start from discussing some generalities of the orthogonal SO(4) group. There are six generators in this group that can be labeled by a pair of numbers $\{i, j\}$, where $i, j = 1, 2, 3, 4$, and in listing these six generators we can agree to take $i < j$. The generator T^{ij} corresponds to rotation in the plane $\{i, j\}$. The generators with $i > j$ are not independent; we can use the property of antisymmetry,

$$T^{ij} = -T^{ji}, \qquad \text{if } i > j. \tag{PS.162}$$

The SO(4) group is real. Therefore, unlike Exercise 4.2.2 on page 169 (see also page 656) it is convenient to use real antisymmetric generators, rather than Hermitean, namely,

$$\left(T^{ij}\right)_{ab} = \left(-\delta_{ia}\delta_{bj} + \delta_{ib}\delta_{ja}\right), \qquad i < j. \tag{PS.163}$$

The antisymmetry in $i \leftrightarrow j$ or $a \leftrightarrow b$ is obvious. In the above convention, instead of (PS.130), the element of SO(4) group is written [5] as follows:

---

[5] Sign conventions in ω_{ij} are to be specified. In terms of notation introduced in (PS.165) one can write O(4) group element as $G = \exp\left\{\vec{v}\left(\omega_v \vec{\mathcal{V}} + \omega_a \vec{\mathcal{A}}\right)\right\}$, cf. Eqs. (PS.130)–(PS.132).

$$G = \exp\left(\frac{1}{2}\omega_{ij}T^{ij}\right) = G = \exp\left(\sum_{i<j}\omega_{ij}\,T^{ij}\right), \tag{PS.164}$$

where ω^{ij} is a 4×4 antisymmetric real matrix of parameters while the generator matrices are defined in (PS.163). In the explicit form we have

$$T^{23} \equiv \mathcal{V}_1 = \begin{bmatrix} 0 & 0 & 0 & 0 \\ 0 & 0 & -1 & 0 \\ 0 & 1 & 0 & 0 \\ 0 & 0 & 0 & 0 \end{bmatrix}, \quad T^{13} \equiv -\mathcal{V}_2 = \begin{bmatrix} 0 & 0 & -1 & 0 \\ 0 & 0 & 0 & 0 \\ 1 & 0 & 0 & 0 \\ 0 & 0 & 0 & 0 \end{bmatrix},$$

$$T^{12} \equiv \mathcal{V}_3 = \begin{bmatrix} 0 & -1 & 0 & 0 \\ 1 & 0 & 0 & 0 \\ 0 & 0 & 0 & 0 \\ 0 & 0 & 0 & 0 \end{bmatrix}, \tag{PS.165}$$

$$T^{14} \equiv \mathcal{A}_1 = \begin{bmatrix} 0 & 0 & 0 & -1 \\ 0 & 0 & 0 & 0 \\ 0 & 0 & 0 & 0 \\ 1 & 0 & 0 & 0 \end{bmatrix}, \quad T^{24} \equiv \mathcal{A}_2 = \begin{bmatrix} 0 & 0 & 0 & 0 \\ 0 & 0 & 0 & -1 \\ 0 & 0 & 0 & 0 \\ 0 & 1 & 0 & 0 \end{bmatrix},$$

$$T^{34} \equiv \mathcal{A}_3 = \begin{bmatrix} 0 & 0 & 0 & 0 \\ 0 & 0 & 0 & 0 \\ 0 & 0 & 0 & -1 \\ 0 & 0 & 1 & 0 \end{bmatrix}. \tag{PS.166}$$

The labeling of the $\mathcal{V}$, $\mathcal{A}$ generators is as follows:

$$\mathcal{V}_\ell = \sum_{j<k}\varepsilon_{\ell jk}T^{jk}, \quad \mathcal{A}_j = T^{j4}, \quad j,k = 1,2,3. \tag{PS.167}$$

The action of the generators on 4-vectors $\hat{Q}_i$ is

$$\left(\delta^{Vp}\hat{Q}\right)_j = \omega_v\left(\mathcal{V}_p\right)_{jk}\hat{Q}_k,$$

$$\left(\delta_{Ap}\hat{Q}\right)_j = \omega_a\left(\mathcal{A}_p\right)_{jk}\hat{Q}_k, \tag{PS.168}$$

for instance,

$$\left(\delta^{V_2}\hat{Q}\right)_1 = \omega_v\hat{Q}_3, \quad \left(\delta^{V_2}\hat{Q}\right)_3 = -\omega_v\hat{Q}_1, \tag{PS.169}$$

and so on.

It is easy to check that

$$[\mathcal{V}_i, \mathcal{V}_j] = \varepsilon_{ijk}\mathcal{V}_k, \quad [\mathcal{A}_i, \mathcal{A}_j] = \varepsilon_{ijk}\mathcal{V}_k, \quad [\mathcal{V}_i, \mathcal{A}_j] = \varepsilon_{ijk}\mathcal{A}_k. \tag{PS.170}$$

Alternatively, one can rewrite Eq. (PS.170) in a factorized form,

$$[(\mathcal{V}_i + \mathcal{A}_i), (\mathcal{V}_j + \mathcal{A}_j)] = 2\varepsilon_{ijk}(\mathcal{V}_k + \mathcal{A}_k),$$

$$[(\mathcal{V}_i - \mathcal{A}_i), (\mathcal{V}_j - \mathcal{A}_j)] = 2\varepsilon_{ijk}(\mathcal{V}_k - \mathcal{A}_k),$$

$$[(\mathcal{V}_i + \mathcal{A}_i), (\mathcal{V}_j - \mathcal{A}_j)] = 0. \tag{PS.171}$$

The above algebra implies that

$$SO(4) = SO(3) \times SO(3).$$

Now, let us define similar generators acting on Euclidean Dirac spinors. First, the Euclidean γ matrices as defined in Eq. (5.27) are

$$\hat{\gamma}_4 = \begin{pmatrix} 0 & 1 \\ 1 & 0 \end{pmatrix}, \quad \hat{\gamma}_j = \begin{pmatrix} 0 & -i\sigma_j \\ i\sigma_j & 0 \end{pmatrix}. \tag{PS.172}$$

Next, we will define the generators (with $a = 1, 2, 3$),

$$\hat{\Sigma}_a = \sum_{j<k} \frac{1}{2} \varepsilon_{ajk} [\hat{\gamma}_j, \hat{\gamma}_k] = i \begin{pmatrix} \sigma_a & 0 \\ 0 & \sigma_a \end{pmatrix}, \quad j = 1, 2, 3; \ k = 1, 2, 3;$$

$$\hat{\Sigma}_{a4} = \frac{1}{2} [\hat{\gamma}_a, \hat{\gamma}_4] = -i \begin{pmatrix} \sigma_a & 0 \\ 0 & -\sigma_a \end{pmatrix}. \tag{PS.173}$$

Using the 't Hooft symbols from (5.43) one can write both equations in (PS.173) in a compact way, namely,

$$\hat{\Sigma}_a = \frac{1}{4} \eta_{a\mu\nu} [\hat{\gamma}_\mu, \hat{\gamma}_\nu]. \tag{PS.174}$$

Next, we observe $\delta(\hat{\psi}^\dagger \hat{\psi}) = 0$ and $\hat{\psi}^\dagger \hat{\psi}$ is O(4) scalar. To warm up let us ask what happens with the vector combination $\hat{Q}_\mu = \hat{\psi}^\dagger \hat{\gamma}_\mu \hat{\psi}$. According to (PS.168), (PS.160)

$$\delta\hat{Q}_\mu \equiv \delta(\hat{\psi}^\dagger \hat{\gamma}_\mu \hat{\psi}) \propto \hat{\psi}^\dagger \frac{1}{2} [\hat{\gamma}_\mu, \hat{\Sigma}_{a(a4)}] \hat{\psi}$$

$$= \begin{cases} \varepsilon_{ajk} \hat{Q}_k, \\ \delta_{aj} \hat{Q}_4, \quad (\mu = j = 1, 2, 3), \end{cases} \tag{PS.175}$$

$$\text{and} \begin{cases} 0, \\ -\hat{Q}_a, \quad (\mu = 4). \end{cases} \tag{PS.176}$$

The first lines in the above vectorial infinitesimal transformations (PS.175) and (PS.176) correspond to the action of $\hat{\Sigma}_a$ while the second lines to $\hat{\Sigma}_{a4}$. The transformation laws above coincide with those following from Eqs. (PS.165) and (PS.166) for the SO(4) generators in the vectorial representation.

The fact that the combination

$$\hat{Q}_{\mu\nu} = \hat{\psi}^\dagger [\hat{\gamma}_\mu, \hat{\gamma}_\nu] \hat{\psi} \tag{PS.177}$$

is transformed as an *antisymmetric tensor* can be readily verified by virtue of the well-known formula

$$[\hat{\Sigma}_a, \hat{\gamma}_\mu \hat{\gamma}_\nu] = [\hat{\Sigma}_a, \hat{\gamma}_\mu] \hat{\gamma}_\nu + \hat{\gamma}_\mu [\hat{\Sigma}_a, \hat{\gamma}_\nu]. \tag{PS.178}$$

and Eqs. (PS.175) and (PS.176). Alternatively, one can note that the commutators $[\hat{\gamma}_\mu, \hat{\gamma}_\nu]$ coincide with the matrices $\hat{\Sigma}_a$ and $\hat{\Sigma}_{a4}$.

Exercise 5.4.1, page 221

To warm up, let us reproduce Eq. (5.93) for anti-instanton. Using various definitions and invoking relations for the Pauli matrices τ collected on pages xvi and xvii we derive

$$
\begin{aligned}
N_{mn,p\dot{q}} &= 2\bar{\eta}_{a\mu\nu}x_\nu(\tau^a)_{mn}(\tau^-_\mu)_{p\dot{q}} \\
&= 2\bar{\eta}_{a4\nu}x_\nu(\tau^a)_{mn}(\tau^-_4)_{p\dot{q}} + 2\bar{\eta}_{a\ell\nu}x_\nu(\tau^a)_{mn}(\tau^\ell)_{p\dot{q}} \\
&= 2i(\vec{x}\vec{\tau})_{mn}\delta_{p\dot{q}} - 2x_4(\tau^a)_{mn}(\tau^a)_{p\dot{q}} + 2\epsilon_{a\ell j}x_j(\tau^a)_{mn}(\tau^\ell)_{p\dot{q}} \\
&= 2i(\vec{x}\vec{\tau})_{mn}\delta_{p\dot{q}} - 2x_4(2\delta_{m\dot{q}}\delta_{np} - \delta_{mn}\delta_{p\dot{q}}) \\
&\quad - 2i\Big[(\vec{x}\vec{\tau})_{mn}\delta_{p\dot{q}} - 2(\vec{x}\vec{\tau})_{m\dot{q}}\delta_{np} + (\vec{x}\vec{\tau})_{p\dot{q}}\delta_{mn}\Big] \\
&= 2i\left\{2(x\tau^-)_{m\dot{q}}\delta_{np} - (x\tau^-)_{p\dot{q}}\delta_{mn}\right\}.
\end{aligned}
\tag{PS.179}
$$

On the other hand, Eq. (5.93) reduces to

$$
\begin{aligned}
&2i\left\{\delta_{pn}(x\tau^-)_{m\dot{q}} - (\delta_{mn}\delta_{ps} - \delta_{ms}\delta_{pn})(x\tau^-)_{s\dot{q}}\right\} \\
&= 2i\left\{2ix_4\delta_{pn}\delta_{m\dot{q}} - ix_4\delta_{mn}\delta_{p\dot{q}} + 2\delta_{pn}(\vec{x}\vec{\tau})_{m\dot{q}} - \delta_{mn}(\vec{x}\vec{\tau})_{p\dot{q}}\right\} \\
&= 2i\left\{2\delta_{pn}(x\tau^-)_{m\dot{q}} - \delta_{mn}(x\tau^-)_{p\dot{q}}\right\}.
\end{aligned}
\tag{PS.180}
$$

The results in Eqs. (PS.179) and (PS.180) coincide.

To pass from anti-instanton to instanton we must replace

$$
\bar{\eta} \to \eta, \quad \tau^- \to \tau^+,
$$

namely,

$$
\left[N_{\dot{m}\dot{n},p q}\right]_{\text{inst}} = 2\eta_{a\mu\nu}x_\nu(\tau^a)_{\dot{m}\dot{n}}(\tau^+_\mu)_{pq}.
\tag{PS.181}
$$

Note that all undotted indices become dotted and vice versa. Performing the same steps as in Eqs. (PS.179) and (PS.180) we readily arrive at

$$
\left[N_{\dot{m}\dot{n},p q}\right]_{\text{inst}} = 2i\Big[\delta_{\dot{p}\dot{n}}(x\tau^+)_{\dot{m}q} - \varepsilon_{\dot{m}\dot{p}}\,\varepsilon_{\dot{n}\dot{s}}(x\tau^+)_{\dot{s}q}\Big].
\tag{PS.182}
$$

Now the undotted q index goes through unaltered. Thus, if in the anti-instanton solution SU(2) color is entangled with one of the factors $SU(2) \in SO(4)$, in the instanton solution it is another $SU(2) \in SO(4)$ that becomes entangled with color.

Exercise 5.4.2, page 221

As a warm-up exercise let us determine the 4-vector $\hat{v}$. Since any rotation matrix M can be written as

$$
M = \exp\left(\frac{i\tau^a\omega^a}{2}\right) = \cos\frac{\omega}{2} + i\vec{n}\vec{\tau}\sin\frac{\omega}{2} = i\hat{v}_\mu\tau^-_\mu
$$

(here $\omega = |\vec{\omega}|$ and $\vec{n}$ is the unit vector in the direction of $\vec{\omega}$), we determine that

$$
\vec{v} = \vec{n}\sin\frac{\omega}{2}, \qquad \hat{v}_4 = -\cos\frac{\omega}{2},
$$

implying that $\hat{v}^2 = 1$. Let us choose the reference frame in which $\vec{R} = 0$ and only $R_4 \neq 0$. One can always do that. Then

$$\eta_{a\alpha\beta}\bar{\eta}_{ba\gamma}R_\beta R_\gamma = -\delta_{ab}R_4^2.$$

One should also use the facts that

$$O^{ab} = \tfrac{1}{2}\mathrm{Tr}\left(\tau^b\hat{v}_\mu\tau_\mu^+\tau^a\hat{v}_\nu\tau_\nu^-\right)$$

and

$$\tau^a\tau_\mu^+\tau^a = -\tau_\mu^+ + s_\mu, \qquad s_\mu = \begin{cases} 0, & \text{for } \mu = 1, 2, 3, \\ -4i, & \text{for } \mu = 4. \end{cases}$$

Now assembling all these expressions one arrives at

$$O^{ab}\eta_{a\alpha\beta}\bar{\eta}_{ba\gamma}R_\beta R_\gamma = \hat{v}^2 R_4^2 - 4\hat{v}_4^2 R_4^2 \rightarrow \hat{v}^2 R^2 - 4\left(\hat{v}R\right)^2.$$

Exercise 5.4.3, page 222

For generic values of the imaginary time τ we have

$$U(\tau, x) = \exp\left(\int_{-\infty}^{\tau} \frac{i\vec{x}\vec{\tau}}{\tau^2 + \vec{x}^2 + \rho^2}d\tau\right)U(t = -\infty, x_i)$$

$$= \exp\left(i\,\vec{x}\vec{\tau}\int_{-\infty}^{\tau}\frac{1}{\tau^2 + \lambda^2}d\tau\right)U(t = -\infty, x_i), \qquad \text{(PS.183)}$$

where

$$\lambda^2 = \vec{x}^2 + \rho^2.$$

Performing integration over τ in (PS.183) we obtain

$$U(\tau, \vec{x}) = \exp\left\{-i\,\vec{x}\vec{\tau}\,\frac{1}{\lambda}\left(\frac{\pi}{2} + \arctan\frac{\tau}{\lambda}\right)\right\}. \qquad \text{(PS.184)}$$

In the limit $\tau \rightarrow \infty$,

$$U(\tau = +\infty, \vec{x}) = \exp\left(-i\pi\frac{\vec{x}\vec{\tau}}{\sqrt{\vec{x}^2 + \rho^2}}\right)U(t = -\infty, \vec{x}). \qquad \text{(PS.185)}$$

If we choose $U = 1$ in the distant past, in the distant future

$$U(\tau = +\infty, \vec{x}) \equiv U(\vec{x}) = \exp\left(-i\pi\frac{\vec{x}\vec{\tau}}{\sqrt{\vec{x}^2 + \rho^2}}\right). \qquad \text{(PS.186)}$$

Correspondingly, the instanton solution in the $A_0 = 0$ gauge is

$$A_i(\tau, x_i) = \begin{cases} 0, & \text{at } \tau = -\infty, \\ iU^\dagger(\vec{x})\partial_i U(\vec{x}) & \text{at } \tau = \infty. \end{cases} \qquad \text{(PS.187)}$$

Exercise 5.4.4, page 222

In this problem for notational convenience we will use non-canonically normalized A_μ. Then the kinetic term of the gauge field and covariant derivative are

$$-\frac{1}{4g^2}(G_{\mu\nu}^2)^2 \quad \text{and} \quad \mathcal{D}_\mu = \partial_\mu - iA_\mu, \qquad\qquad \text{(PS.188)}$$

respectively.

Our initial data are

$$X = v\left(\frac{x^2}{x^2 + \rho^2}\right)^{1/2} S_1 \qquad\qquad \text{(PS.189)}$$

and

$$A_\mu = \frac{x^2}{x^2 + \rho^2} i S_1 \partial_\mu S_1^\dagger, \qquad\qquad \text{(PS.190)}$$

where

$$S_1 = \frac{i\tau_\mu^+ x_\mu}{\sqrt{x^2}}, \qquad \tau_\mu^+ = (\tau^a, -i). \qquad\qquad \text{(PS.191)}$$

By brute force calculation we obtain

$$\partial_\mu \partial_\mu X = -\frac{3v(x^2 + 2\rho^2)}{\sqrt{(x^2 + \rho^2)^5}} i\tau_\mu^+ x_\mu, \qquad\qquad \text{(PS.192)}$$

$$i\tau_\mu^+ x_\mu = \left\{ \begin{array}{cc} (t + iz) & (y + ix) \\ (ix - y) & (t - iz) \end{array} \right\}.$$

Similarly,

$$-2iA_\mu \partial_\mu X = \frac{6v}{\sqrt{(x^2 + \rho^2)^3}} i\tau_\mu^+ x_\mu \qquad\qquad \text{(PS.193)}$$

and

$$-i\left(\partial_\mu A_\mu\right) X = 0. \qquad\qquad \text{(PS.194)}$$

Finally, the last term takes the form

$$-A_\mu A_\mu X = -\frac{3v\, x^2}{\sqrt{(x^2 + \rho^2)^5}} i\tau_\mu^+ x_\mu. \qquad\qquad \text{(PS.195)}$$

Assembling everything together we arrive at

$$D_\mu^2 X = \partial_\mu \partial^\mu X - 2iA_\mu \partial^\mu X - i\left(\partial_\mu A^\mu\right) X - A_\mu A^\mu X$$

$$= 0, \qquad\qquad \text{(PS.196)}$$

q.e.d.

Exercise 5.4.5, page 222

The Chern–Simons charge is defined in Eqs. (5.12) and (5.13). The *ansatz* (5.179) is in the $A_0 = 0$ *regular gauge*, i.e. at $r \to \infty$ the vector potential falls off as $A_i \sim n_i/r$ where $n_i = x_i/r$. Slow convergence makes it inconvenient to work with this *ansatz*. We will first find a gauge transformation that makes manifest a $1/r^2$ fall-off of the gauge field. To this end, following Klinkhamer and Manton, we will perform the gauge transformation with the matrix

$$U(x) = \exp\left(\frac{i}{2}\,\Theta(r)\,\tau_i n_i\right), \tag{PS.197}$$

where $\Theta(\infty) = \pi$ and $\Theta(0) = 0$. Under this gauge transformation the gauge field (5.189) takes the form

$$A_i^a = \frac{1}{g}\left[\frac{(1 - 2f(gvr))\cos\Theta(r) - 1}{r^2}\,\epsilon_{iab}x_b\right.$$

$$\left. + \frac{(1 - 2f(gvr))\sin\Theta(r)}{r^3}\left(\delta_{ia}r^2 - x_i x_a\right) + \frac{d\Theta(r)}{dr}\frac{x_i x_a}{r^2}\right]. \tag{PS.198}$$

Now, Eq. (PS.198) must be inserted in the Chern–Simons current

$$K^\mu = 2\varepsilon^{\mu\nu\alpha\beta}\left(A_\nu^a \partial_\alpha A_\beta^a + \frac{g}{3}\varepsilon_{abc}A_\nu^a A_\alpha^b A_\beta^c\right)$$

$$= 2\varepsilon^{\mu\nu\alpha\beta}\left[A_\nu^a\left(\frac{F_{\alpha\beta}^a - g\varepsilon_{abc}A_\alpha^b A_\beta^c}{2}\right) + \frac{g}{3}\varepsilon_{abc}A_\nu^a A_\alpha^b A_\beta^c\right]$$

$$= \varepsilon^{\mu\nu\alpha\beta}\left(A_\nu^a F_{\alpha\beta}^a - \frac{g}{3}\varepsilon_{abc}A_\nu^a A_\alpha^b A_\beta^c\right), \tag{PS.199}$$

implying

$$K^0 = -\frac{4}{g^2 r^2}\left\{2(\sin\Theta(r))f'(gvr)\right.$$

$$\left. + \left[1 + \cos\Theta(r)(-1 + 2f(gvr))\right]\Theta'(r)\right\}, \tag{PS.200}$$

where the prime denotes differentiation over r. The topological charge of the sphaleron is then

$$\mathcal{K} = \frac{g^2}{32\pi^2}\int K^0 d^3 x$$

$$= \frac{4\pi g^2}{32\pi^2}\int_0^\infty \frac{4}{g^2 r^2}\left[2(\sin\Theta(r))f'(gvr) + (1 + \cos\Theta(r)(-1 + 2f(gvr))\Theta'(r)\right]r^2 dr$$

$$= \frac{1}{2\pi}\int_0^\infty\left[2(\sin\Theta(r))f'(gvr) + \Theta'(r) + (2\cos\Theta(r)f(gvr) - (\cos\Theta(r))\Theta'(r)\right]dr$$

$$= \frac{1}{2\pi}\int_0^\infty\left\{2\frac{d}{dr}\left[\sin\Theta(r)f(gvr)\right] + \Theta'(r) - \frac{d}{dr}\sin\Theta(r)\right\}dr$$

$$= \frac{1}{2\pi}\left[2\sin(\Theta(r))f(gvr) + \Theta(r) - \sin\Theta(r)\right]_0^\infty = \frac{1}{2\pi}\times\pi = \frac{1}{2}, \tag{PS.201}$$

concluding the proof that the topological charge of the sphaleron is semi-integer. In passing from the third to fourth line in Eq. (PS.201) we integrated by parts and in the last line used the boundary values of $\Theta(r)$.

Exercise 5.7.1, page 240

The baryon charge in the Skyrme model is defined in Eq. (4.135), where a matrix $U \in SU(2)$. The topological current (PS.199) can be presented as

$$K^\mu = 2\epsilon^{\mu\nu\alpha\beta}\text{Tr}\left(A_\nu F_{\alpha\beta} + \frac{2i}{3}g\, A_\nu A_\alpha A_\beta\right), \qquad (PS.202)$$

which gives a topological charge

$$\mathcal{K} = \frac{g^2}{16\pi^2}\int d^3x\,\epsilon^{ijk}\text{Tr}\left(A_i F_{jk} + \frac{2i}{3}g\, A_i A_j A_k\right). \qquad (PS.203)$$

Now, if we choose the field $g A_i$ as a *pure gauge*

$$g A_i = i\, U^\dagger \partial_i U \qquad (PS.204)$$

with the same matrix U as in (4.128) and $A_0 = 0$ we obtain

$$F_{ij} = 0,$$

$$\mathcal{K} = \frac{g^2}{16\pi^2}\int d^3x\,\epsilon^{ijk}\text{Tr}\left(A_i F_{jk} + \frac{2i}{3}g\, A_i A_j A_k\right)$$

$$= \frac{\epsilon^{ijk}}{24\pi^2}\int d^3x\,\text{Tr}\left(U^\dagger\partial_i U\right)\left(U^\dagger\partial_j U\right)\left(U^\dagger\partial_k U\right), \qquad (PS.205)$$

to be compared with Eq. (4.135) for the baryon number in the Skyrme model. The first equality in (PS.205) immediately follows from the representation

$$F_{ij} = \partial_i A_j - \partial_j A_i - ig\left(A_i A_j - A_j A_i\right) \qquad (PS.206)$$

and (PS.204).

Chapter 6

Exercise 6.2.1, page 256

From Eqs. (6.1) and (6.2) we can read off the radius r or the target space sphere S^2. Let us normalize the kinetic term canonically by passing from the field $\vec{S}$ to $\vec{\tilde{S}}$,

$$\vec{\tilde{S}} = \frac{1}{g}\vec{S}.$$

Then, the kinetic term is normalized canonically, while the constraint on $\vec{\tilde{S}}$ takes the form

$$\vec{\tilde{S}}\vec{\tilde{S}} = \frac{1}{g^2}, \qquad (PS.207)$$

implying that the radius of the target space sphere $r = 1/g$. As is well-known, the scalar curvature $\mathcal{R}$ for the two-dimensional sphere of radius r reduces to

$$\mathcal{R} \equiv \frac{2}{r^2} = 2g^2. \tag{PS.208}$$

Equation (PS.208) will be useful later for verification.

Let us turn to the Kähler formalism. The Kähler metric in CP(1) model has only one component (see Eq. (6.9)),

$$G \equiv G_{1\bar{1}} = \frac{2}{g^2}\frac{1}{(1+\phi\bar{\phi})^2}, \quad G^{-1} \equiv G^{\bar{1}1} = \frac{g^2}{2}(1+\phi\bar{\phi})^2. \tag{PS.209}$$

Moreover, in the Kähler formalism, in CP(1) there is a single Christoffel symbol with all holomorphic indices and its complex cojugated,

$$\Gamma \equiv \Gamma^1_{11} = -\frac{2\bar{\phi}}{1+\phi\bar{\phi}}, \quad \bar{\Gamma} \equiv \Gamma^{\bar{1}}_{\bar{1}\bar{1}} = -\frac{2\phi}{1+\phi\bar{\phi}}. \tag{PS.210}$$

We then calculate the Riemann and Ricci tensors according to the standard rules of Kähler geometry (specified for CP(1)),

$$R^{\bar{1}}_{\ \bar{1}1\bar{1}} = \frac{\partial}{\partial\phi}\,\Gamma^{\bar{1}}_{\bar{1}\bar{1}} = -\frac{2}{(1+\phi\bar{\phi})^2} = -R_{1\bar{1}}, \tag{PS.211}$$

and complex conjugated for $R^1_{\ 1\bar{1}1}$. Note that $R_{1\bar{1}} = g^2 G_{1\bar{1}}$. Finally, using the metric with the upper components we obtain for the scalar curvature

$$\mathcal{R} = 2\,G^{\bar{1}1}R_{1\bar{1}} = 2g^2, \tag{PS.212}$$

to be compared with Eq. (PS.208).

It is instructive to perform the same calculation passing from the complex coordinates (PS.209) to the same model in real coordinates,

$$x^1 = \mathrm{Re}\,\phi, \quad x^2 = \mathrm{Im}\,\phi. \tag{PS.213}$$

The subsequent calculation is noticeably more labor-consuming than in the Kähler formalism; that is why comparison makes sense for pedagogical purposes.

A subtle point here is as follows: while the canonically normalized kinetic term for complex fields is $\partial\varphi\partial\bar{\varphi}$, in real coordinates it is $\frac{1}{2}\partial\varphi\partial\varphi$. Taking into account this transition factor $\frac{1}{2}$, we obtain from (PS.209) the metric in real coordinates,

$$G_{ij} = \frac{4}{g^2}\frac{1}{\left[(1+(x^1)^2+(x^2)^2\right]^2}\delta_{ij}, \quad i,j = 1,2. \tag{PS.214}$$

In what follows, we will use a shorthand

$$\chi = 1 + (x^1)^2 + (x^2)^2. \tag{PS.215}$$

We first find eight Christoffel symbols,

$$\Gamma^1_{11} = -\frac{2x^1}{\chi}, \qquad \Gamma^1_{22} = \frac{2x^1}{\chi},$$

$$\Gamma^1_{12} = -\frac{2x^2}{\chi}, \qquad \Gamma^1_{22} = -\frac{2x^2}{\chi},$$

$$\Gamma^2_{11} = \frac{2x^2}{\chi}, \qquad \Gamma^2_{22} = -\frac{2x^2}{\chi},$$

$$\Gamma^2_{12} = -\frac{2x^1}{\chi}, \qquad \Gamma^2_{21} = -\frac{2x^1}{\chi}. \tag{PS.216}$$

Out of 16 Riemann tensor components only four are nonvanishing, namely,

$$R^1{}_{221} = -\frac{4}{\chi^2}, \qquad R^1{}_{212} = \frac{4}{\chi^2},$$

$$R^2{}_{121} = \frac{4}{\chi^2}, \qquad R^2{}_{112} = -\frac{4}{\chi^2}. \tag{PS.217}$$

Equation (PS.217) implies the following expression for the Ricci tensor:

$$R_{km} = R^i{}_{kim} = \frac{4}{\chi^2} \delta_{km}. \tag{PS.218}$$

Note that $R_{ij} = g^2 G_{ij}$. Finally, we can obtain the scalar curvature

$$\mathcal{R} = G^{mk} R_{km} = \left(\frac{g^2}{4}\chi^2 \delta_{mk}\right)\left(\frac{4}{\chi^2}\delta_{km}\right) = 2g^2. \tag{PS.219}$$

The first bracket in the above formula presents $\left(G^{mk}\right)$. Equation (PS.219) is to be compared with (PS.208) or (PS.212).

Exercise 6.2.2, page 257

The vector K^μ is called the *Chern–Simons current*,

$$K^\mu = \varepsilon^{\mu\nu}\left(\frac{\bar{\phi}\partial_\nu\phi}{1 + \phi\bar{\phi}}\right). \tag{PS.220}$$

By direct calculation we arrive at

$$\partial_\mu K^\mu = \varepsilon^{\mu\nu}\partial_\mu\left(\frac{\bar{\phi}\partial_\nu\phi}{1 + \phi\bar{\phi}}\right)$$

$$= \frac{\varepsilon^{\mu\nu}\partial_\mu\bar{\phi}\partial_\nu\phi}{1 + \bar{\phi}\phi} - \frac{\varepsilon^{\mu\nu}\bar{\phi}\phi\partial_\mu\bar{\phi}\partial_\nu\phi}{(1 + \bar{\phi}\phi)^2} = \frac{\varepsilon^{\mu\nu}\partial_\mu\bar{\phi}\partial_\nu\phi}{(1 + \phi\bar{\phi})^2}. \tag{PS.221}$$

The Chern–Simons current is obviously invariant under the linear $U(1)$ transformation $\phi \to e^{-i\alpha}\phi$. However, it is *not* invariant under nonlinear transformations (6.11)

$$\phi \to \phi + \epsilon + \bar{\epsilon}\phi^2, \qquad \bar{\phi} \to \bar{\phi} + \bar{\epsilon} + \epsilon\bar{\phi}^2. \tag{PS.222}$$

Indeed,

$$\delta(K^\mu) = \varepsilon^{\mu\nu}\,\bar{\varepsilon}\,\partial_\nu\phi. \tag{PS.223}$$

At the same time, $\partial_\mu K^\mu$ is invariant under all three transformations, linear and non-linear.

Exercise 6.2.3, page 257

Geometry of CP(2) can be summarized as follows. If we define

$$\chi = 1 + \sum_{i=1}^{2}\bar{\phi}^i\phi^i, \tag{PS.224}$$

then

$$\mathcal{K} = \frac{2}{g^2}\log\left(1 + \bar{\phi}^1\phi^1 + \bar{\phi}^2\phi^2\right) = \log\chi. \tag{PS.225}$$

For the metric and its inverse we have

$$\begin{pmatrix} G_{1\bar{1}} & G_{1\bar{2}} \\ G_{2\bar{1}} & G_{2\bar{2}} \end{pmatrix} = \frac{2}{g^2}\frac{1}{\chi^2}\left(\begin{array}{c|c} 1 + \bar{\phi}^2\phi^2 & -\bar{\phi}^1\phi^2 \\ \hline -\bar{\phi}^2\phi^1 & 1 + \bar{\phi}^1\phi^1 \end{array} \right) \tag{PS.226}$$

and

$$\begin{pmatrix} G^{\bar{1}1} & G^{\bar{1}2} \\ G^{\bar{2}1} & G^{\bar{2}2} \end{pmatrix} = \frac{g^2}{2}\chi\left(\begin{array}{c|c} 1 + \bar{\phi}^1\phi^1 & \bar{\phi}^1\phi^2 \\ \hline \bar{\phi}^2\phi^1 & 1 + \bar{\phi}^2\phi^2 \end{array} \right). \tag{PS.227}$$

The only non-vanishing Christoffel symbols are

$$\Gamma^1_{11} = -2\frac{\bar{\phi}^1}{\chi}, \quad \Gamma^2_{22} = -2\frac{\bar{\phi}^2}{\chi}, \tag{PS.228}$$

$$\Gamma^1_{12} = \Gamma^1_{21} = -\frac{\bar{\phi}^2}{\chi}, \tag{PS.229}$$

$$\Gamma^2_{12} = \Gamma^2_{21} = -\frac{\bar{\phi}^1}{\chi}, \tag{PS.230}$$

and the complex conjugated of the above. It is not difficult to derive the elements of the Riemann tensor,

$$R^{\bar{1}}{}_{\bar{1}1\bar{1}} = -2\frac{1 + \bar{\phi}^2\phi^2}{\chi^2}, \qquad R^{\bar{1}}{}_{\bar{1}2\bar{1}} = 2\frac{\bar{\phi}^2\phi^1}{\chi^2}, \tag{PS.231}$$

$$R^{\bar{1}}{}_{\bar{1}1\bar{2}} = R^{\bar{1}}{}_{\bar{2}1\bar{1}} = \frac{\bar{\phi}^1\phi^2}{\chi^2}, \qquad R^{\bar{1}}{}_{\bar{1}2\bar{2}} = R^{\bar{1}}{}_{\bar{2}2\bar{1}} = -\frac{1 + \bar{\phi}^1\phi^1}{\chi^2}, \tag{PS.232}$$

$$R^{\bar{1}}{}_{\bar{2}1\bar{2}} = R^{\bar{1}}{}_{\bar{2}2\bar{2}} = 0, \tag{PS.233}$$

$$R^{\bar{2}}{}_{\bar{1}1\bar{1}} = R^{\bar{2}}{}_{\bar{1}2\bar{1}} = 0, \tag{PS.234}$$

$$R^{\bar{2}}_{\bar{1}1\bar{2}} = R^{\bar{2}}_{\bar{2}1\bar{1}} = -\frac{1 + \bar{\phi}^2 \phi^2}{\chi^2}, \qquad R^{\bar{2}}_{\bar{1}2\bar{2}} = R^{\bar{2}}_{\bar{2}2\bar{1}} = \frac{\bar{\phi}^2 \phi^1}{\chi^2} \tag{PS.235}$$

$$R^{\bar{2}}_{\bar{2}1\bar{2}} = 2\frac{\bar{\phi}^1 \phi^2}{\chi^2}, \qquad R^{\bar{2}}_{\bar{2}2\bar{2}} = -2\frac{1 + \bar{\phi}^1 \phi^1}{\chi^2}. \tag{PS.236}$$

Using the general definitions at the beginning of Section 6.1 we derive the Ricci tensor,

$$R_{1\bar{1}} = -\left(R^{\bar{1}}_{\bar{1}1\bar{1}} + R^{\bar{2}}_{\bar{2}1\bar{1}}\right) = 3\frac{1 + \bar{\phi}^2 \phi^2}{\chi^2},$$

$$R_{1\bar{2}} = -\left(R^{\bar{1}}_{\bar{1}1\bar{2}} + R^{\bar{2}}_{\bar{2}1\bar{2}}\right) = -3\frac{\bar{\phi}^1 \phi^2}{\chi^2}, \quad R_{2\bar{1}} = -\left(R^{\bar{1}}_{\bar{1}2\bar{1}} + R^{\bar{2}}_{\bar{2}2\bar{1}}\right) = -3\frac{\bar{\phi}^2 \phi^1}{\chi^2},$$

$$R_{2\bar{2}} = -\left(R^{\bar{1}}_{\bar{1}2\bar{2}} + R^{\bar{2}}_{\bar{2}2\bar{2}}\right) = 3\frac{1 + \bar{\phi}^1 \phi^1}{\chi^2}. \tag{PS.237}$$

This is to be compared with

$$R_{i\bar{j}} = -\partial_{\bar{j}}\partial_i \log(\mathrm{Det}) = 3\,\partial_{\bar{j}}\partial_i \log \chi = 3\,\frac{2}{g^2}G_{i\bar{j}}. \tag{PS.238}$$

Above,

$$\mathrm{Det} \equiv \det\{G_{i\bar{j}}\} \to \frac{1}{\chi^3}. \tag{PS.239}$$

The scalar curvature $\mathcal{R}$ is

$$\mathcal{R} = 2\,G^{\bar{j}i}\,R_{i\bar{j}} = \frac{2}{g^2}2\cdot 3\cdot 2 \longrightarrow 2N(N-1). \tag{PS.240}$$

The general equation (10.339) valid in CP($N - 1$) models then leads to

$$R_{i\bar{j}k\bar{\ell}} = -\frac{2}{g^2}\left(G_{i\bar{j}}G_{k\bar{\ell}} + G_{i\bar{\ell}}G_{k\bar{j}}\right) \tag{PS.241}$$

is also satisfied.

Exercise 6.3.1, page 266

The action to be analyzed is

$$S = \frac{2}{g^2}\int d^2x \frac{1}{(1 + \bar{\phi}\phi)^2}\,\partial_\mu \bar{\phi}\partial^\mu \phi, \tag{PS.242}$$

The topological term is omitted since it is a total derivative (see Problem 6.2.2 on page 671) and therefore does not contribute to the equations of motion. Equations of motion are determined by varying the action,

$$\delta S/\delta\phi = 0, \quad \delta S/\delta\bar{\phi} = 0. \tag{PS.243}$$

In normalization we used in (PS.242) the overall factor $2/g^2$ can be dropped in the subsequent derivation. We then obtain

$$\delta S \propto \int d^2x\,\delta\bar{\phi}\left\{-\frac{\partial^2 \phi}{(1 + \bar{\phi}\phi)^2} + 2\partial_\mu \phi\,\partial^\mu \phi\,\frac{\bar{\phi}}{(1 + \bar{\phi}\phi)^3} + \mathrm{H.c.}\right\} = 0, \tag{PS.244}$$

where integration by parts is performed in the first term.

As a result we arrive at

$$\partial^2 \phi - 2 \frac{\bar{\phi}\, \partial\phi\, \partial\phi}{(1 + \phi\bar{\phi})} = 0 \tag{PS.245}$$

plus a conjugated expression for $\bar{\phi}$. Note that (PS.245) can be presented as

$$\left(\partial + \Gamma\partial\phi\right)\partial\phi = 0, \tag{PS.246}$$

where Γ is given in (PS.210).

In the second part of this exercise we will derive equations of motion for an equivalent representation of the model given in (6.1) and (6.2). To this end, to implement the constraint 6.1 we will introduce a Lagrange multiplier λ in the following form:

$$S = \frac{1}{2g^2} \int d^2x [\partial S^a \partial S^a - \lambda (S^a S^a - 1)]. \tag{PS.247}$$

Then

$$\delta S = \frac{1}{g^2} \int d^2x [\partial \delta S^a \partial S^a - \lambda \delta S^a S^a]. \tag{PS.248}$$

Integrating by parts in the first term we obtain

$$\partial^2 S^a + \lambda S^a = 0. \tag{PS.249}$$

Multiplying both sides of the above equation by S^a we find λ (given Eq. (5.242)),

$$\lambda = -S^a \partial^2 S^a, \tag{PS.250}$$

implying the following final equation of motion:

$$\left(\delta^{ab} - S^a S^b\right) \partial^2 S^b = 0. \tag{PS.251}$$

Exercise 6.3.2, page 266

Supersymmetric extension of CP(1) model is discussed in detail in Section 10.12.3.4 on page 478. For our purposes we will need the following:

$$\mathcal{L} = G \left[\partial_\mu \phi^\dagger\, \partial^\mu \phi + i\bar{\psi}\gamma^\mu \left(\partial_\mu + \Gamma\partial_\mu\phi\right)\psi + \cdots \right]$$

$$= G \left[\partial_\mu \phi^\dagger\, \partial^\mu \phi + i\bar{\psi}\gamma^\mu \partial_\mu \psi - i\frac{2}{1 + \phi^\dagger\phi}\, \phi^\dagger \partial_\mu \phi\, \bar{\psi}\gamma^\mu \psi + \cdots \right], \tag{PS.252}$$

where

$$G = G_{1\bar{1}} = \frac{2}{g^2} \frac{1}{(1 + \phi^\dagger\phi)^2}, \tag{PS.253}$$

the dots stand for a four-fermion term irrelevant for our present purposes, and ψ is a Dirac spinor.

The overall structure of the ϕ dependence is completely fixed by geometry of CP(1) target space. Therefore, to determine renormalization of $1/g^2$ in G it is sufficient to analyze the vicinity of the origin in the target space, i.e. $\phi \approx 0$. In other words, we will trace the renormalization of the structure $\partial_\mu \phi^\dagger\, \partial^\mu \phi$ in $\mathcal{L}$ using the last term in (PS.252) as the interaction vertex. Then the only two-loop graph which has exactly the needed structure $\partial_\mu \phi^\dagger \, \partial^\mu \phi$ in front of it is obtained from the vertex

$$-i\,\frac{2}{1+\phi^\dagger\phi}\,\phi^\dagger\partial_\mu\phi\,\bar{\psi}\gamma^\mu\psi \tag{PS.254}$$

and its complex conjugated. It is shown in Fig. S.7. It has three propagators: two fermion and one boson. The denominator in (PS.254) can be ignored while the ϕ fields without derivatives in the numerator in (PS.254) and its complex conjugated are convoluted into the boson propagator.

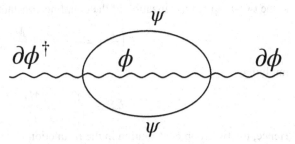

Fig. S.7   Two-loop contribution in the β function of supersymmetric CP(1) model.

For pedagogical purposes we split the calculation in two parts. First we evaluate the fermion loop depicted in Fig. S.8. In CP(1) this is the same famous loop which appears in two-dimensional Schwinger model, see Fig. S.8.

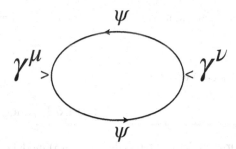

Fig. S.8   Fermion loop subdiagram.

Using

$$-i\,\frac{1}{2\pi}\frac{x_\mu\gamma^\mu}{x^2} \overset{\text{Fourier}}{\longleftrightarrow} \frac{i\partial_\mu\gamma^\mu}{p^2} \tag{PS.255}$$

for the fermion propagator we obtain for the graph in Fig. S.8

$$\frac{1}{2\pi^2}\frac{1}{x^2}\left(g^{\mu\nu}-\frac{2x^\mu x^\nu}{x^2}\right) \overset{\text{Fourier}}{\longleftrightarrow} \frac{i}{\pi}\left(\frac{p^\mu p^\nu}{p^2}-g^{\mu\nu}\right). \tag{PS.256}$$

Next, we return to the diagram of Fig. S.7. The extra factors to be taken into account are i/p^2 for the boson propagator, $(2/g^2)^2$ for two vertices, $(g^2/2)^3$ for three propagators, -2^2 reflecting the factor of 2 in (PS.254) and -2 in its complex conjugated, and the overall factor $-i$. The final result for the diagram in Fig. S.7 is

$$(\partial_\mu \phi^\dagger \, \partial^\mu \phi) \, \frac{g^2}{2\pi^2} \, \log \frac{M_0}{\mu}. \tag{PS.257}$$

As was explained in Section 6.3.5, this contribution must be canceled by the sum of all purely bosonic two-loop graphs. Hence, the two-loop renormalization in non-supersymmetric (*purely bosonic*) CP(1) model is

$$\delta \mathcal{L} = \ldots - (\partial_\mu \phi^\dagger \, \partial^\mu \phi) \, \frac{g^2}{2\pi^2} \, \log \frac{M_0}{\mu} + \cdots, \tag{PS.258}$$

where the dots stand for irrelevant contributions, M_0 is the UV cutoff and μ is the normalization point. Comparing $\delta \mathcal{L}$ above with $G(\partial_\mu \phi^\dagger \, \partial^\mu \phi)$ in (PS.252) we arrive at the two-loop renormalization of the coupling constant in CP(1),

$$\delta \left(\frac{2}{g^2} \right) = -\frac{g^2}{2\pi^2} \, \log \frac{M_0}{\mu} \tag{PS.259}$$

implying

$$\delta g^2 = \frac{g^6}{4\pi^2} \, \log \frac{M_0}{\mu}. \tag{PS.260}$$

Hence, the two-loop contribution in the β function is

$$\beta_2 = \frac{\partial (\delta g^2)}{\partial \log \mu} = -\frac{g^6}{4\pi^2}. \tag{PS.261}$$

This expression coincides with the last term in (6.73) at $N = 2$. Equation (PS.261) can be readily extended to arbitrary N by invoking the general statement[6] that

$$\beta_2 \propto R_{i\bar{k}l\bar{m}} R_{\bar{j}}^{\bar{k}l\bar{m}} \propto 2R_{i\bar{j}} \propto N G_{i\bar{j}}.$$

in non-supersymmetric CP($N - 1$) model. This implies that in CP($N - 1$)

$$\beta_2 = -\frac{g^6 N}{8\pi^2}, \tag{PS.262}$$

in full accord with Eq. (6.73).

Exercise 6.3.3, page 266

To warm up, let us first analyze the O(3) model (6.1), (6.2) using the most familiar spherical angles coordinates,

$$S^3 = r \cos \theta, \quad S^2 = r \sin \theta \sin \phi, \quad S^1 = r \sin \theta \cos \phi, \tag{PS.263}$$

where the radius of the target space sphere $r = 1/g$, θ and φ are the polar and azimuthal angles, respectively, and

$$\mathcal{L} = \frac{1}{2g^2} \partial \vec{S} \partial \vec{S} = \frac{1}{2g^2} \left[(\partial \theta)^2 + \sin^2 \theta (\partial \varphi)^2 \right]. \tag{PS.264}$$

The Lagrangian (PS.264) depends on two independent fields $\theta(x)$ and $\varphi(x)$. Now let us represent both fields as the sum of a background plus a small quantum correction,

$$\theta = \theta_0 + g q_0, \quad \varphi = \varphi_0 + g q_1 \tag{PS.265}$$

---

[6] See e.g. S. V. Ketov, *Quantum Non-Linear Sigma-Models*, (Springer, New York, 2000) or S. J. Graham, Three-loop β function for the bosonic nonlinear sigma model, *Phys. Lett.* B 197, 543 (1987).

with

$$\theta_0 = \frac{\pi}{2}, \tag{PS.266}$$

and $\varphi_0(x)$ remains unspecified. The only requirement is that φ_0 depends on x adiabatically.

Then the Lagrangian above takes the form

$$\mathcal{L} = \frac{1}{2g^2}\left[g^2(\partial q_0)^2 + g^2(\partial q_1)^2 + (1 - g^2 q_0^2)\left((\partial\varphi_0)^2 + 2g\partial\varphi_0\partial q_1\right)\right]$$

$$= \frac{1}{2g^2}(\partial\varphi_0)^2 + \frac{1}{2}\left[(\partial q_0)^2 + (\partial q_1)^2 - q_0^2(\partial\varphi_0)^2\right]. \tag{PS.267}$$

In the preceding expression we omitted all cubic and higher order terms in q irrelevant for one-loop calculation. We also omitted the linear in q term as required in the background field method. The quantum parts of the fields enter in the second line in (PS.267) quadratically. In the given background q_1 is sterile, and the only graph to be considered is the q_0 tadpole diagram in Fig. S.9. A straightforward calculation of this tadpole yields

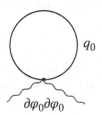

Fig. S.9 One-loop diagram determining the coupling constant renormalization in O(N) models.

$$\mathcal{L} + \delta_{1\text{-loop}}\mathcal{L} = \left(\frac{1}{2g^2} - \frac{1}{4\pi}\log\frac{M_0}{\mu}\right)(\partial\varphi_0)^2, \tag{PS.268}$$

implying the following one-loop coefficient in the beta function:

$$\beta = \frac{\partial g^2(\mu)}{\partial\log\mu} = \beta_1 g^4 + \cdots, \qquad \beta_1 = -\frac{1}{2\pi}. \tag{PS.269}$$

Needless to say, this result coincides with the one-loop beta function in (6.62) for CP(1).

Now we can turn to the general O(n) case, $n \geq 3$. The vector $\vec{S}$ has n components, while the constraint $\vec{S}\vec{S} = r^2 = 1/g^2$ allows us to parametrize the sphere S^{n-1} of dimension $n - 1$ by $n - 1$ angular variables

$$\theta_1, \ \theta_2, \ \theta_3, \ldots, \ \theta_{n-1}. \tag{PS.270}$$

In the previous exercise $n = 3$ we had $\theta_1, \to \theta$ and $\theta_2 \to \varphi$. The metric in the spherical parametrization is well-known, therefore we can write the full Lagrangian

$$\mathcal{L}\big|_{O(n)} = \frac{1}{2g^2}\Big[(\partial\theta_1)^2 + \sin^2\theta_1(\partial\theta_2)^2 + \sin^2\theta_1\sin^2\theta_2(\partial\theta_3)^2$$

$$+ \cdots + \sin^2\theta_1\sin^2\theta_2\ldots\sin^2\theta_{k-1}(\partial\theta_k)^2 + \cdots + \Big(\prod_{\ell=1}^{n-2}\sin^2\theta_\ell\Big)(\partial\theta_{n-1})^2\Big].$$

$$(PS.271)$$

As before, we will choose the following background field:

$$\theta_\ell = \frac{\pi}{2}, \text{ for all } \ell \in [1, n-2], \quad \theta_{n-1} = \varphi_0(x) + g q_{n-1}(x), \qquad (PS.272)$$

assuming $\varphi_0(x)$ to depend on x adiabatically. Moreover,

$$\theta_\ell(x) = \frac{\pi}{2} + q_\ell(x) \text{ for all } \ell \in [1, n-2]. \qquad (PS.273)$$

Then in the quadratic in q approximation (PS.271) reduces to (cf. Eq. (PS.267))

$$\mathcal{L}\big|_{O(n)} \to \frac{1}{2g^2}(\partial\varphi_0)^2 + \frac{1}{2}\Big[\sum_{\ell=1}^{n-1}(\partial q_\ell)^2 - \sum_{\ell=1}^{n-2}(\partial q_\ell)^2(\partial\varphi_0)^2\Big]. \qquad (PS.274)$$

In calculating the one-loop renormalization the only difference with (PS.267) is that now we have $n-2$ tadpole diagrams depicted in Fig. S.9. Correspondingly, for the $O(n)$ models, Eq. (PS.269) takes the form

$$\beta_1\big|_{O(n)} = -\frac{n-2}{2\pi}. \qquad (PS.275)$$

Note that at $n = 2$ the β function vanishes (to all orders in perturbation theory) in accord with the general statement that the $O(2)$ model reduces to a single free field.

Exercise 6.4.1, page 269

The topological charge Q in CP(1) model is defined in Eqs. (6.80) and (6.81). At first, to warm up we will calculate Q using Eqs. (6.47) and the one-instanton solution (6.49). Without loss of generality we can put $b = 0$ in this solution. Then

$$Q = \frac{1}{\pi}\int d^2x \left|\frac{\partial\phi}{\partial z}\right|^2 \frac{1}{(1 + \bar{\phi}\phi)^2} = \frac{1}{\pi}\int d^2x \frac{|a|^2}{(|z|^2 + |a|^2)^2}$$

$$\to \frac{1}{\pi}\int d^2x \frac{1}{(x^2 + 1)^2} = \int_0^\infty dr^2 \frac{1}{(r^2 + 1)^2} = 1. \qquad (PS.276)$$

In the second line we first rescaled the variable z and then passed to the radial variable $r^2 = |x|^2$. As was expected $Q = 1$.

Now, for reasons that will become clear shortly we will redo this calculation starting from (6.81). As we know from Exercise 6.2.2 we can rewrite (6.81) as

$$Q = -\frac{i}{2\pi}\int d^2x\, \varepsilon_{\mu\nu}\frac{\partial_\mu\bar{\phi}\partial_\nu\phi}{(1 + \bar{\phi}\phi)^2} = -\frac{i}{2\pi}\int d^2x\, \partial_\mu K_\mu,$$

$$K_\mu = \varepsilon_{\mu\nu}\frac{\bar{\phi}\,\partial_\nu\phi}{1 + \bar{\phi}\phi}. \qquad (PS.277)$$

Being a full divergence the last integral in the first line reduces to a "surface integral" (in fact, in the case at hand the "surface integral" is the length of a circle). If the circle radius is r, at large r the product $\vec{K}\vec{n}$ falls off as $1/r^3$ so that the "surface integral" tends to zero at $r \to \infty$. Here $\vec{n}$ is the unit vector perpendicular to the circle at a given point on the circle, $\vec{n} = -\{\cos\theta, \sin\theta\}$; see Fig. S.10.

However, near the origin $\vec{K}\vec{n} \sim \frac{1}{r}$ and therefore the "surface integral" does not vanish. Thus, Q reduces to

$$Q = -\frac{i}{2\pi}\left(\frac{i}{r}\right)2\pi r = 1. \tag{PS.278}$$

The origin of various factors in (PS.278) is as follows. The first factor is the overall factor in front of the integral in (PS.277). The last factor is the circle length. Finally, the factor in the parentheses is

$$\vec{K}\vec{n} = -(K_1\cos\theta + K_2\sin\theta) = -i\mathcal{J}e^{i\theta}, \tag{PS.279}$$

$$\mathcal{J} = \frac{\bar{\phi}(\partial\phi/\partial z)}{1 + \bar{\phi}\phi} = -\frac{1}{z}\left(\frac{|a|^2}{|z|^2 + |a|^2}\right)_{|z|\to 0} \to -\frac{1}{z}. \tag{PS.280}$$

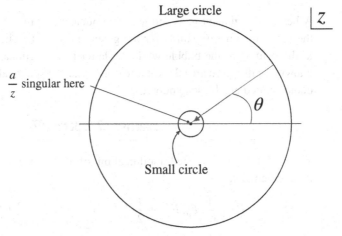

Fig. S.10 Complex z plane used in calculation in Exercise 6.4.1. Inside the small circle we see a perpendicular unit vector. The large circle contribution vanishes due to the fact that $\mathcal{J}$ vanishes at large distances from the origin (the instanton center) fast enough. The small circle contribution is given in (PS.278).

Why is this second method is more suitable for the topological charge calculation in the k-instanton solution? The cricial point is that the corresponding integral is saturated near the position of the instanton center. For k instantons we will have k independent contributions coming from the domains arbitrarily close to the instanton centers. In other words, our calculation is localized on the centers, each producing unity in Q. Thus, the k-instanton solution (6.86) will obviously yield $Q = k$.

Exercise 6.5.1, page 273

The proof is similar to that presented in Section 6.5.1. Some details are different. Now we have two conserved charges, Q and $Q^\dagger$, complex conjugated to each other.

Correspondingly, the fields ϕ and χ by some complex fields, e.g. $\phi \neq \phi^\dagger$. Note that χ and $\chi^\dagger$ may or may not be different. It does not matter whether $\chi \neq \chi^\dagger$ or $\chi = \chi^\dagger$. What is important is that Eq. (6.92) is satisfied, while Eq. (6.91) takes the form

$$\chi(x) = [Q^\dagger, \phi(x)], \quad \chi^\dagger(x) = -[Q, \phi^\dagger(x)].$$

Correspondingly, the correlation function (6.96) must be replaced by

$$\Pi^\mu(q) = -i \int e^{iqx}\, d^D x\, \langle \text{vac}|T\left\{J^{\dagger\,\mu}(x),\, \phi(0)\right\}|\text{vac}\rangle,$$

$$\Pi^{\dagger\,\mu}(q) = i \int e^{iqx}\, d^D x\, \langle \text{vac}|T\left\{J^\mu(x),\, \phi^\dagger(0)\right\}|\text{vac}\rangle. \tag{PS.281}$$

The subsequent steps are the same as in (6.97) and (6.98).

Chapter 7

Exercise 7.1.1, page 284

A heuristic argument is as follows. The nucleation process is obviously facilitated the larger the energy gain is for the given energy loss. The former is proportional to the volume of the bubble while the latter is proportional to the area. As is well known, for the given area the volume is maximal for spherical bubbles. This is clearly demonstrated e.g. by soap bubbles.

Exercise 7.2.1, page 297

To perform the requested modification of (7.51) we can proceed as follows. A modified action is

$$S = \int d^4x \left[\frac{1}{4g^2} F^a_{\mu\nu} F_a^{\mu\nu} + \frac{1}{2}(D_\mu \phi^a)^2 + \frac{1}{2}(D_\mu \chi^a)^2 + V(\chi, \phi)\right], \tag{PS.282}$$

where both fields ϕ^a and χ^a are in the adjoint representation of SU(2). Their interaction potential can be chosen as

$$V(\chi, \phi) = \tilde{\lambda}\left(\phi^a \phi^a - V^2\right)^2 + \frac{\lambda}{4}\left(\chi^a \chi^a - 2v^2\right)^2 + \gamma(\phi^a \chi^a)^2 \tag{PS.283}$$

with

$$V \gg v, \quad \lambda \sim \gamma \sim g^2, \quad \tilde{\lambda} \ll e^2, \text{ but } \tilde{\lambda}V^2 \gg v^2. \tag{PS.284}$$

At a high-energy scale V the field ϕ^3 develops a vacuum expectation value, $\phi^3 = V$ breaking SU(2)$\to$ U(1). In this vacuum $\phi^1_{\text{vac}} = \phi^2_{\text{vac}} = 0$, which can be viewed as a gauge condition while $\phi^3 = V$ ensures that $\chi^3_{\text{vac}} = 0$ by virtue of the last term in (PS.283). As usual, excitations of ϕ^1 and ϕ^2 are eaten up by very heavy $W^\pm$ bosons, with mass $\sim gV$. The mass of the χ^3 excitation is of the same order provided (PS.284) is satisfied. At energies lower than gV we remain with a massless photon A^3 and $\chi^{1,2}$ fields. The latter can be grouped in a single complex field X with charge 1,

$$X = \frac{1}{\sqrt{2}} \left(\chi^1 + i\chi^2 \right). \tag{PS.285}$$

The low-energy Lagrangian reduces to

$$\mathcal{L}_{\text{eff}} = \frac{1}{4g^2} F^3_{\mu\nu} F^3_{\mu\nu} + |(\partial_\mu - iA_\mu)X|^2 + \lambda \left(X^\dagger X - v^2 \right)^2. \tag{PS.286}$$

From this point on, the scenario of the flux tube formation is the same as was discussed in Section 7.2.3, with one significant exception. The field X which develops an expectation value at the scale v

$$|X| = v,$$

has the electric charge 1 rather than that of q considered in Section 7.2.3. The electric charge of q is 1/2. This seemingly insignificant distinction has an important consequence. Indeed, the magnetic flux of the winding-1 string for charge-1/2 is 4π; see Eqs. (3.16) and (3.17). The 't-Hooft–Polyakov monopole has the same magnetic flux. Therefore, this string can break through a monopole-antimonopole pair creation.

However, if a string is formed due to winding of charge-1 field, as is the case now, it carries flux 2π. Hence its breaking cannot be induced by this mechanism.

If we consider a charge-1 winding-2 string (i.e., composite), it does carry the 4π magnetic flux and therefore can break along the same lines as in Section 7.2.3. Some details will be different.

The monopole mass M_M will not change up to small corrections $O(v/V)$. As we know from Section 4.1.3, page 129, for critical monopoles

$$M_M = \frac{4\pi V}{g}.$$

Away from the point of criticality M_M slowly grows (Section 4.1.6). The constraint on $\tilde{\lambda}$ in (PS.284) guarantees that we are not too far away. Therefore, we can accept the avbove value.

As for the string tension, with the constraint (PS.284), i.e., $\lambda \sim g^2$, the tension T is close to critical. Equation (3.25) then implies that

$$T = \begin{cases} 2\pi v^2 & \text{charge } \frac{1}{2}, \\ 4\pi v^2 & \text{charge } 1. \end{cases} \tag{PS.287}$$

The breaking decay rates in the above two cases are given by the following formulas:

$$-\ln \Gamma_{\text{breaking}} = \begin{cases} \frac{8\pi^2}{g^2} \frac{V^2}{v^2}, & \text{charge } \frac{1}{2}, \text{ winding } 1, \\ \frac{4\pi^2}{g^2} \frac{V^2}{v^2}, & \text{charge } 1, \text{ winding } 2. \end{cases} \tag{PS.288}$$

The string tension T can be readily calculated away from the critical value of parameters, in the limit $m_X/m_\gamma \gg 1$ or $\lambda \gg e^2$. This is the so-called Abrikosov limit; see (3.35). The result of this calculation is as follows. The critical string tensions in (PS.287) must be multiplied by $\ln (m_X/m_\gamma)$.

Chapter 8

Exercise 8.1.1, page 316

The regularized currents are

$$j^\mu_{\text{reg}} = \bar{\psi}(t, x + \epsilon)\gamma^\mu \psi(t, x) \exp\left(i \int_x^{x+\epsilon} A_i \, dx\right),$$

$$(j^\mu)^5_{\text{reg}} = \bar{\psi}(t, x + \epsilon)\gamma^\mu \gamma^5 \psi(, x) \exp\left(i \int_x^{x+\epsilon} A_i \, dx\right). \tag{PS.289}$$

We need to show that they are invariant under the gauge transformations

$$\psi \rightarrow e^{i\alpha(t,x)}\psi, \quad A_\mu \rightarrow A_\mu + \partial_\mu \alpha(t, x) \tag{PS.290}$$

with an arbitrary phase $\alpha(t, x)$. The gauge transformed point-split currents can be written as

$$j^\mu_{\text{g.t.}} = e^{i\alpha(t,x) - i\alpha(t,x+\epsilon)} \bar{\psi}(x + \epsilon)\gamma^\mu \psi(x) \exp\left[i \int_x^{x+\epsilon} (A_1 + \partial_x \alpha(x)) \, dx\right]$$

$$= e^{-i\alpha(t,x+\epsilon) + i\alpha(t,x)} \exp\left(i\alpha\Big|_x^{x+\epsilon}\right) j^\mu_{\text{reg}} = j^\mu_{\text{reg}}. \tag{PS.291}$$

The same derivation applies to $j^{\mu 5}_{\text{reg}}$.

Exercise 8.2.1, page 323

The Lagrangian of $\mathcal{N} = 2$ two-dimensional CP(1) model is given in Eq. (10.321). The mass deformation we will need for the Pauli–Villars regularization is presented in Eq. (10.334).

Since *super*symmetry has not yet been discussed, we will forget about it for a while and will analyze the Lagrangian

$$\mathcal{L} = \frac{2}{g^2 \chi^2}\left[\partial_\mu \phi^\dagger \partial^\mu \phi + i\bar{\psi}\gamma^\mu \partial_\mu \psi - \frac{2i}{\chi}\phi^\dagger \partial_\mu \phi \bar{\psi}\gamma^\mu \psi + \frac{1}{\chi^2}(\bar{\psi}\psi)^2\right] \tag{PS.292}$$

per se. Here

$$\chi = 1 + \phi^\dagger \phi. \tag{PS.293}$$

• Equations of motion

First, we will derive Noether's theorem and the Noether current we need to analyze. Instead of doing it in full generality we will focus on linearly realized symmetries, for example on U(1).

We will start from the action

$$S = \int d^d x \, \mathcal{L}(\phi^i, \partial_\mu \phi^i), \tag{PS.294}$$

where $\{\phi^i(x)\}$ is a generic set of fields in the theory at hand, both bosonic and fermionic, d is the space-time dimension. The Euler–Lagrange equations of motion are derived by varying S with respect to all fields and requiring $\delta S = 0$, i.e., extremality of the action. The ordering of the fermion field is important. In quantum filed theory the equations of motion obtained in this way are to be viewed as quantum *operator* equations. Now,

$$
\begin{aligned}
\delta S &= \int d^d x \left\{ \frac{\delta \mathcal{L}}{\delta \phi^i} \delta \phi^i + \frac{\delta \mathcal{L}}{\delta \phi^i_{,\mu}} \delta \partial_\mu \phi^i \right\} \\
&= \int d^d x \left\{ \frac{\delta \mathcal{L}}{\delta \phi^i} - \partial_\mu \left(\frac{\delta \mathcal{L}}{\delta \phi^i_{,\mu}} \right) \right\} \delta \phi^i = 0,
\end{aligned}
\tag{PS.295}
$$

where $\phi^i_{,\mu} \equiv \partial_\mu \phi^i$ and variations $\delta \phi^i$ are arbitrary except a single constraint, $\delta \phi^i(x) \to 0$ at $|x| \to \infty$. This constraint allows one to perform integration by parts in (PS.295). Since the variations $\delta \phi^i$ are arbitrary, Eq. (PS.295) implies

$$
\frac{\delta \mathcal{L}}{\delta \phi^i} - \partial_\mu \left(\frac{\delta \mathcal{L}}{\delta \phi^i_{,\mu}} \right), \qquad \forall i.
\tag{PS.296}
$$

• Noether's theorem

If the action under discussion has a continuous *global* symmetry it automatically vanishes for certain limited variations of $\{\phi^i(x)\}$. Let us denote these deformations as follows:

$$
\phi^i(x) \to \phi^i(x) + \epsilon^i(x),
\tag{PS.297}
$$

and the corresponding variation of action as δS_ϵ. Then

$$
\begin{aligned}
0 = \delta S_\epsilon &= \int d^d x \left\{ \frac{\delta \mathcal{L}}{\delta \phi^i} \epsilon^i + \frac{\delta \mathcal{L}}{\delta \phi^i_{,\mu}} \partial_\mu \epsilon^i \right\} \\
&= \int d^d x \left\{ \partial_\mu \left(\frac{\delta \mathcal{L}}{\delta \phi^i_{,\mu}} \epsilon^i \right) + \left[\frac{\delta \mathcal{L}}{\delta \phi^i} - \partial_\mu \left(\frac{\delta \mathcal{L}}{\delta \phi^i_{,\mu}} \right) \right] \epsilon^i + \right\}.
\end{aligned}
\tag{PS.298}
$$

The preceding equality is valid for every set $\{\phi^i(x)\}$. However, if we use the equation of motion (PS.296) we come to the conclusion that

$$
0 = \int d^d x \, \partial_\mu \left(\frac{\delta \mathcal{L}}{\delta \phi^i_{,\mu}} \epsilon^i \right) = 0 = \int d^d x \, \partial_\mu J^\mu_{\epsilon^i},
\tag{PS.299}
$$

implying

$$
\partial_\mu J^\mu_{\epsilon^i}, \quad \text{where} \quad J^\mu_{\epsilon^i} = \frac{\delta \mathcal{L}}{\delta \phi^i_{,\mu}} \epsilon^i.
\tag{PS.300}
$$

The currents $J^\mu_{\epsilon^i}$ are referred to as the Noether currents. We see that the Noether currents are obtained from kinetic terms upon substituting appropriate variations ϵ^i.

• Classically conserved U(1) currents in (PS.292)

One of the obvious global U(1) symmetries of the theory (PS.292) is

$$
\begin{aligned}
\psi &\to e^{i\alpha} \psi, \quad \delta \psi = i\alpha \psi, \quad \epsilon = i\alpha \psi, \\
\bar{\psi} &\to e^{-i\alpha} \bar{\psi}, \quad \delta \bar{\psi} = -i\alpha \bar{\psi}.
\end{aligned}
\tag{PS.301}
$$

Using (PS.301) and (PS.292) we immediately identify the corresponding Noether current,

$$J_V^\mu = G\,\bar\psi\gamma^\mu\psi, \tag{PS.302}$$

where the subscript V tells us that this is the vector current and the metric

$$G = \frac{2}{g^2\chi^2}. \tag{PS.303}$$

Another classically conserved current can be obtained as follows. We write the Lagrangian in term of left and right components,

$$\psi_L = \frac{1}{2}(1-\gamma^5)\psi, \qquad \psi_R = \frac{1}{2}(1+\gamma^5)\psi, \tag{PS.304}$$

where $\gamma^5 = \mathrm{diag}\{-1,1\}$; see Section 10.2.2. It is easy to check that the Lagrangian (PS.292) is invariant under the global *chiral* rotations

$$\psi_L \to e^{i\alpha}\psi_L, \ \ \psi_R \to e^{-i\alpha}\psi_R \tag{PS.305}$$

The kinetic term has the form

$$\mathcal{L}_{\text{kin ferm}} = G\,i\left[\psi_L^\dagger(\partial_0 - \partial_x)\psi_L + \psi_R^\dagger(\partial_0 + \partial_x)\psi_R\right]. \tag{PS.306}$$

Using (PS.300) we arrive at

$$J_A^\mu = G\,\bar\psi\gamma^\mu\gamma^5\psi. \tag{PS.307}$$

The subscript A marks the axial-vector current.

• Verifying classical conservation

We need to write down relevant equations of motion following from (PS.292). They are

$$i\gamma^\mu D_\mu\psi + \frac{2}{\chi^2}(\bar\psi\psi)\psi = 0, \quad -i\gamma^\mu\bar D_\mu\bar\psi + \frac{2}{\chi^2}(\bar\psi\psi)\bar\psi = 0. \tag{PS.308}$$

Here

$$D_\mu = \partial_\mu + \Gamma\partial_\mu\phi, \tag{PS.309}$$

where the Christoffel symbol is defined in Section 10.12.3.4 (see also Section 10.6.7).

Let us calculate the divergence of the current (PS.302). Using the identity

$$\partial_\mu G = G(\Gamma\partial_\mu\phi + \bar\Gamma\partial_\mu\phi^\dagger), \tag{PS.310}$$

we arrive at

$$\partial_\mu J_V^\mu = G(\bar\psi\gamma^\mu D_\mu\psi + \bar\psi\,\overleftarrow{D}_\mu\,\gamma^\mu\psi). \tag{PS.311}$$

Substituting Eq. (PS.308) in (PS.311) we conclude that indeed $\partial_\mu J_V^\mu = 0$.

Now, let us carry out the same verification for J_A^μ. Then

$$\begin{aligned}
\partial_\mu J_A^\mu &= G(\bar\psi\gamma^\mu\gamma^5 D_\mu\psi + \bar\psi\,\overleftarrow{D}_\mu\,\gamma^\mu\gamma^5\psi) \\
&= G(-\bar\psi\gamma^5\gamma^\mu D_\mu\psi + \bar\psi\,\overleftarrow{D}_\mu\,\gamma^\mu\gamma^5\psi) \\
&= -i\frac{4}{\chi^2}(\bar\psi\psi)(\bar\psi\gamma^5\psi) = 0
\end{aligned} \tag{PS.312}$$

due to the fact that the product $(\bar\psi\psi)(\bar\psi\gamma^5\psi)$ vanishes for anti-commuting fields.

From other similar examples we may expect that while $\partial_\mu J_V^\mu$ continues to vanish at the quantum level, this is not the case for the axial current because of the quantum anomaly.

Let us try to guess the form of this anomaly before actual calculation. The fact that we know the topological term will prompt us the answer. The current J_A^μ has mass dimension 1, implying that the divergence $\partial_\mu J_A^\mu$ must have mass dimension 2. Also it must be a pseudoscalar built from the bosonic fields and be invariant in the target space. Then the answer is unique:

$$\partial_\mu J_A^\mu \propto \frac{1}{\chi^2} \left(\partial_\mu \phi^\dagger\right)\left(\partial_\nu \phi\right)\varepsilon^{\mu\nu}. \tag{PS.313}$$

All we need to do is to calculate the one-loop coefficient. To this end we will use the Pauli–Villars regularisation similar to that presented in Section 8.1.7. As we know from this section in two dimensions relevant diagrams are "diangle" rather than triangle; see Fig. 8.4b.

Following Eq. (10.334) I will introduce a massive regulator field R with the same interactions as ψ. I copy a part of the Lagrangian (10.334) (with replacement $\psi \to R$) relevant for our current calculation:

$$\mathcal{L}_R = G\left[i\bar{R}\gamma^\mu \partial_\mu R + \mathcal{A}_\mu \bar{R}\gamma^\mu R - M_R \bar{R}R\right], \tag{PS.314}$$

$$\mathcal{A}_\mu = -\frac{2i}{\chi}\phi^\dagger \partial_\mu \phi, \quad M_R = \frac{1-\phi^\dagger\phi}{\chi}M_0. \tag{PS.315}$$

At the very end, we will tend $M_R \to \infty$. The concrete expression for M_R is not important since it cancels from the anomalous contribution. Do not forget that the R loop requires an extra minus compared to the same loop with the ψ field.

If we choose the background field ϕ in the same way as in Section 6.3.2 (see Eq. (6.47)) then the metric G is x independent and it can be absorbed in the definition of the ψ, R fields. The subsequent calculation is the same as in the Schwinger model; see Section 8.1.7 and, in particular. Eq. (8.43). As a result, we obtain

$$\partial_\mu J_A^\mu = -\frac{1}{\pi}\partial_\mu \mathcal{A}_\nu \,\varepsilon^{\mu\nu} = \frac{2i}{\pi}\partial_\mu\left(\frac{1}{\chi}\phi^\dagger \partial_\nu \phi\right)\varepsilon^{\mu\nu}$$

$$= \frac{1}{\chi^2}\frac{2i}{\pi}\partial_\mu \phi^\dagger \partial_\nu \phi \,\varepsilon^{\mu\nu}. \tag{PS.316}$$

Thus, we confirm Eq. (PS.313) and establish the proportionality coefficient $2i/\pi$.

Exercise 8.2.2, page 323

- **Gauge potential for constant $F_{\mu\nu}$**

If $F_{\alpha\beta}$ is space-time independent the easiest way to prove that the gauge potential can be chosen as

$$A_\mu(x) = \frac{1}{2}x^\alpha F_{\alpha\mu} \tag{PS.317}$$

is to start from the Taylor expansion

$$A_\mu(x) = A_\mu(0) + x^\alpha C_{\alpha\mu} + x^\alpha x^\beta C_{\alpha\beta\mu} + \cdots. \tag{PS.318}$$

where the coefficients C with two, three, or more indices are constants. Calculating $F_{\alpha\mu}(x)$ we immediately observe that x independence of $F_{\alpha\mu}$ requires all coefficients C with three or more indices to vanish, implying

$$A_\mu(x) = A_\mu(0) + x^\alpha C_{\alpha\mu}, \quad F_{\mu\nu} = C_{\mu\nu} - C_{\nu\mu}. \tag{PS.319}$$

It is convenient to impose the following gauge condition from the very beginning:

$$x^\mu A_\mu(x) = 0, \tag{PS.320}$$

which is known as the *Fock–Schwinger gauge*. Combining the last two equations we conclude that in this gauge

$$x^\mu A_\mu(0) + x^\mu x^\alpha C_{\alpha\mu} = 0, \tag{PS.321}$$

which leads in turn to

$$A_\mu(0) = 0, \quad C_{\alpha\mu} = -C_{\mu\alpha}. \tag{PS.322}$$

Substituting these results in (PS.319) we arrive at (PS.317).

This expansion can be readily generalized to the case of x dependent $F_{\mu\nu}$ see, chapter II in V. A. Novikov, M. A. Shifman, A. I. Vainshtein, and V. I. Zakharov, *Fortsch. Phys.* **32**, 585 (1984).

• **Massless fermion propagator in a constant $F_{\mu\nu}$**

The defining equation for the fermion Green's function in the coordinate space is

$$\left(i\frac{\partial}{\partial x^\mu}\gamma^\mu + A_\mu(x)\gamma^\mu\right) S(x,0) = \delta^{(4)}(x), \tag{PS.323}$$

where $A_\mu(x)$ is given in (PS.317). Next we split S in two parts, the free fermion propagator S_0 and a correction S_1 of the first order in $F_{\mu\nu}$,

$$S = S_0 + S_1, \quad S_0 = \frac{1}{2\pi^2}\frac{x_\mu\gamma^\mu}{x^4} \tag{PS.324}$$

while S_1 is due to propagation in the external field which is assumed to be small. The Green's function is normalized as follows:

$$S(x,y) = -i\langle T\{\psi(x)\bar\psi(y)\}\rangle. \tag{PS.325}$$

Substituting (PS.324) in (PS.323) we obtain

$$i\partial_\mu\gamma^\mu S_1(x,0) = -\frac{1}{4\pi^2}x^\alpha F_{\alpha\mu}\gamma^\mu\frac{x_\alpha\gamma^\alpha}{x^4}. \tag{PS.326}$$

The solution of the above equation is

$$S_1(x,0) = -\frac{1}{8\pi^2}\frac{x^\alpha}{x^2}\tilde{F}_{\alpha\beta}\gamma^\beta\gamma^5 \tag{PS.327}$$

plus irrelevant terms. Note that the choice of the gauge condition (PS.320) breaks the Lorentz invariance. It is restored only in the final answer for gauge-invariant quantities. For further details see the paper by Novikov *et al.* quoted after Eq. (PS.322).

Exercise 8.3.1, page 327

If N is large and we wish to saturate the anomaly by a baryon loop, we would expect each quark loop to be suppressed by a power of $1/N$ entailing an exponential suppression of the type e^{-N}. This cannot realize the linear dependence in N seen in the anomaly calculation, which indicates that the symmetry is spontaneously broken.

Exercise 8.5.1, page 334

Applying $CS_{\text{shift}}C$ to $\Psi(a)$ we arrive at

$$CS_{\text{shift}}C\,\Psi(a) = \Psi(a - \pi)$$

$$= -\Psi(a + \pi) = -S_{\text{shift}}\Psi(a), \tag{PS.328}$$

where the Bloch condition

$$\Psi(a + 2\pi) = e^{i\theta}\,\Psi(a) \tag{PS.329}$$

with $\theta = \pi$ is used in the second equality. Since $C^2 = 1$, from (PS.328) we conclude that

$$S_{\text{shift}}C\,\Psi(a) = -CS_{\text{shift}}\Psi(a) \tag{PS.330}$$

or

$$S_{\text{shift}}C = -CS_{\text{shift}}. \tag{PS.331}$$

The fact that $S_{\text{shift}}C \neq CS_{\text{shift}}$ precludes the ground state wave function $\Psi(a)$ from being simultaneously the eigenfunction for the both operators, S_{shift} and C. This means that there are two *degenerate* ground states. For instance, we can choose them [7] as the eigenfunctions for S_{shift},

$$S_{\text{shift}}\Psi^{1,2}(a) \equiv \Psi^{1,2}(a + \pi) = \pm i\,\Psi^{1,2}(a). \tag{PS.332}$$

Then the C symmetry is spontaneously broken, and transforms Ψ^1 into Ψ^2 and *vice versa*,

$$C\Psi^{1,2} = \Psi^{2,1}, \tag{PS.333}$$

so that the C-eigenfunctions are $\sqrt{\frac{1}{2}}(\Psi^1 \pm \Psi^2)$, with the corresponding eigenvalues ± 1. Alternatively, one could start from the C-eigenstates $(\Psi^1 \pm \Psi^2)$. Then S_{shift} acting on $(\Psi^1 + \Psi^2)$ will produce $(\Psi^1 - \Psi^2)$ and vice versa.

---

[7] The eigenvalues $\pm i$ in (PS.332) follow from the fact that S_{shift}^2 corresponds to the shift $a \to a + 2\pi$.

Chapter 9

Exercise 9.4.1, page 374

Our task is to find the energy eigenvalues for the Schrödinger equation

$$H\psi_k(x) = E_k\psi_n(x),$$

$$H = 2M - \frac{1}{M}\frac{d^2}{dx^2} + \frac{12\pi M^2}{N}x. \tag{PS.334}$$

Here $M/2$ is the reduced mass for the $\bar{n}n$ pair and x is defined in the interval $x \in [0, \infty]$. The fact that x cannot be negative follows from Fig. 9.32. The vacua far to the left and far to the right must be the lowest energy density states (i.e., vac 1 in the notation of Fig. 9.32) and, therefore, $\bar{n}$ cannot be put to the right of n. Choosing the origin on the x axis at the $\bar{n}n$ center of mass we find the coordinate of n at $x/2$ while that of $\bar{n}$ at $-x/2$. Then the distance between n and $\bar{n}$ is $x = x_n - x_{\bar{n}}$. The boundary conditions on the wave functions are

$$\psi_n(0) = 0, \quad \psi_n(x \to \infty) \to 0. \tag{PS.335}$$

It is convenient to introduce a dimensionless coordinate and rescaled energy eigenvalues, namely,

$$y = M\left(12\pi N^{-1}\right)^{1/3} x \quad \text{and} \quad \mathcal{E}_k = \frac{E_k - 2M}{M}\left(\frac{N}{12\pi}\right)^{2/3}. \tag{PS.336}$$

Then Eq. (PS.334) takes the form

$$\left(-\frac{d^2}{dy^2} + y\right)\psi_k(y) = \mathcal{E}_k\psi_k(y). \tag{PS.337}$$

The solution to the above equation is provided by the *Airy* function. To satisfy the boundary conditions at $y \in [0, \infty]$ we should choose

$$\psi_k(y) = \text{Ai}(y - \rho_k), \quad k = 0, 1, 2, 3\ldots \tag{PS.338}$$

where $\text{Ai}(y)$ is the Airy function (i.e., $-\text{Ai}''(y) + y\text{Ai}(y) = 0$). Moreover, $-\rho_k$ is the k-th root of the Airy function (all of them are real and negative, so $\rho_k > 0$). Equation (PS.338) implies that automatically,

$$\psi_k(y = 0) = 0 \quad \text{and} \quad \psi_k(y \to \infty) \to 0$$

and the rescaled energy eigenvalues are

$$\mathcal{E}_k = \rho_k, \quad \rho_0 \approx 2.34, \quad \rho_1 \approx 4.09, \quad \rho_2 \approx 5.52, \ldots, \tag{PS.339}$$

and

$$E_k = 2M + \left(\frac{12\pi}{N}\right)^{2/3} M\mathcal{E}_k. \tag{PS.340}$$

At large values of k (i.e., $k \gg 1$) we have[8]

$$\rho_k \approx \left(\frac{3\pi}{2}k\right)^{2/3}. \tag{PS.341}$$

Thus, we arrive at

$$E_k = 2M + \left(18\pi^2\right)^{2/3} \left(\frac{k}{N}\right)^{2/3} M. \tag{PS.342}$$

Exercise 9.4.2, page 374

Decay of $k = 1$ false vacuum into the $k = 0$ true ground state (see Fig. 9.33) proceeds via nucleation of a kink–antikink pair. What happens is the reverse of Fig. 9.32 on page 373, i.e., vac 1 $\leftrightarrow$ vac 2. In other words, the true vacuum is inside the kink pair and the false vacuum is outside the pair. To calculate the decay rate in the quasiclassical approximation $N \gg 1$ we will need the contents of Sections 7.1, 9.4.2, 9.4.4. The difference in the energy densities between vac 2 and vac 1 can be inferred from Eq. (9.72),

$$\mathcal{E} = \mathcal{E}_{\text{vac 2}} - \mathcal{E}_{\text{vac 1}} = \frac{12\pi M^2}{N}, \tag{PS.343}$$

while the kink mass M replaces the parameter μ in Eqs. (7.8) and (7.9) or the parameter T at the bottom of Eq. (7.20). The the critical action is

$$S_* = \frac{\pi M^2}{\mathcal{E}} = \frac{N}{12}, \qquad S_* \to \infty \text{ at } N \to \infty. \tag{PS.344}$$

Since the critical action is large, the quasiclassical approximation is applicable. Thus, the exponential factor in $d\Gamma$ is $e^{-N/12}$.

One can also establish the pre-exponential factor.[*] That yields the decay rate per unit length per unit interval of time

$$d\Gamma = \frac{3e^2}{2\pi} \frac{1}{N} M^2 \exp\left(-\frac{N}{12}\right). \tag{PS.345}$$

In the pre-exponent of the preceding formula $e \approx 2.718\ldots$.

Exercise 9.5.1, page 387

The Coleman theorem discussed in Section 6.5.2 (also known as Coleman–Mermin–Wagner theorem) does not preclude *sterile* massless particles. If a particle has no interactions, infrared effects cannot shift the position of the pole in its propagator from $p^2 = 0$ to $p^2 \neq 0$. Sterile particles can be neither emitted nor absorbed, therefore they cannot transfer any physical signal. As we know, at $N \to \infty$ all mesons (including pion) become sterile.

An interesting question is what happens at large but finite N. The answer to this question was given in Ian Affleck, *Nucl. Phys.* B **265** [FS15], 448 (1986). At large but finite N the axial U(1) symmetry remains *unbroken*. The two-point function of

---

[8] Equation (PS.341) could also be obtained through quasiclassical quantization. A more exact formula valid even at $k \sim 1$ is $\rho_k \approx \left(\frac{9\pi}{8} + \frac{3\pi}{2}k\right)^{2/3}$.

[*] See A. Monin, M. Voloshin, Phys. Rev. D **78**, 065048, 2008.

the appropriate order parameter at large separation x scales as $\frac{1}{|x|^{1/N}} \rightarrow 0$ at $|x| \rightarrow \infty$. The latter expression still exhibits the absence of the mass gap. Thus, there is a massless excitation, but it is not a *Goldstone* boson unless $N = \infty$. For a brief discussion see Section 6.5.2.

Exercise 9.5.2, page 387

The equation we consider is

$$\mathcal{E}\psi(z) = \left[\pi\lambda|z| - i\gamma^5\partial_z + \gamma^0(m + \Sigma)\right]\psi(z). \tag{PS.346}$$

In the limit $m \gg \sqrt{\lambda}$ we can ignore the term containing Σ. Then, in matrix notation we have ($2D$ γ matrices are defined on page 375, Eq. (9.79))

$$\begin{pmatrix} \mathcal{E} - \pi\lambda|z| - i\partial_z & im \\ -im & \mathcal{E} - \pi\lambda|z| + i\partial_z \end{pmatrix} \begin{pmatrix} \psi_1 \\ \psi_2 \end{pmatrix} = 0. \tag{PS.347}$$

which in turn entails

$$(\mathcal{E} - \pi\lambda|z| - i\partial_z)\psi_1 + im\psi_2 = 0,$$

$$-im\psi_1 + (\mathcal{E} - \pi\lambda|z| + i\partial_z)\psi_2 = 0. \tag{PS.348}$$

The second equation in (PS.348) can be used to express ψ_1 in terms of ψ_2. In this way we eliminate ψ_1 and are left with the equation for a single component $\psi_2(z)$.

Next, we must take into account that in the non-relativistic limit

$$\mathcal{E}^2 = (m + E)^2 \longrightarrow m^2 + 2mE,$$

and other terms of order

$$O\left(\frac{\lambda}{m}\right) \quad \text{and} \quad O\left(\frac{E}{m}\right)$$

must be neglected provided they are non-leading, e.g. in the expression

$$2\pi\lambda|z|(m + E) \longrightarrow 2\pi m\lambda|z|.$$

In this way we straightforwardly recover Eq. (9.98).

Exercise 9.6.1, page 392

Using the standard prescriptions of *dimensional regularization* (e.g. M. Peskin and D. Schroeder, *An Introduction to Quantum Field Theory*, (Addison-Wesley, 1995), pages 249–252) we observe that the leading (logarithmic, or $1/\varepsilon$) divergence cancels and arrive at

$$i\,\Pi^{\mu\nu} = \int_0^1 dx \left\{ -4ie^2 \frac{1}{(4\pi)^{D/2}} \Gamma\left(2 - \frac{D}{2}\right) \left(g^{\mu\nu}p^2 - p^\mu p^\nu\right) \right.$$

$$\left. \times x(1 - x) \left(\frac{1}{m^2 - x(1 - x)p^2}\right)^{2 - \frac{D}{2}} \right\}$$

$$\stackrel{D=2}{\longrightarrow} -i\,\frac{e^2}{\pi}\left(g^{\mu\nu}p^2 - p^\mu p^\nu\right)\int_0^1 dx \frac{x(1 - x)}{m^2 - x(1 - x)p^2}. \tag{PS.349}$$

The limiting cases are

$$i\,\Pi^{\mu\nu} = \begin{cases} \text{If } |p^2| \gg m^2 : & i\,\frac{e^2}{\pi}\left(g^{\mu\nu} - \frac{p^\mu p^\nu}{p^2}\right), \text{ cf. Eq. (9.149);} \\[2mm] \text{If } |p^2| \ll m^2 : & -i\,\frac{e^2}{6\pi}\left(\frac{g^{\mu\nu}p^2 - p^\mu p^\nu}{m^2}\right). \end{cases} \tag{PS.350}$$

Transition between the two regimes occurs at $|p^2| \sim m^2$.

Chapter 10

Exercise 10.2.1, page 427

To solve this problem we will need to use Eq. (10.34). Moreover,

$$U = \exp\left(-\frac{1}{2}\omega_{\mu\nu}\sigma^{\mu\nu}\right) = 1 - \frac{1}{2}\omega_{\mu\nu}\sigma^{\mu\nu} + \cdots \tag{PS.351}$$

Start from

$$\omega_{\mu\nu}\sigma^{\mu\nu} = 2\omega_{0i}\sigma^{0i} + \sum_{i,j=1}^{3}\omega_{ij}\sigma^{ij}$$

$$= -\omega_{0i}\sigma^i - \frac{i}{2}\,\omega_{ij}\varepsilon^{ijk}\sigma^k. \tag{PS.352}$$

Comparing Eqs. (PS.351), (PS.352) with (10.4) and (10.5) we arrive at

$$\omega^{0i} = (\mathbf{n}')^i\,\phi, \quad \omega^{ij} = \theta\varepsilon^{ijk}n^k \tag{PS.353}$$

where Eqs. (10.24) and (10.34) are used. The matrix ω^{ij} being antisymmetric is also *real*.

Exercise 10.2.2, page 427

$$\Sigma^{\mu\nu} = \frac{1}{4}\,[\gamma^\mu, \gamma^\nu] = \begin{pmatrix} \sigma^{\mu\nu} & 0 \\ 0 & \bar{\sigma}^{\mu\nu} \end{pmatrix}. \tag{PS.354}$$

The easiest way to derive Eq. (PS.354) is to use Eqs. (10.3), (10.7), (10.35), (PS.353), and the Weyl representation for the γ matrices; see (10.40).

Exercise 10.2.3, page 427

If a real Majorana spinor exists, then $\psi^* = \psi$. This property should be maintained under all Lorentz transformations. However, $\Sigma^{\mu\nu}$ (see Eq. (PS.354)) is in general complex, and thus some Lorentz transformations would ruin the equality $\psi^* = \psi$. For the Majorana property to be preserved, we require $\Sigma^{\mu\nu}$ to be purely real, which entails in turn that the γ matrices must be either purely real or purely imaginary. If we add the condition that the Dirac equation with a *real* wave function remains intact under complex conjugation we are left only with the latter option.

Consider first the mass term

$$\mathcal{L}_m = -\frac{m}{2}\left(\eta^\alpha \eta_\alpha + \bar\eta_{\dot\alpha}\bar\eta^{\dot\alpha}\right). \tag{PS.355}$$

In the Majorana basis (Section 10.2.1) we must be able to rewrite it as

$$\mathcal{L}_m = -\frac{m}{2}\left(\bar\lambda\lambda\right)_M = -\frac{m}{2}\left(\lambda^T \gamma^0 \lambda\right)_M. \tag{PS.356}$$

Comparison of the last two expressions will allow us to determine γ^0 in the Majorana representation. Although this is a straigtforward algebraic calculation it is rather tedious and, instead, we will use $(\gamma^0)_M$ from Exercise 10.2.4 and verify the coincidence of (PS.355) and (PS.356), namely,

$$\left(\frac{1}{2\sqrt2}\right)^2 \left[(a+a^*)^T, \ (b+b^*)^T\right]\begin{pmatrix} 0 & \sigma^2 \\ \sigma^2 & 0 \end{pmatrix}\begin{pmatrix} a+a^* \\ b+b^* \end{pmatrix}$$

$$= \frac{1}{8}\left[\eta(1+i)^2(1-\sigma^2)^2\eta - \eta(1+i)^2(1+\sigma^2)^2\eta + \text{H.c.}\right] = \eta^\alpha \eta_\alpha + \text{H.c.} \tag{PS.357}$$

The spinors a and b must be taken from Eqs. (10.52) and (10.53).

Next, we will compare the kinetic terms,

$$\mathcal{L}_{\text{kin}} = i\bar\eta_{\dot\beta}\,(\bar\sigma^\mu)^{\dot\beta\alpha}\,\partial_\mu \eta_\alpha \quad \text{vs.} \quad \frac{i}{2}(\lambda^T\gamma^0\gamma^\mu \partial_\mu \lambda)_M. \tag{PS.358}$$

We start from

$$\frac{i}{2}(\lambda^T\partial_t\lambda)_M = \frac{i}{2}\frac{1}{8}8(\eta\partial_t\bar\eta + \bar\eta\partial_t\eta) \to i\bar\eta_{\dot\beta}(\bar\sigma^0)^{\dot\beta\alpha}\,\partial_t\eta_\alpha. \tag{PS.359}$$

For other components in the kinetic term the proof of equality is analogous.

Exercise 10.2.4, page 427

The *Weyl* (chiral) representation for the γ matrices is presented in Eqs. (10.40) and (10.44). In the *Dirac* (standard) representation we have

$$\gamma^0 = \begin{pmatrix} 1 & 0 \\ 0 & -1 \end{pmatrix}, \quad \gamma^k = \begin{pmatrix} 0 & -\sigma^k \\ \sigma^k & 0 \end{pmatrix}, \quad \gamma^5 = \begin{pmatrix} 0 & 1 \\ 1 & 0 \end{pmatrix}. \tag{PS.360}$$

The set of the γ matrices in the *Majorana* representation is

$$\gamma^0 = \begin{pmatrix} 0 & \sigma^2 \\ \sigma^2 & 0 \end{pmatrix}, \quad \gamma^1 = \begin{pmatrix} -i\sigma^3 & 0 \\ 0 & -i\sigma^3 \end{pmatrix}, \quad \gamma^2 = \begin{pmatrix} 0 & \sigma^2 \\ -\sigma^2 & 0 \end{pmatrix},$$

$$\gamma^3 = \begin{pmatrix} i\sigma^1 & 0 \\ 0 & i\sigma^1 \end{pmatrix}, \quad \gamma^5 = \begin{pmatrix} \sigma^2 & 0 \\ 0 & -\sigma^2 \end{pmatrix}. \tag{PS.361}$$

The unitary transformations between these sets are

$$(\gamma^\mu)_W = \left(U\gamma^\mu U^\dagger\right)_D, \quad U_{W\leftarrow D} = U_{D\leftarrow W} = \frac{1}{\sqrt2}\begin{pmatrix} 1 & 1 \\ 1 & -1 \end{pmatrix};$$

$$(\gamma^\mu)_M = \left(U\gamma^\mu U^\dagger\right)_D, \quad U_{M\leftarrow D} = \frac{1}{\sqrt2}e^{i\pi/4}\begin{pmatrix} 1 & \sigma^2 \\ \sigma^2 & -1 \end{pmatrix}. \tag{PS.362}$$

Exercise 10.4.1, page 436

The relevant Jacobi identities of the supersymmetry algebra are

$$[[Q_\alpha, P_\mu], P_\nu] + [[P_\mu, P_\nu], Q_\alpha] + [[P_\nu, Q_\alpha], P_\mu] = 0,$$

$$[[P_\rho, P_\mu], P_\nu] + [[P_\mu, P_\nu], P_\rho] + [[P_\nu, P_\rho], P_\mu] = 0,$$

$$[[M_{\mu\nu}, M_{\rho\sigma}], M_{\tau\gamma}] + [[M_{\rho\sigma}, M_{\tau\gamma}], M_{\mu\nu}] + [[M_{\tau\gamma}, M_{\mu\nu}], M_{\rho\sigma}] = 0,$$

$$[\{Q_\alpha, Q_\beta\}, P_\mu] = \{Q_\alpha, [Q_\beta, P_\mu]\} + \{Q_\beta, [Q_\alpha, P_\mu]\}. \tag{PS.363}$$

The above equations can be supplemented by the standard Poincaré algebra (see Eq. (1.75) on page 36),

$$[P_\mu, P_\nu] = 0, \quad [M_{\mu\nu}, P_\lambda] = -i \left(g_{\mu\lambda} P_\nu - g_{\nu\lambda} P_\mu \right),$$

$$[M_{\mu\nu}, M_{\rho\sigma}] = -i \left(g_{\mu\rho} M_{\nu\sigma} - g_{\nu\rho} M_{\mu\sigma} + g_{\mu\sigma} M_{\rho\nu} - g_{\nu\sigma} M_{\rho\mu} \right). \tag{PS.364}$$

Now, we can reduce the first Jacobi identity to

$$[[Q_\alpha, P_\mu], P_\nu] + [[P_\nu, Q_\alpha], P_\mu] = 0. \tag{PS.365}$$

Let us assume for a short while that

$$[P_\mu, Q_\alpha] = c(\sigma_\mu)_{\alpha\dot\beta} \bar{Q}^{\dot\beta}, \quad [P_\mu, \bar{Q}^{\dot\beta}] = c'(\bar\sigma_\mu)^{\dot\beta\alpha} Q_\alpha \tag{PS.366}$$

where c, c' are numerical constants. Substituting (PS.366) in (PS.365) we arrive at

$$[[Q_\alpha, P_\mu], P_\nu] + [[P_\nu, Q_\alpha], P_\mu] = -c(\sigma_\mu)_{\alpha\dot\beta}[\bar{Q}^{\dot\beta}, P_\nu] + c(\sigma_\nu)_{\alpha\dot\beta}[\bar{Q}^{\dot\beta}, P_\mu]$$

$$= c\left((\sigma_\mu\bar\sigma_\nu)_\alpha^\beta - (\sigma_\nu\bar\sigma_\mu)_\alpha^\beta\right)Q_\beta = c(\sigma_{\mu\nu})_\alpha^\beta Q_\beta = c[M_{\mu\nu}, Q_\alpha]. \tag{PS.367}$$

Comparing with (PS.365) we conclude that $c = 0$. By the same token one can check that $c' = 0$ too.

Exercise 10.4.2, page 436

In the Majorana representation we have four real supercharges instead of Q and $\bar{Q}$. Let us denote them as $\{q_1, q_2, q_3, q_4\}$ with all q_i real. Using Eqs. (10.52) and (10.53) we can write

$$q_1 = \frac{1}{2\sqrt{2}}\left\{(1+i)(Q_1 - iQ_2) + (1-i)(\bar{Q}_1 + i\bar{Q}_2)\right\},$$

$$q_2 = \frac{1}{2\sqrt{2}}\left\{(1+i)(Q_2 + iQ_1) + (1-i)(\bar{Q}_2 - i\bar{Q}_1)\right\},$$

$$q_3 = -\frac{1}{2\sqrt{2}}\left\{(1+i)(Q_1 + iQ_2) + (1-i)(\bar{Q}_1 - i\bar{Q}_2)\right\},$$

$$q_4 = -\frac{1}{2\sqrt{2}}\left\{(1+i)(Q_2 - iQ_1) + (1-i)(\bar{Q}_2 + i\bar{Q}_1)\right\}. \tag{PS.368}$$

After a straightforward algebra, we conclude that the superalgebra (10.74), (10.75) takes the form

$$\{q_i, q_j\} = 2 \, (\gamma^\mu \gamma^0)_{ij} \, P_\mu. \tag{PS.369}$$

Exercise 10.5.1, page 441

We have to show that

$$\bar{D}_{\dot\alpha} x_L^\mu = D_\alpha x_R^\mu = 0, \tag{PS.370}$$

where

$$D_\alpha = \frac{\partial}{\partial \theta^\alpha} - i \bar\theta^{\dot\alpha} \partial_{\alpha\dot\alpha}, \quad \bar{D}_{\dot\alpha} = -\frac{\partial}{\partial \bar\theta^{\dot\alpha}} + i \theta^\alpha \partial_{\alpha\dot\alpha} \tag{PS.371}$$

and

$$x_L^\mu = x^\mu - i \theta^\alpha (\sigma^\mu)_{\alpha\dot\alpha} \bar\theta^{\dot\alpha}, \quad x_R^\mu = x^\mu + i \theta^\alpha (\sigma^\mu)_{\alpha\dot\alpha} \bar\theta^{\dot\alpha}. \tag{PS.372}$$

We have

$$\bar{D}_{\dot\alpha} x_L^\mu = \left(-\frac{\partial}{\partial \bar\theta^{\dot\alpha}} + i \theta^\alpha \partial_{\alpha\dot\alpha} \right) \left[x^\mu - i \theta^\alpha (\sigma^\mu)_{\alpha\dot\alpha} \bar\theta^{\dot\alpha} \right]$$

$$= i \theta^\alpha \partial_{\alpha\dot\alpha} x^\mu + i \frac{\partial}{\partial \bar\theta^{\dot\alpha}} \left(\theta^\alpha (\sigma^\mu)_{\alpha\dot\beta} \bar\theta^{\dot\beta} \right) = i \theta^\alpha \sigma^\nu_{\alpha\dot\alpha} \partial_\nu x^\mu + i \frac{\partial}{\partial \bar\theta^{\dot\alpha}} \left(\theta^\alpha (\sigma^\mu)_{\alpha\dot\beta} \bar\theta^{\dot\beta} \right)$$

$$= i \theta^\alpha (\sigma^\mu)_{\alpha\dot\alpha} - i \theta^\alpha (\sigma^\mu)_{\alpha\dot\alpha} = 0, \tag{PS.373}$$

where the minus sign in the last line reflects the Grassmann anticommuting,

$$\frac{\partial}{\partial \bar\theta} \, \theta = -\theta \frac{\partial}{\partial \bar\theta}. \tag{PS.374}$$

In the same manner,

$$D_\alpha x_R^\mu = \left(\frac{\partial}{\partial \theta^\alpha} - i \bar\theta^{\dot\alpha} \partial_{\alpha\dot\alpha} \right) \left[x^\mu + i \theta^\alpha (\sigma^\mu)_{\alpha\dot\alpha} \bar\theta^{\dot\alpha} \right]$$

$$= -i \bar\theta^{\dot\alpha} (\sigma^\nu)_{\alpha\dot\alpha} \partial_\nu (x^\mu) + \frac{\partial}{\partial \theta^\alpha} \left(i \theta^\beta (\sigma^\mu)_{\beta\dot\alpha} \bar\theta^{\dot\alpha} \right)$$

$$= -i \bar\theta^{\dot\alpha} (\sigma^\mu)_{\alpha\dot\alpha} + i (\sigma^\mu)_{\alpha\dot\alpha} \bar\theta^{\dot\alpha} = 0. \tag{PS.375}$$

Consider now

$$D_\alpha x_L^\mu = \left(\frac{\partial}{\partial \theta^\alpha} - i \bar\theta^{\dot\alpha} \partial_{\alpha\dot\alpha} \right) \left[x^\mu - i \theta^\alpha (\sigma^\mu)_{\alpha\dot\alpha} \bar\theta^{\dot\alpha} \right]$$

$$= -i \frac{\partial}{\partial \theta^\alpha} \left[\theta^\beta (\sigma^\mu)_{\beta\dot\alpha} \bar\theta^{\dot\alpha} \right] - i \bar\theta^{\dot\alpha} (\sigma^\nu)_{\alpha\dot\alpha} \partial_\nu x^\mu$$

$$= -i (\sigma^\mu)_{\alpha\dot\alpha} \bar\theta^{\dot\alpha} - i (\sigma^\mu)_{\alpha\dot\alpha} \bar\theta^{\dot\alpha} = -2i (\sigma^\mu)_{\alpha\dot\alpha} \bar\theta^{\dot\alpha}, \tag{PS.376}$$

and similarly

$$\bar{D}_{\dot\alpha} x_R^\mu = 2i \, \theta^\alpha (\sigma^\mu)_{\alpha\dot\alpha}. \tag{PS.377}$$

Exercise 10.5.2, page 441

In fact, we need to prove only the first two equalities in (10.122). The second pair of equalities is obtained by complex conjugation. We start from

$$\{D_\alpha, Q_\beta\} = \left\{\frac{\partial}{\partial\theta^\alpha} - i\bar{\theta}^{\dot{\alpha}}\partial_{\alpha\dot{\alpha}}, -i\frac{\partial}{\partial\theta^\beta} + \bar{\theta}^{\dot{\beta}}\partial_{\beta\dot{\beta}}\right\}$$

$$= \left(-i\left\{\frac{\partial}{\partial\theta^\alpha}, \frac{\partial}{\partial\theta^\beta}\right\} + \left\{\frac{\partial}{\partial\theta^\alpha}, \bar{\theta}^{\dot{\beta}}(\sigma^\nu)_{\beta\dot{\beta}}\partial_\nu\right\}\right.$$ (PS.378)

$$\left. - \left\{\bar{\theta}^{\dot{\alpha}}(\sigma^\nu)_{\alpha\dot{\alpha}}\partial_\nu, \frac{\partial}{\partial\theta^\beta}\right\} - i\left\{\bar{\theta}^{\dot{\alpha}}(\sigma^\nu)_{\alpha\dot{\alpha}}\partial_\nu, \bar{\theta}^{\dot{\beta}}(\sigma^\mu)_{\beta\dot{\beta}}\partial_\mu\right\}\right).$$

Each of the four terms above clearly vanishes because of anticommutation of the Grassmann numbers. Analogously,

$$\{D_\alpha, \bar{Q}_{\dot{\beta}}\} = \left\{\frac{\partial}{\partial\theta^\alpha} - i\bar{\theta}^{\dot{\alpha}}\partial_{\alpha\dot{\alpha}}, i\frac{\partial}{\partial\bar{\theta}^{\dot{\beta}}} - \theta^\beta\partial_{\beta\dot{\beta}}\right\}$$

$$= \left(i\left\{\frac{\partial}{\partial\theta^\alpha}, \frac{\partial}{\partial\bar{\theta}^{\dot{\beta}}}\right\} - \left\{\frac{\partial}{\partial\theta^\alpha}, \theta^\beta(\sigma^\nu)_{\beta\dot{\beta}}\partial_\nu\right\}\right.$$

$$\left. + \left\{\bar{\theta}^{\dot{\alpha}}(\sigma^\nu)_{\alpha\dot{\alpha}}\partial_\nu, \frac{\partial}{\partial\bar{\theta}^{\dot{\beta}}}\right\} + i\left\{\bar{\theta}^{\dot{\alpha}}(\sigma^\nu)_{\alpha\dot{\alpha}}\partial_\nu, \theta^\beta(\sigma^\mu)_{\beta\dot{\beta}}\partial_\mu\right\}\right)$$

$$= -(\sigma^\nu)_{\alpha\dot{\beta}}\partial_\nu + (\sigma^\nu)_{\alpha\dot{\beta}}\partial_\nu = 0.$$ (PS.379)

Only the cross terms in the second and the third lines contribute to the fourth line.

Exercise 10.5.3, page 441

The general vector superfield has the form

$$V(x, \theta, \bar{\theta}) = C + i\theta\chi - i\bar{\theta}\bar{\chi} + \frac{i}{\sqrt{2}}\theta^2 M - \frac{i}{\sqrt{2}}\bar{\theta}^2\bar{M}$$

$$- 2\theta^\alpha\bar{\theta}^{\dot{\alpha}}v_{\alpha\dot{\alpha}} + \left[2i\theta^2\bar{\theta}_{\dot{\alpha}}\left(\bar{\lambda} - \frac{i}{4}\partial^{\dot{\alpha}\alpha}\chi_\alpha\right) + \text{H.c.}\right]$$

$$+ \theta^2\bar{\theta}^2\left(D - \frac{1}{4}\partial^2 C\right),$$ (PS.380)

(see Eq. (10.110)). Then, using (10.105) we obtain

$$\delta C = i(\epsilon\chi - \bar{\epsilon}\bar{\chi}),$$

$$\delta\chi_\alpha = \sqrt{2}M\epsilon_\alpha + 2i v_{\alpha\dot{\alpha}}\bar{\epsilon}^{\dot{\alpha}} - (\partial_{\alpha\dot{\alpha}}C)\bar{\epsilon}^{\dot{\alpha}},$$

$$\delta M = 2\sqrt{2}\bar{\epsilon}_{\dot{\alpha}}\bar{\lambda}^{\dot{\alpha}} - i\sqrt{2}\bar{\epsilon}_{\dot{\alpha}}\partial^{\alpha\dot{\alpha}}\chi_\alpha,$$

$$\delta v_{\alpha\dot{\alpha}} = \left[-\frac{1}{2}\epsilon^\beta(\partial_{\beta\dot{\alpha}}\chi_\alpha) + \frac{1}{2}\epsilon_\alpha\partial_{\beta\dot{\alpha}}\chi^\beta - 2i\epsilon_\alpha\bar{\lambda}_{\dot{\alpha}}\right] + \text{H.c.},$$

$$\delta\lambda_\alpha = i\,\epsilon_\alpha D + \frac{1}{2}\,\epsilon_\beta\,\partial^{\dot\alpha\beta} v_{\alpha\dot\alpha} - \frac{1}{2}\epsilon_\beta\,\partial_{\alpha\dot\beta}\,v^{\dot\beta\beta},$$

$$\delta D = \epsilon^\alpha\partial_{\alpha\dot\alpha}\bar\lambda^{\dot\alpha} + \text{H.c.} \tag{PS.381}$$

Exercise 10.6.1, page 458

We can begin from the Lagrangian

$$\mathcal{L} = \partial_\mu\bar\phi\partial^\mu\phi - |\tilde m|^2\phi^2 + i\bar\psi\partial_\mu\bar\sigma^\mu\psi - \frac{\tilde m}{2}\psi^2 - \frac{\bar{\tilde m}}{2}\bar\psi^2, \tag{PS.382}$$

where $\tilde m$ is a complex mass parameter and ψ is a Weyl spinor. By writing $m = me^{i\alpha}$ we can absorb the phase $e^{i\alpha}$ in the definition of the fields ψ and $\bar\psi$. Then Eq. (PS.382) reduces to

$$\mathcal{L} = \partial_\mu\bar\phi\partial^\mu\phi - m^2\phi^2 + i\bar\psi\partial_\mu\bar\sigma^\mu\psi - \frac{m}{2}\psi^2 - \frac{m}{2}\bar\psi^2, \tag{PS.383}$$

and we can use Eqs. (PS.355), (PS.356), and (PS.358) to pass to the Majorana representation,

$$\mathcal{L} = \partial_\mu\bar\phi\partial^\mu\phi - m^2\phi^2 + \frac{i}{2}\,\bar\lambda\partial_\mu\gamma^\mu\lambda - \frac{m}{2}\bar\lambda\lambda, \tag{PS.384}$$

which implies, in turn, that both masses, boson and fermion, equal m.

Exercise 10.6.2, page 458

To obtain the desired result one needs to expand the potentials—Kähler and superpotential—around the scalar values of the superfield Φ. Expanding first the Kähler potential, we arrive at

$$\mathcal{K}\big(\Phi^i,\bar\Phi^j\big) \to \left[(\Phi^i - \phi^i)(\bar\Phi^j - \bar\phi^j)\frac{\partial^2\mathcal{K}}{\partial\bar\phi^i\partial\bar\phi^j}\right]$$

$$+ \frac{1}{2}(\Phi^i - \phi^i)(\bar\Phi^j - \bar\phi^j)(\bar\Phi^k - \bar\phi^k)\frac{\partial^3 K}{\partial\phi^i\partial\bar\phi^j\partial\bar\phi^k} + \text{H.c.}$$

$$+ \frac{1}{4}(\Phi^i - \phi^i)(\Phi^j - \phi^j)(\bar\Phi^k - \bar\phi^k)(\bar\Phi^m - \bar\phi^m)\frac{\partial^4\mathcal{K}}{\partial\phi^i\partial\phi^j\partial\bar\phi^k\partial\bar\phi^m}. \tag{PS.385}$$

To obtain the Lagrangian we must isolate the $\theta^2\bar\theta^2$ terms in (PS.385) – these are the terms needed to saturate the integral over superspace. We obtain

$$\star_1 (\Phi^i - \phi^i)(\bar\Phi^j - \bar\phi^j) \to \theta^2\bar\theta^2 F^i\bar F^j + (\theta\sigma^\mu\bar\theta)(\theta\sigma^\nu\bar\theta)\partial_\mu\phi^i\partial_\nu\bar\phi^j$$

$$+ [\sqrt{2}\theta(-i\theta\sigma^\mu\bar\theta)\partial_\mu\psi\bar\theta\bar\psi + \text{H.c.}],$$

$$\star_2 (\Phi^i - \phi^i)(\bar\Phi^j - \bar\phi^j)(\bar\Phi^k - \bar\phi^k) + \text{H.c.}$$

$$\to \theta^2\bar\theta^2 F^i\bar\psi^j\bar\psi^k + 2i(\theta\psi^i)(\theta\sigma^\mu\bar\theta)\partial_\mu\bar\phi^j(\bar\theta\bar\psi^k) + \text{H.c.},$$

$$\star_3 (\Phi^i - \phi^i)(\Phi^j - \phi^j)(\bar\Phi^k - \bar\phi^k)(\bar\Phi^m - \bar\phi^m) \to \theta^2\bar\theta^2\psi\psi\bar\psi\bar\psi. \tag{PS.386}$$

Combining (PS.385) and (PS.386) we obtain that (PS.386$\star_1$) produces

$$\Delta_1 = G_{i\bar{j}}(\partial_\mu \phi \partial^\mu \bar{\phi} + i\bar{\psi}^{\bar{j}} \bar{\sigma}^\mu \partial_\mu \psi^i + F^i \bar{F}^{\bar{j}}), \tag{PS.387}$$

Equations (PS.386)$\star_2$ and (PS.390) generate

$$\Delta_2 = \left(-\frac{1}{2} F^i \Gamma^{\bar{j}}_{\bar{k}\bar{m}} \bar{\psi}^{\bar{k}} \bar{\psi}^{\bar{m}} + \bar{F}^{\bar{j}} G^{i\bar{k}} \frac{\partial \bar{W}}{\partial \bar{\phi}^{\bar{k}}} + \text{H.c.} \right) + iG_{i\bar{j}} \bar{\psi}^{\bar{j}} \bar{\sigma}^\mu \partial_\mu \phi^k \Gamma^i_{kl} \psi^l, \tag{PS.388}$$

while (PS.386$\star_3$) produces

$$\Delta_3 = \frac{\partial^2 G_{i\bar{j}}}{\partial \phi^k \partial \bar{\phi}^{\bar{l}}} \psi^i \psi^k \bar{\psi}^{\bar{j}} \bar{\psi}^{\bar{l}}. \tag{PS.389}$$

For the derivation of the previous expressions we expanded the superportential

$$\mathcal{W}(\Phi^i) \rightarrow (\Phi^i - \phi^i) \frac{\partial \mathcal{W}}{\partial \phi^i} + \frac{1}{2}(\Phi^i - \phi^i)(\Phi^j - \phi^j) \frac{\partial^2 \mathcal{W}}{\partial \phi^i \partial \phi^j}$$

$$\rightarrow \theta^2 F^i \frac{\partial \mathcal{W}}{\partial \phi^i} - \theta^2 \psi^i \psi^j \frac{\partial^2 \mathcal{W}}{\partial \phi^i \partial \phi^j}. \tag{PS.390}$$

Now, we assemble all terms $\Delta_{1,2,3}$ and (PS.390) with appropriate coefficients shown in (PS.385) and eliminate the auxiliary F terms by virtue of the equations of motion. Using the definitions in Eqs. (10.168)–(10.170) we arrive at the Lagrangian (10.165).

The reader is advised to verify all numerical coefficients and the emergence of the covariant derivative and the Riemann tensor in the second line of (10.165).

Exercise 10.6.3, page 458

The supersymmetry transformations in (PS.381) and (10.185) look different because the latter is given in the Wess–Zumino gauge while in the former the supergauge is not imposed.

Exercise 10.6.4, page 458

This exercise is studied in the text in appendix section 10.27.2 and we include the solution here for completeness. We have

$$G = \frac{\xi}{4} \frac{1}{\sqrt{1 + \phi\bar{\phi}}}, \tag{PS.391}$$

and the general equation for the two-sheet hyperboloid with U(1) symmetry is

$$\frac{z^2}{b} - \frac{a}{b}(x^2 + y^2) = 1, \tag{PS.392}$$

where a and b are positive parameters, x, y, z are real coordinates and SO(2) symmetry acts in the x, y plane. Alternatively we can write

$$z^2 - a\left(x^2 + y^2\right) = b. \tag{PS.393}$$

Switching to polar coordinates we have

$$z(r) = \sqrt{b + ar^2}, \qquad r = \sqrt{x^2 + y^2}, \tag{PS.394}$$

and therefore

$$z \to \begin{cases} \sqrt{b} + \frac{ar^2}{2\sqrt{b}} + \ldots, & r \to 0, \\ \sqrt{ar} + \frac{b}{2\sqrt{ar}} + \ldots, & r \to \infty. \end{cases} \tag{PS.395}$$

Parameterizing $\phi = \rho e^{i\alpha}$ we rewrite the metric (PS.391) as follows:

$$Gd\phi d\bar{\phi} = \frac{d\phi d\bar{\phi}}{\sqrt{1 + \phi\bar{\phi}}} = \frac{d\rho^2}{\sqrt{1 + \rho^2}} + \frac{\rho^2 d\alpha^2}{\sqrt{1 + \rho^2}}. \tag{PS.396}$$

The interval on the surface of the hyperboloid under discussion is

$$ds^2 = (dz(r))^2 + dr^2 + r^2 d\alpha^2$$

$$= dr^2 \left[1 + \left(\frac{dz}{dr} \right)^2 \right] + r^2 d\alpha^2. \tag{PS.397}$$

Comparing (PS.396) and (PS.397) one obtains from the angular part

$$r^2 = \frac{\rho^2}{\sqrt{1 + \rho^2}}. \tag{PS.398}$$

Comparison of the radial parts yields

$$\frac{1}{\sqrt{1 + \rho^2}} = \left(\frac{dr}{d\rho} \right)^2 \left[1 + \left(\frac{dz}{dr} \right)^2 \right], \tag{PS.399}$$

implying the following equation for $z(r)$:

$$\frac{dz(r)}{dr} = \frac{\rho \sqrt{1 + \frac{3}{4}\rho^2}}{1 + \frac{\rho^2}{2}},$$

$$\rho^2 = \frac{1}{2} \left(r^4 + \sqrt{r^8 + 4r^4} \right). \tag{PS.400}$$

The limiting cases of small and large r are obvious,

$$z(r) \to \begin{cases} \frac{1}{2}r^2 + O(r^4), & r \to 0, \\ \sqrt{3}\, r + O\left(\frac{1}{r^3} \right), & r \to \infty, \end{cases} \tag{PS.401}$$

up to additive constants of integration.

Exercise 10.9.1, page 466

First of all, remember, in this problem we nullify the superpotential' i.e., we will deal with the $m = 0$ theory. Supersymmetric QED with the vanishing superpotential has a flat direction; see Section 10.6.10. We start from the parametrization

$$q = \sqrt{\xi}\, e^{i\alpha} \cosh \rho, \qquad \tilde{q} = \sqrt{\xi}\, e^{i\alpha} \sinh \rho.$$

Then

$$q\tilde{q} = \frac{\xi}{2} e^{2i\alpha} \sinh 2\rho \equiv \frac{\xi}{2} \varphi, \quad \varphi = e^{2i\alpha} \sinh 2\rho, \quad \bar{\varphi}\varphi = \sinh^2 2\rho.$$

The relevant mass term in the Lagrangian is

$$\Delta\mathcal{L}_m = \left[i\sqrt{2}\left(\lambda\psi\right)\bar{q} + \text{H.c.} \right] - \left[i\sqrt{2}\left(\lambda\tilde{\psi}\right)\bar{\tilde{q}} + \text{H.c.} \right]. \tag{PS.402}$$

In Eq. (PS.402) we must rescale the field λ,

$$\lambda \to e\,\lambda,$$

in order to make its kinetic term canonically normalized. Then, substituting the above expressions for q and $\tilde{q}$ we get

$$\Delta\mathcal{L}_m \longrightarrow e\, i\,\sqrt{2}\, e^{-i\alpha} \left[\left(\lambda\psi\right)\sqrt{\xi}\cosh\rho - \left(\lambda\tilde{\psi}\right)\sqrt{\xi}\sinh\rho \right] + \text{H.c.} \tag{PS.403}$$

The phase factor $i\,e^{-i\alpha}$ can be absorbed in ψ, $\tilde{\psi}$. We will omit it hereafter.

Let us introduce

$$\eta = \frac{\psi\cosh\rho - \tilde{\psi}\sinh\rho}{\sqrt{\cosh 2\rho}}, \qquad \tilde{\eta} = \frac{\psi\sinh\rho + \tilde{\psi}\cosh\rho}{\sqrt{\cosh 2\rho}}. \tag{PS.404}$$

Then the kinetic terms of λ, η, and $\tilde{\eta}$ are canonic, i.e.

$$\bar{\lambda}\mathcal{D}\lambda + \bar{\eta}\mathcal{D}\eta + \bar{\tilde{\eta}}\mathcal{D}\tilde{\eta},$$

while the mass term becomes

$$\Delta\mathcal{L}_m = e\sqrt{2\xi}\sqrt{\cosh 2\rho}\left(\lambda\eta + \bar{\lambda}\bar{\eta}\right). \tag{PS.405}$$

Note that the $\tilde{\eta}$ field stays *massless*. Inspecting Eq. (PS.405) we observe that the diagonal combinations are

$$\lambda_{\pm} = \frac{\lambda\pm\eta}{\sqrt{2}}, \qquad \lambda\eta \equiv \frac{1}{4}\left(\lambda_+^2 - \lambda_-^2\right) \tag{PS.406}$$

while the mass for these diagonal combination is

$$e\sqrt{2\xi}\sqrt{\cosh 2\rho} = e\sqrt{2\xi}\left(1 + \bar{\varphi}\varphi\right)^{1/4}. \tag{PS.407}$$

Supersymmetry is unbroken in this case – the boson masses will be the same. In particular, the mass in (PS.407) will determine the masses of the Higgsed photon and a real scalar field from the photon supermultiplet; cf. Section 10.9 and in particular Eq. (10.233).

Exercise 10.10.1, page 470

First, let us analyze the fermion mass matrix. We consider the case $\xi > m^2/e^2$; hence the vacuum fields are $\tilde{q} = 0$ and $q = \left(\xi - \frac{m^2}{e^2}\right)^{1/2}$, where the parameter m is assumed to be real and positive.

The fermion mass term can be extracted from Eq. (10.191). To ensure that the kinetic term of λ is normalized canonically we must replace $\lambda \to e\lambda$. In fact, it is convenient to absorb the phase factor i into λ too,[9] so that the replacement is as follows:

$$\lambda \to -i\,e\,\lambda. \tag{PS.408}$$

With this substitution the fermion mass term in the Lagrangian takes the form

$$\Delta\mathcal{L}_m = i\sqrt{2}e\,(\lambda\psi)\sqrt{\xi - \frac{m^2}{e^2}} + m(\tilde{\psi}\psi). \tag{PS.409}$$

This mass term can be represented as the following matrix:

$$\left(\lambda,\ \psi,\ \tilde{\psi}\right)\begin{pmatrix} 0 & \frac{e}{\sqrt{2}}\sqrt{\xi - \frac{m^2}{e^2}} & 0 \\ \frac{e}{\sqrt{2}}\sqrt{\xi - \frac{m^2}{e^2}} & 0 & \frac{m}{2} \\ 0 & \frac{m}{2} & 0 \end{pmatrix}\begin{pmatrix} \lambda \\ \psi \\ \tilde{\psi} \end{pmatrix}. \tag{PS.410}$$

This mass matrix can be easily diagonalized. It has one zero eigenvalue and two nonvanishing eigenvalues. The massless Goldstino corresponding to the vanishing eigenvalue is

$$\left(\frac{2e^2\xi}{m^2} - 1\right)^{-1/2}\left[\lambda - \left(\frac{2e^2\xi}{m^2} - 2\right)^{1/2}\tilde{\psi}\right], \tag{PS.411}$$

where we included the normalization factor in front of the square brackets. Two diagonal combinations with nonvanishing mass terms are

$$\frac{1}{\sqrt{2}}\frac{1}{\sqrt{\xi - \frac{m^2}{2e^2}}}\left(\sqrt{\xi - \frac{m^2}{e^2}}\,\lambda \pm \sqrt{\xi - \frac{m^2}{2e^2}}\,\psi + \frac{m}{\sqrt{2}e}\tilde{\psi}\right), \tag{PS.412}$$

where the corresponding mass terms are

$$\pm\frac{e}{\sqrt{2}}\sqrt{\xi - \frac{m^2}{2e^2}}. \tag{PS.413}$$

- In the case at hand, supersymmetry is broken and therefore the fermion masses do not coincide with those of the boson states. Now we will examine the boson masses. Since in the vacuum $\tilde{q} = 0$, fluctuations of this field coincide with the field itself. For the field q we write

$$q = \sqrt{\xi - \frac{m^2}{e^2}} + \delta q, \tag{PS.414}$$

(δq is real!). We start from the scalar potential in (10.190) and expand it (up to quadratic terms) in δq and $\tilde{q}$,

$$V(\delta q, \tilde{q}) = \mathcal{E}_{\mathrm{vac}} + 2e^2\left(\sqrt{\xi - \frac{m^2}{e^2}}\right)(\delta q)^2 + 2m^2|\tilde{q}|^2 + \cdots, \tag{PS.415}$$

---

[9] If this purely imaginary factor were not absorbed, the derivation of the fermion mass spectrum would be somewhat more complicated; the mass matrix to be considered would have to include not only λ, ψ and $\tilde{\psi}$ but also complex conjugated of these fields. It would be 6 by 6, rather than 3 by 3. Needless to say, the final answer will be the same.

where the ellipses denote cubic and higher-order terms. From (PS.415) we see that the mass of the δq quantum is (this is a real field)

$$m(\delta q) = \sqrt{2}\, e \sqrt{\xi - \frac{m^2}{2e^2}}, \tag{PS.416}$$

while that of the complex field $\tilde{q}$ is

$$m(\tilde{q}) = \sqrt{2}\, m. \tag{PS.417}$$

The small-ξ case, Eq. (10.256), is solvable along the same lines.

Exercise 10.12.1, page 483

In this exercise the reader is asked to show that the Lagrangian of CP(1)

$$\mathcal{L} = G\left[\partial_\mu \phi^\dagger \partial^\mu \phi - |m|^2 \phi^\dagger \phi + i\bar{\psi}\gamma^\mu \partial_\mu \psi - \frac{1 - \phi^\dagger \phi}{\chi}\bar{\psi}\mu\psi \right.$$

$$\left. - \frac{2i}{\chi}\phi^\dagger \partial_\mu \phi \bar{\psi}\gamma^\mu \psi + \frac{1}{\chi^2}(\bar{\psi}\psi)^2 \right], \qquad \chi = 1 + \phi^\dagger \phi \tag{PS.418}$$

is equivalent to that of O(3) in Eq. (10.331) under the following map,

$$\phi = \frac{S^1 + iS^2}{1 + S^3}, \quad \psi = \frac{\chi^1 + i\chi^2}{1 + S^3} - \frac{S^1 + iS^2}{(1 + S^3)^2}\chi^3, \tag{PS.419}$$

where χ^a are Majorana spinors and

$$G = \frac{2}{g^2\chi^2} \quad \text{and} \quad \mu = \frac{1}{2}m(1 + \gamma^5) + \frac{1}{2}\bar{m}(1 - \gamma^5).$$

| Below we will assume m to be real. |
Note first that under this map

$$\chi = \frac{2}{(1 + S^3)}. \tag{PS.420}$$

In the following we will also use the constraints

$$S^a S^a = 1, \quad S^a \chi^a = 0. \tag{PS.421}$$

Let us begin with the scalar derivative term

$$\partial_\mu \phi^\dagger = -\frac{1}{(1 + S^3)^2}\partial_\mu S^3 (S^1 - iS^2) + \frac{1}{(1 + S^3)}\left(\partial_\mu S^1 - i\partial_\mu S^2\right). \tag{PS.422}$$

After simple algebra we obtain

$$\partial_\mu \phi^\dagger \partial^\mu \phi = \frac{\partial_\mu S^a \partial^\mu S^a}{(1 + S^3)^2} \tag{PS.423}$$

and hence

$$G\partial_\mu \phi^\dagger \partial^\mu \phi = \frac{2}{g^2 \chi^2}\partial_\mu \phi^\dagger \partial^\mu \phi = \frac{1}{2g^2}\partial_\mu S^a \partial^\mu S^a. \tag{PS.424}$$

Next, let us study

$$\phi^\dagger \partial\phi = -\frac{1}{(1 + S^3)^2}\left[\partial S^3 - i\left(S^1 \partial S^2 - S^2 \partial S^1\right)\right]. \tag{PS.425}$$

Now we address the fermion terms. It is important that χ^i is a Majorana spinor, i.e., $(\chi^1)^2 = (\chi^2)^2 = (\chi^3)^2 = 0$. Also note that the product $\gamma^0 \gamma^1 = \gamma^5 = -\sigma_3$ is diagonal. Therefore, the spinorial indices of χ^i need not be explicitly shown. Using the constraints (PS.421) we arrive at

$$\psi^\dagger \Gamma \psi = 2i \frac{\chi^1 \Gamma \chi^2}{\left(1 + S^3\right)^2}, \qquad \Gamma = 1 \text{ or } \gamma^5, \tag{PS.426}$$

implying

$$-G \frac{2i}{\chi} \phi^\dagger \partial_\mu \phi \bar{\psi} \gamma^\mu \psi = -\frac{1}{g^2(1 + S^3)} \chi^1 \Gamma \chi^2 \left(\partial S^3 - i(S^1 \partial S^2 - S_2 \partial S^1)\right). \tag{PS.427}$$

Moreover,

$$G i \bar{\psi} \gamma^\mu \partial_\mu \psi = \frac{1}{2g^2} \bar{\chi}^a i \gamma^\mu \partial_\mu \chi^a + G \frac{2i}{\chi} \phi^\dagger \partial_\mu \phi \bar{\psi} \gamma^\mu \psi. \tag{PS.428}$$

Assuming that the parameter m is *real* correspondence between the mass terms follows rather trivially. For instance,

$$\bar{\psi}^\dagger \gamma^0 \psi = -\frac{1}{(1 + S^3)^2} \left(\bar{\chi}^a \chi^a\right). \tag{PS.429}$$

Establishing the correspondence between the four-fermion terms is somewhat labor consuming. We can simplify our task if we take into account the fact that the overall structures of these terms are uniquely fixed by CP(1) and O(3) symmetries, correspondingly. All we have to do is to establish the overall transfer coefficient between them at some point in the target space, say, near the north pole of the sphere,

$$S^3 \approx 1, \quad \chi^3 \approx 0. \tag{PS.430}$$

Assembling all derivations above we arrive at (10.311).

The second part of this problem is the proof of correspondence between the conserved currents in both representations of the model, (10.330) and (10.336). In the $O(3)$ representation we have

$$J^\mu = \frac{1}{g^2} \left[(\partial_\lambda S^a) \gamma^\lambda \gamma^\mu \chi^a - im \gamma^\mu \chi^3 \right],$$

$$\tilde{J}^\mu = \frac{1}{g^2} \left[\epsilon^{abc} S^a (\partial_\lambda S^b) \gamma^\lambda \gamma^\mu \chi^c - m\epsilon^{3ab} S^a \gamma^\mu \chi^b \right], \tag{PS.431}$$

while in the $CP(1)$ representation

$$J^\mu_\alpha = \sqrt{2} G \left[(\partial_\nu \phi^\dagger) \gamma^\nu \gamma^\mu \psi + im \phi^\dagger \gamma^\mu \psi \right]_\alpha. \tag{PS.432}$$

Performing calculation along the same lines as above one can easily prove this correspondence. Note that in passing from CP(1) to O(3) the overall normalization of the supercurrents is changed.

Exercise 10.12.2, page 483

The easiest way to implement the constraints in (10.288) is to add them in (10.293) with Lagrange multipliers. Then

$$\mathcal{L} = \frac{1}{2g^2} \left\{ \partial_\mu S^a \partial_\mu S^a + i \bar{\chi}^a \partial \!\!\!/ \chi^a + \frac{1}{4} (\bar{\chi}^a \chi^a)^2 + \lambda (S^a S^a - 1) + S^a \bar{\eta} \chi^a \right\}, \tag{PS.433}$$

implying

$$-\partial^2 S^a + \lambda S^a + \frac{1}{2}\bar{\eta}\chi^a = 0,$$

$$i\partial\chi^a + \frac{1}{2}(\bar{\chi}\chi)\chi^a + \frac{1}{2}S^a\eta = 0,$$

$$\eta = -2iS^b\partial\chi^b. \tag{PS.434}$$

Note also that

$$\bar{\chi}_i^a\chi_j^a = \frac{1}{2}(\bar{\chi}^a\chi^a)\,\delta_{ij}, \tag{PS.435}$$

where i, $j = 1, 2$ are spinorial indices. Differentiating the supercurrent (10.294) we arrive at

$$
\begin{aligned}
g^2\,\partial_\mu J_i^\mu &= \left(\partial^2 S^a\right)\chi_i^a + (\partial_a S^a)\,(\gamma^a\partial\chi^a)_i \\
&= \frac{1}{2}(\bar{\chi}^a\eta)\chi_i^a + \frac{i}{2}\,(\partial_a S^a)\,(\gamma^a\chi^a)_i\,(\bar{\chi}\chi) \\
&= -i(\bar{\chi}^a S^b\partial\chi^b)\chi_i^a + \frac{i}{2}\,(\partial_a S^a)\,(\gamma^a\chi^a)_i\,(\bar{\chi}\chi) \\
&= \frac{i}{2}\left(S^b\partial\chi^b\right)_i(\bar{\chi}\chi) + \frac{i}{2}\,(\partial_a S^a)\,(\gamma^a\chi^a)_i\,(\bar{\chi}\chi) = 0.
\end{aligned}
\tag{PS.436}
$$

The expression on the last line is proportional to the full derivative of $S^a\chi^a$.

Exercises 10.12.3–10.12.6, page 483, 484

In these four exercises we will acquaint the reader with a technique alternative to that used previously, namely, representation of the S fields in terms of the spherical angles ϑ and φ,

$$S^3 = \cos\vartheta, \quad S^1 = \rho\cos\varphi, \quad S^2 = \rho\sin\varphi, \tag{PS.437}$$

where ρ is *defined* as

$$\rho \equiv \sin\vartheta.$$

Then

$$\phi = \frac{\rho e^{i\vartheta}}{1 + S^3}. \tag{PS.438}$$

Our starting point is the result obtained in Exercise 6.2.2, pages 257 and 671,

$$K^\mu = \varepsilon^{\mu\nu}\left(\frac{\bar{\phi}\partial_\nu\phi}{1 + \bar{\phi}\phi}\right). \tag{PS.439}$$

Combining (PS.420), (PS.438), and (PS.439) we arrive at

$$K^\mu = \frac{i}{2}\varepsilon^{\mu\nu}(1 - S^3)\partial_\nu\varphi - \frac{1}{2}\varepsilon^{\mu\nu}\partial_\nu\log(1 + \cos\vartheta),$$

$$\partial_\mu K^\mu = \frac{\varepsilon^{\mu\nu}(\partial_\mu\bar{\phi})(\partial_\nu\phi)}{(1 + \bar{\phi}\phi)} = \frac{i}{2}\varepsilon^{\mu\nu}\sin\vartheta\left(\partial_\mu\vartheta\,\partial_\nu\varphi\right). \tag{PS.440}$$

The second term in the first line trivially vanishes in passing to the second line. The answer to the final question is as follows:

$$\varepsilon^{\mu\nu} \varepsilon^{abc} S^a \partial_\mu S^b \partial_\nu S_c = 2\varepsilon^{\mu\nu} \left(S^3 \partial_\mu S^1 \partial_\nu S^2 + S^1 \partial_\mu S^2 \partial_\nu S^3 + S^2 \partial_\mu S^3 \partial_\nu S^1 \right)$$

$$= 2\varepsilon^{\mu\nu} \left\{ S^3 \left[\rho \partial_\mu \rho \, (\cos\varphi)^2 \, \partial_\nu\varphi - \rho \partial_\nu \rho \, (\sin\varphi)^2 \, \partial_\mu\varphi \right] \right.$$

$$\left. + \left(\partial_\nu S^3 \right) \left[\rho^2 \, (\cos\varphi)^2 \, \partial_\mu\varphi + \rho^2 \, (\sin\varphi)^2 \, \partial_\mu\varphi \right] \right\}$$

$$= 2\varepsilon^{\mu\nu} \left\{ (\cos\vartheta)^2 \sin\vartheta \left(\partial_\mu\vartheta \, \partial_\nu\varphi \right) - \left(\partial_\mu S^3 \right) \rho^2 \partial_\nu\varphi \right\} = 2\varepsilon^{\mu\nu} \sin\vartheta \left(\partial_\mu\vartheta \, \partial_\nu\varphi \right).$$

$$\text{(PS.441)}$$

We see that

$$\varepsilon^{\mu\nu} \varepsilon^{abc} S^a \partial_\mu S^b \partial_\nu S_c = \frac{4}{i} \partial_\mu K^\mu, \qquad \text{(PS.442)}$$

to be compared with Eqs. (10.298) and (10.324).

Exercises 10.12.7, page 484

The only vertex that might be relevant for the one-loop *fermion* contribution in the β function in CP(1) is

$$G \left(-2i\chi^{-1} \phi^\dagger \partial_\mu \phi \right) \bar\psi \gamma^\mu \psi \equiv \mathcal{A}_\mu \bar\psi \gamma^\mu \psi, \qquad \text{(PS.443)}$$

where a shorthand notation $\mathcal{A}_\mu$ is introduced for the fermion-free factor $G(-2i\chi^{-1} \phi^\dagger \partial_\mu \phi)$; see Eq. (10.334). In this convention the relevant fermion loop is the same as in the Schwinger model, which has been already discussed in Chapter 9, see Fig. 9.40a. The result for this loop is given in Eq. (9.149). It does *not* contain any logarithms and, therefore, does not appear in the β function.

Exercise 10.16.1, page 511

In the absence of the anomalies (i.e., $Z = \eta = 0$) Eq. (10.418) implies

$$D^\alpha \mathcal{J}_{\alpha\dot\alpha} = -\frac{4}{g^2} \left[2\bar\theta^{\dot\beta} T^\beta{}_{\dot\alpha\beta\dot\beta} - i\bar\theta^{\dot\beta} \partial^\alpha_{\dot\beta} R_{\alpha\dot\alpha} \right] + \cdots \qquad \text{(PS.444)}$$

Let us examine how these T and R terms appear in $\mathcal{J}_{\alpha\dot\alpha}$. The T term comes from two sources. The θ component of W and $\bar\theta$ component of $\bar W$ produce the bosonic part $G_{\alpha\beta}\bar G_{\dot\alpha\dot\beta}$. The fermion part of T comes from $\lambda(x_L)\bar\lambda(x_R)$, from the expansion $x^\mu_{L,R} = x^\mu \pm iv^\mu$, where

$$v^\mu = \bar\theta_{\dot\beta} (\bar\sigma^\nu)^{\dot\beta\beta} \theta_\beta. \qquad \text{(PS.445)}$$

Performing this expansion in the lowest component of $\mathcal{J}_{\alpha\dot\alpha}$ we arrive at

$$\frac{4i}{g^2} \text{Tr} \left(\lambda_\alpha \overleftrightarrow{\partial_\nu} \bar\lambda_{\dot\alpha} \right) v^\nu = \frac{4i}{g^2} \text{Tr} \left(\lambda_\alpha \overleftrightarrow{\partial^{\dot\beta\beta}} \bar\lambda_{\dot\alpha} \right) \bar\theta_{\dot\beta} \theta_\beta. \qquad \text{(PS.446)}$$

Now, the expression on the right-hand side should be split into two parts: symmetric and antisymmetric under the interchange α and β in $\left(\lambda_\alpha\theta_\beta\right)$ and $\dot\alpha$ and $\dot\beta$ in $\left(\bar\lambda_{\dot\alpha}\bar\theta_{\dot\beta}\right)$,

$$\frac{1}{2}\left[\left(\lambda_\alpha\overset{\leftrightarrow}{\partial^{\beta\dot\beta}}\bar\lambda_{\dot\alpha}\right)\bar\theta_{\dot\beta}\theta_\beta + \left(\lambda_\beta\overset{\leftrightarrow}{\partial^{\dot\beta\alpha}}\bar\lambda_{\dot\beta}\right)\bar\theta_{\dot\alpha}\theta_\alpha\right.$$

$$\left.+ \left(\lambda_\alpha\overset{\leftrightarrow}{\partial^{\dot\beta\alpha}}\bar\lambda_{\dot\alpha}\right)\bar\theta_{\dot\beta}\theta_\beta - \left(\lambda_\beta\overset{\leftrightarrow}{\partial^{\dot\beta\alpha}}\bar\lambda_{\dot\beta}\right)\bar\theta_{\dot\alpha}\theta_\alpha\right].\qquad\text{(PS.447)}$$

The last term is formally proportional to $\varepsilon_{\alpha\beta}\varepsilon_{\dot\alpha\dot\beta}$ but in fact vanishes. As a result we obtain

$$\frac{4i}{g^2}\text{Tr}\left(\lambda_\alpha\overset{\leftrightarrow}{\partial_\nu}\bar\lambda_{\dot\alpha}\right)v^\nu \to \frac{2i}{g^2}\text{Tr}\left[\left(\lambda_\alpha\overset{\leftrightarrow}{\partial^{\beta\dot\beta}}\bar\lambda_{\dot\alpha}\right)\bar\theta_{\dot\beta}\theta_\beta + \left(\lambda_\beta\overset{\leftrightarrow}{\partial^{\dot\beta\alpha}}\bar\lambda_{\dot\beta}\right)\bar\theta_{\dot\alpha}\theta_\alpha\right].$$

$$\text{(PS.448)}$$

The right-hand side above presents the fermion contribution to the energy–momentum tensor.

Finally, let us have a closer look at the R term. We will verify its coefficient using Eq. (PS.444). To this end we must calculate

$$-\frac{4}{g^2}\text{Tr}D^\alpha\left(\lambda_\alpha\bar\lambda_{\dot\alpha}\right) \to -\frac{4}{g^2}\text{Tr}\,\varepsilon^{\alpha\beta}\left(-i\bar\theta^{\dot\beta}\partial_{\beta\dot\beta}\lambda_\alpha\bar\lambda_{\dot\alpha}\right) = -i\bar\theta^{\dot\beta}\partial^\alpha_{\ \dot\beta}R_{\alpha\dot\alpha}.\qquad\text{(PS.449)}$$

Coincidence between Eqs. (PS.444) and (PS.449) is perfect.

Next, we pass to the second part of the problem, namely, rewriting Eq. (10.418) in vector/tensor notation. To this end we must convolute both sides with $\frac{1}{2}(\bar\sigma^\mu)^{\dot\alpha\alpha}$ and use a number of the equations following from Section 10.2.1, for instance,

$$\frac{1}{2}\left(\bar\sigma^\rho\right)^{\dot\alpha\alpha}\left(\sigma^\mu\right)_{\alpha\dot\alpha} = g^{\rho\mu},\qquad \left(\sigma_\mu\right)_{\alpha\dot\alpha}\left(\bar\sigma^\mu\right)^{\dot\beta\beta} = 2\delta^\beta_\alpha\delta^{\dot\alpha}_{\dot\beta},$$

$$\left(\bar\sigma^\rho\right)^{\dot\beta\alpha}\left(\sigma^\mu\right)_{\alpha\dot\alpha} = g^{\rho\mu}\delta^{\dot\beta}_{\dot\alpha} + 2\left(\bar\sigma^{\rho\mu}\right)^{\dot\beta}_{\dot\alpha},\qquad\text{(PS.450)}$$

etc. In this way we arrive at

$$\mathcal{J}^\mu = R^\mu - \left[i\theta^\beta J^\mu_\beta - \frac{1}{\sqrt{2}}\bar\eta_{\dot\beta}\left(\bar\sigma^\mu\right)^{\dot\beta\alpha}\theta_\alpha + \text{H.c.}\right] + 2v_\nu\left[T^{\mu\nu} + g^{\mu\nu}\,\text{Re}\,Z\right]$$

$$+ \frac{1}{4}\varepsilon^{\mu\nu\alpha\beta}v_\nu\partial_\alpha R_\beta + \cdots,\qquad\text{(PS.451)}$$

where v^μ is given in (PS.445).

Exercise 10.16.2, page 511

Let us prove the following assertion: if

$$D^\alpha\mathcal{J}_{\alpha\dot\alpha} = \bar D_{\dot\alpha}D^2 Y\qquad\text{(PS.452)}$$

for some chiral Y, then

$$D^\alpha\left[\mathcal{J}_{\alpha\dot\alpha} + 4i\partial_{\alpha\dot\alpha}(Y - \bar Y)\right] = 0.\qquad\text{(PS.453)}$$

Proof: First note that $\bar{Y}$ can be omitted since $D^\alpha \bar{Y} = 0$. Next, observe that

$$\bar{D}_{\dot\alpha} D^2 = \bar{D}_{\dot\alpha} D^\beta D_\beta = \left(\{\bar{D}_{\dot\alpha} D^\beta\} - D^\beta \bar{D}_{\dot\alpha} \right) D_\beta \to -2i\partial_{\beta\dot\alpha} D^\beta - D^\beta \{\bar{D}_{\dot\alpha} D_\beta\},$$

where in the last term we took into account the fact that $\bar{D}_{\dot\alpha} Y = 0$. Summarizing, we arrive at

$$D^\alpha \mathcal{J}_{\alpha\dot\alpha} = \bar{D}_{\dot\alpha} D^2 Y = -4i D^\alpha \partial_{\alpha\dot\alpha} Y, \qquad (PS.454)$$

which is equivalent to (PS.453).

Exercise 10.16.3, page 511

In this exercise we will limit ourselves to a few examples. For instance,

$$\left(\mathcal{J}^R \right)^\mu \Big|_{\theta^2\bar\theta} \propto \theta^2 \bar\theta_{\dot\alpha} \, \partial^{\dot\alpha\alpha} J_\alpha^\mu, \qquad \left(\mathcal{J}^R \right)^\mu \Big|_{\theta^2\bar\theta^2} \propto \theta^2 \bar\theta^2 \, \partial^2 R^\mu. \qquad (PS.455)$$

Now, we pass to the Konishi current defined in Eq. (10.444). Its lowest component is given in Eq. (10.442). After some straightforward but rather tedious algebra we obtain

$$\left(\mathcal{J}^f \right)^\mu \Big|_\theta \propto (\sigma^{\mu\nu})^\alpha_\beta \, \partial_\nu (\psi^\beta \bar\phi), \qquad (PS.456)$$

and so on.

Exercise 10.16.4, page 511

First, we rewrite the second term in (10.458) as follows:

$$-D^\alpha \bar{D}_{\dot\alpha} D_\alpha = -\frac{1}{2} \Big(D^\alpha \{\bar{D}_{\dot\alpha} D_\alpha\} - D^2 \bar{D}_{\dot\alpha}$$

$$+ \{D^\alpha \bar{D}_{\dot\alpha}\} D_\alpha - \bar{D}_{\dot\alpha} D^\alpha D_\alpha \Big) = \frac{1}{2} \left(D^2 \bar{D}_{\dot\alpha} + \bar{D}_{\dot\alpha} D^2 \right). \qquad (PS.457)$$

Adding the right-hand side in (PS.457) to the first term in (10.458) we complete the proof,

$$D^\alpha \left(D_\alpha \bar{D}_{\dot\alpha} - \bar{D}_{\dot\alpha} D_\alpha \right) = \frac{3}{2} D^2 \bar{D}_{\dot\alpha} + \frac{1}{2} \bar{D}_{\dot\alpha} D^2. \qquad (PS.458)$$

Exercise 10.16.5, page 511

The theory we discussed in Section 10.16.7.1 has the following Lagrangian:

$$\mathcal{L} = \int d^2\theta \left(\frac{1}{4g^2} W^{a\alpha} W_\alpha^a + V_k X_1 V_l \epsilon^{kl} + \text{H.c.} \right)$$

$$+ \sum_{i=1}^2 \int d^4\theta \left(\bar{V}_i e^V V_i + \bar{X}_i e^V X_i \right). \qquad (PS.459)$$

As we know from Section 10.7, this "geometric" global U(1) symmetry simultaneously rotates both the fields in the Lagrangian and Grassmann coordinates in a concerted manner. In particular, the charges of the matter fields under this transformation are universal,

$$V_i, \ X_i \to e^{2i\alpha/3} \{V_i, \ X_i\}. \tag{PS.460}$$

The superpotential in (PS.459) has R charge $+2$ which is necessary for the invariance of the Lagrangian under $U(1)_R$.

Additional global $U(1)$ symmetries are related to matter fields; their $U(1)$ charges need not be universal – distinct flavors may have different charges as long the $U(1)$ charge of the superpotential remains intact. The corresponding assignment of flavor charges need not be unique, generally speaking.

Let us assume that the additional flavor charges are set according to the following transformations:

$$V_i \to e^{i\alpha/3} V_i, \qquad X_1 \to e^{-2i\alpha/3} X_1, \tag{PS.461}$$

or, alternatively,

$$q(V) = \frac{1}{3}, \quad q(X_1) = -\frac{2}{3}. \tag{PS.462}$$

Then the total $U(1)$ charges of the superfields V_i and X_1 acquire the values indicated in Table 10.5 (page 509). The superfield X_2 so far has an undetermined charge because it does not appear in the superpotential. We will determine it later. The corresponding $U(1)$ Noether currents are the lowest components of the hypercurrent. The R current – the lowest component of the geometric hypercurrent (10.439) – is

$$R_\mu = -\frac{1}{g^2} \lambda^a \sigma_\mu \bar{\lambda}^a + \frac{1}{3} \sum_f \left(\psi_f \sigma_\mu \bar{\psi}_f - 2i\phi_f \overleftrightarrow{D}_\mu \bar{\phi}_f \right). \tag{PS.463}$$

The flavor currents – the lowest components of the Konishi currents (10.444) – in this exercise have the form

$$R_\mu^f = q_f \left(-\psi_f \sigma_\mu \bar{\psi}_f - i\phi_f \overleftrightarrow{D}_\mu \bar{\phi}_f \right), \tag{PS.464}$$

where q_f indicates the $U(1)$ charge of the corresponding superfield.

Warning: We modified the Konishi current (10.444) by overall factors q_f to take into account the fact that the *flavor* $U(1)$ symmetries are characterized by individual charge assignments.

Let us calculate the divergence of the above currents. This calculation will allow us to find the flavor charge for the X_2 superfield by requiring cancellation of anomalies. For the geometric current corresponding to (PS.463), (PS.464) we have

$$\partial^{\alpha\dot\alpha} J_{\alpha\dot\alpha} = -\frac{1}{3} i D^2 \left[\left(3\mathcal{W} - \sum_f Q_f \frac{\partial \mathcal{W}}{\partial Q_f} \right) \right.$$

$$\left. - \frac{3T_G - \sum_f T(R_f)}{16\pi^2} \text{Tr} W^2 \right] + \text{H.c.} \tag{PS.465}$$

and

$$\partial^{\alpha\dot\alpha} J_{\alpha\dot\alpha}^f = i(q_f) D^2 \left[\frac{1}{2} Q_f \frac{\partial \mathcal{W}}{\partial Q_f} + \frac{T(R_f)}{16\pi^2} \text{Tr} W^2 \right] + \text{H.c.} \tag{PS.466}$$

Given the superpotential (10.466) we observe that

$$\frac{\partial W}{\partial V_m^{\gamma}} = \frac{\partial}{\partial V_m^{\gamma}} \left(V_k^{\alpha}(X_1)_{\alpha\beta} V_l^{\beta} \varepsilon^{kl} \right) = 2(X_1)_{\gamma\beta} V_l^{\beta} \varepsilon^{ml} \qquad \text{(PS.467)}$$

due to antisymmetry in color of the superfield X_1. Therefore,

$$V_m^{\gamma} \frac{\partial W}{\partial V_m^{\gamma}} = 2W, \qquad \text{(PS.468)}$$

and adding the contribution from X_1 we find

$$\sum_f Q_f \frac{\partial W}{\partial Q_f} = 3W. \qquad \text{(PS.469)}$$

This had to be expected, of course, because of the cubic nature of the superpotential.

Thus, the classical part of the geometric hypercurrent vanishes. Now, let us examine the sum of the flavor currents,

$$\partial^{\alpha\dot\alpha} \sum_f J_{\alpha\dot\alpha}^f = \frac{iD^2}{16\pi^2} \sum_f q_f T(R_f) \text{Tr} W^2 + \frac{iD^2}{2} \sum_f q_f Q_f \frac{\partial W}{\partial Q_f} + \text{H.c.} \qquad \text{(PS.470)}$$

The coefficient from the anomaly in the preceding relation is

$$\sum_f q_f T(R_f) = \frac{1}{3} \left(\frac{1}{2} \right) \times 2 - \frac{2}{3} \left(\frac{3}{2} \right) + q(X_2) \frac{3}{2} = -\frac{11}{3} \qquad \text{(PS.471)}$$

provided

$$q(X_2) = -2 \qquad \text{(PS.472)}$$

(the overall U(1) charge of X_2 is then $-4/3$, as indicated in Table 10.5). The second term in (PS.470) automatically vanishes under our choice of the flavor charges in (PS.462).

Why do we need to have $-11/3$ in the right-hand side of (PS.471)? The anomaly part for the geometric current (PS.463) is

$$\partial^{\alpha\dot\alpha} J_{\alpha\dot\alpha} = \frac{1}{3} \frac{iD^2}{16\pi^2} \left[3T_G - \sum_f T(R_f) \right] \text{Tr} W^2 + \text{H.c.} \qquad \text{(PS.473)}$$

In the model at hand the first coefficient of the β function is

$$3T_G - \sum_f T(R_f) = 11. \qquad \text{(PS.474)}$$

Therefore, introducing the R hypercurrent as

$$\tilde{\mathcal{J}}_{\alpha\dot\alpha} = \mathcal{J}_{\alpha\dot\alpha} + \sum_f J_{\alpha\dot\alpha}^f, \qquad \text{(PS.475)}$$

we achieve our goal since the lowest component of $\tilde{\mathcal{J}}_{\alpha\dot\alpha}$ is conserved,[10]

$$\partial^{\mu} \tilde{R}_{\mu} = \partial^{\mu} \left(R_{\mu} + \frac{1}{3} \left(R_{\mu}^{V_1} + R_{\mu}^{V_2} \right) - \frac{2}{3} R_{\mu}^{X_1} - 2R_{\mu}^{X_2} \right) = 0. \qquad \text{(PS.476)}$$

---

[10] The R_{μ}^{Qi} currents below are defined with $q_f = 1$.

The terms containing 11/3 in (PS.473) and $-11/3$ in (PS.470) cancel each other. It is worth stressing again that the Konishi currents we deal with in this exercise differ from (10.444) by flavor charges q_f; cf. Eq. (PS.470).

We see here the magic of choosing a cubic superpotential with a missing flavor in it. Such choice guarantees simultaneous cancellation of both, quantum anomalies and the classical part in $\partial^\mu \tilde{R}_\mu$.

It is obvious that $\tilde{\chi}_{\dot\alpha} = 0$ for $\tilde{J}_{\alpha\dot\alpha}$ found above.

Exercise 10.16.6, page 511

The simplest case corresponds to pure SQED (no matter) with the Fayet–Iliopoulos term (Section 10.16.9). Then we have

$$\mathcal{L} = \frac{1}{4e^2} \left(\int d^2\theta\, W_\alpha W^\alpha + \text{H.c.} \right) - \xi \int d^4\theta\, V. \qquad (\text{PS.477})$$

In components this action reads

$$S = -\frac{1}{4e^2} F_{\mu\nu} F^{\mu\nu} + \frac{1}{e^2} \bar\lambda_{\dot\alpha} i \partial^{\alpha\dot\alpha} \lambda_\alpha + \frac{1}{2e^2} D^2 - \xi D. \qquad (\text{PS.478})$$

The equations of motion are

$$D = e^2\xi, \quad \partial_\mu D = 0, \quad \partial^{\dot\alpha\alpha} \lambda_\alpha = 0, \quad \bar\lambda_{\dot\alpha} \overleftarrow{\partial}{}^{\dot\alpha\alpha} = 0,$$

$$\partial^\mu F_{\mu\nu} = \partial^\mu \tilde{F}_{\mu\nu} = 0, \text{ or, alternativelly, } \partial^{\dot\alpha\alpha} F_{\alpha\beta} = 0. \qquad (\text{PS.479})$$

The latter spinorial relation contains all four Maxwell equations.

The geometric hypercurrent has the form

$$J_{\alpha\dot\alpha} = \frac{2}{e^2} \left\{ \lambda_{\dot\alpha} \lambda_\alpha + \left[\lambda_\alpha \left(-iD\varepsilon_{\dot\alpha\dot\beta} - F_{\dot\alpha\dot\beta} \right) \bar\theta^{\dot\beta} + i\bar\theta^2 \lambda_\alpha \partial_{\beta\dot\alpha} \lambda^\beta \right] + \text{H.c.} \right\}$$

$$+ \frac{4\xi}{3} \left[A_{\alpha\dot\alpha} + \left(2i\bar\theta_{\dot\alpha} \lambda_\alpha + \text{H.c.} \right) \right] + \cdots \qquad (\text{PS.480})$$

Then, omitting terms vanishing through the equations of motion we arrive at

$$\partial^{\dot\alpha\alpha} J_{\alpha\dot\alpha} = \frac{4\xi}{3} \partial^{\dot\alpha\alpha} A_{\alpha\dot\alpha}. \qquad (\text{PS.481})$$

Let us compare this result with $-\frac{i}{2} \left(D^2 X - \bar{D}^2 \bar{X} \right)$; cf. Eq. (10.410),

$$-\frac{i}{2} \left(D^2 X - \bar{D}^2 \bar{X} \right) = -\frac{i\xi}{12} D^2 \bar{D}^2 V + \text{H.c.} = \frac{16}{12} \xi \partial^{\dot\alpha\alpha} A_{\alpha\dot\alpha}, \qquad (\text{PS.482})$$

where we used Eqs. (10.114), (10.180), and (10.474). In calculating $D^2 \bar{D}^2 V$ one should take into account that out of four spinorial derivatives in front of V three should be represented by $\partial/\partial\theta$ or $\partial/\partial\bar\theta$ while one spinorial derivative should be represented by $\theta\theta$ or $\bar\theta\partial$. The Hermitean conjugated term doubles the result.

The coincidence between (PS.481) and (PS.482) proves the validity of Eq. (10.410) in supersymmetric Maxwell theory with the Fayet–Iliopoulos term plus the condition (10.474). Note that Eqs. (PS.481) and (PS.482) contain no higher components. This proves that the supercurrent component in (PS.480) is strictly conserved which entails in turn conservation of the energy–momentum tensor.

Exercise 10.19.1, page 541

The number of the fermion zero modes depends on the values of $T(V)$, $T(X)$ and $T_{SU(5)}$. In the model at hand we have two quintets (each produces one zero mode), two antidecuplets (each produces three zero modes), and ten gluino's zero modes. Altogether we have $8 + 10$ chiral zero modes. The effective 't Hooft vertex has the form $X^6 V^2 \lambda^{10}$ contracted in color indices in such a way that its overall structure is color-singlet.

Exercise 10.25.1, page 561

The energy per unit area of a wall configuration is given in Eq. (10.648). Assume that ΔW has a phase $\alpha \neq 0$, i.e.,

$$\Delta W = |\Delta W| \, e^{i\alpha}, \qquad \Delta W = W_{z=\infty} - W_{z=-\infty} \qquad \text{(PS.483)}$$

and Eq. (10.652) cannot be satisfied. Then we note that Eq. (10.648) can be presented as

$$\mathcal{E} = \int_{-\infty}^{\infty} dz \left(\partial_z \phi - e^{i\alpha} \bar{W}' \right) \left(\partial_z \bar{\phi} - e^{-i\alpha} W' \right)$$
$$+ e^{i\alpha} \Delta \bar{W} + e^{-i\alpha} \Delta W \qquad \text{(PS.484)}$$

for *arbitrary values of* α, without assuming that ΔW is real. The strongest bound

$$\mathcal{E} \geq 2\text{Re} \left\{ e^{-i\alpha} \left(\Delta W \right) \right\} \qquad \text{(PS.485)}$$

is achieved provided $e^{-i\alpha} (\Delta W)$ is real. Then the right-hand side in (PS.485) reduces to $2 |\Delta W|$. The BPS saturation implies

$$\partial_z \phi = e^{i\alpha} \bar{W}' . \qquad \text{(PS.486)}$$

If a BPS saturated wall or kink solution exists then

$$\mathcal{E}_{\text{wall/kink}} = 2 |\Delta W| . \qquad \text{(PS.487)}$$

Chapter 11

Exercise 11.5.1, page 625

Let us denote the lowest components of the superfields $\mathcal{A}$ and $\mathcal{A}^a$ as a and a^a. Then, in components, after integration over the auxiliary F and D fields, we obtain

$$\mathcal{L}_{\text{bos}} = -\frac{1}{4g_2^2} F_{\mu\nu}^a F^{\mu\nu,a} - \frac{1}{4g_1^2} F_{\mu\nu} F^{\mu\nu} + \frac{1}{g_2^2} \left| \mathcal{D}_\mu a^a \right|^2 + \frac{1}{g_1^2} \left| \mathcal{D}_\mu a \right|^2$$

$$+ \left| \mathcal{D}_\mu q^A \right|^2 + \left| \mathcal{D}_\mu \tilde{\bar{q}}^A \right|^2 - V(q^A, \tilde{q}_A, a, a^a), \qquad \text{(PS.488)}$$

where the scalar potential has the form

$$V(q^A, \tilde{q}_A, a_\mu, a_\mu) = \frac{g_2^2}{2} \left(\frac{i}{g_2^2} f^{abc} \bar{a}^b a^c + \bar{q}_A T^a q^A - \tilde{q}_A T^a \bar{\tilde{q}}^A \right)^2$$

$$+ \frac{g_1^2}{8} \left(\bar{q}_A q^A - \tilde{q}_A \bar{\tilde{q}}^A - 2\xi \right)^2 + 2g_2^2 \left| \tilde{q}_A T^a q^A \right|^2 + \frac{g_1^2}{2} \left| \tilde{q}_A q^A \right|^2$$

$$+ \frac{1}{2} \left\{ \left| (a + 2T^a a^a) q^A \right|^2 + \left| (a + 2T^a a^a) \tilde{q}^A \right|^2 \right\}. \qquad \text{(PS.489)}$$

The fermion part of the Lagrangian is

$$\mathcal{L}_{\text{ferm}} = \frac{i}{g_2^2} \bar{\lambda}_f^a D\lambda^{af} + \frac{i}{g_1^2} \bar{\lambda}_f D\lambda^f + \text{Tr}\left(\bar{\psi} i D\psi \right) + \text{Tr}\left(\tilde{\psi} i D\tilde{\psi} \right)$$

$$+ \frac{1}{\sqrt{2}} \varepsilon^{abc} \bar{a}^a \lambda_f^b \lambda^{fc} + \frac{1}{\sqrt{2}} \varepsilon^{abc} a^a \bar{\lambda}_f^b \bar{\lambda}^{fc}$$

$$+ \frac{i}{\sqrt{2}} \text{Tr}\left[\bar{q}_f (\lambda^f \psi) + (\tilde{\psi} \lambda_f) q^f + (\bar{\psi} \bar{\lambda}_f) q^f + \bar{q}^f (\bar{\lambda}_f \tilde{\psi}) \right]$$

$$+ \frac{i}{\sqrt{2}} \text{Tr}\left[\bar{q}_f \tau^a (\lambda^{af} \psi) + (\tilde{\psi} \lambda_f^a) \tau^a q^f + (\bar{\psi} \bar{\lambda}_f^a) \tau^a q^f + \bar{q}^f \tau^a (\bar{\lambda}_f^a \tilde{\psi}) \right]$$

$$+ \frac{i}{\sqrt{2}} \text{Tr}\left[\tilde{\psi}(a + a^a \tau^a) \psi \right] + \frac{i}{\sqrt{2}} \text{Tr}\left[\bar{\psi}(a + a^a \tau^a) \bar{\tilde{\psi}} \right]. \qquad \text{(PS.490)}$$

Here f is the index of the global SU(2)$_R$ symmetry. Namely, we combine two gluino fields of this theory (one from the vector $\mathcal{N} = 1$ supermultiplet and the other from its scalar superpartner; see Eq. (10.496)) in a single doublet,

$$\left(\lambda^f \right)^{\alpha a} = \begin{pmatrix} \lambda^{\alpha a} \\ \chi^{\alpha a} \end{pmatrix}, \quad f = 1, 2. \qquad \text{(PS.491)}$$

To make SU(2)$_R$ symmetry fully explicit, we form similar doublets for the matter fields, namely,

$$q^f = \begin{pmatrix} q \\ \bar{\tilde{q}} \end{pmatrix}, \quad \bar{q}_f = \{ \bar{q}, \tilde{q} \}, \quad f = 1, 2, \qquad \text{(PS.492)}$$

where each of the fields $q, \tilde{q}, \bar{q}, \bar{\tilde{q}}$, as well as their fermion superpartners $\psi, \tilde{\psi}$, are in fact, two-by-two matrices, for instance,

$$q \rightarrow \left\{ q^{iA} \right\} = \begin{pmatrix} q^{11} & q^{12} \\ q^{21} & q^{22} \end{pmatrix}, \qquad \text{(PS.493)}$$

and so on. That is why Tr signs appear in Eq. (PS.490).

Index

Printed in the United States
by Baker & Taylor Publisher Services